Handbook of Water and Wastewater Treatment Plant Operations

The *Handbook of Water and Wastewater Treatment Plant Operations* is the first thorough resource manual developed exclusively for water and wastewater plant operators. Now regarded as an industry standard, this fifth edition has been updated throughout, and it explains the material in easy-to-understand language. It also provides real-world case studies and operating scenarios, as well as problem-solving practice sets for each scenario.

Key features:

- Updates the material to reflect the developments in the field.
- Includes new math operations with solutions, as well as over 250 new sample questions.
- Adds updated coverage of energy conservation measures with applicable case studies.
- Enables users to properly operate water and wastewater plants and suggests troubleshooting procedures for returning a plant to optimal operation levels.
- Prepares operators for licensure exams.

Handbook of Water and Wastewater Treatment Plant Operations

Fifth Edition

Frank R. Spellman

CRC Press
Taylor & Francis Group
Boca Raton London New York

CRC Press is an imprint of the
Taylor & Francis Group, an **informa** business

Designed cover image: Shutterstock

Fifth edition published 2025
by CRC Press
2385 NW Executive Center Drive, Suite 320, Boca Raton FL 33431

and by CRC Press
4 Park Square, Milton Park, Abingdon, Oxon, OX14 4RN

CRC Press is an imprint of Taylor & Francis Group, LLC

© 2025 Frank R. Spellman

First edition published by Willan 2008

Fourth edition published by CRC Press 2020

Library of Congress Cataloging-in-Publication Data
Names: Spellman, Frank R., author.
Title: Handbook of water/wastewater treatment plant operations / Frank R. Spellman.
Description: 5th edition. | Boca Raton, FL ; Abingdon, Oxon : CRC Press, 2025. |
Includes bibliographical references and index. |
Summary: "The Handbook of Water and Wastewater Treatment Plant Operations is the first thorough resource manual developed exclusively for water and wastewater plant operators. Now regarded as an industry standard, this fifth edition has been updated throughout, and explains the material in easy-to-understand language. It also provides real-world case studies and operating scenarios, as well as problem-solving practice sets for each scenario"– Provided by publisher.
Identifiers: LCCN 2024036322 | ISBN 9781032948324 (hardback) | ISBN 9781032843728 (paperback) |
ISBN 9781003581901 (ebook)
Subjects: LCSH: Water treatment plants–Handbooks, manuals, etc. |
Sewage disposal plants–Handbooks, manuals, etc. | Water–Purification–Handbooks, manuals, etc. |
Sewage–Purification–Handbooks, manuals, etc.
Classification: LCC TD434 .S64 2025 |
DDC 628.1/62–dc23/eng/20240822
LC record available at https://lccn.loc.gov/2024036322

ISBN: 978-1-032-94832-4 (hbk)
ISBN: 978-1-032-84372-8 (pbk)
ISBN: 978-1-003-58190-1 (ebk)

DOI: 10.1201/9781003581901

Typeset in Times
by codeMantra

Contents

Foreword xx
Preface to fifth edition xxi
Author xxiii

PART I Water and Wastewater Operations: An Overview

Chapter 1 Introduction 3

Setting the Stage 3
Publicly Owned Treatment Works: Cash Cows or Cash Dogs? 4
The Bottom Line 5
Note 5

Chapter 2 Sick Water 6

Introduction 6
The Paradigm Shift 8
  A Transformational Change 8
Multiple-Barrier Concept 9
  Multiple-Barrier Approach: Wastewater Operations 9
The Bottom Line 10
References 10

Chapter 3 The Challenges 11

Introduction 11
  Compliance with New, Changing, and Existing Regulations 11
  Maintaining Infrastructure 12
  Privatization and/or Re-engineering 13
  Benchmark It! 14
  Upgrading Security 19
  Technical vs. Professional Management 20
  Energy Conservation Measures and Sustainability 20
  Autonomous Operations 22
Chapter Review Questions 23
  Thought-Provoking Questions (Answers will Vary): 23
Note 23
References 23

Chapter 4 Water/Wastewater Operators 24

Introduction 24
Setting the Record Straight 25
The Digital World 25
Plant Operators as Emergency Responders 26
Operator Duties and Working Conditions 27
Operator Certification and Licensure 28
The Bottom Line 30
Chapter Review Questions 30
  Thought-Provoking Question 30
References 31

Chapter 5   Plant Security 32

Introduction 32
Security Hardware/Devices 32
    Physical Asset Monitoring and Control Devices... 32
    Water Monitoring Devices 46
    Communication and Integration.. 51
    Cyber Protection Devices.. 53
SCADA... 54
    What Is SCADA?... 55
    SCADA Applications in Water and Wastewater Systems 56
    SCADA Vulnerabilities 56
    The Increasing Risk... 57
    Adoption of Technologies with Known Vulnerabilities... 57
    Cyber Threats to Control Systems 58
    Securing Control Systems 59
    Steps To Improve SCADA Security 59
The Bottom Line 63
Chapter Review Questions 63
    Thought-Provoking Question 63
References 63

Chapter 6   Water/Wastewater References, Models & Terminology 64

Setting the Stage.. 64
Treatment Process Models 64
Additional Wastewater Treatment Models 65
Reference 66
References 69
Key Terms, Acronyms, Abbreviations in Water/Wastewater Operations 69
    Definitions 69
Acronyms and Abbreviations.. 75
Chapter Review Questions 77
    Part A... 77
    Part B... 77

PART II   Water/Wastewater Operations: Math, Physics & Technical Aspects

Chapter 7   Water/Wastewater Math Operations 81

Introduction 81
Calculation Steps... 82
Equivalents, Formulae, and Symbols.. 82
    Formulae.. 82
Basic Water/Wastewater Math Operations.. 83
    Arithmetic Average (or Arithmetic Mean) and Median.. 83
    Units and Conversions 84
    Force, Pressure, and Head 86
    Flow 88
    Flow Calculations 89
    Detention Time 91
    Hydraulic Detention Time in Hours.. 92
    Chemical Dosage Calculations.. 92
    Percent Removal 95

Population Equivalent (PE) or Unit Loading Factor 95
Specific Gravity 95
Percent Volatile Matter Reduction in Sludge (Biosolids) 95
Chemical Coagulation and Sedimentation 96
Filtration 96
Water Distribution System Calculations 97
Complex Conversions 100
Applied Math Operations 101
Mass Balance and Measuring Plant Performance 101
Mass Balance for Settling Tanks 102
Mass Balance Using BOD Removal 102
Measuring Plant Performance 103
Water Treatment Math Concepts 103
Water Sources and Storage Calculations 103
Water Source Calculations 104
Water Storage Calculations 106
Copper Sulfate Dosing 106
Coagulation and Flocculation Calculations 107
Sedimentation Calculations 112
Filtration Calculations 117
Flow Rate through a Filter (gpm) 118
Filtration Rate 119
Backwash Rate 120
Backwash Rise Rate 121
Required Depth of Backwash Water Tank (ft) 122
Backwash Pumping Rate, gpm 122
Percent Product Water Used for Backwatering 123
Percent Mud Ball Volume 123
Filter Bed Expansion 123
Water Chlorination Calculations 124
Chlorine Disinfection 124
Determining Chlorine Dosage (Feed Rate) 124
Calculating Hypochlorite Solution Feed Rate 127
Calculating Percent Strength of Solutions 128
Calculating Percent Strength Using Dry Hypochlorite 128
Chemical Use Calculations 128
Fluoridation 129
Percent Fluoride Ion in a Compound 131
Fluoride Feed Rate 132
Water Softening 135
Recarbonation Calculation 139
Water Treatment Capacity 140
Treatment Time Calculation (Until Regeneration Required) 141
Salt and Brine Required for Regeneration 141
Wastewater Math Concepts 142
Preliminary Treatment Calculations 142
Screening 142
Grit Removal 143
Primary Treatment Calculations 145
Trickling Filter Process 147
Recirculation Flow 148
Rotating Biological Contactors (RBCs) 148
Soluble BOD 149
Activated Biosolids 150
Chemical Dosing 158
Biosolids Math 165

Biosolids Digestion/Stabilization 168
    Aerobic Digestion Process Control Calculations 168
    Biosolids Dewatering.. 170
    Biosolids to Compost.. 177
Water/Wastewater Laboratory Calculations.. 178
    Faucet Flow Estimation... 178
    Service Line Flushing Time 178
    Composite Sampling Calculation 179
    Biochemical Oxygen Demand Calculations 180
    Moles and Molarity 181
    Moles 181
Settleability (Activated Biosolids)... 182
    Settleable Solids 182
    Biosolids Total Solids, Fixed Solids and Volatile Solids...................................... 183
    Wastewater Suspended Solids and Volatile Suspended Solids.............................. 183
Chapter Review Problems 185
    General Math Operations 185
    General Wastewater Treatment Problems 186
Back to the Beginning... 195
The Bottom Line 195
References 195

Chapter 8   Water Hydraulics 196

Introduction 196
What Is Water Hydraulics?.. 196
Basic Concepts 196
    Stevin's Law... 197
Density and Specific Gravity 197
Force and Pressure 198
    Hydrostatic Pressure 199
    Effects of Water under Pressure 200
Head 200
    Static Head... 201
    Friction Head 201
    Velocity Head 201
    Total Dynamic Head (Total System Head).. 201
    Pressure and Head 201
    Head and Pressure 201
Flow and Discharge Rates: Water in Motion 201
    Area and Velocity 202
    Pressure and Velocity 203
Piezometric Surface and Bernoulli's Theorem 203
    Conservation of Energy 203
    Energy Head 203
    Piezometric Surface... 203
    Head Loss 204
    Hydraulic Grade Line (HGL) 204
    Bernoulli's Theorem 205
    Bernoulli's Equation 205
Well and Wet Well Hydraulics.. 206
    Well Hydraulics 207
    Wet Well Hydraulics.. 207
Friction Head Loss 207
    Flow in Pipelines 208
    Major Head Loss 209
    Calculating Major Head Loss 209
    C Factor 210

Slope 210
Minor Head Loss 210
Basic Piping Hydraulics..210
Piping Networks 211
Open-Channel Flow 212
Characteristics of Open Channel Flow 212
Parameters Used in Open-Channel Flow 212
Flow Measurement 215
Flow Measurement: The Old-Fashioned Way 215
Basis of Traditional Flow Measurement...216
Flow Measuring Devices..216
Operating Principle 216
Types of Differential Pressure Flowmeters..217
Magnetic Flowmeters 218
Ultrasonic Flowmeters 219
Velocity Flowmeters 219
Open-Channel Flow Measurement..220
The Bottom Line 223
Chapter Review Questions 223
References 223

Chapter 9   Hydraulic Machines: Pumps 225

Introduction 225
Basic Pumping Calculations...226
Velocity of a Fluid through a Pipeline..226
Pressure-Velocity Relationship..226
Static Head..227
Static Suction Head 227
Static Suction Lift...227
Static Discharge Head 227
Friction Head 228
Velocity Head 228
Total Head 229
Conversation of Pressure Head..229
Horsepower...229
Specific Speed 230
Affinity Laws—Centrifugal Pumps..230
Net Positive Suction Head (NPSH) 231
Pumps in Series and Parallel 233
Centrifugal Pumps 233
Description 233
Terminology 234
Pump Theory...235
Pump Characteristics..235
Advantages and Disadvantages of the Centrifugal Pump...236
Centrifugal Pump Applications 237
Pump Control Systems 237
Electrode Control Systems 239
Other Control Systems 239
Centrifugal Pump Modifications 241
Turbine Pumps..242
Positive Displacement Pumps 243
Piston Pump or Reciprocating Pump 243
Diaphragm Pump 243
Peristalic Pumps 243
Chapter Review Questions 244
References 244

Chapter 10 Water/Wastewater Conveyance 245

    Delivering the Lifeblood of Civilization 245
    Conveyance Systems 245
    Definitions 246
    Fluids vs. Liquids 248
    Let's Talk About Pipe 248
    Maintaining Fluid Flow in Piping Systems 248
        Scaling 249
        Piping System Maintenance 249
    Piping System Accessories 250
    Piping System: Temperature Effects and Insulation 251
    Metallic Piping 251
        Piping Materials 251
        Piping: The Basics 251
        Types of Piping Systems 253
        Metallic Piping Materials 253
        Characteristics of Metallic Materials 253
        Maintenance Characteristics of Metallic Piping 254
        Joining Metallic Pipe 256
    Non-metallic Piping 257
        Nonmetallic Piping Materials 258
        Tubing 261
        Tubing vs. Piping: The Difference 261
        Advantages of Tubing 263
        Mechanical Advantages of Tubing 263
        Connecting Tubing 263
        Bending Tubing 265
        Types of Tubing 265
        Typical Tubing Applications 266
    Industrial Hoses 266
        Hose Nomenclature 266
        Factors Governing Hose Selection 267
        Standards, Codes, and Sizes 268
        Hose Couplings 270
        Hose Maintenance 271
    Pipe & Tube Fittings 271
        Fittings 271
        Functions of Fittings 272
        Types of Connection 273
        Tubing Fittings and Connections 273
    Valves 274
        Valve Construction 275
        Types of Valves 276
        Valve Operators 279
        Valve Maintenance 279
    Piping Systems: Protective Devices 279
        Applications 279
        Strainers 280
        Filters 280
        Traps 280
    Piping Ancillaries 282
        Gauges 282
        Pressure Gauges 283
        Temperature Gauges 284
        Vacuum Breakers 285
        Accumulators 285

Air Receivers 285
Heat Exchangers 286
Chapter Review Questions 286
References 287

PART III Characteristics of Water

Chapter 11 Basic Water Chemistry 291
Introduction 291
Chemistry Concepts and Definitions 291
Concepts 291
Definitions 292
Chemistry Fundamentals 293
Matter 293
Compound Substances 295
Water Solutions 296
Water Quality Constituents 296
Solids 297
Turbidity 297
Color 297
Dissolved Oxygen (DO) 297
Metals 298
Organic Matter 298
Inorganic Matter 298
Acids 299
Bases 299
Salts 299
pH 300
Common Water Measurements 301
Alkalinity 301
Water Temperature 301
Specific Conductance 301
Hardness 301
Odor Control (Wastewater Treatment) 302
Water Treatment Chemicals 302
Disinfection 302
Coagulation 302
Taste and Odor Removal 303
Water Softening 303
Chemical Precipitation 303
Ion Exchange Softening 303
Recarbonation 304
Scale and Corrosion Control 304
Drinking Water Parameters: Chemical 304
Organics 304
Synthetic Organic Chemicals (SOCs) 305
Volatile Organic Chemicals (VOCs) 305
Total Dissolved Solids (TDS) 305
Fluorides 306
Heavy Metals 306
Nutrients 306
The Bottom Line 307
Chapter Review Questions 307
Note 308
References 308

Chapter 12 Water Microbiology 309

  Introduction 309
  Microbiology: What Is It? 309
  Water/Wastewater Microorganisms 309
  Key Terms and Definitions..310
  Microorganism Classification and Differentiation..310
    Classification 310
    Differentiation 312
  The Cell 312
    Structure of the Bacterial Cell..312
    Plasma Membrane (Cytoplasmic Membrane) 313
  Bacteria..313
    Bacterial Growth Factors...314
    Destruction of Bacteria 315
    Waterborne Bacteria 315
  Protozoa...315
  Microscopic Crustaceans 316
  Viruses 317
  Algae 317
  Fungi..318
  Nematodes and Flatworms (Worms)...318
  Water Treatment and Microbiological Processes 319
    Pathogenic Protozoa 320
    Giardia..320
    Cryptosporidium...324
    Cyclospora 327
    Helminths 328
  Wastewater Treatment and Biological Processes 328
    Aerobic Process 328
    Anaerobic Process 328
    Anoxic Process 329
    Photosynthesis 329
    Growth Cycles 329
  Biogeochemical Cycles 329
    Carbon Cycle 329
    Nitrogen Cycle 330
    Sulfur Cycle 331
    Phosphorus Cycle 332
  Chapter Review Questions 332
  References 333

Chapter 13 Water Ecology 334

  Introduction 334
  What Is Ecology? 334
  Why Is Ecology Important? 335
  Why Study Ecology?..335
    Leaf Processing in Streams 336
  History of Ecology 337
    Example Ecosystem: Agroecosystem Model 338
  Levels of Organization...339
  Ecosystems 339
  Energy Flow in the Ecosystem..340
  Food Chain Efficiency...341
  Ecological Pyramids 342
  Productivity 342

Population Ecology 343
Stream Genesis and Structure 346
   Water-Flow in a Stream 347
   Stream Water Discharge 347
   Transport of Material 348
   Characteristics of Stream Channels 349
   Stream Profiles 349
   Sinuosity 349
   Bars, Riffles, and Pools 350
   The Floodplain 350
   Adaptation to Stream Current 352
   Types of Adaptive Changes 353
   Specific Adaptations 353
Benthic Life 354
   Benthic Plants and Animals 354
Benthic Macroinvertebrates 354
   Identification of Benthic Macroinvertebrates 355
   Macroinvertebrates and the Food Web 356
   Benthic Macroinvertebrates in Running Wasters 356
   Macroinvertebrate Glossary 356
   Insect Macroinvertebrates 357
   Non-insect Macroinvertebrates 364
Summary of Key Terms 365
The Bottom Line 365
Chapter Review Questions 365
References 365

Chapter 14 Water Quality 367

Introduction 367
The Water Cycle 367
Water Quality Standards 369
   Clean Water Act 369
   Safe Drinking Water Act 370
Water Quality Characteristics of Water and Wastewater 372
   Physical Characteristics of Water/Wastewater 372
   Chemical Characteristics of Water 375
   Chemical Characteristics of Wastewater 379
   Biological Characteristics of Water and Wastewater 381
The Bottom Line 382
Chapter Review Questions 382
Note 382
References 382

Chapter 15 Biomonitoring, Monitoring, Sampling, and Testing 383

Introduction 383
What Is Biomonitoring? 383
   Advantages of Using Periphyton 383
   Advantages of Using Fish 383
   Advantages of Using Macroinvertebrates 384
Periphyton Protocols 384
Fish Protocols 385
Macroinvertebrate Protocols 385
   Determining Incremental Change in Aquatic Ecosystems 385
   The Biotic Index 387
   Metrics within the Benthic Macroinvertebrates 388

Biological Condition Gradient...388
Biological Sampling in Streams...388
Biological Sampling: Planning..389
Sampling Stations..390
Sampling Frequency and Notes...391
Macroinvertebrate Sampling Equipment..391
Macroinvertebrate Sampling in Rocky-Bottom Streams 392
Macroinvertebrate Sampling in Muddy-Bottom Streams...396
Muddy-Bottom Stream Habitat Assessment 398
Post-Sampling Routine..399
Sampling Devices..399
Secchi Disk...401
The Bottom Line on Biological Sampling...402
Drinking Water Quality Monitoring..403
Is the Water Good or Bad?..403
State Water Quality Standards Programs...404
Designing a Water Quality Monitoring Program...404
General Preparation and Sampling Considerations 405
Cleaning Procedures 405
Method A: General Preparation of Sampling Containers 405
Method B: Acid Wash Procedures 405
Sample Types...405
Collecting Samples from a Stream..405
Sample Preservation and Storage..406
Standardization of Methods..407
Test Methods for Drinking Water & Wastewater..407
Titrimetric Methods 407
Colorimetric Methods 407
Visual Methods...407
Electronic Methods...407
Dissolved Oxygen Testing 407
Biochemical Oxygen Demand Testing 411
Temperature Measurement... 413
Hardness Measurement 413
pH Measurement 413
Turbidity Measurement.. 414
Orthophosphate Measurement... 415
Nitrates Measurement.. 417
Solids Measurement 418
Conductivity Testing... 421
Total Alkalinity 422
Fecal Coliform Bacteria Testing... 422
Apparent Color Testing/Analysis... 428
Odor Analysis of Water 429
Chlorine Residual Testing/Analysis 430
Fluorides 431
The Bottom Line 431
Chapter Review Questions 432
References 432

PART IV Water and Water Treatment

Chapter 16 Potable Water Source 435

    Introduction 435
        Key Terms and Definitions 435
    Hydrologic Cycle 436
    Sources of Water 436
    Surface Water 437
        Advantages and Disadvantages of Surface Water 437
        Surface Water Hydrology 437
        Raw Water Storage 438
        Surface Water Intakes 438
        Surface Water Screens 439
        Surface Water Quality 439
    Groundwater..439
        Groundwater Quality 441
    Groundwater under the Direct Influence of Surface Water 441
    Surface Water Quality and Treatment Requirements 441
        Stage 1 D/DBP Rule 442
        Interim Enhanced Surface Water Treatment (IESWT) Rule 442
        Regulatory Deadlines 442
    Public Water System Quality Requirements 442
    Well Systems 442
        Well Site Requirements 443
        Type of Wells 443
        Well Components 444
        Well Evaluation 445
        Well Pumps..446
        Routine Operation and Recordkeeping Requirements.....................................446
        Well Maintenance 447
        Troubleshooting Well Problems 447
        Well Abandonment 447
    The Bottom Line 448
    Chapter Review Questions 448

Chapter 17 Water Treatment Operations 449

    Introduction 449
    Waterworks Operator
    (a.k.a. The Fluid Mechanic) 449
    Purpose of Water Treatment..450
    Stages of Water Treatment 450
    Pretreatment 451
        Aeration 451
        Screening...451
        Chemical Addition...452
        Chemical Feeders 453
        Iron and Manganese Removal 455
        Hardness Treatment...456
        Corrosion Control..457

Coagulation 460
Mixing and Flocculation 462
Sedimentation..462
Filtration...463
    Types of Filter Technologies..463
    Common Filter Problems 465
    Filtration and Compliance with Turbidity Requirements..466
Disinfection 471
    Need for Disinfection in Water Treatment 472
    Pathogens of Primary Concern 475
    Recent Waterborne Disease Outbreaks 476
    Mechanism of Pathogen Inactivation 477
    Other Uses of Disinfectants in Water Treatment...477
    Types of Disinfection Byproducts and Disinfection Residuals 480
    Disinfection Byproduct Formation..480
    Disinfection Strategy Selection 484
    CT Factor...484
    Disinfectant Residual Regulatory Requirements 484
    Summary of Current National Disinfection Practices 485
Summary of Methods of Disinfection...485
    Chlorination...485
    Hypochlorination...489
Arsenic and Other Removal from Drinking Water...497
    Arsenic Exposure 497
    Arsenic Removal Technologies 498
The Bottom Line 503
Chapter Review Questions 503
Note 505
References 505

PART V   Wastewater and Wastewater Treatment

Chapter 18 Wastewater Treatment Operations 511

Introduction 511
Wastewater Operators (aka Fluid Mechanics)...511
Wastewater Treatment Process: The Model 511
Wastewater Terminology and Definitions 511
Measuring Plant Performance 514
    Plant Performance/Efficiency...514
Unit Process Performance and Efficiency...515
Percent Volatile Matter Reduction in Sludge 515
    Hydraulic Detention Time 515
Wastewater Sources and Characteristics...516
    Wastewater Sources...516
    Wastewater Characteristics...517
Wastewater Collection Systems...518
    Gravity Collection System...519
    Force Main Collection System 519
    Vacuum System 519
    Pumping Stations...519
    Wet-Well Pumping Stations...519
    Pumping Station Wet Well Calculations 519
Preliminary Treatment 520
    Screening...520
    Shredding...522

Grit Removal 522
Preaeration 525
Chemical Addition 525
Equalization 525
Aerated Systems 525
Cyclone Degritter 525
Preliminary Treatment Sampling and Testing 525
Other Preliminary Treatment Process Control Calculations 526
Primary Treatment (Sedimentation) 527
Process Description 527
Types of Sedimentation Tanks 528
Operator Observations, Process Problems & Troubleshooting 528
Process Control Calculations 529
Problem Analysis 531
Effluent from Settling Tanks 531
Secondary Treatment 531
Treatment Ponds 532
Troubleshooting Wastewater Ponds 541
Trickling Filters 541
Rotating Biological Contractors (RBCs) 554
Activated Sludge 557
Activated Sludge Terminology 558
Activated Sludge Process: Equipment 560
Overview of Activated Sludge Process 560
Factors Affecting Operation of the Activated Sludge Process 560
Growth Curve 561
Activated Sludge Formation 561
Activated Sludge: Performance-Controlling Factors 561
Activated Sludge Modifications 562
Activated Sludge Process Control Parameters 565
Activated Sludge Operational Control Levels 566
Visual Indicators for Influent or Aeration Tank 567
Final Settling Tank (Clarifier) Observation 568
Process Control Testing and Sampling 568
Symptom 1 573
Symptom 2 573
Symptom 3 573
Symptom 4 574
Symptom 5 574
Symptom 6 574
Symptom 7 574
Symptom 8 574
Symptom 9 574
Symptom 10 574
Solids Concentration: Secondary Clarifier 581
Activated Sludge Process Recordkeeping Requirements 581
Disinfection of Wastewater 582
Chlorine Disinfection 582
Symptom 1 584
Symptom 2 584
Symptom 3 585
Symptom 4 585
Symptom 5 585
Symptom 6 585
Symptom 7 585
Symptom 8 585
Symptom 9 585

Symptom 10 585
Symptom 11 585
Symptom 12 586
Symptom 13 586
Work: Chemical Handling—Chlorine 586
Practice 586
Ultraviolet Irradiation 589
Ozonation 592
Bromine Chloride 593
No Disinfection 593
Advanced Wastewater Treatment 593
Chemical Treatment 594
Operation, Observation, and Troubleshooting Procedures 594
Microscreening 595
Filtration 595
Membrane Bioreactors (MBRs) 597
Biological Nitrification 597
Biological Denitrification 597
Carbon Adsorption 598
Land Application 599
Biological Nutrient Removal (BNR) 601
Enhanced Biological Nutrient Removal (EBNR) 605
0.5 MGD Capacity Plant 605
1.5-MGD Capacity Plant 605
1.55-MGD Capacity Plant 605
2 MGD Capacity Plant 606
2.6-MGD Capacity Plant 606
3-MGD Capacity Plant 606
4.8-MGD Capacity Plant 606
5-MGD Capacity Plant 606
24-MGD Capacity Plant 606
39-MGD Capacity Plant 607
42-MGD Capacity Plant 607
54-MGD Capacity Plant 607
67-MGD Capacity Plant 607
Solids (Sludge/Biosolids) Handling 607
Sludge: Background Information 608
Sources of Sludge 608
Sludge Characteristics 608
Sludge Pumping Calculations 610
Sludge Thickening 612
Operational Observations, Problems, and Troubleshooting Procedures 613
Indicators of Poor Process Performance 613
Dissolved Air Flotation Thickener 614
Sludge Stabilization 615
Step 1: Calculate Volatile Matter 616
Step 2: Calculate Moisture Reduction 616
Symptom 1 618
Symptom 2 618
Rotary Vacuum Filtration 635
Pressure Filtration 637
Centrifugation 639
Sludge Incineration 640
Operational Problems 641
Land Application of Biosolids 643
Process Control: Sampling and Testing 643
Process Control Calculations 643

Disposal Cost 643

Plant Available Nitrogen (PAN) 643

Application Rate Based on Crop Nitrogen Requirement 644

Metals Loading 644

Maximum Allowable Applications Based upon Metals Loading 644

Site Life Based on Metals Loading 645

Permits, Records, and Reports 645

Definitions 645

NPDES Permits 646

Sampling and Testing 646

Reporting Calculations 647

Chapter Review Questions 648

References 650

Chapter 19 Practice Examination 653

Exam 653

Appendix A: Answers to Chapter Review Questions/Problems 667

Appendix B: Formulae 685

Index 687

Foreword

It is very difficult to make an accurate prediction, especially about the future.

—Niels Bohr

A rational person would have difficulty arguing against Bohr's view concerning the difficulty of making accurate predictions. It must be said, however, that rational people are also capable of recognizing that, in a few instances, there are exceptions to every rule. More specifically, as a result of certain "actions" that occur, knowledgeable and observant individuals can make fairly accurate predictions about the potential future consequences of such actions. As a case in point, consider the actions of humans that pollute the air we breathe, the water we drink, and the land we live on and gain our sustenance from. It is probably safe to predict that when contamination and destruction of our life-sustaining environment occur daily, certain fairly accurate predictions can be made about the future consequences. Through the observance and/or awareness of certain "actions," a sense of foreboding concerning the dire consequences of humans destroying the environment motivated a group of concerned and enlightened individuals to organize the first Earth Day celebration in 1970.

The organizers of the first Earth Day celebration were concerned about certain "actions" of humans. To a degree, the organizers' concern was driven by obvious "predictors" of a quality of life anticipated for the future that was not promising. Some of these organizers and other concerned individuals made dire predictions based on what they had observed, what they had read, or what they had heard. For example, they might have observed or learned about certain rivers within the United States that were so oil soaked that they actually burned. Moreover, others had observed, heard, or read about skies above great metropolitan areas that were red with soot. Others had breathed air that they could actually see. Still, others had observed lakes choked with algae; lakes that were dying. Then, there were those mountains they had observed, read about, or heard about. These were mountains unlike the Alps or Rocky Mountains, however. Instead, these were mountains of trash, garbage, refuse, discarded materials, and other waste products. As with all things that disgust humans, the same fate awaited these mountains of waste and filth, that is, they became unbearable for humans to live with and were torn down. After teardown, the entire unsightly, stinking mess was deposited into rivers, lakes, streams, oceans, or landfills. Sometimes, these mountains of unwanted waste were torn down and piled up again on barges that were towed from port to port with no place to land; no one wanted these floating mountains of waste: Not In My Backyard (NIMBY). All these observations, of course, were indicators of what was occurring environmentally in the here and now; moreover, they were "actions" portending what was in store for the inhabitants of Earth and for the future generations to come. Thus, these present indicators of environmental problems became reliable predictors of greater environmental problems ahead in the future.

Along with the organizers and participants of that first Earth Day celebration in 1970, other citizens were concerned about their futures and the futures of their loved ones. Although Niels Bohr was correct in his statement about the difficulty in making accurate predictions, especially about those in the future, in 1970 it was clear to many concerned individuals that if corrective actions were not quickly taken to protect and preserve the Earth's environment, then there would be no need to worry about making future predictions, that is, there would be no future to predict.

To say that we face huge environmental challenges today, as was the case in 1970, is to make an accurate statement. While it is true that since that first Earth Day celebration in 1970 progress has been made in restoring the Earth's environment, it is also true that there is still a long way to go before the "predictors" or "indicators" of the future consequences of the ongoing damage to the Earth's environment are less salient than they are today. It should come as no surprise to anyone that it can be said with a great deal of accuracy that the quality of life here on Earth is directly connected to our "actions."

It should be pointed out that not all the news concerning human-made waste and its disposal is of the doom-and-gloom variety. For example, it is noteworthy to consider the steps that have been taken in recent years to clean up our air and our lakes and to properly dispose of our waste using earth-friendly disposal techniques. One such clean-up step, the cleaning of the water we drink and dispose of, is described in this text.

Note: This work, as with all the other author's works, is written in the author's trademark conversational style; this is simply the case to ensure that there is no failure to communicate. Because the author has been involved with writing examinations for water/wastewater licensure, it is my recommendation that operators and prospective operators take training courses before exams. These courses cover a broad range of knowledge relevant to the specific exam level. This book is often used as a textbook for training. If you work hard at learning the concepts in this text, you will be better prepared to sit for licensure exams.

Preface to fifth edition

This proven operational and study guide has been hailed as the flagship, best-selling work on water and wastewater treatment operations. It is used by plant personnel and administrators as a general information source and troubleshooting guide and as a study aid for operator licensure certification examinations. This fifth edition has been condensed to emphasize important content and updated based on recommendations from users of the first four volumes. This updated volume includes virtually all of the information needed by plant personnel and administrators of water and wastewater operations and conveyance operations, as well as preparation for operator licensure examinations. As in the previous editions, the new fifth edition includes the addition of the following:

- Updates on current issues facing the water and wastewater treatment industries
- Math operations with solutions
- Expanded water operations sample questions—255 multiple-choice sample questions have been added (again, as recommended by previous users of earlier editions of this text)
- Discussion of oxidation ditches
- Discussion of BNR processes
- Discussion of water quality reports
- Update on energy conservation measures with applicable case studies
- Discussion of membrane bioreactors (MBR)
- Discussion of variable frequency controls
- Expanded discussion of ultraviolet disinfection
- Water reuse

Wastewater and water treatment are established environmental technologies that, for the most part, have been somewhat static in technique and operation—that is, water in, water treated, and water out. The treated water out can be and is often used for several different purposes—many of which are talked about in this edition. For example, since the advent of the computer age, "technology it's a changin" … this is the new anthem of change for the times, and this includes water and wastewater operations. Moreover, trends in the science of water and wastewater treatment have shifted for other reasons. The factors driving this shift are incorporated into this new edition; they include the following:

- **Reclaimed Water**: Recycling water can be less expensive, less energy-intensive, and better for the environment than other water treatment methods, such as desalinization.
- **Advanced Filtration and Disinfection Technologies:** Newer filtration systems can filter more complex contaminants, and chlorine is being replaced by ozone, UV, and activated carbon technologies.
- **Smart Meters:** They are being installed at an increasing rate to conserve water.
- **The Internet of Things (IoT):** This technology is a network of computers and devices that collect and share large amounts of data. It provides real-time data on the state of systems and processes (e.g., it indicates leaks in systems and quality parameters in water—and so much more). IoT is technology on the move; it is a transformative technology. The incorporation of IoT into water and wastewater treatment is concerned with predicting water quality. Overall, IoT contributes to efficiency, accuracy, improving safety, and overall unit process management. IoT's integration with artificial intelligence (AI), machine learning (ML), and SCADA systems enhances overall performance and sustainability.
- **Rapid Infiltration Basins (RIBs) (aka Soil Aquifer Treatment):** These are earthen basins designed to promote rapid infiltration and dispersal of treated effluent into the subsurface.
- **Zero Liquid Discharge (ZLD):** This system is capable of eliminating water discharge to local watersheds.

Note: Water/Wastewater Operators by Another Name: After working with and studying water and wastewater operations for decades, I have come to totally respect water/wastewater managers and licensed operators for their significant knowledge and ability to properly manage, operate, and maintain their facilities. What these operators do is manage the fluids: water and wastewater. Even though it is true that water and wastewater are fluids, it is also true that gas phases and plasma are fluids. However, it is the one liquid part of fluids—water and wastewater—that is the focus of this work.

What does all this mean?

Good question.

I have come to realize that licensed water and wastewater managers and operators are actually licensed fluid mechanics. Now, it is true that they are not referred to as fluid mechanics, but in my mind they are.

Consider that a typical fluid water mechanic's job description may be listed and briefly stated as follows:

> Under broad guidance, the plant fluid mechanic is responsible for operating, maintaining, controlling, adjusting, and repairing waterworks and wastewater unit processes/equipment under general supervision. Performing other assignments as required.

The Bottom Line: In the author's opinion, fluid mechanics are polymathic—Jack and Jills of all trades and masters of ONE.

Okay, moving on with this new edition. As with the earlier editions, the fifth edition is designed to assist utility administrators, managers and directors, lawyers, water and wastewater plant managers, plant operators, conveyance personnel, and maintenance operators to oversee, manage, gain knowledge of, and/or to operate their treatment works in a successful and compliant manner. In this edition, several key point notations are included. An example of a key point is:

Key Point: 1 HP = 33,000 ft-b/min and is also expressed as 1 HP = 746 Watts

These key points are reinforcers for imprinting key parameters and other information into the user's memory bank; they also highlight essential information for preparing for licensure exams. Again, this new edition (as was the purpose of the other editions) is designed to aid and assist all personnel preparing for all levels of water and wastewater operator licensure, and as an examination aid, the final chapter provides a 255-question practice examination that may assist operators in passing licensure examinations.

The Bottom Line: State licensure water and wastewater treatment examinations are designed to evaluate the ability of an examinee to operate a water/wastewater treatment facility throughout the United States—if properly used, this guide will aid in proper preparation to sit for and to score high on state examinations.

Frank R. Spellman
Norfolk, VA.

TO THE READER

In reading this text, you are going to spend some time following water on its travels.
Even after being held in bondage, sometimes for eons, eventually water moves.
Do you have any idea where this water has been?
Where is this water going?
What changes it has undergone, during all the long ages water has lain on and under the face of the Earth?
Sometimes we can look at this water…analyze this water…test this water to find out where it has been.
Water, because it is the universal solvent, tends to pick up materials through which it flows.
When this happens, we must sometimes treat the water before we consume it.
Whether this is the case or not, water continues its endless cycle.
And for us this is the best of news.
So, again, do you have any idea where water has been?
More importantly, where the water is going?

—*F.R. Spellman, 2003*

If we could first know where we are and wither we are tending, we could better judge what we do and how to do it…

—*Abraham Lincoln*

Author

Frank R. Spellman is a retired U.S. Naval Officer with 26 years of active duty, a retired Environmental Safety & Health Manager for a large Wastewater Sanitation District in Virginia, and a retired Assistant Professor of Environmental Health at Old Dominion University, Norfolk, VA. The author/co-author of more than 160 books, Spellman consults on environmental matters with the U.S. Department of Justice and various law firms and environmental entities throughout the globe. Spellman holds a BA in Public Administration, a BS in Business Management, an MBA, and an MS/PhD in Environmental Engineering. In 2011/2012, he traced and documented the ancient water distribution system at Machu Picchu, Peru, and surveyed several drinking water resources in Coco and Amazonia, Ecuador. He also studied and surveyed two separate potable water supplies in the Galapagos Islands.

Photo of the author in Upper Amazon Jungle, Ecuador (2013).

Part I

Water and Wastewater Operations

An Overview

1 Introduction

SETTING THE STAGE

For many years, I conducted training and investigations in environmental health, water/wastewater treatment, safety, and accident/fatality investigations for the U.S. Department of Justice. I also performed pre-OSHA audits and, upon request, pre-EPA inspections (audits) at water and wastewater treatment plants. In actual classroom training sessions, I spent years teaching upper-level undergraduates and graduate students in environmental health, water and wastewater treatment operations, plant safety, and environmental courses in nursing programs at Old Dominion University. I also taught short courses in water and wastewater treatment at Virginia Tech for more than 20 years. During these activities, I often wondered about the answers to one of my recurring questions: How do the environmental health students, water and wastewater professionals, medical students, administrators, and regulators I taught or whose facilities I inspected judge or evaluate their professions as compared to other important occupations and operations?

To answer this question, I decided to conduct a non-scientific survey (a sample) to query the respondents, that is, to find out how they ranked their chosen professions compared to other professions.

So, what I did was develop a survey form, listing what I felt—and still feel—are the most important occupations in the United States at present. Now, I admit that the following occupations and their priority, importance, significance, status, and standing were/are personally biased (I still use the survey form shown in Exhibit A), that is, the results are absolutely based on my views of the most important occupations of the past and today (2005–2024). I ranked water/wastewater and sanitation occupations/activities at the top of my survey form.

I get thirsty for fresh, clean, safe drinking water. Don't you?

Exhibit A lists the author's choices in order (again, a personally biased view, of course) of the important occupations that need to be included in any such survey form.

I traditionally administered this blank selection list to both the water/wastewater short course attendees at Virginia Tech (1989–2006) and my Environmental Health students at Old Dominion University (1999–2009).

Exhibit B lists the rankings by the Short Course Attendees at Virginia Tech during the 2009 session.

Exhibit C lists the rankings by Old Dominion University Environmental Health students during the Fall semester of 2009.

EXHIBIT B

10 Most Important Occupations—Selected by Virginia Tech Short Course Attendees (153 Total Respondents)

Occupations	Selected Ranking
Water/Wastewater	97
Sanitation workers	50
Farmers, ranchers, farm laborers	4
Nurse/Home care professional	
Teachers	
Firefighters	
Transportation professions	
Lawmakers	
Construction workers	
Telecommunications workers–techies	
None of the above	2

EXHIBIT C

10 Most Important Occupations—Selected by Old Dominion University Students (111 Total Respondents)

Occupations	Selected Ranking
Water/Wastewater	33
Sanitation workers	28
Farmers, ranchers, farm laborers	4
Nurse/Home care professional	11
Teachers	10
Firefighters	
Transportation professions	
Lawmakers	
Construction workers	
Telecommunications workers–techies	19
None of the above	6

EXHIBIT A

10 Most Important Occupations (Ranked 1 to 10 in order of Importance) Author's Rankings

Occupations	Selected Ranking
Water/Wastewater	1
Sanitation workers	2
Farmers, ranchers, farm laborers	3
Nurse/Home care professional	4
Firefighters	5
Transportation professions	6
Construction workers	7
Teachers	8
Lawmakers	9
Telecommunications workers–techies	10

DOI: 10.1201/9781003581901-2

Okay, what do the exhibits show, illustrate, and disclose about the participants' opinions? Let's start with Exhibit A—the author's selections. As the author, I chose water/wastewater and sanitation as my top and two personal choices for the most important occupations. My selections are biased, of course, because as an advocate for both water/wastewater treatment personnel and sanitation workers, I believe they occupy the most important positions. I based my choices on the need for clean, safe, abundant, and readily available water and sanitary procedures (i.e., proper collection and disposal of waste) to maintain a healthy environment. We need to tackle waste everywhere (this is especially the case with plastic waste and, even more importantly, with microplastics in drinking water).

Now, looking at Exhibit B, there was only one surprise in the respondents' rankings of the most important occupations. Almost the entire group I surveyed at Virginia Tech were operators, managers, directors, superintendents, and regulators in or related to water/wastewater operations. Thus, it is no surprise that many classify water/wastewater and sanitation activities as their top choices. The only surprise to me was the choice made by two people who attended my training sessions who did not feel that any of their choices were listed on the exhibit. I still wonder what they considered the most important occupation—a head-scratcher, for sure.

The results shown in Exhibit C, except for the six who indicated "none of the above," did not surprise me. My ODU classes consisted of students from various backgrounds: active-duty military, former military, nurses, city/state/federal regulators, foreign exchange students, and other students interested in the subject matter and the potential careers available to them upon graduation with degrees (BS, MS, PhD candidates) in environmental health, nursing, or various environmental occupations (knowing that their degree would give them close to 100% employment placement). Moreover, in this particular major, there was no worry about filling my classrooms—standing room only in most cases (don't tell the Fire Marshal).

Key Point: A cross-connection is a point in a plumbing system where a nonpotable substance can come into contact with the potable drinking water supply (BMI, 1999).[1]

PUBLICLY OWNED TREATMENT WORKS: CASH COWS OR CASH DOGS?

Water treatment operations are commonly classified as cash cows and wastewater treatment operations are cash dogs. Water and wastewater treatment facilities are usually owned, operated, and managed by the community (the municipality) where they are located. The increasing use of reclaimed/reused water has turned wastewater treatment plants from cash drains to cash cows—industries and municipalities are now paying for reclaimed/recycled/reused water. While some treatment facilities are privately owned, the majority of Water Treatment Plants (WTPs)

and Wastewater Treatment Plants (WWTPs) are POTWs—Publicly Owned Treatment Works (i.e., owned by local government agencies).

These publicly owned facilities are managed and operated on-site by professionals in the field. On-site management, however, is usually controlled (overseen) by a board of elected, appointed, or hired directors/commissioners, who set policy, determine budgets, plan for expansion or upgrades, hold decision-making power for large purchases, set rates for ratepayers, and, in general, control the overall direction of the operation.

When final decisions on matters that affect plant performance are in the hands of, for example, a board of directors comprised of elected and appointed city officials, their knowledge of the science, engineering, and hands-on problems that on-site staff must solve can range from "everything" to "none." Matters that are of critical importance to on-site management may mean little to the Board. The Board of Directors may also be responsible for other city services and have an agenda (politics) that encompasses more than just the water or wastewater facility. Thus, decisions that affect on-site management can be affected by political and financial concerns that have little to do with the successful operation of a WTP or POTW.

Need to Know Operational/Exam Topics: In water/wastewater operations, possessing knowledge in organizational structure, financial management, planning, reporting, public relations, communications, vulnerability/security, and contingency plans is essential for managing the plant site and sitting for various state licensure examinations.

Finances and funding are always of concern, no matter how small or large, well-supported or under-funded the municipality. Publicly owned treatment works are generally funded from a combination of sources. These include local taxes, state, and federal monies (including grants and matching funds for upgrades), as well as usage fees for water and wastewater customers. In smaller communities, in fact, the water/wastewater plants may be the only city services that generate income. This is especially true in water treatment and delivery, which are commonly viewed as the "cash cow" of city services. As a cash cow, the water treatment works generate cash in excess of the amount needed to maintain the treatment works. These treatment works are "milked" continuously with as little investment as possible. Funds generated by the facility do not always stay with the facility. Funds can be reassigned to support other city services—and when facility upgrade time comes, funding for renovations can be problematic. On the other end of the spectrum, spent water (wastewater), treated in a POTW, is often regarded as one of the "cash dogs" of city services. Typically, these units generate only enough money to sustain operations. This is the case, of course, because managers and oversight boards or commissions are fearful, for political reasons, of charging ratepayers "too much" for treatment services. Some progress has been made, however, in marketing and selling treated wastewater for reuse in industrial cooling applications and some irrigation projects. Also, pipe-to-pipe connections for potable reuse of treated wastewater to drinking water quality are becoming more prominent in many areas where clean, safe water accessibility is declining; thus, in these regions, the reused purified water is made available at a cost. Moreover, wastewater solids have been reused as soil amendments; also in addition, ash from incinerated biosolids has been used as a major ingredient in forming cement revetment blocks used in areas susceptible to heavy erosion from river and sea inlets and outlets.

THE BOTTOM LINE

I deliberately included this section in this new edition of the book, which has become the standard, popular reference for water/wastewater personnel, for several reasons, including preparing for State Licensure examinations for water/wastewater operators and plant managers. As mentioned, for a long time, those involved in water/wastewater viewed drinking water operations as a cash cow—the wastewater utility can and does charge a fee for a safe water supply. On the other hand, in many locations, wastewater treatment was (and still is) classed as a money-down toilet operation—a cash dog. In the past, wastewater went down the drain, down the toilet, down the sink, was dumped in the outhouse, in the backyard, or some places in the street. However, wastewater-treated effluent and solids can be turned into that "green stuff" (no, not trees and/or flowers); instead, into the almighty dollar. This book explains how "down the toilet" is out of flushes, so to speak, and how it is now or how it can be turned "green" in many different ways.

The Real Bottom Line: I have told my latest group of wastewater students and lecture attendees that wastewater treatment plant operators are no longer classified as "sewer rats" but instead as trained, licensed professional fluid mechanics who work in a critical field.

NOTE

1 BMI. Backflow Management Incorporated. (1999). Safe Drinking Water For Everyone Through an Active Cross-Connection Control Program. Portland, OR.

2 Sick Water

INTRODUCTION

Water and wastewater operations are hidden functions. Treatment plants are usually isolated; the piping systems are typically buried underground; the customers and rate-payers hardly think about either—that is, until the rude awakening: until the tap runs dry, the toilet backs up, or the bill is received in the mail or online. Although not often thought of as a commodity (or, for that matter, not thought about at all), water is a commodity—a very valuable and absolutely vital commodity. And even though it is valuable and absolutely vital, we consume the water, waste it, discard it, pollute it, poison it, flush it, and relentlessly modify the hydrological cycles (natural and urban cycles) with total disregard for the consequences: too many people, too little water, and water in the wrong places and the wrong amounts. "The human population is burgeoning, but water demand is increasing twice as fast" (De Villiers 2000). It is the author's position and view that, with the passage of time, potable water will become even more valuable. Moreover, with the passage of even more time, potable water will be even more valuable than we might ever imagine—possibly (likely) comparable in pricing, gallon for gallon, to what we pay for gasoline or even more. From urban growth to infectious disease and newly identified contaminants in water (aka "forever" chemicals—perfluoroalkyl and polyfluoroalkyl chemicals, collectively known as PFAS), greater demands are being placed on our planet's water supply (and other natural resources). As the global population continues to grow, people will place greater and greater demands on our water supply. The fact is—simply—profoundly, without a doubt in the author's mind—water is the "new oil."

Earth was originally allotted a finite amount of water; we have no more or no less than that original allotment today—they are not making any more of it. Thus, it logically follows that, to sustain life as we know it, we must do everything we can to preserve and protect our water supply. Moreover, we must also purify and reuse the water we currently waste (i.e., wastewater).

Key Point: The National Primary Drinking Water Regulations (NPDWR) establish legally enforceable primary standards and treatment techniques for public water systems.

DID YOU KNOW?

More than 50% of Americans drink bottled water occasionally or as their primary source of drinking water—an astounding fact given the high quality and low cost of U.S. tap water.

The term "*Sick Water*" was coined by the United Nations (UN) in a 2010 press release addressing the need to recognize that it is time to arrest the global tide of sick water. The gist of the UN's report pointed out that transforming waste from a major health and environmental hazard into a clean, safe, and economically attractive resource is emerging as a key challenge in the twenty-first century. As a practitioner of environmental health, I certainly support the UN's view on this important topic.

However, when I discuss sick water in the context of this text and many others I have authored or co-authored on the topic, I go a few steps further than the UN in describing the real essence and tragic implications of water that makes people or animals sick, or worse—at least, in my opinion.

Water that is sick is actually a filthy medium, wastewater—a cocktail of fertilizer runoff and sewage disposal alongside animal, industrial, agricultural, and other wastes. In addition to these listed wastes of concern, other wastes are beginning to garner widespread attention; they certainly have garnered our attention in our research on the potential problems related to these so-called "other" wastes.

What are these other wastes? Any waste or product we dispose of in our waters, whether we flush it down the toilet, pour it down the sink or bathtub, or down the drain of a worksite deep sink. Consider the following examples of "pollutants" we discharge to our wastewater treatment plants or septic tanks that we do not often consider as waste products but, in reality, are waste products.

*****

Each morning, a family of four wakes up and prepares for the workday for the two parents and school for the two teenagers. Fortunately, this family has three upstairs bathrooms to accommodate each other's needs for morning natural waste disposal, showering and soap usage, cosmetic application, hair treatments, vitamins, sunscreen, fragrances, and prescribed medications. In addition, the overnight deposit of cat and dog waste is routinely picked up and flushed down the toilet. Let's fashion a short inventory list of what this family of four has disposed of or applied to themselves as they prepare for their day outside the home:

- Toilet-flushed animal waste
- Prescription and over-the-counter therapeutic drugs
- Veterinary drugs
- Fragrances
- Soap
- Shampoo, conditioner, and other hair treatment products

DOI: 10.1201/9781003581901-3

- Body lotion, deodorant, and body powder
- Cosmetics
- Sunscreen products
- Diagnostic agents
- Nutraceuticals (e.g., vitamins, medical foods, functional foods, etc.)

Even though these bioactive substances have been around for decades, today we group all of them (the exception being animal wastes), substances, and/or products under the title of pharmaceuticals and personal care products called "PPCPs").

I pointed to the human activities of the family of four in contributing PPCPs to the environment, but other sources of PPCPs should also be recognized. For example, residues from pharmaceutical manufacturing, residues from hospitals, illicit drug disposal [i.e., police knock on the door, and the frightened user flushes the illicit drugs down the toilet (along with $100 bills, weapons, self-aborted fetuses, dealers' phone numbers, etc.) and into the wastewater stream], veterinary drug use, especially antibiotics, steroids, and agribusiness are all contributors to PPCPs in the environment

With regard to personal deposits of PPCPs into the environment and the local wastewater supply, let's return to the family of four for now. After having applied or taken in the various substances mentioned earlier, the four individuals add these products, PPCPs, to the environment through excretion (the elimination of waste material from the body), when bathing later after returning home, and possibly through the disposal of any unwanted medications into sewers and trash. How many of us have found old medical prescriptions in the family medicine chest and decided they were no longer needed? How many of us have grabbed such unwanted medications and simply disposed of them with a single toilet flush? Many of these medications, for example, antibiotics, are not normally found in the environment. And then some dump deep-fat fryer grease down the toilet—please, do not do that!

Earlier, I stated that wastewater is a cocktail of fertilizer runoff, sewage disposal, alongside animal, industrial, agricultural, and other wastes. When we factor in the addition of PPCPs to this cocktail, we can analogously state that we are simply adding a mixer to the mix.

The questions about our mixed waste cocktail are obvious: Does the disposal of antibiotics and/or other medications into the local wastewater treatment system cause problems for anyone or anything else downstream, so to speak? When we ingest locally treated water, are we also ingesting flushed-down-the-toilet or -drain antibiotics, other medications, illicit drugs, animal excretions, cosmetics, vitamins, vaginal cleaning products, sunscreen products, diagnostic agents, crankcase oil, grease, oil, fats, and veterinary drugs each time we drink a glass of tap water?

Well, Joni, bar the door—we certainly hope not. But hope is not always fact.

The jury is still out on these questions. Simply, we do not know what we do not know about the fate of PPCPs or their impact on the environment once they enter our wastewater treatment systems, the water cycle, and eventually our drinking water supply systems. This is the case even though some PPCPs are easily broken down and processed by the human body or degraded quickly in the environment. Moreover, we have known for some time—since the time of the mythical hero Hercules, arguably the world's first environmental engineer, when he was ordered to perform his fifth labor by Eurystheus to clean up King Augeas' stables. Hercules, faced literally with a mountain of horse and cattle waste piled high in the stable area, had to devise some method to dispose of the waste, and so he did. He diverted a couple of river streams into the stable area so that all the animal waste could simply be deposited into the river: out of sight, out of mind. The waste simply flowed downstream—someone else's problem. Hercules understood the principal point in pollution control technology that is pertinent to this very day: **dilution is the solution to pollution**.

As applied today, the fly in the ointment in Hercules' "dilution is the solution to pollution" approach is modern PPCPs. Although he was able to dispose of animal waste into a running water system where eventually the water's self-purification process would clean the stream, he did not have to deal with today's personal pharmaceuticals and hormones that are given to many types of livestock to enhance their health and growth.

The simple truth is that studies have shown that pharmaceuticals are present in our nation's waterbodies. Further research suggests that certain drugs may cause ecological harm. The EPA and other research agencies are committed to investigating this topic and developing strategies to help protect the health of both the environment and the public. To date, scientists have found no evidence of adverse human health effects from PPCPs in the environment. Moreover, others might argue that even if PPCPs were present today or in ancient (and mythical) times, the amount present in local water systems would represent only a small fraction (ppt—parts per trillion, 10^{-12}) of the total volume of water. Critics would be quick to point out that when we are speaking of parts per trillion (ppt), we are referring to a proportion equivalent to one-twentieth of a drop of water diluted in an Olympic-size swimming pool. I remember one student in my environmental health class who stated that he did not think the water should be termed "sick water" because it was evident to him that if the water contained so many medications, how could it be sick? Instead, he suggested it might be termed "getting well water"—making anyone who drinks it well, cured, no longer sick, etc.

It is important to point out that the term "sick water" can be applied not only to PPCP-contaminated water but also to any filthy, dirty, contaminated, vomit-filled, polluted, pathogen-filled drinking water source. The fact, is dirty or sick water causes more deaths worldwide than all forms of violence, including wars. The United Nations also points out that dirty or sick water is a key factor in the rise of de-oxygenated dead zones that have been emerging in seas and oceans across the globe.

DID YOU KNOW?

More than 50% of Americans drink bottled water occasionally or as their primary source of drinking water—an astounding fact given the high quality and low cost of U.S. tap water.

THE PARADIGM SHIFT

Historically, the purpose of water supply systems has been to provide pleasant drinking water that is free of disease organisms and toxic substances. In addition, the purpose of wastewater treatment has been to protect the health and well-being of our communities. Water/wastewater treatment operations have accomplished this goal by (1) preventing disease and nuisance conditions, (2) avoiding contamination of water supplies and navigable waters, (3) maintaining clean water for the survival of fish, bathing, and recreation, and (4) generally conserving water quality for future use.

The purpose of water supply systems and wastewater treatment processes has not changed. However, primarily because of new regulations that include (1) protecting against protozoan and virus contamination, (2) implementing the multiple barrier approach to microbial control, (3) new requirements of the Ground Water Disinfection Rule (GWDR), the Total Coliform Rule (TCR), and Distribution System (DS), and the Lead and Copper (Pd/Cu) rule, (4) regulations for trihalomethanes (THMs) and Disinfection By-Products (DBPs), and (5) new requirements to remove even more nutrients (nitrogen and phosphorus) from wastewater effluent, the paradigm has shifted. We will discuss this important shift momentarily, but first, it is important to abide by Voltaire's advice: "If you wish to converse with me, please define your terms."

For those not familiar with the term "paradigm," it can be defined in the following ways. A **paradigm** is the consensus of the scientific community: "concrete problem solutions that the profession has come to accept" (Holyningen-Huene, 1993). Thomas Kuhn coined the term "paradigm" and outlined it in terms of the scientific process. He felt that "one sense of paradigm, is global, embracing all the shared commitments of a scientific group; the other isolates a particularly important sort of commitment and is thus a subset of the first" (Holyningen-Huene, 1993). The concept of a paradigm has two general levels. The first is the encompassing whole, the summation of parts. It consists of the theories, laws, rules, models, concepts, and definitions that go into a generally accepted fundamental theory of science. Such a paradigm is "global" in character. The other level of a paradigm is that it can also be just one of these laws, theories, models, etc., that combine to

formulate a "global" paradigm. These have the property of being "local." For instance, Galileo's theory that the earth rotated around the sun became a paradigm in itself, namely, a generally accepted law in astronomy. Yet, on the other hand, his theory, combined with other "local" paradigms in areas such as religion and politics, transformed culture. A paradigm can also be defined as a pattern or point of view that determines what is seen as reality. I use this definition in this text.

A **paradigm shift** is defined as a major change in the way things are thought about, especially scientifically. Once a problem can no longer be solved within the existing paradigm, new laws and theories emerge and form a new paradigm, overthrowing the old one if it is accepted. Paradigm shifts are the "occasional, discontinuous, revolutionary changes in tacitly shared points of view and preconceptions" (Daly 1980). Simply put, a paradigm shift represents "a profound change in the thoughts, perceptions, and values that form a particular vision of reality" (Capra, 1982). For our purposes, I use the term "paradigm shift" to mean a change in the way things are understood and done.

A TRANSFORMATIONAL CHANGE

In water supply systems, the historical focus or traditional approach has been to control turbidity, iron manganese, taste odor, color, and coliforms. New regulations have introduced a new focus, leading to a paradigm shift. Today, the traditional approach is no longer sufficient. Providing acceptable water has become more sophisticated and costly. To meet the requirements of the new paradigm, a systems approach must be employed. In this approach, all components are interrelated; what affects one impacts the others. The focus has shifted to multiple requirements (i.e., new regulations require the process to be modified or the plant to be upgraded).

To illustrate the paradigm shift in the operation of water supply systems, let us look back at the traditional approach to disinfection. Disinfection was originally used in water to destroy harmful organisms. Currently, disinfection is still used in water to destroy harmful organisms, but it is now only one part of the **multiple barrier approach**. Moreover, disinfection has traditionally been used to treat coliforms only. Due to the paradigm shift, disinfection is now (and in the future will be) used against coliforms, *Legionella, Giardia, Cryptosporidium,* and others. (Note: To effectively remove the protozoans Giardia and Cryptosporidium, filtration is required; disinfection is not effective against the oocysts of Cryptosporidium.) Another example of traditional versus current practices is seen in the traditional approach to particulate removal in water to lessen turbidity and improve aesthetics. Current practice still aims to decrease turbidity to improve aesthetics, but now microbial removal plus disinfection is practical.

Another significant factor that contributed to the paradigm shift in water supply systems was the introduction of the Surface Water Treatment Rule (SWTR) in 1989. The

SWTR requires water treatment plants to achieve 99.9% (3 log) removal or activation/inactivation of Giardia and 99.99% (4 log) removal or inactivation of viruses. The SWTR applies to all surface waters and ground waters under direct influence (GWUDI).

As mentioned earlier, the removal of excess nutrients such as nitrogen and phosphorus in wastewater effluent is now receiving more attention from regulators (e.g., USEPA) and others. One of the major concerns is the appearance of dead zones in various water bodies (i.e., excess nutrients cause oxygen-consuming algae to grow, creating oxygen-deficient dead zones).

MULTIPLE-BARRIER CONCEPT

On August 6, 1996, during the Safe Drinking Water Act Reauthorization signing ceremony, President Bill Clinton stated:

> A fundamental promise we must make to our people is that the food they eat and the water they drink are safe.

No rational person could doubt the importance of the promise made in this statement.

The Safe Drinking Water Act (SWDA), passed in 1974 and amended in 1986 and (as stated above) reauthorized in 1996, gives the United States Environmental Protection Agency (USEPA) the authority to set drinking water standards. This document is important for many reasons but is even more significant as it describes how the USEPA establishes these standards.

Drinking water standards are regulations that the USEPA sets to control the level of contaminants in the nation's drinking water. These standards are part of the Safe Drinking Water Act's **"multiple barrier approach"** to drinking water protection. As shown in Figure 2.1, the multiple-barrier approach includes the following elements:

1. **Assessing and Protecting Drinking Water Sources:** This means doing everything possible to prevent microbes and other contaminants from entering water supplies. Minimizing human and animal activity around our watersheds is one part of this barrier.

2. **Optimizing Treatment Processes:** This provides a second barrier. This usually means filtering and disinfecting the water. It also means ensuring that the people responsible for our water are properly trained certified, and knowledgeable about the public health issues involved.

3. **Ensuring the Integrity of Distribution Systems:** This consists of maintaining the quality of water as it moves through the system on its way to the customer's tap.

4. **Effecting Correct Cross-Connection Control Procedures:** This is a critical fourth element in the barrier approach. It is vital because the greatest potential hazard in water distribution systems is associated with cross-connections to nonpotable waters. There are many connections between potable and nonpotable systems—every drain in a hospital constitutes such a connection—but cross-connections are those through which backflow can occur (Angele 1974).

5. **Continuous Monitoring and Testing of the Water before It Reaches the Tap:** Monitoring water quality is a critical element in the barrier approach. It should include specific procedures to follow should potable water ever fail to meet quality standards.

With the involvement of the USEPA, local governments, drinking water utilities, and citizens, these multiple barriers ensure that tap water in the United States and its territories is safe to drink. Simply put, in the Multiple-Barrier Concept, we employ a holistic approach to water management that begins at the source and continues through treatment, disinfection, and distribution.

MULTIPLE-BARRIER APPROACH: WASTEWATER OPERATIONS

Not shown in Figure 2.1 is the fate of the used water. What happens to the wastewater produced? Wastewater is treated via the multiple-barrier treatment train, which is a combination of unit processes used in the system. The primary mission of the wastewater treatment plant (and the operator/practitioner) is to treat the waste stream to a level of purity acceptable for returning it to the environment or for immediate reuse (i.e., at present, reuse in applications such as irrigation of golf courses, etc.).

Water and wastewater professionals maintain a continuous urban water cycle daily. B.D. Jones (1980) summed this up as follows:

> Delivering services is the primary function of municipal government. It occupies the vast bulk of the time and effort of most city employees, is the source of most contacts that citizens have with local governments, occasionally

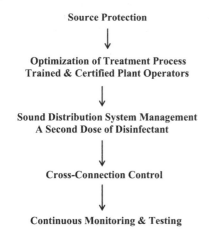

FIGURE 2.1 Multiple-barrier approach.

becomes the subject of heated controversy and is often surrounded by myth and misinformation. Yet, service delivery remains the "hidden function" of local government.

DID YOU KNOW?

Artificially generated water cycles, or the urban water cycles, consists of (1) source (surface or groundwater), (2) water treatment and distribution, (3) use and reuse, and (4) wastewater treatment and disposition, as well as the connection of the cycle to the surrounding hydrological basins.

In the *Handbook of Water and Wastewater Treatment Plant Operations*, 5th ed., the focus is on sanitary (or environmental) services (excluding solid waste disposal)—water and wastewater treatment—because they have been and remain indispensable for the functioning and growth of cities. Water is the most important life-sustaining product on Earth, next to air. Yet, it is the service delivery (and all that it entails) that remains a "hidden function" of local government (Jones 1980). This "hidden function" is what this text is all about. The discussion is presented in a completely new and unique dual manner—in what can be called the new paradigm shift in water management and the concept of the multiple barrier approach. Essentially, in blunt, plain English, the *Handbook* takes the "hidden" part **out** of the services delivered by water and wastewater professionals.

Water service professionals provide water for typical urban domestic and commercial uses, eliminate wastes, protect public health and safety, and help control many forms of pollution. Wastewater service professionals treat the urban waste stream to remove pollutants before discharging the effluent into the environment. Water and wastewater treatment services from the urban circulatory system; the hidden circulatory system. In addition, like the human circulatory system, the urban circulatory system is less effective if the flow is not maintained. In a practical sense, we must keep both systems plaque-free and free-flowing.

Maintaining flow is what water and wastewater operations are all about. This seems easy enough: Water has been flowing literally for eons, emerging from mud, rocks, silt, and the very soul of moving water, to carve a path, to pick up its load, and to cargo its way to the open arms of a waiting sea where the marriage is consummated.

This is not to say that water and wastewater operations are without problems and challenges. After surviving the Y2K fiasco, were you surrounded by dysfunctional managers running about helter-skelter, waiting until midnight, as I was? The dawn of the twenty-first century brought with it, for many of us, aspirations of good things ahead in the constant struggle to provide quality food and water for humanity. However, the only way we can hope to accomplish this is to stay on the cutting edge of technology and to face all challenges head-on. With regard to water and wastewater treatment operations, some of these challenges are addressed in Chapter 3.

THE BOTTOM LINE

With regard to Sick Water, according to the United Nations (2010), globally, two million tons of sewage, industrial, and agricultural waste are discharged into the world's waterways, and at least 1.8 million children under five-years-old die every year from water related diseases [from Sick Water], or one every 20 seconds. More people die as a result of polluted water than are killed by all forms of violence, including wars.

The White Knight to the rescue is water and wastewater treatment operations and the licensed professionals (the Fluid Mechanics) who operate and manage the treatment facilities.

REFERENCES

Angele, F.J., Sr., 1974. *Cross Connections and Backflow Protection*, 2nd ed. Denver: American Water Association.

Capra, F., 1982. *The Turning Point: Science, Society and the Rising Culture*. New York: Simon & Schuster, p. 30.

Daly, H.E., (ed.), 1980. Introduction to the Steady-State Economic, in *Ecology, Ethics: Essays toward a Steady State Economy*. New York: W.H. Freeman & Company.

De Villiers, M., 2000. *Water: The Fate of Our Most Precious Resource*. Boston: Mariner books.

Holyningen-Huene, P., 1993. *Reconstructing Scientific Revolutions*. Chicago: University of Chicago, p. 134.

Jones, B.D., 1980. *Service Delivery in the City: Citizens Demand and Bureaucratic Rules*. New York: Longman, p. 2.

United Nations Environment Programme, & United Nations Human Settlements Programme (2010). *Sick Water: The Central Role of Wastewater Management and Sustainable Development—A Rapid Response Assessment*. Accessed 03/20/24 @ https://wedoc.unep.org/20.500.11822/9156.

3 The Challenges

INTRODUCTION

Challenges and associated problems come and go, shifting from century to century, decade to decade, year to year, operation to operation, and from site to site. They range from the problems caused by natural forces (storms, earthquakes, fires, floods, and droughts) to those caused by social forces, currently including terrorism. In general, eight areas are of concern to many water and wastewater management personnel, including:

1. Complying with regulations and coping with new and changing regulations
2. Maintaining infrastructure
3. Privatization and/or re-engineering
4. Benchmark It!
5. Upgrading security
6. Technical versus professional management
7. Energy conservation measures and sustainability
8. Autonomous operations

COMPLIANCE WITH NEW, CHANGING, AND EXISTING REGULATIONS

Adapting the workforce to the challenges of meeting changing regulations and standards for both water and wastewater treatment is a major concern. As mentioned, drinking water standards are regulations that the USEPA sets to control the level of contaminants in the nation's drinking water. These standards are part of the Safe Drinking Water Act's (SDWA's) multiple-barrier approach to drinking water protection. There are two categories of drinking water standards:

1. **A National Primary Drinking Water Regulation (Primary Standard):** This is a legally enforceable standard that applies to public water systems. Primary standards protect drinking water quality by limiting the levels of specific contaminants that can adversely affect public health and are known or anticipated to occur in water. They take the form of Maximum Contaminant Levels or Treatment Techniques.
2. **A National Secondary Drinking Water Regulation (Secondary Standard):** This is a non-enforceable guideline regarding contaminants that may cause cosmetic effects (such as skin or tooth discoloration) or aesthetic effects (such as taste, odor, or color) in drinking water. The USEPA recommends secondary standards to water systems but does not require systems

to comply. However, states may choose to adopt them as enforceable standards. This information focuses on national primary standards.

Drinking water standards apply to public water systems, which provide water for human consumption through at least 15 service connections or regularly serve at least 25 individuals. Public water systems include municipal water companies, homeowner associations, schools, businesses, campgrounds, and shopping malls.

More recent requirements, such as the Clean Water Act Amendments that went into effect in February 2001, require water treatment plants to meet tougher standards, presenting new challenges for treatment facilities while offering some possible solutions for meeting the new standards. These regulations allow communities to upgrade existing treatment systems, replacing aging and outdated infrastructure with new process systems. Their purpose is to ensure that facilities can filter out higher levels of impurities from drinking water, thus reducing the health risk from bacteria, protozoa, and viruses, and that they can decrease turbidity and reduce concentrations of chlorine by-products in drinking water.

As a Point of Interest: The most recent CWA Amendment was prompted by a U.S. Supreme Court ruling in the case of Sackett v. EPA (2012 1) and (2023 II). This act limited the EPA's regulate wetlands unless they have a significant nexus to major waterways.

With regard to wastewater collection and treatment, the National Pollution Discharge Elimination System (NPDES) program established by the Clean Water Act issues permits that control wastewater treatment plant discharges. Meeting permits is always a concern for wastewater treatment managers because the effluent discharged into water bodies affects those downstream of the release point. Individual point source dischargers must use the best available technology (BAT) to control the levels of pollution in the effluent they discharge into streams. As systems age and BAT changes, meeting permits with existing equipment and unit processes becomes increasingly difficult.

Important Note: In April 2024, the USEPA issued the first-ever national, legally enforceable drinking water standard to protect communities from exposure to harmful PFAS. This new water standard is designed to protect public health by establishing legally enforceable levels for several PFAS known to occur individually and as mixtures in drinking water. This rule sets limits for five individual PFAS: PFOA, PROS, PFNA, PFHxS, and HFPO-DA (also known as: "GenX Chemicals"). The intention of this rule is to reduce exposure to PFAS, which, in turn, will prevent thousands of premature deaths, and tens of thousands of serious illnesses, including cancers and heart impacts (especially in adults).

DOI: 10.1201/9781003581901-4

DID YOU KNOW?

In 2021, USEPA determined that revised effluent limitations guidelines (ELGs) and pretreatment standards are warranted for:

- The Organic Chemicals, Plastics, and Synthetic Fibers category to address per- and polyfluoroalkyl substances (PFAS), also known as "forever" chemicals. This category consists of more than 12,000 human-made chemical compounds used in a wide range of everyday products, including take-out containers, drinking straws, outdoor clothing, medical devices, food packaging, cosmetics, and most products advertised as being greaseproof, fire retardant, non-sticking, and waterproof.
- The Metal Finishing category to address PFAS discharges from chromium electroplating facilities.
- The Meat and Poultry Products category to address nutrient discharges (USEPA 2021).

MAINTAINING INFRASTRUCTURE

During the 1950s and 1960s, the U.S. government encouraged the prevention of pollution by providing funds for the construction of municipal wastewater treatment plants, water pollution research, and technical training and assistance. New processes were developed to treat sewage, analyze wastewater, and evaluate the effects of pollution on the environment. Despite these efforts, however, the expanding population and industrial and economic growth caused pollution and health difficulties to increase.

In response to the need to make a coordinated effort to protect the environment, the National Environmental Policy Act (NEPA) was signed into law on January 1, 1970. In December of that year, a new independent body, the USEPA was created to bring under one roof all the pollution-control programs related to air, water, and solid wastes. In 1972, the Water Pollution Control Act Amendments expanded the role of the federal government in water pollution control and significantly increased federal funding for the construction of wastewater treatment plants.

Many of the wastewater treatment plants in operation today are the result of federal grants made over the years. For example, because of the 1977 Clean Water Act Amendment to the Federal Water Pollution Control Act of 1972 and the 1987 Clean Water Act reauthorization bill, funding for wastewater treatment plants was provided. Many large sanitation districts, with their multiple plant operations, and even a larger number of single plant operations in smaller communities in operation today are a result of these early environmental laws. Because of these laws, the Federal Government provided grants of several hundred million dollars to finance the construction of wastewater treatment facilities throughout the country.

Many of these locally or federally funded treatment plants are aging; based on my experience, I rate some as dinosaurs. The point is that many facilities are facing problems caused by aging equipment, facilities, and infrastructure. Complicating the problems associated with natural aging is the increasing pressure on inadequate older systems to meet the demands of increased population and urban growth. Facilities built in the 1960s and 1970s are now 30–40 years old, and not only are they showing signs of wear and tear, but they simply were also not designed to handle the level of growth that has occurred in many municipalities.

Regulations often necessitate a need to upgrade. By matching funds or providing federal money to cover some of the costs, municipalities can take advantage of a window of opportunity to improve their facilities at a lower direct cost to the community. Those federal dollars, of course, come with strings attached; they must be spent on specific projects in specific areas. On the other hand, many times new regulatory requirements are imposed without the financial assistance needed to implement them. When this occurs, the local community either ignores the new requirements (until caught and forced to comply) or face the situation by implementing local tax hikes to pay the cost of compliance.

An example of how a change in regulations can force the issue is demonstrated by the demands made by OSHA and the USEPA in their Process Safety Management (PSM)/Risk Management Planning (RMP) regulations (29 CFR 1910.119—OSHA). These regulations put the use of elemental chlorine (and other listed hazardous materials) under scrutiny. As a result of these regulations, plant managers throughout the country are forced to choose which side of a double-edged sword cuts their way the most. One edge calls for full compliance with the regulations (analogous to stuffing the regulation through the eye of a needle). The other edge calls for substitution, meaning replacing elemental chlorine with a non-listed chemical (e.g., hypochlorite) or a physical disinfectant (e.g., ultraviolet irradiation, UV)—either way, it is a very costly undertaking.

- **Note:** Many of us who have worked in water and wastewater treatment for years characterize PSM and RMP as the elemental chlorine killer. You have probably heard the old saying: "If you can't do away with something in one way, then regulate it to death."
- **Note:** Changes resulting from regulatory pressure sometimes necessitate replacing or changing existing equipment, increasing chemical costs (e.g., substituting hypochlorite for chlorine typically increases costs threefold), and could easily lead to increased energy and personnel costs. Equipment conditions, new technology, and financial concerns are all considerations when upgrades or new processes are chosen. In addition, the safety of the

process must be considered, of course, because of the demands made by the USEPA and OSHA. The potential harm to workers, the community, and the environment is all under study, as are the possible long-term effects of chlorination on the human population.

PRIVATIZATION AND/OR RE-ENGINEERING

As mentioned, water and wastewater treatment operations are undergoing a new paradigm shift. I explained that this shift focuses on the holistic approach to treating water. The shift is, however, more inclusive. It also includes thinking outside the box. To remain efficient and therefore competitive in the real world of operations, water and wastewater facilities have either bought into the new paradigm shift or been forcibly "shifted" to doing other things (often these "other" things have little to do with water/wastewater operations) (Johnson & Moore 2002).

Experience has shown that few words conjure up more fear among plant managers than "privatization" or "re-engineering." *Privatization* means allowing private enterprises to compete with the government in providing public services, such as water and wastewater operations. Privatization is often proposed as a solution to the numerous woes facing water and wastewater utilities, including corruption, inefficiencies (dysfunctional management becoming a universal malady), and the lack of capital for needed service improvements and infrastructure upgrades and maintenance. Existing management, on the other hand, can accomplish re-engineering internally, or it can be used (and usually is) during the privatization process. *Re-engineering* is the systematic transformation of an existing system into a new form to realize quality improvements in operation, system capability, functionality, performance, or evolvability at a lower cost, schedule, or risk to the customer. A current example of re-engineering in water and wastewater operations is the current shift from manual operations to digital operations (i.e., computerized autonomous systems).

Many on-site managers consider privatization and/or re-engineering schemes threatening. In the worst-case scenario, a private contractor could bid the entire staff out of their jobs. In the best case, privatization and/or re-engineering are often a real threat that forces on-site managers into workforce cuts, improving efficiency and cutting costs. At the same time, on-site managers work to ensure that the community receives safe drinking water and that the facility meets standards and permits, while operating with fewer workers—and without injury to workers, the facility, or the environment.

Local officials should take a hard look at privatization and re-engineering for several reasons:

1. **Decaying Infrastructures:** Many water and wastewater operations include infrastructures that date back to the early 1900s. The most recent systems were built with federal funds during the 1970s, and even these now need upgrading or replacement. The USEPA recently estimated that the nation's 75,000+ drinking water systems alone will require more than $100 billion in investments over the next 20 years. Wastewater systems will require a similar level of investment.

2. **Mandates:** The federal government has reduced its contributions to local water and wastewater systems over the past 30 years, while at the same time imposing stricter water quality and effluent standards under the Clean Water Act and Safe Drinking Water Act. Moreover, as previously mentioned, new unfunded mandated safety regulations, such as OSHA's Process Safety Management and USEPA's Risk Management Planning, are expensive to implement using local revenues sources or state revolving loan funds.

3. **Hidden Function:** Earlier, we stated that much of the work of water and wastewater treatment is a "hidden function." Because of this lack of visibility, it is often difficult for local officials to commit to making the necessary investments in community water and wastewater systems. Simply put, local politicians lack the political will—water pipes and interceptors are not visible and are not perceived as immediately critical for adequate funding. Thus, it is easier for elected officials to ignore them in favor of expenditures of more visible services, such as police and fire. Additionally, raising water and sewage rates to cover operations and maintenance is not always affected because it is an unpopular move for elected officials to make. This means that water and sewer rates do not adequately cover the actual cost of providing services in many municipalities.

In many locations throughout the United States, water and wastewater services are the largest expenditures facing local governments today. (This is certainly the case for municipalities struggling to implement the latest stormwater and nutrient reduction requirements). Thus, this area presents a great opportunity for cost savings. Through privatization, water and wastewater companies can take advantage of advanced technology, more flexible management practices, and streamlined procurement and construction practices to lower costs and make critical improvements more quickly.

With regard to privatization, the view taken in this text is that ownership of water resources, treatment plants, and wastewater operations should be maintained by the public (local government entities) to prevent a Tragedy of the Commons-like event [i.e., free access and unrestricted demand for water (or other natural resources) ultimately structurally dooms the resource through over-exploitation by private interests]. However, because management is also a "hidden function" of many public service operations (e.g., water and wastewater operations), privatization may be a better alternative to prevent creating a home for dysfunctional managers and ROAD Gangers (**R**etired **O**n **A**ctive **D**uty clan members).

The Bottom Line: Water and wastewater are commodities the quantity and quality of which are much too important to leave at the whims of some public authorities.

BENCHMARK IT![1]

Note: Based on personal experience and my own opinion, benchmarking can be a double-edged sword: one edge wielded by a dysfunctional manager and the other by a champion of a functional organization. The dysfunctional edge is primarily used as a self-defense mechanism; namely, by public service utility managers out of their own need for self-preservation (to retain their lucrative positions—to survive their personal in-the-bunker mode). For example, it is their own self-preservation that is the primary reason that many utility directors work against the trend to privatize water, wastewater, and other public operations; they realize that privatization usually results in more efficient and less costly operations and that their positions would be at risk. Usually, the real work to prevent privatization and save unneeded jobs, procedures, methods, and work practices, etc., is delegated to individual managers in charge of specific operations because they usually (usually is a very BIG word in this case) know what they are doing and also have a stake in ensuring that their relatively secure careers are not affected by privatization. It can be easily seen that working against privatization by these "local" managers is in their own self-interest and in the interest of their workers because their jobs may be at stake.

The question is, of course, how does one go about preventing his/her water and wastewater operation from being privatized? The answer is rather straightforward and clear: Use the functional edge of the sword by increasing efficiency and reducing the cost of operations. In the real world, this is easier said than done—but is not impossible. For example, the facilities under properly implemented and managed Total Quality Management (TQM) experience a much easier process. The main advantage a properly installed and managed TQM offers the plant manager is a variety of tools to help plan, develop, and implement water and wastewater operational efficiency measures. These tools include self-assessments, statistical process control, International Organization for Standards (ISO) 9000 and 14,000, process analysis, quality circle, common sense, and benchmarking (see Figure 3.1).

In this text, benchmarking is presented to illustrate how it is used to ensure energy efficiency and sustainability and, in general, form in the Rachel's Creek case study. Cobblers are credited with coining the term *benchmarking*. They used the term to measure people's feet for shoes. They would place the person's foot on a "bench" and mark it out to make the pattern for the shoes. Benchmarking is still

Start---Plan---Research---Observe---Analysis---Adapt

FIGURE 3.1 Benchmarking process.

used to measure but now specifically gauges performance based on specific indicators such as cost per unit of measure, productivity per unit of measure, cycle time of some value per unit of measure, or defects per unit of measure.

It is interesting to note that there is no specific benchmarking process that has been universally adopted; this is the case because of its wide appeal and acceptance. Accordingly, benchmarking manifests itself via various methodologies. Robert Camp (1989) wrote in one of the earliest books on benchmarking and developed a 12-stage approach to benchmarking. Camp's 12-stage methodology consists of:

1. Select the subject
2. Define the process
3. Identify potential partners
4. Identify data sources
5. Collect data and select partners
6. Determine the gap
7. Establish process differences
8. Target future performance
9. Communicate
10. Adjust the goal
11. Implement
12. Review and recalibrate

With regard to improving energy efficiency and sustainability in drinking water and wastewater treatment operations, benchmarking is simply defined (in this text) as the process of comparing the energy usage of a drinking water or wastewater treatment operation to similar operations. Local utilities of similar size and design are excellent points of comparison. Broadening the search, one can find several resources discussing the "typical" energy consumption across the United States for a water or wastewater utility of a particular size and design.

Keep in mind that in drinking water and wastewater treatment utilities (and other utilities and industries), benchmarking is often used by management personnel to increase efficiency and ensure the sustainability of energy resources, but it is also used, as mentioned earlier, to ensure their own self-preservation (i.e., to retain their lucrative positions). With self-preservation as their motive, benchmarking is used as a tool to compare operations with best-in-class facilities or operations to improve performance and avoid the current (and ongoing) trend to privatize water, wastewater, and other public operations.

Earlier, Camp's (1989) 12-stage benchmarking process was discussed. In the pursuit of energy efficiency and sustainability in drinking water and wastewater treatment operations, Camp's 12-stages can be simplified into the six-step process shown in Figure 3.1.

In energy efficiency and sustainability projects, before the benchmarking tool is used, an Energy Team should be formed and assigned the task of studying how to implement energy-saving strategies and how to ensure sustainability in the long run. Keep in mind that forming a "team" is not the same as fashioning a silver bullet—the team is only as good as its leadership and members. Benchmarking is a process for rigorously measuring your performance against best-in-class operations and using the analysis to meet and exceed that standard; thus, those involved in the benchmarking process should be the best of the best (Spellman 2009).

What Benchmarking Is

1. Benchmarking versus best practices gives water and wastewater operations a way to evaluate overall operations.
   a. how effective
   b. how cost-effective
2. Benchmarking shows plants both how well their operations stack up and how well those operations are implemented.
3. Benchmarking is an objective-setting process.
4. Benchmarking is a new way of doing business.
5. Benchmarking forces an external view to ensure the correctness of objective setting.
6. Benchmarking forces internal alignment to achieve plant goals.
7. Benchmarking promotes teamwork by directing attention to those practices necessary to remain competitive.

Potential Results of Benchmarking

Benchmarking may indicate the direction of required change rather than specific metrics: costs must be reduced; customer satisfaction must be increased; return on assets must be increased; maintenance must be improved; and operational practices must be enhanced. Best practices are then translated into operational units of measure.

Targets

Consideration of available resources converts benchmark findings into targets. A target represents what can realistically be accomplished in a given time frame and shows progress toward benchmark practices and metrics. The quantification of precise targets should be based on achieving the benchmark.

Note: Benchmarking can be performance-based, process-based, or strategic-based and can compare financial or operational performance measures, methods or practices, or strategic choices.

Benchmarking: The Process

When forming a benchmarking team, the goal should be to provide a benchmark that evaluates and compares privatized and re-engineered water and wastewater treatment operations to your operation to be more efficient, remain competitive, and make continual improvements. It is important to point out that benchmarking is more than simply setting a performance reference or comparison; it is a way to facilitate learning for continual improvements. The key to the learning process is looking outside one's own plant to other plants that have discovered better ways to achieve improved performance.

Benchmarking Steps

As shown in Figure 3.1, the benchmarking process consists of five major steps:

1. **Planning:** Managers must select a process (or processes) to be benchmarked. A benchmarking team should be formed and the process of benchmarking must be thoroughly understood and documented. The performance measures for the process should be established (i.e., cost, time, and quality).
2. **Research:** Information on the best-in-class performer must be determined through research. The information can be derived from the industry's network, industry experts, industry and trade associations, publications, public information, and other award-winning operations.
3. **Observation:** The observation step is a study of the benchmarking subject's performance level, processes, and practices that have achieved those levels, along with other enabling factors.
4. **Analysis:** In this phase, comparisons in performance levels among facilities are determined. The root causes for the performance gaps are studied. To make accurate and appropriate comparisons, the comparison data must be sorted, controlled for quality, and normalized.
5. **Adaptation:** This phase involves putting what is learned throughout the benchmarking process into action. The findings of the benchmarking study must be communicated to gain acceptance, functional goals must be established, and a plan must be developed. Progress should be monitored, and corrections to the process should be made as needed.

Note: Benchmarking should be interactive. It should also recalibrate performance measures and improve the process itself.

DID YOU KNOW?

Benchmarking can be useful, but no two utilities are ever the same. You'll have some characteristics that affect your relative performance and are beyond a utility's control.or:

Collection of Baseline Data and Tracking Energy Use

Using the five-stage benchmarking procedure detailed above, your benchmarking team identifies, locates, and assembles baseline data that can help determine what is needed to improve your energy performance. Keep in mind that the data you collect will be compared to similar operations in the benchmarking process. The point is that it is important to collect data that is comparable—like oranges to oranges, apples to apples, grapes to grapes, and so forth. It does little good, makes no sense, and wastes time and money to collect nomenclature data from equipment, machinery, and operations that are not comparable to the utility or utilities to which your data will be compared.

The first step is to determine what data you already have available. At a minimum, aim to have one full year of monthly data for consumption of electricity, natural gas, and other fuels—if you can get three years of data, that's even better. However, if you don't have data going this far back, use what you have or can easily collect. In addition, if you can gather the data at daily or hourly intervals, you may be able to identify a wider range of energy opportunities (USEPA 2008).

Here are several data elements to document and track for your utility to review energy improvement opportunities.

- **Water and/or Wastewater Flows** are key to determining your energy performance per gallon treated. For drinking water, the distance traveled and the number of pumps are also key factors.
- **Electricity DATA** includes overall electricity consumption (kWh) as well as peak demand (kW) and load profiles, if available.
- **Other Energy Data** includes purchases of diesel fuel, natural gas, or other energy sources including renewables.
- **Design Specifications** can help you identify how much energy a given process or piece of equipment should be using.
- **Operating Schedules** for intermittent processes will help you make sense of your load profile and possibly plan an energy-saving or cost-saving alternative.

Along with making sure that the data you collect is comparable—i.e., apples to apples—keep in mind that energy units may vary. If you are comparing apples to apples, are you comparing bushel to bushel, pound to pound, or quantity to quantity? In an energy efficiency and sustainability benchmarking study comparison, for example, captured methane or purchased natural gas may be measured in 100 cubic feet (ccf) or Metric Million British Thermal Units (MMBTTU). Develop a table like Table 3.1A shown below to document and track your data needs (USEPA 2008)

REMEMBER

Keep units consistent!

Consider any other quantities that you want to measure. Is there anything you would add to Table 3.1A? Chances are good that you will add quantities and that is why it is labeled Table 3.1A with other renditions to follow. Let's get back to unit selection for your tables. Make sure you select units that your Energy Team is comfortable with and that your

TABLE 3.1A
Data Needs

Data Need	Units
Wastewater flow	MGD
Electricity consumption	kWh
Peak demand	kW
Methane capture (applies to plants that digest biosolids)	MMBTU
Microturbine generation	kWh
Natural gas consumed	MMBTU
Fuel oil consumed	Gallons
Diesel fuel consumed	Gallons
Design specifications	N/A
Operating schedules	N/A
Grease trap waste collected (future renewable fuel source)	Gallons
Other (based on your operation)	TBD

TABLE 3.1B
Data Needs

Data Need	Units	Desired Frequency of Data
Wastewater flow	MGD	Daily
Electricity consumption	kWh	Hourly if possible or daily if not
Peak demand	kW	Monthly
Methane capture	MMBTU	Monthly
Microturbine generation	kWh	Monthly
Natural gas consumed	MMBTU	Monthly
Fuel oil consumed	Gallons	Monthly
Diesel fuel consumed	Gallons	Monthly
Design specifications	N/A	N/A
Operating schedules	N/A	N/A

data is typically available in. If the data is reported using the wrong units, you may encounter conflicting or confusing results.

Keep in mind that units by themselves are not that informative; to be placed in proper context, they need to be associated with an interval of time. Therefore, for the next step, simply expand Table 3.1A by adding another column titled "Desired Frequency of Data."

Remember, while knowing your utility's energy consumption per month is useful, knowing it in kWh per day is better. With hourly consumption data, you can develop a "load profile," which breaks down your energy demand during the day. If your load profile is relatively flat, or if your energy demand is greater during the off-peak hours (overnight and early morning) than during the peak hours (daytime and early evening), your utility may qualify for special pricing plans from your energy provider.

Typically, water and wastewater treatment operations have a predictable diurnal variation (i.e., fluctuations that occur throughout each day). Usage is heaviest during the early morning, lags during the afternoon, has a second, less intensive peak in the early evening, and hits the lowest point overnight. Normally, energy use for water and wastewater treatment operations could be expected to follow a pattern of water flows. However, this effect can be delayed by travel time from the source, through the collection system, to the plant, or by storage tanks within the distribution system to customers. A larger system will have varying travel times, whereas a smaller system will have lower variability. Moreover, this effect can be totally eliminated if the plant has equalization tanks (USEPA 2008).

If your utilities pay a great deal of money for peak demand charges, you might consider the capital investment in an equalization tank. Demand charges can be significant for wastewater utilities, as they are generally about 25% of the utility's electricity bill (WEF 1997).

The next step is to determine how you will collect baseline data. Energy data is recorded by your energy provider (e.g., electric utility, natural gas utility, or heating oil and diesel oil companies). A monthly energy bill contains the total consumption for that month, as well as the peak demand. In some cases, your local utilities will record the demand on every meter at 15-minute intervals throughout the year. Similar data may be available if you have a system at your utility that monitors energy performance. Sources of energy data include the following:

- **Monthly Energy Bills** vary in detail but all contain most essential elements.
- **The Energy Provider** may be able to provide more detailed information.
- **An Energy Management Program** (e.g., Supervisory Control and Data Acquisition—SCADA) automatically tracks energy data, often with sub-meters to identify the load on individual components. If such a system is in place at your utility, you will have a large and detailed data set on hand.

The Baseline Audit

The energy audit is an essential step in energy conservation and management efforts. Your drinking water or wastewater operation may have had an energy audit or program review conducted at some point. If so, find the final report and have your Energy Team review it. While reviewing, consider the following questions: How long did the process take? Who participated in it—your team, the electric utility, independent contractors? What measures were suggested to improve energy efficiency? What measures were actually implemented, and did they meet expectations? Were there lessons learned from the process that should be applied to future audits? In addition, if your facility's previous energy audit recommended measures, determine if they are still viable.

In many cases, electrical utilities offer audits as part of their energy conservation programs. Independent energy service companies also provide these services. An outside review from an electric utility or an engineering company can provide useful input, but it is important to ensure that any third party is familiar with water and wastewater systems.

Some energy audits focus on specific types of equipment such as lighting, Heating Ventilation Air Conditioning (HVAC), or pumps. Others look at the processes used and take a more systematic approach. Audits focused on individual components, as well as in-depth process audits, will include testing equipment. For example, in conducting the baseline energy audit, the Energy Team may compare the nameplate efficiency of a motor or pump to its actual efficiency.

In a process approach, a preliminary walk-through or walk-around audit is often used as a first step to determine if there are likely to be opportunities to save energy. If such opportunities exist, then a detailed process audit is conducted. This may include auditing the performance of the individual components as well as considering how they work together as a whole. Much like an environmental management system's initial assessment, which reviews the current status of regulatory requirements, training, communication, operating conditions, and current practices and processes, a preliminary energy audit or energy program review will provide your utility with a baseline of what your energy consumption is at that point in time.

Once you have collected your utility's baseline data and tracked monthly and annual energy use, there are two additional steps to completing your energy assessment or baseline energy audit: conduct a field investigation and create an equipment inventory and distribution of demand and energy (USEPA 2008).

The Field Investigation

The field investigation is the heart of an energy audit. It will include obtaining information for an equipment inventory, discussing process operations with the individuals responsible for each operation, discussing the impact of specific energy conservation ideas, soliciting ideas from your Energy Team, and identifying the energy profiles of individual system components. The Electric Power Research Institute (EPRI) recommends evaluating how each process or piece of equipment could otherwise be used. For example, it might be possible for a given system to be replaced or complemented for normal operation by one of lower capacity; to run fewer hours; to run during off-peak hours; to employ a variable speed drive; and/or to be replaced by a newer or more efficient system. Depending on the situation, one or more of these changes might be appropriate.

Create Equipment Inventory and Distribution of Demand and Energy

This is a record of your operation's equipment, including equipment names, nameplate horsepower (if applicable), hours of operation per year, measured power consumption, and total kilowatt-hours (kWh) of electrical consumption per year. Other criteria, such as age, may also be included. In addition, different data may be appropriate for other types of systems, such as methane-fired heat and power systems.

You may find that you already have much of this information in your maintenance management system (if applicable). A detailed approach for developing an equipment inventory and identifying the energy demand of each piece of equipment is provided in the 1997 book *Energy Conservation in Wastewater Treatment Facilities: Manual of Practice No. MFD-2, Water Environment Federation*. The basics are presented here, but readers are encouraged to review the WEF (1997) *Manual of Practice* for a more thorough explanation.

Example drinking water and/or wastewater treatment operations equipment inventories and the relevant energy data to collect could include the following:

Motors and Related Equipment

- Start at each motor control center (MCC) and itemize each piece of equipment in order as listed on the MCC.
- Itemize all electric meters on MCCs and local control panels.
- Have a qualified electrician check the power draw of each major piece of equipment.

Pumps

- From the equipment manufacturer's literature, determine the pump's power ratio (this may be expressed in kW/MGD).
- Multiply horsepower by 0.746 to obtain kilowatts.
- Compare the manufacturer's data with field-obtained data.

Aeration Equipment

- Power draw of aeration equipment is difficult to estimate and should be measured.

DID YOU KNOW?

Some utilities will have an inherently higher or lower energy demand due to factors beyond their control. For example, larger plants will, in general, have a lower energy demand per million gallons treated due to economies of scale. A plant that is large relative to its typical load will have a higher energy demand per million gallons treated. Some secondary treatment processes require greater energy consumption than others. Still, benchmarking allows a rough estimate of the utility's relative energy performance. Benchmarking of individual components is also useful. A survey of one's peers may identify what level of performance can realistically be expected from, say, a combined heat and power system or a specific model of methane-fueled microturbine (USEPA 2008).

- Measure aspects related to biochemical oxygen demand (BOD) loading, foot-to-microorganism ratio, and oxygen-transfer efficiency (OTE). Note that OTE levels depend on the type and condition of aeration equipment. Actual OTE levels are often considerably lower than described in the literature or manufacturers' materials.

Case Study 3.1 Benchmarking: A General Example

To gain a better understanding of the general benchmarking process used in water and wastewater treatment plant operations, the following example is provided. (It is in outline and summary form only, as a discussion of a full-blown study is beyond the scope of this text).

RACHEL'S CREEK SANITATION DISTRICT

INTRODUCTION

In January 2007, Rachel's Creek Sanitation District formed a benchmarking team to provide a benchmark that evaluates and compares privatized and re-engineered wastewater treatment operations to Rachel's Creek operations to be more efficient and remain competitive. After three months of evaluating wastewater facilities using the benchmarking tool, our benchmarking is complete. This report summarizes our findings and should serve as a benchmark by which to compare and evaluate Rachel's Creek Sanitation District operations.

FACILITIES

41-wastewater treatment plants throughout the United States.

TARGET AREAS

The benchmarking team focused on the following target areas for comparison:

1. Re-engineering
2. Organization
3. Operations and Maintenance
   a. Contractual services
   b. Materials and supplies
   c. Sampling and data collection
   d. Maintenance
4. Operational Directives
5. Utilities—Energy Consumption
6. Chemicals
7. Technology
8. Permits
   a. Water quality
   b. Solids quality
   c. Air quality
   d. Odor quality
9. Safety
10. Training and Development

11. Process
12. Communication
13. Public Relations
14. Reuse
15. Support Services
    a. Pretreatment
    b. Collection systems
    c. Procurement
    d. Finance and administration
    e. Laboratory
    f. Human resources

SUMMARY OF FINDINGS

Our overall evaluation of Rachel's Creek Sanitation District, compared to our benchmarking targets, is positive; we are in good standing compared to the 41 target facilities we benchmarked against. In the area of safety, we compare quite favorably—only plant 34, with its own full-time safety manager, appeared to perform better than we do. We were very competitive with the privatized plants in our usage of chemicals and far ahead of many public plants. We were also competitive in the use of energy. Our survey of what other plants are doing to cut energy costs showed that we clearly identified areas for improvement, and our current efforts to further reduce energy costs are on track. We were far ahead in the optimization of our unit processes and were leaders in the area of odor control.

We also found areas that we need to improve. To the Rachel's Creek employee, re-engineering applies only to the treatment department and has been limited to cutting staff while plant practices and organizational practices are outdated and inefficient. Under the re-engineering section of this report, we have provided a summary of re-engineering efforts at the re-engineered plants visited. The experiences of these plants can be used to improve our own re-engineering effort. Next is our organization and staffing levels. A private company could reduce the entire treatment department staff by about 18%–24%. The 18%–24% is based on the number of employees and not costs. In the organization section of this report, organizational models and their staffing levels are provided as guidelines for improving our organization and determining optimum staffing levels. The last big area that we need to improve is in the way we accomplish the work we perform. Our people are not used efficiently because of outdated and inefficient policies and work practices. Methods to improve the way we do work are found throughout this report. We noted that efficient work practices used by private companies allow plants to operate with small staff.

Overall, Rachel's Creek Sanitation District's treatment plants are much better than other public service plants. Although some plants may have better equipment, better technology, and cleaner effluents, the costs of labor and materials are much higher than ours. Several of the public plants were in bad condition. Contrary to popular belief, the privately operated plants had good to excellent operations. These plants met permits, complied with safety regulations,

maintained plant equipment, and kept the plant clean. Due to their efficiency and low staff, we felt that most privately operated plants were better than we are. We agreed that this needs to be changed. Using what we learned during our benchmarking effort, we can be just as efficient as a privately operated plant and still maintain our standards of quality (Spellman 2009).

DID YOU KNOW?

Growing algae in wastewater will soak up nutrients at the wastewater plant, thus helping the receiving water body, which could suffer from excessive nutrients discharged by such treatment plants.

UPGRADING SECURITY

Since 9/11, we have heard it said by many of our teachers, neighbors, security specialists, and students (and many others) that there is controversy about the definition of the politically-charged word *terrorism*. Terrorism, like pollution, is a judgment call. For example, regarding the definition of pollution, if two neighbors live next door to an air-polluting facility, one neighbor who has no personal connection with the polluting plant is likely to label the plant's output as pollution. The other neighbor who is an employee of the plant may see the plant's pollution as dollar bills—dollars that are his or her livelihood. I have heard workers who dive into ponds full of raw sewage to find a leak in an effluent pipe say that the sewage in the pond and its associated odor is money in the bank ... and long-term employment because toilets and their function are not likely to vanish anytime soon. Why a money-making enterprise? Because there are few people around who would join them in diving into trenches filled with raw sewage to do their work. On the terrorism front, when someone deliberately spikes a tree to prevent loggers from cutting it down, the tree-spiker might feel that he or she is a patriot, just a knight in shining green armor, and definitely not a terrorist. On the other hand, the logger who has to take the tree down and puts his or her life and limbs at risk in taking down the spiked tree has little doubt in his or her mind about what to call the tree-spiker, and what the tree-spiker is called certainly has nothing to do with patriotism. Thus, what we are saying here in the example presented, pollution versus terrorism, is that along with attempting to define pollution, trying to define terrorism may be a judgment call, especially in the view of the terrorists.

Because water is vital to us all, water treatment plants are potential targets for terrorists; these sites are obviously top targets for some. However, in wastewater treatment plants, you might wonder why security is a concern when we all know that no one in their right mind would fly an airplane into a tank full of raw sewage. Beyond the fact that terrorists are **not** in their right mind, this is not the reason they target tanks full of raw sewage. Instead of targeting raw sewage tanks, wastewater treatment plants use chemicals in the treatment process, including toxic chlorine for disinfection. So, it is not the tank full of raw sewage that is the potential target; rather, it is the tons of chlorine that may be stored at the plant.

The point is because of 9/11 and other such attacks, water and wastewater treatment plants have upgraded security.

TECHNICAL VS. PROFESSIONAL MANAGEMENT

Water treatment operations management is directed toward providing water of the right quality, in the right quantity, at the right place, at the right time, and at the right price to meet various demands. Wastewater treatment management is focused on providing treatment of incoming raw influent (no matter what the quantity), at the right time, to meet regulatory requirements, and at the right price to meet various requirements. The techniques of management are manifold in both water resource management and wastewater treatment operations. In water treatment operations, for example, management techniques may include (Mather 1984):

> Storage to detain surplus water available at one time of the year for use later, transportation facilities to move water from one place to another, manipulation of the pricing structure for water to reduce demand, use of changes in legal systems to make better use of the supplies available, introduction to techniques to make more water available through watershed management, cloud seeding desalination of saline or brackish water, or area-wide educational programs to teach conservation or reuse of water.

Many of the management techniques employed in water treatment operations are also employed in wastewater treatment. In addition, wastewater treatment operations employ management techniques that may include upgrading existing systems for nutrient removal, reusing process residuals in an environmentally friendly manner, and implementing area-wide educational programs to teach proper domestic and industrial waste disposal practices.

Whether managing a waterworks or a wastewater treatment plant, the expertise of the manager must include being a well-rounded, highly skilled individual. No one questions the need to incorporating these highly trained practitioners—well-versed in the disciplines and practices of sanitary engineering, biology, chemistry, hydrology, environmental science, safety principles, accounting, auditing, technical aspects, energy conservation, security needs, and operations—into both professions. Based on years of experience in the water and wastewater profession and personal interactions with high-level public service managers, however, engineers, biologists, chemists, and others with no formal management training and no proven leadership expertise are often hindered (limited) in their ability to solve the complex management problems currently facing both industries. I admit my biased view in this regard because my experience in public service has, unfortunately, exposed me to more dysfunctional than functional managers.

So what is dysfunctional management? How is it defined? Well, we have all encountered one or more exposures to dysfunctional managers in our working careers; thus, there is no need to discuss this matter any further here.

ENERGY CONSERVATION MEASURES AND SUSTAINABILITY

Several long-term economic, social, and environmental trends [(Elkington's (aka Godfather of Sustainability) (1999) so-called Triple Bottom Line] evolve around us. Many of these long-term trends are developing because of us and specifically for us or simply to sustain us. Many of these long-term trends follow general courses and can be described by the jargon of the day; that is, they can be alluded to or specified by a specific buzzword or buzzwords common in usage today. We frequently hear these buzzwords used in general conversation (especially in abbreviated texting form). Buzzwords such as empowerment, outside the box, streamline, wellness, synergy, generation X, face time, exit strategy, LOL, clear goal, and so on are just part of our daily vernacular.

In this section, the popular buzzword we are concerned with, *sustainability*, is often used in business. However, in water and wastewater treatment, sustainability is much more than just a buzzword; it is a way of life (or should be). Numerous definitions of sustainability are overwhelming, vague, and/or indistinct. For our purposes, there is a long definition and a short definition of sustainability. The long definition ensures that water and wastewater treatment operations occur indefinitely without negative impact. The short definition is the capacity of water and wastewater operations to endure. Whether we define long or short fashion, what does sustainability mean in the real world of water and wastewater treatment operations?

We defined sustainability in both long and short terms. However, sustainability in water and wastewater treatment operations can be characterized in broader or all-encompassing terms than those simple definitions. As mentioned, using the Triple Bottom Line scenario, in regard to sustainability, the environmental, economic, and social aspects of water and wastewater treatment operations can more specifically define today's and tomorrow's needs.

Infrastructure is another term used in this text; it can be used to describe water and wastewater operation as a whole or can identify several individual or separate elements of water and wastewater treatment operations (unit processes). For example, in wastewater operations, we can devote extensive coverage of wastewater collection and interceptor systems, lift or pumping stations, influent screening, grit removal, primary clarification, aeration, secondary clarification, disinfection, outfalling, and a whole range of solids handling unit processes. On an individual basis, each of these unit processes can be described as an

TABLE 3.2
2021   Report Card for American Infrastructure

Infrastructure	Grade
Bridges	C
Dams	D
Drinking water	C–
Energy	C–
Hazardous waste	D+
Rail	B
Roads	D
Schools	D+
Wastewater	D+

America's Infrastructure GPA: C–

Source: Modified from American Society of Civil Engineers (2012). *Report Card for American Infrastructure 2021.* Accessed 01/04/2024 @ http://www.infrastrucutrereportcard.org/.

integral infrastructure component of the process. Or, holistically, we simply could group each unit process as one, as a whole, combining all wastewater treatment plant unit processes as "the" operational infrastructure. We could do the same for water treatment operations. For example, as individual water treatment infrastructure components, fundamental systems, or unit processes, we could list source water intake, pretreatment, screening, coagulation and mixing, flocculation, settling and biosolids processing, filtering, disinfection, storage, and distribution systems. Otherwise, we could simply describe water treatment plant operations as the infrastructure.

How one chooses to define infrastructure is not important. What is important is how to maintain and manage infrastructure most efficiently and economically possible to ensure its sustainability. This is no easy task. Consider, for example, the 2021 Report Card for American Infrastructure produced by the American Society of Civil Engineers shown in Table 3.2.

Not only must water and wastewater treatment managers maintain and operate aging and often underfunded infrastructure, but they must also comply with stringent environmental regulations and keep stakeholders and ratepayers satisfied with operations and rates. Moreover, managers must incorporate economic considerations into every decision. For example, they must meet regulatory standards for the quality of treated drinking water and discharged wastewater effluent. They must also plan for future upgrades or retrofits that will enable the facility to meet future water quality and effluent regulatory standards. Finally, and most importantly, managers must optimize the use of manpower, chemicals and electricity to achieve sustainability.

The EPA (2012) points out that *sustainable development* can be defined as meeting the needs of the present generation without compromising the ability of future generations to meet theirs. The current U.S. population benefits from the investments made over the past several decades to build the nation's water/wastewater infrastructure.

Practices that encourage water and wastewater sector utilities and their customers to address existing needs, so that future generations will not be left to address the approaching wave of needs resulting from aging water and wastewater infrastructure, must continuously be promoted by sector professionals. To be on a sustainable path, investments need to result in efficient infrastructure and infrastructure systems and be at a pace and level that allow the water and wastewater sectors to provide the desired levels of service over the long term.

Sounds easy enough: the water/wastewater manager simply needs to put his or her operation on a sustainable path; moreover, he or she can simply accomplish this by investing, right? Well, investing what? Investing in what? Investing how much? These are questions that obviously require answers. Before moving on with this discussion it is important first to discuss plant infrastructure basics (focusing primarily on wastewater infrastructure and, in particular, on piping systems) and secondly to discuss funding (the cash cow versus cash dog syndrome).

During the 1950s and 1960s, the U.S. government encouraged the prevention of pollution by providing funds for the construction of municipal wastewater treatment plants, water pollution research, and technical training and assistance. New processes were developed to treat sewage, analyze wastewater, and evaluate the effects of pollution on the environment. Despite these efforts, however, the expanding population, industrial, and economic growth caused pollution and health difficulties to increase.

DID YOU KNOW?

Looking at water distribution piping only, EPA's 2000 survey on community water systems found that in systems serving more than 100,000 people, about 40% of drinking water pipes are over 40 years old. However, it is important to note that age, in and of itself, does not necessarily point to problems. If a system is well maintained, it can operate for a long time (EPA 2012).

In response to the need for a coordinated effort to protect the environment, the National Environmental Policy Act (NEPA) was signed into law on January 1, 1970. In December of that year, a new independent body, the USEPA, was created to bring all of the pollution-control programs related to air, water, and solid wastes under one roof. In 1972, the Water Pollution Control Act Amendments expanded the role of the federal government in water pollution control and significantly increased federal funding for the construction of wastewater treatment plants.

Many of the wastewater treatment plants in operation today are the result of federal grants made over the years. For example, because of the 1977 Clean Water Act Amendment to the Federal Water Pollution Control Act of

1972 and the 1987 Clean Water Act reauthorization bill, funding for wastewater treatment plants was provided.

Many large sanitation districts, with their multiple plant operations, and an even larger number of single plant operations in smaller communities in operation today are a result of these early environmental laws. Because of these laws, the Federal Government provided grants of several hundred million dollars to finance the construction of wastewater treatment facilities throughout the country.

Many of these locally or federally funded treatment plants are aging; based on experience, I can rate some as dinosaurs. The point is, many facilities are facing problems caused by aging equipment, facilities, and infrastructure. Complicating the problems associated with natural aging is the increasing pressure on inadequate older systems to meet the demands of increased population and urban growth. Facilities built in the 1960s and 1970s are now 40 to 50+ years old, and not only are they showing signs of wear and tear, but they also simply were not designed to handle the level of growth that has occurred in many municipalities.

Regulations often necessitate the need for upgrades. By matching funds or providing federal money to cover some of the costs, municipalities can take advantage of a window of opportunity to improve their facilities at a lower direct cost to the community. Those federal dollars, of course, come with strings attached; they must be spent on specific projects in specific areas. On the other hand, many times new regulatory requirements are put in place without the financial assistance needed to implement them. When this occurs, either the local community ignores the new requirements (until caught and forced to comply) or they face the situation and implement changes through local tax hikes or rate-payer hikes to cover the cost of compliance.

An example of how a change in regulations can force the issue is demonstrated by the demands made by OSHA and USEPA in their Process Safety Management (PSM)/Risk Management Planning (RMP) regulations. These regulations put the use of elemental chlorine (and other listed hazardous materials) under close scrutiny. Moreover, because of these regulations, plant managers throughout the country are forced to choose which side of a double-edged sword cuts their way the most. One edge calls for full compliance with the regulations (analogous to stuffing the regulation through the eye of a needle). The other edge calls for substitution—that is, replacing elemental chlorine (probably the EPA's motive in the first place; see the note) with a non-listed hazardous chemical (e.g., hypochlorite) or a physical disinfectant (e.g., ultraviolet irradiation, UV)—either way, a very costly undertaking (Spellman 2009).

- **Note:** Many of us who have worked in water and wastewater treatment for years characterize PSM and RMP as the elemental chlorine "killer." You have probably heard the old saying: "If you can't do away with something in one way, then regulate it to death."

- **Note:** Changes resulting from regulatory pressure sometimes mean replacing or changing existing equipment, increased chemical costs (e.g., substituting hypochlorite for chlorine typically increases costs threefold), and could easily involve increased energy and personnel costs. Equipment conditions, new technology, and financial concerns are all considerations when upgrades or new processes are chosen. In addition, the safety of the process must be considered, of course, due to the demands made by USEPA and OSHA. The potential harm to workers, the community, and the environment are all under study, as are the possible long-term effects of chlorination on the human population.

Note that water and wastewater treatment plants typically have a useful life of 20–50 years before they require expansion or rehabilitation. Collection, interceptor, and distribution pipes have life cycles that can range from 15 to 100 years, depending on the type of material and where they are laid. Long-term corrosion reduces a pipe's carrying capacity, requiring increasing investments in power and pumping. When water or wastewater pipes age to the point of failure, the result can be contamination of drinking water, the release of wastewater into surface waters or basements, and high costs to both replace the pipes and repair any resulting damage. With pipes, the material used and proper installation can be a greater indicator of failure than age.

A 2002 EPA report referenced a Water Infrastructure Gap Analysis that compared current spending trends at the nation's drinking water and wastewater treatment facilities to the expenses they can expect to incur for both capital and operations and maintenance costs. The "gap" is the difference between projected and needed spending and was found to be over $500 billion over a 20-year period. This important 2002 EPA gap analysis study is just as pertinent today as it was 10 years ago. Moreover, the author draws upon many of the tenets presented in the EPA analysis in formulating many of the basic points and ideas presented herein.

Autonomous Operations

During the mid-1980s through the 1990s, while I was inspecting water and wastewater treatment plants or training plant operators on-site throughout the country, I began to see an increasing number of treatment plants incorporating automation into plant operations. These automated activities began with the installation, operation, and monitoring of automatic samplers. These samplers were, and are, controlled by computers via the advancement of the Internet. The Internet (actually the Internet of Things—IoT) was and is a huge game changer, so to speak, in the operation of water and wastewater treatment plants. Most of these autonomous operations were and are installed in the larger plants I visited in larger metropolitan areas.

This was, and is, the case because the larger entities have the finances to make the changeover to autonomous operations with automatic samplers and data collection inputted to a designated plant computer. I noted that manual sampling operations were also conducted in plants with automatic sampling capabilities. Manual sampling is normally used for collecting grab samples and for immediate in situ field analyses. Also, manual sampling was and is used in lieu of automatic equipment over extended periods of time for composite sampling, especially when it is necessary to evaluate unusual waste stream conditions and verify the automatic sampler results—basically ensuring that $2+2=4$, and so forth, in measuring and providing accurate automatic measurements.

DID YOU KNOW?

An Atmospheric Vacuum Breaker (AVB) is intended to be used in a short-term (less than 12 hours) operation to protect against backsiphonage only.

CHAPTER REVIEW QUESTIONS

Answers to chapter review questions are found in Appendix A.

3.1 Define paradigm as used in this text.
3.2 Define paradigm shift as used in this text.
3.3 List five elements of the multiple-barrier approach.
3.4 "Water service delivery remains one of the 'hidden functions' of local government." Explain.
3.5 _____ Drinking Water Standards are not enforceable.
3.6 Explain the difference between privatization and re-engineering.
3.7 Define benchmarking.
3.8 List the five benchmarking steps.

THOUGHT-PROVOKING QUESTIONS (ANSWERS WILL VARY):

3.9 Given the assignment and the proper resources, how would you clean up Chesapeake Bay?
3.10 Is the use of renewable energy sources in water and wastewater treatment plants a feasible option?
3.11 Of all the current issues facing water and wastewater treatment operations discussed in this chapter, which one do you feel is the most urgent to resolve?

NOTE

1 Adapted from Spellman, F.R. (2013). *Water & Wastewater Infrastructure: Energy Efficiency and Sustainability.* Boca Raton, FL: CRC Press.

REFERENCES

Camp, R., 1989. *The Search for Industry Best Practices that Lead 2 Superior Performance.* Boca Raton, FL: Productivity Press (CRC Press).
Elkington, J., 1999. *Cannibals with Forks.* New York: Wiley.
EPA. 2012. *Frequently Asked Questions: Water Infrastructure and Sustainability.* Accessed 04/01/2023 @ Https://water.ep.gov/infrastrure/sustain/si_faqs.cfm.
Johnson, R. and Moore, A., 2002. Policy Brief 17 Opening the Floodgates: Why Water Privatization Will Continue. *Reason Public Institute.* [www rppi.org.pbrief17].
Mather, J.R., 1984. *Water Resources: Distribution, Use, and Management.* New York: John Wiley & Sons.
Spellman, F.R., 2009. *Handbook of Water and Wastewater Treatment Plant Operations*, 2nd ed. Boca Raton, FL: CRC Press.
USEPA. 2008. *Ensuring a Sustainable Future: An Energy Management Guidebook for Wastewater and water Utilities.* Washington, DC: United States Environmental Protection Agency.
USEPA. 2021. EPA Announces Plans for New Wastewater Regulations, Including First Limits for PFAS, Updated Limits for Nutrients. Accessed 3/29/24 @ https://epa.gov/newsreleases/search/press_office/headquaters-226129.
WEF. 1997. *Energy Conservation in Wastewater Treatment Facilities.* Manual of Practice No. MFD-2. Alexandria, VA: Water Environment Federation.

4 Water/Wastewater Operators

If you are short of water, the choices are stark: conservation, treatment and reuse of wastewater, technological invention, or the politics of violence.

It is not technology (engineering and science) that will mitigate the looming water crisis. Instead, providing safe drinking water to the masses can only be accomplished through politics and management. The irony is that it is both politics and management that are at the heart of the water crisis. Perhaps the most important person at a water [and wastewater] treatment plant is the plant operator because that person is responsible for treating the water to meet or exceed the Federal Safe Drinking Water Act [and Clean Water Act] stands and public expectations.

–D.S. Sarai (2006)

INTRODUCTION

To begin our discussion of water and wastewater operators, it is important to point out a few significant factors:

- Employment as a water and wastewater operator is concentrated in local government and private water supply and sanitary services companies.
- Postsecondary training is increasingly an asset as the number of regulated contaminants grows and treatment unit processes become more complex.
- Because of advances in digital control via computers, remote samplers and other autonomous operations, operators need to be well-versed in computer skills.
- Operators must pass examinations certifying that they are capable of overseeing various treatment processes.
- Plants operate 24/7; therefore, plant operators must be willing to work shifts.
- Operators have a relatively high incidence of on-the-job injuries.

To properly operate a water treatment and distribution system or a wastewater treatment and collection system usually requires a team of highly skilled personnel filling a variety of job classifications. Typical positions include plant manager/plant superintendent, plant engineer, chief operator, lead operator, operator, maintenance operator, distribution and/or interceptor system technicians, assistant operators, laboratory professionals, electricians, instrument technicians, recycling manager, safety and health manager, accountants, and clerical personnel, to name just a few.

Beyond the distinct job classification titles, over the years those operating water and wastewater plants have been called by a variety of titles. These include water jockey, practitioner of water, purveyor of water, sewer rat, or just plain water or wastewater operator. Based on my experience, I have come up with a title that perhaps more closely characterizes what the water/wastewater operator really is: a Jack or Jill of all trades. This characterization seems fitting when you consider the knowledge and skills required of operators to properly perform their assigned duties. Moreover, operating the plant or distribution/collection system is one thing; taking samples, operating equipment, monitoring conditions, determining settings for chemical feed systems and high-pressure pumps, performing laboratory tests, and recording the results in the plant's daily operating log is another.

It is, however, the non-typical functions, the diverse functions, and the off-the-wall functions that cause us to describe operators as Jacks or Jills of all trades. For example, in addition to their normal, routine, daily operating duties, operators may be called upon to make emergency repairs to systems (e.g., making a welding repair to a vital piece of machinery to keep the plant or unit process on line); perform material handling operations; make chemical additions to process flow; respond to hazardous materials emergencies; make confined space entries; perform site landscaping duties; and carry out several other assorted functions. Remember, the plant operator's job is to keep the plant running and to meet permit requirements. Keeping the plant running, the flow flowing, and meeting permit requirements—no matter what—requires not only talent but also the performance of a wide range of functions, many of which are not called for in written job descriptions.

DID YOU KNOW?

Water and wastewater treatment plant and system operators held about 124,800 jobs in 2023. Almost four in five operators worked for local governments. Others worked primarily for private water, sewage, and other systems, utilities, as well as for private waste treatment, disposal, and waste management services companies. Private firms are increasingly providing operation and management services to local governments on a contract basis. Along with steady increases in new positions, there is also a loss of about 8,000 positions each year due to retirement, other employment opportunities, and automation (BLS 2023).

DOI: 10.1201/9781003581901-5

SETTING THE RECORD STRAIGHT

Based on experience, I have found that most people either have a preconceived notion as to what water and wastewater operations are all about, or they have nary a clue. On the one hand, most of us understand that clean water is essential for everyday life. Moreover, we have at least a vague concept that water treatment plants and water operators treat water to make it safe for consumption. On the other hand, when it comes to wastewater treatment and system operations, many of us have an ingrained image of toilets being flushed and a sewer system managed and run by a bunch of sewer rats. Others give wastewater, its treatment and the folks who treat it, no thought at all (that is, unless they are irate ratepayers upset at a back-flushed toilet or the cost of wastewater service).

Typically, the average person has other misconceptions about water and wastewater operations. For example, very few people can identify the exact source of their drinking water supply. Is it pumped from wells, rivers, or streams to water treatment plants? Similarly, where is it treated and distributed to customers? The average person is clueless about the ultimate fate of wastewater. Once the toilet is flushed, it is out of sight, out of mind, and that is that.

Beyond the few functions we have pointed out to this point, what exactly is it that water and wastewater operators, the 124,000+ Jacks or Jills of all trades in the United States do? Operators in both water and wastewater treatment systems control unit processes and equipment to remove or destroy harmful materials, chemical compounds, and microorganisms from the water. They also control pumps, valves, and other processing equipment (including a wide array of computerized systems in a growing number of systems) to convey the water or wastewater through the various treatment processes (unit processes) and dispose of (or reuse) the removed solids (waste materials: sludge or biosolids). Operators also read, interpret, and adjust meters and gauges to make sure plant equipment and processes are working properly. They operate chemical-feeding devices, take samples of the water or wastewater, perform chemical and biological laboratory analyses, and adjust the amount of chemicals, such as chlorine, in the water/wastestream. They use a variety of instruments to sample and measure water quality, as well as common hand and power tools to make repairs and adjustments. Operators also make minor repairs to valves, pumps, and basic electrical equipment (Note: electrical work should only be accomplished by qualified personnel) and other equipment.

As mentioned, water and wastewater system operators increasingly rely on computers to help monitor equipment, store sampling results, make process-control decisions, schedule and record maintenance activities, and produce reports. As pointed out earlier, computer-operated automatic sampling devices are beginning to gain widespread acceptance and use in both industries, especially at larger facilities. When a system malfunction occurs, operators may use system computers to determine the cause and the solution to the problem.

THE DIGITAL WORLD

At many water/wastewater treatment plants, operators are required to perform skilled treatment plant operations work and to monitor, operate, adjust, and regulate a computer-based treatment process. In addition, the operator is also required to operate and monitor electrical, mechanical, and electronic processing and security equipment through central and remote terminal locations in a solids processing, water purification, or wastewater treatment plant. In those treatment facilities that are not completely or partially automated and computer-controlled, computers are used in other applications, such as clerical tasks and a computer maintenance management system (CMMS). The operator must be qualified to operate and navigate such computer systems.

Typical examples of the computer-literate operator's work (for illustrative purposes only) are provided as follows:

- Monitors, adjusts, starts, and stops automated water treatment processes and emergency response systems to maintain a safe and efficient water treatment operation; monitors treatment plant processing equipment and systems to identify malfunctions and their probable causes, following prescribed procedures; places equipment in or out of service or redirects processes around failed equipment; following prescribed procedures, monitors and starts process-related equipment, such as boilers, to maintain process and permit objectives; refers difficult equipment maintenance problems and malfunctions to a supervisor; monitors the system through a process integrated control terminal or remote station terminal to ensure control devices are making proper treatment adjustments; operates the central control terminal keyboard to perform backup adjustments to such treatment processes as influent and effluent pumping, chemical feed, sedimentation, and disinfection; monitors specific treatment processes and security systems at assigned remote plant stations; observes and reviews terminal screen display of graphs, grids, charts and digital readouts to determine process efficiency; responds to visual and audible alarms and indicators that signal deviations from normal treatment processes and chemical hazards; identifies false alarms and other indicators that do not require immediate response; alerts remote control locations to respond to alarms indicating trouble in that area; performs alarm investigations.
- Switches over to semi-automatic or manual control when the computer control system is not properly controlling the treatment process; off-scans

a malfunctioning field sensor point and inserts data obtained from the field to maintain computer control; controls automated mechanical and electrical treatment processes through the computer keyboard when computer programs have failed; performs field tours to take readings when problems cannot be corrected through the computer keyboard; makes regular field tours of the plant to observe physical conditions; manually controls processes when necessary.

- Determines and changes the amount of chemicals to be added for the amount of water, wastewater or biosolids to be treated; takes periodic samples of treated residuals, biosolids processing products and byproducts, clean water, or wastewater for laboratory analysis; receives, stores, handles, and applies chemicals and other supplies needed for the operation of the assigned station; maintains inventory records of supplies on hand and quantities used; prepares and submits daily shift operational reports; records daily activities in the plant operation log, computer database, or from a computer terminal; changes chemical feed tanks, chlorine cylinders, and feed systems; flushes clogged feed and sampling lines.
- Notes any malfunctioning equipment; makes minor adjustments when required; reports major malfunctions to a higher-level operator and enters maintenance and related task information into a computerized maintenance management system (CMMS) and processes work requests for skilled maintenance personnel.
- Performs routine mechanical maintenance, such as packing valves, adjusting belts, and replacing shear pins and air filters; lubricates equipment by applying grease and adding oil, changes and cleans strainers; drains condensate from pressure vessels, gearboxes. and drip traps; performs minor electrical maintenance, such as replacing bulbs and resetting low-voltage circuit switches; prepares equipment for maintenance crews by unblocking pipelines and pumps, and isolating and draining tanks; checks equipment as part of a preventive and predictive maintenance program; reports more complex mechanical and electrical problems to supervisors.
- Responds, in a safe manner, to chlorine leaks and chemical spills in compliance with OSHA's HAZWOPER (29 CFR 1910.120) requirements and with plant-specific emergency response procedures; participates in chlorine and other chemical emergency response drills.
- Prepares operational and maintenance reports as required, including flow and treatment information; changes charts and maintains recording equipment; utilizes system and other software packages to generate reports, charts, and graphs of flow and treatment status and trends; maintains workplace housekeeping.

DID YOU KNOW?

What enables computer operations in water and wastewater treatment plants is the incorporation of algorithms. An *algorithm* is a specific mathematical calculation procedure, a computable set of steps to achieve a desired result. More specifically, according to Cormen et al. (2002), "an algorithm is any well-defined computational procedure that takes some value, or set of values, as input and produces some value, or set of values, as output." In other words, an algorithm is a recipe for an automated solution to a problem. A computer model may contain several algorithms.

Algorithms should not be confused with computations. While an algorithm is a systematic method for solving problems, and computer science is the study of algorithms (although algorithms were developed and used long before any device resembling a modern computer was available), the act of executing an algorithm—that is, systematically manipulating data—is called *computation*.

PLANT OPERATORS AS EMERGENCY RESPONDERS

Occasionally, operators must work under emergency conditions. Sometimes these emergency conditions are operational and not necessarily life-threatening. A good example occurs during a storm event when there may be a temporary loss of electrical power and large amounts of liquid waste flow into sewers, exceeding a plant's treatment capacity. Emergencies can also be caused by conditions inside a plant, such as oxygen deficiency within a confined space or exposure to toxic and/or explosive off-gases, such as hydrogen sulfide and methane. To handle these conditions, operators are trained to make an emergency management response and use special safety equipment and procedures to protect co-workers, public health, the facility, and the environment. During emergencies, operators may work under extreme pressure to correct problems as quickly as possible. These periods may create dangerous working conditions, and operators must be extremely careful and cautious.

Operators who must aggressively respond to hazardous chemical leaks or spills (e.g., enter a chlorine gas-filled room and install chlorine repair kit B on a damaged 1-ton cylinder to stop the leak) must possess a HAZMAT Emergency Response Technician 24-hour certification. Additionally, many facilities where elemental chlorine is used for disinfection, odor control, or other process applications require operators to possess an appropriate certified pesticide applicator training completion certificate. Because of OSHA's specific confined space requirement, whereby a standby rescue team for entrants must be available, many plants require operators to hold and maintain CPR/1st Aid certification.

Note: It is important to point out that many wastewater facilities have substituted elemental chlorine with sodium or calcium hypochlorite, ozone, or ultraviolet irradiation because of the stringent requirements of OSHA's Process Safety Management Standard (29 CFR 1910.119) and USEPA's Risk Management Program. This is not the case in most water treatment operations, however. In water treatment systems, elemental chlorine is still employed because it provides a chlorine residual that is important in maintaining safe drinking water supplies, especially throughout lengthy distribution systems.

OPERATOR DUTIES AND WORKING CONDITIONS

The specific duties of plant operators depend on the type and size of the plant. In smaller plants, one operator may control all machinery, perform sampling and lab analyses, keep records, handle customer complaints, troubleshoot and make repairs, or perform routine maintenance. In some locations, operators may handle both water treatment and wastewater treatment operations. On the other hand, in larger plants with many employees, operators may be more specialized and only monitor one unit process (e.g., a solids-handling operator who operates and monitors an incinerator). Along with treatment operators, plant staffing may include environmentalists, biologists, chemists, engineers, laboratory technicians, maintenance operators, supervisors, clerical help, IT personnel, and various assistants.

In the United States, notwithstanding a certain amount of downsizing brought on by privatization activities, employment opportunities for water/wastewater operators have increased in number. The number of operators has increased because of the ongoing construction of new water/wastewater and solids-handling facilities. In addition, operator jobs have increased due to water pollution standards that have become increasingly stringent since the adoption of two major federal environmental regulations: the Clean Water Act of 1972 (and subsequent amendments), which implemented a national system of regulation on the discharge of pollutants, and the Safe Drinking Water Act of 1974, which established standards for drinking water.

Operators are often hired in industrial facilities to monitor or pretreat wastes before discharge to municipal treatment plants. These wastes must meet certain minimum standards to ensure that they have been adequately pretreated and will not damage municipal treatment facilities. Municipal water treatment plants also must meet stringent drinking water standards. This often means that additional qualified staff members must be hired to monitor and treat/remove specific contaminants. Complicating the problem is the fact that the list of contaminants regulated by these

regulations has grown over time. For example, the 1996 Safe Drinking Water Act Amendments include standards for monitoring Giardia and Cryptosporidium, two biological organisms (protozoans) that cause health problems. Operators must be familiar with the guidelines established by federal regulations and how they affect their plant. In addition to federal regulations, operators must be aware of any guidelines imposed by the state or locality in which the treatment process operates.

Another unique factor related to water/wastewater operators is their working conditions. Water and wastewater treatment plant operators work indoors and outdoors in all kinds of weather. Operators' work is physically demanding and often performed in unclean locations (hence, the emanation of the descriptive but inappropriate title, "sewer rat"). They are exposed to slippery walkways, vapors, odors, heat, dust, and noise from motors, pumps, engines, and generators. They work with hazardous chemicals. In water and wastewater plants, operators may be exposed to many bacterial and viral conditions. As mentioned, dangerous gases such as methane and hydrogen sulfide could be present, so they need to use proper safety gear.

Operators generally work a 5-day, 40-hour week. However, many treatment plants operate 24/7, and operators may have to work nights, weekends, holidays, or rotating shifts. Some overtime is occasionally required in emergencies.

Over the years, statistical reports have provided historical evidence showing that the water/wastewater industry is an extremely unsafe occupational field. This less-than-stellar safety performance has continued to deteriorate, even in the age of the Occupational Safety and Health Act (OHS Act 1970).

The question is, why is the water/wastewater treatment industry's on-the-job injury rate so high? Several reasons help explain this high injury rate. First, all of the major classifications of hazards exist at water/wastewater treatment plants (with the typical exception of radioactivity):

- oxygen deficiency
- physical injuries
- toxic gases and vapors
- infections
- fire
- explosion
- electrocution

Along with all the major classifications of hazards, other factors contribute to the high incidence of injury in the water/wastewater industry. Some of these can be attributed to

- complex treatment systems
- shift work
- new employees
- liberal workers' compensation laws
- absence of safety laws
- absence of safe work practices and safety programs

Experience has shown that a lack of well-managed safety programs and safe work practices are major factors contributing to the water/wastewater industry's high incidence of on-the-job injuries (Spellman 2001).

OPERATOR CERTIFICATION AND LICENSURE

A high school diploma or its equivalent is usually required as the entry-level credential to become a water or wastewater treatment plant operator-in-training. Operators need mechanical aptitude and should be competent in basic mathematics, chemistry, and biology. They must have the ability to apply data to formulas for treatment requirements, flow levels, and concentration levels. Some basic familiarity with computers is also necessary due to the present trend toward computer-controlled equipment and more sophisticated instrumentation. Certain operator positions—particularly in larger cities—are covered by civil service regulations. Applicants for these positions may be required to pass a written examination testing mathematics skills, mechanical aptitude, and general intelligence.

Because treatment operations are becoming more complex, completion of an associate's degree or a 1-year certificate program in water quality and wastewater treatment technology is highly recommended. These credentials increase an applicant's chances for both employment and promotion. Advanced training programs are offered throughout the country. They provide good general thorough advanced training on water and wastewater treatment processes, as well as basic preparation for becoming a licensed operator. They also offer a wide range of computer training courses.

New water and wastewater operators-in-training typically start as attendants or assistants and learn the practical aspects of their job under the direction of an experienced operator. They learn by observing, show-and-tell, and performing routine tasks such as recording meter readings, taking samples of liquid waste and sludge, and performing simple maintenance and repair work on pumps, electrical motors, valves, and other plant or system equipment. Larger treatment plants generally combine this on-the-job training (OJT) with formal classroom or self-paced study programs. Some large sanitation districts operate their own 3–4-year apprenticeship schools. In some of these programs, each year of apprenticeship school completed not only prepares the operator for the next level of certification or licensure but also satisfies a requirement for advancement to the next higher pay grade.

The Safe Drinking Water Act (SDWA) Amendments of 1996, enforced by the USEPA, specify national minimum standards for certification (licensure) and recertification of operators of community and non-transient, non-community water systems. As a result, operators must pass an examination to certify that they are capable of overseeing water/wastewater treatment operations. There are different levels of certification depending on the operator's experience and training. Higher certification levels qualify the operator for a wider variety of treatment processes. Certification requirements vary by state and by the size of treatment plants. Although relocation may require becoming certified in a new location, many states accept other states' certifications.

In an attempt to ensure the currency of training and qualifications and to improve operators' skills and knowledge, most state drinking water and water pollution control agencies offer ongoing training courses. These courses cover principles of treatment processes and process control methods, laboratory practices, maintenance procedures, management skills, collection system operation, general safe work practices, chlorination procedures, sedimentation, biological treatment, sludge/biosolids treatment, biosolids land application and disposal, and flow measurements. Correspondence courses covering both water/wastewater operations and preparation for state licensure examinations are provided by various state and local agencies. Many employers provide tuition assistance for formal college training.

Whether received from formal or informal sources, training provided for or obtained by water and wastewater operators must include coverage of very specific subject/topic areas. Though much of their training is similar or the same, Tables 4.1 and 4.2 list many of the specific specialized topics that waterworks and wastewater operators are expected to have a fundamental knowledge of.

Note: It is important to note that both water and wastewater operators must have fundamental knowledge of basic science and math operations.

Note: For many water/wastewater operators, crossover training or overlapping training is common practice and highly recommended. With the advent of computers, artificial intelligence (AI), the Internet of Things (IoT), privatization, budget cuts, and new regulations, the water/wastewater operator—the fluid mechanic who is a Jack or Jill of all trades—is equipped to survive change including downsizing due to autonomous operations.

Although Tables 4.1 and 4.2 list many of the specific specialized topics that waterworks and wastewater operators are expected to have a fundamental knowledge of, operator licensure examinations require more specific knowledge of a wide range of topics. Many of these specific examination topics are listed in Table 4.3.

TABLE 4.1
Specialized Topics for Waterworks Operators

Chemical addition	Hydraulics – math
Chemical feeders	Laboratory practices
Chemical feeders – math	Measuring and control
Clarification	Piping and valves
Coagulation-flocculation	Public health
Corrosion control	Pumps
Disinfection	Recordkeeping
Disinfection – math	General science
Basic electricity and controls	Electric motors
Filtration	Finances
Filtration – math	Storage
Fluoridation	Leak detection
Fluoridation – math	Hydrants
General safe work practices	Cross connection control and backflow
Bacteriology	Stream ecology

TABLE 4.2

Specialized Topics for Wastewater Operators

Wastewater math	Fecal coliform testing
Troubleshooting techniques	Recordkeeping
Preliminary treatment	Flow measurement
Sedimentation	Sludge dewatering
Ponds	Drying beds
Trickling filters	Centrifuges
Rotating biological contactors	Vacuum filtration
Activated sludge	Pressure filtration
Chemical treatment	Sludge incineration
Disinfection	Land application of biosolids
Solids thickening	Laboratory procedures
Solids stabilization	General safety

TABLE 4.3

Need to Know Licensure Examination Specific Topics

Examination Topics Included in Many State Licensure Examinations

Pollutants	Ammonia
	pH
	Chlorine residual
	Fecal coliform
	Phosphorus
	Nonfilterable reside (e.g., total suspended solids—TSS)
	BOD/CBOD
	Heavy metals
	Toxicity
	Nitrites, nitrates, TKN
	Volatile organics
	Total priority pollutants
	COD
Operation & maintenance	Pumps (types, capacity, head relationships, control)
	O & M manual
	Corrective maintenance
	Preventive maintenance
	Collection system
	Infiltration/Inflow
	Electrical usage
	Blower performance
	Recordkeeping
	Piping system hydraulics
Treatment theory	Aerated lagoons
	Aerobic digestion
	Biosolids disposal
	Comminutors
	Disinfection (chlorination, dechlorination, UV)
	Activated sludge (biosolids)
	Screens
	Grit removal

(Continued)

TABLE 4.3 (*Continued*)

Need to Know Licensure Examination Specific Topics

Examination Topics Included in Many State Licensure Examinations

Pollutants	Ammonia
	Final clarifiers
	Sand drying beds
	Slow sand filters
	Flow measurement
	Primary tanks
	Nitrification
	Rapid and slow sand filters
	Anerobic digestion
	Biosolids disposal
	Sludge thickening
	Sludge dewatering
	Alkaline stabilization
Sampling procedures/ Handling	Grab sampling
	Chlorine residual
	Dissolved oxygen (DO)
	Oil & grease
	CBOD/BOD
	Suspended solids
	Fecal coliform
	Composite sampling (flow proportional)
	Sample holding times and conditions
	pH
	Ammonia
	Phosphorus
	Heavy metals
	Volatile organics
	Toxicity
Laboratory	Laboratory procedures
	Testing requirements and procedures (chemical/biological)
	Biomonitoring
	Microscopic
Regulations	State water board requirements
	NPDES permit
	Local limits
	Reporting
	TMDL
	Water quality standard
Math	Hydraulic loading
	Organic loading
	Surface overflow rate
	Weir overflow rate
	Hydraulic detention time
	Suspended solids test
	CBOD/BOD test
	Flowrate, velocity, area
	Concentration, mass, flowrate
	Aeration requirements

(Continued)

TABLE 4.3 (Continued)

Need to Know Licensure Examination Specific Topics

Examination Topics Included in Many State Licensure Examinations

Pollutants	Ammonia
	Percent removals
	Volumes and areas
	Percentages
	Costs
	SVI
	Unit conversions
	Population equivalents
	Weighted averages
	MCRT/Sludge age
	% Solids, sludge flowrates, specific gravity, mass
	Geometric means
	Mass balances
	Chlorine demand, dosage, residual
	Volatile solids reduction
	Solids loading rate
	Solids detention time
	Pressure/Head
	F:M
	Chemical concentrations
	Energy value of digester gas
	Energy requirements of heating sludge
Safety	Chemical
	Confined space entry
	Lockout/Tagout
	Electrical
	OSHA standards
	First aid/CPR
	Personal protective equipment
	Housekeeping
	Slips, trips, and falls
	Process safety management (PSM)
	Risk management plan (RMP)
	Bloodborne pathogens

THE BOTTOM LINE

The Digital Age, with its quiver containing artificial intelligence (AI), the Internet of Things (IOT), and other advancements in basic autonomous technology that enable treatment plant operation without being controlled "directly" by humans, has the potential to revolutionize water and wastewater treatment operations. I have heard it said during plant training sessions that autonomous technology is needed to address the current challenges faced by the water industries. These challenges include optimizing processes and

predicting water quality to enhance energy management and resource recovery. Autonomous technology such as AI offers numerous benefits that improve efficiency, water quality, and environmental sustainability. While it is true that there are challenges and limitations to overcome, successful applications of autonomous technology in water/wastewater treatment demonstrate its significant potential. With further research, development, collaboration, and incorporation of anti-hacking technologies, autonomous technologies can transform water/wastewater treatment into more efficient, cost-effective, and environmentally friendly processes.

I have also heard plant personnel complain that autonomous technology is designed to replace them. Employees are a huge expense in many industries, and managers who are looking for various means to reduce costs are encouraging the implementation of AI and other digital technologies to lower costs by replacing workers with technology. When I brought this subject up in one of my plant training sessions, a young licensed operator spoke up and said, "Who are they going to call when something goes wrong… a robot, Ghostbusters, or divine intervention?" Now, as you might guess, this comment got a lot of laughs from the more than 100 attendees in the class. But when you think about it, there are many in the industry asking the same questions.

Not all the news is bad, however. Studies estimate potential savings of 15%–30% that are "readily achievable" in water and wastewater treatment plants, with substantial financial returns in the thousands of dollars and within payback periods of only a few months to a few years—all accomplished without calling on Ghostbusters or anyone else.

Key Point: A sanitary survey is an analysis conducted to determine all possible water quality impacts to a proposed well.

CHAPTER REVIEW QUESTIONS

4.1   Briefly explain the causal factors behind the high incidence of on-the-job injuries for water/wastewater operators.

4.2   Why is computer literacy so important in operating a modern water/wastewater treatment system?

4.3   Define CMMS.

4.4   What are the necessary training requirements for HAZMAT responders?

4.5   Specify the national minimum standard for certification (licensure) and recertification for water/wastewater operators.

THOUGHT-PROVOKING QUESTION

4.6   Do you think autonomous technology is a good thing or a bad thing? Explain.

REFERENCES

BLS. 2023. *Occupational Outlook Handbook: Water and Wastewater Treatment Plant System Operators.* Accessed 01/30/24 @ https://www.bls.gov.ooh/productions/water-and-wastewater-treatment/plant-and-system-operations.htm.

Carmen et al. 2002. *Introduction to Algorithm*, 4th ed. Cambridge: MIT Press.

Sarai, D.S., 2006. *Water Treatment Made Simple for Operators.* New York: John Wiley & Sons.

Spellman, F.R., 2001. *Safe Work Practices for Wastewater Treatment Plants.* Boca Raton, FL: CRC Press.

5 Plant Security

INTRODUCTION

According to USEPA (2004), there are approximately 160,000 public water systems (PWSs) in the United States, each of which regularly supplies drinking water to at least 25 persons or 15 service connections. PWSs serve 84% of the total U.S. population, while the remainder is served primarily by private wells. PWSs are divided into community water systems (CWSs) and non-community water systems (NCWSs). Examples of CWSs include municipal water systems that serve mobile home parks or residential developments. Examples of NCWSs include schools, factories, churches, commercial campgrounds, hotels, and restaurants.

DID YOU KNOW?

As of 2003, community water systems serve by far the largest proportion of the U.S. population—273 million out of a total population of 290 million (USEPA 2004).

Because drinking water is consumed directly, health effects associated with contamination have long been a major concern. In addition, interruption or cessation of the drinking water supply can disrupt society, impacting human health and critical activities such as fire protection. Although they have no clue as to its true economic value and its future worth, the general public correctly perceives drinking water as central to the life of an individual and society. However, the general public knows even less about the importance of wastewater treatment and the fate of its end product.

Wastewater treatment is important for preventing disease and protecting the environment. Wastewater is treated by publicly owned treatment works (POTWs) and by private facilities such as industrial plants. There are approximately 2.3 million miles of distribution system pipes and approximately 16,255 POTWs in the United States. POTWs serve 75% of the total U.S. population. POTWs with existing flows of less than 1 MGD are considered small; they number approximately 13,057 systems. To determine the population served, 1 MGD equals approximately 10,000 persons served.

Disruption of a wastewater treatment system or service can cause loss of life, economic impacts, and severe public health incidents. If structural damage occurs, wastewater systems can become vulnerable to inadequate treatment. The public is much less sensitive to wastewater as an area of vulnerability than it is to drinking water; however, wastewater systems provide opportunities for terrorist threats.

Federal and state agencies have long been active in addressing these risks and threats to water and wastewater utilities through regulations, technical assistance, research, and outreach programs. As a result, an extensive system of regulations governing maximum contaminant levels of 90 conventional contaminants (most established by the EPA), construction and operating standards (implemented mostly by the states), monitoring, emergency response planning, training, research, and education has been developed to better protect the nation's drinking water supply and receiving waters. Since the events of 9/11, the EPA has been designated as the sector-specific agency responsible for infrastructure protection activities for the nation's drinking water and wastewater system. The EPA is utilizing its position within the water sector and working with its stakeholders to provide information to help protect the nation's drinking water supply from terrorism or other intentional acts.

SECURITY HARDWARE/DEVICES

Keep in mind that when it comes to making "anything" absolutely secure from intrusion or attack, there is inherently, or otherwise, no silver bullet. However, careful preplanning and installation of security hardware and/or devices and products can significantly affect the plant's ability to weather the storm, so to speak. The USEPA (2005) groups the water/wastewater infrastructure security devices or products described below into four general categories:

- Physical asset monitoring and control devices
- Water monitoring devices
- Communication/integration
- Cyber protection devices

PHYSICAL ASSET MONITORING AND CONTROL DEVICES

Aboveground, Outdoor Equipment Enclosures

Water and wastewater systems consist of multiple components spread over a wide area and typically include a centralized treatment plant, as well as distribution or collection system components that are distributed at multiple locations throughout the community. However, in recent years, distribution and collection system designers have favored placing critical equipment—especially assets that require regular use and maintenance—aboveground.

One of the primary reasons for doing so is that locating this equipment above ground eliminates the safety risks associated with confined space entry, which is often required for maintaining equipment located below ground.

DOI: 10.1201/9781003581901-6

In addition, space restrictions often limit the amount of equipment that can be located inside, and there are concerns that some types of equipment (such as backflow prevention devices) can, under certain circumstances, discharge water that could flood pits, vaults, or equipment rooms. Therefore, many pieces of critical equipment are located outdoors and above ground. Many different system components can be installed outdoors and above ground. Examples of these types of components could include:

- Backflow prevention devices
- Air release and control valves
- Pressure vacuum breakers
- Pumps and motors
- Chemical storage and feed equipment
- Meters
- Sampling equipment
- Instrumentation

Much of this equipment is installed in remote locations and/or in areas where the public can access it.

One of the most effective security measures for protecting aboveground equipment is to place it inside a building. When/where this is not possible, enclosing the equipment or parts of the equipment using some sort of commercial or homemade add-on structure may help to prevent tampering with the equipment. Equipment enclosures can generally be categorized into one of four main configurations, which include:

- One-piece, drop-over enclosures
- Hinged or removable top enclosures
- Sectional enclosures
- Shelters with access locks

Other security features that can be implemented on aboveground outdoor equipment enclosures include locks, mounting brackets, tamper-resistant doors, and exterior lighting.

Alarms

An *alarm system* is a type of electronic monitoring system that is used to detect and respond to specific types of events—such as unauthorized access to an asset or a possible fire. In water and wastewater systems, alarms are also used to alert operators when process operating or monitoring conditions go out of preset parameters (i.e., process alarms). These types of alarms are primarily integrated with process monitoring and reporting systems (i.e., SCADA systems). Note that this discussion does not focus on alarm systems that are not related to a utility's processes.

Alarm systems can be integrated with fire detection systems, IDSs, access control systems, or closed circuit television (CCTV) systems, such that these systems automatically respond when the alarm is triggered. For example, a smoke detector alarm can be set up to automatically notify the fire department when smoke is detected, or an intrusion alarm can automatically trigger cameras to turn on in a remote location so that personnel can monitor that location.

An alarm system consists of sensors that detect different types of events, an arming station that is used to turn the system on and off, a control panel that receives information, processes it, and transmits the alarm, and an annunciator that generates a visual and/or audible response to the alarm. When a sensor is tripped it sends a signal to a control panel, which triggers a visual or audible alarm and/or notifies a central monitoring station. A more complete description of each of the components of an alarm system is provided below.

Detection devices (also called *sensors*) are designed to detect a specific type of event (such as smoke, intrusion, etc.). Depending on the type of event they are designed to detect, sensors can be located inside or outside the facility or other assets. When an event is detected, the sensors use some type of communication method (such as wireless radio transmitters, conductors, or cables) to send signals to the control panel to generate the alarm. For example, a smoke detector sends a signal to a control panel when it detects smoke.

An *arming station*, which is the main user interface with the security system, allows the user to arm (turn on), disarm (turn off), and communicate with the system. How a specific system is armed will depend on how it is used. For example, while IDSs can be armed for continuous operation (24 hours/day), they are usually armed and disarmed according to the work schedule at a specific location so that personnel going about their daily activities do not set off the alarms. In contrast, fire protection systems are typically armed 24 hours a day.

The *control panel* receives information from the sensors and sends it to an appropriate location, such as a central operations station or a 24-hour monitoring facility. Once the alarm signal is received at the central monitoring location, personnel monitoring for alarms can respond (such as by sending security teams to investigate or by dispatching the fire department).

The *annunciator* responds to the detection of an event by emitting a signal, which may be visual, audible, electronic, or a combination of these. For example, fire alarm signals are always connected to audible annunciators, whereas intrusion alarms may not be.

Alarms can be reported locally, remotely, or both. A *local alarm* emits a signal at the location of the event (typically using a bell or siren). A "local only" alarm emits a signal at the location of the event but does not transmit the alarm signal to any other location (i.e., it does not transmit the alarm to a central monitoring location). Typically, the purpose of a "local only" alarm is to frighten away intruders and possibly attract the attention of someone who might notify the proper authorities. Since no signal is sent to a central monitoring location, personnel can only respond to a local alarm if they are in the area and can hear and/or see the alarm signal.

Fire alarm systems must have local alarms, including both audible and visual signals. Most fire alarm signal and response requirements are codified in the National Fire Alarm Code, National Fire Protection Association (NFPA) 72. NFPA 72 discusses the application, installation, performance, and maintenance of protective signaling systems and their components. In contrast to fire alarms, which require a local signal when the fire is detected, many IDSs do not have a local alert device, as monitoring personnel do not wish to inform potential intruders that they have been detected. Instead, these systems silently alert monitoring personnel that an intrusion has been detected, thus allowing monitoring personnel to respond.

In contrast to systems that transmit "local only" alarms when sensors are triggered, systems can also transmit signals to a *central location*, such as a control room or guard post at the utility, or to a police or fire station. Most fire/smoke alarms are set up to signal both at the location of the event and at a fire station or central monitoring station. Many insurance companies require facilities to install certified systems that include alarm communication to a central station. For example, systems certified by Underwriters Laboratory (UL) require the alarm to be reported to a central monitoring station.

The main differences between alarm systems lie in the types of event detection devices used in different systems. **Intrusion sensors**, for example, consist of two main categories: perimeter sensors and interior (space) sensors. *Perimeter intrusion sensors* are typically applied on fences, doors, walls, windows, etc., and are designed to detect an intruder before he/she accesses a protected asset (i.e., perimeter intrusion sensors are used to detect intruders attempting to enter through a door, window, etc.). In contrast, *interior intrusion sensors* are designed to detect an intruder who has already accessed the protected asset (i.e., interior intrusion sensors are used to detect intruders once they are already within a protected room or building). These two types of detection devices can be complementary, and they are often used together to enhance security for an asset. For example, a typical intrusion alarm system might employ a perimeter glass-break detector that protects against intruders accessing a room through a window, as well as an ultrasonic interior sensor that detects intruders that have entered the room without using the window.

Fire detection/Fire alarm systems consist of different types of fire detection devices and fire alarm systems. These systems may detect fire, heat, smoke, or a combination of any of these. For example, a typical fire alarm system might consist of heat sensors, which are located throughout a facility and detect high temperatures or a certain change in temperature over a fixed time period. A different system might be outfitted with both smoke and heat detection devices.

When a sensor in an alarm system detects an event, it must communicate an alarm signal. The two basic types of alarm communication systems are hardwired and wireless. Hardwired systems rely on a wire that runs from the control panel to each of the detection devices and annunciators. Wireless systems transmit signals from a transmitter to a receiver through the air—primarily using radio or other waves. Hardwired systems are usually lower-cost, more reliable (as they are not affected by terrain or environmental factors), and significantly easier to troubleshoot than wireless systems. However, a major disadvantage of hardwired systems is that it may not be possible to hardwire all locations (for example, it may be difficult to hardwire remote locations). In addition, running wires to the required locations can be both time-consuming and costly. The major advantage of using wireless systems is that they can often be installed in areas where hardwired systems are not feasible. However, wireless components can be much more expensive compared to hardwired systems. In the past, it has been difficult to perform self-diagnostics on wireless systems to confirm that proper communication with the controller. Presently, the majority of wireless systems incorporate supervising circuitry, which allows the subscriber to know immediately if there is a problem with the system (such as a broken detection device or a low battery), or if a protected door or window has been left open.

Backflow Prevention Devices

As their name suggests, backflow-prevention devices are designed to prevent backflow, which is the reversal of the normal and intended direction of water flow in a water system. Backflow is a potential problem in a water system because it can spread contaminated water back through a distribution system. For example, backflow at uncontrolled cross-connections (which are any actual or potential connections between the public water supply and a source of contamination) can allow pollutants or contaminants to enter the potable water system. More specifically, backflow from private plumbing systems, industrial areas, hospitals, and other hazardous contaminant-containing systems into public water mains and wells poses serious public health risks and security problems. Cross-contamination from private plumbing systems can contain biological hazards (such as bacteria or viruses) or toxic substances that can contaminate and sicken an entire population in the event of backflow. The majority of historical incidences of backflow have been accidental, but growing concern that contaminants could be intentionally backfed into a system is prompting increased awareness for private homes, businesses, industries, and areas most vulnerable to intentional strikes. Therefore, backflow prevention is a major tool for protecting water systems.

Backflow may occur under two types of conditions: backpressure and backsiphonage. *Backpressure* is the reverse of the normal flow direction within a piping system, resulting from the downstream pressure being higher than the supply pressure. These reductions in supply pressure occur whenever the amount of water being used exceeds the amount of water supplied, such as during water line flushing, firefighting, or breaks in water mains. *Backsiphonage* is the reversal of normal flow direction within a piping system caused by negative pressure in the supply piping (i.e., the reversal of normal flow in a system caused by a vacuum or partial vacuum within the water supply piping). Backsiphonage can occur where there is high velocity in a pipeline; when there is a line repair or break that is lower than a service point; or when there is lowered main pressure due to high water withdrawal rates, such as during firefighting or water main flushing.

To prevent backflow, various types of backflow preventers are appropriate for use. The primary types of backflow preventers are:

- Air Gap Drains
- Double Check Valves
- Reduced Pressure Principle Assemblies
- Pressure Vacuum Breakers

Barriers

Active Security Barriers (Crash Barriers)

Active security barriers (also known as crash barriers) are large structures placed in roadways at entrance and exit points to protect facilities and control vehicle access to these areas. These barriers are placed perpendicular to traffic to block the roadway so the only way for traffic to pass is for the barrier to be moved out of the roadway. These types of barriers are typically constructed from sturdy materials, such as concrete or steel, so that vehicles cannot penetrate through them. They are also designed at a certain height off the roadway to prevent vehicles from going over them.

The key difference between active security barriers, which include wedges, crash beams, gates, retractable bollards, and portable barricades, and passive security barriers, which include non-movable bollards, jersey barriers, and planters, is that active security barriers are designed to be easily raised, lowered, or moved out of the roadway to allow authorized vehicles to pass. Many of these barriers are designed to open and close automatically (e.g., mechanized gates, hydraulic wedge barriers), while others are easy to operate manually (e.g., swing crash beams, manual gates). In contrast to active barriers, passive barriers are permanent and non-movable, typically used to protect the perimeter of a facility, such as sidewalks and other areas that do not require vehicular traffic to pass. Several major types of active security barriers, such as wedge barriers, crash beams, gates, bollards, and portable/removable barricades, are described below.

Wedge barriers are plated, rectangular steel buttresses approximately 2–3 feet high that can be raised and lowered from the roadway. When they are in the open position, they are flush with the roadway, and vehicles can pass over them. However, when they are in the closed (armed) position, they project up from the road at a 45° angle, with the upper end pointing toward the oncoming vehicle and the base of the barrier away from the vehicle. Generally, wedge barriers are constructed from heavy-gauge steel or concrete containing an impact-dampening iron rebar core that is strong and resistant to breaking or cracking, thereby allowing

them to withstand the impact from a vehicle attempting to crash through them. In addition, both these materials help transfer the energy impact over the barrier's entire volume, thus helping to prevent the barrier from being sheared off its base. Because the barrier is angled away from traffic, the force of any vehicle impacting the barrier is distributed over the entire surface and is not concentrated at the base, which helps prevent the barrier from breaking off at the base. Finally, the angle of the barrier helps hang up any vehicles attempting to drive over it.

Wedge barriers can be fixed or portable. Fixed wedge barriers can be mounted on the surface of the roadway ("surface-mounted wedges"), in a shallow mount on the road's surface, or completely below the road surface. Surface-mounted wedge barricades operate by rising from a flat position on the surface of the roadway, while shallow-mount wedge barriers rise from their resting position just below the road surface. In contrast, below-surface wedge barriers operate by rising from beneath the road surface. Both the shallow-mounted and surface-mounted barriers require little or no excavation and thus do not interfere with buried utilities. All three barrier mounting types project above the road surface and block traffic when they are raised into the armed position. Once they are disarmed and lowered, they are flush with the road, thereby allowing traffic to pass. Portable wedge barriers are moved into place on wheels that are removed after the barrier has been set into place.

Installing rising wedge barriers requires preparation of the road surface. Installing surface-mounted wedges does not require that the road be excavated; however, the road surface must be intact and strong enough to allow the bolts anchoring the wedge to the road surface to attach properly. Shallow-mount and below-surface wedge barricades require the excavation of a pit that is large enough to accommodate the wedge structure, as well as any arming/disarming mechanisms. Generally, the bottom of the excavation pit is lined with gravel to allow for drainage. Areas not sheltered from rain or surface runoff can install a gravity drain or self-priming pump.

Crash beam barriers consist of aluminum beams that can be opened or closed across the roadway. While there are several different crash beam designs, every crash beam system consists of an aluminum beam that is supported on each side by a solid footing or buttress, which is typically constructed from concrete, steel, or some other strong material. Beams typically contain an interior steel cable (typically at least 1 inch in diameter) to give the beam added strength and rigidity. The beam is connected by a heavy-duty hinge or other mechanism to one of the footings so that it can swing or rotate out of the roadway when it is open and can swing back across the road when it is in the closed (armed) position, blocking the road and inhibiting access by unauthorized vehicles. The non-hinged end of the beam can be locked into its footing, thus providing anchoring for the beam on both sides of the road and increasing the beam's resistance to any vehicles attempting to penetrate through it. In addition, if the crash beam is hit by a vehicle, the aluminum beam transfers the impact energy to the interior cable, which in turn transfers the impact energy through the footings and into their foundation, thereby minimizing the chance that the impact will snap the beam and allow the intruding vehicle to pass through.

Crash beam barriers can employ drop-arm, cantilever, or swing beam designs. Drop-arm crash beams operate by raising and lowering the beam vertically across the road. Cantilever crash beams are projecting structures that are opened and closed by extending the beam from the hinge buttress to the receiving buttress located on the opposite side of the road. In the swing beam design, the beam is hinged to the buttress such that it swings horizontally across the road. Generally, swing beam and cantilever designs are used at locations where a vertical lift beam is impractical. For example, the swing beam or cantilever designs are utilized at entrances and exits with overhangs, trees, or buildings that would physically block the operation of the drop-arm beam design. Installing any of these crash beam barriers involves the excavation of a pit approximately 48 inches deep for both the hinge and the receiver footings. Due to the depth of excavation, the site should be inspected for underground utilities before digging begins.

In contrast to wedge barriers and crash beams, which are typically installed separately from a fence line, *gates* are often integrated units of a perimeter fence or wall around a facility. Gates are basically movable pieces of fencing that can be opened and closed across a road. When the gate is in the closed (armed) position, the leaves of the gate lock into steel buttresses that are embedded in a concrete foundation located on both sides of the roadway, thereby blocking access to the roadway. Generally, gate barricades are constructed from a combination of heavy-gauge steel and aluminum that can absorb an impact from vehicles attempting to ram through them. Any remaining impact energy not absorbed by the gate material is transferred to the steel buttresses and their concrete foundation.

Gates can utilize a cantilever, linear, or swing design. Cantilever gates are projecting structures that operate by extending the gate from the hinge footing across the roadway to the receiver footing. A linear gate is designed to slide across the road on tracks via a rack and pinion drive mechanism. Swing gates are hinged so that they can swing horizontally across the road. Installation of the cantilever, linear, or swing gate designs described above involves the excavation of a pit approximately 48 inches deep for both the hinge and receiver footings to which the gates are attached. Due to the depth of excavation, the site should be inspected for underground utilities before digging begins.

Bollards are vertical barriers at least 3 feet tall and 1–2 feet in diameter that are typically set 4–5 feet apart from each other to block vehicles from passing between them. Bollards can either be fixed in place, removable, or retractable. Fixed and removable bollards are passive barriers typically used along building perimeters or on sidewalks to prevent vehicles from accessing them while allowing

pedestrians to pass. In contrast to passive bollards, retractable bollards are active security barriers that can easily be raised and lowered to allow vehicles to pass between them. Thus, they can be used in driveways or on roads to control vehicular access. When the bollards are raised, they project above the road surface and block the roadway; when they are lowered, they sit flush with the road surface, thus allowing traffic to pass over them. Retractable bollards are typically constructed from steel or other materials that have a low weight-to-volume ratio so they require low power to raise and lower. Steel is also more resistant to breaking than a more brittle material, such as concrete, and is better able to withstand direct vehicular impact without breaking apart.

Retractable bollards are installed in a trench dug across a roadway—typically at an entrance or gate. Installing retractable bollards requires preparing the road surface. Depending on the vendor, bollards can be installed either in a continuous slab of concrete or in individual excavations with concrete poured in place. The required excavation for a bollard is typically slightly wider and slightly deeper than the bollard height when extended aboveground. The bottom of the excavation is typically lined with gravel to allow for drainage. The bollards are then connected to a control panel that controls the raising and lowering of the bollards. Installation typically requires mechanical, electrical, and concrete work; if utility personnel with these skills are available, then the utility can install the bollards themselves.

Portable/removable barriers, which can include removable crash beams and wedge barriers, are mobile obstacles that can be moved in and out of position on a roadway. For example, a crash beam may be completely removed and stored off-site when it is not needed. An additional example would be wedge barriers that are equipped with wheels that can be removed after the barricade is towed into place.

When portable barricades are needed, they can be moved into position rapidly. To provide them with added strength and stability, they are typically anchored to buttress boxes that are located on either side of the road. These buttress boxes, which may or may not be permanent, are usually filled with sand, water, cement, gravel, or concrete to make them heavy and aid in stabilizing the portable barrier. In addition, these buttresses can help dissipate any impact energy from vehicles crashing into the barrier itself.

Because these barriers are not anchored into the roadway, they do not require excavation or other related construction for installation. In contrast, they can be assembled and made operational in a short period of time. The primary shortcoming of this type of design is that these barriers may move if they are hit by vehicles. Therefore, it is important to carefully assess the placement and anchoring of these types of barriers to ensure that they can withstand the types of impacts that may be anticipated at that location.

Because the primary threat to active security barriers is that vehicles will attempt to crash through them, their most important attributes are their size, strength, and crash resistance. Other important features for an active security barrier are the mechanisms by which the barrier is raised and lowered to allow authorized vehicle entry, as well as other factors such as weather resistance and safety features.

Passive Security Barriers

One of the most basic threats facing any facility is from intruders accessing the facility is from intruders accessing the facility with the intention of causing damage to its assets. These threats may include intruders entering the facility, as well as intruders attacking the facility from outside without actually entering it (i.e., detonating a bomb near enough to the facility to cause damage within its boundaries).

Security barriers are one of the most effective ways to counter the threat of intruders accessing a facility or its perimeter. Security barriers are large, heavy structures used to control access through a perimeter by either vehicles or personnel. They can be used in many ways depending on how and where they are located at the facility. For example, security barriers can be used on or along driveways or roads to direct traffic to a checkpoint (i.e., a facility may install jersey barriers in a road to direct traffic in a certain direction). Other types of security barriers (crash beams, gates) can be installed at the checkpoint so that guards can regulate which vehicles can access the facility. Finally, other security barriers (i.e., bollards or security planters) can be used along the facility perimeter to establish a protective buffer area between the facility and approaching vehicles. Establishing such a protective buffer can help mitigate the effects of the type of bomb blast described above, both by potentially absorbing some of the blast and by increasing the "stand-off" distance between the blast and the facility (the force of an explosion is reduced as the shock wave travels further from the source, and thus the farther the explosion is from the target, the less effective it will be in damaging the target).

Security barriers can be either "active" or "passive." "Active" barriers, which include gates, retractable bollards, wedge barriers, and crash barriers, are readily movable and are typically used in areas where they must be moved often to allow vehicles to pass—such as in roadways at entrances and exits to a facility. In contrast to active security barriers, "passive" security barriers, which include jersey barriers, bollards, and security planters, are not designed to be moved regularly and are typically used in areas where access is not required or allowed—such as along building perimeters or in traffic control areas. Passive security barriers are typically large, heavy structures, usually several feet high, and are designed so that even heavy-duty vehicles cannot go over or through them. Therefore, they can be placed in a roadway parallel to the flow of traffic to direct traffic in a certain direction (such as to a guardhouse, gate, or some other sort of checkpoint), or perpendicular to traffic to prevent vehicles from using a road or approaching a building or area.

Biometric Security Systems

Biometrics involves measuring the unique physical characteristics or traits of the human body. Any aspect of the body that is measurably different from person to person—for example, fingerprints or eye characteristics—can serve as a unique biometric identifier for that individual. Biometric systems recognizing fingerprints, palm shape, eyes, face, voice, and signature comprise the bulk of current biometric systems that recognize other biological features that do exist. Biometric security systems use biometric technology combined with some type of locking mechanism to control access to specific assets. To access an asset controlled by a biometric security system, an individual's biometric trait must be matched with an existing profile stored in a database. If there is a match between the two, the locking mechanism (which could be a physical lock, such as at a doorway, an electronic lock, such as at a computer terminal, or some other type of lock) is disengaged, and the individual is given access to the asset. A biometric security system is typically comprised of the following components:

- A sensor, which measures/records a biometric characteristic or trait.
- A control panel, which serves as the connection point between various system components. The control panel communicates information back and forth between the sensor and the host computer and controls access to the asset by engaging or disengaging the system lock based on internal logic and information from the host computer.
- A host computer, which processes and stores the biometric trait in a database.
- Specialized software, which compares an individual image taken by the sensor with a stored profile or profiles.
- A locking mechanism, which is controlled by the biometric system.
- A power source to power the system.

Biometric Hand and Finger Geometry Recognition

Hand and finger geometry recognition is the process of identifying an individual through the unique "geometry" (shape, thickness, length, width, etc.) of that individual's hand or fingers. Hand geometry recognition has been employed since the early 1980s and is among the most widely used biometric technologies for controlling access to important assets. It is easy to install and use and is appropriate for use in any location requiring the use of two-finger highly-accurate, non-intrusive biometric security. For example, it is currently used in numerous workplaces, daycare facilities, hospitals, universities, airports, and power plants.

A newer option within hand geometry recognition technology is finger geometry recognition (not to be confused with fingerprint recognition). Finger geometry recognition relies on the same scanning methods and technologies as hand geometry recognition, but the scanner only scans two of the user's fingers, as opposed to his entire hand. Finger geometry recognition has been in commercial use since the mid-1990s and is mainly used in time and attendance applications (i.e., to track when individuals have entered and exited a location). To date, the only large-scale commercial use of two-finger geometry for controlling access is at Disney World, where the season pass holders use the geometry of their index and middle fingers to gain access to the facilities.

Hand and finger geometry recognition systems can be used in several types of applications, including access control and time and attendance tracking. While time and attendance tracking can be used for security, it is primarily used for operational and payroll purposes (i.e., clocking in and clocking out). In contrast, access control applications are more likely to be security-related. Biometric systems are widely used for access control and can be used on various assets, including entryways, computers, and vehicles, etc. However, because of their size, hand/finger recognition systems are primarily used in entryway access control applications.

Iris Recognition

The iris, which is the colored or pigmented area of the eye surrounded by the sclera (the white portion of the eye), is a muscular membrane that controls the amount of light entering the eye by contracting or expanding the pupil (the dark center of the eye). The dense, unique patterns of connective tissue in the human iris were first noted in 1936, but it was not until 1994, when algorithms for iris recognition were created and patented, that commercial applications using biometric iris recognition began to be used extensively. There are now two vendors producing iris recognition technology: both the original developer of these algorithms and a second company, which has developed and patented a different set of algorithms for iris recognition.

The iris is an ideal characteristic for identifying individuals because it is formed *in utero*, and its unique patterns stabilize around eight months after birth. No two irises are alike, neither an individual's right nor left iris, nor the irises of identical twins. The iris is protected by the cornea (the clear covering over the eye), and therefore it is not subject to the aging or physical changes (and potential variation) that are common to some other biometric measures, such as the hand, fingerprints, and face. Although some limited changes can occur naturally over time, these changes generally occur in the iris's melanin and therefore affect only the eye's color, not its unique patterns. In addition, because iris scanning uses only black-and-white images, color changes would not affect the scan. Thus, barring specific injuries or certain surgeries directly affecting the iris, the iris's unique patterns remain relatively unchanged over an individual's lifetime.

Iris recognition systems employ a monochromatic, or black-and-white, video camera that uses both visible and near-infrared light to capture video of an individual's iris. Video is used rather than still photography as an extra security procedure. The video confirms the normal continuous fluctuations of the pupil as the eye focuses, ensuring that the scan is of a living human being and not a photograph or some other attempted hoax. A high-resolution image of the iris is then captured or extracted from the video using a device often referred to as a *frame grabber*. The unique characteristics identified in this image are then converted into a numeric code, which is stored as a template for that user.

Card Identification/Access/Tracking Systems

A card reader system is an electronic identification system used to identify a card and then perform an action associated with that card. Depending on the system, the card may identify where a person is or where they were at a certain time, or it may authorize another action, such as disengaging a lock. For example, a security guard may use their card at card readers located throughout a facility to indicate that they have checked a certain location at a certain time. The reader will store the information and/or send it to a central location, where it can be checked later to ensure that the guard has patrolled the area. Other card reader systems can be associated with a lock so that the cardholder must have their card read and accepted by the reader before the lock disengages. A complete card reader system typically consists of the following components:

- Access cards that the user carries.
- Card readers, which read the card signals and send the information to control units.
- Control units, which control the response of the card reader to the card.
- A power source.

Numerous card reader systems are available. The primary differences between card reader systems are different in how data is encoded on the cards, how data are transferred between the card and the card reader, and the types of applications for which they are best suited. However, all card systems are similar in the way that the card reader and control unit interact to respond to the card.

While card readers are similar in the way that the card reader and control unit interact to control access, they differ in the way data is encoded on the cards and the way these data are transferred between the card and the card reader. Several types of technologies are available for card reader systems. These include:

- Proximity
- Wiegand
- Smartcard
- Magnetic Stripe
- Bar Code
- Infrared
- Barium Ferrite
- Hollerith
- Mixed Technologies

The determination of the level of security rate (low, moderate, or high) is based on the level of technology a given card reader system has and how simple it is to duplicate that technology, thus bypassing the security. Vulnerability ratings are based on whether the card reader can be easily damaged due to frequent use or difficult working conditions (i.e., weather conditions if the reader is located outside). Often this is influenced by the number of moving parts in the system—the more moving parts, the greater the system's potential susceptibility to damage. The life cycle rating is based on the durability of a given card reader system over its entire operational period. Systems requiring frequent physical contact between the reader and the card often have a shorter life cycle due to the wear and tear to which the equipment is exposed. For many card reader systems, the vulnerability rating and life cycle ratings have a reciprocal relationship. For instance, if a given system has a high vulnerability rating, it will almost always have a shorter life cycle.

Card reader technology can be implemented for facilities of any size and with any number of users. However, because individual systems vary in the complexity of their technology and in the level of security they can provide to a facility, individual users must determine the appropriate system for their needs. Some important features to consider when selecting a card reader system include:

- What level of technological sophistication and security does the card system have?
- How large is the facility, and what are its security needs?
- How frequently will the card system be used? For systems that will experience a high frequency of use it is important to consider a system that has a longer life cycle and lower vulnerability rating, thus making it more cost-effective to implement.
- Under what conditions will the system be used? (Will it be installed on the interior or exterior of buildings? Does it require light or humidity controls?) Most card reader systems can operate under normal environmental conditions, and therefore this would be a mitigating factor only in extreme conditions.
- What are the system costs?

Fences

A fence is a physical barrier that can be set up around the perimeter of an asset. Fences often consist of individual pieces (such as pickets in a wooden fence or sections of a wrought iron fence) that are fastened together. Individual sections of the fence are fastened together using posts, which are sunk into the ground to provide stability and strength for the sections of the fence hung between them. Gates are installed between individual sections of the fence to allow access inside the fenced area.

Fences are often used as decorative architectural features to separate physical spaces from each other. They may also be used to physically mark the location of a boundary (such as a fence installed along a proper line). However, a fence can also serve as an effective means for physically delaying intruders from gaining access to a water or wastewater asset. For example, many utilities install fences around their primary facilities, remote pump stations, hazardous materials storage areas, or sensitive areas within a facility. Access to the area can be controlled through security at gates or doors in the fence (for example, by posting a guard at the gate or by locking it). To gain access to the asset, unauthorized persons would either have to go around or through the fence.

Fences are often compared with walls when determining the appropriate system for perimeter security. While both fences and walls can provide adequate perimeter security, fences are often easier and less expensive to install than walls. However, they do not usually provide the same physical strength that walls do. In addition, many types of fences have gaps between the individual pieces that make up the fence (i.e., the spaces between chain links in a chain-link fence or the space between pickets in a picket fence). Thus, many types of fences allow the interior of the fenced area to be seen, which may allow intruders to gather important information about the locations or defenses of vulnerable areas within the facility.

Numerous types of materials are used to construct fences, including chain-link, iron, aluminum, wood, and wire. Some types of fences, such as split rails or pickets, may not be appropriate for security purposes because they are traditionally low and are not physically strong. Potential intruders may be able to easily defeat these fences either by jumping or climbing over them or by breaking through them. For example, the rails in a split fence may be easily broken.

Important security attributes of a fence include the height to which it can be constructed, the strength of the material comprising the fence, the method and strength of attaching the individual sections of the fence together at the posts, and the fence's ability to restrict the view of the assets inside the fence. Additional considerations should include the ease of installing the fence and the ease of removing and reusing sections of the fence.

Some fences can include additional measures to delay or even detect potential intruders. Such measures may include the addition of barbed wire, razor wire, or other deterrents at the top of the fence. Barbed wire is sometimes employed at the base of fences as well, which can impede a would-be intruder's progress in reaching the fence. Fences may also be fitted with security cameras to provide visual surveillance of the perimeter. Finally, some facilities have installed motion sensors along their fences to detect movement. Several manufacturers have combined these multiple perimeter security features into one product, offering alarms and other security features.

The correct implementation of a fence can make it a much more effective security measure. Security experts recommend the following when a facility constructs a fence:

- The fence should be at least 7–9 feet high.
- Any outriggers, such as barbed wire, that are affixed on top of the fence should be angled out and away from the facility, not toward it. This will make climbing the fence more difficult and will prevent ladders from being placed against the fence.
- Other types of hardware can increase the security of the fence. This can include installing concertina wire along the fence (this can be done in front of the fence or at the top) or adding intrusion sensors, cameras, or other hardware to the fence.
- All undergrowth should be cleared for several feet (typically 6 feet) on both sides of the fence. This will allow for a clearer view of the fence by any patrols in the area.
- Any trees with limbs or branches hanging over the fence should be trimmed so that intruders cannot use them to climb the fence. Also, it should be noted that fallen trees can damage fences, so management of trees around them can be important. This can be especially important in areas where the fence goes through remote locations.
- Fences that do not block the view from outside to inside allow patrols to see inside the fenced area without having to enter the facility.
- "No Trespassing" signs posted along fence can be a valuable tool in prosecuting any intruders who claim that the fence was broken and that they did not enter through the fence illegally. Adding signs that highlight the local ordinances against trespassing can further persuade simple troublemakers to illegally jump or climb the fence.

Films for Glass Shatter Protection

Most water and wastewater utilities have numerous windows on the outside of buildings, indoors, and interior offices. In addition, many facilities have glass doors or other glass structures, such as glass walls or display cases. These glass objects are potentially vulnerable to shattering when heavy objects are thrown or launched at them, when explosions occur nearby, or during high winds (for exterior glass). If the glass shatters, intruders may potentially enter an area. In addition, shattered glass projected into a room from an explosion or an object being thrown through a door or window can injure and potentially incapacitate personnel in the room. Materials that prevent glass from shattering can help maintain the integrity of the door, window, or other glass objects, delaying an intruder from gaining access. These materials can also prevent flying glass, thus reducing potential injuries.

Materials designed to prevent glass from shattering include specialized films and coatings. These materials can be applied to existing glass objects to improve their strength and ability to resist shattering. The films have been tested against many scenarios that could result in glass breakage, including penetration by blunt objects, bullets, high winds, and simulated explosions. Thus, the films are tested against both simulated weather scenarios (which could include both the high winds themselves and the force of objects blown into the glass), as well as more criminal/terrorist scenarios where the glass is subjected to explosives or bullets. Many vendors provide information on the results of these types of tests, and potential users can compare different product lines to determine which products best suit their needs.

The primary attributes of films for shatter protection are:

- The materials from which the film is made
- The adhesive that bonds the film to the glass surface
- The thickness of the film

Fire Hydrant Locks

Fire hydrants are installed at strategic locations throughout a community's water distribution system to supply water for firefighting. However, because there are many hydrants in a system, and they are often located in residential neighborhoods, industrial districts, and other areas where they cannot be easily observed or guarded, they are potentially vulnerable to unauthorized access. Many municipalities, states, and EPA regions have recognized this potential vulnerability and have instituted programs to lock hydrants. For example, EPA Region 1 has included locking hydrants as number 7 on its "Drinking Water Security and Emergency Preparedness" Top Ten List for small groundwater suppliers.

A "hydrant lock" is a physical security device designed to prevent unauthorized access to the water supply through a hydrant. They can also ensure water and water pressure availability to firefighters and prevent water theft and associated lost water revenue. These locks have been successfully used in numerous municipalities and various climates and weather conditions.

Fire hydrant locks are basically steel covers or caps that are locked in place over the operating nut of a fire hydrant. The lock prevents unauthorized persons from accessing the operating nut and opening the fire hydrant valve. The lock also makes it more difficult to remove the bolts from the hydrant and access the system that way. Additionally, hydrant locks shield the valve from being broken off. Should a vandal attempt to breach the hydrant lock by force and succeed in breaking it, the vandal will only succeed in bending the operating valve. If the hydrant's operating valve is bent, the hydrant will not be operational, but the water asset remains protected and inaccessible to vandals. However, the entire hydrant will need to be replaced.

Hydrant locks are designed so that the hydrants can be operated by special "key wrenches" without removing the lock. These specialized wrenches are generally distributed to the fire department, public works department, and other authorized persons so they can access the hydrants as needed. A municipality generally keeps an inventory of wrenches and their serial numbers so that the location of all wrenches is known. These operating key wrenches may only be purchased by registered lock owners.

The most important features of hydrant locks are their strength and the security of their locking systems. The locks must be strong enough that they cannot be broken off. Hydrant locks are constructed from stainless or alloyed steel. Stainless steel locks are stronger and ideal for all climates; however, they are more expensive than alloy locks. The locking mechanisms for each fire hydrant locking system ensure that the hydrant can only be operated by authorized personnel who have the specialized key to operate the hydrant.

Hatch Security

A hatch is basically a door installed on a horizontal plane (such as in a floor, a paved lot, or a ceiling), instead of on a vertical plane (such as in a building wall). Hatches are usually used to provide access to assets that are either located underground (such as hatches to basements or underground storage areas) or to assets located above ceilings (such as emergency roof exits). At water and wastewater facilities, hatches are typically used to provide access to underground vaults containing pumps, valves, or piping, or to the interior of water tanks or covered reservoirs. Securing a hatch by locking it or upgrading materials to give the hatch added strength can help delay unauthorized access to any asset behind the hatch. Like all doors, a hatch consists of a frame anchored to the horizontal structure, a door or doors, hinges connecting the door/doors to the frame, and a latching or locking mechanism that keeps the hatch door/doors closed.

It should be noted that improving hatch security is straightforward, and hatches with upgraded security features can be newly installed or retrofitted for existing applications. Many municipalities already have specifications for hatch security at their water and wastewater utility assets.

Depending on the application, the primary security-related attributes of a hatch are the strength of the door and frame, its resistance to the elements and corrosion, its ability to be sealed against water or gas, and its locking features. Hatches must be both strong and lightweight so they can withstand typical static loads (such as people or vehicles walking or driving over them) while still being easy to open. In addition, because hatches are typically installed in outdoor locations, they are usually designed from corrosion-resistant metal that can withstand the elements. Therefore, hatches are typically constructed from high-gauge steel or lightweight aluminum.

The hatch locking mechanism is perhaps the most important part of hatch security. There are several locks that can be implemented for hatches, including:

- Slam locks (internal locks that are located within the hatch frame)
- Recessed cylinder locks
- Bolt locks
- Padlocks

Intrusion Sensors

An exterior intrusion sensor is a detection device used in an outdoor environment to detect intrusions into a protected area. These devices are designed to detect an intruder and then communicate an alarm signal to an alarm system. The alarm system can respond to the intrusion in various ways, such as by triggering an audible or visual alarm signal or by sending an electronic signal to a central monitoring location that notifies security personnel of the intrusion. Intrusion sensors can be used to protect many kinds of assets. Intrusion sensors that protect physical space are classified according to whether they protect indoor or "interior" space (i.e., an entire building or room within a building) or outdoor or "exterior" space (i.e., a fence line or perimeter). Interior intrusion sensors are designed to protect the interior space of a facility by detecting an intruder who is attempting to enter or who has already entered a room or building. In contrast, exterior intrusion sensors are designed to detect an intrusion into a protected outdoor/exterior area. Exterior protected areas are typically arranged as zones or exclusion areas so that the intruder is detected early in the intrusion attempt before gaining access to more valuable assets (e.g., a building located within the protected area). Early detection creates additional time for security forces to respond to the alarm.

Buried Exterior Intrusion Sensors

Buried sensors are electronic devices designed to detect potential intruders. The sensors are buried along the perimeters of sensitive assets and can detect intruder activity both above and below ground. Some of these systems are composed of individual stand-alone sensor units, while others consist of buried cables.

Ladder Access Control

Water and wastewater utilities have several assets that are raised above ground level, including raised water tanks, raised chemical tanks, raised piping systems and roof access points into buildings. In addition, communications equipment, antennas, or other electronic devices may be placed on top of these raised assets. Typically, these assets are reached by ladders permanently anchored to them. For example, raised water tanks are typically accessed by ladders bolted to one of the legs of the tank. Controlling access to these raised assets by controlling access to the ladder can increase security at a water or wastewater utility.

A typical ladder access control system consists of some type of cover that is locked or secured over the ladder. The cover can be a casing that surrounds most of the ladder, or a door or shield that covers only part of it. In either case, several rungs of the ladder (the number of rungs depends on the size of the cover) are made inaccessible by the cover, and these rungs can only be accessed by opening or removing the cover. The cover is locked to allow only authorized personnel to open or remove it and use the ladder. Ladder access controls are usually installed several feet above ground level, and they usually extend several feet up the ladder so that they cannot be circumvented by someone accessing the ladder above the control system. The important features of ladder access control are the size and strength of the cover and its ability to lock or otherwise be secured from unauthorized access.

The covers are constructed from aluminum or some type of steel. This should provide adequate protection from being pierced or cut through. The metals are corrosion-resistant, so they will not corrode or become fragile from extreme weather conditions in outdoor applications. The bolts used to install each of these systems are galvanized steel. In addition, the bolts for each cover are installed on the inside of the unit so they cannot be removed from the outside.

Locks

A lock is a type of physical security device that can be used to delay or prevent a door, window, manhole, filing cabinet drawer, or some other physical feature from being opened, moved, or operated. Locks typically operate by connecting two pieces—such as by connecting a door to a door jamb or a manhole to its casement. Every lock has two modes—engaged (or "locked") and disengaged (or "opened"). When a lock is disengaged, the asset on which the lock is installed can be accessed by anyone; but when the lock is engaged, access to the locked asset is restrained.

Locks are excellent security features because they are designed to function in many ways and work on different types of assets. Locks can also provide different levels of security depending on how they are designed and implemented. The security provided by a lock depends on several factors, including its ability to withstand physical damage (i.e., can it be cut off, broken, or otherwise physically disabled) as well as its requirements for supervision or operation (i.e., combinations may need to be changed frequently so that they are not compromised and the locks remain secure). While there is no single definition of the "security" of a lock, locks are often described as minimum, medium, or maximum security. Minimum security locks are those that can be easily disengaged (or "picked") without the correct key or code, or those that can be easily disabled (such as small padlocks that can be cut with bolt cutters). Higher security locks are more complex and, thus, more difficult to pick or sturdier and more resistant to physical damage.

Many locks, such as door locks, only need to be unlocked from one side. For example, most door locks need a key to be unlocked only from the outside. A person opens such devices, called single-cylinder locks, from the inside by pushing a button or by turning a knob or handle. Double-cylinder locks require a key to be locked or unlocked from both sides.

Manhole Intrusion Sensors

Manholes are located at strategic locations throughout most municipal water, wastewater, and other underground utility systems. They are designed to provide access to these underground utilities and, therefore, are potential entry points to a system. For example, manholes in water or wastewater systems may provide access to sewer lines or vaults containing on/off or pressure-reducing water valves.

Because many utilities run under other infrastructure (roads, buildings), manholes also provide potential access points to critical infrastructure as well as water and wastewater assets. In addition, since the portion of the system to which manholes provide entry is primarily located underground access to a system through a manhole increases the chance that an intruder will not be seen. Therefore, protecting manholes can be a critical component of safeguarding an entire community.

The various methods for protecting manholes are designed to prevent unauthorized personnel from physically accessing them and to detect attempts at unauthorized access. A manhole intrusion sensor is a physical security device designed to detect unauthorized access to the utility through a manhole. Monitoring a manhole that provides access to a water or wastewater system can mitigate two distinct types of threats. First, monitoring a manhole may detect the access of unauthorized personnel to water or wastewater systems or assets through the manhole. Second, monitoring manholes may also allow for the detection of hazardous substances into the water system.

Several different technologies have been used to develop manhole intrusion sensors, including mechanical systems, magnetic systems, and fiber optic and infrared sensors. Some of these intrusion sensors have been specifically designed for manholes, while others consist of standard off-the-shelf intrusion sensors implemented in systems specifically designed for application in a manhole.

Manhole Locks

A manhole lock is a physical security device designed to delay unauthorized access to the utility through a manhole. Locking a manhole that provides access to a water or wastewater system can mitigate two distinct types of threats. First, locking a manhole may delay the access of unauthorized personnel to water or wastewater systems through it. Second, locking manholes may also prevent the introduction of hazardous substances into the wastewater or stormwater system.

Radiation Detection Equipment for Monitoring Personnel and Packages

A major potential threat facing water and wastewater facilities is contamination by radioactive substances. Radioactive substances brought on-site at a facility could be used to contaminate the facility, thereby preventing workers from safely entering the facility to perform necessary water treatment tasks. In addition, radioactive substances brought on-site at a water treatment plant could be discharged into the water source or the distribution system, contaminating the downstream water supply. Therefore, the detection of radioactive substances being brought on-site can be an important security enhancement.

Various radionuclides have unique properties, and different equipment is required to detect different types of radiation. However, it is impractical and potentially unnecessary to monitor for specific radionuclides being brought on-site. Instead, for security purposes, it may be more use-ful to monitor for gross radiation as an indicator of unsafe substances.

To protect against these radioactive materials being brought on-site, a facility may set up monitoring sites out-fitted with radiation detection instrumentation at entrances. Depending on the specific types of equipment chosen, this equipment would detect radiation emitted from peo-ple, packages, or other objects being brought through an entrance.

One of the primary differences between the different types of detection equipment is how the equipment reads radiation. Radiation may either be detected by direct mea-surement or through sampling. Direct radiation measure-ment involves measuring radiation using an external probe on the detection instrumentation. Some direct measurement equipment detects radiation emitted into the air around the monitored object. Because this equipment detects radiation in the air, it does not require physical contact with the moni-tored object. Direct methods for detecting radiation include using a walk-through portal-type monitor that detects ele-vated radiation levels on a person or in a package, or using a hand-held detector, which is moved or swept over indi-vidual objects to locate a radioactive source.

Some types of radiation, such as alpha or low-energy beta radiation, have a short range and are easily shielded by various materials. These types of radiation cannot be measured through direct measurement. Instead, they must be measured through sampling. Sampling involves wiping the surface to be tested with a special filter cloth and then reading the cloth in a specialized counter. For example, spe-cialized smear counters measure alpha and low-energy beta radiation.

Reservoir Covers

Reservoirs are used to store raw or untreated water. They can be located underground (buried), at ground level, or on an elevated surface. Reservoirs can vary significantly in size; small reservoirs can hold as little as 1,000 gallons, while larger reservoirs may hold many millions of gallons. Reservoirs can be either natural or man-made. Natural res-ervoirs can include lakes or other contained water bodies, while man-made reservoirs usually consist of engineered structures, such as tanks or other impoundment structures. In addition to the water containment structure itself, reser-voir systems may also include associated water treatment and distribution equipment, including intakes, pumps, pump houses, piping systems, and chemical treatment and storage areas.

Drinking water reservoirs are of particular concern because they are potentially vulnerable to contamination of the stored water, either through direct contamination of the storage area or through infiltration of the equipment, piping, or chemicals associated with the reservoir. For example, many drinking water reservoirs are designed as aboveground, open-air structures, making them vulner-able to airborne deposition, bird and animal waste, human

activities, and dissipation of chlorine or other treatment chemicals. However, one of the most serious potential threats to the system is direct contamination of the stored water through the dumping of contaminants into the reser-voir. Utilities have taken various measures to mitigate this type of threat, including fencing off the reservoir, installing cameras to monitor intruders, and monitoring for changes in water quality. Another option for enhancing security is covering the reservoir with some type of manufactured cover to prevent intruders from gaining physical access to the stored water. Implementing a reservoir cover may or may not be practical, depending on the size of the reservoir (for example, covers are not typically used on natural reser-voirs because they are too large for the cover to be techni-cally feasible and cost-effective). This section will focus on drinking water reservoir covers, where and how they are typically implemented, and how they can be used to reduce the threat of contamination of the stored water. While cov-ers can enhance the reservoir's security, it should be noted that covering a reservoir typically changes its operational requirements. For example, vents must be installed in the cover to ensure gas exchange between the stored water and the atmosphere.

A reservoir cover is a structure installed on or over the surface of the reservoir to minimize water quality degrada-tion. The three basic design types for reservoir covers are:

- Floating
- Fixed
- Air-supported

A variety of materials are used in manufacturing a cover, including reinforced concrete, steel, aluminum, polypro-pylene, chlorosulfonated polyethylene, and ethylene inter-polymer alloys. Several factors affect a reservoir cover's effectiveness and its ability to protect the stored water. These factors include:

- The location, size, and shape of the reservoir.
- The ability to lay/support a foundation (for example, footing, soil, and geotechnical support conditions).
- The length of time the reservoir can be removed from service for cover installation or maintenance.
- Aesthetic considerations.
- Economic factors, such as capital and mainte-nance costs.

It may not be practical, for example, to install a fixed cover over a reservoir if the reservoir is too large or if the local soil conditions cannot support a foundation. A floating or air-supported cover may be more appropriate for these types of applications.

In addition to the practical considerations for the instal-lation of these types of covers, several operations and main-tenance (O&M) concerns affect the utility of a cover for specific applications. These include how different cover materials will withstand local climatic conditions, what types of cleaning and maintenance will be required for each particular type of cover, and how these factors will affect the cover's lifespan and its ability to be repaired when damaged.

The primary feature affecting the security of a reservoir cover is its ability to maintain its integrity. Any type of cover, no matter what its construction material, will provide good protection from contamination by rainwater or atmospheric deposition, as well as from intruders attempting to access the stored water with the intent of causing intentional contamination. The covers are large and heavy, making it difficult to circumvent them to gain access to the reservoir. At the very least, it would take a determined intruder, as opposed to a vandal, to defeat the cover.

Side-Hinged Door Security

Doorways are the main access points to a facility or rooms within a building. They are used on the exterior or the interior of buildings to provide privacy and security for the areas behind them. Different types of doorway security systems may be installed in different doorways, depending on the needs or requirements of the buildings or rooms. For example, exterior doorways tend to have heavier doors to withstand the elements and provide some security for the entrance of the building. Interior doorways in office areas may have lighter doors that are primarily designed to provide privacy rather than security. Therefore, these doors may be made of glass or lightweight wood. Doorways in industrial areas may have sturdier doors than other interior doorways and may be designed to provide protection or security for the areas behind the doorway. For example, fireproof doors may be installed in chemical storage areas or in other areas where there is a danger of fire. Because they are the main entries into a facility or a room, doorways are often prime targets for unauthorized entry into a facility or asset. Therefore, securing doorways may be a major step in providing security at a facility. A doorway includes four main components:

- The door, which blocks the entrance. The primary threat to the actual door is breaking or piercing through it. Therefore, the primary security features of doors are their strength and resistance to various physical threats, such as fire or explosions.
- The door frame, which connects the door to the wall. The primary threat to a door frame is that the door can be pried away from it. Therefore, the primary security feature of a door frame is its resistance to prying.
- The hinges, which connect the door to the door frame. The primary threat to door hinges is that they can be removed or broken, allowing intruders to remove the entire door. Therefore, security hinges are designed to be resistant to breaking. They may also be designed to minimize the threat of removal from the door.

- The lock, which connects the door to the door frame. The use of the lock is controlled through various security features, such as keys and combinations, so that only authorized personnel can open it and go through the door. Locks may also incorporate other security features, such as software or other systems to track the overall use of the door or to track individuals using it.

Each of these components is integral in providing security for a doorway, and upgrading the security of only one of these components while leaving the others unprotected may not increase the overall security of the doorway. For example, many facilities upgrade door locks as a basic step in increasing security. However, if the facilities do not also focus on increasing security for the door hinges or the door frame, the door may remain vulnerable to being removed from its frame, thereby defeating the increased security of the door lock.

The primary attribute of the security of a door is its strength. Many security doors are 4–20-gauge hollow metal doors consisting of steel plates over a hollow cavity reinforced with steel stiffeners to give the door extra stiffness and rigidity. This increases resistance to blunt force used to try to penetrate the door. The space between the stiffeners may be filled with specialized materials to provide fire, blast, or bullet resistance. The Window and Door Manufacturers Association has developed a series of performance attributes for doors. These include:

- Structural Resistance
- Forced Entry Resistance
- Hinge Style Screw Resistance
- Split Resistance
- Hinge Resistance
- Security Rating
- Fire Resistance
- Bullet Resistance
- Blast Resistance

The first five bullets provide information on a door's resistance to standard physical breaking and prying attacks. These tests are used to evaluate the strength of the door and the resistance of the hinges and frame in a standardized way. For example, the Rack Load Test simulates a prying attack on a corner of the door. A test panel is restrained at one end, and a third corner is supported. Loads are applied and measured at the fourth corner. The Door Impact Test simulates a battering attack on a door and frame using impacts of 200-foot pounds by a steel pendulum. The door must remain fully operable after the test. It should be noted that door glazing is also rated for resistance to shattering, etc. Manufacturers will be able to provide security ratings for these features of a door as well.

Door frames are an integral part of doorway security because they anchor the door to the wall. Door frames are typically constructed from wood or steel, and they are installed such that they extend for several inches over the doorway that has been cut into the wall. For added security, frames can be designed to have varying degrees of overlap with or wrapping over the underlying wall. This can make prying the frame from the wall more difficult. A frame formed from a continuous piece of metal (as opposed to a frame constructed from individual metal pieces) will prevent prying between pieces of the frame.

Many security doors can be retrofitted into existing frames; however, many security door installations include replacing the door frame as well as the door itself. For example, bullet resistance per Underwriters Laboratories (UL) 752 requires the resistance of the door and frame assembly, and thus replacing the door only would not meet UL 752 requirements.

Valve Lockout Devices

Valves are utilized as control elements in water and wastewater process piping networks. They regulate the flow of both liquids and gases by opening, closing, or obstructing a flow passageway. Valves are typically located where flow control is necessary. They can be positioned in-line or at pipeline and tank entrance and exit points. They serve multiple purposes in a process pipe network, including:

- Redirecting and throttling flow
- Preventing backflow
- Shutting off flow to a pipeline or tank (for isolation purposes)
- Releasing pressure
- Draining extraneous liquid from pipelines or tanks
- Introducing chemicals into the process network
- As access points for sampling process water

Valves are located at critical junctures throughout water and wastewater systems, both on-site at treatment facilities and off-site within water distribution and wastewater collection systems. They may be located either above ground or below ground. Because many valves are located within the community, it is critical to protect against valve tampering. For example, tampering with a pressure relief valve could result in a pressure buildup and potential explosion in the piping network. On a larger scale, the addition of a pathogen or chemical to the water distribution system through an unprotected valve could result in the release of that contaminant to the general population.

Various security products are available to protect aboveground versus belowground valves. For example, valve lockout devices can be purchased to protect valves and valve controls located aboveground. Vaults containing underground valves can be locked to prevent access to these valves. Valve-specific lockout devices are available in a variety of colors, which can be useful in distinguishing different valves. For example, different colored lockouts can be used to indicate the type of liquid passing through the valve (i.e., treated, untreated, potable, chemical) or to identify the party responsible for maintaining the lockout.

Implementing a system of different-colored locks on operating valves can increase system security by reducing the likelihood of an operator inadvertently opening the wrong valve and causing a problem in the system.

Vent Security

Vents are installed in aboveground, covered water reservoirs and in underground reservoirs to allow ventilation of the stored water. Specifically, vents permit the passage of air that is displaced from or drawn into the reservoir as the water level rises and falls due to system demands. Small reservoirs may require only one vent, whereas larger reservoirs may have multiple vents throughout the system.

The specific vent design for any given application will vary depending on the design of the reservoir, but every vent consists of an open-air connection between the reservoir and the outside environment. Although these air exchange vents are an integral part of covered or underground reservoirs, they also represent a potential security threat. Vent security can be improved by making the vents tamper-resistant or adding other security features, such as security screens or security covers, which can enhance the security of the entire water system. Many municipalities already have specifications for vent security at their water assets. These specifications typically include the following requirements:

- Vent openings are to be angled down or shielded to minimize the entrance of surface and/or rainwater into the vent through the opening.
- Vent designs are to include features to exclude insects, birds, animals, and dust.
- Corrosion-resistant materials are to be used to construct the vents.

Some states have adopted more specific requirements for added vent security at their water utility assets. For example, the State of Utah's Department of Environmental Quality, Division of Drinking Water, and Division of Administrative Rules (DAR) provide specific requirements for public drinking water storage tanks. The rules for drinking water storage tanks as they apply to venting are outlined in Utah R309-545-15: "Venting," and include the following requirements:

- Drinking water storage tank vents must have an open discharge on buried structures.
- The vents must be located 24–36 inches above the earthen covering.
- The vents must be located and sized to avoid blockage during winter conditions.

In a second example, Washington State's "Drinking Water Tech Tips: Sanitary Protection of Reservoirs" document states that vents must be protected to prevent the water supply from being contaminated. The document indicates that non-corrodible No. 4 mesh may be used to screen vents on elevated tanks. It continues to state that the vent opening for storage facilities located underground or at ground level should be 24–36 inches above the roof or ground and that it must be protected with a 24-inch mesh non-corrodible screen. New Mexico's Administrative Code also specifies that vents must be covered with No. 24 mesh (NMAC Title 20, Chapter 7, Subpart I, 208.E). Washington and New Mexico, as well as many other municipalities, require vents to be screened using a non-corrodible mesh to minimize the entry of insects, other animals, and rain-borne contamination into the vents. When selecting the appropriate mesh size, it is important to identify the smallest mesh size that meets both the strength and durability requirements for that application.

Visual Surveillance Monitoring

Visual surveillance is used to detect threats through continuous observation of important or vulnerable areas of an asset. The observations can also be recorded for later review or use (for example, in court proceedings). Visual surveillance systems can be used to monitor various parts of collection, distribution, or treatment systems, including the perimeter of a facility, outlying pumping stations, or entry or access points into specific buildings. These systems are also useful in recording individuals who enter or leave a facility, thereby helping to identify unauthorized access. Images can be transmitted live to a monitoring station, where they can be monitored in real time, or they can be recorded and reviewed later. Many facilities have found that a combination of electronic surveillance and security guards provides an effective means of facility security. Visual surveillance is provided through a closed-circuit television (CCTV) system, in which capture, transmission, and reception of an image are localized within a closed "circuit." This is different from other broadcast images, such as over-the-air television, which are broadcast over the air to any receiver within range. At a minimum, a CCTV system consists of:

- One or more cameras.
- A monitor for viewing the images.
- A system for transmitting the images from the camera to the monitor.

WATER MONITORING DEVICES

Note: Adapted from Spellman, F.R., *Water Infrastructure Protection and Homeland Security*, Government Institutes Press, Lanham, MD, 2007.

Proper security preparation comes down to a three-legged approach: Detect, Delay, and Respond. The third leg of security, Detect, is discussed in this section. Specifically, this section deals with the monitoring of water samples to detect toxicity and/or contamination. Many of the major monitoring tools that can be used to identify anomalies in process streams or finished water that may represent potential threats are discussed, including:

- Sensors for monitoring chemical, biological, and radiological contamination
- Chemical sensor—Arsenic measurement system
- Chemical sensor for toxicity (adapted BOD analyzer)
- Chemical sensor—Total organic carbon analyzer
- Chemical sensor—Chlorine measurement system
- Chemical sensor—Portable cyanide analyzer
- Portable field monitors to measure VOCs
- Radiation detection equipment
- Radiation detection equipment for monitoring water assets
- Toxicity monitoring/toxicity meters

Water quality monitoring sensor equipment may be used to monitor key elements of water or wastewater treatment processes (such as influent water quality, treatment processes, or effluent water quality) to identify anomalies that may indicate threats to the system. Some sensors, such as those for biological organisms or radiological contaminants, measure potential contamination directly, while others, particularly some chemical monitoring systems, measure "surrogate" parameters that may indicate problems in the system but do not identify sources of contamination directly. In addition, sensors can provide more accurate control of critical components in water and wastewater systems and may provide a means of early warning so that the potential effects of certain types of attacks can be mitigated. One advantage of using chemical and biological sensors to monitor for potential threats to water and wastewater systems is that many utilities already employ sensors to monitor potable water (raw or finished) or influent/effluent for Safe Drinking Water Act (SDWA) or Clean Water Act (CWA) water quality compliance or process control.

Chemical sensors that can be used to identify potential threats to water and wastewater systems include inorganic monitors (e.g., chlorine analyzer), organic monitors (e.g., total organic carbon analyzer), and toxicity meters. Radiological meters can be used to measure concentrations of several different radioactive species. Monitors that use biological species can be used as sentinels for the presence of contaminants of concern, such as toxins. At present, biological monitors are not in widespread use, and very few bio-monitors are used by drinking water utilities in the United States.

Monitoring can be conducted using either portable or fixed-location sensors. Fixed-location sensors are usually used as part of a continuous, on-line monitoring system. Continuous monitoring has the advantage of enabling immediate notification when there is an upset. However, the sampling points are fixed, and only certain points in the system can be monitored. In addition, the number of monitoring locations needed to capture the physical, chemical, and biological complexity of a system can be prohibitive. The use of portable sensors can overcome this problem of monitoring many points in the system. Portable sensors can be used to analyze grab samples at any point in the system, but they have the disadvantage of providing measurements only at one point in time.

Sensors for Monitoring Chemical, Biological, and Radiological Contamination

Toxicity tests measure water toxicity by monitoring adverse biological effects on test organisms. Toxicity tests have traditionally been used to monitor wastewater effluent streams for National Pollutant Discharge Elimination System (NPDES) permit compliance or to test water samples for toxicity. However, this technology can also monitor drinking water distribution systems or other water/wastewater streams for toxicity. Currently, several types of bio-sensors and toxicity tests are being adapted for use in the water/wastewater security field. The keys to using bio-monitoring or bio-sensors for drinking water or other water/wastewater asset security are rapid response and the ability to use the monitor at critical locations in the system, such as in water distribution systems downstream of pump stations or before the biological process in a wastewater treatment plant. While several different organisms can be used to monitor for toxicity (including bacteria, invertebrates, and fish), bacteria-based bio-sensors are ideal for use as early warning screening tools for drinking water security because bacteria usually respond to toxics in a matter of minutes. In contrast, toxicity screening methods that use higher-level organisms such as fish may take several days to produce a measurable result. Bacteria-based biosensors have recently been incorporated into portable instruments, making rapid response and field-testing practical. These portable meters detect decreases in biological activity (e.g., decreases in bacterial luminescence), which are highly correlated with increased levels of toxicity.

At present, few utilities are using biologically-based toxicity monitors to monitor water/wastewater assets for toxicity, and very few products are commercially available. Several new approaches to the rapid monitoring of microorganisms for security purposes (e.g., microbial source tracking) have been identified. However, most of these methods are still in the research and development phase.

Chemical Sensors: Arsenic Measurement System

Arsenic is an inorganic toxin that occurs naturally in soils. It can enter water supplies from many sources, including erosion of natural deposits, runoff from orchards, runoff from glass and electronics production wastes, or leaching from products treated with arsenic, such as wood. Synthetic organic arsenic is also used in fertilizers. Arsenic toxicity is primarily associated with inorganic arsenic ingestion, which has been linked to cancerous health effects, including cancer of the bladder, lungs, skin, kidney, nasal passages, liver, and prostate. Arsenic ingestion has also been linked to non-cancerous cardiovascular, pulmonary, immunological, neurological, and endocrine problems. According to the USEPA's Safe Drinking Water Act (SDWA) Arsenic Rule, inorganic arsenic can exert toxic effects after acute

(short-term) or chronic (long-term) exposure. Toxicological data for acute exposure, typically given as an LD50 value (the dose that would be lethal to 50% of the test subjects), suggests that the LD50 of arsenic ranges from 1to 4mg of arsenic per kilogram (mg/kg) of body weight. This dose would correspond to a lethal dose range of 70–280mg for 50% of adults weighing 70kg. At non-lethal, but high, acute doses, inorganic arsenic can cause gastroenterological effects, shock, neuritis (continuous pain), and vascular effects in humans. The USEPA has set a maximum contaminant level goal of 0 for arsenic in drinking water; the current enforceable maximum contaminant level (MCL) is 0.050 mg/L. As of January 23, 2006, the enforceable MCL for arsenic will be 0.010 mg/L.

The SDWA requires arsenic monitoring for public water systems. The Arsenic Rule indicates that surface water systems must collect one sample annually, while groundwater systems must collect one sample in each compliance period (once every three years). Samples are collected at entry points to the distribution system, and analysis is done in the lab using one of several USEPA-approved methods, including Inductively Coupled Plasma Mass Spectrometry (ICP-MS, USEPA 200.8) and several atomic absorption (AA) methods. However, several different technologies, including colorimetric test kits and portable chemical sensors, are currently available for monitoring inorganic arsenic concentrations in the field. These technologies can provide a quick estimate of arsenic concentrations in a water sample. Thus, these technologies may be useful for spot-checking different parts of a drinking water system (for example, reservoirs and isolated areas of distribution systems) to ensure that the water is not contaminated with arsenic.

Chemical Sensor: Adapted BOD Analyzer

One manufacturer has adapted a BOD analyzer to measure oxygen consumption as a surrogate for general toxicity. The critical element in the analyzer is the bioreactor, which is used to continuously measure the respiration of the biomass under stable conditions. As the toxicity of the sample increases, oxygen consumption decreases. An alarm can be programmed to sound if oxygen reaches a minimum concentration (i.e., if the sample is strongly toxic). The operator must then interpret the results as a measure of toxicity. Note that, present, it is difficult to directly define the sensitivity and/or the detection limit of toxicity measurement devices because limited data are available regarding the specific correlation between decreased oxygen consumption and increased toxicity of the sample.

Chemical Sensor: Total Organic Carbon Analyzer

Total organic carbon (TOC) analysis is a well-defined and commonly used methodology that measures the carbon content of dissolved and particulate organic matter present in water. Many water utilities monitor TOC to determine raw water quality or to evaluate the effectiveness of processes designed to remove organic carbon. Some wastewater utilities also employ TOC analysis to monitor the efficiency of the treatment process. In addition to these uses for TOC monitoring, measuring changes in TOC concentrations can be an effective "surrogate" for detecting contamination from organic compounds (e.g., petrochemicals, solvents, pesticides). Thus, while TOC analysis does not provide specific information about the nature of the threat, identifying changes in TOC can be a good indicator of potential threats to a system. TOC analysis includes inorganic carbon removal, oxidation of the organic carbon into CO_2, and quantification of the CO_2. The primary differences between different online TOC analyzers lie in the methods used for oxidation and CO_2 quantification.

The oxidation step can be either high or low temperature. The determination of the appropriate analytical method (and thus the appropriate analyzer) is based on the expected characteristics of the wastewater sample (TOC concentrations and the individual components making up the TOC fraction). In general, high-temperature (combustion) analyzers achieve more complete oxidation of the carbon fraction than low-temperature (wet chemistry/UV) analyzers. This can be important both for distinguishing different fractions of organics in a sample and for achieving a precise measurement of the organic content of the sample. Three different methods are available for detecting and quantifying carbon dioxide produced in the oxidation step of a TOC analyzer. There are:

- Nondispersive infrared (NDIR) detector
- Colorimetric methods
- Aqueous conductivity methods

The most common detector that online TOC analyzers use for source water and drinking water analysis is the nondispersive infrared detector.

Although the differences in analytical methods employed by different TOC analyzers may be important for compliance or process monitoring, high levels of precision and the ability to distinguish specific organic fractions in a sample may not be required for detecting a potential chemical threat. Instead, significant deviations from normal TOC concentrations may provide the best indication of a chemical threat to the system.

The detection limit for organic carbon depends on the measurement technique used (high or low temperature) and the type of analyzer. Since TOC concentrations serve as simply surrogates that can indicate potential problems in a system, gross changes in these concentrations are the best indicators of potential threats. Therefore, high-sensitivity probes may not be required for security purposes. However, the following detection limits can be expected:

- High temperature method (between 680°C and 950°C or higher in a few special cases, providing the best possible oxidation): =1 mg/L carbon.
- Low temperature method (below 100°C, with limited oxidation potential): =0.2 mg/L carbon

The response time of a TOC analyzer may vary depending on the manufacturer's specifications, but it usually takes from 5 to 15 minutes to achieve a stable, accurate reading.

Chemical Sensors: Chlorine Measurement System

Residual chlorine is one of the most sensitive and useful indicator parameters for monitoring water distribution systems. All water distribution systems monitor residual chlorine concentrations as part of their Safe Drinking Water Act (SDWA) requirements, and procedures for monitoring chlorine concentrations are well-established and accurate. Chlorine monitoring assures proper residual levels at all points in the system, helps pace rechlorination when needed, and quickly and reliably signals any unexpected increase in disinfectant demand. A significant decline or loss of residual chlorine could indicate potential threats to the system. Several key points regarding residual chlorine monitoring for security purposes are provided below:

- Chlorine residuals can be measured using continuous online monitors at fixed points in the system, by taking grab samples at any point in the system, or by taking grab samples at any point in the system and using chlorine test kits or portable sensors to determine chlorine concentrations.
- Correct placement of residual chlorine monitoring points within a system is crucial for the early detection of potential threats. For example, while dead ends and low-pressure zones are common trouble spots that may show low residual chlorine concentrations, these zones are generally not of great concern for water security because system hydraulics limit the circulation of contaminants in these areas o.
- Monitoring points and procedures for SDWA compliance versus system security purposes may differ. Utilities must determine the best balance between using online, fixed monitoring systems and portable sensors/test kits to meet both SDWA compliance and security needs.

Various portable and online chlorine monitors are commercially available. These range from sophisticated on-line chlorine monitoring systems to portable electrode sensors and colorimetric test kits. Online systems can be equipped with control, signal, and alarm systems that notify the operator of low chlorine concentrations, and some may be tied into feedback loops that automatically adjust chlorine concentrations in the system. In contrast, the use of portable sensors or colorimetric test kits requires technicians to take a sample and read the results. The technician then initiates the required actions based on the test results.

Several measurement methods are currently available to measure chlorine in water samples, including:

- N, N-diethyl-p-phenylenediamine (DPD) colorimetric method

- Iodometric method
- Amperometric electrodes
- Polarographic membrane sensors

It should be noted that there can be differences in the specific type of analyte, the range, and the accuracy of these different measurement methods. In addition, these different methods have different operations and maintenance requirements. For example, DPD systems require periodic replenishment of buffers, whereas polarographic systems do not. Users may want to consider these requirements when choosing the appropriate sensor for their system.

Chemical Sensors: Portable Cyanide Analyzer

Portable cyanide detection systems are designed for field use to evaluate potential cyanide contamination of a water asset. These detection systems use one of two distinct analytical methods—either a colorimetric method or an ion-selective method—to provide a quick, accurate cyanide measurement that does not require laboratory evaluation. Aqueous cyanide chemistry can be complex. Various factors, including the water asset's pH and redox potential, can affect the toxicity of cyanide in that asset. While personnel using these cyanide detection devices do not need advanced knowledge of cyanide chemistry to successfully screen a water asset for cyanide, understanding aqueous cyanide chemistry can help users interpret whether the asset's cyanide concentration represents a potential threat. Therefore, a summary of aqueous cyanide chemistry, including a discussion of cyanide toxicity, is provided below.

Cyanide (CN-) is a toxic carbon-nitrogen organic compound that is the functional portion of the lethal gas hydrogen cyanide (HCN). The toxicity of aqueous cyanide varies depending on its form. At near-neutral pH, "free cyanide" (commonly designated as "CN-," although it is actually defined as the total of HCN and CN-) is the predominant cyanide form in water. Free cyanide is potentially toxic in its aqueous form, although the primary concern regarding aqueous cyanide is that it could volatilize. Free cyanide is not highly volatile (it is less volatile than most VOCs, but its volatility increases as the pH decreases below 8). However, when free cyanide does volatilize, it volatilizes in its highly toxic gaseous form (gaseous HCN). As a general rule, metal-cyanide complexes are much less toxic than free cyanide because they do not volatilize unless the pH is low.

Analyses for cyanide in public water systems are often conducted in certified labs using various USEPA-approved methods, such as the preliminary distillation procedure with subsequent analysis by colorimetric, ion-selective electrode, or flow injection methods. Lab analyses using these methods require careful sample preservation and pretreatment procedures and are generally expensive and time-consuming. Using these methods, several cyanide fractions are typically defined:

- **Total cyanide:** includes free cyanide (CN-+HCN) and all metal- complexed cyanide.

- **Weak Acid Dissociable (WAD) Cyanide:** includes free cyanide (CN-+HCN) and weak cyanide complexes that could be potentially toxic by hydrolysis to free cyanide in the pH range 4.5–6.0.
- **Amendable Cyanide:** includes free cyanide (CN-+HCN) and weak cyanide complexes that can release free cyanide at high pH (11–12) (this fraction gets its name because it includes measurement of cyanide from complexes that are "amendable" to oxidation by chlorine at high pH). To measure "Amendable Cyanide," the sample is split into two fractions. One of the fractions is analyzed for "Total Cyanide" as described above. The other fraction is treated with high levels of chlorine for approximately 1 hour, dechlorinated, and distilled according to the "Total Cyanide" method above. "Amendable Cyanide" is determined by the difference in the cyanide concentrations in these two fractions.
- **Soluble Cyanide:** measures only soluble cyanide. Soluble cyanide is measured by using the preliminary filtration step, followed by "Total Cyanide" analysis described above.

As discussed, these different methods yield various cyanide measurements which may or may not provide a complete picture of the sample's potential toxicity. For example, the "Total Cyanide" method includes cyanide complexed with metals, some of which will not contribute to cyanide toxicity unless the pH is out of the normal range. In contrast, the "WAD Cyanide" measurement includes metal-complexed cyanide that could become free cyanide at low pH, and "Amendable Cyanide" measurements include metal-complexed cyanide that could become free cyanide at high pH. Personnel using these kits should therefore be aware of the potential differences in actual cyanide toxicity and the cyanide measured in the sample under different environmental conditions.

Ingestion of aqueous cyanide can result in numerous adverse health effects and may be lethal. The USEPA's Maximum Contaminant Level (MCL) for cyanide in drinking water is 0.2 µg/L (0.2 parts per million, or ppm). This MCL is based on free cyanide analysis per the "Amendable Cyanide" method described above (the USEPA has recognized that very stable metal-cyanide complexes, such as iron-cyanide complexes are non-toxic [unless exposed to significant UV radiation], and these fractions are therefore not considered when defining cyanide toxicity). Ingestion of free cyanide at concentrations in excess of this MCL causes both acute effects (e.g., rapid breathing, tremors, and neurological symptoms) and chronic effects (e.g., weight loss, thyroid effects, and nerve damage). Under the current primary drinking water standards, public water systems are required to monitor their systems to minimize public exposure to cyanide levels in excess of the MCL.

Hydrogen cyanide gas is also toxic, and the Office of Safety and Health Administration (OSHA) has set a permissible exposure limit (PEL) of 10 ppmv for HCN inhalation. HCN has a strong, bitter, almond-like smell and an odor threshold of approximately 1 ppmv. Considering that HCN is relatively non-volatile (see above), a slight cyanide odor emanating from a water sample suggests very high aqueous cyanide concentrations—greater than 10–50 mg/L, which is in the range of a lethal or near-lethal dose with the ingestion of one pint of water.

Portable Field Monitors to Measure VOCs

Volatile organic compounds (VOCs) are a group of highly utilized chemicals that have widespread applications, including use as fuel components, solvents, and cleaning and liquefying agents in degreasers, polishes, and dry-cleaning solutions. VOCs are also used in herbicides and insecticides for agricultural applications. Laboratory-based methods for analyzing VOCs are well established; however, analyzing VOCs in the lab is time-consuming—obtaining a result may require several hours to several weeks depending on the specific method. Faster, commercially available methods for analyzing VOCs quickly in the field include the use of portable gas chromatographs (GC), mass spectrometers (MS), or gas chromatographs/mass spectrometers (GC/MS), all of which can be used to obtain VOC concentration results within minutes. These instruments can be useful in the rapid confirmation of the presence of VOCs in an asset or for monitoring an asset regularly. In addition, portable VOC analyzers can analyze a wide range of VOCs, such as toxic industrial chemicals (TICs), chemical warfare agents (CWAs), drugs, explosives, and aromatic compounds. There are several easy-to-use, portable VOC analyzers currently on the market that are effective in evaluating VOC concentrations in the field. These instruments utilize gas chromatography, mass spectrometry, or a combination of both methods to provide near laboratory-quality analysis for VOCs.

Radiation Detection Equipment

Radioactive substances (radionuclides) are known health hazards that emit energetic waves and/or particles, which can cause both carcinogenic and non-carcinogenic health effects. Radionuclides pose unique threats to source water supplies, as well as water treatment, storage, or distribution systems because radiation emitted from radionuclides in water systems can affect individuals through several pathways—by direct contact with, ingestion, inhalation, or external exposure to the contaminated water. While radiation can naturally occur in some cases due to the decay of some minerals, intentional and unintentional releases of man-made radionuclides into water systems also present a realistic threat.

Threats to water and wastewater facilities from radioactive contamination could involve two major scenarios. First, the facility or its assets could be contaminated, preventing workers from accessing and operating the facility/assets. Second, at drinking water facilities, the water supply could be contaminated, leading to tainted water being distributed

to users downstream. These two scenarios require different threat reduction strategies. The first scenario requires monitoring for radioactive substances being brought on-site, while the second requires monitoring water assets for radioactive contamination. Although the effects of radioactive contamination are basically the same in both scenarios, each threat requires different types of radiation monitoring and equipment.

Radiation Detection Equipment for Monitoring Water Assets

Most water systems are required to monitor for radioactivity and certain radionuclides and to meet Maximum Contaminant Levels (MCLs) for these contaminants to comply with the Safe Drinking Water Act (SDWA). Currently, the USEPA requires drinking water to meet MCLs for beta/photon emitters (including gamma radiation), alpha particles, combined radium 226/228, and uranium. However, this monitoring is required only at entry points into the system. In addition, after the initial sampling requirements, only one sample is required every 3–9 years, depending on the contaminant type and initial concentrations. While this is adequate to monitor long-term protection from overall radioactivity and specific radionuclides in drinking water, it may not be adequate to identify short-term spikes in radioactivity, such as from spills, accidents, or intentional releases. In addition, compliance with the SDWA requires analyzing water samples in a laboratory, resulting in a delay in receiving results. In contrast, security monitoring is more effective when results can be obtained quickly in the field. Moreover, monitoring for security purposes does not necessarily require identifying the specific radionuclides causing the contamination. Thus, for security purposes, it may be more appropriate to monitor for non-radionuclide-specific radiation using either portable field meters, which can be used as necessary to evaluate grab samples, or online systems, which can provide continuous monitoring of a system.

Ideally, measuring radioactivity in water assets in the field would involve minimal sampling and sample preparation. However, the physical properties of specific types of radiation combined with the physical properties of water, make evaluating radioactivity in water assets in the field somewhat difficult. For example, alpha particles can only travel short distances and they cannot penetrate most physical objects. Therefore, instruments designed to evaluate alpha emissions must be specially designed to capture emissions at a short distance from the source, and they must not block alpha emissions from entering the detector. Gamma radiation does not have the same physical properties and can thus be measured using different detectors.

Measuring different types of radiation is further complicated by the relationship between the radiation's intrinsic properties and the medium in which the radiation is being measured. For example, gas-flow proportional counters are typically used to evaluate gross alpha and beta radiation from smooth, solid surfaces. However, because water is not a smooth surface and alpha and beta emissions are relatively short range and can be attenuated within the water, these types of counters are not appropriate for measuring alpha and beta activity in water. An appropriate method for measuring alpha and beta radiation in water is by using a liquid scintillation counter. However, this requires mixing an aliquot of water with a liquid scintillation "cocktail." The liquid scintillation counter is a large, sensitive piece of equipment, so it is not appropriate for field use. Therefore, measurements for alpha and beta radiation from water assets are not typically made in the field.

Unlike the problems associated with measuring alpha and beta activity in water in the field, the properties of gamma radiation allow it to be measured relatively well in water samples in the field. The standard instrumentation used to measure gamma radiation from water samples in the field is a sodium iodide (NaI) scintillator.

Although the devices outlined above are the most commonly used for evaluating total alpha, beta, and gamma radiation, other methods and devices can be used. In addition, local conditions (i.e., temperature, humidity) or the properties of the specific radionuclides emitting the radiation may make other types of devices or methods more optimal to achieve the goals of the survey than the devices noted above. Therefore, experts or individual vendors should be consulted to determine the appropriate measurement device for any specific application.

An additional factor to consider when developing a program to monitor for radioactive contamination in water assets is whether to take regular grab samples or sample continuously. For example, portable sensors can be used to analyze grab samples at any point in the system, but they have the disadvantage of providing measurements only at one point in time. On the other hand, fixed-location sensors are usually used as part of a continuous, online monitoring system. These systems continuously monitor a water asset and can be outfitted with an alarm system to alert operators if radiation increases above a certain threshold. However, the sampling points are fixed, and only certain points in the system can be monitored. In addition, the number of monitoring locations needed to capture the physical and radioactive complexity of a system can be prohibitive.

Toxicity Monitoring/Toxicity Meters

Toxicity measurement devices measure general toxicity to biological organisms, and the detection of toxicity in any water/wastewater asset can indicate a potential threat, either to the treatment process (in the case of influent toxicity), to human health (in the case of finished drinking water toxicity) or to the environment (in the case of effluent toxicity). Currently, whole effluent toxicity tests (WET tests), in which effluent samples are tested against test organisms, are required for many National Pollutant Discharge Elimination System (NPDES) discharge permits. The WET tests are used as a complement to the effluent limits on physical and chemical parameters to assess the overall effects of the discharge on living organisms or aquatic biota. Toxicity tests may also be used to monitor wastewater influent streams for

potential hazardous contamination, such as organic heavy metals (arsenic, mercury, lead, chromium, and copper) that might upset the treatment process.

The ability to get feedback on sample toxicity from short-term toxicity tests or toxicity "meters" can be valuable in estimating the overall toxicity of a sample. Online real-time toxicity monitoring is still under active research and development. However, several portable toxicity measurement devices are commercially available. These can generally be divided into categories based on the different ways they measure toxicity:

- Meters measuring direct biological activity (e.g., luminescent bacteria) and correlating decreases in this direct biological activity with increased toxicity
- Meters measuring oxygen consumption and correlating decreases in oxygen consumption with increased toxicity.

COMMUNICATION AND INTEGRATION

This section discusses the devices necessary for communication and integration of water and wastewater system operations, such as electronic controllers, two-way radios, and wireless data communications. Electronic controllers are used to automatically activate equipment (such as lights, surveillance cameras, audible alarms, or locks) when they are triggered. Triggering could occur in response to a variety of scenarios, including the tripping of an alarm or a motion sensor, the breaking of a window or a glass door, variations in vibration sensor readings, or simply input from a timer. Two-way wireless radios allow two or more users who have their radios tuned to the same frequency to communicate instantaneously with each other without the radios being physically linked together with wires or cables. Wireless data communications devices are used to enable the transmission of data between computer systems and/or between a SCADA server and its sensing devices, without individual components being physically linked together via wires or cables. In water and wastewater utilities, these devices are often used to link remote monitoring stations (i.e., SCADA components) or portable computers (i.e., laptops) to computer networks without using physical wiring connections.

Electronic Controllers

An electronic controller is a piece of electronic equipment that receives incoming electric signals and uses preprogrammed logic to generate electronic output signals based on the incoming signals. While electronic controllers can be implemented for any application that involves inputs and outputs (for example, control of a piece of machinery in a factory), in a security application, these controllers essentially act as the system's "brain" and can respond to specific security-related inputs with preprogrammed output responses. These systems combine the control of electronic circuitry with a logic function such that circuits are opened and closed (and thus equipment is turned on and off) through some preprogrammed logic. The basic principle behind the operation of an electronic controller is that it receives electronic inputs from sensors or any device generating an electrical signal (for example, electrical signals from motion sensors) and then uses preprogrammed logic to produce electrical outputs (for example, these outputs could turn on the power to a surveillance camera or an audible alarm). Thus, these systems automatically generate a preprogrammed logical response to a preprogrammed input scenario.

The three major types of electronic controllers are timers, electromechanical relays, and programmable logic controllers (PLCs), which are often called "digital relays." Each of these types of controllers is discussed in more detail below. Timers use internal signals/inputs (in contrast to externally generated inputs) and generate electronic output signals at certain times. More specifically, timers control electric current flow to any application to which they are connected and can turn the current on or off on a schedule pre-specified by the user. The typical timer range (the amount of time that can be programmed to elapse before the timer activates linked equipment) is from 0.2 seconds to 10 hours, although some of the more advanced timers have ranges of up to 60 hours. Timers are useful in fixed applications that don't require frequent schedule changes. For example, a timer can be used to turn on the lights in a room or building at a certain time every day. Timers are usually connected to their own power supply (usually 120–240 V).

In contrast to timers, which have internal triggers based on a regular schedule, electromechanical relays and PLCs have both external inputs and external outputs. However, PLCs are more flexible and powerful than electromechanical relays, and thus this section focuses primarily on PLCs as the predominant technology for security-related electronic control applications. Electromechanical relays are simple devices that use a magnetic field to control a switch. The voltage applied to the relay's input coil creates a magnetic field, which attracts an internal metal switch. This causes the relay's contacts to touch, closing the switch and completing the electrical circuit. This activates any linked equipment. These systems are often used for high-voltage applications, such as in some automotive and other manufacturing processes.

Two-Way Radios

Two-way radios, as discussed here, are limited to direct unit-to-unit radio communication, either via single unit-to-unit transmission and reception or via multiple hand-held units to a base station radio contact and distribution system. Radio frequency spectrum limitations apply to all hand-held units and are directed by the FCC. This also distinguishes a hand-held unit from a base station or base station unit (such as those used by an amateur (ham) radio operator), which operate under different wavelength parameters.

Two-way radios allow a user to contact another user or group of users instantly on the same frequency and to transmit voice or data without the need for wires. They use "half-duplex" communications, meaning communication can be either transmitted or received; it cannot transmit and receive simultaneously. In other words, only one person may talk while other personnel with radio(s) can only listen. To talk, the user depresses the talk button and speaks into the radio. The audio then transmits the voice wirelessly to the receiving radios. When the speaker has finished speaking and the channel has cleared, users on any of the receiving radios can transmit, either to answer the first transmission or to begin a new conversation. In addition to carrying voice data, many types of wireless radios also allow the transmission of digital data, and these radios may be interfaced with computer networks that can use or track this data. For example, some two-way radios can send information such as global positioning system (GPS) data or the ID of the radio. Some two-way radios can also send data through a SCADA system.

Wireless radios broadcast voice or data communications over the airwaves from the transmitter to the receiver. While this can be an advantage in that the signal emanates in all directions and does not need a direct physical connection to be received by the receiver, it can also make the communications vulnerable to being blocked, intercepted, or otherwise altered. However, security features are available to ensure that the communications are not tampered with.

Wireless Data Communications

A wireless data communication system consists of two components: a "Wireless Access Point" (WAP), and a "Wireless Network Interface Card" (sometimes also referred to as a "Client"), which work together to complete the communications link. These wireless systems can link electronic devices, computers, and computer systems together using radio waves, thus eliminating the need for these individual components to be directly connected through physical wires. While wireless data communications have widespread applications in water and wastewater systems, they also have limitations. First, wireless data connections are limited by the distance between components (radio waves scatter over a long distance and cannot be received efficiently unless special directional antennas are used). Second, these devices only function if the individual components are in a direct line of sight with each other, since radio waves are affected by interference from physical obstructions. However, in some cases, repeater units can be used to amplify and retransmit wireless signals to circumvent these problems. The two components of wireless devices are discussed in more detail below.

The wireless access point provides the wireless data communication service. It usually consists of a housing (which is constructed from plastic or metal depending on the environment in which it will be used) containing a circuit board, flash memory that holds software, one of two external ports to connect to existing wired networks, a wireless radio transmitter/receiver, and one or more antenna connections. Typically, the WAP requires a one-time user configuration to allow the device to interact with the local area network (LAN). This configuration is usually done via a web-driven software application that is accessed via a computer.

The wireless network interface card, or client, is a piece of hardware that is plugged into a computer and enables that computer to make a wireless network connection. The card consists of a transmitter, functional circuitry, and a receiver for the wireless signal, all of which work together to enable communication between the computer, its wireless transmitter/receiver, and its antenna connection. Wireless cards are installed in a computer through a variety of connections, including USB adapters, Laptop CardBus (PCMCIA), or Desktop Peripheral (PCI) cards. As with the WAP, software is loaded onto the user's computer, allowing configuration of the card so that it can operate over the wireless network.

Two of the primary applications for wireless data communications systems are to enable mobile or remote connections to a LAN and to establish wireless communications links between SCADA remote telemetry units (RTUs) and sensors in the field. Wireless card connections are usually used for LAN access from mobile computers. Wireless cards can also be incorporated into RTUs to allow them to communicate with sensing devices that are located remotely.

CYBER PROTECTION DEVICES

The USEPA, on March 3, 2023, released a memorandum stressing the need for states to assess cybersecurity risks at drinking water systems to protect our public drinking water. To this point, it is important to note that various cyber protection devices are currently available for use in protecting utility computer systems. These protection devices include anti-virus and pest eradication software, firewalls, and network intrusion hardware/software. These products are discussed in this section.

Anti-Virus and Pest Eradication Software

Anti-virus programs are designed to detect, delay, and respond to programs or pieces of code that are specifically designed to harm computers. These programs are known as "malware." Malware can include computer viruses, worms, and Trojan horse programs (programs that appear to be benign but have hidden harmful effects). Pest eradication tools are designed to detect, delay, and respond to "spyware" (strategies that websites use to track user behavior, such as by sending "cookies" to the user's computer) and hacker tools that track keystrokes (keystroke loggers) or passwords (password crackers).

Viruses and pests can enter a computer system through the Internet or through infected floppy disks or CDs. They can also be placed onto a system by insiders. Some of these programs, such as viruses and worms, then move within a computer's drives and files or between computers if the computers are networked to each other. This malware can deliberately damage files, utilize memory and network capacity, crash application programs, and initiate transmissions of sensitive information from a PC. While the specific mechanisms of these programs differ, they can infect files and even the basic operating program of the computer's firmware/hardware.

The most important features of an anti-virus program are its ability to identify potential malware and alert a user before infection occurs, as well as its ability to respond to a virus already resident on a system. Most of these programs provide a log so that the user can see what viruses have been detected and where they were detected. After detecting a virus, the anti-virus software may delete the virus automatically, or it may prompt the user to delete the virus. Some programs will also fix files or programs damaged by the virus.

Various sources of information are available to inform the general public and computer system operators about new viruses being detected. Since anti-virus programs use signatures (or snippets of code or data) to detect the presence of a virus, periodic updates are required to identify new threats. Many anti-virus software providers offer free upgrades that can detect and respond to the latest viruses.

Firewalls

A firewall is an electronic barrier designed to keep computer hackers, intruders, or insiders from accessing specific data files and information on a utility's computer network or other electronic/computer systems. Firewalls operate by evaluating and filtering information coming through a public network (such as the Internet) into the utility's computer or other electronic systems. This evaluation can include identifying the source or destination addresses and ports, and allowing or denying access based on this identification. Two methods are used by firewalls to limit access to the utility's computers or other electronic systems from the public network:

- The firewall may deny all traffic unless it meets certain criteria.
- The firewall may allow all traffic through unless it meets certain criteria.

A simple example of the first method is to screen requests to ensure that they come from an acceptable (i.e., previously identified) domain name and Internet Protocol address. Firewalls may also use more complex rules that analyze the application data to determine if the traffic should be allowed through. For example, the firewall may require user authentication (i.e., use of a password) to access the system. How a firewall determines what traffic to let through depends on which network layer it operates at and how it is configured. Firewalls may be a piece of hardware, a software program, or an appliance card that contains both.

Advanced features that can be incorporated into firewalls allow for the tracking of attempts to log-on to the local area network system. For example, a report of successful and unsuccessful log-in attempts may be generated for the computer specialist to analyze. For systems with mobile users, firewalls allow remote access to the private network through secure log-on procedures and authentication certificates. Most firewalls have a graphical user interface for managing the firewall. In addition, new Ethernet firewall cards that fit in the slot of an individual computer provide additional layers of defense (like encryption and permit/deny) for individual computer transmissions to the network interface function. The cost of these new cards is only slightly higher than that of traditional network interface cards.

Network Intrusion Hardware and Software

Network intrusion detection and prevention systems are software- and hardware-based programs designed to detect unauthorized attacks on a computer network system. Whereas other applications, such as firewalls and anti-virus software, share similar objectives with network intrusion systems, network intrusion systems provide a deeper layer of protection beyond the capabilities of these other systems because they evaluate patterns of computer activity rather than specific files. It is worth noting that attacks may come from either outside or within the system (i.e., from an insider), and that network intrusion detection systems may be more applicable for detecting patterns of suspicious activity from inside a facility (i.e., accessing sensitive data) than other information technology solutions. Network intrusion detection systems employ a variety of mechanisms to evaluate potential threats. The types of search and detection mechanisms depend on the level of sophistication of the system. Some of the available detection methods include:

- **Protocol Analysis:** Protocol analysis is the process of capturing, decoding, and interpreting electronic traffic. The protocol analysis method of network intrusion detection involves the analysis of data captured during transactions between two or more systems or devices, and the evaluation of this data to identify unusual activity and potential problems. Once a problem is isolated and recorded, problems or potential threats can be linked to pieces of hardware or software. Sophisticated protocol analysis will also provide statistics and trend information on the captured traffic.
- **Traffic Anomaly Detection:** Traffic anomaly detection identifies potentially threatening activity by comparing incoming traffic to "normal" traffic patterns and identifying deviations. It does this by comparing user characteristics against thresholds and triggers defined by the network administrator. This method is designed to detect attacks that span multiple connections rather than a single session.
- **Network Honeypot:** This method establishes non-existent services in order to identify potential hackers. A network honeypot impersonates

services that don't exist by sending fake information to people scanning the network. It identifies the attacker when they attempt to connect to the service. There is no reason for legitimate traffic to access these resources because they don't exist; therefore, any attempt to access them constitutes an attack.
- **Anti-intrusion Detection System Evasion Techniques:** These methods are designed to detect attackers who may be trying to evade intrusion detection system scanning. They include methods called IP defragmentation, TCP stream reassembly, and deobfuscation.

These detection systems are automated, but they can only indicate patterns of activity, and a computer administrator, or other experienced individual must interpret these activities to determine whether or not they are potentially harmful. Monitoring the logs generated by these systems can be time-consuming, and there may be a learning curve to determine a baseline of "normal" traffic patterns from which to distinguish potentially suspicious activity.

SCADA

In Queensland, Australia, on April 23, 2000, police stopped a car on the road and found a stolen computer and radio inside. Using commercially available technology, a disgruntled former employee had turned his vehicle into a pirate command center of sewage treatment along Australia's Sunshine Coast.

The former employee's arrest solved a mystery that had troubled the Maroochy Shire wastewater system for two months. Somehow the system was leaking hundreds of thousands of gallons of putrid sewage into parks, rivers and the manicured grounds of a Hyatt Regency hotel—marine life died, the creek water turned black and the stench was unbearable for residents. Until the former employee's capture—during his 46th successful intrusion—the utility's managers did not know why.

Specialists study this case of cyber-terrorism because it is the only one known in which someone used a digital control system deliberately to cause harm. The former employee's intrusion shows how easy it is to break in—and how restrained he was with his power.

To sabotage the system, the former employee set the software on his laptop to identify itself as a pumping station, and then suppressed all alarms. The former employee was the "central control station" during his intrusions, with unlimited command of 300 SCADA nodes governing sewage and drinking water alike.

–Gellman (2002)

The Bottom Line: As serious as the former employee's intrusions were, they pale in comparison to what he could have done to the freshwater system—he could have done anything he liked.

In 2000, the Federal Bureau of Investigation (FBI) identified and listed threats to critical infrastructure. These threats

TABLE 5.1
Threats to Critical Infrastructure Observed by the FBI

Threat	Description
Criminal groups	There is an increased use of cyber intrusions by criminal groups who attack systems for monetary gain.
Foreign intelligence services	Foreign intelligence services use cyber tools as part of their information-gathering and espionage activities.
Hackers	Hackers sometimes crack into networks for the thrill of the challenge or bragging rights in the hacker community. While remote cracking once required a fair amount of skill or computer knowledge, hackers can now download attack scripts and protocols from the Internet and launch them against victim sites. Thus, while attack tools have become more sophisticated, they have also become easier to use.
Hacktivists	Hacktivism refers to politically motivated attacks on publicly accessible web pages or e-mail servers. These groups and individuals overload e-mail servers and hack into websites to send a political message.
Information warfare	Several nations are aggressively working to develop information warfare doctrine, programs, and capabilities. Such capabilities enable a single entity to have a significant and serious impact by disrupting the supply, communications, and economic infrastructures that support military power—impacts that, according to the Director of Central Intelligence, can affect the daily lives of Americans across the country.
Inside threat	The disgruntled organizational insider is a principal source of computer crimes. Insiders may not need a great deal of knowledge about computer intrusions because their knowledge of a victim system often allows them to gain unrestricted access to cause damage to the system or steal system data. The insider threat also includes outsourcing vendors.
Virus writers	Virus writers are posing an increasingly serious threat. Several destructive computer viruses and "worms" have harmed files and hard drives, including the Melissa Macro Virus, the Explore.Zip worm, the CIH (Chernobyl) Virus, Nimda, and Code Red.

are listed and described in Table 5.1. In the past few years, especially since 9/11, it has been somewhat routine for us to pick up a newspaper or magazine or view a television news program where a major topic of discussion is cybersecurity or the lack thereof. Many of the cyber intrusion incidents we read or hear about have added new terms or new uses for old terms to our vocabulary. For example, old terms such as Trojan Horse, worms, and viruses have taken on new connotations regarding cybersecurity issues. Relatively new terms such as scanners, Windows NT hacking tools, ICQ hacking tools, mail bombs, sniffers, logic bombs, nukers, dots, backdoor Trojans, key loggers, hackers' Swiss knives, password crackers, and BIOS crackers are now commonly encountered.

Not all relatively new and universally recognizable cyber terms have sinister connotations or meanings, of course. Consider, for example, the following digital terms: backup, binary, bit, byte, CD-ROM, CPU, database, e-mail, HTML, icon, memory, cyberspace, modem, monitor, network, RAM, Wi-Fi (wireless fidelity), record, software, and World Wide Web—none of these terms normally generate thoughts of terrorism in most of us. There is, however, one digital term, SCADA, that most people have not heard of. This is not the case, however, for those who work with the nation's critical infrastructure, including water and wastewater. SCADA, or Supervisory Control And Data Acquisition System (also sometimes referred to as Digital Control Systems or Process Control Systems), plays an important role in computer-based control systems. Many water and wastewater systems use computer-based systems to remotely control sensitive processes and system equipment that were previously controlled manually. These systems (commonly

known as SCADA) allow a water and wastewater utility to collect data from sensors and control equipment located at remote sites. Common water and wastewater system sensors measure elements such as fluid level, temperature, pressure, water purity, water clarity, and pipeline flow rates. Common water/wastewater system equipment includes valves, pumps, and mixers for mixing chemicals in the water supply.

WHAT IS SCADA?

Simply put, SCADA is a computer-based system that remotely controls processes previously controlled manually. SCADA allows an operator using a central computer to supervise (control and monitor) multiple networked computers at remote locations. Each remote computer can control mechanical processes (pumps, valves, etc.) and collect data from sensors at its remote location. Thus the phrase: Supervisory Control and Data Acquisition, refers to **SCADA**. The central computer is called the Master Terminal Unit, or MTU. The operator interfaces with the MTU using software called Human Machine Interface, or HMI. The remote computer is called a Programmable Logic Controller (PLC) or Remote Terminal Unit (RTU). The RTU activates a relay (or switch) that turns mechanical equipment "on" and "off." It also collects data from sensors.

Initially, utilities ran wires, also known as hardwires or landlines, from the central computer (MTU) to the remote computers (RTUs). Because remote locations can be hundreds of miles from the central location, utilities began to use public phone lines and modems, leased telephone company lines, and radio and microwave communication.

More recently, they have also begun to use satellite links, the Internet, and newly developed wireless technologies.

Because the SCADA systems' sensors provide valuable information, many utilities established "connections" between their SCADA systems and their business systems. This allowed utility management and other staff access to valuable statistics, such as water usage. When utilities later connected their systems to the Internet, they were able to provide stakeholders with water/wastewater statistics on the utility web pages.

SCADA Applications in Water and Wastewater Systems

As stated above, SCADA systems can be designed to measure a variety of equipment operating conditions and parameters, volumes and flow rates, or water quality parameters, and to respond to changes in those parameters either by alerting operators or by modifying system operation through a feedback loop system without having personnel physically visit each process or piece of equipment daily to check it and/or ensure that it is functioning properly. SCADA systems can also be used to automate certain functions so that they can be performed without the need to be initiated by an operator (e.g., injecting chlorine in response to periodic low chlorine levels in a distribution system or turning on a pump in response to low water levels in a storage tank). As described above, in addition to process equipment, SCADA systems can also integrate specific security alarms and equipment, such as cameras, motion sensors, lights, and data from card reading systems, thereby providing a clear picture of what is happening in areas throughout a facility. Finally, SCADA systems also provide constant, real-time data on processes, equipment, location access, etc., allowing the necessary response to be made quickly. This can be extremely useful during emergency conditions, such as when distribution mains break or when potentially disruptive BOD spikes appear in wastewater influent.

Because these systems can monitor multiple processes, equipment, and infrastructure, they provide quick notification of or response to problems or upsets. SCADA systems typically provide the first line of detection for atypical or abnormal conditions. For example, a SCADA system connected to sensors that measure specific water quality parameters is measured outside of a specific range. A real-time customized operator interface screen could display and control critical systems monitoring parameters.

The system could transmit warning signals back to the operators, such as by initiating a call to a personal pager. This might allow the operators to initiate actions to prevent contamination and disruption of the water supply. Further automation of the system could ensure that it initiates measures to rectify the problem. Preprogrammed control functions (e.g., shutting a valve, controlling flow, increasing chlorination, or adding other chemicals) can be triggered and operated based on SCADA utility.

SCADA Vulnerabilities

According to the USEPA (2005), SCADA networks were developed with little attention paid to security, making the security of these systems often weak. Studies have found that, while technological advancements introduced vulnerabilities, many water and wastewater utilities have spent little time securing their SCADA networks. As a result, many SCADA networks may be susceptible to attacks and misuse. Remote monitoring and supervisory control of processes began to develop in the early 1960s and adopted many technological advancements. The advent of minicomputers made it possible to automate a vast number of once manually operated switches. Advancements in radio technology reduced the communication costs associated with installing and maintaining buried cables in remote areas. SCADA systems continued to adopt new communication methods including satellite and cellular. As the price of computers and communications dropped, it became economically feasible to distribute operations and expand SCADA networks to include even smaller facilities.

Advances in information technology and the necessity for improved efficiency have resulted in increasingly automated and interlinked infrastructures creating new vulnerabilities due to equipment failure, human error, weather and other natural causes, and physical and cyber-attacks. Some areas and examples of possible SCADA vulnerabilities include:

- **Human:** People can be tricked or corrupted and commit errors.
- **Communications:** Messages can be fabricated, intercepted, changed, deleted, or blocked.
- **Hardware:** Security features are not easily adapted to small self-contained units with limited power supplies.
- **Physical:** Intruders can break into a facility to steal or damage SCADA equipment.
- **Natural:** Tornadoes, floods, earthquakes, and other natural disasters can damage equipment and connections.
- **Software:** Programs can be poorly written.

A survey found that many water utilities were doing little to secure their SCADA network vulnerabilities (Ezell 1998); for example, many respondents reported that they had remote access, which can allow an unauthorized person to access the system without being physically present. More than 60% of the respondents believed that their systems were not safe from unauthorized access and use. Twenty percent of the respondents even reported known attempts at, or successful unauthorized access to, their system. Yet 22 of 43 respondents reported that they do not spend any time ensuring their network is safe, and 18 of 43 respondents reported that they spend less than 10% of their time ensuring network safety.

SCADA system computers and their connections are susceptible to different types of information system attacks and misuse, such as system penetration and unauthorized access to information. The Computer Security Institute and Federal Bureau of Investigation conduct an annual Computer Crime and Security Survey (FBI 2004). The survey reported on ten types of attacks or misuse and reported that viruses and denial of service had the greatest negative economic impact. The same study also found that 15% of the respondents reported abuse of wireless networks, which can be a SCADA component. On average, respondents from all sectors did not believe that their organization invested enough in security awareness. Utilities, as a group, reported a lower average computer security expenditure/investment per employee than many other sectors, such as transportation, telecommunications, and finance.

Sandia National Laboratories' *Common Vulnerabilities in Critical Infrastructure Control Systems* described some of the common problems it has identified in the following five categories (Stamp et al. 2003):

1. **System Data:** Important data attributes for security include availability, authenticity, integrity, and confidentiality. Data should be categorized according to its sensitivity, and ownership and responsibility must be assigned. However, SCADA data is often not classified at all, making it difficult to identify where security precautions are appropriate.
2. **Security Administration:** Vulnerabilities emerge because many systems lack a properly structured security policy, equipment and system implementation guides, configuration management, training, and enforcement and compliance auditing.
3. **Architecture:** Many common practices negatively affect SCADA security. For example, while it is convenient to use SCADA capabilities for other purposes such as fire and security systems, these practices create single points of failure. Additionally, the connection of SCADA networks to other automation systems and business networks introduces multiple entry points for potential adversaries.
4. **Network (Including Communication Links):** Legacy systems' hardware and software have very limited security capabilities, and the vulnerabilities of contemporary systems (based on modern information technology) are publicized. Wireless and shared links are susceptible to eavesdropping and data manipulation.
5. **Platforms:** Many platform vulnerabilities exist, including default configurations retained, poor password practices, shared accounts, inadequate protection for hardware, and nonexistent security monitoring controls. In most cases, important security patches are not installed, often due to concerns about negatively impacting system operation; in some cases, technicians are contractually forbidden from updating systems by their vendor agreements.

The following incident helps illustrate some of the risks associated with SCADA vulnerabilities.

- During the course of conducting a vulnerability assessment, a contractor stated that personnel from his company penetrated the information system of a utility within minutes. Contractor personnel drove to a remote substation and noticed a wireless network antenna. Without leaving their vehicle, they plugged in their wireless radios and connected to the network within 5 minutes. Within 20 minutes, they had mapped the network, including SCADA equipment, and accessed the business network and data.

This illustrates what a cybersecurity advisor from Sandia National Laboratories, specialized in SCADA, stated—that utilities are moving to wireless communication without understanding the added risks.

THE INCREASING RISK

According to GAO (2003), historically, security concerns about control systems (including SCADA) were related primarily to protecting against physical attacks and misuse of refining and processing sites or distribution and holding facilities. However, there has been a growing recognition that control systems are now vulnerable to cyber-attacks from numerous sources, including hostile governments, terrorist groups, disgruntled employees, and other malicious intruders. In addition to the control system vulnerabilities mentioned earlier, several factors have contributed to the escalation of risk to control systems, including (1) the adoption of standardized technologies with known vulnerabilities, (2) the connectivity of control systems to other networks, (3) constraints on the implementation of existing security technologies and practices, (4) insecure remote connections, and (5) the widespread availability of technical information about control systems.

ADOPTION OF TECHNOLOGIES WITH KNOWN VULNERABILITIES

When a technology is not well known, not widely used, not understood, or not publicized, it is difficult to penetrate and thus disable it. Historically, proprietary hardware, software, and network protocols made it difficult to understand how control systems operated—and therefore how to hack into them. Today, however, to reduce costs and improve performance, organizations have been transitioning from proprietary systems to less expensive, standardized technologies such as Microsoft's Windows and Unix-like operating systems, along with the common networking protocols used by the Internet. These widely used standardized technologies have commonly known vulnerabilities, and sophisticated and effective exploitation tools are widely available and relatively easy to use. As a consequence, both the number of people with the knowledge to wage attacks and the number of systems subject to attack have increased. Additionally, common communication protocols and the emerging use of Extensible Markup Language (commonly referred to as XML) can make it easier for a hacker to interpret the content of communications among the components of a control system.

Control systems are often connected to other networks—enterprises often integrate their control systems with their enterprise networks. This increased connectivity has significant advantages, including providing decision-makers with access to real-time information and allowing engineers to monitor and control the process control system from different points on the enterprise network. In addition, enterprise networks are often connected to strategic partners' networks and the Internet. Further, control systems are increasingly using wide area networks and the Internet to transmit data to remote or local stations and individual devices. This convergence of control networks with public and enterprise networks potentially exposes the control systems to additional security vulnerabilities. Unless appropriate security controls are deployed in both the enterprise network and the control system network, enterprise security breaches can impact the operation of control systems. According to industry experts, the use of existing security technologies, strong user authentication, and patch management practices, are generally not implemented in control systems because these systems operate in real time, are typically not designed with cybersecurity in mind, and usually have limited processing capabilities.

Existing security technologies such as authorization, authentication, encryption, intrusion detection, and filtering of network traffic and communications require more bandwidth, processing power, and memory than control system components typically have. Because controller stations are generally designed to perform specific tasks, they use low-cost, resource-constrained microprocessors. In fact, some devices in the electrical industry still use the Intel 8088 processor, introduced in 1978. Consequently, it is difficult to install existing security technologies without seriously degrading the performance of the control system.

Further, complex passwords and other strong password practices are not always used to prevent unauthorized access to control systems, partly because they could hinder rapid response to safety procedures during an emergency. As a result, experts report that weak passwords—those that are easy to guess, shared among users, and infrequently changed—are common in control systems. This includes the use of default passwords or, in some cases, no password at all.

In addition, although modern control systems are based on standard operating systems, they are typically customized to support control system applications. Consequently, vendor-provided software patches are generally either incompatible or cannot be implemented without compromising service shutting down "always-on" systems, or affecting interdependent operations.

Potential vulnerabilities in control systems are exacerbated by insecure connections. Organizations often leave access links—such as dial-up modems to equipment and control information—open for remote diagnostics, maintenance, and examination of system status. Such links may not be protected with authentication or encryption, which increases the risk that hackers could use these insecure connections to break into remotely controlled systems. Also, control systems often use wireless communication systems, which are especially vulnerable to attack, or leased lines that pass through commercial telecommunications facilities. Without encryption to protect data as it flows through these insecure connections or authentication mechanisms to limit access, there is limited protection for the integrity of the information being transmitted.

Public information about infrastructures and control systems is available to potential hackers and intruders. The availability of this infrastructure and vulnerability data was demonstrated by a university graduate student, whose dissertation reportedly mapped every business and industrial sector in the American economy to the fiber-optic network that connects them—using publicly available material from the Internet, none of which was classified. Many of the electric utility officials who were interviewed for the National Security Telecommunications Advisory Committee's Information Assurance Task Force's Electric Power Risk Assessment expressed concern over the amount of information about their infrastructure that is readily available to the public.

In the electric power industry, open sources of information—such as product data and educational videotapes from engineering associations—can be used to understand the basics of the electrical grid. Other publicly available information—including filings from the Federal Energy Regulatory Commission (FERC), industry publications, maps, and material available on the Internet—is sufficient to allow someone to identify the most heavily loaded transmission lines and the most critical substations in the power grid.

In addition, significant information on control systems is publicly available—including design and maintenance documents, technical standards for the interconnection of control systems and RTUs, and standards for communication among control devices—all of which could assist hackers in understanding the systems and how to attack them. Moreover, there are numerous former employees, vendor, support contractors, and other end users of the same equipment worldwide with inside knowledge of the operation of control systems.

CYBER THREATS TO CONTROL SYSTEMS

There is a consensus—and increasing concern—among government officials and control systems experts about potential cyber threats to the systems that govern our critical infrastructures. As components of control systems increasingly make critical decisions once made by humans, the potential impact of a cyber-threat becomes more devastating. Such threats could come from numerous sources, ranging from hostile governments and terrorist groups to disgruntled employees and other malicious intruders. Based on interviews and discussions with representatives throughout the electric power industry, the Information Assurance Task Force of the National Security Telecommunications Advisory Committee concluded that an organization with sufficient resources, such as a foreign intelligence service or a well-supported terrorist group, could conduct a structured attack on the electric power grid electronically, with a high degree of anonymity and without having to set foot in the target nation.

In July 2002, the National Infrastructure Protection Center (NIPC) reported that the potential for compound cyber and physical attacks, referred to as "swarming attacks," is an emerging threat to U.S. critical infrastructure. As NIPC reports, the effects of a swarming attack include slowing or complicating the response to a physical attack. For instance, a cyber-attack that disabled the water supply or the electrical system in conjunction with a physical attack could deny emergency services the necessary resources to manage the consequences—such as controlling fires, coordinating actions, and generating light.

Control systems, such as SCADA, can be vulnerable to cyber-attacks. Entities or individuals with malicious intent might take one or more of the following actions to successfully attack control systems:

- Disrupt the operation of control systems by delaying or blocking the flow of information through control networks, thereby denying the availability of the networks to control system operations.
- Make unauthorized changes to programmed instructions in PLCs, RTUs, or DCS controllers, change alarm thresholds, or issue unauthorized commands to control equipment, which could potentially result in damage to equipment (if tolerances are exceeded), a premature shutdown of processes (such as prematurely shutting down transmission lines), or even the disabling of control equipment.
- Send false information to control system operators either to disguise unauthorized changes or to initiate inappropriate actions by system operators.
- Modify the control system software, producing unpredictable results.
- Interfere with the operation of safety systems.

In addition, in control systems covering a wide geographic area, the remote sites are often unstaffed and may not be physically monitored. If such remote systems are physically breached, the attackers could establish a cyber connection to the control network.

SECURING CONTROL SYSTEMS

Several challenges must be addressed to effectively secure control systems against cyberthreats. These challenges include: (1) the limitations of current security technologies in securing control systems; (2) the perception that securing control systems may not be economically justifiable; and

(3) the conflicting priorities within organizations regarding the security of control systems. A significant challenge in effectively securing control systems is the lack of specialized security technologies for these systems. The computing resources in control systems that are needed to perform security functions tend to be quite limited, making it very challenging to use security technologies within control system networks without severely hindering performance. Securing control systems may not be perceived as economically justifiable. Experts and industry representatives have indicated that organizations may be reluctant to spend more money to secure control systems. Hardening the security of control systems would require industries to expend more resources, including acquiring more personnel, providing training for personnel, and potentially prematurely replacing current systems that typically have a lifespan of about 20 years. Finally, several experts and industry representatives indicated that the responsibility for securing control systems typically includes two separate groups: IT security personnel and control system engineers and operators. IT security personnel tend to focus on securing enterprise systems, while control system engineers and operators are more concerned with the reliable performance of their control systems. As a result, these two groups do not always fully understand each other's requirements and collaborate to implement secure control systems.

STEPS TO IMPROVE SCADA SECURITY

The President's Critical Infrastructure Protection Board and the Department of Energy (DOE) have developed the steps outlined below to help organizations improve the security of their SCADA networks. DOE (2001) points out that these steps are not meant to be prescriptive or all-inclusive. However, they address essential actions to be considered to improve the protection of SCADA networks. The steps are divided into two categories: specific actions to improve implementation and actions to establish essential underlying management processes and policies (DOE 2001).

21 Steps to Increase SCADA Security

The following steps focus on specific actions to be taken to increase the security of SCADA networks:

1. **Identify All Connections to SCADA Networks.**
   Conduct a thorough risk analysis to assess the risk and necessity of each connection to the SCADA network. Develop a comprehensive understanding of all connections to the SCADA network and how well those connections are protected. Identify and evaluate the following types of connections:
   - Internal local area and wide area networks, including business networks
   - The Internet
   - Wireless network devices, including satellite uplinks

- Modem or dial-up connections
- Connections to business partners, vendors, or regulatory agencies

2. **Disconnect Unnecessary Connections to the SCADA Network.**

   To ensure the highest degree of security for SCADA systems, isolate the SCADA network from other network connections to as great a degree as possible. Any connection to another network introduces security risks, particularly if the connection creates a pathway from or to the Internet. Although direct connections with other networks may allow important information to be passed efficiently and conveniently, insecure connections are simply not worth the risk; isolation of the SCADA network must be a primary goal to provide needed protection. Strategies such as the utilization of "demilitarized zones" (DMZs) and data warehousing can facilitate the secure transfer of data from the SCADA network to business networks. However, they must be designed and implemented properly to avoid the introduction of additional risk through improper configuration.

3. **Evaluate and Strengthen the Security of Any Remaining Connections to the SCADA Networks.**

   Conduct penetration testing or vulnerability analysis of any remaining connections to the SCADA network to evaluate the protection posture associated with these pathways. Use this information in conjunction with risk management processes to develop a robust protection strategy for any pathways to the SCADA network. Since the SCADA network is only as secure as its weakest connecting point, it is essential to implement firewalls, intrusion detection systems (IDSs), and other appropriate security measures at each point of entry. Configure firewall rules to prohibit access from and to the SCADA network and be as specific as possible when permitting approved connections. For example, an Independent System Operator (ISO) should not be granted "blanket" network access simply because there is a need for a connection to certain components of the SCADA system. Strategically place IDSs at each entry point to alert security personnel of potential breaches of network security. Organization management must understand and accept responsibility for risks associated with any connection to the SCADA network.

4. **Harden SCADA Networks by Removing or Disabling Unnecessary Services.**

   SCADA control servers built on commercial or open-source operating systems can be exposed to attacks through default network services. To the greatest degree possible, remove or disable unused services and network daemons to reduce the risk of direct attack. This is particularly important when SCADA networks are interconnected with other networks. Do not permit a service or feature on a SCADA network unless a thorough risk assessment of the consequences of allowing the service/feature shows that the benefits far outweigh the potential for vulnerability exploitation. Examples of services to remove from SCADA networks include automated meter reading/remote billing systems, email services, and Internet access. An example of a feature to disable is remote maintenance to numerous secure configurations, such as the National Security Agency's series of security guides. Additionally, work closely with SCADA vendors to identify secure configurations and coordinate any and all changes to operational systems to ensure that removing or disabling services does not cause downtime, interruption of service, or loss of support.

5. **Do Not Rely on Proprietary Protocols to Protect Your System.**

   Some SCADA systems use unique, proprietary protocols for communications between field devices and servers. Often, the security of SCADA systems is based solely on the secrecy of these protocols. Unfortunately, obscure protocols provide very little "real" security. Do not rely on proprietary protocols or factor default configuration settings to protect your system. Additionally, demand that vendors disclose any backdoors or vendor interfaces to your SCADA systems and expect them to provide systems that are capable of being secured.

6. **Implement the Security Features Provided by Device and System Vendors.**

   Older SCADA systems (most systems in use) have no security features whatsoever, so SCADA system owners must insist that their system vendor implement security features in the form of product patches or upgrades. Some newer SCADA devices are shipped with basic security features, but these are usually disabled to ensure ease of installation.

   Analyze each SCADA device to determine whether security features are present. Additionally, factory default security settings (such as in computer network firewalls) are often set to provide maximum usability but minimal security. Set all security features to provide maximum security only after a thorough risk assessment of the consequences of reducing the security level.

7. **Establish Strong Controls over Any Medium That is Used as a Backdoor into the SCADA Network.**

   Where backdoors or vendor connections exist in SCADA systems, strong authentication must be implemented to ensure secure communications. Modems, wireless, and wired networks used for communications and maintenance represent a significant vulnerability to the SCADA network and remote sites. Successful "war dialing" or "war driving" attacks could allow an attacker to bypass all other controls and gain direct access to the SCADA network or resources. To minimize the risk of such attacks, disable inbound access and replace it with some type of callback system.

8. **Implement Internal and External Intrusion Detection Systems and Establish 24-Hour-a-Day Incident Monitoring.**

To effectively respond to cyber-attacks, establish an intrusion detection strategy that includes alerting network administrators of malicious network activity originating from internal or external sources. Intrusion detection system monitoring is essential 24 hours a day; this capability can be easily set up through a pager. Additionally, incident response procedures must be in place to allow an effective response to any attack. To complement network monitoring, enable logging on all systems and audit system logs daily for detecting suspicious activity as soon as possible.

9. **Perform Technical Audits of SCADA Devices and Networks, and Any Other Connected Networks, to Identify Security Concerns.**

Technical audits of SCADA devices and networks are critical to ongoing security effectiveness. Many commercial and open-source security tools are available that allow system administrators to conduct audits of their systems/networks to identify active services, patch levels, and common vulnerabilities. The use of these tools will not solve systemic problems but will eliminate the "paths of least resistance" that an attacker could exploit. Analyze identified vulnerabilities to determine their significance and take corrective actions as appropriate. Track corrective actions and analyze this information to identify trends. Additionally, retest systems after corrective actions are taken to ensure that vulnerabilities have been eliminated. Actively scan non-production environments to identify and address potential problems.

10. **Conduct Physical Security Surveys and Assess All Remote Sites Connected to the SCADA Network to Evaluate Their Security.**

Any location that has a connection to the SCADA network is a target, especially unmanned or unguarded remote sites. Conduct a physical security survey and inventory access points at each facility that has a connection to the SCADA system. Identify and assess any sources of information including remote telephone/computer network/fiber optic cables that could be tapped; radio and microwave links that are exploitable computer terminals

that could be accessed; and wireless local area network access points. Identify and eliminate single points of failure. The security of the site must be adequate to detect or prevent unauthorized access. Do not allow "live" network access points at remote, unguarded sites simply for convenience.

11. **Establish SCADA "Red Teams" to Identify and Evaluate Possible Attack Scenarios.**

Establish a "Red Team" to identify potential attack scenarios and evaluate system vulnerabilities. Use a variety of people who can provide insight into the weaknesses of the overall network, SCADA system, physical systems, and security controls. Those who work on the system every day have valuable insight into the vulnerabilities of your SCADA network and should be consulted when identifying potential attack scenarios and possible consequences. Additionally, ensure that the risk from a malicious insider is fully evaluated, as this represents one of the greatest threats to an organization. Feed information resulting from the "Red Team" evaluation into risk management processes to assess the information and establish appropriate protection strategies.

The following steps focus on management actions to establish an effective cybersecurity program:

12. **Clearly Define Cyber Security Roles, Responsibilities, and Authorities for Managers, System Administrators, and Users.**

Organization personnel need to understand the specific expectations associated with protecting information technology resources through the definition of clear and logical roles and responsibilities. In addition, key personnel need to be given sufficient authority to carry out their assigned responsibilities. Too often, good cyber security is left up to the initiative of the individual, which usually leads to inconsistent implementations and ineffective security. Establish a cyber security organizational structure that defines roles and responsibilities and clearly identifies how cybersecurity issues are escalated and who is notified in an emergency.

13. **Document Network Architecture and Identify Systems That Serve Critical Functions or Contain Sensitive Information That Require Additional Levels of Protection.**

Develop and document a robust information security architecture as part of a process to establish an effective protection strategy. It is essential that organizations design their network with security in mind and continue to have a strong understanding of their network architecture throughout its lifecycle. Of particular importance is an in-depth understanding of the functions that the systems perform and the sensitivity of the stored information. Without this understanding, risk cannot be properly assessed and protection strategies may not be sufficient. Documenting the information security architecture and its components is critical to understanding the overall protection strategy and identifying single points of failure.

14. **Establish a Rigorous, Ongoing Risk Management Process.**

A thorough understanding of the risks to network computing resources from denial-of-service attacks and the vulnerability of sensitive information to compromise is essential for an effective cybersecurity program. Risk assessments, based on this technical understanding, are critical for formulating effective strategies to mitigate vulnerabilities and preserve the integrity of computing resources. Initially, perform a baseline risk analysis based on the current threat assessment to develop a network protection strategy. Due to rapidly changing technology and the emergence of new threats daily, an ongoing risk assessment process is also needed so that routine changes can be made to the protection strategy to ensure it remains effective. Fundamental to risk management is the identification of residual risk with a network protection strategy in place and the acceptance of that risk by management.

15. **Establish a Network Protection Strategy Based on the Principle of Defense-in-Depth.**

A fundamental principle that must be part of any network protection strategy is defense-in-depth. Defense-in-depth must be considered early in the design phase of the development process and must be an integral consideration in all technical decision-making associated with the network. Utilize technical and administrative controls to mitigate threats from identified risks to as great a degree as possible at all levels of the network. Single points of failure must be avoided, and cybersecurity defenses must be layered to limit and contain the impact of any security incidents. Additionally, each layer must be protected against other systems at the same layer. For example, to protect against the internal threats, restrict users to access only those resources necessary to perform their job functions.

16. **Clearly Identity Cyber Security Requirements.**

Organizations and companies need structured security programs with mandated requirements to establish expectations and allow personnel to be held accountable. Formalized policies and procedures are typically used to establish and institutionalize a cybersecurity program. A formal program is essential for establishing a consistent, standards-based approach to cybersecurity throughout an organization and eliminates sole dependence on individual initiative. Policies and procedures also inform employees of their specific cybersecurity

responsibilities and the consequences of failing to meet those responsibilities. They provide guidance regarding actions to be taken during a cyber security incident and promote efficient and effective actions during a time of crisis. As part of identifying cyber security requirements, include user agreements and notification and warning banners. Establish requirements to minimize the threat from malicious insiders, including the need to conduct background checks and limit network privileges to those absolutely necessary.

17. **Establish Effective Configuration Management Processes.**

A fundamental management process needed to maintain a secure network is configuration management. Configuration management needs to cover both hardware configurations and software configurations. Changes to hardware or software can easily introduce vulnerabilities that undermine network security. Processes are required to evaluate and control any change to ensure that the network remains secure. Configuration management begins with well-tested and documented security baselines for your various systems.

18. **Conduct Routine Self-Assessments.**

Robust performance evaluation processes are needed to provide organizations with feedback on the effectiveness of cyber security policy and technical implementation. A sign of a mature organization is one that is able to self-identify issues, conduct root-cause analyses, and implement effective corrective actions that address individual and systemic problems. Self-assessment processes that are normally part of an effective cyber security program include routine scanning for vulnerabilities, automated auditing of the network, and self-assessments of organizational and individual performance.

19. **Establish System Backups and Disaster Recovery Plans.**

Establish a disaster recovery plan that allows for rapid recovery from any emergency (including a cyber-attack). System backups are an essential part of any plan and allow for the rapid reconstruction of the network. Routinely exercise disaster recovery plans to ensure that they work and that personnel are familiar with them. Make appropriate changes to disaster recovery plans based on lessons learned from exercises.

20. **Senior Organizational Leadership Should Establish Expectations for Cyber Security Performance and Hold Individuals Accountable for Their Performance.**

Effective cybersecurity performance requires commitment and leadership from senior managers in the organization. Senior management must establish an expectation for strong cybersecurity and communicate this to their subordinate managers throughout the organization. It is also essential that senior organizational leadership establish a structure for the implementation of a cybersecurity program. This structure will promote consistent implementation and the ability to sustain a strong cybersecurity program. It is then important for individuals to be held accountable for their performance as it relates to cybersecurity. This includes managers, system administrators, technicians, and users/operators.

21. **Establish Policies and Conduct Training to Minimize the Likelihood That Organizational Personnel will Inadvertently Disclose Sensitive Information Regarding SCADA System Design, Operations, or Security Controls.**

Release data related to the SCADA network only on a strict, need-to-know basis, and only to persons explicitly authorized to receive such information. "Social engineering," the gathering of information about a computer or computer network via questions to naïve users, is often the first step in a malicious attack on computer networks. The more information revealed about a computer or computer network, the more vulnerable the computer/network becomes. Never divulge data related to a SCADA network, including the names and contact information of the system operators/administrators, computer operating systems, and/or physical and logical locations of computers and network systems over the telephone or to personnel unless they are explicitly authorized to receive such information. Any requests for information from unknown persons should be sent to a central network security location for verification and fulfillment. People can be a weak link in an otherwise secure network. Conduct training and information awareness campaigns to ensure that personnel remain diligent in guarding sensitive network information, particularly their passwords.

THE BOTTOM LINE

Again, when it comes to the security of our nation and even of water and wastewater treatment facilities, few have summed it up better than Governor Ridge (Henry 2002).

Now, obviously, the further removed we get from September 11, I think the natural tendency is to let down our guard. Unfortunately, we cannot do that. The government will continue to do everything we can to find and stop those who seek to harm us. And I believe we owe it to the American people to remind them that they must be vigilant, as well.

CHAPTER REVIEW QUESTIONS

THOUGHT-PROVOKING QUESTION

5.1 Do you feel that water and/or wastewater facilities are realistic targets for terrorism? Why? (Answers will vary.)

5.2 Are we more vulnerable to homegrown terrorists than foreign terrorists? Explain. (Answers will vary.)

REFERENCES

DOE. 2001. *21 Steps to Improve Cyber Security of SCADA Networks.* Washington, DC: Department of Energy.

Ezell, B.C., 1998. *Risks of Cyber Attack to Supervisory Control and Data Acquisition.* Charlottesville, VA: University of Virginia.

FBI. 2004. *Ninth Annual Computer Crime and Security Survey.* Washington, DC:: Computer Crime Institute and Federal Bureau of Investigations.

GAO. 2003. *Critical Infrastructure Protection: Challenges in Securing Control System.* Washington, DC: United States General Accounting Office.

Gellman, B., 2002. Cyber-Attacks by Al Qaeda Feared: Terrorists at Threshold of Using Internet as Tool of Bloodshed, Experts Say. *Washington Post,* June 27; p. A01.

Henry, K., 2002. New Face of Security. *Government Security.* April; pp. 30–31.

Stamp, J. et al., 2003. *Common Vulnerabilities in Critical Infrastructure Control Systems,* 2nd ed. White Sands: Sandia National Laboratories.

USEPA. 2004. *Water Security: Basic Information.* Accessed 09/30/23 @ https://cfpub.epa.gov/safewater/watersecurity/basicinformation.cfm.

USEPA. 2005. EPA Needs to Determine What Barriers Prevent Water Systems from Securing Known SCADA Vulnerabilities, in *Final Briefing Report,* Harris, J., (ed.). Washington, DC: USEPA.

6 Water/Wastewater References, Models & Terminology

SETTING THE STAGE

This handbook is a compilation and summary of information available from many expert sources. While I have attempted to cover all aspects of water and wastewater treatment system operation, let me point out that no single handbook contains *all* the information or all the answers. Moreover, because of the physical limits of any written text, some topics are only given cursory exposure and limited coverage. For those individuals seeking a more in-depth treatment of specific topics relevant to water and wastewater treatment system operations, we recommend consulting one or more of the references listed in Table 6.1, and any of the many other outstanding references referred to throughout this text.

- **Note:** Technomic Publishing Company originally published many of the texts in Table 6.1. Technomic is now part of CRC/Lewis Publishers; the listed Technomic texts are available from CRC Press.

TABLE 6.1
Recommended Reference Material

1   *Small Water System O and M*, Kerri, K. et al. California State University, Sacramento, CA.

2   *Water Distribution System O and M*, Kerri, K. et al. California State University, Sacramento, CA.

3   *Water Treatment Plant Operation*, Vol. 1 and 2. Kerri, K. et al. California State University, Sacramento, CA.

4   Basic Mathematics, #3014-g. Atlanta: Centers for Disease Control.

5   *Waterborne Disease Control.* Atlanta: Centers for Disease Control.

6   Water Fluoridation, #3017-G. Atlanta: Centers for Disease Control.

7   *Introduction to Water Sources and Transmission*—Volume 1. Denver: American Water Works Association.

8   *Introduction to Water Treatment*—Volume 2. Denver: American Water Works Association.

9   *Introduction to Water Distribution*—Volume 3. Denver: American Water Works Association.

10 *Introduction to Water Quality Analysis*—Volume 4. Denver: American Water Works Association.

11 *Reference Handbook: Basic Science Concepts and Applications.* Denver: American Water Works Association.

12 *Handbook of Water Analysis*, 2nd ed., HACH Chemical Company, P.O. Box 389, Loveland, CO., 1992.

(Continued)

TABLE 6.1 (*Continued*)
Recommended Reference Material

13 *Methods for Chemical Analysis of Water and Wastes*, U.S. Environmental Protection Agency, Environmental Monitoring Systems Laboratory-Cincinnati (ESSL-CL), EPA-6000/4-79-020, Revised March 1983 and 1979 (where applicable).

14 *Standard Methods for the Examination of Water and Wastewater*, American Public Health Association, Washington, D.C., current edition.

15 *Basic Math Concepts: For Water and Wastewater Plant Operators.* Price, J. K. Lancaster, PA: Technomic Publishing Company, 1991.

16 *Spellman's Standard Handbook for Wastewater Operators, Vol. 1, 2, & 3.* Spellman, F. R., Lancaster, PA: Technomic Publishing Company, 199-2000.

17 *The Handbook for Waterworks Operator Certification, Vols. 1, 2, & 3.* Spellman, F. R., & Drinan, J. Lancaster, PA: Technomic Publishing Company, 2001.

18 *Fundamentals for the Water & Wastewater Maintenance Operator Series: Electricity, Electronics, Pumping, Water Hydraulics, Piping and Valves, & Blueprint Reading.* Spellman, F. R. & Drinan, J. Lancaster, PA: Technomic Publishing Company (Distributed by CRC Press, Boca Raton, FL), 2000-2002.

19 *Wastewater Treatment Plants: Planning, Design, and Operation*, 2nd ed. Qasim, S. R. Lancaster, PA: Technomic Publishing Company, Inc., 1999.

20 *Simplified Wastewater Treatment Plant Operations.* Haller, E. Lancaster, PA: Technomic Publishing Company, 1999.

21 *Operation of Wastewater Treatment Plants, A Field Study Program, Volume I*, 4th ed., Kerri, K., et al. California State University, Sacramento, CA.

22 *Operation of Wastewater Treatment Plants, A Field Study Program, Volume II*, 4th ed., Kerri, K., et al. California State University, Sacramento, CA.

TREATMENT PROCESS MODELS

Figure 6.1 illustrates a basic schematic of the water treatment process. Other unit processes used in the treatment of water (fluoridation, for example) are not represented in Figure 6.1; however, we discuss many of the other unit processes in detail within the handbook. Figure 6.2 shows a basic schematic or model of a wastewater treatment process that provides primary and secondary treatment using the *activated sludge* process. In secondary treatment (which provides BOD removal beyond what is achievable

DOI: 10.1201/9781003581901-7

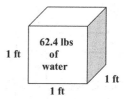

FIGURE 6.1 Unit processes water treatment.

by simple sedimentation), three approaches are commonly

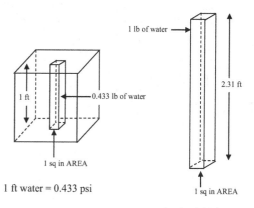

1 ft water = 0.433 psi

1 psi = 2.31 ft water

FIGURE 6.2 Schematic of an example wastewater treatment facility providing primary and secondary treatment using the activated sludge process.

used: trickling filter, activated sludge, and oxidation ponds. We discuss these systems in detail later in the text. We also discuss BNR (biological nutrient removal) and standard tertiary or advanced wastewater treatment.

The purpose of the models shown in Figures 6.1 and 6.2 is to allow readers to visually follow the water and wastewater treatment process step-by-step as they are presented in this text. The figures help the reader understand how all the various unit processes sequentially follow and tie into each other. This format simply provides a pictorial presentation along with pertinent written information, enhancing the learning process.

ADDITIONAL WASTEWATER TREATMENT MODELS

To demonstrate that there are more wastewater treatment models available and actually being used in the industry than just the so-called conventional unit processes shown in Figure 6.2, seven different models are described and discussed in this section.

SIDE BAR 6.1 GREEN BAY METROPOLITAN SEWERAGE DISTRICT

The Green Bay (Wisconsin) Metropolitan Sewerage District's (GBMSD) De Pere Wastewater Treatment Plant is an 8.0 MGD (average daily flow) two-stage activated sludge plant with biological phosphorus removal and tertiary effluent filtration.

TABLE 6.2
Profile of De Pere WWTF Influent Data (Y2009)

Parameter	Daily Average
Flow (MGD)	8
BOD (lbs/day)	29,070
TSS (lbs/day)	18,587
Ammonia-N (lbs/day)	Not monitored
Phosphorus (lb/day)	307.5

The Green Bay (WI) Metropolitan Sewerage District's (GBMSD) De Pere Wastewater Treatment Facility (WWTF) serves the City of De Pere, portions of the village of Ashwaubenon, and portions of the Towns of Lawrence, Belleview, and Hobart. GBMSD acquired ownership of the De Pere WWTF from the City of De Pere on January 1, 2008.

The original circa mid-1930s plant (a primary treatment facility with biosolids digestion) was upgraded in 1964 to an activated biosolids process, with chlorination for disinfection. In the later 1970s, there was a major upgrade to the facility (which represents the current operational scheme), including a two-stage activated biosolids process with biological phosphorus removal, tertiary filtration (gravity sand filters), solids dewatering with incineration, and liquid chlorine disinfection. Influent data for the De Pere WWTF are presented in Table 6.2.

In 1997, additional upgrades to the facility were initiated, beginning with UV disinfection replacing the liquid chlorine system. The chlorine disinfection system is currently maintained as a backup system. Several other major upgrades included the replacement of the coarse influent screen with fine screens (1998–1999), renovation of the multi-media tertiary filtration system to a signal media U.S. Filter Multiwash air scour system (1999–2000), and a solids handling upgrade that included the installation of two gravity belt thickeners (replacing dissolved air flotation) and the addition of two filter presses (2001–2002).

Figure 6.3 presents the process flow diagram for the GBMSD—De Pere WWTF, a two-stage activated biosolids treatment plant (online 1978 to present) with tertiary filtration and design flows as follows:

Average Dry—8.5 MGD
Design Flow—14.2 MGD
Maximum Hourly Dry—23.8 MGD
Maximum Hourly Wet—30.0 MGD

Influent to the plant undergoes fine screening and is subsequently pumped to preliminary treatment (grit removal followed by grease removal, utilizing two 50 feet×50 feet clarifiers with grease/scum collection). The influent pump station consists of four 150 hp, 10 MGD pumps. Screenings are disposed of in a landfill. Grit,

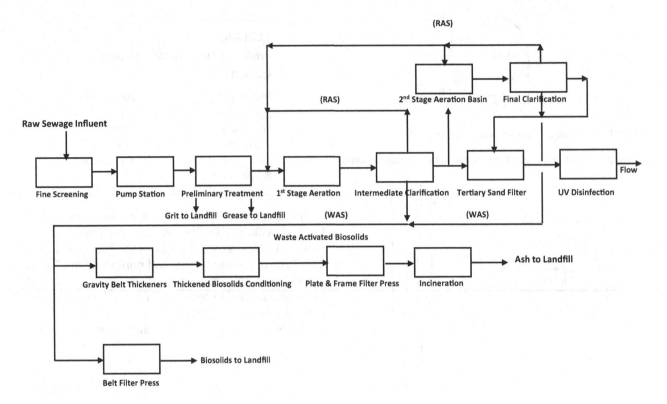

FIGURE 6.3 De Pere WWT process flow diagram (biological treatment). (EPA 832-R-10-00. 2010. *Evaluation of Energy Conservation Measures.* Washington, DC: Environmental Protection Agency.)

oil, and grease removed in preliminary treatment units are also disposed of in a landfill.

Biological treatment is conducted in two serial stages, each with a 1.1 MGD anaerobic zone (for phosphorus removal) followed by a 2.2 MGD aeration zone. Approximately 100% of the mixed liquor suspended solids from the aeration zone are recycled to the anoxic zone. Aeration is provided by six (each) 6,000 standard cubic feet per minute (scfm), 330 hp turbo blowers for the first stage aeration process, and there are (each) 4,000 scfm, 250 hp multi-stage centrifugal blowers for the second stage aeration process.

The first stage of biological treatment is followed by clarification [two each, 100-foot diameter, 13.7-foot side water depth clarifiers (one online for each aeration basin)]. Clarifier effluent from the first stage biological treatment process can be further treated in the second stage treatment process. However, all wastewater is currently treated only in the first stage biological process. The second stage of biological treatment is not currently utilized since it is not required to achieve discharge compliance. Biological treatment is followed by their 125-foot diameter, 10.9-foot side water depth clarifiers. Clarifier effluent is polished by tertiary sand filtration and disinfected using UV prior to discharge. During periods of high flow, UV disinfection is supplemented by disinfection with liquid chlorine.

Clarifier underflow (WAS—waste activated sludge) from biological treatment undergoes one of two dewatering processes. Approximately 75% of the WAS undergoes thickening (two each, 2 m gravity belt thickeners), chemical conditioning (lime and ferric chloride), dewatering (two each, 1.5 m × 2 m plate and frame filter presses), and incineration (18.75-foot diameter, 7 hearth, 7,500 lb/hour multiple hearth incinerator). The incinerator ash is disposed of in a landfill. The balance of the WAS is chemically conditioned with polymer and dewatered in two 2 m belt filter presses. The dewatered sludge is disposed of in a landfill. The filtrate from sludge thickening and dewatering operations is returned to the first stage biological treatment.

The most recent upgrade (2003–2004) replaced the facility's first stage treatment centrifugal blowers with high-speed, magnetic turbo blowers, the first installation of this new, energy-efficient technology in the country. Because the second stage aeration process is currently not utilized, only the first stage process blowers were replaced under the ECM project.

REFERENCE

Shumaker, G., 2007. High Speed Technology Brings Low Costs, *Water and Wastes Digest*, August.

SIDE BAR 6.2 SHEBOYGAN REGIONAL WASTEWATER TREATMENT PLANT

The Sheboygan Regional Wastewater Treatment Plant in Sheboygan, MI, is an 11.8 MGD (average daily flow) activated biosolids plant with biological

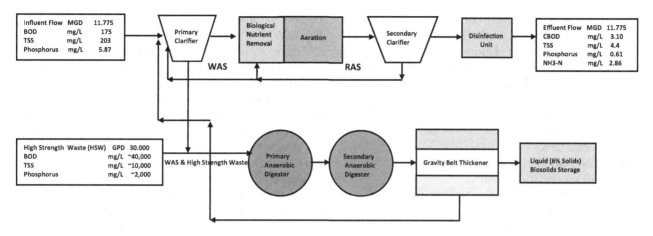

FIGURE 6.4 Sheboygan wastewater treatment plant process flow diagram. (EPA. 2010. Adapted from *Evaluation of Energy Conservation Measures*. EPA 832-R-10-005. Washington, DC: United States Environmental Protection Agency.)

TABLE 6.3

Profile of Sheboygan WWTP Influent Data (Y2009)

Parameter	Daily Average
Flow (MGD)	11.78
BOD (mg/L)	175
TSS (mg/L)	203
Ammonia-N (mg/L)	Not monitored
Phosphorus (mg/L)	5.7

phosphorus removal. The Sheboygan Regional Wastewater Treatment Plant (WWTP) serves approximately 68,000 residential customers in the cities of Sheboygan and Sheboygan Falls, the Village of Kohler, and the Town(s) of Sheboygan, Sheboygan Falls, and Wilson. The plant was originally constructed in 1982 as a conventional activated biosolids plant using turbine aerators with sparger rings. In 1990, the plant was upgraded to include a fine bubble diffused air system with positive displacement blowers. From 1997 through 1999, additional improvements were made to the facility to implement biological nutrient removal and to the bar screens, grit removal facilities, biosolids storage tanks, and the primary and secondary clarifiers. The plant currently operates as an 18.4 MGD biological nutrient removal plant with fine screens, grit removal, primary clarification, biological nutrient removal, activated biosolids thickening, and liquid (6% solids) biosolids storage. Table 6.3 provides average daily influent data for the plant. Figure 6.4 provides a process flow diagram of the plant treatment scheme.

Influent to the plant goes through two automatic self-cleaning fine screens. A 20-foot diameter cyclone separator removes grit before the wastewater enters primary clarification. Primary clarification is provided by four primary clarifiers. Secondary biological treatment

is conducted in six basins. The first two basins are anaerobic to provide phosphorus removal and are configured with baffles in an "N" pattern. The remaining four basins are currently aerated using two Turblex blowers. Following aeration, secondary clarification is provided by four clarifiers. Return activated sludge (RAS) from the clarifiers is sent to the anaerobic zone. A portion of the RAS is conveyed upstream of the primary clarifier. Plant effluent is disinfected with chlorine and is then dechlorinated before discharge into Lake Michigan.

The combined primary and secondary biosolids underflow from the primary clarifier (waste sludge—biosolids) is sent to three primary anaerobic digesters. From the primary digesters, the biosolids flow to a single secondary anaerobic digester. Methane from the digesters is used to provide heat to the digesters as well as fuel for ten 30 kW microturbines that provide electricity to the plant. Two belt thickeners (one at 2 m and one at 3 m) increase the solids content of the digested biosolids from 2.5% to 6% solids. Digested, thickened biosolids are held in two storage tanks before being land-applied.

SIDE BAR 6.3 BIG GULCH WASTEWATER TREATMENT PLANT

The Big Gulch WWTP, owned and operated by the Mukilteo Water and Wastewater District (Washington), is a 1.5 MGD (average daily flow) oxidation ditch plant operating two parallel oxidation ditches. Ditch A treats approximately 40% of the plant flow, and Ditch B treats approximately 60% of the flow.

The Big Gulch WWTP provides wastewater treatment services for 22,455 people residing in portions of the City of Mukilteo and Snohomish County (Washington). Originally constructed in 1970, the

WWTP consisted of a coarse bar screen and a single oxidation ditch using brush rotor aerators, followed by a secondary clarifier and chlorine disinfection. Between 1989 and 1991, the Big Gulch WWTP underwent significant upgrades including the following:

- New headworks with a grit removal channel
- Influent screw pumps
- Selector tank
- Second oxidation ditch
- Third secondary clarifier
- Aerobic biosolids holding tanks with a rotary drum thickener
- Biosolids return piping
- Scum and waste activated biosolids pumps
- Biosolids pumps
- Biosolids dewatering belt filter press
- Chlorine contact chamber

Subsequent to the 1991 facility upgrade, the following upgrades to the treatment plant were implemented

- Influent screening (perforated-plate fine screens)
- Submersible mixers (in the oxidation ditches)
- UV disinfection (replacing chlorine disinfection)

To address the need for additional oxidation ditch aeration capacity to handle intermittent increases in BOD loading, the aeration system in both ditches was upgraded with fine bubble diffusers and automatically controlled turbo blowers. Influent data for the Big Gulch WWTP are presented in Table 6.4.

Figure 6.5 presents the process flow diagram for the Big Gulch WWTP, an activated biosolids treatment plant with UV disinfection.

Influent to the plant passes through a perforated-plate mechanical fine screen (rated capacity of 6.5 MGD) into a gravity grit channel. Effluent from the grit removal system is returned to the headworks and grit is sent to the dumpsters.

Degritted influent, combined with return activated sludge (RAS) from the secondary clarifiers and filtrate from the sludge dewatering belt filter press, is lifted to the selector mixing basin using two influent screw lift pumps (3.83 MGD capacity each). Selector mixing basin effluent is conveyed to the oxidation ditches via overflow channels equipped with adjustable weir gates to distribute the flow to the ditches (40% to Oxidation Ditch A and 60% to Oxidation Ditch B). The two oxidation ditches, operating in parallel and providing a combined 18-hour hydraulic residence time, are followed by three secondary clarifiers.

Effluent from the secondary clarifiers is conveyed to the UV disinfection system. The UV system consists of 96 lamps and provides 35 mJ/cm² (energy) at a peak flow of 8.7 MGD (based on 60% UV transmittance). The UV disinfection system produces an effluent with fecal coliform counts below the facility's permit limit of 200 colonies/100 mL (monthly average).

Waste activated biosolids and scum from the secondary clarifiers are transferred via a rotary lobe pump to a pair of two-cell aerobic biosolids holding tanks for aerobic digestion, producing Class B biosolids. In 2006, the aerobic biosolids digestion system was upgraded with fine-bubble air diffusers and positive displacement blowers. Biosolids are thickened through either settling in the aerobic biosolids holding tanks or through rotary drum thickening. In 2007, the rotary drum biosolids thickener was installed to increase digestion capacity. Digested biosolids are dewatered using a gravity belt dewatering press, and the dewatered biosolids are transported for land application.

TABLE 6.4
Profile of Big Gulch WWTP Influent Data (Y2004–Y2010)

Parameter	Average	Minimum	Maximum
Flow (MGD)	1.68	1.21	2.40
CBOD (mg/L)	217	116	462
TSS (mg/L)	255	131	398

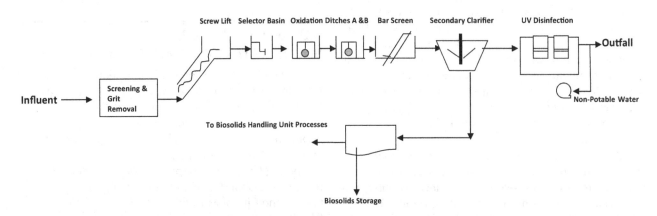

FIGURE 6.5 Simplified schematic flow diagram of Big Gulch WWTP.

REFERENCES

Bridges, T.G., 2010. Phone Conversation—Plant Manager, Big Gulch WWTP. Mukilteo, Washington. As reported *In* EPA (2010) Evaluation of Energy Conservation Measures, EPA 832-R-10-005. Washington, DC: Environmental Protection Agency.

Gray & Osborne, Inc., (2008). *Wastewater Treatment Plant Capacity Study and Engineering Report*. Seattle, WA: Gray & Osborne, Inc.

SIDE BAR 6.4 CITY OF BARTLETT WASTEWATER TREATMENT PLANT

The City of Bartlett, TN, WWTP is a 1.0 MGD (average daily flow) secondary facility utilizing two mechanically aerated oxidation ditches to provide secondary treatment. Each of the aeration basins is equipped with three rotor aerators. The City of Bartlett's Wastewater Treatment Plant (WWTP) #1, located in West Tennessee near Memphis, serves approximately 24,000 residential customers and one school. One hundred percent of the plant influent is domestic wastewater. The facility was originally commissioned in 1994 as a 0.5 MGD aerated lagoon and has undergone three major expansions (in 1993, 2003, and 2005) to meet the city's growing population. In 1993, the facility was upgraded to a secondary treatment facility (one oxidation ditch and secondary clarification). In 2003, the facility was upgraded with solids handling (aerobic digester and belt filter press). In 2005, a second oxidation ditch was added. Influent data for the City of Bartlett WWTP #1 is presented in Table 6.5.

Plant influent undergoes mechanical screening followed by biological treatment in two mechanically aerated oxidation ditches. Each oxidation ditch is equipped with three 60 HP rotor aerators. Oxidation ditch effluent undergoes secondary clarification followed by UV disinfection prior to discharge into the Loosahatchie River. Waste biosolids from the secondary clarifiers undergo aerobic digestion. Digested biosolids are dewatered in a belt filter press and then land-applied as an agricultural soil amendment and fertilizer.

TABLE 6.5
Profile of the City of Bartlett WWTP #1 Influent Data (Y2009)

Parameter	Daily Average
Flow (mg/L)	1.0
BOD (mg/L)	130
TSS (mg/L)	180
Ammonia-N (mg/L)	Not monitored
TKN (mg/L)	41
Phosphorus (mg/L)	6

KEY TERMS, ACRONYMS, ABBREVIATIONS IN WATER/WASTEWATER OPERATIONS

To learn about water/wastewater treatment operations (or any other technology, for that matter), you must master the language associated with the technology. Each technology has its own terms, acronyms, and abbreviations, each with their accompanying definitions. Many terms and acronyms used in water/wastewater treatment are unique, while others combine words from many different technologies and professions. One thing is certain: water/wastewater operators without a clear understanding of the relevant terms and acronyms are ill-equipped to perform their duties effectively.

Usually, a handbook or text like this one includes a glossary of terms at the end. In this *Handbook*, a glossary is included; however, definitions of many frequently used terms are presented right up front. Experience shows that an early introduction to keywords and definitions benefits readers. An upfront introduction to key terms, acronyms, and abbreviations facilitates a more orderly, logical, and systematic learning activity. Terms not defined in this section will be defined as they appear in the text and in the glossary. A short quiz on many of the following terms follows at the end of this chapter.

DEFINITIONS

Absorb: to take in. Many things absorb water.

Acid Rain: the acidic rainfall that results when rain combines with sulfur oxides emitted from combustion of fossil fuels (coal, for example).

Acre-Feet (Acre-Foot): an expression of water quantity. One acre-foot will cover one acre of ground one foot deep. An acre-foot contains 43,560 cubic feet, 1,233 m^3, or 325,829 gallons (U.S). It is also abbreviated as ac-ft.

Activated Carbon: derived from vegetable or animal materials by roasting in a vacuum furnace. Its porous nature gives it a very high surface area per unit mass—as much as 1,000 m^2/g, which is 10 million times the surface area of 1 g of water in an open container. Used in adsorption (see definition), activated carbon adsorbs substances that are not, or are only slightly, adsorbed by other methods.

Activated Sludge: the solids formed when microorganisms are used to treat wastewater in the activated sludge treatment process. It includes organisms, accumulated food materials, and waste products from the aerobic decomposition process.

Advanced Wastewater Treatment: treatment technology to produce an extremely high-quality discharge.

Adsorption: the adhesion of a substance to the surface of a solid or liquid. Adsorption is often used to extract pollutants by causing them to attach to adsorbents such as activated carbon or silica gel. *Hydrophobic* (water-repulsing adsorbents) are used to extract oil from waterways during oil spills.

Aeration: the process of bubbling air through a solution, sometimes cleaning water of impurities through exposure to air.

Aerobic: conditions in which free, elemental oxygen is present. Also used to describe organisms, biological activity, or treatment processes that require free oxygen.

Agglomeration: floc particles colliding and gathering into a larger, settleable mass.

Air Gap: the air space between the free-flowing discharge end of a supply pipe and an unpressurized receiving vessel.

Algae Bloom: a phenomenon whereby excessive nutrients within a river, stream, or lake cause an explosion of plant life that results in the depletion of the oxygen in the water needed by fish and other aquatic life. Algal bloom is usually the result of urban runoff (of lawn fertilizers, etc.). The potential tragedy is that of a "fish kill," where the stream life dies in one mass execution.

Alum: aluminum sulfate; a standard coagulant used in water treatment.

Ambient: the expected natural conditions that occur in water unaffected or uninfluenced by human activities.

Anaerobic: conditions in which no oxygen (free or combined) is available. Also used to describe organisms, biological activity, or treatment processes that function in the absence of oxygen.

Anoxic: conditions in which no free, elemental oxygen is present. The only source of oxygen is combined oxygen, such as that found in nitrate compounds. Also used to describe the biological activity of treatment processes that function only in the presence of combined oxygen.

Aquifer: a water-bearing stratum of permeable rock, sand, or gravel.

Aquifer System: a heterogeneous body of introduced permeable and less permeable material that acts as a water-yielding hydraulic unit of regional extent

Artesian Water: a well tapping a confined or artesian aquifer in which the static water level stands above the top of the aquifer. The term is sometimes used to include all wells tapping confined water. Wells with water levels above the water table are said to have a positive artesian head (pressure), and those with water levels below the water table, have negative artesian head.

Average Monthly Discharge Limitation: the highest allowable discharge over a calendar month.

Average Weekly Discharge Limitation: the highest allowable discharge over a calendar week.

Backflow: reversal of flow when the pressure in a service connection exceeds the pressure in the distribution main.

Backwash: fluidizing filter media with water, air, or a combination of the two so that individual grains can be cleaned of the material that has accumulated during the filter run.

Bacteria: any of several one-celled organisms, some of which cause disease.

Bar Screen: a series of bars formed into a grid used to screen out large debris from the influent flow.

Base: a substance that has a pH value between 7 and 14.

Basin: a groundwater reservoir defined by the overlying land surface and underlying aquifers that contain water stored in the reservoir.

Beneficial Use of Water: the use of water for any beneficial purpose. Such uses include domestic use, irrigation, recreation, fish and wildlife, fire protection, navigation, power, industrial use, etc. The benefit varies from one location to another and by custom. What constitutes beneficial use is often defined by statute or court decisions.

Biochemical oxygen demand (BOD₅): the oxygen used in meeting the metabolic needs of aerobic microorganisms in water rich in organic matter.

Biosolids: from *Merriam-Webster's Collegiate Dictionary, Tenth Ed.* (1998): biosolids *n* (1977) solid organic matter recovered from a sewage treatment process and used especially as fertilizer [or soil amendment] – usually used in plural.

- **Note:** In this text, *biosolids* is used in many places (activated sludge being the exception) to replace the standard term sludge. The authors view the term sludge as an ugly four-letter word inappropriate to use to describe biosolids. Biosolids is a product that can be reused; it has some value. Because biosolids have value, they certainly should not be classified as a "waste" product – and when biosolids for beneficial reuse are addressed, it is made clear that they are not.

Biota: all the species of plants and animals indigenous to a certain area.

Boiling Point: the temperature at which a liquid boils. The temperature at which the vapor pressure of a liquid equals the pressure on its surface. If the pressure of the liquid varies, the actual boiling point varies. The boiling point of water is 212°F or 100°C.

Breakpoint: point at which chlorine dosage satisfies chlorine demand.

Breakthrough: in filtering, when unwanted materials start to pass through the filter.

Buffer: a substance or solution that resists changes in pH.

Calcium Carbonate: compound principally responsible for hardness.

Calcium Hardness: portion of total hardness caused by calcium compounds.

Carbonaceous Biochemical Oxygen Demand (CBOD₅): the amount of biochemical oxygen demand that can be attributed to carbonaceous material.

Carbonate Hardness: caused primarily by compounds containing carbonate.

Chemical Oxygen Demand (COD): the amount of chemically oxidizable materials present in the wastewater.

Chlorination: disinfection of water using chlorine as the oxidizing agent.

Clarifier: a device designed to permit solids to settle or rise and be separated from the flow. Also known as a settling tank or sedimentation basin.

Coagulation: the neutralization of the charges of colloidal matter.

Coliform: a type of bacteria used to indicate possible human or animal contamination of water.

Combined Sewer: a collection system that carries both wastewater and stormwater flows.

Comminution: a process to shred solids into smaller, less harmful particles.

Composite Sample: a combination of individual samples taken in proportion to flow.

Connate Water: pressurized water trapped in the pore spaces of sedimentary rock at the time it was deposited. It is usually highly mineralized.

Consumptive Use: (1) the quantity of water absorbed by crops and transpired or used directly in the building of plant tissue, together with the water evaporated from the cropped area; (2) the quantity of water transpired and evaporated from a cropped area or the normal loss of water from the soil by evaporation and plant transpiration; (3) the quantity of water discharged to the atmosphere or incorporated in the products of the process in connection with vegetative growth, food processing, or an industrial process.

Contamination (Water): damage to the quality of water sources by sewage, industrial waste, or other materials.

Cross-Connection: a connection between a storm drain system and a sanitary collection system; a connection between two sections of a collection system to handle anticipated overloads of one system; or a connection between drinking (potable) water and an unsafe water supply or sanitary collection system.

Daily Discharge: the discharge of a pollutant measured during a calendar day or any 24-hour period that reasonably represents a calendar day for the purposes of sampling. Limitations expressed as weight are total mass (weight) discharged over the day. Limitations expressed in other units are the average measurement of the day.

Daily Maximum Discharge: the highest allowable values for a daily discharge.

Darcy's Law: an equation for the computation of the quantity of water flowing through porous media. Darcy's Law assumes that the flow is laminar and that inertia can be neglected. The law states that the rate of viscous flow of homogeneous fluids through isotropic porous media is proportional to and in the direction of, the hydraulic gradient.

Detention Time: the theoretical time water remains in a tank at a given flow rate.

De-Watering: the removal or separation of a portion of water present in a sludge or slurry.

Diffusion: the process by which both ionic and molecular species dissolved in water move from areas of higher concentration to areas of lower concentration.

Discharge Monitoring Report (DMR): the monthly report required by the treatment plant's National Pollutant Discharge Elimination System (NPDES) discharge permit.

Disinfection: water treatment process that kills pathogenic organisms.

Disinfection By-products (DBPs): chemical compounds formed by the reaction of disinfectant with organic compounds in water.

Dissolved Oxygen (DO): the amount of oxygen dissolved in water or sewage. Concentrations of less than 5 parts per million (ppm) can limit aquatic life or cause offensive odors. Excessive organic matter present in water because of inadequate waste treatment and runoff from agricultural or urban land generally causes low DO.

Dissolved Solids: the total amount of dissolved inorganic material contained in water or wastes. Excessive dissolved solids make water unsuitable for drinking or industrial uses.

Domestic Consumption (Use): water used for household purposes such as washing, food preparation, and showers. The quantity (or quantity per capita) of water consumed in a municipality or district for domestic uses or purposes during a given period, sometimes encompasses all uses, including the quantity wasted, lost, or otherwise unaccounted for.

Drawdown: lowering the water level by pumping. It is measured in feet for a given quantity of water pumped during a specified period, or after the pumping level has become constant.

Drinking Water Standards: established by state agencies, the U.S. Public Health Service, and the Environmental Protection Agency (EPA) for drinking water in the United States.

Effluent: something that flows out, usually a polluting gas or liquid discharge.

Effluent Limitation: any restriction imposed by the regulatory agency on quantities, discharge rates, or concentrations of pollutants discharged from point sources into state waters.

Energy: in scientific terms, the ability or capacity to do work. Various forms of energy include kinetic, potential, thermal, nuclear, rotational, and electromagnetic. One form of energy may be changed to another, as when coal is burned to produce steam to drive a turbine, which produces electric energy.

Erosion: the wearing away of the land surface by wind, water, ice, or other geologic agents. Erosion occurs naturally from weather or runoff but is often intensified by human land use practices.

Eutrophication: the process of enrichment of water bodies by nutrients. Eutrophication of a lake normally contributes to its slow evolution into a bog or marsh and ultimately to dry land. Eutrophication may be accelerated by human activities, thereby speeding up the aging process.

Evaporation: the process by which water becomes vapor at a temperature below the boiling point.

Facultative: organisms that can survive and function in the presence or absence of free, elemental oxygen.

Fecal Coliform: the portion of the coliform bacteria group that is present in the intestinal tracts and feces of warm-blooded animals.

Field Capacity: the capacity of soil to hold water. It is measured as the ratio of the weight of water retained by the soil to the weight of the dry soil.

Filtration: the mechanical process that removes particulate matter by separating water from solid material, usually by passing it through sand.

Floc: solids that join to form larger particles that will settle better.

Flocculation: a slow mixing process in which particles are brought into contact, with the intent of promoting their agglomeration.

Flume: a flow rate measurement device.

Fluoridation: chemical addition to water to reduce the incidence of dental caries in children.

Food-to-Microorganisms Ratio (F/M): an activated sludge process control calculation based upon the amount of food (BOD_5 or COD) available per pound of mixed liquor volatile suspended solids.

Force Main: a pipe that carries wastewater under pressure from the discharge side of a pump to a point of gravity flows downstream.

Grab Sample: an individual sample collected at a randomly selected time.

Graywater: water that has been used for showering, clothes washing, and faucet uses. The kitchen sink and toilet water are excluded. This water has excellent potential for reuse as irrigation for yards.

Grit: heavy inorganic solids, such as sand, gravel, eggshells, or metal filings.

Groundwater: the supply of fresh water found beneath the Earth's surface (usually in aquifers) often used for supplying wells and springs. Because groundwater is a major source of drinking water, concern is growing over areas where leaching agricultural or industrial pollutants or substances from leaking underground storage tanks (USTs) are contaminating groundwater.

Groundwater Hydrology: the branch of hydrology that deals with groundwater; its occurrence and movements, its replenishment and depletion, the properties of rocks that control groundwater movement and storage, and the methods of investigation and use of groundwater.

Groundwater Recharge: the inflow to a groundwater reservoir.

Groundwater Runoff: a portion of runoff that has passed into the ground, has become groundwater, and has been discharged into a stream channel as spring or seepage water.

Hardness: the concentration of calcium and magnesium salts in water.

Headloss: the amount of energy used by water in moving from one point to another.

Heavy Metals: metallic elements with high atomic weights, e.g., mercury, chromium, cadmium, arsenic, and lead. They can damage living things at low concentrations and tend to accumulate in the food chain.

Holding Pond: a small basin or pond designed to hold sediment-laden or contaminated water until it can be treated to meet water quality standards or used in some other way.

Hydraulic Cleaning: *cleaning pipe with water under enough pressure to produce high water velocities.*

Hydraulic Gradient: a measure of the change in groundwater head over a given distance.

Hydraulic Head: the height above a specific datum (generally sea level) that water will rise in a well.

Hydrologic Cycle (Water Cycle): the cycle of water movement from the atmosphere to the earth and back to the atmosphere through various processes. These processes include precipitation, infiltration, percolation, storage, evaporation, transpiration, and condensation.

Hydrology: the science dealing with the properties, distribution, and circulation of water.

Impoundment: a body of water such as a pond, confined by a dam, dike, floodgate, or other barrier, and used to collect and store water for future use.

Industrial Wastewater: wastes associated with industrial manufacturing processes.

Infiltration: the gradual downward flow of water from the surface into soil material.

Infiltration/Inflow: extraneous flows in sewers; simply, inflow is water discharged into sewer pipes or service connections from such sources as foundation drains, roof leaders, cellar and yard area drains, cooling water from air conditioners, and other clean-water discharges from commercial and industrial establishments. Defined by Metcalf & Eddy as follows:

- **Infiltration:** water entering the collection system through cracks, joints, or breaks.
- **Steady Inflow:** water discharged from cellar and foundation drains, cooling water discharges, and drains from springs and swampy areas. This type of inflow is steady and is identified, and measured along with infiltration.
- **Direct Flow:** those types of inflow that have a direct stormwater runoff connection to the sanitary sewer and cause an almost immediate increase in wastewater flows. Possible sources are roof leaders, yard and areaway drains, manhole covers, cross-connections from storm drains and catch basins, and combined sewers.
- **Total Inflow:** the sum of the direct inflow at any point in the system plus any flow discharged from the system upstream through overflows, pumping station bypasses, and the like.
- **Delayed Inflow:** stormwater that may require several days or more to drain through the sewer system. This category can include the discharge of sump pumps from cellar drainage as well as the slowed entry of surface water through manholes in ponded areas.

Influent: wastewater entering a tank, channel, or treatment process.

Inorganic Chemical/Compounds: chemical substances of mineral origin, not of carbon structure. These include metals such as lead, iron (ferric chloride), and cadmium.

Ion Exchange Process: used to remove hardness from water.

Jar Test: laboratory procedure used to estimate proper coagulant dosage.

Langelier Saturation Index (L.I.): a numerical index that indicates whether calcium carbonate will be deposited or dissolved in a distribution system.

Leaching: the process by which soluble materials in the soil such as nutrients, pesticide chemicals, or contaminants, are washed into a lower layer of soil or are dissolved and carried away by water.

License: a certificate issued by the State Board of Waterworks/Wastewater Works Operators authorizing the holder to perform the duties of a wastewater treatment plant operator.

Lift Station: a wastewater pumping station designed to "lift" the wastewater to a higher elevation. A lift station normally employs pumps or other mechanical devices to pump the wastewater and discharges it into a pressure pipe called a force main.

Maximum Contaminant Level (MCL): an enforceable standard for the protection of human health.

Mean Cell Residence Time (MCRT): the average length of time a mixed liquor suspended solids particle remains in the activated sludge process. This may also be known as sludge retention time.

Mechanical Cleaning: clearing pipe by using equipment (bucket machines, power rodders, or hand rods) that scrapes, cuts, pulls, or pushes the material out of the pipe.

Membrane Process: a process that draws a measured volume of water through a filter membrane with small enough openings to take out contaminants.

Metering Pump: a chemical solution feed pump that adds a measured amount of solution with each stroke or rotation of the pump.

Mixed Liquor: the suspended solids concentration of the mixed liquor.

Mixed Liquor Volatile Suspended Solids (MLVSS): the concentration of organic matter in the mixed liquor suspended solids.

Milligrams/Liter (mg/L): a measure of concentration equivalent to parts per million (ppm).

Nephelometric Turbidity Unit (NTU): indicates the amount of turbidity in a water sample.

Nitrogenous Oxygen Demand (NOD): a measure of the amount of oxygen required to biologically oxidize nitrogen compounds under specified conditions of time and temperature.

Nonpoint Source (NPS) Pollution: forms of pollution caused by sediment, nutrients, organic and toxic substances originating from land use activities that are carried to lakes and streams by surface runoff. Nonpoint source pollution occurs when the rate of materials entering these water bodies exceeds natural levels.

NPDES Permit: National Pollutant Discharge Elimination System permit, which authorizes the discharge of treated wastes and specifies the conditions that must be met for discharge.

Nutrients: substances required to support living organisms. Usually refers to nitrogen, phosphorus, iron, and other trace metals.

Organic Chemicals/Compounds: animal or plant-produced substances containing mainly carbon, hydrogen, and oxygen, such as benzene and toluene.

Parts per Million (ppm): the number of parts by weight of a substance per million parts of water. This unit is commonly used to represent pollutant concentrations. Large concentrations are expressed in percentages.

Pathogenic: disease causing. A pathogenic organism is capable of causing illness.

Percolation: the movement of water through the subsurface soil layers, usually continuing downward to the groundwater or water table reservoirs.

pH: a way of expressing both acidity and alkalinity on a scale of 0–14, with 7 representing neutrality; numbers less than 7 indicate increasing acidity and numbers greater than 7 indicate increasing alkalinity.

Photosynthesis: a process in green plants in which water, carbon dioxide, and sunlight combine to form sugar.

Piezometric Surface: an imaginary surface that coincides with the hydrostatic pressure level of water in an aquifer.

Point Source Pollution: a type of water pollution resulting from discharges into receiving waters from easily identifiable points. Common point sources of pollution are discharges from factories and municipal sewage treatment plants.

Pollution: the alteration of the physical, thermal, chemical, or biological quality of, or the contamination of, any water in the state that renders the water harmful, detrimental, or injurious to humans, animal life, vegetation, property, or to public health, safety, or welfare, or impairs the usefulness or the public enjoyment of the water for any lawful or reasonable purpose.

Porosity: that part of a rock that contains pore spaces without regard to size, shape, interconnection, or arrangement of openings. It is expressed as a percentage of the total volume occupied by spaces.

Potable Water: water satisfactorily safe for drinking purposes from the standpoint of its chemical, physical, and biological characteristics.

Precipitate: a deposit on the earth of hail, rain, mist, sleet, or snow. The common process by which atmospheric water becomes surface or subsurface water.

Precipitation: is also commonly used to designate the quantity of water precipitated.

Preventive Maintenance (PM): regularly scheduled servicing of machinery or other equipment using appropriate tools, tests, and lubricants. This type of maintenance can prolong the useful life of equipment and machinery and increase its efficiency by detecting and correcting problems before they cause a breakdown of the equipment.

Purveyor: an agency or person that supplies potable water.

Radon: a radioactive, colorless, odorless gas that occurs naturally in the earth. When trapped in buildings, concentrations build up and can cause health hazards such as lung cancer.

Recharge: the addition of water into a groundwater system.

Reservoir: a pond, lake, tank, or basin (natural or human-made) where water is collected and used for storage. Large bodies of groundwater are called groundwater reservoirs; water behind a dam is also called a reservoir of water.

Reverse Osmosis: a process in which almost pure water is passed through a semipermeable membrane.

Return Activated Sludge Solids (RASS): the concentration of suspended solids in the sludge flow being returned from the settling tank to the head of the aeration tank.

River Basin: a term used to designate the area drained by a river and its tributaries.

Sanitary Wastewater: wastes discharged from residences and from commercial, institutional, and similar facilities that include both sewage and industrial wastes.

Schmutzdecke: a layer of solids and biological growth that forms on top of a slow sand filter, allowing the filter to remove turbidity effectively without chemical coagulation.

Scum: the mixture of floatable solids and water removed from the surface of the settling tank.

Sediment: transported and deposited particles derived from rocks, soil, or biological material.

Sedimentation: a process that reduces the velocity of water in basins so that suspended material can settle out by gravity.

Seepage: the appearance and disappearance of water at the ground surface. Seepage designates the movement of water in saturated material. It differs from percolation, which is predominantly the movement of water in unsaturated material.

Septic Tanks: used to hold domestic wastes when a sewer line is not available to carry it to a treatment plant. The wastes are piped to underground tanks directly from a home or homes. Bacteria in the wastes decompose some of the organic matter, the sludge settles on the bottom of the tank, and the effluent flows out of the tank into the ground through drains.

Settleability: a process control test used to evaluate the settling characteristics of the activated sludge. Readings taken at 30–60 minutes are used to calculate the settled sludge volume (SSV) and the sludge volume index (SVI).

Settled Sludge Volume (SSV): the volume (in percent) occupied by an activated sludge sample after 30–60 minutes of settling. Normally written as SSV with a subscript to indicate the time of the reading used for calculation (SSV_{60} or SSV_{30}).

Sludge: the mixture of settleable solids and water removed from the bottom of the settling tank.

Sludge Retention Time (SRT): see mean cell residence time.

Sludge Volume Index (SVI): a process control calculation used to evaluate the settling quality of the activated sludge. Requires the SSV_{30} and mixed liquor suspended solids test results to calculate.

Soil Moisture (Soil Water): water diffused in the soil. It is found in the upper part of the zone of aeration from which water is discharged by transpiration from plants or by soil evaporation.

Specific Heat: the heat capacity of a material per unit mass. The amount of heat (in calories) required to raise the temperature of 1 g of a substance by 1°C; the specific heat of water is 1 calorie.

Storm Sewer: a collection system designed to carry only stormwater runoff.

Stormwater: runoff resulting from rainfall and snowmelt.

Stream: a general term for a body of flowing water. In hydrology, the term is generally applied to the water flowing in a natural channel as distinct from a canal. More generally, it is applied to the water flowing in any channel, natural or artificial. Some types of streams: (1) Ephemeral: a stream that flows only in direct response to precipitation and whose channel is at all times above the water table. (2) Intermittent or Seasonal: a stream that flows only at certain times of the year when it receives water from springs, rainfall, or from surface sources such as melting snow. (3) Perennial: a stream that flows continuously. (4) Gaining: a stream or reach of a stream that receives water from the zone of saturation; an effluent stream. (5) Insulated: a stream or reach of a stream that neither contributes water to the zone of saturation nor receives water from it. It is separated from the zones of saturation by an impermeable bed. (6) Losing: a stream or reach of a stream that contributes water to the zone of saturation; an influent stream. (7) Perched: a perched stream is either a losing stream or an insulated stream that is separated from the underlying groundwater by a zone of aeration.

Supernatant: the liquid standing above a sediment or precipitate.

Surface Water: lakes, bays, ponds, impounding reservoirs, springs, rivers, streams, creeks, estuaries, wetlands, marshes, inlets, canals, gulfs inside the territorial limits of the state, all other bodies of surface water, natural or artificial, inland or coastal, fresh or salt, navigable or non-navigable, and including the beds and banks of all watercourses and bodies of surface water, that are wholly or partially inside or bordering the state or subject to the jurisdiction of the state. Waters in treatment systems authorized by state or federal law, regulation, or permit, and created for water treatment is not considered water in the state.

Surface Tension: the free energy produced in a liquid surface by the unbalanced inward pull exerted by molecules underlying the layer of surface molecules.

Thermal Pollution: the degradation of water quality by the introduction of a heated effluent, primarily from the discharge of cooling waters from industrial processes (particularly from electrical power generation); waste heat eventually results from virtually every energy conversion.

Titrant: a solution of known strength or concentration used in titration.

Titration: a process in which a solution of known strength (titrant) is added to a certain volume of a treated sample containing an indicator. A color change shows when the reaction is complete.

Titrator: an instrument (usually a calibrated cylinder or tube-form) used in titration to measure the amount of titrant being added to the sample.

Total Dissolved Solids: the amount of material (inorganic salts and small amounts of organic material) dissolved in water, commonly expressed as a concentration in milligrams per liter.

Total Suspended Solids (TSS): total suspended solids in water, commonly expressed as a concentration in milligrams per liter.

Toxicity: the occurrence of lethal or sublethal adverse effects on representative sensitive organisms due to exposure to toxic materials. Adverse effects caused by conditions of temperature, dissolved oxygen, or nontoxic dissolved substances are excluded from the definition of toxicity.

Transpiration: the process by which water vapor escapes from the living plant—principally the leaves—and enters the atmosphere.

Vaporization: the change of a substance from a liquid or solid state to a gaseous state.

VOC (Volatile Organic Compound): any organic compound that participates in atmospheric photochemical reactions except for those designated by the USEPA Administrator as having negligible photochemical reactivity.

Wastewater: the water supply of a community after it has been soiled by use.

Waste Activated Sludge Solids (WASS): the concentration of suspended solids in the sludge being removed from the activated sludge process.

Water Cycle: the process by which water travels in a sequence from the air (condensation) to the earth (precipitation) and returns to the atmosphere (evaporation). It is also referred to as the hydrologic cycle.

Water Quality Standard: a plan for water quality management containing four major elements: water use, criteria to protect uses, implementation plans and enforcement plans. An anti-degradation statement is sometimes prepared to protect existing high-quality waters.

Water Quality: a term used to describe the chemical, physical, and biological characteristics of water with respect to its suitability for a particular use.

Water Supply: any quantity of available water.

Waterborne Disease: a disease caused by a microorganism that is carried from one person or animal to another by water.

Watershed: the area of land that contributes surface runoff to a given point in a drainage system.

Weir: a device used to measure wastewater flow.

Zone of Aeration: a region in the earth above the water table. Water in the zone of aeration is under atmospheric pressure and would not flow into a well.

Zoogleal Slime: the biological slime that forms on fixed film treatment devices. It contains a wide variety of organisms essential to the treatment process.

ACRONYMS AND ABBREVIATIONS

°C	Degrees Centigrade or Celsius
°F	Degrees Fahrenheit
µ	Micron
µg	Microgram
µm	Micrometer
A-C	Alternating Current
ACEE	American Council for an Energy Efficient Economy
Al3	Aluminum Sulfate (or Alum)
Amp	Amperes
ANAMMOX	Anaerobic Ammonia Oxidation
APPA	American Public Power Association
ASCE	American Society of Civil Engineers
A/O	Anoxic/Oxic
A2/O	Anaerobic/Anoxic/Oxic
AS	Activated Sludge
ATM	Atmosphere
AT 3	Aeration Tank 3 process
ASE	Alliance to Save Energy
AWWA	American Water Works Association
BABE	Bio-Augmentation Batch Enhanced
BAF	Biological Aerated Filter
BAR	Bioaugmentation Reaeration
BASIN	Biofilm Activated Sludge Innovative Nitrification
BEP	Best Efficiency Point
bhp	Brake Horsepower
BNR	Biological Nutrient Removal
BOD	Biochemical Oxygen Demand
BOD-to-TKN	Biochemical Oxygen Demand-to-Total Kjeldahl Nitrogen Ratio
BOD-to-TP	Biochemical Oxygen Demand-to-Total Phosphorus Ratio
BPR	Biological Phosphorus Removal
CANON	Completely Autotrophic Nitrogen Removal over Nitrate
CAS	Cyclic Activated Sludge
CBOD	Carbonaceous Biochemical Oxygen Demand
CCCSD	Central Contra Costa Sanitary District
CEC	California Energy Commission
CEE	Consortium for Energy Efficiency
CFO	Cost Flow Opportunity
CFM	Cubic Feet per Minute
CFS	Cubic Feet per Second
CHP	Combined Heat and Power

Ci	Curie
COD	Chemical Oxygen Demand
COV	Coefficient of Variation
CP	Central Plant
CWSRF	Clean Water State Revolving Fund
DAF	Dissolved-Air Flotation Unit
DCS	Distributed Control System
DO	Dissolved Oxygen
DOE	Department of Energy
DON	Dissolved Organic Nitrogen
DSIRE	Database of State Incentives for Renewables and Efficiency
EBPR	Enhanced Biological Phosphorus Removal
ECM	Energy Conservation Measure
ENR	Engineering News-Record
EPA	Environmental Protection Agency
EPACT	Energy Policy Act
EPC	Energy Performance Contracting
EPRI	Electric Power Research Institute
ESCO	Energy Services Company
FeCl3	Ferric Chloride
FFS	Fixed Film System
GAO	Glycogen Accumulating Organism
GBMSD	Green Bay (Wisconsin) Metropolitan Sewerage District
GPD	Gallons per Day
GPM	Gallons per Minute
HCO_3-	Bicarbonate
H_2CO_3	Carbonic Acid
HDWK	Headworks
HP	Horsepower
HPC	Heterotrophic Plant Count; an analytic method that measures common bacteria in water. The lower the concentration of bacteria, the better.
HRT	Hydraulic Retention Time
Hz	Hertz
I&I	Inflow and Infiltration
IFAS	Integrated Fixed-Film Activated Sludge
I&C	Instrumentation and Control
IOA	International Ozone Association
IUVA	International Ultraviolet Association
kW	Kilowatt hour
kWh/year	Kilowatt-Hours per Year
LPHO	Low Pressure High Output
MBR	Membrane Bioreactor
M	Mega
M	Million
MG	Million Gallons
MGD	Million Gallons per Day
Mg/L	Milligrams per Liter (equivalent to parts per million)
MLE	Modified Ludzack-Ettinger Process
MLSS	Mixed Liquor Suspended Solids

MPN	Most Probable Number
MW	Molecular Weight
N	Nitrogen
NAESCO	National Association of Energy Service Companies
NEMA	National Electrical Manufacturers Association
NH4	Ammonium
NH4-N	Ammonia Nitrogen
NL	No Limit
NPDES	National Pollutant Discharge Elimination System
NYSERDA	New York State Energy Research and Development Authority
ORP	Oxidation-Reduction Potential
O&M	Operation and Maintenance
Pa	Pascal
PAO	Phosphate Accumulating Organisms
PG&E	Pacific Gas and Electric
PID	Phased Isolation Ditch
PLC	Programmable Logic Controller
$PO_{4}{}^{3-}$	Phosphate
POTWs	Publicly Owned Treatment Works
PSAT	Pump System Assessment Tool
psi	Pounds per Square Inch
psig	Pounds per Square Inch Gauge
RAS	Return Activated Sludge
rpm	Revolutions per Minute
SCADA	Supervisory Control and Data Acquisition
SCFM	Standard Cubic Feet per Minute
SBR	Sequencing Batch Reactor
SRT	Solids Retention Time
TDH	Total Dynamic Head
TKL	Total Kjeldahl Nitrogen
TMDL	Total Maximum Daily Load
TN	Total Nitrogen
TP	Total Phosphorus
TSS	Total Suspended Solids
TVA	Tennessee Valley Authority
UV	Ultraviolet Light
UVT	UV Transmittance
VFD	Variable Frequency Drive
VSS	Volatile Suspended Solids
W	Watt
WAS	Waste Activated Sludge
WEF	Water Environment Federation
WEFTEC	Water Environment Federation Technical Exhibition & Conference
WERF	Water Environment Research Foundation
WMARSS	Waco Metropolitan Area Regional Sewer System
WPCP	Water Pollution Control Plant
WRF	Water Research Foundation
WSU	Washington State University
WWTP	Wastewater Treatment Plant

CHAPTER REVIEW QUESTIONS

6.1 **Matching Exercise:** Match the definitions listed in Part A with the terms listed in Part B by placing the correct letter in the blank.

Note: After completing this exercise, check your answers with those provided in Appendix A.

PART A

1. A nonchemical turbidity removal layer in a slow sand filter. _____
2. Region in earth (soil) above the water table. _____
3. Compound associated with photochemical reaction. _____
4. Oxygen is used in water-rich inorganic matter. _____
5. A stream that receives water from the zone of saturation. _____
6. The addition of water into a groundwater system. _____
7. The natural water cycle. _____
8. Present in intestinal tracts and feces of animals and humans. _____
9. Discharge from a factory or municipal sewage treatment plant. _____
10. Common to fixed film treatment devices. _____
11. Identified water that is safe to drink. _____
12. The capacity of soil to hold water. _____
13. Used to measure acidity and alkalinity. _____
14. Rain mixed with sulfur oxides. _____
15. Enrichment of water bodies by nutrients. _____
16. A solution of known strength of concentration. _____
17. Water lost by foliage. _____
18. Another name for a wastewater pumping station. _____
19. Plants and animals indigenous to an area. _____
20. The amount of oxygen dissolved in water. _____
21. A stream that flows continuously. _____
22. A result of excessive nutrients within a water body. _____
23. Change in groundwater head over a given distance. _____
24. Water trapped in sedimentary rocks. _____
25. Heat capacity of a material per unit mass. _____
26. A compound derived from material that once lived. _____

PART B

a. pH
b. algae bloom
c. zone of aeration
d. hydrological cycle
e. point source pollution
f. perennial
g. organic
h. connate water
i. fecal coliform
j. BOD
k. field capacity
l. transpiration
m. biota
n. specific heat
o. Schmutzdecke
p. recharge
q. Zoogleal slime
r. eutrophication
s. gaining
t. VOC
u. potable
v. acid rain
w. titrant
x. lift station
y. DO
z. hydraulic gradient

Part II

Water/Wastewater Operations

Math, Physics & Technical Aspects

7 Water/Wastewater Math Operations

To operate a waterworks and/or a wastewater treatment plant, and to pass the examination for an operator's license, you must know how to perform certain mathematical operations. However, do not panic, as Price points out "Those who have difficulty in math often do not lack the ability for mathematical calculation, they merely have not learned, or have been taught, the 'language of math'."

—(Price, 1991)

INTRODUCTION

Without the ability to perform mathematical calculations, it would be difficult for operators to properly operate water/wastewater treatment systems. In reality, most of the calculations operators need to perform are not difficult. Generally, math ability through basic algebra is all that is needed. Experience has shown that skill with math operations used in water/wastewater treatment system operations is an acquired skill that is enhanced and strengthened with practice. In this chapter, we assume the reader is well grounded in basic math principles; thus, we do not cover basic operations such as addition, subtraction, multiplication, and division. However, based on recommendations provided by a large number of previous users, we have included more than 400 additional, real-world practice problems and solutions in this new edition.

Note: Keep in mind that mathematics is a *language*—a universal language. Mathematical symbols have the same meaning to people speaking many different languages throughout the globe. The key to learning mathematics is to learn the language—the symbols, definitions, and terms of mathematics, which allow you to understand the concepts necessary to perform the operations.

Key Question: Why is there so much emphasis on math in water/wastewater treatment operations? This is a common question I am often asked by trainees whenever I teach water/wastewater treatment principles and practices in short courses for pre-licensure preparation. My standard reply is that without math, the operator is unable to optimize treatment processes, maintain efficiency, and protect public health. The ability to calculate chemical dosages, flow rates, or contact times and a sound understanding of math are vital for successful water and wastewater treatment operations. Operators must be able to calculate volume, chemical dosage, chlorine contact time, surface area, velocity, flow rates, pump efficiency, area, and volume conversions, and pressure calculations in filters, pumps, and pipelines, as well as recordkeeping, compliance, and reporting.

Key Demonstration: After I explain why math is so important in water/wastewater operations, I write the following on the chalkboard or flip chart—

Convert $4\,\mathrm{ft}^3/\mathrm{sec}$ to MGD

Then I ask the trainees if any of them can make this conversion. Usually, about 30% of the trainees can make the conversion—these are licensed operators who are attempting to sit for higher levels of licensure, meaning they have a number of years of practical treatment experience.

Anyway, I use the chalkboard or flip chart to demonstrate the conversion process.

$$\frac{4\,\mathrm{ft}^3}{\mathrm{sec}} \times \frac{50\,\mathrm{sec}}{\mathrm{min}} \times \frac{60\,\mathrm{min}}{\mathrm{h}} \times \frac{24\,\mathrm{h}}{\mathrm{day}} = 345{,}600 \times \frac{\mathrm{ft}^3}{\mathrm{day}}$$

$$\frac{345{,}600\,\mathrm{ft}^3}{\mathrm{day}} \times \frac{7.48\,\mathrm{gal}}{\mathrm{ft}^3} = \frac{2{,}585{,}088\,\mathrm{gal}}{\mathrm{day}}$$

$$\frac{2{,}585{,}088\,\mathrm{gal}}{\mathrm{day}} \times \frac{1\,\mathrm{MG}}{1{,}000{,}000\,\mathrm{gal}} = 2.6\,\mathrm{MGD}\,(\text{rounded})$$

After demonstrating this conversion, I tell the trainees a couple of things. First, in this particular instance, we need to look at the units so we know what steps need to be taken to determine the answer in the required units—we went from seconds to minutes to hours to days to cubic feet to gallons in the series of steps shown here. Finally, I tell the trainees that knowing how to make conversions like this one enables them to not only properly operate the plant but also that this knowledge, ability, and skill will aid them in passing operator licensure examinations.

After demonstrating this basic conversion process, and before beginning the math refresher training, I provide another example of the type of math operation needed in water/wastewater operations and commonly found on licensure exams. I make the point that the trainees need to solve math operations such as:

Question: How long will it take water to pass through a rectangular basin at a flow rate of 1.2 MGD, if the basin is 90 ft long, 40 ft wide, and 12 ft deep?

I then ask the trainees to determine the answer by working the problem on paper without the aid of a computer or pocket calculator. I am never surprised when only a few in the class can solve this problem, and when I check their solutions, I find that a fewer number are able to obtain the correct answer.

For those scratching their heads or just sitting in their chairs, all of whom are clueless, I simply state that if they

DOI: 10.1201/9781003581901-9

want to be licensed water or wastewater operators, they must be able to solve problems like this one.

And then I provide the **bottom line**: "The good news is that by the end of this training session all of you will know how to solve this and other math operations that will aid you in practice and in working such problems on licensure exams ... and you will find this question number 1 on this session's final examination."

CALCULATION STEPS

As with all math operations, many methods can be successfully used to solve water/wastewater system problems. Standard steps/methods of problem-solving are provided in the following:

1. If appropriate, make a drawing of the information in the problem.
2. Place the given data on the drawing.
3. Answer "What is the question?" This is the first thing you should ask as you begin to solve the problem, as well as asking yourself, "What are they really looking for?" Writing down exactly what is being looked for is always smart. Sometimes the answer has more than one unknown. For instance, you may need to find x and then find y.
4. If the calculation calls for an equation, write it down.
5. Fill in the data in the equation; look to see what is missing.
6. Rearrange or transpose the equation, if necessary.
7. Use a calculator when necessary.
8. Always write down the answer.
9. Check any solution obtained.

EQUIVALENTS, FORMULAE, AND SYMBOLS

In order to work mathematical operations to solve problems (for practical application or for taking licensure examinations), it is essential to understand the language, equivalents, symbols, and terminology used. Because this handbook is designed for use in practical on-the-job applications, equivalents, formulas, and symbols are included as a ready reference in Table 7.1.

FORMULAE

$$SVI = \frac{volume}{concentration} \times 100$$

$$Q = A \times V$$

$$Detention\ time = \frac{volume}{Q}$$

$$Volume = L \times W \times D$$

TABLE 7.1
Equivalents, Formulae, and Symbols

Equivalents

12 in	= 1 ft
36 in	= 1 yard
144 in²	= 1 ft²
9 ft²	= 1 yard²
43,560 ft²	= 1 acre
1 ft³	= 1,728 in³
1 ft³ of water	= 7.48 gal
1 ft³ of water weighs	= 62.4 lbs
1 gal of water weighs	= 8.34 lb
1 liter	= 1.000 mL
1 g	= 1.000 mg
1 million gal/day	= 694 gal/min
	= 1.545 ft³/sec
Average BOD/capita/day	= 0.17 lb
Average SS/capita/day	= 0.20
Average daily flow	= assume 100 gal/capita/day

Symbols

A	= Area
V	= Velocity
t	= time
SVI	= Sludge volume index
Vol	= Volume
#	= pounds (lbs)
eff	= effluent
W	= width
D	= depth
L	= length
Q	= flow
r	= radius
π	= pi (3.14)
WAS	= Waste activated sludge
RAS	= Return activated sludge
MLSS	= Mixed liquor suspended solids
MLVSS	= Mixed liquor volatile suspended. Solids

$$Area = W \times L$$

$$Circular\ area = \pi \times D^2 \left(=0.785 \times D^2\right)$$

$$Circumference = \pi D$$

$$Hydraulic\ loading\ rate = \frac{Q}{A}$$

$$Sludge\ age = \frac{\#MLSS\ in\ Aeration\ Tank}{\#SS\ in\ primary\ eff/day}$$

$$MCRT = \frac{\#SS\ in\ secondary\ system\ (aeration\ tank + sec.\ clarifier)}{\#WAS/day + SS\ in\ eff/day}$$

$$\text{Organic loading rate} = \frac{\#\text{BOD/day}}{\text{volume}}$$

BASIC WATER/WASTEWATER MATH OPERATIONS

ARITHMETIC AVERAGE (OR ARITHMETIC MEAN) AND MEDIAN

During the day-to-day operation of a wastewater treatment plant, considerable mathematical data is collected. The data, if properly evaluated, can provide useful information for trend analysis and indicate how well the plant or unit process is operating. However, because there may be much variation in the data, it is often difficult to determine trends in performance. The *arithmetic average* refers to a statistical calculation used to describe a series of numbers, such as test results. By calculating an average, a group of data is represented by a single number. This number may be considered typical of the group. The *arithmetic mean* is the most commonly used measurement of average value.

Note: When evaluating information based on averages, remember that the "average" reflects the general nature of the group and does not necessarily reflect any one element of that group.

The arithmetic average is calculated by dividing the sum of all of the available data points (test results) by the number of test results.

$$\frac{\text{Test } 1 + \text{Test } 2 + \text{Test } 3 + \cdots + \text{Test } N}{\text{Number of Tests Performed (N)}} \quad (7.1)$$

Example 7.1

Problem: Effluent BOD test results for the treatment plant during the month of September are shown below:

Test 1	20 mg/L
Test 2	31 mg/L
Test 3	22 mg/L
Test 4	15 mg/L

What is the average effluent BOD for the month of September?

Solution:

$$\text{Average} = \frac{20 \text{ mg/L} + 31 \text{ mg/L} + 22 \text{ mg/L} + 15 \text{ mg/L}}{4}$$

$$= 22 \text{ mg/L}$$

Example 7.2

Problem: For the primary influent flow, the following composite-sampled solids concentrations were recorded for the week:

Monday	300 mg/L SS
Tuesday	312 mg/L SS
Wednesday	315 mg/L SS
Thursday	320 mg/L SS
Friday	311 mg/L SS
Saturday	320 mg/L SS
Sunday	310 mg/L SS
Total	2,188 mg/L SS

Solution:

$$\text{Average SS} = \frac{\text{Sum of All Measurements}}{\text{Number of Measurements Used}}$$

$$= \frac{2,188 \text{ mg/L SS}}{7}$$

$$= 312.6 \text{ mg/L SS}$$

Example 7.3

Problem: A waterworks operator takes a chlorine residual measurement every day. We show part of the operating log in Table 7.2. Find the mean.

Solution:

Add up the seven chlorine residual readings.

$$0.9 + 1.0 + 1.2 + 1.3 + 1.4 + 1.1 + 0.9 = 7.8$$

Next, divide by the number of measurements (in this case, seven).

$$7.8 \div 7 = 1.11$$

The mean chlorine residual for the week was 1.11 mg/L.

Definition: The *median* is defined as the value of the central item when the data are arrayed by size. First, arrange all of the readings in either ascending or descending order, then find the middle value.

TABLE 7.2
Daily Chlorine Residual Results

Day	Chlorine Residual (mg/L)
Monday	0.9
Tuesday	1.0
Wednesday	1.2
Thursday	1.3
Friday	1.4
Saturday	1.1
Sunday	0.9

Example 7.4

Problem: In our chlorine residual example, what is the median?

Solution:

Arrange the values in ascending order:

| 0.9 | 0.9 | 1.0 | 1.1 | 1.2 | 1.3 | 1.4 |

The middle value is the fourth one (1.1); therefore, the median chlorine residual is 1.1 mg/L. (Usually, the median will be a different value than the mean.) If the data contains an even number of values, you must add one more step, since no middle value is present. You must find the two values in the middle and then find the mean of those two values.

Example 7.5

Problem: A water system has four wells with the following capacities: 115 gpm, 100 gpm, 125 gpm, and 90 gpm. What are the mean and the median pumping capacities?

Solution:

The mean is

$$\frac{115 \text{ gpm} + 100 \text{ gpm} + 125 \text{ gpm} + 90 \text{ gpm}}{4} = 107.5 \text{ gpm}$$

To find the median, arrange the values in order:

| 90 gpm | 100 gpm | 115 gpm | 125 gpm |

With four values, there is no single middle value, so we must take the mean of the two middle values:

$$\frac{100 \text{ gpm} + 115 \text{ gpm}}{2} = 107.5 \text{ gpm}$$

Note: At times, determining what the original numbers were like is difficult (if not impossible) when dealing only with averages.

Example 7.6

Problem: A water system has four storage tanks. Three of them have a capacity of 100,000 gal each, while the fourth has a capacity of 1 million gallons. What is the mean capacity of the storage tanks?

Solution:

The mean capacity of the storage tanks is

$$\frac{100,000 + 100,000 + 100,000 + 1,000,000}{4} = 325,000 \text{ gal}$$

Note: Notice that no tank in Example 7.6 has a capacity anywhere close to the mean. The median capacity requires us to take the mean of the two middle values; since they are both 100,000 gal, the median is 100,000 gal. Although

three of the tanks have the same capacity as the median, this data offers no indication that one of these tanks holds a million gallons—information that could be important for the operator to know.

UNITS AND CONVERSIONS

Most of the calculations made in water/wastewater operations involve using *units*. While the number tells us how many, the units tell us what we have. Examples of units include inches, feet, square feet, cubic feet, gallons, pounds, milliliters, milligrams per liter, pounds per square inch, miles per hour, and so on. *Conversions* are a process of changing the units of a number to make the number usable in a specific instance. Multiplying or dividing a number to change the units of that number accomplishes conversions. Common conversions in water/wastewater operations are:

- Gallons per minute (gpm) to cubic feet per second (cfs)
- Million gallons to acre-feet
- Cubic feet to acre-feet
- Cubic feet of water to weight
- Cubic feet of water to gallons
- Gallons of water to weight
- Gallons per minute (gpm) to million gallons per day (MGD)
- Pounds per square inch (psi) to feet of head (the measure of the pressure of water expressed as height of water in feet); 1 psi = 2.31 ft of head.

In many instances, the conversion factor cannot be derived—it must be known. Therefore, we use tables such as the one below (Table 7.3) to determine the common conversions.

Note: Conversion factors are used to change measurements or calculated values from one unit of measure to another. In making the conversion from one unit to another, you must know two things:

TABLE 7.3
Common Conversions

Linear Measurements	Weight
1 in = 2.54 cm	1 ft³ of water = 62.4 lbs
1 ft = 30.5 cm	1 gal = 8.34 lbs
1 m = 100 cm 3.281 ft = 39.4 in	1 lb = 453.6 g
1 acre = 43,560 ft²	1 kg = 1,000 g = 2.2 lbs
1 yard = 3 ft	1% = 10,000 mg/L
Volume	**Pressure**
1 gal = 3.78 L	1 ft of head = 0.433 psi
1 ft³ = 7.48 gal	1 psi = 2.31 ft of head
1 L = 1,000 mL	**Flow**
1 acre ft = 43,560 ft³	1 cfs = 448 gpm
1 gal = 32 cups	1 gpm = 1,440 gpd
1 lb = 16 oz dry wt.	

1. The exact number that relates the two units
2. Whether to multiply or divide by that number

Most operators memorize some standard conversions. This happens because of using the conversions, not because of attempting to memorize them.

Temperature Conversions

An example of a type of conversion typical in water and wastewater operations is provided in this section on temperature conversions.

Note: Operators should keep in mind that temperature conversions are only a small part of the many conversions that must be made in real-world systems operations.

Most water and wastewater operators are familiar with the formulas used for Fahrenheit and Celsius temperature conversions:

$$°C = 5/9(°F - 32°) \qquad (7.2)$$

$$°F = 5/9(°C + 32°) \qquad (7.3)$$

These conversions are not difficult to perform. The difficulty arises when we must recall these formulas from memory. Probably the easiest way to recall these important formulas is to remember three basic steps for both Fahrenheit and Celsius conversions:

1. Add 40°
2. Multiply by the appropriate fraction (5/9 or 9/5)
3. Subtract 40°

Obviously, the only variable in this method is the choice of 5/9 or 9/5 in the multiplication step. To make the proper choice, you must be familiar with two scales. On the Fahrenheit scale, the freezing point of water is 32°, and 0° on the Celsius scale. The boiling point of water is 212° on the Fahrenheit scale and 100° on the Celsius scale.

Why is it important? Note, for example, that at the same temperature, higher numbers are associated with the Fahrenheit scale and lower numbers with the Celsius scale. This important relationship helps you decide whether to multiply by 5/9 or 9/5. Let us look at a few conversion problems to see how the three-step process works.

Example 7.7

Problem: Convert 220°F to Celsius. Using the three-step process, we proceed as follows:

Solution:

(1) Step 1: add 40°F

$$220°$$
$$+40°$$
$$\overline{260°}$$

(2) Step 2: 260°F must be multiplied by either 5/9 or 9/5. Since the conversion is to the *Celsius* scale, you will be moving to a number *smaller* than 260. Through reason and observation, obviously we see that multiplying 260 by 9/5 would almost be the same as multiplying by 2, which would double 260, rather than make it smaller. On the other hand, multiplying by 5/9 is about the same as multiplying by 1/2, which would cut 260 in half. Since in this problem you wish to move to a smaller number, you should multiply by 5/9:

$$(5/9)(260°) = 144.4°C$$

(3) Step 3: Now subtract 40°C

$$144.4°C$$
$$-40.0°C$$
$$\overline{104.4°C}$$

Thus, 220°F = 104.4°C

Example 7.8

Problem: Convert 22°C to Fahrenheit

(1) Step 1: add 40°F

$$22°F$$
$$+40°F$$
$$\overline{62°F}$$

Because we are converting from Celsius to Fahrenheit, you are moving from a smaller to a larger number, and you should use 9/5 in the multiplication:

(2) Step 2:

$$(9/5)(62°) = 112°$$

(3) Step 3: Subtract 40°

$$112°$$
$$-40°$$
$$\overline{72°}$$

Thus, 22°C = 72°F

Obviously, knowing how to make these temperature conversion calculations is useful. However, in practical (real-world) operations, you may wish to use a temperature conversion table.

Milligrams per Liter (Parts per Million)

One of the most common terms for concentration is *milligrams per liter (mg/L)*. If a mass of 15 mg of oxygen is dissolved in a volume of 1 L of water, the concentration of that solution is expressed simply as 15 mg/L. Very dilute solutions are more conveniently expressed in terms of *micrograms per liter (μg/L)*. For example, a concentration of 0.005 mg/L is preferably written as its equivalent 5 μg/L. Since 1,000 μg = 1 mg, simply move the decimal point three places to the right when converting from mg/L to μg/L. Move the decimal three places to the left when converting from μg/L to mg/L. For example, a concentration of 1,250 μg/L is equivalent to 1.25 mg/L.

A liter of water has a mass of 1 kg. But 1 kg is equivalent to 1,000 g or 1,000,000 mg. Therefore, if we dissolve 1 mg of a substance in 1 liter of water, we can say that there is 1 mg of solute per 1 million mg of water—or in other words, *one part per million* (ppm).

Note: For comparative purposes, we like to say that 1 ppm is analogous to a full shot glass of water sitting at the bottom of a full standard swimming pool.

Neglecting the small change in the density of water as substances are dissolved in it, we can say that, in general, a concentration of 1 milligram per liter is equivalent to one part per million: *1 mg/L = 1 ppm*. Conversions are very simple; for example, a concentration of 18.5 mg/L is identical to 18.5 ppm. The expression *mg/L* is preferred over *ppm*, just as the expression μg/L is preferred over its equivalent of *ppb*. However, both types of units are still used, and the waterworks operator should be familiar with both of them.

Area and Volume

Water and wastewater operators are often required to calculate surface areas and volumes. *Area* is a calculation of the surface of an object. For example, the length and the width of a water tank can be measured, but the surface area of the water in the tank must be calculated. An area is found by multiplying two length measurements, so the result is a square measurement. For example, when multiplying feet by feet, we get square feet, which is abbreviated ft². *Volume* is the calculation of the space inside a three-dimensional object and is calculated by multiplying three length measurements, or an area by a length measurement. The result is a cubic measurement, such as cubic feet (abbreviated ft³).

Force, Pressure, and Head

Force, pressure, and head are important parameters in water and wastewater operations. Before we study calculations involving the relationship between force, pressure, and head, we must first define these terms:

- **Force:** the push exerted by water on any confining surface. Force can be expressed in pounds, tons, grams, or kilograms.
- **Pressure:** the force per unit area. The most common way of expressing pressure is in pounds per square inch (psi).
- **Head:** the vertical distance or height of water above a reference point. The head is usually expressed in feet. In the case of water, head and pressure are related.

Figure 7.1 illustrates these terms. A cubical container measuring 1 ft on each side can hold 1 ft³ of water. A basic fact of science states that 1 ft³ of water weighs 62.4 lb. The force acting on the bottom of the container would be 62.4 lb. The pressure acting on the bottom of the container would be 62.4 lb/ft². The area of the bottom in square inches is:

$$1 \text{ ft}^2 = 12 \text{ in} \times 12 \text{ in} = 144 \text{ in}^2 \qquad (7.4)$$

Therefore, the pressure in pounds per square inch (psi) is:

$$\frac{62.4\text{-lb/ft}^2}{1 \text{ ft}^2} = \frac{62.4 \text{ lb/ft}^2}{144 \text{ in}^2/\text{ft}^2} = 0.433 \text{ lb/in}^2 \left(\text{psi}\right) \qquad (7.5)$$

If we use the bottom of the container as our reference point, the head would be 1 ft. From this, we can see that 1 ft of head is equal to 0.433 psi. Figure 7.2 illustrates some other important relationships between pressure and head.

Note: In water/wastewater operations, 0.433 psi is an important parameter.

Note: When we say 'head' in water/wastewater science, we are talking about "pressure."

Note: Force acts in a particular direction. Water in a tank exerts force down on the bottom and out of the sides. Pressure, however, acts in all directions. A marble at a water depth of 1 ft would have 0.433 psi of pressure acting inward on all sides.

Key Point: 1 ft of head = 0.433 psi.

This is a valuable parameter that should be committed to memory. You should also know the relationship between pressure and feet of head; in other words, how many feet of head 1 psi represents. This is determined by dividing 1 by 0.433.

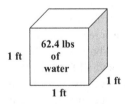

FIGURE 7.1   One cubic foot of water weighs 62.4 lb.

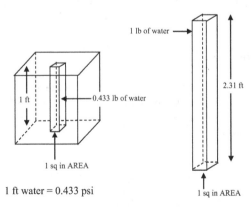

FIGURE 7.2   The relationship between pressure and head.

$$\text{Feet of head} = \frac{1 \text{ ft}}{0.433 \text{ psi}} = 2.31 \text{ ft/psi}$$

What we are saying here is that if a pressure gauge were reading 12 psi, the height of the water necessary to represent this pressure would be 12 psi × 2.31 ft/psi = 27.7 ft.

Again, the key points: 1 ft = 0.433 psi

1 psi = 2.31 ft

Having two conversion methods for the same thing is often confusing. Thus, memorizing one and staying with it is best. The most accurate conversion is: 1 ft = 0.433 psi; this is the standard conversion used throughout this handbook.

Example 7.9

Problem: Convert 50 psi to ft of head.

Solution:

$$50 \frac{\text{psi}}{1} \times \frac{\text{ft}}{0.433 \text{ psi}} = 115.5 \text{ ft}$$

Problem: Convert 50 ft to psi.

Solution:

$$50 \frac{\text{ft}}{1} \times \frac{0.433 \text{ psi}}{\text{ft}} = 21.7 \text{ psi}$$

As the above examples demonstrate, when attempting to convert psi to ft, we divide by 0.433; when attempting to convert ft to psi, we multiply by 0.433. The above process can be most helpful in clearing up the confusion about whether to multiply or divide. Another way, however, may be more beneficial and easier for many operators to use. Notice that the relationship between psi and feet is almost 2–1. It takes slightly more than 2 ft to make 1 psi. Therefore, when looking at a problem where the data is in pressure and the result should be in feet, the answer will be at least twice as large as the starting number. For example, if the pressure were 25 psi, we intuitively know that the head is over 50 ft. Therefore, we must divide it by 0.433 to obtain the correct answer.

Example 7.10

Problem: Convert a pressure of 55 psi to ft of head

Solution:

$$55 \frac{\text{psi}}{1} \times \frac{1 \text{ ft}}{0.433 \text{ psi}} = 127 \text{ ft}$$

Example 7.11

Problem: Convert 14 psi to ft

Solution:

$$14 \frac{\text{psi}}{1} \times \frac{1 \text{ ft}}{0.433 \text{ psi}} = 32.3 \text{ ft}$$

Example 7.12

Problem: Between the top of a reservoir and the watering point, the elevation is 115 ft. What will the static pressure be at the watering point?

Solution:

$$115 \frac{\text{ft}}{1} \times \frac{0.433 \text{ psi}}{1 \text{ ft}} = 49.8 \text{ psi}$$

Using the preceding information, we can develop Equations (7.6) and (7.7) for calculating pressure and head.

$$\text{Pressure (psi)} = 0.433 \times \text{Head (ft)} \qquad (7.16)$$

$$\text{Head (ft)} = 2.31 \times \text{Pressure (psi)} \qquad (7.7)$$

Example 7.13

Problem: Find the pressure (psi) in a 12-ft-deep tank at a point 15 ft below the water surface.

Solution:

$$\text{Pressure (psi)} = 0.433 \times 5 \text{ ft}$$

$$= 2.17 \text{ psi (rounded)}$$

Example 7.14

Problem: A pressure gauge at the bottom of a tank reads 12.2 psi. How deep is the water in the tank?

Solution:

$$\text{Head (ft)} = 2.31 \times 12.2 \text{ psi}$$

$$= 28.2 \text{ ft (rounded)}$$

Example 7.15

Problem: What is the pressure (static pressure) 4 miles beneath the ocean surface?

Solution:

Change miles to ft, then to psi.

$$5,380 \text{ ft/mile} \times 4 = 21,120 \text{ ft}$$

$$\frac{21,120 \text{ ft}}{2.31 \text{ ft/psi}} = 9,143 \text{ psi (rounded)}$$

Example 7.16

Problem: A 150 ft diameter cylindrical tank contains 2.0 MG of water. What is the water depth? At what pressure would a gauge at the bottom read in psi?

Solution:

- Step 1: Change MG to ft³.

$$\frac{2{,}000{,}000 \text{ gal}}{7.48} = 267{,}380 \text{ ft}^3$$

- Step 2: Using volume, solve for depth.

$$\text{Volume} = 0.785 \times \text{D}^2 \times \text{depth}$$

$$267{,}380 \text{ ft}^3 = 0.785 \times (150)^2 \times \text{depth}$$

$$\text{Depth} = 15.1 \text{ ft}$$

Example 7.17

Problem: The pressure in a pipe is 70 psi. What is the pressure in feet of water? What is the pressure in psf?

Solution:

- Step 1: Convert pressure to feet of water.

$$70 \text{ psi} \times 2.31 \text{ ft/psi} = 161.7 \text{ ft of water}$$

- Step 2: Convert psi to psf.

$$70 \text{ psi} \times 144 \text{ in}^2/\text{ft}^2 = 10{,}080 \text{ psf}$$

Example 7.18

Problem: The pressure in a pipeline is 6,476 psf. What is the head on the pipe?

Solution:

$$\text{Head on pipe} = \text{feet of pressure}$$

$$\text{Pressure} = \text{Weight} \times \text{Height}$$

$$6{,}476 \text{ psf} = 62.4 \text{ lb/ft}^3 \times \text{Height}$$

$$\text{Height} = 104 \text{ ft (rounded)}$$

FLOW

Flow is expressed in many different terms (English system of measurements). The most common flow terms are:

- Gallons per minute (gpm)
- Cubic feet per second (cfs)
- Gallons per day (gpd)
- Million gallons per day (MGD)

In converting flow rates, the most common flow conversions are: 1 cfs = 448 gpm and 1 gpm to 1,440 gpd. To convert gallons per day to MGD, divide the gpd by 1,000,000. For instance, convert 150,000 gal to MGD:

$$\frac{150{,}000 \text{ gpd}}{1{,}000{,}000} = 0.150 \text{ MGD}$$

In some instances, flow is given in MGD but needed in gpm. To make the conversion (MGD to gpm) requires two steps.

Step 1: convert the gpd by multiplying by 1,000,000
Step 2: convert to gpm by dividing by the number of minutes in a day (1,440 min/day)

Example 7.19

Problem: Convert 0.135 MGD to gpm.

Solution:

Convert the flow in MGD to gpd:

$$0.135 \text{ MGD} \times 1{,}000{,}000 = 135{,}000 \text{ gpd}$$

Convert to gpm by dividing by the number of minutes in a day (24 h/day × 60 min/h) = 1,440 min/day.

$$\frac{135{,}000 \text{ gpd}}{1{,}440 \text{ min/day}} = 93.8 \text{ or } 94 \text{ gpm}$$

To determine flow through a pipeline, channel, or stream, we use the following equation:

$$Q = VA \qquad (7.8)$$

where:
    Q = cubic feet per second (cfs)
    V = velocity in feet per second (ft/sec)
    A = area in square feet (ft²)

Example 7.20

Problem: Find the flow in cfs in an 8-in line if the velocity is 3 ft/sec.

Solution:

Step 1: Determine the cross-sectional area of the line in square feet. Start by converting the diameter of the pipe to inches.
Step 2: The diameter is 8 in; therefore, the radius is 4 in. 4 in is 4/12 of a foot or 0.33 ft.
Step 3: Find the area in square feet.

$$A = \pi r^2 \qquad (7.9)$$

$$A = \pi (0.33 \text{ ft})^2$$

$$A = \pi \times 0.109 \text{ ft}^2$$

$$A = 0.342 \text{ ft}^2$$

Step 4: Q = VA

$$Q = 3 \text{ ft/sec} \times 0.342 \text{ ft}^2$$

$$Q = 1.03 \text{ cfs}$$

Example 7.21

Problem: Find the flow in gpm when the total flow for the day is 75,000 gpd.

Solution:

$$\frac{75{,}000 \text{ gpd}}{1{,}440 \text{ min/day}} = 52 \text{ gpm}$$

Example 7.22

Problem: Find the flow in gpm when the flow is 0.45 cfs.

Solution:

$$0.45\frac{\text{cfs}}{1} \times \frac{448 \text{ gpd}}{1 \text{ cfs}} = 52 \text{ gpm}$$

FLOW CALCULATIONS

In water and wastewater treatment, one of the major concerns of the operator is not only to maintain flow but also to measure it. Normally, flow measurements are determined by metering devices. These devices measure water flow at a particular moment (instantaneous flow) or over a specified time (total flow). Instantaneous flow can also be determined mathematically. In this section, we discuss how to mathematically determine instantaneous and average flow rates and how to make flow conversions.

Instantaneous Flow Rates

In determining instantaneous flow rates through channels, tanks, and pipelines, we can use Q=AV (Equation 7.8).

Key Point: It is important to remember that when using an equation such as Q=AV, the units on the left side of the equation must match th units on the right side of the equation (A and V) with respect to volume (cubic feet or gallons) and time (seconds, minutes, hours, or days).

Example 7.23

Problem: A channel 4 ft wide has water flowing to a depth of 2 ft. If the velocity through the channel is 2 feet per second (fps), what is the cubic feet per second (cfs) flow rate through the channel?

Solution:

$$Q, \text{cfs} = (A)(V, \text{fps})$$

$$= (4 \text{ ft})(2 \text{ ft})(2 \text{ fps})$$

$$= 16 \text{ cfs}$$

Instantaneous Flow Into and Out of a Rectangular Tank

One of the primary flow measurements water/wastewater operators are commonly required to calculate is flow through a tank. This measurement can be determined using the Q=AV equation. For example, if the discharge valve to a tank were closed, the water level would begin to rise. If you time how fast the water rises, this would give you an indication of the velocity of flow into the tank. This information can be "plugged" into Q=VA (Equation 7.8) to determine the flow rate through the tank. Let us look at an example.

Example 7.24

Problem: A tank is 8 ft wide and 12 ft long. With the discharge valve closed, the influent to the tank causes the water level to rise 1.5 ft in 1 min. What is the gpm flow into the tank?

Solution:

Calculate the cfm flow rate:

$$Q, \text{cfm} = (A)(V, \text{fpm})$$

$$= (8 \text{ ft})(12 \text{ ft})(1.5 \text{ fpm})$$

$$= 144 \text{ cfm}$$

Then, convert cfm flow rate to gpm flow rate:

$$(144 \text{ cfm})(7.48 \text{ gal/ft}^3) = 1{,}077 \text{ gpm}$$

How do we compute the flow rate from a tank when the influent valve is closed and the discharge pump remains on, lowering the wastewater level in the tank? First, we time the rate of this drop in the wastewater level to calculate the velocity of flow from the tank. Then, we use the Q=AV equation to determine the flow rate out of the tank, as illustrated in Example 7.25.

Example 7.25

Problem: A tank is 9 ft wide and 11 ft long. The influent valve to the tank is closed and the water level drops 2.5 ft in 2 min. What is the gpm flow from the tank?

Solution:

Drop rate = 2.5 ft/2 min = 1.25 ft/min
First, calculate the cfm flow rate:

$$Q, \text{cfm} = (A)(V, \text{fpm})$$

$$= (9 \text{ ft})(11 \text{ ft})(1.25 \text{ fpm})$$

$$= 124 \text{ cfm}$$

Convert cfm flow rate to gpm flow rate:

$$(124 \text{ cfm})(7.48 \text{ gal/ft}^3) = 928 \text{ gpm}$$

Flow Rate into a Cylindrical Tank

We can use the same basic method to determine the flow rate when the tank is cylindrical in shape, as shown in Example 7.26.

Example 7.26

Problem: The discharge valve to a 25-ft diameter cylindrical tank is closed. If the water rises at a rate of 12 in. in 4 min, what is the gpm flow into the tank?

Solution:

$$Rise = 12 \text{ in} = 1 \text{ ft}$$
$$= 1 \text{ ft/4 min}$$
$$= 0.25 \text{ ft/min}$$

First, calculate the cfm flow into the tank:

$$Q, \text{cfm} = (A)(V, \text{fpm})$$
$$= (0.785)(25 \text{ ft})(25 \text{ ft})(0.25 \text{ ft/min})$$
$$= 123 \text{ cfm}$$

Then, convert cfm flow rate to gpm flow rate:

$$(123 \text{ cfm})(7.48 \text{ gal/ft}^3) = 920 \text{ gpm}$$

Flow through a Full Pipeline

Flow through pipelines is of considerable interest to water distribution operators and wastewater collection workers. The flow rate can be calculated using the Q=AV (Equation 7.8). The cross-sectional area of a round pipe is a circle, so the area (A) is represented by $(0.785) \times (\text{Diameter})^2$.

Note: To avoid errors in terms, it is prudent to express pipe diameters in feet.

Example 7.27

Problem: The flow through an 8-in diameter pipeline is moving at a velocity of 4 ft/sec. What is the cfs flow rate through the full pipeline?
Convert 8 in to feet

$$8 \text{ in/12 in} = 0.67 \text{ ft}$$

Calculate the cfs flow rate:

$$Q, \text{cfs} = (A)(V, \text{fps})$$
$$= (0.785)(0.67 \text{ ft})(0.67 \text{ ft})(4 \text{ fps})$$
$$= 1.4 \text{ cfs}$$

Velocity Calculations

To determine the velocity of flow in a channel or pipeline, we use the Q=AV equation. However, to use the equation correctly we must transpose it. We simply input the information given into the equation and then transpose for the unknown (V in this case), as illustrated in Example 7.28 for channels and Example 7.29 for pipelines.

Example 7.28

Problem: A channel has a rectangular cross section. The channel is 5 ft wide with wastewater flowing to a depth of 2 ft. If the flow rate through the channel is 8,500 gpm, what is the velocity of the wastewater in the channel (ft/sec)?

Solution:

Convert gpm to cfs: $\dfrac{8,500 \text{ gpm}}{(7.48 \text{ gal})(60 \text{ sec})} = 18.9 \text{ cfs}$

Calculate the velocity:

$$Q, \text{cfs} = (A)(V, \text{fps})$$

$$\text{Velocity (the unknown, fps)} = \frac{18.9}{(5)(2)} = 1.89 \text{ fps}$$

Example 7.29

Problem: A full 8-in. diameter pipe delivers 250 gpm. What is the velocity of flow in the pipeline (ft/sec)?

Solution:

Convert: 8 in/12 into feet=0.67 ft
Convert: gpm to cfs flow:

$$\frac{250 \text{ gpm}}{(7.18 \text{ gal/ft}^3)(60 \text{ sec/min})} = 0.56 \text{ cfs}$$

Calculate the velocity:

$$Q = AV$$

$$0.56 \text{ cfs} = (0.785)(0.67 \text{ ft})(0.67 \text{ ft})(\text{unknown, Vel, fps})$$

$$\text{Velocity} = \frac{0.56 \text{ cfs}}{(0.785)(0.67)(0.67)} = 1.6 \text{ fps}$$

Average Flow Rate Calculations

Flow rates in water/wastewater systems vary considerably during the course of a day, week, month, or year. Therefore, when computing flow rates for trend analysis or other purposes, an average flow rate is used to determine the typical flow rate.

Example 7.30

Problem: The following flows were recorded for the week:

Monday	8.2 MGD
Tuesday	8.0 MGD
Wednesday	7.3 MGD
Thursday	7.6 MGD
Friday	8.2 MGD
Saturday	8.9 MGD
Sunday	7.7 MGD

What was the average daily flow rate for the week?

$$\text{Average Daily Flow} = \frac{\text{Total of all Sample Flows}}{\text{Number of Days}} \quad (7.10)$$

$$= \frac{55.9 \text{ MGD}}{7 \text{ Days}} = 8.0 \text{ MGD}$$

Flow Conversion Calculations

One of the tasks involving calculations that the wastewater operator is typically called on to perform is converting one expression of flow to another. The ability to do this is also a necessity for those preparing for licensure examinations. Probably the easiest way to accomplish flow conversions is to employ the box method illustrated in Figure 7.3. When using the box method, it is important to remember that moving from a smaller box to a larger box requires multiplication by the indicated factor. Moving from a larger box to a smaller box requires division by the factor indicated. From Figure 7.3 it should be obvious that memorizing the nine boxes and the units in each box is not that difficult. The values of 60, 1,440, 7.48, and 8.34 are not that difficult to remember either; it is a matter of remembering the exact placement of the units and the values. Once this is accomplished, you will have obtained a powerful tool that will enable you to make flow conversions in a relatively easy manner.

DETENTION TIME

Detention time is the length of time water is retained in a vessel or basin, or the period from the time the water enters a settling basin until it flows out the other end. To calculate the detention period of a basin, the volume of the basin must first be obtained. Using a basin that is 70 ft long, 25 ft wide, and 12 ft deep, the volume would be:

$$V = L \times W \times D$$

$$V = 70 \text{ ft} \times 25 \text{ ft} \times 12 \text{ ft}$$

$$V = 21,000 \text{ ft}^3$$

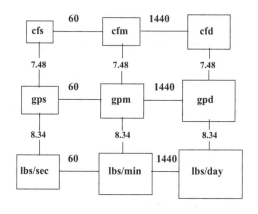

FLOW CONVERSIONS USING THE BOX METHOD*

cfs = cubic feet per second
cfm = cubic feet per minute
cfd = cubic feet per day

gps = gallons per second
gpm = gallons per minute
gpd = gallons per day

*The factors shown in the diagram have the following units associated with them: 60 sec/min, 1440 min/day, 7.48 gal/cu ft, and 8.34 lbs/gal.

FIGURE 7.3 Flow Conversions using the box method. (Adapted from Price, 1991, p. 32.)

$$\text{Gallons} = V \times 7.48 \text{ gal/ft}^3$$

$$\text{Gallons} = 21,000 \times 7.48 = 157,080 \text{ gal}$$

If we assume that the plant filters 300 gpm, $157,080 \div 300 = 524$ min (rounded), or roughly 9 h of detention time. Stated another way, the detention time is the length of time theoretically required for the coagulated water to flow through the basin. If chlorine were added to the water as it entered the basin, the chlorine contact time would be 9 h. To determine the CT used to assess the effectiveness of chlorine, we must calculate detention time.

Key point: True detention time is the "T" portion of the CT value.

Note: Detention time is also important when evaluating the sedimentation and flocculation basins of a water treatment plant.

Detention time is expressed in units of time (obviously). The most common units are seconds, minutes, hours, and days. The simplest way to calculate detention time is to divide the volume of the container by the flow rate into the container. The theoretical detention time of a container is the same as the amount of time it would take to fill the container if it were empty. For volume, the most common units used are gallons. However, on occasion, cubic feet may also be used. Time units will correspond to whatever units are used to express the flow. For example, if the flow is in gpm, the detention time will be in days. If the result gives the detention time in the wrong time units, simply convert to the appropriate units.

Example 7.31

Problem: The reservoir for the community is 110,000 gal. The well will produce 60 gpm. What is the detention time in the reservoir in hours?

Solution:

$$DT = \frac{110,000\text{-gal}}{60 \text{ gal/min}} = 1,834 \text{ min or } \frac{1,834 \text{ min}}{60 \text{ min/h}} = 30.6 \text{ h}$$

Example 7.32

Problem: Find the detention time in a 55,000-gal reservoir if the flow rate is 75 gpm.

Solution:

$$DT = \frac{55,000\text{-gal}}{75 \text{ gal/min}} = 734 \text{ min or } \frac{634 \text{ min}}{60 \text{ min/h}} = 12 \text{ h}$$

Example 7.33

Problem: If the fuel consumption to the boiler is 30 gal/day, how many days will the 1,000-gal tank last?

Solution:

$$\text{Days} = \frac{1,000 \text{ gal}}{30 \text{ gal/day}} = 33.3 \text{ days}$$

Hydraulic Detention Time

The term detention time or hydraulic detention time (HDT) refers to the average length of time (theoretical time) that a drop of water, wastewater, or suspended particles remains in a tank or channel. It is calculated by dividing the volume of water/wastewater in the tank by the flow rate through the tank. The units of flow rate used in the calculation depend on whether the detention time is to be calculated in seconds, minutes, hours, or days. Detention time is used in conjunction with various treatment processes, including sedimentation and coagulation-flocculation. Generally, in practice, detention time is associated with the amount of time required for a tank to empty. The range of detention time varies with the process. For example, in a tank used for sedimentation, detention time is commonly measured in minutes. The calculation methods used to determine detention time are illustrated in the following sections.

Hydraulic Detention Time in Days

Note: The general hydraulic detention time calculation is:

$$\text{Hydraulic Detention Time} = \frac{\text{Tank Volume}}{\text{Flow Rate}} \quad (7.11)$$

This general formula is then modified based upon the information provided or available and the "normal" range of detention times for the unit being evaluated.

$$\text{HDT, Days} = \frac{\text{Tank Volume, ft}^3 \times 7.48 \text{ gal/ft}^3}{\text{Flow, gal/day}} \quad (7.12)$$

Example 7.34

Problem: An anaerobic digester has a volume of 2,200,000 gal. What is the detention time in days when the influent flow rate is 0.06 MGD?

Solution:

$$\text{D.T., Days} = \frac{2,200,000 \text{ gal}}{0.06 \text{ MGD} \times 1,000,000 \text{ gal/MG}} = 37 \text{ days}$$

HYDRAULIC DETENTION TIME IN HOURS

$$\text{HDT (h)} = \frac{\text{Tank Volume, ft}^3 \times 7.48 \text{ gal/ft}^3 \times 24 \text{ h/day}}{\text{Flow, gal/day}}$$

$$(7.13)$$

Example 7.35

Problem: A settling tank has a volume of 40,000 ft³. What is the detention time in hours when the flow is 4.35 MGD?

Solution:

$$\text{HDT (h)} = \frac{40,000 \text{ ft}^3 \times 7.48 \text{ gal/ft}^3 \times 24 \text{ h/day}}{4.35 \text{ MGD} \times 1,000,000 \text{ gal/MG}}$$

$$\text{HDT (h)} = 1.7 \text{ h}$$

Hydraulic Detention Time in Minutes

$$\text{HDT, (min)} = \frac{\text{Tank Volume, ft}^3 \times 7.48 \text{ gal/ft}^3 \times 1,440 \text{ min/day}}{\text{Flow, gal/day}}$$

$$(7.14)$$

Example 7.36

Problem: A grit channel has a volume of 1,240 ft³. What is the detention time in minutes when the flow rate is 4.1 MGD?

Solution:

$$\text{D.T., (min)} = \frac{1,240 \text{ ft}^3 \times 7.48 \text{ gal/ft}^3 \times 1,440 \text{ min/day}}{4,100,000 \text{ gal/day}} = 3.26 \text{ min}$$

Note: The tank volume and the flow rate must be in the same dimensions before calculating the hydraulic detention time.

CHEMICAL DOSAGE CALCULATIONS

Chemicals are used extensively in wastewater treatment plant operations. Wastewater treatment plant operators add chemicals to various unit processes for slime-growth control, corrosion control, odor control, grease removal, BOD reduction, pH control, sludge-bulking control, ammonia oxidation, bacterial reduction, and other reasons. In order to apply any chemical dose correctly, it is important to be able to make certain dosage calculations. One of the most frequently used calculations in wastewater mathematics is the conversion of milligrams per liter (mg/L) concentration to pounds per day (lb/day) or pounds (lb) dosage or loading. The general types of mg/L to lb/day or lb calculations are for chemical dosage, BOD, COD, or SS loading/removal, pounds of solids under aeration, and WAS pumping rate. These calculations are usually made using either of the following equations:

$$(\text{mg/L})(\text{MGD flow})(8.34 \text{ lbs/gal}) = \text{lbs/day} \quad (7.15)$$

$$(\text{mg/L})(\text{MG volume})(8.34 \text{ lbs/gal}) = \text{lbs} \quad (7.16)$$

Note: If mg/L concentration represents a concentration in a flow, then million gallons per day (MGD) flow is used as the second factor. However, if the concentration pertains to a tank or pipeline volume, then million gallons (MG) volume is used as the second factor.

Dosage Formula Pie Chart

In converting pounds (lb) or mg/L, million gallons (MG) and 8.34 are key parameters. The pie chart shown in Figure 7.4 and the steps listed below can help find lb or mg/L.

Step 1: Determine what unit the question is asking you to find (lb or mg/L).

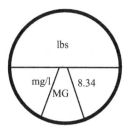

FIGURE 7.4 Dosage formula pie chart.

Step 2: Physically cover or hide the area of the chart containing the desired unit. Write the desired unit down alone on one side of the "equals sign" to begin the necessary equation (e.g., lb =).

Step 3: Look at the remaining uncovered areas of the circle. These *exactly* represent the other side of the "equals sign" in the necessary equation. If the unit above the center line is not covered, your equation will have a top (numerator) and a bottom (denominator), just like the pie chart. Everything above the center line goes in the numerator, or the top of the equation, and everything below the center line goes in the denominator, or on the bottom of the equation. Remember that all units below the line are *always multiplied together*. For example, if you are asked to find the dosage in mg/L, you would cover mg/L in the pie chart and write it down on one side of the "equals sign" to start your equation, like this:

$$mg/L =$$

The remaining portions of the pie chart are lb on top, divided by MGD×8.34 on the bottom, and would be written like this:

$$mg/L = \frac{lbs}{MGD \times 8.34}$$

If the area above the center line is covered, the right side of your equation will be made up of only the units below the center line. Remember that all units below the line are always multiplied together.

If you are asked to find the number of pounds needed, you would cover lb in the pie chart and write it down on one side of the "equals sign" to start your equation, like this

$$lb =$$

All of the remaining areas of the pie chart are together on one line (below the center line of the circle), multiplied together on the other side of the "equals sign," and written like this:

$$lbs = mg/L \times MGD \times 8.34$$

Chlorine Dosage

Chlorine is a powerful oxidizer commonly used in water treatment for purification and in wastewater treatment for disinfection, odor control, bulking control, and other applications. When chlorine is added to a unit process, we want to ensure that a measured amount is added. The amount of chemical added or required is expressed in two ways:

- milligrams per liter (mg/L)
- pounds per day (lb/day)

To convert from mg/L (or ppm) concentration to lb/day, we use Equation (7.17).

$$(mg/L) \times (MGD) \times (8.34) = 1b/day \qquad (7.17)$$

Note: In previous years, it was normal practice to use the expression parts per million (ppm) as an expression of concentration, since 1 mg/L = 1 ppm. However, current practice is to use mg/L as the preferred expression of concentration.

Example 7.37

Problem: Determine the chlorinator setting (lb/day) needed to treat a flow of 8 MGD with a chlorine dose of 6 mg/L.

Solution:

$$(mg/L)(MGD)(8.34) = lbs/day$$

$$(6 \text{ mg/L})(8 \text{ MGD})(8.34 \text{ lbs/gal}) = lbs/day$$

$$= 400 \text{ lbs/day}$$

Example 7.38

Problem: What should the chlorinator setting be (lb/day) to treat a flow of 3 MGD if the chlorine demand is 12 mg/L and a chlorine residual of 2 mg/L is desired?

Note: The chlorine demand is the amount of chlorine used in reacting with various components of the wastewater such as harmful organisms and other organic and inorganic substances. When the chlorine demand has been satisfied, these reactions stop.

$$(mg/L) \times (MGD) \times (8.34) = lbs/day$$

To find the unknown value (lb/day), we must first determine the chlorine dose. To do this, we will use Equation (7.18).

$$\text{Chlorine Dose, mg/L} = \text{Chlorine Demand, mg/L}$$

$$+ \text{Chlorine Residual, mg/L}$$

$$= 12 \text{ mg/L} + 2 \text{ mg/L}$$

$$= 14 \text{ mg/L} \qquad (7.18)$$

Then, we can make the mg/L to lb/day calculation:

$$(12 \text{ mg/L}) \times (3 \text{ MGD}) \times (8.34 \text{ lbs/gal}) = 300 \text{ lbs/day}$$

Hypochlorite Dosage

At many wastewater facilities, sodium hypochlorite or calcium hypochlorite is used instead of chlorine. The reasons for substituting hypochlorite for chlorine vary. However, with the passage of stricter hazardous chemical regulations under OSHA and the USEPA, many facilities are deciding to substitute the hazardous chemical chlorine with non-hazardous hypochlorite. Obviously, the potential liability involved in using deadly chlorine is also a factor in the decision to substitute it with a less toxic chemical substance.

For whatever reason the wastewater treatment plant decides to substitute chlorine for hypochlorite, there are differences between the two chemicals that the wastewater operator needs to be aware of. Chlorine is a hazardous material. Chlorine gas is used in wastewater treatment applications at 100% available chlorine. This is an important consideration to keep in mind when making or setting chlorine feed rates. For example, if the chlorine demand and residual require 100-lb/day of chlorine, the chlorinator setting would be just that100-lb/24 h. Hypochlorite is less hazardous than chlorine; it is similar to strong bleach and comes in two forms: dry calcium hypochlorite (often referred to as HTH) and liquid sodium hypochlorite. Calcium hypochlorite contains about 65% available chlorine, while sodium hypochlorite contains about 12%–15% available chlorine (in industrial strengths).

Note: Because neither type of hypochlorite is 100% pure chlorine, more lb/day must be fed into the system to obtain the same amount of chlorine for disinfection. This is an important economic consideration for those facilities thinking about substituting hypochlorite for chlorine. Some studies indicate that such a switch can increase operating costs overall by up to three times the cost of using chlorine.

To calculate the lb/day of hypochlorite required, a two-step calculation is required:

$$\text{Step (1)} \quad (\text{mg/L})(\text{MGD})(8.34) = \text{lb/day} \qquad (7.19)$$

$$\text{Step (2)} \quad \frac{\text{Chlorine, lb/day}}{\% \text{ Available}} \times 100 = \text{Hypochlorite, lb/day}$$

$$(7.20)$$

Example 7.39

Problem: A total chlorine dosage of 10 mg/L is required to treat a particular wastewater. If the flow is 1.4 MGD and the hypochlorite has 65% available chlorine, how many lb/day of hypochlorite will be required?

Solution:

Step 1: Calculate the lb/day of chlorine required using the mg/L to lb/day equation:

$$(\text{mg/L}) \times (\text{MGD}) \times (8.34) = \text{lb/day}$$

$$(10 \text{ mg/L}) \times (1.4 \text{ MGD}) \times (8.34 \text{ lbs/gal}) = 117 \text{ lb/day}$$

Step 2: Calculate the lb/day of hypochlorite required. Since only 65% of the hypochlorite is chlorine, more than 117 lb/day will be required:

$$\frac{117 \text{ lb/day Chlorine}}{65 \text{ Avail. Chlorine}} \times = 100 = 180 \text{ lb/day Hypochlorite}$$

Example 7.40

Problem: A wastewater flow of 840,000 gpd requires a chlorine dose of 20 mg/L. If sodium hypochlorite (15% available chlorine) is to be used, how many lb/day of sodium hypochlorite are required? How many gal/day of sodium hypochlorite is this?

Solution:

Step (1) Calculate the lb/day of chlorine required:

$$(\text{mg/L}) \times (\text{MGD}) \times (8.34) = \text{lb/day}$$

$$(20 \text{ mg/L}) \times (0.84 \text{ MGD}) \times (8.34 \text{ lb/gal}) = 140 \text{ lb/day Chlorine}$$

Step (2) Calculate the lb/day of sodium hypochlorite:

$$\frac{140 \text{ lb/day Chlorine}}{15 \text{ Avail. Chlorine}} \times = 100 = 933 \text{ lb/day Hypochlorite}$$

Step (3) Calculate the gal/day of sodium hypochlorite:

$$\frac{933 \text{ lb/day}}{8.34 \text{ lb/gal}} = 112 \text{ gal/day Sodium Hypochlorite}$$

Example 7.41

Problem: How many pounds of chlorine gas are necessary to 5,000,000 gal of wastewater at a dosage of 2 mg/L?

Solution:

Step (1) Calculate the pounds of chlorine required.

$$(V \times 10^6 \text{ gal}) = \text{Chlorine conc. (mg/L)} \times 8.34 = \text{lb chlorine}$$

Step (2) Substitute

$$(5 \times 10^6 \text{ gal}) \times 2 \text{ mg/L} \times 8.34 = 83 \text{ lb chlorine}$$

PERCENT REMOVAL

Percent removal is used throughout the wastewater treatment process to express or evaluate the performance of the plant and individual treatment unit processes. The results can be used to determine if the plant is performing as expected or in troubleshooting unit operations by comparing the results with those listed in the plant's *O & M manual* (Operations and Maintenance Manual). It can be used with either concentration or quantities.

For concentrations use:

$$\% \text{ Removal} = \frac{\left[\text{Influent Conc.} - \text{Effluent Conc.}\right] \times 100}{\text{Influent Conc.}}$$

(7.21)

For quantities use:

$$\% \text{ Removal} = \frac{\left[\text{Inf. Quantity} - \text{Eff. Quantity}\right] \times 100}{\text{Influent Quantity}}$$

(7.22)

Note: The calculation used for determining the performance (percent removal) for a digester is different from that used for performance (percent removal) for other processes, such as some process residuals or biosolids treatment processes. Ensure that the correct formula is selected.

Example 7.42

Problem: The plant influent contains 259 mg/L BOD_5, and the plant effluent contains 17 mg/L BOD_5. What is the % BOD_5 removal?

Solution:

$$\% \text{ Removal} = \frac{\left[(259 \text{ mg/L} - 17 \text{ mg/L}) \times 100\right]}{259 \text{ mg/L}} = 93.4\%$$

POPULATION EQUIVALENT (PE) OR UNIT LOADING FACTOR

When it is impossible to conduct a wastewater characterization study and other data are unavailable, population equivalent or unit per capita loading factors are used to estimate the total waste loadings to be treated. If the BOD contribution of a discharger is known, the loading placed upon the wastewater treatment system in terms of the equivalent number of people can be determined. The BOD contribution of a person is normally assumed to be 0.17 lb BOD/day.

$$\text{P.E., people} = \frac{BOD_5 \text{ Contribution, lb/day}}{0.17 \text{ lb } BOD_5/\text{day/person}} \quad (7.23)$$

Example 7.43

Problem: A new industry wishes to connect to the city's collection system. The industrial discharge will contain an average BOD concentration of 349 mg/L, and the average daily flow will be 50,000 gpd. What is the population equivalent of the industrial discharge?

Solution:

First, convert the flow rate to million gallons per day:

$$\text{Flow} = \frac{50,000 \text{ gpd}}{1,000,000 \text{ gal/MG}} = 0.050 \text{ MGD}$$

Next, calculate the population equivalent:

$$\text{P. E., people} = \frac{349 \text{ mg/L} \times 0.050 \text{ MGD} \times 8.34 \text{ lb/mg/L/MG}}{0.17 \text{ lb BOD/person/day}}$$

$$= 856 \text{ people/day}$$

SPECIFIC GRAVITY

Specific gravity is the ratio of the density of a substance to that of a standard material under standard conditions of temperature and pressure. The standard material for gases is air, and for liquids and solids, it is water. Specific gravity can be used to calculate the weight of a gallon of liquid chemical:

$$\text{Chemical, lb/gal} = \text{Water, lb/gal} \times \text{specific gravity} \quad (7.24)$$

Example 7.44

Problem: The label of the chemical states that the contents of the bottle have a specific gravity of 1.4515. What is the weight of 1 gal of solution?

Solution:

$$\text{Weight (lb/gal)} = 1.4515 \times 8.34 \text{ lb/gal}$$

$$= 12.1 \text{ lb/gal}$$

PERCENT VOLATILE MATTER REDUCTION IN SLUDGE (BIOSOLIDS)

The calculation used to determine *percent volatile matter reduction* is complicated because of the changes occurring during sludge digestion.

$$\% \text{ V.M. Reduction} = \frac{\left(\% \text{ VM}_{\text{in}} - \% \text{ VM}_{\text{out}}\right) \times 100}{\left[\% \text{ VM}_{\text{in}} - \left(\% \text{ V.M}_{\text{in}} \times \% \text{ V.M}_{\text{out}}\right)\right]}$$

(7.25)

where VM = Volatile Matter

Example 7.45

Problem: Determine the % volatile matter reduction for the digester using the digester data provided below.

Raw Sludge Volatile Matter 72%
Digested Sludge Volatile Matter 51%

$$\% \text{ Volatile Matter Reduction} = \frac{(0.72 - 0.51) \times 100}{[0.72 - (0.72 \times 0.51)]} = 59\%$$

CHEMICAL COAGULATION AND SEDIMENTATION

Chemical *coagulation* consists of treating the water with certain chemicals to bring non-settleable particles together into larger, heavier masses of solid material (called *floc*), which are then relatively easy to remove.

Calculating Feed Rate

The following equation is used to calculate the feed rate of chemicals used in coagulation:

Chemical Feed Rate (lb/day) = Dose (mg/L)

$$\times \text{Flow (MGD)} \times 8.34 \quad (7.26)$$

Example 7.46

Problem: A water treatment plant operates at a rate of 5 MGD. The dosage of alum is 40 ppm (or mg/L). How many pounds of alum are used a day?

Solution:

Chemical Feed Rate = Dose (mg/L) × Flow (MGD) × (8.34)
= (40 mg/L) × (5 MGD) × (8.34)
= 1,668 lb/day of alum

Calculating Solution Strength

Use the following procedure to calculate solution strength.

Example 7.47

Problem: Eight pounds of alum are added to 115 lb of water. What is the solution strength?

Solution:

$$70\% = \frac{8}{(8 + 115)} \times 100 = 6.5\% \text{ Solution}$$

We use this same concept in determining other solution strengths.

Example 7.48

Problem: Twenty-five pounds of alum are added to 90 lb of water. What is the solution strength?

Solution:

$$\frac{25}{(25 + 90)} \times 100 = 22\% \text{ Solution}$$

In the previous examples, we added pounds of chemicals to pounds of water. Recall that 1 gal of water = 8.34 lb. By multiplying the number of gallons by the 8.34 factor, we can find pounds.

Example 7.49

Problem: Forty pounds of soda ash is added to 65 gal of water. What is the solution strength?

Solution:

Units must be consistent, so convert gallons water to pounds water.

$$65\text{-gal} \times 8.34 \text{ lb/gal} = 542.7 \text{ lb water}$$

$$\frac{40 \text{ lb}}{542.7 \text{ lb} + 40 \text{ lb}} \times 100 = 6.9\% \text{ Solution}$$

FILTRATION

In waterworks operation (and to an increasing degree in wastewater treatment), the rate of flow through filters is an important operational parameter. While the flow rate can be controlled by various means or may proceed at a variable declining rate, the important point is that with flow suspended matter continuously builds up within the filter bed, affecting the rate of filtration.

Calculating the Rate of Filtration

Example 7.50

Problem: A filter box is 20 ft × 30 ft (including the sand area). If the influent valve is shut, the water drops 3.0 in/min. What is the rate of filtration in MGD?

Solution:

Given:

Filter Box = 20 ft × 30 ft

Water drops = 3.0 in/min

Find the volume of water passing through the filter:

Volume = Area × Height

Area = Width × Length

Note: The best way to perform calculations of this type is systematic, breaking down the problem into what is given and what is to be found.

- Step 1: Determine area:

$$\text{Area} = 20 \text{ ft} \times 30 \text{ ft} = 600 \text{ ft}^2$$

Convert 3.0 in into feet:

$$(3 \text{ in}) \div (12 \text{ in/ft}) = 0.25 \text{ ft}$$

Determine the volume of water passing through the filter in 1 min.

$$\text{Volume} = 600 \text{ ft}^2 \times 0.25 \text{ ft} = 150 \text{ ft}^3$$

- Step 2: Convert cubic feet to gallons.

$$150 \text{ ft}^3 \times 0.48 \text{ gal/ft}^3 = 1,122 \text{ gpm}$$

- Step 3: The problem asks for the rate of filtration in MGD. To find MGD, multiply the number of gallons per minute by the number of minutes per day.

$$1,122 \text{ gal/min} \times 1,440 \text{ min/day} = 1.62 \text{ MGD}$$

Filter Backwash

In filter backwashing, one of the most important operational parameters to be determined is the amount of water in gallons required for each backwash. This amount depends on the design of the filter and the quality of the water being filtered. The actual washing typically lasts 5–10 min and uses amounts to 1%–5% of the flow produced.

Example 7.51

Problem: A filter has the following dimensions:

$$\text{Length} = 30 \text{ ft}$$

$$\text{Width} = 20 \text{ ft}$$

$$\text{Depth of filter media} = 24 \text{ in}$$

Assuming a backwash rate of 15 gal/ft²/min is recommended, and 10 min of backwash is required, calculate the amount of water in gallons required for each backwash.

Solution:

Given:

$$\text{Length} = 30 \text{ ft}$$

$$\text{Width} = 20 \text{ ft}$$

$$\text{Depth of filter media} = 24 \text{ in}$$

$$\text{Rate} = 15 \text{ gal/ft}^2/\text{min}$$

$$\text{Backwash time} = 10 \text{ min}$$

Find the amount of water in gallons required:

- Step 1: Area of filter = 30 ft × 20 ft = 600 ft²
- Step 2: Gallons of water used per square foot of filter = 15 gal/ft²/min × 10 min = 15 gal/ft²
- Step 3: Gallons required = 150 gal/ft² × 600 ft² = 90,000 gal required for backwash

WATER DISTRIBUTION SYSTEM CALCULATIONS

After water is adequately treated, it must be conveyed or distributed to the customer for domestic, commercial, industrial, and fire-fighting applications. Water distribution systems should be capable of meeting the demands placed on them at all times and at satisfactory pressures. Waterworks operators responsible for water distribution must be able to perform basic calculations for both practical and licensure purposes; such calculations deal with water velocity, rate of water flow, water storage tanks, and water disinfection.

Water Flow Velocity

The velocity of a particle (any particle) is the speed at which it is moving. Velocity is expressed by indicating the length of travel and how long it takes to cover the distance. Velocity can be expressed in almost any distance and time units.

$$\text{Velocity} = \frac{\text{Distance Traveled}}{\text{Time}} \quad (7.27)$$

Note that the water flow that enters the pipe (any pipe) is the same flow that exits the pipe (under steady flow conditions). Water flow is continuous. Water is incompressible; it cannot accumulate inside. The flow at any given point is the same flow at any other given point in the pipeline; therefore, a given flow volume may not change (it shouldn't), but the velocity of the water may change. At any given flow, velocity is dependent upon the cross-sectional area of the pipe or conduit. Velocity (the speed at which the flow is traveling) is an important parameter. Recall that when dealing with the velocity of flow, the most basic hydraulic equation is:

$$Q = AV$$

where:

Q = flow

A = area [cross-sectional area of conduit

$$-(0.785) \times (\text{diameter})^2]$$

V = velocity

Example 7.52

Problem: A flow of 2 MGD occurs in a 10-in diameter conduit. What is the water velocity?

Solution:

Change MGD to cfs, inches to feet.

$$Q = AV$$

$$2 \times 1.55 = 0.785 \times 0.83^2 \, V$$

$$3.1 = 0.785 \times 0.69 \, V$$

$$V = 5.7 \text{ ft/sec}$$

Example 7.53

Problem: A 24-in-diameter pipe carries water at a velocity of 140 ft/min. What is the flow rate (gpm)?

Solution:

Change ft/min to ft/sec and inches to feet, then solve for flow.

$$Q = AV$$

$$Q = 0.785 \times 2^2 \times 2.3$$

$$Q = 7.2 \text{ cfs}$$

$$Q = 7.2 \text{ cfs} \times 7.48 \text{ ft}^3 \times 60 \text{ min}$$

$$Q = 3,231 \text{ gpm}$$

Example 7.54

Problem: If water travels 700 ft in 5 min, what is the velocity?

Solution:

$$\text{Velocity} = \frac{\text{Distance Traveled}}{\text{Time}}$$

$$= \frac{700 \text{ ft}}{5 \text{ min}}$$

$$= 140 \text{ ft/min}$$

Example 7.55

Problem: Flow in a 6-in pipe is 400 gpm. What is the average velocity?

Solution:

- Step 1:

$$\text{Area} = (0.785)(\text{diameter})^2$$

Convert 6 inch to feet

by dividing by 12: 6/12 = 0.5 or 0.5 ft

$$= 0.785(0.5)^2$$

$$= 0.785(0.25)$$

$$= 0.196 \text{ ft}^2 \text{ (rounded)}$$

- Step 2:

$$\text{Flow (cfs)} = \text{Flow(gal/min)} \times \text{ft}^3/7.48\text{-gal} \times 1 \text{ min}/60 \text{ sec}$$

$$\text{Flow (cfs)} = 400 \text{ gal/min} \times \text{ft}^3/7.48\text{-gal} \times 1 \text{ min}/60 \text{ sec}$$

$$\text{Flow (cfs)} = 400 \text{ ft}^3/448.3 \text{ sec}$$

$$Q = 0.89 \text{ cfs}$$

- Step 3:

$$\text{Velocity (ft/sec)} = \frac{\text{Flow} \left(\text{ft}^3/\text{sec} \right)}{\text{Area} \left(\text{ft}^2 \right)}$$

$$= \frac{0.89 \text{ ft}^3/\text{sec}}{0.196 \text{ ft}^2}$$

$$V = 4.5 \text{ ft/sec}$$

Example 7.56

Problem: Flow in a 2.0-ft wide rectangular channel is 1.2 ft deep and measures 11.0 cfs. What is the average velocity?

Solution:

- Step 1:

$$\text{Transpose } Q = VA \text{ to } V = Q/A$$

Given:

$$Q = \text{Rate of flow} = 11.0 \text{ cfs}$$

$$A = \text{Area in square feet}$$

$$2.0 \text{ ft wide}$$

$$1.2 \text{ ft deep}$$

Find: Average Velocity

- Step 2:

$$\text{Area} = (\text{Width}) \times (\text{Depth})$$

$$= 2.0 \text{ ft} \times 1.2 \text{ ft}$$

$$A = 2.4 \text{ ft}^2$$

- Step 3:

$$\text{Velocity (ft/sec)} = \frac{\text{Flow} \left(\text{ft}^3/\text{sec} \right)}{\text{Area} \left(\text{ft}^2 \right)}$$

$$\text{Velocity (V)} = \frac{11.0 \text{ ft}^3/\text{sec}}{2.4 \text{ ft}^2}$$

$$= 4.6 \text{ ft/sec (rounded)}$$

Storage Tank Calculations

Water is stored at a waterworks to provide allowance for differences in water production rates and high-lift pump discharge to the distribution system. Water within the distribution system may be stored in elevated tanks, standpipes, covered reservoirs, and/or underground basins. The waterworks operator should be familiar with the basic storage tank calculations illustrated in the following example.

Example 7.57

Problem: A cylindrical tank is 120 ft high and 25 ft in diameter. How many gallons of water will it contain?

Solution:

Given:

Height = 120 ft

Diameter = 25 ft

Cylindrical Shape

Find: Total gallons of water contained in the tank

- Step 1: Find the volume in cubic feet.

$$\text{Volume} = 0.785 \times (\text{Diameter})^2 \times (\text{Height})$$
$$= 0.785(25 \text{ ft})^2(120 \text{ ft})$$
$$= 0.785(625 \text{ ft}^2)(120 \text{ ft})$$
$$= 58,875 \text{ ft}^3$$

- Step 2: Find the number of gallons of water the cylindrical tank will contain.

$$= 58,875 \text{ ft}^3 \times 7.48 \text{ gal/ft}^3$$
$$= 440,385 \text{ gal}$$

Distribution System Disinfection Calculations

Delivering a clean, pathogen-free product to the customer is what water treatment operation is all about. Before being placed in service, all facilities and appurtenances associated with the treatment and distribution of water must be disinfected because water may become tainted anywhere in the system. In the examples that follow, we demonstrate how to perform the necessary calculations for this procedure.

Example 7.58

Problem: A waterworks has a tank containing water that needs to be disinfected using HTH with 70% available chlorine. The tank is 100 ft high and 25 ft in diameter. The dose to use is 50 ppm. How many pounds of HTH are needed?

Solution:

Given:

Height = 100 ft

Diameter = 25 ft

Chlorine dose = 50 ppm

Available chlorine = 70%

Find pounds of HTH:

- Step 1: Find the volume of the tank.

$$\text{Volume} = (3.14) \times (r^2) \times (H)$$
$$r = \frac{\text{Diameter}}{2} = \frac{25 \text{ ft}}{2} = 12.5 \text{ ft}$$
$$\text{Volume} = 3.14(12.5)^2(100)$$
$$\text{Volume} = 3.14(156.25)(100)$$
$$\text{Volume} = 3.14(15,625)$$
$$\text{Volume} = 49,062.5 \text{ ft}^3$$

- Step 2: Convert cubic feet to million gallons (MG).

$$49,062.5 \text{ ft}^3 \times \frac{7.48 \text{ gal}}{\text{ft}^3} = \frac{\text{MG}}{1,000,000 \text{ gal}} = 0.367 \text{ MG}$$

Chemical Wt (lbs) = Chem. Dose (mg/L)
$$\times \text{Water Vol. (MG)} \times 8.34$$

- Step 3: Chlorine = 50 mg/L × 0.367 MG × 8.34

$$\text{Chlorine} = 153 \text{ lbs (available)}$$

Note: The fundamental concept to keep in mind when computing hypochlorite calculations is that once we determine how many pounds of chlorine will be required for disinfection, we will always need more pounds of hypochlorite compared to elemental chlorine.

- Step 4

$$\text{Hypochlorite} = \frac{\text{Available Chlorine}}{\text{Chlorine Fraction}}$$
$$\text{Hypochlorite} = \frac{153 \text{ lbs}}{0.7}$$
$$= 218.6 \text{ lbs HTH required (rounded)}$$

Example 7.59

Problem: When treating 4,000 ft of 8-in water line by applying enough chlorine for an 80-ppm dosage, how many pounds with hypochlorite of 70% available chlorine are required?

Solution:

Given:

Length = 4,000 ft

Available chlorine = 70%

Diameter = 8 in

Chlorine Dose = 80 ppm

Find the pound of hypochlorite required.

- Step 1: Find the volume of the pipe.
  Change 8 in to ft by dividing by 12.

$$Diameter = \frac{8 \text{ in}}{12 \text{ in/ft}} = 0.66 \text{ ft} = 0.70 \text{ ft (rounded)}$$

To get the radius: $r = \frac{Diameter}{2} = \frac{0.70 \text{ ft}}{2} = 0.35 \text{ ft}$

The determine volume:

$$Volume = (3.14) \times (r^2) \times (H)$$

$$Volume = 3.14 \times (0.35)^2 \times 4,000 \text{ ft}$$

$$Volume = 3.14 \times (0.1225) \times 4,000 \text{ ft}$$

$$Volume = 1,538.6 \text{ ft}^3$$

- Step 2: Convert cubic ft to million gallons (MG).

$$1,538.6 \text{ ft}^3 \times \frac{7.48 \text{ gal/ft}^3}{\text{ft}^3} \times \frac{MG}{1,000,000 \text{ gal}} = 0.0115 \text{ MG}$$

- Step 3: Determine available chlorine:

$$Chlorine = 80 \text{ mg/L} \times 0.0115 \text{ MG} \times 8.34$$

$$Chlorine = 7.67 \text{ lbs (available)}$$

- Step 4: Determine hypochlorite (HTH) required:

$$\frac{7.67 \text{ lb Chlorine}}{0.7} = 11 \text{ lb of hypochlorite (rounded)}$$

COMPLEX CONVERSIONS

Water and wastewater operators use complex conversions, for example, converting laboratory test results to other units of measure that can be used to adjust or control the treatment process. Conversions such as these require several measurements (i.e., concentration, flow rate, tank volume, etc.) and an appropriate conversion factor. The most widely used conversions are discussed in the following sections.

Concentration to Quantity

Concentration (milligrams/Liter) to pounds

Pounds = Concentration (mg/L) × tank volume (MG)

$$\times 8.34 \text{ lb/MG/mg/L}$$

(7.28)

Example 7.60

Problem:
  Given:
    MLSS = 2,580 mg/L
    Aeration Tank Vol. = 0.90 MG
  What is the concentration in pounds?

Solution:

Pounds = 2,580 mg/L × 0.90 MG × 8.34 lbs/MG/mg/L

= 19,366 lbs

Concentration (milligrams/liter) to pounds/day

Pounds/Day = Concentration, mg/L × Flow, MGD

$$\times 8.34 \text{ lb/MG/mg/L}$$
(7.29)

Example 7.61

Problem:
  Given:
    Effluent BOD$_5$ = 23 mg/L
    Effluent Flow = 4.85 MGD
  What is the concentration in pounds per day?

Solution:

Pounds/Day = 23 mg/L × 4.85 MGD × .34 lbs/MG/mg/L

= 930 lbs/day

Concentration (milligrams/liter) to kilograms/day

Kilograms/Day = Concentration, mg/L × Flow, MGD

$$\times 3.785 \text{ lbs/MG/mg/L}$$
(7.30)

Example 7.62

Problem:
  Given:

    Effluent TSS = 29 mg/L

    Effluent Flow = 11.5 MGD

What is the concentration in kilograms per day?

Solution:

Kilograms/Day = 29 mg/L × 11.5 MGD × 3.785 lbs/MG/mg/L

= 1,263 kg/day

Concentration (milligrams/kilogram) to pounds/ton

Pounds/ton = Concentration, mg/kg × 0.002 lbs/ton/mg/kg   (7.31)

Example 7.63

Problem: Given that biosolids contain 0.97 mg/kg of lead, how many pounds of lead are being applied per acre if the current application rate is 5 dry tons of solids per acre?

Solution:

Pounds/acre = 0.97 mg/kg × 5 tons/acre × 0.002 lbs/ton/mg/kg

= 0.0097 lbs/acre

Quantity to Concentration

Pounds to Concentration (milligrams/Liter)

$$\text{Concentration, mg/L} = \frac{\text{Quantity, lbs}}{\text{Volume, MG} \times 8.34 \text{ lb/mg/L/MG}}$$
(7.32)

Example 7.64

Problem: The aeration tank contains 73,529 lb of solids. The volume of the tank is 3.20 MG. What is the concentration of solids in the aeration tank in milligrams/liter?

Solution:

$$\text{Concentration, mg/L} = \frac{73,529 \text{ lb}}{3.20 \text{ MG} \times 8.34 \text{ lb/mg/L/MG}}$$

Concentration, mg/L = 2,755.1 (2,755) mg/L

Pounds/Day to concentration (milligrams/liter)

$$\text{Concentration, mg/L} = \frac{\text{Quantity, lb/day}}{\text{Flow, MGD} \times 8.34 \text{ lb/mg/L/MG}}$$
(7.33)

Example 7.65

Problem: What is the chlorine dose in milligrams/liter when 490 lb/day of chlorine is added to an effluent flow of 11.0 MGD?

Solution:

$$\text{Dose, mg/L} = \frac{490 \text{ lb/day}}{11.0 \text{ MGD} \times 8.34 \text{ lb/mg/L/MG}}$$

Dose, mg/L = 5.34 mg/L

Kilograms/day to concentration (milligrams/liter)

$$\text{Concentration, mg/L} = \frac{\text{Quantity, kg/Day}}{\text{Flow, MGD} \times 3.785 \text{ kg/mg/L/MG}}$$
(7.34)

Quantity to Volume or Flow Rate

Pounds to Tank Volume (million gallons)

$$\text{Volume, MG} = \frac{\text{Quantity, lb}}{\text{Concentration, mg/L} \times 8.34 \text{ lb/mg/L/MG}}$$
(7.35)

Pounds/day to flow (million gallons/day)

$$\text{Flow, MGD} = \frac{\text{Quantity, lb/day}}{\text{Concentration, mg/L} \times 8.34 \text{ lb/mg/L/MG}}$$
(7.36)

Example 7.66

Problem: You must remove 8,485 lb of solids from the activated sludge process. The waste-activated sludge solids concentration is 5,636 mg/L. How many million gallons must be removed?

Solution:

$$\text{Flow, MGD} = \frac{8,485 \text{ lb/day}}{5,636 \text{ mg/L} \times 8.34 \text{ lb/MG/mg/L}}$$

Flow, MGD = 0.181 MGD

Kilograms/day to flow (million gallons/day)

$$\text{Flow, MGD} = \frac{\text{Quantity, kg/day}}{\text{Concentration, mg/L} \times 3.785 \text{ kg/MG/mg/L}}$$
(7.37)

APPLIED MATH OPERATIONS

MASS BALANCE AND MEASURING PLANT PERFORMANCE

The simplest way to express the fundamental engineering principle of *mass balance* is to say, "Everything has to go somewhere." More precisely, the *law of conservation of mass* states that when chemical reactions take place, matter is neither created nor destroyed. This important concept allows us to track materials—pollutants, microorganisms, chemicals, and other materials—from one place to another. The concept of mass balance plays an important role in treatment plant operations (especially wastewater treatment), where we assume a balance exists between the material entering and leaving the treatment plant or a treatment process: "what comes in must equal what goes out." This concept is very helpful in evaluating biological systems, sampling and testing procedures, and many other unit processes within the treatment system. In the following sections, we illustrate how the mass balance concept is used to determine the quantity of solids entering and leaving settling tanks and the mass balance using BOD removal.

MASS BALANCE FOR SETTLING TANKS

The mass balance for the settling tank calculates the quantity of solids entering and leaving the unit.

Key Point: The two numbers (in—influent and out—effluent) must be within 10%–15% of each other to be considered acceptable. Larger discrepancies may indicate sampling errors, increasing solids levels in the unit, or undetected solids discharge in the tank effluent.

To get a better feel for how the mass balance for settling tanks procedure is formatted for actual use, consider the steps used in the computation of Example 7.65 below.

Step 1: Solids In=Pounds of Influent Suspended Solids

Step 2: Pounds of Effluent Suspended Solids

Step 3: Biosolids Solids Out=Pounds of Biosolids Solids Pumped Per Day

Step 4: Solids In—(Solids Out+Biosolids Solids Pumped)

Example 7.67

Problem: The settling tank receives a daily flow of 4.20 MGD. The influent contains 252 mg/L suspended solids, and the unit effluent contains 140 mg/L suspended solids. The biosolids pump operates 10 min/h and removes biosolids at the rate of 40 gpm. The biosolids are 4.2% solids. Determine if the mass balance for solids removal is within the acceptable 10%–15% range.

Solution:

Step 1: Solids In = 252 mg/L × 4.20 MGD × 8.34 = 8,827 lb/day

Step 2: Solids Out: 140 mg/L × 4.20 MGD × 8.34 4,904 lb/day

$$\text{Biosolids Solids} = 10\,\frac{\text{min}}{\text{h}} \times 24\,\frac{\text{h}}{\text{day}} \times 40\ \text{gpm}$$

Step 3:   × 8.34 × 0.042

                              = 3,363 lb/day

           Balance = 8,827 lb/day – (4,904 lb/day

Step 4:   + 3,363 lb/day)

                        = 560 lb or 6.3%

MASS BALANCE USING BOD REMOVAL

The amount of BOD removed by a treatment process is directly related to the quantity of solids the process will

TABLE 7.4
General Conversion Rates

Process Type	Conversion Factor (lb Solids/lb BOD Removal)
Primary treatment	1.7
Trickling filters	1.0
Rotating biological contactors	1.0
Activated biosolids with/primary	0.7
Activated biosolids without/primary	
Conventional	0.85
Extended air	0.65
Contact stabilization	1.0
Step feed	0.85
Oxidation ditch	0.65

generate. Because the actual amount of solids generated will vary with operational conditions and design, exact figures must be determined on a case-by-case basis. However, research has produced general conversion rates for many of the common treatment processes. These values are given in Table 7.4 and can be used if plant-specific information is unavailable. Using these factors, the mass balance procedure determines the amount of solids the process is anticipated to produce. This is compared with the actual biosolids production to determine the accuracy of the sampling and/or the potential for solids buildup in the system or unrecorded solids discharges.

Step 1: BOD_{in} = Influent BOD × Flow × 8.34

Step 2: BOD_{Out} = Effluent BOD × Flow × 8.34.

Step 3: BOD Pound Removed = $\text{BOD}_{in} - \text{BOD}_{out}$

Step 4: Solids Generated, lb = BOD Removed, lb × Factor

Step 5: Solids Removed = Sludge Pumped, gpd × % Solids × 8.34

Step 6: Effluent Solids, mg/L × Flow, MGD × 8.34

Example 7.68

Problem: A conventional activated biosolids system with primary treatment is operating at the levels listed below. Does the mass balance for the activated biosolids system indicate that a problem exists?

Plant influent BOD	250 mg/L
Primary effluent BOD	166 mg/L
Activated biosolids system effluent BOD	25 mg/L
Activated biosolids system effluent TSS	19 mg/L
Plant flow	11.40 MGD
Waste concentration	6,795 mg/L
Waste flow	0.15 MGD

Solution:

BOD_{in} = 166 mg/L × 11.40 MGD × 8.34 = 15,783 lb/day

$$BOD_{out} = 25 \text{ mg/L} \times 11.40 \text{ MGD} \times 8.34 = 2,377 \text{ lb/day}$$

$$BOD \text{ Removed} = 15,783 \text{ lb/day} - 2,377 \text{ lb/day} = 13,406 \text{ lb/day}$$

$$\text{Solids Produced} = 13,406 \text{ lb/day} \times 0.7 \text{ lb Solids/lb BOD}$$
$$= 9,384 \text{ lb Solids/day}$$

$$\text{Solids Removed} = 6,795 \text{ mg/L} \times 0.15 \text{ MGD} \times 8.34 = 8,501 \text{ lb/day}$$

$$\text{Difference} = 9,384 \text{ lb/day} - 8,501 \text{ lb/day} = 883 \text{ lb/day or } 9.4\%$$

These results are within the acceptable range.

Key Point: We have demonstrated two ways in which mass balance can be used. However, it is important to note that the mass balance concept can be used for all aspects of wastewater and solids treatment. In each case, the calculations must consider all of the sources of material entering the process and all of the methods available for the removal of solids.

MEASURING PLANT PERFORMANCE

To evaluate how well a plant or unit process is performing, **performance efficiency** or **percent (%) removal** is used. The results obtained can be compared with those listed in the plant's operation and maintenance (O & M) manual to determine if the facility is performing as expected. In this section, sample calculations often used to measure plant performance/efficiency are presented. The *efficiency* of a unit process is its effectiveness in removing various constituents from the wastewater or water. Suspended solids and BOD removal are therefore the most common calculations of unit process efficiency. In wastewater treatment, the efficiency of a sedimentation basin may be affected by factors such as the types of solids in the wastewater, the temperature of the wastewater, and the age of the solids. Typical removal efficiencies for a primary sedimentation basin are as follows:

* Settleable Solids----------90%–99%
* Suspended Solids---------40%–60%
* Total Solids----------------10%–15%
* BOD----------------------20%–50%

Plant Performance/Efficiency

Key Point: The calculation used for determining the performance (percent removal) for a digester is different from that used for performance (percent removal) for other processes. Care must be taken to select the correct formula.

$$\% \text{ Removal} = \frac{[\text{Influent Concentration} - \text{Effluent Concentration}] \times 100}{\text{Influent Concentration}}$$
$$(7.38)$$

Example 7.69

Problem: The influent BOD_5 is 247 mg/L, and the plant effluent BOD is 17 mg/L. What is the percent removal?

$$\% \text{ Removal} = \frac{(247 \text{ mg/L} - 17 \text{ mg/L}) \times 100}{247 \text{ mg/L}} = 93\%$$

Unit Process Performance/Efficiency

The concentration entering the unit and the concentration leaving the unit (i.e., primary, secondary, etc.) are used to determine the unit's performance.

$$\% \text{ Removal} = \frac{[\text{Influent Concentration} - \text{Effluent Concentration}] \times 100}{\text{Influent Concentration}}$$
$$(7.39)$$

Example 7.70

Problem: The primary influent BOD is 235 mg/L, and the primary effluent BOD is 169 mg/L. What is the percent removal?

Solution:

$$\% \text{ Removal} = \frac{(235 \text{ mg/L} - 169 \text{ mg/L}) \times 100}{235 \text{ mg/L}} = 28\%$$

Percent Volatile Matter Reduction in Sludge

The calculation used to determine *percent volatile matter reduction* is more complicated because of the changes occurring during biosolids digestion.

$$\% \text{ VM Reduction} = \frac{(\% \text{ VM}_{in} - \% \text{ VM}_{out}) \times 100}{[\% \text{ VM}_{in} - (\% \text{ VM}_{in} \times \% \text{ VM}_{out})]}$$
$$(7.40)$$

Example 7.71

Problem: Using the digester data provided below, determine the % Volatile Matter Reduction for the digester.
   Data:

Raw Biosolids Volatile Matter	74%
Digested Biosolids Volatile Matter	54%

$$\% \text{ Volatile Matter Resolution} = \frac{(0.74 - 0.54) \times 100}{[0.74 - (0.74 \times 0.54)]}$$
$$= 59\%$$

WATER TREATMENT MATH CONCEPTS

WATER SOURCES AND STORAGE CALCULATIONS

Approximately 40 million cubic miles of water cover or reside within the Earth. The oceans contain about 97% of all water on Earth. The other 3% is freshwater: (1) snow and ice on the surface of Earth contain about 2.25% of the water; (2) usable ground water is approximately 0.3%; and (3) surface freshwater is less than 0.5%. In the United States, for example, average rainfall is approximately 2.6 ft (a volume of 5,900 km³). Of this amount, approximately 71% evaporates (about 4,200 km³), and 29% goes to stream flow (about 1,700 km³).

Beneficial freshwater uses include manufacturing, food production, domestic and public needs, recreation, hydroelectric power production, and flood control. Stream flow withdrawn annually is about 7.5% (440 km³). Irrigation and industry use almost half of this amount (3.4% or 200 km³/year). Municipalities use only about 0.6% (35 km³/year) of this amount. Historically, in the United States, water usage is increasing (as might be expected). For example, in 1900, 40 billion gallons of freshwater were used. In 1975, the total increased to 455 billion gallons. Projected use in 2000 is about 720 billion gallons.

The primary sources of freshwater include the following:

- Captured and stored rainfall in cisterns and water jars
- Groundwater from springs, artesian wells, and drilled or dug wells
- Surface water from lakes, rivers, and streams
- Desalinized seawater or brackish groundwater
- Reclaimed wastewater

WATER SOURCE CALCULATIONS

Water source calculations covered in this section apply to wells and pond/lake storage capacity. Specific well calculations discussed include well drawdown, well yield, specific yield, well casing disinfection, and deep-well turbine pump capacity.

Well Drawdown

Drawdown is the drop in the level of water in a well when water is being pumped. Drawdown is usually measured in feet or meters. One of the most important reasons for measuring drawdown is to ensure that the source water is adequate and not being depleted. The data collected to calculate drawdown can indicate if the water supply is slowly declining. Early detection can give the system time to explore alternative sources, establish conservation measures, or obtain any special funding that may be needed to secure a new water source. Well drawdown is the difference between the pumping water level and the static water level.

$$\text{Drawdown, ft} = \text{Pumping Water Level, ft} - \text{Static Water Level, ft} \quad (7.41)$$

Example 7.72

Problem: The static water level for a well is 70 ft. If the pumping water level is 90 ft, what is the drawdown?

Solution:

$$\text{Drawdown, ft} = \text{Pumping Water Level, ft} - \text{Static Water Level, ft}$$

$$= 90 \text{ ft} - 70 \text{ ft}$$

$$= 20 \text{ ft}$$

Example 7.73

Problem: The static water level of a well is 122 ft. The pumping water level is determined using the sounding line. The air pressure applied to the sounding line is 4.0 psi, and the length of the sounding line is 180 ft. What is the drawdown?

Solution:

First, calculate the water depth in the sounding line and the pumping water level:

1. Water depth in sounding line = (4.0 psi)(2.31 ft/psi)

$$= 9.2 \text{ ft}$$

2. Pumping water level = 180 ft − 9.2 ft = 170.8 ft

Then, calculate drawdown as usual:

$$\text{Drawdown, ft} = \text{Pumping Water Level, ft} - \text{Static Water Level, ft}$$

$$= 170.8 \text{ ft} - 122 \text{ ft}$$

$$= 48.8 \text{ ft}$$

Well Yield

Well yield is the volume of water per unit of time that is produced from well pumping. Usually, well yield is measured in terms of gallons per minute (gpm) or gallons per hour (gph). Sometimes, large flows are measured in cubic feet per second (cfs). Well yield is determined using the following equation.

$$\text{Well Yield, gpm} = \frac{\text{Gallons Produced}}{\text{Duration of Test, min}} \quad (7.42)$$

Example 7.74

Problem: Once the drawdown level of a well stabilized, it was determined that the well produced 400 gal during a 5-min test.

Solution:

$$\text{Well Yield, gpm} = \frac{\text{Gallons Produced}}{\text{Duration of Test, min}}$$

$$= \frac{400 \text{ gal}}{5 \text{ min}}$$

$$= 80 \text{ gpm}$$

Example 7.75

Problem: During a 5-min test for well yield, a total of 780 gal is removed from the well. What is the well yield in gpm? In gph?

Solution:

$$\text{Well Yield, gpm} = \frac{\text{Gallons Produced}}{\text{Duration of Test, min}}$$

$$= \frac{780 \text{ gal}}{5\text{-min}}$$

$$= 156 \text{ gpm}$$

Then convert gpm flow to gph flow:

$$(156 \text{ gal/min})(60/\text{h}) = 9,360 \text{ gph}$$

Specific Yield

Specific yield is the discharge capacity of the well per foot of drawdown. The specific yield may range from 1 gpm/ft drawdown to more than 100 gpm/ft drawdown for a properly developed well. Specific yield is calculated using the Equation (7.43).

$$\text{Specific Yield, gpm/ft} = \frac{\text{Well Yield, gpm}}{\text{Drawdown, ft}} \quad (7.43)$$

Example 7.76

Problem: A well produces 260 gpm. If the drawdown for the well is 22 ft, what is the specific yield in gpm/ft of drawdown?

Solution:

$$\text{Specific Yield, gpm/ft} = \frac{\text{Well Yield, gpm}}{\text{Drawdown, ft}}$$

$$= \frac{260 \text{ gpm}}{22 \text{ ft}}$$

$$= 11.8 \text{ gpm/ft}$$

Example 7.77

Problem: The yield for a particular well is 310 gpm. If the drawdown for this well is 30 ft, what is the specific yield in gpm/ft of drawdown?

Solution:

$$\text{Specific Yield, gpm/ft} = \frac{\text{Well Yield, gpm}}{\text{Drawdown, ft}}$$

$$= \frac{310 \text{ gpm}}{30 \text{ ft}}$$

$$= 10.3 \text{ gpm/ft}$$

Well Casing Disinfection

A new, cleaned, or repaired well normally contains contamination that may remain for weeks unless the well is thoroughly disinfected. This may be accomplished using ordinary bleach in a concentration of 100 parts per million (ppm) of chlorine. The amount of disinfectant required is determined by the amount of water in the well. The following equation is used to calculate the pounds of chlorine required for disinfection:

$$\text{Chlorine, lb} = (\text{Chlorine, mg/L}) \times (\text{Casing Vol., MG}) \times (8.34 \text{ lb/gal}) \quad (7.44)$$

Example 7.78

Problem: A new well is to be disinfected with chlorine at a dosage of 50 mg/L. If the well casing diameter is 8 in and the length of the water-filled casing is 110 ft, how many pounds of chlorine will be required?

Solution:

First, calculate the volume of the water-filled casing:

$$(0.785)(0.67)(67)(110 \text{ ft})(7.48 \text{ gal/ft}^3) = 290 \text{ gal}$$

Then, determine the pounds of chlorine required using the mg/L to lb equation:

$$\text{Chlorine, lb} = (\text{chlorine, mg/L}) \text{ Volume, MG})(8.34 \text{ lb/gal})$$

$$(50 \text{ mg/L})(0.000290 \text{ MG})(8.34 \text{ lb/gal}) = 0.12 \text{ lb Chlorine}$$

Deep-Well Turbine Pump Calculations

The deep well turbine pump is used for high-capacity deep wells. The pump, usually consisting of more than one stage of a centrifugal pump, is fastened to a pipe called the pump column; the pump is located in the water. The pump is driven from the surface through a shaft running inside the pump column. The water is discharged from the pump up through the pump column to the surface. The pump may be driven by a vertical shaft electric motor at the top of the well or by some other power source, usually through a right-angle gear drive located at the top of the well. A modern version of the deep well turbine pump is the submersible type pump, in which the pump, along with a close-coupled electric motor built as a single unit, is located below water level in the well. The motor is built to operate submerged in water.

Vertical Turbine Pump Calculations

The calculations pertaining to well pumps include head, horsepower, and efficiency calculations. *Discharge head* is measured at the pressure gauge located close to the pump discharge flange. The pressure (psi) can be converted to feet of head using the equation:

$$\text{Discharge Head, ft} = (\text{press, psi})(2.31 \text{ ft/psi}) \quad (7.45)$$

Total pumping head (*field head*) is a measure of the lift below the discharge head pumping water level (*discharge head*). Total pumping head is calculated as follows:

$$\text{Pumping Head, ft} = \text{Pumping Water Level, ft} + \text{Discharge Head, ft}$$
(7.46)

Example 7.79

Problem: The pressure gauge reading at a pump discharge head is 4.1 psi. What is this discharge head expressed in feet?

Solution:

$$(4.1 \text{ psi})(2.31 \text{ ft/psi}) = 9.5 \text{ ft}$$

Example 7.80

Problem: The static water level of a pump is 100 ft. The well drawdown is 26 ft. If the gauge reading at the pump discharge head is 3.7 psi, what is the total pumping head?

Solution:

$$\text{Total pumping head, ft} = \text{Pumping water level, ft}$$
$$+ \text{discharge head, ft}$$
$$= (100 \text{ ft} + 26 \text{ ft}) + (3.7 \text{ psi}) (2.31 \text{ ft/psi})$$
$$= 126 \text{ ft} + 8.5 \text{ ft}$$
$$= 134.5 \text{ ft}$$

WATER STORAGE CALCULATIONS

Water storage facilities for water distribution systems are required primarily to provide for fluctuating demands of water usage (to provide a sufficient amount of water to average or equalize daily demands on the water supply system). In addition, other functions of water storage facilities include increasing operating convenience, leveling pumping requirements (to keep pumps from running 24 h a day), decreasing power costs, providing water during power source or pump failure, providing large quantities of water to meet fire demands, providing surge relief (to reduce the surge associated with stopping and starting pumps), increasing detention time (to provide chlorine contact time and satisfy the desired CT (contact time) value requirements), and blending water sources. The storage capacity, in gallons, of a reservoir, pond, or small lake can be estimated using Equation (7.47).

$$\text{Capacity, gal} = (\text{Ave. Length, ft}) \times (\text{Ave. Width, ft})$$
$$\times (\text{Ave. Depth, ft})$$
$$\times (7.48 \text{ gal/ft}^3)$$
(7.47)

Example 7.81

Problem: A pond has an average length of 250 ft, an average width of 110 ft, and an estimated average depth of 15 ft. What is the estimated volume of the pond in gallons?

Solution:

$$\text{Volume (gal)} = (\text{Ave. Length, ft}) \times (\text{Ave. Width, ft})$$
$$\times (\text{Ave. Depth, ft})$$
$$\times (7.48 \text{ gal/ft}^3)$$
$$= (250 \text{ ft})(110 \text{ ft})(15 \text{ ft})(7.48 \text{ gal/ft}^3)$$
$$= 3,085,500 \text{ gal}$$

Example 7.82

Problem: A small lake has an average length of 300 ft and an average width of 95 ft. If the maximum depth of the lake is 22 ft, what is the estimated volume of the lake in gallons?

Note: For small ponds and lakes, the average depth is generally about 0.4 times the greatest depth. Therefore, to estimate the average depth, measure the greatest depth, then multiply that number by 0.4.

Solution:

First, the average depth of the lake must be estimated:

$$\text{Estimated Aver. Depth, ft} = (\text{Greatest Depth, ft}) \times (0.4 \text{ Depth, ft})$$
$$= (22 \text{ ft}) \times (0.4 \text{ ft})$$
$$= 8.8 \text{ ft}$$

Then, the lake volume can be determined:

$$\text{Volume, gal} = (\text{Aver. Length, ft}) \times (\text{Aver. Width, ft})$$
$$\times (\text{Aver. Depth, ft}) \times (7.48 \text{ gal/ft}^3)$$
$$= (300 \text{ ft}) \times (95 \text{ ft}) \times (8.8 \text{ ft}) \times (7.48 \text{ ft}^3)$$
$$= 1,875,984 \text{ gal}$$

COPPER SULFATE DOSING

Algal control is perhaps the most common in situ treatment of lakes, ponds, and reservoirs by the application of copper sulfate—the copper ions in the water kill the algae. Copper sulfate application methods and dosages will vary depending on the specific surface water body being treated. The desired copper sulfate dosage may be expressed in mg/L of copper sulfate, lb of copper sulfate per acre-ft, or lb of copper sulfate acre. For a dose expressed as mg/L of copper, the following equation is used to calculate the lb of copper sulfate required:

$$\text{Copper Sulfate, lbs} = \frac{\text{Copper (mg/L)(Volume, MG)(8.34 lb/gal)}}{\dfrac{\%\text{ Available copper}}{100}} \tag{7.48}$$

Example 7.83

Problem: For algae control in a small pond, a dosage of 0.5-mg/L of copper is desired. The pond has a volume of 15 MG. How many pounds of copper sulfate will be required? (Note that copper sulfate contains 25% available copper.)

Solution:

$$\text{Copper Sulfate, lbs} = \frac{(\text{mg/L Copper})(\text{Volume MG})(8.34 \text{ lb/gal})}{\dfrac{\%\text{ Available copper}}{100}}$$

$$= \frac{(0.5 \text{ mg/L})(15 \text{ MG})(8.34 \text{ lb/gal})}{\dfrac{25}{100}}$$

$$= 250 \text{ lb Copper Sulfate}$$

For calculating pound of copper sulfate/acre-ft, use the following equation (assume the desired copper sulfate dosage is 0.9 lb/acre-ft):

$$\text{Copper Sulfate, lbs} = \frac{(0.9 \text{ lb Copper Sulfate}) (\text{acre-ft})}{1 \text{ acre-ft}} \tag{7.49}$$

Example 7.84

Problem: A pond has a volume of 35 acre-ft. If the desired copper sulfate dose is 0.9 lb/acre-ft, how many pounds of copper sulfate will be required?

Solution:

$$\text{Copper Sulfate, lb} = \frac{(0.9 \text{ lb Copper Sulfate}) (\text{acre-ft})}{1 \text{ acre-ft}}$$

$$\frac{0.9 \text{ lb Copper Sulfate}}{1 \text{ acre-ft}} = \frac{x \text{ lb Copper Sulfate}}{35 \text{ acre-ft}}$$

Then, solve for x:

$$(0.9)(35) = x$$

$$31.5 \text{ lb copper sulfate}$$

The desired copper sulfate dosage may also be expressed in terms of pounds of copper sulfate per acre. The following equation is used to determine pounds of copper sulfate (assume a desired dose of 5.2 lb/acre):

$$\text{Copper Sulfate, lbs} = \frac{5.2 \text{ lb copper sulfate} \times \text{acres}}{1 \text{ acre}} \tag{7.50}$$

Example 7.85

Problem: A small lake has a surface area of 6.0 acres. If the desired copper sulfate dose is 5.2 lb/acre, how many pounds of copper sulfate are required?

Solution:

$$\text{Copper Sulfate, lb} = \frac{5.2 \text{ lb copper sulfate} \times 6 \text{ acres}}{1 \text{ acre}}$$

$$= 31.2 \text{ lb copper sulfate}$$

COAGULATION AND FLOCCULATION CALCULATIONS

Coagulation

Following screening and the other pretreatment processes, the next unit process in a conventional water treatment system is a mixer where the first chemicals are added in what is known as coagulation. The exception to this situation occurs in small systems using groundwater, where chlorine or other taste and odor control measures are introduced at the intake and are the extent of treatment. The term *coagulation* refers to the series of chemical and mechanical operations by which coagulants are applied and made effective. These operations are comprised of two distinct phases: (1) rapid mixing to disperse coagulant chemicals by violent agitation into the water being treated and (2) flocculation to agglomerate small particles into well-defined floc by gentle agitation for a much longer time. The coagulant must be added to the raw water and perfectly distributed into the liquid; such uniformity of chemical treatment is achieved through rapid agitation or mixing.

Coagulation results from adding salts of iron or aluminum to the water. Common coagulants (salts) are as follows (coagulation is the reaction between one of these salts and water):

- alum—aluminum sulfate
- sodium aluminate
- ferric sulfate
- ferrous sulfate
- ferric chloride
- polymers

Flocculation

Flocculation follows coagulation in the conventional water treatment process. *Flocculation* is the physical process of slowly mixing the coagulated water to increase the probability of particle collision. Through experience, we find that effective mixing reduces the required amount of chemicals and greatly improves the sedimentation process, resulting in longer filter runs and higher-quality finished water. The goal of flocculation is to form a uniform, feather-like material similar to snowflakes—a dense, tenacious floc that entraps fine, suspended, and colloidal particles and carries them down rapidly in the settling basin. To increase the speed of floc formation and the strength and weight of the floc, polymers are often added.

Coagulation and Flocculation Calculations

Proper operation of the coagulation and flocculation unit processes requires calculations to determine chamber or basin volume, chemical feed calibration, chemical feeder settings, and detention time.

Chamber and Basin Volume Calculations

To determine the volume of a square, rectangular chamber, or basin, we use Equation (7.51) or Equation (7.52).

$$\text{Volume, ft}^3 = \text{length (ft)} \times \text{width (ft)} \times \text{depth (ft)} \quad (7.51)$$

$$\text{Volume, gal} = \text{length (ft)} \times \text{width (ft)} \times \text{depth (ft)} \times 7.48 \text{ (gal/ft}^3) \quad (7.52)$$

Example 7.86

Problem: A flash mix chamber is 4 ft² with water to a depth of 3 ft. What is the volume of water (in gallons) in the chamber?

Solution:

$$\text{Volume, gal} = (\text{length, ft})(\text{width, ft})(\text{depth, ft})(7.48 \text{ gal/ft}^3)$$
$$= (4 \text{ ft})(4 \text{ ft})(3 \text{ ft})(7.48 \text{ gal/ft}^3)$$
$$= 359 \text{ gal}$$

Example 7.87

Problem: A flocculation basin is 40 ft long, 12 ft wide, and has water to a depth of 9 ft. What is the volume of water (in gallons) in the basin?

Solution:

$$\text{Volume, gal} = (\text{length, ft})(\text{width, ft})(\text{depth, ft})(7.48 \text{ gal/ft}^3)$$
$$= (40 \text{ ft})(12 \text{ ft})(9 \text{ ft})(7.48 \text{ gal/ft}^3)$$
$$= 32,314 \text{ gal}$$

Example 7.88

Problem: A flocculation basin is 50 ft long, 22 ft wide, and contains water to a depth of 11 ft, 6 in. How many gallons of water are in the tank?

Solution:

First, convert the 6-in portion of the depth measurement to feet:

$$\frac{6 \text{ in}}{12 \text{ in/ft}} = 0.5 \text{ ft}$$

Then, calculate basin volume:

$$\text{Volume, ft} = (\text{length, ft})(\text{width, ft})(\text{depth, ft}),(7.48 \text{ gal/ft}^3)$$
$$= (50 \text{ ft})(22 \text{ ft})(11.5 \text{ ft})(7.48 \text{ gal/ft}^3)$$
$$= 94,622 \text{ gal}$$

Detention Time

Because coagulation reactions are rapid, detention time for flash mixers is measured in seconds, whereas the detention time for flocculation basins is generally between 5 and 30 min. The equation used to calculate detention time is shown below.

$$\text{Detention Time, min} = \frac{\text{Volume of Tank, gal}}{\text{Flow Rate, gpm}} \quad (7.53)$$

Example 7.89

Problem: The flow to a flocculation basin that is 50 ft long, 12 ft wide, and 10 ft deep is 2,100 gpm. What is the detention time in the tank, in minutes?

Solution:

$$\text{Tank Volume, gal} = (50 \text{ ft})(12 \text{ ft})(10 \text{ ft})(7.48 \text{ gal/ft}^3)$$
$$= 44,880 \text{ gal}$$

$$\text{Detention Time, min} = \frac{\text{Volume of Tank, gal}}{\text{Flow Rate, gpm}}$$
$$= \frac{44,880 \text{ gal}}{2100 \text{ gpm}}$$
$$= 21.4 \text{ min}$$

Example 7.90

Problem: A flash mix chamber is 6 ft long, 4 ft wide, and has water to a depth of 3 ft. If the flow to the flash mix chamber is 6 MGD, what is the chamber detention time in seconds (assuming that the flow is steady and continuous)?

Solution:

First, convert the flow rate from gpd to gps so that the time units will match:

$$\frac{6,000,000}{(1,440 \text{ min/day})(60 \text{ sec/min})} = 69 \text{ gps}$$

Then, calculate detention time:

$$\text{Detention Time, sec} = \frac{\text{Volume of Tank, gal}}{\text{Flow Rate, gps}}$$
$$= \frac{(6 \text{ ft})(4 \text{ ft})(3 \text{ ft})(7.48 \text{ gal/ft}^3)}{69 \text{ gps}}$$
$$= 7.8 \text{ sec}$$

Determining Dry Chemical Feeder Setting, Pound/Day

When adding (dosing) chemicals to the water flow, a measured amount of chemical is called for. The amount of chemical required depends on factors such as the type of chemical used, the reason for dosing, and the flow rate being treated. To convert from mg/L to lb/day, the following equation is used:

Chemical added, lb/day = (Chemical, mg/L)

$$(Flow, MGD)(8.34 \text{ lb/gal}) \quad (7.54)$$

Example 7.91

Problem: Jar tests indicate that the best alum dose for water is 8 mg/L. If the flow to be treated is 2,100,000 gpd, what should the lb/day setting be on the dry alum feeder?

Solution:

Setting lb/day = (Chemical, mg/L)(Flow, MGD)(8.34 lb/gal)

$$= \left(8 \text{ mg/L}\right)\left(2.10 \text{ MGD}\right)\left(8.34 \text{ lb/gal}\right)$$

$$= 140 \text{ lb/day}$$

Example 7.92

Problem: Determine the desired lb/day setting on a dry chemical feeder if jar tests indicate an optimum polymer dose of 12 mg/L and the flow to be treated is 4.15 MGD.

Solution:

Setting, lb/day = (12 mg/L)(4.15 MGD)(8.34 lb/gal)

$$= 415 \text{ lb/day}$$

Determining Chemical Solution Feeder Setting, gpd

When solution concentration is expressed as pound chemical per gallon solution, the required feed rate can be determined using the following equations:

Chemical, lb/day = (Chemical, mg/L)(Flow, MGD)(8.34 lb/gal)

$$(7.55)$$

Then, convert the lb/day dry chemical to gpd solution

$$\text{Solution, gpd} = \frac{\text{Chemical, lb/day}}{\text{lb Chemical/gal Solution}} \quad (7.56)$$

Example 7.93

Problem: Jar tests indicate that the best alum dose for water is 7 mg/L. The flow to be treated is 1.52 MGD. Determine the gpd setting for the alum solution feeder if the liquid alum contains 5.36 lb of alum per gallon of solution.

Solution:

First, calculate the lb/day of dry alum required, using the mg/L to lb/day equation:

Dry alum, lb/day = (mg/L)(Flow, MGD)(8.34 lb/gal)

$$= (7 \text{ mg/L})(1.52 \text{ MGD})(8.34 \text{ lb/gal})$$

$$= 89 \text{ lb/day}$$

Then, calculate gpd solution required.

$$\text{Alum Solution, gpd} = \frac{89 \text{ lb/day}}{5.36 \text{ lb alum/gal solution}}$$

$$= 16.6 \text{ gpd}$$

Determining Chemical Solution Feeder Setting, mL/min

Some solution chemical feeders dispense chemicals as milliliters per minute (mL/min). To calculate the mL/min solution required, use the following procedure:

$$\text{Solution, mL/min} = \frac{(\text{gpd})(3785 \text{ mL/gal})}{1,440 \text{ min/day}} \quad (7.57)$$

Example 7.94

Problem: The desired solution feed rate was calculated to be 9 gpd. What is this feed rate expressed as mL/min?

Solution:

$$\text{Solution (mL/min)} = \frac{(\text{gpd})(3,785 \text{ mL/gal})}{1,440 \text{ min/day}}$$

$$= \frac{(9 \text{ gpd}) \times (3,785 \text{ mL/gal})}{1,440 \text{ min/day}}$$

$$= 24 \text{ mL/min feed rate}$$

Example 7.95

Problem: The desired solution feed rate has been calculated to be 25 gpd. What is this feed rate expressed as mL/min?

Solution:

$$\text{mL/min} = \frac{(\text{gpd}) \ (3,785 \text{ mL/gal})}{1,440 \text{ min/day}}$$

$$= \frac{(25 \text{ gpd})(3,785 \text{ mL/gal})}{1,440 \text{ min/day}}$$

$$= 65.7 \text{ mL/min Feed Rate}$$

Sometimes we need to know the mL/min solution feed rate, but we may not know the gpd solution feed rate. In such cases, calculate the gpd solution feed rate first using the following equation:

$$\text{gpd} = \frac{(\text{Chemical, mg/L})(\text{Flow, MGD})(8.34 \text{ lb/gal})}{\text{Chemical, lb/Solution, gal}} \quad (7.58)$$

Determining Percent of Solutions

The strength of a solution is a measure of the amount of chemical solute dissolved in the solution. We use the following equation to determine the percent strength of solution using the following equation:

$$\% \text{ Strength} = \frac{\text{Chemical, lb}}{\text{Water, lbs + Chemical, lb}} \times 100 \quad (7.59)$$

Example 7.96

Problem: If a total of 10 oz of dry polymer is added to 15 gal of water, what is the percent strength (by weight) of the polymer solution?

Solution:

Before calculating percent strength, the ounces chemical must be converted to lb chemical:

$$\frac{10 \text{ oz.}}{16 \text{ oz./lb}} = 0.625 \text{ lb chemical}$$

Now, calculate the percent strength:

$$\% \text{ Strength} = \frac{\text{Chemical, lb}}{\text{Water, lb + Chemical, lb}} \times 100$$

$$= \frac{0.625 \text{ lb chemical}}{(15 \text{ gal}) (8.34 \text{ lb/gal}) + 0.625 \text{ lb}} \times 100$$

$$= \frac{0.625 \text{ lb Chemical}}{125.7 \text{ lb Solution}} \times 100$$

$$= 0.5\%$$

Example 7.97

Problem: If 90 g (1 g = 0.0022 lb) of dry polymer is dissolved in 6 gal of water, what percent strength is the solution?

Solution:

First, convert grams of chemical to pounds: 90 g polymer × 0.0022 lb:

$$(90 \text{ g polymer})(0.0022 \text{ lb/g}) = 0.198 \text{ lb Polymer}$$

Now, calculate the percent strength of the solution:

$$\% \text{ Strength} = \frac{\text{lb Polymer}}{\text{lb Water + lb Polymer}} \times 100$$

$$= \frac{0.198 \text{ lb Polymer}}{(6 \text{ gal}) (8.34 \text{ lb/gal}) + 0.198 \text{ lb}} \times 100$$

$$= 4\%$$

Determining Percent Strength of Liquid Solutions

When using liquid chemicals to make up solutions (e.g., liquid polymer), a different calculation is required, as shown below:

$$\text{Liq. Poly., lb} \frac{\text{Liquid Poly (\% Strength)}}{100}$$

$$\quad (7.60)$$

$$= \text{Poly. Sol., lb} \frac{\text{Poly. Sol. (\% Strength)}}{100}$$

Example 7.98

Problem: A 12% liquid polymer is to be used in making up a polymer solution. How many lb of liquid polymer should be mixed with water to produce 120 lb of a 0.5% polymer solution?

Solution:

$$\frac{(\text{Liq. Poly., lb}) \ (\text{Liq. Poly. \% Strength})}{100}$$

$$= \frac{(\text{Poly Sol., lbs}) \ (\text{Poly. Sol. \% Strength})}{100}$$

$$\frac{(\text{x lb}) \ (12)}{100} = \frac{(120 \text{ lb})(0.5)}{100}$$

$$x = \frac{(120)(0.005)}{0.12}$$

$$x = 5 \text{ lb}$$

Determining % Strength of Mixed Solutions

The percent strength of the solution mixture is determined using the following equation:

$$\% \text{ Strength of Mix.} = \frac{\dfrac{(\text{Sol. 1, lb})(\% \text{ Strength of Sol.1})}{100} + \dfrac{(\text{Sol. 2, lb})(\% \text{ Strength of Sol 2})}{100}}{\text{lb Solution 1} \ + \ \text{lb Solution 2}}$$

$$\times 100$$

$$\quad (7.61)$$

Example 7.99

Problem: If 12 lb of a 10% strength solution is mixed with 40 lb of a 1% strength solution, what is the percent strength of the solution mixture?

Solution:

$$\text{Chemical Feed Rate, lb/min} = \frac{\text{Chemical Applied, lb}}{\text{Length of Application, min}}$$

$$\% \text{ Strength of Mix.} = \frac{\dfrac{(\text{Sol 1, lb})(\% \text{ Strength, Sol 1})}{100} + \dfrac{(\text{Sol 2, lb})(\% \text{ Strength, Sol 2})}{100}}{\text{lbs solution 1} \ + \ \text{lbs Solution 2}}$$

$$= \frac{(12 \text{ lb})(0.1) + (40 \text{ lb})(0.01)}{12 \text{ lb} + 40 \text{ lb}} \times 100$$

$$= \frac{1.2 \text{ lb} + 0.40}{52 \text{ lb}} \times 100$$

$$= 3.1\%$$

Dry Chemical Feeder Calibration

Occasionally we need to perform a calibration calculation to compare the actual chemical feed rate with the feed rate indicated by the instrumentation. To calculate the actual feed rate for a dry chemical feeder, place a container under the feeder, weigh the container when empty, and then weigh the container again after a specified length of time (e.g., 30 min). The actual chemical feed rate can be calculated using the following equation:

$$\text{Chemical Feed Rate, lb/min} = \frac{\text{Chemical Applied, lb}}{\text{Length of Application, min}}$$
(7.62)

If desired, the chemical feed rate can be converted to lbs/day:

$$\text{Feed Rate, lb/day} = \text{Feed Rate, lb/min} \times 1,440 \text{ min/day}$$
(7.63)

Example 7.100

Problem: Calculate the actual chemical feed rate in lb/day if a container is placed under a chemical feeder and a total of 2 lb is collected during a 30-min period.

Solution:

First, calculate the lb/min feed rate:

$$\text{Chemical Feed Rate, lb/min} = \frac{\text{Chemical Applied, lb}}{\text{Length of Application (min)}}$$

$$= \frac{2 \text{ lb}}{30 \text{ min}}$$

$$= 0.06 \text{ lb/min Feed Rate}$$

Then, calculate the lb/day feed rate:

$$\text{Feed Rate, lb/day} = (0.06 \text{ lb/min})(1,440 \text{ min/day})$$

$$= 86.4 \text{ lbs/day Feed Rate}$$

Example 7.101

Problem: Calculate the actual chemical feed rate in lb/day if a container is placed under a chemical feeder and a total of 1.6 lb is collected during a 20-min period.

Solution:

First, calculate the lb/min feed rate:

$$\text{Chemical Feed Rate, lb/min} = \frac{\text{Chemical Applied, lb}}{\text{Length of Application, min}}$$

$$= \frac{1.6 \text{ lb}}{20 \text{ min}}$$

$$= 0.08 \text{ lb/min feed rate}$$

Then, calculate the lb/day feed rate:

$$\text{Feed Rate, lb/day} = (0.08 \text{ lb/min})(1,440 \text{ min/day})$$

$$= 115 \text{ lb/day Feed Rate}$$

Chemical Solution Feeder Calibration

As with other calibration calculations, the actual solution chemical feed rate is determined and then compared with the feed rate indicated by the instrumentation. To calculate the actual solution chemical feed rate, first express the solution feed rate in MGD. Once the MGD solution flow rate has been calculated, use the mg/L equation to determine the chemical dosage in lb/day. If the solution feed is expressed as mL/min, first convert the mL/min flow rate to gpd flow rate.

$$\text{gpd} = \frac{(\text{mL/min})(1,440 \text{ min/day})}{3,785 \text{ mL/gal}}$$
(7.64)

Then, calculate the chemical dosage, lbs/day.

$$\text{Chemical, lb/day} = (\text{mg/L Chemical})(\text{MGD Flow})(8.34 \text{ lb/day})$$
(7.65)

Example 7.102

Problem: A calibration test is conducted for a solution chemical feeder. During a 5-min test, the pump delivered 940 mg/L of the 1.20% polymer solution (assuming the polymer solution weighs 8.34 lb/gal). What is the polymer dosage rate in lb/day?

Solution:

The flow rate must be expressed as MGD. Therefore, the mL/min solution flow rate must first be converted to gpd and then to MGD. The mL/min flow rate is calculated as:

$$\frac{940 \text{ mL}}{5 \text{ min}} = 188 \text{ mL/min}$$

Next, convert the mL/min flow rate to gpd flow rate:

$$\frac{(188 \text{ mL/min})(1,440 \text{ min/day})}{3,785 \text{ mL/gal}} = 72 \text{ gpd flow rate}$$

Then, calculate the lb/day polymer feed rate:

$$(12,000 \text{ mg/L})(0.000072 \text{ MGD})(8.34 \text{ lb/day})$$

$$= 7.2 \text{ lb/day Polymer}$$

Example 7.103

Problem: A calibration test is conducted for a solution chemical feeder. During a 24-h period, the solution feeder delivers a total of 100 gal of solution. The polymer solution is a 1.2-% solution. What is the lb/day feed rate? (Assume the polymer solution weighs 8.34 lb/gal.)

Solution:

The solution feed rate is 100 gallons per day, or 100 gpd. Expressed as MGD, this is 0.000100 MGD. Use the mg/L to lb/day equation to calculate the actual feed rate, lb/day:

$$\text{lb/day Chemical} = (\text{Chemical, mg/L})(\text{Flow, MGD})(8.34 \text{ lb/day})$$

$$= (12,000 \text{ mg/L})(0.000100 \text{ MGD})(8.34 \text{ lb/day})$$

$$= 10 \text{ lb/day Polymer}$$

The actual pumping rates can be determined by calculating the volume pumped during a specified time frame. For example, if 60 gal are pumped during a 10-min test, the average pumping rate during the test is 6 gpm. The actual volume pumped is indicated by the drop in tank level. By using the following equation, we can determine the flow rate in gpm.

$$\text{Flow Rate, gpm} = \frac{(0.785)(D^2)(\text{Drop in Level, ft})(7.48 \text{ gal/ft}^3)}{\text{Duration of Test, min}}$$

$$(7.66)$$

Example 7.104

Problem: A pumping rate calibration test is conducted for a 15-min period. The liquid level in the 4-ft diameter solution tank is measured before and after the test. If the level drops 0.5 ft during the 15-min test, what is the pumping rate in gpm?

$$\text{Flow Rate, gpm} = \frac{(0.785)(D^2)(\text{Drop, ft})(7.48 \text{ gal/ft}^3)}{\text{Duration of Test, min}}$$

$$= \frac{(0.785)(4 \text{ ft})(0.5 \text{ ft})(7.48 \text{ gal/ft}^3)}{15 \text{ min}}$$

$$= 3.1 \text{ gpm Pumping Rate}$$

Determining Chemical Usage

One of the primary functions performed by water operators is the recording of data. The lb/day or gpd chemical use is part of this data. From this data, the average daily use of chemicals and solutions can be determined. This information is important in forecasting expected chemical use, comparing it with chemicals in inventory, and determining when additional chemicals will be required. To determine average chemical use, we use Equation (7.66) (lb/day) or Equation (7.67) (gpd)

$$\text{Average Use, lb/day} = \frac{\text{Total Chemical Used, lb}}{\text{Number of days}} \quad (7.67)$$

or

$$\text{Average Use, gpd} = \frac{\text{Total Chemical Used, gal}}{\text{Number of Days}} \quad (7.68)$$

Then, we can calculate days' supply in inventory:

$$\text{Days' Supply in Inventory} = \frac{\text{Total Chemical in Inventory, lb}}{\text{Average Use, lbs/day}}$$

$$(7.69)$$

or

$$\text{Days' Supply in Inventory} = \frac{\text{Total Chemical in Inventory, gal}}{\text{Average Use, gpd}}$$

$$(7.70)$$

Example 7.105

Problem: The chemical used each day during the week is given below. Based on this data, what was the average lb/day of chemical use during the week?

Monday—88 lb/day	Friday—96 lb/day
Tuesday—93 lb/day	Saturday—92 lb/day
Wednesday—91 lb/day	Sunday—86 lb/day
Thursday—88 lb/day	

Solution:

$$\text{Average Use, lb/day} = \frac{\text{Total Chemical Used, lb}}{\text{Number of Days}}$$

$$= \frac{634 \text{ lb}}{7 \text{ days}}$$

$$= 90.6 \text{ lb/day Average Use}$$

Example 7.106

Problem: The average chemical use at a plant is 77 lb/day. If the chemical inventory is 2,800 lb, how many days' supply is this?

Solution:

$$\text{Days' Supply in Inventory} = \frac{\text{Total Chemical in Inventory, lb}}{\text{Average Use, lb/day}}$$

$$= \frac{2,800 \text{ lb in Inventory}}{77 \text{ lb/day Average Use}}$$

$$= 36.4 \text{ days' Supply in Inventory}$$

Sedimentation Calculations

Sedimentation, the solid-liquid separation by gravity, is one of the most basic processes in water and wastewater treatment. In water treatment, plain sedimentation, such as the use of a pre-sedimentation basin for grit removal and a sedimentation basin following coagulation-flocculation is the most commonly used approach. The two common tank shapes of sedimentation tanks are rectangular and cylindrical. The equations for calculating the volume of each type of tank are shown below.

Calculating Tank Volume

For rectangular sedimentation basins, we use Equation (7.71).

$$\text{Volume, gal} = (\text{length, ft})(\text{width, ft})(\text{depth, ft})(7.48 \text{ gal/ft}^3) \quad (7.71)$$

For circular clarifiers, we use Equation (7.72).

$$\text{Volume, gal} = (0.785)\left(D^2\right)(\text{depth, ft})(7.48 \text{ gal/ft}^3) \quad (7.72)$$

Example 7.107

Problem: A sedimentation basin is 25 ft wide by 80 ft long and contains water to a depth of 14 ft. What is the volume of water in the basin in gallons?

Solution:

$$\text{Volume, gal} = (\text{length, ft})(\text{width, ft})(\text{depth, ft})\left(7.48 \text{ gal/ft}^3\right)$$

$$= (80 \text{ ft})(25 \text{ ft})(14 \text{ ft})\left(7.48 \text{ gal/ft}^3\right)$$

$$= 209,440 \text{ gal}$$

Example 7.108

Problem: A sedimentation basin is 24 ft wide and 75 ft long. When the basin contains 140,000 gal, what would the water depth be?

Solution:

$$\text{Volume, gal} = (\text{length, ft})(\text{width, ft})(\text{depth, ft})\left(7.48 \text{ gal/ft}^3\right)$$

$$140,000 \text{ gal} = (75 \text{ ft})(24 \text{ ft})(x \text{ ft})\left(7.48 \text{ gal/ft}^3\right)$$

$$x \text{ ft} = \frac{140,000}{(75)(24)(7.48)}$$

$$x \text{ ft} = 10.4 \text{ ft}$$

Detention Time

Detention time for clarifiers varies from 1 to 3 h. The equations used to calculate detention time are shown below.

Basic detention time equation

$$\text{Detention Time, h} = \frac{\text{Volume of Tank, gal}}{\text{Flow Rate, gph}} \quad (7.73)$$

Rectangular sedimentation basin equation:

$$\text{Detention Time, h} = \frac{(\text{Length, ft})(\text{Width, ft})(\text{Depth, ft})\left(7.48 \text{ gal/ft}^3\right)}{\text{Flow Rate, gph}} \quad (7.74)$$

Circular basin equation:

$$\text{Detention Time, h} = \frac{(0.785)\left(D^2\right)(\text{Depth, ft})\left(7.48 \text{ gal/ft}^3\right)}{\text{Flow Rate, gph}} \quad (7.75)$$

Example 7.109

Problem: A sedimentation tank has a volume of 137,000 gal. If the flow to the tank is 121,000 gph, what is the detention time in the tank in hours?

Solution:

$$\text{Detention Time, h} = \frac{\text{Volume of Tank, gal}}{\text{Flow Rate, gph}}$$

$$= \frac{137,000 \text{ gal}}{121,000 \text{ gph}}$$

$$= 1.1 \text{ h}$$

Example 7.110

Problem: A sedimentation basin is 60 ft long, 22 ft wide, and has water to a depth of 10 ft. If the flow to the basin is 1,500,000 gpd, what is the sedimentation basin detention time?

Solution:

First, convert the flow rate from gpd to gph so that time units will match. (1,500,000 gpd ÷ 24 h/day = 62,500 gph). Then, calculate the detention time.

$$\text{Detention Time, h} = \frac{\text{Volume of Tank, gal}}{\text{Flow Rate, gph}}$$

$$= \frac{(60 \text{ ft})(22 \text{ ft})(10 \text{ ft})\left(7.48 \text{ gal/ft}^3\right)}{62,500 \text{ gph}}$$

$$= 1.6 \text{ h}$$

Surface Overflow Rate

The surface loading rate—similar to the hydraulic loading rate (flow per unit area)—is used to determine loading on sedimentation basins and circular clarifiers. The hydraulic loading rate, however, measures the total water entering the process, whereas the surface overflow rate measures only the water overflowing the process (plant flow only).

Note: Surface overflow rate calculations do not include recirculated flows. Other terms used synonymously with surface overflow rate are Surface Loading Rate and Surface Settling Rate.

The surface overflow rate is determined using the following equation:

$$\text{Surface Overflow Rate} = \frac{\text{Flow, gpm}}{\text{Area, ft}^2} \qquad (7.76)$$

Example 7.111

Problem: A circular clarifier has a diameter of 80 ft. If the flow to the clarifier is 1,800 gpm, what is the surface overflow rate in gpm/ft²?

Solution:

$$\text{Surface Overflow Rate} = \frac{\text{Flow, gpm}}{\text{Area, ft}^2}$$

$$= \frac{1,800 \text{ gpm}}{(0.785)(80 \text{ ft})(80 \text{ ft})}$$

$$= 0.36 \text{ gpm/ft}^2$$

Example 7.112

Problem: A sedimentation basin 70-ft by 25 ft receives a flow of 1,000 gpm. What is the surface overflow rate in gpm/ft²?

Solution:

$$\text{Surface Overflow Rate} = \frac{\text{Flow, gpm}}{\text{Area, ft}^2}$$

$$= \frac{1,000 \text{ gpm}}{(70 \text{ ft})(25 \text{ ft})}$$

$$= 0.6 \text{ gpm/ft}^2$$

Mean Flow Velocity

The measure of the average velocity of the water as it travels through a rectangular sedimentation basin is known as **mean flow velocity**. Mean flow velocity is calculated using Equation (7.77).

$$Q \text{ (Flow), ft}^3/\text{min} = A \text{ (Cross-Sectional Area), ft}^2 \\ \times V \text{ (Volume) ft/min} \qquad (7.77)$$

$$(Q = A \times V)$$

Example 7.113

Problem: A sedimentation basin 60 ft long and 18 ft wide has water to a depth of 12 ft. When the flow through the basin is 900,000 gpd, what is the mean flow velocity in the basin in ft/min?

Solution:

Because velocity is desired in ft/min, the flow rate in the Q=AV equation must be expressed in ft³/min (cfm):

$$\frac{900,000 \text{ gpd}}{(1,440 \text{ min/day}) (7.48 \text{ gal/ft}^3)} = 84 \text{ cfm}$$

Then, use the Q=AV equation to calculate velocity:

$$Q = AV$$

$$84 \text{ cfm} = (18 \text{ ft})(12 \text{ ft})(x \text{ fpm})$$

$$x = \frac{84}{(18)(12)}$$

$$= 0.4 \text{ fpm}$$

Example 7.114

Problem: A rectangular sedimentation basin 50 ft long and 20 ft wide has a water depth of 9 ft. If the flow to the basin is 1,880,000 gpd, what is the mean flow velocity in ft/min?

Solution:

Because velocity is desired in ft/min, the flow rate in the Q=AV equation must be expressed in ft3/min (cfm):

$$\frac{1,880,000 \text{ gpd}}{(1,440 \text{ min/day}) (7.48 \text{ gal/ft}^3)} = 175 \text{ cfm}$$

The use the Q=AV equation to calculate velocity:

$$Q = AV$$

$$175 \text{ cfm} = (20 \text{ ft})(9 \text{ ft})(x \text{ fpm})$$

$$x = \frac{175 \text{ cfm}}{(20)(9)}$$

$$x = 0.97 \text{ fpm}$$

Weir Loading Rate (Weir Overflow Rate)

The weir loading rate (weir overflow rate) is the amount of water leaving the settling tank per linear foot of the weir. The result of this calculation can be compared with the design. Normally, the weir overflow rates of 10,000–20,000 gal/day/ft are used in the design of a settling tank. Typically, the weir loading rate is a measure of the gallons per minute (gpm) flow over each foot (ft) of the weir. The weir loading rate is determined using the following equation:

$$\text{Weir Loading Rate, gpm/ft} = \frac{\text{Flow, gpm}}{\text{Weir Length, ft}}$$

Example 7.115

Problem: A rectangular sedimentation basin has a total of 115 ft of weir. What is the weir loading rate in gpm/ft when the flow is 1,110,000 gpd?

Solution:

$$\frac{1,110,000 \text{ gpd}}{1,440 \text{ min/day}} = 771 \text{ gpm}$$

$$\text{Weir Loading Rate} = \frac{\text{Flow, gpm}}{\text{Weir Length, ft}}$$

$$= \frac{771 \text{ gpm}}{115 \text{ ft}}$$

$$= 6.7 \text{ gpm/ft}$$

Example 7.116

Problem: A circular clarifier receives a flow of 3.55 MGD. If the diameter of the weir is 90 ft, what is the weir loading rate in gpm/ft?

Solution:

$$\frac{3,550,000 \text{ gpd}}{1,440 \text{ min/day}} = 2,465 \text{ gpm}$$

$$\text{Feet of weir} = (3.14)(90 \text{ ft})$$

$$= 283 \text{ ft}$$

$$\text{Weir Loading Rate} = \frac{\text{Flow, gpm}}{\text{Weir Length, ft}}$$

$$= \frac{2,465 \text{ gpm}}{283 \text{ ft}}$$

$$= 8.7 \text{ gpm/ft}$$

Percent Settled Biosolids

The percent settled biosolids test (volume over volume test, or V/V test) is conducted by collecting a 100-mL slurry sample from the solids contact unit and allowing it to settle for 10 min. After 10 min, the volume of settled biosolids at the bottom of the 100-mL graduated cylinder is measured and recorded. The equation used to calculate percent settled biosolids is shown below.

$$\% \text{ Settled Biosolids} = \frac{\text{Settled Biosolids Volume, mL}}{\text{Total Sample Volume, mL}} \times 100$$

$$(7.78)$$

Example 7.117

Problem: A 100-mL sample of slurry from a solids contact unit is placed in a graduated cylinder and allowed to sit for 10 min. The settled biosolids at the bottom of the graduated cylinder after 10 min are 22 mL. What is the percent of settled biosolids in the sample?

Solution:

$$\% \text{ Settled Biosolids} = \frac{\text{Settled Biosolids, mL}}{\text{Total Sample, mL}} \times 100$$

$$= \frac{22 \text{ mL}}{100 \text{ mL}} \times 100$$

$$= 19 \% \text{ Settled Biosolids}$$

Example 7.118

Problem: A 100-mL sample of slurry from a solids contact unit is placed in a graduated cylinder. After 10 min, a total of 21 mL of biosolids settled at the bottom of the cylinder. What is the percent of settled biosolids in the sample?

$$\text{Total Alk. Required, mg/L} = \text{Alk. Reacting with Alum, mg/L}$$

$$+ \text{Alk. in the Water, mg/L}$$

Determining Lime Dosage (mg/L)

During the alum dosage process, lime is sometimes added to provide adequate alkalinity (HCO_3^-) in the solids contact clarification process for the coagulation and precipitation of solids. To determine the lime dose required, in mg/L, three steps are necessary.

In **Step 1**, the total alkalinity required is calculated. The total alkalinity required to react with the alum to be added and to provide proper precipitation is determined using the following equation:

$$\text{Total Alk. Required, mg/L} = \text{Alk. Reacting with Alum, mg/L}$$

$$+ \text{Alk. in the Water, mg/L}$$

$$\uparrow$$

$$\left(1 \text{ mg/L alum reacts w/0.45 mg/L Alk.}\right)$$

$$(7.79)$$

Example 7.119

Problem: Raw water requires an alum dose of 45 mg/L, as determined by jar testing. If a residual 30-mg/L alkalinity must be present in the water to ensure complete precipitation of alum added, what is the total alkalinity required in mg/L?

Solution:

First, calculate the alkalinity that will react with 45 mg/L alum:

$$\frac{0.45 \text{ mg/L Alk.}}{1 \text{ mg/L Alum}} = \frac{x \text{ mg/L Alk}}{45 \text{ mg/L Alum}}$$

$$(0.45)(45) = x$$

$$= 20.25 \text{ mg/L Alk.}$$

Then, calculate the total alkalinity required:

Total Alkalinity Req., mg/L = Alk to React w/Alum, mg/L

+Residual Alk, mg/L

= 20.25 mg/L + 30 mg/L

= 50.25 mg/L

Example 7.120

Problem: Jar tests indicate that 36 mg/L alum is optimum for a particular raw water. If a residual 30 mg/L alkalinity must be present to promote complete precipitation of the alum added, what is the total alkalinity required, in mg/L?

Solution:

First, calculate the alkalinity that will react with 36 mg/L alum:

$$\frac{0.45 \text{ mg/L Alk.}}{1 \text{ mg/L Alum}} = \frac{x \text{ mg/L Alk.}}{x \text{ mg/L Alk.36 mg/L Alum}}$$

$$(0.45)(36) = x$$

$$= 16.2$$

Then, calculate the total alkalinity required:

Total Alk. Required, mg/L = 16.2 mg/L + 30 mg/L

= 46.2 mg/L

In **Step 2,** we make a comparison between the required alkalinity and alkalinity already in the raw water to determine how much mg/L alkalinity should be added to the water. The equation used to make this calculation is shown below:

Added Alkalinity = Tot. Alk. Req'd, mg/L

− Alk. Present in Water, mg/L   (7.80)

Example 7.121

Problem: A total of 44 mg/L alkalinity is required to react with alum and ensure proper precipitation. If the raw water has an alkalinity of 30 mg/L as bicarbonate, how many mg/L of alkalinity should be added to the water?

Solution:

Alk. to be added, mg/L = Total Alk. Req'd, mg/L

−Alk. Present in the Water, mg/L

= 44 mg/L − 30 mg/L

= 14 mg/L Alkalinity to be added

In **Step 3,** after determining the amount of alkalinity to be added to the water, we calculate how much lime

(the source of alkalinity) needs to be added. We accomplish this by using the ratio shown in Example 7.119.

Example 7.122

Problem: It has been calculated that 16 mg/L of alkalinity must be added to raw water. How much mg/L of lime will be required to provide this amount of alkalinity? (1 mg/L of alum reacts with 0.45 mg/L and 1 mg/L of alum reacts with 0.35 mg/L of lime.)

Solution:

First, determine the mg/L of lime required by using a proportion that relates bicarbonate alkalinity to lime:

$$\frac{0.45 \text{ mg/L Alk.}}{0.35 \text{ mg/L Lime}} = \frac{16 \text{ mg/L Alk.}}{x \text{ mg/L Lime}}$$

Then, cross-multiply:

$$0.45x = (16)(0.35)$$

$$x = \frac{(16)(0.35)}{0.45}$$

$$x = 12.4 \text{ mg/L Lime}$$

In Example 7.123, we use all three steps to determine the lime dosage (mg/L) required.

Example 7.123

Problem: Given the following data, calculate the lime dose required in mg/L:
    Alum dose required (determined by jar tests)—52 mg/L
    Residual alkalinity required for precipitation—30 mg/L
    1 mg/L alum reacts with 0.35 mg/L lime
    1 mg/L alum reacts with 0.45 mg/L alkalinity
    Raw water alkalinity—36 mg/L

Solution:

To calculate the total alkalinity required, you must first calculate the alkalinity that will react with 52 mg/L of alum:

$$\frac{0.45 \text{ mg/L Alk.}}{1 \text{ mg/L Alum}} = \frac{x \text{ mg/L Alk.}}{52 \text{ mg/L Alum}}$$

$$(0.45)(52) = x$$

$$23.4 \text{ mg/L Alk.} = x$$

The total alkalinity requirement can now be determined:

Total Alk. Required, mg/L = Alk. to React w/Alum, mg/L

+Residual Alk, mg/L

= 23.4 mg/L + 30 mg/L

= 53.4 mg/L Total Alkalinity Required

Next, calculate how much alkalinity must be **added** to the water:

Alk. to be Added, mg/L = Total Alk Required, mg/L

$$-\text{Alk. Present, mg/L}$$

$$= 53.4 \text{ mg/L} - 36 \text{ mg/L}$$

$$= 17.4 \text{ mg/L Alk to be added to the Water}$$

Finally, calculate the lime required to provide this additional alkalinity:

$$\frac{0.45 \text{ mg/L Alk.}}{0.35 \text{ mg/L Lime}} = \frac{17.4 \text{ mg/L Alk.}}{x \text{ mg/L Lime}}$$

$$0.45 \, x = (17.4)(0.35)$$

$$x = \frac{(17.4)(0.35)}{0.45}$$

$$x = 13.5 \text{ mg/L Lime}$$

Determining Lime Dosage (lb/day)

After the lime dose has been determined in terms of mg/L, it is a fairly simple matter to calculate the lime dose in lb/day, which is one of the most common calculations in water and wastewater treatment. To convert from mg/L to lb/day lime dose, we use the following equation:

$$\text{Lime, lb/day} = (\text{Lime, mg/L})(\text{Flow, MGD})(8.34 \text{ lb/gal})$$

(7.81)

Example 7.124

Problem: The lime dose for raw water has been calculated to be 15.2 mg/L. If the flow to be treated is 2.4 MGD, how many lb/day of lime will be required?

Solution:

$$
\begin{aligned}
\text{Lime, lb/day} &= (\text{Lime, mg/L})(\text{Flow, MGD})(8.34 \text{ lb/gal}) \\
&= (15.2 \text{ mg/L})(2.4 \text{ MGD})(8.34 \text{ lb/gal}) \\
&= 304 \text{ lb/day Lime}
\end{aligned}
$$

Example 7.125

Problem: The flow to a solids contact clarifier is 2,650,000 gpd. If the lime dose required is determined to be 12.6 mg/L, how many lb/day of lime will be required?

Solution:

$$
\begin{aligned}
\text{Lime, lb/day} &= (\text{Lime, mg/L})(\text{Flow, MGD})(8.34 \text{ lb/gal}) \\
&= (12.6 \text{ mg/L})(2.65 \text{ MGD})(8.34 \text{ lb/gal}) \\
&= 278 \text{ lb/day Lime}
\end{aligned}
$$

Key Point: Turbidity is the result of suspended solids casing cloudiness or haziness of the water.

Determining Lime Dosage (g/min)

In converting from mg/L lime to grams/min (g/min) lime, use Equation (7.82).

Key Point: 1 lb = 453.6 g.

$$
\begin{aligned}
\text{Lime, lb/day} &= (\text{Lime, mg/L})(\text{Flow, MGD})(8.34 \text{ lb/gal}) \\
&= (12.6 \text{ mg/L})(2.65 \text{ MGD})(8.34 \text{ lb/gal}) \\
&= 278 \text{ lb/day Lime}
\end{aligned}
$$

(7.82)

Example 7.126

Problem: A total of 275 lb/day of lime will be required to raise the alkalinity of the water passing through a solids-contact clarification process. How many g/min lime does this represent?

Solution:

$$
\begin{aligned}
\text{Lime, g/min} &= \frac{(\text{lb/day})(453.6 \text{ g/lb})}{1,440 \text{ min/day}} \\
&= \frac{(275 \text{ lb/day})(453.6 \text{ g/lb})}{1,440 \text{ min/day}} \\
&= 86.6 \text{ g/min Lime}
\end{aligned}
$$

Example 7.127

Problem: A lime dose of 150 lb/day is required for a solids-contact clarification process. How many g/min of lime does this represent?

Solution:

$$
\begin{aligned}
\text{Lime, g/min} &= \frac{(\text{lb/day})(453.6 \text{ g/lb})}{1,440 \text{ min/day}} \\
&= \frac{(150 \text{ lb/day})(453.6 \text{ g/lb})}{1,440 \text{ min/day}} \\
&= 47.3 \text{ g/min Lime}
\end{aligned}
$$

FILTRATION CALCULATIONS

Water filtration is a physical process of separating suspended and colloidal particles from waste by passing the water through granular material. The process of filtration involves straining, settling, and adsorption. As floc passes into the filter, the spaces between the filter grains become clogged, reducing the openings and increasing removal. Some material is removed merely because it settles on a media grain. One of the most important processes is the adsorption of the floc onto the surface of individual filter grains. In addition to removing silt and sediment, floc, algae, insect larvae, and any other large elements, filtration also contributes to the removal of bacteria and protozoans such as Giardia lamblia and Cryptosporidium. Some filtration processes are also used for iron and manganese removal.

The *Surface Water Treatment Rule* (SWTR) specifies four filtration technologies, although SWTR also allows the use of alternative filtration technologies, e.g., cartridge filters. These include slow sand filtration, rapid sand filtration, pressure filtration, diatomaceous earth filtration, and direct filtration. Of these, all but rapid sand filtration is commonly employed in small water systems that use filtration. Each type of filtration system has advantages and disadvantages. Regardless of the type of filter, however, filtration involves the processes of *straining* (where particles are captured in the small spaces between filter media grains), *sedimentation* (where the particles land on top of the grains and stay there), and *adsorption* (where a chemical attraction occurs between the particles and the surface of the media grains).

FLOW RATE THROUGH A FILTER (GPM)

The flow rate in gpm through a filter can be determined by simply converting the gpd flow rate, as indicated on the flow meter. The flow rate (gpm) can be calculated by taking the meter flow rate (gpd) and dividing it by 1,440 min/day, as shown in Equation (7.83).

$$\text{Flow Rate, gpm} = \frac{\text{Flow Rate, gpd}}{1,440 \text{ min/day}} \quad (7.83)$$

Example 7.128

Problem: The flow rate through a filter is 4.25 MGD. What is this flow rate expressed in gpm?

Solution:

$$\begin{aligned}\text{Flow Rate, gpm} &= \frac{4.25 \text{ MGD}}{1,440 \text{ min/day}} \\ &= \frac{4,250,000 \text{ gpd}}{1,440 \text{ min/day}} \\ &= 2,951 \text{ gpm}\end{aligned}$$

Example 7.129

Problem: During a 70-h filter run, a total of 22.4 million gallons of water are filtered. What is the average flow rate through the filter in gpm during this filter run?

Solution:

$$\begin{aligned}\text{Flow Rate, gpm} &= \frac{\text{Total Gallons Produced}}{\text{Filter Run, min}} \\ &= \frac{22,400,000 \text{ gal}}{(70 \text{ h})(60 \text{ min/h})} \\ &= 5,333 \text{ gpm}\end{aligned}$$

Example 7.130

Problem: At an average flow rate of 4,000 gpm, how long would a filter run (in hours) be required to produce 25 MG of filtered water?

Solution:

Write the equation as usual, filling in known data:

$$\text{Flow Rate (gpm)} = \frac{\text{Total Gallons Produced}}{\text{Filter Run, min}}$$

$$4,000 \text{ gpm} = \frac{25,000,000 \text{ gal}}{(x \text{ h})(60 \text{ min/h})}$$

Then, solve for x:

$$= \frac{25,000,000 \text{ gal}}{(4,000)(60)}$$

$$= 104 \text{ h}$$

Example 7.131

Problem: A filter box is 20 ft×30 ft (including the sand area). If the influent valve is shut, the water drops 3.0 in/min. What is the rate of filtration in MGD?

Solution:

Given:

$$\text{Filter Box} = 20 \text{ ft} \times 30 \text{ ft}$$

$$\text{Water drops} = 3.0 \text{ in/min}$$

Find the volume of water passing through the filter.

$$\text{Volume} = \text{Area} \times \text{Height}$$

$$\text{Area} = \text{Width} \times \text{Length}$$

Note: The best way to perform calculations for this type of problem is step-by-step, breaking down the problem into what is given and what is to be found.

Step 1
$$\text{Area} = 20 \text{ ft} \times 30 \text{ ft} = 600 \text{ ft}^2$$

Convert 3.0 inches into feet
Divide 3.0 by 12 to find feet.

$$3.0/12 = 0.25 \text{ ft}$$

$$\text{Volume} = 600 \text{ ft}^2 \times 0.25 \text{ ft}$$
$$= 150 \text{ ft}^3 \text{ of water passing through}$$
$$\text{the filter in 1 min}$$

Step 2
Convert cubic feet to gallons.
$$150 \text{ ft}^3 \times 7.48 \text{ gal/ft}^3 = 1,122 \text{ gal/min}$$

Step 3

The problem asks for the rate of filtration in MGD. To find MGD, multiply the number of gallons per minute by the number of minutes per day.

$$1,122 \text{ gal/min} \times 1,440 \text{ min/day} = 1.62 \text{ N}$$

Example 7.132

Problem: The influent valve to a filter is closed for 5 min. During this time, the water level in the filter drops 0.8 ft (10 in). If the filter is 45 ft long and 15 ft wide, what is the gpm flow rate through the filter? The water drop equals 0.16 ft/min.

Solution:

First, calculate the cfm flow rate using the Q=AV equation:

$$
\begin{aligned}
Q, \text{cfm} &= (\text{Length, ft})(\text{Width, ft})(\text{Drop Velocity, ft/min}) \\
&= (45 \text{ ft})(15 \text{ ft})(0.16 \text{ ft/min}) \\
&= 108 \text{ cfm}
\end{aligned}
$$

Then, convert the cfm flow rate to gpm flow rate:

$$(108 \text{ cfm})(7.48 \text{ gal/ft}^3) = 808 \text{ gpm}$$

FILTRATION RATE

One measure of filter production is the filtration rate (generally ranging from 2 to 10 gpm/ft²). Along with the filter run time, it provides valuable information for the operation of filters. It is the gallons per minute of water filtered through each square foot of filter area. The filtration rate is determined using Equation (7.84).

$$\text{Filtration Rate, gpm/ft}^2 = \frac{\text{Flow Rate, gpm}}{\text{Filter Surface Area, ft}^2} \quad (7.84)$$

Example 7.133

Problem: A filter 18 ft by 22 ft receives a flow of 1,750 gpm. What is the filtration rate in gpm/ft²?

Solution:

$$
\begin{aligned}
\text{Filtration Rate} &= \frac{\text{Flow Rate, gpm}}{\text{Filter Surface Area, ft}^2} \\
&= \frac{1,750 \text{ gpm}}{(18 \text{ ft})(22 \text{ ft})} \\
&= 4.4 \text{ gpm/ft}^2
\end{aligned}
$$

Example 7.134

Problem: A filter 28 ft long and 18 ft wide treats a flow of 3.5 MGD. What is the filtration rate in gpm/ft²?

Solution:

$$\text{Flow Rate} = \frac{3,500,000 \text{ gpd}}{1440 \text{ min/day}} = 2,431 \text{ gpm}$$

$$
\begin{aligned}
\text{Filtration Rate, gpm/ft}^2 &= \frac{\text{Flow Rate, gpm}}{\text{Filter Surface Area, ft}^2} \\
&= \frac{2,431 \text{ gpm}}{(28 \text{ ft})(18 \text{ ft})} \\
&= 4.8 \text{ gpm/ft}^2
\end{aligned}
$$

Example 7.135

Problem: A filter 45 ft long and 20 ft wide produces a total of 18 MG during a 76-h filter run. What is the average filtration rate in gpm/ft² for this filter run?

Solution:

First, calculate the gpm flow rate through the filter:

$$
\begin{aligned}
\text{Flow Rate, gpm} &= \frac{\text{Total Gallons Produced}}{\text{Filter Run, min}} \\
&= \frac{18,000,000 \text{ gal}}{(76 \text{ h})(60 \text{ min/h})} \\
&= 3,947 \text{ gpm}
\end{aligned}
$$

Then, calculate the filtration rate:

$$
\begin{aligned}
\text{Filtration Rate} &= \frac{\text{Flow Rate, gpm}}{\text{Filter Area, ft}^2} \\
&= \frac{3,947 \text{ gpm}}{(45 \text{ ft})(20 \text{ ft})} \\
&= 4.4 \text{ gpm/ft}^2
\end{aligned}
$$

Example 7.136

Problem: A filter is 40 ft long and 20 ft wide. During a test of flow rate, the influent valve to the filter is closed for 6 min. The water level drop during this period is 16 in. What is the filtration rate for the filter in gpm/ft²?

Solution:

First, calculate gpm flow rate, using the Q=AV equation:

$$Q, \text{gpm} = (\text{Length, ft})(\text{Width, ft})$$

$$(\text{Drop Velocity, ft/min})(7.48 \text{ gal/ft}^3)$$

$$= \frac{(40 \text{ ft})(20 \text{ ft})(1.33 \text{ ft})(7.48 \text{ gal/ft}^3)}{6 \text{ min}}$$

$$= 1,316 \text{ gpm}$$

Then, calculate the filtration rate:

$$
\begin{aligned}
\text{Filtration Rate} &= \frac{\text{Flow Rate, gpm}}{\text{Filter Area, ft}^2} \\
&= \frac{1,316 \text{ gpm}}{(40 \text{ ft})(20 \text{ ft})} \\
&= 1.6 \text{ gpm/ft}^2
\end{aligned}
$$

Unit Filter Run Volume (UFRV)

The unit filter run volume (UFRV) calculation indicates the total gallons passing through each square foot of filter surface area during an entire filter run. This calculation is used to compare and evaluate filter runs. UFRVs are usually at least 5,000 gal/ft² and generally in the range of 10,000 gpd/ft². The UFRV value will begin to decline as the performance of the filter begins to deteriorate. The equation to be used in these calculations is shown below.

$$UFRV = \frac{\text{Total Gallons Filtered}}{\text{Filter Surface Area, ft}^2} \qquad (7.85)$$

Example 7.137

Problem: The total water filtered during a filter run (between backwashes) is 2,220,000 gal. If the filter is 18 ft by 18 ft, what is the unit filter run volume (UFRV) in gal/ft²?

Solution:

$$\begin{aligned} UFRV &= \frac{\text{Total Gallons Filtered}}{\text{Filter Surface Area, ft}^2} \\ &= \frac{2{,}220{,}000 \text{ gal}}{(18 \text{ ft})(18 \text{ ft})} \\ &= 6{,}852 \text{ gal/ft}^2 \end{aligned}$$

Key Point: When sampling for copper and lead, the tap must remain off for 6 h.

Example 7.138

Problem: The total water filtered during a filter run is 4,850,000 gal. If the filter is 28 ft by 18 ft, what is the unit filter run volume in gal/ft²?

Solution:

$$\begin{aligned} UFRV &= \frac{\text{Total Gallons Filtered}}{\text{Filter Surface Area, ft}^2} \\ &= \frac{4{,}850{,}000 \text{ gal}}{(28 \text{ ft})(18 \text{ ft})} \\ &= 9{,}623 \text{ gal/ft}^2 \end{aligned}$$

Equation (7.85) can be modified, as shown in Equation (7.86), to calculate the unit filter run volume given the filtration rate and filter run data.

$$UFRV = (\text{Filtration Rate, gpm, ft}^2)(\text{Filter Run Time, min}) \qquad (7.86)$$

Example 7.139

Problem: The average filtration rate for a filter was determined to be 2.0 gpm/ft². If the filter run time was 4,250 min, what was the unit filter run volume in gal/ft²?

Solution:

$$\begin{aligned} UFRV &= (\text{Filtration Rate, gpm/ft}^2)(\text{Filter Run Time, min}) \\ &= 8{,}500 \text{ gal/ft}^2 \end{aligned}$$

The problem indicates that, at an average filtration rate of 2.0 gal entering each square foot of the filter each minute, the total gallons entering during the total filter run is 4,250 times that amount.

Example 7.140

Problem: The average filtration rate during a particular filter run was determined to be 3.2 gpm/ft². If the filter run time was 61.0 h, what was the UFRV in gal/ft² for the filter run?

Solution:

$$\begin{aligned} UFRV &= (\text{Filtration Rate, gpm/ft}^2)(\text{Filter Run, h})(60 \text{ min/h}) \\ &= (3.2 \text{ gpm/ft}^2)(61.0 \text{ h})(60 \text{ min/h}) \\ &= 11{,}712 \text{ gal/ft}^2 \end{aligned}$$

BACKWASH RATE

In filter backwashing, one of the most important operational parameters to be determined is the amount of water in gallons required for each backwash. This amount depends on the design of the filter and the quality of the water being filtered. The actual washing typically lasts 5–10 min and amounts to 1%–5% of the flow produced.

Example 7.141

Problem: A filter has the following dimensions:

$$\text{Length} = 30 \text{ ft}$$

$$\text{Width} = 20 \text{ ft}$$

$$\text{Width} = 20 \text{ ft}$$

Assuming a backwash rate of 15 gal/ft²/min is recommended, and 10 min of backwash is required, calculate the amount of water in gallons required for each backwash.
Given:

$$\text{Length} = 30 \text{ ft}$$

$$\text{Width} = 20 \text{ ft}$$

$$\text{Depth of filter media} = 24 \text{ in}$$

$$\text{Rate} = 15 \text{ gal/ft}^2/\text{min}$$

Solution:

Find the amount of water in gallons required:

Step 1: Area of filter = 30 ft × 20 ft = 600 ft²
Step 2: Gallons of water used per square foot of filter
= 15 gal/ft²/min × 10 min = 150 gal/ft²
Step 3: Gallons required = 150 gal/ft² × 600 ft²
= 90,000 gal required for backwash

Typically, backwash rates will range from 10 to 25 gpm/ft². The backwash rate is determined by using Equation (7.87).

$$\text{Backwash} = \frac{\text{Flow Rate, gpm}}{\text{Filter Area, ft}^2} \quad (7.87)$$

Example 7.142

Problem:
A filter 30 ft by 10 ft has a backwash rate of 3,120 gpm. What is the backwash rate in gpm/ft²?

Solution:

$$\text{Backwash Rate, gpm/ft}^2 = \frac{\text{Flow Rate, gpm}}{\text{Filter Area, ft}^2}$$
$$= \frac{3120 \text{ gpm}}{(30 \text{ ft})(10 \text{ ft})}$$
$$= 10.4 \text{ gpm/ft}^2$$

Example 7.143

Problem: A filter 20 ft long and 20 ft wide has a backwash flow rate of 4.85 MGD. What is the filter backwash rate in gpm/ft²?

Solution:

$$\text{Backwash Rate} = \frac{\text{Flow Rate, gpm}}{\text{Filter Area, ft}^2}$$
$$= \frac{4,850,000 \text{ gpd}}{1,440 \text{ min/day}}$$
$$= 3,368 \text{ gpm}$$
$$= \frac{3,368 \text{ gpm}}{(20 \text{ ft})(20 \text{ ft})}$$
$$= 8.42 \text{ gpm/ft}^2$$

BACKWASH RISE RATE

Backwash rate is occasionally measured as the upward velocity of the water during backwashing, expressed as in/min rise. To convert from gpm/ft² backwash rate to in/min rise rate, use either Equation (7.88) or Equation (7.89).

$$\text{Backwash Rate, in/min} = \frac{\left(\text{Backwash Rate, gpm/ft}^2\right)(12 \text{ in/ft})}{7.48 \text{ gal/ft}^2}$$
$$(7.88)$$

$$\text{Backwash Rate, in/min} = \left(\text{Backwash Rate, gpm/ft}^2\right)(1.6) \quad (7.89)$$

Example 7.144

Problem: A filter has a backwash rate of 16 gpm/ft². What is this backwash rate expressed as in/min rise rate?

Solution:

$$\text{Backwash Rate, in/min} = \frac{\left(\text{Backwash Rate, gpm/ft}^2\right)(12 \text{ in/ft})}{7.48 \text{ gal/ft}^3}$$
$$= \frac{\left(16 \text{ gpm/ft}^2\right)(12 \text{ in/ft})}{7.48 \text{ gal/ft}^3}$$
$$= 25.7 \text{ in/min}$$

Example 7.145

Problem: A filter 22 ft long and 12 ft wide has a backwash rate of 3,260 gpm. What is this backwash rate expressed as in/min rise?

Solution:

First, calculate the backwash rate as gpm/ft²:

$$\text{Backwash Rate} = \frac{\text{Flow Rate, gpm}}{\text{Filter Area, ft}^2}$$
$$= \frac{3,260 \text{ gpm}}{(22 \text{ ft})(12 \text{ ft})}$$
$$= 12.3 \text{ gpm/ft}^2$$

Then, convert gpm/sq ft to in./min rise rate:

$$= \frac{\left(12.3 \text{ gpm/ft}^2\right)(12 \text{ in/ft})}{7.48 \text{ gal/ft}^3}$$
$$= 19.7 \text{ in/min}$$

Volume of Backwash Water Required, gal

To determine the volume of water required for backwashing, we must know both the desired backwash flow rate (gpm) and the duration of backwash (min):

$$\text{Backwash Water Vol., gal} = \left(\text{Backwash, gpm}\right)$$
$$\left(\text{Duration of Backwash, min}\right)$$
$$(7.90)$$

Example 7.146

Problem: For a backwash flow rate of 9,000 gpm and a total backwash time of 8 min, how many gallons of water will be required for backwashing?

Solution:

$$\text{Backwash Water Vol., gal} = (\text{Backwash, gpm})$$
$$(\text{Duration of Backwash, min})$$
$$= (9,000 \text{ gpm}) \ (8 \text{ min})$$
$$= 72,000 \text{ gal}$$

Example 7.147

Problem: How many gallons of water would be required to provide a backwash flow rate of 4,850 gpm for a total of 5 min?

$$\text{Backwash Water Vol., gal} = (\text{Backwash, gpm})$$
$$(\text{Duration of Backwash, min})$$
$$= (4,850 \text{ gpm}) \ (7 \text{ min})$$
$$= 33,950 \text{ gal}$$

REQUIRED DEPTH OF BACKWASH WATER TANK (FT)

The required depth of water in the backwash water tank is determined from the volume of water required for backwashing. To make this calculation, simply use Equation (7.91).

$$\text{Volume, gal} = (0.785)(D^2)(\text{Depth, ft})(7.48 \text{ gal/ft}^3)$$
$$(7.91)$$

Example 7.148

Problem: The volume of water required for backwashing has been calculated to be 85,000 gal. What is the required depth of water in the backwash water tank to provide this amount of water if the diameter of the tank is 60 ft?

Solution:

Use the volume equation for a cylindrical tank, fill in the known data, and then solve for x:

$$\text{Volume, gal} = (0.785)(D^2)(\text{Depth, ft})(7.48 \text{ gal/ft}^3)$$
$$85,000 \text{ gal} = (0.785)(60 \text{ ft})(60 \text{ ft})(x \text{ ft})(7.48 \text{ gal/ft}^3)$$
$$= \frac{85,000}{(0.785)(60)(60)(7.48)}$$
$$x = 4 \text{ ft}$$

Example 7.149

Problem: A total of 66,000 gal of water will be required for backwashing a filter at a rate of 8,000 gpm for a 9-min period. What depth of water is required in the backwash tank if it has a diameter of 50 ft?

Solution:

Use the volume equation for cylindrical tanks:

$$\text{Volume, gal} = (0.785)(D^2)(\text{Depth, ft})(7.48 \text{ gal/ft}^3)$$
$$66,000 \text{ gal} = (0.785)(50 \text{ ft})(50 \text{ ft})(x \text{ ft})(7.48 \text{ gal/ft}^3)$$
$$x = \frac{66,000}{(0.785)(50)(50)(7.48)}$$
$$x = 4.5 \text{ ft}$$

BACKWASH PUMPING RATE, GPM

The desired backwash pumping rate (gpm) for a filter depends on the desired backwash rate in gpm/ft^2, and the ft^2 area of the filter. The backwash pumping rate, gpm, can be determined by using Equation (7.92).

$$\text{Backwash Pumping Rate, gpm} = (\text{Desired Backwash}$$
$$\text{Rate, gpm/ft}^2)$$
$$(\text{Filter Area, ft}^2) \quad (7.92)$$

Example 7.150

Problem: A filter is 25 ft long and 20 ft wide. If the desired backwash rate is 22 gpm/ft^2, what backwash pumping rate (gpm) will be required?

Solution:

The desired backwash flow through each square foot of the filter area is 20 gpm. Therefore, he total gpm flow through the filter is 20 gpm times the entire square foot area of the filter:

$$\text{Backwash Pump. Rate, gpm} = (\text{Desired Backwash Rate, gpm/ft}^2)$$
$$(\text{Filter Area, ft}^2)$$
$$= (20 \text{ gpm/ft}^2)(25 \text{ ft})(20 \text{ ft})$$
$$= 10,000 \text{ gpm}$$

Example 7.151

Problem: The desired backwash pumping rate for a filter is 12 gpm/ft^2. If the filter is 20 ft long and 20 ft wide, what backwash pumping rate (gpm) will be required?

Solution:

$$\text{Backwash Pumping Rate, gpm} = (\text{Desired Backwash Rate, ft}^2)$$
$$(\text{Filter Area, ft}^2)$$
$$= (12 \text{ gpm/ft}^2)(20 \text{ ft})(20 \text{ ft})$$
$$= 4,800 \text{ gpm}$$

Percent Product Water Used for Backwatering

Along with measuring the filtration rate and filter run time, another aspect of filter operation monitored for filter performance is the percent of product water used for backwashing. The equation for the percent of product water used for backwashing calculations is shown below.

$$\text{Backwash Water, \%} = \frac{\text{Backwash Water, gal}}{\text{Water Filtered, gal}} \times 100 \quad (7.93)$$

Example 7.152

Problem: A total of 18,100,000 gal of water were filtered during a filter run. If 74,000 gal of this product water were used for backwashing, what percent of the product water was used for backwashing?

Solution:

$$
\begin{aligned}
\text{Backwash Water, \%} &= \frac{\text{Backwash Water, gal}}{\text{Water Filtered, gal}} \times 100 \\
&= \frac{74,000 \text{ gal}}{18,100,000 \text{ gal}} \times 100 \\
&= 0.4\,\%
\end{aligned}
$$

Example 7.153

Problem: A total of 11,400,000 gal of water are filtered during a filter run. If 48,500 gal of product water are used for backwashing, what percent of the product water is used for backwashing?

Solution:

$$
\begin{aligned}
\text{Backwash Water, \%} &= \frac{\text{Backwash Water, gal}}{\text{Water Filtered, gal}} \times 100 \\
&= \frac{48,500 \text{ gal}}{11,400,000 \text{ gal}} \times 100 \\
&= 0.43\,\% \text{ Backwash Water}
\end{aligned}
$$

Percent Mud Ball Volume

Mud balls are heavier deposits of solids near the top surface of the medium that break into pieces during backwash, resulting in spherical accretions (usually less than 12 in. in diameter) of floc and sand. The presence of mud balls in the filter media is checked periodically. The principal objection to mudballs is that they diminish the effective filter area.

Example 7.154

Problem: A 3,350 mL sample of filter media was taken for mud ball evaluation. The volume of water in the graduated cylinder rose from 500 to 525 mL when mud balls were placed in the cylinder. What is the percent mud ball volume of the sample?

Solution:

First, determine the volume of mud balls in the sample:

$$525 \text{ mL} - 500 \text{ mL} = 25 \text{ mL}$$

Then, calculate the percent mud ball volume:

$$
\begin{aligned}
\text{\% Mud Ball Volume} &= \frac{\text{Mud Ball Volume, mL}}{\text{Total Sample, Volume, mL}} \times 100 \\
&= \frac{25 \text{ mL}}{3,350 \text{ mL}} \times 100 \\
&= 0.75\%
\end{aligned}
$$

Example 7.155

Problem: A filter is tested for the presence of mud balls. The mud ball sample has a total volume of 680 mL. Five samples were taken from the filter. When the mud balls were placed in 500 mL of water, the water level rose to 565 mL. What is the percent mud ball volume of the sample?

Solution:

$$\text{\% Mud Ball Volume} = \frac{\text{Mud Ball Volume, mL}}{\text{Total Sample Vol., mL}} \times 100$$

The mud ball volume is the volume the water rose:

$$565 \text{ mL} - 500 \text{ mL} = 65 \text{ mL}$$

Because five samples of media were taken, the total sample volume is five times the sample volume:

$$(5) \times (680 \text{ mL}) = 3,400 \text{ mL}$$

$$
\begin{aligned}
\text{\% Mud Ball Volume} &= \frac{65 \text{ mL}}{3,400 \text{ mL}} \times 100 \\
&= 1.9\%
\end{aligned}
$$

Filter Bed Expansion

In addition to backwash rate, it is also important to expand the filter media during the wash to maximize the removal of particles held in the filter or by the media; that is, the efficiency of the filter wash operation depends on the expansion of the sand bed. Bed expansion is determined by measuring the distance from the top of the unexpanded media to a reference point (e.g., the top of the filter wall) and from the top of the expanded media to the same reference. A proper backwash rate should expand the filter by 20%–25%. Percent bed expansion is given by dividing the bed expansion by the total depth of expandable media (i.e., media depth less support gravels) and multiplying by 100 as follows:

Expanded Measurement = Depth to top of media during backwash (in)

Unexpanded Measurement = Depth to top of media before backwash (in)

Bed Exp. = Unexpanded measurement (in) − Expanded measurement (in)

$$\text{Bed Expansion}, \% = \frac{\text{Bed expansion measurement (in)}}{\text{Total depth of expandable media (in)}} \times 100$$

(7.94)

Example 7.156

Problem: The backwashing practices for a filter with 30 in of anthracite and sand are being evaluated. While at rest, the distance from the top of the media to the concrete floor surrounding the top of the filter is measured to be 41 in. After the backwash has been started and the maximum backwash rate is achieved, a probe containing a white disk is slowly lowered into the filter bed until anthracite is observed on the disk. The distance from the expanded media to the concrete floor is measured to be 34 in. What is the percent bed expansion?

Solution:

Given:

Unexpanded measurement = 41 in

Expanded measurement = 34.5 in

Bed expansion = 6.5

Bed expansion (percent) = (6.5 in/30 in) × 100 = 22%

WATER CHLORINATION CALCULATIONS

Chlorine is the most commonly used substance for the disinfection of water in the United States. The addition of chlorine or chlorine compounds to water is called chlorination. Chlorination is considered to be the single most important process for preventing the spread of waterborne diseases.

CHLORINE DISINFECTION

Chlorine deactivates microorganisms through several mechanisms, assuring that it can destroy most biological contaminants:

- It causes damage to the cell wall.
- It alters the permeability of the cell (the ability to pass water in and out through the cell wall).
- It alters the cell protoplasm.
- It inhibits the enzyme activity of the cell, making it unable to use its food to produce energy.
- It inhibits cell reproduction.

Chlorine is available in several different forms: (1) as pure elemental gaseous chlorine (a greenish-yellow gas possessing a pungent and irritating odor that is heavier than air, nonflammable, and nonexplosive); when released to the atmosphere, this form is toxic and corrosive; (2) as solid calcium hypochlorite (in tablets or granules); or (3) as a liquid sodium hypochlorite solution (in various strengths).

The strengths of one form of chlorine over the others for a given water system depend on the amount of water to be treated, the configuration of the water system, the local availability of the chemicals, and the skill of the operator. One of the major advantages of using chlorine is the effective residual that it produces. A residual indicates that disinfection is complete, and that the system has an acceptable bacteriological quality. Maintaining a residual in the distribution system helps to prevent the regrowth of microorganisms that were injured but not killed during the initial disinfection stage.

DETERMINING CHLORINE DOSAGE (FEED RATE)

The expressions milligrams per liter (mg/L) and pounds per day (lbs/day) are most often used to describe the amount of chlorine added or required. Equation (7.95) can be used to calculate either mg/L or lbs/day chlorine dosage.

$$\text{Chlorine Feed Rate, lb/day} = (\text{Chlorine, mg/L})(\text{flow, MGD})$$

$$(8.34, \text{lb/gal})$$

(7.95)

Example 7.157

Problem: Determine the chlorinator setting (lb/day) needed to treat a flow of 4 MGD with a chlorine dose of 5 mg/L.

Solution:

$$\text{Chlorine, lb/day} = (\text{Chlorine, mg/L})$$
$$(\text{Flow, MGD})(8.34 \text{ lb/gal})$$
$$= (5 \text{ mg/L})(4 \text{ MGD})(8.34 \text{ lb/gal})$$
$$= 167 \text{ lb/day}$$

Example 7.158

Problem: A pipeline 12 in. in diameter and 1,400 ft long is to be treated with a chlorine dose of 48 mg/L. How many lbs of chlorine will this require?

Solution:

First, determine the gallon volume of the pipeline:

$$\text{Volume, gal} = (0.785)(D^2)(\text{Length, ft})(7.48 \text{ gal/ft}^3)$$
$$= (0.785)(1 \text{ ft})(1 \text{ ft})(1,400 \text{ ft})(7.48 \text{ gal/ft}^3)$$
$$= 8,221 \text{ gal}$$

Now, calculate the lb of chlorine required

$$\text{lbs Chlorine} = (\text{Chlorine, mg/L})(\text{MG Volume})(8.34 \text{ lb/gal})$$
$$= (48 \text{ mg/L})(0.008221 \text{ MG})(8.34 \text{ lbs/gal})$$
$$= 3.3 \text{ lbs}$$

Example 7.159

Problem: A chlorinator setting is 30 lb/24 h. If the flow being chlorinated is 1.25 MGD, what is the chlorine dosage expressed as mg/L?

Solution:

$$\text{Chlorine, lb/day} = (\text{Chlorine, mg/L})(\text{flow, MGD})(8.34\ \text{lb/gal})$$

$$30\ \text{lb/day} = (x\ \text{mg/L})(\text{flow, MGD})(8.34\ \text{lb/gal})$$

$$x = \frac{30}{(1.25)(8.34)}$$

$$x = 2.9\ \text{mg/L}$$

Example 7.160

Problem: A flow of 1,600 gpm is to be chlorinated. At a chlorinator setting of 48 lb/24 h, what would be the chlorine dosage in mg/L?

Solution:

Convert the gpm flow rate to MGD flow rate:

$$(1,600\ \text{gpm})(1,440\ \text{min/day}) = 2,304,000\ \text{gpd}$$
$$= 2.304\ \text{MGD}$$

Now, calculate the chlorine dosage in mg/L:

$$\text{Chlorine, lb/day} = (\text{Chlorine, mg/L})$$
$$(\text{Flow, MGD}) \times 8.34\ \text{lb/gal}$$

$$(x\ \text{mg/L})(2.304\ \text{MGD})(8.34\ \text{lb/gal}) = 48\ \text{lb/day}$$

$$x = \frac{48}{(2.304)(8.34)}$$

$$x = 2.5\ \text{mg/L}$$

Calculating Chlorine Dose, Demand, and Residual

Common terms used in chlorination include the following:

- **Chlorine Dose:** The amount of chlorine added to the system. It can be determined by adding the desired residual for the finished water to the chlorine demand of the untreated water. Dosage can be either milligrams per liter (mg/L) or pounds per day (lbs/day). The most common is mg/L.

Chlorine Dose, mg/L = Chlorine Demand, mg/L

+Chlorine, mg/L Residual, mg/L

- **Chlorine Demand:** The amount of chlorine used by iron, manganese, turbidity, algae, and microorganisms in the water. Because the reaction between chlorine and microorganisms is not instantaneous, demand is relative to time. For instance, the demand 5 min after applying chlorine will be less than the demand after 20 min. Demand, like dosage is expressed in mg/L. The chlorine demand is as follows:

Chlorine Demand = Chlorine dose − Chlorine Residual

- **Chlorine Residual:** The amount of chlorine (determined by testing) remaining after the demand is satisfied. Residual, like demand, is based on time. The longer the time after dosage, the lower the residual will be, until all of the demand has been satisfied. Residual, like dosage and demand, is expressed in mg/L. The presence of a *free residual* of at least 0.2–0.4 ppm usually provides a high degree of assurance that the disinfection of the water is complete. *Combined residual* is the result of combining free chlorine with nitrogen compounds. Combined residuals are also called chloramines. *Total chlorine residual* is the mathematical combination of free and combined residuals. Total residual can be determined directly with standard chlorine residual test kits.

The following examples illustrate the calculation of chlorine dose, demand, and residual using Equation (7.96).

$$\text{Chlorine dose, mg/L} = \text{Chlorine demand, mg/L}$$
$$+\text{Chlorine Residual, mg/L} \quad (7.96)$$

Example 7.161

Problem: A water sample is tested and found to have a chlorine demand of 1.7 mg/L. If the desired chlorine residual is 0.9 mg/L, what is the desired chlorine dose in mg/L?

Solution:

$$\text{Chlorine dose, mg/L} = \text{Chlorine demand, mg/L}$$
$$+\text{Chlorine residual, mg/L}$$
$$= 1.7\ \text{mg/L} + 0.9\ \text{mg/L}$$
$$= 2.6\ \text{mg/L Chlorine dose}$$

Example 7.162

Problem: The chlorine dosage for water is 2.7 mg/L. If the chlorine residual after 30 min of contact time is found to be 0.7 mg/L, what is the chlorine demand expressed in mg/L?

Solution:

$$\text{Chlorine dose, mg/L} = \text{Chlorine demand, mg/L}$$
$$+\text{Chlorine residual, mg/L}$$
$$2.7\ \text{mg/L} = x\ \text{mg/L} + 0.6\ \text{mg/L}$$
$$2.7\ \text{mg/L} - 0.7\ \text{mg/L} = x\ \text{mg/L}$$
$$x\ \text{Chlorine Demand, mg/L} = 2.0\ \text{mg/L}$$

Example 7.163

Problem: What should the chlorinator setting be (lb/day) to treat a flow of 2.35 MGD if the chlorine demand is 3.2 mg/L and a chlorine residual of 0.9 mg/L is desired?

Solution:

Determine the chlorine dosage in mg/L:

$$\text{Chlorine dose, mg/L} = \begin{array}{l}\text{Chlorine demand, mg/L}\\+\text{Chlorine residual, mg/L}\end{array}$$
$$= 3.2 \text{ mg/L} + 0.9 \text{ mg/L}$$
$$= 4.1 \text{ mg/L}$$

Calculate the chlorine dosage (feed rate) in lb/day:

$$\text{Chlorine, lb/day} = \frac{(\text{Chlorine, mg/L})(\text{Flow, MGD})}{(8.34 \text{ lb/gal})}$$
$$= (4.1 \text{ mg/L})(2.35 \text{ MGD})(8.34 \text{ lb/gal})$$
$$= 80.4 \text{ lb/day Chlorine}$$

To calculate the actual increase in chlorine residual that would result from an increase in chlorine dose, we use the mg/L to lb/day equation as shown below.

$$\text{Increase in Chl., lb/day} = (\text{Expected Increase, mg/L})$$
$$(\text{Flow, MGD})(8.34 \text{ lb/gal}) \quad (7.97)$$

Key Point: The actual increase in residual is simply a comparison of the new and old residual data.

Example 7.164

Problem: A chlorinator setting is increased by 2 lb/day. The chlorine residual before the increased dosage was 0.2 mg/L. After the increased chlorine dose, the chlorine residual was 0.5 mg/L. The average flow rate being chlorinated is 1.25 MGD. Is the water being chlorinated beyond the breakpoint?

Solution:

Calculate the expected increase in chlorine residual. Use the mg/L to lb/day equation:

$$\text{Lbs/day Increase} = (\text{mg/L Increase})(\text{Flow, MGD})(8.34 \text{ lb/gal})$$
$$2 \text{ lb/day} = (x \text{ mg/L})(1.25 \text{ MGD})(8.34 \text{ lb/gal})$$
$$x = \frac{2}{(1.25)(8.34)}$$
$$x = 0.19 \text{ mg/L}$$

The actual increase in residual chlorine is:

$$0.5 \text{ mg/L} - 0.19 \text{ mg/L} = 0.31 \text{ mg/L}$$

Example 7.165

Problem: A chlorinator setting of 18 lb chlorine per 24 h results in a chlorine residual of 0.3 mg/L. The chlorinator setting is increased to 22 lb/24 h. The chlorine residual increased to 0.4 mg/L at this new dosage rate. The average flow being treated is 1.4 MGD. On the basis of this data, is the water being chlorinated past the breakpoint?

Solution:

Calculate the expected increase in chlorine residual:

$$\text{Lbs/day Increase} = (\text{mg/L Increase})(\text{Flow, MGD})(8.34 \text{ lb/gal})$$
$$4 \text{ lb/day} = (x \text{ mg/L})(1.4 \text{ MGD})(8.34 \text{ lb/gal})$$
$$x = \frac{4}{(1.4 \text{ MGD})(8.34)}$$
$$x = 0.34 \text{ mg/L}$$

The actual increase in residual:

$$0.4 \text{ mg/L} - 0.3 \text{ mg/L} = 0.1 \text{ mg/L}$$

Calculating Dry Hypochlorite Rate

The most commonly used dry hypochlorite, calcium hypochlorite, contains about 65%–70% available chlorine, depending on the brand. Because hypochlorites are not 100% pure chlorine, more lbs/day must be fed into the system to obtain the same amount of chlorine for disinfection. The equation used to calculate the lbs/day of hypochlorite needed can be found using Equation (7.98).

$$\text{Hypochlorite, lb/day} = \frac{\text{lb/day Chlorine}}{\frac{\% \text{ Available Chlorine}}{100}} \quad (7.98)$$

Example 7.166

Problem: A chlorine dosage of 110 lb/day is required to disinfect a flow of 1,550,000 gpd. If the calcium hypochlorite to be used contains 65% available chlorine, how many lb/day of hypochlorite will be required for disinfection?

Solution:

Because only 65% of the hypochlorite is chlorine, more than 110 lb of hypochlorite will be required:

$$\text{Hypochlorite, lbs/day} = \frac{\text{lb/day Chlorine}}{\frac{\% \text{ Available Chlorine}}{100}}$$
$$= \frac{110 \text{ lb/day}}{\frac{65}{100}}$$
$$= \frac{110}{0.65}$$
$$= 169 \text{ lb/day Hypochlorite}$$

Example 7.167

Problem: A water flow of 900,000 gpd requires a chlorine dose of 3.1 mg/L. If calcium hypochlorite (65% available chlorine) is to be used, how many lbs/day of hypochlorite are required?

Solution:

Calculate the lb/day of chlorine required:

$$\text{Chlorine, lb/day} = \frac{(\text{Chlorine, mg/L})(\text{Flow, MGD})}{(8.34 \text{ lb/gal})}$$

$$= (3.1 \text{ mg/L})(0.90 \text{ MGD})(8.34 \text{ lb/gal})$$

$$= 23 \text{ lb/day}$$

Calculate the lb/day of hypochlorite:

$$\text{Hypochlorite, lb/day} = \frac{\text{Chlorine, lb/day}}{\frac{\% \text{ Available Chlorine}}{100}}$$

$$= \frac{23 \text{ lb/day Chlorine}}{0.65 \text{ Available Chlorine}}$$

$$= 35 \text{ lb/Available day Hypochlorite}$$

Example 7.168

Problem: A tank contains 550,000 gal of water and is to receive a chlorine dose of 2.0 mg/L. How many pounds of calcium hypochlorite (65% available chlorine) will be required?

Solution:

$$\text{Hypochlorite, lbs} = \frac{\frac{(\text{mg/L Chlorine})(\text{MG Volume})}{(8.34 \text{ lbs/gal})}}{\frac{\% \text{ Available Chlorine}}{100}}$$

$$= \frac{\frac{(2.0 \text{ mg/L})(0.550 \text{ MG})(8.34 \text{ lbs/gal})}{65}}{100}$$

$$= \frac{9.2 \text{ lbs}}{0.65}$$

$$= 14.2 \text{ lb Hypochlorite}$$

Example 7.169

Problem: A total of 40 lb of calcium hypochlorite (65% available chlorine) is used in a day. If the flow rate treated is 1,100,000 gpd, what is the chlorine dosage in mg/L?

Solution:

Calculate the lb/day chlorine dosage:

$$\text{Hypochlorite, lb/day} = \frac{\text{Chlorine, lb/day}}{\frac{\% \text{ Available Chlorine}}{100}}$$

$$40 \text{ lb/day Hypochlorite} = \frac{x \text{ lb/day Chlorine}}{0.65}$$

$$(0.65)(40) = x$$

$$26 \text{ lb/day Chlorine} = x$$

Then, calculate mg/L chlorine using the mg/L to lb/day equation and filling in the known information:

$$26 \text{ lb/day Chlorine} = (x \text{ mg/L Chlorine})(1.10 \text{ MGD})(8.34 \text{ lb/gal})$$

$$x = \frac{26 \text{ lb/day}}{(1.10 \text{ MGD})(8.34 \text{ lb/gal})}$$

$$= 2.8 \text{ mg/L Chlorine}$$

Example 7.170

Problem: A flow of 2,550,000 gpd is disinfected with calcium hypochlorite (65% available chlorine). If 50 lb of hypochlorite is used in a 24-h period, what is the mg/L chlorine dosage?

Solution:

Calculate the lb/day chlorine dosage:

$$50 \text{ lb/day Hypochlorite} = \frac{x \text{ lb/day Chlorine}}{0.65}$$

$$x = 32.5 \text{ Chlorine}$$

Calculate mg/L chlorine:

$$(x \text{ mg/L Chlorine})(2.55 \text{ MGD})(8.34 \text{ lb/gal}) = 32.5 \text{ lb/day}$$

$$x = 1.5 \text{ mg/L Chlorine}$$

CALCULATING HYPOCHLORITE SOLUTION FEED RATE

Liquid hypochlorite (i.e., sodium hypochlorite) is supplied as a clear, greenish-yellow liquid in strengths from 5.25% to 16% available chlorine. Often referred to as "bleach," it is, in fact, used for bleaching—common household bleach is a solution of sodium hypochlorite containing 5.25% available chlorine. When calculating gallons per day (gpd) of liquid hypochlorite, the lb/day of hypochlorite required must be converted to gpd of hypochlorite required. This conversion is accomplished using Equation (7.99).

$$\text{Hypochlorite, gpd} = \frac{\text{Hypochlorite, lb/day}}{8.34 \text{ lb/gal}} \quad (7.99)$$

Example 7.171

Problem: A total of 50 lb/day sodium hypochlorite is required for disinfection of a 1.5-MGD flow. How many gallons per day of hypochlorite is this?

Solution:

Because lb/day of hypochlorite has already been calculated, we simply convert lbs/day to gpd of hypochlorite required:

$$\text{Hypochlorite, gpd} = \frac{\text{Hypochlorite, lb/day}}{8.34 \text{ lb/gal}}$$
$$= \frac{50 \text{ lb/day}}{8.34 \text{ lb/gal}}$$
$$= 6.0 \text{ gpd Hypochlorite}$$

Example 7.172

Problem: A hypochlorinator is used to disinfect the water pumped from a well. The hypochlorite solution contains 3% available chlorine. A chlorine dose of 1.3 mg/L is required for adequate disinfection throughout the system. If the flow being treated is 0.5 MGD, how many gpd of the hypochlorite solution will be required?

Solution:

Calculate the lb/day of chlorine required:

$$(1.3 \text{ mg/L})(0.5 \text{ MGD})(8.34 \text{ lb/gal}) = 5.4 \text{ lb/day Chlorine}$$

Calculate the lb/day of hypochlorite solution required:

$$\text{Hypochlorite, lb/day} = \frac{5.4 \text{ lb/day Chlorine}}{0.03}$$
$$= 180 \text{ lb/day Hypochlorite}$$

Calculate the gpd of hypochlorite solution required:

$$= \frac{180 \text{ lb/day}}{8.34 \text{ lb/gal}}$$
$$= 21.6 \text{ gpd Hypochlorite}$$

CALCULATING PERCENT STRENGTH OF SOLUTIONS

If a teaspoon of salt is dropped into a glass of water it gradually disappears. The salt dissolves in the water. A microscopic examination of the water would not show the salt. Only examination at the molecular level, which is not easily done, would show salt and water molecules intimately mixed. If we taste the liquid, of course, we would know that the salt is there. And we could recover the salt by evaporating the water. In a solution, the molecules of the salt, the **solute,** are homogeneously dispersed among the molecules of water, the **solvent**. This mixture of salt and water is homogeneous on a molecular level. Such a homogeneous mixture is called a **solution**. The composition of a solution can be varied within certain limits. The three common states of matter are gas, liquid, and solid. In this discussion, of course, we are only concerned, at the moment, with solids (calcium hypochlorite) and liquid (sodium hypochlorite) states.

CALCULATING PERCENT STRENGTH USING DRY HYPOCHLORITE

To calculate the percent strength of a chlorine solution, we use Equation (7.100).

$$\% \text{ Chlorine Strength} = \frac{\dfrac{(\text{Hypochlorite, lb})(\% \text{ Available Chlorine})}{100}}{\text{Water, lb} + \dfrac{(\text{Hypochlorite, lb})(\% \text{ Available Chlorine})}{100}} \times 100$$

$$(7.100)$$

Example 7.173

Problem: If a total of 72 oz. of calcium hypochlorite (65% available chlorine) is added to 15 gal of water, what is the percent chlorine strength (by weight) of the solution?

Solution:

Convert the ounces of hypochlorite to lb of hypochlorite:

$$\frac{72 \text{ oz.}}{16 \text{ oz./lb}} = 4.5 \text{ lb chemical}$$

$$\% \text{ Chlorine Strength} = \frac{\dfrac{(\text{Hypochlorite, lb})(\% \text{ Available Chlorine})}{100}}{\text{Water, lb} + \dfrac{(\text{Hypochlorite, lb})(\% \text{ Available Chlorine})}{100}} \times 100$$

$$= \frac{(4.5 \text{ lb})(0.65)}{(15 \text{ gal})(8.34 \text{ lb/gal}) + (4 \text{ lb})(0.65)} \times 100$$

$$= \frac{2.9 \text{ lb}}{125.1 \text{ lb} + 2.9 \text{ lb}} \times 100$$

$$= \frac{(2.9)(100)}{126}$$

$$= 2.3 \text{ Chlorine Strength}$$

CHEMICAL USE CALCULATIONS

In typical plant operations, chemical use is recorded each day. Such data provide a record of daily use from which the average daily use of the chemical or solution can be calculated. To calculate the average use in pounds per day (lb/day), we use Equation (7.101). To calculate the average use in gallons per day (gpd), we use Equation (7.102).

$$\text{Average Use, lb/day} = \frac{\text{Total Chemical Used, lb}}{\text{Number of Days}} \quad (7.101)$$

$$\text{Average Use, gpd} = \frac{\text{Total Chemicals Used, gal}}{\text{Number of Days}} \quad (7.102)$$

To calculate the day's supply in inventory, we use Equation (7.103) or Equation (7.104).

$$\text{Day's Supply in Inventory} = \frac{\text{Total Chemical in Inventory, lb}}{\text{Average Use, lb/day}} \quad (7.103)$$

$$\text{Day's Supply in Inventory} = \frac{\text{Total Chemical in Inventory, gal}}{\text{Average Use, gpd}} \quad (7.104)$$

Example 7.174

Problem: The pounds (lbs) of calcium hypochlorite used for each day during the week are given below. Based on this data, what was the average lb/day of hypochlorite chemical use during the week?

Monday—50 lb/day	Friday—56 lb/day
Tuesday—55 lb/day	Saturday—51 lb/day
Wednesday—51 lb/day	Sunday—48 lb/day
Thursday—46 lb/day	

Solution:

$$\text{Average Use, lb/day} = \frac{\text{Total Chemical Used, lb}}{\text{Number of Days}}$$
$$= \frac{357}{7}$$
$$= 51 \text{ lb/day Average Use}$$

Example 7.175

Problem: The average calcium hypochlorite use at a plant is 40 lb/day. If the chemical inventory in stock is 1,100 lb, how many days' supply is this?

$$\text{Days' Supply in Inventory} = \frac{\text{Total Chemical in Inventory, lb}}{\text{Average Use, lb/day}}$$
$$\text{Days' Supply in Inventory} = \frac{1,100 \text{ lb in Inventory}}{40 \text{ lb/day Average Use}}$$
$$= 27.5 \text{ days' Supply in Inventory}$$

FLUORIDATION

Note: The key terms used in this chapter are defined as follows:

- **Fluoride:** Found in many waters, it is also added to many water systems to reduce tooth decay.
  - **Dental Carries:** Tooth decay.
  - **Dental Fluorosis:** Result of excessive fluoride content in drinking water, causing mottled or discolored teeth.

Water Fluoridation

As of 1989, fluoridation in the United States was being practiced in approximately 8,000+ communities serving more than 126 million people. Residents of over 1,800 additional communities, serving more than nine million people, consumed water that contained at least 0.7-mg/L fluoride from natural sources. Some key facts about fluoride:

- Fluoride is seldom found in appreciable quantities in surface waters and appears in groundwater in only a few geographical regions.
- Fluoride is sometimes found in a few types of igneous or sedimentary rocks.
- Fluoride is toxic to humans in large quantities; it is also toxic to some animals
  - Based on human experience, fluoride, used in small concentrations (about 1.0 mg/L in drinking water), can be beneficial.

Fluoride Compounds

Theoretically, any compound that forms fluoride ions in a water solution can be used for adjusting the fluoride content of a water supply. However, there are several practical considerations involved in selecting compounds:

- The compound must have sufficient solubility to permit its use in routine water plant practice
- The cation to which the fluoride ion is attached must not have any undesirable characteristics
- The material should be relatively inexpensive and readily available in grades of size and purity suitable for its intended use.

Caution: Fluoride chemicals, like chlorine, caustic soda, and many other chemicals used in water treatment can constitute a safety hazard for the water plant operator unless proper handling precautions are observed. It is essential that the operator be aware of the hazards associated with each individual chemical prior to its use.

The three commonly used fluoride chemicals should meet the American Water Works Association (AWWA) standards for use in water fluoridation: sodium fluoride (B701-90), sodium fluorosilicate (B702-90), and fluorosilicic acid (B703-90).

Sodium Fluoride

The first fluoride compound used in water fluoridation was *sodium fluoride*. It was selected based on the above criteria and also because its toxicity and physiological effects had been so thoroughly studied. It has become the reference standard used in measuring fluoride concentration. Other compounds came into use, but sodium fluoride is still widely used because of its unique physical characteristics. Sodium fluoride (NaF) is a white, odorless material available either as a powder or in the form of crystals of various sizes. It is a salt that in the past was manufactured by adding sulfuric acid to fluorspar and then neutralizing the mixture with sodium carbonate. Neutralizing fluorosilicic acid with caustic soda (NaOH) now produces it. Approximately 19 lbs of sodium fluoride will add 1 ppm of fluoride to 1 million gallons of water. Sodium fluoride's solubility is practically constant at 4.0 g/100 mL in water at temperatures generally encountered in water treatment practice (see Table 7.5).

Sodium Fluorosilicate

Fluorosilicic acid can readily be converted into various salts, and one of these, *sodium fluorosilicate* (Na_2SiF_6), also known as sodium silicofluoride is widely used as a chemical for water fluoridation. As with most fluorosilicates, it is generally obtained as a by-product from the manufacture of phosphate fertilizers. Phosphate rock is ground up and treated with sulfuric acid, thus forming a gas by-product. This gas reacts with water and forms fluorosilicic acid. When neutralized with sodium carbonate, sodium fluorosilicate will precipitate out. The conversion of fluorosilicic acid to a dry material containing a high percentage of available fluoride results in a compound that has most of the advantages of the acid, with few of its disadvantages. Once it was shown that fluorosilicates form fluoride ions in water solution as readily as simple fluoride compounds and that there is no difference in the physiological effect, fluorosilicates were rapidly accepted for water fluoridation and, in many cases, have displaced the use of sodium fluoride, except in saturators. Sodium fluorosilicate is a white, odorless crystalline powder. Its solubility varies, see Table 7.5. Approximately 14 lbs of sodium fluorosilicate will add 1 ppm of fluoride to 1 million gallons of water.

Fluorosilicic Acid

Fluorosilicic acid (H_2SiF_6), also known as hydrofluorosilicic or silicofluoric acid is a 20%–35% aqueous solution with a formula weight of 144.08. It is a straw-colored, transparent, fuming, corrosive liquid having a pungent odor and an irritating action on the skin. Solutions of 20%–35% fluorosilicic acid exhibit a low pH (1.2), and at a concentration of 1 ppm can slightly depress the pH of poorly buffered potable waters. It must be handled with great care because it will cause a "delayed burn" on skin tissue. The specific gravity and density of fluorosilicic acid are given in Table 7.6. Approximately 46 lbs (4.4 gal) of 23% acid is required to add 1 ppm of fluoride to 1 million gallons of

TABLE 7.5
Solubility of Fluoride Chemicals

Chemical	Temperature	Solubility (g per 100 mL of H_2O)
Sodium fluoride	0.0	4.00
	15.0	4.03
	20.0	4.05
	25.0	4.10
	100.0	5.00
Sodium fluorosilicate	0.0	0.44
	25.0	0.76
	37.8	0.98
	65.6	1.52
	100.0	2.45
Fluorosilicic acid	Infinite at all temperatures	

Source: Reeves (1994, p. 21).

TABLE 7.6
Properties of Fluorosilicic Acid

Acid (%)[a]	Specific Gravity (s.g.)	Density (lbs/gal)
0 (water)	1.000	8.345
10	1.0831	9.041
20	1.167	9.739
23	1.191	9.938
25	1.208	10.080
30	1.250	10.431
35	1.291	10.773

Source: Reeves (1994, p. 21).

Note: Actual densities and specific gravities will be slightly higher when distilled water is not used. Add approximately 0.2 lb/gal to density depending on impurities.

[a] Based on the other percentage being distilled water.

water. Two different processes, resulting in products with differing characteristics, manufacture fluorosilicic acid. The largest production of the acid is a by-product of phosphate fertilizer manufacture. Phosphate rock is ground up and treated with sulfuric acid, forming a gas by-product. Hydrofluoric acid (HF) is an extremely corrosive material. Its presence in fluorosilicic acid, whether from intentional addition, i.e., "fortified" acid, or from normal production processes demands careful handling.

Optimal Fluoride Levels

The recommended optimal fluoride concentrations for fluoridated water supply systems are given in Table 7.7. These levels are based on the annual average of the maximum daily air temperature in the area of the involved school or community. In areas where the mean temperature is not shown in Table 7.7, the optimal fluoride level can be determined by the following formula.

TABLE 7.7
Recommended Optimal Fluoride Level

Annual Ave. of Max. Daily Air Temps[a] (°F)	Recommended Fluoride Concentrations		Recommended Control Range			
	Community (ppm)	School[b] (ppm)	Community Systems		School Systems	
			0.1 Below	0.5 Above	20% Low	20% High
40.0–53.7	1.2	5.4	1.1	1.7	4.3	6.5
53.8–58.3	1.1	5.0	1.0	1.6	4.0	6.0
58.4–63.8	1.0	4.5	0.9	1.5	3.6	5.4
63.9–70.6	0.9	4.1	0.8	1.4	3.3	4.9
70.7–79.2	0.8	3.6	0.7	1.3	2.9	4.3
79.3–90.5	0.7	3.2	0.6	1.2	2.6	3.8

Source: Reeves (1994, p. 21).
[a] Based on temperature data obtained for a minimum of 5 years.
[b] Based on 4.5 times the optimal fluoride level for communities.

$$\text{ppm fluoride} = \frac{0.34}{E} \quad (7.105)$$

where

E = the estimated average daily water consumption for children up to 10 years of age in ounces of water per pound of body weight. E is obtained from the formula:

$$E = 0.038 + 0.0062 \text{ average max. daily air temp. (°F)} \quad (7.106)$$

In Table 7.7, the recommended control range is shifted to the high side of the optimal fluoride level for two reasons:

1. It has become obvious that many water plant operators try to maintain the fluoride level in their community at the lowest level possible. The result is that the actual fluoride level in the water will vary around the lowest value in the range instead of around the optimal level.
2. Some studies have shown that sub-optimal fluorides are relatively ineffective in actually preventing dental caries. Even a drop of 0.2 ppm below optimal levels can significantly reduce dental benefits.

Important Point: In water fluoridation, underfeeding is a much more serious problem than overfeeding.

Fluoridation Process Calculations

In this section, important process calculations are discussed.

PERCENT FLUORIDE ION IN A COMPOUND

When calculating the percent fluoride ion present in a compound, we need to know the chemical formula for the compound (e.g., NaF) and the atomic weight of each element in the compound. The first step is to calculate the molecular weight of each element in the compound (Number of atoms × atomic weight = molecular weight). Then, we calculate the percent fluoride in the compound using Equation (7.107).

$$\% \text{ Fluoride in Compound} = \frac{\text{Molecular Weight of Fluoride}}{\text{Molecular Weight of Compound}} \times 100 \quad (7.107)$$

Important Point: Available fluoride ion concentration is abbreviated as AFI in the calculations that follow.

Example 7.176

Problem: Given the following data, calculate the percent fluoride in sodium fluoride (NaF).
Given:

Element	No. of Atoms	Atomic Weight	Molecular Weight
Na	1	22.997	22.997
F	1	19.00	19.00
		Molecular Weight of NaF	41.997

Solution:

Calculate the percent fluoride in NaF.

$$\% \text{ F in NaF} = \frac{\text{Molecular Weight of F}}{\text{Molecular Weight of NaF}} \times 100$$
$$= \frac{19.00}{41.997} \times 100$$
$$= 45.2\%$$

Key Point: The molecular weight of hydrofluorosilicic acid (H_2SiF_6) is 79.1% and sodium silicofluoride (Na_2SiF_6).

Fluoride Feed Rate

Adjusting the fluoride level in a water supply to an optimal level is accomplished by adding the proper concentration of a fluoride chemical at a consistent rate. To calculate the fluoride feed rate for any fluoridation feeder in terms of pounds of fluoride to be fed per day, it is necessary to determine:

- Dosage
- Maximum pumping rate (capacity)
- Chemical purity
- Available fluoride ion concentration

The fluoride feed rate formula is a general equation used to calculate the concentration of a chemical added to water. It will be used for all fluoride chemicals except sodium fluoride when used in a saturator.

Important Point: mg/L is equal to ppm.

The fluoride feed rate (the amount of chemical required to raise the fluoride content to the optimal level) can be calculated as follows:

$$\text{Fluoride Fd. Rt. (lb/day)} = \frac{\text{Dosage(mg/L)} \times \text{cap. (MGD)} \times 8.34 \text{ lb/gal}}{\text{AFI} \times \text{chemical purity}}$$

(7.108)

If the capacity is in MGD, the fluoride feed rate will be in pounds per day. If the capacity is in gpm, the feed rate will be in pounds per minute if a factor of 1 million is included in the denominator.

$$\text{Fluoride Fd. Rt. (lb/min)} = \frac{\text{Dosage(mg/L)} \times \text{cap. (gpm)} \times 8.34 \text{ lb/gal}}{1,000,000 \times \text{AFI} \times \text{chemical purity}}$$

(7.109)

Example 7.177

Problem: A water plant produces 2,000 gpm, and the city wants to add 1.1 mg/L of fluoride. What would the fluoride feed rate be?

Solution:

$$2,000 \text{ gpm} \times 1,440 \text{ min/day} = 2,880,000 \text{ gpd}$$

$$2,880,000 \text{ gpd} \div 1,000,000 = 2.88 \text{ MGD}$$

$$\text{Fluoride Feed Rate (lb/day)} = \frac{1.1 \text{ mg/L} \times 2.88 \text{ MGD} \times 8.34 \text{ lb/gal}}{0.607 \times 0.985}$$

$$\text{Fluoride Feed Rate, lb/day} = 44.2 \text{ lb/day}$$

The fluoride feed rate is 44.2 lb/day. Some feed rates from equipment design data sheets are given in grams/minute. To convert to grams/minute, divide by 1,440 min/day and multiply by 454 g/lb.

$$\text{Fluoride Fd. Rt. (g/min)} = 44.2 \text{ lb/day}, 1,440 \text{ min/day} \times 454 \text{ g/lb}$$

$$\text{Fluoride Feed Rate} = 13.9 \text{ g/min.}$$

Example 7.178

Problem: If it is known that the plant rate is 4,000 gpm and the dosage needed is 0.8 mg/L, what is the fluoride feed rate in milliliters per minute for 23% fluorosilicic acid?

Solution:

$$1,000,000 = 10^6$$

$$\text{Fluoride Feed Rate (lb/min)} = \frac{\text{Dosage(mg/L)} \times \text{cap.(gpm)} \times 8.34 \text{ lb/gal}}{10^6 \times \text{AFI} \times \text{chemical purity}}$$

$$\text{Fluoride Feed Rate (lb/min)} = \frac{0.8 \text{ mg/L} \times 4,000 \text{ gpm} \times 8.34 \text{ lb/gal}}{10^6 \times 0.79 \times 0.23}$$

$$\text{Fluoride Feed Rate} = 0.147 \text{ lb/min}$$

Note: A gallon of 23% fluorosilicic acid weighs 10 lb, and there are 3,785 mL/gal. The following formula can be used to convert the feed rate to mL/min:

$$\text{Fluoride Feed Rate (mL/min)} = \frac{0.147 \text{ lb/min,}}{10 \text{ lb/gal} \times 3,785 \text{ mL/gal}}$$

$$\text{Fluoride Feed Rate} = 55.6 \text{ mL/min}$$

Example 7.179

Problem: If a small water plant wishes to use sodium fluoride in a dry feeder, and the water plant has a capacity (flow) of 180 gpm, what would be the fluoride feed rate? Assume 0.1 mg/L natural fluoride and 1.0 mg/L is desired in the drinking water.

Important Point: The Centers for Disease Control (CDC) recommends against using sodium fluoride in a dry feeder.

Solution:

$$1,000,000 = 10^6$$

$$\text{Fluoride Feed Rate (lb/min)} = \frac{\begin{array}{c}\text{dosage}(\text{mg/L})\\ \times \text{cap. (gpm)} \times 8.34 \text{ lb/gal}\end{array}}{10^6 \times \text{AFI} \times \text{chemical purity}}$$

$$\text{Fluoride Feed Rate (lb/min)} = \frac{\begin{array}{c}(1.0 - 0.1)\ \text{mg/L}\\ \times 180\ \text{gpm} \times 8.34 \text{ lb/gal}\end{array}}{10^6 \times 0.45 \times 0.98}$$

$$\text{Fluoride Feed Rate} = 0.003 \text{ lb/min or } 0.18 \text{ lb/h}$$

Thus, sodium fluoride can be fed at a rate of 0.18 lbs/h to obtain 1.0 mg/L of fluoride in the water.

Fluoride Feed Rates for Saturator

A sodium fluoride saturator is unique in that the strength of the saturated solution formed is always 18,000 ppm. This is because sodium fluoride has a solubility that is practically constant at 4.0 g/100 mL of water at temperatures generally encountered in water treatment. This means that each liter of solution contains 18,000 mg of fluoride ion (40,000 mg/L times the percent available fluoride [45%] equals 18,000 mg/L). This simplifies calculations because it eliminates the need for weighing the chemicals. A meter on the water inlet of the saturator provides this volume. All that is needed is the volume of solution added to the water for calculated dosage.

$$\text{Fluoride Feed Rate (gpm)} = \frac{\text{Cap. (gpm)} \times \text{dosage (mg/L)}}{18,000 \text{ mg/L}}$$

$$(7.110)$$

The fluoride feed rate will have the same units as the capacity. If the capacity is in gallons per minute (gpm), the feed rate will be in gpm as well. If the capacity is in gallons per day (gpd), the feed rate will be in gpd.

Note: For the mathematician, the following derivation is given.

$$\text{Fluoride Feed Rate (lb/min)} = \frac{\begin{array}{c}\text{Dosage (mg/L)}\\ \times \text{cap. (gpm)} \times 8.34 \text{ lb/gal}\end{array}}{10^6 \times \text{AFI} \times \text{chemical purity}}$$

$$(7.111)$$

To change the Fluoride Feed Rate from pounds of dry feed to gallons of solution, divide by the concentration of sodium fluoride and the density of the solution (water).

Note: The chemical purity of the sodium fluoride in solution will be 4% × 8.34 lb/gal.

$$\text{Fluoride Feed Rate (gal/min)} = \frac{\begin{array}{c}\text{cap. (gpm)} \times \text{dosage (mg/L)}\\ \times 8.34 \text{ lb/gal}\end{array}}{10^6 \times \text{AFI} \times \text{chemical purity}}$$

$$\text{Fluoride Feed Rate (gal/min)} = \frac{\begin{array}{c}\text{cap. (gpm)} \times \text{dosage (mg/L)}\\ \times 8.34 \text{ lb/gal}\end{array}}{10^6 \times 0.45 \times 4\% \times 8.34 \text{ lb/gal}}$$

$$\text{Fluoride Feed Rate (gal/min)} = \frac{\text{cap. (gpm)} \times \text{dosage (mg/L)}}{10^6 \times 0.45 \times 0.04}$$

$$\text{Fluoride Feed Rate (gpm)} = \frac{\text{cap. (gpm)} \times \text{dosage (mg/L)}}{18,000 \text{ mg/L}}$$

$$(7.112)$$

Example 7.180 Feed Rate for Saturator

Problem: A water plant produces 1.0 MGD and has less than 0.1 mg/L of natural fluoride. What would the fluoride feed rate be to obtain 1.0 mg/L in the water?

Solution:

$$\text{Fluoride Feed Rate (gpd)} = \frac{\text{capacity (gpd)} \times \text{dosage (mg/L)}}{18,000 \text{ mg/L}}$$

$$\text{Fluoride Feed Rate (gpd)} = \frac{1,000,000 \text{ gpd} \times 1.0 \text{ mg/L}}{18,000 \text{ mg/L}}$$

$$\text{Fluoride Feed Rate} = 55.6 \text{ gpd}$$

Thus, it takes approximately 56 gal of saturated solution to treat 1 MG of water at a dose of 1.0 mg/L.

Calculated Dosages

Some states require that records be kept regarding the amount of chemical used and that the theoretical concentration of the chemical in the water be determined mathematically. To find the theoretical concentration of fluoride, the calculated dosage must be determined. Adding the calculated dosage to the natural fluoride level in the water supply will yield the theoretical concentration of fluoride in the water. This number, the theoretical concentration, is calculated as a safety precaution to help ensure that an overfeed or accident does not occur. It is also an aid in solving troubleshooting problems. If the theoretical concentration is significantly higher or lower than the measured concentration, steps should be taken to determine the discrepancy. The fluoride feed rate formula can be changed to find the calculated dosage as follows:

$$\text{Dosage (mg/L)} = \frac{\text{Fluoride Feed Rate (lb/day)}}{\text{capacity (MGD)} \times 8.34 \text{ lb/gal}} \times \text{AFI} \times \text{chemical purity} \quad (7.113)$$

When the fluoride feed rate is changed to fluoride fed and the capacity is changed to the actual daily production of water in the water system, the dosage becomes the calculated dosage. The units remain the same, except that fluoride feed goes from lb/day to lbs and actual production goes from MGD to MG (million gallons) (the "day" units cancel).

Note: The amount of fluoride fed (lb) will be determined over a time period [day, week, month, etc.], and the actual production will be determined over the same period.

$$\text{Cal.Dosage (mg/L)} = \frac{\text{Fluoride fed (lb)}}{\text{Actual production (MG)}} \times \text{AFI} \times \text{chemical purity} \\ \times 8.34 \text{ lb/gal} \quad (7.114)$$

The numerator of the equation gives the pounds of fluoride ion added to the water while the denominator gives million pounds of water treated. Pounds of fluoride divided by million pounds of water equals ppm or mg/L. The formula for the calculated dosage for the saturator is as follows:

$$\text{Calculated Dosage (mg/L)} = \frac{\text{Sol. fed (gal)} \times 18,000 \text{ mg/L}}{\text{Actual production (gal)}} \quad (7.115)$$

Determining the calculated dosage for an unsaturated sodium fluoride solution is based on the particular strength of the solution. For example, a 2% strength solution is equal to 8,550 mg/L. The percent strength is based on the pounds of sodium fluoride dissolved in a certain amount of water. For example, find the percent solution if 6.5 lbs of sodium fluoride are dissolved in 45 gal of water:

$$45\text{-gal} \times 8.34 \text{ lb/gal} = 375 \text{ lbs. of water}$$

$$\frac{6.5 \text{ lb NaF}}{375 \text{ lbs H}_2\text{O}} = 1.7\% \text{ NaF solution}$$

This means that 6.5 lb of fluoride chemical dissolved in 45 gal of water will yield a 1.7% solution. To find the solution concentration of an unknown sodium fluoride solution, use the following formula:

$$\text{Solution Concentration} = \frac{18,000 \text{ mg/L}}{4\%} \times \text{solution strength (\%)} \quad (7.116)$$

Example 7.181

Problem: Assume that 6.5 lb of NaF is dissolved in 45 gal of water, as previously given. What would be the solution concentration? The solution strength is 1.7% (see above).

Solution:

$$\text{Solution Concentration} = \frac{18,000 \text{ mg/L}}{4\%} \times \text{solution strength (\%)}$$

$$\text{Solution Concentration} = \frac{18,000 \text{ mg/L} \times 1.7\%}{4\%}$$

$$\text{Solution Concentration} = 7,650 \text{ mg/L}$$

Note: The calculated dosage formula for an unsaturated sodium fluoride solution is:

$$\text{Cal.Dosage (mg/L)} = \frac{\text{Sol. fed (gal)} \times \text{sol. conc. (mg/L)}}{\text{Actual production (gal)}}$$

Caution: The CDC recommends against the use of unsaturated sodium fluoride solution in water fluoridation.

Calculated Dosage Problems

Example 7.182 Sodium Fluorosilicate Dosage

Problem: A plant uses 65 lb of sodium fluorosilicate in treating 5,540,000 gal of water in 1 day. What is the calculated dosage?

Solution:

$$\text{Calculated Dosage (mg/L)} = \frac{\text{Fluoride fed (lb)}}{\text{Actual production (MG)}} \times \text{AFI} \times \text{purity} \\ \times 8.34 \text{ lb/gal}$$

$$\text{Calculated Dosage (mg/L)} = \frac{65 \text{ lbs} \times 0.607 \times 0.985}{5,540 \text{ MG} \times 8.34 \text{ lb/gal}}$$

$$\text{Calculated Dosage} = 0.84 \text{ mg/L}$$

Example 7.183 Fluorosilicic Acid Dosage

Problem: A plant uses 43 lb of fluorosilicic acid in treating 1,226,000 gal of water. Assume the acid is 23% pure. What is the calculated dosage?

Solution:

$$\text{Calculated Dosage (mg/L)} = \frac{\text{Fluoride fed (lb)} \times \text{AFI} \times \text{purity}}{\text{Actual production (MG)}}$$

$$\times \ 8.34 \ \text{lb/gal}$$

$$\text{Calculated Dosage (mg/L)} = \frac{43 \ \text{lbs} \times 0.792 \times 0.23}{1.226 \ \text{MG} \times 8.34 \ \text{lb/gal}}$$

$$\text{Calculated Dosage} = 0.77 \ \text{mg/L}$$

Note: The calculated dosage is 0.77 mg/L. If the natural fluoride level is added to this dosage, it should equal the actual fluoride level in the drinking water.

Example 7.184 Sodium Fluoride (Dry) Dosage

Problem: A water plant feeds sodium fluoride in a dry feeder. They use 5.5 lb of the chemical to fluoridate 240,000 gal of water. What is the calculated dosage?

Solution:

$$\text{Calculated Dosage (mg/L)} = \frac{\text{Fluoride fed (lb)} \times \text{AFI} \times \text{purity}}{\text{Actual production (MG)}}$$

$$\times 8.34 \ \text{lb/gal}$$

$$\text{Calculated Dosage (mg/L)} = \frac{5.5 \ \text{lb} \times 0.45 \times 0.98}{0.24 \ \text{MG} \times 8.34 \ \text{lb/gal}}$$

$$\text{Calculated Dosage} = 1.2 \ \text{mg/L}$$

Example 7.185 Sodium Fluoride— Saturator Dosage

Problem: A plant uses 10 gal of sodium fluoride from its saturator in treating 200,000 gal of water. What is the calculated dosage?

Solution:

$$\text{Calculated Dosage (mg/L)} = \frac{\text{Solution fed (gal)} \times 18,000 \ \text{mg/L}}{\text{Actual production (gal)}}$$

$$\text{Calculated Dosage (mg/L)} = \frac{10 \ \text{gal} \times 18,000 \ \text{mg/L}}{200,000 \ \text{gal}}$$

$$\text{Calculated Dosage} = 0.9 \ \text{mg/L}$$

Example 7.186 Sodium Fluoride— Unsaturated Solution Dosage

Problem: A water plant adds 93 gpd of a 2% solution of sodium fluoride to fluoridate 800,000 gpd. What is the calculated dosage?

Solution:

$$\text{Solution Concentration (mg/L)} = \frac{18,000 \ \text{mg/L} \times \text{sol. strength (\%)}}{4\%}$$

$$\text{Solution Concentration (mg/L)} = \frac{18,000 \ \text{mg/L} \times 0.02}{0.04}$$

$$\text{Solution Concentration} = 9,000 \ \text{mg/L}$$

$$\text{Calculated Dosage (mg/L)} = \frac{\text{solution fed (gal)} \times \text{sol. conc. (mg/L)}}{\text{actual production (gal)}}$$

$$\text{Calculated Dosage (mg/L)} = \frac{93\text{-gal} \times 9,000 \ \text{mg/L}}{800,000 \ \text{gal}}$$

$$\text{Calculated Dosage} = 1.05 \ \text{mg/L}$$

WATER SOFTENING

Hardness in water is caused by the presence of certain positively charged metallic ions in solution. The most common of these hardness-causing ions are calcium and magnesium; others include iron, strontium, and barium. The two primary constituents of water that determine hardness are calcium and magnesium. If the concentrations of these elements in the water are known, the total hardness of the water can be calculated. To make this calculation, the equivalent weights of calcium, magnesium, and calcium carbonate must be known; the equivalent weights are given below.

Equivalent Weights

Calcium, Ca	20.04
Magnesium, Mg	12.15
Calcium Carbonate $CaCO_3$	50.045

Calculating Calcium Hardness, as $CaCO_3$

The hardness (in mg/L as $CaCO_3$) for any given metallic ion is calculated using Equation (7.117).

$$\frac{\text{Calcium Hardness, mg/L as } CaCO_3}{\text{Equivalent Weight of } CaCO_3} = \frac{\text{Calcium, mg/L}}{\text{Equivalent Weight of Calcium}}$$

$$(7.117)$$

Example 7.187

Problem: A water sample has a calcium content of 51 mg/L. What is this calcium hardness expressed as $CaCO_3$?

Solution:

Calcium Hardness,

$$\frac{\text{mg/L as CaCO}_3}{\begin{array}{c}\text{Equivalent Weight}\\ \text{of CaCO}_3\end{array}} = \frac{\text{Calcium, mg/L}}{\begin{array}{c}\text{Equivalent Weight}\\ \text{of Calcium}\end{array}}$$

$$\frac{\text{x mg/L}}{50.045} = \frac{51 \text{ mg/L}}{20.04}$$

$$\text{x} = \frac{(51)(50.045)}{20.45}$$

$$\text{x} = 124.8 \text{ mg/L Ca as CaCO}_3$$

Example 7.188

Problem: The calcium content of a water sample is 26 mg/L. What is this calcium hardness expressed as $CaCO_3$?

Solution:

Calcium Hardness,

$$\frac{\text{mg/L as CaCO}_3}{\begin{array}{c}\text{Equivalent Weight}\\ \text{of CaCO}_3\end{array}} = \frac{\text{Calcium, mg/L}}{\begin{array}{c}\text{Equivalent Weight}\\ \text{of Calcium}\end{array}}$$

$$\frac{\text{x mg/L}}{50.045} = \frac{26 \text{ mg/L}}{20.04}$$

$$\text{x} = \frac{(26)(50.045)}{20.04}$$

$$\text{x} = 64.9 \text{ mg/L Ca as CaCO}_3$$

Calculating Magnesium Hardness, as $CaCO_3$

To calculate magnesium harness, we use Equation (7.118).

Magnesium Hardness,

$$\frac{\text{m/L as CaCO}_3}{\begin{array}{c}\text{Equivalent Weight}\\ \text{of CaCO}_3\end{array}} = \frac{\text{Magnesium, mg/L}}{\begin{array}{c}\text{Equivalent Weight}\\ \text{of Magnesium}\end{array}} \quad (7.118)$$

Example 7.189

Problem: A sample of water contains 24-mg/L magnesium. Express this magnesium hardness as $CaCO_3$.

Solution:

Magnesium Hardness,

$$\frac{\text{m/L as CaCO}_3}{\begin{array}{c}\text{Equivalent Weight}\\ \text{of CaCO}_3\end{array}} = \frac{\text{Magnesium, mg/L}}{\begin{array}{c}\text{Equivalent Weight}\\ \text{of Magnesium}\end{array}}$$

$$\frac{\text{x mg/L}}{50.045} = \frac{24 \text{ mg/L}}{12.15}$$

$$\text{x} = \frac{(24)(50.045)}{12.15}$$

$$\text{x} = 98.9 \text{ mg/L}$$

Example 7.190

Problem: The magnesium content of a water sample is 16 mg/L. Express this magnesium hardness as $CaCO_3$.

Solution:

Magnesium Hardness,

$$\frac{\text{mg/L as CaCO}_3}{\begin{array}{c}\text{Equivalent Weight}\\ \text{of CaCO}_3\end{array}} = \frac{\text{Magnesium, mg/L}}{\begin{array}{c}\text{Equivalent Weight}\\ \text{of Magnesium}\end{array}}$$

$$\frac{\text{x mg/L}}{50.045} = \frac{16 \text{ mg/L}}{12.15}$$

$$\text{x} = \frac{(16)(50.045)}{12.15}$$

$$\text{x} = 65.9 \text{ mg/L Mg as CaCO}_3$$

Calculating Total Hardness

Calcium and magnesium ions are the two constituents that are the primary causes of hardness in water. To find total hardness, we simply add the concentrations of calcium and magnesium ions, expressed in terms of calcium carbonate, $CaCO_3$, using Equation (7.119).

Total Hardness,

$$\text{mg/L as CaCO}_3 = \text{Cal. Hard., mg/L as CaCO}_3$$

$$+ \text{Mg. Hardness., mg/L as CaCO}_3 \quad (7.119)$$

Example 7.191

Problem: A sample of water has calcium content of 70 mg/L as $CaCO_3$ and magnesium content of 90 mg/L as $CaCO_3$.

Solution:

$$\begin{array}{ll} \text{Total Hardness,} & \text{Calcium Hard., mg/L} \\ \text{mg/L as CaCO}_3 = & + \text{Magnesium Hard., mg/L} \\ = & 70 \text{ mg/L} + 90 \text{ mg/L} \\ = & 160 \text{ mg/L as CaCO}_3 \end{array}$$

Example 7.192

Problem: Determine the total hardness as $CaCO_3$ of a sample of water that has calcium content of 28 mg/L and magnesium content of 9 mg/L.

Solution:

Express calcium and magnesium in terms of $CaCO_3$:

Calcium Hardness,

$$\frac{\text{mg/L as } CaCO_3}{\begin{array}{c}\text{Equivalent Weight}\\ \text{of } CaCO_3\end{array}} = \frac{\text{Calcium, mg/L}}{\begin{array}{c}\text{Equivalent Weight}\\ \text{of Calcium}\end{array}}$$

$$\frac{x \text{ mg/L}}{50.045} = \frac{28 \text{ mg/L}}{20.04}$$

$$x = 69.9 \text{ mg/L Mg as } CaCO_3$$

Magnesium Hardness,

$$\frac{\text{mg/L as } CaCO_3}{\begin{array}{c}\text{Equivalent Weight}\\ \text{of } CaCO_3\end{array}} = \frac{\text{Magnesium, mg/L}}{\begin{array}{c}\text{Equivalent Weight}\\ \text{of Magnesium}\end{array}}$$

$$\frac{x \text{ mg/L}}{50.045} = \frac{9 \text{ mg/L}}{12.15}$$

$$x = 37.1 \text{ mg/L Mg as } CaCO_3$$

Now, total hardness can be calculated:

$$\begin{array}{c}\text{Total Hardness,}\\ \text{mg/L as } CaCO_3\end{array} = \begin{array}{l}\text{Cal.Hardness, mg/L +}\\ \text{Mag. Hardness, mg/L}\end{array}$$

$$= 69.9 \text{ mg/L} + 37.1 \text{ mg/L}$$

$$= 107 \text{ mg/L as } CaCO_3$$

Calculating Carbonate and Noncarbonate Hardness

As mentioned, total hardness is comprised of calcium and magnesium hardness. Once total hardness has been calculated, it is sometimes used to determine another expression of hardness—carbonate and noncarbonate. When hardness is numerically greater than the sum of bicarbonate and carbonate alkalinity, the amount of hardness equivalent to the total alkalinity (both in units of mg $CaCO_3$/L) is called **carbonate hardness**; the amount of hardness in excess of this is called **noncarbonate hardness**. When the hardness is numerically equal to or less than the sum of carbonate and noncarbonate alkalinity, all hardness is carbonate hardness, and noncarbonate hardness is absent. Again, total hardness is comprised of carbonate hardness and noncarbonate hardness:

$$\text{Total Hardness} = \text{Carbonate Hardness} + \text{Noncarbonate Hardness}$$

$$(7.120)$$

When the alkalinity (as $CaCO_3$) is greater than the total hardness, all the hardness is carbonate hardness:

Total Hardness, mg/L as $CaCO_3$ = Carbonate Hardness,

$$\text{mg/L as } CaCO_3 \quad (7.121)$$

When the alkalinity (as $CaCO_3$) is less than the total hardness, the alkalinity represents carbonate hardness, and the remaining hardness is noncarbonate hardness:

Total Hardness,

mg/L as $CaCO_3$ = Carb. Hard., mg/L as $CaCO_3$

$$+\text{Noncarb. Hard., mg/L as } CaCO_3 \quad (7.122)$$

When carbonate hardness is represented by the alkalinity, we use Equation (7.123).

Total Hardness,

mg/L as $CaCO_3$ = Alk., mg/L as $CaCO_3$

$$+\text{Noncarb. Hardness, mg/L as } CaCO_3 \quad (7.123)$$

Example 7.193

Problem: A water sample contains 110 mg/L alkalinity as $CaCO_3$ and 105 mg/L total hardness as $CaCO_3$. What is the carbonate and noncarbonate hardness of the sample?

Solution:

Because the alkalinity is greater than the total hardness, all the hardness is carbonate hardness:

Total Hardness, mg/L as $CaCO_3$ = Carbonate Hardness,

$$\text{mg/L as } CaCO_3$$

$$105 \text{ mg/L as } CaCO_3 = \text{Carbonate Hardness}$$

No noncarbonate hardness is present in this water.

Example 7.194

Problem: The alkalinity of a water sample is 80 mg/L as $CaCO_3$. If the total hardness of the water sample is 112 mg/L as $CaCO_3$, what are the carbonate and noncarbonate hardnesses in mg/L as $CaCO_3$?

Solution:

Alkalinity is less than total hardness; therefore, both carbonate and noncarbonate hardnesses will be present in the hardness of the sample.

Total Hard., mg/L $CaCO_3$ = Carb. Hard., mg/L as $CaCO_3$

$$+\text{Noncarb. Hard., mg/L as } CaCO_3$$

$$112 \text{ mg/L} = 80 \text{ mg/L} - x \text{ mg/L}$$

$$112 \text{ mg/L} - 80 \text{ mg/L} = x \text{ mg/L}$$

$$32 \text{ mg/L noncarbonate hardness} = x$$

Alkalinity Determination

Alkalinity measures the acid-neutralizing capacity of a water sample. It is an aggregate property of the water sample and can be interpreted in terms of specific substances only when a complete chemical composition of the sample is also performed. The alkalinity of surface waters is primarily due to the carbonate, bicarbonate, and hydroxide content and is often interpreted in terms of the concentrations of these constituents. The higher the alkalinity, the greater the capacity of the water to neutralize acids; conversely, the lower the alkalinity, the less the neutralizing capacity. To detect the different types of alkalinity, the water is tested for phenolphthalein and total alkalinity using Equations (7.124) and (7.125).

Phenolphthalein

$$\text{Alkalinity mg/L as } CaCO_3 = \frac{(A)(N)(50,000)}{\text{ML of Sample}} \quad (7.124)$$

$$\text{Total Alkalinity mg/L as } CaCO_3 = \frac{(B)(N)(50,000)}{\text{ML of Sample}} \quad (7.125)$$

where

A = mL titrant used to pH 8.3
B = total mL of titrant used to titrate to pH 4.5
N = normality of the acid ($0.02\,N\ H_2SO_4$ for this alkalinity test)
50,000 = a conversion factor to change the normality into units of $CaCO_3$

Example 7.195

Problem: A 100-mL water sample is tested for phenolphthalein alkalinity. If 1.3 mL of titrant is used to reach pH 8.3 and the sulfuric acid solution has a normality of 0.02 N, what is the phenolphthalein alkalinity of the water?

Solution:

$$\begin{aligned}
\text{Phenolphthalein Alkalinity,} \\
\text{mg/L as } CaCO_3
\end{aligned} = \frac{(A)(N)(50,000)}{\text{mL of Sample}}$$

$$= \frac{(1.3\ \text{mL})\ (0.02\ N)\ (50,000)}{100\ \text{mL}}$$

$$= \frac{13\ \text{mg/L as } CaCO_3}{\text{Phenolphthalein Alk.}}$$

Example 7.196

Problem: A 100-mL sample of water is tested for alkalinity. The normality of the sulfuric acid used for titration is 0.02 N. If 0 mL is used to reach pH 8.3, and 7.6 mL titrant is used to reach pH 4.5, what are the phenolphthalein and total alkalinity of the sample?

Solution:

$$\begin{aligned}
\text{Phenolphthalein} \\
\text{Alk. mg/L as } CaCO_3
\end{aligned} = \frac{(0\ \text{mL})(0.02\ N)(50,000)}{100\ \text{mL}}$$

$$= 0\ \text{mg/L}$$

$$\begin{aligned}
\text{Total Alkalinity,} \\
\text{mg/L as } CaCO_3
\end{aligned} = \frac{(7.6\ \text{mL})(0.02\ N)(50,000)}{100\ \text{mL}}$$

$$= 76\ \text{mg/L}$$

Calculation for Removal of Noncarbonate Hardness

Soda ash is used for the precipitation and removal of noncarbonate hardness. To calculate the required soda ash dosage, we use, in combination, Equations (7.126) and (7.127).

$$\begin{aligned}
\text{Tot. Hardness,} \\
\text{mg/L as } CaCO_3 = \text{Carb. Hardness, mg/L as } CaCO_3 \\
+ \text{Noncarb. Hard., mg/L as } CaCO_3
\end{aligned} \quad (7.126)$$

$$\begin{aligned}
\text{Soda Ash } (Na_2CO_3)\ \text{Fd., mg/L} = (\text{Noncarb.})\ \text{Hard.,} \\
\text{mg/L as } CaCO_3\ (106/100)
\end{aligned}$$

$$(7.127)$$

Example 7.197

Problem: A water sample has a total hardness of 250 mg/L as $CaCO_3$ and a total alkalinity of 180 mg/L. What soda ash dosage (mg/L) is required to remove the noncarbonate hardness?

Solution:

Calculate the noncarbonate hardness:

$$\begin{aligned}
\text{Total Hardness, mg/L as } CaCO_3 = \text{Carb. Hard.,} \\
\text{mg/L as } CaCO_3 \\
+ \text{Noncarb. Hard., mg/L as } CaCO_3
\end{aligned}$$

$$250\ \text{mg/L} - 180\ \text{mg/L} = x\ \text{mg/L}$$

$$70\ \text{mg/L} = x$$

Calculate the soda ash required:

$$\begin{aligned}
\text{Soda Ash, mg/L} &= \frac{(\text{Noncarbonate Hardness}),}{\text{mg/L as } CaCO_3\,(106)/100} \\
&= (70\ \text{mg/L})(106)/100 \\
&= 74.2\ \text{mg/L soda ash}
\end{aligned}$$

Example 7.198

Problem: Calculate the soda ash required (in mg/L) to soften water if the water has a total hardness of 192 mg/L and a total alkalinity of 103 mg/L.

Solution:

Determine noncarbonate hardness:

$$192\ \text{mg/L} = 103\ \text{mg/L} + x\ \text{mg/L}$$

$$192\ \text{mg/L} - 103\ \text{mg/L} = x$$

$$89\ \text{mg/L} = x$$

Calculate the soda ash required:

$$\text{Soda Ash, mg/L} = \frac{(\text{Noncarbonate}) \text{ Hardness,}}{\text{mg/L as } CaCO_3 \ (106)/100}$$

$$= (89 \text{ mg/L})(106)/100$$

$$= 94 \text{ mg/L soda ash}$$

RECARBONATION CALCULATION

Recarbonation involves the reintroduction of carbon dioxide into the water, either during or after lime softening, lowering the pH of the water to about 10.4. After the addition of soda ash, recarbonation lowers the pH of the water to about 9.8, promoting better precipitation of calcium carbonate and magnesium hydroxide. Equations (7.128) and (7.129) are used to estimate the carbon dioxide dosage.

$$\text{Excess Lime, mg/L} = (A + B + C + D)(0.15) \qquad (7.128)$$

$$\text{Total } CO_2 \text{ Dosage, mg/L} = \left[Ca(OH)_2 \text{ Excess, mg/L} \right](44)/74 + \left(Mg^{+2} \right) \text{Residual mg/L} \ (44)/24.3 \qquad (7.129)$$

Example 7.199

Problem: The A, B, C, and D factors of the excess lime equation have been calculated as follows: A=14 mg/L; B=126 mg/L; C=O; D=66 mg/L. If the residual magnesium is 5 mg/L, what is the carbon dioxide (in mg/L) required for recarbonation?

Solution:

Calculate the excess lime concentration:

$$\begin{aligned}\text{Excess Lime (mg/L)} &= (A + B + C + D) \times 0.15 \\ &= (14 \text{ mg/L} + 126 \text{ mg/L} + 0 + 66 \text{ mg/L})(0.15) \\ &= 31 \text{ mg/L}\end{aligned}$$

Determine the required carbon dioxide dosage:

$$\begin{aligned}\text{Total } CO_2 \text{ Dosage, mg/L} &= \frac{(31 \text{ mg/L})(44)/74}{+(5 \text{ mg/L})(44)/24.3} \\ &= 18 \text{ mg/L} + 9 \text{ mg/L} \\ &= 27 \text{ mg/L } CO_2\end{aligned}$$

Example 7.200

Problem: The A, B, C, and D Factors of the excess lime equation have been calculated as: A=10 mg/L;

B=87 mg/L; C=0; D =111 mg/L. If the residual magnesium is 5 mg/L, what carbon dioxide dosage would be required for recarbonation?

Solution:

The excess lime is:

$$\begin{aligned}\text{Excess Lime, mg/L} &= (A + B + C + D) \ (0.15) \\ &= \left(\begin{matrix} 10 \text{ mg/L} + 87 \text{ mg/L} \\ + 0 + 111 \text{ mg/L} \end{matrix} \right)(0.15) \\ &= (208)(0.15) \\ &= 31 \text{ mg/L}\end{aligned}$$

The required carbon dioxide dosage for recarbonation is:

$$\begin{aligned}\text{Total } CO_2 \text{ Dosage, mg/L} &= \frac{(31 \text{ mg/L})(44)/74}{+(5 \text{ mg/L})(44)/24.3} \\ &= 18 \text{ mg/L} + 9 \text{ mg/L} \\ &= 27 \text{ mg/L } CO_2\end{aligned}$$

Calculating Feed Rates

The appropriate chemical dosage for various unit processes is typically determined by lab or pilot scale testing (e.g., jar testing, pilot plant), only monitoring, and historical experience. Once the chemical dosage is determined, the feed rate can be calculated using Equation (7.130). Once the chemical feed rate is known, this value must be translated into a chemical feeder setting.

$$\text{Feed Rate (lbs/day)} = \text{Flow Rate, MGD} \times \text{Chem. Does,}$$
$$\text{mg/L} \times 8.34 \text{ lb/gal} \qquad (7.130)$$

To calculate the lbs/min chemical required we use Equation (7.131).

$$\text{Chemical, lb/min} = \frac{\text{Chemical, lb/day}}{1,440 \text{ min/day}} \qquad (7.131)$$

Example 7.201

Problem: Jar tests indicate that the optimal lime dosage is 200 mg/L. If the flow to be treated is 4.0 MGD, what should the chemical feeder setting be in lb/day and lb/min?

Solution:

Calculate the lb/day feed rate:

$$\begin{aligned}\text{Feed Rate, lb/day} &= \frac{(\text{Flow Rate, MGD})}{(\text{Chemical Dose, mg/L})(8.34 \text{ lb/gal})} \\ &= (200 \text{ mg/L})(4.0 \text{ MGD})(8.34 \text{ lb/gal}) \\ &= 6,672 \text{ lb/day}\end{aligned}$$

Convert this feed rate to lb/min:

$$= \frac{6{,}672 \text{ lb/day}}{1{,}440 \text{ min/day}}$$

$$= 4.6 \text{ lb/min}$$

Example 7.202

Problem: What should the lime dosage setting be in lbs/day and lbs/h if the optimal lime dosage has been determined to be 125 mg/L and the flow to be treated is 1.1 MGD?

Solution:

The lb/day feed rate for lime is:

$$\text{Lime, lb/day} = \left(\text{Lime, mg/L}\right)\left(\text{Flow, MGD}\right)\left(8.34 \text{ lb/gal}\right)$$

$$= \left(125 \text{ mg/L}\right)\left(1.1 \text{ MGD}\right)\left(8.34 \text{ lb/day}\right)$$

$$= 1{,}147 \text{ lb/day}$$

Convert this to lb/min feed rate, as follows:

$$= \frac{1{,}147 \text{ lb/day}}{24 \text{ h/day}}$$

$$= 48 \text{ lb/h}$$

Ion Exchange Capacity

An ion exchange softener is a common alternative to the use of lime and soda ash for softening water. Natural water sources contain dissolved minerals that dissociate in water to form charged particles called *ions*. Of main concern are the positively charged ions of calcium, magnesium, and sodium, while bicarbonate, sulfate, and chloride are the normal negatively charged ions of concern. An ion exchange medium, called *resin*, is a material that will exchange a hardness-causing ion for another one that does not cause hardness, hold the new ion temporarily, and then release it when a regenerating solution is poured over the resin. The removal capacity of an exchange resin is generally reported as grains of hardness removal per cubic foot of resin. To calculate the removal capacity of the softener, we use Equation (7.132).

$$\text{Exchange Capacity., grains} = \left(\text{Removal Cap., grains/ft}^3\right)$$

$$\left(\text{Media Vol., ft}^3\right)$$

$$(7.132)$$

Example 7.203

The hardness removal capacity of an exchange resin is 24,000 grains/ft³. If the softener contains a total of 70 ft³ of resin, what is the total exchange capacity (grains) of the softener?

$$\text{Exchange Cap., grains} = \frac{\left(\text{Removal Cap., grains/ft}^3\right)}{\left(\text{Media Vol., ft}^3\right)}$$

$$= \left(22{,}000 \text{ grains/ft}^3\right)\left(70 \text{ ft}^3\right)$$

$$= 1{,}540{,}000 \text{ grains}$$

Example 7.204

Problem: An ion exchange water softener has a diameter of 7 ft. The depth of resin is 5 ft. If the resin has a removal capacity of 22 kg/ft³, what is the total exchange capacity of the softener (in grains)?

Solution:

Before the exchange capacity of a softener can be calculated, the ft³ resin volume must be known:

$$\text{Vol., ft}^3 = (0.785)\left(D^2\right)(\text{Depth, ft})$$

$$= (0.785)(7 \text{ ft})(7 \text{ ft})(5 \text{ ft})$$

$$= 192 \text{ ft}^3$$

Calculate the exchange capacity of the softener:

$$\text{Exchange Cap., grains} = \frac{\left(\text{Removal Cap., grains/ft}\right)}{\left(\text{Media Vol., ft}^3\right)}$$

$$= \left(22{,}000 \text{ grains/ft}^3\right)\left(192 \text{ ft}^3\right)$$

$$= 4{,}224{,}000 \text{ grains}$$

WATER TREATMENT CAPACITY

To calculate when the resin must be regenerated (based on the volume of water treated), we know the softener's exchange capacity and the water's hardness. Equation (7.133) is used for this calculation

$$\text{Water Treatment Cap., gal} = \frac{\text{Exchange Capacity, grains}}{\text{Hardness, grains/gal}}$$

$$(7.133)$$

Example 7.205

Problem: An ion-exchange softener has an exchange capacity of 2,445,000 grains. If the hardness of the water to be treated is 18.6 grains/gal, how many gallons of water can be treated before regeneration of the resin is required?

Solution:

$$\text{Water Treatment Capacity, gal} = \frac{\text{Exchange Capacity, grains}}{\text{Hardness, grains/gal}}$$

$$= \frac{2{,}455{,}000 \text{ grains}}{18.6 \text{ gpg}}$$

$$= 131{,}989 \text{ gal}$$

Example 7.206

Problem: An ion exchange softener has an exchange capacity of 5,500,000 grains. If the hardness of the water to be treated is 14.8 grains/gal, how many gallons of water can be treated before resin regeneration is required?

Solution:

$$\text{Water Treatment Capacity, gal} = \frac{\text{Exchange Capacity, grains}}{\text{Hardness, grains/gal}}$$

$$= \frac{5,500,000 \text{ grains}}{14.8 \text{ g/gal}}$$

$$= 371,622 \text{ gal}$$

Example 7.207

Problem: The hardness removal capacity of an ion exchange resin is 25 kilograins/ft³. The softener contains a total of 160 ft³ of resin. If the water to be treated contains 14.0 g/gal hardness, how many gallons of water can be treated before regeneration of the resin is required?

Solution:

Both the water hardness and the exchange capacity of the softener must be determined before the gallons of water can be calculated.

$$\text{Exchange Capacity, grains} = \frac{\left(\text{Removal Capacity, grains/ft}^3\right)}{\left(\text{Media Volume, ft}^3\right)}$$

$$= \left(25,000 \text{ grains/ft}^3\right)\left(160 \text{ ft}^3\right)$$

$$= 4,000,000 \text{ grains}$$

Calculate the gallons of water treated:

$$\text{Water Treatment Capacity, gal} = \frac{4,000,000 \text{ grains}}{14.0 \text{ g/gal}}$$

$$= 285,714 \text{ gal}$$

TREATMENT TIME CALCULATION (UNTIL REGENERATION REQUIRED)

After calculating the total number of gallons of water to be treated (before regeneration), we can also calculate the operating time required to treat that amount of water. Equation (7.134) is used to make this calculation.

$$\text{Operating Time, h} = \frac{\text{Water Treated, gal}}{\text{Flow Rate, gph}} \quad (7.134)$$

Example 7.208

Problem: An ion exchange softener can treat a total of 642,000 gal before regeneration is required. If the flow rate treated is 25,000 gph, how many hours of operation are there before regeneration is required?

Solution:

$$\text{Operating Time, h} = \frac{\text{Water Treated, gal}}{\text{Flow Rate, gph}}$$

$$= \frac{642,000 \text{ gal}}{25,000 \text{ gph}}$$

$$= 25.7 \text{ h of operation before regeneration}$$

Example 7.209

Problem: An ion exchange softener can treat a total of 820,000 gal of water before regeneration of the resin is required. If the water is to be treated at a rate of 32,000 gph, how many hours of operation are there until regeneration is required?

Solution:

$$\text{Operating Time, h} = \frac{\text{Water Treatment, gal}}{\text{Flow Rate, gph}}$$

$$= \frac{820,000 \text{ gal}}{32,000 \text{ gph}}$$

$$= 25.6 \text{ h before regeneration}$$

SALT AND BRINE REQUIRED FOR REGENERATION

When calcium and magnesium ions replace the sodium ions in the ion exchange resin, the resin can no longer remove the hardness ions from the water. When this occurs, pumping a concentrated solution (10%–14% sodium chloride solution) onto the resin must regenerate the resin. When the resin is completely recharged with sodium ions, it is ready for softening again. Typically, the salt dosage required to prepare the brine solution ranges from 5 to 15 lb salt/ft³ resin. Equation (7.135) is used to calculate the salt required (pounds, lbs), and Equation (7.136) is used to calculate brine (gallons).

$$\text{Salt Required, lb} = \left(\text{Salt Req., lb/kgrains. Rem.}\right)$$

$$\left(\text{Hard. Rem., kgrains}\right) \quad (7.135)$$

$$\text{Brine, gal} = \frac{\text{Salt Required, lb}}{\text{Brine Solution, lb salt/gal brine}} \quad (7.136)$$

To determine the brine solution, lbs salt/gal brine factor used in Equation (7.136), we must refer to the salt solutions table shown below.

Salt Solutions		
% NaCl	lbs NaCl/gal	lbs NaCl/ft³
10	0.874	6.69
11	0.990	7.41
12	1.09	8.14
13	1.19	8.83
14	1.29	9.63
15	1.39	10.4

Example 7.210

Problem: An ion exchange softener removes 1,310,000 grains of hardness from the water before the resin must be regenerated. If 0.3 lb salt is required for each kilograin removed, how many pounds of salt will be required for preparing the brine to be used in resin regeneration?

Solution:

$$\text{Salt Required, lb} = \frac{\left(\text{Salt Required, lb/1,000 grains}\right)}{\left(\text{Hardness Removed, kg}\right)}$$

$$= \frac{\left(0.3 \text{ lb salt/kilograins removed}\right)}{\left(1,310 \text{ kilograins}\right)}$$

$$= 393 \text{ lb salt required}$$

Example 7.211

Problem: A total of 430 lb salt will be required to regenerate an ion exchange softener. If the brine solution is to be a 12% (see Salt Solutions table to determine the lb salt/gal brine for a 12% brine solution) brine solution, how many gallons brine will be required?

Solution:

$$\text{Brine, gal} = \frac{\text{Salt Required, lb}}{\text{Brine Solution, lb salt/gal brine}}$$

$$= \frac{430 \text{ lb salt}}{1.09 \text{ lb salt/gal brine}}$$

$$= 394 \text{ gal of 12\% brine}$$

Thus, it takes 430 lb salt to make up a total of 394-gal brine, which will result in the desired 12% brine solution.

WASTEWATER MATH CONCEPTS

PRELIMINARY TREATMENT CALCULATIONS

The initial stage of treatment in the wastewater treatment process (following collection and influent pumping) is *preliminary treatment*. Process selection is normally based on the expected characteristics of the influent flow. Raw influents entering the treatment plant may contain many kinds of materials (trash), and preliminary treatment protects downstream plant equipment by removing these materials, which could cause clogs, jams, or excessive wear on plant machinery. In addition, the removal of various materials at the beginning of the treatment train saves valuable space within the treatment plant.

Two of the processes used in preliminary treatment include screening and grit removal. However, preliminary treatment may also include other processes, each designed to remove a specific type of material that presents a potential problem for downstream unit treatment processes.

These processes include shredding, flow measurement, pre-aeration, chemical addition, and flow equalization. Except in extreme cases, plant design will not include all of these items. In this chapter, we focus on and describe typical calculations used in two of these processes: screening and grit removal.

SCREENING

Screening removes large solids, such as rags, cans, rocks, branches, leaves, and roots, from the flow before it moves on to downstream processes.

Screenings Removal Calculations

Wastewater operators responsible for screenings disposal are typically required to keep a record of the amount of screenings removed from the flow. To maintain accurate screenings records, the volume of screenings withdrawn must be determined. Two methods are commonly used to calculate the volume of screenings withdrawn:

$$\text{Screenings Removed, ft}^3/\text{day} = \frac{\text{Screenings, ft}^3}{\text{days}} \quad (7.137)$$

$$\text{Screenings Removed, ft}^3/\text{MG} = \frac{\text{Screenings, ft}^3}{\text{Flow, MG}} \quad (7.138)$$

Example 7.212

Problem: A total of 65 gal of screenings are removed from the wastewater flow during a 24-h period. What is the screenings removal reported as ft³/day?

Solution:

First, convert gallon screenings to cubic feet.

$$\frac{65 \text{ gal}}{7.48 \text{ gal/ft}^3} = 8.7 \text{ ft}^3 \text{screenings}$$

Then, calculate screenings removed as cu ft/day:

$$\text{Screenings Removed } \left(\text{ft}^3/\text{day}\right) = \frac{8.7 \text{ ft}^3}{1 \text{ day}} = 8.7 \text{ ft}^3/\text{day}$$

Example 7.213

Problem: For 1 week, a total of 310 gal of screenings were removed from the wastewater screens. What is the average removal in ft³/day?

Solution:

First, gallon screenings must be converted to ft³ screenings:

$$\frac{310 \text{ gal}}{7.48 \text{ gal/ft}^3} = 41.4 \text{ ft}^3 \text{ screenings}$$

Then, calculate the screenings removed as ft³/day.

$$\text{Screenings Removed, ft}^3/\text{day} = \frac{41.4 \text{ ft}^3}{7.48 \text{ gal/ft}^3} = 5.5 \text{ ft}^3/\text{day}$$

Screenings Pit Capacity Calculations

Recall that detention time may be considered the time required for flow to pass through a basin or tank or the time required to fill a basin or tank at a given flow rate. In screenings pit capacity problems, the time required to fill a screenings pit is calculated. The equation used in screenings pit capacity problems is given below:

$$\text{Screenings Pit Fill Time, days} = \frac{\text{Volume of Pit, ft}^3}{\text{Screenings Removed, ft}^3/\text{day}}$$

(7.139)

Example 7.214

Problem: A screenings pit has a capacity of 500 ft³. (The pit is actually larger than 500 ft³ to accommodate soil for covering.) If an average of 3.4 ft³ of screenings is removed daily from the wastewater flow, in how many days will the pit be full?

Solution:

$$\begin{aligned}
\text{Screenings Pit Fill Time, days} &= \frac{\text{Volume of Pit, ft}^3}{\text{Screenings Removed, ft}^3/\text{day}} \\
&= \frac{500 \text{ ft}^3}{3.4 \text{ ft}^3/\text{day}} \\
&= 147.1 \text{ days}
\end{aligned}$$

Example 7.215

Problem: A plant has been averaging a screenings removal of 2 ft³/MG. If the average daily flow is 1.8 MGD, how many days will it take to fill the pit with an available capacity of 125 ft³?

Solution:

The filling rate must first be expressed as ft³/day:

$$\frac{(2 \text{ ft}^3)(1.8 \text{ MGD})}{\text{MG}} = 3.6 \text{ ft}^3/\text{day}$$

$$\text{Screenings Pit Fill Time, days} = \frac{125 \text{ ft}^3}{3.6 \text{ ft}^3/\text{day}}$$

$$= 34.7 \text{ days}$$

Example 7.216

Problem: A screenings pit has a capacity of 12 yd³ available for screenings. If the plant removes an average of 2.4 ft³ of screenings per day, in how many days will the pit be filled?

Solution:

Because the filling rate is expressed as ft³/day, the volume must be expressed in ft³:

$$12 \text{ yd}^3 \left(27 \text{ ft}^3/\text{yd}^3\right) = 324 \text{ ft}^3$$

Now calculate fill time:

$$\begin{aligned}
\text{Screenings Pit Fill Time, days} &= \frac{\text{Volume of Pit, ft}^3}{\text{Screenings Removed, ft}^3/\text{day}} \\
&= \frac{324 \text{ ft}^3}{2.4 \text{ ft}^3/\text{day}} \\
&= 135 \text{ days}
\end{aligned}$$

GRIT REMOVAL

The purpose of *grit removal* is to remove inorganic solids (sand, gravel, clay, egg shells, coffee grounds, metal filings, seeds, and other similar materials) that could cause excessive mechanical wear. Several processes or devices are used for grit removal, all based on the fact that grit is heavier than the organic solids, which should be kept in suspension for treatment in subsequent unit processes. Grit removal may be accomplished in grit chambers or by the centrifugal separation of biosolids. Processes use gravity/velocity, aeration, or centrifugal force to separate the solids from the wastewater.

Grit Removal Calculations

Wastewater systems typically average 1–15 ft³ of grit per million gallons of flow (sanitary systems: 1–4 ft³/MG; combined wastewater systems: 4–15 ft³/MG of flow), with higher ranges during storm events. Generally, grit is disposed of in sanitary landfills. Because of this process, for planning purposes, operators must keep accurate records of grit removal. Most often, the data are reported as cubic feet of grit removed per million gallons of flow:

$$\text{Grit Removed, ft}^3/\text{MG} = \frac{\text{Grit Volume, ft}^3}{\text{Flow, MG}}$$

(7.140)

Over a given period, the average grit removal rate at a plant (at least a seasonal average) can be determined and used for planning purposes. Typically, grit removal is calculated in cubic yards because excavation is normally expressed in terms of cubic yards.

$$\text{Cubic yd}^3 = \frac{\text{Total grit, ft}^3}{27 \text{ ft}^3/\text{yd}^3}$$

(7.141)

Example 7.217

Problem: A treatment plant removes 10 ft³ of grit in 1 day. How many ft³ of grit are removed per million gallons if the plant flow was 9 MGD?

Solution:

$$\text{Grit Removed, ft}^3/\text{MG} = \frac{\text{Grit Volume, ft}^3}{\text{Flow, MG}}$$

$$= \frac{10 \text{ ft}^3}{9 \text{ MG}} = 1.1 \text{ ft}^3/\text{MG}$$

Example 7.218

Problem: The total daily grit removed for a plant is 250 gal. If the plant flow is 12.2 MGD, how many cubic feet of grit are removed per MG flow?

Solution:

First, convert the gallon grit removed to ft³:

$$\frac{250 \text{ gal}}{7.48 \text{ gal/ft}^3} = 33 \text{ ft}^3$$

Next, complete the calculation of ft³/MG:

$$\text{Grit Removed, ft}^3/\text{MG} = \frac{\text{Grit Vol. ft}^3}{\text{Flow, MG}}$$

$$= \frac{33 \text{ ft}^3}{12.2 \text{ MGD}} = 2.7 \text{ ft}^3/\text{MGD}$$

Example 7.219

Problem: The monthly average grit removal is 2.5 ft³/MG. If the monthly average flow is 2,500,000 gpd, how many yd³ must be available for grit disposal if the disposal pit is to have a 90-day capacity?

Solution:

First, calculate the grit generated each day:

$$\frac{(2.5 \text{ ft}^3)}{\text{MG}}(2.5 \text{ MGD}) = 6.25 \text{ ft}^3 \text{each day}$$

The ft³ grit generated for 90 days would be:

$$\frac{(6.25 \text{ ft})}{\text{day}}(90 \text{ days}) = 562.5 \text{ ft}^3$$

Convert ft³ to yd³ grit:

$$\frac{562.5 \text{ ft}^3}{27 \text{ ft}^3/\text{yd}^3} = 21 \text{yd}^3$$

Grit Channel Velocity Calculation

The optimum velocity in sewers is approximately 2 fps at peak flow because this velocity normally prevents solids from settling in the lines. However, when the flow reaches the grit channel, the velocity should decrease to about 1 fps to permit the heavy inorganic solids to settle. In the example calculations that follow, we describe how the velocity of the flow in a channel can be determined by the float and stopwatch method and by channel dimensions.

Example 7.220 Velocity by Float and Stopwatch

$$\text{Velocity, fps} = \frac{\text{Distance Traveled (ft)}}{\text{Time required (sec)}} \quad (7.142)$$

Problem: A float takes 30 sec to travel 37 ft in a grit channel. What is the velocity of the flow in the channel?

Solution:

$$\text{Velocity (fps)} = \frac{37 \text{ ft}}{30 \text{ sec}} = 1.2 \text{ fps}$$

Example 7.221 Velocity by Flow and Channel Dimensions

This calculation can be used for a single channel or tank or multiple channels or tanks with the same dimensions and equal flow. If the flow through each unit of the unit dimensions is unequal, the velocity for each channel or tank must be computed individually.

$$\text{Velocity, fps} = \frac{\text{Flow, MGD} \times 1.55 \text{ cfs/MGD}}{\text{Channels in Service} \times \text{Channel Width, ft} \times \text{Water Depth, ft}} \quad (7.143)$$

Problem: The plant is currently using two grit channels. Each channel is 3 ft wide and has a water depth of 1.3 ft. What is the velocity when the influent flow rate is 4.0 MGD?

Solution:

$$\text{Velocity, fps} = \frac{4.0 \text{ MGD} \times 1.55 \text{ cfs/MGD}}{2 \text{ Channels} \times 3 \text{ ft} \times 1.3 \text{ ft}}$$

$$\text{Velocity, fps} = \frac{6.2 \text{ cfs}}{7.8 \text{ ft}^2} = 0.79 \text{ fps}$$

Key Point: Because 0.79 is within the 0.7–1.4 level, the operator of this unit would not make any adjustments.

Key Point: The channel dimensions must always be in feet. Convert inches to feet by dividing by 12 in/ft.

Example 7.222 Required Settling Time

This calculation can be used to determine the time required for a particle to travel from the surface of the liquid to the bottom at a given settling velocity. To compute the settling time, the settling velocity in fps must be provided or determined by an experiment in a laboratory.

$$\text{Settling Time (sec)} = \frac{\text{Liquid Depth in ft}}{\text{Settling, Velocity, fps}} \quad (7.144)$$

Problem: The plant's grit channel is designed to remove sand, which has a settling velocity of 0.080 fps.

The channel is currently operating at a depth of 2.3 ft. How many seconds will it take for a sand particle to reach the bottom of the channel?

Solution:

$$\text{Settling Time, sec} = \frac{2.3 \text{ ft}}{0.080 \text{ fps}} = 28.7 \text{ sec}$$

Example 7.223 Required Channel Length

This calculation can be used to determine the length of the channel required to remove an object with a specified settling velocity.

$$\text{Required Channel Length} = \frac{\text{Channel Depth, ft} \times \text{Flow Velocity, fps}}{0.080 \text{ fps}}$$

$$(7.145)$$

Problem: The plant's grit channel is designed to remove sand, which has a settling velocity of 0.080 fps. The channel is currently operating at a depth of 3 ft. The calculated velocity of flow through the channel is 0.85 fps. The channel is 36 ft long. Is the channel long enough to remove the desired sand particle size?

Solution:

$$\text{Required Channel Length} = \frac{3 \text{ ft} \times 0.85 \text{ fps}}{0.080 \text{ fps}} = 31.6 \text{ ft}$$

Yes, the channel is long enough to ensure all the sand will be removed.

PRIMARY TREATMENT CALCULATIONS

Primary treatment (primary sedimentation or clarification) should remove both settleable organic and floatable solids. Poor solids removal during this step of treatment may cause organic overloading of the biological treatment processes following primary treatment. Normally, each primary clarification unit can be expected to remove 90%–95% of settleable solids, 40%–60% of the total suspended solids, and 25%–35% of BOD.

Process Control Calculations

As with many other wastewater treatment plant unit processes, several process control calculations may be helpful in evaluating the performance of the primary treatment process. Process control calculations are used in the sedimentation process to determine:

- Surface loading rate (surface settling rate)
- Weir overflow rate (weir loading rate)
- BOD and SS removed, lb/day
- Percent removal
- Hydraulic detention time
- Biosolids pumping
- Percent total solids (% TS)

In the following sections, we will take a closer look at a few of these process control calculations and example problems.

Key Point: The calculations presented in the following sections allow you to determine values for each function performed. Again, keep in mind that an optimally operated primary clarifier should have values in an expected range. Recall that the expected range for % removal in a primary clarifier is

- settleable solids   90%–95%
- suspended solids   40%–60%
- BOD   25%–35%

The expected range of hydraulic detention time for a primary clarifier is 1–3 h. The expected range of surface loading/settling rate for a primary clarifier is 600–1,200 gpd/ft² (ballpark estimate). The expected range of weir overflow rate for a primary clarifier is 10,000–20,000 gpd/ft.

Surface Loading Rate (Surface Settling Rate/Surface Overflow Rate)

Surface loading rate is the number of gallons of wastewater passing over 1 ft² of tank per day. This can be used to compare actual conditions with design. Plant designs generally use a surface loading rate of 300–1,200 gal/day/ft².

$$\text{Surface Loading Rate, gpd/ft}^2 = \frac{\text{gal/day}}{\text{Surface Tank Area, ft}^2}$$

$$(7.146)$$

Example 7.224

Problem: The circular settling tank has a diameter of 120 ft. If the flow to the unit is 4.5 MGD, what is the surface loading rate in gal/day/ft²?

Solution:

$$\text{Surface Loading Rate} = \frac{4.5 \text{ MGD} \times 1,000,000 \text{ gal/MGD}}{0.785 \times 120 \text{ ft} \times 120 \text{ ft}}$$

$$= 398 \text{ gpd/ft}^2$$

Example 7.225

Problem: A circular clarifier has a diameter of 50 ft. If the primary effluent flow is 2,150,000 gpd, what is the surface overflow rate in gpd/ft²?

Solution:

Key Point

Remember that Area = $(0.785)(50 \text{ ft})(50 \text{ ft})$

$$\text{Surface Overflow Rate} = \frac{\text{Flow, gpd}}{\text{Area, ft}^2}$$

$$= \frac{2,150,000}{(0.785)(50\ \text{ft})(50\ \text{ft})} = 1,096\ \text{gpd/ft}^2$$

Example 7.226

Problem: A sedimentation basin 90 ft by 20 ft receives a flow of 1.5 MGD. What is the surface overflow rate in gpd/ft²?

$$\text{Surface Overflow Rate} = \frac{\text{Flow, gpd}}{\text{Area, ft}^2}$$

$$= \frac{1,500,000\ \text{gpd}}{(90\ \text{ft})(20\ \text{ft})}$$

$$= 833\ \text{gpd/ft}^2$$

Weir Overflow Rate (Weir Loading Rate)

A weir is a device used to measure wastewater flow. *Weir overflow rate (weir loading rate)* is the amount of water leaving the settling tank per gallons of flow per foot (flow (gpm)/weir length, feet). The result of this calculation can be compared with design. Normally, weir overflow rates of 10,000–20,000 gpd/ft are used in the design of a settling tank.

$$\text{Weir Overflow Rate, gpd/ft} = \frac{\text{Flow, gal/day}}{\text{Weir Length, ft}} \qquad (7.147)$$

Key Point: To calculate weir circumference, use total feet of weir = 3.14 × weir diameter (ft).

Example 7.227

Problem: The circular settling tank is 80 ft in diameter and has a weir along its circumference. The effluent flow rate is 2.75 MGD. What is the weir overflow rate in gallons per day per foot?

Solution:

$$\text{Weir Overflow Rate, gpd/ft} = \frac{2.75\ \text{MGD} \times 1,000,000\ \text{gal}}{3.14 \times 80\ \text{ft}}$$

$$= 10,947\ \text{gal/day/ft}$$

Key Point: Notice that 10,947 gal/day/ft is above the recommended minimum of 10,000.

Example 7.228

Problem: A rectangular clarifier has a total of 70 ft of weir. What is the weir overflow rate in gpd/ft when the flow is 1,055,000 gpd?

Solution:

$$\text{Weir Overflow Rate} = \frac{\text{Flow, gpd}}{\text{Weir Length, ft}}$$

$$= \frac{1,055,000\ \text{gpd}}{70\ \text{ft}} = 15,071\ \text{gpd}$$

BOD and Suspended Solids Removed (lb/day)

To calculate the pounds of BOD or suspended solids removed each day, we need to know the mg/L BOD or SS removed and the plant flow. Then, we can use the mg/L to lb/day equation.

$$\text{SS Removed} = \text{mg/L} \times \text{MGD} \times 8.34\ \text{lb/gal} \qquad (7.148)$$

Example 7.229

Problem: If 120 mg/L suspended solids are removed by a primary clarifier, how many lb/day suspended solids are removed when the flow is 6,250,000 gpd?

Solution:

$$\text{SS Removed} = 120\ \text{mg/L} \times 6.25\ \text{MGD} \times 8.34\ \text{lb/gal} = 6,255\ \text{lb/day}$$

Example 7.230

Problem: The flow to a secondary clarifier is 1.6 MGD. If the influent BOD concentration is 200 mg/L and the effluent BOD concentration is 70 mg/L, how many pounds of BOD are removed daily?

$$\text{lb/day BOD removed} = 200\ \text{mg/L} - 70\ \text{mg/L} = 130\ \text{mg/L}$$

After calculating mg/L BOD removed, calculate lb/day BOD removed.

$$\text{BOD removed, lb/day} = (130\ \text{mg/L})(1.6\ \text{MGD})(8.34\ \text{lb/gal})$$

$$= 1,735\ \text{lb/day}$$

Example 7.231

Problem: If 120 mg/L of suspended solids are removed by a trickling filter, how many lb/day of suspended solids are removed when the flow is 4.0 MGD?

Solution:

$$(\text{mg/L})(\text{MGD flow}) \times (8.34\ \text{lb/gal}) = \text{lb/day}$$

$$(120\ \text{mg/L})(4.0\ \text{MGD}) \times (8.34\ \text{lb/gal}) = 4,003\ \text{lb SS/day}$$

Example 7.232

Problem: The 3,500,000 gpd influent flow to a trickling filter has a BOD content of 185 mg/L. If the trickling filter effluent has a BOD content of 66 mg/L, how many pounds of BOD are removed daily?

Solution:

$$(\text{mg/L})(\text{MGD flow})(8.34\ \text{lb/gal}) = \text{lb/day removed}$$

$$185\ \text{mg/L} - 66\ \text{mg/L} = 119\ \text{mg/L}$$

$$(119\ \text{mg/L})(3.5\ \text{MGD})(8.34\ \text{lb/gal}) = 3,474\ \text{lb/day removed}$$

TRICKLING FILTER PROCESS

The *trickling filter process* is one of the oldest forms of dependable biological treatment for wastewater. By its very nature, the trickling filter has advantages over other unit processes. For example, it is a very economical and dependable process for treating wastewater before discharge. Capable of withstanding periodic shock loading, its energy demands are low because aeration is a natural process.

Trickling filter operation involves spraying wastewater over a solid media such as rock, plastic, or redwood slats (or laths). As the wastewater trickles over the surface of the media, the growth of microorganisms (bacteria, protozoa, fungi, algae, helminths or worms, and larvae) develops. This growth is visible as a shiny slime very similar to the slime found on rocks in a stream. As wastewater passes over this slime, the slime adsorbs the organic (food) matter. This organic matter is used as food by the microorganisms. At the same time, air moving through the open spaces in the filter transfers oxygen to the wastewater. This oxygen is then transferred to the slime to keep the outer layer aerobic. As the microorganisms use the food and oxygen, they produce more organisms, carbon dioxide, sulfates, nitrates, and other stable by-products; these materials are then discarded from the slime back into the wastewater flow and carried out of the filter.

Trickling Filter Process Calculations

Several calculations are useful in the operation of trickling filters: these include hydraulic loading, organic loading, biochemical oxygen demand (BOD), and suspended solids (SS) removal. Each type of trickling filter is designed to operate with specific loading levels, vary greatly depending on the filter classification. To operate the filter properly, the filter loading must be within the specified levels. The three main loading parameters for the trickling filter are hydraulic loading, organic loading, and recirculation ratio.

Hydraulic Loading

Calculating the *hydraulic loading rate* is important in accounting for both the primary effluent and the recirculated trickling filter effluent. These are combined before being applied to the filter surface. The hydraulic loading rate is calculated based on the filter surface area. The normal hydraulic loading rate ranges for standard rate and high-rate trickling filters are:

Standard Rate—25–100 gpd/ft² or 1–40 MGD/acre
High Rate—100–1,000 gpd/ft² or 4–40 MGD/acre

Key Point: If the hydraulic loading rate for a particular trickling filter is too low, septic conditions will begin to develop.

Example 7.233

Problem: A trickling filter 80 ft in diameter is operated with a primary effluent of 0.588 MGD and a recirculated effluent flow rate of 0.660 MGD. Calculate the hydraulic loading rate on the filter in units gpd/ft².

Solution:

The primary effluent and recirculated trickling filter effluent are applied together across the surface of the filter; therefore
0.588 MGD + 0.660 MGD = 1.248 MGD = 1,248,000 gpd.

$$\text{Circular surface area} = 0.785 \times (\text{diameter})^2$$
$$= 0.785 \times (80 \text{ ft})^2$$
$$= 5,024 \text{ ft}^2$$

$$\frac{1,248,000 \text{ gpd}}{5,024 \text{ ft}^2} = 248.4 \text{ gpd/ft}^2$$

Example 7.234

Problem: A trickling filter 80 ft in diameter treats a primary effluent flow of 550,000 gpd. If the recirculated flow to the clarifier is 0.2 MGD, what is the hydraulic loading on the trickling filter?

Solution:

$$\text{Hydraulic loading rate} = \frac{\text{Total Flow, gpd}}{\text{Area, ft}^2}$$
$$= \frac{750,000 \text{ gpd total flow}}{(0.785)(80 \text{ ft})(80 \text{ ft})}$$
$$= 149 \text{ gpd/ft}^2$$

Example 7.235

Problem: A high-rate trickling filter receives a daily flow of 1.8 MGD. What is the dynamic loading rate in MGD/acre if the filter is 90 ft in diameter and 5 ft deep?

Solution:

$$(0.785)(90 \text{ ft})(90 \text{ ft}) = 6,359 \text{ ft}^2$$

$$\frac{6,359 \text{ ft}^2}{43,560 \text{ ft}^2/\text{acre}} = 0.146 \text{ acres}$$

$$\text{Hydraulic Loading Rate} = \frac{1.8 \text{ MGD}}{0.146 \text{ acres}} = 12.3 \text{ MGD/acre}$$

Key Point: When the hydraulic loading rate is expressed as MGD per acre, it still is an expression of gallon flow over the surface area of the trickling filter.

Organic Loading Rate

Trickling filters are sometimes classified by the *organic loading rate* applied. The organic loading rate is expressed as a certain amount of BOD applied to a certain volume of media. In other words, the organic loading is defined as the pounds of BOD_5 or Chemical Oxygen Demand (COD) applied per day per 1,000 ft³ of media—a measure of the amount of food being applied to the filter slime. To calculate the organic loading on the trickling filter, two things must be known: the pounds of BOD or COD being applied to the filter media per day and the volume of the filter media in $1,000\,ft^3$ units. The BOD and COD contribution of the recirculated flow is not included in the organic loading.

Example 7.236

Problem: A trickling filter, 60 ft in diameter, receives a primary effluent flow rate of 0.440 MGD. Calculate the organic loading rate in units of pounds of BOD applied per day per $1,000\,ft^3$ of media volume. The primary effluent BOD concentration is 80 mg/L. The media depth is 9 ft.

Solution:

$$0.440\ \text{MGD} \times 80\ \text{mg/L} \times 8.34\ \text{lb/gal} = 293.6\ \text{lb of BOD applied/day}$$

$$\text{Surface Area} = 0.785 \times (60)^2 = 2,826\ ft^2$$

$$\text{Area} \times \text{Depth} \times \text{Volume}$$

$$2,826\ ft^2 \times 9\ ft = 25,434\ ft^3\ \left(\text{TF Volume}\right)$$

Key Point: To determine the pounds of BOD per $1,000\,ft^3$ in a volume of thousands of cubic feet, we must set up the equation as shown below.

$$\frac{293.6\ \text{lb BOD/day}}{25,434\ ft^3} \times \frac{1,000}{1,000}$$

Regrouping the numbers and the units together:

$$\frac{293.6\ \text{lb BOD/day} \times 1,000}{25,434\ ft^3} \times \frac{\text{lb BOD/day}}{1,000\ ft^3} = 11.5\ \frac{\text{lb BOD/day}}{1,000\ ft^3}$$

RECIRCULATION FLOW

Recirculation in trickling filters involves the return of filter effluent back to the head of the trickling filter. It can level flow variations and assist in solving operational problems, such as ponding, filter flies, and odors. The operator must check the rate of recirculation to ensure that it is within design specifications. Rates above design specifications indicate hydraulic overloading; rates under design specifications indicate hydraulic underloading. The **trickling filter recirculation ratio** is the ratio of the recirculated trickling filter flow to the primary effluent flow. The trickling filter recirculation ratio may range from 0.5:1 (0.5) to 5:1 (5). However, the ratio is often found to be 1:1 or 2:1.

$$\text{Recirculation} = \frac{\text{Recirculated Flow, MGD}}{\text{Primary Effluent Flow, MGD}} \qquad (7.149)$$

Example 7.237

Problem: A treatment plant receives a flow of 3.2 MGD. If the trickling filter effluent is recirculated at a rate of 4.50 MGD, what is the recirculation ratio?

Solution:

$$
\begin{aligned}
\text{Recirculation Ratio} &= \frac{\text{Recirculated Flow, MGD}}{\text{Primary Effluent Flow, MGD}}\\[4pt]
&= \frac{4.5\ \text{MGD}}{3.2\ \text{MGD}}\\[4pt]
&= 1.4\ \text{Recirculation Ratio}
\end{aligned}
$$

Example 7.238

Problem: A trickling filter receives a primary effluent flow of 5 MGD. If the recirculated flow is 4.6 MGD, what is the recirculation ratio?

Solution:

$$
\begin{aligned}
\text{Recirculation Ratio} &= \frac{\text{Recirculated Flow, MGD}}{\text{Primary Effluent Flow, MGD}}\\[4pt]
&= \frac{4.6\ \text{MGD}}{5\ \text{MGD}}\\[4pt]
&= 0.92\ \text{Recirculation Ratio}
\end{aligned}
$$

ROTATING BIOLOGICAL CONTACTORS (RBCs)

The *rotating biological contactor (RBC)* is a variation of the attached growth idea provided by the trickling filter. Still relying on microorganisms that grow on the surface of a medium, the RBC is instead a *fixed film* biological treatment device. The basic biological process, however, is similar to that occurring in trickling filters. An RBC consists of a series of closely spaced (mounted side by side), circular, plastic synthetic disks, typically about 11.5 ft in diameter. Attached to a rotating horizontal shaft, approximately 40% of each disk is submerged in a tank that contains the wastewater to be treated. As the RBC rotates, the attached biomass film (zoogleal slime) that grows on the surface of the disks moves into and out of the wastewater. While submerged in the wastewater, the microorganisms absorb organics; while they are rotated out of the wastewater, they are supplied with needed oxygen for aerobic decomposition. As the zoogleal slime re-enters the wastewater, excess solids, and waste products are stripped off the media as *sloughings*. These sloughings are transported with the wastewater flow to a settling tank for removal.

Several process control calculations may be useful in the operation of an RBC. These include soluble BOD, total media area, organic loading rate, and hydraulic loading. Settling tank calculations and biosolids pumping calculations may be helpful for the evaluation and control of the settling tank following the RBC.

Hydraulic Loading Rate

The manufacturer normally specifies the RBC media surface area, and the hydraulic loading rate is based on the media surface area, usually in square feet (ft^2). Hydraulic loading is expressed in terms of gallons of flow per day per square foot of media. This calculation can be helpful in evaluating the current operating status of the RBC. Comparison with design specifications can determine if the unit is hydraulically overloaded or underloaded. Hydraulic loading on an RBC can range from 1 to 3 gpd/ft^2.

Example 7.239

Problem: An RBC treats a primary effluent flow rate of 0.244 MGD. What is the hydraulic loading rate in gpd/ft^2 if the media surface area is 92,600 ft^2?

Solution:

$$\frac{244,000 \text{ gpd}}{92,000 \text{ ft}^2} = 2.63 \text{ gpd/ft}^2$$

Example 7.240

Problem: An RBC treats a flow of 3.5 MGD. The manufacturer's data indicate a media surface area of 750,000 ft^2. What is the hydraulic loading rate on the RBC?

Solution:

$$\text{Hydraulic Loading Rate} = \frac{\text{Flow, gpd}}{\text{Media Area, ft}^2}$$
$$= \frac{3,500,000 \text{ gpd}}{750,000 \text{ ft}^2} = 4.7 \text{ ft}^2$$

Example 7.241

Problem: A rotating biological contactor treats a primary effluent flow of 1,350,000 gpd. The manufacturer's data indicate that the media surface area is 600,000 ft^2. What is the hydraulic loading rate on the filter?

Solution:

$$\text{Hydraulic Loading Rate} = \frac{\text{Flow, gpd}}{\text{Area, ft}^2}$$
$$= \frac{1,350,000 \text{ gpd}}{600,000 \text{ ft}^2} = 2.3 \text{ ft}^2$$

SOLUBLE BOD

The *soluble BOD* concentration of the RBC influent can be determined experimentally in the laboratory, or it can be estimated using the suspended solids concentration and the "K" factor. The "K" factor is used to approximate the BOD (particulate BOD) contributed by the suspended matter. The K factor must be provided or determined experimentally in the laboratory. The K factor for domestic wastes is normally in the range of 0.5–0.7.

$$\text{Soluble BOD}_5 = \text{Total BOD}_5$$
$$- \left(\text{K Factor} \times \text{Total Suspended Solids} \right) \quad (7.150)$$

Example 7.242

Problem: The suspended solids concentration of wastewater is 250 mg/L. If the amount of K-value at the plant is 0.6, what is the estimated particulate biochemical oxygen demand (BOD) concentration of the wastewater?

Solution:

$$\left(250 \text{ mg/L} \right)\left(0.6 \right) = 150 \text{ mg/L Particulate BOD}$$

Key Point: The K-value of 0.6 indicates that about 60% of the suspended solids are organic suspended solids (particulate BOD).

Example 7.243

Problem: A rotating biological contactor receives a flow of 2.2 MGD with a BOD content of 170 mg/L and a suspended solids (SS) concentration of 140 mg/L. If the K-value is 0.7, how many pounds of soluble BOD enter the RBC daily?

Solution:

$$\text{Total BOD} = \text{Particulate BOD} + \text{Soluble BOD}$$
$$170 \text{ mg/L} = (140 \text{ mg/L})(0.7) + x \text{ mg/L}$$
$$170 \text{ mg/L} = 98 \text{ mg/L} + x \text{ mg/L}$$
$$170 \text{ mg/L} - 98 \text{ mg/L} = x$$
$$x = 72 \text{ mg/L Soluble BOD}$$

Now, lb/day soluble BOD may be determined:

$$(\text{mg/L})(\text{Soluble BOD})(\text{MGD Flow})(8.34 \text{ lb/gal})$$
$$= \text{lb/day}$$
$$72 \text{ mg/L} \times 2.2 \text{ MGD} \times 8.34 \text{ lb/gal}$$
$$= 1,321 \text{ lb/day}$$

Organic Loading Rate

The *organic loading rate* can be expressed as the total BOD loading in pounds per day per 1,000 ft^2 of media. The actual values can then be compared with plant design specifications to determine the current operating condition of the system.

$$\text{Organic Loading Rate} = \frac{\text{Sol. BOD} \times \text{Flow, MGD} \times 8.34 \text{ lb/gal}}{\text{Media Area, 1,000 ft}^2}$$

$$(7.151)$$

Example 7.244

Problem: A rotating biological contactor (RBC) has a media surface area of 500,000 ft² and receives a flow of 1,000,000 gpd. If the soluble BOD concentration of the primary effluent is 160 mg/L, what is the organic loading on the RBC in lb/day/1,000 ft²?

Solution:

$$\text{Organic Loading Rate} = \frac{\text{Sol. BOD, lb/day}}{\text{Media Area, 1,000 ft}}$$

$$= \frac{(160 \text{ mg/L})(1.0 \text{ MGD})(8.34 \text{ lb/gal})}{500 \times 1,000 \text{ ft}^2}$$

$$= \frac{2.7 \text{ lb/day Sol. BOD}}{1,000 \text{ ft}^2}$$

Example 7.245

Problem: The wastewater flow to an RBC is 3,000,000 gpd. The wastewater has a soluble BOD concentration of 120 mg/L. The RBC consists of six shafts (each 110,000 ft²), with two shafts comprising the first stage of the system. What is the organic loading rate in lb/day/1,000 ft² on the first stage of the system?

Solution:

$$\text{Organic Loading Rate} = \frac{\text{Sol. BOD, lb/day}}{\text{Media Area, 1,000 ft}^2}$$

$$= \frac{(120 \text{ mg/L}) \times (3.0 \text{ MGD})(8.34 \text{ lb/gal})}{220 \qquad 1,000 \text{ ft}^2}$$

$$= 13.6 \text{ lb Sol. BOD/day/1,000 ft}^2$$

Total Media Area

Several process control calculations for the RBC use the total surface area of all the stages within the train. As was the case with the soluble BOD calculation, plant design information or information supplied by the unit manufacturer must provide the individual stage areas (or the total train area) because the physical determination of this would be extremely difficult.

$$\text{Total Area} = \text{1st stage Area} + \text{2nd Stage Area}$$
$$+ \cdots + \text{nth Stage Area}$$

$$(7.152)$$

ACTIVATED BIOSOLIDS

The *activated biosolids process* is a man-made process that mimics the natural self-purification process that takes place in streams. In essence, we can state that the activated biosolids treatment process is a "stream in a container." In wastewater treatment, activated biosolids processes are used for both secondary treatment and complete aerobic treatment without primary sedimentation. Activated biosolids refer to biological treatment systems that use a suspended growth of organisms to remove BOD and suspended solids.

The basic components of an activated biosolids sewage treatment system include an aeration tank and a secondary basin, settling basin, or clarifier. Primary effluent is mixed with settled solids recycled from the secondary clarifier and is then introduced into the aeration tank. Compressed air is injected continuously into the mixture through porous diffusers located at the bottom of the tank, usually along one side.

Wastewater is fed continuously into an aerated tank, where the microorganisms metabolize and biologically flocculate the organics. Microorganisms (activated biosolids) are settled from the aerated mixed liquor under quiescent conditions in the final clarifier and are returned to the aeration tank. Left uncontrolled, the number of organisms would eventually become too great; therefore, some must periodically be removed (wasted). A portion of the concentrated solids from the bottom of the settling tank must be removed from the process (waste activated sludge or WAS). Clear supernatant from the final settling tank is the plant effluent.

As with other wastewater treatment unit processes, process control calculations are important tools used by the operator to control and optimize process operations. In this chapter, we review many of the most frequently used activated biosolids process calculations.

Moving Averages

When performing process control calculations, the use of a 7-day *moving average* is recommended. The moving average is a mathematical method to level the impact of any one test result. The moving average is determined by adding the test results collected during the past 7 days and dividing them by the number of tests.

$$\text{Moving Average} = \frac{\text{Test 1} + \text{Test 2} + \text{Test 3} + \cdots \text{Test 6} + \text{Test 7}}{\text{\# of Tests Performed during the 7 Days}}$$

$$(7.153)$$

Example 7.246

Problem: Calculate the 7-day moving average for days 7, 8, and 9.

Day	MLSS	Day	MLSS
1	3,340	6	2,780
2	2,480	7	2,476
3	2,398	8	2,756
4	2,480	9	2,655
5	2,558	10	2,396

Solution:

(1) Moving Ave., Day 7 $= \dfrac{\begin{array}{l}3{,}340+2{,}480+2{,}398+2{,}480\\+2{,}558+2{,}780+2{,}476\end{array}}{7}$

$= 2{,}645$

(2) Moving Ave., Day 7 $= \dfrac{\begin{array}{l}2{,}480+2{,}398+2{,}480+2{,}558\\+2{,}780+2{,}476+2{,}756\end{array}}{7}$

$= 2{,}561$

(3) Moving Ave., Day 7 $= \dfrac{\begin{array}{l}2{,}398+2{,}480+2{,}558+2{,}780\\+2{,}476+2{,}756+2{,}655\end{array}}{7}$

$= 2{,}586$

BOD or COD Loading

When calculating BOD, COD, or SS loading on an aeration process (or any other treatment process), the loading on the process is usually calculated as lbs/day. The following equation is used:

$$\text{BOD, COD, or SS Loading, lb/day} = (\text{mg/L})(\text{MGD})(8.34 \text{ lb/gal})$$
(7.154)

Example 7.247

Problem: The BOD concentration of the wastewater entering an aerator is 210 mg/L. If the flow to the aerator is 1,550,000 gpd, what is the lb/day BOD loading?

Solution:

BOD, lb/day = (BOD, mg/L)(Flow, MGD)(8.34 lb/gal)
$= (210 \text{ mg/L})(1.55 \text{ MGD})(8.34 \text{ lb/gal})$
$= 2{,}715 \text{ lb/day}$

Example 7.248

Problem: The flow to an aeration tank is 2,750 gpm. If the BOD concentration of the wastewater is 140 mg/L, how many pounds of BOD are applied to the aeration tank daily?

Solution:

First, convert the gpm flow to gpd flow:

$$(2{,}750 \text{ gpm})(1{,}440 \text{ min/day}) = 3{,}960{,}000 \text{ gpd}$$

Then, calculate lb/day BOD:

BOD, lb/day = (BOD, mg/L)(Flow, MGD)(8.34 lb/gal)
$= (140, \text{mg/L})(3.96 \text{ MGD})(8.34 \text{ lb/day})$
$= 4{,}624 \text{ lb/day}$

Solids Inventory

In the activated biosolids process, it is important to control the amount of solids under aeration. The suspended solids in an aeration tank are called Mixed Liquor Suspended Solids (MLSS). To calculate the pounds of solids in the aeration tank, we need to know the mg/L MLSS concentration and the aeration tank volume. Then, lbs of MLSS can be calculated as follows:

$$\text{lb MLSS} = (\text{MLSS, mg/L})(\text{MG})(8.34 \text{ lb/gal})$$
(7.155)

Example 6.249

Problem: If the mixed liquor suspended solids concentration is 1,200 mg/L, and the aeration tank has a volume of 550,000 gal, how many pounds of suspended solids are in the aeration tank?

Solution:

lbs = (mg/L)(MG Volume)(8.34 lb/gal)
$= (1{,}200 \text{ mg/L})(0.550 \text{ MG})(8.34 \text{ lbs/gal})$
$= 5{,}504 \text{ lbs MLSS}$

Food-to-Microorganism Ratio (F/M Ratio)

The food-to-microorganism ratio (F/M ratio) is a process control method/calculation based on maintaining a specified balance between available food materials (BOD or COD) in the aeration tank influent and the aeration tank mixed liquor volatile suspended solids (MLVSS) concentration. The chemical oxygen demand (COD) test is sometimes used because the results are available in a relatively short period of time. To calculate the F/M ratio, the following information is required:

- Aeration tank influent flow rate, MGD
- Aeration tank influent BOD or COD, mg/L
- Aeration tank MLVSS, mg/L
- Aeration tank volume, MG

$$\text{F/M Ratio} = \frac{\text{Primary Eff. COD/BOD mg/L} \times \text{Flow MGD} \times 8.34 \text{ lb/mg/L/MG}}{\text{MLVSS mg/L} \times \text{Aerator Vol., MG} \times 8.34 \text{ lb/mg/L/MG}}$$
(7.156)

The typical F/M ratio for the activated biosolids process is shown in the following:

Process	lb BOD / lb MLVSS	lb COD / lb MLVSS
Conventional	0.2–0.4	0.5–1.0
Contact stabilization	0.2–0.6	0.5–1.0
Extended aeration	0.05–0.15	0.2–0.5
Pure oxygen	0.25–1.0	0.5–2.0

Example 7.250

Problem: The aeration tank influent BOD is 145 mg/L, and the aeration tank influent flow rate is 1.6 MGD. What is the F/M ratio if the MLVSS is 2,300 mg/L and the aeration tank volume is 1.8 MG?

Solution:

$$\text{F/M ratio} = \frac{145\ \text{mg/L} \times 1.6\ \text{MGD} \times 8.34\ \text{lb/mg/L/MG}}{2,300\ \text{mg/L} \times 1.8\ \text{MGD} \times 8.34\ \text{lb/mg/L/MG}}$$
$$= 0.0.6\ \text{lb BOD/lb MLVSS}$$

Key Point: If the MLVSS concentration is not available, it can be calculated if the % volatile matter (% VM) of the mixed liquor suspended solids (MLSS) is known.

$$\text{MLVSS} = \text{MLSS} \times \% \text{ (decimal) Volatile Matter (VM)}$$
(7.157)

Key Point: The "F" value in the F/M ratio for computing loading to an activated biosolids process can be either BOD or COD. Remember, the reason for biosolids production in the activated biosolids process is to convert BOD to bacteria. One advantage of using COD over BOD for the analysis of organic load is that COD is more accurate.

Example 7.251

Problem: The aeration tank contains 2,885 mg/L of MLSS. Lab tests indicate that the MLSS is 66% volatile matter. What is the MLVSS concentration in the aeration tank?

Solution:

$$\text{MLVSS, mg/L} = 2,885\ \text{mg/L} \times 0.66 = 1,904\ \text{mg/L}$$

Required MLVSS Quantity (Pounds)

The pounds of MLVSS required in the aeration tank to achieve the optimum F/M ration can be determined from the average influent food (BOD or COD) and the desired F/M ratio:

$$\text{MLVSS, lb} = \frac{\text{Primary Effluent BOD or COD} \times \text{Flow, MGD} \times 8.34}{\text{Desired F/M Ratio}}$$
(7.158)

The required pounds of MLVSS determined by this calculation can then be converted to a concentration value by

$$\text{MLVSS, mg/L} = \frac{\text{Desired MLVSS (lb)}}{[\text{Aeration Volume, MG} \times 8.34]}$$ (7.159)
$$= 16,124\ \text{lb MLVSS}$$

Example 7.252

Problem: The aeration tank influent flow is 4.0 MGD, and the influent COD is 145 mg/L. The aeration tank volume is 0.65 MG. The desired F/M ratio is 0.3 lb COD/lb MLVSS. How many pounds of MLVSS must be maintained in the aeration tank to achieve the desired F/M ratio?

Solution:

Determine the required concentration of MLVSS in the aeration tank.

$$\text{MLVSS} = \frac{145\ \text{mg/L} \times 4.0\ \text{MGD} \times 8.34\ \text{lb/gal}}{0.3\ \text{lb COD/lb MLVSS}} = 16,124\ \text{lb MLVSS}$$

$$\text{MLVSS, mg/L} = \frac{16,124\ \text{lb MLVSS}}{[0.65\ \text{MG} \times 8.34]} 2,974\ \text{mg/L MLVSS}$$

Calculating Waste Rates Using F/M Ratio

Maintaining the desired F/M ratio is accomplished by controlling the MLVSS level in the aeration tank. This may be accomplished by adjusting return rates; however, the most practical method is through proper control of the waste rate.

$$\text{Waste Vol. Solids, lb/day} = \text{Actual MLVSS, lb}$$
$$- \text{Desired MLVSS, lb}$$ (7.160)

If the desired MLVSS is greater than the actual MLVSS, wasting is stopped until the desired level is achieved. Practical considerations require that the required waste quantity be converted to a required volume of waste per day. This is accomplished by converting the waste pounds to flow rate in million gallons per day or gallons per minute.

$$\text{Waste, MGD} = \frac{\text{Waste Volatile, lb/day}}{[\text{Waste Volatile Conc., mg/L} \times 8.34\ \text{lb/gal}]}$$
(7.161)

$$\text{Waste, gpm} = \frac{\text{Waste, MGD} \times 1,000,000\ \text{gpd/MGD}}{1,440\ \text{min/day}}$$ (7.162)

Key Point: When the F/M ratio is used for process control, the volatile content of the waste activated sludge should be determined.

Example 7.253

Problem: Given the following information, determine the required waste rate in gallons per minute to maintain an F/M ratio of 0.17 lb COD/lb MLVSS.

Primary effluent COD	140 mg/L
Primary effluent flow	2.2 MGD
MLVSS, mg/L	3,549 mg/L
Aeration tank volume	0.75 MG
Waste volatile concentrations	4,440 mg/L (volatile solids)

Solution:

$$\text{Actual MLVSS, lb} = 3{,}549 \text{ mg/L} \times 0.75 \text{ MG} \times 8.34 = 22{,}199 \text{ lb}$$

$$\text{Required MLVSS, lb} = \frac{140 \text{ mg/L} \times 2.2 \text{ MGD} \times 8.34}{0.17 \text{ lb COD/lb MLVSS}}$$

$$15{,}110 \text{ lb MLVSS}$$

$$\text{Waste, lb/day} = 22{,}199 \text{ lb} - 15{,}110 \text{ lb} = 7{,}089 \text{ lb}$$

$$\text{Waste, MGD} = \frac{7{,}089 \text{ lb/day}}{4{,}440 \text{ mg/L} \times 8.34} = 0.19 \text{ MGD}$$

$$\text{Waste, gpm} = \frac{0.19 \text{ MGD} \times 1{,}000{,}000 \text{ gpd/MGD}}{1{,}440 \text{ min/day}} = 132 \text{ gpm}$$

Gould Biosolids Age

Biosolids age refers to the average number of days a particle of suspended solids remains under aeration. It is a calculation used to maintain the proper amount of activated biosolids in the aeration tank. This calculation is sometimes referred to as Gould Biosolids Age so that it is not confused with similar calculations such as Solids Retention Time (or Mean Cell Residence Time). When considering sludge age, in effect we are asking, "How many days of suspended solids are in the aeration tank?" For example, if 3,000 lb of suspended solids enter the aeration tank daily and the aeration tank contains 12,000 lb of suspended solids, then there are 4 days of solids in the aeration tank—a sludge age of 4 days.

$$\text{Sludge Age, days} = \frac{\text{SS in Tank, lb}}{\text{SS Added, lb/day}} \quad (7.163)$$

Example 7.254

Problem: A total of 2,740 lb/day of suspended solids enter an aeration tank in the primary effluent flow. If the aeration tank has a total of 13,800 lb of mixed liquor suspended solids, what is the biosolids age in the aeration tank?

Solution:

$$\text{Sludge Age, days} = \frac{\text{MLSS, lb}}{\text{SS Added, lb/day}}$$

$$= \frac{13{,}800 \text{ lb}}{2{,}740 \text{ lb/day}}$$

$$= 5.0 \text{ days}$$

Mean Cell Residence Time (MCRT)

Mean cell residence time (MCRT), sometimes called *sludge retention time*, is another process control calculation used for activated biosolids systems. MCRT represents the average length of time an activated biosolids particle remains in the activated biosolids system. It can also be defined as the length of time required at the current removal rate to remove all the solids in the system.

$$\text{Mean Cell Residence Time, days} = \frac{\left[\text{MLSS mg/L} \times \binom{\text{Aeration Vol.}}{+\text{Clarifier Vol.}}\right] \times 8.34 \text{ lb/mg/L/MG}}{[\text{WAS, mg/L} \times \text{WAS flow} \times 8.34) + (\text{TSS out} \times \text{flow out} \times 8.34)]}$$

(7.164)

Key Point: MCRT can be calculated using only the aeration tank solids inventory. When comparing plant operational levels to reference materials, you must determine which calculation the reference manual uses to obtain its example values. Other methods are available to determine the clarifier solids concentrations. However, the simplest method assumes that the average suspended solids concentration is equal to the aeration tank's solids concentration.

Example 7.255

Problem: Given the following data, what is the MCRT?

Aerator volume	=1,000,000 gal
Final clarifier	= 600,000 gal
Flow	= 5.0 MGD
Waste rate	= 0.085 MGD
MLSS mg/L	= 2,500 mg/L
Waste mg/L	= 6,400 mg/L
Effluent TSS	= 14 mg/L

Solution:

$$\text{MRCT} = \frac{\left[2{,}500 \text{ mg/L} \times (1.0 \text{ MG} + 0.60 \text{ MG}) \times 8.34\right]}{\left[(6{,}4000 \text{ mg/L} \times 0.085 \text{ MGD} \times 8.34) + (14 \text{ mg/L} \times 5.0 \text{ mgd} \times .34)\right]}$$

$$= 6.5 \text{ days}$$

Waste Quantities/Requirements

MCRT for process control requires the determination of the optimum range for MCRT values. This is accomplished by comparing the effluent quality with MCRT values. When the optimum MCRT is established, the quantity of solids to be removed (wasted) is determined by

$$\text{Waste, lb/day} = \frac{\text{MLSS} \times (\text{Aer., MG} + \text{Clarifier, MG}) \times 8.34}{\text{Desired MCRT}}$$

$$- [\text{TSS}_{\text{out}} \times \text{Flow} \times 8.34]$$

(7.165)

Example 7.256

$$\frac{3,400 \text{ mg/L} \times (1.4 \text{ MG} + 0.50 \text{ MG}) \times 8.34}{8.6 \text{ days}}$$

$$- [10 \text{ mg/L} \times 5.0 \text{ MGD} \times 8.34]$$

Waste Quality, lb/day = 5,848 lb

Waste Rate in Million Gallons/Day

When the quantity of solids to be removed from the system is known, the desired waste rate in million gallons per day can be determined. The unit used to express the rate (MGD, gpd, and gpm) is a function of the volume of waste to be removed and the design of the equipment.

$$\text{Waste, MGD} = \frac{\text{Waste Pounds/day}}{\text{WAS Concentration, mg/L} \times 8.34} \quad (7.166)$$

$$\text{Waste, gpm} = \frac{\text{Waste MGD} \times 1,000,000 \text{ gpd/MGD}}{1.440 \text{ min/day}} \quad (7.167)$$

Example 7.257

Problem: Given the following data, determine the required waste rate to maintain an MCRT of 8.8 days.

MLSS, mg/L	2,500 mg/L
Aeration volume	1.20 MG
Clarifier volume	0.20 MG
Effluent TSS	11 mg/L
Effluent flow	5.0 MGD
Waste concentration	6,000 mg/L

Solution:

$$\text{Waste, lb/day} = \frac{2,500 \text{ mg/L} \times (1.20 + 0.20) \times 8.34}{8.8 \text{ days}}$$
$$- [11 \text{ mg/L} \times 5.0 \text{ MGD} \times 8.34]$$
$$= 3,317 \text{ lb/day} - 459 \text{ lb/day}$$
$$= 2,858 \text{ lb/day}$$

$$\text{Waste, lb/day} = \frac{2,858 \text{ lb/day}}{[6,000 \text{ mg/L} \times 8.34]} = 0.057 \text{ MGD}$$

$$\text{Waste, gpm} = \frac{0.057 \text{ MGD} \times 1,000,000 \text{ gpd/MGD}}{1,440 \text{ min/day}} = 40 \text{ gpm}$$

Estimating Return Rates from SSV_{60}

Many methods are available for estimating the proper return biosolids rate. A simple method described in the *Operation of Wastewater Treatment Plants, Field Study*

Programs (1986), developed by California State University, Sacramento, uses the 60-min percent settled sludge volume. The $\%SSV_{60}$ test results can provide an approximation of the appropriate return-activated biosolids rate. This calculation assumes that the SSV_{60} results are representative of the actual settling occurring in the clarifier. If this is true, the return rate in percent should be approximately equal to the SSV_{60}. To determine the approximate return rate in million gallons per day (MGD), the influent flow rate, the current return rate, and the SSV_{60} must be known. The results of this calculation can then be adjusted based on sampling and visual observations to develop the optimum return biosolids rate.

Key Point: The $\% SSV_{60}$ must be converted to a decimal percent, and the total flow rate (wastewater flow and current return rate in million gallons per day must be used).

$$\text{Est. Return Rate, MGD} = \left(\begin{array}{l} \text{Influent Flow, MGD} \\ + \text{Current Return Flow, MGD} \end{array} \right) \times \%SSV_{60}$$

$$\text{RAS Rate, GPM} = \frac{\text{Return, Biosolids Rate, gpd}}{1,440 \text{ min/day}} \quad (7.168)$$

where it is assumed that:

- $\%SSV_{60}$ is representative
- the return rate in percent equals $\%SSV_{60}$
- the actual return rate is normally set slightly higher to ensure organisms are returned to the aeration tank as quickly as possible. The rate of return must be adequately controlled to prevent the following:
  - aeration and settling hydraulic overloads
  - low MLSS levels in the aerator
  - organic overloading of aeration
  - septic return-activated biosolids
  - solids loss due to excessive biosolids blanket depth

Example 7.258

Problem: The influent flow rate is 5.0 MGD, and the current return-activated sludge flow rate is 1.8 MGD. The SSV_{60} is 37%. Based on this information, what should the return biosolids rate be in million gallons per day (MGD)?

Solution:

$$\text{Return, MGD} = (5.0 \text{ MGD} + 1.8 \text{ MGD}) \times 0.37 = 2.5 \text{ MGD}$$

Sludge Volume Index (SVI)

Sludge volume index (SVI) is a measure (an indicator) of the settling quality (a quality indicator) of the activated biosolids. As the SVI increases, the biosolids settle slower, do not compact as well, and are likely to result in an increase in effluent suspended solids. As the SVI decreases, the biosolids become denser, settling is more rapid, and the biosolids age. SVI is the volume in milliliters occupied by 1 g of activated biosolids. For the settled biosolids volume (mL/L) and the mixed liquor suspended solids (MLSS) calculation, mg/L is required. The proper SVI range for any plant must be determined by comparing SVI values with plant effluent quality.

$$\text{Sludge Volume Index (SVI)} = \frac{\text{SSV, mL/L} \times 1{,}000}{\text{MLSS, mg/L}} \qquad (7.169)$$

Example 7.259

Problem: The SSV_{30} is 365 ml/L, and the MLSS is 2,365 mg/L. What is the SVI?

Solution:

$$\text{Sludge Volume Index (SVI)} = \frac{365 \text{ mL/L} \times 1{,}000}{2{,}365 \text{ mg/L}} = 154.3$$

SVI equals 154.3.
What does this mean? It means that the system is operating normally with good settling and low effluent turbidity. How do we know this? We know this because we compare the 154.3 results with the parameters listed below to obtain the expected condition (the result).

SVI	Expected Conditions (Indicates)
Less than 100	Old biosolids—possible pin floc
	Effluent turbidity increasing
100–250	Normal operation—good settling
	Low effluent turbidity
Greater than 250	Bulking biosolids—poor settling
	High effluent turbidity

Mass Balance: Settling Tank Suspended Solids

Solids are produced whenever biological processes are used to remove organic matter from wastewater. The mass balance for anaerobic biological processes must consider both the solids removed by physical settling processes and the solids produced by the biological conversion of soluble organic matter to insoluble suspended matter. Research has shown that the amount of solids produced per pound of BOD removed can be predicted based on the type of process being used. Although the exact amount of solids produced can vary from plant to plant, research has developed a series of K factors that can be used to estimate the solids production for plants using a particular treatment process. These average factors provide a simple method to evaluate the effectiveness of a facility's process control program. The mass balance also provides an excellent mechanism to evaluate the validity of process control and effluent monitoring data generated. Recall that average K factors are listed in pounds of solids produced per pound of BOD removed for selected processes:

Mass Balance Calculation

$$\text{BOD in, lb} = \text{BOD, mg/L} \times \text{Flow, MGD} \times 8.34$$

$$\text{BOD out, lb} = \text{BOD, mg/L} \times \text{Flow, MGD} \times 8.34$$

$$\text{Solids Produced, lb/day} = [\text{BOD in. lb} - \text{BOD out, lb}] \times \text{K}$$

$$\text{TSS out, lb/day} = \text{TSS out, mg/L} \times \text{Flow, MGD} \times 8.34$$

$$\text{Waste, lb/day} = \text{Waste, mg/L} \times \text{Flow, MGD} \times 8.34$$

$$\text{Solids Removed, lb/day} = \text{TSS out, lb/day} + \text{Waste, lb/day}$$

$$\% \text{ Mass Balance} = \frac{(\text{Solids Produced} - \text{Solids Removed}) \times 100}{\text{Solids Produced}} \qquad (7.170)$$

Biosolids Waste Based Upon Mass Balance

$$\text{Waste Rate, MGD} = \frac{\text{Solids Produced, lb/day}}{(\text{Waste Concentration} \times 8.34)} \qquad (7.171)$$

Example 7.260

Problem: Given the following data, determine the mass balance of the biological process and the appropriate waste rate to maintain current operating conditions.

Process:	Extended Aeration (No Primary)
Influent flow	1.1 MGD
BOD	220 mg/L
TSS	240 mg/L
Effluent flow	1.5 MGD
BOD	18 mg/L
TSS	22 mg/L
Waste flow	24,000 gpd
TSS	8,710 mg/L

Solution:

$$\text{BOD in} = 220 \text{ mg/L} \times 1.1 \text{ MGD} \times 8.34 = 2{,}018 \text{ lb/day}$$

$$\text{BOD out} = 18 \text{ mg/L} \times 1.1 \text{ MGD} \times 8.34 = 165 \text{ lb/day}$$

$$\text{BOD Removed} = 2{,}018 \text{ lb/day} - 165 \text{ lb/day} = 1{,}853 \text{ lb/day}$$

$$\text{Solids Produced} = 1{,}853 \text{ lb/day} \times 0.65 \text{ lb/lb BOD}$$

$$= 1{,}204 \text{ lb solids/day}$$

$$\text{Solids Out, lb/day} = 22 \text{ mg/L} \times 1.1 \text{ MGD} \times 8.34 = 202 \text{ lb/day}$$

$$\text{Sludge Out, lb/day} = 8{,}710 \text{ mg/L} \times 0.024 \text{ MGD} \times 8.34 = 1{,}743 \text{ lb/day}$$

$$\text{Solids Removed, lb/day} = (202 \text{ lb/day} + 1{,}743 \text{ lb/day}) = 1{,}945 \text{ lb/day}$$

$$\text{Mass Balance} = \frac{(1{,}204 \text{ lb Solids/day} - 1{,}945 \text{ lb/day}) \times 100}{1{,}204 \text{ lb/day}} = 62\%$$

The mass balance indicates:

1. The sampling point(s), collection methods, and/or laboratory testing procedures are producing non-representative results.
2. The process is removing significantly more solids than required. Additional testing should be performed to isolate the specific cause of the imbalance.

To assist in the evaluation, the waste rate based on the mass balance information can be calculated.

$$\text{Waste, GPD} = \frac{\text{Solids Produced, lb/day}}{(\text{Waste TSS, mg/L} \times 8.34)} \quad (7.172)$$

$$\text{Waste, GPD} = \frac{1{,}204 \text{ lb/day} \times 1{,}000{,}000}{8{,}710 \text{ mg/L} \times 8.34} = 16{,}575 \text{ gpd}$$

Oxidation Ditch Detention Time

Oxidation ditch systems may be used where the treatment of wastewater is amenable to aerobic biological treatment and the plant design capacities generally do not exceed 1.0 mgd. The oxidation ditch is a form of aeration basin where the wastewater is mixed with return biosolids. It is essentially a modification of a completely mixed activated biosolids system used to treat wastewater from small communities. This system can be classified as an extended aeration process and is considered to be a low-loading rate system. Such treatment facilities can remove 90% or more of influent BOD. Oxygen requirements will generally depend on the maximum diurnal organic loading, degree of treatment, and suspended solids concentration to be maintained in the aerated channel mixed liquor suspended solids (MLSS). Detention time is the length of time required for a given flow rate to pass through a tank. Although detention time is not normally calculated for aeration basins, it is calculated for oxidation ditches.

Key Point: When calculating detention time, the time and volume units used in the equation must be consistent with each other.

$$\text{Detention Time, h} = \frac{\text{Vol. of Oxidation Ditch, gal}}{\text{Flow Rate, gph}} \quad (7.173)$$

Example 7.261

Problem: An oxidation ditch has a volume of 160,000 gal. If the flow to the oxidation ditch is 185,000 gpd, what is the detention time in hours?

Solution:

Because detention time is desired in hours, the flow must be expressed as gph:

$$\frac{185{,}000 \text{ gpd}}{24 \text{ h/day}} = 7{,}708 \text{ gph}$$

Now, calculate the detention time:

$$\begin{aligned} \text{Detention Time, h} &= \frac{\text{Vol. of Oxidation Ditch, gal}}{\text{Flow Rate, gph}} \\ &= \frac{160{,}000 \text{ gal}}{7{,}708 \text{ gph}} \\ &= 20.8 \text{ h} \end{aligned}$$

Treatment Ponds

The primary goals of wastewater treatment ponds focus on the simplicity and flexibility of operation, protection of the water environment, and protection of public health. Moreover, ponds are relatively easy to build and manage; they accommodate large fluctuations in flow and can also provide treatment that approaches conventional systems (producing a highly purified effluent) at much lower cost. It is the cost (the economics) that drives many managers to decide on the pond option for treatment. The actual degree of treatment provided in a pond depends on the type and number of ponds used. Ponds can be used as the sole type of treatment, or they can be used in conjunction with other forms of wastewater treatment—that is, other treatment processes followed by a pond or a pond followed by other treatment processes. Ponds can be classified based on their location in the system, the type of waste they receive, and the main biological process occurring in the pond.

Before we discuss pond process control calculations, it is important first to describe the calculations for determining the area, volume, and flow rate parameters that are crucial for making treatment pond calculations.

Pond Area in Inches

$$\text{Area, acres} = \frac{\text{Area, ft}^2}{43{,}560 \text{ ft}^2/\text{acre}} \quad (7.174)$$

Pond Volume in Acre-Feet

$$\text{Volume, acre-ft} = \frac{\text{Volume, ft}^3}{43{,}560 \text{ ft}^2/\text{acre-ft}} \quad (7.175)$$

Flow Rate in Acre-Feet/Day

$$\text{Flow, acre-ft/day} = \text{flow, MGD} \times 3{,}069 \text{ acre-ft/MG} \quad (7.176)$$

Key Point: Acre-feet (ac-ft) is a unit that can cause confusion, especially for those not familiar with pond or lagoon operations. One acre-foot is the volume of a box with a 1-acre top and 1 ft of depth—but the top doesn't have to be an even number of acres in size to use acre-feet.

Determining Flow Rate in Acre-Inches Day

$$\text{Flow, acre-in/day} = \text{flow, MGD} \times 36.8 \text{ acre-in} \quad (7.177)$$

Although there are no recommended process control calculations for the treatment pond, several calculations may help evaluate process performance or identify causes of poor performance. These include hydraulic detention time, BOD loading, organic loading rate, BOD removal efficiency, population loading, and hydraulic loading rate. In the following, we provide a few calculations that might be helpful in pond performance evaluation and the identification of causes of poor performance, along with other calculations and/or equations that may be useful.

Hydraulic Detention Time (Days)

$$\text{Hydraulic detention time, days} = \frac{\text{Pond volume, acre-ft}}{\text{Influent flow, acre-ft/day}} \quad (7.178)$$

Key Point: Normally, hydraulic detention time ranges from 30 to 120 days for stabilization ponds.

Example 7.262

Problem: A stabilization pond has a volume of 54.5 acre-ft. What is the detention time on days when the flow is 0.35 MGD?

Solution:

$$\text{Flow, acre-ft/day} = 0.35 \text{ MGD} \times 3.069 \text{ acre-ft/MG}$$

$$= 1.07 \text{ acre-ft/day}$$

$$\text{DT days} = \frac{54.5 \text{ acre/ft}}{1.07 \text{ acre-ft/day}}$$

BOD Loading

When calculating BOD loading on a wastewater treatment pond, the following equation is used:

$$\text{lb/day} = (\text{BOD, mg/L})(\text{flow, MGD})(8.34 \text{ lb/gal}) \quad (7.179)$$

Example 7.263

Problem: Calculate the BOD loading (lb/day) on a pond if the influent flow is 0.3 MGD with a BOD of 200 mg/L.

Solution:

$$\begin{aligned}
\text{lbs/day} &= (\text{BOD, mg/L})(\text{flow, MGD})(8.34 \text{ lb/gal}) \\
&= (200 \text{ mg/L})(0.3 \text{ MGD})(8.34 \text{ lb/gal}) \\
&= 500 \text{ lb/day BOD}
\end{aligned}$$

Organic Loading Rate

Organic loading can be expressed as pounds of BOD per acre per day (most common), pounds of BOD per acre foot per day, or people per acre per day.

$$\text{Organic Loading, lb BOD/acre/day}$$

$$= \frac{\text{BOD, mg/L Influent Flow, MGD} \times 8.34}{\text{Pond area, acres}} \quad (7.180)$$

Key Point: The normal range is 10–50 lbs BOD per day per acre.

Example 7.264

Problem: A wastewater treatment pond has an average width of 370 ft and an average length of 730 ft. The influent flow rate to the pond is 0.10 MGD with a BOD concentration of 165 mg/L. What is the organic loading rate to the pond in pounds per day per acre (lb/day/acre)?

Solution:

$$730 \text{ ft} \times 370 \text{ ft} \times \frac{1 \text{ acre}}{43{,}560 \text{ ft}^2} = 6.2 \text{ acre}$$

$$0.10 \text{ MGD} \times 165 \text{ mg/L} \times 8.34 \text{ lb/gal} = 138 \text{ lb/day}$$

$$\frac{138 \text{ lb/day}}{6.2 \text{ acre}} = 22.2 \text{ lb/day/acre}$$

BOD Removal Efficiency

As mentioned, the efficiency of any treatment process is its effectiveness in removing various constituents from the water or wastewater. BOD removal efficiency is, therefore, a measure of the effectiveness of the wastewater treatment pond in removing BOD from the wastewater.

$$\% \text{ BOD Removed} = \frac{\text{BOD Removed, mg/L}}{\text{BOD Total, mg/L}} \times 100 \quad (7.181)$$

Example 7.265

Problem: The BOD entering a wastewater treatment pond is 194 mg/L. If the BOD in the pond effluent is 45 mg/L, what is the BOD removal efficiency of the pond?

$$\% \text{ BOD Removed} = \frac{\text{BOD Removal, mg/L}}{\text{BOD Total, mg/L}} \times 100$$

$$= \frac{149 \text{ mg/L}}{194 \text{ mg/L}} \times 100$$

$$= 77\%$$

Population Loading

Population loading, people/acre/day

$$= \frac{\text{BOD, mg/L Infl.flow, MGD} \times 8.34}{\text{Pond area, acres}} \quad (7.182)$$

$$\text{Hydraulic Loading, in/day} = \frac{\text{Influent flow, acre-in/day}}{\text{Pond area, acres}} \quad (7.183)$$

CHEMICAL DOSING

Chemicals are used extensively in wastewater treatment (and water treatment) operations. Plant operators add chemicals to various unit processes for slime growth control, corrosion control, odor control, grease removal, BOD reduction, pH control, biosolids bulking control, ammonia oxidation, bacterial reduction, and other reasons. To apply any chemical dose correctly, it is important to make certain dosage calculations. One of the most frequently used calculations in wastewater/water mathematics is the dosage or loading. The general types of mg/L to lb/day or lb calculations include chemical dosage, BOD, COD, SS loading/removal, pounds of solids under aeration, and WAS pumping rate. These calculations are usually made using either Equation (7.184) or Equation (7.185).

$$(\text{Chemical, mg/L})(\text{MGD flow})(8.34 \text{ lb/gal}) = \text{lb/day}$$
$$(7.184)$$

$$(\text{Chemical, mg/L})(\text{MG volume})(8.34 \text{ lb/gal}) = \text{lb} \quad (7.185)$$

Key Point: If mg/L concentration represents a concentration in a flow, then a million gallons per day (MGD) flow is used as the second factor. However, if the concentration pertains to a tank or pipeline volume, then million gallons (MG) volume is used as the second factor.

Key Point: Typically, especially in the past, the expression parts per million (ppm) was used as an expression of concentration because 1 mg/L = 1 ppm. However, current practice is to use mg/L as the preferred expression of concentration.

Chemical Feed Rate

In chemical dosing, a measured amount of chemicals is added to the wastewater (or water). The amount of chemical required depends on the type of chemical used, the reason for dosing, and the flow rate being treated. The two expressions most often used to describe the amount of chemical added or required are:

- milligrams per liter (mg/L)
- pounds per day (lbs/day)

A milligram per liter is a measure of concentration. As shown below, if a concentration of 5 mg/L is desired, then a total of 15-mg of chemical would be required to treat 3 L:

$$\frac{5 \text{ mg} \times 3}{\text{L} \times 3} = \frac{15 \text{ mg}}{3 \text{ L}}$$

The amount of chemical required, therefore, depends on two factors:

- the desired concentration (mg/L)
- the amount of wastewater to be treated (normally expressed as MGD).

To convert from mg/L to lb/day, Equation (7.266) is used:

Example 7.266

Problem: Determine the chlorinator setting (lb/day) needed to treat a flow of 5 MGD with a chemical dose of 3 mg/L.

Solution:

Chemical, lb/day = Chemical, mg/L × flow, MGD × 8.34 lb/gal

$$= 3 \text{ mg/L} \times 5 \text{ MGD} \times 8.34 \text{ lb/gal}$$

$$= 125 \text{ lb/day}$$

Example 7.267

Problem: The desired dosage for a dry polymer is 10 mg/L. If the flow to be treated is 2,100,000 gpd, how many lb/day of polymer will be required?

Solution:

Polymer, lb/day = Polymer, mg/L × flow, MGD × 8.34 lb/day

$$= 10 \text{ mg/L Polymer} \times (2.10 \text{ MGD})(8.34 \text{ lb/day})$$

$$= 175 \text{ lb/day Polymer}$$

Key Point: To calculate chemical dose for tanks or pipelines, a modified equation must be used. Instead of MGD flow, MG volume is used:

$$\text{lb Chemical} = \text{Chemical, mg/L} \times \text{Tank Volume, MG}$$

$$\times 8.34 \text{ lb/gal}$$

$$(7.186)$$

Example 7.268

Problem: To neutralize a sour digester, 1 lb of lime is added for every pound of volatile acids in the digester biosolids. If the digester contains 300,000 gal of biosolids with a volatile acid (VA) level of 2,200 mg/L, how many pounds of lime should be added?

Solution:

Because the volatile acid concentration is 2,200 mg/L the lime concentration should also be 2,200 mg/L:

$$\text{lb lime required} = \text{lime, mg/L} \times \text{digester volume, MG} \times 8.34\ \text{lb/gal}$$

$$= (2{,}200\ \text{mg/L})(0.30\ \text{MG})(8.34\ \text{lb/gal})$$

$$= 5{,}504\ \text{lb lime}$$

Chlorine Dose, Demand, and Residual

Chlorine is a powerful oxidizer that is commonly used in wastewater and water treatment for disinfection, in wastewater treatment for odor control, bulking control, and other applications. When chlorine is added to a unit process, we want to ensure that a measured amount is added, obviously. The chlorine dose depends on two considerations: the chlorine demand and the desired chlorine residual:

$$\text{Chlorine Dose} = \text{Chlorine Demand} + \text{Chlorine Residual}$$

$$(7.187)$$

Chlorine Dose

In describing the amount of chemical added or required, we apply Equation (7.188):

$$\text{lb/day} = \text{Chemical, mg/L} \times \text{MGD} \times 8.34\ \text{lb/day} \quad (7.188)$$

Example 7.269

Problem: Determine the chlorinator setting (lb/day) needed to treat a flow of 8 MGD with a chlorine dose of 6 mg/L.

Solution:

$$(\text{mg/L})(\text{MGD})(8.34) = \text{lb/day}$$

$$(6\ \text{mg/L})(8\ \text{MGD})(8.34\ \text{lb/gal}) = \text{lb/day}$$

$$= 400\ \text{lb/day}$$

Chlorine Demand

The chlorine demand is the amount of chlorine used in reacting with various components of the water, such as harmful organisms and other organic and inorganic substances. When the chlorine demand has been satisfied, these reactions cease.

Example 7.270

Problem: The chlorine dosage for a secondary effluent is 6 mg/L. If the chlorine residual after 30 min of contact time is found to be 0.5 mg/L, what is the chlorine demand expressed in mg/L?

Solution:

$$\text{Chlorine Dose} = \text{Chlorine demand} + \text{Chlorine Residual}$$

$$6\ \text{mg/L} = x\ \text{mg/L} + 0.5\ \text{mg/L}$$

$$6\ \text{mg/L} - 0.5\ \text{mg/L} = x\ \text{mg/L}$$

$$x = 5.5\ \text{mg/L Chlorine Demand}$$

Chlorine Residual

Chlorine residual is the amount of chlorine remaining after the demand has been satisfied.

Example 7.271

Problem: What should the chlorinator setting (lb/day) be to treat a flow of 3.9 MGD if the chlorine demand is 8 mg/L and a chlorine residual of 2 mg/L is desired?

Solution:

First, calculate the chlorine dosage in mg/L:

$$\text{Chlorine Dose} = \text{Chlorine demand} + \text{Chlorine Residual}$$

$$= 8\ \text{mg/L} + 2\ \text{mg/L}$$

$$= 10\ \text{mg/L}$$

Then, calculate the chlorine dosage (feed rate) in lb/day:

$$(\text{Chlorine, mg/L})(\text{MGD flow})(8.34\ \text{lb/gal}) = \text{lb/day Chlorine}$$

$$(10\ \text{mg/L})(3.9\ \text{MGD})(8.34\ \text{lb/gal}) = 325\ \text{lb/day Chlorine}$$

Hypochlorite Dosage

Hypochlorite is less hazardous than chlorine; therefore, it is often used as a substitute chemical for elemental chlorine. Hypochlorite is similar to strong bleach and comes in two forms: dry calcium hypochlorite (often referred to as HTH) and liquid sodium hypochlorite. Calcium hypochlorite contains about 65% available chlorine, while sodium hypochlorite contains about 12%–15% available chlorine (in industrial strengths).

Key Point: Because either type of hypochlorite is not 100% pure chlorine, more lb/day must be fed into the system to obtain the same amount of chlorine for disinfection. This is an important economic consideration for those facilities thinking about substituting hypochlorite for chlorine. Some studies indicate that such a substitution can increase operating costs overall by up to three times the cost of using chlorine.

To calculate the lb/day of hypochlorite required, a two-step calculation is necessary:

Step (1) $mg/L(MGD)(8.34) = lb/day$

Step (2) $\dfrac{\text{Chlorine, lb/day}}{\dfrac{\% \text{ available}}{100}} = \text{hypochlorite, lb/day}$ (7.189)

Example 7.272

Problem: A total chlorine dosage of 10 mg/L is required to treat a particular wastewater. If the flow is 1.4 MGD and the hypochlorite has 65% available chlorine, how many lb/day of hypochlorite will be required?

Solution:

Step 1: Calculate the lb/day chlorine required using the mg/L to lb/day equation:

$$(mg/L)(MGD)(8.34) = lb/day$$

$$(10 \text{ mg/L})(1.4 \text{ MGD})(8.34 \text{ lb/gal}) = 117 \text{ lb/day}$$

Step 2: Calculate the lb/day hypochlorite required. Because only 65% of the hypochlorite is chlorine, more than 117 lb/day will be required:

$$\dfrac{117 \text{ lb/day chlorine}}{\dfrac{65 \text{ available chlorine}}{100}} = 180 \text{ lb/day hypochlorite}$$

Example 7.273

Problem: A wastewater flow of 840,000 gpd requires a chlorine dose of 20 mg/L. If sodium hypochlorite (15% available chlorine) is to be used, how many lb/day of sodium hypochlorite are required? How many gal/day of sodium hypochlorite is this?

Solution:

Calculate the lb/day chlorine required:

$$(mg/L)(MGD)(8.34) = lb/day$$

$$(20 \text{ mg/L})(0.84 \text{ MGD})(8.34 \text{ lb/gal}) = 140 \text{ lb/day chlorine}$$

Calculate the lb/day sodium hypochlorite:

$$\dfrac{140 \text{ lb/day chlorine}}{\dfrac{15 \text{ available chlorine}}{100}} = 933 \text{ lb/day hypochlorite}$$

140 lb/day chlorine
    Calculate the gal/day sodium hypochlorite:

$$\dfrac{933 \text{ lb/day}}{8.34 \text{ lb/gal}} = 112 \text{ gal/day sodium hypochlorite}$$

Example 7.274

Problem: How many pounds of chlorine gas are necessary to treat 5,000,000 gal of wastewater at a dosage of 2 mg/L?

Solution:

Step 1: Calculate the pounds of chlorine required.

$$V, 10^6 \text{ gal} = \text{chlorine concentration } (mg/L) \times 8.34 = lb \text{ chlorine}$$

Step 2: Substitute $\dfrac{5 \times 10^6\text{-gal} \times 2 \text{ mg/L} \times 8.34}{= 83 \text{ lb chlorine}}$

Chemical Solutions

A *water solution* is a homogeneous liquid consisting of the *solvent* (the substance that dissolves another substance) and the *solute* (the substance that dissolves in the solvent). Water is the solvent. The solute (whatever it may be) may dissolve up to a certain point. This is called its *solubility*—that is, the solubility of the solute in a particular solvent (water) at a particular temperature and pressure. Remember, in chemical solutions, the substance being dissolved is called the solute, and the liquid present in the greatest amount in a solution (and that does the dissolving) is called the solvent. We should also be familiar with another term, *concentration*—the amount of solute dissolved in a given amount of solvent. Concentration is measured as:

$$\begin{aligned} \% \text{ Strength} &= \dfrac{\text{Wt. of solute}}{\text{Wt. of solution}} \times 100 \\ &= \dfrac{\text{Wt of solute}}{\text{Wt. of solute} + \text{solvent}} \times 100 \end{aligned}$$

(7.190)

Example 7.275

Problem: If 30 lb of chemical is added to 400 lb of water, what is the percent strength (by weight) of the solution?

Solution:

$$\begin{aligned} \% \text{ Strength} &= \dfrac{\dfrac{30 \text{ lb solute}}{400 \text{ lb water}}}{\times 100} = \dfrac{30 \text{ lb solute}}{30 \text{ lb solute} + 400 \text{ lb water}} \times 100 \\ &= \dfrac{30 \text{ lb solute}}{430 \text{ lb solute/water}} \times 100 \\ \% \text{ Strength} &= 7.0\% \end{aligned}$$

Important to making accurate computations of chemical strength is a complete understanding of the dimensional units involved. For example, it is important to understand exactly what *milligrams per liter* (mg/L) signify.

$$\text{Milligrams per Liter } (mg/L) = \dfrac{\text{Milligrams of Solute}}{\text{Liters of Solution}}$$

(7.191)

Another important dimensional unit commonly used when dealing with chemical solutions is *parts per million* (ppm).

$$\text{Parts per Million } (ppm) = \dfrac{\text{Parts of Solute}}{\text{Million Parts of Solution}}$$

(7.192)

Key Point: "Parts" is usually a weight measurement; for example:

$$8 \text{ ppm} = \frac{8 \text{ lb solids}}{1,000,000 \text{ lb solution}}$$

$$8 \text{ ppm} = \frac{8 \text{ mg solids}}{1,000,000 \text{ mg solution}}$$

Solution Chemical Feeder Setting, GPD

Calculating GPD feeder setting depends on how the solution concentration is expressed, lb/gal or percent. If the solution strength is expressed as lb/gal, use the following equation:

$$(\text{Chemical, mg/L})$$

$$\text{Solution, gpd} = \frac{(\text{Flow, MGD})(8.34, \text{lb/gal})}{\text{lb Chemical Solution}} \quad (7.193)$$

In water and wastewater operations, a standard trial-and-error method known as jar testing is conducted to determine the optimum chemical dosage. Jar testing has been the accepted bench testing procedure for many years. After jar testing results are analyzed to determine the best chemical dosage, the following example problems demonstrate how the actual calculations are made.

Example 7.276

Problem: Jar tests indicate that the best liquid alum dose for water is 8 mg/L. The flow to be treated is 1.85 MGD. Determine the gpd setting for the liquid alum chemical feeder if the liquid alum contains 5.30 lb of alum per gallon of solution.

Solution:

First, calculate the lb/day of dry alum required using the mg/L to lb/day equation:

$$
\begin{aligned}
\text{lb/day} &= (\text{dose, mg/L})(\text{flow, MGD})(8.34, \text{lb/gal}) \\
&= (8 \text{ mg/L})(1.85 \text{ MGD})(8.34 \text{ lb/gal}) \\
&= 123 \text{ lb/day dry alum}
\end{aligned}
$$

Then, calculate gpd solution required.

$$\text{Alum Solution, gpd} = \frac{123 \text{ lb/day alum}}{5.30 \text{ lb alum/gal solution}}$$

The feeder setting, then is 23 gpd alum solution. If the solution strength is expressed as a percent, we use the following equation:

$$(\text{Chem., mg/L})$$

$$(\text{Flow Treated, MGD})(8.34 \text{ lb/gal}) = (\text{Sol., mg/L})(\text{Sol. Flow, MGD})$$

$$(8.34, \text{lb/gal})$$

$$(7.194)$$

Example 7.277

Problem: The flow to the plant is 3.40 MGD. Jar testing indicates that the optimum alum dose is 10 mg/L. What should the gpd setting be for the solution feeder if the alum solution is a 52% solution?

Solution:

A solution concentration of 52% is equivalent to 520,000 mg/L:

$$\text{Desired Dose, lb/day} = \text{Actual Dose, lb/day}$$

$$(\text{Chemical, mg/L})$$

$$(\text{Flow Treated, MGD})(8.34, \text{lb/gal}) = (\text{Sol, mg/L})(\text{Sol. Flow, MGD})$$

$$(8.34 \text{ lb/gal})$$

$$(10 \text{ mg/L})(3.40 \text{ MGD})(8.34 \text{ lb/gal}) = (520,00 \text{ mg/L})$$

$$(\text{x MGD})(8.34 \text{ lb/gal})$$

$$x = \frac{(10)(3.40)(8.34)}{(520,000)(8.34)}$$

$$x = 0.0000653 \text{ MGD}$$

This can be expressed as gpd flow:

$$0.0000653 \text{ MGD} = 65.3 \text{ gpd flow}$$

Chemical Feed Pump: Percent Stroke Setting

Chemical feed pumps are generally positive displacement pumps (also called "piston" pumps). This type of pump displaces, or pushes out, a volume of chemical equal to the volume of the piston. The length of the piston, called the stroke, can be lengthened or shortened to increase or decrease the amount of chemical delivered by the pump. As mentioned, each stroke of a piston pump "displaces" or pushes out chemical. To calculate the percent stroke setting, use the following equation:

$$\% \text{ Stroke Setting} = \frac{\text{Required Feed, gpd}}{\text{Maximum Feed, gpd}} \quad (7.195)$$

Example 7.278

Problem: The required chemical pumping rate has been calculated as 8 gpm. If the maximum pumping rate is 90 gpm, what should the percent stroke setting be?

Solution:

The percent stroke setting is based on the ratio of the gpm required to the total possible gpm:

$$\% \text{ Stroke Setting} = \frac{\text{Required Feed, gpd}}{\text{Maximum Feed, gpd}} \times 100$$

$$= \frac{8 \text{ gpm}}{90 \text{ gpm}} \times 100$$

$$= 8.9\%$$

Chemical Solution Feeder Setting, mL/min

Some chemical solution feeders dispense chemicals as milliliters per minute (mL/min). To calculate the mL/min solution required, use the following equation:

$$\text{Solution, mL/min} = \frac{(\text{gpd})(3,785 \text{ mL/gal})}{1,440 \text{ min/day}} \quad (7.196)$$

Example 7.279

Problem: The desired solution feed rate was calculated to be 7 gpd. What is this feed rate expressed as mL/min?

Solution:

Since the gpd flow has already been determined, the mL/min flow rate can be calculated directly:

$$\text{Feed rate, mL/min} = \frac{(\text{gpd})(3,785 \text{ mL/gal})}{1,440 \text{ min/day}}$$
$$= \frac{(7 \text{ gpd})(3,785 \text{ mL/gal})}{1,440 \text{ min/day}}$$
$$= 18 \text{ mL/min feed rate}$$

Chemical Feed Calibration

Routinely, to ensure accuracy, we need to compare the actual chemical feed rate with the feed rate indicated by the instrumentation. To accomplish this, we use calibration calculations. To calculate the actual chemical feed rate for a dry chemical feed, place a container under the feeder, weigh the container when empty, and then weigh the container again after a specified length of time, such as 30 min. The actual chemical feed rate can then be determined as:

$$\text{Chemical Feed Rate, lb/min} = \frac{\text{Chemical Applied, lb}}{\text{Length of Application, min}} \quad (7.197)$$

Example 7.280

Problem: Calculate the actual chemical feed rate, lb/day, if a container is placed under a chemical feeder and a total of 2.2 lb is collected during a 30-min period.

Solution:

First, calculate the lb/min feed rate:

$$\text{Chemical Feed Rate, lb/min} = \frac{\text{Chemical Applied, lb}}{\text{Length of Applications, min}}$$
$$= \frac{2.2 \text{ lb}}{30 \text{ min}}$$
$$= 0.07 \text{ lb/min Feed Rate}$$

Then, calculate the lb/day feed rate:

$$\text{Chemical Feed Rate, lb/day} = (0.07 \text{ lb/min})(1,440 \text{ min/day})$$
$$= 101 \text{ lb/day Feed Rate}$$

Example 7.281

Problem: A chemical feeder is to be calibrated. The container used to collect the chemical is placed under the chemical feeder and weighed (0.35 lb). After 30 min, the weight of the container and chemical is found to be 2.2 lb. Based on this test, what is the actual chemical feed rate, in lb/day?

Key Point: The chemical applied is the weight of the container and chemical minus the weight of the empty container.

Solution:

First, calculate the lb/min feed rate:

$$\text{Chemical Feed Rate, lb/min} = \frac{\text{Chemical Applied, lb}}{\text{Length of Application, min}}$$
$$= \frac{2.2 \text{ lb} - 0.35 \text{ lb}}{30 \text{ min}}$$
$$= \frac{1.85 \text{ lb}}{30 \text{ min}}$$
$$= 0.062 \text{ lb/min Feed Rate}$$

Chemical Applied, lb
Then, calculate the lbs/day feed rate:

$$(0.062 \text{ lb/min})(1,440 \text{ min/day}) = 89. \text{ lb/day Feed Rate}$$

When the chemical feeder is for a solution, the calibration calculation is slightly more difficult than that for a dry chemical feeder. As with other calibration calculations, the actual chemical feed rate is determined and then compared with the feed rate indicated by the instrumentation. The calculations used for solution feeder calibration are as follows:

$$\text{Flow rate, gpd} = \frac{(\text{mL/min})(1,440 \text{ min/day})}{3,785 \text{ mL/gal}} = \text{gpd} \quad (7.198)$$

Then, calculate chemical dosage, lb/day:

$$\text{Chemical, lb/day} = (\text{Chemical, mg/L})$$
$$(\text{Flow, MGD})(8.34 \text{ lb/day}) \quad (7.199)$$

Example 7.282

Problem: A calibration test is conducted for a chemical solution feeder. For 5 min, the solution feeder delivers a total of 700 mL. The polymer solution is a 1.3% solution. What is the lbs/day feed rate? (Assume the polymer solution weighs 8.34 lb/gal.)

Solution:

The mL/min flow rate is calculated as:

$$\frac{700 \text{ mL}}{5 \text{ min}} = 140 \text{ mL/min}$$

Then, convert mL/min flow rate to gpd flow rate:

$$\frac{(140 \text{ mL/min})(1{,}440 \text{ min/day})}{3{,}785 \text{ mL/gal}} = 53 \text{ gpd flow rate}$$

Now, calculate the lb/day feed rate:

$$(\text{Chemical, mg/L})(\text{Flow, MGD})(8.34 \text{ lb/day}) = \text{Chemical, lb/day}$$

$$(13{,}000 \text{ mg/L})(0.000053 \text{ MGD})(8.34 \text{ lb/day}) = 5.7 \text{ lb/day polymer}$$

Actual pumping rates can be determined by calculating the volume pumped during a specified timeframe. For example, if 120 gal are pumped during a 15-min test, the average pumping rate during the test is 8 gpm. The gallons pumped can be determined by measuring the drop in tank level during the timed test.

$$\text{Flow, gpm} = \frac{\text{Volume Pumped, gal}}{\text{Duration of Test, min}} \qquad (7.200)$$

Then the actual flow rate (gpm) is calculated using:

$$\text{Flow Rate, gpm} = \frac{\dfrac{(0.785)(D^2)(\text{Drop in Level, ft})}{(7.48 \text{ gal/ft}^3)}}{\text{Duration of Test, min}}$$

$$(7.201)$$

Example 7.283

Problem: A pumping rate calibration test is conducted for 5 min. The liquid level in the 4-ft diameter solution tank is measured before and after the test. If the level drops 0.4 ft during the 5-min test, what is the pumping rate in gpm?

Solution:

$$\begin{aligned}
\text{Flow Rate, gpm} &= \frac{(0.785)(D^2)(\text{Drop, ft})(7.48 \text{ gal/ft}^3)}{\text{Duration of Test, min}} \\[2mm]
&= \frac{(0.785)(4 \text{ ft})(4 \text{ ft})(0.4 \text{ ft})(7.48 \text{ gal/ft}^3)}{5 \text{ min}}
\end{aligned}$$

$$\text{Pumping Rate} = 38 \text{ gpm}$$

Average Use Calculations

During a typical shift, operators log in or record several parameter readings. The data collected is important for monitoring plant operation, providing information on how to best optimize plant or unit process operations. One of the important parameters monitored each shift or each day is the actual use of chemicals. From the recorded chemical use data, expected chemical use can be forecasted. This data is also important for inventory control; that is, a determination can be made when additional chemical supplies will be required. In determining average chemical use, we first must determine the average chemical use:

$$\text{Average Use, lb/day} = \frac{\text{Total Chemical Used, lb}}{\text{Number of Days}} \qquad (7.202)$$

or

$$\text{Average Use, gpd} = \frac{\text{Total Chemical Used, gal}}{\text{Number of Days}} \qquad (7.203)$$

Then, calculate the day's supply in inventory:

$$\text{Day's Supply in Inventory} = \frac{\text{Total Chemical in Inventory, lb}}{\text{Average Use, lb/day}}$$

$$(7.204)$$

or

$$\text{Day's Supply in Inventory} = \frac{\text{Total Chemical in Inventory, gal}}{\text{Average Use, gpd}}$$

$$(7.205)$$

Example 7.284

Problem: The chemical used for each day during the week is given below. Based on this data, what was the average lb/day chemical use during the week?

Monday—92 lb/day	Friday—96 lb/day
Tuesday—94 lb/day	Saturday—92 lb/day
Wednesday—92 lb/day	Sunday—88 lb/day
Thursday—88 lb/day	

Solution:

$$\text{Average Use, lb/day} = \frac{\text{Total Chemical Used, lb}}{\text{Number of Days}}$$

$$= \frac{642 \text{ lb}}{7 \text{ days}}$$

$$\text{Average Use} = 91.7 \text{ lb/day}$$

Example 7.285

Problem: The average chemical used at a plant is 83 lb/day. If the chemical inventory in stock is 2,600 lb, how many days' supply is this?

Solution:

$$\begin{aligned}
\text{Days' Supply in Inventory} &= \frac{\text{Total Chemical in Inventory, lb}}{\text{Average Use, lb/day}} \\[2mm]
&= \frac{2{,}600 \text{ lb in Inventory}}{83 \text{ lb/day Average Use}} \\[2mm]
&= 31.3 \text{ days' supply}
\end{aligned}$$

Process Residuals: Biosolids Production and Pumping Calculations

The wastewater unit treatment processes remove solids and biochemical oxygen demand (BOD) from the waste stream before the liquid effluent is discharged to its receiving waters. What remains to be disposed of is a mixture of solids and wastes, called *process residuals*—more commonly referred to as biosolids (or sludge).

Key Point: Sludge is the commonly accepted name for wastewater residual solids. However, if wastewater sludge is used for beneficial reuse (i.e., as a soil amendment or fertilizer), it is commonly called biosolids. I choose to refer to process residuals as biosolids in this text.

The most costly and complex aspect of wastewater treatment can be the collection, processing, and disposal of biosolids. This is the case because the quantity of biosolids produced may be as high as 2% of the original volume of wastewater, depending somewhat on the treatment process being used. Because the 2% biosolids can be as much as 97% water content, and because the cost of disposal will be related to the volume of biosolids being processed, one of the primary purposes or goals (along with stabilizing it so it is no longer objectionable or environmentally damaging) of biosolids treatment is to separate as much of the water from the solids as possible.

Primary and Secondary Solids Production Calculations

It is important to point out that when making calculations pertaining to solids and biosolids, the term *solids* refers to dry solids and the term *biosolids* refers to the solids and water. The solids produced during primary treatment depend on the solids that settle in or are removed by the primary clarifier. In making primary clarifier solids production calculations, we use the mg/L to lb/day equation shown below:

$$\text{SS removed (lb/day)} = (\text{SS Removed, mg/L})$$

$$(\text{Flow, MGD})(8.34 \text{ lb/gal}) \quad (7.206)$$

Primary Clarifier Solids Production Calculations

Example 7.286

Problem: A primary clarifier receives a flow of 1.80 MGD with suspended solids concentrations of 340 mg/L. If the clarifier effluent has a suspended solids concentration of 180 mg/L, how many pounds of solids are generated daily?

Solution:

$$\text{SS, lb/day Removed} = (\text{SS Removed, mg/L})$$

$$(\text{Flow, MGD})(8.34 \text{ lb/gal})$$

$$= (160 \text{ mg/L})(1.80 \text{ MGD})(8.34 \text{ lb/gal})$$

$$\text{Solids} = 2,402 \text{ lb/day}$$

Example 7.287

Problem: The suspended solids content of the primary influent is 350 mg/L, and the primary effluent is 202 mg/L. How many pounds of solids are produced during a day when the flow is 4,150,000 gpd?

Solution:

$$\text{SS, lb/day Removed} = (\text{SS Removed, mg/L})$$

$$(\text{Flow, MGD})(8.34 \text{ lb/gal})$$

$$= (148 \text{ mg/L})(4.15 \text{ MGD})(8.34 \text{ lb/gal})$$

$$\text{Solids Removed} = 5,122 \text{ lb/day}$$

Secondary Clarifier Solids Production Calculation

Solids produced during secondary treatment depend on many factors, including the amount of organic matter removed by the system and the growth rate of the bacteria. Because the precise calculation of biosolids production is complex, we use a rough estimate method of solids production that uses an estimated growth rate (unknown) value. We use the BOD removed lbs/day equation shown below.

$$\text{BOD Removed, lb/day} = (\text{BOD Removed, mg/L})$$

$$(\text{Flow, MGD})(8.34 \text{ lb/day})$$

$$(7.207)$$

Example 7.288

Problem: The 1.5-MGD influent to the secondary system has a BOD concentration of 174 mg/L. The secondary effluent contains 22 mg/L BOD. If the bacteria growth rate, unknown *x-value* for this plant, is 0.40 lb SS/lb BOD removed, how many pounds of dry biosolids solids are produced each day by the secondary system?

Solution:

$$\begin{aligned}
\text{BOD Removed, lb/day} &= \frac{(\text{BOD, mg/L})(\text{Flow, MGD})}{(8.34 \text{ lb/gal})} \\
&= (152 \text{ mg/L})(1.5 \text{ MGD})8.34 \text{ lb/gal} \\
&= 1,902 \text{ lb/day}
\end{aligned}$$

Then, use the unknown *x-value* to determine lb/day solids produced.

$$\frac{0.44 \text{ lb SS Produced}}{1 \text{ lb BOD Removed}} = \frac{x \text{ lb SS Produced}}{1,902 \text{ lb/day BOD Removed}}$$

$$\frac{(0.44)(1902)}{1} = x$$

$$837 \text{ lb/day Solids Produced} = x$$

Key Point: Typically, for every pound of food consumed (BOD removed) by the bacteria, between 0.3 and 0.7 lb of new bacteria cells are produced; these are solids to be removed from the system.

BIOSOLIDS MATH

Percent Solids

Biosolids are composed of water and solids. The vast majority of biosolids is water, usually in the range of 93%–97%. To determine the solids content of biosolids, a sample is dried overnight in an oven at 103°F–105°F. The solids that remain after drying represent the total solids content of the biosolids. Solids content may be expressed as a percent or as mg/L. Either of two equations is used to calculate percent solids.

$$\% \text{ Solids} = \frac{\text{Total Solids, g}}{\text{Biosolids Sample, g}} \times 100 \quad (7.208)$$

$$\% \text{ Solids} = \frac{\text{Solids, lb/day}}{\text{Biosolids, lb/day}} \times 100 \quad (7.209)$$

Example 7.289

Problem: The total weight of a biosolids sample (sample only, not the dish) is 22 g. If the weight of the solids after drying is 0.77 g, what is the percent total solids of the biosolids?

Solution:

$$
\begin{aligned}
\% \text{ Solids} &= \frac{\text{Total Solids (g)}}{\text{Biosolids Sample (g)}} \times 100 \\
&= \frac{0.77 \text{ g}}{22 \text{ g}} \times 100 \\
&= 3.5\%
\end{aligned}
$$

Biosolids Pumping

While on shift, wastewater operators are often required to make various process control calculations. An important calculation involves biosolids pumping. The biosolids pumping calculations the operator may be required to make are covered in this section.

Estimating Daily Biosolids Production

The calculation for estimating the required biosolids pumping rate provides a method to establish an initial pumping rate or to evaluate the adequacy of the current withdrawal rate.

$$\text{Est. pump rate} = \frac{(\text{Influ. TSS Conc.} - \text{Effluent TSS Conc.}) \times \text{Flow} \times 8.34}{\% \text{ Solids in Sludge} \times 8.34 \times 1{,}440 \text{ min/day}}$$

$$(7.210)$$

Example 7.290

Problem: The biosolids withdrawn from the primary settling tank contain 1.4% solids. The unit influent contains 285 mg/L TSS, and the effluent contains 140 mg/L TSS. If the influent flow rate is 5.55 MGD, what is the estimated biosolids withdrawal rate in gallons per minute (assuming the pump operates continuously)?

Solution:

$$\text{Biosolids Rate, gpm} = \frac{(285 \text{ mg/L} - 140 \text{ mg/L}) \times 5.55 \times 8.34}{0.014 \times 8.34 \times 1{,}440 \text{ min/day}} = 40 \text{ gpm}$$

Biosolids Production in Pounds per Million Gallons

A common method of expressing biosolids production is in pounds of biosolids per million gallons of wastewater treated.

$$\text{Biosolids, lb/MG} = \frac{\text{Total Biosolids Production, lb}}{\text{Total Wastewater Flow, MG}} \quad (7.211)$$

Example 7.291

Problem: Records show that the plant has produced 85,000 gal of biosolids during the past 30 days. The average daily flow for this period was 1.2 MGD. What was the plant's biosolids production in pounds per million gallons?

Solution:

$$\text{Biosolids, lb/MG} = \frac{85{,}000\text{-gal} \times 8.34 \text{ lb/gal}}{1.2 \text{ MGD} \times 30 \text{ days}} = 19{,}692 \text{ lb/MG}$$

Biosolids Production in Wet Tons/Year

Biosolids production can also be expressed in terms of the amount of biosolids (water and solids) produced per year. This is normally expressed in wet tons per year.

$$\text{Biosolids, Wet Tons/year} = \frac{\text{Biosolids Prod., lb/MG} \times \text{Ave. Daily Flow, MGD} \times 365 \text{ days/year}}{2{,}000 \text{ lb/ton}} \quad (7.212)$$

Example 7.292

Problem: The plant is currently producing biosolids at the rate of 16,500 lb/MG. The current average daily wastewater flow rate is 1.5 MGD. What will be the total amount of biosolids produced per year in wet tons per year?

Solution:

$$
\begin{aligned}
\text{Biosolids, Wet Tons/year} &= \frac{16{,}500 \text{ lb/MG} \times 1.5 \text{ MGD} \times 365 \text{ days/year}}{2{,}000 \text{ lb/ton}} \\
&= 4{,}517 \text{ Wet Tons/year}
\end{aligned}
$$

Biosolids Pumping Time

The biosolids pumping time is the total time the pump operates during a 24-h period in minutes.

Pump Operating Time = Time/Cycle, min

$$\times \text{Frequency, cycles/day} \qquad (7.213)$$

Note: The following information is used for Examples 7.293–7.294.

Frequency	24 times/day
Pump rate	120 gpm
Solids	3.70%
Volatile matter	66%

Example 7.293

Problem: What is the pump operating time?

Solution:

Pump Operating Time = 15 min/h × 24 (cycles)/day = 360 min/day

Biosolids Pumped/Day in Gallons

Biosolids, gpd = Operating Time, min/day

$$\times \text{Pump Rate, gpm} \qquad (7.214)$$

Example 7.294

Problem: What are the biosolids pumped per day in gallons?

Solution:

Biosolids, gpd = 360 min/day × 120 gpm = 43,200 gpd

Biosolids Pumped/Day in Pounds

Sludge, lb/day = Gallons of Biosolids Pumped × 8.34 lb/gal

$$(7.215)$$

Example 7.295

Problem: What are the biosolids pumped per day in pounds?

Solution:

What are the biosolids pumped per day in pounds?

Biosolids, lb/day = 43,200 gal/day × 8.34 lb/gal = 360,000 lb/day

Solids Pumped per Day in Pounds

Solids Pumped, lb/day = Biosolids Pumped, lb/day

$$\times \% \text{ Solids} \qquad (7.216)$$

Example 7.296

Problem: What are the solids pumped per day?

Solution:

Solids Pumped lb/day = 360,300 lb/day × 0.0370 = 13,331 lb/day

Volatile Matter Pumped per Day in Pounds

Vol. Matter (lb/day) = Solids Pumped, lb/day

$$\times \% \text{ Volatile Matter} \qquad (7.217)$$

Example 7.297

Problem: What is the volatile matter in pounds per day?

Solution:

Volatile Matter, lb/day = 13,331 lb/day × 0.66 = 8,798 lb/day

Biosolids Thickening Calculations

Biosolids thickening (or concentration) is a unit process used to increase the solids content of the biosolids by removing a portion of the liquid fraction. In other words, biosolids thickening is all about volume reduction. By increasing the solids content, more economical treatment of the biosolids can be achieved. Biosolids thickening processes include the following:

- Gravity thickeners
- Flotation thickeners
- Solids concentrators

Biosolids thickening calculations are based on the concept that the solids in the primary or secondary biosolids are equal to the solids in the thickened biosolids. The solids are the same. It is primarily water that has been removed to thicken the biosolids and result in higher percent solids. In these unthickened biosolids, the solids might represent 1% or 4% of the total pounds of biosolids. But when some of the water is removed, those same amount of solids might represent 5%–7% of the total pounds of biosolids.

Key Point: The key to biosolids thickening calculations is that solids remain constant.

Gravity/Dissolved Air Flotation Thickener Calculations

As mentioned, biosolids thickening calculations are based on the concept that the solids in the primary or secondary biosolids are equal to the solids in the thickened biosolids. That is, assuming a negligible amount of solids are lost in the thickener overflow, the solids are the same. Note that water is removed to thicken the biosolids, resulting in a higher percent solids.

Estimating Daily Biosolids Production

The calculation for estimating the required biosolids pumping rate provides a method to establish an initial pumping rate or to evaluate the adequacy of the current pump rate.

$$\text{Est. Pump Rate} = \frac{\left(\begin{array}{c}\text{Influent TSS Conc.}\\ -\text{Eff. TSS Conc.}\end{array}\right) \times \text{Flow} \times 8.34}{\% \text{ Solids in Biosolids}}$$

$$\times 8.34 \times 1{,}440 \text{ min/day} \qquad (7.218)$$

Example 7.298

Problem: The biosolids withdrawn from the primary settling tank contain 1.5% solids. The unit influent contains 280 mg/L TSS, and the effluent contains 141 mg/L TSS. If the influent flow rate is 5.55 MGD, what is the estimated biosolids withdrawal rate in gallons per minute (assuming the pump operates continuously)?

Solution:

$$\text{Biosolids Withdrawal Rate, gpm} = \frac{\left(280 \text{ mg/L} - 141 \text{ mg/L}\right)}{0.015 \times 8.34 \times 1{,}440 \text{ min/day}}$$

$$= 36 \text{ gpm}$$

Surface Loading Rate (gal/day/ft²)

Surface loading rate (surface settling rate) is the hydraulic loading—the amount of biosolids applied per square foot of a gravity thickener.

$$\text{Surface Loading, gal/day/ft}^2 = \frac{\begin{array}{c}\text{Biosolids Applied to}\\ \text{the Thickener, gpd}\end{array}}{\text{Thickener Area, ft}^2} \qquad (7.219)$$

Example 7.299

Problem: A 70-ft diameter gravity thickener receives 32,000 gpd of biosolids. What is the surface loading in gallons per square foot per day?

Solution:

$$\text{Surface Loading} = \frac{32{,}000 \text{ gpd}}{0.785 \times 70 \text{ ft} \times 70 \text{ ft}} = 8.32 \text{ gpd/ft}^2$$

Solids Loading Rate, lb/day/ft²

The solids loading rate is the pounds of solids per day being applied to 1 ft² of tank surface area. The calculation uses the surface area of the bottom of the tank. It assumes the floor of the tank is flat and has the same dimensions as the surface.

$$\text{Surf. Loading Rt., lb/day/ft}^2 = \frac{\begin{array}{c}\% \text{ Biosolids solids}\\ \times \text{biosolids flow, gpd}\\ \times 8.34 \text{ lb/gal}\end{array}}{\text{Thickener Area, ft}^2} \qquad (7.220)$$

Example 7.300

Problem: The thickener influent contains 1.6% solids. The influent flow rate is 39,000 gpd. The thickener is 50 ft in diameter and 10 ft deep. What is the solid loading in pounds per day?

Solution:

$$\text{Solids Loading Rate, lb/day/ft}^2 = \frac{\begin{array}{c}0.016 \times 39{,}000 \text{ gpd}\\ \times 8.34 \text{ lb/gal}\end{array}}{0.785 \times 50 \text{ ft} \times 50 \text{ ft}}$$

$$= 2.7 \text{ lb/ft}^2$$

Concentration Factor (CF)

The concentration factor (CF) represents the increase in concentration resulting from the thickener—it is a means of determining the effectiveness of the gravity thickening process.

$$\text{CF} = \frac{\text{Thickened Biosolids Concentration, \%}}{\text{Influent Biosolids Concentration, \%}} \qquad (7.221)$$

Example 7.301

Problem: The influent biosolids contain 3.5% solids. The thickened biosolids solids concentration is 7.7%. What is the concentration factor?

Solution:

$$\text{CF} = \frac{7.7\%}{3.5\%} = 2.2$$

Air-to-Solids Ratio

The air-to-solids ratio is the ratio between the pounds of solids entering the thickener and the pounds of air being applied.

$$\text{Air/Solids Ratio} = \frac{\text{Air Flow ft}^3/\text{min} \times 0.0785 \text{ lb/ft}^3}{\begin{array}{c}\text{Biosolids Flow, gpm} \times \% \text{ Solids}\\ \times 8.34 \text{ lb/gal}\end{array}} \qquad (7.222)$$

Example 7.302

Problem: The biosolids pumped to the thickener are 0.85% solids. The airflow is 13 cfm. What is the air-to-solids ratio if the current biosolids flow rate entering the unit is 50 gpm?

Solution:

$$\text{Air:Solids Ratio} = \frac{13\ \text{cfm} \times 0.075\ \text{lb/ft}^3}{50\ \text{gpm} \times 0.0085 \times 8.34\ \text{lb/gal}} = 0.28$$

Recycle Flow in Percent

The amount of recycle flow is expressed as a percent.

$$\text{Recycle \%} = \frac{\text{Recycle Flow Rate, gpm} \times 100}{\text{Sludge Flow, gpm}} = 175\%$$

(7.223)

Example 7.303

Problem: The sludge flow to the thickener is 80 gpm. The recycle flow rate is 140 gpm. What is the % recycle?

Solution:

$$\text{\% Recycle} = \frac{140\ \text{gpm} \times 100}{80\ \text{gpm}} = 175\%$$

Centrifuge Thickening Calculations

A centrifuge exerts a force on the biosolids thousands of times greater than gravity. Sometimes polymer is added to the influent of the centrifuge to help thicken the solids. The two most important factors that affect the centrifuge are the volume of the biosolids put into the unit (gpm) and the pounds of solids put in. The water that is removed is called centrate. Normally, hydraulic loading is measured as flow rate per unit of area. However, because of the variety of sizes and designs, hydraulic loading to centrifuges does not include area considerations and is expressed only as gallons per hour. The equations to be used if the flow rate to the centrifuge is given as gallons per day or gallons per minute are:

$$\text{Hydraulic Loading, gph} = \frac{\text{Flow, gpd}}{24\ \text{h/day}} \qquad (7.224)$$

$$\text{Hydraulic Loading, gph} = \frac{(\text{gpm flow})(60\ \text{min})}{\text{h}} \qquad (7.225)$$

Example 7.304

Problem: A centrifuge receives a waste-activated biosolids flow of 40 gpm. What is the hydraulic loading on the unit in gal/h?

Solution:

$$
\begin{aligned}
\text{Hydraulic Loading, gph} &= \frac{(\text{gpm flow})(60\ \text{min})}{\text{h}} \\
&= \frac{(40\ \text{gpm})(60\ \text{min})}{\text{h}} \\
&= 2{,}400\ \text{gph}
\end{aligned}
$$

Example 7.305

Problem: A centrifuge receives 48,600 gal of biosolids daily. The biosolids concentration before thickening is 0.9%. How many pounds of solids are received each day?

Solution:

$$\frac{48{,}600\ \text{gal}}{\text{day}} \times \frac{8.34\ \text{lb}}{\text{gal}} \times \frac{0.9}{100} = 3{,}648\ \text{lb/day}$$

BIOSOLIDS DIGESTION/STABILIZATION

A major problem in designing wastewater treatment plants is the disposal of biosolids into the environment without causing damage or nuisance. Untreated biosolids are even more difficult to dispose of. Untreated raw biosolids must be stabilized to minimize disposal problems. In many cases, the term *stabilization* is considered synonymous with digestion.

Key Point: The *stabilization* of organic matter is accomplished biologically using a variety of organisms. The microorganisms convert the colloidal and dissolved organic matter into various gases and protoplasm. Because protoplasm has a specific gravity slightly higher than that of water, it can be removed from the treated liquid by gravity.

Biosolids digestion is a process in which biochemical decomposition of the organic solids occurs; in the decomposition process, the organics are converted into simpler and more stable substances. Digestion also reduces the total mass or weight of biosolids solids, destroys pathogens, and makes it easier to dry or dewater the biosolids. Well-digested biosolids have the appearance and characteristics of rich potting soil.

Biosolids may be digested under aerobic or anaerobic conditions. Most large municipal wastewater treatment plants use anaerobic digestion, while aerobic digestion finds application primarily in small, package-activated biosolids treatment systems.

Aerobic Digestion Process Control Calculations

The purpose of *aerobic digestion* is to stabilize organic matter, reduce volume, and eliminate pathogenic organisms. Aerobic digestion is similar to the activated biosolids process. Biosolids are aerated for 20 days or more. Volatile solids are reduced by biological activity.

Volatile Solids Loading, lb/ft³/day

Volatile solids (organic matter) loading for the aerobic digester is expressed in pounds of volatile solids entering the digester per day per cubic foot of digester capacity.

Volatile Solids

$$\text{Loading, lb/day/ft}^3 = \frac{\text{Volatile Solids Added, lb/day}}{\text{Digester Volume, ft}^3} \qquad (7.226)$$

Example 7.306

Problem: The aerobic digester is 20 ft in diameter and has an operating depth of 20 ft. The biosolids that are added to the digester daily contain 1,500 lb of volatile solids. What are the volatile solids loading in pounds per day per cubic foot?

Solution:

$$\text{Vol. Solids Loading, lb/day/ft}^3 = \frac{1,500 \text{ lb/day}}{0.785 \times 20 \text{ ft} \times 20 \text{ ft} \times 20 \text{ ft}}$$

$$= 0.24 \text{ lb/day/ft}^3$$

Digestion Time, Days

The theoretical time the biosolids remain in the aerobic digester:

$$\text{Digestion Time, Days} = \frac{\text{Digester volume (gal)}}{\text{Biosolids added (gpd)}} \quad (7.227)$$

Example 7.307

Problem: The digester volume is 240,000 gal. Biosolids are added to the digester at the rate of 15,000 gpd. What is the digestion time in days?

Solution:

$$\text{Digestion Time, Days} = \frac{240,000 \text{ gal}}{15,000 \text{ gpd}} = 16 \text{ days}$$

pH Adjustment

In many instances, the pH of the aerobic digester will fall below the levels required for good biological activity. When this occurs, the operator must perform a laboratory test to determine the amount of alkalinity required to raise the pH to the desired level. The results of the lab test must then be converted to the actual quantity required by the digester.

$$\text{Chemical Required, lb} = \frac{\substack{\text{Chem. Used in Lab Test, mg} \\ \times \text{Dig. Vol.} \times 3.785}}{\substack{\text{Sample Vol., l} \\ \times 454 \text{ g/lb} \times 1,000 \text{ mg/g}}} \quad (7.228)$$

Example 7.308

Problem: 240 mg of lime will increase the pH of a 1-L sample of the aerobic digester contents to pH 7.1. The digester volume is 240,000 gal. How many pounds of lime will be required to increase the digester's pH to 7.3?

Solution:

$$\text{Chemical Required, lb} = \frac{240 \text{ mg} \times 240,000\text{-gal} \times 3.785 \text{ L/gal}}{1 \text{ L} \times 454 \text{ g/lb} \times 1,000 \text{ mg/g}}$$

Anaerobic Digestion Process Control Calculations

The purpose of *anaerobic digestion* is the same as aerobic digestion: to stabilize organic matter, reduce volume, and eliminate pathogenic organisms. Equipment used in anaerobic digestion includes an anaerobic digester of either the floating or fixed cover type. These include biosolids pumps for biosolids addition and withdrawal, as well as heating equipment such as heat exchangers, heaters, pumps, and mixing equipment for recirculation. Typical ancillaries include gas storage, cleaning equipment, and safety equipment such as vacuum relief and pressure relief devices, flame traps, and explosion-proof electrical equipment. In the anaerobic process, biosolids enter the sealed digester where organic matter decomposes anaerobically. Anaerobic digestion is a two-stage process:

1. Sugars, starches, and carbohydrates are converted to volatile acids, carbon dioxide, and hydrogen sulfide.
2. Volatile acids are converted to methane gas.

Key anaerobic digestion process control calculations are covered in the sections that follow.

Required Seed Volume in Gallons

$$\text{Seed Volume (gal)} = \text{Digester Volume, gal} \times \% \text{ Seed} \quad (7.229)$$

Example 7.309

Problem: The new digester requires 25% seed to achieve normal operation within the allotted time. If the digester volume is 280,000 gal, how many gallons of seed material will be required?

Solution:

$$\text{Seed Volume} = 280,000 \times 0.25 = 70,000 \text{ gal}$$

Volatile Acids to Alkalinity Ratio

The volatile acids to alkalinity ratio can be used to control anaerobic digester.

$$\text{Ratio} = \frac{\text{Volatile Acids Concentration}}{\text{Alkalinity Concentration}} \quad (7.230)$$

Example 7.310

Problem: The digester contains 240 mg/L of volatile acids and 1,840 mg/L of alkalinity. What is the volatile acids/alkalinity ratio?

Solution:

$$\text{Ratio} = \frac{240 \text{ mg/L}}{1,840 \text{ mg/L}} = 0.13$$

Key Point: Increases in the ratio normally indicate a potential change in the operating condition of the digester.

Biosolids Retention Time (BRT)

The length of time the biosolids remain in the digester.

$$BRT = \frac{Digester\ Volume\ in\ Gallons}{Biosolids\ Volume\ added\ per\ day,\ gpd} \quad (7.231)$$

Example 7.311

Problem: Biosolids are added to a 520,000-gal digester at the rate of 12,600 gal/day. What is the biosolids retention time?

Solution:

$$BRT = \frac{520,000\ gal}{12,600\ gpd} = 41.3\ days$$

Estimated Gas Production in Cubic Feet/Day

The rate of gas production is normally expressed as the volume of gas (ft^3) produced per pound of volatile matter destroyed. The total cubic feet of gas a digester will produce per day can be calculated by

$$Gas\ Production,\ ft^3/day = Vol.\ Matter\ In,\ lb/day$$
$$\times\%\ Vol.\ Mat.\ Reduction$$
$$\times Prod.\ Rate\ ft^3/lb$$
$$(7.232)$$

Key Point: Multiplying the volatile matter added to the digester per day by the % of volatile matter reduction (in decimal form) gives the amount of volatile matter being destroyed by the digestion process per day.

Example 7.312

Problem: The digester reduces 11,500 lb of volatile matter per day. Currently, the volatile matter reduction achieved by the digester is 55%. The rate of gas production is 11.2 ft^3 of gas per pound of volatile matter destroyed.

Solution:

$$Gas\ Prod. = 11,500\ lb/day \times 0.55 \times 11.2\ ft^3/lb = 70,840\ ft^3/day$$

Percent Volatile Matter Reduction

Due to the changes occurring during biosolids digestion, the calculation used to determine the percent of volatile matter reduction is more complicated.

$$\%\ Red. = \frac{\left(\%\ Vol.\ Matter_{in} - \%\ Vol.\ Matter_{out}\right) \times 100}{\left[\%\ Vol.\ Matter_{in} - \left(\%\ Vol.\ Matter_{in} \times \%\ Vol.\ Matter_{out}\right)\right]} \quad (7.233)$$

Example 7.313

Problem: Using the digester data provided here, determine the % of volatile matter reduction for the digester: raw biosolids volatile matter, 71%; digested biosolids volatile matter, 54%.

Solution:

$$\%\ Volatile\ Matter\ Reduction = \frac{0.71 - 0.54}{\left[0.71 - (0.71 \times 0.54)\right]} = 52\%$$

Percent Moisture Reduction in Digested Biosolids

$$\%\ Moisture\ Reduction = \frac{\left(\%\ Moisture_{in} - \%\ Moisture_{out}\right) \times 100}{\left[\%\ Moisture_{in} - \left(\%\ Moisture_{in} \times \%\ Moisture_{out}\right)\right]} \quad (7.234)$$

Key Point: % Moisture = 100% − Percent Solids

Example 7.314

Problem: Using the digester data provided below, determine the % of moisture reduction and % of volatile matter reduction for the digester.

Raw biosolids	% Solids	9%
	% Moisture	91% (100%−9%)
Digested biosolids	% Solids	15%
	% Moisture	85% (100%−15%)

Solution:

$$\%\ Moisture\ Reduction = \frac{(0.91 - 0.85) \times 100}{\left[0.91 - (0.91 \times 0.85)\right]} \times 44\%$$

Biosolids Dewatering

The process of removing enough water from liquid biosolids to change its consistency to that of a damp solid is called *biosolids dewatering*. Although the process is also called *biosolids drying*, the "dry" or dewatered biosolids may still contain a significant amount of water, often as much as 70%. At moisture contents of 70% or less, the biosolids no longer behave as a liquid and can be handled manually or mechanically. Several methods are available to dewater biosolids. The particular types of dewatering techniques/devices used best describe the actual processes used to remove water from biosolids and change their form from a liquid to a damp solid. The commonly used techniques/devices include the following:

- Filter presses
- Vacuum filtration
- Sand drying beds

Key Point: Centrifugation is also used in the dewatering process. However, in this text, we concentrate on those unit processes listed above that are traditionally used for biosolids dewatering.

Note that an ideal dewatering operation would capture all of the biosolids at minimum cost, and the resultant dry biosolids solids or cake would be capable of being handled without causing unnecessary problems. Process reliability, ease of operation, and compatibility with the plant environment would also be optimized.

Pressure Filtration

In *pressure filtration*, the liquid is forced through the filter media by positive pressure. Several types of presses are available, but the most commonly used are plate and frame presses and belt presses.

Plate and Frame Press Calculations

The *plate and frame press* consists of vertical plates held in a frame that are pressed together between a fixed and moving end. A cloth filter medium is mounted on the face of each individual plate. The press is closed, and biosolids are pumped into the press at pressures of up to 225 psi, passing through feed holes in the trays along the length of the press. Filter presses usually require a precoat material, such as incinerator ash or diatomaceous earth, to aid in solids retention on the cloth and to allow easier release of the cake. Performance factors for plate and frame presses include feed biosolids characteristics, type and amount of chemical conditioning, operating pressures, and the type and amount of precoat. Filter press calculations (and other dewatering calculations) typically used in wastewater solids handling operations include solids loading rate, net filter yield, hydraulic loading rate, biosolids feed rate, solids loading rate, flocculant feed rate, flocculant dosage, total suspended solids, and percent solids recovery.

Solids Loading Rate

The solids loading rate is a measure of the lbs/hr of solids applied per square foot of plate area, as shown in Equation (7.235).

$$\text{Sol. Loading Rate } \left(\text{lbs/h/ft}^2\right) = \frac{\left(\text{Biosolids, gph}\right)\left(8.34, \text{lbs/gal}\right)\left(\%\text{ Sol./100}\right)}{\text{Plate Area }\left(\text{ft}^2\right)}$$
(7.235)

Key Point: The solids loading rate measures the lb/h of solids applied to each ft^2 of plate surface area. However, this does not reflect the time when biosolids feeding to the press is stopped.

Net Filter Yield

Operated in batch mode, biosolids are fed to the plate and frame filter press until the space between the plates is completely filled with solids. The biosolids flow to the press is then stopped, and the plates are separated, allowing the biosolids cake to fall into a hopper or conveyor below. The *net filter yield*, measured in lbs/h/ft^2, reflects the run time as well as the downtime of the plate and frame filter press. To calculate the net filter yield, simply multiply the solids loading rate (in lbs/h/ft^2) by the ratio of filter run time to total cycle time as follows:

$$\text{N. F. Y.} = \frac{\frac{\left(\text{Biosolids, gph}\right)\left(8.34\text{ lb/gal}\right)}{\left(\%\text{ Sol/100}\right)}}{\text{Plate Area, ft}^2} \times \frac{\text{Filter Run Time}}{\text{Total Cycle Time}}$$
(7.235)

Example 7.315

Problem: A plate and frame filter press receives a flow of 660 gal of biosolids during a 2-h period. The solids concentration of the biosolids is 3.3%. The surface area of the plate is 110 ft^2. If the down time for biosolids cake discharge is 20 min, what is the net filter yield in lbs/h/ft^2?

Solution:

First, calculate solids loading rate and then multiply that number by the corrected time factor:

$$\begin{aligned}\text{Solids Loading Rate} &= \frac{\frac{\left(\text{Biosolids, gph}\right)\left(8.34\text{ lbs/gal}\right)}{\left(\%\text{ Sol./100}\right)}}{\text{Plate Area, ft}^2} \\ &= \frac{\left(330\text{ gph}\right)\left(8.34\text{ lb/gal}\right)\left(3.3/100\right)}{100\text{ ft}^2} \\ &= 0.83\text{ lbs/h/ft}^2\end{aligned}$$

Next, calculate the net filter yield, using the corrected time factor:

$$\text{Net Filter Yield, lbs/h/ft}^2 = \frac{\left(0.83\text{ lb/h/ft}^2\right)\left(2\text{ h}\right)}{2.33\text{ h}}$$
$$= 0.71\text{ lbs/h/ft}^2$$

Belt Filter Press Calculations

The *belt filter press* consists of two porous belts. The biosolids are sandwiched between the two porous belts. The belts are pulled tight together as they are passed around a series of rollers to squeeze water out of the biosolids. Polymer is added to the biosolids just before it gets to the unit. The biosolids are then distributed across one of the belts to allow for some of the water to drain by gravity. The belts are then put together with the biosolids in between.

Hydraulic Loading Rate

Hydraulic loading for belt filters is a measure of gpm flow per foot of belt width.

$$\text{Hydraulic Loading Rate, gpm/ft} = \frac{\text{Flow, gpm}}{\text{Belt Width, ft}} \quad (7.236)$$

Example 7.316

Problem: A 6-ft wide belt press receives a flow of 110 gpm of primary biosolids. What is the hydraulic loading rate in gpm/ft?

Solution:

$$\begin{aligned}\text{Hydraulic Loading Rate, gpm/ft} &= \frac{\text{Flow, gpm}}{\text{Belt Width, ft}} \\ &= \frac{110 \text{ gpm}}{6 \text{ ft}} \\ &= 18.3 \text{ gpm/ft}\end{aligned}$$

Example 7.317

Problem: A belt filter press 5 ft wide receives a primary biosolids flow of 150 gpm. What is the hydraulic loading rate in gpm/ft²?

Solution:

$$\begin{aligned}\text{Hydraulic Loading Rate, gpm/ft} &= \frac{\text{Flow, gpm}}{\text{Belt Width, ft}} \\ &= \frac{150 \text{ gpm}}{5 \text{ ft}} \\ &= 30 \text{ gpm/ft}\end{aligned}$$

Biosolids Feed Rate

The biosolids feed rate to the belt filter press depends on several factors, including the biosolids, lb/day, that must be dewatered, the maximum solids feed rate, lbs/h, that will produce an acceptable cake dryness, and the number of hours per day the belt press is in operation. The equation used to calculate the biosolids feed rate is:

$$\text{Biosolids Feed Rate, lb/h} = \frac{\text{Biosolids to be dewatered, lb/day}}{\text{Operating Time, h/day}} \quad (7.237)$$

Example 7.318

Problem: The amount of biosolids to be dewatered by the belt filter press is 20,600 lb/day. If the belt filter press is to be operated 10 h each day, what should the biosolids feed rate in lbs/h be to the press?

Solution:

$$\begin{aligned}\text{Biosolids Feed Rate, lb/h} &= \frac{\text{Biosolids to be dewatered, lb/day}}{\text{Operating Time, h/day}} \\ &= \frac{20,600 \text{ lb/day}}{10 \text{ h/day}} \\ &= 2,060 \text{ lb/h}\end{aligned}$$

Solids Loading Rate

The solids loading rate may be expressed as lbs/h or tons/h. In either case, the calculation is based on the biosolids flow (or feed) to the belt press and percent of the mg/L concentration of total suspended solids (TSS) in the biosolids. The equation used to calculate the solids loading rate is

$$\text{Sol. Load. Rate, lb/h} = (\text{Feed, gpm})(60 \text{ min/h})$$

$$(8.34 \text{ lb/gal})(\% \text{ TSS}/100) \quad (7.238)$$

Example 7.319

Problem: The biosolids feed to a belt filter press is 120 gpm. If the total suspended solids concentration of the feed is 4%, what is the solids loading rate in lb/h?

Solution:

$$\text{Sol. Load. Rate, lb/h} = (\text{Feed, gpm})(60 \text{ min/h})$$

$$(8.34 \text{ lb/gal})(\% \text{ TSS})/100$$

$$= (120 \text{ gpm})(60 \text{ min/h})(8.34 \text{ lb/gal})(4/100)$$

$$= 2,402 \text{ lb/h}$$

Flocculant Feed Rate

The flocculant feed rate may be calculated like all other mg/L to lb/day calculations, and then converted to pounds hour feed rate, as follows:

$$\text{Flocculant Feed, lb/day} = \frac{(\text{Floc., mg/L})(\text{Feed Rate, MGD})(8.34 \text{ lb/gal})}{24 \text{ h/day}}$$

$$(7.239)$$

Example 7.320

Problem: The flocculant concentration for a belt filter press is 1% (10,000 mg/L). If the flocculant feed rate is 3 gpm, what is the flocculant feed rate in lb/h?

Solution:

First, calculate the lb/day flocculant using the mg/L to lbs/day calculation. Note that the gpm feed flow must be expressed as MGD feed flow:

$$\frac{(3 \text{ gpm})(1,440 \text{ min/day})}{1,000,000} = 0.00432 \text{ MGD}$$

$$\text{Flocculant Feed, lb/day} = (\text{mg/L Floc.})$$

$$(\text{Feed Rate, MGD})(8.34 \text{ lb/gal})$$

$$= (10,000 \text{ mg/L})$$

$$(0.00432 \text{ MGD})(8.34 \text{ lb/gal})$$

$$= 360 \text{ lb/day}$$

Then, convert lb/day flocculant to lb/hr:

$$= \frac{360 \text{ lb/day}}{24 \text{ h/day}} = 15 \text{ lb/h}$$

Flocculant Dosage

Once the solids loading rate (ton/h) and flocculant feed rate (lb/h) have been calculated, the flocculant dose in lb/ton can be determined. The equation used to determine flocculant dosage is

$$\text{Flocculant Dosage, lb/ton} = \frac{\text{Flocculant, lb/h}}{\text{Solids Treated, ton/h}} \quad (7.240)$$

Example 7.321

Problem: A belt filter has solids loading rate of 3,100 lb/h and a flocculant feed rate of 12 lb/h. Calculate the flocculant dose in lb per ton of solids treated?

Solution:

First, convert lb/h solids loading to ton/h solids loading:

$$\frac{3,100 \text{ lb/h}}{2,000 \text{ lb/ton}} = 1.55 \text{ ton/h}$$

Now, calculate lb flocculant per ton of solids treated:

$$\begin{aligned}
\text{Flocculant Dosage, lb/ton} &= \frac{\text{Flocculant, lb/h}}{\text{Solids Treated, ton/h}} \\
&= \frac{12 \text{ lb/h}}{1.55 \text{ ton/h}} \\
&= 7.8 \text{ lb/ton}
\end{aligned}$$

Total Suspended Solids

The feed biosolids solids are comprised of two types of solids: suspended solids and dissolved solids. *Suspended solids* will not pass through a glass fiber filter pad. Suspended solids can be further classified as total suspended solids (TSS), volatile suspended solids, and fixed suspended solids. They can also be separated into three components based on settling characteristics: settleable solids, floatable solids, and colloidal solids. Total suspended solids in wastewater are normally in the range of 100–350 mg/L. *Dissolved solids*, which pass through a glass fiber filter pad, can also be classified as total dissolved solids (TDS), volatile dissolved solids, and fixed dissolved solids. Total dissolved solids normally range from 250 to 850 mg/L.

Two lab tests can be used to estimate the total suspended solids concentration of the feed biosolids to the filter press: the *total residue test* (measures both suspended and dissolved solids concentrations) and the *total filterable residue test* (measures only the dissolved solids concentration). By subtracting the total filterable residue from the total residue, the result is the total non-filterable residue (total suspended solids), as shown in Equation (7.241).

Total Res., mg/L – Total Filterable Residue, mg/L =

Total Non-Filterable Res., mg/L   (7.241)

Example 7.322

Lab tests indicate that the total residue portion of a feed biosolids sample is 22,000 mg/L. The total filterable residue is 720 mg/L. On this basis, what is the estimated total suspended solids concentration of the biosolids sample?

$$\begin{aligned}
&\text{Total Residue, mg/L} \\
&-\text{Total Filterable Res., mg/L} \\
&22,000 \text{ mg/L} - 720 \text{ mg/L}
\end{aligned}
\begin{aligned}
&= \text{Total Non-} \\
&\quad \text{Filterable Res., mg/L} \\
&= 21,280 \text{ mg/L Total SS}
\end{aligned}$$

Rotary Vacuum Filter Dewatering Calculations

The *rotary vacuum filter* is a device used to separate solid material from liquid. The vacuum filter consists of a large drum with large holes in it covered with a filter cloth. The drum is partially submerged and rotates through a vat of conditioned biosolids. Capable of excellent solids capture and high-quality supernatant/filtrate, solids concentrations of 15%–40% can be achieved.

Filter Loading

The filter loading for vacuum filters is a measure of lbs/h of solids applied per square foot of drum surface area. The equation used for this calculation is shown below.

$$\text{Filter Loading, lb/h/ft}^2 = \frac{\text{Solids to Filter, lb/h}}{\text{Surface Area, ft}^2} \quad (7.242)$$

Example 7.323

Problem: Digested biosolids are applied to a vacuum filter at a rate of 70 gpm, with a solids concentration of 3%. If the vacuum filter has a surface area of 300 ft², what is the filter loading in lb/h/ft²?

Solution:

$$\begin{aligned}
\text{Filter Loading, lb/h/ft}^2 &= \frac{(\text{Biosolids, gpm})(60 \text{ min/h})}{(8.34 \text{ lb/gal})(\% \text{ Sol./100})} \Big/ \text{Surface Area, ft}^2 \\
&= \frac{(70 \text{ gpm})(60 \text{ min/h})(8.34 \text{ lb/gal})(3/100)}{300 \text{ ft}^2} \\
&= 3.5 \text{ lb/h/ft}^2
\end{aligned}$$

Filter Yield

One of the most common measures of vacuum filter performance is filter yield. It is the lb/h of dry solids in the dewatered biosolids (cake) discharged per square foot of filter area. It can be calculated using Equation (7.243).

$$\text{Filter Yield, lbs/h/ft}^2 = \frac{(\text{Wet Cake Flow, lb/h})\dfrac{(\% \text{ Solids in Cake})}{100}}{\text{Filter Area, ft}^2} \quad (7.243)$$

Example 7.324

Problem: The wet cake flow from a vacuum filter is 9,000 lb/h. If the filter area is 300 ft² and the percent solids in the cake is 25%, what is the filter yield in lb/h/ft²?

Solution:

$$\text{Filter Area, ft}^2 = \frac{(\text{Wet Cake Flow, lb/h})\dfrac{(\% \text{ Solids in Cake})}{100}}{\text{Filter Area, ft}^2}$$

$$= \frac{(9{,}000 \text{ lb/h})\dfrac{(25)}{100}}{300 \text{ ft}^2}$$

$$= 7.5 \text{ lb/h/ft}^2$$

Vacuum Filter Operating Time

Example 7.325

Problem: A total of 4,000 lb/day of primary biosolids solids are to be processed by a vacuum filter. The vacuum filter yield is 2.2 lb/h/ft². The solids recovery is 95%. If the area of the filter is 210 ft², how many hours per day must the vacuum filter remain in operation to process these solids?

Solution:

$$\text{Filter Yield, lb/h/ft}^2 = \frac{\dfrac{\text{Solids. to filter, lb/day}}{\text{Filter operation., lb/day}}}{\text{Filter Area, ft}^2} \frac{(\% \text{ Recovery})}{100}$$

$$2.2\text{-lb/h/ft}^2 = \frac{\dfrac{4{,}000 \text{ lb/day}}{\text{x h/day Oper.}}}{210 \text{ ft}^2} \frac{(95)}{100}$$

$$2.2\text{-lb/h/ft}^2 = \frac{(4{,}000 \text{ lb/day})}{\text{x h/day}} \frac{(1)}{210 \text{ ft}^2} \frac{(95)}{100}$$

$$\text{x} = \frac{(4{,}000)(1)(95)}{(2.2)(210)(100)}$$

$$\text{x} = 8.2 \text{ h/day}$$

Percent Solids Recovery

As mentioned, the function of the vacuum filtration process is to separate the solids from the liquids in the biosolids being processed. Therefore, the percent of feed solids "recovered" (sometimes referred to as the percent solids captured) is a measure of the efficiency of the process. Equation (7.244) is used to determine the percent solids recovery.

$$\% \text{ Sol Rec.} = \frac{(\text{Wet Cake Flow, lb/h})\dfrac{(\% \text{ Sol. In Cake})}{100}}{(\text{Biosolids Feed, lb/h})\dfrac{(\% \text{ Sol. In Feed})}{100}} \times 100$$

$$(7.244)$$

Example 7.326

Problem: The biosolids feed to a vacuum is 3,400 lb/day, with a solids content of 5.1%. If the wet cake flow is 600 lb/h with a 25% solids content, what is the percent solids recovery?

Solution:

$$\% \text{ Sol. Rec.} = \frac{(\text{Wet Cake Flow, lb/h})\dfrac{(\% \text{ Sol. In Cake})}{100}}{(\text{Biosolids Feed, lb/h})\dfrac{(\% \text{ Sol. In Feed})}{100}} \times 100$$

$$= \frac{(600 \text{ lb/h})\dfrac{(25)}{100}}{(3{,}400 \text{ lb/h})\dfrac{(5.1)}{100}} \times 100$$

$$= \frac{150 \text{ lb/h}}{173 \text{ lb/h}} \times 100$$

$$\% \text{ Sol. Rec.} = 87\%$$

Sand Drying Beds

Drying beds are generally used for dewatering well-digested biosolids. Biosolids drying beds consist of a perforated or open-joint drainage system in a support medium, usually gravel or wire mesh. Drying beds are usually separated into workable sections by wood, concrete, or other materials. They may be enclosed or open to the weather. They may rely entirely on natural drainage and evaporation processes or may use a vacuum to assist the operation. *Sand drying beds* are the oldest biosolids dewatering technique and consist of 6–12 in of coarse sand underlain by layers of graded gravel, ranging from 1/8 to ¼ in at the top and ¾ to 1-1/2 in at the bottom. The total gravel thickness is typically about 1 ft. Graded natural earth (4–6 in) usually makes up the bottom with a web of drain tile placed on 20–30-ft centers. Sidewalls and partitions between bed sections are usually made of wooden planks or concrete and extend about 14 in above the sand surface. Typically, three calculations are used to monitor sand drying bed performance: total biosolids applied, solids loading rate, and biosolids withdrawal to drying beds.

Total Biosolids Applied

The total gallons of biosolids applied to sand drying beds may be calculated using the dimensions of the bed and the depth of biosolids applied, as shown by Equation (7.245).

$$\text{Volume, gal} = (\text{length, ft})(\text{width, ft})$$

$$(\text{depth, ft})(7.48, \text{gal/ft}^3) \qquad (7.245)$$

Example 7.327

Problem: A drying bed is 220 ft long and 20 ft wide. If biosolids are applied to a depth of 4 in, how many gallons of biosolids are applied to the drying bed?

Solution:

$$\text{Volume, gal} = (l)(w)(d)(7.48 \text{ gal/ft}^3)$$

$$= (220 \text{ ft})(20 \text{ ft})(0.33 \text{ ft})(7.48 \text{ gal/ft}^3)$$

$$= 10{,}861 \text{ gal}$$

Solids Loading Rate

The biosolids loading rate may be expressed as lb/year/ft². The loading rate depends on biosolids applied per application, lb, percent solids concentration, cycle length, and square feet of sand bed area. The equation for the biosolids loading rate is given below.

$$\text{Sol. Load. Rate, lb/year/ft}^2 = \frac{\dfrac{\text{Biosolids applied (lb)}}{\text{Days of Application}}(365 \text{ days/year})\dfrac{(\% \text{ Solids})}{100}}{(\text{length, ft})(\text{width, ft})}$$

$$(7.246)$$

Example 7.328

Problem: A biosolids bed is 210 ft long and 25 ft wide. A total of 172,500 lb of biosolids are applied each application of the sand drying bed. The biosolids has a solids content of 5%. If the drying and removal cycle requires 21 days, what is the solids loading rate in lb/year/ft²?

Solution:

$$
\begin{aligned}
\text{Solids. Load. Rate,} \\
\text{lb/year/ft}^2
\end{aligned}
= \frac{\dfrac{\text{lb Biosolids Applied}}{\text{Days of Application}}(365)\dfrac{(\% \text{ Solids})}{100}}{\text{Bed Area, ft}^2}
$$

$$= \frac{\dfrac{(172{,}500 \text{ lb})}{21 \text{ days}}(365 \text{ days/year})\dfrac{(5)}{100}}{(210 \text{ ft})(25 \text{ ft})}$$

$$= 37.5 \text{ lb/year/ft}^2$$

Biosolids Withdrawal to Drying Beds

Pumping digested biosolids to drying beds is one method among many for dewatering biosolids, thus making the dried biosolids useful as a soil conditioner. Depending on the climate of a region, the drying bed depth may range from 8 to 18 in. Therefore, the area covered by these drying beds may be substantial. For this reason, the use of drying beds is more common in smaller plants than in larger plants. When calculating biosolids withdrawal to drying beds, use Equation (7.247):

$$\text{Biosolids Withdrawn, ft}^3 = (0.785)(D^2)(\text{Drawdown, ft})$$

$$(7.247)$$

Example 7.329

Problem:
Biosolids are withdrawn from a digester that has a diameter of 40 ft. If the biosolids are drawn down 2 ft, how many ft³ will be sent to the drying beds?

Solution:

$$\text{Biosolids Withdrawal, ft}^3 = (0.785)(D^2)(\text{ft drop})$$

$$= (0.785)(40 \text{ ft})(40 \text{ ft})(2 \text{ ft})$$

$$= 2{,}512 \text{ ft}^3 \text{ withdrawn}$$

Biosolids Disposal

In the disposal of biosolids, land application, in one form or another, has become not only a necessity (because of the banning of ocean dumping in the United States in 1992 and the shortage of landfill space since then) but also quite popular as a beneficial reuse practice. That is, beneficial reuse means that the biosolids are disposed of in an environmentally sound manner by recycling nutrients and soil conditions. Biosolids are being applied throughout the United States to agricultural and forest lands. For use in land applications, the biosolids must meet certain conditions. Biosolids must comply with state and federal biosolids management/disposal regulations and must also be free of materials dangerous to human health (i.e., toxicity, pathogenic organisms, etc.) and/or dangerous to the environment (i.e., toxicity, pesticides, heavy metals, etc.). Biosolids are land applied by direct injection, by application and incorporation (plowing in), or by composting.

Land application of biosolids requires precise control to avoid problems. The use of process control calculations is part of the overall process control process. Calculations include determining disposal cost, plant available nitrogen (PAN), application rate (dry tons and wet tons/acre), metals loading rates, maximum allowable applications based on metals loading, and site life based on metals loading.

Disposal Cost

The cost of disposal of biosolids can be determined by:

Cost = Wet Tons Biosolids Produced/Year

$$\times\% \text{ Solids} \times \text{Cost/dry ton} \qquad (7.248)$$

Example 7.330

Problem: The treatment system produces 1,925 wet tons of biosolids for disposal each year. The biosolids are 18% solids. A contractor disposes of the biosolids for $28.00 per dry ton. What is the annual cost for biosolids disposal?

Solution:

Cost = 1,925 wet tons/year × 0.18 × $2,800/dry ton = $9,702

Plant Available Nitrogen (PAN)

One factor considered when land applying biosolids is the amount of nitrogen in the biosolids available to the plants grown on the site. This includes ammonia nitrogen and organic nitrogen. The organic nitrogen must be mineralized for plant consumption. Only a portion of the organic nitrogen is mineralized per year. The mineralization factor (f^1) is assumed to be 0.20. The amount of ammonia nitrogen available is directly related to the time elapsed between applying the biosolids and incorporating (plowing) the biosolids into the soil. Volatilization rates are presented in the example below.

$$\text{PAN, lb/dry ton} = [(\text{Or. Nit., mg/kg} \times F^1)$$
$$+ (\text{Amm. Nit., mg/kg} \times V_1)]$$
$$\times 0.002 \text{ lb/dry ton} \qquad (7.249)$$

where:

F^1 = Mineral rate for organic nitrogen (assume 0.20)
V_1 = Volatilization rate ammonia nitrogen
V_1 = 1.00 if biosolids are injected
V_1 = 0.85 if biosolids are plowed in within 24 h
V_1 = 0.70 if biosolids are plowed in within 7 days

Example 7.331

Problem: The biosolids contain 21,000 mg/kg of organic nitrogen and 10,500 mg/kg of ammonia nitrogen. The biosolids are incorporated into the soil within 24 h after application. What is the plant available nitrogen (PAN) per dry ton of solids?

Solution:

$$\text{PAN, lb/dry ton} = [(21,000 \text{ mg/kg} \times 0.20) + (10,500 \times 0.85)]$$
$$\times 0.002$$
$$= 26.3 \text{ lb PAN/dry ton}$$

Application Rate Based on Crop Nitrogen Requirement

In most cases, the application rate of domestic biosolids to crop lands will be controlled by the amount of nitrogen the crop requires. The biosolids application rate based on the nitrogen requirement is determined by the following:

1. Using an agricultural handbook to determine the nitrogen requirement of the crop to be grown.
2. Determining the amount of biosolids in dry tons required to provide this much nitrogen.

$$\text{Dry tons/acre} = \frac{\text{Plant Nitrogen Requirement, lb/acre}}{\text{Plant Available Nitrogen, lb/dry ton}}$$
$$(7.250)$$

Example 7.332

Problem: The crop to be planted on the land application site requires 150 lb of nitrogen per acre. What is the required biosolids application rate if the PAN of the biosolids is 30 lb/dry ton?

Solution:

$$\text{Dry tons/acre} = \frac{150 \text{ lb nitrogen nitrogen/acre}}{30 \text{ lb/dry ton}} = 5 \text{ dry tons/acre}$$

Metals Loading

When biosolids are land applied, metal concentrations are closely monitored, and the loading on land application sites is calculated.

$$\text{Loading, lb/acre} = \text{Metal Conc., mg/kg}$$
$$\times 0.002 \text{ lb/dry ton}$$
$$\times \text{Appl. Rate, dry tons/acre} \qquad (7.251)$$

Example 7.333

Problem: The biosolids contain 14 mg/kg of lead. Biosolids are currently being applied to the site at a rate of 11 dry tons/acre. What is the metal loading rate for lead in pounds per acre?

Solution:

$$\text{Loading Rate, lb/acre} = 14 \text{ mg/kg} \times 0.002 \text{ lb/dry ton}$$
$$\times 11 \text{ dry tons} = 0.31 \text{ lb/acre}$$

Maximum Allowable Applications Based upon Metals Loading

If metals are present, they may limit the total number of applications a site can receive. Metal loadings are normally expressed in terms of the maximum total amount of metal that can be applied to a site during its use.

$$\text{Applications} = \frac{\begin{array}{c}\text{Max. Allowable Cumulative}\\ \text{Load for the Metal, lb/Ac}\end{array}}{\text{Metal Loading, lb/acre/application}} \quad (7.252)$$

Example 7.334

Problem: The maximum allowable cumulative lead loading is 48 lb/acre. Based on the current loading of 0.35 lb/acre, how many applications of biosolids can be made to this site?

Solution:

$$\text{Applications} = \frac{48.0 \text{ lb/acre}}{0.35 \text{ lb/acre}} = 137 \text{ applications}$$

Site Life Based on Metals Loading

The maximum number of applications based on metal loading and the number of applications per year can be used to determine the maximum site life.

$$\text{Site Life, years} = \frac{\begin{array}{c}\text{Maximum Allowable}\\ \text{Applications}\end{array}}{\begin{array}{c}\text{Number of Applications}\\ \text{Planned/Year}\end{array}} \quad (7.253)$$

Example 7.335

Problem: Biosolids are currently applied to a site twice annually. Based on the lead content of the biosolids, the maximum number of applications is determined to be 135 applications. Based on the lead loading and the application rate, how many years can this site be used?

Source:

$$\text{Site Life} = \frac{135 \text{ applications}}{2 \text{ applications/year}} = 68 \text{ year}$$

Key Point: When more than one metal is present, the calculations must be performed for each metal. The site life would then be the lowest value generated by these calculations.

BIOSOLIDS TO COMPOST

The purpose of composting biosolids is to stabilize the organic matter, reduce volume, eliminate pathogenic organisms, and produce a product that can be used as a soil amendment or conditioner. Composting is a biological process. In a composting operation, dewatered solids are usually mixed with a bulking agent (i.e., hardwood chips) and stored until biological stabilization occurs. The composting mixture is ventilated during storage to provide sufficient oxygen for oxidation and to prevent odors. After the solids are stabilized, they are separated from the bulking agent. The composted solids are then stored for curing and are applied to farmlands or other beneficial uses. The expected performance of the composting operation for both percent volatile matter reduction and percent moisture reduction ranges from 40% to 60%.

Performance factors related to biosolids composting include moisture content, temperature, pH, nutrient availability, and aeration. The biosolids must contain sufficient moisture to support biological activity. If the moisture level is too low (less than 40%), biological activity will be reduced or stopped. At the same time, if the moisture level exceeds approximately 60%, it will prevent sufficient airflow through the mixture. The composting process operates best when the temperature is maintained within an operating range of 130°F–140°F—biological activities provide enough heat to increase the temperature well above this range. Forced air ventilation or mixing is used to remove heat and maintain the desired operating temperature range. The temperature of the composting solids, when maintained at the required levels, will be sufficient to remove pathogenic organisms. The influent pH can affect the performance of the process if it is extreme (less than 6.0 or greater than 11.0). The pH during composting may have some impact on biological activity but does not appear to be a major factor. Composted biosolids generally have a pH in the range of 6.8–7.5. The critical nutrient in the composting process is nitrogen. The process works best when the ratio of nitrogen to carbon is in the range of 26–30 carbon to one nitrogen. Above this ratio, composting is slowed. Below this ratio, the nitrogen content of the final product may be less attractive than compost. Aeration is essential to provide oxygen to the process and to control the temperature. In forced air processes, some means of odor control should be included in the design of the aeration system.

Pertinent composting process control calculations include the determination of the percent of moisture of the compost mixture and compost site capacity. An important consideration in compost operation is the solids processing capability (fill time), lb/day, or lb/week. Equation (7.254) is used to calculate site capacity.

$$\text{Fill Time, days} = \frac{\text{Total Available Capacity, yd}^3}{\dfrac{\text{Wet Compost, lb/day}}{\text{Compost Bulk Density, lbs/yd}^3}}$$

$$(7.254)$$

Example 7.336

Problem: A composting facility has an available capacity of 7,600 yd³. If the composting cycle is 21 days, how many lb/day of wet compost can this facility process? Assume a compost bulk density of 900 lbs/yd³.

Solution:

$$\text{Fill Time, days} = \frac{\text{Total Available Capacity, yd}^3}{\dfrac{\text{Wet Compost, lb/day}}{\text{Compost Bulk Density, lb/yd}^3}}$$

$$21 \text{ days} = \frac{7,600 \text{ yd}^3}{\dfrac{x \text{ lb/day}}{900 \text{ lb/yd}^3}}$$

$$21 \text{ days} = \frac{\left(7,600 \text{ yd}^3\right)\left(900 \text{ lb/yd}^3\right)}{x \text{ lb/day}}$$

$$x \text{ lb/day} = \frac{\left(7,600 \text{ yd}^3\right)\left(900 \text{ lb/yd}^3\right)}{21 \text{ days}}$$

$$x = 325,714 \text{ lb/day}$$

WATER/WASTEWATER LABORATORY CALCULATIONS

Waterworks and wastewater treatment plants are sized to meet current needs, as well as future requirements. No matter the size of the treatment plant, some space or area within the plant is designated as the lab area, which can range from closet-sized to fully equipped and staffed environmental laboratories. Water and wastewater laboratories usually perform several different tests. Lab test results give the operator the information necessary to operate the treatment facility optimally. Laboratory testing usually includes service line flushing time, solution concentration, pH, COD, total phosphorus, fecal coliform count, chlorine residual, and BOD (seeded) test, to name a few. The standard reference for performing wastewater testing is contained in *Standard Methods for the Examination of Water & Wastewater* (APHA, 2012).

In this section, the focus is on standard water/wastewater lab tests that involve various calculations. Specifically, the focus is on calculations used to determine proportioning factors for composite sampling, flow from a faucet estimation, service line flushing time, solution concentration, biochemical oxygen demand (BOD), molarity and moles, normality, settleability, settleable solids, biosolids total, fixed and volatile solids, suspended solids and volatile suspended solids, and biosolids volume index and biosolids density index.

FAUCET FLOW ESTIMATION

Occasionally, the waterworks sampler must take water samples from a customer's residence. The sample is usually taken from the customer's front yard faucet in small water systems. A convenient flow rate for taking water samples is about 0.5 gpm. To estimate the flow from a faucet, use a 1-gal container and record the time it takes to fill the container. To calculate the flow in gpm, insert the recorded information into Equation (7.255).

$$\text{Flow, gpm} = \frac{\text{Volume, gal}}{\text{Time, min}} \qquad (7.255)$$

Example 7.337

Problem: The flow from a faucet filled up the gallon container in 48 sec. What was the gpm flow rate from the faucet? Because the flow rate is desired in minutes, the time should also be expressed as minutes:

$$\frac{48 \text{ sec}}{60 \text{ sec/min}} = 0.80 \text{ min}$$

Solution:

Calculate the flow rate from the faucet:

$$\text{Flow, gpm} = \frac{\text{Volume, gal}}{\text{Time, min}}$$
$$= \frac{1 \text{ gal}}{0.80 \text{ min}}$$
$$= 1.25 \text{ gpm}$$

Example 7.338

Problem: The flow from a faucet filled up the gallon container in 55 sec. What was the gpm flow rate from the faucet?

Solution:

$$\frac{55 \text{ sec}}{60 \text{ sec/min}} = 0.92 \text{ min}$$

Calculate the flow rate:

$$\text{Flow, gpm} = \frac{\text{Volume, gal}}{\text{Time, min}}$$
$$= \frac{1 \text{ gal}}{0.92 \text{ min}}$$
$$= 1.1 \text{ gpm}$$

SERVICE LINE FLUSHING TIME

To determine the quality of potable water delivered to the consumer, a sample is taken from the customer's outside faucet—water that is typical of the water delivered. To obtain an accurate indication of the system's water quality, this sample must be representative. Further, to ensure that the sample taken is typical of the water delivered, the service line must be flushed twice. Equation (7.256) is in calculating flushing time.

$$\text{Flushing Time, min} = \frac{(0.785)\left(D^2\right)(\text{Length, ft})}{\left(7.48 \text{ gal/ft}^3\right)(2)}{\text{Flow Rate, gpm}} \qquad (7.256)$$

Example 7.339

Problem: How long (minutes) will it take to flush a 40-ft length of ½-in diameter service line if the flow through the line is 0.5 gpm?

Solution:

Calculate the diameter of the pump in feet.

$$\frac{(0.50)}{12 \text{ in/ft}} = 0.04 \text{ ft}$$

Calculate the flushing time:

$$\text{Flushing Time, min} = \frac{\dfrac{(0.785)\left(D^2\right)(\text{Length, ft})}{\left(7.48 \text{ gal/ft}^3\right)(2)}}{\text{Flow Rate, gpm}}$$

$$= \frac{\dfrac{(0.785)(0.04 \text{ ft})(0.04 \text{ ft})(40 \text{ ft})}{\left(7.48 \text{ gal/ft}^3\right)(2)}}{0.5 \text{ gpm}}$$

$$= 1.5 \text{ min}$$

Example 7.340

Problem: At a flow rate of 0.5 gpm, how long (minutes and seconds) will it take to flush a 60-ft length of ¾-in service line?

Solution:

$$3/4\text{-in diameter} = 0.06 \text{ ft}$$

$$\text{Flushing Time, min} = \frac{\dfrac{(0.785)\left(D^2\right)(\text{Length, ft})}{\left(7.48 \text{ gal/ft}^3\right)(2)}}{\text{Flow Rate, gpm}}$$

$$= \frac{\dfrac{(0.785)(0.06 \text{ ft})(0.06 \text{ ft})(60 \text{ ft})}{\left(7.48 \text{ gal/ft}^3\right)(2)}}{0.5 \text{ gpm}}$$

$$= 5.1 \text{ min}$$

Convert the fractional part of a minute (0.1) to seconds:

$$(0.1 \text{ min})(60 \text{ sec/min}) = 6 \text{ sec}$$

$$= 5 \text{ min, 6 sec}$$

Composite Sampling Calculation

When preparing oven-baked food, a cook pays close attention to setting the correct oven temperature, usually setting the temperature correctly and then moving on to some other chore. The oven thermostat ensures that the oven-baked food is cooked at the correct temperature, and that is that. Unlike the cook, in water and wastewater treatment plant operations, the operator does not have the luxury of setting a plant parameter and then walking away and forgetting about it. To optimize plant operations, various adjustments to unit processes must be made on an ongoing basis.

The operator makes unit process adjustments based on local knowledge (experience) and lab test results. However, before lab tests can be performed, samples must be taken. There are two basic types of samples: grab samples and composite samples. The type of sample taken depends on the specific test, the reason the sample is being collected, and the requirements in the plant discharge permit.

A *grab sample* is a discrete sample collected at one time and one location. It is primarily used for any parameter whose concentration can change quickly (e.g., dissolved oxygen, pH, temperature, total chlorine residual), and it is representative only of the conditions at the time of collection.

A *composite sample* consists of a series of individual grab samples taken at specified time intervals and in proportion to flow. The individual grab samples are mixed together in proportion to the flow rate at the time the sample was collected to form the composite sample. The composite sample represents the character of the water/wastewater over a period of time. Because knowledge of the procedure used in processing composite samples is important (a basic requirement) to the water/wastewater operator, the actual procedure used is covered in this section:

- Determine the total amount of sample required for all tests to be performed on the composite sample.
- Determine the treatment system's average daily flow.

Key Point: The average daily flow can be determined using several months of data, which will provide a more representative value.

- Calculate a proportioning factor.

$$\text{Prop. Factor (PF)} = \frac{\dfrac{\text{Total Sample Volume}}{\text{Required, mm}}}{\text{\# of Samples to be Calculated}} \quad (7.257)$$
$$\times \text{ Ave. Daily Flow, MGD}$$

Key Point: Round the proportioning factor to the nearest 50 units (i.e., 50, 100, 150, etc.) to simplify the calculation of the sample volume.

- Collect the individual samples in accordance with the schedule (once/hour, once/15 min, etc.).
- Determine the flow rate at the time the sample was collected.
- Calculate the specific amount to add to the composite container.

$$\text{Required Volume, mL} = \text{Flow}^T \times \text{PF} \quad (7.258)$$

T = Time sample was collected

Mix the individual samples thoroughly, measure the required volume, and add them to the composite storage container.

Keep the composite sample refrigerated throughout the collection period.

Example 7.341

Problem: The effluent testing will require 3,825 mL of sample. The average daily flow is 4.25 MGD. Using the flows given below, calculate the amount of sample to be added at each of the times shown:

Time	Flow, MGD
8 A.M.	3.88
9 A.M.	4.10
10 A.M.	5.05
11 A.M.	5.25
12 Noon	3.80
1 P.M.	3.65
2 P.M.	3.20
3 P.M.	3.45
4 P.M.	4.10

Solution:

$$\text{Proportioning Factor (PF)} = \frac{3,825 \text{ mL}}{9 \text{ Samples} \times 4.25 \text{ MGD}}$$

$$= 110$$

Volume$_{8 \text{ AM}}$ = 3.88 × 100 = 388 (400) mL
Volume$_{9 \text{ AM}}$ = 4.10 × 100 = 410 (410) mL
Volume$_{10 \text{ AM}}$ = 5.05 × 100 = 505 (500) mL
Volume$_{11 \text{ AM}}$ = 5.25 × 100 = 525 (530) mL
Volume$_{12 \text{ N}}$ = 3.80 × 100 = 380 (380) mL
Volume$_{1 \text{ PM}}$ = 3.65 × 100 = 365 (370) mL
Volume$_{2 \text{ PM}}$ = 3.20 × 100 = 320 (320) mL
Volume$_{3 \text{ PM}}$ = 3.45 × 100 = 345 (350) mL
Volume$_{4 \text{ PM}}$ = 4.10 × 100 = 410 (410) mL

BIOCHEMICAL OXYGEN DEMAND CALCULATIONS

Biochemical oxygen demand (BOD$_5$) measures the amount of organic matter that can be biologically oxidized under controlled conditions (5 days at 20°C in the dark). Several criteria are used when selecting which BOD$_5$ dilutions to be used for calculating test results. Consult a laboratory testing reference manual (such as *Standard Methods*) for this information. There are two basic calculations for BOD$_5$. The first is used for samples that have not been seeded. The second must be used whenever BOD$_5$ samples are seeded. Both methods are introduced, and examples are provided below.

BOD$_5$ (Unseeded)

$$\text{BOD}_5(\text{Unseeded}) = \frac{(\text{DO}_{start}, \text{mg/L} - \text{DO}_{final}, \text{mg/L}) \times 300 \text{ mL}}{\text{Sample Volume, mL}}$$

(7.259)

Example 7.342

Problem: The BOD$_5$ test is competed. Bottle 1 of the test had dissolved oxygen (DO) of 7.1 mg/L at the start of the test. After 5 days, bottle 1 had a DO of 2.9 mg/L. Bottle 1 contained 120 mg/L of sample.

Solution:

$$\text{BOD}_5(\text{Unseeded}) = \frac{(7.1 \text{ mg/L} - 2.9 \text{ mg/L}) \times 300 \text{ mL}}{120 \text{ mL}} = 10.5 \text{ mg/L}$$

BOD$_5$ (Seeded)

If the BOD$_5$ sample has been exposed to conditions that could reduce the number of healthy, active organisms, the sample must be seeded with organisms. Seeding requires the use of a correction factor to remove the BOD$_5$ contribution of the seed material.

$$\text{Seed Correction} = \frac{\text{Seed Material BOD}_5 \times \text{Seed in Dilution, mL}}{300 \text{ mL}}$$

(7.260)

$$\text{BOD}_5(\text{Seeded}) = \frac{\left[\left(\text{DO}_{start}, \text{mg/L} - \text{DO}_{final}, \text{mg/L}\right) - \text{Seed Corr.}\right] \times 300}{\text{Sample Volume, mL}}$$

(7.261)

BOD 7-Day Moving Average

Because the BOD characteristics of wastewater varies from day to day, even hour to hour, operational control of the treatment system is most often accomplished based on trends in data rather than individual data points. The BOD 7-day moving average is a calculation of the BOD trend.

Key Point: The 7-day moving average is called a moving average because a new average is calculated each day by adding the new day's value and the six previous days' values.

$$\text{7-day Average BOD} = \frac{\begin{array}{c}\text{BOD Day 1} + \text{BOD Day 2} \\ + \text{BOD Day 3} + \text{BOD Day 4} \\ + \text{BOD Day 5} + \text{BOD Day 6} \\ + \text{BOD Day 7}\end{array}}{7}$$

(7.262)

Example 7.343

Problem: Given the following primary effluent BOD test results, calculate the 7-day average.

June 1—200 mg/L	June 5—222 mg/L
June 2—210 mg/L	June 6—214 mg/L
June 3—204 mg/L	June 7—218 mg/L
June 4—205 mg/L	

$$\text{7-day average BOD} = \frac{200 + 210 + 204 + 205 + 222 + 214 + 218}{7}$$

$$= 210 \text{ mg/L}$$

MOLES AND MOLARITY

Chemists have defined a very useful unit called the *mole*. Moles and molarity, a concentration term based on the mole, have many important applications in water/wastewater operations. A mole is defined as a gram molecular weight; that is, the molecular weight expressed as grams. For example, a mole of water is 18 g of water, and a mole of glucose is 180 g of glucose. A mole of any compound always contains the same number of molecules. The number of molecules in a mole is called Avogadro's number and has a value of 6.022×10^{23}.

Interesting Point: How big is Avogadro's number? An Avogadro's number of soft drink cans would cover the surface of the Earth to a depth of over 200 miles.

Key Point: Molecular weight is the weight of one molecule. It is calculated by adding the weights of all the atoms present in one molecule. The units are atomic mass units (amu). A mole is a gram molecular weight, that is, the molecular weight expressed in grams. The molecular weight is the weight of one molecule in daltons. All moles contain the same number of molecules, Avogadro's number, equal to 6.022×10^{23}. The reason all moles have the same number of molecules is that the value of the mole is proportional to the molecular weight.

MOLES

As mentioned, a mole is a quantity of a compound equal in weight to its formula weight. For example, the formula weight for water can be determined using the Periodic Table of Elements:

$$\text{Hydrogen} \ (1.008) \times 2 = 2.016$$

$$+ \text{Oxygen} = 16.000$$

$$\text{Formula weight of } H_2O = \overline{18.016}$$

Because the formula weight of water is 18.016, a mole is 18.016 units of weight. A *gram-mole* is 18.016 g of water.

A *pound-mole* is 18.016 lb of water. For our purposes in this text, the term "mole" will be understood to mean "gram-mole." The equation used to determine moles is shown below.

$$\text{Moles} = \frac{\text{Grams of chemical}}{\text{Formula weight of chemical}} \quad (7.263)$$

Example 7.344

Problem: The atomic weight of a certain chemical is 66. If 35 g of the chemical is used in making up a 1-L solution, how many moles are used?

$$\text{Moles} = \frac{\text{Grams of chemical}}{\text{Formula weight}}$$

$$= \frac{66 \text{ g}}{35 \text{ g/mol}}$$

$$= 1.9 \text{ mol}$$

The molarity of a solution is calculated by taking the moles of solute and dividing it by the liters of solution. The molarity of a solution is calculated by taking the moles of solute and dividing by the liters of solution:

$$\text{Molarity} = \frac{\text{Moles of solute}}{\text{Liters of solution}} \quad (7.264)$$

Example 7.345

Problem: What is the molarity of 2 mol of solute dissolved in 1 L of solvent?

Solution:

$$\text{Molarity} = \frac{2 \text{ mol}}{1 \text{ L}} = 2 \text{ M}$$

Key Point: Measurement in moles is a measurement of the amount of a substance. Measurement in molarity is a measurement of the concentration of a substance—the amount (moles) per unit volume (liters).

Normality

As mentioned, the *molarity* of a solution refers to its concentration (the solute dissolved in the solution). The *normality* of a solution refers to the number of *equivalents* of solute per liter of solution. The definition of a chemical equivalent depends on the substance or type of chemical reaction under consideration. Because the concept of equivalents is based on the "reacting power" of an element or compound, it follows that a specific number of equivalents of one substance will react with the same number of equivalents of another substance. When the concept of equivalents is taken into consideration, it is less likely that chemicals will be wasted as excess amounts. Keeping in mind that normality is a measure of the reacting power of a solution (i.e., 1 equivalent of a substance reacts with 1 equivalent of another substance), we use the following equation to determine normality.

$$\text{Normality} = \frac{\text{No. of Equivalents of Solute}}{\text{Liters of Solution}} \quad (7.265)$$

Example 7.346

Problem: If 2.0 equivalents of a chemical are dissolved in 1.5 L of solution, what is the normality of the solution?

Solution:

$$
\begin{aligned}
\text{Normality} &= \frac{\text{No of Equivalents of Solute}}{\text{Liters of Solution}} \\
&= \frac{2.0 \text{ Equivalents}}{1.5 \text{ L}} \\
&= 1.33 \text{ N}
\end{aligned}
$$

Example 7.347

Problem: An 800-mL solution contains 1.6 equivalents of a chemical. What is the normality of the solution?

Solution:

First, convert 800 mL to liters:

$$\frac{800 \text{ mL}}{1,000 \text{ mL}} = 100$$

Then, calculate the normality of the solution:

$$
\begin{aligned}
\text{Normality} &= \frac{\text{No. of Equivalents of Solute}}{\text{Liters of Solution}} \\
&= \frac{1.6 \text{ Equivalents}}{0.8 \text{ L}} \\
&= 2 \text{ N}
\end{aligned}
$$

SETTLEABILITY (ACTIVATED BIOSOLIDS)

The settleability test is a test of the quality of the activated biosolids—or activated sludge solids (Mixed Liquor Suspended Solids). Settled biosolids volume (SBV)—or settled sludge volume (SSV)—is determined at specified times during sample testing. Thirty- and 60-min observations are used for control. Subscripts (SBV_{30} or SSV_{30} and SBV_{60} or SSV_{60}) indicate settling time. A sample of activated biosolids is taken from the aeration tank, poured into a 2,000-mL graduated, and allowed to settle for 30 or 60 min. The settling characteristics of the biosolids in the graduate give a general indication of the settling of the MLSS in the final clarifier. From the settleability test the percent settleable solids can be calculated using the following equation.

$$\% \text{ Settleable Solids} = \frac{\text{mL Settled Solids}}{2,000\text{-mL Sample}} \times 100 \quad (7.266)$$

Example 7.348

Problem: The settleability test is conducted on a sample of MLSS. What is the percent settleable solids if 420 mL settle in the 2,000-mL graduate?

Solution:

$$
\begin{aligned}
\% \text{ Settleable Solids} &= \frac{\text{mL Settled Solids}}{2,000\text{- mL Sample}} \times 100 \\
&= \frac{420 \text{ mL}}{2,000 \text{ mL}} \times 100 \\
&= 21\%
\end{aligned}
$$

Example 7.349

Problem: A 2,000-mL sample of activated biosolids is tested for settleability. If the settled solids are measured as 410 mL, what is the percent settled solids?

Solution:

$$
\begin{aligned}
\% \text{ Settleable Solids} &= \frac{\text{mL Settled Solids}}{2,000\text{-mL Sample}} \times 100 \\
&= \frac{410 \text{ mL}}{2,000 \text{ mL}} \times 100 \\
&= 20.5\%
\end{aligned}
$$

SETTLEABLE SOLIDS

The settleable solids test is an easy, quantitative method to measure sediment found in wastewater. An Imhoff cone is filled with 1 L of sample wastewater, stirred, and allowed to settle for 60 min. The settleable solids test, unlike the settleability test, is conducted on samples from sedimentation tanks or clarifier influent and effluent to determine the % removal of settleable solids. The percent settleable solids is determined by

$$\% \text{ Settleable Solids Removed} = \frac{\text{Set. Solids Removed, mL/L}}{\text{Set. Solids in Influent, mL/L}} \times 100$$

$$(7.267)$$

Example 7.350

Problem: Calculate the percent removal of settleable solids if the settleable solids of the sedimentation tank influent is 15 mL/L and the settleable solids of the effluent are 0.4 mL/L.

Solution:

First, subtract 0.4 mL/L from 15.0 = 14.6 mL/L removed settleable solids.

Next, insert parameters into Equation (7.267).

$$
\begin{aligned}
\% \text{ Set. Sol. Removed} &= \frac{14.6 \text{ mL/L}}{15.0 \text{ mL/L}} \times 100 \\
&= 97\%
\end{aligned}
$$

Example 7.351

Problem: Calculate the percent removal of settleable solids if the settleable solids of the sedimentation tank influent is 13 mL/L and the settleable solids of the effluent are 0.5 mL/L.

Solution:

First, determine the removed settleable solids.

$$\% \text{ Set. Sol. Removed} = \frac{12.5 \text{ mL/L}}{13.0 \text{ mL/L}} \times 100$$

$$= 96\%$$

BIOSOLIDS TOTAL SOLIDS, FIXED SOLIDS AND VOLATILE SOLIDS

Wastewater consists of both water and solids. The *total solids* may be further classified as either *volatile solids* (organics) or *fixed solids* (inorganics). Normally, total solids and volatile solids are expressed as percentages; whereas suspended solids are generally expressed as mg/L. To calculate either percentages or mg/L concentrations, certain concepts must be understood:

- **Total Solids:** The residue left in the vessel after the evaporation of liquid from a sample and subsequent drying in an oven at 103°C–105°C.
- **Fixed Solids:** The residue left in the vessel after a sample is ignited (heated to dryness at 550°C).
- **Volatile Solids:** The weight loss after a sample is ignited (heated to dryness at 550°C). Determinations of fixed and volatile solids do not distinguish precisely between inorganic and organic matter because the loss on ignition is not confined to organic matter. It includes losses due to decomposition or volatilization of some mineral salts.

Key Point: When the word *biosolids* is used, it may be understood to mean a semi-liquid mass composed of solids and water. The term *solids*, however, is used to mean dry solids after the evaporation of water.

Percent total solids and volatile solids are calculated as follows:

$$\% \text{ Total Solids} = \frac{\text{Total solids weight}}{\text{Biosolids sample weight}} \times 100 \qquad (7.268)$$

$$\% \text{ Volatile Solids} = \frac{\text{Volatile solids weight}}{\text{Total solids weight}} \times 100 \qquad (7.269)$$

Example 7.352

Problem: Given the information below determine (1) the percent solids in the sample and (2) the percent of volatile solids in the biosolids sample:

	Biosolids	After Drying	After Burning (Ash) Sample
Weight of sample and dish	73.43 g	24.88	22.98
Weight of dish (tare weight)	22.28 g	22.28	22.28

To calculate the % total solids, the grams total solids (solids after drying) and grams biosolids sample must be determined:

Total Solids	Biosolids Sample
24.88 g Total solids & dish	73.43 g Biosolids & dish
−22.28 g Weight of dish	−22.28 g Dish
2.60 g Total solids	51.15 g Biosolids

$$
\begin{aligned}
\% \text{ Total Solids} &= \frac{\text{Weight of total solids}}{\text{Weight of biosolids sample}} \times 100 \\[6pt]
&= \frac{2.60 \text{ g}}{51.15 \text{ g}} \times 100 \\[6pt]
&= 5\% \text{ Total Solids}
\end{aligned}
$$

To calculate the % volatile solids, the grams total solids and grams volatile solids must be determined. Because total solids have already been calculated (above), only volatile solids must be calculated:

Volatile Solids

24.88 g Sample and Dish <u>Before</u> Burning

22.98 g Sample and Dish <u>After</u> Burning

1.90 g Solids Lost in Burning

$$
\begin{aligned}
\% \text{ Volatile Solids} &= \frac{\text{Weight of Volatile Solids}}{\text{Weight of Total Sample}} \times 100 \\[6pt]
&= \frac{1.90 \text{ g}}{2.60 \text{ g}} \times 100 \\[6pt]
&= 73\% \text{ Volatile Solids}
\end{aligned}
$$

WASTEWATER SUSPENDED SOLIDS AND VOLATILE SUSPENDED SOLIDS

Total suspended solids (TSS) are the amount of filterable solids in a wastewater sample. Samples are filtered through a glass fiber filter. The filters are dried and weighed to determine the amount of total suspended solids in mg/L of the sample. *Volatile suspended solids* (VSS) are those solids lost on ignition (heating to 500°C.). They are useful to the treatment plant operator because they give a rough approximation of the amount of organic matter present in the solid fraction of wastewater, activated biosolids, and industrial wastes. With the exception of the required drying time, the suspended solids and volatile suspended solids tests of wastewater are similar to those of the total and volatile solids performed for biosolids.

Key Point: The total and volatile solids of biosolids are generally expressed as percentages, by weight. The biosolids samples are 100 mL and are unfiltered. Calculation of suspended solids and volatile suspended solids is demonstrated in the example below.

Example 7.353

Problem: Given the following information regarding a primary effluent sample, calculate: (1) the mg/L suspended solids and (2) the percent volatile suspended solids of the sample.

	After Drying (Before Burning)	After Burning (Ash)
Weight of sample and dish	24.6268 g	24.6232 g
Weight of dish (tare wt.)	24.6222 g	24.6222 g

Sample Volume = 50 mL

Solution:

To calculate the milligrams of suspended solids per liter of sample (mg/L), you must first determine grams of suspended solids:

$$24.6268 \text{ g Dish and Suspended Solids}$$

$$-24.6222 \text{ g Dish}$$

$$\overline{00.0046 \text{ g Suspended Solids}}$$

Next, we calculate mg/L of suspended solids [using a multiplication factor of 20 (this number will vary with sample volume) to make the denominator equal to 1 L (1,000 mL)]:

$$\frac{0.0046 \text{ g SS}}{50 \text{ mL}} \times \frac{1,000 \text{ mg}}{1 \text{ g}} \times \frac{20}{20} \times \frac{92 \text{ mg}}{1,000 \text{ mL}} = 92 \text{ mg/L SS}$$

To calculate the percent of volatile suspended solids, we must know the weight of both total suspended solids and volatile suspended solids.

$$24.6268 \text{ g Dish \& SS \underline{Before} Burning}$$

$$-24.6234 \text{ g Dish \& SS \underline{After} Burning}$$

$$\overline{0.0034 \text{ g Solids Lost in Burning}}$$

$$
\begin{aligned}
\% \text{ VSS} &= \frac{\text{Wt. of Volatile Solids}}{\text{Wt. of Suspended Solids}} \times 100 \\
&= \frac{0.0034 \text{ g VSS}}{0.0046 \text{ g}} \times 100 \\
&= 70\% \text{ VSS}
\end{aligned}
$$

Biosolids Volume Index (BVI) and Biosolids Density Index (BDI)

Two variables are used to measure the settling characteristics of activated biosolids and determine what the return

biosolids pumping rate should be. These are the *volume of the biosolids* (BVI) and the *density of the biosolids* (BDI) indices.

$$
\begin{aligned}
\text{BVI} &= \frac{\% \text{ MLSS volume after 30 minutes}}{\% \text{ MLSS mg/L MLSS}} \\
&= \text{mL settled biosolids} \times 1,000
\end{aligned}
\tag{7.270}
$$

$$
\text{BDI} = \frac{\text{MLSS (\%)}}{\% \text{ volume MLSS after } 30 \text{ minutes settling}} \times 100
\tag{7.271}
$$

These indices relate the weight of biosolids to the volume the biosolids occupy. They show how well the liquid-solids separation part of the activated biosolids system is performing its function on the biological floc that has been produced and is to be settled out and returned to the aeration tanks or wasted. The better the liquid-solids separation is, the smaller the volume occupied by the settled biosolids will be, and the lower the pumping rate required to keep the solids in circulation.

Example 7.354

Problem: The settleability test indicates that after 30 min, 220 mL of biosolids settles in the 1-L graduated cylinder. If the mixed liquor suspended solids (MLSS) concentration in the aeration tank is 2,400 mg/L, what is the biosolids volume?

Solution:

$$
\text{BVI} = \frac{\text{Volume (determined by settleability test)}}{\text{Density (determined by the MLSS conc.)}}
$$

$$
\text{BVI} = \frac{220 \text{ mL/L}}{2,400 \text{ mg/L}}
$$

$$
= \frac{220 \text{ mL}}{2,400 \text{ mg}} (\text{convert milligrams to grams}) \frac{220 \text{ mL}}{2.4 \text{ g}}
$$

$$
= 92
$$

The biosolids density index (BDI) is also a method of measuring the settling quality of activated biosolids; however, like the BVI parameter, may or may not provide a true picture of the quality of the biosolids in question unless compared with other relevant processes parameters. It differs from BVI in that the higher the BDI value, the better the settling quality of the aerated mixed liquor. Similarly, the lower the BDI, the poorer the settling quality of the mixed liquor. BDI is the concentration in percent solids that the activated biosolids will assume after settling for 30 min. BDI will range from 2.00 to 1.33, and biosolids with values of one or more are generally considered to have good settling characteristics. In calculating the biosolids density index (BDI), we simply inverted the numerators and denominators and multiplied by 100.

Example 7.355

Problem: The MLSS concentration in the aeration tank is 2,500 mg/L. If the activated biosolids settleability test indicates 225 mL settled in the 1-L graduated cylinder, what is the biosolids density index?

$$BDI = \frac{Density \; (determined \; by \; the \; MLSS \; concentration)}{Volume \; (Determined \; by \; the \; settleability \; test)} \times 100$$

$$BDI = \frac{2,500 \; mg}{225 \; mL} \times 100 (convert \; milligrams \; to \; grams)$$

$$= \frac{2.5 \; g}{225 \; mL} \times 100$$

$$= 1.11 \; mL \; settle \; in \; the \; 2,000\text{-}mL \; graduate?$$

CHAPTER REVIEW PROBLEMS

GENERAL MATH OPERATIONS

7.1 The diameter of a tank is 70 ft. If the water depth is 25 ft, what is the volume of water in the tank, in gallons?

7.2 A tank is 60 ft in length, 20 ft wide, and 10 ft deep. Calculate the cubic feet volume of the tank.

7.3 A tank 20 ft wide and 60 ft long is filled with water to a depth of 12 ft. What is the volume of the water in the tank (in gal)?

7.4 What is the volume of water in a tank, in gallons, if the tank is 20 ft wide, 40 ft long, and contains water to a depth of 12 ft?

7.5 A tank has a diameter of 60 ft and a depth of 12 ft. Calculate the volume of water in the tank in gallons.

7.6 What is the volume of water in a tank, in gallons, if the tank is 20 ft wide, 50 ft long, and contains water to a depth of 16 ft?

7.7 A rectangular channel is 340 ft in length, 4 ft in depth, and 6 ft wide. What is the volume of water, in cubic feet?

7.8 A replacement section of 10 in pipe is to be sandblasted before it is put into service. If the length of the pipeline is 1,600 ft, how many gallons of water will be needed to fill the pipeline?

7.9 A trapezoidal channel is 800 ft in length, 10 ft wide at the top, and 5 ft wide at the bottom with a distance of 4 ft from the top edge to the bottom along the sides. Calculate the volume in gallon.

7.10 A section of 8-in diameter pipeline is to be filled with treated water for distribution. If the pipeline is 2,250 ft in length, how many gallons of water will be distributed?

7.11 A channel is 1,200 ft in length, carries water 4 ft in depth, and is 5 ft wide. What is the volume of water in gallons?

7.12 A pipe trench is to be excavated that is 4 ft wide, 4 ft deep, and 1,200 ft long. What is the volume of the trench in cubic yards?

7.13 A trench is to be excavated that is 3 ft wide, 4 ft deep, and 500 yards long. What is the cubic yard volume of the trench?

7.14 A trench is 300 yards long, 3 ft wide, and 3 ft deep. What is the cubic feet volume of the trench?

7.15 A rectangular trench is 700 ft long, 6.5 ft wide, and 3.5 ft deep. What is the cubic feet volume of the trench?

7.16 The diameter of a tank is 90 ft. If the water depth in the tank is 25 ft, what is the volume of water in the tank, in gallons?

7.17 A tank is 80 ft long, 20 ft wide, and 16 ft deep. What is the cubic feet volume of the tank?

7.18 How many gallons of water will it take to fill an 8 in diameter pipe that is 4,000 ft in length?

7.19 A trench is 400 yards long, 3 ft wide, and 3 ft deep. What is the cubic feet volume of the trench?

7.20 A trench is to be excavated. If the trench is 3 ft wide, 4 ft deep, and 1,200 ft long, what is the cubic yard volume of the trench?

7.21 A tank is 30 ft wide and 80 ft long. If the tank contains water to a depth of 12 ft, how many gallons of water are in the tank?

7.22 What is the volume of water (in gallons) contained in a 3,000-ft section of the channel if the channel is 8 ft wide and the water depth is 3.5 ft?

7.23 A tank has a diameter of 70 ft and a depth of 19 ft. What is the volume of water in the tank, in gallons?

7.24 If a tank is 25 ft in diameter and 30 ft deep, how many gallons of water will it hold?

7.25 A channel 44 in wide has water flowing to a depth of 2.4 ft. If the velocity of the water is 2.5 fps, what is the cfm flow in the channel?

7.26 A tank is 20 ft long and 12 ft wide. With the discharge valve closed, the influent to the tank causes the water level to rise 0.8 ft in 1 min. What is the gpm flow to the tank?

7.27 A trapezoidal channel is 4 ft wide at the bottom and 6 ft wide at the water surface. The water depth is 40 in. If the flow velocity through the channel is 130 ft/min, what is the cfm flow rate through the channel?

7.28 An 8-in diameter pipeline has water flowing at a velocity of 2.4 fps. What is the gpm flow rate through the pipeline? Assume the pipe is flowing full.

7.29 A pump discharges into a 3-ft diameter container. If the water level in the container rises 28 in in 30 sec, what is the gpm flow into the container?

7.30 A 10-in diameter pipeline has water flowing at a velocity of 3.1 fps. What is the gpm flow rate through the pipeline if the water is flowing at a depth of 5 in?

7.31   A channel has a rectangular cross-section. The channel is 6 ft wide with water flowing to a depth of 2.6 ft. If the flow rate through the channel is 14,200 gpm, what is the velocity of the water in the channel (ft/sec)?

7.32   An 8-in diameter pipe flowing fully delivers 584 gpm. What is the velocity of flow in the pipeline (ft/sec)?

7.33   A special dye is used to estimate the velocity of flow in an interceptor line. The dye is injected into the water at one pumping station, and the travel time to the first manhole 550 ft away is noted. The dye first appears at the downstream manhole in 195 sec. The dye continues to be visible until the total elapsed time is 221 sec. What is the ft/sec velocity of flow through the pipeline?

7.34   The velocity in a 10-in diameter pipeline is 2.4 ft/sec. If the 10-in pipeline flows into an 18-in diameter pipeline, what is the velocity in the 8-in pipeline in ft/sec?

7.35   A float travels 500 ft in a channel in 1 min and 32 sec. What is the estimated velocity in the channel (ft/sec)?

7.36   The velocity in an 8-in diameter pipe is 3.2 ft/sec. If the flow then travels through a 10-in diameter section of the pipeline, what is the ft/sec velocity in the 10-in pipeline?

7.37   The following flows were recorded for the week:
Monday—4.8 MGD
Tuesday—5.1 MGD
Wednesday—5.2 MGD
Thursday—5.4 MGD
Friday—4.8 MGD
Saturday—5.2 MGD
Sunday—4.8 MGD
What was the average daily flow rate for the week?

7.38   The totalizer reading for September was 121.4 MG. What was the average daily flow (ADF) for September?

7.39   Convert 0.165 MGD to gpm.

7.40   The total flow for 1 day at a plant was 3,335,000 gal. What was the average gpm flow for that day?

7.41   Express a flow of 8 cfs in terms of gpm.

7.42   What is 35 gps expressed as gpd?

7.43   Convert a flow of 4,570,000 gpd to cfm.

7.44   What is 6.6 MGD expressed as cfs?

7.45   Express 445,875 cfd as gpm.

7.46   Convert 2,450 gpm to gpd.

7.47   A channel has a rectangular cross-section. The channel is 6 ft wide with water flowing to a depth of 2.5 ft. If the flow rate through the channel is 14,800 gpm, what is the velocity of the water in the channel (ft/sec)?

7.48   A channel 55 in wide has water flowing to a depth of 3.4 ft. If the velocity of the water is 3.6 fps, what is the cfm flow in the channel?

7.49   The following flows were recorded for June, July, and August: June—102.4 MG; July—126.8 MG; August—144.4 MG. What was the average daily flow for these 3 months?

7.50   A tank is 12 ft by 12 ft. With the discharge valve closed, the influent to the tank causes the water level to rise 8-in in 1 min. What is the gpm flow to the tank?

7.51   An 8-in diameter pipe flowing full delivers 510 gpm. What is the ft/sec velocity of flow in the pipeline?

7.52   Express a flow of 10 cfs in terms of gpm.

7.53   The totalizer reading for December was 134.6 MG. What was the average daily flow (ADF) for September?

7.54   What is 5.2 MGD expressed as cfs?

7.55   A pump discharges into a 3-ft diameter container. If the water level in the container rises 20 in in 30 sec, what is the gpm flow into the container?

7.56   Convert a flow of 1,825,000 gpd to cfm.

7.57   Six-inch diameter pipeline has water flowing at a velocity of 2.9 fps. What is the gpm flow rate through the pipeline?

7.58   The velocity in a 10-in pipeline is 2.6 ft/sec. If the 10-in pipeline flows into an 8-in diameter pipeline, what is the ft/sec velocity in the 8-in pipeline?

7.59   Convert 2,225 gpm to gpd.

7.60   The total flow for 1 day at a plant was 5,350,000 gal. What was the average gpm flow for that day?

GENERAL WASTEWATER TREATMENT PROBLEMS

7.61   Determine the chlorinator setting (lbs/day) needed to treat a flow of 5.5 MGD with a chlorine dose of 2.5 mg/L.

7.62   To dechlorinate wastewater, sulfur dioxide is to be applied at a level of 4 mg/L more than the chlorine residual. What should the sulfonator feed rate be (lbs/day) for a flow of 4.2 MGD with a chlorine residual of 3.1 mg/L?

7.63   What should the chlorinator setting be (lbs/day) to treat a flow of 4.8 MGD if the chlorine demand is 8.8 mg/L and a chlorine residual of 3 mg/L is desired?

7.64   A total chlorine dosage of 10 mg/L is required to treat the water in a unit process. If the flow is 1.8 MGD and the hypochlorite has 65% available chlorine, how many lb/day of hypochlorite will be required?

7.65   A storage tank is to be disinfected with 60 mg/L of chlorine. If the tank holds 86,000 gal, how many pounds of chlorine (gas) will be needed?

7.66   To neutralize a sour digester, 1 lb of lime is to be added for every pound of volatile acids in the digester liquor. If the digester contains

225,000 gal of sludge with a volatile acid (VA) level of 2,220 mg/L, how many pounds of lime should be added?

7.67 A flow of 0.83 MGD requires a chlorine dosage of 8 mg/L. If the hypochlorite has 65% available chlorine, how many lb/day of hypochlorite will be required?

7.68 The suspended solids concentration of the wastewater entering the primary system is 450 mg/L. If the plant flow is 1,840,000 gpd, how many lbs/day of suspended solids enter the primary system?

7.69 Calculate the BOD loading (lbs/day) on a stream if the secondary effluent flow is 2.90 MGD and the BOD of the secondary effluent is 25 mg/L.

7.70 The daily flow to a trickling filter is 5,450,000 gpd. If the BOD content of the trickling filter influent is 260 mg/L, how many lbs/day of BOD enter the trickling filter?

7.71 The flow to an aeration tank is 2,540 gpm. If the COD concentration of the water is 144 mg/L, how many pounds of COD are applied to the aeration tank daily?

7.72 The daily flow to a trickling filter is 2,300 gpm with a BOD concentration of 290 mg/L. How many lbs of BOD are applied to the trickling filter daily?

7.73 If a primary clarifier removes 152 mg/L suspended solids, how many lbs/day of suspended solids are removed when the flow is 5.7 MGD?

7.74 The flow to a primary clarifier is 1.92 MGD. If the influent to the clarifier has a suspended solids concentration of 310 mg/L and the primary effluent has 122 mg/L SS, how many lbs/day of suspended solids are removed by the clarifier?

7.75 The flow to a primary clarifier is 1.88 MGD. If the influent to the clarifier has a suspended solids concentration of 305 mg/L and the primary effluent has 121 mg/L SS, how many lbs/day of suspended solids are removed by the clarifier?

7.76 The flow to a trickling filter is 4,880,000 gpd. If the primary effluent has a BOD concentration of 150 mg/L and the trickling filter effluent has a BOD concentration of 25 mg/L, how many pounds of BOD are removed daily?

7.77 A primary clarifier receives a flow of 2.13 MGD with a suspended solids concentration of 367 mg/L. If the clarifier effluent has a suspended solids concentration of 162 mg/L, how many pounds of suspended solids are removed daily?

7.78 The flow to the trickling filter is 4,200,000 gpd with a BOD concentration of 210 mg/L. If the trickling filter effluent has a BOD concentration of 95 mg/L, how many lbs/day of BOD do the trickling filter remove?

7.79 The aeration tank has a volume of 400,000 gal. If the mixed liquor suspended solids concentration is 2,230 mg/L, how many pounds of suspended solids are in the aerator?

7.80 The aeration tank of a conventional activated sludge plant has a mixed liquor volatile suspended solids concentration of 1,890 mg/L. If the aeration tank is 115 ft long, 40 ft wide, and has wastewater to a depth of 12 ft, how many pounds of MLVSS are under aeration?

7.81 The volume of an oxidation ditch is 23,800 ft³. If the MLVSS concentration is 3,125 mg/L, how many pounds of volatile solids are under aeration?

7.82 An aeration tank is 110 ft long and 40 ft wide. The operating depth is 16 ft. If the mixed liquor suspended solids concentration is 2,250 mg/L, how many pounds of mixed liquor suspended solids are under aeration?

7.83. An aeration tank is 105 ft long and 50 wide. The depth of wastewater in the tank is 16 ft. If the tank contains an MLSS concentration of 2,910 mg/L, how many lbs of MLSS are under aeration?

7.84 The WAS-suspended solids concentration is 6,150 mg/L. If 5,200 lb/day solids are to be wasted, what must the WAS pumping rate be in MGD?

7.85 The WAS-suspended solids concentration is 6,200 mg/L. If 4,500 lb/day solids are to be wasted, (1) what must the WAS pumping rate be, in MGD? (2) What is this rate expressed in gpm?

7.86 It has been determined that 6,070 lb/day of solids must be removed from the secondary system. If the RAS SS concentration is 6,600 mg/L, what must be the WAS pumping rate, in gpm?

7.87 The RAS suspended solids concentration is 6,350 mg/L. If a total of 7,350 lb/day solids are to be wasted, what should the WAS pumping rate be in gpm?

7.88. A total of 5,750 lb/day of solids must be removed from the secondary system. If the RAS SS concentration is 7,240 mg/L, what must be the WAS pumping rate in gpm?

7.89 Determine the chlorinator setting (lbs/day) required to treat a flow of 3,650,000 gpd with a chlorine dose of 2.5 mg/L.

7.90 Calculate the BOD loading (lbs/day) on a stream if the secondary effluent flow is 2.10 MGD and the BOD of the secondary effluent is 17 mg/L.

7.91 The flow to a primary clarifier is 4.8 MGD. If the influent to the clarifier has a suspended solids concentration of 310 mg/L and the primary effluent suspended solids concentration is 120 mg/L, how many lbs/day of suspended solids are removed by the clarifier?

7.92 What should the chlorinator setting be (lbs/day) to treat a flow of 5.5 MGD if the chlorine demand is 7.7 mg/L and a chlorine residual of 2 mg/L is desired?

7.93 The suspended solids concentration of the wastewater entering the primary system is 305 mg/L. If the plant flow is 3.5 MGD, how many lbs/day of suspended solids enter the primary system?

7.94 A total chlorine dosage of 10 mg/L is required to treat water in a unit process. If the flow is 3.1 MGD and the hypochlorite has 65% available chlorine, how many lbs/day of hypochlorite will be required?

7.95 A primary clarifier receives a flow of 3.44 MGD with a suspended solids concentration of 350 mg/L. If the clarifier effluent has a suspended solids concentration of 140 mg/L, how many pounds of suspended solids are removed daily?

7.96 A storage tank is to be disinfected with 60 mg/L of chlorine. If the tank holds 90,000 gal, how many pounds of chlorine gas will be needed?

7.97 An aeration tank is 110 ft long and 45 ft wide. This operating depth is 14 ft. If the mixed liquor suspended solids concentration is 2,720 mg/L, how many pounds of mixed liquor suspended solids are under aeration?

7.98 The WAS-suspended solids concentration is 5,870 mg/L. If 5,480 lbs/day of solids are to be wasted, what must the WAS pumping rate be, in MGD?

7.99 The flow to an aeration tank is 2,300 gpm. If the COD concentration of the water is 120 mg/L, how many pounds of COD enter the aeration tank daily?

7.100 The daily flow to a trickling filter is 2,210 gpm. If the BOD concentration of the trickling filter influent is 240 mg/L, how many lbs/day of BOD are applied to the trickling filter?

7.101 The 1.7-MGD influent to the secondary system has a BOD concentration of 220 mg/L. The secondary effluent contains 24 mg/L BOD. How many pounds of BOD are removed each day by the secondary system?

7.102 The chlorine feed rate at a plant is 330 lbs/day. If the flow is 5,300,000 gpd, what is this dosage expressed in mg/L?

7.103 It has been determined that 6,150 lbs/day of solids must be removed from the secondary system. If the RAS SS concentration is 5,810 mg/L, what must be the WAS pumping rate, in gpm?

7.104 A trickling filter 100 ft in diameter treats a primary effluent flow of 2.5 MGD. If the recirculated flow to the clarifier is 0.9 MGD, what is the hydraulic loading on the trickling in gpd/ft^2?

7.105 The flow to a 90-ft diameter trickling filter is 2,850,000 gpd. The recirculated flow is 1,675,000 gpd. At this flow rate, what is the hydraulic loading rate in gpd/ft^2?

7.106 A rotating biological contactor treats a flow of 3.8 MGD. The manufacturer's data indicate a media surface area of 870,000 ft^2. What is the hydraulic loading rate on the RBC in gpd/ft^2?

7.107 A pond receives a flow of 2,100,000 gpd. If the surface area of the pond is 16 acres, what is the hydraulic loading in in/day?

7.108 What is the hydraulic loading rate in gpd/ft^2 to a 90-ft diameter trickling filter if the primary effluent flow to the trickling filter is 3,880,000 gpd, and the recirculated flow is 1,400,000 gpd?

7.109 A 20-acre pond receives a flow of 4.4 acre-ft/day. What is the hydraulic loading on the pond in in/day?

7.110 A sedimentation tank 70 ft by 25 ft receives a flow of 2.05 MGD. What is the surface overflow rate in gpd/ft^2?

7.111 A circular clarifier has a diameter of 60 ft. If the primary effluent flow is 2.44 MGD, what is the surface overflow rate in gpd/ft^2?

7.112 A sedimentation tank is 110 ft long and 50 ft wide. If the flow to the tank is 3.45 MGD, what is the surface overflow rate in gpd/ft^2?

7.113 The primary effluent flow to a clarifier is 1.66 MGD. If the sedimentation tank is 25 and 70 ft long, what is the surface overflow rate of the clarifier in gpd/ft^2?

7.114 The flow to a circular clarifier is 2.66 MGD. If the diameter of the clarifier is 70 ft, what is the surface overflow rate in gpd/ft^2?

7.115 A filter 40 ft by 20 ft receives a flow of 2,230 gpm. What is the filtration rate in gpm/ft^2?

7.116 A filter 40 ft by 25 ft receives a flow rate of 3,100 gpm. What is the filtration rate in gpm/ft^2?

7.117 A filter 26 ft by 60 ft receives a flow of 2,500 gpm. What is the filtration rate in gpm/ft^2?

7.118 A filter 40 ft by 20 ft treats a flow of 2.2 MGD. What is the filtration rate in gpm/ft^2?

7.119 A filter has a surface area of 880 ft^2. If the flow treated is 2,850 gpm, what is the filtration rate?

7.120 A filter 14 ft by 14 ft has a backwash flow rate of 4,750 gpm. What is the filter backwash rate in gpm/ft^2?

7.121 A filter 20 ft by 20 ft has a backwash flow rate of 4,900 gpm. What is the filter backwash rate in gpm/ft^2?

7.122 A filter is 25 ft by 15 ft. If the backwash flow rate is 3,400 gpm, what is the filter backwash rate in gpm/ft^2?

7.123 A filter 25 ft by 30 ft backwashes at a rate of 3,300 gpm. What is this backwash rate expressed as gpm/ft^2?

7.124 The backwash flow rate for a filter is 3,800 gpm. If the filter is 15 ft by 20 ft, what is the backwash rate expressed as gm/ft?

7.125 The total water filtered during a filter is 3,770,000 gal. If the filter is 15 ft by 30 ft, what is the unit filter run volume (UFRV) in gal/ft^2?

7.126 The total water filtered during a filter run (between backwashes) is 1,860,000 gal. If the filter is 20 ft by 15 ft, what is the UFRV in gal/ft^2?

7.127 A filter 25 ft by 20 ft filters a total of 3.88 MG during the filter run. What is the unit filter run volume in gal/ft^2?

7.128 The total water filtered between backwashes is 1,410,200 gal. If the length of the filter is 20 ft and the width is 14 ft, what is the unit filter run volume in gal/ft^2?

7.129 A filter is 30 ft by 20 ft. If the total water filtered between backwashes is 5,425,000 gal, what is the UFRV in gal/ft^2?

7.130 A rectangular clarifier has a total of 163 ft of weir. What is the weir overflow rate in gpd/ft when the flow is 1,410,000 gpd?

7.131 A circular clarifier receives a flow of 2.12 MGD. If the diameter of the weir is 60 ft, what is the weir overflow rate in gpd/ft?

7.132 A rectangular clarifier has a total of 240 ft of weir. What is the weir overflow rate in gpd/ft when the flow is 2.7 MGD?

7.133 The flow rate to a clarifier is 1,400 gpm. If the diameter of the weir is 80 ft, what is the weir overflow rate in gpd/ft?

7.134 A rectangular sedimentation basin has a total weir length of 189 ft. If the flow to the basin is 4.01 MGD, what is the weir loading rate in gpm/ft?

7.135 A trickling filter 80 ft in diameter with a media depth of 5 ft receives a flow of 2,450,000 gpd. If the BOD concentration of the primary effluent is 210 mg/L, what is the organic loading on the trickling filter in lbs BOD/day/1,000 ft^3?

7.136 The flow to a 3.5 acre wastewater pond is 120,000 gpd. The influent BOD concentration is 170 mg/L. What is the organic loading to the pond in lbs BOD/day/acre?

7.137 An 85-ft diameter trickling filter with a media depth of 6 ft receives a primary effluent flow of 2,850,000 gpd with a BOD of 120 mg/L. What is the organic loading on the trickling filter in lbs BOD/day/1,000 ft^3?

7.138 A rotating biological contactor (RBC) receives a flow of 2.20 MGD. If the soluble BOD of the influent wastewater to the RBC is 140 mg/L and the surface area of the media is 900,000 ft^2, what is the organic loading rate in lbs BOD/day/1,000 ft^2?

7.139 A 90-ft diameter trickling filter with a media depth of 4 ft receives a primary effluent flow of 3.5 MGD. If the BOD concentration of the wastewater flow to the trickling filter is 150 mg/L, what is the organic loading rate in lbs BOD/day/1,000 ft^3?

7.140 An activated sludge aeration tank receives a primary effluent flow of 3,420,000 gpd with a BOD

of 200 mg/L. The mixed liquor volatile suspended solids are 1,875 mg/L and the aeration tank volume is 420,000 gal. What is the current F/M ratio?

7.141 The volume of an aeration tank is 280,000 gal. The mixed liquor suspended solids is 1,710 mg/L. If the aeration tank receives a primary effluent flow of 3,240,000 gpd with a BOD of 190 mg/L, what is the F/M ratio?

7.142 The desired F/M ratio at a particular activated sludge plant is 0.9 lbs COD/1 lb mixed liquor volatile suspended solids. If the 2.25-MGD primary effluent flow has a COD of 151 mg/L how many lbs of MLVSS should be maintained?

7.143 An activated sludge plant receives a flow of 2,100,000 gpd with a COD concentration of 160 mg/L. The aeration tank volume is 255,000 gal, and the MLVSS is 1,900 mg/L. What is the current F/M ratio?

7.144 The flow to an aeration tank is 3,110,000 gpd, with a BOD content of 180 mg/L. If the aeration tank is 110 ft long, 50 ft wide, has wastewater to a depth of 16 ft, and the desired F/M ratio is 0.5, what is the desired MLVSS concentration (mg/L) in the aeration tank?

7.145 A secondary clarifier is 70 ft in diameter and receives a combined primary effluent and return activated sludge (RAS) flow of 3.60 MGD. If the MLSS concentration in the aerator is 2,650 mg/L, what is the solids loading rate on the secondary clarifier in lbs/day/ft^2?

7.146 A secondary clarifier, 80 ft in diameter, receives a primary effluent flow of 3.10 MGD and a return sludge flow of 1.15 MGD. If the MLSS concentration is 2,825 mg/L, what is the solids loading rate on the clarifier in lbs/day/ft^2?

7.147 The desired solids loading rate for a 60-ft diameter clarifier is 26-lbs/day/ft^2. If the total flow to the clarifier is 3,610,000 gpd, what is the desired MLSS concentration?

7.148 A secondary clarifier 60-ft in diameter receives a primary effluent flow of 2,550,000 gpd and a return sludge flow of 800,000 gpd. If the MLSS concentration is 2,210 mg/L, what is the solids loading rate on the clarifier in lbs/day/ft^2?

7.149 The desired solids loading rate for a 60-ft diameter clarifier is 20 lbs/day/ft^2. If the total flow to the clarifier is 3,110,000 gpd, what is the desired MLSS concentration?

7.150 A digester receives a total of 12,110 lbs/day of volatile solids. If the digester volume is 33,100 ft^3, what is the digester loading in the volatile solids added/day/ft^3?

7.151 A digester 60 ft in diameter with a water depth of 25 ft receives 124,000 lbs/day of raw sludge. If the sludge contains 6.5% solids with 70%

volatile matter, what is the digester loading in lbs of volatile solids added/day/ft³?

7.152 A digester 50 ft in diameter with a liquid level of 20 ft receives 141,000 lbs/day sludge with 6% total solids and 71% volatile solids. What is the digester loading in lbs volatile solids added/day/ft³?

7.153 A digester 40 ft in diameter with a liquid level of 16 ft receives 21,200 gpd sludge with 5.5% solids and 69% volatile solids. What is the digester loading in lbs/volatile solids/day/ft³? Assume the sludge weighs 8.34 lbs/gal.

7.154 A digester 50 ft in diameter with a liquid level of 20 ft receives 22,000 gpd sludge with 5.3% total solids and 70% volatile solids. What is the digester loading in the volatile solids/day/ft³? Assume the sludge weighs 8.6 lbs/gal.

7.155 A total of 2,050 lbs/day of volatile solids are pumped into a digester. The digester sludge contains a total of 32,400 lbs of volatile solids. What are the volatile solids loading on the digester in lbs volatile solids in digester?

7.156 A digester contains a total of 174,600 lbs of sludge that has a total solids content of 6.1% and volatile solids of 65%. If 620 lbs/day volatile solids are added to the digester, what is the volatile solids loading on the digester in lbs volatile solids added/day/lb volatile solids in the digester?

7.157 A total of 63,200-lbs/day sludge is pumped to an 115,000-gal digester. The sludge being pumped into the digester has a total solids content of 5.5% and a volatile solids content of 73%. The sludge in the digester has a solids content of 6.6% with a 59% volatile solids content. What are the volatile solids loading on the digester in lbs volatile solids added/day/lb VS in the digester? Assume the sludge in the digester weighs 8.34 lbs/gal.

7.158 A total of 110,000 gal of digested sludge is in a digester. The digested sludge contains 5.9% total solids and 58% volatile solids. If the desired volatile solids loading ratio is 0.08 lbs volatile solids added/day/lb volatile solids under digestion, what are the desired lbs volatile/day to enter the digester? Assume the sludge in the digester weighs 8.34 lbs/gal.

7.159 A total of 7,900 gpd sludge is pumped to the digester. The sludge has 4.8% solids with a volatile solids content of 73%. If the desired volatile solids loading ratio is 0.06 lbs volatile solids added/day/lb volatile solids under digestion, how many lbs volatile solids should be in the digester for this volatile solids load? Assume the sludge pumped to the digester weighs 8.34 lbs/gal.

7.160 A 5.3-acre wastewater pond serves a population of 1,733. What is the population loading on the pond in persons per acre?

7.161 A wastewater pound serves a population of 4,112. If the pond is 10 acres, what is the population loading on the pond?

7.162 A 381,000-gpd wastewater flow has a BOD concentration 1,765 mg/L. Using an average of 0.2 lbs/day BOD/person, what is the population equivalent of this wastewater flow?

7.163 A wastewater pond is designed to serve a population of 6,000. If the desired population loading is 420 persons per acre, how many acres of pond will be required?

7.164 A 100,000-gpd wastewater flow has a BOD content of 2,210 mg/L. Using an average of 0.2 lbs/day BOD/person, what is the population equivalent of this flow?

7.165 A circular clarifier has a diameter of 80 ft. If the primary effluent flow is 2.25 MGD, what is the surface overflow rate in gpd/ft²?

7.166 A filter has a square foot area of 190 ft². If the flow rate to the filter is 2,960 gpm, what is this filter backwash rate expressed as gpm/ft²?

7.167 The flow rate to a circular clarifier is 2,100,000 gpd. If the diameter of the weir is 80 ft, what is the weir overflow rate in gpd/ft?

7.168 A trickling filter, 90 ft in diameter, treats a primary effluent flow of 2.8 MGD. If the recirculated flow to the clarifier is 0.5 MGD, what is the hydraulic loading on the trickling filter in gpd/ft²?

7.169 The desired F/M ratio at an activated sludge plant is 0.7 lbs BOD/day/lb mixed liquor volatile suspended solids. If the 2.1-MGD primary effluent flow has a BOD of 161 mg/L, how many lbs of MLVSS should be maintained in the aeration tank?

7.170 A digester contains a total of 182,000 lbs of sludge that has a total solids content of 6.4% and volatile solids of 67%. If 500 lbs/day volatile solids are added to the digester, what is the volatile solids loading on the digester in lbs/day volatile solids added/lb volatile solids in the digester?

7.171 A secondary clarifier is 80 ft in diameter and receives a combined primary effluent and return activated sludge (RAS) flow of 3.58 MGD. If the MLSS concentration in the aerator is 2,760 mg/L, what is the solids loading rate on the secondary clarifier in lbs/day/ft²?

7.172 A digester, 70 ft in diameter with a water depth of 21 ft, receives 115,000 lbs/day of raw sludge. If the sludge contains 7.1% solids with 70% volatile solids, what is the digester loading in lbs volatile solids added/day/ft³ volume?

7.173 A 25-acre pond receives a flow of 4.15 acre-ft/day. What is the hydraulic loading on the pond in in/day?

7.174 The flow to an aeration tank is 3,335,000 gpd with a BOD content of 174 mg/L. If the aeration tank is 80 ft long, 40 ft wide, has wastewater to a depth of 12 ft, and the desired F/M ratio is 0.5, what is the desired MLVSS concentration (mg/L) in the aeration tank?

7.175 A sedimentation tank 80 ft by 25 ft receives a flow of 2.0 MGD. What is the surface overflow rate in gpd/da/ft?

7.176 The total water filtered during a filter run (between backwashes) is 1,785,000 gal. If the filter is 25 ft by 20 ft, what is the unit filter run volume (UFRV) in gal/ft^2?

7.177 The volume of an aeration tank is 310,000 gal. The mixed liquor volatile suspended solids are 1,920 mg/L. If the aeration tank receives a primary effluent flow of 2,690,000 gpd with a COD of 150 mg/L, what is the F/M ratio?

7.178 A total of 24,500 gal of digested sludge is in a digester. The digested sludge contains 5.5% solids and 56% volatile solids. To maintain a desired volatile loading ratio of 0.09 lbs volatile solids added/day/lb volatile solids under digestion, what is the desired lbs of volatile solids/day loading to the digester?

7.179 The flow to a filter is 4.44 MGD. If the filter is 40 ft by 30 ft, what is the filter-loading rate in gpm/ft^2?

7.180 An 80-ft diameter trickling filter with a media depth of 4 ft receives a primary effluent flow of 3.3 MGD with a BOD concentration of 115 mg/L. What is the organic loading on the filter in lbs BOD/day/1,000 ft^3?

7.181 A circular clarifier receives a flow of 2.56 MGD. If the diameter of the weir is 80 ft, what is the weir overflow rate in gpd/ft?

7.182 A 5.5-acre wastewater pond serves a population of 1,900. What is the population loading on the pond (people/acre)?

7.183 A rotating biological contactor (RBC) receives a flow of 2.44 MGD. If the soluble BOD of the influent wastewater to the RBC is 140 mg/L and the surface area of the media is 750,000 ft^2, what is the organic loading rate in lbs soluble BOD/day/1,000 ft^2?

7.184 A filter 40 ft by 30 ft treats a flow of 4.15 MGD. What is the filter-loading rate in gpm/ft^2?

7.185 A flocculation basin is 8 ft deep, 16 ft wide, and 30 ft long. If the flow through the basin is 1.45 MGD, what is the detention time in minutes?

7.186 The flow to a sedimentation tank 80 ft long, 20 ft wide, and 12 ft deep is 1.8 MGD. What is the detention time in the tank in hours?

7.187 A basin 3 ft by 4 ft, is to be filled to the 3 ft level. If the flow to the tank is 6 gpm, how long will it take to fill the tank (in hours)?

7.188 The flow rate to a circular clarifier is 5.20 MGD. If the clarifier is 80 ft in diameter with water to a depth of 10 ft, what are the detention time in hours?

7.189 A waste treatment pond is operated at a depth of 6 ft. The average width of the pond is 500 ft and the average length is 600 ft. If the flow to the pond is 222,500 gpd, what is the detention time in days?

7.190 An aeration tank has a total of 12,300 lbs of mixed liquor-suspended solids. If a total of 2,750 lbs/day of suspended solids enter the aerator with the primary effluent flow, what is the sludge age in the aeration tank?

7.191 An aeration tank is 110 ft long, and 30 ft wide with wastewater to a depth of 20 ft. The mixed liquor suspended solids concentration is 2,820 mg/L. If the primary effluent flow is 988,000 gpd with a suspended solids concentration of 132 mg/L, what is the sludge age in the aeration tank?

7.192 An aeration tank contains 200,000 gal of wastewater. The MLSS is 2,850 mg/L. If the primary effluent flow is 1.52 MGD with a suspended solids concentration of 84 mg/L, what is the sludge age?

7.193 The 2.10-MGD primary effluent flow to an aeration tank has a suspended solids concentration of 80 mg/L. The aeration tank volume is 205,000 gal. If a sludge age of 6 days is desired, what is the desired MLSS concentration?

7.194 A sludge age of 5.5 days is desired. Assume 1,610 lbs/day of suspended solids enter the aeration tank in the primary effluent. To maintain the desired sludge age, how many lbs of MLSS must be maintained in the aeration tank?

7.195 An aeration tank has a volume of 320,000 gal. The final clarifier has a volume of 180,000 gal. The MLSS concentration in the aeration tank is 3,300 mg/L. If a total of 1,610 lbs/day of suspended solids are wasted and 340 lbs/day suspended solids are in the secondary effluent, what is the solids retention time for the activated sludge system? Use the solids retention equation that uses combined aeration tank and final clarifier volumes to estimate system solids.

7.196 Determine the solids retention time (SRT) given the data below. Use the solids retention time equation that uses the combined aeration tank and final clarifier volumes to estimate system solids.

7.197 Calculate the solids retention time given the data below. Use the SRT equation that uses the combined aeration tank and final clarifier volumes to estimate system solids.

Aeration Tank	MLSS—2,550 mg/L
Volume—1.4 MG	
Final Clarifier	
Volume—0.4 MG	WAS SS—6,240 mg/L
Population Equivalent	Sec. Effluent
Flow—2.8 MGD	SS—20 mg/L
WAS Pumping	
Rate—85,000 gpd	

Aeration tank	MLSS—
volume—1.5 MG	2,408 mg/L
Final clarifier	WAS
volume—0.4 MG	SS—6,320 mg/L
Population equivalent	Secondary
flow—2.85 MGD	effluent
	SS—25 mg/L
WAS pumping	
rate—71,200 gpd	

7.198 The volume of an aeration tank is 800,000 gal and the final clarifier is 170,000 gal. The desired solids retention time (SRT) for the plant is 8 days. The primary effluent flow is 2.6 MGD and the WAS pumping rate is 32,000 gpd. If the WAS SS concentration is 6,340 mg/L and the secondary effluent SS concentration is 20 mg/L, what is the MLSS concentration in mg/L?

7.199 The flow to a sedimentation tank 75 ft long, 30 ft wide, and 14 ft deep is 1,640,000 gpd. What is the detention time in the tank in hours?

7.200 An aeration tank has a total of 12,600 lbs of mixed liquor-suspended solids. If a total of 2,820 lbs/day of suspended solids enter the aeration tank in the primary effluent flow, what is the sludge age in the aeration tank?

7.201 An aeration tank has a volume of 310,000 gal. The final clarifier has a volume of 170,000 gal. The MLSS concentration in the aeration tank is 3,120 mg/L. If a total of 1,640 lbs/day of suspended solids are wasted and 320 lbs/day of suspended solids are in the secondary effluent, what is the solids retention time for the activated sludge system?

7.202 The flow through a flocculation basin is 1.82 MGD. If the basin is 40 ft long, 20 ft wide, and 10 ft deep, what is the detention time in minutes?

7.203 Determine the solids retention time given the data below.

Aeration	MLSS—2,810 mg/L
volume—220,000 gal	
Final clarifier	WAS
volume—115,000 gal	SS—6,100 mg/L
Population equivalent	Sec. Effluent
flow—2,400,000 gpd	SS—18 mg/L
WAS pumping	
rate—18,900 gpd	

7.204 The mixed liquor suspended solids concentration in an aeration tank is 3,250 mg/L. The aeration tank contains 330,000 gal. If the primary effluent flow is 2,350,000 gpd with suspended solids concentrations of 100 mg/L, what is the sludge age?

7.205 Calculate the solids retention time given the following data:

7.206 An aeration tank is 80 ft long and 25 ft wide, with wastewater to a depth of 10 ft. The mixed liquor suspended solids concentration is 2,610 mg/L. If the influent flow to the aeration tank is 920,000 gpd with a suspended solids concentration of 140 mg/L, what is the sludge age in the aeration tank?

7.207 A tank 6 ft in diameter is to be filled to the 4 ft level. If the flow to the tank is 12 gpm, how long will it take to fill the tank (in minutes)?

7.208 A sludge age of 6 days is desired. The suspended solids concentration of the 2.14-MGD influent flow to the aeration tank is 140 mg/L. To maintain the desired sludge age, how many pounds of MLSS must be maintained in the aeration tank?

7.209 The average width of a pond is 400 ft, and the average length is 440 ft. The depth is 6 ft. If the flow to the pond is 200,000 gpd, what is the detention time in days?

7.210 The volume of an aeration tank is 480,000 gal, and the volume of the final clarifier is 160,000 gal. The desired solids retention time for the plant is 7 days. The primary effluent flow is 2,920,000 gpd, and the WAS pumping rate is 34,000 gpd. If the WAS SS concentration is 6,310 mg/L and the secondary effluent SS concentration is 12 mg/L, what is the desired MLSS concentration in mg/L?

7.211 The suspended solids concentration entering a trickling filter is 110 mg/L. If the suspended solids concentration in the trickling filter effluent is 21 mg/L, what is the suspended solids removal efficiency of the trickling filter?

7.212 The BOD concentration of the raw wastewater at an activated sludge plant is 230 mg/L. If the BOD concentration of the final effluent is 14 mg/L, what is the overall efficiency of the plant in BOD removal?

7.213 The influent flow to a waste treatment pond has a BOD content of 260 mg/L. If the pond effluent has a BOD content of 60 mg/L, what is the BOD removal efficiency of the pond?

7.214 The suspended solids concentration of the primary clarifier influent is 310 mg/L. If the suspended solids concentration of the primary effluent is 135 mg/L, what is the suspended solids removal efficiency?

7.215 A total of 3,700 gal of sludge is pumped into a digester. If the sludge has a 4.9% solids content, how many lbs/day of solids are pumped into the digester? Assume the sludge weighs 8.34 lbs/gal.

7.216 The total weight of a sludge sample is 12.87 g (sample only, not the dish). If the weight of the solids after drying is 0.87 g, what is the percent total solids of the sludge?

7.217 A total of 1,450 lbs/day of suspended solids are removed from a primary clarifier and pumped into a sludge thickener. If the sludge has a solids content of 3.3%, how many lbs/day sludge is this?

7.218 It is anticipated that 258 lbs/day of suspended solids will be pumped from the primary clarifier of a new plant. If the primary clarifier sludge has a solids content of 4.4%, how many gpd sludges will be pumped from the clarifier? Assume a sludge weight of 8.34 lbs/gal.

7.219 A total of 291,000 lbs/day of sludge is pumped from a primary clarifier to a sludge thickener. If the total solids content of the sludge is 3.6%, how many lbs/day of total solids are sent to the thickener?

7.220 A primary sludge flow of 3,100 gpd with a solids content of 4.4% is mixed with a thickened secondary sludge flow of 4,100 gpd that has a solids content of 3.6%. What is the percent solids content of the mixed sludge flow? Assume the density of both sludges is 8.34 lbs/gal.

7.221 Primary and thickened secondary sludges are to be mixed and sent to the digester. The 8,100-gpd primary sludge has a solids content of 5.1%, and the 7,000-gpd thickened secondary sludge has a solids content of 4.1%. What would be the percent solids content of the mixed sludge? Assume the density of both sludges is 8.34 lbs/gal.

7.222 A 4,750-gpd primary sludge has a solids content of 4.7%. The 5,250-gpd thickened secondary sludge has a solids content of 3.5%. If the sludges were blended, what would be the percent solids content of the mixed sludge? Assume the density of both sludges is 8.34 lbs/gal.

7.223 A primary sludge flow of 8,925 gpd with a solids content of 4.0% is mixed with a thickened secondary sludge flow of 11,340 gpd with 6.6% solids content. What is the percent solids of the combined sludge flow? Assume the density of both sludges is 8.34 lbs/gal.

7.224 If 3,250 lbs/day of solids with a volatile solids content of 65% are sent to the digester, how many lbs/day of volatile solids are sent to the digester?

7.225 A total of 4,120 gpd of sludge is to be pumped into the digester. If the sludge has a 7% solids content with 70% volatile solids, how many lbs/day of volatile solids are pumped into the digester? Assume the sludge weighs 8.34 lbs/gal.

7.226 The static water level for a well is 91 ft. If the pumping water level is 98 ft, what is the well drawdown?

7.227 The static water level for a well is 110 ft. The pumping water level is 125 ft. What is the well drawdown?

7.228 Before the pump is started, the water level is measured at 144 ft. The pump is then started. If the pumping water level is determined to be 161 ft, what is the well drawdown?

7.229 The static water level of a well is 86 ft. The pumping water level is determined using the sounding line. The air pressure applied to the sounding line is 3.7 psi and the length of the sounding line is 112 ft. What is the drawdown?

7.230 A sounding line is used to determine the static water level for a well. The air pressure applied is 4.6 psi, and the length of the sounding line is 150 ft. If the pumping water level is 171 ft, what is the drawdown?

7.231 If the well yield is 300 gpm and the drawdown is measured to be 20 ft, what is the specific capacity?

7.232 During a 5-min well yield test, a total of 420 gal were pumped from the well. What is the well yield in gpm?

7.233 Once the drawdown of a well stabilized, it was determined that the well produced 810 gal during a 5-min pumping test. What is the well yield in gpm?

7.234 During a test for well yield, a total of 856 gal were pumped from the well. If the well yield test lasted 5 min, what was the well yield in gpm? In gph?

7.235 A bailer is used to determine the approximate yield of a well. The bailer is 12 ft long and has a diameter of 12 in. If the bailer is placed in the well and removed a total of 12 times during a 5-min test, what is the well yield in gpm?

7.236 During a 5-min well yield test, a total of 750 gal of water were pumped from the well. At this yield, if the pump is operated a total of 10 h each day, how many gallons of water are pumped daily?

7.237 The discharge capacity of a well is 200 gpm. If the drawdown is 28 ft, what is the specific yield in gpm/ft of drawdown?

7.238 A well produces 620 gpm. If the drawdown for the well is 21 ft, what is the specific yield in gpm/ft of drawdown?

7.239 A well yields 1,100 gpm. If the drawdown is 41.3 ft, what is the specific yield in gpm/ft of drawdown?

7.240 The specific yield of a well is listed as 33.4 gpm/ft. If the drawdown for the well is 42.8 ft, what is the well yield in gpm?

7.241 A new well is to be disinfected with chlorine at a dosage of 40 mg/L. If the well casing diameter is 6 in and the length of the water-filled casing is 140 ft, how many pounds of chlorine will be required?

7.242 A new well with a casing diameter of 12 in is to be disinfected. The desired chlorine dosage is 40 mg/L. If the casing is 190 ft long and the water level in the well is 81 ft from the top of the well, how many pounds of chlorine will be required?

7.243 An existing well has a total casing length of 210 ft. The top 180 ft of casing has a 12-in diameter, and the bottom 40 ft of the casing has an 8-in diameter. The water level is 71 ft from the top of the well. How many pounds of chlorine will be required if a chlorine dosage of 110 mg/L is desired?

7.244 The water-filled casing of a well has a volume of 540 gal. If 0.48 lbs of chlorine were used in disinfection, what was the chlorine dosage in mg/L?

7.245 A total of 0.09 lbs of chlorine is required for the disinfection of a well. If sodium hypochlorite (5.25% available chlorine) is to be used, how many fluid ounces of sodium hypochlorite are required?

7.246 A new well is to be disinfected with calcium hypochlorite (65% available chlorine). The well casing diameter is 6 in and the length of the water-filled casing is 120 ft. If the desired chlorine dosage is 50 g/L, how many ounces (dry measure) of calcium hypochlorite will be required?

7.247 How many pounds of chloride of lime (25% available chlorine) will be required to disinfect a well if the casing is 18-in in diameter and 200 ft long with a water level at 95 ft from the top of the well? The desired chlorine dosage is 100 mg/L.

7.248 The water-filled casing of a well has a volume of 240 gal. How many fluid ounces of sodium hypochlorite (5.25% available chlorine) are required to disinfect the well if a chlorine concentration of 60 mg/L is desired?

7.249 The pressure gauge reading at a pump discharge head is 4.0 psi. What is this discharge head expressed in feet?

7.250 The static water level of a well is 94 ft. The well drawdown is 24 ft. If the gauge reading at the pump discharge head is 3.6 psi, what is the field head?

7.251 A pond has an average length of 400 ft, an average width of 110 ft, and an estimated average depth of 14 ft. What is the estimated volume of the pond in gallons?

7.252 A pond has an average length of 400 ft and an average width of 110 ft. If the maximum depth of the pond is 30 ft, what is the estimated volume of the pond in gallon?

7.253 A pond has an average length of 200 ft, an average width of 80 ft, and an average depth of 12 ft. What is the acre-feet volume of the pond?

7.254 A small pond has an average length of 320 ft, an average width of 170 ft, and a maximum depth of 16 ft. What is the acre-feet volume of the pond?

7.255 For algae control in a reservoir, a dosage of 0.5 mg/L copper is desired. The reservoir has a volume of 20 MG. How many pounds of copper sulfate pentahydrate (25% available copper) will be required?

7.256 The static water level for a well is 93.5 ft. If the pumping water level is 131.5 ft, what is the drawdown?

7.257 During a 5-min well yield test, a total of 707 gal was pumped from the well. What is the good yield in gpm? In gph?

7.258 A bailer is used to determine the approximate yield of a well. The bailer is 12 ft long and has a diameter of 12 in. If the bailer is placed in the well and removed a total of eight times during a 5-min test, what is the well yield in gpm?

7.259 The static water level in a well is 141 ft. The pumping water level is determined using the sounding line. The air pressure applied to the sounding line is 3.5 psi, and the length of the sounding line is 167 ft. What is the drawdown?

7.260 A well produces 610 gpm. If the drawdown for the well is 28 ft, what is the specific yield in gpm/ft of drawdown?

7.261 A new well is to be disinfected with a chlorine dose of 55 mg/L. If the well casing diameter is 6 in and the length of the water-filled casing is 150 ft, how many pounds of chlorine will be required?

7.262 During a 5-min well yield test, a total of 780 gal of water was pumped from the well. At this yield, if the pump is operated for a total of 8 h each day, how many gallons of water are pumped daily?

7.263 The water-filled casing of a well has a volume of 610 gal. If 0.47 lbs of chlorine was used for disinfection, what was the chlorine dosage in mg/L?

7.264 An existing well has a total casing length of 230 ft. The top 170 ft of the casing has a 12-in diameter and the bottom 45 ft of the casing has an 8-in diameter. The water level is 81 ft from the top of the well. How many pounds of chlorine will be required if a chlorine dosage of 100 mg/L is desired?

7.265 A total of 0.3 lbs of chlorine is required for the disinfection of a well. If sodium hypochlorite is to be used (5.25% available chlorine), how many fluid ounces of sodium hypochlorite is required?

7.266 A flash mix chamber is 4 ft wide and 5 ft long, with water to a depth of 3 ft. What is the gallon volume of water in the flash mix chamber?

7.267   A flocculation basin is 50 ft long, and 20 ft wide, with water to a depth of 8 ft. What is the volume of water in the basin (in gallons)?

7.268   A flocculation basin is 40 ft long, 16 ft wide, with water to a depth of 8 ft. How many gallons of water are in the basin?

7.269   A flash mix chamber is 5 ft² with water to a depth of 42 in. What is the volume of water in the flash mixing chamber (in gallons)?

7.270   A flocculation basin is 25 ft wide, 40 ft long, and contains water to a depth of 9 ft 2 in. What is the volume of water (in gallons) in the flocculation basin?

BACK TO THE BEGINNING

At the beginning of this chapter, the importance of the math operations that water and wastewater operators are required to perform, both in plant operations and on licensure examinations, was explained, along with the following question:

Question: How long will it take water to pass through a rectangular basin at a flow rate of 1.2 MGD, if the basin is 90 ft long, 40 ft wide, and 12 ft deep?

After finishing this chapter, the reader should have no problem calculating the answer to this question.

Answer:

$$\text{Detention Time} = \frac{\text{Volume}}{\text{Flowrate}}$$

$$22°F$$
$$+ 40°F$$
$$\overline{62°F}$$

$$= \frac{90 \text{ ft} \times 40 \text{ ft} \times 12 \text{ ft}}{1.2 \text{ MGD}}$$

$$= \frac{43,200 \text{ ft}^3}{1.2 \text{ MGD}}$$

$$= \frac{43,200 \text{ ft}^3 \times 7.48 \text{ gal/ft}^3}{1.2 \text{ MGD}}$$

$$= \frac{323,136 \text{ gal}}{1.2 \text{ MGD}}$$

$$= \frac{323,136 \text{ gal}}{1.2 \text{ MGD}} \times \frac{1 \text{ MGD}}{1,000,000 \text{ gal/day}} = 0.27 \text{ day}$$

$$= 0.27 \text{ day} \times \frac{24 \text{ h}}{\text{day}}$$

$$= 6.5 \text{ h (rounded)}$$

THE BOTTOM LINE

For water and wastewater operators (aka licensed Fluid Mechanics) to optimize treatment operations, they need to be able to calculate flow rates, chemical dosages, and treatment efficiency. Moreover, the operators can't perform proper mixing and distribution of chemicals without being able to perform basic math operations. An important (lawfully required) operational function is to record data for accurate regulatory compliance. Math skills are also an important tool in the operator's toolbox to measure and record parameters such as pH levels, turbidity, and chemical concentrations. The real bottom line is that these measurements directly impact the quality of treated water and environmental compliance.

REFERENCES

APHA. 2012. *Standard Methods for Examination of Water & Wastewater*, 22nd ed. Washington, DC: American Public Health Association.

Price, J.M., 1991. *Applied Math for Wastewater Plant Operators*. Boca Raton, FL: CRC Press.

Reeves, T.G., 1994. *Water Fluoridation: A Manual for Water Plant Operators*. Atlanta: U.S. Department of Health and Human Services.

8 Water Hydraulics

INTRODUCTION

Beginning students of water hydraulics and its principles often come to the subject matter with certain misgivings. For example, water and wastewater operators quickly learn on the job that their primary operational/maintenance concerns involve a daily routine of monitoring, sampling, laboratory testing, operation, and process maintenance. How does water hydraulics relate to daily operations? The hydraulic functions of the treatment process have already been designed into the plant. Why learn water hydraulics at all?

> Simply put, while having hydraulic control of the plant is obviously essential to the treatment process, maintaining and ensuring continued hydraulic control is also essential. No water/wastewater facility (and/or distribution collection system) can operate without proper hydraulic control. The operator must know what hydraulic control is and what it entrails to know how to ensure proper hydraulic control. Moreover, in order to understand the basics of piping and pumping systems, water and wastewater maintenance operators must have a fundamental knowledge of basic water hydraulics.
>
> *Spellman & Drinan (2001)*

Note: The practice and study of water hydraulics is not new. Even in medieval times, water hydraulics was not new. "Medieval Europe had inherited a highly developed range of Roman hydraulic components" (Magnusson, 2001). The basic conveyance technology, based on low-pressure systems of pipes and channels, was already established. In studying "modern" water hydraulics, it is important to remember that the science of water hydraulics is the direct result of two immediate and enduring problems: "The acquisition of freshwater and access to continuous strip of land with a suitable gradient between the source and the destination" (Magnusson, 2001).

WHAT IS WATER HYDRAULICS?

The word "hydraulic" is derived from the Greek words "hydro" (meaning water) and "aulis" (meaning pipe). Originally, the term hydraulics referred only to the study of water at rest and in motion (the flow of water in pipes or channels). Today, it is taken to mean the flow of any "liquid" in a system.

What is a liquid? In terms of hydraulics, a liquid can be either oil or water. In fluid power systems used in modern industrial equipment, the hydraulic liquid of choice is oil. Some common examples of hydraulic fluid power systems include automobile braking and power steering systems, hydraulic elevators, and hydraulic jacks or lifts. Probably the most familiar hydraulic fluid power systems in water/wastewater operations are used in dump trucks, front-end loaders, graders, and earth-moving and excavation equipment. In this text, we are concerned with liquid water.

Many find the study of water hydraulics difficult and puzzling (especially the licensure examination questions), but we know it is neither mysterious nor difficult. It is the function or output of practical applications of the basic principles of water physics. Because water/wastewater treatment is based on the principles of water hydraulics, concise, real-world training is necessary for operators who must operate the plant and for those sitting for state licensure/certification examinations.

BASIC CONCEPTS

Air Pressure (@ Sea Level = 14.7 pounds per square in^{ch} (psi)

This relationship shown above is important because our study of hydraulics begins with air. A blanket of air, many miles thick, surrounds the Earth. The weight of this blanket on a given square inch of the Earth's surface will vary according to the thickness of the atmospheric blanket above that point. As shown above, at sea level, the pressure exerted is 14.7 pounds per square inch (psi). On a mountaintop, air pressure decreases because the blanket is not as thick.

1 ft^3 H$_2$O = 62.4 lb

The relationship shown above is also important: both cubic feet and pounds are used to describe a volume of water. There is a defined relationship between these two methods of measurement. The specific weight of water is defined relative to a cubic foot. One cubic foot of water weighs 62.4 lbs. This relationship is true only at a temperature of 4°C and at a pressure of one atmosphere, conditions referred to as *standard temperature and pressure* (STP). Note that 1 atmosphere = 14.7 lb/in^2 at sea level and 1 ft^3 of water contains 7.48 gal. The weight varies so little that, for practical purposes, this weight is used for temperatures ranging from 0°C to 100°C. One cubic inch of water weighs 0.0362 lb. Water 1 ft deep will exert a pressure of 0.43 lb/in^2 on the bottom area (12 in × 0.0362 lb/in^3). A column of water 2 ft high exerts 0.86 psi (2 ft × 0.43 psi/ft), and one 55 ft high exerts 23.65 psi (55 ft × 0.43 psi/ft). A column of water 2.31 ft high will exert 1.0 psi (2.31 ft × 0.43 psi/ft). To produce a pressure of 50 psi requires a 115.5-ft water column (50 psi × 2.31 ft/psi).

Remember the important points being made here:

1. 1 ft^3 H$_2$O = 62.4 lb
2. A column of water 2.31 ft high will exert 1.0 psi

DOI: 10.1201/9781003581901-10

Another relationship is also important:

$$1 \text{ gal } H_2O = 8.34 \text{ lb}$$

At standard temperature and pressure, 1 ft^3 of water contains 7.48 gal. With these two relationships, we can determine the weight of 1 gal of water. This is accomplished by:

Weight of gal of water = 62.4 lb ÷ 7.48 gal = 8.34 lb/gal

Thus,

$$1 \text{ gal } H_2O = 8.34 \text{ lb}$$

Note: Further, this information allows cubic feet to be converted to gallons by simply multiplying the number of cubic feet by 7.48 gal/ft^3.

Example 8.1

Problem: Find the number of gallons in a reservoir that has a volume of 855.5 ft^3.

Solution:

$$855.5 \text{ ft}^3 \times 7.48 \text{ gal/ft}^3 = 6,399 \text{ gal (rounded)}$$

Note: As mentioned earlier, the term *head* is used to designate water pressure in terms of the height of a column of water in feet. For example, a 10-ft column of water exerts 4.3 psi. This can be called 4.3-psi pressure or 10 ft of head.

STEVIN'S LAW

Stevin's law (also known as the hydrostatic pressure law) deals with water at rest. Specifically, the law states: "The pressure at any point in a fluid at rest depends on the distance measured vertically to the free surface and the density of the fluid." Basically, in simpler terms, this describes how water pressure varies with depth in a fluid. Stated as a formula, this becomes

$$p = w \times h \qquad (8.1)$$

where

p = pressure in pounds per square foot (psf)
w = density in pounds per cubic foot (lb/ft^3)
h = vertical distance in feet

Example 8.2

Problem: What is the pressure at a point 18 ft below the surface of a reservoir?

Solution:

To calculate this, we must know that the density of the water (w) is 62.4 lb/ft^3.

$$p = w \times h$$

$$= 62.4 \text{ lb/ft}^3 \times 18 \text{ ft}$$

$$= 1,123 \text{ lb/ft}^2 \text{ or } 1,123 \text{ psf}$$

Water/wastewater operators generally measure pressure in pounds per square inch rather than pounds per square foot. To convert, divide by 144 in^2/ft^2 (12 in × 12 in = 144 in^2):

$$p = \frac{1,123 \text{ psf}}{144 \text{ in}^2/\text{ft}^2} = 7.8 \text{ lb/in}^2 \text{ or psi (rounded)}$$

Key Point: The weight of the immersed volume must be equal to the weight of the same volume of liquid.

Key Point: Stevin's law applies to any fluid, not just water.

DENSITY AND SPECIFIC GRAVITY

Table 8.1 shows the relationship between temperature, specific weight, and density of water. When we say that iron is heavier than aluminum, we mean that iron has greater density than aluminum. In practice, what we are really saying is that a given volume of iron is heavier than the same volume of aluminum.

TABLE 8.1
Water Properties (Temperature, Specific Weight, and Density)

Temperature (°F)	Specific Weight (lb/ft^3)	Density (Slugs/ft^3)	Temperature (°F)	Specific Weight (lb/ft^3)	Density (Slugs/ft^3)
32	62.4	1.94	130	61.5	1.91
40	62.4	1.94	140	61.4	1.91
50	62.4	1.94	150	61.2	1.90
60	62.4	1.94	160	61.0	1.90
70	62.3	1.94	170	60.8	1.89
80	62.2	1.93	180	60.6	1.88
90	62.1	1.93	190	60.4	1.88
100	62.0	1.93	200	60.1	1.87
110	61.9	1.92	210	59.8	1.86
120	61.7	1.92			

Note: What is density? *Density* is the *mass per unit volume* of a substance.

Suppose you had a tub of lard and a large box of cold cereal, each having a mass of 600 g. The density of the cereal would be much less than the density of the lard because the cereal occupies a much larger volume than the lard occupies. The density of an object can be calculated by using the formula:

$$\text{Density} = \frac{\text{Mass}}{\text{Volume}} \qquad (8.2)$$

In water and wastewater treatment, perhaps the most common measures of density are pounds per cubic foot (lb/ft^3) and pounds per gallon (lb/gal):

- 1 ft^3 of water weights 62.4 lb—Density = 62.4 lb/ft^3
- One gallon of water weighs 8.34 lb—Density = 8.34 lb/gal

The density of a dry material, such as cereal, lime, soda, and sand, is usually expressed in pounds per cubic foot. The density of a liquid, such as liquid alum, liquid chlorine, or water, can be expressed either as pounds per cubic foot or as pounds per gallon. The density of a gas, such as chlorine gas, methane, carbon dioxide, or air, is usually expressed in pounds per cubic foot.

As shown in Table 8.1, the density of a substance like water changes slightly as the temperature of the substance changes. This occurs because substances usually increase in volume (size—they expand) as they become warmer. Because of this expansion with warming, the same weight is spread over a larger volume, so the density is lower when a substance is warm than when it is cold.

Note: What is specific gravity? Specific gravity is the weight (or density) of a substance compared to the weight (or density) of an equal volume of water. The specific gravity of water is 1.

This relationship is easily seen when a cubic foot of water, which weighs 62.4 lb, is compared to a cubic foot of aluminum, which weighs 178 lb. Aluminum is 2.8 times heavier than water.

It is not difficult to find the specific gravity of a piece of metal. All you have to do is weigh the metal in the air, then weigh it underwater. Its loss of weight is the weight of an equal volume of water. To find the specific gravity, divide the weight of the metal by its loss of weight in water.

$$\text{Specific Gravity} = \frac{\text{Weight of Substance}}{\text{Weight of Equal Volume of Water}}$$

$$(8.3)$$

Example 8.3

Problem: Suppose a piece of metal weighs 150 lb in air and 85 lb underwater. What is the specific gravity?

Solution:

Step 1: 150 lb subtract 85 lb = 65 lb loss of weight in water

Step 2:

$$\text{Specific gravity} = \frac{150}{65} = 2.3$$

Note: In a calculation of specific gravity, it is *essential* that the densities be expressed in the same units.

As stated earlier, the specific gravity of water is 1, which serves as the standard reference against which all other liquid or solid substances are compared. Specifically, any object with a specific gravity greater than 1 will sink in water (e.g., rocks, steel, iron, grit, floc, sludge). Substances with a specific gravity of less than 1 will float (e.g., wood, scum, and gasoline). Considering the total weight and volume of a ship, its specific gravity is less than 1; therefore, it can float.

The most common use of specific gravity in water/wastewater treatment operations is in gallons-to-pounds conversions. In many cases, the liquids being handled have a specific gravity of 1.00 or very close to 1.00 (between 0.98 and 1.02), so 1.00 may be used in the calculations without introducing significant error. However, in calculations involving a liquid with a specific gravity of less than 0.98 or greater than 1.02, the conversions from gallons to pounds must consider specific gravity. The technique is illustrated in the following example.

Example 8.4

Problem: There is 1,455 gal of a certain liquid in a basin. If the specific gravity of the liquid is 0.94, how many pounds of liquid are in the basin?

Solution:

Normally, for conversion from gallons to pounds, we would use the factor 8.34 lb/gal (the density of water) if the substance's specific gravity were between 0.98 and 1.02. However, in this instance, the substance has a specific gravity outside this range, so the 8.34 factor must be adjusted by multiplying 8.34 lb/gal by the specific gravity to obtain the adjusted factor:

Step 1: (8.34 lb/gal) (0.94) = 7.84 lb/gal (rounded)
Step 2: Then convert 1,455 gal to pounds using the corrected factor:
(1,455 gal)(7.84 lb/gal) = 11,407 lb (rounded)

FORCE AND PRESSURE

Water exerts force and pressure against the walls of its container, whether it is stored in a tank or flowing in a pipeline. Force and pressure are different, although they are closely related. *Force* is the push or pull influence that causes motion. In the English system, force and weight are often used in the same way. The weight of a cubic foot of water is 62.4 lb. The force exerted on the bottom of a 1-ft^3 is 62.4 lb.

If we stack two cubes on top of one another, the force on the bottom will be 124.8 lbs. *Pressure* is a force per unit of area. In equation form, this can be expressed as:

$$P = \frac{F}{A} \qquad (8.4)$$

where

P = pressure
F = force
A = area over which the force is distributed

Earlier, we pointed out that pounds per square inch (lbs/in² or psi) and pounds per square foot (lbs/ft²) are common expressions of pressure. The pressure on the bottom of the cube is 62.4 lb/ft². It is normal to express pressure in pounds per square inch (psi). This can be easily accomplished by determining the weight of 1 in² of 1-ft³. If we have a cube that is 12 in on each side, the number of square inches on the bottom surface of the cube is $12 \times 12 = 144$ in². Dividing the weight by the number of square inches determines the weight of each square inch.

$$psi = \frac{62.4 \, lb/ft}{144 \, in^2} = 0.433 \, psi/ft$$

This is the weight of a column of water 1 in² and 1 ft tall. If the column of water were 2 ft tall, the pressure would be 2 ft × 0.433 psi/ft = 0.866.

Note: 1 ft of water = 0.433 psi

With the above information, the feet of the head can be converted to psi by multiplying the feet of the head times 0.433 psi/ft.

Example 8.5

Problem: A tank is mounted at a height of 90 ft. Find the pressure at the bottom of the tank.

Solution:

$$90 \, ft \times 0.433 \, psi/ft = 39 \, psi \, (rounded)$$

Note: To convert psi to feet, divide the psi by 0.433 psi/ft.

Example 8.6

Problem: Find the height of water in a tank if the pressure at the bottom of the tank is 22 psi.

Solution:

$$\text{Height in feet} = \frac{22 \, psi}{0.433 \, psi/ft} = 51 \, ft \, (rounded)$$

Important Point: One of the problems encountered in a hydraulic system is storing the liquid. Unlike air, which is readily compressible and capable of being stored in large quantities in relatively small containers, a liquid such as water cannot be compressed. Therefore, it is not possible to store a large amount of water in a small tank, as 62.4 lb of water occupies a volume of 1 ft³, regardless of the pressure applied to it.

HYDROSTATIC PRESSURE

Figure 8.1 shows a number of differently shaped, connected, open containers of water. Note that the water level is the same in each container, regardless of the shape or size of the container. This occurs because pressure is developed within the water (or any other liquid) by the weight of the water above. If the water level in any one container were to be momentarily higher than that in any of the other containers, the higher pressure at the bottom of this container would cause some water to flow into the container with the lower liquid level. In addition, the pressure of the water at any level (such as Line T) is the same in each of the containers. Pressure increases because of the weight of the water. The farther down from the surface, the more pressure is created. This illustrates that the weight, not the volume, of water contained in a vessel determines the pressure at the bottom of the vessel.

Nathanson (1997) pointed out some very important principles that always apply to hydrostatic pressure:

1. The pressure depends only on the depth of water above the point in question (not on the water surface area).
2. The pressure increases in direct proportion to the depth.

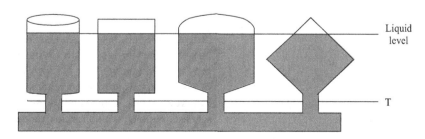

FIGURE 8.1 Hydrostatic pressure.

3. The pressure in a continuous volume of water is the same at all points that are at the same depth.
4. The pressure at any point in the water acts in all directions at the same depth.

Effects of Water under Pressure

Hauser (1995) points out that water under pressure and in motion can exert tremendous forces inside a pipeline. One of these forces, called hydraulic shock or *water hammer*, is the momentary increase in pressure that occurs when there is a sudden change in direction or velocity of the water. When a rapidly closing valve suddenly stops water flowing in a pipeline, pressure energy is transferred to the valve and pipe wall. Shock waves are set up within the system. Waves of pressure move in a horizontal yo-yo fashion—back and forth—against any solid obstacles in the system. Neither the water nor the pipe will compress to absorb the shock, which may result in damage to pipes, valves, and shaking of loose fittings.

Another effect of water under pressure is called thrust. *Thrust* is the force that water exerts on a pipeline as it rounds a bend. As shown in Figure 8.2, thrust usually acts perpendicular (at 90°) to the inside surface it pushes against. It affects not only bends, but also reducers, dead ends, and tees. Uncontrolled, the thrust can cause movement in the fitting or pipeline, which will lead to the separation of the pipe coupling from both sections of the pipeline or at some other nearby coupling upstream or downstream of the fitting.

Key Point: When we dive into the ocean and down every 33 ft (10.06 m), the pressure increases by 1 atm [i.e., 1 atm equals 1,013 millibars or 760 mm (29.92 in) of mercury].

Two types of devices commonly used to control thrust in larger pipelines are thrust blocks and thrust anchors. A *thrust block* is a mass of concrete cast in place onto the pipe and around the outside bend of the turn. An example is shown in Figure 8.3. These are used for pipes with tees or elbows that turn left or right or slant upward. The thrust is transferred to the soil through the larger bearing surface of the block. A *thrust anchor* is a massive block of concrete, often a cube, cast in place below the fitting to be anchored (see Figure 8.4). As shown in Figure 8.4, imbedded steel shackle rods anchor the fitting to the concrete block,

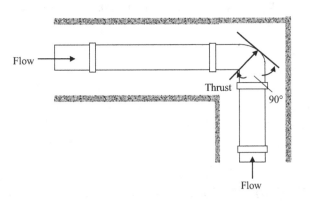

FIGURE 8.2 Direction of thrust in a pipe in a trench (viewed from above).

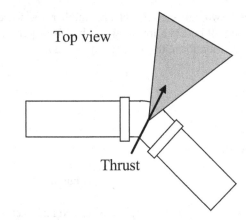

FIGURE 8.3 Thrust block.

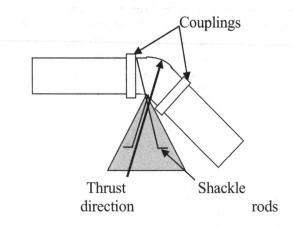

FIGURE 8.4 Thrust anchor.

effectively resisting upward thrust. The size and shape of a thrust control device depend on pipe size, type of fitting, water pressure, water hammer, and soil type.

HEAD

Head is defined as the vertical distance the water/wastewater must be lifted from the supply tank to the discharge, or as the height a column of water would rise due to the pressure at its base. A perfect vacuum plus atmospheric pressure of 14.7 psi would lift the water 34 ft. When the top of the sealed tube is opened to the atmosphere and the reservoir is enclosed, the pressure in the reservoir is increased; thus, the water will rise in the tube. Because atmospheric pressure is essentially universal, we usually ignore the first 14.7 psi of actual pressure measurements and measure only the difference between the water pressure and the atmospheric pressure; we call this *gauge pressure*. For example, water in an open reservoir is subjected to the 14.7 psi of atmospheric pressure; subtracting this 14.7 psi leaves a gauge pressure of 0 psi, indicating that the water would rise 0 ft above the reservoir surface. If the gauge pressure in a water main were 120 psi, the water would rise in a tube connected to the main:

$$120\,\text{psi} \times 2.31\,\text{ft/psi} = 277\,\text{ft (rounded)}$$

The *total head* includes the vertical distance the liquid must be lifted (static head), the loss due to friction (friction head), and the energy required to maintain the desired velocity (velocity head).

Total head = Static head + friction head + velocity head

(8.5)

STATIC HEAD

The static head is the actual **vertical** distance the liquid must be lifted.

Static head = Discharge elevation − supply elevation   (8.6)

Example 8.7

Problem: The supply tank is located at an elevation of 118 ft. The discharge point is at an elevation of 215 ft. What is the static head in feet?

Solution:

Static Head, ft = 215 ft − 118 ft = 97 ft

FRICTION HEAD

Friction head is the equivalent distance of the energy that must be supplied to overcome friction. Engineering references include tables that show the equivalent vertical distance for various sizes and distances for each component.

Friction head, ft = Energy losses due to friction   (8.7)

VELOCITY HEAD

Velocity head is the equivalent distance of the energy consumed in achieving and maintaining the desired velocity in the system.

Velocity head, ft = Energy losses to maintain velocity

(8.8)

TOTAL DYNAMIC HEAD (TOTAL SYSTEM HEAD)

Total head = Static head + friction head + velocity head

(8.9)

PRESSURE AND HEAD

The pressure exerted by water and/or wastewater is directly proportional to its depth or head in the pipe, tank, or channel. If the pressure is known, the equivalent head can be calculated.

Head, ft = Pressure, psi × 2.31 ft/psi   (8.10)

Example 8.8

Problem: The pressure gauge on the discharge line from the influent pump reads 72.3 psi. What is the equivalent head in feet?

Solution:

Head, ft = 72.3 × 2.31 ft/psi = 167 ft

HEAD AND PRESSURE

If the head is known, the equivalent pressure can be calculated by:

$$\text{Pressure, psi} = \frac{\text{Head, ft}}{2.31 \, \text{ft/psi}} \quad (8.11)$$

Example 8.9

Problem: A tank is 22 ft deep. What is the pressure in psi at the bottom of the tank when it is filled with water?

Solution:

$$\text{Pressure, psi} = \frac{22 \, \text{ft}}{2.31 \, \text{ft/psi}} = 9.52 \, \text{psi (rounded)}$$

FLOW AND DISCHARGE RATES: WATER IN MOTION

The study of fluid flow is much more complicated than that of fluids at rest, but it is important to understand these principles because the water in a waterworks and distribution system, as well as in a wastewater treatment plant and collection system is nearly always in motion. *Discharge* (or flow) is the quantity of water passing a given point in a pipe or channel during a given period. Stated another way for open channels, the flow rate through an open channel is directly related to the velocity of the liquid and the cross-sectional area of the liquid in the channel.

$$Q = A \times V \quad (8.12)$$

where

Q = Flow – discharge in cubic feet per second (cfs)
A = Cross-sectional area of the pipe or channel (ft²)
V = water velocity in feet per second (fps or ft/sec)

Example 8.10

Problem: The channel is 6 ft wide, and the water depth is 3 ft. The velocity in the channel is 4 fps. What is the discharge or flow rate in cubic feet per second?

Solution:

$$\text{Flow, cfs} = 6\,\text{ft} \times 3\,\text{ft} \times 4\,\text{ft/sec} = 72\,\text{cfs}$$

Discharge or flow can be recorded as gallons/day (gpd), gallons/minute (gpm), or cubic feet (cfs). Flows treated by many waterworks or wastewater treatment plants are large and often referred to in million gallons per day (MGD). The discharge or flow rate can be converted from cfs to other units such as gallons per minute (gpm) or million gallons per day (MGD) using appropriate conversion factors.

Example 8.11

Problem: A 12-in diameter pipe has water flowing through it at 10 ft/sec. What is the discharge in (a) cfs, (b) gpm, and (c) MGD?

Solution:
Before we can use the basic formula, we must determine the area (A) of the pipe. The formula for the area of a circle is:

$$\text{Area (A)} = \pi \times \frac{D^2}{4} = \pi \times r^2 \qquad (8.13)$$

Note: π is the constant value 3.14159 or simply 3.14.
Where:

> D = diameter of the circle in feet
> r = radius of the circle in feet

Therefore, the area of the pipe is:

$$A = \pi \frac{D^2}{4} = 3.14 \times \frac{(1\,\text{ft})^2}{4} = 0.785\,\text{ft}^2$$

a. Now, we can determine the discharge in cfs:

$$Q = V \times A = 10\,\text{ft/sec} \times 0.785\,\text{ft}^2 = 7.85\,\text{ft}^3/\text{sec or cfs}$$

b. We need to know that 1 cfs is 449 gpm, so 7.85 cfs × 449 gpm/cfs = 3,525 gpm. (rounded)
c. 1 million gallons per day is 1.55 cfs, so:

$$\frac{7.85\,\text{cfs}}{1.55\,\dfrac{\text{cfs}}{\text{MGD}}} = 5.06\,\text{MGD}$$

Note: Flow may be *laminar* (streamline, see Figure 8.5) or *turbulent* (see Figure 8.6). Laminar flow occurs at extremely low velocities. The water moves in straight parallel lines, called streamlines or laminae, which slide upon each other

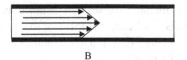

FIGURE 8.5 Laminar (streamline) flow.

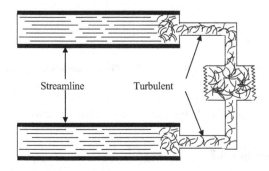

FIGURE 8.6 Turbulent flow.

as they travel, rather than mixing. Normal pipe flow is turbulent flow, which occurs because of friction encountered on the inside of the pipe. The outer layers of flow are thrown into the inner layers; the result is that all the layers mix and move in different directions and at different velocities. However, the direction of flow is forward.

Note: Flow may be steady or unsteady. For our purposes, we consider steady-state flow only; meaning most of the hydraulic calculations in this manual assume steady-state flow.

AREA AND VELOCITY

The *law of continuity* states that the discharge at each point in a pipe or channel is the same as the discharge at any other point (if water does not leave or enter the pipe or channel). That is, under the assumption of steady-state flow, the flow that enters the pipe or channel is the same flow that exits the pipe or channel. In equation form, this becomes

$$Q_1 = Q_2 \quad \text{or} \quad A_1 V_1 = A_2 V_2 \qquad (8.14)$$

Note: In regard to the area/velocity relationship, Equation (8.14) also makes clear that for a given flow rate, the velocity of the liquid varies indirectly with changes in the cross-sectional area of the channel or pipe. This principle provides the basis for many of the flow measurement devices used in open channels (weirs, flumes, and nozzles).

Example 8.12

Problem: A pipe 12-in in diameter is connected to a 6-in-diameter pipe. The velocity of the water in the 12-in pipe is 3 fps. What is the velocity in the 6-in pipe?

Solution:
Using the equation $A_1 V_1 = A_2 V_2$, we need to determine the area of each pipe:

$$12 \text{ in: } A = \pi \times \frac{D^2}{4}$$

$$= 3.14 \times \frac{(1 \text{ ft})^2}{4}$$

$$= 0.785 \text{ ft}^2$$

$$6 \text{ in: } A = 3.14 \times \frac{(0.5)^2}{4}$$

$$= 0.196 \text{ ft}^2$$

The continuity equation now becomes:

$$\left(0.785 \text{ ft}^2\right) \times \left(3 \frac{\text{ft}}{\text{sec}}\right) = \left(0.196 \text{ ft}^2\right) \times V_2$$

Solving for V_2

$$V_2 = \frac{\left(0.785 \text{ ft}^2\right) \times (3 \text{ ft/sec})}{\left(0.196 \text{ ft}^2\right)}$$

$$= 12 \text{ ft/sec or fps}$$

PRESSURE AND VELOCITY

In a closed pipe flowing full (under pressure), the pressure is indirectly related to the velocity of the liquid. This principle, when combined with the principle discussed in the previous section, forms the basis for several flow measurement devices (Venturi meters and rotameters), as well as the injectors used for dissolving chlorine into water and for introducing chlorine, sulfur dioxide, and/or other chemicals into wastewater.

$$\text{Velocity }_1 \times \text{Pressure }_1 = \text{Velocity }_2 \times \text{Pressure }_2$$
$$\text{or} \qquad (8.15)$$
$$V_1 P_1 = V_2 P_2$$

PIEZOMETRIC SURFACE AND BERNOULLI'S THEOREM

To keep the systems in your plant operating properly and efficiently, you must understand the basics of hydraulics—the laws of force, motion, and others. As stated previously, most applications of hydraulics in water/wastewater treatment systems involve water in motion—in pipes under pressure or in open channels under the force of gravity. The volume of water flowing past any given point in the pipe or channel per unit time is called the *flow rate* or *discharge rate*—or just *flow*. The *continuity of flow* and the *continuity equation* have been discussed (i.e., Equation 8.15). Along with the continuity of flow principle and continuity equation, the law of conservation of energy, piezometric surface, and Bernoulli's theorem (or principle) are also important to our study of water hydraulics.

CONSERVATION OF ENERGY

Many of the principles of physics are important to the study of hydraulics. When applied to problems involving the flow of water, few principles of physical science are more important and useful than the *Law of Conservation of Energy*. Simply put, the Law of Conservation of Energy states that energy can neither be created nor destroyed, but it can be converted from one form to another. In a given closed system, the total energy is constant.

ENERGY HEAD

In hydraulic systems, two types of energy (kinetic and potential) and three forms of mechanical energy (potential energy due to elevation, potential energy due to pressure, and kinetic energy due to velocity) exist. Energy is measured in foot-pounds (ft-lbs). It is convenient to express hydraulic energy in terms of *energy head*, in feet of water. This is equivalent to foot-pounds per pound of water (ft-lbs/lb = ft).

PIEZOMETRIC SURFACE

As mentioned earlier, we have seen that when a vertical tube, open at the top, is installed onto a vessel of water, the water will rise in the tube to the water level in the tank. The water level to which the water rises in the tube is called the *piezometric surface*. That is, the piezometric surface is an imaginary surface that coincides with the level to which water in a system would rise in a *piezometer* (an instrument used to measure pressure).

The surface of water that is in contact with the atmosphere is known as the *free water surface*. Many important hydraulic measurements are based on the difference in height between the free water surface and some point in the water system. The piezometric surface is used to locate this free water surface in a vessel where it cannot be observed directly.

To understand how a piezometer measures pressure, consider the following example: If a clear, see-through pipe is connected to the side of a clear glass or plastic vessel, the water will rise in the pipe to indicate the level of the water in the vessel. Such a see-through pipe, the piezometer, allows you to see the level of the top of the water in the pipe; this is the piezometric surface.

In practice, a piezometer is connected to the side of a tank or pipeline. If the water-containing vessel is not under pressure (as is the case in Figure 8.7), the piezometric surface will be the same as the free water surface in the vessel, just as it would be if a drinking straw (the piezometer) were left standing in a glass of water.

When a tank and pipeline system is pressurized, as they often are, the pressure will cause the piezometric surface to rise above the level of the water in the tank. The greater the pressure, the higher the piezometric surface

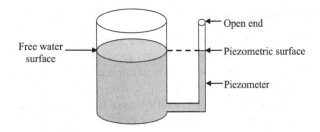

FIGURE 8.7 A container not under pressure where the piezometric surface is the same as the free water surface in the vessel.

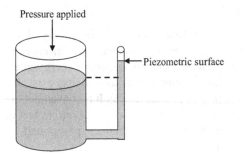

FIGURE 8.8 A container under pressure where the piezometric surface is above the level of the water in the tank.

(see Figure 8.8). Increased pressure in a water pipeline system is usually obtained by elevating the water tank.

Note: In practice, piezometers are not installed on water towers because water towers are hundreds of feet high, nor on pipelines. Instead, pressure gauges are used to record pressure in feet of water or psi.

Water only rises to the water level of the main body of water when it is at rest (static or standing water). The situation is quite different when water is flowing. Consider, for example, an elevated storage tank feeding a distribution system pipeline. When the system is at rest, with all valves closed, all the piezometric surfaces are the same height as the free water surface in storage. On the other hand, when the valves are opened and the water begins to flow, the piezometric surface changes. This is an important point because as water continues to flow down a pipeline, less pressure is

exerted. This happens because some pressure is lost (used up) to keep the water moving over the interior surface of the pipe (friction). The pressure that is lost is called *head loss*.

HEAD LOSS

Head loss is best explained by example. Figure 8.9 shows an elevated storage tank feeding a distribution system pipeline. When the valve is closed (Figure 8.9a), all the piezometric surfaces are at the same height as the free water surface in storage. When the valve opens and water begins to flow (Figure 8.9b), the piezometric surfaces *drop*. The farther along the pipeline, the lower the piezometric surface becomes because some of the pressure is used up to keep the water moving over the rough interior surface of the pipe. Thus, pressure is lost and is no longer available to push water up in a piezometer; this is the head loss.

HYDRAULIC GRADE LINE (HGL)

When the valve shown in Figure 8.9 is opened, flow begins, with a corresponding energy loss due to friction. The pressures along the pipeline can measure this loss. In Figure 8.9b, the difference in pressure heads between sections 1, 2, and 3 can be seen in the piezometer tubes attached to the pipe. A line connecting the water surface in the tank with the water levels at section 1, 2, and 3 shows the pattern of continuous pressure loss along the pipeline. This is called the *Hydraulic Grade Line (HGL)* or *Hydraulic Gradient* of the system.

Note: It is important to point out that in a static water system, the HGL is always horizontal. The HGL is a very useful graphical aid when analyzing pipe flow problems.

Note: During the early design phase of a treatment plant, it is important to establish the hydraulic grade line across the plant because both the proper selection of the plant site elevation and the suitability of the site depend on this consideration. Typically, most conventional water treatment plants require 16–17 ft of headloss across the plant.

Key Point: Changes in the piezometric surface occur when water is flowing.

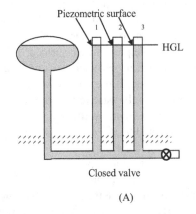

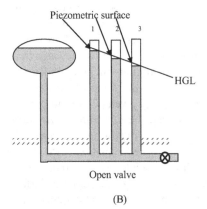

Closed valve	Open valve
(A)	(B)

FIGURE 8.9 Head loss and piezometric surface changes when water is flowing.

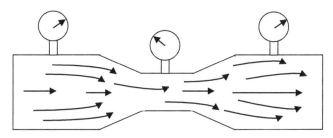

FIGURE 8.10 Bernoulli's principle.

BERNOULLI'S THEOREM

Nathanson (1997) noted that Swiss physicist and mathematician Samuel Bernoulli developed the calculation for the total energy relationship from point to point in a steady-state fluid system in the 1700s. Before discussing Bernoulli's energy equation, it is important to understand the basic principle behind it. Water (and any other hydraulic fluid) in a hydraulic system possesses two types of energy—kinetic and potential. *Kinetic energy* is present when the water is in motion; the faster the water moves, the more kinetic energy is used. *Potential energy* results from the water pressure. The *total energy* of the water is the sum of the kinetic and potential energy. Bernoulli's principle states that the total energy of the water (fluid) always remains constant. Therefore, when the water flow in a system increases, the pressure must decrease. When water starts to flow in a hydraulic system, the pressure drops. When the flow stops, the pressure rises again. The pressure gauges shown in Figure 8.10 indicate this balance more clearly.

Note: This discussion of Bernoulli's equation ignores friction losses from point to point in a fluid system employing steady-state flow.

Note: By using Bernoulli's equation, we make assumptions regarding flow conditions (laminar, steady flow). We can use the principle of conservation of energy to find the relationship between elevation, velocity, and pressure in a pipe at any given point.

BERNOULLI'S EQUATION

In a hydraulic system, the total energy head is equal to the sum of three individual energy heads. This can be expressed as

Total Head = Elevation Head + Pressure Head + Velocity Head

where

> elevation head = pressure due to the elevation of the water
> pressure head = the height of a column of water that a given hydrostatic pressure in a system could support
> velocity head = energy present due to the velocity of the water

This can be expressed mathematically as

$$E = z + \frac{p}{w} + \frac{v^2}{2g} \tag{8.16}$$

where

> E = total energy head
> z = height of the water above a reference plane, ft
> p = pressure, psi
> w = unit weight of water, 62.4 lb/ft³
> v = flow velocity, ft/sec
> g = acceleration due to gravity, 32.2 ft/sec²

Consider the constriction in the section of the pipe shown in Figure 8.11.

We know, based on the law of energy conservation, that the total energy head at section A, E_1, must equal the total energy head at section B, E_2, and using Equation 8.16, we get Bernoulli's equation.

$$z_A = \frac{P_A}{w} + \frac{v_A^2}{2g} = z_B + \frac{P_B}{w} + \frac{V_B^2}{2g} \tag{8.17}$$

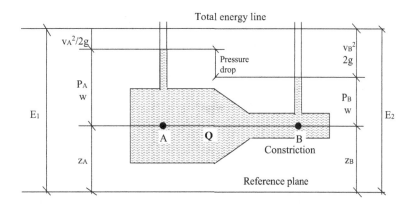

FIGURE 8.11 The law of conservation. Since the velocity and kinetic energy of the water flowing in the constricted section must increase, the potential energy may decrease. This is observed as a pressure drop in the constriction. (Adapted from Nathanson, 1997, p. 29.)

The pipeline system shown in Figure 8.11 is horizontal; therefore, we can simplify Bernoulli's equation because $z_A = z_B$. Since, they are equal, the elevation heads cancel out from both sides, leaving:

$$\frac{P_A}{W} + \frac{v_A^2}{2g} + \frac{P_B}{w} + \frac{v_B^2}{2g} \qquad (8.18)$$

As water passes through the constricted section of the pipe (section B), we know from the continuity of flow that the velocity at section B must be greater than the velocity at section A due to the smaller flow area at section B. This means that the velocity head in the system increases as the water flows into the constricted section. However, the total energy must remain constant. For this to occur, the pressure head, and therefore the pressure, must drop. In effect, pressure energy is converted into kinetic energy in the constriction.

The fact that the pressure in the narrower pipe section (constriction) is less than the pressure in the bigger section seems to defy common sense. However, it does follow logically from continuity of flow and conservation of energy. The fact that there is a pressure difference allows the measurement of the flow rate in the closed pipe.

Example 8.13

Problem: In Figure 8.11, the diameter at Section A is 8 in and at Section B, it is 4 in. The flow rate through the pipe is 3.0 cfs, and the pressure at Section A is 100 psi. What is the pressure in the constriction at Section B?

Solution:
Compute the flow area at each section, as follows:

$$A_A = \frac{\pi(0.666\,\text{ft})^2}{4} = 0.349\,\text{ft}^2 \text{ (rounded)}$$

$$A_B = \frac{\pi(0.333\,\text{ft})^2}{4} = 0.087\,\text{ft}^2$$

From $Q = A \times V$ or $V = Q/A$, we get

$$V_A = \frac{3.0\,\text{ft}^3/\text{sec}}{0.349\,\text{ft}^2} = 8.6\,\text{ft/sec (rounded)}$$

$$V_B = \frac{3.0\,\text{ft}^3/\text{sec}}{0.087\,\text{ft}^2} = 34.5\,\text{ft/sec (rounded)}$$

And we get:

$$\frac{100 \times 144}{62.4} + \frac{8.6^2}{2 \times 32.2} = \frac{p_B \times 144}{62.4} + \frac{34.5^2}{2 \times 32.2}$$

Note: The pressures are multiplied by 144 in²/ft² to convert from psi to lb/ft² to be consistent with the units for w; the energy head terms are in feet of head.

Continuing, we get

$$231 + 1.15 = 2.3p_B + 18.5$$

and

$$p_B = \frac{232.2 - 18.5}{2.3} = \frac{213.7}{2.3} = 93\,\text{psi (rounded)}$$

WELL AND WET WELL HYDRAULICS

When the source of water for a water distribution system is from a groundwater supply, knowledge of well hydraulics is important to the operator. Basic well hydraulics terms are presented and defined, and they are related pictorially (see Figure 8.12). Wet wells, which are important in both water and wastewater operations are also discussed.

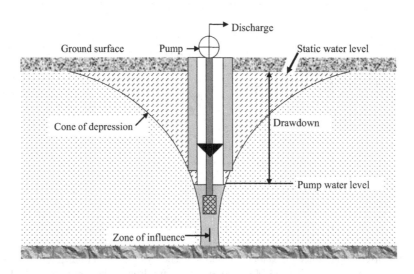

FIGURE 8.12 Hydraulic characteristics of a well.

Well Hydraulics

- **Static Water Level:** the water level in a well when no water is being taken from the groundwater source (i.e., the water level when the pump is off; see Figure 8.12). The static water level is normally measured as the distance from the ground surface to the water surface. This is an important parameter because it is used to measure changes in the water table.
- **Pumping Water Level:** the water level when the pump is off. When water is pumped out of a well, the water level usually drops below the level in the surrounding aquifer and eventually stabilizes at a lower level; this is the pumping level (see Figure 8.12).
- **Drawdown:** the difference, or the drop, between the Static Water Level and the Pumping Water Level, measured in feet. Simply, it is the distance the water level drops once pumping begins (see Figure 8.12).
- **Cone of Depression:** in unconfined aquifers, there is a flow of water in the aquifer from all directions toward the well during pumping. The free water surface in the aquifer then takes the shape of an inverted cone or curved funnel line. The curve of the line extends from the Pumping Water Level to the Static Water Level at the outside edge of the Zone (or Radius) of Influence (see Figure 8.12).

Note: The shape and size of the cone of depression depends on the relationship between the pumping rate and the rate at which water can move toward the well. If the rate is high, the cone will be shallow and its growth will stabilize. If the rate is low, the cone will be sharp and continue to grow in size.

- **Zone (or Radius) of Influence:** the distance between the pump shaft and the outermost area affected by drawdown (see Figure 8.12). The distance depends on the porosity of the soil and other factors. This parameter becomes important in well fields with many pumps. If wells are set too close together, the zones of influence will overlap, increasing the drawdown in all wells. Pumps should obviously be spaced apart to prevent this from happening.

Two important parameters not shown in Figure 8.12 are well yield and specific capacity:

1. *Well yield* is the rate of water withdrawal that a well can supply over a long period, or the maximum pumping rate that can be achieved without increasing the drawdown. The yield of small wells is usually measured in gallons per minute (liters per minute) or gallons per hour (liters per hour). For large wells, it may be measured in cubic feet per second (cubic meters per second).

2. *Specific capacity* is the pumping rate per foot of drawdown (gpm/ft), or

$$\text{Specific capacity} = \text{Well yield} \div \text{drawdown} \quad (8.19)$$

Example 8.14

Problem: If the well yield is 300 gpm, and the drawdown is measured to be 20 ft, what is the specific capacity?

Solution:

$$\text{Specific Capacity} = 300 \div 20$$

$$\text{Specific Capacity} = 15 \, \text{gpm/ft of drawdown}$$

Specific capacity is one of the most important concepts in well operation and testing. The calculation should be made frequently in the monitoring of well operation. A sudden drop in specific capacity indicates problems such as pump malfunction, screen plugging, or other problems that can be serious. Such problems should be identified and corrected as soon as possible.

Wet Well Hydraulics

Water pumped from a wet well by a pump set above the water surface exhibits the same phenomena as groundwater wells. In operation, a slight depression of the water surface forms right at the intake line (drawdown), but in this case, it is minimal because there is free water at the pump entrance at all times (at least there should be). The most important consideration in wet well operations is to ensure that the suction line is submerged far enough below the surface so that air entrained by the active movement of the water at this section is not able to enter the pump. Because water or wastewater flow is not always constant or at the same level, variable speed pumps are commonly used in wet well operations, or several pumps are installed for single or combined operations. In many cases, pumping is accomplished in an on/off mode. Control of pump operation is in response to the water level in the well. Level control devices such as mercury switches are used to sense high and low levels in the well and transmit the signal to pumps for action.

FRICTION HEAD LOSS

Materials or substances capable of flowing cannot flow freely. Nothing flows without encountering some type of resistance. Consider electricity: the flow of free electrons in a conductor. Whatever type of conductor is used (i.e., copper, aluminum, silver, etc.) offers some resistance. In hydraulics, the flow of water/wastewater is analogous to the flow of electricity. Within a pipe or open channel, for instance, flowing water, like electron flow in a conductor, encounters resistance. However, resistance to the flow of water is generally termed *friction loss* (or, more appropriately, head loss).

FLOW IN PIPELINES

The problem of waste/wastewater flow in pipelines—the prediction of flow rate through pipes of given characteristics, the calculation of energy conversions therein, and so forth—is encountered in many applications of water/wastewater operations and practice. Although the subject of pipe flow embraces only those problems in which pipes flow full (as in water lines), we also address pipes that flow partially full (wastewater lines, normally treated as open channels) in this section.

Also discussed is the solution of practical pipe flow problems resulting from the application of the energy principle; the equation of continuity; and the principle and equation of water resistance. Resistance to flow in pipes is not only the result of long reaches of pipe but is also caused by pipe fittings, such as bends and valves, which dissipate energy by producing relatively large-scale turbulence.

To gain an understanding of what friction head loss is all about, it is necessary to review a few terms presented earlier in the text and to introduce some new terms pertinent to the subject (Lindeburg, 1986):

- **Laminar and Turbulent Flow:** Laminar flow is ideal flow; that is, water particles move along straight, parallel paths in layers or streamlines. Moreover, in laminar flow, there is no turbulence in the water and no friction loss. This is not typical of normal pipe flow because the water velocity is too great but is typical of groundwater flow. Turbulent flow (characterized as "normal" for a typical water system) occurs when water particles move in a haphazard fashion and continually cross each other in all directions resulting in pressure losses along a length of pipe.
- **Hydraulic Grade Line (HGL):** Recall that the hydraulic grade line (HGL) (shown in Figure 8.13) is a line connecting two points to which the liquid would rise at various places along any pipe or open channel if piezometers were inserted in the liquid. It is a measure of the pressure head available at these various points.

Note: When water flows in an open channel, the HGL coincides with the profile of the water surface.

- **Energy Grade Line:** The total energy of flow in any section concerning some datum (i.e., a reference line, surface, or point) is the sum of the elevation head z, the pressure head y, and the velocity head $V^2/2g$. Figure 8.13 shows the energy grade line or energy gradient, which represents the energy from section to section. In the absence of frictional losses, the energy grade line remains horizontal, although the relative distribution of energy may vary between the elevation, pressure, and velocity heads. In all real systems, however, losses of energy occur because of resistance to flow, and the resulting energy grade line is sloped (i.e., the energy grade line is the slope of the specific energy line).
- **Specific Energy (E):** Sometimes called specific head, it is the sum of the pressure head y and the velocity head $V^2/2g$. The specific energy concept is especially useful in analyzing flow in open channels.
- **Steady Flow:** Occurs when the discharge or rate of flow at any cross-section is constant.
- **Uniform and Non-uniform Flow:** Uniform flow occurs when the depth, cross-sectional area, and other elements of flow are substantially constant from section to section. Non-uniform flow occurs when the slope, cross-sectional area, and velocity change from section to section. The flow through a Venturi section used for measuring flow is a good example.
- **Varied Flow:** Flow in a channel is considered varied if the depth of flow changes along the length of the channel. The flow may be gradually varied or rapidly varied (i.e., when the depth of flow changes abruptly), as shown in Figure 8.14.
- **Slope (Gradient):** The head loss per foot of channel.

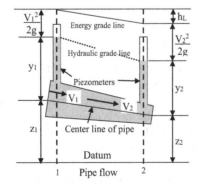
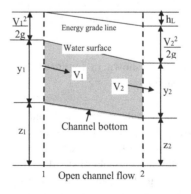

FIGURE 8.13 Comparison of pipe flow and open-channel flow. (Adapted from Metcalf & Eddy, 1981, p. 11.)

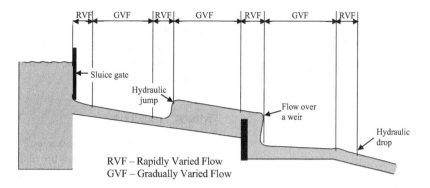

RVF – Rapidly Varied Flow
GVF – Gradually Varied Flow

FIGURE 8.14 Varied flow.

MAJOR HEAD LOSS

Major head loss consists of pressure decreases along the length of the pipe caused by friction created as water encounters the surfaces of the pipe. It typically accounts for most of the pressure drop in a pressurized or dynamic water system. The components that contribute to major head loss are roughness, length, diameter, and velocity:

1. **Roughness**

    Even when new, the interior surfaces of pipes are rough. The roughness varies, of course, depending on pipe material, corrosion (tuberculation and pitting), and age. Because normal flow in a water pipe is turbulent, the turbulence increases with pipe roughness, which, in turn, causes pressure to drop over the length of the pipe.
2. **Pipe Length**

    With every foot of pipe length, friction losses occur. The longer the pipe, the more head loss. Friction loss because of pipe length must be factored into head loss calculations.
3. **Pipe Diameter**

    Generally, small-diameter pipes have more head loss than large-diameter pipes. This is the case because, in large-diameter pipes, less of the water actually touches the interior surfaces of the pipe (encountering less friction) than in a small-diameter pipe.
4. **Water Velocity**

    Turbulence in a water pipe is directly proportional to the speed (or velocity) of the flow. Thus, the velocity head also contributes to head loss.

Note: For the same diameter pipe, when flow increases, head loss increases.

CALCULATING MAJOR HEAD LOSS

Darcy, Weisbach, and others developed the first practical equation used to determine pipe friction in about 1850. The equation or formula now known as the *Darcy-Weisbach* equation for circular pipes (valid for both laminar and turbulent flow in a pipe) is:

$$h_f = f \frac{LV^2}{D2g} \tag{8.20}$$

In terms of the flow rate Q, the equation becomes:

$$h_f = \frac{8fLQ^2}{\pi^2 g D^5} \tag{8.21}$$

where

h_f = head loss, (ft)
f = coefficient of friction
L = length of pipe, (ft)
V = mean velocity, (ft/sec)
D = diameter of pipe, (ft)
g = acceleration due to gravity, (32.2 ft/sec²)
Q = flow rate, (ft³/sec)

The Darcy-Weisbach formula was meant to apply to the flow of any fluid and into this friction factor was incorporated the degree of roughness and an element called the *Reynold's Number*, which was based on the viscosity of the fluid and the degree of turbulence of flow. The Darcy-Weisbach formula is used primarily for determining head loss calculations in pipes. For making this determination in open channels, the *Manning Equation* was developed during the later part of the nineteenth century. Later, this equation was used for both open channels and closed conduits.

In the early 1900s, a more practical equation, the *Hazen-Williams* equation, was developed for use in making calculations related to water pipes and wastewater force mains:

$$Q = 0.435 \times CD^{2.63} \times S^{0.54} \tag{8.22}$$

where

Q = flow rate, (ft³/sec)
C = coefficient of roughness (C decreases with roughness)
D = hydraulic radius R, (ft)
S = slope of energy grade line, (ft/ft)

TABLE 8.2
C Factors (Lindeburg, 1986)

Type of Pipe	C Factor
Asbestos cement	140
Brass	140
Brick sewer	100
Cast iron	
10 years old	110
20 years old	90
Ductile iron, (cement lined)	140
Concrete or concrete lined	
Smooth, steel forms	140
Wooden forms	120
Rough	110
Copper	140
Fire hose (rubber lined)	135
Galvanized iron	120
Glass	140
Lead	130
Masonry conduit	130
Plastic	150
Steel	
Coal-tar enamel lined	150
New unlined	140
Riveted	110
Tin	130
Vitrified	120
Wood stave	120

C FACTOR

The *C factor,* as used in the Hazen-Williams formula, designates the coefficient of roughness. C does not vary appreciably with velocity, and by comparing pipe types and ages, it includes only the concept of roughness, ignoring fluid viscosity and Reynold's Number. Based on experience (experimentation), accepted tables of C factors have been established for pipe (see Table 8.2). Generally, the C factor decreases by one with each year of pipe age. Flow for a newly designed system is often calculated with a C factor of 100, based on averaging it over the life of the pipe system.

Note: A high C factor means a smooth pipe. A low C factor means a rough pipe.

Note: An alternative to calculating the Hazen-Williams formula, called an alignment chart, has become quite popular for fieldwork. The alignment chart can be used with reasonable accuracy.

SLOPE

The slope is defined as the head loss per foot. In open channels, where the water flows by gravity, slope is the amount of incline of the pipe, and is calculated as feet of drop per foot of pipe length (ft/ft). The slope is designed to be just enough to overcome frictional losses so that the velocity remains constant, the water keeps flowing, and solids do not settle in the conduit. In piped systems, where pressure loss for every foot of pipe is experienced, the slope is not provided by slanting the pipe but instead by pressure added to overcome friction.

MINOR HEAD LOSS

In addition to the head loss caused by friction between the fluid and the pipe wall, losses are also caused by turbulence created by obstructions (i.e., valves and fittings of all types) in the line, changes in direction, and changes in flow area.

Note: In practice, if minor head loss is less than 5% of the total head loss, it is usually ignored.

BASIC PIPING HYDRAULICS

Water, regardless of the source, is conveyed to the waterworks for treatment and distributed to the users. Conveyance from the source to the point of treatment occurs through aqueducts, pipelines, or open channels, but the treated water is normally distributed in pressurized closed conduits. After use, regardless of the purpose, the water becomes wastewater, which must be disposed of somehow; however, it almost always ends up being conveyed back to a treatment facility before being discharged to some water body, beginning the cycle again. We call this the urban water cycle because it provides a human-generated imitation of the natural water cycle. Unlike the natural water cycle, however, without pipes, the cycle would be non-existent or, at the very least, short-circuited.

For use as water mains in a distribution system, pipes must be strong and durable in order to resist applied forces and corrosion. The pipe is subjected to internal pressure from the water and external pressure from the weight of the backfill (soil) and vehicles above it. The pipe may also have to withstand water hammer. Damage due to corrosion or rusting may occur internally because of the water quality or externally because of the nature of the soil conditions.

Pipes used in a wastewater system must be strong and durable to resist the abrasive and corrosive properties of the wastewater. Like water pipes, wastewater pipes must also be able to withstand stresses caused by the soil backfill material and the effects of vehicles passing above the pipeline. Joints between wastewater collection/interceptor pipe sections should be flexible but tight enough to prevent excessive leakage, either of sewage out of the pipe or groundwater into the pipe. Of course, pipes must be constructed to withstand the expected conditions of exposure, and pipe configuration systems for water distribution and/or wastewater collection and interceptor systems must be properly designed and installed in terms of water hydraulics. Because the water and wastewater operator should have a basic knowledge of water hydraulics related to commonly used standard piping configurations, piping basics are briefly discussed in this section.

Piping Networks

It would be far less costly and make for more efficient operation if municipal water and wastewater systems were built with separate single-pipe networks extending from the treatment plant to user's residence, or from user's sink or bathtub drain to the local wastewater treatment plant. Unfortunately, this ideal single-pipe scenario is not practical for real-world applications. Instead of a single piping system, a network of pipes is laid under the streets. Each of these piping networks is composed of different materials that vary (sometimes considerably) in diameter, length, and age. These networks range in complexity to varying degrees, and each of these joined-together pipes contributes energy losses to the system.

Water and wastewater flow networks may consist of pipes arranged in series, parallel, or some complicated combination. In any case, an evaluation of friction losses for the flows is based on energy conservation principles applied to the flow junction points. Methods of computation depend on the particular piping configuration. In general, however, they involve establishing a sufficient number of simultaneous equations or employing a friction loss formula where the friction coefficient depends only on the roughness of the pipe (e.g., Hazen-Williams Equation—Equation 8.22).

Note: Demonstrating the procedure for making these complex computations is beyond the scope of this text. We only present the operator's "need to know" aspects of complex or compound piping systems in this text.

When two pipes of different sizes or roughness's are connected in series (see Figure 8.15), head loss for a given discharge, or discharge for a given head loss, may be calculated by applying the appropriate equation between the bonding points, considering all losses in the interval. Thus, head losses are cumulative. Series pipes may be treated as a single pipe of constant diameter to simplify the calculation of friction losses. The approach involves determining an "equivalent length" of a constant diameter pipe that has the same friction loss and discharge characteristics as the actual series pipe system. In addition, the application of the continuity equation to the solution allows the head loss to be expressed in terms of only one pipe size.

Note: In addition to the head loss caused by friction between the water and the pipe wall, losses are also caused by minor losses: obstructions in the line, changes in direction, and changes in flow area. In practice, the method of equivalent length is often used to determine these losses. The method of equivalent length uses a table to convert each valve or fitting into an equivalent length of straight pipe.

The solution of pipes in series is fairly straightforward and uncomplicated. Moreover, when making calculations involving pipes in series, remember these two important basic operational tenets:

1. The same flow passes through all pipes connected in series.
2. The total head loss is the sum of the head losses of all the component pipes.

In some operations involving series networks where the flow is given and the total head loss is unknown, we can use the Hazen-Williams formula to solve for the slope and the head loss of each pipe as if they were separate pipes. Adding up the head losses to get the total head loss is then a simple matter.

Other series network calculations may not be as simple to solve using the Hazen-Williams Equation. For example, one problem we may face is what diameter to use with varying-sized pipes connected in a series combination. Moreover, the head loss is applied to both pipes (or other multiples), and it is not known how much loss originates from each one; thus, determining the slope would be difficult—but not impossible.

In such cases, the equivalent pipe theory, as mentioned earlier, can be used. Again, one single "Equivalent Pipe" is created that will carry the correct flow. This is practical because the head loss through it is the same as that in the actual system. The equivalent pipe can have any C factor and diameter, as long as those same dimensions are maintained through to the end. Keep in mind that the equivalent pipe must have the correct length so that it will allow the correct flow through, which yields the correct head loss (the given head loss) (Lindeburg, 1986).

Two or more pipes are connected (as in Figure 8.16) so that flow is first divided among the pipes and is then rejoined to comprise a parallel pipe system. A parallel pipe system is a common method for increasing the capacity of an existing line. Determining flows in pipes arranged in parallel is also made by applying energy conservation principles—specifically, energy losses through all pipes connecting common junction points must be equal. Each leg of the parallel network is treated as a series piping system and converted to a single equivalent-length pipe. The friction losses through the equivalent-length parallel pipes are then considered

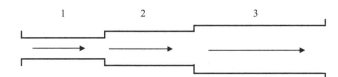

FIGURE 8.15 Pipes in series.

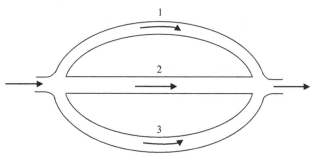

FIGURE 8.16 Pipes in parallel.

equal, and the respective flows are determined by proportional distribution.

Note: Computations used to determine friction losses in parallel combinations may be accomplished using a simultaneous solution approach for a parallel system that has only two branches. However, if the parallel system has three or more branches, a modified procedure using the Hazen-Williams loss formula is easier.

Key Point: It may be helpful to remember the difference between pipes in series and pipes in parallel by considering that in series the same discharge passes through all the pipes, as shown below:

$$Q_1 = Q_2 = Q_3$$

And in parallel as shown below:

$$Q = Q_1 + Q_2 + Q_3$$

OPEN-CHANNEL FLOW

Water is transported over long distances through aqueducts to locations where it is to be used and/or treated. The selection of an aqueduct type rests on factors such as topography, head availability, climate, construction practices, economics, and water quality protection. Along with pipes and tunnels, aqueducts may also include or be solely composed of open channels (Lindeburg, 1986). In this section, we deal with water passage in open channels, which allow part of the water to be exposed to the atmosphere. This type of channel—an open-flow channel—includes natural waterways (streams and rivers), irrigation channels, navigation channels, sewers, canals, culverts, flumes, and pipes flowing under the influence of gravity.

CHARACTERISTICS OF OPEN CHANNEL FLOW

McGhee (1991) pointed out that basic hydraulic principles apply in free surface flow or open channel flow (with water depth constant), although there is no pressure to act as the driving force. Velocity head is the only natural energy this water possesses, and at normal water velocities, this is a small value ($V^2/2g$). Several parameters can be (and often are) used to describe open-channel flow. However, we begin our discussion with a few characteristics, including laminar or turbulent, uniform, or varied, and subcritical, critical, or supercritical.

Laminar and Turbulent Flow

Laminar and *turbulent* flow in open channels is analogous to that in closed pressurized conduits (i.e., pipes). It is important to point out, however, that flow in open channels is usually turbulent. In addition, there is no important circumstance in which laminar flow occurs in open channels in either water or wastewater unit processes or structures.

Uniform and Varied Flow

Flow can be a function of time and location. If the flow quantity is invariant, it is said to be steady. *Uniform* flow is flow in which the depth, width, and velocity remain constant along a channel. That is, if the flow cross-section does not depend on the location along the channel, the flow is said to be uniform. *Varied* or *non-uniform* flow involves a change in these parameters, with a change in one producing a change in the others. Most circumstances of open-channel flow in water and wastewater systems involve varied flow. The concept of uniform flow is valuable, however, in that it defines a limit that the varied flow may be considered to be approaching in many cases.

Note: Uniform channel construction does not ensure uniform flow.

Critical Flow

Critical flow (i.e., flow at the critical depth and velocity) defines a state of flow between two flow regimes. Critical flow coincides with minimum specific energy for a given discharge and maximum discharge for a given specific energy. It occurs in flow measurement devices at or near free discharges and establishes controls in open-channel flow. Critical flow occurs frequently in water/wastewater systems and is very important in their operation and design.

Note: Critical flow minimizes specific energy and maximizes discharge.

PARAMETERS USED IN OPEN-CHANNEL FLOW

The three primary parameters used in open channel flow are *hydraulic radius, hydraulic depth*, and *slope* (S).

Hydraulic Radius

The *hydraulic radius* is the ratio of the area in flow to the wetted perimeter.

$$r_H = \frac{A}{P} \tag{8.23}$$

where

r_H = hydraulic radius
A = the cross-sectional area of the water
P = wetted perimeter

Why is hydraulic radius important? Good question.

Probably the best way to answer this question is by illustration. Consider, for example, that in open channels it is of primary importance to maintain the proper velocity. This is the case, of course, because if velocity is not maintained then flow stops (theoretically). To maintain velocity at a constant level, the channel slope must be adequate to overcome friction losses. As with other flows, the calculation of head loss at a given flow is necessary, and the Hazen-Williams Equation is useful ($Q = 0.435 \times C \times d^{2.63} \times S^{0.54}$). Keep in mind that the concept of slope has not changed. The difference?

We are now measuring, or calculating for, the physical slope of a channel (ft/ft), equivalent to head loss.

The preceding seems logical and makes sense—but there is a problem. The problem is with the diameter. In conduits that are not circular (grit chambers, contact basins, streams, and rivers) or in pipes that are only partially full (drains, wastewater gravity mains, sewers, etc.), where the cross-sectional area of the water is not circular, there is no diameter. Without a diameter, what do we do? Another good question.

Because we do not have a diameter in situations where the cross-sectional area of the water is not circular, we must use another parameter to designate the size of the cross-section and the amount of it that contacts the sides of the conduit. This is where the hydraulic radius (r_H) comes in. The hydraulic radius is a measure of the efficiency with which the conduit can transmit water. Its value depends on pipe size and the amount of fullness. Simply put, we use the hydraulic radius to measure how much of the water is in contact with the sides of the channel or how much of the water is not in contact with the sides (see Figure 8.17).

Note: For a circular channel flowing either full or half-full, the hydraulic radius is (D/4). Hydraulic radii of other channel shapes are easily calculated from the basic definition.

Hydraulic Depth

The *hydraulic depth* is the ratio of the area in flow to the width of the channel at the fluid surface (note that another name for hydraulic depth is the *hydraulic mean depth* and *hydraulic radius*):

$$d_H = \frac{A}{w} \tag{8.24}$$

where

d_H=hydraulic depth
A=area in flow
w=width of the channel at the fluid surface

Slope, S

The *slope* (S) in open channel equations is the slope of the energy line. If the flow is uniform, the slope of the energy line will parallel the water surface and channel bottom.

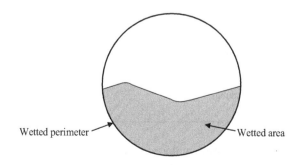

FIGURE 8.17   Hydraulic radius.

In general, the slope can be calculated from the Bernoulli equation as the energy loss per unit length of channel:

$$S = \frac{Dh}{Dl} \tag{8.25}$$

As mentioned, the calculation for head loss at a given flow is typically accomplished by using the Hazen-Williams Equation. In addition, in open-channel flow problems, although the concept of slope has not changed, the problem arises with the diameter. Again, in pipes that are only partially full where the cross-sectional area of the water is not circular, there is no diameter. Thus, the hydraulic radius is used for these non-circular areas. In the original version of the Hazen-Williams Equation, the hydraulic radius was incorporated. Moreover, similar versions developed by Chezy (pronounced "Shay-zee") and Manning, among others, incorporated the hydraulic radius. For use in open channels, Manning's formula has become the most commonly used:

$$Q = \frac{1.5}{n} A \times R^{0.66} \times s^{0.5} \tag{8.26}$$

where

Q=channel discharge capacity (ft³/sec)
1.5=constant
n=channel roughness coefficient
A=cross-sectional flow area (ft²)
R=hydraulic radius of the channel (ft)
S=slope of the channel bottom, dimensionless

The hydraulic radius of a channel is defined as the ratio of the flow area to the wetted perimeter P. In formula form, R=A/P. The new component is n (the roughness coefficient), which depends on the material and age of a pipe or lined channel and topographic features for a natural streambed. It approximates roughness in open channels and can range from a value of 0.01 for a smooth clay pipe to 0.1 for a small natural stream. The value of n commonly assumed for concrete pipes or lined channels is 0.013. n values decrease as the channels get smoother (see Table 8.3). The following example illustrates the application of Manning's formula for a channel with a rectangular cross-section.

Note: Manning's roughness coefficient, usually denoted (n), is a crucial parameter used in hydraulic calculations for open channels.

Example 8.15

Problem: A rectangular drainage channel is 3 ft wide and is lined with concrete, as illustrated in Figure 8.18. The bottom of the channel drops in elevation at 0.5/100 ft. What is the discharge in the channel when the water depth is 2 ft?

Solution:
    Assume n=0.013

TABLE 8.3
Manning Roughness Coefficient, n

Type of Conduit	N	Type of Conduit	N
		Pipe	
Cast iron, coated	0.012–0.014	Cast iron, uncoated	0.013–0.015
Wrought iron, galvanized	0.015–0.017	Wrought iron, black	0.012–0.015
Steel, riveted, and spiral	0.015–0.017	Corrugated	0.021–0.026
Wood stave	0.012–0.013	Cement surface	0.010–0.013
Concrete	0.012–0.017	Vitrified	0.013–0.015
Clay, drainage tile	0.012–0.014		
		Lined Channels	
Metal, smooth semicircular	0.011–0.015	Metal, corrugated	0.023–0.025
Wood, planed	0.010–0.015	Wood, unplaned	0.011–0.015
Cement lined	0.010–0.013	Concrete	0.014–0.016
Cement rubble	0.017–0.030	Grass	------0.020
		Unlined Channels	
Earth: straight and uniform	0.017–0.025	Earth: dredged	0.025–0.033
Earth: winding	0.023–0.030	Earth: stony	0.025–0.040
Rock: smooth and uniform	0.025–0.035	Rock: jagged and irregular	0.035–0.045

Referring to Figure 8.19, we see the cross-sectional flow area A=3 ft×2 ft=6 ft², and the wetted perimeter P=2 ft+3 ft+2 ft=7 ft. The hydraulic radius R=A/P=6 ft²/7 ft=0.86 ft. The slope S=0.5/100=0.005. Applying Manning's formula, we get:

$$Q = \frac{2.0}{0.013} \times 6 \times 0.86^{0.66} \times 0.005^{0.5}$$

$$Q = 59\,cfs$$

To this point, we have set the stage for explaining (in the simplest possible way) what open-channel flow is—what it is all about. Thus, now that we have explained the necessary foundational material and important concepts, we are ready to explain open-channel flow in a manner that it will be easily understood.

We stated that when water flows in a pipe or channel with a **free surface** exposed to the atmosphere, it is called *open-channel flow*. We also know that gravity provides the motive force, the constant push, while friction resists the motion and causes energy expenditure. River and stream flow is open channel flow. Flow in sanitary sewers and stormwater drains is open-channel flow, except in force mains where the water is pumped under pressure.

The key to solving stormwater and/or sanitary sewer routine problems is a condition known as steady uniform flow; that is, we assume steady uniform flow. Steady flow, of course, means that the discharge is constant over time. Uniform flow means that the slope of the water surface and the cross-sectional flow area are also constant. It is common practice to call a length of channel, pipeline, or stream with a relatively constant slope and cross section a *reach* (Nathanson, 1997).

Under steady uniform flow conditions, the slope of the water surface, is the same as the slope of the channel bottom. The hydraulic grade line (HGL) lies along the water surface, and as in pressure flow in pipes, the HGL slopes downward in the direction of flow. Energy loss is evident as the water surface elevation drops. Figure 8.19 illustrates a typical profile view of uniform steady flow. The slope of the water surface represents the rate of energy loss.

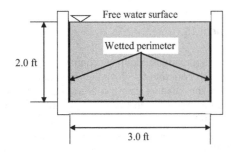

FIGURE 8.18   Illustration of Example 8.15.

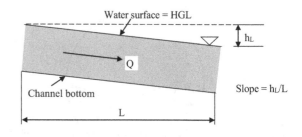

FIGURE 8.19   Steady uniform open-channel flow, where the slope of the water surface (or HGL) is equal to the slope of the channel bottom.

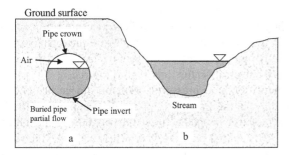

FIGURE 8.20 Open-channel flow, whether in a surface stream or an underground pipe. (Adapted from Nathanson, 1997, p. 35.)

Note: The rate of energy loss (see Figure 8.19) may be expressed as the ratio of the drop in elevation of the surface in the reach to the length of the reach.

Figure 8.20 shows typical cross-sections of open-channel flow. In Figure 8.20, the pipe is only partially filled with water, and there is a free surface at atmospheric pressure. This is still an open-channel flow, although the pipe is a closed underground conduit. Remember, the important point is that gravity, not a pump, is moving the water.

FLOW MEASUREMENT

Although it is clear that maintaining water and wastewater flow is at the heart of any treatment process, it is equally clear that the measurement of flow is essential to ensuring the proper operation of a water and wastewater treatment system. Few knowledgeable operators would argue with this statement. Hauser (1995) asks: "Why measure flow?" She explains, "The most vital activities in the operation of water and wastewater treatment plants are dependent on a knowledge of how much water is being processed." In the statement above, Hauser makes clear that flow measurement is not only important but also routine in water/wastewater operations. Routine, yes, but also the most important variable measured in a treatment plant. Hauser also pointed out that there are several reasons to measure flow in a treatment plant. The American Water Works Association (AWWA, 1995) lists several additional reasons to measure flow:

- The flow rate through the treatment processes needs to be controlled so that it matches distribution system use.
- It is important to determine the proper feed rate of chemicals added in the processes.
- The detention times through the treatment processes must be calculated. This is particularly applicable to surface water plants that must meet C×T values required by the Surface Water Treatment Rule.
- Flow measurement allows operators to maintain a record of water furnished to the distribution system for periodic comparison with the total water metered to customers. This provides a measure of "water accounted for" or, conversely (as pointed out earlier by Hauser), the amount of water wasted, leaked, or otherwise not paid for; that is, lost water.

- Flow measurement allows operators to determine the efficiency of pumps. Pumps that are not delivering their designed flow rate are probably not operating at maximum efficiency, and so power is being wasted.
- For well systems, it is very important to maintain records of the volume of water pumped and the hours of operation for each well. The periodic computation of well pumping rates can identify problems such as worn pump impellers and blocked well screens.
- Reports that must be furnished to the state by most water systems must include records of raw and finished water pumpage.
- Wastewater generated by a treatment system must also be measured and recorded.
- Individual meters are often required for the proper operation of individual pieces of equipment. For example, the makeup water to a fluoride saturator is always metered to assist in tracking the fluoride feed rate.

Note: Simply put, the measurement of flow is essential for the operation, process control, and recordkeeping of water and wastewater treatment plants.

All of the uses just discussed create an obvious need for a variety of flow-measuring devices, often with different capabilities. In this section, we will discuss many of the major flow-measuring devices currently used in water/wastewater operations.

FLOW MEASUREMENT: THE OLD-FASHIONED WAY

An approximate but very simple method to determine open-channel flow has been used for many years. The procedure involves measuring the velocity of a floating object moving in a straight uniform reach of the channel or stream. If the cross-sectional dimensions of the channel are known and the depth of flow is measured, the flow area can be computed. From the relationship Q=A×V, the discharge Q can be estimated. In preliminary fieldwork, this simple procedure is useful in obtaining a ballpark estimate for the flow rate but is not suitable for routine measurements.

Example 8.16

Problem: A floating object is placed on the surface of water flowing in a drainage ditch and is observed to travel a distance of 20 m downstream in 30 sec. The ditch is 2 m wide, and the average depth of flow is estimated to be 0.5 m. Estimate the discharge under these conditions.

Solution:
The flow velocity is computed as distance over time, or

$$V = D/T = 20\,m/30\,sec = 0.67\,m/sec$$

The channel area is A=2 m×0.5 m=1.0 m². The discharge Q=A×V=1.0 m²×0.66 m²=0.66 m³/sec.

BASIS OF TRADITIONAL FLOW MEASUREMENT

Flow measurement can be based on flow rate or flow amount. *Flow rate* is measured in gallons per minute (gpm), million gallons per day (MGD), or cubic feet per second (cfs). Water/wastewater operations need flow rate meters to determine process variables within the treatment plant, in wastewater collection, and potable water distribution. Typically, flow rate meters used are pressure differential meters, magnetic meters, and ultrasonic meters. Flow rate meters are designed for metering flow in a closed pipe or open channel flow.

Flow amount is measured in either gallons (gal) or cubic feet (ft^3). Typically, a totalizer is used, which sums up the gallons or cubic feet that pass through the meter,. Most service meters are of this type and are used in private, commercial, and industrial activities where the total amount of flow measured is used to determine customer billing. In wastewater treatment, where sampling operations are important, automatic composite sampling units—flow proportioned to grab a sample every so many gallons—are used. Totalizer meters can be of the velocity (propeller or turbine), positive displacement, or compound types. In addition, weirs and flumes are used extensively for measuring flow in wastewater treatment plants because they are not affected (to a degree) by dirty water or floating solids.

FLOW MEASURING DEVICES

In recent decades, flow measurement technology has evolved rapidly from the "old-fashioned way" of measuring flow, discussed earlier, to the use of simple practical measuring devices to much more sophisticated devices. Physical phenomena discovered centuries ago have been the starting point for many of the viable flowmeter designs used today. Moreover, the recent technology explosion has enabled flowmeters to handle many more applications than could have been imagined centuries ago. Before selecting a particular type of flow measurement device, Kawamura (2000) recommends consideration of several questions:

1. Is liquid or gas flow being measured?
2. Is the flow occurring in a pipe or in an open channel?
3. What is the magnitude of the flow rate?
4. What is the range of flow variation?
5. Is the liquid being measured clean, or does it contain suspended solids or air bubbles?
6. What is the accuracy requirement?
7. What is the allowable head loss by the flow meter?
8. Is the flow corrosive?
9. What types of flow meters are available in the region?
10. What types of post-installation services are available in the area?

Differential Pressure Flowmeters

For many years, *differential pressure* flowmeters have been the most widely applied flow-measuring devices for water flow in pipes that require accurate measurement at a reasonable cost. The differential pressure type of flowmeter makes up the largest segment of the total flow measurement devices currently being used. Differential pressure-producing meters currently on the market are the venturi, Dall type, Herschel venturi, universal venturi, and venturi inserts.

The differential pressure-producing device has a flow restriction in the line that causes a differential pressure or "head" to be developed between the two measurement locations. Differential pressure flowmeters are also known as head meters, and of all the head meters, the orifice flowmeter is the most widely applied device. The advantages of differential pressure flowmeters include:

- Simple construction
- Relatively inexpensive
- No moving parts
- Transmitting instruments are external
- Low maintenance
- Wide application of flowing fluids; suitable for measuring both gas and liquid flow
- Ease of instrument and range selection
- Extensive product experience and performance database

Disadvantages include:

- Flow rate is a nonlinear function of the differential pressure
- Low flow rate rangeability with normal instrumentation

OPERATING PRINCIPLE

Differential pressure flowmeters operate on the principle of measuring pressure at two points in the flow, which indicates the rate of flow passing by. The difference in pressure between the two measurement locations of the flowmeter is the result of the change in flow velocities. Simply put, there is a set relationship between the flow rate and volume, so the meter instrumentation automatically translates the differential pressure into a volume of flow. The volume of flow rate through the cross-sectional area is given by:

$$Q = A \times v(\text{average})$$

where:

Q = the volumetric flow rate
A = flow in the cross-sectional area
v = the average fluid velocity

Differential pressure flowmeters operate on the principle of developing a differential pressure across a restriction that can be related to the fluid flow rate.

Note: Optimum measurement accuracy is maintained when the flowmeter is calibrated, installed in accordance with standards and codes of practice, and when the transmitting instruments are periodically calibrated.

TYPES OF DIFFERENTIAL PRESSURE FLOWMETERS

The most commonly used differential pressure flowmeter types used in water/wastewater treatment are:

1. Orifice
2. Venturi
3. Nozzle
4. Pitot-static tube

Orifice

An orifice is an opening with a closed perimeter through which water flows. The most commonly applied *orifice* is a thin, **concentric**, flat metal plate with an opening in the plate (see Figure 8.21), installed perpendicular to the flowing stream in a circular conduit or pipe. Typically, a sharp-edged hole is bored in the center of the orifice plate. As the flowing water passes through the orifice, the restriction causes an increase in velocity. A concurrent decrease in pressure occurs as potential energy (static pressure) is converted into kinetic energy (velocity). As the water leaves the orifice, its velocity decreases, and its pressure increases as kinetic energy is converted back into potential energy according to the laws of conservation of energy. However, there is always some permanent pressure loss due to friction, and this loss is a function of the ratio of the diameter of the orifice bore (d) to the pipe diameter (D).

For dirty water applications (i.e., wastewater), a concentric orifice plate will eventually have impaired performance due to dirt buildup on the plate. Instead, eccentric or segmental orifice plates (see Figure 8.22) are often used. Measurements are typically less accurate than those obtained from the concentric orifice plate. Eccentric or segmental orifices are rarely applied in current practice.

The orifice differential pressure flowmeter is the lowest-cost differential flowmeter, is easy to install, and has no moving parts. However, it also has a high permanent head loss (ranging from 40% to 90%), resulting in higher pumping costs. Its accuracy is ±2% for a flow range of 4:1, and it is susceptible to wear or damage.

Note: Orifice meters are not recommended for permanent installation to measure wastewater flow; solids in the water easily catch on the orifice, throwing off accuracy. For installation, it is necessary to have ten diameters of straight pipe ahead of the orifice meter to create a smooth flow pattern and five diameters of straight pipe on the discharge side.

Venturi

A *Venturi* is a restriction with a relatively long passage with smooth entry and exit (see Figure 8.23). It has a long life expectancy, simplicity of construction, and relatively high-pressure recovery (i.e., it produces less permanent pressure loss than a similarly sized orifice). However, it is more expensive, is not linear with flow rate, and is the largest and heaviest differential pressure flowmeter. It is often used in wastewater flows since the smooth entry allows foreign material to be swept through instead of building up, as it would in front of an orifice. The accuracy of this type of flowmeter is ±1% for a flow range of 10:1. The head loss across a Venturi flowmeter is relatively small, ranging from 3% to 10% of the differential, depending on the ratio of the throat diameter to the inlet diameter (also known as beta ratio).

Nozzle

Flow nozzles (flow tubes) have a smooth entry and sharp exit. For the same differential pressure, the permanent pressure loss of a nozzle is of the same order as that of an orifice, but it can handle wastewater and abrasive fluids better than an orifice can. Note that, for the same line size and flow rate, the differential pressure at the nozzle is lower (head loss ranges from 10% to 20% of the differential) than the differential pressure for an orifice; hence, the total pressure loss is lower than that of an orifice. Nozzles are primarily used in steam service because of their rigidity, which makes them dimensionally more stable at high temperatures and velocities than orifices.

Note: A useful characteristic of nozzles is that they reach a critical flow condition, which is a point at which further reduction in downstream pressure does not produce a greater velocity through the nozzle. When operated in this mode, nozzles are very predictable and repeatable.

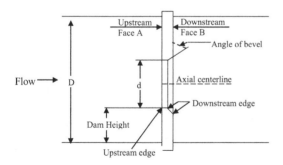

FIGURE 8.21 Orifice plate.

Concentric   Eccentric   Segmental

FIGURE 8.22 Orifice plates.

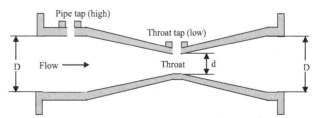

FIGURE 8.23 Venturi tube.

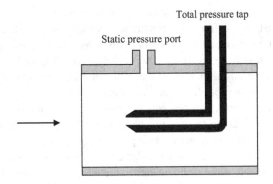

FIGURE 8.24 Pitot tube inside a pipe.

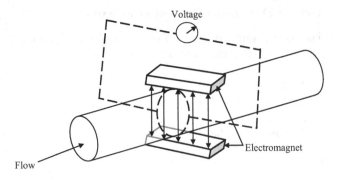

FIGURE 8.25 Magnetic flowmeter.

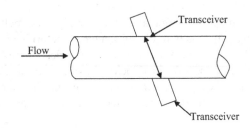

FIGURE 8.26 Time-of-flight ultrasonic flowmeter.

Pitot Tube

A *Pitot tube* is a point velocity-measuring device (see Figure 8.24). It has an impact port; as fluid hits the port, its velocity is reduced to zero and kinetic energy (velocity) is converted to potential energy (pressure head). The pressure at the impact port is the sum of the static pressure and the velocity head. The pressure at the impact port is also known as stagnation pressure or total pressure. The pressure difference between the impact pressure and the static pressure measured at the same point is the velocity head. The flow rate is the product of the measured velocity and the cross-sectional area at the point of measurement. Note that the Pitot tube has negligible permanent pressure drop in the line, but the impact port must be located in the pipe where the measured velocity is equal to the average velocity of the flowing water through the cross section.

MAGNETIC FLOWMETERS

Magnetic flowmeters are relatively new to the water/wastewater industry (USEPA, 1991). They are volumetric flow devices designed to measure the flow of electrically conductive liquids in a closed pipe. They measure the flow rate based on the voltage created between two electrodes (in accordance with Faraday's Law of Electromagnetic Induction) as the water passes through an electromagnetic field (see Figure 8.25). The induced voltage is proportional to the flow rate, depending on magnetic field strength (constant), the distance between electrodes (constant), and velocity of flowing water (variable). Properties of the magnetic flowmeter include (1) minimal head loss (no obstruction with line size meter); (2) no effect on flow profile; (3) suitable for size range between 0.1 in and 120 in; (4) have an accuracy rating of form 0.5%–2% of flow rate; and (5) it measures forward or reverses flow. The advantages of magnetic flowmeters include:

- Obstruction less flow
- Minimal head loss
- Wide range of sizes
- Bi-directional flow measurement
- Variations in density, viscosity, pressure, and temperature yield negligible effects.
- Can be used for wastewater
- No moving parts

Disadvantages include:

- The metered liquid must be conductive (but you wouldn't use this type of meter on clean fluids anyway)
- They are bulky, expensive in smaller sizes, and may require periodic calibration to correct the drifting of the signal

The combination of the magnetic flowmeter and the transmitter is considered a system. A typical system, schematically illustrated in Figure 8.26, shows a transmitter mounted remotely from the magnetic flowmeter. Some systems are available with transmitters mounted integral to the magnetic flowmeter. Each device is individually calibrated during the manufacturing process, and the accuracy statement of the magnetic flowmeter includes both pieces of equipment. One is not sold or used without the other. It is also interesting to note that since 1983 almost every manufacturer now offers a microprocessor-based transmitter.

Regarding minimum piping straight run requirements, magnetic flowmeters are quite forgiving of piping configurations. The downstream side of the magnetic flowmeter is much less critical than the upstream side. Essentially, all that is required of the downstream side is that sufficient backpressure is provided to keep the magnetic flowmeter full of liquid during flow measurement. Two diameters downstream should be acceptable (Mills, 1991).

Note: Magnetic flowmeters are designed to measure conductive liquids only. If air or gas is mixed with the liquid, the output becomes unpredictable.

ULTRASONIC FLOWMETERS

Ultrasonic flowmeters use an electronic transducer to send a beam of ultrasonic sound waves through the water to another transducer on the opposite side of the unit. The velocity of the sound beam varies with the liquid flow rate, so the beam can be electronically translated to indicate flow volume. The accuracy is ±1% for a flow velocity ranging from 1 to 25 ft/sec, but the meter reading is greatly affected by a change in the fluid composition. Two types of ultrasonic flowmeters are in general use for closed pipe flow measurements. The first (time of flight or transit time) usually uses pulse transmission and is for clean liquids, while the second (Doppler) usually uses continuous wave transmission and is for dirty liquids.

Time of Flight Ultrasonic Flowmeters

Time-of-flight flowmeters make use of the difference in the time it takes for a sonic pulse to travel a fixed distance, first in the direction of flow and then against the flow (Brown, 1991). This is accomplished by opposing transceivers positioned on a diagonal path across the meter spool as shown in Figure 8.26. Each transmits and receives ultrasonic pulses with and against the flow. The fluid velocity is directly proportional to the time difference of pulse travel. The time-of-flight ultrasonic flowmeter operates with minimal head loss, has an accuracy range of 1%–2.5% of full scale, and can be mounted as integral spool piece transducers or as externally mountable clamp-ons. They can measure flow accurately when properly installed and applied. The advantages of time-of-flight ultrasonic flowmeters include:

- No obstruction to flow
- Minimal head loss
- Clamp-ons
  - can be portable
  - no interruption of flow
- No moving parts
- Linear over a wide range
- Wide range of pipe sizes
- Bidirectional flow measurement

The disadvantages include:

- Sensitive to solids or bubble content
  - interfere with sound pulses
- Sensitive to flow disturbances
- Alignment of transducers is critical
- Clamp-on—pipe walls must freely pass ultrasonic pulses

Doppler Type Ultrasonic Flowmeters

Doppler ultrasonic flowmeters make use of the Doppler frequency shift caused by sound scattered or reflected from moving particles in the flow path. Doppler meters are not considered to be as accurate as time-of-flight flowmeters. However, they are very convenient to use and are generally

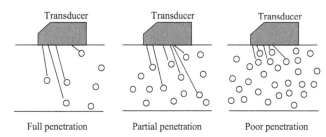

FIGURE 8.27 Particle concentration effect. The more particles there are, the more the error.

more popular and less expensive than time-of-flight flowmeters. In operation, a propagated ultrasonic beam is interrupted by particles in the moving fluid and reflected toward a receiver. The difference between propagated and reflected frequencies is directly proportional to the fluid flow rate. Ultrasonic Doppler flowmeters feature minimal head loss with an accuracy of 2%–5% of full scale. They can be either of the integral spool piece transducer type or externally mountable clamp-ons. The advantages of the Doppler ultrasonic flowmeter include:

- No obstruction or interruption of flow
- Minimal head loss
- Clamp-on
  - can be portable
  - no interruption of flow
- No moving parts
- Linear over a wide range
- Wide range of pipe sizes
- Low installation and operating costs
- Bidirectional flow measurement

The disadvantages include:

- Minimum concentration and size of solids or bubbles for reliable operation (see Figure 8.27)
- Minimum speed to maintain suspension
- Limited to sonically conductive pipe (clamp-on type)

VELOCITY FLOWMETERS

Velocity or *turbine* flowmeters use a propeller or turbine to measure the velocity of the flow passing through the device (see Figure 8.28). The velocity is then translated into a volumetric amount by the meter register. Sizes exist from a variety of manufacturers to cover the flow range from 0.001 to over 25,000 gpm for liquid service. End connections are available to meet various piping systems. The flowmeters are typically manufactured from stainless steel but are also available in a wide variety of materials, including plastic. Velocity meters are applicable to all clean fluids. Velocity meters are particularly well suited for measuring intermediate flow rates in clean water (Oliver, 1991). The advantages of the velocity meter include:

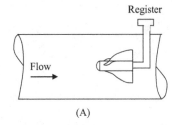

FIGURE 8.28   (a) Propeller meter; (b) turbine meter.

- Accuracy
- Composed of corrosion-resistant materials
- Long-term stability
- Liquid or gas operation
- Wide operating range
- Low-pressure drop
- Wide temperature and pressure limits
- High shock capability
- Wide variety of electronics available

As shown in Figure 8.28, a turbine flowmeter consists of a rotor mounted on a bearing and shaft in a housing. The fluid to be measured is passed through the housing, causing the rotor to spin with a rotational speed proportional to the velocity of the flowing fluid within the meter. A device to measure the speed of the rotor is employed to make the actual flow measurement. The sensor can be a mechanically gear-driven shaft to a meter or an electronic sensor that detects the passage of each rotor blade generating a pulse. The rotational speed of the sensor shaft and the frequency of the pulse are proportional to the volumetric flow rate through the meter.

Positive-Displacement Flowmeters

Positive-displacement flowmeters are most commonly used for customer metering; they have long been used to measure liquid products. These meters are very reliable and accurate for low flow rates because they measure the exact quantity of water passing through them. Positive-displacement flowmeters are frequently used for measuring small flows in a treatment plant because of their accuracy. Repair or replacement is easy since they are so common in the distribution system (Barnes, 1991).

In essence, a positive-displacement flowmeter is a hydraulic motor with high volumetric efficiency that absorbs a small amount of energy from the flowing stream. This energy is used to overcome internal friction in driving the flowmeter and its accessories and is reflected as a pressure drop across the flowmeter. Pressure drop is regarded as unavoidable and must be minimized. It is the pressure drop across the internals of a positive-displacement flowmeter that actually creates a hydraulically unbalanced rotor, which causes rotation. A positive-displacement flowmeter continuously divides the flowing stream into known volumetric segments, isolates the segments momentarily, and

returns them to the flowing stream while counting the number of displacements.

A positive-displacement flowmeter can be broken down into three basic components: the external housing, the measuring unit, and the counter drive train. The external housing is the pressure vessel that contains the product being measured. The measuring unit is a precision metering element made up of the measuring chamber and the displacement mechanism. The most common displacement mechanisms include the oscillating piston, sliding vane, oval gear, trirotor, birotor, and nutating disc types (see Figure 8.29). The counter drive train is used to transmit the internal motion of the measuring unit into a usable output signal. Many positive-displacement flowmeters use a mechanical gear train that requires a rotary shaft seal or packing gland where the shaft penetrates the external housing.

The positive-displacement flowmeter can offer excellent accuracy, repeatability, and reliability in many applications. The positive-displacement flowmeter has satisfied many needs in the past and should play a vital role in serving future needs as required.

OPEN-CHANNEL FLOW MEASUREMENT

The majority of industrial liquid flows are carried in closed conduits that flow full and under pressure. However, this is not the case for high-volume flows of liquids in waterworks, sanitary, and stormwater systems, which are commonly carried in open channels. Low system heads and high volumetric flow rates characterize flow in open channels (Grant, 1991). The most commonly used method of measuring the rate of flow in open-channel configurations is that of hydraulic structures. In this method, flow in an open channel is measured by inserting a hydraulic structure into the channel, which changes the level of liquid in or near the structure. By selecting the shape and dimensions of the hydraulic structure, the rate of flow through or over the restriction will be related to the liquid level in a known manner. Thus, the flow rate through the open channel can be derived from a single measurement of the liquid level in or near the structure. The hydraulic structures used in measuring flow in open channels are known as primary measuring devices and may be divided into two broad categories: weirs and flumes, which are covered in the following subsections.

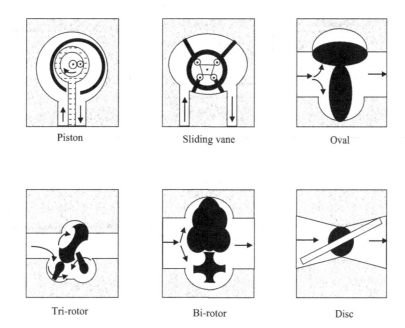

FIGURE 8.29   Six common positive-displacement meter principles.

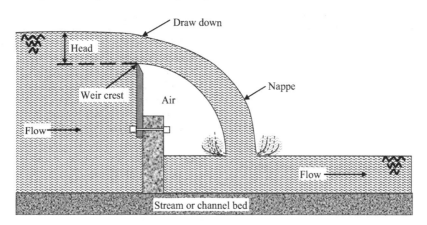

FIGURE 8.30   Side view of a weir.

Weirs

The *weir* is a widely used device to measure open-channel flow. As can be seen in Figure 8.30, a weir is simply a dam or obstruction placed in the channel so that water backs up behind it and then flows over it. The sharp crest or edge allows the water to spring clear of the weir plate and fall freely in the form of a *nappe*. As Nathanson (1997) points out, when the nappe discharges freely into the air, there is a hydraulic relationship between the height or depth of water flowing over the weir crest and the flow rate. This height, the vertical distance between the crest and the water surface, is called the head on the weir; it can be measured directly with a meter or yardstick or automatically by float-operated recording devices. Two common weirs, rectangular and triangular, are shown in Figure 8.31.

Rectangular weirs are commonly used for large flows (see Figure 8.31a). The formula used to make rectangular weir computations is:

$$Q = 3.33 \times L \times h^{1.5} \qquad (8.27)$$

where

Q = flow
L = width of the weir
h = head on weir (measured from the edge of the weir in contact with the water, up to the water surface)

Example 8.17

Problem: A weir 4 ft high extends 15 ft across a rectangular channel in which 80 cfs are flowing. What is the depth just upstream from the weir?

Solution:

$$Q = 3.33 \times L \times h^{1.5}$$

$$80 = 3.33 \times 15 h^{1.5}$$

h = 1.4 ft (w/calculator: $1.6 \, \text{INV} \, y^{\times 1.5} = 1.36$ or   1.4)

4 ft height of weir + 1.4 ft head of water = 5.4 ft depth

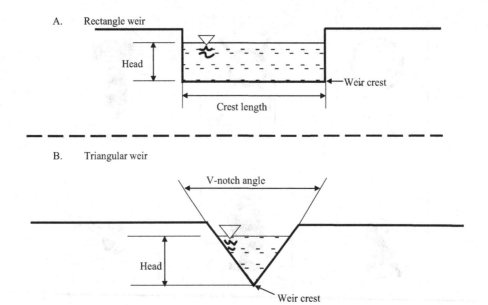

FIGURE 8.31   (a) Rectangular weir; (b) triangular V-notch weir.

Triangular weirs, also called V-notch weirs, can have notch angles ranging from 22.5° to 90°, but right-angle notches are the most common (see Figure 8.31b).

The formula used to make V-notch (90°) weir calculations is

$$Q = 2.5 \times h^{2.5} \qquad (8.28)$$

where

Q = flow
h = head on weir (measured from bottom of notch to water surface)

Example 8.18

Problem: What should be the minimum weir height for measuring a flow of 1,200 gpm with a 90° V-notch weir if the flow is now moving at 4 ft/sec in a 2.5 ft wide rectangular channel?

Solution:

$$\frac{1,200 \text{ gpm}}{60 \text{ sec/min} \times 7.48 \text{ gal/ft}^3} = 2.67 \text{ cfs}$$

$$Q = A \times V$$

$$2.67 = 2.5 \times d \times 4$$

$$d = 0.27 \text{ ft}$$

$$Q = 2.5 \times h^{2.5}$$

$$2.67 = 2.5 \times h^{2.5}$$

$$h = 1.03 \text{ (calculator: } 1.06 \text{ INV } y^{\times 2.5} = 1.026 \text{ or } 1.03 \text{)}$$

0.27 ft (original depth) + 1.03(head on weir) = 1.3 ft

It is important to point out that weirs, aside from being operated within their flow limits, must also be operated within the available system head. In addition, the operation of the weir is sensitive to the approach velocity of the water, often necessitating a stilling basin or pound upstream of the weir. Weirs are unsuitable for water that carries excessive solid materials or silt, which deposit in the approach channel behind the weir and destroy the conditions required for accurate discharge measurements.

Note: Accurate flow rate measurements with a weir cannot be expected unless the proper conditions and dimensions are maintained.

Flumes

A *flume* is a specially shaped constricted section in an open channel (similar to the Venturi tube in a pressure conduit). The special shape of the flume (see Figure 8.32) restricts the channel area and/or changes the channel slope, resulting in increased velocity and a change in the level of the liquid flowing through the flume. The flume restricts the flow and then expands it in a definite fashion. The flow rate through

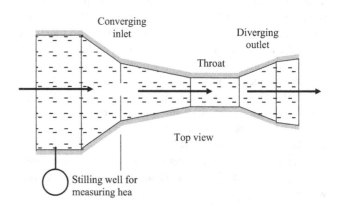

FIGURE 8.32   Parshall flume.

the flume may be determined by measuring the head on the flume at a single point, usually at some distance downstream from the inlet.

Flumes can be categorized as belonging to one of three general families, depending on the state of flow induced: subcritical, critical, or supercritical. Typically, flumes that induce a critical or supercritical state of flow are most commonly used. This is because when critical or supercritical flow occurs in a channel, one head measurement can indicate the discharge rate if it is made far enough upstream so that the flow depth is not affected by the drawdown of the water surface as it achieves or passes through a critical state of flow. For critical or supercritical states of flow, a definitive head-discharge relationship can be established and measured, based on a single head reading. Thus, the most commonly encountered flumes are designed to pass the flow from subcritical through critical near the point of measurement.

The most common flume used for a permanent wastewater flow-metering installation is called the *Parshall flume*, shown in Figure 8.32. Formulas for flow through Parshall flumes differ depending on throat width. The formula below can be used for widths of 1–8 ft and applies to a medium range of flows.

$$Q = 4 \times W \times H_a^{1.52} \times W^{0.026} \qquad (8.29)$$

where

Q = flow
H_a = depth in stilling well upstream
W = width of throat

Note: Parshall flumes are low maintenance items.

THE BOTTOM LINE

Water and wastewater operators play a crucial role in maintaining safe and efficient water supply and treatment systems. Understanding basic hydraulics is essential for several reasons:

- System operation and maintenance
- Flow control and distribution
- Process optimization
- Safety and public health

CHAPTER REVIEW QUESTIONS

8.1 What should be the minimum weir height for measuring a flow of 900 gpm with a 90° V-notch weir, if the flow is now moving at 3 ft/sec in a 2-ft wide rectangular channel?

8.2 A 90° V-notch weir is to be installed in a 30-in diameter sewer to measure 600 gpm. What head should be expected?

8.3 For dirty water operations, a _____ or _____ orifice plate should be used.

8.4 A _____ has a smooth entry and sharp exit.

8.5 _____ sends a beam of ultrasonic sound waves through the water to another transducer on the opposite side of the unit.

8.6 Find the number of gallons in a storage tank that has a volume of 660 ft³.

8.7 Suppose a rock weighs 160 lbs in air and 125 lbs under water. What is the specific gravity?

8.8 A 110-ft diameter cylindrical tank contains 1.6 MG water. What is the water depth?

8.9 The pressure in a pipeline is 6,400 psf. What is the head on the pipe?

8.10 The pressure on a surface is 35 psig. If the surface area is 1.6 ft², what is the force (lb) exerted on the surface?

8.11 Bernoulli's principle states that the total energy of a hydraulic fluid is _____ _____.

8.12 What is *pressure head*?

8.13 What is a *hydraulic grade line*?

8.14 A flow of 1,500 gpm takes place in a 12-in pipe. Calculate the velocity head.

8.15 Water flows at 5.00 mL/sec in a 4-in line under a pressure of 110 psi. What is the pressure head (ft of water)?

8.16 In question 8.15, what is the velocity head in the line?

8.17 What is velocity head in a 6-in pipe connected to a 1-ft pipe, if the flow in the larger pipe is 1.46 cfs?

8.18 What is *velocity head*?

8.19 What is *suction lift*?

8.20 Explain *energy grade line*.

REFERENCES

AWWA. 1995. *Basic Science Concepts and Applications: Principles and Practices of Water Supply Operations*, 2nd ed. Denver: American Water Works Association.

Barnes, R.G., 1991. Positive Displacement Flowmeters for Liquid Measurement, in *Flow Measurement,* Spitzer, D.W., (ed.). Research Triangle Park, NC: Instrument Society of America.

Brown, A.E., 1991. Ultrasonic Flowmeters, in *Flow Measurement,* Spitzer, D.W., (ed.). Research Triangle Park, NC: Instrument Society of America.

Grant, D.M., 1991. Open Channel *Flow Measurement,* in Flow Measurement, Spitzer, D.W., (ed.). Research Triangle Park, NC: Instrument Society of America.

Hauser, B.A., 1995. *Practical Hydraulics Handbook*, 2nd ed. Boca Raton, FL: Lewis Publishers.

Kawamura, S., 2000. *Integrated Design and Operation of Water Treatment Facilities*, 2nd ed. New York: John Wiley & Sons.

Lindeburg, M.R., 1986. *Civil Engineering Reference Manual*, 4th ed. San Carlos, CA: Professional Publications, Inc.

Magnusson, R.J., 2001. *Water Technology in the Middle Ages*. Baltimore, MD: The John Hopkins University Press.

McGhee, T.J., 1991. *Water Supply and Sewerage*, 2nd ed. New York: McGraw-Hill.

Metcalf & Eddy. 1981. *Wastewater Engineering: Collection and Pumping of Wastewater*. New York: McGraw-Hill.

Mills, R.C., 1991. Magnetic Flowmeters, in Flow Measurement, Spitzer, D.W., (ed.). Research Triangle Park, NC: Instrument Society of America.

Nathanson, J.A., 1997. *Basic Environmental Technology: Water Supply Waste Management, and Pollution Control*, 2nd ed. Upper Saddle River, NJ: Prentice Hall.

Oliver, P.D., 1991. Turbine Flowmeters, in *Flow Measurement,* Spitzer, D.W., (ed.). Research Triangle Park, NC: Instrument Society of America.

Spellman, F.R., and Drinan, J., 2001. *Water Hydraulics*. Boca Raton, FL: CRC Press.

USEPA. 1991. *Flow Instrumentation: A Practical Workshop on Making Them Work*. Sacramento, CA: Water & Wastewater Instrumentation Testing Association.

9 Hydraulic Machines
Pumps

A hydraulic machine, or pump, is a device that raises, compresses, or transfers fluids.

INTRODUCTION

Garay (1990) points out that "few engineered artifacts are as essential as pumps in the development of the culture which our western civilization enjoys." This statement is germane to any discussion about pumps simply because humans have always needed to move water from one place to another against the forces of nature. As the need for potable water increases, the need to pump water from distant locations to where it is most needed is also increasing.

Initially, humans relied on one of the primary forces of nature—gravity—to assist them in moving water from one place to another. Gravity only works, of course, if the water is moved downhill on a sloping grade. Humans soon discovered that if they accumulated water behind the water source (e.g., behind a barricade, levee, or dam), pressure would move the water further. But when pressure is dissipated by various losses (e.g., friction loss), and when water in low-lying areas is needed in higher areas, the energy needed to move that water must be created. Simply put, some type of pump is needed.

In 287 B.C., Archimedes (Greek mathematician and physicist) invented the screw pump (see Figure 9.1). The Roman emperor Nero, around 100 A.D., is often credited with the development of the piston pump. In operation, the piston pump displaces volume after volume of water with each stroke. The piston pump has two basic problems: (1) its size limits capacity and (2) it is a high-energy consumer. It was not until the nineteenth century before pumping technology took a leap forward from its rudimentary beginnings. The first fully functional centrifugal pumps were developed in the 1800s. Centrifugal pumps can move great quantities of water with much smaller units than the pumps previously in use.

The pump is a type of hydraulic machine. Pumps convert mechanical energy into fluid energy. Whether it is taken from groundwater or a surface water body, or from one unit treatment process to another, or to the storage tank for eventual final delivery through various sizes and types of pipes to the customer, *pumps* are the usual source of energy necessary for the conveyance of water. Again, the only exception may be, of course, where the source of energy is supplied entirely by gravity. Waterworks and wastewater maintenance operators must therefore be familiar with pumps, pump characteristics, pump operation, and maintenance.

There are three general requirements of pump and motor combinations. These requirements are (1) reliability (2) adequacy and (3) economy. **Reliability** is generally obtained by installing in duplicate the very best equipment

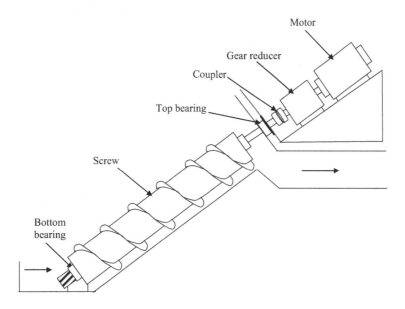

FIGURE 9.1 Archimedes' screw pump.

DOI: 10.1201/9781003581901-11

available and by the use of an auxiliary power source. **Adequacy** is obtained by securing liberal sizes of pumping equipment. **Economics** can be achieved by considering the life and depreciation, first cost, standby charges, interest and operating costs.

—*Texas Manual (Texas Utilities Association, 1988)*

Over the past several years, it has become more evident that many waterworks and wastewater facilities have been unable to meet their optimum supply and/or treatment requirements for one of three reasons:

1. Untrained operations and maintenance staff
2. Poor plant maintenance
3. Improper plant design

BASIC PUMPING CALCULATIONS

Calculations, calculations, calculations, and more calculations! Indeed, we can't get away from them—not in water/wastewater treatment and collection/distribution operations, nor in licensure certification examinations, nor real life. Basic calculations are a fact of life that the water/wastewater maintenance operator soon learns, and hopefully learns well enough to use as required to operate a water/wastewater facility correctly. In the following sections, basic calculations used frequently in water hydraulics and pumping applications are discussed. The basic calculations that the water and wastewater maintenance operator may be required to know for operational and certification purposes are discussed. In addition, calculations for pump specific speed, suction specific speed, affinity formulas, and other advanced calculations are also covered in this section, although at a higher technical level.

VELOCITY OF A FLUID THROUGH A PIPELINE

The speed or velocity of a fluid flowing through a channel or pipeline is related to the cross-sectional area of the pipeline and the quantity of water moving through it. For example, if the diameter of a pipeline is reduced, then the velocity of the water in the pipeline must increase to allow the same amount of water to pass through.

$$\text{Velocity (V) fps} = \frac{\text{Flow (Q) cfs}}{\text{Cross-Sectional Area (A) ft}^2} \quad (9.1)$$

Example 9.1

Problem: If the flow through a 2-ft-diameter pipe is 9 MGD, the velocity is:

Solution:

$$\text{Velocity, fps} = \frac{9\,\text{MGD} \times 1.55\,\text{cfs/MGD}}{0.785 \times 2\,\text{ft} \times 2\,\text{ft}}$$

$$V = \frac{14\,\text{cfs}}{3.14\,\text{ft}^2}$$

$$V = 4.5\,\text{fps (rounded)}$$

Example 9.2

Problem: If the same 9-MGD flow used in Example 9.1 is transferred to a pipe with a 1-ft diameter, what would the velocity be?

Solution:

$$\text{Velocity, fps} = \frac{9\,\text{MGD} \times 1.55\,\text{cfs/MGD}}{0.785 \times 1\,\text{ft} \times 1\,\text{ft}}$$

$$V = \frac{14\,\text{cfs}}{0.785\,\text{ft}^2}$$

$$V = 17.8\,\text{fps (rounded)}$$

Based upon these sample problems, you can see that if the cross-sectional area is decreased, the velocity of the flow must be increased. Mathematically, we can say that the velocity and cross-sectional area are inversely proportional when the amount of flow (Q) is constant.

$$\text{Area}_1 \times \text{Velocity}_1 = \text{Area}_2 \times \text{Velocity}_2 \quad (9.2)$$

Note: The concept just explained is extremely important in the operation of a centrifugal pump and will be discussed later.

PRESSURE-VELOCITY RELATIONSHIP

A relationship similar to that of velocity and cross-sectional area exists for velocity and pressure. As the velocity of flow in a full pipe increases, the pressure of the liquid decreases. This relationship is:

$$\text{Pressure}_1 \times \text{Velocity}_1 = \text{Pressure}_2 \times \text{Velocity}_2 \quad (9.3)$$

Example 9.3

Problem: If the flow in a pipe has a velocity of 3 fps and a pressure of 4 psi and the velocity of the flow increases to 4 fps, what will the pressure be?

Solution:

$$P_1 \times V_1 = P_2 \times V$$

$$4\,\text{psi} \times 3\,\text{fps} = P_2 \times 4\,\text{fp}$$

Rearranging:

$$P_2 = \frac{4\,\text{psi} \times 3\,\text{fps}}{4\,\text{fps}}$$

$$P_2 = \frac{12\,\text{psi}}{4}$$

$$P_2 = 3\,\text{psi}$$

Again, this is another important hydraulics principle that is crucial to the operation of a centrifugal pump.

STATIC HEAD

Pressure at a given point originates from the height or depth of water above it. It is this pressure, or *head*, that gives the water energy and causes it to flow. By definition, *static head* is the vertical distance the liquid travels from the supply tank to the discharge point. This relationship is shown as:

$$\text{Static Head, ft} = \text{Discharge Level, ft} - \text{Supply Level, ft} \tag{9.4}$$

In many cases, it is desirable to separate the static head into two separate parts: (1) the portion that occurs before the pump (suction head or suction lift) and (2) the portion that occurs after the pump (discharge head). When this is done, the center (or datum) of the pump becomes the reference point.

STATIC SUCTION HEAD

Static suction head refers to the situation when the supply is located above the pump datum:

$$\text{Static Suction Head, ft} = \text{Supply Level, ft} - \text{Pump Level, ft} \tag{9.5}$$

STATIC SUCTION LIFT

Static suction lift refers to when the supply is located below the pump datum:

$$\text{Static Suction Lift, ft} = \text{Pump Level, ft} - \text{Supply Level, ft} \tag{9.6}$$

STATIC DISCHARGE HEAD

$$\text{Static Discharge Head, ft} = \text{Discharge Level, ft} - \text{Pump datum, ft} \tag{9.7}$$

If the total static head is to be determined after calculating the static suction head or lift and the static discharge head individually, two separate calculations can be used, depending on whether there is a suction head or a suction lift.

For Suction Head:

$$\text{Total Static Head} = \text{Static Discharge Head, ft} - \text{Static Suction Lift, ft} \tag{9.8}$$

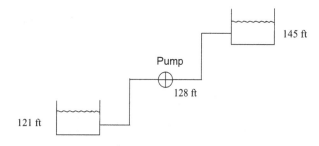

FIGURE 9.2 Illustration for Example 9.4.

For Suction Lift:

$$\text{Total Static Head, ft} = \text{Static Discharge Head, ft} + \text{Static Suction Lift, ft} \tag{9.9}$$

Example 9.4

Problem: Refer to Figure 9.2.

Solution:

Step 1: Static Suction Lift, ft = Pump Level, ft

−Supply Level, ft

$$\text{Static Suction Lift, ft} = 128\,\text{ft} - 121\,\text{ft}$$

$$= 7\,\text{ft}$$

Step 2: Static Discharge Head, ft = Discharge Level, ft

+Static Suction Lift, ft

$$= 145\,\text{ft} - 128\,\text{ft}$$

$$= 17\,\text{ft}$$

Step 3: Total Static Head, ft = Static Discharge Head, ft

+Static Suction Lift

$$= 17\,\text{ft} + 7\,\text{ft}$$

$$= 24\,\text{ft}$$

- or -

Total Static Head, ft = Discharge Level, ft − Supply Level, ft

$$= 145\,\text{ft} - 119\,\text{ft}$$

$$= 24\,\text{ft}$$

Example 9.5

Problem: See Figure 9.3

Solution:

Step 1: Static Suction Head, ft = Supply Level, ft

−Pump Level, ft

$$= 124\,\text{ft} - 117\,\text{ft}$$

$$= 7\,\text{ft}$$

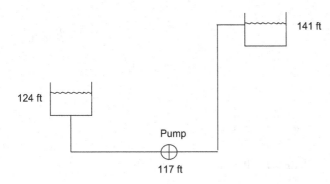

FIGURE 9.3   Illustration for Example 9.5.

Step 2: Static Discharge Head, ft = Discharge Level, ft

$$-\text{Pump Level, ft}$$

$$= 141\,\text{ft} - 117\,\text{ft}$$

$$= 24\,\text{ft}$$

Step 3: Total Static Head, ft = Static Discharge Head, ft

$$-\text{Static Suction Head}$$

$$= 24\,\text{ft} - 7\,\text{ft}$$

$$= 17\,\text{ft}$$

or

Total Static Head, ft = Discharge Level, ft − Supply Level, ft

$$= 141\,\text{ft} - 124\,\text{ft}$$

$$= 17\,\text{ft}$$

FRICTION HEAD

Various formulae calculate friction losses. Hazen-Williams wrote one of the most common formulas for smooth steel pipe. Usually, we do not need to calculate the friction losses, as handbooks such as the *Hydraulic Institute Pipe Friction Manual* have tabulated these long ago. This important manual also shows velocities in different pipe diameters at varying flows, as well as the resistance coefficient (K) for valves and fittings (Wahren, 1997). *Friction head* (in feet) is the amount of energy used to overcome resistance to the flow of liquids through the system. It is affected by the length and diameter of the pipe, the roughness of the pipe, and the velocity head. It is also affected by the physical construction of the piping system. The number and types of elbows, valves, T's, etc., will greatly influence the friction head for the system. These must be converted to their equivalent length of pipe and included in the calculation.

Friction Head, ft = Roughness Factor, f

$$\times \frac{\text{Length}}{\text{Diameter}} \times \frac{\text{Velocity}^2}{2_g} \quad (9.10)$$

The *roughness factor (f)* varies with length and diameter, as well as the condition of the pipe and the material from which it is constructed; it normally ranges from 0.01 to 0.04.

Important Point: For centrifugal pumps, good engineering practice is to try to keep velocities in the suction pipe to 3 ft/sec or less. Discharge velocities higher than 11 ft/sec may cause turbulent flow and/or erosion in the pump casing.

Example 9.6

Problem: What is the friction head in a system that uses 150 ft of 6-in diameter pipe when the velocity is 3 fps? The system's valving is equivalent to an additional 75 ft of pipe. Reference material indicates a roughness factor (f) of 0.025 for this particular pipe and flow rate.

Solution:

Friction Head, ft = Roughness factor

$$\times \frac{\text{Length}}{\text{Diameter}} \times \frac{\text{Velocity}^2}{2_g}$$

$$\text{Friction Head, ft} = 0.025 \times \frac{(150\,\text{ft} + 75\,\text{ft})}{0.5\,\text{ft}} \times \frac{(3\,\text{fps})^2}{2 \times 32\,\text{ft/sec}}$$

$$\text{Friction Head, ft} = 0.025 \times \frac{225\,\text{ft}}{0.5\,\text{ft}} \times \frac{9\,\text{ft}^2/\text{sec}^2}{64\,\text{ft/sec}^2}$$

$$\text{Friction Head, ft} = 0.025 \times 450 \times 0.140\,\text{ft}$$

$$\text{Friction Head, ft} = 1.58\,\text{ft}$$

It is also possible to compute friction head using tables. Friction head can also be determined on both the suction side and the discharge side of the pump. In each case, it is necessary to determine:

1. The length of the pipe
2. The diameter of the pipe
3. Velocity
4. Pipe equivalent of valves, elbows, T's, etc.

VELOCITY HEAD

Velocity head is the amount of head or energy required to maintain a stated velocity in the suction and discharge lines. The design of most pumps makes the total velocity head for the pumping system zero.

Note: Velocity head only changes from one point to another in a pipeline if the diameter of the pipe changes.

Velocity head and total velocity head are determined by:

$$\text{Velocity Head, ft} = \frac{(\text{Velocity})^2}{2g} \quad (9.11)$$

Total Velocity Head, ft = Velocity Head Discharge, ft

$$- \text{Velocity Head Suction, ft}$$

$$(9.12)$$

Example 9.7

Problem: What is the velocity head for a system that has a velocity of 5 fps?

Solution:

$$\text{Velocity Head, ft} = \frac{(\text{Velocity})^2}{2 \times \text{Acceleration due to gravity}}$$

$$\text{Velocity Head, ft} = \frac{(5\,\text{fps})^2}{2 \times 32\,\text{ft/sec}^2}$$

$$\text{Velocity Head, ft} = \frac{25\,\text{ft}^2/\text{sec}^2}{64\,\text{ft}^2/\text{sec}^2}$$

$$\text{Velocity Head, ft} = 0.39\,\text{ft}$$

Note: There is no velocity head in a static system. The water is not moving.

TOTAL HEAD

Total head is the sum of the static, friction, and velocity heads.

$$\text{Total Head, ft} = \text{Static Head, ft} + \text{Friction Head, ft}$$

$$+ \text{Velocity Head, ft}$$

$$(9.13)$$

CONVERSATION OF PRESSURE HEAD

Pressure is directly related to head. If the liquid in a container subjected to a given pressure is released into a vertical tube, the water will rise 2.31 ft for every pound per square inch of pressure. To convert pressure to head in feet:

$$\text{Head, ft} = \text{Pressure, psi} \times 2.31\,\text{ft/psi} \qquad (9.14)$$

This calculation can be very useful in cases where the liquid is moved through another line that is under pressure. Since the liquid must overcome the pressure in the line it is entering, the pump must supply this additional head.

Example 9.8

Problem: A pump discharges to a pipe full of liquid under a pressure of 20 psi. The pump and piping system has a total head of 97 ft. How much additional head must the pump supply to overcome the line pressure?

Solution:

$$\text{Head, ft} = \text{Pressure, psi} \times 2.31\,\text{ft/psi}$$

$$= 20\,\text{psi} \times 2.31\,\text{ft/psi}$$

$$= 46\,\text{ft (rounded)}$$

Note: The pump must supply an additional head of 46 ft to overcome the internal pressure of the line.

HORSEPOWER

The unit of work is *foot-pound*; defined as the amount of work required to lift a 1-lb object 1 ft off the ground (ft-lb). For practical purposes, we consider the amount of work being done. It is more valuable to work faster; that is, for economic reasons, we consider the rate at which work is being done (i.e., power or foot pound/second). At some point, the horse was determined to be the ideal work animal; it could move 550 lbs 1 ft, in 1 sec, which is considered equivalent to 1 hp.

$$550\,\text{ft/lb/sec} = 1\,\text{Horsepower (hp)}$$

or

$$33,000\,\text{ft/lb/min} = 1\,\text{Horsepower (hp)}$$

A pump performs work while it pushes a certain amount of water at a given pressure. The two basic terms for horsepower are (1) hydraulic horsepower and (2) *brake horsepower*.

Hydraulic (Water) Horsepower (WHP)

A pump has power because it does work. A pump lifts water (which has weight) a given distance in a specific amount of time (ft/lb/min). One hydraulic (water) horsepower (WHP) provides the necessary power to lift the water to the required height; it equals the following:

- 550 ft-lb/sec
- 33,000 ft-lb/min
- 2,545 British thermal units per hour (Btu/h)
- 0.746 kW
- 1.014 metric hp

To calculate the hydraulic horsepower (WHP) using flow in gpm and head in feet, use the following formula for centrifugal pumps:

$$\text{WHP} = \frac{\text{flow (in gpm)} \times \text{head (in ft)} \times \text{specific gravity}}{3,960}$$

$$(9.15)$$

Note: 3,960 is derived by dividing 33,000 ft-lb by 8.34 lb/gal = 3,960.

Brake Horsepower (BHP)

A water pump does not operate alone. It is driven by the motor, and electrical energy drives the motor. Brake horsepower is the horsepower applied to the pump. A pump's brake horsepower (BHP) equals its hydraulic horsepower divided by the pump's efficiency. Note that neither the pump nor its prime mover (motor) is 100% efficient. There are friction losses within both of these units, and it will take **more** horsepower applied to the pump to get the required amount of horsepower to move the water, and even more horsepower applied to the motor to get the job done (Hauser, 1993). Thus, the formula for BHP is:

$$BHP = \frac{flow\ (gpm) \times head\ (ft) \times specific\ gravity}{3,960 \times efficiency} \quad (9.16)$$

Important Points: (1) *Water Horsepower* (whp) is the power necessary to lift the water to the required height; (2) *Brake Horsepower* (bhp) is the horsepower applied to the pump; (3) *Motor Horsepower* (hp) is the horsepower applied to the motor; and (4) *Efficiency* is the power produced by the unit divided by the power used in operating the unit.

SPECIFIC SPEED

The capacity of the flow rate of a centrifugal pump is governed by the impeller thickness (Lindeburg, 1986). For a given impeller diameter, the deeper the vanes, the greater the capacity of the pump.

Each desired flow rate or desired discharge head will have one optimum impeller design. The impeller best for developing a high discharge pressure will have different proportions from an impeller designed to produce a high flow rate. The quantitative index of this optimization is called *specific speed* (n_s). The higher the specific speed of a pump, the higher its efficiency.

The specific speed of an impeller is its speed when pumping 1 gpm of water at a differential head of 1 ft. The following formula is used to determine specific speed (where *H* is at the best efficiency point):

$$N_s = \frac{rpm \times Q^{0.5}}{H^{0.75}} \quad (9.17)$$

where:

rpm = revolutions per minute
Q = flow (in gpm)
H = head (in ft)

Pump specific speeds vary between pumps. Although no absolute rule sets the specific speed for different kinds of centrifugal pumps, the following rule of thumb for N_s can be used:

- Volute, diffuser, and vertical turbine = 500–5,000
- Mixed flow = 5,000–10,000
- Propeller pumps = 9,000–15,000

Suction Specific Speed

Suction specific speed (n_{ss}), also an impeller design characteristic, is an index of the suction characteristics of the impeller (i.e., the suction capacities of the pump) (Wahren, 1997). For practical purposes, n_{ss} ranges from about 3,000 to 15,000. The limit for the use of suction specific speed impellers in water is approximately 11,000. The following equation expresses n_{ss}:

$$n_{ss} = \frac{rpm \times Q^{0.5}}{NPSHR^{0.75}} \quad (9.18)$$

where

rpm = revolutions per minute
Q = flow in gpm
NPSHR = net positive suction head required

Ideally, n_{ss} should be approximately 7,900 for single suction pumps and 11,200 for double suction pumps. (The value of Q in Equation (9.18) should be halved for double suction pumps.)

AFFINITY LAWS—CENTRIFUGAL PUMPS

Most parameters (impeller diameter, speed, and flow rate) determining a pump's performance can vary. If the impeller diameter is held constant and the speed is varied, the following ratios are maintained with no change in efficiency (because of inexact results, some deviations may occur in the calculations):

$$Q_2/Q_1 = D_2/D_1 \quad (9.19)$$

$$H_2/H_1 = (D_2/D_1)^2 \quad (9.20)$$

$$BHP_2/BHP_1 = (D_2/D_1)^3 \quad (9.21)$$

where:

Q = flow
H_1 = head before change
H_2 = head after change
DHP = brake horsepower
D_1 = impeller diameter before change
D_2 = impeller diameter after change

The relationships for speed (N) changes are as follows:

$$Q_2/Q_1 = N_2/N_1 \quad (9.22)$$

$$H_2/H_1 = (N_2/N_1)^2 \quad (9.23)$$

$$BHP_2/BHP_1 = (N_2/N_1)^3 \quad (9.24)$$

where:

N_1 = initial rpm
N_2 = changed rpm

Example 9.9

Problem: Change an 8-in-diameter impeller for a 9-in diameter impeller, and find the new flow (Q), head (H), and brake horsepower (BHP) where the 8-in diameter data are:

$$Q_1 = 340 \, \text{rpm}$$

$$H_1 = 110 \, \text{ft}$$

$$BHP_1 = 10$$

Solution:

Data for the 9-in impeller diameter data will be as follows:

$$Q_2 = 340 \times 9/8 = 383 \, \text{gpm}$$

$$H_2 = 110 \times (9/8)^2 = 139 \, \text{ft}$$

$$BHP_2 = 10 \times (9/8)^3 = 14$$

NET POSITIVE SUCTION HEAD (NPSH)

Earlier, we referred to the *net positive suction head required* (NPSHR); also important in pumping technology is the *net positive suction head* (NPSH) (Lindeburg, 1986; Wahren, 1997). NPSH is different from both suction head and suction pressure. This important point tends to be confusing to those first introduced to the term and to pumping technology in general. When an impeller in a centrifugal pump spins, the motion creates a partial vacuum in the impeller's eye. The NPSH is the height of the column of liquid that will fill this partial vacuum without allowing the liquid's vapor pressure to drop below its flash point; that is, this is the NPSH required (NPSHR) for the pump to function properly.

The Hydraulic Institute (1994) defines NPSH as "the total suction head in feet of liquid absolute determined at the suction nozzle and referred to datum less the vapor pressure of the liquid in feet absolute." This defines the NPSH *available* (NPSHA) for the pump. (Note that NPSHA is the actual water energy at the inlet.) The important point is that a pump will run satisfactorily if the NPSHA equals or exceeds the NPSHR. Most authorities recommend that the NPSHA be at least 2 ft absolute or 10% larger than the NPSHR, whichever number is larger.

Note: With regard to NPSHR, contrary to popular belief, water is not sucked into a pump. A positive head (normally atmospheric pressure) must push the water into the impeller (i.e., flood the impeller). NPSHR is the minimum water energy required at the inlet by the pump for satisfactory operation. The pump manufacturer usually specifies NPSHR.

It is important to point out that if NPSHA is less than NPSHR, the water will cavitate. *Cavitation* is the vaporization of fluid within the casing or suction line. If the water pressure is less than the vapor pressure, pockets of vapor will form. As vapor pockets reach the surface of the impeller, the local high water pressure will collapse them, causing noise, vibration, and possible structural damage to the pump.

Calculating NPSHA

In the following two examples, we demonstrate how to calculate NPSH for two real-world situations: (1) determining NPSHA for an open-top water tank or a municipal water storage tank with a roof and correctly sized vent, and (2) determining NPSHA for a suction lift from an open reservoir.

Atmospheric Tank

The following calculation may be used for an open-top water tank or a municipal water storage tank with a roof and correctly sized vent as shown in Figures 9.4 and 9.5. The formula for calculating NPSHA is:

$$NPSHA = P_a + h - P_y - h_e - h_f \qquad (9.25)$$

where:

P_v = vapor pressure in absolute of water at a given temperature
P_a = atmospheric pressure in absolute or pressure of gases against the surface of water
h = weight of liquid column from surface of the water to center of pump suction nozzle in feet absolute
h_e = entrance losses in feet absolute
h_f = friction losses in suction line in feet absolute

Example 9.10

Problem: Given the following, find the NPSHA.

Liquid = water
Temperature (t) = 60°F
Specific gravity = 1.0
P_v = 0.256 psia (0.6 ft)
h_e = 0.4 ft
h = 15 ft
h_f = 2 ft
P_a = 14.7 psia (34 ft)

Solution:

NPSHA = 34 ft + 15 ft − 0.6 ft − 0.4 ft − 2 ft
NPSHA = 46 ft

NPSHA: Suction Lift from Open Reservoir
See Figure 9.6.

Example 9.11

Problem: Find the NPSHA, where:

Liquid = water
Temperature (t) = 60°F

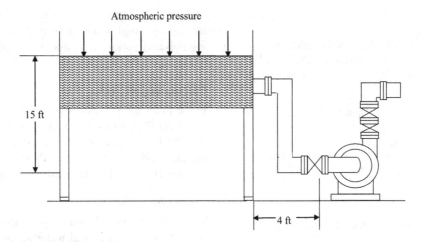

FIGURE 9.4   Open atmospheric tank.

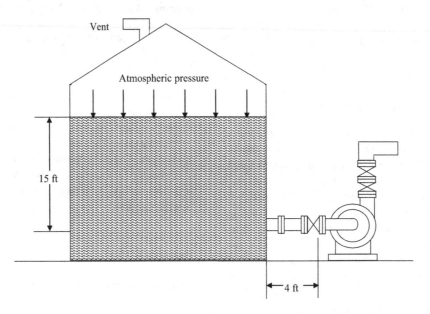

FIGURE 9.5   Roofed water storage tank.

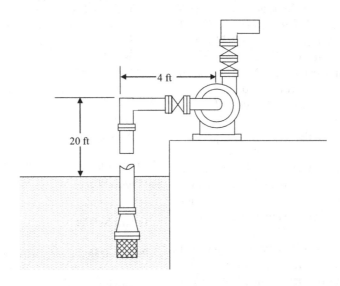

FIGURE 9.6   Suction lift from open reservoir.

Specific gravity = 1.0
P_v = 0.256 psia (0.6 ft)
Q = 120 gpm
h_e = 0.4 ft
h_f = 2 ft
h = −20 ft
P_a = 14.7 psia (34 ft)

Solution:

NPSHA = 34 ft − 20 ft − 0.6 ft − 0.4 ft − 2 ft
NPSHA = 11 ft

PUMPS IN SERIES AND PARALLEL

Parallel operation is obtained by having two pumps discharging into a common header. This type of connection is advantageous when the system demand varies greatly. An advantage of operating pumps in parallel is that when two pumps are online, one can be shut down during low demand. This allows the remaining pump to operate close to its optimum efficiency. *Series* operation is achieved by having one pump discharge into the suction of the next. This arrangement is used primarily to increase the discharge head, although a small increase in capacity also results.

CENTRIFUGAL PUMPS

The *centrifugal pump* (and its modifications) is the most widely used type of pumping equipment in water/wastewater operations. This type of pump is capable of moving high volumes of water/wastewater (and other liquids) in a relatively efficient manner. The centrifugal pump is very dependable, has relatively low maintenance requirements, and can be constructed from a wide variety of construction materials. It is considered one of the most dependable systems available for water transfer.

DESCRIPTION

The centrifugal pump consists of a rotating element (*impeller*) sealed in a casing (*volute*). The rotating element is connected to a drive unit (*motor/engine*) that supplies the energy to spin the rotating element. As the impeller spins inside the volute casing, an area of low pressure is created in the center of the impeller. This low pressure allows the atmospheric pressure on the liquid in the supply tank to force the liquid up to the impeller. Because the pump will not operate if no low-pressure zone is created at the center of the impeller, it is important that the casing be sealed to prevent air from entering the casing. To ensure the casing is airtight, the pump employs some type of seal (*mechanical* or *conventional packing*) assembly at the point where the shaft enters the casing. This seal also includes lubrication, provided by water, grease, or oil, to prevent excessive wear.

From a hydraulic standpoint, note the energy changes that occur in the moving water. As water enters the casing, the spinning action of the impeller imparts (transfers) energy to the water. This energy is transferred to the water in the form of increased speed or velocity. The liquid is thrown outward by the impeller into the volute casing where the design of the casing allows the velocity of the liquid to be reduced, which, in turn, converts the velocity energy (*velocity head*) to pressure energy (*pressure head*). The process by which this change occurs is described later. The liquid then travels out of the pump through the pump discharge. The major components of the centrifugal pump are shown in Figure 9.7.

Key Point: A centrifugal pump is a pumping mechanism whose rapidly spinning impeller imparts a high velocity to the water that enters, and then converts that velocity to pressure upon exit.

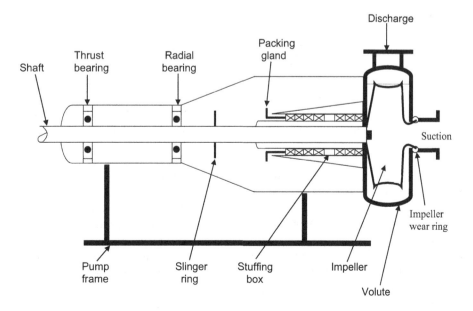

FIGURE 9.7 Major components of a centrifugal pump.

TERMINOLOGY

To understand centrifugal pumps and their operation, we must understand the terminology associated with centrifugal pumps:

- **Base Plate:** The foundation under a pump. It usually extends far enough to support the drive unit. The base plate is often referred to as the pump frame.
- **Bearings:** Devices used to reduce friction and allow the shaft to rotate easily. Bearings may be a sleeve, roller, or ball.
  - **Thrust Bearing:** In a single suction pump, it is the bearing located nearest the motor, farthest from the impeller. It takes up the major thrust of the shaft, which is opposite from the discharge direction.
  - **Radial (Line) Bearing:** In a single suction pump, it is the one closest to the pump. It rides free in its own section and takes up-and-down stresses.

Note: In most cases where the pump and motor are constructed on a common shaft (no coupling), the bearings will be part of the motor assembly.

- **Casing:** The housing surrounding the rotating element of the pump. In the majority of centrifugal pumps, this casing can also be called the **volute**.
  - **Split Casing:** A pump casing that is manufactured in two pieces fastened together using bolts. Split casing pumps may be vertically (perpendicular to the shaft direction) split or horizontally (parallel to the shaft direction) split.
  - **Coupling:** A device to join the pump shaft to the motor shaft. If the pump and motor are constructed on a common shaft, it is called a *close-coupled* arrangement.
  - **Extended Shaft:** For a pump constructed on one shaft that must be connected to the motor by a coupling.
  - **Frame:** The housing that supports the pump bearing assemblies. In an end suction pump, it may also support the pump casing and the rotating element.
  - **Impeller:** The rotating element in the pump that actually transfers the energy from the drive unit to the liquid. Depending on the pump application, the impeller may be open, semi-open, or closed. It may also be single or double suction.
  - **Impeller Eye:** The center of the impeller, the area that is subject to lower pressures due to the rapid movement of the liquid to the outer edge of the casing.

- **Priming:** Filling the casing and impeller with liquid. If this area is not completely full of liquid, the centrifugal pump will not pump efficiently.
- **Seals:** Devices used to stop the leakage of air into the inside of the casing around the shaft.
- **Packing:** The material placed around the pump shaft to seal the shaft opening in the casing and prevent air leakage into the casing.
- **Stuffing Box:** The assembly is located around the shaft at the rear of the casing. It holds the packing and lantern ring.
- **Lantern Ring:** Also known as the *seal cage,* it is positioned between the rings of packing in the stuffing box to allow the introduction of a lubricant (water, oil, or grease) onto the surface of the shaft to reduce friction between the packing and the rotating shaft.
- **Gland:** Also known as the *packing gland,* it is a metal assembly designed to apply even pressure to the packing to compress it tightly around the shaft.
- **Mechanical Seal:** A device consisting of a stationary element, a rotating element, and a spring to supply force to hold the two elements together. Mechanical seals may be either single or double units.
- **Shaft:** The rigid steel rod that transmits energy from the motor to the pump impeller. Shafts may be either vertical or horizontal.
- **Shaft Sleeve:** A piece of metal tubing placed over the shaft to protect it as it passes through the packing or seal area. In some cases, the sleeve may also help to position the impeller on the shaft.
- **Shut-off Head:** The head or pressure at which the centrifugal pump will stop discharging. It is also the pressure developed by the pump when it is operated against a closed discharge valve. This is also known as the *cut-off head.*
- **Shroud:** The metal plate that is used to either support the impeller vanes (open or semi-open impeller) or to enclose the vanes of the impeller (closed impeller).
- **Slinger Ring:** A device to prevent pumped liquids from traveling along the shaft and entering the bearing assembly. A slinger ring is also called a *deflector.*
- **Wearing Rings:** Devices that are installed on stationary or moving parts within the pump casing to protect the casing and the impeller from wear due to the movement of liquid through points of small clearances.
- **Impeller Ring:** A wearing ring installed directly on the impeller.

- **Casing Ring:** A wearing ring installed in the casing of the pump. A casing ring is also known as the suction head ring.
- **Stuffing Box Cover Ring:** A wearing ring installed at the impeller in an end suction pump to maintain the impeller clearances and prevent casing wear.

PUMP THEORY

The *volute-cased centrifugal pump* (see Figure 9.8) provides the pumping action necessary to transfer liquids from one point to another. First, the drive unit (usually an electric motor) supplies energy to the pump impeller to make it spin. This energy is then transferred to the water by the impeller. The vanes of the impeller spin the liquid toward the outer edge of the impeller at a high rate of speed or velocity. This action is very similar to that which would occur when a bucket full of water with a small hole in the bottom is attached to a rope and spun. When sitting still, the water in the bucket will drain out slowly. However, when the bucket is spinning, the water will be forced through the hole at a much higher rate of speed.

Centrifugal pumps may be single-stage, having a single impeller, or they may be multiple-stage, having several impellers through which the fluid flows in series. Each impeller in the series increases the pressure of the fluid at the pump discharge. Pumps may have 30 or more stages in extreme cases. In centrifugal pumps, a correlation of pump capacity, head, and speed at optimum efficiency is used to classify the pump impellers with respect to their specific geometry. This correlation is called *specific speed* and is an important parameter for analyzing pump performance (Garay, 1990).

The volute of the pump is designed to convert velocity energy to pressure energy. As a given volume of water moves from one cross-sectional area to another within the volute casing, the velocity or speed of the water changes proportionately. The volute casing has a cross-sectional area that is extremely small at the point in the case that is farthest from

the discharge (see Figure 9.8). This area increases continuously to the discharge. As this area increases, the velocity of the water passing through it decreases as it moves around the volute casing to the discharge point.

As the velocity of the water decreases, the velocity head decreases, and the energy is converted to pressure head. There is a direct relationship between the velocity of the water and the pressure it exerts. Therefore, as the velocity of the water decreases, the excess energy is converted to additional pressure (pressure head). This pressure head supplies the energy to move the water through the discharge piping.

PUMP CHARACTERISTICS

The centrifugal pump operates on the principle of energy transfer and, therefore, has certain definite characteristics that make it unique. The type and size of the impeller limit the amount of energy that can be transferred to the water, as well as the characteristics of the material being pumped and the total head of the system through which the water is moving. For any given centrifugal pump, there is a definite relationship between these factors, along with head (capacity), efficiency, and brake horsepower.

Head (Capacity)

As might be expected, the capacity of a centrifugal pump is directly related to the total head of the system. If the total head on the system is increased, the volume of the discharge will be reduced proportionately. As the head of the system increases, the capacity of the pump will decrease proportionately until the discharge stops. The head at which the discharge no longer occurs is known as the *cut-off head*. As pointed out earlier, the total head includes a certain amount of energy to overcome the friction of the system. This friction head can be greatly affected by the size and configuration of the piping and the condition of the system's valves. If the control valves on the system are partially closed, the friction head can increase dramatically. When this happens, the total head increases, and the capacity or volume discharged by the pump decreases. In many cases, this method is employed to reduce the discharge of a centrifugal pump. It should be noted, however, that this increases the load on the pump and drive system, causing additional energy requirements and wear.

The total closure of the discharge control valve increases the friction head to the point where all the energy supplied by the pump is consumed by the friction head and is not converted to pressure head. Consequently, the pump exceeds its cut-off head, and the pump discharge is reduced to zero. Again, it is important to note that, although the operation of a centrifugal pump against a closed discharge may not be hazardous (as it can be with other types of pumps), it should be avoided because of the excessive load placed on the drive unit and pump. Our experience has shown that, on occasion, the pump can produce pressure higher than what the pump discharge piping can withstand. Whenever this occurs, the

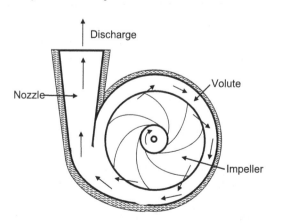

FIGURE 9.8 Cross-sectional diagram showing the features of a centrifugal pump.

discharge piping may be severely damaged by the operation of the pump against a closed or plugged discharge.

Efficiency

Every centrifugal pump operates with varying degrees of efficiency across its entire capacity and head ranges. An important factor in selecting a centrifugal pump is to select a unit that will perform near its maximum efficiency in the expected application.

Brake Horsepower Requirements

In addition to the head capacity and efficiency factors, most pump literature includes a graph showing the amount of energy in horsepower that must be supplied to the pump to achieve optimal performance.

ADVANTAGES AND DISADVANTAGES OF THE CENTRIFUGAL PUMP

The primary reason why centrifugal pumps have become one of the most widely used pumps is that they offer several advantages:

- **Construction:** The pump consists of a single rotating element and a simple casing, which can be constructed using a wide assortment of materials. If the fluids to be pumped are highly corrosive, the pump parts that are exposed to the fluid can be made of lead or other materials that are not likely to corrode. If the fluid being pumped is highly abrasive, the internal parts can be made of abrasion-resistant material or coated with a protective material. Also, the simple design of a centrifugal pump allows it to be constructed in a variety of sizes and configurations. No other pump currently available has the range of capacities or applications available through the use of the centrifugal pump.
- **Operation:** "Simple and quiet" best describes the operation of a centrifugal pump. An operator-in-training with a minimal amount of experience may be capable of operating facilities that use centrifugal-type pumps. Even when improperly operated, the centrifugal pump's rugged construction allows it to operate (in most cases) without major damage.
- **Maintenance:** The amount of wear on a centrifugal pump's moving parts is reduced, and its operating life is extended because its moving parts are not required to be constructed to very close tolerances.
- **Pressure is Self-Limited:** Because of the nature of its pumping action, the centrifugal pump will not exceed a predetermined maximum pressure. Thus, if the discharge valve is suddenly closed, the pump cannot generate additional pressure that might result in damage to the system or potentially create a hazardous working condition. The power

supplied to the impeller will only generate a specified amount of head (pressure). If a major portion of this head or pressure is consumed in overcoming friction or lost as heat energy, the pump will have a decreased capacity.
- **Adaptable to High- Speed Drive Systems:** Centrifugal pumps can make use of high-speed, high-efficiency motors. In situations where the pump is selected to match a specific operating condition that remains relatively constant, the pump drive unit can be used without the need for expensive speed reducers.
- **Small Space Requirements:** For most pumping capacities, the amount of space required for the installation of the centrifugal-type pump is much less than that of any other type of pump.
- **Fewer Moving Parts:** The rotary rather than reciprocating motion employed in centrifugal pumps reduces space and maintenance requirements due to the fewer number of moving parts required.

Although the centrifugal pump is one of the most widely used pumps, it does have a few disadvantages:

- **Additional Equipment Needed for Priming:** The centrifugal pump can be installed in a manner that makes it self-priming, but it is not capable of drawing water to the pump impeller unless the pump casing and impeller are filled with water. This can cause problems because, if the water in the casing drains out, the pump will cease pumping until it is refilled. Therefore, it is normally necessary to start a centrifugal pump with the discharge valve closed. The valve is then gradually opened to its proper operating level. Starting the pump against a closed discharge valve is not hazardous provided the valve is not left closed for extended periods.
- **Air Leaks Affect Pump Performance:** Air leaks on the suction side of the pump can cause reduced pumping capacity in several ways. If the leak is not serious enough to result in a total loss of prime, the pump may operate at a reduced head or capacity due to air mixing with the water. This causes the water to be lighter than normal and reduces the efficiency of the energy transfer process.
- **Narrow Range of Efficiency:** Centrifugal pump efficiency is directly related to the head capacity of the pump. The highest performance efficiency is available for only a very small section of the head-capacity range. When the pump is operated outside this optimum range, the efficiency may be greatly reduced.
- **Pump May Run Backwards:** If a centrifugal pump is stopped without closing the discharge line, it may run backwards because the pump does not have any built-in mechanism to prevent flow

from moving through the pump in the opposite direction (i.e., from the discharge side to the suction side). If the discharge valve is not closed or the system does not contain the proper check valves, the flow that was pumped from the supply tank to the discharge point will immediately flow back to the supply tank when the pump shuts off. This results in increased power consumption due to the frequent startup of the pump to transfer the same liquid from supply to discharge.

Note: It is sometimes difficult to tell whether a centrifugal pump is running forward or backwards because it appears and sounds like it is operating normally when running in reverse.

- **Pump Speed Is Difficult to Adjust:** Centrifugal pump speed cannot usually be adjusted without the use of additional equipment, such as speed-reducing or speed-increasing gears or special drive units. Because the speed of the pump is directly related to the discharge capacity of the pump, the primary method available to adjust the output of the pump, other than a valve on the discharge line, is to adjust the speed of the impeller. Unlike some other types of pumps, the delivery of the centrifugal pump cannot be adjusted by changing some operating parameters of the pump.

CENTRIFUGAL PUMP APPLICATIONS

The centrifugal pump is probably the most widely used pump available at this time because of its simplicity of design and wide-ranging diversity (it can be adjusted to suit a multitude of applications). Proper selection of the pump components (impeller, casing, etc.) and construction materials can produce a centrifugal pump capable of transporting not only water but also other materials ranging from material/chemical slurries to air (centrifugal blowers). Attempting to list all of the various applications for the centrifugal pump would exceed the limitations of this guidebook. Therefore, our discussion of pump applications is limited to those that frequently occur in water/wastewater operations.

- **Large Volume Pumping:** In water/wastewater operations, the primary use of centrifugal pumps is large volume pumping. In large volume pumping, generally low-speed, moderate-head, vertically shafted pumps are used. Centrifugal pumps are well suited for water/wastewater system operations because they can be used in conditions where high volumes are required and a change in flow is not a problem. As the discharge pressure on a centrifugal pump is increased, the quantity of water/wastewater pumped is reduced. Also, centrifugal pumps can be operated for short periods with the discharge valve closed.

- **Non-Clog Pumping:** These specifically designed centrifugal pumps use closed impellers with, at most, two to three vanes. They are usually designed to pass solids or trash up to 3 in in diameter.
- **Dry Pit Pump:** Depending on the application, this pump may be either a large volume pump or a non-clog pump. It is located in a dry pit that shares a common wall with the wet well. This pump is normally placed in such a position to ensure that the liquid level in the wet well is sufficient to maintain the pump's prime.
- **Wet Pit or Submersible Pump:** This type of pump is usually a non-clog type that can be submerged, with its motor, directly in the wet well. In a few instances, the pump may be submerged in the wet well while the motor remains above the water level. In these cases, the pump is connected to the motor by an extended shaft.
- **Underground Pump Stations:** Utilizing a wet well-dry well design, the pumps are located in an underground facility. Wastes are collected in a separate wet well, then pumped upward and discharged into another collector line or manhole. This system normally uses a non-clog type pump and is designed to add sufficient head to the water/waste flow to allow gravity to move the flow to the plant or the next pump station.
- **Recycle or Recirculation Pumps:** Because the liquids being transferred by the recycle or recirculation pump normally do not contain any large solids, the use of the non-clog type centrifugal pump is not always required. A standard centrifugal pump may be used to recycle trickling filter effluent, return activated sludge, or digester supernatant.
- **Service Water Pumps:** The wastewater plant effluent may be used for many purposes, such as cleaning tanks, watering lawns, providing the water to operate the chlorination system, and backwashing filters. Because the plant effluent used for these purposes is normally clean, the centrifugal pumps used closely parallel the units used for potable water. In many cases, the double suction, closed impeller, or turbine-type pump will be used.

PUMP CONTROL SYSTEMS

Pump operations usually control only one variable: flow, pressure or level. All pump control systems have a measuring device that compares a measured value with a desired one. This information relays to a control element that makes the changes. The user may obtain control with manually operated valves or sophisticated microprocessors. Economics dictate the accuracy and complication of a control system.

—Wahren (1997)

Most centrifugal pumps require some form of pump control system. The only exception to this practice is when the plant pumping facilities are designed to operate continuously at a constant rate of discharge. The typical pump control system includes a sensor to determine when the pump should be turned on or off and the electrical/electronic controls to actually start and stop the pump.

The control systems currently available for the centrifugal pump range from a very simple on-off float control to an extremely complex system capable of controlling several pumps in sequence.

In the following sections, we briefly describe the operation of various types of control devices/systems used with centrifugal pumps.

Float Control

Currently, the *float control system* is the simplest of the centrifugal pump controls (see Figure 9.9). In the float control system, the float rides on the water's surface in the well, storage tank, or clear well, attached to the pump controls by a rod with two collars. One collar activates the pump when the liquid level in the well or tank reaches a preset level, and a second collar shuts the pump off when the level in the well reaches a minimum level. This type of control system is simple to operate and relatively inexpensive to install and maintain. However, the system has several disadvantages. It operates at one discharge rate, which can result in (1) extreme variations in the hydraulic loading on succeeding units and (2) long periods of non-operation due to low flow periods or maintenance activities.

Pneumatic Controls

Pneumatic control systems (also called *bubbler tube control system*) are relatively simple systems that can be used to control one or more pumps. The system consists of an

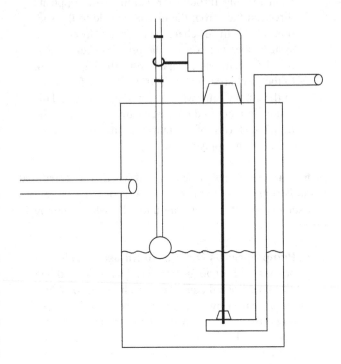

FIGURE 9.9 Float system for pump motor control.

air compressor, a tube extending into the well, clear well, or storage tank/basin, pressure-sensitive switches with varying on/off set points, and a pressure relief valve (see Figure 9.10). The system works on the basic principle of measuring the depth of the water in the well or tank by determining the air pressure that is necessary to just release a bubble from the bottom of the tube (see Figure 9.10); hence, the name *bubbler tube*. The air pressure required to force a bubble out of the tube is determined by the liquid pressure, which is directly related to the depth of the liquid (1 psi = 2.31 ft).

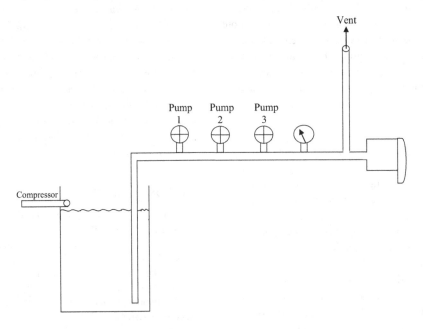

FIGURE 9.10 Pneumatic system for pump motor control.

By installing a pressure switch on the airline to activate the pump starter at a given pressure, the level of the water can be controlled by activating one or more pumps.

Installation of additional pressure switches with slightly different pressure settings allows several pumps to be activated in sequence. For example, the first pressure switch can be adjusted to activate a pump when the level in the wet well/tank is 3.8 ft (1.6 psi) and shut off at 1.7 ft (0.74 psi). If the flow into the pump well/tank varies greatly and additional pumps are available to ensure that the level in the well/tank does not exceed the design capacity, additional pressure switches may be installed. These additional pressure switches are set to activate a second pump when the level in the well/tank reaches a preset level (i.e., 4.5 ft or 1.95 psi) and cut off when the well/tank level is reduced to a preset level (i.e., 2.7 ft or 1.2 psi). If the first pump's capacity is less than the rate of flow into the well/tank, the level of the well/tank continues to rise. Upon reaching the preset level (4 ft), it will activate the second pump. If necessary, a third pump can be added to the system set to activate at a third preset well/tank depth (4.6 ft or 1.99 psi) and cut off at a preset depth (3.0 ft or 1.3 psi).

The pneumatic control system is relatively simple with minimal operation and maintenance requirements. The major operational problem involved with this control system is the clogging of the bubbler tube. If, for some reason, the tube becomes clogged, the pressure in the system can increase and may activate all pumps to run even when the well/tank is low. This can result in excessive power consumption, which, in turn, may damage the pumps.

ELECTRODE CONTROL SYSTEMS

The *Electrode Control System* uses a probe or electrode to control the pump on-and-off cycle. A relatively simple control system consists of two electrodes extended into the clear well, storage tank, or basin. One electrode is designed to activate the pump starter when it is submerged in the water; the second electrode extends deeper into the well/tank and is designed to open the pump circuit when the water drops below the electrode (see Figure 9.11). The major maintenance requirement of this system is keeping the electrodes clean.

Important Point: Because the Electrode Control System uses two separate electrodes, the unit may be locked into an on-cycle or off-cycle, depending on which electrode is involved.

OTHER CONTROL SYSTEMS

Several other systems that use electrical energy are available for controlling the centrifugal pump. These include a *tube-like device* that has several electrical contacts mounted inside (see Figure 9.12). As the water level rises in the clear well, storage tank, or basin, the water rises in the tube, contacting the electrical contacts and activating

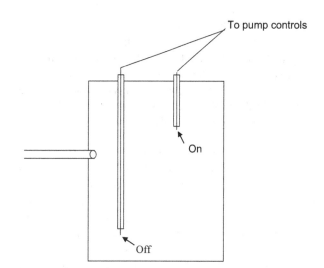

FIGURE 9.11   Electrode system for pump motor control.

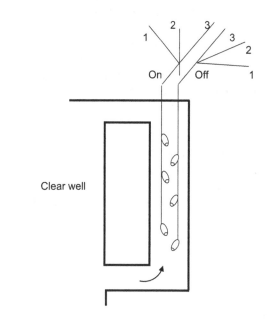

FIGURE 9.12   Electrical contacts for pump motor control.

the motor starter. Again, this system can be used to activate several pumps in series by installing several sets of contact points. As the water level drops in the well/tank, the level in the tube drops below a second contact that deactivates the motor and stops the pumping. Another control system uses a *mercury switch* (or a similar type of switch) enclosed in a protective capsule. Again, two units are required per pump: one switch activates the pump when the liquid level rises, and the second switch shuts the pump off when the level reaches the desired minimum depth.

Electronic Control Systems

Several centrifugal pump control systems are available that use electronic systems for the control of pump operation. A brief description of some of these systems is provided in the sections that follow.

Flow Equalization System

In any multiple pump operation, the flow delivered by each pump will vary due to the basic hydraulic design of the system. To obtain equal loads on each pump when two or more are in operation, the flow equalization system electronically monitors the delivery of each pump and adjusts the speed of the pumps to achieve similar discharge rates for each pump.

Sonar or Other Transmission Types of Controllers

A *sonar* or *low-level radiation system* can be used to control centrifugal pumps. This type of system uses a transmitter and receiver to locate the level of the water in a tank, clear well, or basin. When the level reaches a predetermined set point, the pump is activated and when the level is reduced to a predetermined set point, the pump is shut off. Basically, the system is very similar to a radar unit. The transmitter sends out a beam that travels to the liquid, bounces off the surface, and returns to the receiver. The time required for this is directly proportional to the distance from the liquid to the instrument. The electronic components of the system can be adjusted to activate the pump when the time interval corresponds to a specific depth in the well or tank. The electronic system can also be set to shut off the pump when the time interval corresponds to a preset minimum depth.

Motor Controllers

Several types of controllers are available that start and stop motors. They also protect the motor from overloads and short circuit conditions. Many motor controllers also function to adjust motor speed to increase or decrease the discharge rate for a centrifugal pump. This type of control may use one of the previously described controls to start and stop the pump and, in some cases, adjust the speed of the unit. As the depth of the water in a well or tank increases, the sensor automatically increases the speed of the motor in predetermined steps to the maximum design speed. If the level continues to increase, the sensor may be designed to activate an additional pump.

Protective Instrumentation

Protective instrumentation of some type is normally employed in pump or motor installations. (Note that the information provided in this section applies to the centrifugal pump as well as to many other types of pumps.) Protective instrumentation for centrifugal pumps (or most other types of pumps) is dependent on pump size, application, and the amount of operator supervision. That is, pumps under 500 hp often only come with pressure gauges and temperature indicators. These gauges or transducers may be mounted locally (on the pump itself) or remotely (in suction and discharge lines immediately upstream and downstream of the suction and discharge nozzles). If transducers are employed, readings are typically displayed and taken (or automatically recorded) at a remote operating panel or control center.

Temperature Detectors

Resistance temperature devices (RTDs) and *thermocouples* (see Figure 9.13) (Grimes, 1976) are commonly used as temperature detectors on the pump prime movers (motors) to indicate temperature problems. In some cases, dial thermometers, armored glass-stem thermometers, or bimetallic-actuated temperature indicators are used. Whichever device is employed, it typically monitors temperature variances that may indicate a possible source of trouble. On electric motors, greater than 250 hp, RTD elements are used to monitor temperatures in stator winding coils. Two RTDs per phase is standard. One RTD element is usually installed in the shoe of the loaded area employed on journal bearings in pumps and motors. Normally, tilted-pad thrust bearings have an RTD element in the active as well as the inactive side. RTD elements are used when remote indication, recording, or automatic logging of temperature readings is required. Because of their smaller size, RTDs provide more flexibility in locating the measuring device near the measuring point. When dial thermometers are installed, they monitor oil thrown from bearings. Sometimes, temperature detectors also monitor bearings with water-cooled jackets to warn against water supply failure. Pumps with heavy wall casings may also have casing temperature monitors.

Vibration Monitors

Vibration sensors are available to measure either bearing vibration or shaft vibration direction directly. Direct measurement of shaft vibration is desirable for machines with stiff bearing supports where bearing-cap measurements will be only a fraction of the shaft vibration. Wahren (1997) noted that pumps and motors of 1,000 hp and larger may have the following vibration monitoring equipment:

- Seismic pickup with double set points installed on the pump outboard housing
- Proximators with X–Y vibration probes complete with interconnecting coaxial cables at each radial and thrust journal bearing
- Key phasor with proximator and interconnecting coaxial cables

Supervisory Instrumentation

Supervisory instruments are used to monitor the routine operation of pumps, their prime movers, and their accessories in order to sustain a desired level of reliability and

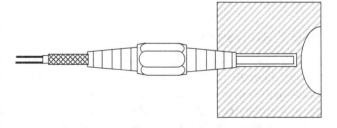

FIGURE 9.13   Thermocouple installation in journal bearing.

performance. Generally, these instruments are not used for accurate performance tests or for automatic control, although they may share connections or functions. Supervisory instruments consist of annunciators and alarms that provide operators with warnings of abnormal conditions that, unless corrected, will cause pump failure. Annunciators used for both alarm and pre-alarm have both visible and audible signals.

CENTRIFUGAL PUMP MODIFICATIONS

The centrifugal pump can be modified to meet the needs of several different applications. If there is a need to produce higher discharge heads, the pump may be modified to include several additional impellers. If the material being pumped contains a large amount of material that could clog the pump, the pump construction may be modified to remove a major portion of the impeller from direct contact with the material being pumped. Although there are numerous modifications of the centrifugal pump available, the scope of this text covers only those that have found wide application in the water distribution and wastewater collection and treatment fields. Modifications to be presented in this section include:

- Submersible pumps
- Recessed impeller or vortex pumps
- Turbine pumps

Submersible Pumps

The *submersible pump* is, as the name suggests, placed directly in the wet well or groundwater well. It uses a water-proof electric motor located below the static level of the wet well/well to drive a series of impellers. In some cases, only the pump is submerged, while in other cases, the entire pump-motor assembly is submerged. Figure 9.14 illustrates this system.

Description

The submersible pump may be either a close-coupled centrifugal pump or an extended shaft centrifugal pump. In a close-coupled system, both the motor and pump are submerged in the liquid being pumped. Seals prevent water and wastewater from entering the inside of the motor, protecting the electric motor in a close-coupled pump from shorts and motor burnout. In the extended shaft system, the pump is submerged, while the motor is mounted above the pump wet well. In this situation, an extended shaft assembly must connect the pump and motor.

Applications

The submersible pump has wide applications in the water/wastewater treatment industry. It generally can be substituted for any application of other types of centrifugal pumps. However, it has found its widest application in distribution or collector system pump stations.

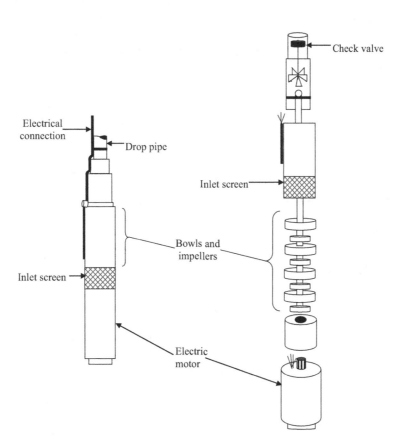

FIGURE 9.14 Submersible pump.

Advantages

In addition to the advantages discussed earlier for a conventional centrifugal pump, the submersible pump has additional advantages:

- Because it is located below the surface of the liquid, there is less chance that the pump will lose its prime, develop air leaks on the suction side of the pump or require initial priming.
- Since the pump or the entire assembly is located in the well/wet well, there are lower costs associated with the construction and operation of this system. It is not necessary to construct a dry well or a large structure to hold the pumping equipment and necessary controls.

Disadvantages

The major disadvantage associated with the submersible pump is the lack of access to the pump or pump and motor. The performance of any maintenance requires either drainage of the wet well, extensive lift equipment to remove the equipment from the wet well, or both. This may be a major factor in determining whether a pump receives the attention it requires. Also, in most cases, all major maintenance on close-coupled submersible pumps must be performed by outside contractors due to the need to re-seal the motor to prevent leakage.

Recessed Impeller or Vortex Pumps

The *recessed impeller* or *vortex pump* uses an impeller that is either partially or wholly recessed into the rear of the casing (see Figure 9.15). The spinning action of the impeller creates a vortex or whirlpool. This whirlpool increases the velocity of the material being pumped. As in other centrifugal pumps, this increased velocity is then converted into increased pressure or head.

Applications

The recessed impeller or vortex pump is widely used in applications where the liquid being pumped contains large

amounts of solids, debris, or slurries that could clog or damage the pump's impeller. It has found increasing use as a sludge pump in facilities that continuously withdraw sludge from their primary clarifiers.

Advantages

The major advantage of this modification is the increased ability to handle materials that would normally clog or damage the pump impeller. Because the majority of the flow does not come into direct contact with the impeller, there is much less potential for problems.

Disadvantages

Because there is less direct contact between the liquid and the impeller, the energy transfer is less efficient. This results in somewhat higher power costs and limits the pump's application to low to moderate capacities. Objects that might clog a conventional centrifugal pump are able to pass through this pump. Although this is very beneficial in reducing pump maintenance requirements, it has, in some situations, allowed material to be passed into a less accessible location before becoming an obstruction. To be effective, the piping and valving must be designed to pass objects of a size equal to that which the pump will discharge.

Turbine Pumps

The turbine pump consists of a motor, drive shaft, a discharge pipe of varying lengths, and one or more impeller-bowl assemblies (see Figure 9.16). It is normally a vertical assembly where water enters at the bottom, passes

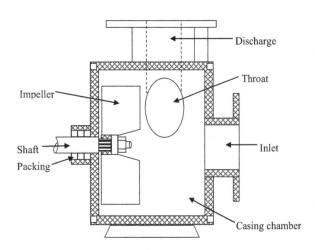

FIGURE 9.15 Schematic of a recessed impeller or vortex pump.

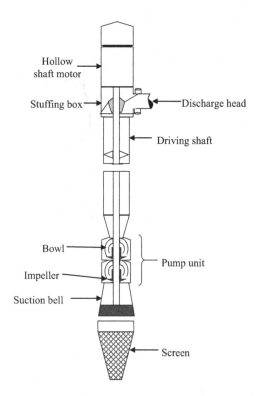

FIGURE 9.16 Vertical turbine pump.

axially through the impeller-bowl assembly where the energy transfer occurs, and then moves upward through additional impeller-bowl assemblies to the discharge pipe. The length of this discharge pipe will vary with the distance from the wet well to the desired point of discharge.

Applications

Due to the construction of the turbine pump, its major applications have traditionally been for pumping relatively clean water. The line shaft turbine pump has been used extensively for drinking water pumping, especially in situations where water is withdrawn from deep wells. The main application in wastewater plants has been pumping plant effluent back into the plant for use as service water.

Advantages

The turbine pump has a major advantage in the amount of head it is capable of producing. By installing additional impeller-bowl assemblies, the pump can achieve even greater production. Moreover, the turbine pump has simple construction, a low noise level, and is adaptable to several drive types—motor, engine, or turbine.

Disadvantages

High initial costs and high repair costs are two of the major disadvantages of turbine pumps. In addition, the presence of large amounts of solids within the liquid being pumped can seriously increase the amount of maintenance the pump requires; consequently, the unit has not found widespread use in any situation other than service water pumping.

POSITIVE DISPLACEMENT PUMPS

Positive displacement pumps force or displace water through the pumping mechanism. Most have a reciprocating element that draws water into the pump chamber on one stroke and pushes it out on the other. Unlike centrifugal pumps, which are meant for low-pressure, high-flow applications, positive displacement pumps can achieve greater pressures but are slower-moving, low-flow pumps. Other positive displacement pumps include the piston pump, diaphragm pump, and peristaltic pumps, which are the focus of our discussion. In the water/wastewater industry, positive displacement pumps are most often found as chemical feed pumps. It is important to remember that positive displacement pumps cannot be operated against a closed discharge valve. As the name indicates, something must be displaced with each stroke of the pump. Closing the discharge valve can cause rupturing of the discharge pipe, the pump head, the valve, or some other component.

PISTON PUMP OR RECIPROCATING PUMP

The *piston or reciprocating pump* is one type of positive displacement pump. This pump works just like the piston in an automobile engine: on the intake stroke, the intake valve opens, filling the cylinder with liquid. As the piston

reverses direction, the intake valve is pushed closed and the discharge valve is pushed open, allowing the liquid to be pushed into the discharge pipe. With the next reversal of the piston, the discharge valve is pulled closed and the intake valve is pulled open, and the cycle repeats. A piston pump is usually equipped with an electric motor and a gear and cam system that drives a plunger connected to the piston. Just like an automobile engine piston, the piston must have packing rings to prevent leakage and must be lubricated to reduce friction. Because the piston is in contact with the liquid being pumped, only high-grade lubricants can be used when pumping materials that will be added to drinking water. The valves must be replaced periodically as well.

DIAPHRAGM PUMP

A *diaphragm pump* is composed of a chamber used to pump the fluid,; a diaphragm that is operated by either electric or mechanical means, and two valve assemblies: a suction valve assembly and a discharge valve assembly (see Figure 9.17). A diaphragm pump is a variation of the piston pump in which the plunger is isolated from the liquid being pumped by a rubber or synthetic diaphragm. As the diaphragm is moved back and forth by the plunger, liquid is pulled into and pushed out of the pump. This arrangement provides better protection against leakage of the liquid being pumped and allows the use of lubricants that otherwise would not be permitted. Care must be taken to assure that diaphragms are replaced before they rupture. Diaphragm pumps are appropriate for discharge pressures up to about 125 psi but do not work well if they must lift liquids more than about 4 ft. Diaphragm pumps are frequently used for chemical feed pumps. By adjusting the frequency of the plunger motion and the length of the stroke, extremely accurate flow rates can be metered. The pump may be driven hydraulically by an electric motor or by an electronic driver in which the plunger is operated by a solenoid. Electronically driven metering pumps are extremely reliable (with few moving parts) and inexpensive.

PERISTALIC PUMPS

Peristaltic pumps (sometimes called tubing pumps) use a series of rollers to compress plastic tubing to move the liquid through the tubing. A rotary gear turns the rollers

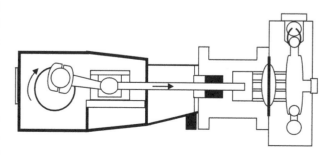

FIGURE 9.17 Diaphragm pump.

at a constant speed to meter the flow. Peristaltic pumps are mainly used as chemical feed pumps. The flow rate is adjusted by changing the speed at which the roller gear rotates (to push the waves faster) or by changing the size of the tubing (so there is more liquid in each wave). As long as the right type of tubing is used, peristaltic pumps can operate at discharge pressures up to 100 psi. Note that the tubing must be resistant to deterioration from the chemical being pumped. The principal item of maintenance is the periodic replacement of the tubing in the pump head. There are no check valves or diaphragms in this type of pump.

CHAPTER REVIEW QUESTIONS

9.1 Applications in which chemicals must be metered under high pressure require high-powered _____ pumps.

9.2 _____ materials are materials that resist any flow-producing force.

9.3 What type of pump is usually used for pumping high-viscosity materials?

9.4 High-powered positive displacement pumps are used to pump chemicals that are under _____ pressure.

9.5 _____ viscosity materials are thick.

9.6 When the _____ of a pump impeller is above the level of the pumped fluid, the condition is called suction lift.

9.7 When a pump is not running, conditions are referred to as _____; when a pump is running the conditions are _____.

9.8 With the _____, the difference in elevation between the suction and discharge liquid levels is called static head.

9.9 Velocity head is expressed mathematically as _____.

9.10 The sum of total static head, head loss, and dynamic head is called _____.

9.11 What are the three basic types of curves used for centrifugal pumps?

9.12 The liquid used to rate pump capacity is _____.

9.13 Because of the reduced amount of air pressure at high altitudes, less _____ is available for the pump.

9.14 With the pump shut off, the difference between the suction and discharge liquid levels is called _____.

9.15 _____ and _____ is the largest single contributing factor to the reduction of pressure at a pump impeller.

9.16 The operation of a centrifugal pump is based on _____.

9.17 The casing of a pump encloses the pump impeller, the shaft, and the _____.

9.18 The _____ is the part of the pump that supplies energy to the fluid.

9.19 If wearing rings are used only on the volute case, we must replace the _____ and _____ at the same time.

9.20 Which part of the end-suction pump directs water flow into and out of the pump?

REFERENCES

Garay, P.N., 1990. *Pump Application Desk Book*. Lilburn, GA: The Fairmont Press, Inc., p. 1.

Hauser, B.A., 1993. *Hydraulics for Operators*. Boca Raton, FL: Lewis Publishers.

Hydraulic Institute Complete Pump Standards, 4th ed. Cleveland: Hydraulic Institute, 1994.

Lindeburg, M.R., 1986. *Civil Engineering Reference Manual*, 4th ed. San Carlos, CA: Professional Publications, Inc.

Wahren, U. 1997. *Practical Introduction to Pumping Technology*. Houston: Gulf Publishing Company.

10 Water/Wastewater Conveyance

The design considerations for the piping system are the function of the specifics of the system. However, all piping systems have a few common issues: The pipe strength must be able to resist internal pressure, handling, and earth and traffic loads; the pipe characteristics must enable the pipe to withstand corrosion and abrasion and expansion and contraction of the pipeline (if the line is exposed to atmospheric conditions); engineers must select the appropriate pipe support, bedding, and backfill conditions; the design must account for the potential for pipe failure at the connection point to the basins due to subsidence of a massive structure; and the composition of the pipe must not give rise to any adverse effects on the health of consumers.

—Kawamura (1999)

In the U.S., water and wastewater conveyance systems are quite extensive. Consider, for example, wastewater conveyance. In the year 2000, the United States operated 21,264 collection and conveyance systems that included both sanitary and combined sewer systems. Publicly owned sewer systems in the country account for 724,000 miles of sewer pipe and privately owned sewer pipe comprises an additional 500,000 miles. Most of our nation's conveyance systems are beginning to show signs of aging, with some systems dating back more than 100 years (USEPA, 2006).

DELIVERING THE LIFEBLOOD OF CIVILIZATION

Conveyance or piping systems resemble veins, arteries, and capillaries. According to Nayyar (2000), "They carry the lifeblood of modern civilization. In a modern city they transport water from the sources of water supply to the points of distribution; convey waste from residential and commercial buildings and other civic facilities to the treatment facility or the point of discharge." Water and wastewater operators must be familiar with piping, piping systems, and the many components that make piping systems function. Operators are directly concerned with various forms of piping, tubing, hose, and the fittings that connect these components to create workable systems.

This chapter covers important, practical information about the piping systems that are a vital part of plant operation, and essential to the success of the total activity. To prevent major system trouble, skilled operators are called upon to perform the important function of preventive maintenance to avoid major breakdowns and must be able to make needed repairs when breakdowns do occur.

A comprehensive knowledge of piping systems and accouterments is essential to maintaining plant operations.

Key Point: While water is transported entirely through a pressurized system, wastewater is preferably conveyed via a gravity flow system.

CONVEYANCE SYSTEMS

With regard to early conveyance systems, the prevailing practice in medieval England was the use of closed pipes. This practice differed from that of the Romans, who generally employed open channels in their long-distance aqueducts and used pipes mainly to distribute water within cities. The English preferred to lay long runs of pipes from the water source to the final destination. The Italians, on the other hand, where ancient aqueduct arches were still visible, seem to have had more of a tendency to follow the Roman tradition of long-distance channel conduits. At least some of the channel aqueducts seem to have fed local distribution systems of lead or earthenware pipes (Magnusson, 2001).

With today's water and wastewater conveyance, not much has changed from the past. Our goal today remains the same: (1) Convey water from the source to the treatment facility to the user, and (2) convey wastewater from the user to treatment and then to the environment. In water and wastewater operations, the term conveyance or *piping system* refers to a complete network of pipes, valves, and other components. For water and wastewater operations in particular, the piping system is all-inclusive; it includes both the network of pipes, valves, and other components that bring the flow (water or wastewater) to the treatment facility, as well as the piping, valves, and other components that distribute treated water to the end-user and/or treated wastewater to the outfall. In short, all piping systems are designed to perform a specific function.

Probably the best way to illustrate the importance of a piping system is to describe many of its applications in water and wastewater operations. In the modern water and wastewater treatment plant, piping systems are critical to successful operation. In water and wastewater operations, fluids and gases are extensively used in processing operations, and they are usually conveyed through pipes. Piping carries water and wastewater into the plant for treatment, fuel oil to heating units, steam to steam services, lubricants to machinery, compressed air to pneumatic service outlets for air-powered tools, and chemicals to unit processes. For water treatment alone, as Kawamura (1999) pointed out, there are "six basic piping systems: (1) raw water and finished waste distribution mains; (2) plant yard piping that connects the unit processes; (3) plant utility, including the fire hydrant lines; (4) chemical

DOI: 10.1201/9781003581901-12

lines; (5) sewer lines; and (6) miscellaneous piping, such as drainage and irrigation lines.

In addition to raw water, treated water, wastewater influent, and treated wastewater effluent, the materials conveyed through piping systems include oils, chemicals, liquefied gases, acids, paints, sludge, and many other others.

Important Point: Because of the wide variety of materials that piping systems can convey, the components of piping systems are made from different materials and are furnished in many sizes to accommodate the requirements of numerous applications. For example, pipes and fittings can be made of stainless steel, many different types of plastic, brass, lead, glass, steel, and cast iron.

Any waterworks or wastewater treatment plant has many piping systems, not just those systems that convey water and wastewater. Along with these systems mentioned earlier, plant piping systems also include those that provide hot and cold water for personnel use. Another system heats the plant, while yet another may be used for air conditioning.

Water and wastewater operators have many responsibilities and basic skills. The typical plant operator is skilled in HVAC systems, chemical feed systems, mechanical equipment operation and repair, and piping system maintenance. However, in this text, only the fluid transfer systems themselves are important. The units that the piping system serves or supplies (such as pumps, unit processes, and machines) are discussed in other chapters.

For water and wastewater operators, a familiar example of a piping system is the network of sodium hypochlorite pipes in treatment plants that use this chemical for disinfection and other purposes. The whole group of components—pipes, fittings, and valves—working together for one purpose makes up a *system*. This particular system has a definite purpose—to carry and distribute sodium hypochlorite, conveying it to the point of application.

Note: This chapter is concerned only with the piping system used to circulate the chemical, not with the hypochlorination equipment itself. Our concern begins where the chemical outlet is connected to the storage tank and continues to the point where the pipe is connected to the point of application. The piping, fittings, and valves of the hypochlorination pipeline (and others) are important to us. Gate, needle, pressure-relief, air-and-vacuum relief, diaphragm, pinch, butterfly, check, rotary, and globe valves, as well as traps, expansion joints, plugs, elbows, tee fittings, couplings, reducers, laterals, caps, and other fittings, help ensure the effective flow of fluids through the lines. As you trace a piping system through your plant site, you will find many of them (see Figure 10.1). They are important because they are directly related to the operation of the system. Piping system maintenance is focused on keeping the system functioning properly, and to function properly, piping systems must be kept closed and leakproof.

Important Point: Figure 10.1 shows a single-line diagram similar to an electrical schematic. It uses symbols to represent all the diagram components. A double-line diagram (not shown here) provides a pictorial view of the pipe, joints, valves, and other major components similar to an electrical wiring diagram and an electrical schematic.

DEFINITIONS

Key terms related to water/wastewater conveyance are listed and defined in this section.

Absolute Pressure: Gauge pressure plus atmospheric pressure.

Alloy: A substance composed of two or more metals.

Anneal: To heat and then cool a metal to make it softer and less brittle.

Annealing: The process of heating and then cooling a metal, usually to make it softer and less brittle.

Asbestos: A fibrous mineral form of magnesium silicate.

Backsiphonage: A condition in which the pressure in the distribution system is less than atmospheric pressure,

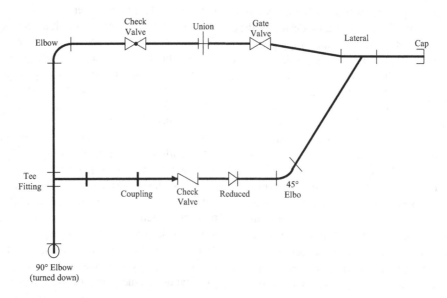

FIGURE 10.1   Various components in a single-line piping diagram.

allowing contamination to enter a water system through a cross-connection.

Bellows: A device that uses a bellow for measuring pressure.

Bimetallic: Made of two different types of metal.

Bourbon Tube: A semicircular tube of elliptical cross-section, used to sense pressure changes.

Brazing: Soldering with a nonferrous alloy that melts at a lower temperature than the metals being joined; also known as hard soldering.

Butterfly Valve: A valve in which a disk rotates on a shaft as the valve opens and closes. In the fully open position, the disk is parallel to the axis of the pipe.

Carcass: The reinforcement layers of a hose, located between the inner tube and the outer cover.

Cast Iron: A generic term for the family of high carbon-silicon-iron casting alloys, including gray, white, malleable, and ductile iron.

Check Valve: A valve designed to open in the direction of normal flow and close with flow reversal. An approved check valve has substantial construction and suitable materials, closes positively, and permits no leakage in a direction opposite to normal flow.

Condensate: Steam that condenses into water within a piping system.

Diaphragm Valve: A valve in which the closing element is a thin, flexible disk, often used in low-pressure systems.

Differential Pressure: The difference between the inlet and outlet pressures in a piping system.

Double-Line Diagram: A pictorial view of pipes, joints, valves, and other major components similar to an electrical wiring diagram.

Ductile: A term applied to a metal that can be fashioned into a new form without breaking.

Expansion Joint: A device that absorbs thermal expansion and contraction in piping systems.

Extruding: The process of shaping metal or plastic by forcing it through a die.

Ferrous: A term applied to a metal that contains iron.

Ferrule: A short bushing used for making a tight connection.

Filter: An accessory fitting used to remove solids from a fluid stream.

Fluids: Any substance that flows.

Flux: A substance used in soldering to prevent the formation of oxides during the soldering operation and to increase the wetting action, allowing solder to flow more freely.

Friable: Readily crumbled by hand.

Gate Valve: A valve in which the closing element consists of a disk that slides across an opening to stop the flow of water.

Gauge Pressure: The amount by which the total absolute pressure exceeds the ambient atmospheric pressure.

Globe Valve: A valve having a round, ball-like shell, and a horizontal disk.

Joint: A connection between two lengths of pipe or between a length of pipe and a fitting.

Laminar: Flow arranged in or consisting of thin layers.

Mandrel: A central core or spindle around which material may be shaped.

Neoprene: A synthetic material that is highly resistant to oil, flame, various chemicals, and weathering.

Metallurgy: The science and study of metals.

Nominal Pipe Size: The thickness given in the product material specifications or standards to which manufacturing tolerances are applied.

Nonferrous: A term applied to a material that does not contain iron.

Piping Systems: A complete network of pipes, valves, and other components.

Ply: One of several thin sheets or layers of material.

Prestressed Concrete: Concrete that has been compressed with wires or rods to reduce or eliminate cracking and tensile forces.

Pressure-Regulating Valve: A valve with a horizontal disk for automatically reducing water pressures in a main to a preset value.

PVC: Polyvinyl chloride plastic pipe.

Schedule: An approximate value of the expression 1,000 P/S, where P is the service pressure and S is the allowable stress, both expressed in pounds per square inch.

Single-Line Diagram: A diagram that uses symbols for all the components.

Soldering: A form of brazing in which nonferrous filler metals having melting temperatures below 800°F (427°C) are used. The filler material is called *solder* and is distributed between surfaces by *capillary action*.

Solenoid: An electrically energized coil of wire surrounding a movable iron case.

Stainless Steel: An alloy steel having unusual corrosion-resisting properties, usually imparted by nickel and chromium.

Strainer: An accessory fitting used to remove large particles of foreign matter from a fluid.

Throttle: Controlling flow through a valve by means of intermediate steps between fully open and fully closed.

Tinning: Covering metal to be soldered with a thin coat of solder to work properly. Overheating or failure to keep the metal clean can cause the point to become covered with oxide. The process of replacing this coat of oxide is called tinning.

Trap: An accessory fitting used to remove condensate from steam lines.

Vacuum Breaker: A mechanical device that allows air into the piping system, thereby preventing backflow that could otherwise be caused by the siphoning action created by a partial vacuum.

Viscosity: The thickness or resistance to flow of a liquid.

Vitrified Clay: Clay that has been treated in a kiln to produce a glazed, watertight surface.

Water Hammer: The concussion of moving water against the sides of a pipe, caused by a sudden change in the rate of flow or stoppage of flow in the line.

FLUIDS VS. LIQUIDS

We use the term *fluids* throughout this text to describe the substance(s) being conveyed through various piping systems from one part of the plant to another. We normally think of pipes conveying some type of liquid substance, which most of us take to mean the same as fluid; however, there is a subtle difference between the two terms. The dictionary definition of *fluid* is any substance that flows, which can mean a liquid or gas (air, oxygen, nitrogen, etc.). Some fluids carried by piping systems include thick, viscous mixtures such as sludge in a semi-fluid state. Although sludge and other such materials might seem more solid (at times) than liquid, they do flow, and are considered fluids. In addition to carrying liquids such as oil, hydraulic fluids, and chemicals, piping systems also carry compressed air and steam, which are considered fluids because they flow.

Important Point: Fluids travel through a piping system at various pressures, temperatures, and speeds.

LET'S TALK ABOUT PIPE

We can say that a means of conveyance can be accomplished by any discernible, confined, and discrete conveyance, such as a ditch, channel, tunnel, conduit, discrete fissure, container, or pipe. Obviously, in our discussion here, we are talking about the conveyance of water and wastewater in a closed conveyance apparatus, that is, we are typically referring to pipe—all kinds of pipe. At present, there are three commonly used types of pipe for water and wastewater conveyance: DIP, PVC, and HDPE pipe.

- **DIP (Ductile Iron Pipe):** DIP is made from ferric scrap metal, typically containing over 98% recycled content (using recycled materials for any purpose is a laudable effort and resource-saving activity). DIP pipe is strong, with a yield strength (i.e., the maximum point of stress at which a material becomes plastic; usually measured in N/m² or Pascals) of 42,000 psi.

  Because of its recycled material content, DIP is highly recyclable. It also allows for the fabrication of larger inside diameters compared to HDPE and PVC pipes. It also can withstand or handle pressures of up to almost 1,600 psi in 12-inch pipes. To determine the maximum internal pressure for thin-walled pipe, we use the Universal Hoop Stress Equation:

$$P = \frac{2(P_w + P_s) - t}{D} \qquad (10.1)$$

  where
  (P) is the pressure (psi).
  (P_w) is the working pressure (psi).
  (P_s) is the surge allowance (e.g., 100 psi or equivalent to a 2-fps velocity change in the pipe).

(t) is the minimum wall thickness to contain the design pressure; it simply designates the net thickness of the pipe.
(D) is the inside diameter of the pipe.

Note: DIP can withstand severe stresses caused externally by shifting ground and heavy loads, as well as internally by water pressure and water hammer. It has a minimum tensile strength of 60,000 psi, a yield strength of 42,000 psi, and a minimum elongation of 10 %. Additionally, it is easy to install, has high impact resistance, demonstrates tremendous bursting strength, and is virtually maintenance-free.

- **PVC (Polyvinyl Chloride Pipe):** PVC is a vinyl polymer that has a high resistance to heat and corrosion, and it is used extensively in plumbing, drainage, water mains, and irrigation. It is a strong pipe but not as flexible as HDPE. PVC can withstand about 100 psi above its pressure class and can handle standard water surges. It is easy to join other PVC connections. However, PVC's Achilles Heel has a tendency to fail and crack under extreme conditions of stress.
- **HDPE (High-Density Polyethylene):** HDPE is a pipe commonly used in industrial applications such as carrying water, wastewater, hazardous chemicals, slurries, and compressed gases—all of which are common in water and wastewater treatment operations. The pipe is resistant to stress, rain, and wind.

MAINTAINING FLUID FLOW IN PIPING SYSTEMS

The primary purpose of any piping system is to maintain the free and smooth flow of fluids through the system. Another purpose is to ensure that the fluids being conveyed are kept in good condition (i.e., free of contamination). Piping systems are purposely designed to ensure the free and smooth flow of fluids throughout the system, but additional components are often included to maintain fluid quality. Piping system filters are one example, and strainers and traps are two others.

It is extremely important to maintain free and smooth flow and fluid quality in piping systems, especially those that feed vital pieces of equipment or machinery. Consider the internal combustion engine, for example. Impurities such as dirt and metal particles can damage internal components and cause excessive wear and eventual breakdown. To help prevent such wear, the oil is continuously run through a filter designed to trap and remove the impurities.

Other piping systems need the same type of protection that the internal combustion engine does, which is why most piping systems include filters, strainers, and traps. These filtering components can prevent damage to valves, fittings, the pipe itself, and downstream equipment or machinery. Chemicals, various types of waste products, paint, and pressurized steam are good examples of potentially damaging fluids. Filters and strainers play an important role in piping systems—protecting both the piping system and the equipment that the system serves.

Scaling

Because sodium and calcium hypochlorite are widely used in water and wastewater treatment operations, problems common in piping systems feeding this chemical are of special concern. In this section, we discuss *scaling* problems that can occur in piping systems convey hypochlorite solutions. To maintain the chlorine in solution (used primarily as a disinfectant), sodium hydroxide (caustic) is used to raise the pH of the hypochlorite; the excess caustic extends its shelf life. A high pH caustic solution raises the pH of the dilution water to over pH 9.0 after it is diluted. The calcium in the dilution water reacts with dissolved CO_2 and forms calcium carbonate. Experience has shown that two-inch pipes have turned into 3/4-inch pipes due to scale buildup. The scale deposition is greatest in areas of turbulence such as pumps, valves, rotameters, and backpressure devices.

If lime (calcium oxide) is added (for alkalinity), plant water used as dilution water will have higher calcium levels and generate more scale. While it is true that softened water will not generate scale, it is also true that it is expensive in large quantities. Many facilities only use softened water on hypochlorite mist odor scrubbers.

Scaling often occurs in solution rotameters, making flow readings impossible and freezing the flow indicator in place. Various valves can freeze up, and pressure-sustaining valves freeze and become plugged. Various small diffuser holes are filled with scale. To slow the rate of scaling, many facilities purchase water from local suppliers to dilute hypochlorite for the RAS and miscellaneous uses. Some facilities have experimented with the system by not adding lime. When they did this, manganese dioxide (black deposits) developed on the rotameter glass, making viewing the float impossible. In many instances, moving the point of hypochlorite addition downstream of the rotameter seemed to solve the problem.

If remedial steps are not taken, scaling from hypochlorite solutions can cause problems. For example, scale buildup can reduce the inside diameter of pipes so much that the actual supply of hypochlorite solution required to properly disinfect water or wastewater is reduced. As a result, the water sent to the customer or discharged to the receiving body may not be properly disinfected. Because of the scale buildup, the treatment system itself will not function as designed and could result in a hazardous situation in which the reduced pipe size increases the pressure level to the point of catastrophic failure. Scaling, corrosion, or other clogging problems in certain piping systems are far from ideal.

Example 10.2

The scale problem can be illustrated by an example. Assume we have a piping system designed to provide chemical feed to a critical plant unit process. If the motive force for the chemical being conveyed is provided by a positive-displacement pump, it delivers a given volume of solution at 70 psi through clean pipe. After clogging takes place, the pump continues trying to force the same volume of chemical through the system at 70 psi, but the pressure drops to 25 psi. Friction causes the pressure drop. The reduction of the inside diameter of the pipe increases the friction between the chemical solution and the inside wall of the pipe.

Important Point: A basic principle in fluid mechanics states that fluid flowing through a pipe is affected by friction—the greater the friction, the greater the loss of pressure.

Important Point: Another principle states that the amount of friction increases as the square of the velocity. (Note that speed and velocity are not the same, but common practice refers to the "velocity" of a fluid.) In short, if the velocity of the fluid doubles, the friction increases to four times what it was before. If the velocity is multiplied by five, the friction is multiplied by 25, and so on.

In Example 10.2, the pressure dropped from 70 to 25 psi because the solution had to run faster to move through the pipe. Since the velocity of the solution pushed by the pump had to increase to levels above what it was when the pipe was clean, the friction increased at a higher rate than before. The friction loss was the reason that a pressure of 25 psi reached the far end of the piping system. The equipment, designed to operate at a pressure of 70 psi, could not work on the 25 psi of pressure being supplied.

Important Point: After reviewing the previous example, you might ask: Why couldn't the pump be slowed down so that the chemical solution could pass more slowly through the system, thus avoiding the effect of increased friction? Lower pressure results as pump speed is reduced, but this causes other problems as well. Pumps that run at a speed other than that for which they are designed do so with a reduction in efficiency.

What is the solution to our pressure loss problem in Example 10.2? Actually, we can solve this problem in two possible ways—either replace the piping or clean it. Replacing the piping or cleaning it sounds simple and straightforward, but it can be complicated. If the pipe is relatively short, no more than 20 to a few 100 feet in length, then we may decide to replace the pipe. What would we do if the pipe were three to five miles or more in length? Cleaning this length of pipe probably makes more sense than replacing its entire length. Each situation is different, requiring remedial choices based on practicality and expense.

Piping System Maintenance

Maintaining a piping system can be an involved process. However, good maintenance practices can extend the life of piping system components and rehabilitation can further prolong their life. The performance of a piping system depends on the ability of the pipe to resist unfavorable conditions and to operate at or near the capacity and efficiency it was designed for. This performance can be checked in several ways: flow measurement, fire flow tests, loss-of-head tests, pressure tests, simultaneous flow and pressure tests, tests for leakage, and chemical and bacteriological water tests. These tests are an important part of system maintenance and should be scheduled as part of the regular operation of the system (AWWA 1996).

Most piping systems are designed with various protective features to minimize wear and catastrophic failure, and therefore reduce the amount of maintenance required. Such protective features include pressure relief valves, blow-off valves, and clean-out plugs.

- **Pressure Relief Valves:** A valve that opens automatically when the fluid pressure reaches a preset limit to relieve the stress on a piping system.
- **Blow-Off Valve:** A valve that can be opened to blow out any foreign material in a pipe.
- **Clean-Out Plug:** A threaded plug that can be removed to allow access to the inside of the pipe for cleaning.

Important Point: Use caution when removing a clean-out plug from a piping system. Before removing the plug, pressure must be cut off and the system bled off residual pressure.

Many piping systems (including water distribution networks and wastewater lines and interceptors) can be cleaned either by running chemical solvents through the lines or by using mechanical clean-out devices.

PIPING SYSTEM ACCESSORIES

Depending on the complexity of the piping system, the number of valves included in a system can range from no more than one in a small, simple system to a large number in very complex systems such as water distribution systems. *Valves* are necessary for both the operation of a piping system and for control of the system and it components. In water and wastewater treatment, this control function is used to manage various unit processes, pumps, and other equipment. Valves also function as protective devices. For example, valves used to protect a piping system may be designed to open automatically to vent fluid out of the pipe when the pressure in the lines becomes too high. In lines that carry liquids, relief valves preset to open at a given pressure are commonly used.

Important Point: Not all valves function as safety valves. For example, hand-operated gate and globe valves primarily function as control valves.

The size and type of valve are selected depending on their intended use. Most valves require periodic inspections to ensure they are operating properly.

Along with valves, piping systems typically include accessories such as pressure and temperature gauges, filters, strainers, and pipe hangers and supports.

- *Pressure gauges* indicate the pressure in the piping system.
- *Temperature gauges* indicate the temperature in the piping system.
- *Filters* and *strainers* are installed in piping systems to help keep fluids clean and free from impurities.
- *Pipe hangers and supports* support piping to keep the lines straight and prevent sagging, especially in long runs. Various types of pipe hangers and supports are shown in Figure 10.2.

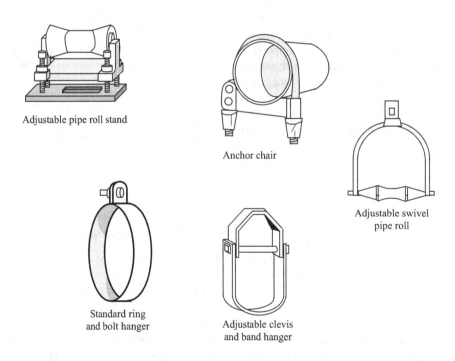

Adjustable pipe roll stand

Anchor chair

Adjustable swivel pipe roll

Standard ring and bolt hanger

Adjustable clevis and band hanger

FIGURE 10.2 Pipe hangers and supports.

PIPING SYSTEM: TEMPERATURE EFFECTS AND INSULATION

Most materials, especially metals, expand as the temperature increases and contract as the temperature decreases. This can be a significant problem in piping systems. To combat this problem, and to allow for expansion and contraction in piping systems, expansion joints must be installed in the line between sections of rigid pipe. An *expansion joint* absorbs thermal expansion and/or thermal movement; as the pipe sections expand or contract with the temperature, the expansion joint expands or compresses accordingly, eliminating stress on the pipes.

Piping system temperature requirements also impact how pipes are insulated. For example, you do not need to wander far in most plant sites to find pipes covered with layers of insulation. Piping insulation amounts to wrapping the pipe in an envelope of insulating material. The thickness of the insulation depends on the application. Under normal circumstances, heat passes from a hot or warm surface to a cold or cooler one. Insulation helps prevent hot fluid from cooling as it passes through the system. For systems conveying cold fluid, insulation helps keep the fluid cold. Materials used for insulation vary and are selected according to the requirements of the application. Various types of insulating materials are also used to protect underground piping against rust and corrosion caused by exposure to water and chemicals in the soil.

METALLIC PIPING

Pipe materials used for transporting water may also be used to collect wastewater. It is more usual, however, to employ less expensive materials since wastewater lines rarely are required to withstand any internal pressure. Iron and steel pipes convey wastewater only under unusual loading conditions or for force mains (interceptor lines) in which the wastewater flow is pressurized (McGhee 1991).

PIPING MATERIALS

Materials selected for piping applications must be chosen with the physical characteristics needed for the intended service in mind. For example, the piping material selected must be suitable for the flow medium and the given operating conditions of temperature and pressure during the intended design life of the product. For long-term service capability, the material's mechanical strength must be appropriate; the piping material must be able to resist operational variables such as thermal or mechanical cycling. Extremes in application temperature must also be considered with respect to material capabilities.

Environmental factors must also be considered. The operating environment surrounding the pipe or piping components affects pipe durability and lifespan. Corrosion, erosion, or a combination of the two can result in degradation of material properties or loss of effective load-carrying cross-section. The nature of the substance contained by the piping is an important factor as well.

Knowledge of the basic characteristics of the metals and nonmetals used for piping provides clues to the uses of the piping materials with which we work in water/wastewater treatment operations. Such knowledge is especially helpful to operators, making their jobs much easier and more interesting. In this section, metallic piping is discussed, along with piping joints, how to join or connect sections of metallic piping, and how to maintain metallic pipe.

PIPING: THE BASICS

Earlier, we pointed out that "piping" includes pipe, flanges, fittings, bolting, gaskets, valves, and the pressure-containing portions of other piping components.

Important Point: According to Nayyar (2000), "A *pipe* is a tube with a round cross section conforming to the dimensional requirements of ASME B36.10M (Welded and Seamless Wrought Steel Pipe) and ASME B36.19M (Stainless Steel Pipe)."

Piping also includes pipe hangers and supports and other accessories necessary to prevent over pressurization and overstressing of the pressure-containing components. From a system viewpoint, a pipe is one element or part of piping. Accordingly, when joined with fittings, valves, and other mechanical devices or equipment, pipe sections are called *piping*.

Pipe Sizes

With time and technological advancements (development of stronger and corrosion-resistant piping materials), pipe sizes have become standardized and are usually expressed in inches or fractions of inches. As a rule, the size of a pipe is given in terms of its outside or inside diameter. Figure 10.3 shows the terminology that applies to a section of pipe. Pipes are designated by diameter. The principal dimensions are as follows:

- Wall thickness
- Length
- Outside diameter (O.D.): which is used to designate pipe greater than 12 inches in diameter
- Inside diameter (I.D.): which is used to designate pipe less than 12 inches in diameter

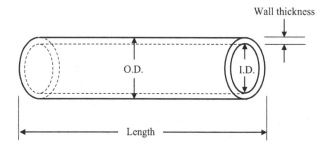

FIGURE 10.3 Pipe terminology.

Important Point: Another important pipe consideration not listed above or shown in Figure 10.3 is weight per foot, which varies according to the pipe material and the pipe's wall thickness.

In the continuing effort to standardize pipe size and wall thickness, the designation *nominal pipe size (NPS)* replaced the iron pipe size designation, and the term *schedule (SCH)* was developed to specify the nominal wall thickness of pipe. The NPS diameter (approximate dimensionless designator of pipe size) is generally somewhat different from its actual diameter. For example, the pipe we refer to as a "3-inch diameter pipe" has an actual O.D. of 3.5 inches, while the actual O.D. of a "12-inch pipe" may be 0.075 inches greater (i.e., 12.750 inches) than the nominal diameter. On the other hand, a pipe 14 inches or greater in diameter has an actual O.D. equal to the nominal size. The inside diameter will depend upon the pipe wall thickness specified by the schedule number.

Important Point: Keep in mind that whether the O.D. is small or large, the dimensions must be within certain tolerances in order to accommodate various fittings.

Pipe Wall Thickness

Original pipe wall thickness designations of STD (standard), XS (extra-strong), and XXS (double extra-strong) are still in use today; however, because this system allowed no variation in wall thickness, and because pipe requirements became more numerous, greater variation was needed. As a result, pipe wall thickness, or *schedule*, today is expressed in numbers (5, 5S, 10, 10S, 20, 20S, 30, 40, 40S, 60, 80, 80S, 100, 120, 140, 160). (Note that you may often hear piping referred to either in terms of its diameter or schedule number.) The most common schedule numbers are 40, 80, 120, and 160. The outside diameter of each pipe size is standardized. Therefore, a particular nominal pipe size will have a different inside diameter depending upon the schedule number specified. For example, a Schedule 40 pipe with a 3-inch nominal diameter (actual O.D. of 3.500 inches) has a wall thickness of 0.216 inches. The same pipe in Schedule 80 (XS) would have a wall thickness of 0.300 inches.

Important Point: A schedule number indicates the approximate value of the expression 1,000 P/S, where P is the service pressure and S is the allowable stress, both expressed in pounds per square inch (psi). The higher the schedule number, the thicker the pipe is.

Important Point: The schedule numbers followed by the letter S are per ASME B36.19M, and they are primarily intended for use with stainless steel pipe (ASME 1996).

Piping Classification

The usual practice is to classify pipe following the pressure-temperature rating system used for classifying flanges. However, due to the increasing variety and complexity of requirements for piping, several engineering societies and standards groups have devised codes, standards, and specifications that meet most applications. By consulting such codes, (e.g., ASTM, Manufacturer's Specifications, NFPA, AWWA, and others), a designer can determine exactly what piping specification should be used for any application.

Important Point: Because pipelines often carry hazardous materials and fluids under high pressure, following a code helps ensure the safety of personnel, equipment, and the piping system itself.

1. **ASTM Ratings:** The American Society for Testing and Materials (ASTM) publishes standards (codes) and specifications, which are used to determine the minimum pipe size and wall thickness to use in a given application.
2. **Manufacturer's Rating:** Pipe manufacturers, because of the proprietary design of pipe, fittings, or joints, often assign a pressure-temperature rating that may form the design basis for the piping system. (In addition, the manufacturer may impose limitations that must be adhered to.)

    Important Point: Under no circumstances shall the manufacturer's rating be exceeded.
3. **NFPA Ratings:** Certain piping systems fall within the jurisdiction of the National Fire Protection Association (NFPA). These pipes are required to be designed and tested to certain required pressures (usually rated for 175 psi, 200 psi, or as specified).
4. **AWWA Ratings:** The American Water Works Association (AWWA) publishes standards and specifications that are used to design and install water pipelines and distribution system piping. The ratings used may be in accordance with the flange ratings of AWWA, or the rating could be based on the rating of the joints used in the piping.
5. **Other Ratings:** Sometimes a piping system may not fall within the above-related rating systems. In this case, the designer may assign a specific rating to the piping system. This is a common practice in classifying or rating piping for mainstream or hot reheat piping of power plants, whose design pressure and design temperature may exceed the pressure-temperature rating of ASME B16.5. In assigning a specific rating to such piping, the rating must be equal to or higher than the design conditions.

Important Point: The rating of all pressure-containing components in the piping system must meet or exceed the specific rating assigned by the designer (Nayyar 2000).

When piping systems are subjected to full-vacuum conditions or submerged in water, they experience both the internal pressure of the flow medium and external pressure. In such instances, piping must be rated for both internal and external pressures at the given temperature. Moreover, if a piping system is designed to handle more than one flow medium during its different modes of operation, it must be assigned a dual rating for two different flow media.

Types of Piping Systems

Piping systems consist of two main categories: process lines and service lines. Process lines convey the flow medium used in a manufacturing process or a treatment process (such as fluid flow in water and/or wastewater treatment). For example, one of the major unit process operations in wastewater treatment is sludge digestion. The sludge is converted from bulky, odorous raw sludge to a relatively inert material that can be rapidly dewatered without obnoxious odors. Because sludge digestion is a unit process operation, the pipes used in the system are called process lines. *Service lines* (or utility lines) carry water, steam, compressed air, air conditioning fluids, and gas. Normally, all of these are part of the plant's general service system, which is composed of service lines. Service lines cool and heat the plant, provide water where it is needed, and carry the air that drives air equipment and tools.

Code for Identification of Pipelines

Under guidelines provided by the American National Standards Institute (ANSI - A 13.1), a code has been established for the identification of pipelines. This code involves the use of nameplates (tags), legends, and colors. The code states that the contents of a piping system shall be identified by a lettered legend giving the name of the contents. In addition, the code requires that information relating to temperature and pressure should be included. Stencils, tape, or markers can be used to accomplish the marking. To identify the characteristic hazards of the contents, color should be used, but its use must be in combination with legends.

Important Point: Not all plants follow the same code recommendations, which can be confusing if you are not familiar with the system used. Standard piping color codes are often used in water and wastewater treatment operations. Plant maintenance operators need to be familiar with the pipe codes used in their plants.

Metallic Piping Materials

In the not-too-distant past, it was not (relatively speaking) that difficult to design certain pipe delivery systems. For example, several hundred years ago (and even more recently in some cases), when it was desirable to convey water from a source to a point of use, the designer was faced with only two issues. First, a source of fresh water had to be found. Next, if the source was found and determined suitable for whatever need required, a means of conveying the water to the point of use was needed.

When designing an early water conveyance system, gravity was the key player. This point is clear when you consider that before the advent of the pump, a motive force to power the pump, and the energy required to provide power to the motive force, gravity was the means by which water was conveyed (with the exception of burdened humans or animals that physically carried the water) from one location to another. Early gravity conveyance systems employed the use of clay pipe, wood pipe, natural gullies or troughs, aqueducts fashioned from stone, and any other means that were suitable or available to convey the water. Some of these earlier pipe or conveyance materials are still in use today. With the advent of modern technology (electricity, the electric motor, the pump and various machines/processes) and the need to convey fluids other than water, there also came the need to develop piping materials that could carry a wide variety of fluids.

Modern waterworks have a number of piping systems made up of different materials. One of the principal materials used in piping systems is metal. Metal pipes may be made of cast iron, stainless steel, brass, copper, and various alloys. As a waterworks and wastewater maintenance operator who works with metal piping, you must be knowledgeable about the characteristics of individual metals as well as the kinds of considerations common to all piping systems. These considerations include the effect of temperature changes, impurities in the line, shifting of pipe supports, corrosion, and water hammer.

Characteristics of Metallic Materials

Metallurgy (the science and study of metals) deals with the extraction of metals from ores and with the combining, treating, and processing of metals into useful materials. Different metals have different characteristics, making them usable in a wide variety of applications. Metals are divided into two types: *ferrous*, which includes iron and iron-base alloys (a metal made up of two or more metals that dissolve into each other when melted together); and *nonferrous*, which includes other metals and alloys.

Important Point: Mixing a metal and a nonmetal (e.g., steel, which is a mixture of iron [a metal] and carbon [a nonmetal]) can also form an alloy.

A *ferrous* metal contains iron (elemental symbol—Fe). Iron is one of the most common metals but is rarely found in nature in its pure form. Comprising about 6% of the Earth's crust, iron ore is actually in the form of iron oxides (Fe_2O_3 or Fe_3O_4). Coke and limestone are used in the reduction of iron ore in a blast furnace, where oxygen is removed from the ore, leaving a mixture of iron and carbon, along with small amounts of other impurities. The end product removed from the furnace is called *pig iron*—an impure form of iron. Sometimes the liquid pig iron is cast from the blast furnace and used directly for metal castings; however, the iron is more often remelted in a furnace to further refine it and adjust its composition (Babcock & Wilcox 1972).

Important Note: Piping is commonly made of wrought iron, cast iron, or steel. The difference among them is largely the amount of carbon that each contains.

Remelted pig iron is known as *cast iron* (meaning the iron possesses carbon in excess of 2% by weight). Cast iron is inferior to steel in malleability, strength, toughness, and ductility (i.e., it is hard and brittle). Cast iron has, however, better fluidity in the molten state and can be cast satisfactorily into complicated shapes.

Steel is an alloy of iron with not more than 2.0% by weight carbon. The most common method of producing steel is to refine pig iron by oxidation or removing impurities and excess carbon—both of which have a higher affinity for oxygen than iron. *Stainless steel* is an alloy of steel and chromium.

Important Note: When piping is made of stainless steel, an "S" after the schedule number identifies it as such.

Various heat treatments can be used to manipulate specific properties of steel, such as hardness and ductility (meaning it can be fashioned into a new form without breaking). One of the most common heat treatments employed in steel processing is annealing. Annealing (sometimes referred to as *stress-relieving*) consists of heating the metal and permitting it to cool gradually to make it softer and less brittle.

Important Point: Steel is one of the most important basic production materials of modern industry.

Nonferrous metals, unlike ferrous metals, do not contain iron. A common example of a nonferrous metal used in piping is brass. Other examples of nonferrous materials used in pipe include polyethylene, polybutylene, polyurethane, and polyvinyl chloride (PVC). Pipes made from these materials are commonly used in low-pressure applications for transporting coarse solids (Snoek and Carney 1981).

In addition to the more commonly used ferrous and nonferrous metals, special pipe materials for specific applications are gaining wider use in industry—even though they are more expensive. One of the most commonly used materials in this category is the aluminum pipe. Aluminum pipe has the advantage of being lightweight and corrosion-resistant with relatively good strength characteristics.

Important Note: Although aluminum is relatively strong, it is important to note that its strength decreases as temperature increases.

Lead is another special pipe material used for certain applications, especially where a high degree of resistance to corrosive materials is desired. Tantalum, titanium, and zirconium piping materials are also highly resistant to corrosives.

Piping systems convey many types of water, including service water, city water, treated or processed water, and distilled water. Service water, used for flushing and cooling purposes, is untreated water that is usually strained but is otherwise raw water taken directly from a source (e.g., lake, river, or deep well). City water is treated potable water. Treated water has been processed to remove various minerals that could cause deterioration or sludge in piping. Distilled water is specially purified.

Important Point: Piping material selection for use in water treatment or distribution operations should be based on commonly accepted piping standards such as those provided by the American Society for Testing Materials (ASTM), American Water Works Association (AWWA), American National Standards Institute (ANSI), American Society of Mechanical Engineers (ASME), and American Petroleum Institute (API).

Cast Iron Pipe

According to the AWWA (1996), "There are more miles of [cast iron pipe] in use today than of any other type. There are many water systems having cast-iron mains that are over 100 years old and still function well in daily use." Cast iron pipe has the advantages of strength, long service life, and reasonable maintenance-free function. The disadvantages include its susceptibility to electrolysis and attack from acidic and alkali soil, as well as its heaviness (Gagliardi and Liberatore 2000).

Ductile Iron Pipe

Ductile-iron pipe resembles cast-iron pipe in appearance and has many of the same characteristics. It differs from cast-iron pipe in that the graphite in the metal is in spheroidal or nodular form, that is, in ball shape rather than in flake form. Ductile-iron pipe is strong, durable, has high flexural strength, and good corrosion resistance, is lighter in weight than cast iron, has a greater carrying capacity for the same external diameter, and is easily tapped. However, ductile-iron pipe is subject to general corrosion if installed unprotected in a corrosive environment (Gigliarda and Liberatore 2000).

Steel Pipe

Steel pipe is sometimes used as large feeder mains in water distribution systems. It is frequently used where there is particularly high pressure or where very large diameter pipe is required. Steel pipe is relatively easy to install, has high tensile strength, lower cost, performs well hydraulically when lined, and is adapted to locations where some movement may occur. However, it is subject to external corrosion from electrolysis in acid or alkali soil, and has poor corrosion-resistance unless properly lined, coated, and wrapped.

Note: The materials from which street wastewater (sewer) pipes are most commonly constructed are vitrified clay pipe, plastic, concrete, and ductile iron pipe. However, it is metallic ductile iron pipe that is most commonly used in wastewater collection—primarily for force mains (interceptor lines, etc.) and for piping in and around buildings. Ductile iron pipe is generally not used for gravity sewer applications, however.

MAINTENANCE CHARACTERISTICS OF METALLIC PIPING

The maintenance required for metallic piping is not only determined in part by the characteristics of the metal (i.e., expansion, flexibility, support, etc.) but also includes the kinds of maintenance common to nonmetallic piping systems as well. The major considerations are:

- Expansion and flexibility
- Pipe support systems
- Valve selection
- Isolation
- Preventing backflow
- Water hammer

- Air binding
- Corrosion effects

Expansion and Flexibility

Because of thermal expansion, water and wastewater systems (which are rigid and laid out in specified lengths) must have adequate flexibility. In water and wastewater systems without adequate flexibility, thermal expansion may lead to failure of piping or anchors. Moreover, it may also lead to joint leakage and excessive loads on appurtenances. The thermal expansion of piping can be controlled using the proper locations of anchors, guides, and snubbers. Where expansion cannot be controlled, flexibility is provided by the use of bends, loops, or expansion joints (Gigliardi and Liberatore 2000).

Important Point: Metals expand or contract according to temperature variations. Over a long run (length of pipe), the effects can cause considerable strain on the lines—damage or failure may result.

Pipe Support Systems

Pipe supports are normally used to carry dead weight and thermal expansion loads. These pipe supports may loosen in time and, therefore, they require periodic inspection. Along with normal expansion and contraction, vibration (water hammer and/or fluids traveling at high speeds and pressures) can cause the supports to loosen.

Valve Selection

Proper valve selection and routine preventive maintenance are critical for the proper operation and maintenance of any piping system. In water or wastewater piping systems, valves are generally used for isolating a section of a water main or wastewater collection line, draining the water and wastewater line, throttling liquid flow, regulating water and wastewater storage levels, controlling water hammer, bleeding off air, or preventing backflow.

Isolation

Various valves are used in piping systems to provide isolation. For instance, gate valves isolate specific areas (valve closed) of the system during repair work or reroute water or wastewater flow (valve open) throughout the distribution or collection system. Service stop valves are commonly used to shut off service lines to individual homes or industries. Butterfly valves are also used for isolation purposes.

Preventing Backflow

Backflow, or reversed flow, could result in contaminated or polluted water entering the potable water system. There are numerous places in a water distribution system where unsafe water may be drawn into the potable water mains if a temporary vacuum occurs in the system. In addition, contaminated water from a higher-pressure source can be forced through a water system connection that is not properly controlled. A typical backflow condition from a recirculated system is illustrated in Figure 10.4.

Important Point: Valves, air gaps, reduced-pressure-zone backflow preventers, vacuum breakers, and barometric loops are often used as backflow prevention devices, depending on the situation.

Water Hammer

In water/wastewater operations specifically involving flow through piping, the term water hammer is often used. The term water hammer (often called surging) is actually a misnomer, as it implies only water and connotes a "hammering" noise. However, it has become a generic term for pressure wave effects in liquids. By definition, a *water hammer* is a pressure (acoustic) wave phenomenon created by relatively sudden changes in liquid velocity. In pipelines, sudden changes in flow (velocity) can occur as a result of (1) pump and valve operation, (2) vapor pocket collapse, or (3) the impact of water following the rapid expulsion of air from a vent or a partially open valve (Marine 1999). A water hammer can damage or destroy piping, valves, fittings, and equipment.

Important Point: When a water hammer occurs, there is little the maintenance operator can do except to repair any damage that results.

Air Binding

Air enters a piping system from several sources, such as the release of air from the water, air carried through vortices into the pump suction, air leaking through joints that may be under negative pressure, and air present in the piping system before it is filled. The problem with air entry or air binding, due to air accumulation in the piping, is that the effective cross-sectional area for water or wastewater flow is reduced. This flow reduction can, in turn, lead to an increase in pumping costs because of the resulting extra head loss.

Corrosion Effects

All metallic pipes are subject to corrosion. Many materials react chemically with metal piping to produce rust, scale, and other oxides. Regarding water treatment processes, when raw water is taken from wells, rivers, or lakes, the water solution is an extremely dilute liquid of mineral salts and gases. The dissolved mineral salts result from water flowing over and through the Earth's layers. The dissolved gases are atmospheric oxygen and carbon dioxide, picked

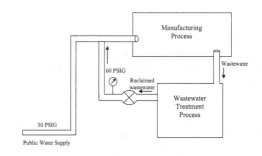

FIGURE 10.4 Backflow from the recirculated system.

up through water-atmosphere contact. Wastewater picks up corrosive materials mainly from industrial processes and/or chemicals added to the wastewater during treatment.

Important Point: Materials such as acids, caustic solutions, and similar solutions are typical causes of pipe corrosion.

Several types of corrosion should be considered in water and wastewater distribution and collection piping systems (AWWA 1996).

1. **Internal Corrosion:** caused by aggressive water flowing through the pipes.
2. **External Corrosion:** caused by the soil's chemical and electrical conditions.
3. **Bimetallic Corrosion:** caused when components made of dissimilar metals are connected.
4. **Stray-Current Corrosion:** caused by uncontrolled DC electrical currents flowing in the soil.

Joining Metallic Pipe

According to Crocker (2000), pipe joint design and selection can significantly impact the initial cost, long-range operating cost, and the performance of the piping system. When determining the type of joint to be used in connecting pipes, certain considerations must be made. For example, initial considerations include material cost, installation labor cost, and degree of leakage integrity required. The operator is also concerned with periodic maintenance requirements, and specific performance requirements.

Metallic piping can be joined or connected in several ways. The method used depends on (1) the nature of the metal sections (ferrous, nonferrous) being joined; (2) the kind of liquid or gas to be carried by the system; (3) pressure and temperature in the line; and (4) access requirements.

A *joint* is defined simply as the connection between elements in a piping system. At present, there are five major types of joints, each used for a specific purpose, for joining metal pipe (see Figure 10.5):

1. Bell-and-spigot joints
2. Screwed or threaded joints
3. Flanged joints
4. Welded joints
5. Soldered joints

Bell-and-Spigot Joints

The bell-and-spigot joint has been around since its development in the late 1780s. The joint is used for connecting lengths of cast iron water and wastewater pipe (gravity flow only). The *bell* is the enlarged section at one end of the pipe, and the plain end is the spigot (see Figure 10.5). The spigot end is placed into the bell, and the joint is sealed. The joint sealing compound is typically made up of lead and oakum. Lead and oakum constitute the prevailing joint sealers for sanitary systems. Bell-and-spigot joints are usually reserved for sanitary sewer systems; they are no longer used in water systems.

Important Point: Bell-and-spigot joints are not used in ductile iron pipes.

Screwed or Threaded Joints

Screwed or threaded joints (see Figure 10.5) are commonly used to join sections of smaller-diameter low-pressure pipe; they are used in low-cost, non-critical applications such as domestic water, industrial cooling, and fire protection systems. Diameters of ferrous or nonferrous pipe joined by threading range from 1/8 inch to 8 inches. Most couplings have threads on the inside surface. The advantages of this type of connection are its relative simplicity, ease of installation (where disassembly and reassembly are necessary to accommodate maintenance needs or process changes), and high leakage integrity at low pressure and temperature where vibration is not encountered. Screwed construction is commonly used with galvanized pipe and fittings for domestic water and drainage applications.

Important Point: Maintenance supervisors must ensure that screwed or threaded joints are used within the limitations imposed by the rules and requirements of the applicable code.

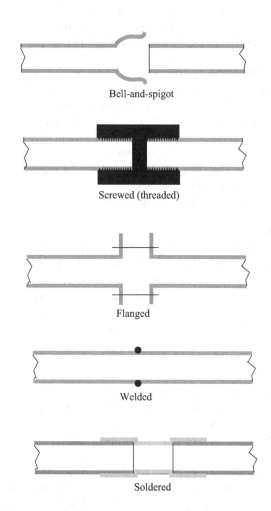

FIGURE 10.5 Common pipe joints.

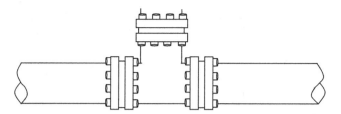

FIGURE 10.6 Flanged assembly.

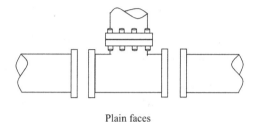

Plain faces

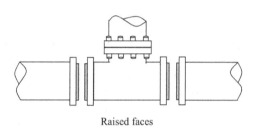

Raised faces

FIGURE 10.7 Flange faces.

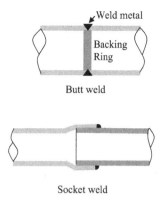

FIGURE 10.8 Two kinds of welding pipe joints.

whereby metal sections to be joined are heated to such a high temperature that they melt and blend. The advantage of welded joints is obvious: the pieces joined become one continuous piece. When properly welded, the joint is as strong as the piping itself. The two basic types of welded joints are (see Figure 10.8):

1. *Butt-welded Joints*, in which the sections to be welded are placed end-to-end, are the most common method of joining pipe used in large industrial piping systems.
2. *Socket-welded Joints*, in which one pipe fits inside the other, with the weld being made on the outside of the lap; they are used in applications where leakage integrity and structural strength are important.

Soldered and Brazed Joints

Soldered and brazed joints are often used to join copper and copper-alloy (non-ferrous metals) piping systems, although brazing of steel and aluminum pipe and tubing is possible. The main difference between *brazing* and welding is the temperatures employed in each process. Brazing is accomplished at far lower temperatures. Brazing, in turn, requires higher temperatures than *soldering*. In both brazing and soldering, the joint is cleaned (using an emery cloth) and then coated with flux, which prevents oxides from forming. The clean, hot joint draws solder or brazing rod (via capillary action) into the joint to form the connection. The parent metal does not melt in brazed or soldered construction.

NON-METALLIC PIPING

Although metal piping is in wide use today, nonmetallic piping (especially clay and cement) is of equal importance. New processes to make them more useful in meeting today's requirements have modified these older materials. Relatively speaking, using metallic piping is a new practice. Originally, all piping was made from clay or wood, and stone soon followed. Open stone channels or aqueducts were used to transport water over long distances. After nearly 2,000 years of service, some of these open channels are still in use today.

Flanged Joints

As shown in Figure 10.6, flanged joints consist of two machined surfaces tightly bolted together with a gasket between them. The flange is a rim or ring at the end of the fitting, which mates with another section. Flanges are joined either by being bolted or welded together. Some flanges have raised faces, while others have plain faces, as shown in Figure 10.7. Steel flanges generally have raised faces, and iron flanges usually have plain or flat faces.

Important Point: A flange with a raised face should never be joined to one with a plain face.

Flanged joints are used extensively in water and wastewater piping systems because of their ease of assembly and disassembly; however, they are expensive. Contributing to the higher cost are the material costs of the flanges themselves and the labor costs for attaching the flanges to the pipe and then bolting the flanges to each other (Crocker 2000). Flanged joints are not normally used for buried pipe due to their lack of flexibility to compensate for ground movement. Instead, flanged joints are primarily used in exposed locations where rigidity, self-restraint, and tightness are required (e.g., inside treatment plants and pumping stations).

Welded Joints

Weld joints are preferred for applications involving high pressures and temperatures. Welding of joints is the process

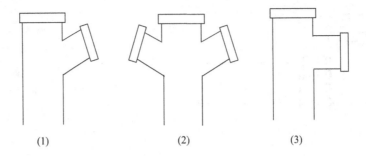

(1)   (2)   (3)

FIGURE 10.9   Section of bell-and-spigot fittings for clay pipe: (a) wye, (b) double wye, and (c) tee.

Common practice today is to use metal piping, though nonmetallic piping is of equal importance and has many applications in water/wastewater operations. Many of the same materials that have been used for centuries (clay, for example) are still used today, but many new piping materials are now available, and the choice depends on the requirements of the planned application. The development of new technological processes has enabled the modification of older materials for new applications in modern facilities and has also brought about the use of new materials for old applications.

In this section, we study nonmetallic piping materials: what they are, and where they are most commonly used. We also describe how to join sections of nonmetallic piping and how to maintain them.

NONMETALLIC PIPING MATERIALS

Nonmetallic piping materials used in water/wastewater applications include clay (wastewater), concrete (water/wastewater), asbestos-cement pipe (water/wastewater), and plastic (water/wastewater). Other nonmetallic piping materials include glass (chemical porcelain pipe) and wood (continuous-strip wooden pipes for carrying water and waste chemicals, used in some areas, especially in the western part of the United States); however, these materials are not discussed in this text due to their limited application in water and wastewater operations.

Important Point: As with metallic piping, nonmetallic piping must be used in accordance with specifications established and codified by a number of engineering societies and standards organizations. These codes were devised to help ensure personnel safety and protection of equipment.

Clay Pipe

Clay pipes are used to carry and/or collect industrial wastes, wastewater, and stormwater (they are not typically used to carry potable water). Clay pipes typically range in size from 4 to 36 inches in diameter and are available in more than one grade and strength. Clay pipe is used in non-pressurized systems. For example, when used in drainpipe applications, liquid flow is solely dependent on gravity; that is, it is used as an open-channel

pipe, whether partially or completely filled. Clay pipe is manufactured in two forms:

* Vitrified (glass-like)
* Unglazed (not glassy)

Important Point: *Vitrified clay pipe* is extremely corrosion proof. It is ideal for many industrial waste and wastewater applications.

Important Point: McGhee (1991) recommended that wyes and tees (see Figure 10.9) should be used for joining various sections of wastewater piping. Failure to provide wyes and tees in common wastewater lines invites builders to break the pipe to make new connections. Obviously, this practice should be avoided, because such breaks are seldom properly sealed and can become a major source of infiltration.

Both vitrified and unglazed clay pipes are made and joined with the same type of bell-and-spigot joint described earlier. The bell-and-spigot shape is shown in Figure 10.10. When joining the sections of clay pipe, both ends of the pipe must first be thoroughly cleaned. The small (spigot) end of the pipe must be properly centered and then seated securely in the large (bell) end. The bell is then packed with fibrous material (usually jute) for solid joints, which is tamped down until about 30% of the space is filled. The joint is then filled with sealing compound. In flexible joint applications, the sealing elements are made from natural or synthetic rubber, or from a plastic-type material.

Drainage and wastewater collection lines designed for gravity flow are laid downgrade at an angle, with the bell ends of the pipe pointing upgrade. The pipe is normally placed in a trench with strong support members (along its small dimension and not on the bell end). Vitrified clay pipe can be placed directly into a trench and covered with soil. However, unglazed clay pipe must be protected against the effects of soil contaminants and ground moisture.

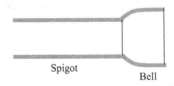

Spigot

Bell

FIGURE 10.10   Bell-and-spigot ends of clay pipe sections.

Concrete Pipe

Concrete is another common pipe material and is sometimes used for sanitary sewers in locations where grades, temperatures, and wastewater characteristics prevent corrosion (ACPA 1987). The pipe provides both high tensile and compressive strength, as well as corrosion resistance. Concrete pipe is generally found in three basic forms: (1) non-reinforced concrete pipe; (2) reinforced concrete, cylinder and non-cylinder pipe; and (3) reinforced and prestressed concrete pressure pipe. With the exception of reinforced and prestressed pressure pipe, most concrete pipe is limited to low-pressure applications. Moreover, almost all concrete piping is used for conveying industrial wastes, wastewater, and stormwater; similarly, some is used for water service connections. Rubber gaskets are used to join sections of many non-reinforced concrete pipes. However, for circular concrete sewer and culvert pipe, flexible, watertight rubber joints are used to join pipe sections.

The general advantages of concrete pipe include the following:

- It is relatively inexpensive to manufacture.
- It can withstand relatively high internal pressure or external loads.
- It has high resistance to corrosion (internal and external).
- Generally, cement pipe, when installed properly, has a very long, trouble-free life.
- Bedding requirements during installation are minimal.

Disadvantages of concrete pipe include:

- It is very heavy, and thus expensive, when shipped long distances.
- Its weight makes special handling equipment necessary.
- Exact pipes and fittings must be laid out in advance for installation (AWWA 1996).

Non-reinforced Concrete Pipe

Non-reinforced concrete pipe, or ordinary concrete pipe, is manufactured in diameters from 4 to 24 inches. As in vitrified clay pipe, nonreinforced concrete pipe is made with bell-and-spigot ends. Non-reinforced concrete pipe is normally used for small wastewater (sewer) lines and culverts.

Reinforced Concrete Pipe

All concrete pipes made in sizes larger than 24 inches are reinforced; however, reinforced pipe can also be obtained in sizes as small as 12 inches. Reinforced concrete pipe is used for water conveyance (cylinder pipe), carrying wastewater, stormwater, and industrial wastes. It is also used in culverts. It is manufactured by wrapping high-tensile strength wire or rods around a steel cylinder that has been lined with cement mortar. Joints are either bell-and-spigot

or tongue-and-groove in sizes up to 30 inches and tongue-and-groove exclusively above that size.

Reinforced and Prestressed Concrete Pipe

When concrete piping is to be used for heavy load, high-pressure applications (up to 600 psi), it is strengthened by reinforcement and pre-stressing. Prestressed concrete pipe is reinforced by steel wire, steel rods, or bars embedded lengthwise in the pipe wall. If a wire is used, it is wound tightly to pre-stress the core and is covered with an outer coating of concrete. Pre-stressing is accomplished by manufacturing the pipe with a permanent built-in compression force.

Asbestos Cement (A-C) Pipe

Prior to 1971, asbestos was known as the "material of a thousand uses" (Coastal Video Communications 1994). It was used for fireproofing (primarily), insulation (secondarily, on furnaces, ducts, boilers, and hot water pipes, for example), soundproofing, as well as a host of other applications. Other applications included its use in the conveyance of water and wastewater. However, while still used in some industrial applications and in many water/wastewater piping applications, asbestos-containing materials (ACM), including asbestos cement (A-C) pipe, are not as widely used as they were before 1971.

Asbestos-containing materials lost favor with regulators and users primarily because of the health risks involved. Asbestos has been found to cause chronic and often fatal lung diseases, including asbestosis and certain forms of lung cancer. Although debatable, there is some evidence that asbestos fibers in water may cause intestinal cancers as well. While it is true that asbestos fibers are found in some natural waters (Bales et al. 1984) and can be leached from asbestos cement pipe by very aggressive waters (i.e., those that dissolve the cement itself; Webber et al. 1989), it is also true that the danger from asbestos exposure is not so much due to the danger of specific products (A-C pipe, for example) as it is to the overall exposure of people involved in the mining, production, installation, and ultimate removal and disposal of asbestos products (AWWA 1996).

A-C pipe is composed of a mixture of Portland cement and asbestos fiber, which is built up on a rotating steel mandrel and then compacted with steel pressure rollers. This pipe has been used for over 70 years in the United States. Because it has a very smooth inner surface, it has excellent hydraulic characteristics (McGhee 1991).

In water and wastewater operations, it is the ultimate removal and disposal of asbestos cement pipe that poses the problem for operators. For example, consider an underground wastewater line break that must be repaired. After locating exactly where the line break is (this is sometimes difficult to accomplish because A-C pipe is not as easily located as conventional pipe), the work crew must first excavate the soil covering the line break, being careful not to cause further damage (A-C pipe is relatively fragile). Once the soil has been removed, exposing the line break, the damaged pipe section must be removed. In some instances, it may be more economical or practical to remove only the damaged portion of the pipe and to install a replacement portion and then girdle it with a clamping mechanism (sometimes referred to as a *saddle clamp*).

To this point in the described repair operation, there is little chance for exposure to personnel from asbestos. This is the case, of course, because in order to be harmful, ACM must release fibers that can be inhaled. The asbestos in undamaged A-C pipe is not *friable* (non-friable asbestos); that is, it cannot be readily reduced to powder form by hand pressure when it is dry. Thus, it poses little or no hazard in this condition. However, if the maintenance crew making the pipe repair must cut, grind, or sand the A-C pipe section under repair, the non-friable asbestos is separated from its bond. This type of repair activity is capable of releasing friable airborne fibers—and herein lies the hazard of working with A-C pipe.

To guard against the hazard of exposure to asbestos fibers, A-C pipe repairs must be accomplished in a safe manner. Operators must avoid any contact with ACM that disturbs its position or arrangement; disturbs its matrix or renders it friable; and generates any visible debris from it.

Important Point: Visibly damaged, degraded, or friable ACM in the vicinity are **always** indicators that surface debris or dust could be contaminated with asbestos. OSHA standards require that we assume that such dust or debris contains asbestos fibers (Coastal Video Communications 1994).

In the A-C pipe repair operation described above, repairs to the A-C pipe require that prescribed USEPA, OSHA, State, and Local guidelines be followed. General EPA/OSHA guidelines, at a minimum, require that trained personnel perform repairs made to the A-C pipe only. The following safe work practice is provided for those who must work with ACMs (Spellman 1996).

Safe Work Practice: A-C Pipe

1. When repairs or modifications are conducted that require cutting, sanding, or grinding on cement pipe containing asbestos, USEPA-trained asbestos workers or supervisors are to be called to the work site *immediately*.
2. Excavation personnel will unearth buried pipe to the point necessary to make repairs or modifications. The immediate work area will then be cleared of personnel as directed by the asbestos-trained supervisor.
3. The on-scene supervisor will direct the asbestos-trained workers as required to accomplish the work task.
4. The work area will be barricaded 20 feet in all directions to prevent unauthorized personnel from entering.
5. Asbestos-trained personnel will wear *all* required Personal Protective Equipment (PPE). Required

PPE shall include Tyvek totally enclosed suits, a 1/2 face respirator equipped with HEPA filters, rubber boots, goggles, gloves, and a hard hat.
6. Supervisor will perform the required air sampling before entry.
7. Air sampling *shall* be conducted using the NIOSH 7400 Protocol.
8. A portable decontamination station will be set up as directed by the supervisor.
9. Workers will enter the restricted area *only* when directed by the supervisors and, using wet methods *only*, will either perform pipe cutting using a rotary cutter assembly or inspect the broken area to be covered with a repair saddle device.
10. After performing the required repair or modifications, workers will encapsulate bitter ends and/or fragmented sections.
11. After encapsulation, the supervisor can authorize entry into the restricted area for other personnel.
12. Broken ACM pipe pieces must be properly disposed of following EPA, State, and Local guidelines.

Important Point: Although exposure to asbestos fibers is dangerous, it is important to note that studies by the USEPA, AWWA, and other groups have concluded that the asbestos in water mains does not generally constitute a health threat to the public (AWWA 1996).

Because A-C pipe is strong and corrosion-resistant, it is widely used for carrying water and wastewater. Standard sizes range from 3 to 36 inches. Though highly resistant to corrosion, A-C pipe should not be used for carrying highly acidic solutions or unusually soft water unless its inner and outer surface walls are specially treated. A-C pipe is preferred for use in many outlying areas, because of its lightweight, which results in greater ease of handling. An asbestos-cement sleeve is used to join A-C pipe. The sleeve's inner diameter (I.D.) is larger than the pipe's outer diameter (O.D.). The ends of the pipes fit snugly into the sleeve and are sealed with a natural or synthetic rubber seal or gasket, which acts as an expansion joint.

Plastic Pipe

Plastic pipe has been used in the United States for about 60 years, and its use is becoming increasingly common. In fact, because of its particular advantages, plastic pipe is replacing both metallic and nonmetallic piping. The advantages of plastic piping include:

- High internal and external corrosion resistance
- Rarely needs to be insulated or painted
- Lightweight
- Ease of joining
- Freedom from rot and rust
- Resistance to burning
- Lower cost
- Long service life
- Easy to maintain

Several types of plastic pipe exist; however, where plastic pipe is commonly used in water and wastewater service, polyvinyl chloride (PVC) is the most common plastic pipe for municipal water distribution systems. PVC is a polymer extruded (shaped by forcing through a die) under heat and pressure into a thermoplastic that is nearly inert when exposed to most acids, fuels, and corrosives. PVC is commonly used to carry cold drinking water because it is nontoxic and will not affect the water's taste or cause odor. The limitations of PVC pipe include its limited temperature range—approximately 150°F–250°F—and low-pressure capability—usually 75–100 psi. Joining sections of plastic pipe is accomplished by welding (solvent, fusion, fillet), threading, and flanges.

Important Point: The strength of plastic piping decreases as the temperature of the materials it carries increases.

TUBING

Piping by another name might be tubing. A logical question might be: "When is a pipe a tube, or a tube a pipe?" However, does it really matter if we call piping or tubing by two distinct, separate, and different names? It depends, of course, on the difference(s) between the two. When we think of piping and tubing, we think of tubular, which infers cylindrical products that are hollow. Does this description help us determine the difference between piping and tubing? No, not really. We need more—a better, more concise description or delineation.

Maybe size will work. It is true that when we normally think of pipe, we think in terms of either metallic or nonmetallic cylindrical products that are hollow and range in nominal size from about 0.5 inch (or less) to several feet in diameter. On the other hand, when we think of tubing, we think of cylindrical, hollow products that are relatively smaller in diameter to than many piping materials.

Maybe application will work. It is true that when we normally think of pipe, we think of any number of possible applications from conveying raw petroleum from field to refinery, to the conveyance of raw water from source to treatment facility, to wastewater discharge point to treatment to outfall, and several others. On the other hand, when we think in terms of tubing applications and products conveyed, the conveyance of compressed air, gases (including liquefied gas), steam, water, lubricating oil, fuel oil, chemicals, fluids in hydraulic systems, and waste products comes to mind.

On the surface, and evidenced by the discussion above, it is apparent that when we attempt to classify or differentiate piping and tubing, our effort is best characterized as somewhat arbitrary, capricious, vague, and/or ambiguous. It appears that piping by any other name is just piping. In reality, however, piping is not tubing, and in the end (so to speak), the difference may come down to determination by end use.

The Bottom Line: It is important to differentiate between piping and tubing because they are different. They

are different in physical characteristics and methods of installation, as well as in their advantages and disadvantages. In this section, these differences become clear.

TUBING VS. PIPING: THE DIFFERENCE

Lohmeir and Avery (2000) pointed out that piping and tubing are considered separate products, even though they are geometrically quite similar. Moreover, the classification of "pipe" or "tube" is determined by end use. As mentioned, many of the differences between piping and tubing are related to physical characteristics, methods of installation, as well as the advantages and disadvantages.

Simply, *tubing* refers to tubular materials (products) made to either an inside (I.D.) or outside diameter (O.D.; expressed in even inches or fractions). Tubing walls are generally much thinner than those of piping; thus, wall thickness in tubing is of particular importance.

Important Point: Wall thickness tolerance in tubing is held so closely that wall thickness is usually given in thousandths of an inch rather than as a fraction of an inch. Sometimes a gauge number is used to indicate the thickness according to a given system.

Tubing of different diameters has different wall thicknesses. An example from *Pipe Properties* and *Tubing Properties* illustrates the difference between piping and tubing. The wall thickness of a commercial type of 8-inch pipe is 0.406 inch. Light-wall 8-inch copper tubing, by contrast, has a wall thickness of 0.050 inch. When we compare these figures, it is clear that tubing has much thinner walls than piping of the same general diameter (Basavaraju 2000).

Important Note: It is important to note that the range between "thick" and "thin" is narrower for tubing than it is for piping.

The list of tubing applications is a lengthy one. Some tubing types can be used not only as conduits for electrical wire but also to convey waste products, compressed air, hydraulic fluids, gases, fuel oil, chemicals, lubricating oil, steam, waters, and other fluids (i.e., both gaseous and liquid).

Tubing is made from both metals and plastics. Metal tubing is designed to be somewhat flexible but also strong. Metallic materials such as copper, aluminum, steel, and stainless steel are used in applications where fluids are carried under high pressure (some types of tubing—e.g., stainless steel—can accommodate very high pressures—>5,000 psi). As the diameter of the tubing increases, the wall thickness increases accordingly (slightly).

Ranging in size from 1/32 to 12 inches in diameter, it is the smaller sizes that are most commonly used. Standard copper tubing ranges from 1/32 to 10 inches in diameter, steel from 3/15 to 10 3/4 inches, aluminum from 1/8 to 12 inches, and special alloy tubing is available up to 8 inches in diameter.

Typically, in terms of initial cost, metal tubing materials are more expensive than iron piping; however, the high initial cost versus the ability to do a particular application as designed (as desired) is a consideration that can't be

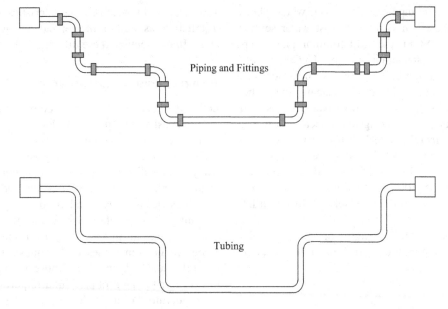

FIGURE 10.11 Tubing eliminates fittings.

overlooked or under-emphasized. Consider, for example, an air compressor. Typically, while in operation, air compressors are mechanical devices that not only produce a lot of noise but also vibrate. Installing a standard rigid metal piping system to such a device might not be practical. Installing tubing that is flexible to the same device, however, may have no detrimental impact on operation whatsoever. An even more telling example is the internal combustion engine. For example, a lawnmower engine, like the air compressor, also vibrates and is used in less than static conditions (i.e., the lawnmower is typically exposed to all kinds of various dynamic stresses). Obviously, we would **not** want the fuel lines (tubing) in such a device to be "hard-wired" with rigid pipe; instead, we would want the fuel lines to be durable but also somewhat flexible. Thus, flexible metal tubing is called for in this application because it will hold up.

Simply put, initial cost can be important. However, considerations such as maintenance requirements, durability, length of life, and ease of installation often favor the use of metallic tubing over the use of metallic pipe.

Although it is true that most metallic tubing materials have relatively thin walls, it is also true that most are quite strong. Small tubing material with thin walls (i.e., soft materials up to approximately 1 inch O.D.) can be bent quite easily by hand. Tubing with larger diameters requires special bending tools. The big advantage of flexible tubing should be obvious: Tubing can be run from one point to another with fewer fittings than if piping were used.

Note: Figure 10.11 shows how the use of tubing can eliminate several pipe fittings.

The advantages of the tubing type of arrangement shown in Figure 10.11 include the following:

- It eliminates 18 potential sources of leaks
- The cost of the eighteen 90° elbow fittings needed for the piping installation is eliminated

- The time needed to cut, gasket, and flange the separate sections of pipe is conserved (obviously, it takes little time to bend tubing into the desired configuration)
- The tubing configuration is much lighter in weight than the separate lengths of pipe and the pipe flanges would have been.

For the configuration shown in Figure 10.12, the weight is considerably less for the copper tubing than for the piping arrangement. Moreover, the single length of tubing bent to follow the same general conveyance route is much easier to install.

It may seem apparent to some readers that many of the weight and handling advantages of tubing compared to piping can be eliminated or at least matched simply by reducing the wall thickness of the piping. It is important to

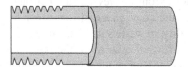

Threaded pipe section

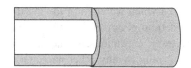

Pipe section without threads

FIGURE 10.12 Pipe wall thickness is important when threading is required.

remember, however, that piping has a thick wall because it often needs to be threaded to make connection(s). If the wall thickness of iron pipe, for example, were made comparable to the thickness of copper tubing and then threaded at connection points, its mechanical integrity would be reduced. The point is that piping must have sufficient wall thickness left after threading to not only provide a tight fit but also to handle the fluid pressure. On the other hand, copper tubing is typically designed for brazed and soldered connections rather than threaded ones. Thus, its wall thickness can be made uniformly thin. This advantage of tubing over iron piping is illustrated in Figure 10.12.

Important Point: The lighter weight of tubing means greater ease of handling, as well as lower shipping costs.

ADVANTAGES OF TUBING

To this point, in regard to design requirements, reliability, and maintenance activities of using tubing instead of piping, we have pointed out several advantages. These advantages can be classified as *mechanical* or *chemical* advantages.

MECHANICAL ADVANTAGES OF TUBING

Probably the major mechanical advantage of using tubing is its relatively small diameter and flexibility, which makes it user-friendly in tight spaces where piping would be difficult to install and maintain (i.e., for the tightening or repair/replacement of fittings). Another mechanical advantage of tubing important to water and wastewater maintenance operators is its ability to absorb shock from *water hammer*. Water hammer can occur whenever fluid flow is started or stopped. In water/wastewater operations, certain fluid flow lines have a frequent on-off cycle. In a conventional piping system, this may produce vibration, which is transmitted along the rigid conduit, shaking joints, valves, and other fittings. The resulting damage usually results in leaks, which, of course, necessitate repairs. In addition, the piping supports can also be damaged. When tubing, with its built-in flexibility, is used in place of conventional iron piping, however, the conduit absorbs most of the vibration and shock. The result is far less wear and tear on the fittings and other appurtenances.

As mentioned, sections of tubing are typically connected by means of soldering, brazing, or welding rather than by threaded joints, although steel tubing is sometimes joined by threading. In addition to the advantages in cost and time savings, avoiding the use of threaded joints precludes other problems. For example, anytime piping is threaded, it is weakened. At the same time, threading is commonly used for most piping systems and usually presents no problem.

Another advantage of tubing over iron piping is the difference in inner-wall surfaces between the two. Specifically, tubing generally has a smoother inner-wall surface than iron piping does. This smoother inner-wall characteristic aids in reducing turbulent flow (wasted energy and decreased pressure) in tubing. Instead, flow in the smoother-walled tubing is more laminar; that is, it has less turbulence. Laminar flow is

characterized as flow in layers—very thin layers. (Somewhat structurally analogous to this liquid laminar flow phenomenon are wood-type products such as kitchen cabinets, many of which are constructed of laminated materials.)

This might be a good time to address laminar flow inside a section of tubing. First, we need to discuss both laminar and turbulent flow to point out the distinct differences between them. Simply, in *laminar* flow, streamlines remain parallel to one another and no mixing occurs between adjacent layers. In *turbulent* flow, mixing occurs across the pipe. The distinction between the two regimes lies in the fact that the shear stress in laminar flow results from viscosity, while that in turbulent flow results from momentum exchanges occurring as a result of the motion of fluid particles from one layer to another (McGhee 1991). Normally, flow is laminar inside tubing. However, if there are irregularities (dents, scratches, or bumps) on the tubing's inner wall, the fluid will be forced across the otherwise smooth surface at a different velocity which causes turbulence. Iron piping has more irregularities along its inner walls, which cause turbulence in the fluid flowing along the conduit. Ultimately, this turbulence can reduce the delivery rate of the piping system considerably.

Chemical Advantages of Tubing

The major chemical advantage of tubing compared to piping comes from the corrosion-resistant properties of the metals used to make the tubing. Against some corrosive fluids, most tubing materials perform very well. Some metals perform better than others, however, depending on the metal and the corrosive nature of the fluid. It is also important to point out that the tubing used must be compatible with the fluid being conveyed. When conveying a liquid stream from one point to another, the last thing we want is for contamination from the tubing to be added to the fluid. Many tubing conveyance systems are designed for use in food-processing operations, for example. If we were conveying raw milk to or from a unit process, we certainly would not want to contaminate the milk. To avoid such contamination, where conditions of particular sanitation are necessary, stainless steel, aluminum, or appropriate plastic tubing must be used.

CONNECTING TUBING

The skill required to properly connect metal or non-metallic tubing can be learned by just about anyone. However, a certain amount of practice and experience is required to ensure that the tubing is properly connected. Moreover, certain tools are required for connecting sections of tubing. The tools used to make either a soldered connection or a compression connection (where joint sections are pressed together) include:

- Hacksaw
- Tube cutter
- Scraper
- Flat file
- Burring tool

- Flaring tool
- Presetting tool for flareless fittings
- Assorted wrenches
- Hammer
- Tube bender

Cutting Tubing

No matter what type of connection you are making (soldered or compressed), it is important to cut the tubing cleanly and squarely. This can be accomplished using a tubing cutter. The use of a tubing cutter is recommended because it provides a much smoother cut than that made with a hacksaw. A typical tubing cutter has a pair of rollers on one side and a cutting wheel on the other. The tubing cutter is turned all the way around the tubing, making a clean cut.

Important Point: When cutting stainless steel tubing, cut the tubing as rapidly and safely as you can, with as few strokes as possible. This is necessary because as stainless steel is cut, it hardens, especially when cut with a hacksaw.

After making the tubing cut, the rough edge of the cut must be smoothed with a *burring* tool to remove the small metal chads, burrs, or whiskers. If a hacksaw is used to cut the tubing, ensure that the rough cut is filed until it is straight and square to the length of the tubing.

Soldering Tubing

Soldering is a form of brazing in which nonferrous filler metals having melting temperatures below 800°F (427°C) are used. The filler metal is called solder (usually a tin-lead alloy, which has a low melting point) and is distributed between surfaces by capillary action. Whether soldering two sections of tubing together or connecting tubing to a fitting, such as an elbow, the soldering operation is the same. Using emery cloth or a wire brush, the two pieces to be soldered must first be cleaned (turned to bright metal). Clean, oxide-free surfaces are necessary to make sound soldered joints. Uniform capillary action is possible only when surfaces are completely free of foreign substances such as dirt, oil, grease, and oxide.

Important Point: During the cleaning process care must be taken to avoid getting the prepared adjoining surfaces too smooth. Surfaces that are too smooth will prevent the filler metal (solder) from effectively wetting the joining areas.

The next step is to ensure that both the tubing outside and the fitting inside are covered with soldering flux and fitted together. When joining two tubing ends, use a sleeve. The purpose of flux is to prevent or inhibit the formation of oxide during the soldering process. The two ends are fitted into the sleeve from opposite sides. Make sure the fit is snug.

Next, heat the joint. First, heat the tubing next to the fitting, then the fitting itself. When the flux begins to spread, solder should be added (this is known as *tinning*). The heat will suck the solder into the space between the tubing and the sleeve. Then heat the fitting on and off and apply more solder until the joint is fully penetrated (Giachino and Weeks 1985).

Important Point: During the soldering operation, it is important to ensure that the heat is applied evenly around the tubing. A continuous line of solder will appear where the fitting and tubing meet at each end of the sleeve. Also, ensure that the joined parts are held so that they do not move. After soldering the connection, wash the connection with hot water to prevent future corrosion.

The heat source normally used for soldering is heated using an oxyacetylene torch or some other high-temperature heat source. Important soldering points to remember include:

1. Always use the recommended flux when soldering.
2. Make sure the parts to be soldered are clean and their surfaces fit closely together.
3. During the soldering process, do not allow the parts to move while the solder is in a liquid state.
4. Be sure the soldering heat is adequate for the soldering job to be done, including the types of metals and the fluxes.
5. Wash the solder work in hot water to stop corrosive action later.

Connecting Flared/Nonflared Joints

In addition to being connected by brazing or soldering, tubing can also be connected by either *flared* or *non-flared* joints. Flaring is accomplished by evenly spreading the end of the tube outward, as shown in Figure 10.13. The accuracy of the angle of the flare is important; it must match the angle of the fitting being connected. The flaring tool is inserted into the squared end of the tubing, and then hammered or impacted into the tube a short distance, spreading the tubing end as required.

Figure 10.14 shows the resulting flared connection. The flared section is inserted into the fitting in such a way that the flared edge of the tube rests against the angled face of

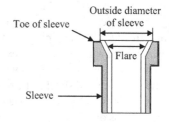

FIGURE 10.13 Flared tubing end.

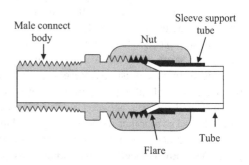

FIGURE 10.14 Flared fitting.

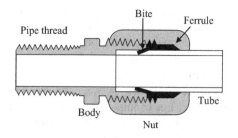

FIGURE 10.15 Flareless fitting.

the male connector body—a sleeve supports the tubing. The nut is tightened firmly on the male connector body, making a firm joint that will not leak, even if the tubing ruptures because of excess pressure. Figure 10.15 shows a flareless fitting. As shown, the plain tube end is inserted into the body of the fitting. Notice that there are two threaded outer sections with a ferrule or bushing located between them. As the threaded members are tightened, the ferrule bites into the tubing, making a tight connection.

BENDING TUBING

A type of tool typically used in water/wastewater maintenance applications for bending tubing. is the hand bender, which is nothing more than a specifically sized spring-type apparatus. Spring-type benders come in several different sizes (the size that fits the particular tubing to be bent is used to bend it). The spring-type tubing bender is slipped over the tubing section to be bent. Then, carefully, the spring and tubing are bent by hand to conform to the angle of bend desired. When using any type of tubing bender, it is important to obtain the desired bend without damaging (flattening, kinking, or wrinkling) the tubing. As mentioned, any distortion of the smooth inner wall of a tubing section causes turbulence in the flow, which lowers the pressure. Figure 10.16 shows three different

kinds of incorrect bends and one correct bend. From the figure, it should be apparent how the incorrect bends constrict the flow, causing turbulence and lower pressure.

TYPES OF TUBING

Common types of metal tubing in industrial service include:

- *Copper* (seamless, fully annealed, furnished in coils or in straight lengths). In water treatment applications, copper tubing has replaced lead and galvanized iron in service line installations because it is flexible, easy to install, corrosion-resistant in most soils, and able to withstand high pressure. It is not sufficiently soluble in most water to be a health hazard, but corrosive water may dissolve enough copper to cause green stains on plumbing fixtures. Copper water service tubing is usually connected by either flare or compression fittings. Copper plumbing is usually connected with solder joints (AWWA 1996).

Important Point: *Annealing* is the process of reheating a metal and then letting it cool slowly. In the production of tubing, annealing is performed to make the tubing softer and less brittle.

- *Aluminum* (seamless, annealed, and suitable for bending and flaring).
- *Steel* (seamless, fully annealed, also available as a welded type, suitable for bending and flaring).
- *Stainless steel* (seamless, fully annealed, also available as a welded type, suitable for bending and flaring).
- *Special alloy* (made for carrying corrosive materials).

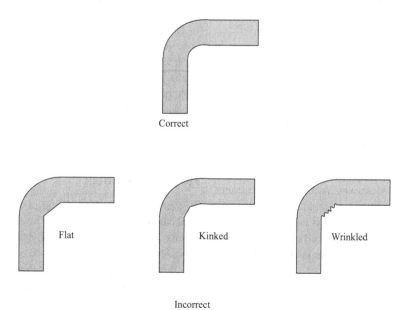

FIGURE 10.16 Correct and incorrect tubing bends.

Like metal piping, metal tubing is made in both welded and seamless styles. *Welded tubing* begins as flat strips of metal that are then rolled and formed into tubing. The seam is then welded. *Seamless* tubing is formed as a long, hot metal ingot and then shaped into a cylindrical shape. The cylinder is then extruded (passed through a die), producing tubing in larger sizes and wall thicknesses. If smaller tubing (with thinner walls and closer tolerances) is desired, the extruded tubing is reworked by drawing it through another die.

Typical Tubing Applications

In a typical water/wastewater operation, tubing is used in unit processes and/or machinery. Heavy-duty tubing is used for carrying gas, oxygen, steam, and oil in many underground services, interior plumbing, and heating and cooling systems throughout the plant site. Steel tubing is used in high-pressure hydraulic systems. Stainless steel tubing is used in many of the chemical systems. In addition, in many plants, aluminum tubing is used as raceways or containers for electrical wires. Plastics have become very important as nonmetallic tubing materials. The four most common types of plastic tubing are Plexiglas (acrylic), polycarbonate, vinyl, and polyethylene (PE). For plant operations, plastic tubing usage is most prevalent where it meets corrosion resistance demands, and the temperatures are within its working range—primarily used in chemical processes. Plastic tubing is connected either by fusing with solvent-cement or by heating. Reducing the plastic ends of the tubing to a soft, molten state and then pressing them together makes fused joints. In the solvent cement method, the ends of the tubing are coated with a solvent that dissolves the plastic. The tube ends are firmly pressed together, and as the plastic hardens, they are securely joined. When heat fused, the tubes are held against a hot plate. When molten, the ends are joined, and the operation is complete.

INDUSTRIAL HOSES

Earlier we described the uses and merits of piping and tubing. This section describes industrial hoses, which are classified as a slightly different tubular product. Their basic function is the same, however, and that is to carry fluids (liquids and gases) from one point to another. The outstanding feature of industrial hoses is their flexibility, which allows them to be used in applications where vibrations would make the use of rigid pipe impossible. Most water/wastewater treatment plants use industrial hoses to convey steam, water, air, and hydraulic fluids over short distances. It is important to point out that each application must be analyzed individually, and an industrial hose must be selected that is compatible with the system specifications.

In this section, we study industrial hoses—what they are, how they are classified and constructed, and the ways in which sections of hose are connected to one another and to piping or tubing. We will also read about the maintenance requirements of industrial hoses and what to look for when we make routine inspections or checks for specific problems.

Industrial hoses, piping, and tubing are all used to convey a variety of materials under a variety of circumstances. Beyond this similar ability to convey a variety of materials, however, there are differences between industrial hoses and piping and tubing. For example, in their construction and advantages, industrial hoses **are** different from piping and tubing. As mentioned, the outstanding advantage of hoses is their flexibility; their ability to bend means that hoses can meet the requirements of numerous applications that can't be met by rigid piping and some tubing systems. Two examples of this flexibility are Camel hose (used in wastewater collection systems to clean out interceptor lines and/ or to remove liquid from excavations where broken lines need repair) and the hose that supplies hydraulic fluids used on many forklifts. Clearly, rigid piping would be impractical to use in both situations.

Industrial hose is not only flexible but also has a dampening effect on vibration. Certain tools used in water and wastewater maintenance activities must vibrate to do their jobs. Probably the best and most familiar such tool is the power hammer or jackhammer. Obviously, the built-in rigidity of piping and tubing would not allow vibrating tools to stand up very long under such conditions. Other commonly used tools and machines in water/wastewater operations have pneumatically or hydraulically driven components. Many of these devices are equipped with moving members that require the air or oil supply to move with them. In such circumstances, of course, rigid piping could not be used.

It is important to note that the flexibility of industrial hose is not the only consideration that must be taken into account when selecting hose over either piping or tubing. That is, hose must be selected according to the potentially damaging conditions of an application. These conditions include the effects of pressure, temperature, and corrosion.

Hose applications range from the lightweight ventilating hose (commonly called "elephant trunk") used to supply fresh air to maintenance operators working in manholes, vaults, or other tight places. In water and wastewater treatment plants, hoses are used to carry water, steam, corrosive chemicals and gases, and hydraulic fluids under high pressure. To meet such service requirements, hoses are manufactured from a number of different materials.

Hose Nomenclature

To gain a fuller understanding of industrial hoses and their applications, it is important to be familiar with the nomenclature or terminology normally associated with industrial hoses. Accordingly, in this section, we explain hose terminology that water/wastewater operators should be familiar with. Figure 10.17 is a cutaway view of a high-pressure air hose of the kind that supplies portable air hammers, drills, and other pneumatic tools commonly used in water/wastewater maintenance operations. The hose is the most common type of reinforced nonmetallic hose in general use. Many of the terms given have already been mentioned. The I.D., which designates the hose size, refers to the inside diameter throughout the length of the hose body unless the hose has enlarged ends. The O.D. is the diameter of the outside wall of the hose.

Important Point: If the ends of an industrial hose are enlarged, as shown in Figure 10.18, the letters E.E. are used (meaning *expanded* or *enlarged end*). Some hoses have enlarged ends to fit a fixed end of piping tightly (e.g., an automobile engine).

As shown in Figure 10.17, the *tube* is the inner section (i.e., the core) of the hose through which the fluid flows. Surrounding the tube is the *reinforcement* material, which provides resistance to pressure—either from the inside or outside. Notice that the hose shown in Figure 10.17 has two layers of reinforcement *braid* (this braid is fashioned from high-strength synthetic cord). The hose is said to be *mandrel-braided* because a spindle or core (the mandrel) is inserted into the tube before the reinforcing materials are put on. The mandrel provides a firm foundation over which the cords are evenly and tightly braided. The *cover* of the hose is an outer protective covering. The hose in Figure 10.17 has a cover of tough, abrasion-resistant material.

The *overall length* is the true length of a straight piece of hose. Hose, which is not too flexible, is formed or molded in a curve (e.g., automobile hose used in heating systems; see Figure 10.19). As shown in Figure 10.19, the *arm* is the section of a curved hose that extends from the end of the hose to the nearest centerline intersection. The *body* is the middle section or sections of the curved hose. Figure 10.20 shows the *bend radius* (i.e., the radius of the bend measured to the centerline) of the curved hose and is designated as the radius R. In a straight hose bent on the job, the radius of the bend is measured to the surface of the hose (i.e., the radius r in Figure 10.20).

Important Point: Much of the nomenclature used above does not apply to non-metallic hose that is not reinforced. However, non-reinforced non-metallic hose is not very common in water/wastewater treatment plant operations.

FACTORS GOVERNING HOSE SELECTION

The amount of *pressure* that a hose will be required to convey is one of the important factors governing hose selection. Typically, the pressure range falls into one of three general groups:

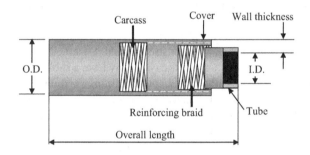

FIGURE 10.17 Common hose nomenclature.

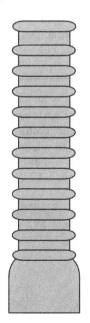

FIGURE 10.18 Expanded-end hose.

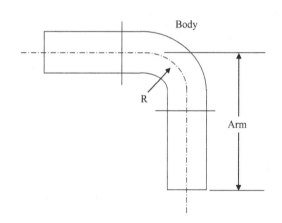

FIGURE 10.19 Bend radius.

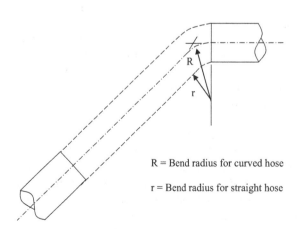

R = Bend radius for curved hose

r = Bend radius for straight hose

FIGURE 10.20 Bend radius: measurement.

- <250 psi (low pressure applications)
- 250–3,000 psi (medium pressure applications)
- 3,000–6,000+ psi (high pressure applications)

Important Point: Note that some manufacturers have their own distinct hose pressure rating schemes; we can't assume that a hose rated as "low-pressure" will automatically be useful at 100 or 200 psi. It may, in fact, be built for pressures not to exceed 50 psi, for example. Therefore, whenever we replace a particular hose, we must ensure that the same type of hose with the same pressure rating as the original hose is used. In high-pressure applications, this precaution is of particular importance.

In addition to the pressure rating of a hose, we must also consider, for some applications, the *vacuum rating* of a hose, which refers to suction hose applications in which the pressure outside the hose is greater than the pressure inside the hose. It is important, obviously, to know the degree of vacuum that can be created before a hose begins to collapse. A drinking straw, for example, collapses rather easily if too much vacuum is applied; thus, it has a low vacuum rating. In contrast, the lower automobile radiator hose (which also works under vacuum) has a relatively higher vacuum rating.

Standards, Codes, and Sizes

Just as they have for piping and tubing, authoritative standards organizations have devised standards and codes for hoses. These standards and codes are safety measures designed to protect personnel and equipment. For example, specifications are provided for working pressures, sizes, and material requirements. The working pressure of a hose, for example, is typically limited to one-fourth or 25% of the amount of pressure needed to burst the hose. Let's look at an example: if we have a hose that has a maximum rated working pressure of 200 psi, it should not rupture until 800 psi is reached, and possibly even then. Thus, the importance of using hoses that meet specified standards or codes is quite evident.

Hose Size

The parameter typically used to designate hose size is its inside diameter (I.D.). In regard to the classification of hose, ordinarily a *dash numbering* system is used. Current practice by most manufacturers is to use the dash system to identify both hose and fittings. In determining the size of a hose, we simply convert the size in 16ths. For example, a hose size of 1/2 inch (a hose with a 1/2-inch I.D.) is the same as 8/16 inch. The numerator of the fraction (the top number, or "8" in this case) is the dash size of the hose. In the same way, a 1 1/2-inch size can be converted to 24/16 inch and is therefore identified as a –24 (pronounced "dash 24") hose. By using the dash system, we can match a hose line to a tubing or piping section and be sure the I.D. of both will be the same. This means, of course, that the non-turbulent flow of fluid will not be interrupted. Based on I.D., hoses range in size from 3/16 inch to as large as 24 inch.

Hose Classifications

Hose is classified in a number of ways; for example, hose can be classified by type of service (hydraulic, pneumatic, and corrosion-resistant), by material, by pressure, and by type of construction. Hose may also be classified by type. The three types include metallic, nonmetallic, and reinforced nonmetallic. Generally, the terminology is the same for each type.

Nonmetallic Hose

Relatively speaking, the use of hose is not a recent development. Hoses, in fact, have been used for one application or another for hundreds of years. Approximately 100 years ago, after new developments in the processing of rubber, layering rubber around mandrels usually made hoses. Later, the mandrel was removed, leaving a flexible rubber hose. However, these flexible hoses tended to collapse easily. Even so, they were an improvement over the earlier types. Manufacturers later added layers of rubberized canvas. This improvement gave hoses more strength and the ability to handle higher pressures. Later, after the development of synthetic materials, manufacturers had more rugged and corrosion-resistant rubber-type materials to work with. Today, neoprene, nitrile rubber, and butyl rubber are commonly used in hoses. However, current manufacturing practice is not to make hoses from a single material. Instead, different materials form layers in the hose, reinforcing it in various ways for strength and resistance to pressure. Hoses manufactured today usually have a rubber-type inner tube or a synthetic (e.g., plastic) lining surrounded by a *carcass* (usually braided) and cover. The requirements of the application determine the type of carcass braiding used. To reinforce a hose, two types of braiding are used: *vertical* braiding and *horizontal* braiding. Vertical braiding strengthens the hose against pressure applied at right angles to the centerline of the hose. Horizontal braiding strengthens the hose along its length, giving it greater resistance to expansion and contraction. Descriptions of the types of nonmetallic hoses follow, with references to their general applications.

Vertical-Braided Hose

Vertical-braided hose has an inner tube of seamless rubber (see Figure 10.21). The reinforcing wrapping (carcass) around the tube is made of one or more layers of braided yarn. This type of hose is usually made in lengths of up to 100 feet with I.D.s of up to 1.5 inches. Considered a small

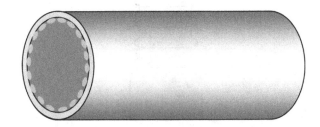

FIGURE 10.21 Vertical-braided hose.

hose, it is used in low-pressure applications to carry fuel oil, acetylene gas and oxygen for welding, water for lawns, gardens, and other household uses, and paint for spraying.

Horizontal-Braided Hose

The horizontal-braided hose is mandrel-built and used to make a hose with an I.D. of up to 3 inches. Used in high-pressure applications, the seamless rubber tube is reinforced by one or more layers of braided fibers or wire. This hose is used to carry propane and butane gas, steam, and for various hydraulic applications that require high working pressures.

Reinforced Horizontal Braided-Wire Hose

In this type of hose, the carcass around the seamless tube is made up of two or more layers of fiber braid with steel wire reinforcement between them. The I.D. may be up to 4 inches. Mechanically very strong, this hose is used where there are high working pressures and/or strong suction (vacuum) forces, such as in chemical transfer and petroleum applications.

Wrapped Hose

Made in diameters up to 24 inches, wrapped hose is primarily used for pressure service rather than suction. The hose is constructed on mandrels, and to close tolerances (see Figure 10.22). It also has a smooth bore, which encourages laminar flow and avoids turbulence. Several plies (layers) of woven cotton or synthetic fabric make up the reinforcement. The tube, selected for its resistance to corrosive fluids, is made from several synthetic rubbers. It is also used in sandblasting applications.

Wire-Reinforced Hose

In this type of hose, wires wound in a spiral around the tube, or inside the carcass, in addition to a number of layers of wrapped fabrics, provide the reinforcement (see Figure 10.23). With I.D.s of 16–24 inches common, this type of hose is used in oil-suction and discharge situations that require special hose ends, maximum suction (without collapsing), special flexing characteristics (must be able to bend in a small radius without collapsing), or a combination of these three requirements.

Wire-Woven Hose

Wire-woven hose (see Figure 10.24) has cords interwoven with wire running spirally around the tube and is highly flexible, lightweight, and resistant to collapse even under suction conditions. This kind of hose is well suited for negative pressure applications.

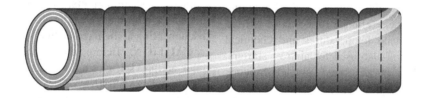

FIGURE 10.22   Wrapped hose.

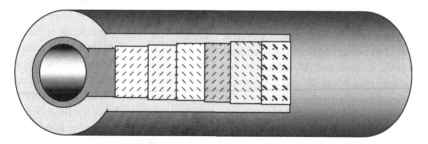

FIGURE 10.23   Wire-reinforced hose.

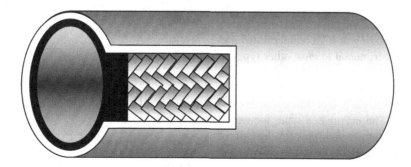

FIGURE 10.24   Wire-woven hose.

Other Types of Nonmetallic Hose

Hoses are also made of other nonmetallic materials, many of them nonreinforced. For example, materials like Teflon®, Dacron®, polyethylene, and nylon have been developed. Dacron remains flexible at very low temperatures, even as low as −200+°C (up to −350°F), nearly the temperature of liquid nitrogen. Consequently, these hoses are used to carry liquefied gas in cryogenic applications. Where corrosive fluids and fluids up to 230+°C (up to 450°F) are to be carried, Teflon is often used. Teflon can also be used at temperatures as low as −55°C (−65°F). Usually sheathed in a flexible, braided metal covering, Teflon hoses are well protected against abrasion and also have added resistance to pressure.

Nylon hoses (small diameter) are commonly used as air hoses, supplying compressed air to small pneumatic tools. The large plastic hoses (up to 24 inches) used to ventilate manholes are made of neoprene-coated materials such as nylon fabric, glass fabric, and cotton duck. The cotton duck variety is for light-duty applications, while the glass fabric type is used with portable heaters and for other applications involving hot air and fumes.

Various hoses made from natural latex, silicone rubber, and pure gum are available. The pure gum hose can safely carry acids, chemicals, and gases. Small hoses of natural latex, which can be sterilized, are used in hospitals, pharmaceuticals, blood, intravenous solutions, food-handling operations, and laboratories. Silicone rubber hose is used in situations where extreme temperatures and chemical reactions are possible. It is also used for aircraft starters, providing compressed air in very large volumes. Silicone rubber hose works successfully over a temperature range from −57°C (−70°F) to 232°C (450°F).

Metallic Hose

The construction of a braided, flexible all-metal hose includes a tube of corrugated bronze. The tube is covered with a woven metallic braid to protect against abrasion and provide increased resistance to pressure. Metal hoses are also available in steel, aluminum, Monel®, stainless steel, and other corrosion-resistant metals in diameters up to 3 inches and lengths of up to 24 inches. In addition to protecting against abrasion and resistance to pressure, flexible metal hoses also dampen vibration. For example, a plant air compressor produces a considerable amount of vibration. The use of flexible hoses in such machines increases their portability and dampens vibrations. Other considerations, such as constant bending at high temperatures and pressures, are extremely detrimental to most other types of hoses.

Other common uses for metallic hoses include serving as steam lines, lubricating lines, gas and oil lines, and exhaust hoses for diesel engines. The corrugated type, for example, is used for high-temperature, high-pressure leak-proof service. Another type of construction is the *interlocked*

flexible metal hose, which is used mainly for low-pressure applications. The standard shop oil can use a flexible hose for its flexible spout. Other metal hoses, with a liner of flexible, corrosion-resistant material, are available in diameters of up to 24 inches.

Another type of metallic hose is used in ductwork. This type of hose is usually made of aluminum, galvanized steel, or stainless steel and is used to protect against corrosive fumes, as well as gases at extreme hot or cold temperatures. This hose is fire-resistant because it usually does not burn.

Hose Couplings

The methods of connecting or coupling hoses vary. Hose couplings may be either permanent or reusable. They can also be manufactured for the obvious advantage of quick-connect or quick-disconnect. Probably the best example of the need for quick-connect is a fire hose—quick-disconnect couplings permit rapid connection between separate lengths of hose and between hose ends and hydrants or nozzles. Another good example of where the quick-connect, quick-disconnect feature is user-friendly is in plant or mobile compressed air systems—a single line may have several uses. Changes involve disconnecting one section and connecting another. In plant shops, for example, compressed air from a single source is used to power pneumatic tools, cleaning units, paint sprayers, and so on. Each unit has a hose that is equipped for rapid connecting and disconnecting at the fixed airline.

CAUTION: Before connections are broken unless quick-acting, self-closing connectors are used, pressure must be released first.

For general low-pressure applications, a coupling like that shown in Figure 10.25 is used. To place this coupling on the hose by hand, first cut the hose to the proper length, then oil the inside of the hose and the outside of the coupling stem. Force the hose over the stem into the protective cap until it seats against the bottom of the cap. No brazing is involved, and the coupling can be used repeatedly. After the coupling has been inserted into the hose, a yoke is placed over it in such a way that its arms are positioned along opposite sides of the hose behind the fitting. The arms are then tightly strapped or banded.

CAUTION: Where the pressure demands are greater, such a coupling can be blown out of the tube. Hose couplings designed to meet high-pressure applications must be used.

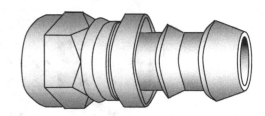

FIGURE 10.25 Low-pressure hose coupling.

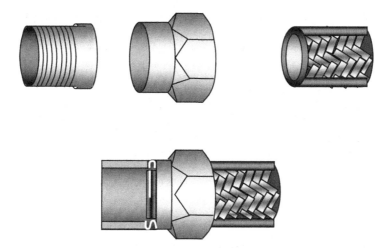

FIGURE 10.26 Coupling installation for an all-metal hose.

A variation of this type uses a clamp that is put over the inner end of the fitting and is then tightly bolted, thus holding the hose firmly. In other cases, a plain clamp is used. Each size clamp is designed for a hose of a specified size (diameter). The clamp slides snugly over the hose and is then crimped tight by means of a special hand tool or air-powered tool.

Couplings for an all-metal hose, described earlier, involve two brazing operations, as shown in Figure 10.26. The sleeve is slipped over the hose end and brazed to it, and then the nipple is brazed to the sleeve.

Important Point: For large hoses of rugged wall construction, it is not possible to insert push-on fittings by hand. Special bench tools are required.

Quick-connect, quick-disconnect hose couplings provide flexibility in many plant process lines where a number of different fluids or dry chemicals from a single source are either to be blended or routed to different vats or other containers. Quick-connect couplings can be used to pump out excavations, manholes, and so forth. They would not be used, however, where highly corrosive materials are involved.

Hose Maintenance

All types of equipment and machinery require proper care and maintenance, including hoses. Depending on the hose type and its application, some require more frequent checking than others. The maintenance procedures required for most hoses are typical and are outlined here as an example. To maintain a hose, we should:

1. Examine for cracks in the cover caused by weather, heat, oil, or usage.
2. Look for a restricted bore because of tube swelling or foreign objects.
3. Look for cover blisters, which permit material pockets to form between the carcass and cover.
4. Look for leaking materials, which are usually caused by improper couplings or faulty fastenings of couplings.

5. Look for corrosion damage to couplings.
6. Look for kinked or otherwise damaged hose.

CAUTION: Because any of the faults listed above can result in a dangerous hose failure, regular inspection is necessary. At the first sign of weakness or failure, replace the hose. System pressure and temperature gauges should be checked regularly. Do not allow the system to operate above design conditions—especially when the hose is a component of the system.

PIPE & TUBE FITTINGS

The term *piping* refers to the overall network of pipes or tubing, fittings, flanges, valves, and other components that comprise a conduit system used to convey fluids. Whether a piping system is used to simply convey fluids from one point to another or to process and condition the fluid, piping components serve an important role in the composition and operation of the system. A system used solely to convey fluids may consist of relatively few components, such as valves and fittings, whereas a complex chemical processing system may consist of a variety of components used to measure, control, condition, and convey the fluids. In this section, the characteristics and functions of various piping and tubing fittings are described (Geiger 2000).

Fittings

The primary function of *fittings* is to connect sections of piping and tubing and to change the direction of flow. Whether used in piping or tubing, fittings are similar in shape and type, even though pipe fittings are usually heavier than tubing fittings. Several methods can be used to connect fittings to piping and tubing systems. However, most tubing is threadless because it does not have the wall thickness needed to carry threads. Most pipes, on the other hand, because they have heavier walls, are threaded.

With regard to changing the direction of flow, the simplest way would be simply to bend the conduit, which, of

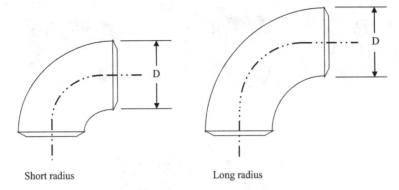

FIGURE 10.27 Short- and long-radius elbows.

course, is not always practical or possible. When piping is bent, it is usually accomplished by the manufacturer in the production process (in larger shops equipped with their pipe-bending machines) but not by the maintenance operator on the job. Tube bending, on the other hand, is a common practice. Generally, a tubing line requires fewer fittings than a pipeline; however, in actual practice, many tube fittings are used.

Important Point: Recall that improperly made bends can restrict fluid flow by changing the shape of the pipe and weakening the pipe wall.

Fittings are basically made from the same materials (and in the same broad ranges of sizes) as piping and tubing, including bronze, steel, cast iron, glass, and plastic. Various established standards are in place to ensure that fittings are made from the proper materials and can withstand the pressures required; they are also made to specific tolerances so that they will properly match the piping or tubing that they join. A fitting stamped "200 lb," for example, is suitable (and safe) for use up to 200 psi.

FUNCTIONS OF FITTINGS

Fittings in piping and tubing systems have five main functions:

- Changing the direction of flow
- Providing branch connections
- Changing the sizes of lines
- Closing lines
- Connecting lines

Changing the Direction of Flow

Usually, a 45° or 90° *elbow* (or "ell") fitting is used to change the direction of flow. Elbows are among the most commonly used fittings in piping and are occasionally used in tubing systems. Two types of 90° elbows are shown in Figure 10.27. From the figure, it is apparent that the *long-radius* fitting (the most preferred elbow) has the more gradual curve of the two. This type of elbow is used in applications where the rate of flow is critical and space presents no problem. The gradual curve minimizes the flow loss caused by turbulence. The *short-radius* elbow (see

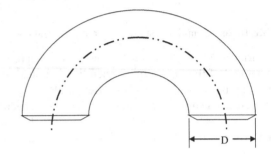

FIGURE 10.28 Long-radius return bend.

Figure 10.27) should not be used in a system made up of long lines that have many changes in direction. Because of the greater frictional loss in the short-radius elbow, heavier, more expensive pumping equipment may be required. Figure 10.28 shows a *return bend* fitting that carries fluid through a 180° ("hairpin") turn. This type of fitting is used for piping in heat exchangers and heater coils. Note that tubing, which can be bent into this form, does not require any fittings in this kind of application.

Providing Branch Connections

Because they are often more than single lines running from one point to another, piping and tubing systems usually have a number of intersections. In fact, many complex piping and tubing systems resemble the layout of a town or city.

Changing the Sizes of Lines

For certain applications, it is important to reduce the volume of fluid flow or to increase flow pressure in a piping or tubing system. To accomplish this, a *reducer* (which reduces a line to a smaller pipe size) is commonly used.

Important Point: Reducing is also sometimes accomplished by means of a bushing inserted into the fitting.

Sealing Lines

Pipe *caps* are used to seal or close off (similar to "corking" a bottle) the end of a pipe or tube. Usually, caps are used in a part of the system that has been dismantled. To seal off openings in fittings, *plugs* are used. Plugs also provide a means of access into the piping or tubing system in case the line becomes clogged.

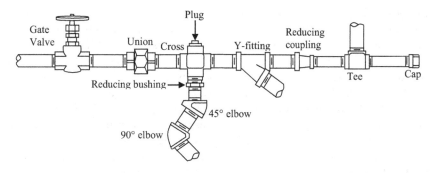

FIGURE 10.29 Diagram of a hypothetical shortened piping system.

Connecting Lines

To connect two lengths of piping or tubing, a coupling or union is used. A *coupling* is simply a threaded sleeve. A *union* is a three-piece device that includes a threaded end, an internally threaded bottom end, and a ring. A union does not change the direction of flow, close off the pipe, or provide for a branch line. Unions make it easy to connect or disconnect pipes without disturbing the position of the pipes. Figure 10.29 is a diagram of a shortened piping system, which illustrates how some fittings are used in a piping system. (Figure 10.29 is only for illustrative purposes; it is unlikely that such a system with so many fittings would actually be used.)

FIGURE 10.30 Flanged fitting.

TYPES OF CONNECTION

Pipe connections may be screwed, flanged, or welded. Each method is widely used, and each has its advantages and disadvantages.

Screwed Fittings

Screwed fittings are joined to the pipe by means of threads. The main advantage of using threaded pipe fittings is that they can be easily replaced. The actual threading of a section of replacement pipe can be accomplished on the job. The threading process itself, however, cuts right into the pipe material and may weaken the pipe in the joint area. The weakest links in a piping system are the connection points. Because threaded joints can be potential problem areas, especially where higher pressures are involved, the threads must be properly cut to ensure the "weakest" link is not further compromised. Typically, the method used to ensure a good seal in a threaded fitting is to coat the threads with a paste dope. Another method is to wind the threads with Teflon® tape.

Flanged Connections

Figure 10.30 shows a flanged fitting. Flanged fittings are forged or cast-iron pipes. The flange is a rim at the end of the fitting, which mates with another section. Pipe sections are also made with flanged ends. Flanges are joined either by being bolted or welded together. The flange faces may be ground and lapped to provide smooth, flat mating surfaces. Obviously, a tight joint must be provided to prevent leakage

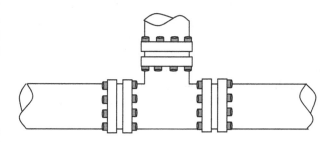

FIGURE 10.31 Flanged joint.

of fluid and pressure. Figure 10.31 shows a typical example of a flanged joint. The mating parts are bolted together with a gasket inserted between their faces to ensure a tight seal. The procedure requires proper alignment of clean parts and tightening of bolts.

Important Point: Some flanges have raised faces, and others have plain faces. Like faces must be matched, a flange with a raised face should never be joined to one with a plain face.

Welded Connections

Currently, because of improvements in piping technology and welding techniques and equipment, the practice of using welded joints is increasing. When properly welded, a piping system forms a continuous system that combines piping, valves, flanges, and other fittings. Along with providing a long leak-proof and maintenance-free life, the smooth joints simplify insulation and take up less room.

TUBING FITTINGS AND CONNECTIONS

Tubing is connected by brazed or welded flange, compression, and flare fittings.

The *welded flange* connection is a reliable means of connecting tubing components. The flange welded to the tube end fits against the end of the fitting. The locknut of the flange is then tightened securely onto the fitting. The *compression fitting* connection uses a ferrule that pinches the tube as the locknut is tightened on the body of the fitting. The *flare fitting* connection uses tubing flared on one end that matches the angle of the fitting. The tube's flared end is butted against the fitting, and a locknut is screwed tightly onto the fitting, sealing the tube connection properly.

Other fittings used for flanged connections include expansion joints and vibration dampeners. *Expansion joints* function to compensate for slight changes in the length of pipe by allowing joined sections of rigid pipe to expand and contract with changes in temperature. They also allow pipe motion, either along the length of the pipe or to the side, as the pipe shifts around slightly after installation. Finally, expansion joints help dampen vibration and noise carried along the pipe from distant equipment (e.g., pumps). One type of expansion joint has a leak-proof tube that extends through the bore and forms the outside surfaces of the flanges. Natural or synthetic rubber compounds are normally used, depending on the application. Other types of expansion joints include metal corrugated types, slip-joint types, and spiral-wound types. In addition, high-temperature lines are usually made with a large bend or loop to allow for expansion. *Vibration dampeners* absorb vibrations that, unless reduced, could shorten the life of the pipe and the service life of the operating equipment. They also eliminate line humming and hammering (water hammer) carried by the pipes.

VALVES

Any water or wastewater operation will have many valves that require attention. Simply as a matter of routine, a fluid mechanic must be able to identify and locate different valves to inspect, adjust, and repair or replace them. For this reason, the operator should be familiar with all valves, especially those that are vital parts of a piping system. A *valve* is defined as any device by which the flow of fluid may be started, stopped, or regulated by a movable part that opens or obstructs passage. As applied in fluid power systems, valves are used for controlling the flow, pressure, and direction of the fluid flow through a piping system. The fluid may be a liquid, a gas, or some loose material in bulk (like a biosolids slurry). Designs of valves vary, but all valves have two features in common: a passageway through which fluid can flow and some kind of movable (usually machined) part that opens and closes the passageway (Globe Valves 1998).

Important Point: It is all but impossible to operate a practical fluid power system without some means of controlling the volume and pressure of the fluid and directing the flow of fluid to the operating units. This is accomplished by incorporating different types of valves.

Whatever type of valve is used in a system, it must be accurate in controlling fluid flow and pressure and the sequence of operation. Leakage between the valve element and the valve seat is reduced to a negligible quantity by precision-machined surfaces, resulting in carefully controlled clearances. This is, of course, one of the very important reasons for minimizing contamination in fluid power systems. Contamination causes valves to stick, plugs small orifices, and causes abrasions of the valve seating surfaces, which result in leakage between the valve element and valve seat when the valve is in the closed position. Any of these can result in inefficient operation or complete stoppage of the equipment. Valves may be controlled manually, electrically, pneumatically, mechanically, hydraulically, or by combinations of two or more of these methods. Factors that determine the method of control include the purpose of the valve, the design and purpose of the system, the location of the valve within the system, and the availability of the source of power.

Valves are made from bronze, cast iron, steel, Monel®, stainless steel, and other metals. They are also made from plastic and glass (see Table 10.1). Special valve trim is used where seating and sealing materials differ from the basic material of construction (see Table 10.2). (*Valve trim* usually

TABLE 10.1
Valves: Materials of Construction

Cast iron	Grey cast iron. Also, referred to as flake graphite iron.
Ductile iron	Malleable iron or spheroidal graphite (nodular) cast iron.
Carbon steel	Steel forgings or steel castings, according to the method of manufacture. Carbon steel valves may also be manufactured by fabrication using wrought steel.
Stainless steel	May also be in the form of forgings, castings, or wrought steels for fabrication.
Copper alloy	Gunmetal, bronze, or brass. Aluminum bronze may also be used.
High duty alloys	Nickel or nickel molybdenum alloys manufactured under various trade names.
Other metals	Pure metals having extreme corrosion resistance, such as titanium or aluminum.
Non-metals	Typically the plastics materials, such as PVC or polypropylene.

TABLE 10.2
Valve Trim

Metal seating	Commonly used in gate and globe valves, particularly in the latter for control applications where seatings may additionally be coated with hard metal.
Soft seating	Commonly used in ball, butterfly, and diaphragm valves. Seatings may be made from a wide variety of elastomers and polymers, including fluorocarbons.
Lined	Valves are usually made of cast iron with an internal lining of elastomer or polymer material. Inorganic materials such as glass, together with metals such as titanium, are used for lining. The lining thickness will depend on the design and the type of material used. In many cases, the valve lining will also form the seating trim.

means those internal parts of a valve controlling the flow and in physical contact with the line fluid.) Valves are made in a full range of sizes that match pipe and tubing sizes. Actual valve size is based upon the internationally agreed definition of nominal size. *Nominal size (DN)* is a numerical designation of size common to all components in a piping system other than components designated by outside diameters. It is a convenient number for reference purposes and is only loosely related to manufacturing dimensions. Valves are made for service at the same pressures and temperatures to which piping and tubing is subject. Valve pressures are based upon the internationally agreed definition of nominal pressure. *Nominal pressure (PN)* is a pressure that is conventionally accepted or used for reference purposes. All equipment of the same nominal size (DN) designated by the same nominal pressure (PN) number must have the same mating dimensions appropriate to the type of end connections. The permissible working pressure depends upon materials, design, and working temperature and should be selected from the relevant, pressure/temperature tables. The pressure rating of many valves is designated under the American (ANSI) class system. The equivalent class rating to PN ratings is based upon international agreement.

Usually, valve end connections are classified as flanged, threaded, or other (see Table 10.3). Valves are also covered by various codes and standards, as are the other components of piping and tubing systems. Many valve manufacturers offer valves with special features. Table 10.4 lists a few of these special features; however, this is not an exhaustive list, and for more details about other features, the manufacturer should be consulted. The various types of valves used in fluid power systems, their classifications, and their applications are discussed in this section.

Valve Construction

Figure 10.32 shows the basic construction and principle of operation of a common valve type. Fluid flows into the valve through the inlet. The fluid flows through passages in the body and past the opened element that closes the valve. It then flows out of the valve through the outlet or discharge. If the closing element is in the closed position, the passageway is blocked and fluid flow is stopped at that point. The closing element keeps the flow blocked until the valve is opened again. Some valves are opened automatically, while manually operated hand wheels control others. Other valves, such as check valves, operate in response to pressure or the direction of flow. To prevent leakage whenever the closing element is in the closed position, a seal is used. In Figure 10.32, the seal consists of a *stuffing box* fitted with packing. The closing element fits against the *seat* in the valve body to keep the valve tightly closed.

TABLE 10.3
Valve End Connections

Flanged	Valves will normally be supplied with flanges conforming to either BS4505 (Equivalent to DIN) or BS 1560 (equivalent to ANSI) according to specifications. Manufacturers may also be able to supply valves with flanges to other standards.
Threaded	Valves will normally be supplied with threads to BS21 (ISO/7), either parallel or taper.
Other	End connections include butt or socket weld ends, as well as wafer valves designed to fit between pipe flanges.

TABLE 10.4
Valve Special Features

High temperature	Valves that are usually able to operate continuously on services above 250°C.
Cryogenic	Valves that will operate continuously on services in the range of −50°C to 196°C.
Bellows sealed	Valves that are glandless designs having a metal bellows for stem sealing.
Actuated	Valves that may be operated by a gearbox, pneumatic or hydraulic cylinder (including diaphragm actuator), or electric motor and gearbox.
Fire tested design	Refers to a valve that has passed a fire test procedure specified in an appropriate inspection standard.

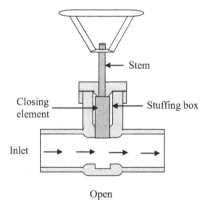

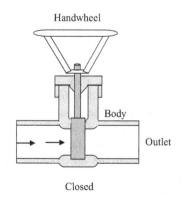

FIGURE 10.32 Basic valve operation.

TYPES OF VALVES

The types of valves covered in this text include:

1. Ball valves
2. Gate valves
3. Globe valves
4. Needle valves
5. Butterfly valves
6. Plug valves
7. Check valves
8. Quick-opening valves
9. Diaphragm valves
10. Regulating valves
11. Relief valves
12. Reducing valves

Each of these valves is designed to control the flow, pressure, and direction of fluid flow, or to serve some other special application. With a few exceptions, these valves are named after the type of internal element that controls the passageway. The exceptions are the check valve, quick-opening valve, regulating valve, relief valve, and reducing valves.

Ball Valves

Ball valves, as the name implies, are stop valves that use a ball to stop or start fluid flow. The ball performs the same function as the disk in other valves. As the valve handle is turned to open the valve, the ball rotates to a point where part or all of the hole through the ball is in line with the valve body inlet and outlet, allowing fluid to flow through the valve. When the ball is rotated so the hole is perpendicular to the flow openings of the valve body, the flow of fluid stops. Most ball valves are the quick-acting type. They require only a 90° turn to either completely open or close the valve. However, many are operated by planetary gears. This type of gearing allows the use of a relatively small handwheel and operating force to operate a large valve. The gearing does, however, increase the operating time for the valve. Some ball valves also contain a swing check located within the ball to give the valve a check valve feature. The two main advantages of using ball valves are: (1) fluid can flow through it in either direction, as desired; and (2) when closed, pressure in the line helps to keep it closed.

Gate Valves

Gate valves are used when a straight-line flow of fluid and minimum flow restriction are needed; they are the most common type of valve found in a water distribution system. Gate valves are so named because the part that either stops or allows flow through the valve acts somewhat like a gate. The gate is usually wedge-shaped. When the valve is wide open, the gate is fully drawn up into the valve bonnet. This leaves an opening for flow through the valve the same size as the pipe in which the valve is installed. For these reasons, the pressure loss (pressure drop) through these types

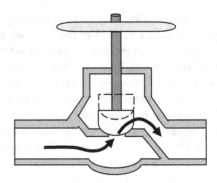

FIGURE 10.33 Globe valve.

of valves is about equal to the loss in a piece of pipe of the same length. Gate valves are not suitable for throttling (means to control the flow as desired, by means of intermediate steps between fully open and fully closed) purposes. The control of flow is difficult because of the valve's design, and the flow of fluid slapping against a partially open gate can cause extensive damage to the valve.

Important Point: Gate valves are well suited to service on equipment in distant locations, where they may remain in the open or closed position for a long time. Generally, gate valves are not installed where they will need to be operated frequently because they require too much time to operate from fully open to closed (AWWA 1996).

Globe Valves

Probably the most common valve type in existence, the globe valve is commonly used for water faucets and other household plumbing. As illustrated in Figure 10.33, the valves have a circular disk (the globe) that presses against the valve seat to close the valve. The disk is the part of the globe valve that controls flow. The disk is attached to the valve stem. As shown in Figure 10.33, fluid flow through a globe valve is at right angles to the direction of flow in the conduits. Globe valves sit very tightly and can be adjusted with fewer turns of the wheel than gate valves; thus, they are preferred for applications that call for frequent opening and closing. On the other hand, globe valves create high head loss when fully open; thus, they are not suited for systems where head loss is critical.

Important Point: The globe valve should never be jammed in the open position. After a valve is fully opened, the hand wheel should be turned toward the closed position approximately one-half turn. Unless this is done, the valve is likely to seize in the open position, making it difficult, if not impossible, to close the valve. Another reason for not leaving globe valves in the fully open position is that it is sometimes difficult to determine whether the valve is open or closed (Globe Valves 1998).

Needle Valves

Although similar in design and operation to the globe valve (a variation of globe valves), the *needle* valve has a closing element in the shape of a long-tapered point, which is at the end of the valve stem. Figure 10.34 shows a cross-sectional

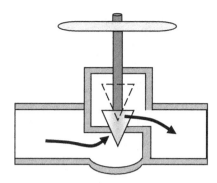

FIGURE 10.34 Common needle valve.

view of a needle valve. In Figure 10.34, the long taper of the valve closing element permits a much smaller seating surface area than that of the globe valve; accordingly, the needle valve is more suitable as a throttle valve. Needle valves are used for very accurate throttling.

Butterfly Valves

Figure 10.35 shows a cross-sectional view of a butterfly valve. The valve itself consists of a body in which a disk (butterfly) rotates on a shaft to open or close the valve. Butterfly valves may be flanged or wafer design, the latter intended for fitting directly between pipeline flanges. In the fully open position, the disk is parallel to the axis of the pipe and the flow of fluid. In the closed position, the disk seals against a rubber gasket-type material bonded either to the valve seat of the body or to the edge of the disk. Because the disk of a butterfly valve stays in the fluid path in the open position, the valve creates more turbulence (higher resistance to flow—equaling a higher pressure loss) than a gate valve. On the other hand, butterfly valves are compact. They can also be used to control flow in either direction. This feature is useful in water treatment plants that periodically backwash to clean filter systems.

Plug Valves

A plug valve (also known as a *cock* or *petcock*) is similar to a ball valve. Plug valves:

1. Offer high-capacity operation with 1/4 turn operation.
2. Use either a cylindrical or conical plug as the closing member.
3. Are directional.
4. Offer moderate vacuum service.
5. Allow flow throttling with interim positioning.
6. Are of simple construction with an O-ring seal.
7. Are not necessarily full on and off.

8. Are easily adapted to automatic control.
9. Can safely handle gases and liquids.

Check Valves

Check valves are usually self-acting and designed to allow the flow of fluid in one direction only. They are commonly used at the discharge of a pump to prevent backflow when the power is turned off. When the direction of the flow is moving in the proper direction, the valve remains open. When the direction of flow reverses, the valve closes automatically due to the fluid pressure against it.

Several types of check valves are used in water/wastewater operations, including:

1. Slanting disk check valves
2. Cushioned swing check valves
3. Rubber flapper swing check valves
4. Double door check valves
5. Ball check valves
6. Foot valves
7. Backflow prevention devices

In each case, pressure from the flow in the proper direction pushes the valve element to an open position. Flow in the reverse direction pushes the valve element to a closed position.

Important Point: Check valves are also commonly referred to as *nonreturn* or *reflux* valves.

Quick-Opening Valves

Quick-opening valves are nothing more than adaptations of some of the valves already described. Modified to provide a quick on/off action, they use a lever device in place of the usual threaded stem and control handle to operate the valve. This type of valve is commonly used in water/wastewater operations where deluge showers and emergency eyewash stations are installed in work areas where chemicals are loaded, transferred, and/or where chemical systems are maintained. They also control the air supply for some emergency alarm horns around chlorine storage areas, for example. Moreover, they are usually used to cut off the flow of gas to a main or individual outlets.

Diaphragm Valves

Diaphragm valves are glandless valves that use a flexible elastomeric diaphragm (a flexible disk) as the closing member and, in addition, affect an external seal. They are well suited for service in applications where tight, accurate closure is important. The tight seal is effective whether the fluid is a gas or a liquid. This tight closure feature makes

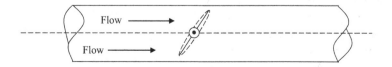

FIGURE 10.35 Cross-section of a butterfly valve.

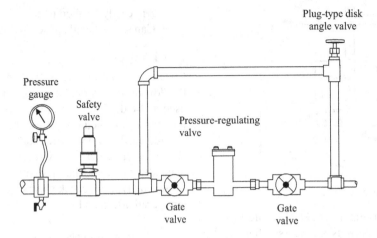

FIGURE 10.36 Pressure-regulating valve system.

these valves useful in vacuum applications. Diaphragm valves operate similarly to globe valves and are usually multi-turn in operation; they are available as weir type and full bore. A common application of diaphragm valves in water or wastewater operations is to control fluid to an elevated tank.

Regulating Valves

As their name implies, *regulating* valves regulate either pressure or temperature in a fluid line, keeping them very close to a preset level. If the demands and conditions of a fluid line always remained steady, no regulating valve would be needed. In the real world, however, ideal conditions do not occur. *Pressure-regulating* valves regulate fluid pressure levels to meet flow demand variations. Flow variations depend on the number of pieces of equipment in operation and the change in demand as pumps and other machines operate. In such fluid line systems, demands are constantly changing. Probably the best example of this situation is seen in the operation of the plant's low-pressure air supply system. For shop use, no more than 30 psi air is usually required (depending on required usage, of course). This air is supplied by the plant's air compressor, which normally operates long enough to fill an accumulator with pressurized air at a set pressure level.

When shop air is required, for whatever reason, compressed air is drawn from the connection point in the shop. The shop connection point is usually connected via a pressure reducer (which sets the pressure at the desired usage level) that, in turn, is fed from the accumulator, where the compressed air is stored. If the user draws a large enough quantity of compressed air from the system (from the accumulator), a sensing device within the accumulator will send a signal to the air compressor to start, which will produce compressed air to recharge the accumulator. In addition to providing service in airlines, pressure-regulating valves are also used in liquid lines. The operating principle is much the same for both types of service. Simply put, the valve is set to monitor the line and to make needed adjustments in response to a signal from a sensing device.

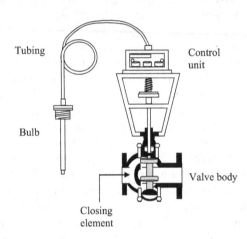

FIGURE 10.37 Temperature-regulating valve assembly.

Temperature-regulating valves (also referred to as *thermostatic control* valves) are closely related to pressure-regulating valves (see Figure 10.36). Their purpose is to monitor the temperature in a line or process solution tank and to regulate it—to raise or lower the temperature as required. In water and wastewater operations, probably the most familiar application where temperature-regulating valves (see Figure 10.37) are used is in heat exchangers. A *heat exchanger-type* water system utilizes a water-to-coolant heat exchanger for heat dissipation. This is an efficient and effective method to dispose of unwanted heat. Heat exchangers are equipped with temperature-regulating valves that automatically modulate the shop process water, limiting usage to just what is required to achieve the desired coolant temperature.

Relief Valves

Some fluid power systems, even when operating normally, may temporarily develop excessive pressure. For example, whenever an unusually strong work resistance is encountered, dangerously high pressure may develop. *Relief* valves are used to control this excess pressure. Such valves are automatic; they start to open at a preset pressure but require a 20% overpressure to open wide. As the pressure increases, the valve continues to open farther until it has reached its maximum travel. As the pressure drops, it starts to close and finally shuts off at about the set pressure. Main system relief valves are generally installed between the pump or pressure source and the first system isolation valve. The valve must be large enough to allow the full output of the hydraulic pump to be delivered back to the reservoir.

Important Point: Relief valves do not maintain flow or pressure at a given amount but prevent pressure from rising above a specific level when the system is temporarily overloaded.

Reducing Valves

Pressure-reducing valves provide a steady pressure into a system that operates at a lower pressure than the supply system. In practice, they are very much like pressure-regulating valves. A pressure-reducing valve reduces pressure by throttling the fluid flow. A reducing valve can normally be set for any desired downstream pressure within the design limits of the valve. Once the valve is set, the reduced pressure will be maintained regardless of changes in supply pressure (as long as the supply pressure is at least as high as the reduced pressure desired) and regardless of the system load, provided the load does not exceed the design capacity of the reducer.

VALVE OPERATORS

In many modern water and wastewater operations, devices called operators or actuators mechanically operate many valves. These devices may be operated by air, electricity, or fluid; that is, pneumatic, hydraulic, and magnetic operators.

Pneumatic and Hydraulic Valve Operators

Pneumatic and hydraulic valve operators are much the same in appearance and work in much the same way. Hydraulic cylinders using either plant water pressure or hydraulic fluid frequently operate valves in treatment plants and pumping stations (AWWA 1996). In a typical pneumatic ball-valve actuator, the cylinder assembly is attached to the ball-valve stem close to the pipe. A piston inside the cylinder can move in either direction. The piston rod is linked to the valve stem, opening or closing the valve, depending on the direction in which the piston is traveling. As a fail-safe feature, some of these valves are spring-loaded. In case of hydraulic or air pressure failure, the valve operator returns the valve to the safe position.

Note: According to Casada (2000), valve operators and positioners usually require more maintenance than the valves themselves.

Magnetic Valve Operators

Magnetic valve operators use electric *solenoids*. A solenoid is a coil of magnetic wire, roughly in the shape of a doughnut. When a bar of iron is inserted as a plunger mechanism inside an energized coil, it moves along the coil because of the magnetic field that is created. If the plunger (the iron bar) is fitted with a spring, it returns to its starting point when the electric current is turned off. Solenoids are used as operators for many different types of valves in water and wastewater operations. For example, in a direct-operating valve, the solenoid plunger is used in place of a valve stem and handwheel. The plunger is connected directly to the disk of a globe valve. As the solenoid coil is energized or de-energized, the plunger rises or falls, operating or closing the valve.

VALVE MAINTENANCE

As with any other mechanical device, effective valve maintenance begins with its correct operation. As an example of incorrect operation, consider the standard household water faucet. As the faucet washers age, they harden and deteriorate. The valve becomes more difficult to operate properly; eventually, the valve begins to leak. A common practice is simply to apply as much force as possible to the faucet handle. Doing so, however, damages the valve stem and the body of the valve. Good maintenance includes preventive maintenance, which, in turn, includes the inspection of valves, correct lubrication of all moving parts, and the replacement of seals or stem packing.

PIPING SYSTEMS: PROTECTIVE DEVICES

Piping systems must be protected from the harmful effects of undesirable impurities (solid particles) entering the fluid stream. Because of the considerable variety of materials carried by piping systems, there is an equal range of choices in protective devices. Such protective devices include strainers, filters, and traps. In this section, we describe the design and function of strainers, filters, and traps. The major maintenance considerations of these protective devices are also explained.

APPLICATIONS

Filters, strainers, and traps are normally thought of in terms of specific components used in specific systems. However, it is important to keep in mind that the basic principles apply in many systems. While the examples used in this chapter include applications found in water/wastewater treatment, collection, and distribution systems, these applications are also found in almost every plant—hot and cold water lines, lubricating lines, pneumatic and hydraulic lines, and steam lines. With regard to steam lines, it is important to point out that in our discussion of traps, their primary application is in steam systems where they remove unwanted air and condensate from the lines.

Important Point: A very large percentage (estimated to be >70%) of all plant facilities in the United States make use of steam in some applications.

Other system applications of piping protective devices include the conveyance of hot and chilled water for heating and air conditioning, as well as lines that convey fluids for various processes. Any foreign contamination in any of these lines can cause potential trouble. Piping systems can become clogged, thereby causing greatly increased friction and lower line pressure. Foreign contaminants (dirt and other particles) can also damage valves, seals, and pumping components.

Important Point: Foreign particles in a high-pressure line can damage a valve by clogging it so that it cannot close tightly. In addition, foreign particles may wear away the closely machined valve parts.

STRAINERS

Strainers, usually wire mesh screens, are used in piping systems to protect equipment sensitive to contamination that may be carried by the fluid. Strainers can be used in pipelines conveying air, gas, oil, steam, water, wastewater, and nearly any other fluid conveyed by pipes. Generally, strainers are installed ahead of valves, pumps, regulators, and traps to protect them against the damaging effects of corrosion products that may become dislodged and conveyed throughout the piping system (Geiger 2000).

A common strainer is shown in Figure 10.38. This type of strainer is generally used upstream of traps, control valves, and instruments. This strainer resembles a lateral branch fitting with the strainer element installed in the branch. The end of the lateral branch is removable to permit servicing of the strainer. In operation, the fluid passes through the strainer screen, which catches most of the contaminants. Then the fluid passes back into the line. Contaminants in the fluid are caught in two ways: either they do not make it through the strainer screen, or they do not make the sharp turn that the fluid must take as it leaves the unit. The bottom of the unit serves as a sump where the solids are collected. A blowout connection may be provided in the end cap to flush the strainer. The blowout plug can be removed, and the pressure in the line can be used to blow the fixture clean.

Important Point: Before removing the blowout plug, the valve system must be locked out or tagged out first.

FILTERS

The purpose of any filter is to reduce or remove impurities or contaminants from a fluid (liquid or gas) to an acceptable or predetermined level. This is accomplished by passing the fluid through some kind of porous barrier. Filter cartridges have replaceable elements made of paper, wire cloth, nylon cloth, or fine-mesh nylon cloth between layers of coarse wire. These materials filter out unwanted contaminants, which collect on the entry side of the filter element. When clogged, the element is replaced. Most filters operate in two ways: (1) they cause the fluid to make sharp changes in direction as it passes through (this is important because the larger particles are too heavy to change direction quickly), or (2) they contain some kind of barrier that will not let larger contaminants pass.

TRAPS

Traps, used in steam processes are automatic valves that release condensate (condensed steam) from a steam space while preventing the loss of live steam. Condensate is undesirable because water produces rust, and water plus steam leads to water hammer. In addition, steam traps remove air and non-condensate from the steam space. The operation of a trap depends on what is called *differential pressure* (or delta-P), measured in psi. Differential pressure is the difference between the inlet and outlet pressures. A trap will not operate correctly at a differential pressure higher than the one for which it was designed.

There are many types of steam traps because there are many different types of applications. Each type of trap has a range of applications for which it is best suited. For example, thermostatic and float-and-thermostatic are the names given to the two general types of traps. *Thermostatic* traps have a corrugated bellows-operating element that is filled with an alcohol mixture that has a boiling point lower than that of water (see Figure 10.39). The bellows contract when in contact with condensate and expand when the steam is present. If a heavy condensate load occurs, the bellows will remain contracted, allowing condensate to flow continuously. As steam builds up, the bellows close. Thus, at times,

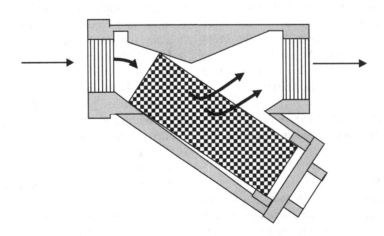

FIGURE 10.38 A common strainer.

the trap acts as a "continuous flow" type, while at other times it acts intermittently as it opens and closes to condensate and stream, or it may remain totally closed (Bandes and Gorelick 2000).

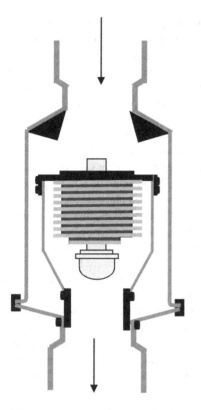

FIGURE 10.39 A thermostatic trap (shown in the open positions).

Important Point: The thermostatic trap is designed to operate at a definite temperature drop of a certain number of degrees below the saturated temperature for the existing steam pressure.

A *float-and-thermostatic* trap is shown in Figure 10.40. It consists of a ball float and a thermostatic bellows element. As condensate flows through the body, the float rises and falls, opening the valve according to the flow rate. The thermostatic element discharges air from the steam lines. They are suitable for heavy and light loads and for both high and low pressure, but they are not recommended where water hammer is a possibility.

Trap Maintenance and Testing

Because they operate under constantly varying pressure and temperature conditions, traps used in steam systems require maintenance. Just as significant, because of these varying conditions, traps can fail. When they do fail, most traps fail in the open mode, which may require the boiler to work harder to perform a task, creating high backpressure in the condensate system. This inhibits the discharge capacities of some traps, which may exceed their rating, causing system inefficiency.

Important Point: While it is true that most traps operate with backpressure, it is also true that they do so only at a percentage of their rating, affecting everything downstream of the failed trap. Steam quality and product can be affected.

A closed trap produces condensate backup into the steam space. The equipment cannot produce the intended heat. Consider, for example, a four-coil dryer with only three operating. In this setup, it will take longer for the dryer to dry a product, which has a negative effect on production.

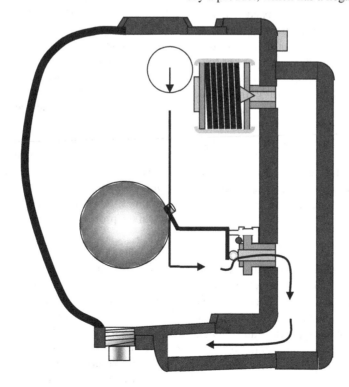

FIGURE 10.40 A float-and-thermostatic trap.

Trap Maintenance

Excluding design problems, two of the most common causes of trap failure are oversizing and dirt. Oversizing causes traps to work too hard. In some cases, this can result in the blowing of live steam. For example, certain trap types can lose their prime due to an abrupt change in pressure, causing the valve to open. Traps tend to accumulate dirt (sludge) that prevents tight closing. The moving parts of the traps are subject to wear. Because the moving parts of traps operate in a mixture of steam and water, or sometimes in a mixture of compressed air and water, they are difficult to lubricate.

Important Point: Dirt (sludge) is generally produced from pipe scale or from over-treating chemicals in a boiler.

Trap maintenance includes periodic cleaning, removing dirt that interferes with valve action, adjusting the mechanical linkage between moving parts and valves, and reseating the valves when necessary. If these steps are not taken, the trap will not operate properly.

Trap Testing

Important Point: A word of caution is advised before testing any steam trap. Inspectors should be familiar with the particular function, review the types of traps, and know the various pressures within the system. This can help ensure inspector safety, avoid misdiagnosis, and allow for proper interpretation of trap conditions.

The three main categories of online trap inspection are visual, thermal, and acoustic. *Visual* inspection depends on a release valve situated downstream of certain traps. A maintenance operator opens these valves to check if the trap discharges condensate or steam. *Thermal* inspection relies on upstream/downstream temperature variations in a trap. It includes pyrometry, infrared, heat bands (wrapped around a trap; they change color as temperature increases), and heat sticks (which melt at various temperatures). *Acoustic* techniques require a maintenance operator to listen to and detect steam trap operations and malfunctions. This method includes various forms of listening devices such as medical stethoscopes, screwdrivers, mechanical stethoscopes, and ultrasonic detection instruments.

Important Point: A simple trap test—just listening to the trap action—tells us how the trap opens and closes. Moreover, if the trap has a bypass line around it, leaky valves will be apparent when the main line to the trap is cut off, forcing all the fluid through the bypass.

PIPING ANCILLARIES

Earlier, we described various devices associated with process piping systems designed to protect the system. In this section, we discuss some of the most widely used ancillaries (or accessories) designed to improve the operation and control of the system. These include pressure and temperature gauges, vacuum breakers, accumulators, receivers, and heat exchangers. We need to know how these ancillary devices work, how to care for them, and, more importantly, how to use them.

GAUGES

To properly operate a system, any system, the operator must know certain things. For example, to operate a plant air compressor, the operator needs to know: (1) how to operate it; (2) how to maintain it; (3) how to monitor its operation; and, in many cases, (4) how to repair it. In short, the operator must understand system parameters and how to monitor them. Simply put, operating parameters refer to those physical indications of system operation. The term, *parameter*, refers to a system's limits or restrictions. For example, let's consider, again, the plant's air compressor. Obviously, it is important to know how the air compressor operates, or at least how to start and place the compressor online properly. However, it is also just as important to determine if the compressor is operating as per design.

Before starting any machine or system, we must first perform a pre-start check to ensure that it has the proper level of lubricating oil, etc. Then, after starting the compressor, we need to determine (observe) if the compressor is actually operating (normally, this is not difficult to discern considering that most air compressor systems make a lot of noise while in operation). Once in operation, our next move is to double-check system line-up to ensure that various valves in the system are correctly positioned (opened or closed). We might even go to a remote plant compressed air service outlet to make sure that the system is producing compressed air. (Keep in mind that some compressed air systems have a supply of compressed air stored in an air receiver; thus, when an air outlet is opened, air pressure might be present even if the compressor is not functioning as per design). On the other hand, instead of using a remote outlet to test for compressed air supply, all we need to do is look at the compressor air pressure gauge. This gauge should indicate that the compressor is producing compressed air.

Gauges are the main devices that provide us with parameter indications needed to determine equipment or system operation. Concerning the air compressor, the parameter we are most concerned about now is air pressure (gauge pressure). Not only is the correct pressure generation by the compressor important, but also the correct pressure in system pipes, tubes, and hoses is essential. Keeping air pressure at the proper level is necessary mainly for four reasons:

1. Safe operation
2. Efficient, economic conveyance of air through the entire system, without waste of energy
3. Delivery of compressed air to all outlet points in the system (the places where the air is to be used) at the required pressure
4. Prevention of too much or too little pressure (either condition can damage the system and become hazardous to personnel).

We pointed out that, before starting the air compressor, certain pre-start checks must be made. This is important for all machinery, equipment, and systems. In the case of our air compressor example, we want to ensure that proper lubricating oil pressure is maintained. This is important, of course, because pressure failure in the lubricating line that serves the compressor can mean inadequate lubrication of bearings and, in turn, expensive mechanical repairs.

Pressure Gauges

As mentioned, many pressure-measuring instruments are called *gauges*. Generally, pressure gauges are located at key points in piping systems. Usually expressed in pounds per square inch (psi), there is a difference between *gauge pressure* (psig) and *absolute pressure* (psia). Simply put, "gauge pressure" refers to the pressure level indicated by the gauge. However, even when the gauge reads zero, it is subject to ambient atmospheric pressure (i.e., 14.7 psi at sea level). When a gauge reads 50 psi, that is 50 lb *gauge pressure* (psig). The true pressure is the 50 lb shown plus the 14.7 lb of atmospheric pressure acting on the gauge. The total "actual" pressure is called the *absolute pressure*: gauge pressure plus atmospheric pressure (50 psi + 14.7 psi = 64.7 psi). It is written 64.7 psi.

Important Point: Pressure in any fluid pushes equally in all directions. The total force on any surface is the psi multiplied by the area in square inches. For example, a fluid under a pressure of 10 psi, pushing against an area of 5 inch^2, produces a total force against that surface of 50 lb (10 × 5).

Spring-Operated Pressure Gauges

Pressure, by definition, must operate against a surface. Thus, the most common method of measuring pressure in a piping system is to have the fluid press against some type of surface—a flexible surface that moves slightly. This movable surface, in turn, is linked mechanically to a gear-lever mechanism that moves the indicator arrow to indicate the pressure on the dial (i.e., a pressure gauge). The surface that the pressure acts against may be a disk or diaphragm, the inner surface of a coiled tube, a set of bellows, or the end of a plunger. No matter the element type, if the mechanism is fitted with a spring that resists the pressure and returns the element (i.e., the indicator pointer) to the zero position when the pressure drops to zero, it is called a *spring-loaded* gauge.

Bourdon-Tube Gauges

Many pressure gauges in use today use a coiled tube as a measuring element called a *Bourdon tube* (named for its inventor, Eugene Bourdon, a French engineer). The Bourdon tube is a device that senses pressure and converts it to displacement. Under pressure, the fluid fills the tube (see Figure 10.41). Since the Bourdon tube's displacement is a function of the pressure applied, it may be mechanically

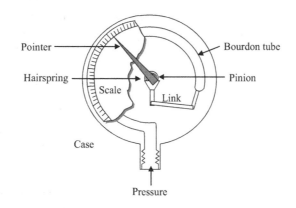

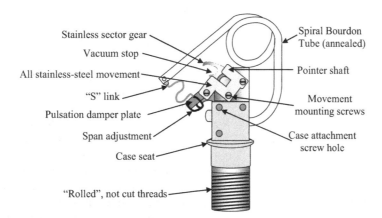

FIGURE 10.41 (Top) Bourdon tube gauge; (bottom) internal components.

amplified and indicated by a pointer. Thus, the pointer position indirectly indicates pressure.

Important Point: The Bourdon tube gauge is available in various tube shapes: helical, C-shaped or curved, and spiral. The size, shape, and material of the tube depend on the pressure range and the type of gauge desired.

Bellows Gauge

Figure 10.42 shows how a simplified *bellows* gauge works. The bellows itself is a convoluted unit that expands and contracts axially with changes in pressure. The pressure to be measured can be applied to either the outside or the inside of the bellows; in practice, most bellows measuring devices have the pressure applied to the outside of the bellows. When pressure is released, the spring returns the bellows and the pointer to the zero position.

Plunger Gauge

Most of us are familiar with the simple tire-pressure gauge. This device is a type of plunger gauge. Figure 10.43 shows a *plunger* gauge used in industrial hydraulic systems. The plunger gauge is a spring-loaded gauge, where pressure from the line acts on the bottom of a cylindrical plunger in the center of the gauge and moves it upward. At full

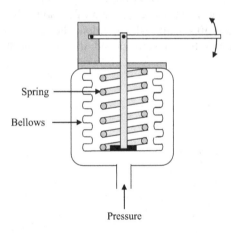

FIGURE 10.42   Bellows gauge.

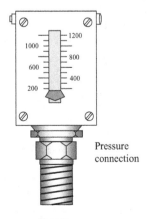

FIGURE 10.43   Plunger gauge.

pressure, the plunger extends above the gauge, indicating the measured pressure. As the pressure drops, the spring contracts to pull the plunger downward back into the body (the zero-reading indication).

Note: Spring-loaded gauges are not extremely accurate, but they are entirely adequate where there is no need for more precise readings.

TEMPERATURE GAUGES

As mentioned, ensuring that system pressures are properly maintained in equipment and piping systems is critical for safe and proper operation. Likewise, ensuring that the temperature of fluids in industrial equipment and piping systems is correct is just as critical. Various temperature-measuring devices are available for measuring the temperature of fluids in industrial systems.

Temperature has been defined in a variety of ways. One example defines temperature as the measure of heat (thermal energy) associated with the movement (kinetic energy) of the molecules of a substance. This definition is based on the theory that molecules of all matter are in continuous motion that is sensed as heat. For our purposes, we define temperature as the degree of hotness or coldness of a substance measured on a definite scale. Temperature is measured when a measuring instrument is brought into contact with the medium being measured (e.g., a thermometer). All temperature-measuring instruments use some change in a material to indicate temperature. Some of the effects that are used to indicate temperature are changes in physical properties and altered physical dimensions (e.g., the change in the length of a material in the form of expansion and contraction).

Several temperature scales have been developed to provide a standard for indicating the temperatures of substances. The most commonly used scales include the Fahrenheit, Celsius, Kelvin, and Rankine temperature scales. The Celsius scale is also called the centigrade scale. The Fahrenheit (°F) and Celsius (°C) scales are based on the freezing point and boiling point of water. The freezing point of a substance is the temperature at which it changes its physical state from a liquid to a solid. The boiling point is the temperature at which a substance changes from a liquid state to a gaseous state.

Note: Thermometers are classified as mechanical temperature sensing devices because they produce some type of mechanical action or movement in response to temperature changes. There are many types of thermometers: liquid, gas, vapor-filled systems, and bimetallic thermometers.

Figure 10.44 shows an *industrial-type thermometer* commonly used for measuring the temperature of fluids in industrial piping systems. This type of measuring instrument is nothing more than a rugged version of the familiar mercury thermometer. The bulb and capillary tube are contained inside a protective metal tube called a *well*. The thermometer is attached to the piping system (vat, tank, or other component) by a union fitting.

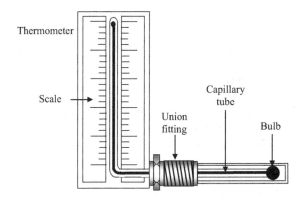

FIGURE 10.44   Industrial thermometer.

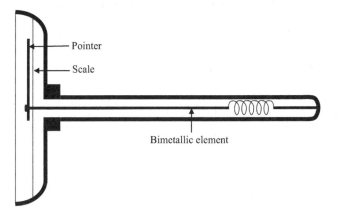

FIGURE 10.45   Bimetallic gauge.

Another common type of temperature gauge is the *bimetallic* gauge shown in Figure 10.45. Bimetallic means that if two materials with different linear coefficients of expansion (i.e., how much a material expands with heat) are bonded together, their rates of expansion will differ as the temperature changes. This will cause the entire assembly to bend in an arc. When the temperature is raised, an arc is formed around the material with the smaller expansion coefficient. The amount of arc is reflected in the movement of the pointer on the gauge. Because two dissimilar materials form the assembly, it is known as a bimetallic element, which is also commonly used in thermostats.

Vacuum Breakers

Another common ancillary device found in pipelines is a *vacuum breaker* (components shown in Figure 10.46). A vacuum breaker is a mechanical device that allows air

into the piping system, thereby preventing backflow that could otherwise be caused by the siphoning action created by a partial vacuum. In other words, a vacuum breaker is designed to admit air into the line whenever a vacuum develops. A vacuum, obviously, is the absence of air. Vacuum in a pipeline can be a serious problem. For example, it can cause fluids to run in the wrong direction, possibly mixing contaminants with purer solutions. In water systems, back-siphonage can occur when a partial vacuum pulls nonpotable liquids back into the supply lines (AWWA 1996). In addition, it can cause the collapse of tubing or equipment.

As illustrated in Figure 10.46, this particular type of vacuum breaker uses a ball that is usually held against a seat by a spring. The ball is contained in a retainer tube mounted inside the piping system or the component being protected. If a vacuum develops, the ball is forced (sucked) down into the retainer tube, where it works against the spring. Air flows into the system to fill the vacuum. In water systems, when air enters the line between a cross-connection and the source of the vacuum, the vacuum will be broken and backsiphonage is prevented (AWWA 1996). The spring then returns the ball to its usual position, which acts to seal the system again.

Accumulators

As mentioned, in a plant compressed air system, a means of storing and delivering air as needed is usually provided. An air receiver normally accomplishes this. In a hydraulic system, an accumulator provides the functions served by an air receiver for an air system. That is, the accumulator (usually a dome-shaped or cylindrical chamber or tank attached to a hydraulic line) in a hydraulic system works to help store and deliver energy as required. Moreover, accumulators help keep pressure in the line smoothed out. For example, if pressure in the line rises suddenly, the accumulator absorbs the rise, preventing shock to the piping. If the pressure in the line drops, the accumulator acts to bring it back up to normal.

Important Point: The primary function of an accumulator in a hydraulic system is to supplement pump flow.

Air Receivers

As shown in Figure 10.47, an air receiver is a tank or cylindrical-type vessel used for a number of purposes. Most importantly, it has the ability to store compressed air.

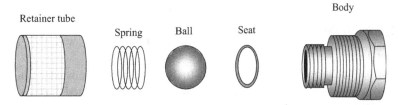

FIGURE 10.46   Vacuum breaker components.

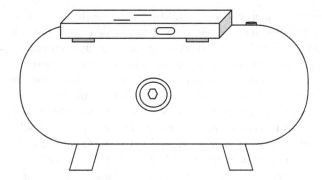

FIGURE 10.47 Air receiver.

Much like accumulators, air receivers cushion shock from sudden pressure rises in an airline. That is, the air receiver serves to absorb the shock of valve closure and load starts, stops, and reversals. There is no liquid in an air receiver; the air compresses as pressure rises. As pressure drops, the air expands to maintain pressure in the line.

Important Note: OSHA standard, 29 CFR 1910.169(a), requires air receivers to be drained. Specifically, the standard states, "a drain pipe and valve shall be installed at the lowest point of every air receiver to provide for the removal of accumulated oil and water" (OSHA 1978). This is an item that should be taken seriously, not only for safety reasons but also because it is a compliance item that OSHA inspectors often check.

HEAT EXCHANGERS

Operating on the principle that heat flows from a warmer body to a cooler one, *heat exchangers* are devices used for adding or removing heat or cold from a liquid or gas. The purpose may be to cool one body or to warm the other; nonetheless, whether used to warm or cool, the principle remains the same. Various designs are used in heat exchangers. The simplest form consists of a tube, or possibly a large coil of tubing, placed inside a larger cylinder. In an oil lubrication system, the purpose of a heat exchanger is to cool the hot oil. However, a heat exchanger system can also be used to heat a process fluid circulating through part of the heat exchanger while steam circulates through its other section.

Final Note: In this section, we have discussed the major ancillary or accessory equipment used in many piping systems. It is important to point out that there are other accessories commonly used in piping systems (e.g., rotary pressure joints, actuators, intensifiers, pneumatic pressure line accessories, and so forth); however, a discussion of these accessories is beyond the scope of this text.

CHAPTER REVIEW QUESTIONS

10.1   What is an expansion joint?
10.2   A _____ is defined as any substance or material that flows.
10.3   Compressed air is considered to be a _____.
10.4   Sections or lengths of pipe are _____ with fittings.
10.5   The _____ of fluids through a pipe is controlled by valves.
10.6   Friction causes _____ _____ in a piping system.
10.7   As friction _____ in a piping system, the output pressure decreases.
10.8   Relief valves are designed to open _____.
10.9   _____ is used to help keep the fluids carried in piping systems hot or cold.
10.10 The major problems in piping systems are caused by _____ and corrosion.
10.11 If the speed of the fluid in a pipe doubles, the friction is _____.
10.12 The most important factor in keeping a piping system operating efficiently is _____
10.13 Pipe sizes above _____ in. are usually designated by outside diameter.
10.14 The difference in _____ numbers represents the difference in the wall _____ of pipes.
10.15 When pipe wall thickness _____, the I.D. decreases.
10.16 A _____ metal contains iron.
10.17 As temperature _____, the viscosity of a liquid decreases.
10.18 Another name for rust is:
10.19 Sections of _____ water pipe are usually connected with a bell-and-spigot joint.
10.20 A ferrous metal always contains _____.
10.21 Asbestos-cement pipe has the advantage of being highly resistant to _____.
10.22 As temperature increases, the strength of plastic pipe _____.
10.23 Name four basic nonmetallic piping materials.
10.24 Vitrified clay pipe is the most _____ pipe available for carrying industrial wastes.
10.25 Cast iron pipe can be lined with _____ to increase its resistance to corrosion.
10.26 A joint made so that the sections of tubing are _____ together is called a compression joint.
10.27 Incorrect tube bends can cause _____ flow and _____ pressure.
10.28 High-pressure hydraulic systems use _____ tubing.
10.29 One process used to join plastic tubing is called _____ welding.
10.30 Compared to pipe, tubing is more _____.
10.31 _____ tubing is most likely used in food-processing applications.
10.32 Before tubing can be bent or flared, it should be _____.
10.33 Plastic tubing is usually joined by _____.
10.34 The materials used most commonly for tubing are _____ and _____.
10.35 Smooth fluid flow is called _____ flow.

10.36 The _____ hose is the most common type of hose in general use.

10.37 The type of hose construction most suitable for maximum suction conditions is _____.

10.38 _____ is the nonmetallic hose best suited for use at extremely low temperatures.

10.39 Each size of hose clamp is designed for hose of a specific _____.

10.40 _____ is the outstanding advantage of hose.

10.41 Applied to a hose, the letters _____ stand for enlarged end.

10.42 Hose is _____ to provide strength and greater resistance to _____.

10.43 Dacron hose remains _____ at extremely low temperatures.

10.44 The _____ fitting allows for a certain amount of pipe movement.

10.45 The _____ fitting helps reduce the effects of water hammer.

10.46 A flange that has a plain face should be joined to a flange that has a _____ face.

10.47 Improperly made _____ restrict fluid flow in a pipeline.

10.48 The designation 200 lb refers to the _____ at which a fitting can safely be used.

10.49 Used to close off an unused outlet in a fitting with a _____.

10.50 A _____ connects two or more pipes of different diameters.

REFERENCES

ACPA. 1987. *Concrete Pipe Design Manual.* Vienna, VA: American Concrete Pipe Association.

ASME. 1996. *ASME b 36.10M. Welded and Seamless Wrought Steel Pipe.* New York: American Society of Mechanical Engineers.

AWWA. 1996. *Water Transmission and Distribution*, 2nd ed. Denver, CO: American Water Works Association.

Babcock & Wilcox, 1972. *Steam: Its Generation and Use.* Cambridge, ON: The Babcock & Wilcox Company.

Bales, R.C., NewKirk, D.D., and Hayward, S.B., 1984. Chrysotile asbestos in California surface waters from upstream rivers through water treatment. *J Am Water Works Assoc*; 76: 66.

Bandes, A., and Gorelick, B., 2000. *Inspect Steam Traps for Efficient System.* Terre Haute, IN: TWI Press.

Basavaraju, C., 2000. Pipe properties, Geiger, E.L., 2000. Tube Properties, in *Piping Handbook*, 7th ed., Nayyar, M.L., (ed.), pp. 211–240. New York: McGraw-Hill.

Casada, D., 2000. *Valve Replacement Savings.* Oak Ridge, TN: Oak Ridge Laboratories.

Coastal Video Communications Corp., 1994. *Asbestos Awareness.* Virginia Beach, VA.

Crocker, S., Jr., 2000. Hierarchy of Design Documents, in *Piping Handbook*, 7th ed. Nayyar, M.L., (ed.), pp. 104–123. New York: McGraw-Hill.

Gagliardi, M.G., and Liberatore, L.J., 2000. Water Piping Systems, in *Piping Handbook,* 7th ed., Nayyar, M.L., (ed.), pp. 143–180. New York: McGraw-Hill.

Geiger, E.L., 2000. Piping Components, in Piping Handbook, 7th ed., Nayyar, M. L., (ed.), pp. 124–142. New York: McGraw-Hill.

Giachino, J.W., and Weeks, W., 1985. *Welding Skills.* Homewood, IL: American Technical Publishers.

Globe Valves. 1998. *Integrated Publishing's* official web page. Accessed 07/06/24 @ http/tpub.com/fluid/ch2c.htm.

Kawamura, S., 1999. *Integrated Design and Operation of Water Treatment Facilities*, 2nd ed. New York: John Wiley & Sons.

Lohmeir, A., and Avery, D.R., 2000. Manufacture of Metallic Pipe, in *Piping Handbook*, 7th ed., Nayyar, M.L., (ed.), pp. 181–190. New York: McGraw-Hill.

Magnusson, R.J., 2001. *Technology in the Middle Ages.* Baltimore: John Hopkins University.

Marine, C.S., 1999. Hydraulic Transient Design for Pipeline Systems, in *Water Distribution Systems Handbook*, Mays, L.W., (ed.). New York: McGraw-Hill.

McGhee, T.J., 1991. *Water Supply and Sewerage*, 6th ed. New York: McGraw-Hill.

Nayyar, M.L., 2000. Introduction to Piping, in *Piping Handbook*, 7th ed., Nayyar, M.L., (ed.), 1–7. New York: McGraw-Hill.

OSHA. 1978. *Drain on Air Receivers.* 29 CFR 1910.169. Washington, DC.

Snoek, P.E., and Carney, J.C., 1981. Pipeline Material Selection for Transport of Abrasive Tailings. *Proceedings, Sixth International Technical Conference on Slurry Transportation,* Las Vegas, NV.

Spellman, F.R., 1996. *Safe Work Practices for Wastewater Treatment Plants.* Boca Raton, FL: CRC Press.

USEPA. 2006. *Emerging Technologies for Conveyance Systems.* Washington, DC: EPA 832-R-06-004.

Webber, J.S., Covey, J.R., and King, M.V. 1989. Asbestos in drinking water is supplied through grossly deteriorated pipe. *J Am Water Works Assoc*; 81:80.

Part III

Characteristics of Water

11 Basic Water Chemistry

Chemical testing can be divided into two types. The first type measures a bulk physical property of the sample, such as volume, temperature, melting point, or mass. These measurements are normally performed with an instrument, and one simply has to calibrate the instrument to perform the test. Most analysis, however, are of the second type, in which a chemical property of the sample is determined that generates information about how much of what is present.

—Smith (1993)

INTRODUCTION

Water is a unique molecule. Although no one has seen a water molecule, we have determined through X-rays that atoms in water are elaborately meshed. Moreover, although it is true that we do not know as much as we need to know about water—our growing knowledge of water is a work in progress—we have determined many things about water. A large amount of our current knowledge comes from studies of water chemistry.

Water chemistry is important because several factors about water to be treated and then distributed or returned to the environment are determined through simple chemical analysis. Probably the most important determination that the water practitioner makes about water is its hardness.

Why chemistry? "I'm not a chemist," you say. But when you add chlorine to water to make it safe to drink or discharge into a receiving body (usually a river or lake), you are a chemist. Chemistry is the study of substances and the changes they undergo. Water specialists and those interested in the study of water must possess a fundamental knowledge of chemistry. Before beginning our discussion of water chemistry, it is important for the reader to have a basic understanding of chemistry concepts and chemical terms. The following section presents a review of chemistry terms, definitions, and concepts.

DID YOU KNOW?

A colorless, odorless, tasteless liquid, water is the only common substance that occurs naturally on the Earth in all three physical states: solid, liquid, and gas. Approximately 73% of the Earth's surface, almost 328 million cubic miles, is covered with water. The human body is 70% water by weight, and water is essential to the life of every living thing (Hauser 1995).

CHEMISTRY CONCEPTS AND DEFINITIONS

Chemistry, like the other sciences, has its own language; thus, to understand chemistry, you must understand the following concepts and key terms.

CONCEPTS

Miscible and Solubility

Substances that are miscible are capable of being mixed in all proportions. Simply put, when two or more substances disperse themselves uniformly in all proportions when brought into contact, they are said to be completely soluble in one another, or completely miscible. The precise chemistry definition is: "homogenous molecular dispersion of two or more substances" (Jost1992). Examples include the following:

- All gases are completely miscible.
- Water and alcohol are completely miscible.
- Water and mercury (in its liquid form) are immiscible liquids.

Between the two extremes of miscibility, there is a range of *solubility*; that is, various substances mix with one another up to a certain proportion. In many environmental situations, a rather small amount of contaminant may be soluble in water in contrast to the complete miscibility of water and alcohol. The amounts are measured in parts per million (ppm).

Suspension, Sediments, Particles, and Solids

Often water carries *solids* or *particles* in suspension. These dispersed particles are much larger than molecules and may comprise millions of molecules. The particles may be suspended in flowing conditions and initially under quiescent conditions, but eventually, gravity causes settling of the particles. The resultant accumulation by settling is often called *sediment* or *biosolids* (sludge) or *residual solids* in wastewater treatment vessels. Between this extreme of readily falling out by gravity and permanent dispersal as a solution at the molecular level, there are intermediate types of dispersion or suspension. Particles can be so finely milled or of such small intrinsic size as to remain in suspension almost indefinitely and, in some respects, similarly to solutions.

Emulsion

Emulsions represent a special case of suspension. As you know, oil and water do not mix. Oil and other hydrocarbons derived from petroleum generally float on water with

DOI: 10.1201/9781003581901-14

negligible solubility in water. In many instances, oils may be dispersed as fine oil droplets (an emulsion) in water and not readily separated by floating because of size or the addition of dispersal-promoting additives. Oil and, in particular, emulsions can prove detrimental to many treatment technologies and must be treated in the early steps of a multi-step treatment train.

Ion

An ion is an electrically charged particle. For example, sodium chloride or table salt forms charged particles on dissolution in water; sodium is positively charged (a cation), and chloride is negatively charged (an anion). Many salts similarly form cations and anions on dissolution in water.

Mass Concentration

Concentration is often expressed in terms of parts per million (ppm) or mg/L. Sometimes, parts per thousand (ppt) or parts per billion (ppb) are also used.

$$ppm = \frac{\text{Mass of substance}}{\text{Mass of solutions}} \qquad (11.1)$$

Because 1 kg of solution with water as a solvent has a volume of approximately 1 L:

$$1\ ppm \approx 1\ mg/L$$

Definitions

Chemistry: The science that deals with the composition and changes in composition of substances. Water is an example of this composition; it is composed of two gases, hydrogen, and oxygen. Water also changes form from liquid to solid to gas but does not necessarily change composition.

Matter: Anything that has weight (mass) and occupies space. Kinds of matter include elements, compounds, and mixtures.

Solids: Substances that maintain a definite size and shape are solids in water that fall into one of the following categories:

- *Dissolved solids* are in solution and pass through a filter. The solution consisting of the dissolved components and water forms a single phase (a homogeneous solution).
- *Colloidal solids* (sols) are uniformly dispersed in solution, but they form a solid phase that is distinct from the water phase.
- *Suspended solids* are also a separate phase from the solution. Some suspended solids are classified as settleable solids. Settleable solids are determined by placing a sample in a cylinder and measuring the amount of solids that have settled after a set amount of time. The size of solids increases going from dissolved solids to suspended solids.

Liquids: A definite volume, but not shape; liquids will fill containers to certain levels and form free level surfaces.

Gases: Of neither definite volume nor shape, gases completely fill any container in which they are placed.

Element: The simplest form of chemical matter. Each element has chemical and physical characteristics different from all other kinds of matter.

Compound: A substance of two or more chemical elements chemically combined. Examples: water (H_2O) is a compound formed by hydrogen and oxygen. Carbon dioxide (CO_2) is composed of carbon and oxygen.

Mixture: A physical, not chemical, intermingling of two or more substances. Sand and salt stirred together form a mixture.

Atom: The smallest particle of an element that can unite chemically with other elements. All the atoms of an element are the same in chemical behavior, although they may differ slightly in weight. Most atoms can combine chemically with other atoms to form molecules.

Molecule: The smallest particle of matter or a compound that possesses the same composition and characteristics as the rest of the substance. A molecule may consist of a single atom, two or more atoms of the same kind, or two or more atoms of different kinds.

Radical: Two or more atoms that unite in a solution and behave chemically as if they were a single atom.

Solvent: The component of a solution that does the dissolving.

Solute: The component of a solution that is dissolved by the solvent.

Ion: An atom or group of atoms that carries a positive or negative electric charge as a result of having lost or gained one or more electrons.

Ionization: The formation of ions by the splitting of molecules or electrolytes in solution. Water molecules are in continuous motion, even at lower temperatures. When two water molecules collide, a hydrogen ion is transferred from one molecule to the other. The water molecule that loses the hydrogen ion becomes a negatively charged hydroxide ion, While the water molecule that gains the hydrogen ion becomes a positively charged hydronium ion. This process is commonly referred to as the self-ionization of water.

Cation: A positively charged ion.

Anion: A negatively charged ion.

Organic: Chemical substances of animal or vegetable origin made of a carbon structure.

Inorganic: Chemical substances of mineral origin.

Solids: As it pertains to water—suspended and dissolved material in water.

Dissolved Solids: The material in water that will pass through a glass fiber filter and remain in an evaporating dish after evaporation of the water.

Suspended Solids: The quantity of material deposited when a quantity of water, sewage, or other liquid is filtered through a glass fiber filter.

Total Solids: The solids in water, sewage, or other liquids; it includes the suspended solids (largely removable by a filter) and filterable solids (those which pass through the filter).

Saturated Solution: The physical state in which a solution will no longer dissolve more of the dissolving substance—solute.

Colloidal: Any substance in a certain state of fine division in which the particles are less than 1 μm in diameter.

Turbidity: A condition in water caused by the presence of suspended matter, resulting in the scattering and absorption of light rays.

Precipitate: A solid substance that can be dissolved but is separated from solution because of a chemical reaction or change in conditions such as pH or temperature.

CHEMISTRY FUNDAMENTALS

Whenever water and wastewater practitioners add a substance to another substance (from adding sugar to a cup of tea to adding chlorine to water to make it safe to drink) they perform chemistry. Water and wastewater operators (as well as many others) are chemists because they work with chemical substances, and it is important for operators to know and understand how those substances react.

MATTER

Going through a day without coming into contact with many kinds of matter would be impossible. Paper, coffee, gasoline, chlorine, rocks, animals, plants, water, and air—all the materials of which the world is made—are all different forms or kinds of matter. Earlier matter was defined as anything that has mass (weight) and occupies space—matter is distinguishable from empty space by its presence. Thus, obviously, the statement about going through a day without coming into contact with "matter" is not only correct, but avoiding some form of matter is virtually impossible. Not all matter is the same, even though we narrowly classify all matter into three groups: solids, liquids, and gases. These three groups are called the *physical states of matter* and are distinguishable from one another by means of two general features: shape and volume.

Important Point: *Mass* is closely related to the concept of *weight*. On Earth, the weight of matter is a measure of the force with which it is pulled by gravity toward the Earth's center. As we leave Earth's surface, the gravitational pull decreases, eventually becoming virtually insignificant, while the weight of matter accordingly reduces to zero. Yet, the matter still possesses the same amount of "mass." Hence, the mass and weight of matter are proportional to each other.

Important Point: Since matter occupies space, a given form of matter is also associated with a definite volume. Space should not be confused with air, since air is itself a form of matter. *Volume* refers to the actual amount of space that a given form of matter occupies.

Solids have a definite, rigid shape with their particles closely packed together and sticking firmly to each other. A solid does not change its shape to fit a container. Put a solid on the ground and it will keep its shape and volume—it will never spontaneously assume a different shape. Solids also possess a definite volume at a given temperature and pressure.

Liquids maintain a constant volume but change shape to fit the shape of their container; they do not possess a characteristic shape. The particles of the liquid move freely over one another but still stick together enough to maintain a constant volume. Consider a glass of water. The liquid water takes the shape of the glass up to the level it occupies. If we pour the water into a drinking glass, the water takes the shape of the glass; if we pour it into a bowl, the water takes the shape of the bowl. Thus, if space is available, any liquid assumes whatever shape its container possesses. Like solids, liquids possess a definite volume at a given temperature and pressure, and they tend to maintain this volume when they are exposed to a change in either of these conditions.

Gases have no definite fixed shape and their volume can be expanded or compressed to fill different sizes of containers. A gas or mixture of gases, like air, can be put into a balloon and will take the shape of the balloon. Particles of gases do not stick together at all and move about freely, filling containers of any shape and size. A gas is also identified by its lack of a characteristic volume. When confined to a container with non-rigid, flexible walls, for example, the volume that a confined gas occupies depends on its temperature and pressure. When confined to a container with rigid walls, however, the volume of the gas is forced to remain constant.

Internal linkages among its units, including between one atom and another, maintain the constant composition associated with a given substance. These linkages are called *chemical bonds*. When a particular process occurs that involves the making and breaking of these bonds, we say that a *chemical reaction* or a *chemical change* has occurred. Let's take a closer look at both chemical and physical changes in matter.

Chemical changes occur when new substances are formed that have entirely different properties and characteristics. When wood burns or iron rusts, a chemical change occurs; the linkages—the chemical bonds—are broken. *Physical changes* occur when matter changes its physical properties, such as size, shape, and density, as well as when it changes its state, i.e., from gas to liquid to solid. When ice melts or when a glass window breaks into pieces, a physical change has occurred.

Content of Matter: The Elements

Matter is composed of pure basic substances. Earth is made up of the fundamental substances of which all matter is composed. These substances that resist attempts to decompose them into simpler forms of matter are called *elements*.

To date, there are more than 100 known elements. They range from simple, lightweight elements to very complex, heavyweight elements. Some of these elements exist in nature in pure form; others are combined. The smallest unit of an element is the *atom*.

The simplest atom possible consists of a nucleus having a single proton with a single electron traveling around it. This is an atom of hydrogen, which has an atomic weight of one because of the single proton. The *atomic weight* of an element is equal to the total number of protons and neutrons in the nucleus of an atom of an element.

To better understand the basic atomic structure and related chemical principles it is useful to compare the atom to our solar system. In our solar system, the sun is the center of everything, whereas the *nucleus* is the center of the atom. The sun has several planets orbiting around it, while the atom has *electrons* orbiting about the nucleus. It is interesting to note that the astrophysicist, who would likely find this analogy overly simplistic, is concerned mostly with activity within the nucleus. This is not the case, however, with the chemist. The chemist deals principally with the activity of the planetary electrons; chemical reactions between atoms or molecules involve only electrons, with no changes in the nuclei.

The nucleus is made up of positively electrically charged *protons* and *neutrons,* which are neutral (no charge). The negatively charged electrons orbiting it balance the positive charge in the nucleus. An electron has negligible mass (less than 0.02% of the mass of a proton), which makes it practical to consider the weight of the atom as the weight of the nucleus.

Atoms are identified by name, atomic number, and atomic weight. The *atomic number*, or *proton number*, is the number of protons in the nucleus of an atom. It is equal to the positive charge on the nucleus. In a neutral atom, it is also equal to the number of electrons surrounding the nucleus. As mentioned, the atomic weight of an atom depends on the number of protons and neutrons in the nucleus, with the electrons having negligible mass. Atoms (elements) receive their names and symbols in interesting ways. The discoverer of the element usually proposes a name for it. Some elements get their symbols from languages other than English. The following is a list of common elements with their common names and the names from which the symbol is derived.

- Chlorine   Cl
- Copper   Cu (*Cuprum*—Latin)
- Hydrogen   H
- Iron   Fe (*Ferrum*—Latin)
- Nitrogen   N
- Oxygen   O
- Phosphorus   P
- Sodium   Na (*Natrium*—Latin)
- Sulfur   S

As shown above, a capital letter or a capital letter and a small letter designate each element. These are called chemical symbols. As is apparent from the above list, most of the time the symbol is easily recognized as an abbreviation of the atom's name, such as O for oxygen.

Typically, we do not find most of the elements as single atoms. They are more often found in combinations of atoms called *molecules.* A molecule is the least common denominator of making a substance what it is. A system of formulae has been devised to show how atoms are combined into molecules. When a chemist writes the symbol for an element, it stands for one atom of the element. A subscript following the symbol indicates the number of atoms in the molecule. O_2 is the chemical formula for an oxygen molecule. It shows that oxygen occurs in molecules consisting of two oxygen atoms. As you know, a molecule of water contains two hydrogen atoms and one oxygen atom, so the formula is H_2O.

Important Point: The chemical formula of the water molecule, H_2O, was defined in 1860 by the Italian scientist Stanislao Cannizzaro.

Some elements have similar chemical properties. For example, a chemical such as bromine (atomic number 35) has chemical properties that are similar to those of the element chlorine (atomic number 17, which most water operators are familiar with) and iodine (atomic number 53).

In 1865, English chemist John Newlands arranged some of the known elements in increasing order of atomic weights. Newlands' arrangement had the lightest element he knew about at the top of his list and the heaviest element at the bottom. He was surprised when he observed that starting from a given element, every eighth element repeated the properties of the given element.

Later, in 1869, Mendeleev, a Russian chemist, published a table of the 63 known elements. In his table, Mendeleev, like Newlands, arranged the elements in increasing order of atomic weights. He also grouped them in eight vertical columns so that the elements with similar chemical properties would be found in one column. It is interesting to note that Mendeleev left blanks in his table. He correctly hypothesized that undiscovered elements existed that would fill in the blanks when they were discovered. Because he knew the chemical properties of the elements above and below the blanks in his table, he was able to predict quite accurately the properties of some of the undiscovered elements.

Today, our modern form of the periodic table is based on the work done by the English scientist Henry Moseley, who was killed during World War I. Following the work of Ernest Rutherford (a New Zealand physicist) and Niels Bohr (a Danish physicist), Moseley used X-ray methods to determine the number of protons in the nucleus of an atom.

The atomic number, or number of protons, of an atom is related to its atomic structure. In turn, atomic structure governs chemical properties. The atomic number of an element is more directly related to its chemical properties than its atomic weight. It is more logical to arrange the periodic table according to atomic numbers than atomic weights. By demonstrating the atomic numbers of elements, Moseley enabled chemists to create a better periodic table.

In the periodic table, each box or section contains the atomic number, symbol, and atomic weight of an element. The numbers down the left side of the box show the arrangement, or configuration, of the electrons in the various shells around the nucleus. For example, the element carbon has an atomic number of 6, its symbol is C, and its atomic weight is 12.01.

In the periodic table, a horizontal row of boxes is called a *period* or *series*. Hydrogen is all by itself because of its special chemical properties. Helium is the only element in the first period. The second period contains lithium, beryllium, boron, carbon, nitrogen, oxygen, fluorine, and neon. Other elements may be identified by looking at the table. A vertical column is called a *group* or *family*. Elements in a group have similar chemical properties. The periodic table is useful because by knowing where an element is located in the table, you can have a general idea of its chemical properties.

As mentioned, for convenience, elements have a specific name and symbol but are often identified by chemical symbol only. The symbols of the elements consist of either one or two letters, with the first letter capitalized. Table 11.1 lists the elements important to the water practitioner (about a third of the 100+ elements) below. Those elements most closely associated with water treatment are marked (*).

TABLE 11.1
Elements and their Symbols

Element	Symbol
Aluminum*	Al
Arsenic	As
Barium	Ba
Cadmium	Ca
Carbon*	C
Calcium	Ca
Chlorine*	Cl
Chromium	Cr
Cobalt	Co
Copper	Cu
Fluoride*	F
Helium	He
Hydrogen*	H
Iodine	I
Iron*	Fe
Lead	Pb
Magnesium*	Mg
Manganese*	Mn
Mercury	Hg
Nitrogen*	N
Nickel	Ni
Oxygen*	O
Phosphorus	P
Potassium	K
Silver	Ag
Sodium*	Na
Sulfur*	S
Zinc	Zn

COMPOUND SUBSTANCES

If we take a pure substance like calcium carbonate (limestone) and heat it, the calcium carbonate ultimately crumbles to a white powder. However, careful examination of the heating process shows that carbon dioxide also evolves from the calcium carbonate. Substances like calcium carbonate that can be broken down into two or more simpler substances are called *compound substances* or simply *compounds*. Heating is a common way of decomposing compounds, but other forms of energy are often used as well.

Chemical elements that make up compounds such as calcium carbonate combine with each other in definite proportions. When atoms of two or more elements are bonded together to form a compound, the resulting particle is called a *molecule*.

Important Point: This law simply means that only a certain number of atoms or radicals of one element will combine with a certain number of atoms or radicals of a different element to form a chemical compound.

Water (H_2O) is a compound. As stated, compounds are chemical substances made up of two or more elements bonded together. Unlike elements, compounds can be separated into simpler substances by chemical changes. Most forms of matter in nature are composed of combinations of the 100 pure elements.

If we have a particle of a compound, for example, a crystal of salt (sodium chloride), and subdivide, subdivide, and subdivide until you get the smallest unit of sodium chloride possible, you would have a molecule. As stated, a molecule (or least common denominator) is the smallest particle of a compound that still has the characteristics of that compound.

Important Point: Because the weights of atoms and molecules are relative and the units are extremely small, the chemist works with units he/she identifies as moles. A mole (symbol mol) is defined as the amount of a substance that contains as many elementary entities (atoms, molecules, and so on) as there are atoms in 12 g of the isotope carbon-12.

Important Point: An isotope of an element is an atom having the same structure as the element—the same electrons orbiting the nucleus and the same protons in the nucleus—but having more or fewer neutrons.

One mole of an element that exists as single atoms weighs as many grams as its atomic number (so one mole of carbon weighs 12 g), and it contains 6.022045×10^{23} atoms, which is *Avogadro's number*.

As stated previously, symbols are used to identify elements. This is a shorthand method for writing the names of the elements. This shorthand method is also used for writing the names of compounds. Symbols used in this manner show the kinds and numbers of different elements in the compound. These shorthand representations of chemical compounds are called chemical *formulas*. For example, the formula for table salt (sodium chloride) is NaCl. The formula shows that one atom of sodium combines with one atom of chlorine to form sodium chloride. Let's look at a more complex formula for the compound sodium carbonate (soda ash): Na_2CO_3. The formula shows that this compound is made up of three elements: sodium, carbon, and oxygen. In addition, there are two atoms of sodium, one atom of carbon, and three atoms of oxygen in each molecule.

When depicting chemical reactions, chemical *equations* are used. The following equation shows a chemical reaction that most water/wastewater operators are familiar with: chlorine gas added to water. It shows the formulas of the molecules that react together and the formulas of the product molecules.

$$Cl_2 + H_2O \rightarrow HOCl + HCl$$

As stated previously, a chemical equation tells what elements and compounds are present before and after a chemical reaction. Sulfuric acid poured over zinc will cause the release of hydrogen and the formation of zinc sulfate. This is shown by the following equation:

$$Zn + H_2SO_4 \rightarrow ZnSO_4 + H_2$$

One atom (also one molecule) of zinc unites with one molecule of sulfuric acid, giving one molecule of zinc sulfate and one molecule (two atoms) of hydrogen. Notice that there is the same number of atoms of each element on each side of the arrow. However, the atoms are combined differently.

Let us look at another example. When hydrogen gas is burned in the air, the oxygen from the air unites with the hydrogen and forms water. The water is the product of burning hydrogen. This can be expressed as an equation.

$$2H_2 + O_2 \rightarrow 2H_2O$$

This equation indicates that two molecules of hydrogen unite with one molecule of oxygen to form two molecules of water.

WATER SOLUTIONS

A *solution* is a condition in which one or more substances are uniformly and evenly mixed or dissolved. A solution has two components: a solvent and a solute. The *solvent* is the component that does the dissolving, while the *solute* is the component that is dissolved. In water solutions, water is the solvent. Water can dissolve many other substances—given enough time, there are not too many solids, liquids, and gases that water cannot dissolve. When water dissolves substances, it creates solutions with many impurities. Generally, a solution is usually transparent and not cloudy. However, a solution may be colored when the solute remains uniformly distributed throughout the solution and does not settle with time.

When molecules dissolve in water, the atoms making up the molecules come apart (dissociate) in the water. This dissociation in water is called *ionization*. When the atoms in the molecules come apart, they do so as charged atoms (both negatively and positively charged) called *ions*.

The positively charged ions are called *cations*, and the negatively charged ions are called *anions*. A good example of ionization occurs when calcium carbonate ionizes:

$$CaCO_3 \leftrightarrow Ca^{++} + CO_3^{-2}$$

Calcium carbonate   Calcium ion   Carbonate ion
(cation)   (anion)

Another good example is the ionization that occurs when table salt (sodium chloride) dissolves in water:

$$NaCl \leftrightarrow Na^+ + Cl^-$$

Sodium chloride   Sodium ion   Chloride ion
(cation)   (anion)

The symbols of some of the common ions found in water are provided below:

Ion	Symbol
Hydrogen	H^+
Sodium	Na^+
Potassium	K^+
Chloride	Cl^-
Bromide	Br^-
Iodide	I^-
Bicarbonate	HCO_3^-

Water dissolves polar substances better than non-polar substances. This makes sense when you consider that water is a polar substance. Polar substances such as mineral acids, bases, and salts are easily dissolved in water, while non-polar substances such as oils, fats, and many organic compounds do not dissolve easily in water.

Water dissolves polar substances better than non-polar substances—only to a point. Polar substances dissolve in water up to a point—only so much solute will dissolve at a given temperature, for example. When that limit is reached, the resulting solution is saturated. When a solution becomes saturated, no more solute can be dissolved. For solids dissolved in water, if the temperature of the solution is increased, the amount of solids (solutes) required to reach saturation increases.

WATER QUALITY CONSTITUENTS

Natural water can contain a number of substances (what we may call impurities) or constituents in water treatment operations. The concentrations of various substances in water, in dissolved, colloidal, or suspended form, are typically low but vary considerably. A hardness value of up to 400 ppm of calcium carbonate, for example, is sometimes tolerated in public supplies, whereas 1 ppm of dissolved iron would be unacceptable. When a particular constituent can affect the health of the water user or the environment, it is called a contaminant or pollutant. These contaminants, of course, are what the water operator works to prevent from entering or to remove from the water supply. In this section, we discuss some of the more common constituents of water.

SOLIDS

Other than gases, all contaminants in water contribute to the solid content. Natural water carries many dissolved and undissolved solids. The undissolved solids are non-polar substances and consist of relatively large particles of materials such as silt that will not dissolve. Classified by their size and state, as well as their chemical characteristics and size distribution, solids can be dispersed in water in both suspended and dissolved forms.

The size of solids in water can be classified as *suspended*, *settleable*, *colloidal*, or *dissolved*. *Total solids* are those suspended and dissolved solids that remain behind when the water is removed by evaporation. Solids are also characterized as volatile or non-volatile.

The distribution of solids is determined by computing the percentage of filterable solids by size range. Solids typically include inorganic solids such as silt and clay from riverbanks, as well as organic matter, such as plant fibers and microorganisms from natural or human-made sources.

Important Point: Though not technically accurate from a chemical point of view because some finely suspended material can actually pass through the filter, suspended solids are defined as those that can be filtered out in the suspended solids laboratory test. The material that passes through the filter is defined as dissolved solids.

As mentioned, colloidal solids are extremely fine suspended solids (particles) of less than 1 μm in diameter; they are so small (though they still make water cloudy) that they will not settle even if allowed to sit quietly for days or weeks.

TURBIDITY

Simply put, turbidity refers to how clear the water is. Water clarity is one of the first characteristics people notice. Turbidity in water is caused by the presence of suspended matter, which results in the scattering and absorption of light rays. The greater the amount of total suspended solids (TSS) in the water, the murkier it appears and the higher the measured turbidity. Thus, in plain English, turbidity is a measure of the light-transmitting properties of water. Natural water that is very clear (low turbidity) allows you to see images at considerable depths. As mentioned, high turbidity water, on the other hand, appears cloudy. Keep in mind that water with low turbidity is not necessarily free of dissolved solids. Dissolved solids do not cause light to be scattered or absorbed; thus, the water looks clear. High turbidity causes problems for the waterworks operator, as the components that cause high turbidity can lead to taste and odor problems and will reduce the effectiveness of disinfection.

COLOR

Color in water can be caused by several contaminants, such as iron, which changes in the presence of oxygen to yellow or red sediments. The color of water can be deceiving. First, color is considered an aesthetic quality of water with no direct health impact. Second, many of the colors associated with water are not true colors but rather the result of colloidal suspension (apparent color). This apparent color can often be attributed to iron and dissolved tannins extracted from decaying plant material. True color is the result of dissolved chemicals (most often organics) that cannot be seen. True color is distinguished from apparent color by filtering the sample.

DISSOLVED OXYGEN (DO)

Although water molecules contain an oxygen atom, this oxygen is not what is needed by aquatic organisms living in our natural waters. A small amount of oxygen, up to about ten molecules of oxygen per million of water, is actually dissolved in water. This dissolved oxygen is breathed by fish and zooplankton and is needed for their survival. Other gases can also be dissolved in water. In addition to oxygen, carbon dioxide, hydrogen sulfide, and nitrogen are examples of gases that dissolve in water. Gases dissolved in water are important. For example, carbon dioxide is important because of the role it plays in pH and alkalinity. Carbon dioxide is released into the water by microorganisms and consumed by aquatic plants. However, dissolved oxygen (DO) in water is of utmost importance, not only because it is crucial for most aquatic organisms but also because also it serves an important indicator of water quality.

Key Point: Simply stated, DO is an important indicator of the overall biological health of a waterbody and is required to support aquatic life, which is why we measure DO.

Like terrestrial life, aquatic organisms need oxygen to survive. As water moves past their breathing apparatus, microscopic bubbles of oxygen gas in the water, called dissolved oxygen (DO), are transferred from the water to their blood. Like any other gas diffusion process, the transfer is efficient only above certain concentrations. In other words, oxygen can be present in the water but at too low a concentration to sustain aquatic life. Oxygen is also needed by virtually all algae and macrophytes, as well as for many chemical reactions that are important to water body functioning.

Rapidly moving water, such as in a mountain stream or large river, tends to contain a lot of dissolved oxygen, while stagnant water contains little. Bacteria in water can consume oxygen as organic matter decays. Thus, excess organic material in our lakes and rivers can cause an oxygen-deficient situation. Aquatic life can struggle in stagnant water with a lot of decaying organic material, especially in summer, when dissolved oxygen levels are at a seasonal low.

Important Point: As mentioned, solutions can become saturated with solute and this is the case with water and oxygen. As with other solutes, the amount of oxygen that can be dissolved at saturation depends on the temperature

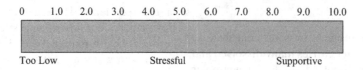

FIGURE 11.1 Range of tolerance for dissolved oxygen in fish—tolerances vary by species. Bottom feeders, crabs, oysters, and worms need minimal amounts of oxygen (1–6 mg/L), while shallow water fish need higher levels (4–15 mg/L) (USEPA 2021; Chesapeake Bay Program 2023).

of the water. In the case of oxygen, the effect is the opposite of other solutes: the higher the temperature, the lower the saturation level; the lower the temperature, the higher the saturation level.

DO is generally measured in the field (for our purposes, the field includes unit processes in waterworks or wastewater treatment plants) along with water temperature, turbidity (clarity), specific conductance, and pH. This information is then assessed against water quality standards to determine whether the water is fit for aquatic life and also meets drinking water standards. With regard to water quality fit for aquatic life, Figure 11.1 is a generalized illustration of how DO affects fish health—keep in mind that sensitivities vary by species.

In Figure 11.1, the range labeled "too low DO" is insufficient to support fish. In the successful range, DO conditions impede spawning and reproduction and limit growth and activity. A higher DO level is needed to support fish spawning, growth, and activity. Note that different levels of DO are required to support aquatic life, depending on the species present and their stages of life (spawning, larvae, etc.).

Key Point: Dissolved oxygen is necessary for more than just fish. It is also necessary for invertebrates, bacteria, and plants. It's all about these organisms' use of oxygen in respiration, similar to organisms on land. Fish and crustaceans obtain oxygen for respiration through their gills, while plant life and phytoplankton require dissolved oxygen for respiration when there is no light for photosynthesis (CMG Garden Notes 2013).

METALS

Metals are elements present in chemical compounds as positive ions or the form of cations (+ ions) in solution. Metals with a density over 5 kg/dm^3 are known as *heavy metals*. Metals are one of the constituents or impurities often carried by water. Although most metals are not harmful at normal levels, a few can cause taste and odor problems in drinking water. In addition, some metals may be toxic to humans, animals, and microorganisms. Most metals enter water as part of compounds that ionize to release the metal as positive ions.

Table 11.2 lists some metals commonly found in water and their potential health hazards.

Important Point: Metals may be found in various chemical and physical forms. These forms, or "species," can be particles, simple organic compounds, organic complexes, or colloids. The dominant form is determined largely by

TABLE 11.2
Common Metals Found in Water

Metal	Health Hazard
Barium	Circulatory system affects and increases blood pressure
Cadmium	Concentration in the liver, kidneys, pancreas and thyroid
Copper	Nervous system damage and kidney effects, toxic to humans
Lead	Same as copper
Mercury	Central nervous system (CNS) disorders
Nickel	CNS disorders
Selenium	CNS disorders
Silver	Turns skin gray
Zinc	Causes taste problems—not a health hazard

the chemical composition of the water, the matrix, and, in particular, the pH.

ORGANIC MATTER

Organic matter or compounds contain the element carbon and are derived from material that was once alive (i.e., plants and animals). Organic compounds include fats, dyes, soaps, rubber products, plastics, wood, fuels, cotton, proteins, and carbohydrates. Organic compounds in water are usually large, nonpolar molecules that do not dissolve well in water. They often provide large amounts of energy to animals and microorganisms.

Important Point: Natural organic matter (NOM) describes the complex mixture of organic material, such as humic and hydrophilic acids, present in all drinking water sources. NOM can cause major problems in water treatment as it reacts with chlorine to form disinfection by-products (DBPs). Many of the disinfection DBPs formed by the reaction of NOM with disinfectants are reported to be toxic and carcinogenic to humans if ingested over an extended period. The removal of NOM and, hence, the reduction of DBPs is a major goal in the treatment of any water source.

INORGANIC MATTER

Inorganic matter or compounds are carbon-free, not derived from living matter, and easily dissolved in water; they are of mineral origin. Inorganics include acids, bases, oxides, salts, etc. Several inorganic components are important in establishing and controlling water quality. Two important inorganic constituents in water are nitrogen and phosphorus.

TABLE 11.3
Relative Strengths of Acids in Water

Perchloric acid	$HClO_4$
Sulfuric acid	H_2SO_4
Hydrochloric acid	HCl
Nitric acid	HNO_3
Phosphoric acid	H_3PO_4
Nitrous acid	HNO_2
Hydrofluoric acid	HF
Acetic acid	CH_3COOH
Carbonic acid	H_2CO_3
Hydrocyanic acid	HCN
Boric acid	H_3BO_3

ACIDS

Lemon juice, vinegar, and sour milk are acidic or contain acid. The common acids used in waterworks operations are hydrochloric acid (HCl), sulfuric acid (H_2SO_4), nitric acid (HNO_3), and carbonic acid (H_2CO_3). Note that, in each of these acids, hydrogen (H) is one of the elements.

Important Point: An acid is a substance that produces hydrogen ions (H^+) when dissolved in water. Hydrogen ions are hydrogen atoms stripped of their electrons. A single hydrogen ion is nothing more than the nucleus of a hydrogen atom.

The relative strengths of acids in water (listed in descending order of strength) are classified in Table 11.3.

Note: Acids and bases become solvated—they loosely bond to water molecules.

BASES

A *base* is a substance that produces hydroxide ions (OH^-) when dissolved in water. Lye or common soap (bitter substances) contains bases. The bases used in waterworks operations are calcium hydroxide ($Ca(OH)_2$), sodium hydroxide ($NaOH$), and potassium hydroxide (KOH). Note that the hydroxyl group (OH) is found in all bases. In addition, bases contain metallic substances, such as sodium (Na), calcium (Ca), magnesium (Mg), and potassium (K). These bases contain the elements that produce alkalinity in water.

SALTS

When acids and bases chemically interact, they neutralize each other. The compound (other than water) that forms from the neutralization of acids and bases is called a *salt*. Salts constitute, by far, the largest group of inorganic compounds. A common salt used in waterworks operations, copper sulfate, is utilized to kill algae in water. This intentional addition of the sulfate, copper sulfate should not be confused with the presence of naturally occurring sulfates in drinking water. For further illustration, consider the 1999 EPA case study on sulfates in drinking water presented below.

HEALTH EFFECTS FROM EXPOSURE TO HIGH LEVELS OF SULFATE IN DRINKING WATER[1]

As mentioned, sulfate is a naturally occurring substance often found in drinking water. Health concerns regarding sulfate in drinking water have been raised due to reports of diarrhea associated with the ingestion of water containing high levels of sulfate. Available data suggest that people acclimate rapidly to the presence of sulfates in their drinking water. However, some groups within the general population may be at greater risk from the laxative effects of sulfate when they experience an abrupt change from drinking water with low sulfate concentrations to drinking water with high sulfate concentrations.

One potentially sensitive population is infants receiving their first bottles containing tap water, whether as water alone or as formula mixed with water. A series of three case histories from Saskatchewan reported by Chien et al. (1968) suggested that infants may experience gastroenteritis, including diarrhea and dehydration, upon their first exposure to water that contains high levels of sulfate. The three infants discussed in the report were symptom-free until their families moved to areas with water supplies that contained high levels of sulfate (650–1,150 mg/L). Interestingly, the infants developed diarrhea when they were given water from these new sources. Stools from two of the infants tested negative for bacterial pathogens, ova (female reproductive cells), and parasites, and the diarrhea subsided when alternative water sources were used. In light of these reports, the authors suggested that sulfate levels are important with respect to their laxative effect on babies. They recommended that water be screened for sulfate content if a sample is submitted for assessment of suitability for infant feeding.

Other groups of people who could potentially be adversely affected by water with high sulfate concentrations are transient populations (i.e., tourists, hunters, students, and other temporary visitors) and people moving to areas with high sulfate concentrations in drinking water from areas with low sulfate concentrations. This concern is based primarily on anecdotal reports rather than on published studies. For example, an analysis of 300 responses to an informal survey conducted by the North Dakota Department of Health suggested that water with sulfate levels ≥750 mg/L was considered laxative by most consumers (Peterson 1951). Peterson noted that a high concentration of magnesium sulfate was even more likely to have a laxative effect on consumers than a high concentration of sodium sulfate. He also observed that reports of a laxative effect of water with a low concentration of sulfate could have come from people new to the area, whereas reports of no laxative effect in areas of high sulfate concentration could have come from people acclimated to sulfate exposure.

In another informal report, Moore (1952) evaluated data collected in North and South Dakota on well water quality. The data from South Dakota included 67 wells with

1,000–2,000 mg/L sulfate and indicated that the water was at least tolerable as drinking water with no apparent extensive physiological effect. The data from North Dakota included information from 248 private drinking water wells. As the concentration of sulfate in these wells increased, more adults reported a laxative effect. For example, for well water containing <200 mg/L sulfate, only 22% of consumers reported that their water had a laxative effect. Water containing high (≥100 mg/L) concentrations of magnesium sulfate affected 62% of consumers. Neither of these reports suggested that the population affected considered the laxative effect of their drinking water to be an adverse health issue.

Experimental studies of the association between exposure to sulfate and subsequent diarrhea have been conducted on pigs and piglets, as well as in human adults. For example, groups of ten artificially-reared (using a mechanical "auto-sow") neonatal piglets were provided diets containing 0, 1,200, 1,600, or 2,000 mg of added inorganic sulfate (as anhydrous sodium sulfate)/L of diet for 28 days (Gomez et al. 1995). Sulfate concentrations of ≥1,800 mg/L in the diet caused persistent but nonpathogenic diarrhea in the piglets. Growth was not affected in any of the exposure groups. A study by Veenhuizen et al. (1991) found no adverse effect on nursery pig performance (e.g., mean weight gain, feed consumption, water consumption, prevalence of diarrhea) when the pigs were fed concentrations of up to 1,800 mg of sodium sulfate, magnesium sulfate, or a combination of sodium and magnesium sulfate/L of diet for 16 or 18 days.

In another study, Veenhuizen (1993) conducted a water quality survey of 54 swine farms in Ohio, in which water samples were analyzed for concentrations of sulfates and total dissolved solids. The study found that sulfate concentrations ranged from 6 to 16,000 mg/L. There was no association between sulfate concentration and the prevalence of diarrhea on the farms.

Heizer et al. (1997) provided four healthy adult subjects with drinking water containing increasing levels of sulfate (0, 40, 600, 800, 1,000, and 1,200 mg/L from sodium sulfate) for six consecutive 2-day periods. In a single-dose study, six other volunteers received water with 0–1,200 mg/L sulfate for two consecutive 6-day periods. In the dose-range study, there was a decrease in mouth-to-anus appearance time (using colored markers) with increasing sulfate concentration. In the single-dose study, there was a significant increase in stool mass for the 6 days of exposure to sulfate compared to the 6 days without exposure. None of the study subjects reported diarrhea.

While the studies mentioned above address the acute effects of sulfate on adult human intestinal function and provide animal data that can be extrapolated to people, it has not been feasible to conduct an experimental study to verify the reported effects of exposure to high levels of sulfate on human infants. Esteban et al. (1997) conducted a field study in 19 South Dakota counties to determine the risk for diarrhea in infants exposed to high levels of sulfate in tap water compared to the risk for diarrhea in those unexposed. In this study, there was no significant association between sulfate ingestion and the incidence of diarrhea for the range of sulfate concentrations studied (mean sulfate level 264 mg/L; range 0–2,787 mg/L). In addition, there was no dose-response or threshold effect, and the results suggested that breast milk has a more significant laxative effect than sulfate in drinking water. However, because the sample size was small (274 infants) and the age of the infants ranged from 6.5 to 30 weeks, it is probable that some of the infants could have been exposed (and become acclimated) to drinking water containing high levels of sulfate prior to being enrolled in the study.

pH

pH is an important indicator of chemical, physical, and biological changes in a waterbody and plays a critical role in chemical processes in natural waters. pH is a measure of the hydrogen ion (H^+) concentration. Solutions range from very acidic (having a high concentration of H^+ ions) to very basic (having a high concentration of OH^- ions). The pH scale ranges from 0 to 14, with 7 being the neutral value. The pH of water is important to the chemical reactions that take place within water, and pH values that are too high or low can inhibit the growth of microorganisms. High pH values are considered basic, while low pH values are considered acidic. Stated another way, low pH values indicate a high level of H^+ concentration, while high pH values indicate a low H^+ concentration. Because of this inverse logarithmic relationship, there is a tenfold difference in H^+ concentration.

Natural water varies in pH depending on its source. Pure water has a neutral pH, with an equal number of H^+ and OH^-. Adding an acid to water causes additional + ions to be released so the H^+ ion concentration goes up and the pH value goes down.

$$HCl \leftrightarrow H^+ + Cl^-$$

To control water coagulation and corrosion, the waterworks operator must test for the hydrogen ion concentration of the water to determine the water's pH. In coagulation testing, as more alum (acid) is added, the pH value lowers. If more lime (alkali) is added, the pH value rises. This relationship should be remembered—if a good floc is formed, the pH should then be determined and maintained at that pH value until the raw water changes.

Pollution can change the pH of water, which in turn can harm animals and plants living in the water. For instance, water coming out of an abandoned coal mine can have a pH of 2, which is very acidic and would definitely affect any fish crazy enough to try to live in it. By using the logarithmic scale, this mine-drainage water would be 100,000 times more acidic than neutral water—so stay out of abandoned mines.

Important Point: Seawater is slightly more basic (the pH value is higher) than most natural freshwater. Neutral water (such as distilled water) has a pH of 7, which is in the middle of being acidic and alkaline. Seawater happens to be slightly alkaline (basic), with a pH of about 8. Most natural water has a pH range of 6–8, although acid rain can have a pH as low as 4.

COMMON WATER MEASUREMENTS

Water and wastewater practitioners and regulators, such as waterworks operators and the U.S. Environmental Protection Agency (USEPA), along with their scientific counterparts at the U.S. Geological Survey (USGS), have been measuring water for decades. Millions of measurements and analyses have been made. Some measurements are taken almost every time water is sampled and investigated, no matter where in the United States the water is being studied. Even these simple measurements can sometimes reveal something important about the water and the environment around it.

The USGS (2006) has noted that the results of a single measurement of water's properties are actually less important than looking at how the properties vary over time. For example, if you take the pH of the river running through your town and find that it is 5.5, you might say, "Wow, the water is acidic!" But a pH of 5.5 might be "normal" for that river. It is similar to how normal body temperature (when not ill) is about 97.5°, but a youngster's normal temperature is "really normal"—right on the 98.6 mark. As with our temperatures, if the pH of your river begins to change, then you might suspect that something is going on somewhere that is affecting the water and possibly the water quality. So, often, the changes in water measurements are more important than the actual measured values.

Up to this point, the important constituents and parameters of turbidity, dissolved oxygen, pH, and others have been discussed; there are others, too. In the following sections, the parameters of alkalinity, water temperature, specific conductance and hardness are discussed.

ALKALINITY

Alkalinity is defined as the capacity of water to accept protons; it can also be defined as a measure of water's ability to neutralize an acid. Bicarbonates, carbonates, and hydrogen ions cause alkalinity and hydrogen compounds in a raw or treated water supply. Bicarbonates are the major components because of carbon dioxide action on basic materials in the soil; borates, silicates, and phosphates may be minor components. The alkalinity of raw water may also contain salts formed from organic acids such as humic acids.

Alkalinity in water acts as a buffer that tends to stabilize and prevent fluctuations in pH. In fact, alkalinity is closely related to pH, but the two must not be confused. Total alkalinity is a measure of the amount of alkaline materials in the water. The alkaline materials act as the buffer to changes in pH. If the alkalinity is too low (below 80 ppm), there can be rapid fluctuations in pH—i.e., there is insufficient buffer for the pH. High alkalinity (above 200 ppm) results in the water being too buffered. Thus, having significant alkalinity in water is usually beneficial because it tends to prevent quick changes in pH, which interfere with the effectiveness of common water treatment processes. Low alkalinity also contributes to water's corrosive tendencies.

Note: When alkalinity is below 80 mg/L, it is considered to be low.

WATER TEMPERATURE

Water temperature influences the majority of physical, biological, chemical, and ecosystem processes in aquatic environments. Moreover, water temperature is not only important to fishermen, but also to industries and even fish and algae. A lot of water is used for cooling purposes in power plants that generate electricity. They need to cool water to start with, and they generally release warmer water back into the environment. The temperature of the released water can affect downstream habitats. Temperature can also affect the ability of water to hold oxygen, as well as the ability of organisms to resist certain pollutants. Increased water temperatures can result in decreased DO available to aquatic life and can increase the solubility of metals and other toxins in the water.

SPECIFIC CONDUCTANCE

Specific conductance is a measure of the ability of water to conduct an electrical current. It is highly dependent on the amount of dissolved solids (such as salt) in the water. Pure water, such as distilled water, will have a very low specific conductance, while seawater will have a high specific conductance. Rainwater often dissolves airborne gases and airborne dust while it is in the air, and thus often has a higher specific conductance than distilled water. Specific conductance is an important water-quality measurement because it gives a good idea of the amount of dissolved material in the water. When electrical wires are attached to a battery and light bulb, and the wires are placed into a beaker of distilled water, the light will not turn on. However, the bulb does light up when the beaker contains saline (salt water). In the saline water, the salt has dissolved, releasing free electrons, and the water will conduct an electric current.

HARDNESS

Hardness may be considered a physical or chemical characteristic or parameter of water. It represents the total concentration of calcium and magnesium ions, reported as calcium carbonate. Simply put, the amount of dissolved calcium and magnesium in water determines its "hardness." Hardness causes soaps and detergents to be less effective

TABLE 11.4
Water Hardness

Classification	mg/L CaCO$_3$
Soft	0–75
Moderately hard	75–150
Hard	150–300
Very hard	Over 300

and contributes to scale formation in pipes and boilers. Hardness is not considered a health hazard; however, water that contains hardness must often be softened by lime precipitation or ion exchange. Hard water can even shorten the life of fabrics and clothes. Low hardness contributes to the corrosive tendencies of water. Hardness and alkalinity often occur together because some compounds can contribute both alkalinity and hardness ions. Hardness is generally classified as shown in Table 11.4.

ODOR CONTROL (WASTEWATER TREATMENT)

There is an old saying in wastewater treatment: "Odor is not a problem until the neighbors complain" (Spellman 1997). Experience has shown that when treatment plant odor is apparent, it is not long before the neighbors do complain. Thus, odor control is an important factor affecting the performance of any wastewater treatment plant, especially regarding public relations.

According to Metcalf & Eddy (1991), in wastewater operations, "The principal sources of odors are from (1) septic wastewater containing hydrogen sulfide and odorous compounds, (2) industrial wastes discharged into the collection system, (3) screenings and unwanted grit, (4) septage handling facilities, (5) scum on primary settling tanks, (6) organically overloaded treatment processes, (7) [biosolids]-thickening tanks, (8) waste gas-burning operations where lower-than-optimum temperatures are used, (9) [biosolids]-conditioning and dewatering facilities, (10) [biosolids] incineration, (11) digested [biosolids] in drying beds or [biosolids]-holding basins, and (12) [biosolids]-composting operations."

Odor control can be accomplished by chemical or physical means. Physical means include utilizing buffer zones between the process operation and the public, making operational changes, controlling discharges to collection systems, using containments, dilution, fresh air, adsorption, activated carbon, scrubbing towers, and other methods. Odor control by chemical means involves scrubbing with various chemicals, chemical oxidation, and chemical precipitation methods. In *scrubbing with chemicals,* odorous gases are passed through specially designed scrubbing towers to remove odors. The commonly used chemical scrubbing solutions are chlorine and potassium permanganate. When hydrogen sulfide concentrations are high, sodium hydroxide is often used. In *chemical oxidation* applications,

the oxidants chlorine, ozone, hydrogen peroxide, and potassium permanganate are used to oxidize the odor compounds. *Chemical precipitation* works to precipitate sulfides from odor compounds using iron and other metallic salts.

WATER TREATMENT CHEMICALS

To operate a water treatment process correctly and safely, water operators need to know the types of chemicals used in the processes, the purpose of each, and the safety precautions required in the use of each. This section briefly discusses chemicals used in:

- Disinfection (also used in wastewater treatment)
- Coagulation
- Taste and odor removal
- Water softening
- Recarbonation
- Ion exchange softening
- Scale and corrosion control

DISINFECTION

In water practice, disinfection is often accomplished using chemicals. The purpose of disinfection is to selectively destroy disease-causing organisms. Chemicals commonly used in disinfection include chlorine and its compounds (the most widely used), ozone, bromide, iodine, hydrogen peroxide, and others. Many factors must be considered when choosing the type of chemical to be used for disinfection. These factors include contact time, intensity and nature of the physical agent, temperature, and the type and number of organisms.

COAGULATION

Chemical coagulation conditions water for further treatment by the removal of:

- Turbidity, color, and bacteria
- Iron and manganese
- Tastes, odors, and organic pollutants

In water treatment, normal sedimentation processes do not always settle out particles efficiently, especially when attempting to remove particles of less than 50 μm in diameter.

In some instances, it is possible to agglomerate (to make or form into a rounded mass) particles into masses or groups. These rounded masses are of increased size and, therefore, increased settling velocities in some instances. For colloidal-sized particles, however, agglomeration is difficult—making it challenging to clarify colloidal particles without special treatment.

Chemical coagulation is usually accomplished by the addition of metallic salts such as aluminum sulfate (alum) or ferric chloride. Alum is the most commonly used coagulant in water treatment and is most effective within pH ranges of 5.0 to 7.5. Sometimes, a polymer is added to alum to help form small flocs together for faster settling. Ferric chloride, effective down to a pH of 4.5, is sometimes used.

In addition to pH, a variety of other factors influence the chemical coagulation process, including:

1. Temperature
2. Influent quality
3. Alkalinity
4. Type and amount of coagulant used
5. Type and length of flocculation
6. Type and length of mixing

TASTE AND ODOR REMOVAL

Although odor can be a problem with wastewater treatment, the taste and odor parameter is only associated with potable water. Either organic or inorganic materials may produce tastes and odors in water. The perceptions of taste and odor are closely related and often confused by water practitioners as well as by consumers. Thus, it is difficult to measure either one precisely. Experience has shown that a substance that produces an odor in water almost invariably imparts a perception of taste as well. However, this is not always the case. Taste is generally attributed to mineral substances in the water. Most of these minerals affect water taste but do not cause odors.

Along with the impact minerals can have on water taste, some other substances or practices can affect both water tastes and odors (e.g., metals, salts from the soil, constituents of wastewater, and end products generated from biological reactions). When water has a distinct taste but no odor, the taste might be the result of inorganic substances. Anyone who has tasted alkaline water has also tasted its biting bitterness. The salts not only give water a salty taste but also contribute to its bitter taste. Other than from natural causes, water can take on a distinctive color or taste, or both, from human contamination of the water. Organic materials can produce both taste and odor in water. Petroleum-based products are probably the prime contributors to both these problems in water.

Biological degradation or decomposition of organics in surface waters also contributes to both taste and odor problems in water. Algae are another issue, as certain species produce oily substances that may result in both taste and odor. Synergy can also work to create taste and odor problems in water; mixing water and chlorine is one example.

With regard to chemically treating water for odor and taste problems, oxidants such as chlorine, chlorine dioxide, ozone, and potassium permanganate can be used. These chemicals are especially effective when water is associated with an earthy or musty odor caused by the nonvolatile metabolic products of actinomycetes and blue-green algae. Tastes and odors associated with dissolved gases and some volatile organic materials are normally removed by oxygen in aeration processes.

WATER SOFTENING

The reduction of hardness, or softening, is a process commonly practiced in water treatment. Chemical precipitation and ion exchange are the most commonly used softening processes. Softening hard water is desired (for domestic users) to reduce the amount of soap used, increase the life of water heaters, and reduce the encrustation of pipes (cementing together the individual filter media grains).

CHEMICAL PRECIPITATION

In chemical precipitation, it is necessary to adjust the pH. To precipitate the two ions most commonly associated with hardness in water, calcium (Ca^{+2}) and magnesium (Mg^{+2}), the pH must be raised to about 9.4 for calcium and about 10.6 for magnesium. To raise the pH to the required levels, lime is added.

Chemical precipitation is accomplished by converting calcium hardness to calcium carbonate and magnesium hardness to magnesium hydroxide. This is normally achieved using the lime-soda ash or the caustic soda process.

The lime-soda ash process reduces the total mineral content of the water, removes suspended solids, removes iron and manganese, and reduces color and bacterial numbers. The process, however, has a few disadvantages. McGhee (1991) points out, for example, that the process produces large quantities of sludge, requires careful operation, and, as stated earlier, if the pH is not properly adjusted, it may create operational problems downstream of the process.

In the caustic soda process, caustic soda reacts with the alkalinity to produce carbonate ions for reduction with calcium. The process works to precipitate calcium carbonate in a fluidized bed of sand grains, steel grit, marble chips, or some other similar dense material. As particles grow in size by deposition of $CaCO_3$, they migrate to the bottom of the fluidized bed, from which they are removed. This process has the advantages of requiring short detention times (about 8 seconds) and producing no sludge.

ION EXCHANGE SOFTENING

Hardness can be removed by ion exchange. In water softening, ion exchange replaces calcium and magnesium with a non-hardness cation, usually sodium. Calcium and magnesium in solution are removed by interchange with sodium within a solid interface (matrix) through which the flow is passed. Similar to a filter, the ion exchanger contains a bed of granular material, a flow distributor, and an effluent vessel that collects the product. The exchange media include greensand (sand or sediment given a dark greenish color by grains of glauconite), aluminum silicates, synthetic siliceous gels, bentonite clay, sulfonated coal, and synthetic organic resins. These media are generally in particle form usually ranging up to a diameter of 0.5 mm. Modern applications more often employ artificial organic resins, which are clear, BB-sized spheres. These resins provide a greater number of exchange sites. Each resin sphere contains sodium ions, which are released into the water in exchange for calcium and magnesium. As long as exchange sites are available, the reaction is virtually instantaneous and complete.

When all the exchange sites have been utilized, hardness begins to appear in the influent (breakthrough). When a breakthrough occurs, this necessitates the regeneration of the medium by contacting it with a concentrated sodium chloride solution.

Ion exchange used in water softening has both advantages and disadvantages. One of its major advantages is that it produces softer water than chemical precipitation. Additionally, ion exchange does not produce the large quantity of sludge encountered in the lime-soda process. One disadvantage is that, although it does not produce sludge, ion exchange does produce concentrated brine. Moreover, the water must be free of turbidity and particulate matter or the resin might function as a filter and become plugged.

RECARBONATION

Recarbonation (stabilization) is the adjustment of the ionic condition of water so that it will neither corrode pipes nor deposit calcium carbonate, producing an encrusting film. During or after the lime-soda ash softening process, this recarbonation is accomplished by the reintroduction of carbon dioxide into the water. Lime softening of hard water supersaturates the water with calcium carbonate and may result in a pH greater than 10. Because of this, pressurized carbon dioxide is bubbled into the water, lowering the pH and removing calcium carbonate. The high pH can also create a bitter taste in drinking water. Recarbonation removes this bitterness.

SCALE AND CORROSION CONTROL

Controlling scale and corrosion is important in water systems. Carbonate and noncarbonate hardness constituents in water cause scale, which forms a chalky-white deposit frequently found on the bottoms of teakettles. When controlled, this scale can be beneficial, forming a protective coating inside tanks and pipelines. Problems arise when the scale is not controlled. Excessive scaling reduces the capacity of pipelines and the efficiency of heat transfer in boilers. Corrosion is the oxidation of unprotected metal surfaces. Of particular concern in water treatment is the corrosion of iron and its alloys (i.e., the formation of rust). Several factors contribute to the corrosion of iron and steel. Alkalinity, pH, DO, and carbon dioxide can all cause corrosion. Along with the corrosion potential of these chemicals, their corrosive tendencies are significantly increased when water temperature and flow are increased.

DRINKING WATER PARAMETERS: CHEMICAL

Water, in any of its forms, also … [has] scant respect for the laws of chemistry.

Most materials act either as acids or bases, settling on either side of a natural reactive divided. Not water. It is one of the few substances that can behave both as an acid and as a base, so that under certain conditions it is capable of reacting chemically with itself. Or with anything else.

Molecules of water are off balance and hard to satisfy. They reach out to interfere with every other molecule they meet, pushing its atoms apart, surrounding them, and putting them into solution. Water is the ultimate solvent, wetting everything, setting other elements free from the rocks, making them available for life. Nothing is safe. There isn't a container strong enough to hold it.

—Watson (1988)

Water chemical parameters are categorized into two basic groups: inorganic and organic chemicals. Both groups enter water through natural processes or pollution.

Note: The solvent capabilities of water are directly related to its chemical parameters.

In this section, we do not examine each organic or inorganic chemical individually; instead, we focus on general chemical parameter categories such as DO, organics (BOD and COD), synthetic organic chemicals (SOCs), volatile organic chemicals (VOCs), total dissolved solids (TDS), fluorides, metals, and nutrients—the major chemical parameters of concern.

ORGANICS

Natural organics contain carbon and consist of biodegradable organic matter such as wastes from biological material processing, human sewage, and animal feces. Microbes aerobically break down these complex organic molecules into simpler, more stable end products. Microbial degradation yields products such as carbon dioxide, water, phosphate, and nitrate. Organic particles in water may harbor harmful bacteria and pathogens. Infection by microorganisms may occur if the water is used for primary contact or as a raw drinking water source. Treated drinking water does not present the same health risks. In a potable drinking water plant, all organics should be removed from the water before disinfection.

Organic chemicals also contain carbon and are substances that come directly from, or are manufactured from, plant or animal matter. Plastics are a good example of organic chemicals made from petroleum, which originally came from plant and animal matter. Some organic chemicals (like those discussed above) released by decaying vegetation, occur naturally and tend not to pose health problems when they get in our drinking water. However, more serious problems arise from the over 100,000 different manufactured or synthetic organic chemicals in commercial use today. These include paints, herbicides, synthetic fertilizers, pesticides, fuels, plastics, dyes, preservatives, flavorings, and pharmaceuticals, to name a few.

Many organic materials are soluble in water, and are toxic, and many are found in public water supplies. According to Tchobanoglous and Schroeder (1987), the presence of organic matter in water is troublesome. Organic matter causes (1) color formation, (2) taste and odor problems, (3) oxygen depletion in streams, (4) interference with the water treatment process, and (5) the formation of halogenated compounds when chlorine is added to disinfect the water.

Remember, organics in natural water systems may come from natural sources or may result from human activities. Generally, the principal source of organic matter in water is from natural sources including decaying leaves, weeds, and trees; the amount of these materials present in natural waters is usually low. Anthropogenic (man-made) sources of organic substances come from pesticides and other synthetic organic compounds.

Again, many organic compounds are soluble in water, and surface waters are more prone to contamination by natural organic compounds than groundwaters. In water, dissolved organics are usually divided into two categories: *biodegradable* and *nonbiodegradable*. Biodegradable (able to break down) materials consist of organics that can be used as nutrients by naturally occurring microorganisms within a reasonable length of time. Alcohols, acids, starches, fats, proteins, esters, and aldehydes are the main constituents of biodegradable materials. They may result from domestic or industrial wastewater discharges or may be end products of the initial microbial decomposition of plant or animal tissue. Biodegradable organics in surface waters cause problems mainly associated with the effects that result from the action of microorganisms. As microbes metabolize organic material, they consume oxygen.

When this process occurs in water, the oxygen consumed is DO. If the oxygen is not continually replaced in the water by artificial means, the DO level will decrease as the organics are decomposed by the microbes. This need for oxygen is called the *biochemical oxygen demand* (BOD): the amount of DO required by bacteria to break down organic materials during the stabilization of decomposable organic matter under aerobic conditions over a 5-day incubation period at 20°C (68°F). This bioassay test measures the oxygen consumed by living organisms using the organic matter contained in the sample and dissolved oxygen in the liquid. The organics are broken down into simpler compounds, and the microbes use the energy released for growth and reproduction. A BOD test is not required for monitoring drinking water.

Note: The more organic material in the water, the higher the BOD exerted by the microbes will be. Note also that some biodegradable organics can cause color, taste, and odor problems.

Nonbiodegradable organics are resistant to biological degradation. The constituents of woody plants are a good example. These constituents, including tannin and lignic

acids, phenols, and cellulose are found in natural water systems and are considered refractory (resistant to biodegradation). Some polysaccharides with exceptionally strong bonds, and benzene (for example, associated with the refining of petroleum) with its ringed structure, are essentially nonbiodegradable.

Certain nonbiodegradable chemicals can react with oxygen dissolved in water. The *chemical oxygen demand* (COD) is a more complete and accurate measurement of the total depletion of dissolved oxygen in water. Greenberg et al. (1999) defines COD as a test that provides a measure of the oxygen equivalent of that portion of the organic matter in a sample that is susceptible to oxidation by a strong chemical oxidant. The procedure is detailed in *Standard Methods*.

Note: COD is not normally used for monitoring water supplies but is often used for evaluating contaminated raw water.

SYNTHETIC ORGANIC CHEMICALS (SOCs)

Synthetic organic chemicals (SOCs) are man-made, and because they don't occur naturally in the environment, they are often toxic to humans. More than 50,000 SOCs are in commercial production including common pesticides, carbon tetrachloride, chloride, dioxin, xylene, phenols, aldicarb, and thousands of other synthetic chemicals. Unfortunately, even though they are so prevalent, little data have been collected on these toxic substances. Determining definitively just how dangerous many of the SOCs are is rather difficult.

VOLATILE ORGANIC CHEMICALS (VOCs)

Volatile organic chemicals (VOCs) are a type of organic chemical that is particularly dangerous. VOCs are absorbed through the skin during contact with water, as in the shower or bath. Hot water exposure allows these chemicals to evaporate rapidly, and they are harmful if inhaled. VOCs can be present in any tap water, regardless of what part of the country one lives in or the water supply source.

TOTAL DISSOLVED SOLIDS (TDS)

Earlier we pointed out that solids in water occur either in solution or in suspension and are distinguished by passing the water sample through a glass-fiber filter. By definition, the *suspended solids* are retained on top of the filter, while the *dissolved solids* pass through the filter with the water. When the filtered portion of the water sample is placed in a small dish and then evaporated, the solids in the water remain as residue in the evaporating dish. This material is called *total dissolved solids*, or TDS. Dissolved solids may be organic or inorganic. Water may come into contact with these substances within the soil, on surfaces, and in the atmosphere. The organic dissolved constituents of water come from the decay products of vegetation, from organic chemicals, and from organic gases. Removing these dissolved minerals, gases, and organic constituents is desirable because they may cause physiological effects and produce aesthetically displeasing color, taste, and odors.

Note: In water distribution systems, a high TDS means high conductivity, resulting in higher ionization in corrosion control. However, high TDS also means more likelihood of a protective coating, which is a positive factor in corrosion control.

FLUORIDES

According to Phyllis J. Mullenix, Ph.D., water fluoridation is not the safe public health measure we have been led to believe. Concerns about uncontrolled dosage, accumulation in the body over time and effects beyond the teeth (including the brain as well as bones) have not been resolved for fluoride. The health of citizens necessitates that all the facts be considered, not just those that are politically expedient (Mullenix 1997).

Most medical authorities would take issue with Ms. Mullenix's view on the efficacy of fluoride in reducing tooth decay. Most authorities seem to hold that a moderate amount of *fluoride* ions (F⁻) in drinking water contributes to good dental health. Fluoride is seldom found in appreciable quantities in surface waters and appears in groundwater in only a few geographical regions, though it is sometimes found in certain types of igneous or sedimentary rocks. Fluoride is toxic to humans in large quantities (the key phrases are "large quantities" or, in Ms. Mullenix's view, "uncontrolled dosages") and is also toxic to some animals.

Fluoride used in small concentrations (about 1.0 mg/L in drinking water) can be beneficial. Experience has shown that drinking water containing a proper amount of fluoride can reduce tooth decay by 65% in children between the ages of 12 and 15. However, when the concentration of fluoride in untreated natural water supplies is excessive, either an alternative water supply must be used, or treatment to reduce the fluoride concentration must be applied, because excessive amounts of fluoride can cause mottled or discolored teeth, a condition known as *dental fluorosis*.

HEAVY METALS

Heavy metals are elements with atomic weights between 63.5 and 200.5 and a specific gravity greater than 4.0. Living organisms require trace amounts of some heavy metals, including cobalt, copper, iron, manganese, molybdenum, vanadium, strontium, and zinc. Excessive levels of essential metals, however, can be detrimental to the organism. Non-essential heavy metals of particular concern to surface water systems include cadmium, chromium, mercury, lead, arsenic, and antimony.

Heavy metals in water are classified as either nontoxic or toxic. Only those metals that are harmful in relatively small amounts are labeled toxic; other metals fall into the nontoxic group. In natural waters (other than in groundwater), sources of metals include dissolution from natural deposits and discharges of domestic, agricultural, or industrial wastes.

All heavy metals exist in surface waters in colloidal, particulate, and dissolved phases, although dissolved concentrations are generally low. The colloidal and particulate metals may be found in (1) hydroxides, oxides, silicates, or sulfides, or (2) adsorbed to clay, silica, or organic matter. The soluble forms are generally ions or unionized organometallic chelates or complexes. The solubility of trace metals in surface waters is predominantly controlled by water pH, the type and concentration of liquids on which the metal could adsorb, the oxidation state of the mineral components, and the redox environment of the system. The behavior of metals in natural waters is a function of the substrate sediment composition, the suspended sediment composition, and the water chemistry. Sediment composed of fine sand and silt will generally have higher levels of adsorbed metals than quartz, feldspar, and detrital carbonate-rich sediment.

The water chemistry of the system controls the rate of adsorption and desorption of metals to and from sediment. Adsorption removes the metal from the water column and stores it in the substrate. Desorption returns the metal to the water column, where recirculation and bioassimilation may take place. Metals may be desorbed from the sediment if the water experiences increases in salinity, decreases in redox potential, or decreases in pH.

Although heavy metals such as iron (Fe) and manganese (Mn) do not cause health problems, they do impart a noticeable bitter taste to drinking water, even at very low concentrations. These metals usually occur in groundwater in solution, and they may cause brown or black stains on laundry and plumbing fixtures.

NUTRIENTS

Elements in water (such as carbon, nitrogen, phosphorus, sulfur, calcium, iron, potassium, manganese, cobalt, and boron—all essential to the growth and reproduction of plants and animals) are called *nutrients* (or biostimulants). Nutrients play a critical role in the biodiversity and healthy functioning of aquatic ecosystems by supporting the growth of aquatic plants and algae that provide food and habitat for fish, shellfish, and smaller organisms (USEPA 2021). The two most important nutrients that concern us in this text are nitrogen and phosphorus. Nitrogen (N_2), an extremely stable gas, is the primary component of the Earth's atmosphere (78%). The nitrogen cycle is composed of four processes. Three of the processes—fixation, ammonification, and nitrification—convert gaseous nitrogen into usable chemical forms. The fourth process—denitrification—converts fixed nitrogen back to the unusable gaseous nitrogen state.

Nitrogen occurs in many forms in the environment and takes part in many biochemical reactions. Major sources of nitrogen include runoff from animal feedlots, fertilizer runoff from agricultural fields, municipal wastewater discharges, and certain bacteria and blue-green algae that obtain nitrogen directly from the atmosphere. Certain forms of acid rain can also contribute nitrogen to surface waters.

Nitrogen in water is commonly found in the form of *nitrate* (NO_3), which indicates that the water may be contaminated with sewage. Nitrates can also enter groundwater from chemical fertilizers used in agricultural areas. Excessive nitrate concentrations in drinking water pose an immediate health threat to infants, both human and animal, and can cause death. The bacteria commonly found in the intestinal tract of infants can convert nitrate to highly toxic nitrites (NO_2). Nitrite can replace oxygen in the bloodstream, resulting in oxygen starvation that causes a bluish discoloration of the infant (known as "blue baby" syndrome).

Note: Lakes and reservoirs usually have less than 2 mg/L of nitrate measured as nitrogen. Higher nitrate levels are found in groundwater ranging up to 20 mg/L, but much higher values are detected in shallow aquifers polluted by sewage and/or excessive use of fertilizers.

Phosphorus (P) is an essential nutrient that contributes to the growth of algae and the eutrophication of lakes, although its presence in drinking water has little effect on health. In aquatic environments, phosphorus is found in the form of phosphate and is considered a limiting nutrient. If all phosphorus is used, plant growth ceases, no matter the amount of nitrogen available. Many bodies of freshwater currently experience influxes of nitrogen and phosphorus from outside sources. The increasing concentration of available phosphorus allows plants to assimilate more nitrogen before the phosphorus is depleted. If sufficient phosphorus is available, high concentrations of nitrates will lead to phytoplankton (algae) and macrophyte (aquatic plant) production.

Major sources of phosphorus include phosphates in detergents, fertilizer and feedlot runoff, and municipal wastewater discharges. The 1976 USEPA Water Quality Standards recommended a phosphorus criterion of 0.10 µg/L (elemental) phosphorus for marine and estuarine waters, but no freshwater criterion.

The biological, physical, and chemical conditions of our water are of enormous concern to us all, as we must live in intimate contact with it. When these parameters shift and change, the changes affect us, often in ways that science cannot yet define. Water pollution is an external element that can significantly affect our water. But what exactly is water pollution? We quickly learn that water pollution doesn't always go straight from source to water. Controlling what goes into our water is difficult because the hydrologic cycle carries water (and whatever it picks up along the way) through all of our environment's media, affecting the biological, physical, and chemical conditions of the water we must drink to live.

DID YOU KNOW?

Note that nutrients affect other water quality parameters, such as:

- **pH:** Daily variation in pH is amplified when nutrients promote increased growth of aquatic primary producers (plants, algae, and cyanobacteria).
- **Organic Carbon:** The nutrients in the water may increase the amount of organic matter.
- **Total Suspended Solids (TSS):** Nutrients increase algal growth depending on the TSS in the water.
- **Turbidity:** Water may be more turbid (cloudy) if algae and aquatic plant growth increase in response to higher nutrient concentrations.
- **Chlorophyll A:** This is an algal and aquatic plant pigment that is useful as an indicator of algal biomass and may increase in response to higher nutrient concentrations.
- **DO:** It may decrease if an algal bloom proceeds to the point where bacterial decay of dead algae consumes oxygen (USEPA 2021).

THE BOTTOM LINE

The chemistry of pure water ensures that it is flavorless and odorless. The purpose of water and wastewater treatment is to ensure that water for human utilization is purified to the extent possible, making it safe and healthy for use.

CHAPTER REVIEW QUESTIONS

11.1 The chemical symbol for sodium is _____.
11.2 The chemical symbol for sulfuric acid is _____.
11.3 Neutrality on the pH scale is _____.
11.4 Is NaOH a salt or a base?
11.5 Chemistry is the study of substances and the _____ they undergo.
11.6 The three stages of matter are _____, _____, and _____.
11.7 A basic substance that cannot be broken down any further without changing the nature of the substance is _____.
11.8 A combination of two or more elements is a _____.
11.9 A table of the basic elements is called the _____ table.
11.10 When a substance is mixed into water to form a solution, the watt is called the _____, and the substance is called the _____.
11.11 Define ion.

11.12 A solid that is less than 1 µm in size is called a
_____.

11.13 The property of water that causes light to be scattered and absorbed is _____.

11.14 What is true color?

11.15 What is the main problem with metals found in water?

11.16 Compounds derived from material that once was alive are called _____ chemicals.

11.17 pH range is from _____ to _____.

11.18 What is alkalinity?

11.19 The two ions that cause hardness are _____ and _____.

11.20 What type of substance produces hydroxide ions (OH^-) in water?

NOTE

1 From USEPA. 1999. *Health Effects from Exposure to High Levels of Sulfate in Drinking Water Study.* EPA 815-R-99-001. Washington, DC.

REFERENCES

Chien, L., Robertson, H., and Gerrard, U., 1968. Infantile gastroenteritis is due to water with high sulfate content. *Can Med Assoc J*; 99:102–104.

Chesapeake Bay Program. 2023. *Dissolved Oxygen.* Accessed 04/08/24 @ https://www.chesapekaebay.net/discover/bayecosystem/dissolvedoxygen.

CMG Garden Notes. 2013. *Plant Growth Factors: Photosynthesis, Respiration, and Transportation.* Accessed 04/07/24 @ https://www.ext.colostate.edu/mg/gardennotes/141.html.

Esteban, E., Rubin, C., McGeehin, M., Flanders, D., Baker, J.J., and Sinks, T.H., 1997. Evaluation of human health effects associated with elevated levels of sulfate in drinking water: a cohort investigation in South Dakota. Manuscript submitted. *Int J Occup Med Environ Heat*; 3:171–176.

Gomez, G.G., Sandler, R.S., and Seal E., Jr., 1995. High levels of inorganic sulfate cause diarrhea in neonatal piglets. *J Nutr*; 125:2325–2322.

Greenberg, A.E., et al., Eds., 1999. *Standard Methods for Examination of Water and Wastewater*, 20th ed. American Public Health Assn.

Hauser, B.A., 1995. *Practical Hydraulics Handbook*, 2nd ed. Boca Raton, FL: Lewis Publishers.

Heizer, W.D., Sandler, R.S., Seal, E., Jr., Murray, S.C., Busby, M.G., Schliebe, B.G., and Pusek, S.N., 1997. Intestinal effects of sulfate on drinking water on normal human subjects. *Dig Dis Sci*; 42(No. 5):10055–1061.

Jost, N.J., 1992. Surface and Ground Water Pollution Control Technology, in *Fundamentals of Environmental Science and Technology*, Knowles, P.-C., (ed.). Rockville, MD: Government Institutes, Inc.

McGhee, T.J., 1991. *Water Supply and Sewerage*, 6th ed. New York: McGraw-Hill, Inc.

Metcalf & Eddy, Inc., 1991. *Wastewater Engineering: Treatment, Disposal, Reuse*, 3rd ed. New York: McGraw-Hill, Inc.

Moore, F.W., (1952). Physiological effects of the consumption of saline drinking water. *A progress report to the 16th Meeting of the Subcommittee on Sanitary Engineering and Environment. Appendix B.* January 1952. Washington, DC: National Academy of Sciences.

Mullenix, P. J., 1997. In a letter sent to Operations and Environmental Committee, City of Calgary Canada.

Peterson, N.L., 1951. Sulfates in drinking water. *Official Bulletin: North Dakota Water and Sewage Works Conference;* 18:11.

Smith, R.K., 1993. *Water and Wastewater Laboratory techniques.* Alexandria, VA: Water Environment Federation.

Spellman, F.R., 1997. *Wastewater Biosolids to Compost.* Boca Raton, FL: CRC Press.

Tchobanoglous, G., and Schroeder, E.D., 1987. *Water Quality.* Reading, MA: Addison-Wesley Publishing Company.

USEPA. 2021. *Nutrients.* Accessed 04/09/24 @ epa.gov/awma/factsheets-water-quality-parameters.

USGS. 2006. *Water Science for Schools: Water Measurements.* Washington, DC: U.S. Geological Survey.

Veenhuizen, M.F., 1993. Association between water sulfate and diarrhea in swine on Ohio farms. *JAVMA*; 8:1255–1260.

Veenhuizen, M.F., Shurson, G.C., and Kohler, E.M., 1991. Effect of concentration and source of sulfate on nursery pig performance and health. *JAVMA*; 201(No. 8):1203–1208.

Watson, L., 1988. *The Water Planet: A Celebration of the Wonder of Water.* New York: Crown Publishers, Inc.

12 Water Microbiology

INTRODUCTION

Microorganisms are significant in water and wastewater because they play a significant role in disease transmission, and they are the primary agents of biological treatment. Thus, water, wastewater, and other water practitioners must have considerable knowledge of the microbiological characteristics of water and wastewater. Simply put, waterworks operators cannot fully comprehend the principles of effective water treatment without knowing the fundamental factors concerning microorganisms and their relationships to one another, their effects on the treatment process, and their impacts on consumers, animals, and the environment.

Water/wastewater operators must know the principal groups of microorganisms found in water supplies (surface and groundwater) and wastewater, as well as those that must be treated (pathogenic organisms) and /or removed or controlled for biological treatment processes. They must be able to identify the organisms used as indicators of pollution and know their significance, and they must know the methods used to enumerate the indicator organisms. This chapter provides microbiology fundamentals specifically targeting the needs of water and wastewater specialists.

Note: To have microbiological activity, the body of water or wastewater must possess the appropriate environmental conditions. The majority of wastewater treatment processes, for example, are designed to operate using an aerobic process. The conditions required for aerobic operation are (1) sufficient free, elemental oxygen, (2) sufficient organic matter (food), (3) sufficient water, (4) enough nitrogen and phosphorus (nutrients) to permit oxidation of the available carbon materials, (5) proper pH (6.5–9.0), and (6) lack of toxic materials.

MICROBIOLOGY: WHAT IS IT?

Biology is generally defined as the study of living organisms (i.e., the study of life). *Microbiology* is a branch of biology that deals with the study of microorganisms so small in size that they must be studied under a microscope. Microorganisms of interest to the water and wastewater operator include bacteria, protozoa, viruses, algae, and others.

Note: The science and study of bacteria is known as bacteriology (discussed later).

As mentioned, waterworks operators' primary concern is how to control microorganisms that cause waterborne diseases—waterborne pathogens—to protect the consumer (human and animal). Wastewater operators have the same microbiological concerns as water operators, but instead of directly purifying water for consumer consumption, their focus is on removing harmful pathogens from the waste stream before outfalling it to the environment. To summarize, each technical occupation, water and wastewater operator, in regard to its reliance on knowledge of microbiological principles, is described in the following (Spellman 1996).

1. *Water Operators* are concerned with water supply and water purification through a treatment process. In treating water, the primary concern is producing potable water that is safe to drink (free of pathogens) with no accompanying offensive characteristics such as foul taste and odor. The treatment operator must possess a wide range of knowledge to correctly examine water for pathogenic microorganisms and to determine the type of treatment necessary to ensure the water quality of the end product, potable water, meets regulatory requirements.

2. *Wastewater operators* are also concerned with water quality. However, they are not as focused as water specialists on the total removal or reduction of most microorganisms. The wastewater treatment process actually benefits from microorganisms that act to degrade organic compounds and, thus, stabilize organic matter in the waste stream. Thus, wastewater operators must be trained to operate the treatment process in a manner that controls the growth of microorganisms and puts them to work. Moreover, to fully understand wastewater treatment, it is necessary to determine which microorganisms are present and how they function to break down components in the wastewater stream. Then, of course, the operator must ensure that before outfalling or dumping treated effluent into the receiving body, the microorganisms that worked so hard to degrade organic waste products, especially the pathogenic microorganisms, are not sent from the plant with effluent as viable organisms.

WATER/WASTEWATER MICROORGANISMS

As mentioned, microorganisms of interest to water and wastewater operators include bacteria, protozoa, rotifers, viruses, algae, fungi, and nematodes. These organisms are the most diverse group of living organisms on earth and occupy important niches in the ecosystem. Their simplicity and minimal survival requirements allow them to exist in diverse situations. Because they are a major health concern, water treatment specialists are mostly concerned with how to control microorganisms that cause *waterborne diseases*. These waterborne diseases are carried by *waterborne pathogens* (i.e., bacteria, viruses, protozoa, etc.). The focus of wastewater operators, on the other hand, is on the millions of organisms that arrive at the plant with the influent. The majority of these organisms are nonpathogenic and beneficial to plant operations. From a microbiological standpoint, the predominant species of microorganisms depend on the characteristics of the influent, environmental conditions, process design, and the mode of plant operation. There are, however, pathogenic organisms that may be present. These include the organisms responsible for diseases such as typhoid, tetanus, hepatitis, dysentery, gastroenteritis, and others.

To understand how to minimize or maximize the growth of microorganisms and control pathogens, one must study the microorganisms' structure and characteristics. In the subsequent sections, we will examine each of the major groups of microorganisms (those important to water/wastewater operators) in relation to their size, shape, types, nutritional needs, and control.

Note: Koren (1991) pointed out that, in a water environment, water is not a medium for the growth of microorganisms but is instead a means of transmission (meaning it serves as a conduit; hence, the name *waterborne*) of the pathogen to the place where an individual can consume it and thereby start the outbreak of disease. This is contrary to the view taken by the average person. That is, when the topic of waterborne disease is brought up, we might mistakenly assume that waterborne diseases are at home in the water. Nothing could be further from the truth. A water-filled ambiance is not the environment in which the pathogenic organism would choose to live, that is, if it had such a choice. The point is that microorganisms do not normally grow, reproduce, languish, and thrive in watery surroundings. Pathogenic microorganisms temporarily residing in water are simply biding their time, going with the flow, waiting for their opportunity to meet up with their unsuspecting host or hosts. To a degree, when the pathogenic microorganism finds its host or hosts, it is finally home or may have found its final resting place (Spellman 1997).

KEY TERMS AND DEFINITIONS

Algae Simple: plants, many microscopic, containing chlorophyll. Freshwater algae are diverse in shape, color, size, and habitat. They are the basic link in the conversion of inorganic constituents in water into organic constituents.

Algal Bloom: sudden spurts of algal growth, which can adversely affect water quality and indicate potentially hazardous changes in local water chemistry.

Anaerobic: able to live and grow in the absence of free oxygen.

Autotrophic Organisms: produce food from inorganic substances.

Bacteria: Single-celled, microscopic living organisms (single-celled microorganisms) that possess rigid cell walls. They may be aerobic, anaerobic, or facultative; they can cause disease; and some are important in pollution control.

Biogeochemical Cycle: the chemical interactions between the atmosphere, hydrosphere, and biosphere.

Coliform Organism: microorganisms found in the intestinal tract of humans and animals. Their presence in water indicates fecal pollution and potentially adverse contamination by pathogens.

Denitrification: the anaerobic biological reduction of nitrate to nitrogen gas.

Fungi: simple plants lacking the ability to produce energy through photosynthesis.

Heterotrophic Organism: organisms that are dependent on organic matter for food.

Prokaryotic Cell: the simple cell type, characterized by the lack of a nuclear membrane and the absence of mitochondria.

Virus: the smallest form of microorganisms capable of causing disease.

MICROORGANISM CLASSIFICATION AND DIFFERENTIATION

The microorganisms we are concerned with are tiny organisms that make up a large and diverse group of free-living forms; they exist either as single cells, cell bunches, or clusters. Found in abundance almost anywhere on earth, the vast majority of microorganisms are not harmful. Many microorganisms, or microbes, occur as single cells (unicellular); others are multicellular; and still others, like viruses, do not have a true cellular appearance. A single microbial cell, for the most part, exhibits the characteristic features common to other biological systems, such as metabolism, reproduction, and growth.

CLASSIFICATION

The Greek scholar and philosopher Aristotle classified animals based on flying, swimming, and walking/crawling/running. For centuries thereafter, scientists simply classified the life forms visible to the naked eye as either animal or plant. We started to have trouble differentiating microorganisms, so this classification had to be changed. The Swedish naturalist Carolus Linnaeus organized much of the current knowledge about living things in 1735. The importance of organizing or classifying organisms cannot be overstated, for without a classification scheme, it would be difficult to establish criteria for identifying organisms and to arrange similar organisms into groups. Probably the most important reason for classifying organisms is to make things less confusing (Wistriech and Lechtman 1980). Linnaeus was quite innovative in the classification of organisms. One of his innovations is still with us today: the *binomial system of nomenclature*. Under the binomial system, all organisms are generally described by a two-word scientific name, the *genus* and *species*. Genus and species are groups that are part of a hierarchy of groups of increasing size, based on their taxonomy. This hierarchy follows:

Kingdom
  Phylum
    Class
      Order
        Family
          Genus
            Species

Using this system, a fruit fly might be classified as:

Animalia
  Arthropoda
    Insecta
      Diptera
        Drosophilidae
          Drosophila
            Melanogaster

This means that this organism is the species *Melanogaster* in the genus *Drosophila*, in the family Drosophilidae, in the order Diptera, in the class Insecta, in the phylum Arthropoda, in the kingdom Animalia.

To further illustrate how the hierarchical system is exemplified by the classification system, the standard classification of the mayfly is provided below:

Kingdom   Animalia
  Phylum   Arthropoda
    Class   Insecta
      Order   Ephermeroptera
        Family   Ephemeridae
          Genus   *Hexagenia*
            Species   *limbata*

Utilizing this hierarchy and Linnaeus's binomial system of nomenclature, the scientific name of any organism (as stated previously) includes both generic and specific names. In the above instances, to uniquely name the species it is necessary to supply the genus and the species, *Drosophila melanogaster* (i.e., the fruit fly) and *Hexagenia limbata*

(mayfly). As shown, the first letter of the generic name is usually capitalized; hence, for example, *E. coli* indicates that *coli* is the species and *Escherichia* (abbreviated to E.) is the genus. The largest, most inclusive category—the kingdom—is plant. The names are always in Latin, so they are usually printed in italics or underlined. Some organisms also have English common names. Microbe names of particular interest in water/wastewater treatment include:

- *Escherichia coli* – a coliform bacterium
- *Salmonella typhi* – the typhoid bacillus
- *Giardia lamblia* – a protozoan
- *Shigella* spp.
- *Vibrio cholerae*
- *Campylobacter*
- *Leptospira* spp.
- *Entamoeba histolytica*
- *Cryptosporidia*

Note: *Escherichia coli* is commonly known simply as *E. coli,* while *Giardia lamblia* is usually referred to by only its genus name, *Giardia.*

Generally, we use a simplified system of microorganism classification in water science, breaking down classification into the kingdoms of animal, plant, and protista. As a rule, the animal and plant kingdoms contain all the multicellular organisms, and the protists contain all single-cell organisms. Along with microorganism classification based on the animal, plant, and protista kingdoms, microorganisms can be further classified as being *eukaryotic* or *prokaryotic* (see Table 12.1).

Note: A eukaryotic organism is characterized by a cellular organization that includes a well-defined nuclear membrane. Prokaryotes have a structural organization that sets them apart from all other organisms. They are simple cells characterized by a nucleus *lacking* a limiting membrane, an endoplasmic reticulum, chloroplasts, and mitochondria. They are remarkably adaptable, existing abundantly in the soil, the sea, and freshwater.

TABLE 12.1
Simplified Classification of Microorganisms

Kingdom	Members	Cell Classification
Animal	Rotifers	
	Crustaceans	
	Worms and larvae	Eukaryotic
Plant	Ferns	
	Mosses	
Protista	Protozoa	
	Algae	
	Fungi	
	Bacteria	Prokaryotic
	Lower algae forms	

DIFFERENTIATION

Differentiation among the higher forms of life is based almost entirely upon morphological (form or structure) differences. However, differentiation (even among the higher forms) is not as easily accomplished as one might expect, because normal variations among individuals of the same species occur frequently. Because of this variation even within a species, securing accurate classification when dealing with single-celled microscopic forms that present virtually no visible structural differences becomes extremely difficult. Under these circumstances, considering physiological, cultural, and chemical differences is necessary, as well as structure and form. Differentiation among the smaller groups of bacteria is based almost wholly upon chemical differences.

THE CELL

The structural and fundamental unit of both plants and animals, no matter how complex, is the cell. Since the nineteenth century, scientists have known that all living things, whether animal or plant, are made up of cells. A typical cell is an entity, isolated from other cells by a membrane or cell wall. The cell membrane contains protoplasm and the nucleus (see Figure 12.1). The protoplasm within the cell is a living mass of viscous, transparent material. Within the protoplasm is a dense spherical mass called the nucleus or nuclear material. In a typical mature plant cell, the cell wall is rigid and composed of nonliving material, while in the typical animal cell, the wall is an elastic living membrane. Cells exist in a very great variety of sizes and shapes, as well as functions. Their average size ranges from bacteria too small to be seen with the light microscope to the largest known single cell, the ostrich egg. Microbial cells also have an extensive size range, some being larger than human cells (Kordon 1992).

Note: The nucleus cannot always be observed in bacteria.

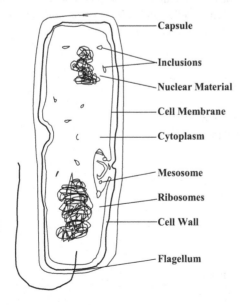

FIGURE 12.1   Bacterial cell.

STRUCTURE OF THE BACTERIAL CELL

The structural form and various components of the bacterial cell are probably best understood by referring to the simplified diagram of a rod-form bacterium shown in Figure 12.1. When studying Figure 12.1, keep in mind that cells of different species may differ greatly, both in structure and chemical composition; for this reason, no typical bacterium exists. Figure 12.1 shows a generalized bacterium used for the discussion that follows. [Not all bacteria have all of the features shown in the figure, and some bacteria have structures not shown in the figure.]

Capsules

Bacterial *capsules* (see Figure 12.1) are organized accumulations of gelatinous materials on cell walls, in contrast to *slime layers* (a water secretion that adheres loosely to the cell wall and commonly diffuses into the cell), which are unorganized accumulations of similar material. The capsule is usually thick enough to be seen under the ordinary light microscope (macrocapsule), while thinner capsules (microcapsules) can be detected only by electron microscopy (Singleton and Sainsbury 1994). The production of capsules is determined largely by genetics as well as environmental conditions and depends on the presence or absence of capsule-degrading enzymes and other growth factors. Varying in composition, capsules are mainly composed of water; the organic contents are made up of complex polysaccharides, nitrogen-containing substances, and polypeptides. Capsules confer several advantages when bacteria grow in their normal habitat. For example, they help to (1) prevent desiccation; (2) resist phagocytosis by host phagocytic cells; (3) prevent infection by bacteriophages; and (4) aid bacterial attachment to tissue surfaces in plant and animal hosts or the surfaces of solid objects in aquatic environments. Capsule formation often correlates with pathogenicity.

Flagella

Many bacteria are motile, and this ability to move independently is usually attributed to a special structure, the *flagella* (singular: flagellum). Depending on the species, a cell may have a single flagellum (see Figure 12.1) (*monotrichous* bacteria; *trichous* means "hair"); one flagellum at each end (*amphitrichous* bacteria; *amphi* means "on both sides"); a tuft of flagella at one or both ends (*lophotrichous* bacteria; *lopho* means "tuft"); or flagella that arise all over the cell surface (*peritrichous* bacteria; *peri* means "around").

A flagellum is a threadlike appendage extending outward from the plasma membrane and cell wall. Flagella are slender, rigid, locomotor structures, about 20 μm across and up to 15–20 μm long. Flagellation patterns are very useful in identifying bacteria and can be seen by light microscopy, but only after being stained with special techniques designed to increase their thickness. The detailed structure of flagella can be seen only in the electron microscope.

Bacterial cells benefit from flagella in several ways. They can increase the concentration of nutrients or decrease the concentration of toxic materials near the bacterial surfaces by causing a change in the flow rate of fluids. They can also disperse flagellated organisms to areas where colony formation can take place. The main benefit of flagella to organisms is their increased ability to flee from areas that might be harmful.

Cell Wall

The main structural component of most prokaryotes is the rigid *cell wall* (see Figure 12.1). Functions of the cell wall include (1) protecting the delicate protoplast from osmotic lysis (bursting); (2) determining a cell's shape; (3) acting as a permeability layer that excludes large molecules and various antibiotics and playing an active role in regulating the cell's intake of ions; and (4) providing solid support for flagella. Cell walls of different species may differ greatly in structure, thickness, and composition. The cell wall accounts for about 20%–40% of the dry weight of a bacterium.

Plasma Membrane (Cytoplasmic Membrane)

Surrounded externally by the cell wall and composed of a lipoprotein complex, the *plasma membrane* or cell membrane, is the critical barrier separating the inside from the outside of the cell (see Figure 12.1). About 7–8 µm thick and comprising 10%–20% of a bacterium's dry weight, the plasma membrane controls the passage of all material into and out of the cell. The inner and outer faces of the plasma membrane are embedded with water-loving (hydrophilic) lips, whereas the interior is hydrophobic. Control of material in the cell is accomplished by screening, as well as by electric charge. The plasma membrane is the site of the surface charge of the bacteria.

In addition to serving as an osmotic barrier that passively regulates the passage of material into and out of the cell, the plasma membrane participates in the active transport of various substances into the bacterial cell. Inside the membrane, many highly reactive chemical groups guide the incoming material to the proper points for further reaction. This active transport system provides bacteria with certain advantages, including the ability to maintain a fairly constant intercellular ionic state in the presence of varying external ionic concentrations. In addition to participating in the uptake of nutrients, the cell membrane transport system also participates in waste excretion and protein secretion.

Cytoplasm

Within a cell and bounded by the cell membrane is a complicated mixture of substances and structures called the *cytoplasm* (see Figure 12.1). The cytoplasm is a water-based fluid containing ribosomes, ions, enzymes, nutrients, storage granules (under certain circumstances), waste products, and various molecules involved in synthesis, energy metabolism, and cell maintenance.

Mesosome

A common intracellular structure found in the bacterial cytoplasm is the mesosome (see Figure 12.1). *Mesosomes* are invaginations of the plasma membrane in the shape of tubules, vesicles, or lamellae. Their exact function is unknown. Currently, many bacteriologists believe that mesosomes are artifacts generated during the fixation of bacteria for electron microscopy.

Nucleoid (Nuclear Body or Region)

The *nuclear region* of the prokaryotic cell is primitive and a striking contrast to that of the eukaryotic cell (see Figure 12.1). Prokaryotic cells lack a distinct nucleus, with the function of the nucleus being carried out by a single, long, double strand of DNA that is efficiently packaged to fit within the nucleoid. The nucleoid is attached to the plasma membrane. A cell can have more than one nucleoid when cell division occurs after the genetic material has been duplicated.

Ribosomes

The bacterial cytoplasm is often packed with ribosomes (see Figure 12.1). *Ribosomes* are minute, rounded bodies made of RNA and are loosely attached to the plasma membrane. Ribosomes are estimated to account for about 40% of a bacterium's dry weight; a single cell may have as many as 10,000 ribosomes. Ribosomes are the site of protein synthesis and are part of the translation process.

Inclusions

Inclusions (or storage granules) are often seen within bacterial cells (see Figure 12.1). Some inclusion bodies are not bound by a membrane and lie free in the cytoplasm. A single-layered membrane about 2–4 µm thick encloses other inclusion bodies. Many bacteria produce polymers that are stored as granules in the cytoplasm.

BACTERIA

The simplest wholly contained life systems are *bacteria or prokaryotes*, which are the most diverse group of microorganisms. As mentioned, they are among the most common microorganisms in water and are primitive, unicellular (single-celled) organisms, possessing no well-defined nucleus and presenting a variety of shapes and nutritional needs. Bacteria contain about 85% water and 15% ash or mineral matter. The ash is largely composed of sulfur, potassium, sodium, calcium, and chlorides, with small amounts of iron, silicon, and magnesium. Bacteria reproduce by binary fission.

Note: *Binary fission* occurs when one organism splits or divides into two or more new organisms.

Bacteria, once called the smallest living organisms (now it is known that smaller forms of matter exhibit many of the characteristics of life), range in size from 0.5 to 2 µm in diameter and about 1–10 µm long.

Note: A *micron* is a metric unit of measurement equal to 1,000th of a millimeter. To visualize the size of bacteria, consider that about 1,000 bacteria lying side-by-side would reach across the head of a straight pin.

Bacteria are categorized into three general groups based on their physical form or shape (though almost every variation has been found; see Table 12.2). The simplest form is the sphere. Spherical-shaped bacteria are called *cocci* (meaning "berries"). They are not necessarily perfectly round, but may be somewhat elongated, flattened on one side, or oval. Rod-shaped bacteria are called *bacilli*. Spiral-shaped bacteria (called spirilla), which have one or more twists and are never straight, make up the third group (see Figure 12.2). Such formations are usually characteristic of a particular genus or species. Within these three groups are many different arrangements. Some exist as single cells; others as pairs, in packets of four or eight, in chains, and in clumps.

Most bacteria require organic food to survive and multiply. Plant and animal material that gets into the water provides the food source for bacteria. Bacteria convert the food to energy and use the energy into make new cells. Some bacteria can use inorganics (e.g., minerals such as iron) as an energy source and exist and multiply even when organics (pollution) are not available.

BACTERIAL GROWTH FACTORS

Several factors affect the rate at which bacteria grow, including temperature, pH, and oxygen levels. The warmer the environment, the faster the rate of growth. Generally, for each increase of 10°C, the growth rate doubles. Heat can also be used to kill bacteria. Most bacteria grow best at neutral pH. Extreme acidic or basic conditions generally inhibit growth, though some bacteria may require acidic conditions and some may require alkaline conditions for growth.

Bacteria are aerobic, anaerobic, or facultative. If *aerobic*, they require free oxygen in the aquatic environment. *Anaerobic* bacteria exist and multiply in environments that lack dissolved oxygen. *Facultative* bacteria (e.g., iron bacteria) can switch from aerobic to anaerobic growth or grow in either anaerobic or aerobic environments.

Under optimum conditions, bacteria grow and reproduce very rapidly. As stated previously, bacteria reproduce by *binary fission*. An important point to consider in connection with bacterial reproduction is the rate at which the process

TABLE 12.2
Forms of Bacteria

	Technical Name		
Form	*Singular*	*Plural*	**Example**
Sphere	Coccus	Cocci	Streptococcus
Rod	Bacillus	Bacilli	Bacillus typhosis
Curved or spiral	Spirillum	Spirilla	Spirillum cholera

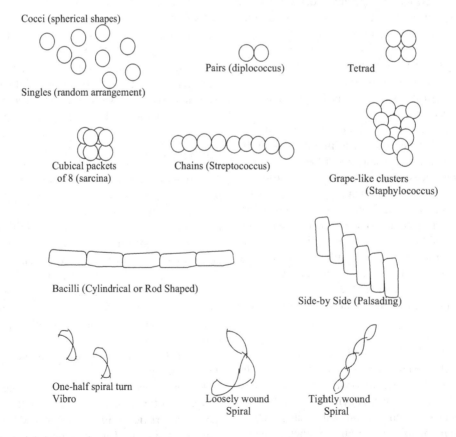

FIGURE 12.2   Bacterial shapes and arrangements.

can take place. The total time required for an organism to reproduce and for the offspring to reach maturity is called *generation time*. Bacteria growing under optimal conditions can double their numbers about every 20–30 minutes. Obviously, this generation time is very short compared with that of higher plants and animals. Bacteria continue to grow at this rapid rate as long as nutrients hold out—even the smallest contamination can result in sizable growth in a very short time.

Note: Even though wastewater can contain bacterial counts in the millions per mL, in wastewater treatment, under controlled conditions, bacteria can help to destroy and to identify pollutants. In such a process, bacteria stabilize organic matter (e.g., activated sludge processes) and thereby assist the treatment process in producing effluent that does not impose an excessive oxygen demand on the receiving body. Coliform bacteria can be used as an indicator of pollution by human or animal wastes.

DESTRUCTION OF BACTERIA

In water and wastewater treatment, the destruction of bacteria is usually called *disinfection*.

Disinfection does not mean that all microbial forms are killed; that would be *sterilization*. However, disinfection does reduce the number of disease-causing organisms to an acceptable number. Growing bacteria are easy to control by disinfection. Some bacteria, however, form spores—survival structures—which are much more difficult to destroy.

Note: Inhibiting the growth of microorganisms is termed antisepsis while destroying them is called *disinfection*.

WATERBORNE BACTERIA

All surface waters contain bacteria. Waterborne bacteria, as we have said, are responsible for infectious epidemic diseases. Bacterial numbers increase significantly during storm events when streams are high. Heavy rainstorms increase stream contamination by washing material from the ground surface into the stream. After the initial washing occurs, few impurities are left to be washed into the stream, which may then carry relatively "clean" water. A river of good quality shows its highest bacterial numbers during rainy periods; however, a much-polluted stream may show the highest numbers during low flows because of the constant influx of pollutants. Water and wastewater operators are primarily concerned with bacterial pathogens responsible for disease. These pathogens enter potential drinking water supplies through fecal contamination and are ingested by humans if the water is not properly treated and disinfected.

Note: Regulations require that owners of all public water supplies collect water samples and deliver them to a certified laboratory for bacteriological examination at least monthly. The number of samples required is usually in accordance with federal standards, which generally require that one sample per month be collected for each 1,000 persons served by the waterworks.

PROTOZOA

Protozoans (or "first animals") are a large group of eucaryotic organisms of more than 50,000 known species belonging to the Kingdom Protista that have adapted a form of the cell to serve as the entire body. In fact, protozoans are one-celled animal-like organisms with complex cellular structures. In the microbial world, protozoans are giants, many times larger than bacteria. They range in size from 4 to 500 μm. The largest ones can almost be seen by the naked eye. They can exist as solitary or independent organisms [for example, the stalked ciliates (see Figure 12.3) such as *Vorticella* sp.], or they can colonize like the sedentary *Carchesium* sp. Protozoa get their name because they employ the same type of feeding strategy as animals. That is, they are heterotrophic, meaning they obtain cellular energy from organic substances such as proteins. Most are harmless, but some are parasitic. Some forms have two life stages: *active trophozoites* (capable of feeding) and *dormant cysts*.

The major groups of protozoans are based on their method of locomotion (motility). For example, the *Mastigophora* are motile by means or one of more *flagella* (the whip-like projections that propel free-swimming organisms—*Giardia lamblia* is a flagellated protozoan); the *Ciliophora* by means of shortened modified flagella called *cilia* (short hair-like structures that beat rapidly and propel them through the water); the *Sarcodina* by means of amoeboid movement (streaming or gliding action—the shape of amoebae change as they stretch, then contract, from place to place); and the *Sporozoa*, which are nonmotile; they have simply swept along, riding the current of the water.

Protozoa consume organics to survive; their favorite food is bacteria. Protozoa are mostly aerobic or facultative concerning oxygen requirements. Toxic materials, pH, and temperature affect protozoan growth rates in the same way they affect bacteria.

Most protozoan life cycles alternate between an active growth phase (*trophozoites*) and a resting stage (*cysts*). Cysts are extremely resistant structures that protect the organism from destruction when it encounters harsh environmental conditions, including chlorination.

Note: Those protozoans not completely resistant to chlorination require higher disinfectant concentrations and longer contact times for disinfection than are normally used in water treatment.

The protozoa and the waterborne diseases associated with them that are of most concern to the waterworks operator are:

- *Entamoeba histolytica:* Amoebic dysentery
- *Giardia lamblia:* Giardiasis
- *Cryptosporidium*: Cryptosporidiosis

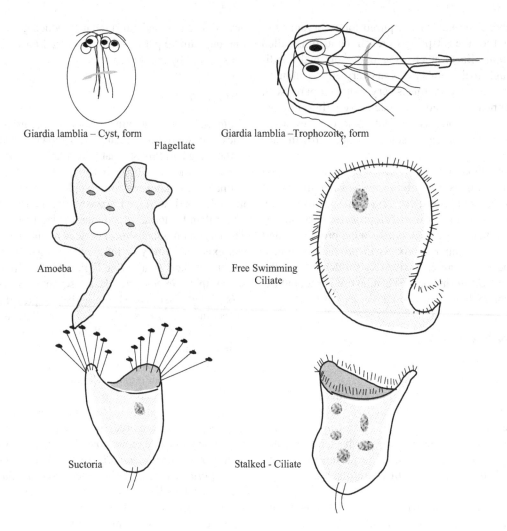

FIGURE 12.3   Protozoa.

In wastewater treatment, protozoa are a critical part of the purification process and can be used to indicate the condition of treatment processes. Protozoa normally associated with wastewater include amoeba, flagellates, free-swimming ciliates, and stalked ciliates.

Amoebae are associated with poor wastewater treatment of a young biosolids mass (see Figure 12.3). They move through wastewater in a streaming or gliding motion. Moving the liquids stored within the cell wall affects this movement. They are normally associated with an effluent high in BODs and suspended solids.

Flagellates (flagellated protozoa) have a single, long hair-like or whip-like projection (flagella) that is used to propel the free-swimming organisms through wastewater and to attract food (see Figure 12.3). Flagellated protozoans are normally associated with poor treatment and young biosolids. When the free-swimming ciliated protozoan is the predominant organism, the plant effluent will contain large amounts of BODand suspended solids.

The *free-swimming ciliated protozoan* uses its tiny, hair-like projections (cilia) to move through the wastewater and attract food (see Figure 12.3). The free-swimming ciliated protozoan is normally associated with a moderate biosolids age and effluent quality. When the free-swimming ciliated protozoan is the predominant organism, the plant effluent will normally be turbid and contain a high amount of suspended solids.

The *stalked ciliated protozoan* attaches itself to the wastewater solids and uses its cilia to attract food (see Figure 12.3). The stalked ciliated protozoan is normally associated with plant effluent that is very clear and contains low amounts of both BOD and suspended solids.

Rotifers make up a well-defined group of the smallest, simplest multicellular microorganisms and are found in nearly all aquatic habitats (see Figure 12.4). Rotifers are a higher life form associated with cleaner waters. Normally found in well-operated wastewater treatment plants, they can be used to indicate the performance of certain types of treatment processes.

MICROSCOPIC CRUSTACEANS

Because they are important members of freshwater zooplankton, microscope *crustaceans* are of interest to water and wastewater operators. These microscopic organisms are characterized by a rigid shell structure. They are

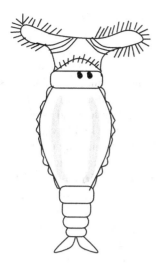

FIGURE 12.4 *Philodina*, a common rotifer.

multicellular animals that have strict aerobes, and as primary producers, they feed on bacteria and algae. They are important as a source of food for fish. Additionally, microscopic crustaceans have been used to clarify algae-laden effluents from oxidation ponds. *Cyclops* and *Daphnia* are two microscopic crustaceans of interest to water and wastewater operators.

VIRUSES

Viruses are very different from other microorganisms. Consider their size relationship, for example. Relative to size, if protozoans are the Goliaths of microorganisms, then viruses are the Davids. More specifically and accurately, viruses are intercellular parasitic particles that are the smallest living infectious materials known—the midgets of the microbial world. Viruses are very simple life forms consisting of a central molecule of genetic material surrounded by a protein shell called a *capsid* and sometimes by a second layer called an *envelope*. They contain no mechanisms to obtain energy or reproduce on their own; thus, to live, viruses must have a host. After invading the cells of their specific host (animal, plant, insect, fish, or even bacteria), they take over the host's cellular machinery and force it to make more viruses. In the process, the host cell is destroyed and hundreds of new viruses are released into the environment. Viruses occur in many shapes, including long slender rods, elaborate irregular shapes, and geometric polyhedrals (see Figure 12.5).

The viruses of most concern to the waterworks operator are the pathogens that cause hepatitis, viral gastroenteritis, and poliomyelitis.

Smaller and different from bacteria, viruses are prevalent in water contaminated with sewage. Detecting viruses in water supplies is a major challenge due to the complexity of non-routine procedures involved, although experience has shown that the normal coliform index can serve as a rough guide for viruses, as it does for bacteria. More attention must be paid to viruses, however, when surface water supplies are used for sewage disposal. Viruses are difficult to destroy by normal disinfection practices, requiring increased disinfectant concentration and contact time for effective destruction.

Note: Viruses that infect bacterial cells cannot infect and replicate within the cells of other organisms. It is possible to utilize specificity to identify bacteria in a procedure called *phage typing*.

ALGAE

You do not have to be a water or wastewater operator to understand that algae can be a nuisance. Many ponds and lakes in the United States are currently undergoing *eutrophication*, the enrichment of an environment with inorganic substances (e.g., phosphorus and nitrogen), causing excessive algae growth and premature aging of the water body. When eutrophication occurs, especially when filamentous algae like *Caldophora* break loose in a pond or lake and wash ashore, algae make their stinking, noxious presence known.

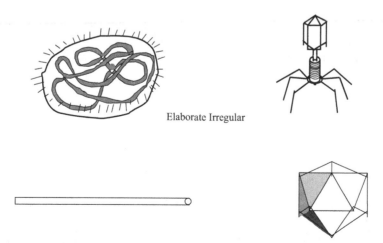

Elaborate Irregular

Long Slender Rod

Geometric Polyhedral

FIGURE 12.5 Virus shapes.

Algae are a form of aquatic plants classified by color (e.g., green algae, blue-green algae, golden-brown algae, etc.). They come in many shapes and sizes (see Figure 12.6). Although not pathogenic, algae cause problems with water/wastewater treatment plant operations. They grow easily on the walls of troughs and basins, and heavy growth can plug intakes and screens. Additionally, some algae release chemicals that give off undesirable tastes and odors. Although usually classified by their color, they are also commonly classified based on their cellular properties or characteristics. Several characteristics are used to classify algae, including (1) cellular organization and cell wall structure; (2) the nature of the chlorophyll(s); (3) the type of motility, if any; (4) the carbon polymers that are produced and stored; and (5) the reproductive structures and methods.

Many algae (in mass) are easily seen by the naked eye, while others are microscopic. They occur in fresh and polluted water, as well as in salt water. Since they are plants, they are capable of using energy from the sun in photosynthesis. They usually grow near the surface of the water because light cannot penetrate very far through the water. Algae are controlled in raw waters with chlorine and potassium permanganate. Algae blooms in raw water reservoirs are often controlled with copper sulfate.

Note: By producing oxygen, which is utilized by other organisms, including animals, algae play an important role in the balance of nature.

FUNGI

Fungi are of relatively minor importance in water/wastewater operations (except for biosolids composting, where they are critical). Fungi, like bacteria, are also extremely diverse. They are multicellular, autotrophic, photosynthetic protists. They grow as filamentous, mold-like forms or as yeast-like (single-celled) organisms. They feed on organic material.

Note: Aquatic fungi grow as parasites on living plants or animals and as saprophytes on those that are dead.

NEMATODES AND FLATWORMS (WORMS)

Along with inhabiting organic mud, worms also inhabit biological slimes; they have been found in activated sludge and in trickling filter slimes (wastewater treatment processes).

Microscopic in size, they range in length from 0.5 to 3 mm and in diameter from 0.01 to 0.05 mm. Most species have a similar appearance. They have a body that is covered by a cuticle, are cylindrical, nonsegmented, and taper at both ends.

These organisms continuously enter wastewater treatment systems, primarily through attachment to soils that reach the plant through inflow and infiltration (I & I). They are present in large, often highly variable numbers, but as strict aerobes, they are found only in aerobic treatment processes where they metabolize solid organic matter.

When nematodes are firmly established in the treatment process, they can promote microfloral activity and decomposition. They crop bacteria in both the activated sludge and trickling filter systems. Their activities in these systems enhance oxygen penetration by tunneling through floc particles and biofilm. In activated sludge processes, they are present in relatively small numbers because the liquefied environment is not a suitable habitat for crawling, which they prefer over the free-swimming mode. In trickling filters, where the fine stationary substratum is suitable to permit crawling and mating, nematodes are quite abundant.

Along with preferring the trickling filter habitat, nematodes play a beneficial role in this habitat; for example, they break loose portions of the biological slime coating the filter bed. This action prevents excessive slime growth and filter clogging. They also aid in keeping slime porous and accessible to oxygen by tunneling through slime. In the activated sludge process, nematodes play important roles as agents of better oxygen diffusion. They accomplish this by tunneling through floc particles. They also act as parameters of operational conditions in the process, such as low dissolved oxygen levels (anoxic conditions) and the presence of toxic wastes.

Environmental conditions have an impact on the growth of nematodes. For example, in anoxic conditions, their swimming and growth are impaired. The most important condition they indicate is when the wastewater strength and composition have changed. Temperature fluctuations directly affect their growth and survival; the population decreases when temperatures increase.

Aquatic flatworms (improperly named because they are not all flat) feed primarily on algae. Because of their aversion to light, they are found in the lower depths of pools.

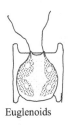
Euglenoids

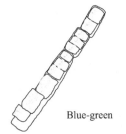
Blue-green

Diatom

FIGURE 12.6   Algae.

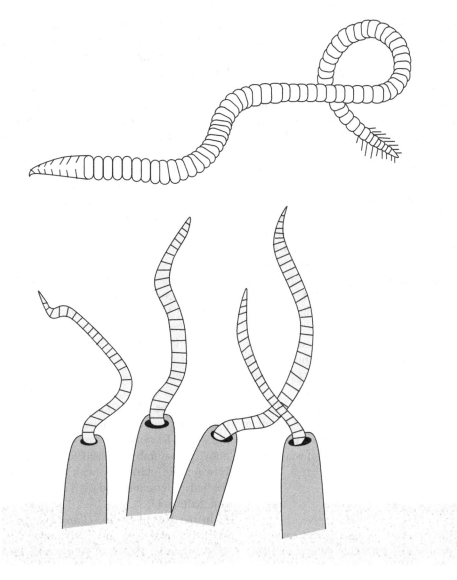

FIGURE 12.7   Tubificid worms.

Two varieties of flatworms are seen in wastewater treatment processes: *microturbellarians* are more round than flat and average about 0.5–5 mm in size, and *macroturbellarians* (planarians) are more flat than round and average about 5–20 mm in body size. Flatworms are very hardy and can survive in wide variations in humidity and temperature. As inhabitants of sewage sludge, they play an important role in sludge stabilization and as bioindicators of process problems. For example, their inactivity or sluggishness might indicate a low dissolved oxygen level or the presence of toxic wastes.

Surface waters grossly polluted with organic matter (especially domestic sewage) have a fauna that is capable of thriving in very low concentrations of oxygen. A few species of tubificid worms dominate this environment. Pennak reported (1989) that the bottoms of severely polluted streams can be literally covered with a "writhing" mass of these tubificids.

The *Tubifex* (commonly known as sludge worms) are small, slender, reddish worms that normally range in length

from 25 to about 50 mm. They are burrowers; their posterior end protrudes to obtain nutrients (see Figure 12.7). When found in streams, *Tubifex* are indicators of pollution.

WATER TREATMENT AND MICROBIOLOGICAL PROCESSES

The primary goal of water treatment is to protect the consumer of potable drinking water from disease. Drinking water safety is a worldwide concern. Drinking untreated or improperly treated water is a major cause of illness in developing countries. Water contains several biological (as well as chemical) contaminants that must be removed efficiently to produce safe drinking water that is also aesthetically pleasing to the consumer. The finished water must be free of microbial pathogens and parasites, turbidity, color, taste, and odor. To achieve this goal, raw surface water or groundwater is subjected to a series of treatment processes that will be described in detail later. Disinfection alone is sufficient if the raw water originates from a protected source.

More commonly, several processes are used to treat water. For example, disinfection may be combined with coagulation, flocculation, and filtration.

As mentioned, several unit processes are used in the water treatment process to produce microbiologically (and chemically) safe drinking water. The extent of treatment depends on the source of raw water; surface waters generally require more treatment than groundwaters. With the exception of disinfection, the other unit processes in the treatment train do not specifically address the destruction or removal of pathogens.

Water treatment unit processes include (1) storage of raw water, (2) prechlorination, (3) coagulation-flocculation, (4) water softening, (5) filtration, and (6) disinfection. Although filtration and disinfection are the primary means of removing contaminants and pathogens from drinking water supplies, in each of these unit processes, reduction or destruction of pathogens is accomplished but is variable and influenced by several factors such as sunlight, sedimentation, and temperature.

The water treatment unit processes mentioned above are important and are described in detail later in this text. For the moment, because of relatively recent events involving pathogenic protozoans causing adverse reactions, including death, to consumers in various locations in the United States (and elsewhere), it is important to turn our attention to these pathogenic protozoans. One thing is certain: these pathogenic protozoans have the full attention of water treatment operators everywhere.

PATHOGENIC PROTOZOA

As mentioned, certain types of protozoans can cause disease. Of particular interest to the drinking water practitioner are *Entamoeba histolytica* (amebic dysentery and amebic hepatitis), *Giardia lamblia* (Giardiasis), *Cryptosporidium* (Cryptosporidiosis), and the emerging *Cyclospora* (Cyclosporiasis). Sewage contamination transports eggs, cysts, and oocysts of parasitic protozoa and helminths (tapeworms, hookworms, etc.) into raw water supplies, leaving water treatment (in particular filtration) and disinfection as the means by which to diminish the danger of contaminated water for the consumer.

To prevent the occurrence of *Giardia* and *Cryptosporidium* spp. in surface water supplies, and to address increasing problems with waterborne diseases, USEPA implemented its Surface Water Treatment Rule (SWTR) in 1989. The rule requires both filtration and disinfection of all surface water supplies as a means of primarily controlling *Giardia* spp. and enteric viruses. Since the implementation of its Surface Water Treatment Rule, USEPA has also recognized that *Cryptosporidium* spp. is an agent of waterborne disease. In 1996, in its next series of surface water regulations, the USEPA included *Cryptosporidium*.

To test the need for and the effectiveness of USEPA's Surface Water Treatment Rule, LeChevallier et al. (1991) conducted a study on the occurrence and distribution of *Giardia* and *Cryptosporidium* organisms in raw water supplies to 66 surface water filter plants. These plants were located in 14 states and one Canadian province. A combined immunofluorescence test indicated that cysts and oocysts were widely dispersed in the aquatic environment. *Giardia* spp. was detected in more than 80% of the samples. *Cryptosporidium* spp. was found in 85% of the sample locations. Considering several variables, *Giardia* or *Cryptosporidium* spp. were detected in 97% of the raw water samples. After evaluating their data, the researchers concluded that the Surface Water Treatment Rule might have to be upgraded (subsequently, it has been) to require additional treatment.

GIARDIA

Giardia (gee-ar-dee-ah) *lamblia* (also known as hiker's/traveler's scourge or disease) is a microscopic parasite that can infect warm-blooded animals and humans. Although *Giardia* was discovered in the nineteenth century, it was not until 1981 that the World Health Organization (WHO) classified *Giardia* as a pathogen. *An outer shell called a cyst that allows it to survive outside the body for long periods protects giardia.* If viable cysts are ingested, *Giardia* can cause the illness known as *Giardiasis*, an intestinal illness that can cause nausea, anorexia, fever, and severe diarrhea. The symptoms last only for several days, and the body can naturally rid itself of the parasite in 1–2 months. However, for individuals with weakened immune systems, the body often cannot rid itself of the parasite without medical treatment.

In the United States, *Giardia* is the most commonly identified pathogen in waterborne disease outbreaks. Contamination of a water supply by *Giardia* can occur in two ways: (1) by the activity of animals in the watershed area of the water supply; or (2) by the introduction of sewage into the water supply. Wild and domestic animals are major contributors to contaminating water supplies. Studies have also shown that, unlike many other pathogens, *Giardia* is not host-specific. In short, *Giardia* cysts excreted by animals can infect and cause illness in humans. Additionally, in several major outbreaks of waterborne diseases, the source of *Giardia* cysts was sewage-contaminated water supplies.

Treating the water supply, however, can effectively control waterborne Giardia. Chlorine and ozone are examples of disinfectants known to effectively kill *Giardia* cysts. Filtration of the water can also effectively trap and remove the parasite from the water supply. The combination of disinfection and filtration is the most effective water treatment process available today for the prevention of *Giardia* contamination.

In drinking water, *Giardia* is regulated under the Surface Water Treatment Rule (SWTR). Although the SWTR does not establish a Maximum Contaminant Level (MCL) for *Giardia,* it specifies treatment requirements to achieve at least 99.9% (3-log) removal and/or inactivation of *Giardia*. This regulation requires that all drinking water systems using surface water or groundwater under the influence of surface water must disinfect and filter the water. The Enhanced Surface Water Treatment Rule (ESWTR), which includes *Cryptosporidium* and further regulates *Giardia,* was established in December 1996.

Giardiasis

Giardiasis is recognized as one of the most frequently occurring waterborne diseases in the United States. *Giardia lamblia* cysts have been discovered in the United States in places as far apart as Estes Park, Colorado (near the Continental Divide); Missoula, Montana; Wilkes-Barre, Scranton, and Hazleton, Pennsylvania; and Pittsfield and Lawrence, Massachusetts, just to name a few (CDC 1995).

Giardiasis is characterized by intestinal symptoms that usually last 1 week or more and may be accompanied by one or more of the following: diarrhea, abdominal cramps, bloating, flatulence, fatigue, and weight loss. Although vomiting and fever are commonly listed as relatively frequent symptoms, people involved in waterborne outbreaks in the United States have not commonly reported them.

While most *Giardia* infections persist only for 1 or 2 months, some people undergo a more chronic phase, which can follow the acute phase or may become manifest without an antecedent acute illness. Loose stools and increased abdominal gassiness with cramping, flatulence, and burping characterize the chronic phase. Fever is not common, but malaise, fatigue, and depression may ensue. For a small number of people, the persistence of infection is associated with the development of marked malabsorption and weight loss (Weller 1985). Similarly, lactose (milk) intolerance can be a problem for some people. This can develop coincidentally with the infection or be aggravated by it, causing an increase in intestinal symptoms after ingestion of milk products.

Some people may have several of these symptoms without evidence of diarrhea or have only sporadic episodes of diarrhea every 3 or 4 days. Still, others may not have any symptoms at all. Therefore, the problem may not be whether you are infected with the parasite or not, but how harmoniously you both can live together or how to get rid of the parasite (either spontaneously or by treatment) when the harmony does not exist or is lost.

Note: Three prescription drugs are available in the United States to treat giardiasis: quinacrine, metronidazole, and furazolidone.

Giardiasis occurs worldwide. In the United States, *Giardia* is the parasite most commonly identified in stool specimens submitted to state laboratories for parasitologic examination. During a 3-year period, approximately 4% of one million stool specimens submitted to state laboratories tested positive for Giardia (CDC 1979). Other surveys have demonstrated *Giardia* prevalence rates ranging from 1% to 20%, depending on the location and the ages of persons studied. Giardiasis ranks among the top 20 infectious diseases causing the greatest morbidity in Africa, Asia, and Latin America; it has been estimated that about two million infections occur annually in these regions (Walsh and Warren 1979). People who are at the highest risk for acquiring *Giardia* infection in the United States can be categorized into five major groups:

1. People in cities whose drinking water originates from streams or rivers and whose water treatment processes do not include filtration, or where filtration is ineffective because of malfunctioning equipment.
2. Hikers/campers/outdoor people.
3. International travelers.
4. Children who attend day-care centers, as well as day-care center staff and parents or siblings of children infected in day-care centers.
5. Homosexual men.

People in categories 1, 2, and 3 have in common the same general source of infection, i.e., they acquire *Giardia* from fecally contaminated drinking water. City residents usually become infected because the municipal water treatment process does not include the filter necessary to physically remove the parasite from the water. The number of people in the United States at risk (i.e., the number who receive municipal drinking water from unfiltered surface water) is estimated to be 20 million. International travelers may also acquire the parasite from improperly treated municipal waters in cities or villages in other parts of the world, particularly in developing countries. In Eurasia, only travelers to Leningrad appear to be at increased risk. In prospective studies, 88% of the United States and 35% of Finnish travelers to Leningrad who had negative stool tests for *Giardia* on departure to the Soviet Union developed symptoms of giardiasis and had positive tests for *Giardia* after they returned home (Brodsky et al. 1974). With the exception of visitors to Leningrad, however, *Giardia* has not been implicated as a major cause of traveler's diarrhea—it has been detected in fewer than 2% of travelers who develop diarrhea. However, hikers and campers risk infection every time they drink untreated raw water from a stream or river. Persons in categories 4 and 5 become exposed through more direct contact with feces or an infected person by exposure to soiled diapers of an infected child (day-care center-associated cases) or through direct or indirect anal-oral sexual practices in the case of homosexual men.

Although community waterborne outbreaks of giardiasis have received the greatest publicity in the United States during the past decade, about half of the *Giardia* cases discussed with the staff of the Centers for Disease Control over a 3-year period had a day-care exposure as the most likely source of infection. Numerous outbreaks of *Giardia* in day-care centers have been reported in recent years. Infection rates for children in day-care center outbreaks range from 21% to 44% in the United States and from 8% to 27% in Canada (Black et al. 1981). The highest infection rates are usually observed in children who wear diapers (1–3 years of age). In a study of 18 randomly selected day-care centers in Atlanta, 10% of diapered children were found to be infected (CDC Unpublished). Transmission from this age group to older children, day-care staff, and household contacts is also common. About 20% of parents caring for an infected child become infected.

Local health officials and managers of water utility companies need to realize that sources of *Giardia* infection other than municipal drinking water exist. Armed with this knowledge, they are less likely to make a quick (and sometimes wrong) assumption that a cluster of recently diagnosed cases in a city is related to municipal drinking water. Of course, drinking water must not be ruled out as a source of infection when a larger-than-expected number of cases is recognized in a community, but the possibility that the cases are associated with a day-care center outbreak, drinking untreated stream water, or international travel should also be entertained.

To understand the finer aspects of *Giardia* transmission and strategies for control, the drinking water practitioner must become familiar with several aspects of the parasite's biology. Two forms of the parasite exist: a *trophozoite* and a *cyst*, both of which are much larger than bacteria (see Figure 12.8). Trophozoites live in the upper small intestine, where they attach to the intestinal wall by means of a disc-shaped suction pad on their ventral surface. Trophozoites actively feed and reproduce at this location. At some time, during the trophozoite's life, it releases its hold on the bowel wall and floats in the fecal stream through the intestine. As it makes this journey, it undergoes a morphologic transformation into an egg-like structure called a cyst. The cyst (about 6–9 nm in diameter×8–12 μm – 1/100 mm – in length) has a thick exterior wall that protects the parasite against the harsh elements that it will encounter outside the body. This cyst form of the parasite is infectious to other people or animals. Most people become infected either directly (by hand-to-mouth transfer of cysts from the feces of an infected individual) or indirectly (by drinking feces-contaminated water). Less common modes of transmission include ingestion of fecally contaminated food and hand-to-mouth transfer of cysts after touching a fecally contaminated surface. After the cyst is swallowed, the trophozoite is liberated through the action of stomach acid and digestive enzymes and becomes established in the small intestine.

Although infection after ingestion of only one *Giardia* cyst is theoretically possible, the minimum number of cysts shown to infect a human under experimental conditions is ten (Rendtorff 1954). Trophozoites divide by binary fission about every 12 hours. What this means in practical terms is that if a person swallowed only a single cyst, reproduction at this rate would result in more than one million parasites 10 days later, and one billion parasites by day 15.

The exact mechanism by which *Giardia* causes illness is not yet well understood but is not necessarily related to the number of organisms present. Nearly all of the symptoms, however, are related to dysfunction of the gastrointestinal tract. The parasite rarely invades other parts of the body, such as the gallbladder or pancreatic ducts. Intestinal infection does not result in permanent damage.

Note: *Giardia* has an incubation period of 1–8 weeks.

Data reported by the CDC indicate that *Giardia* is the most frequently identified cause of diarrheal outbreaks associated with drinking water in the United States. The remainder of this section is devoted specifically to waterborne transmissions of *Giardia*. *Giardia* cysts have been detected in 16% of potable water supplies (lakes, reservoirs, rivers, springs, groundwater) in the United States at an average concentration of 3 cysts per 100 L (Rose et al. 1983). Waterborne epidemics of giardiasis are a relatively frequent occurrence. In 1983, for example, *Giardia* was identified as the cause of diarrhea in 68% of waterborne outbreaks in which the causal agent was identified. From 1965 to 1982, more than 50 waterborne outbreaks were reported (CDC 1983). In 1984, about 250,000 people in Pennsylvania were advised to boil drinking water for 6 months because of *Giardia*-contaminated water.

Many of the municipal waterborne outbreaks of *Giardia* have been subjected to intense study to determine their cause. Several general conclusions can be made from data obtained in those studies. Waterborne transmission of *Giardia* in the United States usually occurs in mountainous regions where community drinking water obtained from clear running streams is chlorinated but not filtered before distribution. Although mountain streams appear to be clean, fecal contamination upstream by human residents or visitors, as well as by *Giardia*-infected animals such as beavers, has been well documented. Water obtained from deep wells is an unlikely source of *Giardia* because of the natural filtration of water as it percolates through the soil to reach underground cisterns. Well-waste sources that pose the greatest risk of fecal contamination are poorly constructed or improperly located ones. A few outbreaks have occurred in towns that included filtration in the water treatment process, where the filtration was not effective in removing *Giardia* cysts because of defects in filter construction, poor maintenance of the filter media, or inadequate pretreatment of the water before filtration. Occasional outbreaks have also occurred because of accidental cross-connections between water and sewage systems.

Important Point: From these data, we conclude that two major ingredients are necessary for waterborne outbreaks. *Giardia* cysts must be present in untreated source water, and the water purification process must either fail to kill or remove *Giardia* cysts from the water.

Although beavers are often blamed for contaminating water with *Giardia* cysts, that they are responsible for introducing the parasite into new areas seems unlikely. Far more likely is that they are also victims: *Giardia* cysts may be carried in untreated human sewage discharged into the water by small-town sewage disposal plants or may originate from cabin toilets that drain directly into streams and rivers. Backpackers, campers, and sports enthusiasts may also deposit *Giardia*-contaminated feces in the environment, which are subsequently washed into streams by rain.

In support of this concept is a growing amount of data that indicate a higher *Giardia* infection rate in beavers living downstream from U.S. National Forest campgrounds compared with beavers living in more remote areas that have a near-zero rate of infection.

Although beavers may be unwitting victims of the *Giardia* story, they still play an important part in the contamination scheme because they can (and probably do) serve as amplifying hosts. An *amplifying host* is easy to infect, serves as a good habitat for the parasite to reproduce, and in the case of *Giardia*, returns millions of cysts to the water for every one ingested. Beavers are especially important in this regard because they tend to defecate in or very near the water, which ensures that most of the *Giardia* cysts excreted are returned to the water.

The microbial quality of water resources and the management of the microbially laden wastes generated by the burgeoning animal agriculture industry are critical local, regional, and national problems. Animal wastes from cattle, hogs, sheep, horses, poultry, and other livestock and commercial animals can contain high concentrations of microorganisms, such as *Giardia,* that are pathogenic to humans.

The contribution of other animals to waterborne outbreaks of *Giardia* is less clear. Muskrats (another semiaquatic animal) have been found in several parts of the United States to have high infection rates (30%–40%) (Frost et al. 1984). Recent studies have shown that muskrats can be infected with *Giardia* cysts from humans and beavers. Occasional *Giardia* infections have been reported in coyotes, deer, elk, cattle, dogs, and cats (but not in horses and sheep) encountered in mountainous regions of the United States. Naturally occurring *Giardia* infections have not been found in most other wild animals (bear, nutria, rabbit, squirrel, badger, marmot, skunk, ferret, porcupine, mink, raccoon, river otter, bobcat, lynx, moose, bighorn sheep) (Frost et al. 1984).

Scientific knowledge about what is required to kill or remove *Giardia* cysts from a contaminated water supply has increased considerably. For example, we know that cysts can survive in cold water (4°C) for at least 2 months, and they are killed instantaneously by boiling water (100°C) (Frost et al. 1984). We do not know how long the cysts will remain viable at other water temperatures (e.g., at 0°C or in a canteen at 15°C–20°C), nor do we know how long the parasite will survive on various environmental surfaces, e.g., under a pine tree, in the sun, on a diaper-changing table, or in carpets in a daycare center.

The effect of chemical disinfection (chlorination, for example) on the viability of *Giardia* cysts is an even more complex issue. The number of waterborne outbreaks of *Giardia* that have occurred in communities where chlorination was employed as a disinfectant process demonstrates that the amount of chlorine used routinely for municipal water treatment is not effective against *Giardia* cysts. These observations have been confirmed in the laboratory under experimental conditions (Jarroll et al. 1979). This does not mean that chlorine does not work at all; it does work under

certain favorable conditions. Without getting too technical, gaining some appreciation of the problem can be achieved by understanding a few of the variables that influence the efficacy of chlorine as a disinfectant:

- **Water pH:** At pH values above 7.5, the disinfectant capability of chlorine is greatly reduced.
- **Water Temperature:** The warmer the water, the higher the efficacy. Chlorine does not work in ice-cold water from mountain streams.
- **Organic Content of the Water:** Mud, decayed vegetation, or other suspended organic debris in water chemically combines with chlorine, making it unavailable as a disinfectant.
- **Chlorine Contact Time:** The longer *Giardia* cysts are exposed to chlorine, the more likely the chemical will kill them.
- **Chlorine Concentration:** The higher the chlorine concentration, the more likely chlorine will kill *Giardia* cysts. Most water treatment facilities try to add enough chlorine to give a free (unbound) chlorine residual at the customer's tap of 0.5 mg/L of water.

These five variables are so closely interrelated that improving one can often compensate for another; for example, if chlorine efficacy is expected to be low because water is obtained from an icy stream, the chlorine contact time, chlorine concentration, or both could be increased. In the case of *Giardia*-contaminated water, producing safe drinking water with a chlorine concentration of 1 mg/L and contact time as short as 10 minutes might be possible—if all the other variables were optimal (i.e., pH of 7.0, water temperature of 25°C, and a total organic content of the water close to zero). On the other hand, if all of these variables were unfavorable (i.e., pH of 7.9, water temperature of 5°C, and high organic content), chlorine concentrations in excess of 8 mg/L with several hours of contact time may not be consistently effective. Because water conditions and water treatment plant operations (especially those related to water retention time, and therefore, to chlorine contact time) vary considerably in different parts of the United States, neither the USEPA nor the CDC has been able to identify a chlorine concentration that would be safe yet effective against *Giardia* cysts under all water conditions. Therefore, the use of chlorine as a preventive measure against waterborne giardiasis generally has been applied under outbreak conditions when the amount of chlorine and contact time have been tailored to fit specific water conditions and the existing operational design of the water utility.

In an outbreak, for example, the local health department and water utility may issue an advisory to boil water, increase the chlorine residual at the consumer's tap from 0.5 to 1 or 2 mg/L, and, if the physical layout and operation of the water treatment facility permit, increase the chlorine contact time. These emergency procedures are intended to reduce the risk of transmission until a filtration device can

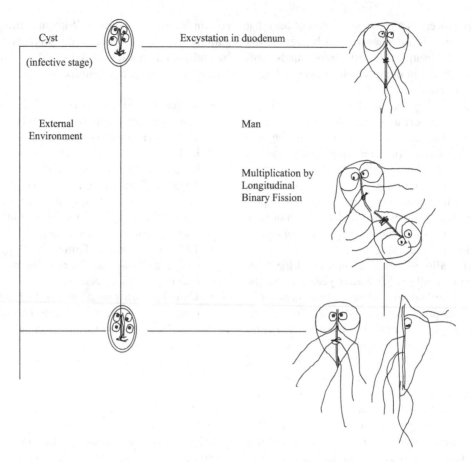

FIGURE 12.8 Life cycle of *Giardia lamblia*.

be installed, or repaired or until an alternative source of safe water (a well, for example) can be made operational.

The long-term solution to the problem of municipal waterborne outbreaks of giardiasis involves improvements in and more widespread use of filters in the municipal water treatment process. The sand filters most commonly used in municipal water treatment today cost millions of dollars to install, which makes them unattractive for many small communities. The pore sizes in these filters are not sufficiently small to remove a *Giardia* cyst ($6–9\,\mu m \times 8–12\,\mu m$). For the sand filter to effectively remove *Giardia* cysts from the water, the water must receive some additional treatment before it reaches the filter. Furthermore, the flow of water through the filter bed must be carefully regulated.

An ideal prefilter treatment for muddy water would include sedimentation (a holding pond where large, suspended particles are allowed to settle out by the action of gravity) followed by flocculation or coagulation (the addition of chemicals such as alum or ammonium to cause microscopic particles to clump together). The sand filter easily removes the large particles resulting from the flocculation/coagulation process, including Giardia cysts bound to other microparticulates. Chlorine is then added to kill the bacteria and viruses that may escape the filtration process. If the water comes from a relatively clear source, chlorine may be added to the water before it reaches the filter.

The successful operation of a complete waterworks operation is a complex process that requires considerable training. Troubleshooting breakdowns or recognizing potential problems in the system before they occur often requires the skills of an engineer. Unfortunately, most small water utilities with water treatment facilities that include filtration cannot afford the services of a full-time engineer. Filter operation or maintenance problems in such systems may not be detected until a *Giardia* outbreak is recognized in the community. The bottom line is that although filtration is the best that water treatment technology has to offer for municipal water systems against waterborne giardiasis, it is not infallible. For municipal water filtration facilities to work properly, they must be properly constructed, operated, and maintained.

Whenever possible, persons outdoors should carry drinking water of known purity with them. When this is not practical, and when water from streams, lakes, ponds, and other outdoor sources must be used, time should be taken to properly disinfect the water before drinking it.

CRYPTOSPORIDIUM

Ernest E. Tyzzer first described the protozoan parasite *Cryptosporidium* in 1907. He frequently found the parasite in the gastric glands of laboratory mice, identified the parasite as a sporozoan of uncertain taxonomic status and named it *Cryptosporidium muris*. Later, in 1910, after a more detailed study, he proposed *Cryptosporidium* as a new genus and *C. muris* as the type species. Amazingly, except for developmental stages, Tyzzer's original description of the life cycle (see Figure 12.9) was later confirmed by electron microscopy. In 1912, Tyzzer described a new species, *Cryptosporidium parvum* (Tyzzer 1912).

For almost 50 years, Tyzzer's discovery of the genus *Cryptosporidium* remained (like himself) relatively obscure because it appeared to be of no medical or economic importance. Slight rumblings of the genus's importance were felt in the medical community when Slavin (1955) wrote about a new species, *Cryptosporidium meleagridis,* associated with illness and death in turkeys. Interest remained slight even when *Cryptosporidium* was found to be associated with bovine diarrhea (Panciera et al, 1971).

Not until 1982 did worldwide interest focus on the study of organisms in the genus *Cryptosporidium*. During this period, the medical community and other interested parties began a full-scale, frantic effort to find out as much as possible about Acquired Immune Deficiency Syndrome (AIDS). The CDC reported that 21 AIDS-infected males from six large cities in the United States had severe protracted diarrhea caused by *Cryptosporidium*. It was in 1993, though, that when the "bug – the pernicious parasite *Cryptosporidium* – made [itself and] Milwaukee famous" (Mayo Foundation 1996).

Note: The *Cryptosporidium* outbreak in Milwaukee caused the deaths of 100 people —the largest episode of waterborne disease in the United States in the 70 years since health officials began tracking such outbreaks.

The massive waterborne outbreak in Milwaukee (more than 400,000 persons developed acute and often prolonged diarrhea or other gastrointestinal symptoms) increased interest in *Cryptosporidium* at an exponential level. The Milwaukee Incident spurred both public interest and the interest of public health agencies, agricultural agencies and groups, environmental agencies and groups, and suppliers of drinking water. This increase in interest level and concern has spurred new studies of *Cryptosporidium* with an emphasis on developing methods for recovery, detection, prevention, and treatment (Fayer et al. 1997).

The USEPA has become particularly interested in this "new" pathogen. For example, in the reexamination of regulations on water treatment and disinfection, the USEPA issued MCLG and CCL for *Cryptosporidium*. The similarity to *Giardia lamblia* and the necessity to provide an efficient conventional water treatment capable of eliminating viruses at the same time forced the USEPA to regulate the surface water supplies in particular. The proposed "Enhanced Surface Water Treatment Rule" (ESWTR) included regulations from watershed protection to the specialized operation of treatment plants (certification of operators and state overview) and effective chlorination. Protection against *Cryptosporidium* included control of waterborne pathogens such as *Giardia* and viruses (De Zuane 1997).

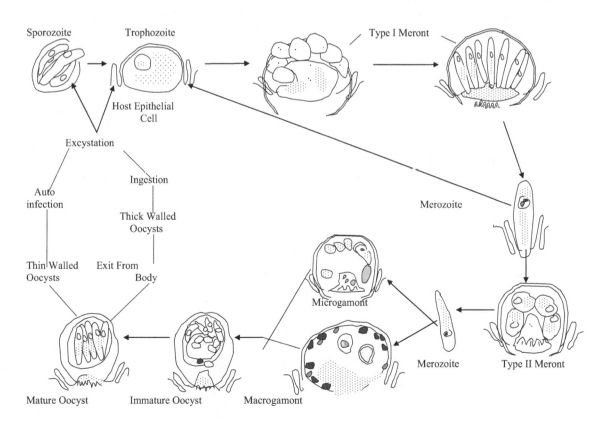

FIGURE 12.9 Life cycle of *Cryptosporidium parvum*.

The Basics of Cryptosporidium

Cryptosporidium (crip-toe-spor-ID-ee-um) is one of several single-celled protozoan genera in the phylum Apicomplexa (all referred to as coccidian). *Cryptosporidium*, along with other genera in the phylum Apircomplexa, develops in the gastrointestinal tract of vertebrates throughout its life cycle—in short, it lives in the intestines of animals and people. This microscopic pathogen causes a disease called *cryptosporidiosis* (crip-toe-spor-id-ee-O-sis). The dormant (inactive) form of *Cryptosporidium*, called an oocyst (O-o-sist), is excreted in the feces (stool) of infected humans and animals. The tough-walled oocysts survive under a wide range of environmental conditions.

Several species of *Cryptosporidium* were incorrectly named after the host in which they were found; subsequent studies have invalidated many species. Now, eight valid species of Cryptosporidium (see Table 12.3) have been named. Upton (1997) reports that *C. muris* infects the gastric glands of laboratory rodents and several other mammalian species but, (even though several texts state otherwise) is not known to infect humans. However, *C. parvum* infects the small intestine of an unusually wide range of mammals, including humans, and is the zoonotic species responsible for human Cryptosporidiosis. In most mammals, *C. parvum* is predominantly a parasite of neonate (newborn) animals. He points out that even though exceptions occur, older animals generally develop poor infections, even when unexposed previously to the parasite. Humans are the only host that can be seriously infected at any time in their lives, and only previous exposure to the parasite results in either full or partial immunity to challenge infections.

Oocysts are present in most surface bodies of water across the United States, many of which supply public drinking water. Oocysts are more prevalent in surface waters when heavy rains increase runoff of wild and domestic animal wastes from the land or when sewage treatment plants are overloaded or break down. Only laboratories with specialized capabilities can detect the presence of *Cryptosporidium* oocysts in water. Unfortunately, present sampling and detection methods are unreliable. Recovering oocysts trapped in the material used to filter water samples is difficult. Once a sample is obtained, however, determining whether the oocyst is alive or whether it is the species *C. parvum* that can infect humans is easily accomplished by looking at the sample under a microscope.

The number of oocysts detected in raw (untreated) water varies with location, sampling time, and laboratory methods. Water treatment plants remove most, but not always all, oocysts. Low numbers of oocysts are sufficient to cause cryptosporidiosis, but the low numbers of oocysts sometimes present in drinking water are not considered a cause for alarm in the public.

Protecting water supplies from *Cryptosporidium* demands multiple barriers. Why? *Cryptosporidium* oocysts have tough walls that can withstand many environmental stresses and are resistant to chemical disinfectants such as chlorine that are traditionally used in municipal drinking water systems.

Physical removal of particles, including oocysts, from water by filtration is an important step in the water treatment process. Typically, water pumped from rivers or lakes into a treatment plant is mixed with coagulants, which help settle out particles suspended in the water. If sand filtration is used, even more particles are removed. Finally, the clarified water is disinfected and piped to customers. Filtration is the only conventional method now in use in the United States for controlling Cryptosporidium.

Ozone is a strong disinfectant that kills protozoa if sufficient doses and contact times are used, but ozone leaves no residual for killing microorganisms in the distribution system, as does chlorine. The high costs of new filtration or ozone treatment plants must be weighed against the benefits of additional treatment. Even well-operated water treatment plants cannot ensure that drinking water will be completely free of *Cryptosporidium* oocysts. Water treatment methods alone cannot solve the problem; watershed protection and monitoring of water quality are critical. As mentioned, watershed protection is another barrier to *Cryptosporidium* in drinking water. Land use controls such as septic system regulations and best management practices to control runoff can help keep human and animal wastes out of water.

Under the Surface Water Treatment Rule of 1989, public water systems must filter surface water sources unless water quality and disinfection requirements are met and a watershed control program is maintained. This rule, however, did not address *Cryptosporidium*. The USEPA has now set standards for turbidity (cloudiness) and coliform bacteria (which indicate that pathogens are probably present) in drinking water. Frequent monitoring must occur to provide officials with early warning of potential problems to enable them to take steps to protect public health. Unfortunately, no water quality indicators can reliably predict the occurrence of cryptosporidiosis. More accurate and rapid assays of oocysts will make it possible to notify residents promptly if their water supply is contaminated with *Cryptosporidium*, averting outbreaks.

The Bottom Line: The collaborative efforts of water utilities, government agencies, health care providers, and individuals are needed to prevent outbreaks of cryptosporidiosis.

TABLE 12.3
Valid Named Species of *Cryptosporidium* (Fayer et al. 1997)

Species	Host
C. baileyi	Chicken
C. felis	Domestic cat
C. meleagridis	Turkey
C. murishouse	House mouse
C. nasorium	Fish
C. parvum	House mouse
C. serpentis	Corn snake
C. wrairi	Guinea pig

Cryptosporidiosis

Juranek (1995) wrote in the journal *Clinical Infectious Diseases:*

> Cryptosporidium *parvum* is an important emerging pathogen in the U.S. and a cause of severe, life-threatening disease in patients with AIDS. No safe and effective form of specific treatment for Cryptosporidiosis has been identified to date. The parasite is transmitted by ingestion of oocysts excreted in the feces of infected humans or animals. The infection can therefore be transmitted from person-to-person, through ingestion of contaminated water (drinking water and water used for recreational purposes) or food, from animal to person, or by contact with fecally contaminated environmental surfaces. Outbreaks associated with all of these modes of transmission have been documented. Patients with human immunodeficiency virus infection should be made more aware of the many ways that *Cryptosporidium* species are transmitted, and they should be given guidance on how to reduce their risk of exposure.

Since the Milwaukee outbreak, concern about the safety of drinking water in the United States has increased, and new attention has been focused on determining and reducing the risk of Cryptosporidiosis from the community and municipal water supplies. Cryptosporidiosis is spread by putting something in the mouth that has been contaminated with the stool of an infected person or animal. In this way, people swallow the *Cryptosporidium* parasite. As previously mentioned, a person can become infected by drinking contaminated water or eating raw or undercooked food contaminated with *Cryptosporidium* oocysts; direct contact with the droppings of infected animals or stools of infected humans; or hand-to-mouth transfer of oocysts from surfaces that may have become contaminated with microscopic amounts of stool from an infected person or animal.

The symptoms may appear 2–10 days after infection by the parasite. Although some persons may not have symptoms, others have watery diarrhea, headache, abdominal cramps, nausea, vomiting, and low-grade fever. These symptoms may lead to weight loss and dehydration. In otherwise healthy persons, these symptoms usually last 1–2 weeks, at which time the immune system can stop the infection. In persons with suppressed immune systems, such as those who have AIDS or who have recently had an organ or bone marrow transplant, the infection may continue and become life-threatening.

Currently, no safe and effective cure for Cryptosporidiosis exists. People with normal immune systems improve without taking antibiotic or antiparasitic medications. The treatment recommended for this diarrheal illness is to drink plenty of fluids and to get extra rest. Physicians may prescribe medication to slow the diarrhea during recovery.

The best way to prevent Cryptosporidiosis is to:

- Avoid water or food that may be contaminated.
- Wash hands after using the toilet and before handling food.

- If you work in a childcare center where you change diapers, be sure to wash your hands thoroughly with plenty of soap and warm water after every diaper change, even if you wear gloves.

During community-wide outbreaks caused by contaminated drinking water, drinking water practitioners should inform the public to boil drinking water for 1 minute to kill the *Cryptosporidium* parasite.

CYCLOSPORA

Cyclospora organisms, which until recently were considered blue-green algae, were discovered at the turn of the century. The first human cases of *Cyclospora* infection were reported in the 1970s. In the early 1980s, *Cyclospora* was recognized as a pathogen in patients with AIDS. We now know that *Cyclospora* is endemic in many parts of the world and appears to be an important cause of traveler's diarrhea. *Cyclospora* are two to three times larger than *Cryptosporidium* but otherwise have similar features. *Cyclospora* diarrheal illness in patients with healthy immune systems can be cured with a week of therapy with trimethoprim-sulfamethoxazole (TMP-SMX).

So, exactly what is *Cyclospora?* In 1998, the CDC described *Cyclospora cayetanensis* as a unicellular parasite previously known as a cyanobacterium-like (blue-green algae-like) or coccidian-like body (CLB). The disease is known as cyclosporiasis. *Cyclospora* infects the small intestine and causes an illness characterized by diarrhea with frequent stools. Other symptoms can include loss of appetite, bloating, gas, stomach cramps, nausea, vomiting, fatigue, muscle aches, and fever. Some individuals infected with Cyclospora may not show symptoms. Since the first known cases of illness caused by *Cyclospora* infection were reported in medical journals in the 1970s, cases have been reported with increased frequency from various countries since the mid-1980s (in part because of the availability of better techniques for detecting the parasite in stool specimens).

Huang et al. (1995) detailed what they believe is the first known outbreak of diarrheal illness associated with *Cyclospora* in the United States. The outbreak, which occurred in 1990, consisted of 21 cases of illness among physicians and others working at a Chicago hospital. Contaminated tap water from a physicians' dormitory at the hospital was the probable source of the organisms. The tap water probably picked up the organism while in a storage tank at the top of the dormitory after the failure of a water pump.

The transmission of *Cyclospora* is not a straightforward process. When infected persons excrete the oocyst state of *Cyclospora* in their feces, the oocysts are not infectious and may require days to weeks to become so (i.e., to sporulate). Therefore, transmission of *Cyclospora* directly from an infected person to someone else is unlikely. However, indirect transmission can occur if an infected person contaminates the environment and the oocysts have sufficient time, under appropriate conditions, to become infectious. For example, *Cyclospora* may be transmitted by ingestion of water or food contaminated with oocysts. Outbreaks linked to contaminated water, as well as outbreaks linked to various types of fresh produce, have been reported in recent years (Herwaldt et al. 1997). How common the various modes of transmission and sources of infection are is not yet known, nor is it known whether animals can be infected and serve as sources of infection for humans.

Note: *Cyclospora* organisms have not yet been grown in tissue cultures or laboratory animal models.

Persons of all ages are at risk for infection. Persons living in or traveling to developing countries may be at increased risk; however, infection can be acquired worldwide, including in the United States. In some countries, infection appears to be seasonal. Based on currently available information, avoiding water or food that may be contaminated with stool is the best way to prevent infection. Reinfection can occur.

Note: De Zuane (1997) points out that pathogenic parasites are not easily removed or eliminated by conventional treatment and disinfection unit processes. This is particularly true for *Giardia lamblia, Cryptosporidium,* and *Cyclospora.* Filtration facilities can be adjusted in terms of depth, prechlorination, filtration rate, and backwashing to become more effective in the removal of cysts. The pretreatment of protected watershed raw water is a major factor in the elimination of pathogenic protozoa.

HELMINTHS

Helminths are parasitic worms that grow and multiply in sewage (biological slimes) and wet soil (mud). They multiply in wastewater treatment plants; as strict aerobes, they have been found in activated sludge and particularly in trickling filters, and therefore appear in large concentrations in treated domestic liquid waste. They enter the skin or are ingested as worms in their many lifecycle phases. Generally, they are not a problem in drinking water supplies in the United States because both their egg and larval forms are large enough to be trapped during conventional water treatment. In addition, most helminths are not waterborne, so the chances of infection are minimized (WHO 1996).

WASTEWATER TREATMENT AND BIOLOGICAL PROCESSES

Uncontrolled bacteria in industrial water systems produce an endless variety of problems including disease, equipment damage, and product damage. Unlike the microbiological problems that can occur in water systems, in wastewater treatment, microbiology can be applied as a beneficial science for the destruction of pollutants in wastewater (Kemmer 1979).

It should be noted that all the biological processes used for the treatment of wastewater (in particular) are derived from or modeled on processes occurring naturally in nature. The processes discussed in the following are typical examples. It should also be noted that "by controlling the environment of microorganisms, the decomposition of wastes is speeded up. Regardless of the type of waste, the biological treatment process consists of controlling the environment required for optimum growth of the microorganism involved" (Metcalf & Eddy 2003).

AEROBIC PROCESS

In *aerobic treatment processes,* organisms use free, elemental oxygen, organic matter, nutrients (nitrogen, phosphorus), and trace metals (iron, etc.) to produce more organisms and stable dissolved and suspended solids, as well as carbon dioxide (see Figure 12.10).

ANAEROBIC PROCESS

The *anaerobic treatment process* consists of two steps, occurs completely in the absence of oxygen, and produces a usable by-product, methane gas. In the first step of the process, facultative microorganisms use the organic matter as food to produce more organisms, volatile (organic) acids, carbon dioxide, hydrogen sulfide, other gases, and some stable solids (see Figure 12.11).

In the second step, anaerobic microorganisms use the volatile acids as their food source. The process produces more organisms, stable solids, and methane gas, which can be used to provide energy for various treatment system components (see Figure 12.12).

Oxygen		More bacteria
Bacteria	⇒	Stable solids
Organic matter		Settleable solids
Nutrients		Carbon dioxide

FIGURE 12.10 Aerobic decomposition.

Facultative Bacteria		More Bacteria
Organic matter	⇒	Volatile solids
Nutrients		Settleable solids
		Hydrogen sulfide

FIGURE 12.11 Anaerobic decomposition—first step.

Anaerobic bacteria		More bacteria
Volatile acids	⇒	Stable solids
Nutrients		Settleable solids
		Methane

FIGURE 12.12 Anaerobic decomposition—second step.

ANOXIC PROCESS

In the *anoxic treatment process* (anoxic means without oxygen), microorganisms use fixed oxygen in nitrate compounds as a source of energy. The process produces more organisms and removes nitrogen from the wastewater by converting it to nitrogen gas, which is released into the air (see Figure 12.13).

PHOTOSYNTHESIS

Green algae use carbon dioxide and nutrients in the presence of sunlight and chlorophyll to produce more algae and oxygen (see Figure 12.14).

GROWTH CYCLES

All organisms follow a basic growth cycle that can be represented as a growth curve. This curve occurs when the environmental conditions required for the particular organism are reached. It is the environmental conditions (i.e., oxygen availability, pH, temperature, presence or absence

Nitrate oxygen		More bacteria
Bacteria		Stable solids
Organic matter	⇒	Settleable solids
Nutrients		Nitrogen

FIGURE 12.13 Anoxic decomposition.

Sun		
Algae	⇒	More algae
Carbon dioxide		Oxygen
Nutrients		

FIGURE 12.14 Photosynthesis.

of nutrients, presence or absence of toxic materials) that determine when a particular group of organisms will predominate. Obviously, this information can be very useful in operating a biological treatment process (see Figure 12.15).

BIOGEOCHEMICAL CYCLES

Several chemicals are essential to life and follow predictable cycles through nature. In these natural cycles or *biogeochemical cycles,* the chemicals are converted from one form to another as they progress through the environment. The water/wastewater operator should be aware of those cycles dealing with the nutrients (e.g., carbon, nitrogen, and sulfur) because they have a major impact on the performance of the plant and may require changes in operation at various times of the year to keep them functioning properly; this is especially the case in wastewater treatment. The microbiology of each cycle deals with the biotransformation and subsequent biological removal of these nutrients in wastewater treatment plants.

Note: Smith categorizes biogeochemical cycles into two types, the *gaseous* and the *sedimentary.* Gaseous cycles include the carbon and nitrogen cycles. The main sink of nutrients in the gaseous cycle is the atmosphere and the ocean. Sedimentary cycles include the sulfur cycle. The main sink for sedimentary cycles is soil and rocks of the earth's crust (Smith 1974).

CARBON CYCLE

Carbon, which is an essential ingredient of all living things, is the basic building block of the large organic molecules necessary for life. Carbon is cycled into food chains from the atmosphere, as shown in Figure 12.16. From Figure 12.16, it can be seen that green plants obtain carbon dioxide (CO_2) from the air and, through photosynthesis, described by Asimov (1989) as the "most important chemical process on Earth," produce the food and oxygen that all organisms live on. Part of the carbon produced remains in

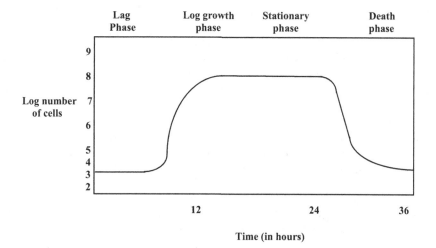

FIGURE 12.15 Microorganism growth curve.

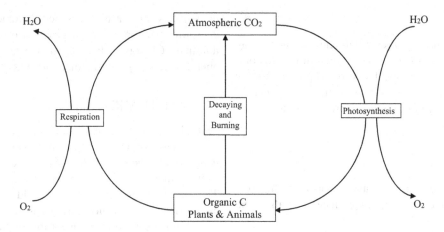

FIGURE 12.16 Carbon cycle.

living matter; the other part is released as CO_2 in cellular respiration. Miller (1988) points out that the carbon dioxide released by cellular respiration in all living organisms is returned to the atmosphere.

Some carbon is contained in buried dead animal and plant materials. Much of these buried materials were transformed into fossil fuels. Fossil fuels, such as coal, oil, and natural gas, contain large amounts of carbon. When fossil fuels are burned, stored carbon combines with oxygen in the air to form carbon dioxide, which enters the atmosphere. In the atmosphere, carbon dioxide acts, as a beneficial heat screen, as it does not allow the radiation of Earth's heat into space. This balance is important. The problem is that as more carbon dioxide from burning is released into the atmosphere, the balance can and is being altered. Odum (1983) warns that the recent increase in the consumption of fossil fuels, "coupled with the decrease in 'removal capacity' of the green belt is beginning to exceed the delicate balance." Massive increases in carbon dioxide in the atmosphere tend to increase the possibility of global warming. The consequences of global warming "would be catastrophic … and the resulting climatic change would be irreversible" (Abrahamson 1988).

NITROGEN CYCLE

Nitrogen is an essential element that all organisms need. In animals, nitrogen is a component of crucial organic molecules such as proteins and DNA and constitutes 1%–3% dry weight of cells. Our atmosphere contains 78% nitrogen by volume, yet it is not a common element on Earth. Although nitrogen is an essential ingredient for plant growth, it is chemically very inactive, and before the vast majority of the biomass can incorporate it, it must be *fixed*. Special nitrogen-fixing bacteria found in soil and water fix nitrogen. Thus, microorganisms play a major role in nitrogen cycling in the environment. These microorganisms (bacteria) can take nitrogen gas from the air and convert it to nitrate. This is called *nitrogen fixation*. Some of these bacteria occur as free-living organisms in the soil, while others live in a

symbiotic relationship (a close relationship between two organisms of different species, where both partners benefit from the association) with plants. An example of a symbiotic relationship related to nitrogen can be seen in the roots of peas. These roots have small swellings along their length, which contain millions of symbiotic bacteria that can take nitrogen gas from the atmosphere and convert it to nitrates that can be used by the plant. Then the plant is plowed into the soil after the growing season to improve the nitrogen content. Price (1984) describes the nitrogen cycle as an example "of a largely complete chemical cycle in ecosystems with little leaching out of the system." Simply, the nitrogen cycle provides various bridges between the atmospheric reservoirs and the biological communities (see Figure 12.17).

Atmospheric nitrogen is fixed either by natural or industrial means. For example, nitrogen is fixed by lightning or by soil bacteria that convert it to ammonia, then to nitrite, and finally to nitrates, which plants can use. Nitrifying bacteria make nitrogen from animal wastes. Denitrifying bacteria convert nitrates back to nitrogen and release it as nitrogen gas.

The logical question now is, "What does all of this have to do with water?" The best way to answer this question is to ask another question: Have you ever dived into a slow-moving stream and had the noxious misfortune to surface right in the middle of an algal bloom? When this happens to you, the first thought that runs through your mind is, "Where is my nose plug?" Why? Because of the horrendous stench, disablement of the olfactory sense is a necessity.

If too much nitrate, for example, enters the water supply—as runoff from fertilizers—it produces an overabundance of algae, called an algal bloom. If this runoff from fertilizer gets into a body of water, algae may grow so profusely that they form a blanket over the surface. This usually happens in summer when the light levels and warm temperatures favor rapid growth.

In their voluminous and authoritative text, *Wastewater Engineering: Treatment, Disposal, & Reuse* (Metcalf &

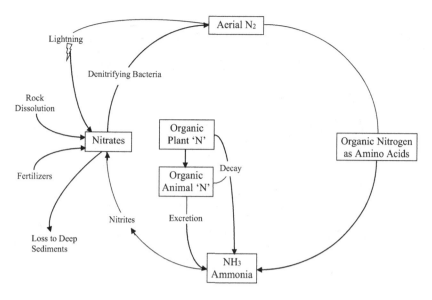

FIGURE 12.17 Nitrogen cycle.

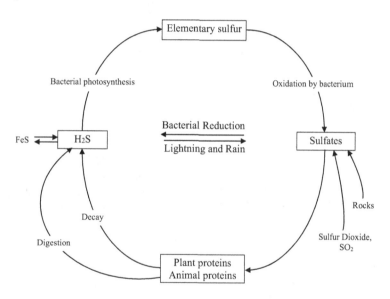

FIGURE 12.18 Sulfur cycle.

Eddy 2003), it is noted that nitrogen is found in wastewater in the form of urea. During wastewater treatment, the urea is transformed into ammonia nitrogen. Since ammonia exerts a BOD and chlorine demand, high quantities of ammonia in wastewater effluents are undesirable. The process of nitrification is utilized to convert ammonia to nitrates. *Nitrification* is a biological process that involves the addition of oxygen to the wastewater. If further treatment is necessary, another biological process called *denitrification* is used. In this process, nitrate is converted into nitrogen gas, which is lost to the atmosphere, as can be seen in Figure 12.17. From the wastewater operator's point of view, nitrogen and phosphorus are both considered limiting factors for productivity. Phosphorus discharged into streams contributes to pollution. Of the two, nitrogen is harder to control but is found in smaller quantities in wastewater.

SULFUR CYCLE

Sulfur, like nitrogen, is characteristic of organic compounds. The *sulfur cycle* is both sedimentary and gaseous (see Figure 12.18). The principal forms of sulfur that are of special significance in water quality management are organic sulfur, hydrogen sulfide, elemental sulfur, and sulfate (Tchobanoglous and Schroeder 1985). Bacteria play a major role in the conversion of sulfur from one form to another. In an anaerobic environment, bacteria break down organic matter, thereby producing hydrogen sulfide with its characteristic rotten-egg odor. A bacterium called *Beggiatoa* converts hydrogen sulfide into elemental sulfur and then into sulfates. Other sulfates are contributed by the dissolving of rocks and some sulfur dioxide. Sulfur is incorporated by plants into proteins. Organisms then consume some of these plants. Many heterotrophic anaerobic bacteria liberate sulfur from proteins, such as hydrogen sulfide.

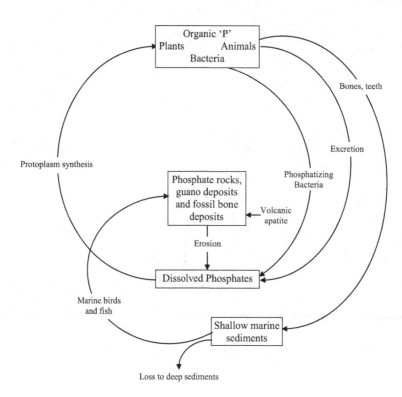

FIGURE 12.19   Phosphorus cycle.

Phosphorus Cycle

Phosphorus is another chemical element common in the structure of living organisms (see Figure 12.19). However, the phosphorus cycle is different from the hydrologic, carbon, and nitrogen cycles because phosphorus is found in sedimentary rock. These massive deposits are gradually eroding to provide phosphorus to ecosystems. A large amount of eroded phosphorus ends up in deep sediments in the oceans and in lesser amounts in shallow sediments. Part of the phosphorus comes to land when marine animals surface. Decomposing plant or animal tissue and animal droppings return organic forms of phosphorus to the water and soil. For example, fish-eating birds play a role in the recovery of phosphorus. The guano deposit, or bird excreta, on the Peruvian coast is an example. Humans have hastened the rate of phosphorus loss through mining and the production of fertilizers, which are washed away and lost. Odum (1983) suggested, however, that there is no immediate cause for concern since the known reserves of phosphate are quite large.

Phosphorus has become very important in water quality studies because it is often found to be a limiting factor. Metcalf & Eddy Inc. (2003) reported that control of phosphorus compounds that enter surface waters and contribute to the growth of algal blooms is of considerable interest and has generated much study. Phosphorus, upon entering a stream, acts as a fertilizer, promoting the growth of undesirable algae populations or algal blooms. As the organic matter decays, DO levels decrease and fish and other aquatic species die.

Although it is true that phosphorus discharged into streams is a contributing factor to stream pollution, it is also true that phosphorus is not the lone factor. Odum (1975) warns against what he calls the one-factor control hypothesis (i.e., one-problem/one-solution syndrome). He notes that environmentalists in the past have focused on one or two items, like phosphorus contamination, and have failed to understand that the strategy for pollution control must involve reducing the input of all enriching and toxic materials.

CHAPTER REVIEW QUESTIONS

12.1 The three major groups of microorganisms that cause disease in water are:

12.2 When does a river of good quality show its highest bacterial numbers?

12.3 Are coliform organisms pathogenic?

12.4 How do bacteria reproduce?

12.5 The three common shapes of bacteria are:

12.6 Three waterborne diseases caused by bacteria are:

12.7 Two protozoa-caused waterborne diseases are:

12.8 When a protozoon is in a resting phase, it is called a _____.

12.9 For a virus to live, it must have a _____.

12.10 What problems do algae cause in drinking water?

REFERENCES

Abrahamson, D.E., Ed., 1988. *The Challenge of Global Warming.* Washington, DC: Island Press.

Asimov, I., 1989. *How Did We Find Out About Photosynthesis?* New York: Walker & Company.

Black, R.E., Dykes, A.C., Anderson, K.E., Wells, J.G., Sinclair, S.P., Gary, G.W., Hatch, M.H., and Ginagaros, E.J., 1981. Handwashing to prevent diarrhea in day-care centers. *Am J Epidmilo;* 113:445–451.

Brodsky, R.E., Spencer, H.C., and Schultz, M.G., 1974. Giardiasis in American travelers to the Soviet Union. *J Infect Dis;* 130:319–323.

CDC. 1979. *Intestinal Parasite Surveillance, Annual Summary 1978.* Atlanta, Georgia: Centers for Disease Control.

CDC. 1983. *Water-Related Disease Outbreaks Surveillance, Annual Summary 1983.* Atlanta, Georgia: Centers for Disease Control.

CDC. 1995. *Giardiasis,* Juranek, D.D. (Ed.). Atlanta, Georgia: Centers for Disease Control.

De Zuane, J., 1997. *Handbook of Drinking Water Quality.* New York: John Wiley & Sons, Inc.

Fayer, R., Speer, C.A., and Dudley, J.P., 1997. *The General Biology Cryptosporidium in Cryptosporidium and Cryptosporidiosis,* Fayer, R., (ed.). Boca Raton, FL: CRC Press.

Frost, F., Plan, B., and Liechty, B., 1984. Giardia prevalence in commercially trapped mammals, *J Environ Health;* 42:245–249.

Herwaldt, F.L., et al., 1997. An outbreaking 1996 of Cyclosporasis associated with imported raspberries. *N Engl J Med;* 336:1548–1556.

Huang, P., Weber, J.T. Sosin, D.M., et al., 1995. Cyclospora. *Ann Intern Med;* 123:401–414.

Jarroll, E.L., Jr., Gingham, A.K, and Meyer, E.A., 1979. Giardia cyst destruction; effectiveness of six small-quantity water disinfection methods. *Am J Trop Med Hygiene;* 29:8–11.

Juranek, D.D., 1995. Cryptosporidium parvum. *Clinical Infect Dis;* 21:S37–S61.

Kemmer, F.N., 1979. *Water: The Universal Solvent,* 2nd ed. Oak Brook, IL: Nalco Chemical Company.

Kordon, C., 1992. *The Language of the Cell.* New York: McGraw-Hill.

Koren, H., 1991. *Handbook of Environmental Health and Safety: Principles and Practices.* Chelsea, MI: Lewis Publishers.

LeChevallier, M.W., Norton, W.D., and Less, R.G., 1991. Occurrences of Giardia and Cryptosporidium spp. in surface water supplies. *Appl Environ Microbiol;* 57:2610–2616.

Mayo Foundation. 1996. *The "Bug" That Made Milwaukee Famous.* Rochester, MN: Mayo Foundation.

Metcalf & Eddy, Inc., 2003. *Wastewater Engineering: Treatment, Disposal and Reuse.* New York: McGraw-Hill, Inc.

Miller, G.T., 1988. *Environmental Science: An Introduction.* Belmont, CA: Wadsworth Publishing Company.

Odum, E.P., 1975. *Ecology: The Link between the Natural and the Social Sciences.* New York: Holt, Rinehart and Winston, Inc.

Odum, E.P., 1983. *Basic Ecology.* Philadelphia, PA: Saunders College Publishing.

Panciera, R.J., Thomassen, R.W., and Garner, R.M., 1971. Cryptosporidial infection in a calf. *Vet Pathol;* 8:479.

Pennak, R.W., 1989. *Fresh-Water Invertebrates of the United States,* 3rd ed. New York: John Wiley & Sons.

Price, P.W., 1984. *Insect Ecology.* New York: John Wiley & Sons.

Rendtorff, R.C., 1954. The experimental transmission of human intestinal protozoan parasites II *Giardia lamblia* cysts given in capsules. *Am J Hygiene;* 59:209–220.

Rose, J.B., Gerb, C.P., and Jakubowski, W., 1983. Survey of potable water supplies for Cryptosporidium and Giardia. *Environ Sci Technol;* 25:1393–1399.

Singleton, P., and Sainsbury, D., 1994. *Dictionary of Microbiology and Molecular Biology,* 2nd ed. New York: John Wiley & Sons.

Slavin, D., 1955. *Cryptosporidium melagridis. J Comp Pathol;* 65:262.

Smith, R.L., 1974. *Ecology and Field Biology.* New York: Harper & Row.

Spellman, F.R., 1996. *Stream Ecology and Self-Purification: An Introduction for Wastewater and Water Specialists.* Boca Raton, FL: CRC Press.

Spellman, F.R., 1997. *Microbiology for Water/Wastewater Operators.* Boca Raton, FL: CRC Press.

Tchobanoglous, G., and Schroeder, E.D., 1985. *Water Quality.* Reading, MA: Addison-Wesley Publishing Company.

Tyzzer, E.E., 1912. Cryptosporidium parvum sp.: a Coccidium found in the small intestine of the common mouse. *Arch Protistenkd;* 26:394 m.

Upton, S.J., 1997. *Basic Biology of Cryptosporidium.* Manhattan, Kansas: Kansas State University.

Walsh, J.D., and Warren, K.S., 1979. Selective primary health care: an interim strategy for disease control in developing countries. *N Engl J Med;* 301:974–976.

Weller, P.F., 1985. Intestinal protozoa: Giardiasis. *Sci Am Med;* 273(3):13–27.

WHO. 1990. *Guidelines for Drinking Water Quality,* 2nd ed., Vol. 2. Austria: World Health Organization.

Wistriech, G.A., and Lechtman, M.D., 1980. *Microbiology,* 3rd ed. New York: Macmillan Publishing Company.

13 Water Ecology

INTRODUCTION

The "control of nature" is a phrase conceived in arrogance, born of the Neanderthal age of biology and the convenience of man.

—*Rachel Carson (1962)*

What is ecology? Why is ecology important? Why study ecology? These are all simple, straightforward questions; however, providing simple, straightforward answers is not that easy, notwithstanding the inherent difficulty of explaining any complex science in simple, straightforward terms, which is the purpose, goal, and mission of this text and this chapter. In short, the task of this chapter is to outline basic information that explains the functions and values of ecology and its interrelationships with other sciences, including ecology's direct impact on our lives. In doing so, the author hopes not only to dispel the common misconception that ecology is too difficult for the average person to understand but also to instill the concept of ecology as an asset that can not only be learned but also cherished.

WHAT IS ECOLOGY?

Ecology can be defined in various ways. For example, ecology, or ecological science, is commonly defined in the literature as the scientific study of the distribution and abundance of living organisms and how these factors are affected by interactions between the organisms and their environment. The term *ecology* was coined in 1866 by the German biologist Haeckel, and it loosely means "the study of the household [of nature]." Odum (1983) explains that the word "ecology" is derived from the Greek oikos, meaning home. Ecology, then, refers to the study of organisms at home. It means the study of an organism in its home. In a broader sense, ecology is the study of the relationships of organisms or groups to their environment.

Important Point: No ecosystem can be studied in isolation. If we were to describe ourselves, our histories, and what made us the way we are, we could not leave the world around us out of our description! So, it is with streams: they are directly tied to the world around them. They take their chemistry from the rocks and dirt beneath them as well as from a great distance around them (Spellman 1996).

Charles Darwin (1998) explained ecology in a famous passage in *The Origin of Species*, a passage that helped establish the science of ecology. According to Darwin, a "web of complex relations" binds all living things in any region, he writes. Adding or subtracting even a single species causes waves of change that race through the web,

"onwards in ever-increasing circles of complexity." The simple act of adding cats to an English village would reduce the number of field mice. Killing mice would benefit the bumblebees, whose nests and honeycombs the mice often devour. Increasing the number of bumblebees would benefit the heartsease and red clover, which are fertilized almost exclusively by bumblebees. So, adding cats to the village could ultimately lead to more flowers. For Darwin, the whole of the Galapagos archipelago illustrates this fundamental lesson. The island volcanoes are much more diverse in their ecology than in their biology. The contrast suggests that in the struggle for existence, species are shaped at least as much by the local flora and fauna as by the local soil and climate. "Why else would the plants and animals differ radically among islands that have the same geological nature, the same height, and climate?" (Darwin 1998).

Probably the best way to understand ecology—to get a really good "feel" for it or to get to the heart of what ecology is all about—is to read the following by Rachel Carson (1962):

> We poison the caddis flies in a stream and the salmon runs dwindle and die.
>
> We poison the gnats in a lake and the poison travels from link to link of the food chain and soon the birds of the lake margins become victims. We spray our elms and the following springs are silent of robin song, not because we sprayed the robins directly but because the poison traveled, step by step, through the now familiar elm leaf-earthworm-robin cycle. These are matters of record, observable, part of the visible world around us. They reflect the web of life—or death—that scientists know as ecology.

As Carson pointed out, what we do to any part of our environment has an impact on other parts. In other words, there is an interrelationship between the parts that make up our environment. Probably the best way to state this interrelationship is to define ecology definitively—that is, to define it as it is used in this text: "Ecology is the science that deals with the specific interactions that exist between organisms and their living and nonliving environments" (Tomera 1989).

When environment is mentioned in the preceding context and as it is discussed throughout this text, it (the environment) includes everything important to the organism in its surroundings. The organism's environment can be divided into four parts:

1. Habitat and distribution—its place to live
2. Other organisms—whether friendly or hostile
3. Food
4. Weather—light, moisture, temperature, soil, etc.

DOI: 10.1201/9781003581901-16

The four major subdivisions of ecology are:

- Behavioral Ecology
- Population Ecology (Autecology)
- Community Ecology (Synecology)
- Ecosystem Ecology

Behavioral ecology is the study of the ecological and evolutionary basis for animal behavior. *Population ecology* (or autecology) is the study of the individual organism or a species. It emphasizes life history, adaptations, and behavior. It is also the study of communities, ecosystems, and the biosphere. An example of autecology would be when biologists spend their entire lifetime studying the ecology of the salmon. *Community ecology* (or synecology), on the other hand, is the study of groups of organisms associated together as a unit and deals with the environmental problems caused by humans. For example, the effect of discharging phosphorus-laden effluent into a stream involves several organisms. The activities of human beings have become a major component of many natural areas. As a result, it is important to realize that the study of ecology must involve people. *Ecosystem ecology* is the study of how energy flow and matter interact with the biotic elements of ecosystems (Odum 1971).

Important Point: Ecology is generally categorized according to complexity; the primary kinds of organisms under study (plant, animal, insect ecology); the biomes principally studied (forest, desert, benthic, grassland, etc.); the climatic or geographic area (e.g., Arctic or tropics); and/or the spatial scale (macro or micro) under consideration.

WHY IS ECOLOGY IMPORTANT?

Ecology, in its true sense, is a holistic discipline that does not dictate what is right or wrong. Instead, ecology is important to life on Earth simply because it makes us aware, to a certain degree, of what life on Earth is all about. Ecology shows us that each living organism has an ongoing and continual relationship with every other element that makes up our environment. Simply put, ecology is all about interrelationships, both intraspecific and interspecific, and how important it is to maintain these relationships to ensure our very survival.

At this point in this discussion, there are literally countless examples that could be used to illustrate the importance of ecology and interrelationships. However, to demonstrate not only the importance of ecology but also to point out that an ecological principle can be a double-edged sword, depending on one's point of view (ecological problems along with pollution can be a judgment call—that is, they are a matter of opinion), a famous parable (*The Keeper of the Spring*), adapted from Peter Marshall's (1950) *Mr. Jones: Meet the Master* is used here to demonstrate that many misconceptions exist in ecology.

The Keeper of the Spring

This is the story of the keeper of the spring. He lived high in the Alps above an Austrian town and had been hired by the town council to clear debris from the mountain springs that fed the stream that flowed through the town. The man did his work well and the village prospered. Graceful swans floated in the stream. The surrounding countryside was irrigated. Several mills used the water for power. Restaurants flourished for townspeople and for a growing number of tourists.

Years went by. One evening, at the town council meeting, someone questioned the money being paid to the keeper of the spring. No one seemed to know who he was or even if he was still on the job high up in the mountains. Before the evening was over, the council decided to dispense with the old man's services.

Weeks went by, and nothing seemed to change. Then autumn came. The trees began to shed their leaves. Branches broke and fell into the pools high up in the mountains. Down below, the villagers began to notice the water becoming darker. A foul odor appeared. The swans disappeared. Also, the tourists. Soon, the disease spread through the town.

When the town council reassembled, they realized that they had made a costly error. They found the old keeper of the spring and hired him back again. Within a few weeks, the stream cleared up, and life returned to the village as they had known it before (Marshall 1950).

After reviewing Marshall's parable and the restoration of the spring, the average person might say to him/herself, "Gee, all is well with the town again." For swans, irrigation, hydropower, and pretty views, residents seem to be pleased that the stream was restored to its "normal" state. The trained ecologist, however, would take a different view of this same stream. The ecologist would go beyond the hype (as portrayed in the popular media, including literature) about what a healthy stream is. For example, the trained ecologist would know that a perfectly clean stream, clear of all terrestrial plant debris (woody debris and leaves), would not be conducive to ensuring diverse, productive invertebrates and fish, would not preserve natural sediment and water regimes, and would not ensure overall stream health (Dolloff and Webster 2000).

WHY STUDY ECOLOGY?

Does anyone need to study ecology to observe, relish, feel, and/or sense the real thing? Now, we could be referring to those clear and sunny days and cool, crisp nights of autumn that provide an almost irresistible lure to those of us (ecologists and non-ecologists alike) who enjoy the outdoors. To take in the splendor and delight of autumn's color display, many head for the hills, the mountains, the countryside, lakes, streams, and recreation areas of the National Forests. The more adventurous horseback ride or backpack through Nature's glory and solitude on trails winding deep into forest tranquility—just being outdoors in those golden days rivals any thrill in life. Even those of us fettered to the chains of city life are often exposed to city streets with those columns of life ablaze in color.

No, one need not study ecology to witness, appreciate, and/or understand the enchantment of autumn's annual color display—summer extinguished in a blaze of color. It is a different story, however, for those involved in trying to understand all of the complicated actions—and even more complicated interactions—involving pigments, sunlight, moisture, chemicals, temperatures, site, hormones, length of daylight, genetic traits, and so on that contribute to a perfect autumn color display (USDA 1999). This is the work of the ecologist—to probe deeper and deeper into the basics of nature, constantly seeking answers. To find the answers, the ecologist must be a synthesis scientist; that is, he or she must be a generalist well versed in botany, zoology, physiology, genetics, and other disciplines like geology, physics, and chemistry.

Earlier, we used Marshall's parable to make the point that a clean stream and other downstream water bodies can be a good thing, depending on one's point of view—pollution is a judgment call. It was also pointed out that this view may not be shared by a trained ecologist, especially a stream ecologist. The ecologist knows, for example, that terrestrial plant debris is not only beneficial but that it is absolutely necessary. Why? Consider the following explanation (Spellman 1996).

A stream has two possible sources of primary energy: (1) instream photosynthesis by algae, mosses, and higher aquatic plants, and imported organic matter from streamside vegetation (e.g., leaves and other parts of plants). Simply put, a significant portion of the food eaten by aquatic organisms grows right in the stream, such as algae, diatoms, nymphs, larvae, and fish. This food that originates from within the stream is called autochthonous (Benfield 1996).

Most food in a stream, however, comes from outside the stream. This is especially the case in small, heavily wooded streams, where there is normally insufficient light to support substantial instream photosynthesis; thus, energy pathways are supported largely by imported energy. A large portion of this imported energy is provided by leaves. Worms drown in floods and get washed in. Leafhoppers and caterpillars fall from trees. Adult mayflies and other insects mate above the stream, lay their eggs in it, and then die in it. All of this food from outside the stream is called allochthonous.

Leaf Processing in Streams

Autumn leaves entering streams are nutrient-poor because trees absorb most of the sugars and amino acids (nutrients)

present in the green leaves (Suberkoop et al. 1976). Leaves falling into streams may be transported short distances but are usually caught by structures in the streambed to form leaf packs. These leaf packs are then processed in place by components of the stream communities in a series of well-documented steps (Figure 13.1) (Peterson and Cummins 1974).

Within 24–48 hours of entering a stream, many of the remaining nutrients in leaves leach into the water. After leaching, leaves are composed mostly of structural materials like non-digestible cellulose and lignin. Within a few days, fungi (especially Hyphomycetes), protozoa, and bacteria process the leaves through microbial processing (see Figure 13.1) (Barlocher and Kendrick 1975). Two weeks later, microbial conditioning leads to structural softening of the leaf and, among some species, fragmentation. The reduction in particle size from whole leaves (coarse particulate organic matter, CPOM) to fine particulate organic matter (FPOM) is accomplished mainly through the feeding activities of a variety of aquatic invertebrates collectively known as "shredders" (Cummins 1974; Cummins and Klug 1979). Shredders (stoneflies, for example) help produce fragments shredded from leaves but not ingested, as well as fecal pellets, which reduce the particle size of organic matter. The particles are then collected (by mayflies, for example) and serve as a food resource for a variety of micro- and macroconsumers. Collectors eat what they want and send even smaller fragments downstream. These tiny fragments may be filtered out of the water by a true fly larva (i.e., a filterer). Leaves may also be fragmented by a combination of microbial activity and physical factors such as current and abrasion (Benfield et al. 1977; Paul et al. 1978).

Leaf-pack processing by all the elements mentioned above (i.e., leaf species, microbial activity, and physical and chemical features of the stream) is important. However, the most important point is that these integrated ecosystem processes convert whole leaves into fine particles, which are then distributed downstream and used as an energy source by various consumers.

The Bottom Line on Allochthonous Material in a Stream: Insects that have fallen into a stream are ready to eat and may join leaves, exuviae (cast-off coverings, such as crab shells or the skins of snakes, etc.), copepods, dead and dying animals, rotifers, bacteria, and dislodged algae and immature insects in their float (drift) downstream to a waiting, hungry mouth.

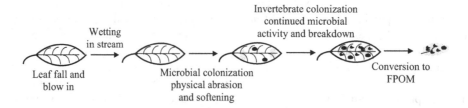

FIGURE 13.1 Leaf processing in streams.

Another important reason to study and learn ecology can be garnered from a simple stream ecology example. Consider the following:

HISTORY OF ECOLOGY

The chronological development of most sciences is clear and direct. Listing the progressive stages in the development of biology, math, chemistry, and physics is a relatively easy, straightforward process. The science of ecology is different. Having only gained prominence in the later part of the twentieth century, ecology is generally considered a new science. However, ecological thinking at some level has been around for a long time, and the principles of ecology have developed gradually, more like a multi-stemmed bush than a tree with a single trunk (Smith 1996).

Smith and Smith (2006) point out that one can argue that ecology goes back to Aristotle or perhaps his friend and associate, Theophrastus, both of whom had an interest in the relations between organisms and the environment and in many species of animals. Theophrastus described interrelationships between animals and between animals and their environment as early as the fourth century B.C. (Ramalay 1940).

Modern ecology has its early roots in plant geography (i.e., plant ecology, which developed earlier than animal ecology) and natural history. The early plant geographers (ecologists) included Carl Ludwig Willdenow (1765–1812) and Friedrich Alexander von Humboldt (1769–1859). Willdenow was one of the first phytogeographers; he was also a mentor to von Humboldt. Willdenow, for whom the perennial vine Willdenow spike moss (*Selaginella willdenowii*) is named, developed the notion, among many others, that plant distribution patterns changed over time. Von Humboldt, considered by many to be the father of ecology, further developed many of the Willdenow notions, including the notion that barriers to plant dispersion were not absolute (Smith 2007).

Another scientist who is considered a founder of plant ecology was Johannes E.B. Warming (1841–1924). Warming studied the tropical vegetation of Brazil, and he is best known for his work on the relations between living plants and their surroundings. He is also recognized for his flagship text on plant ecology, *Plantesamfund* (1895), and for writing *A Handbook of Systematic Botany* (1878).

Meanwhile, other naturalists were assuming important roles in the development of ecology. First and foremost among the naturalists was Charles Darwin. While working on his *Origin of Species*, Darwin came across the writings of Thomas Malthus (1766–1834). Malthus advanced the principle that populations grow in a geometric fashion, doubling at regular intervals until they outstrip the food supply—ultimately resulting in death and thus restraining population growth (Smith and Smith 2006). In his autobiography written in 1876, Darwin wrote:

In October 1838, that is, fifteen months after I had begun my systematic inquiry, I happened to read for amusement Malthus on *Population* and being well prepared to appreciate the struggle for existence which everywhere goes on from long-continued observation of the habits of animals and plants it at once struck me that under these circumstances favorable variations would tend to be preserved, and unfavorable ones to be destroyed. The results of this would be the formation of a new species. Here, then I had at last got a theory by which to work.

During the period when Darwin was formulating his *Origin of Species*, Gregor Mendel (1822–1884) was studying the transmission of characteristics from one generation of pea plants to another. Mendel's and Darwin's work provided the foundation for *population genetics*, the study of evolution and adaptation.

Time marched on, and the preceding work of chemists Lavoisier (who lost his head during the French Revolution) and Horace B. de Saussure, as well as the Austrian geologist Eduard Suess, who proposed the term *biosphere* in 1875, all set the foundations for the advanced work that followed.

Important Point: The Russian geologist Vladimir Vernadsky detailed the idea of the biosphere in 1926.

Several forward strides in animal ecology, independent of plant ecology, were made during the nineteenth century that enabled the twentieth-century scientists R. Hesse, Charles Elton, Charles Adams, and Victor Shelford to refine the discipline. Smith and Smith (2006) point out that many early plant ecologists were "concerned with observing the patterns of organisms in nature, attempting to understand how patterns were formed and maintained by interactions with the physical environment." Instead of looking for patterns, Frederic E. Clements (1874–1945) sought a system for organizing nature. Conducting his studies on vegetation in Nebraska, he postulated that the plant community behaves as a complex organism that grows and develops through stages, resembling the development of an individual organism, to a mature (climax) stage. Clements's (1960) theory of vegetation was significantly criticized by Arthur Tansley, a British ecologist, and others.

Ecology (2007) points out that in 1935, Tansley coined the term *ecosystem*—the interactive system established between a group of living creatures (biocoenosis), and the environment in which they live (biotype). Tansley's ecosystem concept was adopted by the well-known and influential biology educator Eugene P. Odum. Along with his brother Howard Odum, Eugene P. Odum wrote a textbook that, starting in 1952, educated multiple generations of biologists and ecologists in North America (including the author of this text). Eugene P. Odum is often called the "father of modern ecosystem ecology."

A new direction in ecology was given a boost in 1913 when Victor Shelford stressed the interrelationship of plants and animals. He conducted early studies on succession in the Indiana dunes and on experimental *physiological ecology.* Because of his work, ecology became a science of communities. His *Animal Communities in Temperate America* (1913) was one of the first books to treat ecology as a separate science. Eugene P. Odum was one of Shelford's students.

Human ecology began in the 1920s. Around the same time, the study of populations split into two fields: *population ecology* and *evolutionary ecology.* Closely associated with population ecology and evolutionary ecology is *community ecology.* At the same time, physiological ecology arose. Later, natural history observations spawned behavioral ecology (Smith and Smith 2006).

The history of ecology has been tied to advances in biology, physics, and chemistry that have spawned new areas of study in ecology, such as landscape, conservation, restoration, and global ecology. At the same time, ecology was rife with conflicts and opposing camps. Smith (1996) notes that the first major split in ecology was between plant ecology and animal ecology, which even led to a controversy over the term ecology, with botanists dropping the initial "o" from oecology, the spelling in use at the time, and zoologists refusing to use the term ecology at all because of its perceived affiliation with botany. Other historical schisms were between organismal and individualist ecology, holism versus reductionism, and theoretical versus applied ecology (Ecology 2007).

To illustrate one way in which ecosystem classification is used, a real-world example from the United States Department of Agriculture is provided below (USDA 2007).

EXAMPLE ECOSYSTEM: AGROECOSYSTEM MODEL

What are the basic components of agroecosystems? Just like natural ecosystems, they can be thought of as including the processes of primary production, consumption, and decomposition interacting with abiotic environmental components, resulting in energy flow and nutrient cycling. Economic, social, and environmental factors must be added to this primary concept because of the human element that is so closely involved with agroecosystem creation and maintenance.

Agroecosystem Characteristics

Agricultural ecosystems (referred to as agroecosystems) have been described by Odum (1984) as domesticated ecosystems. He states that they are, in many ways, intermediate between natural ecosystems (such as grasslands and forests) and fabricated ecosystems (cities). Agroecosystems are solar-powered (as are natural systems) but differ from natural systems in that:

- Auxiliary energy sources are used to enhance productivity; these sources include processed fuels along with animal and human labor.
- Species diversity is reduced by human management to maximize the yield of specific foodstuffs (plant or animal).

- Dominant plant and animal species are under artificial rather than natural selection.
- Control is external and goal-oriented rather than internal via subsystem feedback, as in natural ecosystems.

Agroecosystems do not arise without human intervention in the landscape. Therefore, the creation of these ecosystems (and maintenance of them as well) is necessarily concerned with the (human) economic goals of production, productivity, and conservation. By definition, agroecosystems are controlled through the management of ecological processes.

Crossley et al. (1984) addressed the potential of agroecosystems as a unifying and clarifying concept for the effective management of managed landscape units. All ecosystems are open, meaning they exchange biotic and abiotic elements with other ecosystems. Agroecosystems are particularly open, with significant exports of primary and secondary production (plant and animal production) as well as increased opportunities for nutrient loss. Modern agroecosystems depend entirely on human intervention; without this intervention, they would not persist. For this reason, they are sometimes referred to as artificial systems, in contrast to natural systems that can sustain themselves without human intervention.

Definitions of agroecosystems often include the entire support base of energy and material subsidies, seeds, chemicals, and even the sociopolitical and economic matrix in which management decisions are made. Crossley et. al. (1984) noted that while this approach is logical, they preferred to designate the individual field as the agroecosystem. This designation aligns with assigning an individual forest catchment or lake as an ecosystem. He envisions the **farm system** as consisting of a set of agroecosystems—fields with similar or different crops—alongside the support mechanisms and socioeconomic factors contributing to their management. Agroecosystems retain most, if not all, of the functional properties of natural ecosystems, including nutrient conservation mechanisms, energy storage and use patterns, and regulation of biotic diversity.

Ecosystem Pattern and Process

Throughout the United States, the landscape consists of patches of natural ecosystems scattered (or embedded) in a matrix of different agroecosystems and fabricated ecosystems. In fact, about three-quarters of the land area of the United States (USDA, 1982) is occupied by agroecosystems.

The pattern created by this interspersion incorporates elements of the variability of structure and separation of functions among the various ecosystems. Pattern variables quantify the structure and relationships between systems; processes imply functional relationships between and within the biotic and abiotic ecosystem components. Within agroecosystems, processes include:

- Enhanced productivity of producers through fertilization
- Improved productivity through selective breeding

- Management of pests with various control methods
- Management of various aspects of the hydrologic cycle
- Land forming

LEVELS OF ORGANIZATION

Odum (1983) suggested that the best way to delimit modern ecology is to consider the concept of *levels of organization*. Levels of organization can be simplified as shown in Figure 13.2. In this relationship, organs form an organism; organisms of a particular species form a population; populations occupying a particular area form a community. Communities, interacting with non-living or abiotic factors, separate into a natural unit to create a stable system known as the *ecosystem (the major ecological unit)*; and the part of the Earth in which the ecosystem operates is known as the *biosphere*. Tomera points out, "every community is influenced by a particular set of abiotic factors" (Tomera 1989). Inorganic substances such as oxygen, carbon dioxide, several other inorganic substances, and some organic substances represent the abiotic part of the ecosystem.

The physical and biological environment in which an organism lives is referred to as its *habitat*. For example, the habitat of two common aquatic insects, the "backswimmer" (Notonecta) and the "water boatman" (Corixa), is the littoral zone of ponds and lakes (shallow, vegetation-choked areas) (see Figure 13.3) (Odum 1983). Within each level of organization of a particular habitat, each organism has a special role. The role the organism plays in the environment is referred to as its *niche*. A niche might be that the organism is food for some other organism or is a predator of other organisms. Odum refers to an organism's niche as its "profession" (Odum 1971). In other words, each organism has a job or role to fulfill in its environment. Although two different species might occupy the same habitat, "niche separation based on food habits" differentiates between the two species (Odum 1983). Comparing the niches of the water backswimmer and the water boatman reveals such niche separation. The backswimmer is an active predator, while the water boatman feeds largely on decaying vegetation (McCafferty 1981).

ECOSYSTEMS

An *ecosystem* denotes an area that includes all organisms therein and their physical environment. The ecosystem is the major ecological unit in nature. Living organisms and their nonliving environment are inseparably interrelated and interact with each other to create a self-regulating and self-maintaining system. To create a self-regulating and self-maintaining system, ecosystems are homeostatic; they resist any change through natural controls. These natural controls are important in ecology, especially because it is people, through their complex activities, who tend to disrupt natural controls.

As stated earlier, the ecosystem encompasses both the living and nonliving factors in a particular environment. The living, or biotic, part of the ecosystem is formed by two components: *autotrophic* and *heterotrophic*. The autotrophic (self-nourishing) component does not require food from its environment but can manufacture food from inorganic substances. For example, some autotrophic components (plants) manufacture the energy needed through photosynthesis. Heterotrophic components, on the other hand, depend on autotrophic components for food.

The non-living, or abiotic, part of the ecosystem is formed by three components: inorganic substances, organic compounds (which link the biotic and abiotic parts), and climate regime. Figure 13.4 is a simplified diagram showing a few of the living and nonliving components of an ecosystem found in a freshwater pond.

An ecosystem is a cyclic mechanism in which biotic and abiotic materials are constantly exchanged through *biogeochemical cycles*. Biogeochemical cycles are defined as follows: *bio* refers to living organisms, and *geo* refers to

Organs → Organism → Population → Communities → Ecosystem → Biosphere

FIGURE 13.2 Levels of organization.

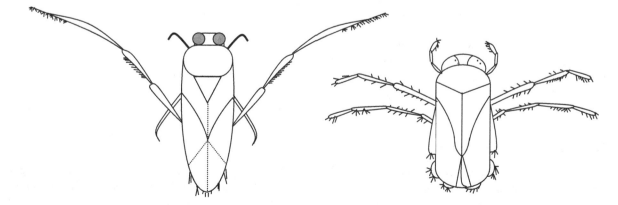

FIGURE 13.3 *Notonecta* (left) and *Corixa* (right). (Adapted from Odum (1983, p. 42).).

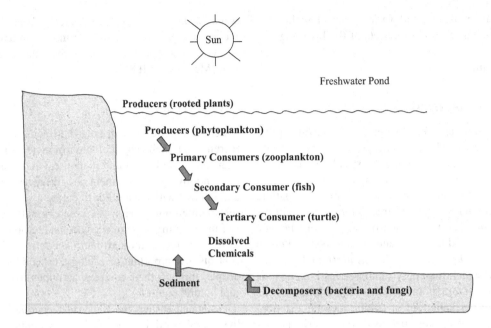

FIGURE 13.4   Major components of a freshwater pond ecosystem.

FIGURE 13.5   Aquatic food chain.

water, air, rocks, or solids. *Chemical* is concerned with the chemical composition of the Earth. Biogeochemical cycles are driven by energy, directly or indirectly, from the sun.

Figure 13.4 depicts an ecosystem where biotic and abiotic materials are constantly exchanged. Producers construct organic substances through photosynthesis and chemosynthesis. Consumers and decomposers use organic matter as their food and convert it into abiotic components. That is, they dissipate energy fixed by producers through food chains. The abiotic part of the pond in Figure 13.4 is formed of inorganic and organic compounds dissolved in sediments, such as carbon, oxygen, nitrogen, sulfur, calcium, hydrogen, and humic acids. Producers, such as rooted plants and phytoplankton, represent the biotic part. Fish, crustaceans, and insect larvae make up the consumers. Mayfly nymphs represent detritivores, which feed on organic detritus. Decomposers make up the final biotic part. They include aquatic bacteria and fungi, which are distributed throughout the pond.

As stated earlier, an ecosystem is a cyclic mechanism. From a functional viewpoint, an ecosystem can be analyzed in terms of several factors. The factors important to this study include the biogeochemical cycles, energy, and food chains.

ENERGY FLOW IN THE ECOSYSTEM

Simply defined, *energy* is the ability or capacity to do work. For an ecosystem to exist, it must have energy. All activities of living organisms involve work, which is the expenditure of energy. This means the degradation of a higher state of energy to a lower state. Two laws govern the flow of energy through an ecosystem: *The First and Second Laws of Thermodynamics*. The first law, sometimes called the *conservation law*, states that energy may not be created or destroyed. The second law states that no energy transformation is 100% efficient. That is, in every energy transformation, some energy is dissipated as heat. The term *entropy* is used as a measure of the nonavailability of energy to a system. Entropy increases with an increase in dissipation. Because of entropy, the input of energy in any system is higher than the output or work done; thus, the resultant efficiency is less than 100%.

The interaction of energy and materials in the ecosystem is important. Energy drives the biogeochemical cycles. Note that energy does not cycle as nutrients do in biogeochemical cycles. For example, when food passes from one organism to another, the energy contained in the food is reduced systematically until all the energy in the system is dissipated as heat. Price (1984) refers to this process as "a *unidirectional flow* of energy through the system, with no possibility for recycling of energy." When water or nutrients are recycled, energy is required. The energy expended in this recycling is not recyclable.

As mentioned, the principal source of energy for any ecosystem is sunlight. Green plants, through the process of photosynthesis, transform the sun's energy into carbohydrates, which are consumed by animals. This transfer of energy, again, is unidirectional—from producers to consumers. Often, this transfer of energy to different organisms is called a *food chain*. Figure 13.5 shows a simple aquatic food chain.

All organisms, alive or dead, are potential sources of food for other organisms. All organisms that share the same general type of food in a food chain are said to be at the same *trophic level* (nourishment or feeding level). Since green plants use sunlight to produce food for animals, they are called the *producers*, or the first trophic level. The herbivores, which eat plants directly, are called the second trophic level or the *primary consumers*. The carnivores are flesh-eating consumers; they include several trophic levels from the third one up. At each transfer, a large amount of energy (about 80%–90%) is lost as heat and waste. Thus, nature normally limits food chains to four or five links. In aquatic ecosystems, however, food chains are commonly longer than those on land. The aquatic food chain is longer because several predatory fish may be feeding on the plant consumers. Even so, the built-in inefficiency of the energy transfer process prevents the development of extremely long food chains.

Only a few simple food chains are found in nature. Most simple food chains are interlocked. This interlocking of food chains forms a *food web*. Most ecosystems support a complex food web. A food web involves animals that do not feed on just one trophic level. For example, humans feed on both plants and animals. An organism in a food web may occupy one or more trophic levels. The trophic level is determined by an organism's role in its particular community, not by its species. Food chains and webs help to explain how energy moves through an ecosystem.

An important trophic level of the food web is comprised of the *decomposers*. The decomposers feed on dead plants or animals and play an important role in recycling nutrients in the ecosystem. Simply put, there is no waste in ecosystems. All organisms, dead or alive, are potential sources of food for other organisms. An example of an aquatic food web is shown in Figure 13.6.

FOOD CHAIN EFFICIENCY

Earlier, we pointed out that energy from the sun is captured (via photosynthesis) by green plants and used to make food. Most of this energy is used to carry on the plant's life activities. The rest of the energy is passed on as food to the next level of the food chain. Nature limits the amount of energy that is accessible to organisms within each food chain. Not all food energy is transferred from one trophic level to the next. Only about 10% (10% rule) of the amount of energy is actually transferred through a food chain. For example, if we apply the 10% rule to the diatoms-copepods-minnows-medium fish-large fish food chain shown in Figure 13.7, we can predict that 1,000 g of diatoms produce 100 g of copepods, which will produce 10 g of minnows, which will produce 1 g of medium fish, which, in turn, will produce 0.1 g of large fish. Thus, only about 10% of the chemical energy available at each trophic level is transferred and stored in usable form at the next level. The other 90% is lost to the environment as low-quality heat in accordance with the second law of thermodynamics.

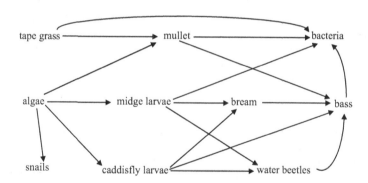

FIGURE 13.6   Aquatic food web.

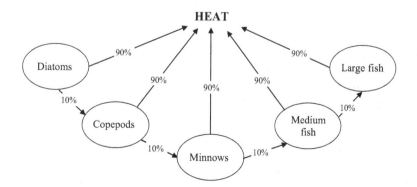

FIGURE 13.7   Simple food chain.

ECOLOGICAL PYRAMIDS

In the food chain, from the producer to the final consumer, it is clear that a particular community in nature often consists of several small organisms associated with a smaller and smaller number of larger organisms. A grassy field, for example, has a larger number of grasses and other small plants, a smaller number of herbivores like rabbits, and an even smaller number of carnivores like foxes. The practical significance of this is that we must have several more producers than consumers.

This pound-for-pound relationship, where it takes more producers than consumers, can be demonstrated graphically by building an *ecological pyramid*. In an ecological pyramid, separate levels represent the number of organisms at various trophic levels in a food chain, or bars are placed one above the other with a base formed by producers and the apex formed by the final consumer. The pyramid shape is formed due to a significant amount of energy loss at each trophic level. The same is true if the corresponding biomass or energy substitutes for numbers. Ecologists generally use three types of ecological pyramids: *pyramids of number, biomass,* and *energy.* Obviously, there will be differences among them. Some generalizations are given as follows:

1. Energy pyramids must always be larger at the base than at the top (because of the Second Law of Thermodynamics, which relates to the dissipation of energy as it moves from one trophic level to another).
2. Likewise, biomass pyramids (in which biomass is used as an indicator of production) are usually pyramid-shaped. This is particularly true of terrestrial systems and aquatic ones dominated by large plants (such as marshes), where consumption by heterotrophs is low and organic matter accumulates over time.

    However, biomass pyramids can sometimes be inverted. This is common in aquatic ecosystems, where the primary producers are microscopic planktonic organisms that multiply very rapidly, have very short life spans, and experience heavy grazing by herbivores. At any single point in time, the amount of biomass in primary producers is less than that in larger, long-lived animals that consume primary producers.
3. Number pyramids can have various shapes (and may not be pyramids at all) depending on the sizes of the organisms that make up the trophic levels. In forests, the primary producers are large trees, and the herbivore level usually consists of insects, so the base of the pyramid is smaller than the herbivore level above it. In grasslands, the number of primary producers (grasses) is much larger than that of the herbivores above (large grazing animals) (Ecosystems Topics 2000).

PRODUCTIVITY

As mentioned, the flow of energy through an ecosystem starts with the fixation of sunlight by plants through photosynthesis. In evaluating an ecosystem, measuring photosynthesis is important. Ecosystems may be classified as highly productive or less productive. Therefore, the study of ecosystems must involve some measure of the productivity of that ecosystem. Primary production is the rate at which the ecosystem's primary producers capture and store a given amount of energy in a specified time interval. In simpler terms, primary productivity is a measure of the rate at which photosynthesis occurs. Four successive steps in the production process are:

1. **Gross Primary Productivity:** the total rate of photosynthesis in an ecosystem during a specified interval.
2. **Net Primary Productivity:** the rate of energy storage in plant tissues in excess of the rate of aerobic respiration by primary producers.
3. **Net Community Productivity:** the rate of storage of organic matter not used.
4. **Secondary Productivity:** the rate of energy storage at consumer levels.

When attempting to comprehend the significance of the term *productivity* as it relates to ecosystems, it is wise to consider an example. Consider the productivity of an agricultural ecosystem, such as a wheat field. Often, its productivity is expressed as the number of bushels produced per acre. This is an example of the harvest method for measuring productivity. For a natural ecosystem, several one-square-meter plots are marked off, and the entire area is harvested and weighed to give an estimate of productivity in grams of biomass per square meter per given time interval. From this method, a measure of net primary production (net yield) can be obtained.

Productivity, both in natural and cultured ecosystems, may vary considerably, not only between types of ecosystems but also within the same ecosystem. Several factors influence year-to-year productivity within an ecosystem. Factors such as temperature, availability of nutrients, fire, animal grazing, and human cultivation activities are directly or indirectly related to the productivity of a particular ecosystem.

Productivity can be measured in several different ways in aquatic ecosystems. For example, the production of oxygen may be used to determine productivity. Oxygen content can be measured in several ways. One method is to measure it in the water every few hours over a period of 24 hours. During daylight, when photosynthesis is occurring, the oxygen concentration should rise. At night, the oxygen level should drop. The oxygen level can be measured using a simple x–y graph. The oxygen level can be plotted on the y-axis, with time plotted on the x-axis, as shown in Figure 13.8.

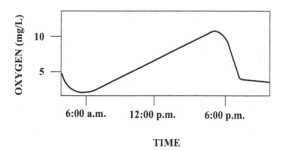

FIGURE 13.8   Diurnal oxygen curve for an aquatic ecosystem.

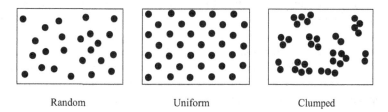

Random   Uniform   Clumped

FIGURE 13.9   Basic patterns of distribution. (Adapted from Odum, 1971, p. 205.)

Another method of measuring oxygen production in aquatic ecosystems is to use light and dark bottles. Biochemical oxygen demand (BOD) bottles (300 mL) are filled with water to a particular height. One of the bottles is tested for the initial dissolved oxygen (DO), and then the other two bottles (one clear, one dark) are suspended in the water at the depth from which they were taken. After a 12-hour period, the bottles are collected, and the DO values for each bottle are recorded. Once the oxygen production is known, productivity in terms of grams/m²/day can be calculated. In the aquatic ecosystem, pollution can have a profound impact on the system's productivity.

POPULATION ECOLOGY

Webster's Third New International Dictionary defines population as "the total number or amount of things, especially within a given area; the organisms inhabiting a particular area or biotype; and a group of interbreeding biotypes that represents the level of organization at which speciation begins." The concept of population is interpreted differently in various sciences. In *human demography*, a population is a set of humans in a given area. In *genetics*, a population is a group of interbreeding individuals of the same species that is isolated from other groups. In *population ecology*, a population is a group of individuals of the same species inhabiting the same area.

If we wanted to study the organisms in a slow-moving stream or pond, we would have two options. We could study each fish, aquatic plant, crustacean, and insect one by one. In that case, we would be studying individuals. It would be easier to do this if the subject were trout, but it would be difficult to separate and study each aquatic plant. The second option would be to study all of the trout, all of the insects of each specific kind, and all of a certain type of aquatic plant in the stream or pond at the time of the study.

When ecologists study a group of the same kind of individuals in a given location at a given time, they are investigating a *population*. When attempting to determine the population of a particular species, it is important to remember that time is a factor. Time is important because populations change, whether at various times during the day, during different seasons, or from year to year.

Population density may change dramatically. For example, if a dam is closed off in a river midway through the spawning season, with no provision allowed for fish movement upstream (such as a fish ladder), it would drastically decrease the density of spawning salmon upstream. Along with the swift and sometimes unpredictable consequences of change, it can be difficult to draw exact boundaries between various populations. The population density or level of a species depends on *natality, mortality, immigration*, and *emigration*. Changes in population density are the result of both births and deaths. The birth rate of a population is called *natality*, and the death rate is called *mortality*. In aquatic populations, two factors besides natality and mortality can affect density. For example, in a run of returning salmon to their spawning grounds, the density could vary as more salmon migrated in or as others left the run for their own spawning grounds. The arrival of new salmon to a population from other places is termed *immigration* (ingress). The departure of salmon from a population is called *emigration* (egress). Thus, natality and immigration increase population density, whereas mortality and emigration decrease it. The net increase in population is the difference between these two sets of factors.

Each organism occupies only those areas that can provide for its requirements, resulting in an irregular distribution. How a particular population is distributed within a given area has a considerable influence on density. As shown in Figure 13.9, organisms in nature may be distributed in three ways. In a random distribution, there is an equal probability of an organism occupying any point in space, and "each individual is independent of the others." In a regular or uniform distribution, organisms are spaced more evenly; they are not distributed by chance. Animals compete with each other and effectively defend a specific territory, excluding other individuals of the same species. In a regular or uniform distribution, the competition between individuals can be quite severe and antagonistic to the point where the spacing generated is quite even. The most common distribution is the contagious or clumped distribution, where organisms are found in groups; this may reflect the heterogeneity of the habitat. Organisms that exhibit a contagious or clumped distribution may develop social hierarchies in order to live together more effectively. Animals within the same species have evolved many symbolic aggressive displays that carry meanings that are not only mutually understood but also prevent injury or death within the same species.

The size of animal populations is constantly changing due to natality, mortality, emigration, and immigration. As mentioned, the population size will increase if the natality and immigration rates are high. On the other hand, it will decrease if the mortality and emigration rates are high. Each population has an upper limit on size, often called the *carrying capacity*. Carrying capacity is the optimum number of individuals of a species that can survive in a specific area over time. Stated differently, the carrying capacity is the maximum number of individuals of a species that can be supported in a bioregion. A pond may be able to support only a dozen frogs, depending on the food resources available for them. If there were thirty frogs in the same pond, at least half of them would probably die because the pond environment would not have enough food for them to survive. Carrying capacity is based on the quantity of food supplies, the physical space available, the degree of predation, and several other environmental factors.

The carrying capacity is of two types: ultimate and environmental. *Ultimate carrying capacity* is the theoretical maximum density; that is, it is the maximum number of individuals of a species in a place that can support itself without rendering the place uninhabitable. The *environmental carrying capacity* is the actual maximum population density that a species maintains in an area. Ultimate carrying capacity is always higher than environmental carrying capacity. Ecologists have concluded that a major factor affecting population stability or persistence is *species diversity*. Species diversity is a measure of the number of species and their relative abundance.

If the stress on an ecosystem is small, the ecosystem can usually adapt quite easily. Moreover, even when severe stress occurs, ecosystems have a way of adapting. Severe environmental changes in an ecosystem can result from natural occurrences such as fires, earthquakes, and floods, as well as from human-induced changes such as land clearing, surface mining, and pollution. One of the most important applications of species diversity is in the evaluation of pollution. Stress of any kind will reduce the species diversity of an ecosystem to a significant degree. In the case of domestic sewage pollution, for example, the stress is caused by a lack of dissolved oxygen (DO) for aquatic organisms.

Ecosystems can and do change; for example, if a fire devastates a forest, it will grow back eventually, because of *ecological succession*. Ecological succession is the observed process of change (a normal occurrence in nature) in the species structure of an ecological community over time. Succession usually occurs in an orderly, predictable manner. It involves the entire system. The science of ecology has developed to such a point that ecologists are now able to predict several years in advance what will occur in a given ecosystem. For example, scientists know that if a burned-out forest region receives light, water, nutrients, and an influx or immigration of animals and seeds, it will eventually develop into another forest through a sequence of steps or stages. Ecologists recognize two types of ecological succession: primary and secondary. The particular type that takes place depends on the conditions at a particular site at the beginning of the process.

Primary succession, sometimes called *bare-rock succession*, occurs on surfaces such as hardened volcanic lava, bare rock, and sand dunes, where no soil exists and where nothing has ever grown before (see Figure 13.10). Obviously, in order to grow, plants need soil. Thus, soil must form on the bare rock before succession can begin. Usually, this soil formation process results from weathering. Atmospheric exposure—weathering, wind, rain, and frost—forms tiny cracks and holes in rock surfaces. Water collects in the rock fissures and slowly dissolves the minerals out of the rock's surface. A pioneer soil layer is formed from the dissolved minerals and supports such plants as lichens. Lichens gradually cover the rock surface and secrete carbonic acid, which dissolves additional minerals from the rock. Eventually, mosses replace the lichens. Organisms called *decomposers* move in and feed on dead lichen and moss. A few small animals, such as mites and spiders, arrive next. The result is what is known as a *pioneer community*. The pioneer community is defined as the first successful integration of plants, animals, and decomposers into a bare-rock community.

After several years, the pioneer community builds up enough organic matter in its soil to support rooted plants like herbs and shrubs. Eventually, the pioneer community is crowded out and replaced by a different environment. This, in turn, works to thicken the upper soil layers. The progression continues through several other stages until a mature or climax ecosystem is developed, several decades later. In bare-rock succession, each stage in the complex succession pattern dooms the stage that existed before it. *Secondary succession* is the most common type of succession. Secondary succession occurs in an area where the natural vegetation has been removed or destroyed but the soil is not destroyed. For example, succession that occurs in abandoned farm fields, known as *old-field succession*, illustrates secondary succession. An example of secondary succession can be seen in the Piedmont region of North Carolina. Early settlers of the area cleared away the native

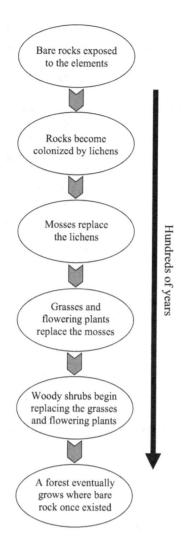

FIGURE 13.10 Bare-rock succession. (Adapted from Tomera, 1989, p. 67.)

oak-hickory forests and cultivated the land. In the ensuing years, the soil became depleted of nutrients, reducing the soil's fertility. As a result, farming ceased in the region a few generations later, and the fields were abandoned. Some 150–200 years after abandonment, the climax oak-hickory forest was restored.

In a stream ecosystem, growth is enhanced by biotic and abiotic factors, including:

- Ability to produce offspring
- Ability to adapt to new environments
- Ability to migrate to new territories
- Ability to compete with species for food and space to live
- Ability to blend into the environment so as not to be eaten
- Ability to find food
- Ability to defend itself from enemies
- Favorable light conditions
- Favorable temperature
- Favorable dissolved oxygen (DO) content
- Sufficient water level

The biotic and abiotic factors in an aquatic ecosystem that reduce growth include:

- Predators
- Disease
- Parasites
- Pollution
- Competition for space and food
- Unfavorable stream conditions (i.e., low water levels)
- Lack of food

With regard to stability in a freshwater ecosystem, the higher the species diversity, the greater the inertia and resilience of the ecosystem. At the same time, when species diversity is high within a stream ecosystem, a population within the stream can become out of control because of an imbalance between growth and reduction factors, while the ecosystem still remains stable. In regard to instability in a freshwater ecosystem, recall that imbalance occurs when growth and reduction factors are out of balance. For example, when sewage is accidentally dumped into a stream, the stream ecosystem, via the self-purification process (discussed later), responds and returns to normal. This process can be described as follows:

- Raw sewage is dumped into the stream.
- Oxygen availability decreases as the detritus food chain breaks down the sewage.
- Some fish die at the pollution site and downstream.
- Sewage is broken down, washed out to sea, and is finally broken down in the ocean.
- Oxygen levels return to normal.
- Fish populations that were depleted are restored as fish about the spill reproduce, and the young occupy the space formerly held by the dead fish.
- Populations all return to "normal."

A shift in balance in a stream's ecosystem (or any ecosystem) similar to the one just described is a common occurrence. In this particular case, the stream responded on its own to the imbalance caused by the sewage and, through the self-purification process returned, to normal. Recall that succession is the method by which an ecosystem either forms or heals itself. Thus, we can say that a type of succession occurred in the polluted stream described above, because, in the end, it healed itself. More importantly, this healing process is a beneficial; otherwise, long ago, there would have been few streams on Earth suitable for much more than the dumping of garbage.

In summary, through research and observation, ecologists have found that succession patterns in different ecosystems usually display common characteristics. First, succession brings about changes in the plant and animal members present. Second, organic matter increases from stage to stage. Finally, as each stage progresses, there is a tendency toward greater stability or persistence. Remember, succession is usually predictable—this is the case unless humans interfere.

STREAM GENESIS AND STRUCTURE

Consider the following: Early in the spring, on a snow- and ice-covered high alpine meadow, the water cycle continues. The cycle's main component, water, has been held in reserve—literally frozen—for the long, dark winter months. But with longer, warmer spring days, the sun is higher, more direct, and of longer duration, and the frozen masses of water respond to the increased warmth. The melt begins with a single drop, then two, then increasingly more. As the snow and ice melt, the drops join a chorus that continues unending; they fall from their ice-bound lip to the bare rock and soil terrain below.

The terrain on which the snowmelt falls is not like glacial till, which is an unconsolidated, heterogeneous mixture of clay, sand, gravel, and boulders, dug out, ground out, and exposed by the force of a huge, slow, and inexorably moving glacier. Instead, this soil and rock ground is exposed to the falling drops of snowmelt because of a combination of wind and tiny, enduring force exerted by drops of water as, over season after season, they collide with the thin soil cover, exposing the intimate bones of the earth.

Gradually, the single drops increase to a small rush—they join to form a splashing, rebounding, helter-skelter cascade of many separate rivulets that trickle, then run, down the face of the granite mountain. At an indented ledge halfway down the mountain slope, a pool forms, whose beauty, clarity, and sweet iciness provide the visitor with an incomprehensible, incomparable gift—a blessing from the Earth.

The mountain pool fills slowly, tranquil under the blue sky, reflecting the pines, snow, and sky around and above it, an open invitation to lie down and drink, and to peer into the glass-clear, deep phantom blue-green eye, so clear that it seems possible to reach down over 50 feet and touch the very bowels of the mountain. The pool has no transition from shallow margin to depth; it is simply deep and pure. As the pool fills with more melt water, we wish to freeze time, to hold this place and this pool in its perfect state forever. It is such a rarity to us in our modern world. However, this cannot be—Mother Nature calls, prodding, urging—and for a brief instant, the water laps in the breeze against the outermost edge of the ridge, and then, a trickle flows over the rim. The giant hand of gravity reaches out and tips the overflowing melt onward, and it continues the downward journey, following the path of least resistance to its next destination, several thousand feet below.

When the overflow, still high in altitude, but with its rock-strewn bed bent downward toward the sea, meets the angled, broken rocks below, it bounces, bursts, and mists its way against the steep, V-shaped walls that form a small valley, carved out over time by water and the forces of the earth. Within the valley confines, the meltwater has grown from drops to rivulets to a small mass of flowing water. It flows through what is at first a narrow opening, gaining strength, speed, and power as the V-shaped valley widens to form a U-shape. The journey continues as the water mass picks up speed and tumbles over massive boulders, then slows again.

At a larger but shallower pool, waters from higher elevations have joined the main body—from the hillsides, crevices, springs, rills, and mountain creeks. At the influent poolside, all appears peaceful, quiet, and restful, but not far away, at the effluent end of the pool, gravity takes control again. The overflow is flung over the jagged lip and cascades downward several hundred feet, where the waterfall again brings its load to a violent, mist-filled meeting.

The water separates and joins repeatedly, forming a deep, furious, wild stream that calms gradually as it continues to flow over lands that are less steep. The waters widen into pools overhung by vegetation and surrounded by tall trees. The pure, crystalline waters have become progressively discolored on their downward journey, stained brown-black with humic acid, and literally filled with suspended sediments; the once-pure stream is now muddy.

The mass divides and flows in different directions over different landscapes. Small streams divert and flow into open country. Different soils work to retain or speed the waters, and in some places, the waters spread out into shallow swamps, bogs, marshes, fens, or mires. Other streams pause long enough to fill deep depressions in the land and form lakes. For a time, the water remains and pauses in its journey to the sea. However, this is only a short-term pause because lakes are only a short-term resting place in the water cycle. The water will eventually move on by evaporation or seepage into groundwater. Other portions of the water mass stay with the main flow, and the speed of flow changes to form a river, which braids its way through the landscape, heading for the sea. As it changes speed and slows, the river bottom changes from rock and stone to silt and clay. Plants begin to grow, stems thicken, and leaves broaden. The river is now full of life and the nutrients needed to sustain it. However, the river courses onward, its destiny met when the flowing rich mass slows its last and finally spills into the sea.

Freshwater systems are divided into two broad categories: running waters (lotic systems) and standing waters (lentic systems). We concentrate on lotic systems (although many of the principles described herein apply to other freshwater surface bodies as well, which are known by common names). Some examples include seeps, springs, brooks, branches, creeks, streams, and rivers. Again, because it is the best term to use in freshwater ecology, it is the stream we are concerned with here. Although there is no standard scientific definition of a stream, it is usually distinguished subjectively as follows: a stream is of intermediate size that can be waded across from one side to the other.

The physical processes involved in the formation of a stream are important to the ecology of the stream because stream channel and flow characteristics directly influence the functioning of the stream's ecosystem and the biota found therein. Thus, in this section, we discuss the pathways of water flow contributing to stream flow; namely, we discuss precipitation inputs as they contribute to the flow. We also discuss stream flow discharge, transport of material, characteristics of stream channels, stream profile, sinuosity, the floodplain, pool-riffle sequences, and depositional features—all of which directly or indirectly impact the ecology of the stream.

WATER-FLOW IN A STREAM

Most elementary students learn early in their education process that water on Earth flows downhill—from land to the sea. However, they may or may not be told that water flows downhill toward the sea by various routes. The route (or pathway) that we are primarily concerned with is the surface water route taken by surface water runoff. Surface runoff depends on various factors. For example, climate, vegetation, topography, geology, soil characteristics, and land-use determine how much surface runoff occurs compared with other pathways.

The primary source (input) of water to total surface runoff, of course, is precipitation. This is the case even though a substantial portion of all precipitation input returns directly to the atmosphere by evapotranspiration. *Evapotranspiration* is a combination process, as the name suggests, whereby water in plant tissue and in the soil evaporates and transpires to water vapor in the atmosphere. A substantial portion of precipitation input returns directly to the atmosphere by evapotranspiration. It is also important to point out that when precipitation occurs, some rainwater is intercepted by vegetation, where it evaporates, never reaching the ground or being absorbed by plants. A large portion of the rainwater that reaches the surface on the ground, in lakes, and in streams also evaporates directly back into the atmosphere.

Although plants display a special adaptation to minimize transpiration, they still lose water to the atmosphere during the exchange of gases necessary for photosynthesis. Notwithstanding the large percentage of precipitation that evaporates, rain or meltwater that reaches the ground surface follows several pathways in reaching a stream channel or groundwater.

Soil can absorb rainfall to its *infiltration capacity* (i.e., to its maximum rate). During a rain event, this capacity decreases. Any rainfall in excess of infiltration capacity accumulates on the surface. When this surface water exceeds the depression storage capacity of the surface, it moves as an irregular sheet of overland flow. In arid areas, overland flow is likely because of the low permeability of the soil. Overland flow is also likely when the surface is frozen and/or when human activities have rendered the land surface less permeable. In humid areas, where infiltration capacities are high, overland flow is rare.

In rain events where the infiltration capacity of the soil is not exceeded, rain penetrates the soil and eventually reaches the groundwater—from which it discharges to the stream slowly and over a long period. This phenomenon helps to explain why stream flow through a dry weather region remains constant; the flow is continuously augmented by groundwater. This type of stream is known as a *perennial stream*, as opposed to an *intermittent* one, because the flow continues during periods of no rainfall.

When a stream courses through a humid region, it is fed water via the water table, which slopes toward the stream channel. Discharge from the water table into the stream accounts for flow during periods without precipitation and explains why this flow increases, even without tributary input, as one proceeds downstream. Such streams are called *gaining* or *effluent*, as opposed to *losing* or *influent streams* that lose water into the ground (see Figure 13.11). The same stream can shift between gaining and losing conditions along its course because of changes in underlying strata and local climate.

STREAM WATER DISCHARGE

The current velocity (speed) of water (driven by gravitational energy) in a channel varies considerably within a stream's cross-section owing to friction with the bottom and

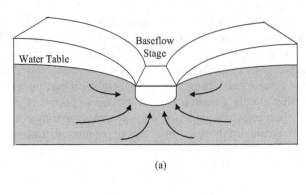

(a)

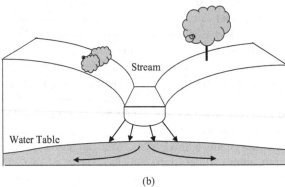

(b)

FIGURE 13.11   (a) Cross-section of a gaining stream. (b) Cross-section of a losing stream.

sides, with sediment, and with the atmosphere, as well as to sinuosity (bending or curving) and obstructions. The highest velocities, obviously, are found where friction is least, generally at or near the surface and near the center of the channel. In deeper streams, current velocity is greatest just below the surface due to friction with the atmosphere; in shallower streams, current velocity is greatest at the surface due to friction with the bed. Velocity decreases as a function of depth, approaching zero at the substrate surface.

Transport of Material

Water flowing in a channel may exhibit *laminar flow* (parallel layers of water shearing over one another vertically) or *turbulent flow* (complex mixing). In streams, laminar flow is uncommon, except at boundaries where flow is very low and in groundwater. Thus, the flow in streams is generally turbulent. Turbulence exerts a shearing force that causes particles to move along the streambed by pushing, rolling, and skipping referred to as *bed load*. This same shear causes turbulent eddies that entrain particles in suspension (called the *suspended* load—particle size under 0.06 mm).

Entrainment is the incorporation of particles when stream velocity exceeds the *entraining velocity* for a particular particle size. The entrained particles in suspension (suspended load) also include fine sediment, primarily clays, silts, and fine sands, which require only low velocities and minor turbulence to remain in suspension. These are referred to as *wash* load (under 0.002 mm). Thus, the suspended load includes the wash load and coarser materials (at lower flows). Together, the suspended load and bed load constitute the *solid load*. It is important to note that in bedrock streams, the bed load will be a lower fraction than in alluvial streams where channels are composed of easily transported material.

A substantial amount of material is also transported as the *dissolved load*. Solutes are generally derived from the chemical weathering of bedrock and soils, and their contribution is greatest in subsurface flow, and in regions of limestone geology. The relative amount of material transported as solute rather than solid load depends on basin characteristics, lithology (i.e., the physical character of rock), and hydrologic pathways. In areas of very high runoff, the contribution of solutes approaches or exceeds the sediment load, whereas in dry regions, sediments make up as much as 90% of the total load.

Deposition occurs when *stream competence* (i.e., the largest particle that can be moved as bedload, and the critical erosion—competent—velocity is the lowest velocity at which a particle resting on the streambed will move) falls below a given velocity. Simply stated, the size of the particle that can be eroded and transported is a function of current velocity.

Sand particles are the most easily eroded. The greater the mass of larger particles (e.g., coarse gravel), the higher the initial current velocities must be for movement. However, smaller particles (silts and clays) require even greater initial velocities because of their cohesiveness and because they present smaller, streamlined surfaces to the flow. Once in transport, particles will continue in motion at somewhat slower velocities than initially required to initiate movement and will settle at still lower velocities.

Particle movement is determined by size, flow conditions and mode of entrainment. Particles over 0.02 mm (medium-coarse sand size) tend to move by rolling or sliding along the channel bed as *traction load*. When sand particles fall out of the flow, they move by *saltation* or repeated bouncing. Particles under 0.06 mm (silt) move as *suspended load* and particles under 0.002 mm (clay) move indefinitely as *wash* load. Unless the supply of sediments becomes depleted, the concentration and the amount of transported solids increase. However, discharge is usually too low throughout most of the year to scrape or scour, shape channels, or move significant quantities of sediment in all but sand-bed streams, which can experience changes more rapidly. During extreme events, the greatest scour occurs, and the amount of material removed increases dramatically.

Sediment inflow into streams can be both increased and decreased because of human activities. For example, poor agricultural practices and deforestation greatly increase erosion. Fabricated structures, such as dams and channel diversions, can, on the other hand, greatly reduce sediment inflow.

Characteristics of Stream Channels

Flowing waters (rivers and streams) determine their own channels, and these channels exhibit relationships attesting to the operation of physical laws—laws that are not yet fully understood. The development of stream channels and entire drainage networks, as well as the existence of various regular patterns in the shape of channels, indicate that streams are in a state of dynamic equilibrium between erosion (sediment loading) and deposition (sediment deposit), governed by common hydraulic processes. However, because channel geometry is four-dimensional, including a long profile, cross-section, depth, and slope profile, and because these dimensions mutually adjust over a time scale as short as years and as long as centuries or more, cause-and-effect relationships are difficult to establish. Other variables that are presumed to interact as the stream achieves its graded state include width and depth, velocity, size of sediment load, bed roughness, and the degree of braiding (sinuosity).

Stream Profiles

Mainly because of gravity, most streams exhibit a downstream decrease in gradient along their length. Beginning at the headwaters, the steep gradient becomes less pronounced as one proceeds downstream, resulting in a concave longitudinal profile. Though diverse geography provides for almost unlimited variation, a lengthy stream that originates in a mountainous area typically comes into existence as a series of springs and rivulets; these coalesce into a fast-flowing, turbulent mountain stream, and the addition of tributaries results in a large, smoothly flowing river that winds through the lowlands to the sea.

When studying a stream system of any length, it becomes readily apparent (almost from the start) that what we are studying is a body of flowing water that varies considerably from place to place along its length. For example, a common variable—the results of which can be readily seen—is whenever discharge increases, causing corresponding changes in the stream's width, depth, and velocity. In addition to physical changes that occur from location to location along a stream's course, there is a legion of biological variables that correlate with stream size and distance downstream. The most apparent and striking changes are in the steepness of slope and in the transition from a shallow stream with large boulders and a stony substrate to a deep stream with a sandy substrate. The particle size of bed material at various locations is also variable along the stream's course. The particle size usually shifts from an abundance of coarser material upstream to mainly finer material in downstream areas.

Sinuosity

Unless forced by humans in the form of heavily regulated and channelized streams, straight channels are uncommon. Stream flow creates distinctive landforms composed of straight (usually in appearance only), meandering, and braided channels, channel networks, and floodplains. Simply put flowing water will follow a sinuous course. The most commonly used measure is the *sinuosity index* (SI). Sinuosity equals 1 in straight channels and more than 1 in sinuous channels. *Meandering* is the natural tendency for alluvial channels and is usually defined as an arbitrarily extreme level of sinuosity, typically an SI greater than 1.5. Many variables affect the degree of sinuosity.

Even in many natural channel sections of a stream course that appear straight, meandering occurs in the line of maximum water or channel depth (known as the *thalweg*). Keep in mind that a stream has to meander; that is how they renew themselves. By meandering, they wash plants and soil from the land into their waters, which serve as nutrients for the plants in the rivers. If rivers are not allowed to meander and are *channelized*, the amount of life they can support will gradually decrease. That means fewer fish, ultimately—and fewer bald eagles, herons, and other fishing birds (Spellman 1996).

Meander flow follows a predictable pattern and causes regular regions of erosion and deposition (see Figure 13.12). The streamlines of maximum velocity and the deepest part of the channel lie close to the outer side of each bend and cross over near the point of inflection between the banks (see Figure 13.12). A huge elevation of water at the outside of a bend causes a helical flow of water toward the opposite bank. In addition, a separation of surface flow causes a back eddy. The result is zones of erosion and deposition, which explains why point bars develop in a downstream direction in depositional zones.

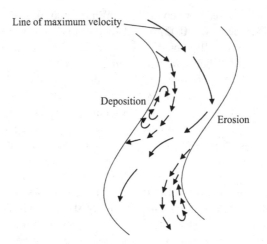

FIGURE 13.12   A meandering reach.

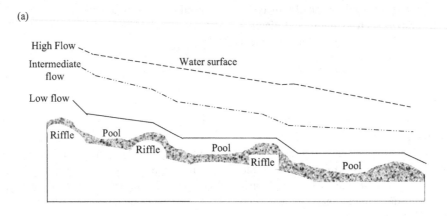

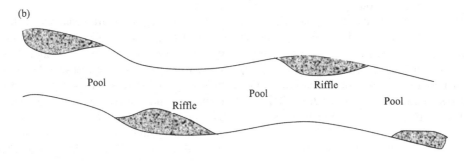

FIGURE 13.13   (a) Longitudinal profile of a riffle-pool sequence; (b) plain view of riffle-pool sequence.

BARS, RIFFLES, AND POOLS

Implicit in the morphology and formation of meanders are *bars*, *riffles*, and *pools*. Bars develop by deposition in slower, less competent flow on either side of the sinuous mainstream. As onward-moving water, depleted of bed load, regains competence, it shears a pool in the meander—reloading the stream for the next bar. Alternating bars migrate to form riffles (see Figure 13.13). As stream flow continues along its course, a pool-riffle sequence is formed. The riffle is a mound or hillock and the pool is a depression.

THE FLOODPLAIN

A stream channel influences the shape of the valley floor through which it courses. The self-formed, self-adjusted flat area near the stream is the *floodplain*, which loosely describes the valley floor prone to periodic inundation during over-bank discharges. What is not commonly known is that valley flooding is a regular and natural behavior of the stream. A stream's aquatic community has several unique characteristics. The aquatic community operates under the same ecological principles as terrestrial ecosystems, but the physical structure of the community is more isolated and exhibits limiting factors that are very different from the limiting factors of a terrestrial ecosystem. Certain materials and conditions are necessary for the growth and reproduction of organisms. If, for instance, a farmer plants wheat in a field containing too little nitrogen, it will stop growing when it has used up the available nitrogen, even if the wheat's requirements for oxygen, water, potassium, and other nutrients are met. In this particular case, nitrogen is said to be the limiting factor. A *limiting factor* is a condition or a substance (the resource in the shortest supply) that limits the presence and success of an organism or a group of organisms in an area.

Even the smallest mountain stream provides an astonishing number of different places for aquatic organisms to live, or *habitats*. If it is a rocky stream, every rock of the substrate provides several different habitats. On the side facing upriver, organisms with special adaptations that are very good at clinging to rocks do well here. On the side that faces downriver, a certain degree of shelter is provided from the current but still allows organisms to hunt for food. The top of a rock, if it contacts air, is a good place for organisms that cannot breathe underwater and need to surface now and then. Underneath the rock is a popular place for organisms that hide to prevent predation. Normal stream life can be compared to that of a "balanced aquarium" (ASTM 1969); that is, nature continuously strives to provide clean, healthy, normal streams. This is accomplished by maintaining the stream's flora and fauna in a balanced state. Nature balances stream life by maintaining both the number and the type of species present in any one part of the stream. Such balance ensures that there is never an overabundance of one species compared to another. Nature structures the stream environment so that both plant and animal life are dependent upon the existence of others within the stream.

As mentioned, lotic (washed) habitats are characterized by continuously running water of current flow. These running water bodies typically have three zones: riffle, run, and pool. The *riffle zone* contains faster-flowing, well-oxygenated water with coarse sediments. In the riffle zone, the velocity of the current is great enough to keep the bottom clear of silt and sludge, thus providing a firm bottom for organisms. This zone contains specialized organisms that are adapted to live in running water. For example, organisms adapted to live in fast streams or rapids (trout) have streamlined bodies, which aid in their respiration and in obtaining food (Smith 1996). Stream organisms that live under rocks to avoid the strong current have flat or streamlined bodies. Others have hooks or suckers to cling to or attach themselves to a firm substrate to avoid the washing-away effect of the strong current. The *run zone* (or intermediate zone) is the slow-moving, relatively shallow part of the stream with moderately low velocities and little or no surface turbulence. The *pool zone* of the stream is usually a deeper water region where the velocity of the water is reduced and silt and other settling solids provide a soft bottom (more homogeneous sediments), which is unfavorable for sensitive bottom-dwellers. Decomposition of some

of these solids causes a lower amount of dissolved oxygen (DO). Some stream organisms spend some of their time in the rapids part of the stream and other times can be found in the pool zone (trout, for example). Trout typically spend about the same amount of time in the rapid zone pursuing food as they do in the pool zone pursuing shelter.

Organisms are sometimes classified based on their mode of life:

1. **Benthos (Mud Dwellers):** The term originates from the Greek word for bottom and broadly includes aquatic organisms living on the bottom or on submerged vegetation. They live under and on rocks and in the sediments. A shallow sandy bottom has sponges, snails, earthworms, and some insects. A deep, muddy bottom will support clams, crayfish, and the nymphs of damselflies, dragonflies, and mayflies. A firm, shallow, rocky bottom has the nymphs of mayflies, stoneflies, and larvae of water beetles.

2. **Periphytons or Aufwuchs:** The first term usually refers to microfloral growth upon substrata (i.e., benthic-attached algae). The second term, *aufwuchs* (pronounced: OWF-vooks; German: "growth upon"), refers to the fuzzy, sort of furry-looking, slimy green coating that attaches or clings to them stems and leaves of rooted plants or other objects projecting above the bottom without penetrating the surface. It consists not only of algae like Chlorophyta but also diatoms, protozoans, bacteria, and fungi.

3. **Planktons (Drifters):** They are small, mostly microscopic plants and animals that are suspended in the water column; movement depends on water currents. They mostly float in the direction of the current. There are two types of plankton: (1) *Phytoplankton* are assemblages of small plants (algae) and have limited locomotion abilities; they are subject to movement and distribution by water movements. (2) *Zooplankton* are animals that are suspended in water and have limited means of locomotion. Examples of zooplankton include crustaceans, protozoans, and rotifers.

4. **Nektons or Pelagic Organisms (Capable of Living in Open Waters):** They are distinct from other plankton in that they are capable of swimming independently of turbulence. They are swimmers that can navigate against the current. Examples of nekton include fish, snakes, diving beetles, newts, turtles, birds, and large crayfish.

5. **Neustons:** They are organisms that float or rest on the surface of the water (never breaking water tension). Some varieties can spread out their legs so that the surface tension of the water is not broken; for example, water striders (see Figure 13.14).

6. **Madricoles:** Organisms that live on rock faces in waterfalls or seepages.

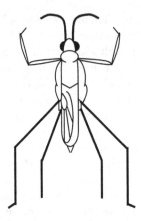

FIGURE 13.14 Water strider. (Adapted from Standard Methods, 15th Edition. Copyright © 1981 by the American Public Health Association, the American Water Works Association, and the Water Pollution Control Federation.)

In a stream, the rocky substrate is the home for many organisms. Thus, we need to know something about the particles that make up the substrate. Namely, we need to know how to measure the particles so we can classify them by size.

Substrate particles are measured with a metric ruler, in centimeters (cm). Because rocks can be long and narrow, we measure them twice: first the width, then the length. By adding the width to the length and dividing by two, we obtain the average size of the rock.

It is important to randomly select the rocks we wish to measure. Otherwise, we would tend to select larger rocks, more colorful rocks, or those with unusual shapes. Instead, we should just reach down and collect the rocks in front of us and within easy reach. Then measure each rock. Upon completion of measurement, each rock should be classified. Ecologists have developed a standard scale (Wentworth scale) for size categories of substrate rock and other mineral materials, along with the different sizes. These are:

Boulder	>256 mm
Cobble	64–256 mm
Pebble	16–64 mm
Gravel	2–16 mm
Sand	0.0625–2 mm
Silt	0.0039–0.0625 mm
Clay	<0.0039 mm

Organisms that live in/on/under rocks or small spaces occupy what is known as a *microhabitat*. Some organisms make their own microhabitats: many caddisflies build a case around themselves and use it for their shelter.

Rocks are not the only physical features of streams where aquatic organisms can be found. For example, fallen logs and branches (commonly referred to as Large Woody Debris, or LWD) provide an excellent place for some aquatic organisms to burrow into and surfaces for others to attach to, as they might to a rock. They also create areas where small detritus, such as leaf litter, can pile up underwater. These piles of leaf litter are excellent shelters for

many organisms, including the large, fiercely predaceous larvae of dobsonflies.

Another important aquatic habitat is found in the matter, or *drift*, that floats along downstream. Drift is important because it is the main source of food for many fish. It may include insects such as mayflies (*Ephemeroptera*), some true flies (*Diptera*), and some stoneflies (*Plecoptera*) and caddisflies (*Trichoptera*). In addition, dead or dying insects and other small organisms, terrestrial insects that fall from trees, leaves, and other matter are common components of drift. Among the crustaceans, amphipods (small crustaceans) and isopods (small crustaceans, including sow bugs and gribbles) have also been reported in the drift.

ADAPTATION TO STREAM CURRENT

The current in streams is the outstanding feature of streams and the major factor limiting the distribution of organisms. The current is determined by the steepness of the bottom gradient, the roughness of the streambed, and the depth and width of the streambed. The current in streams has promoted many special adaptations in stream organisms. Odum (1971) listed these adaptations as follows (see Figure 13.15):

1. **Attachment to a Firm Substrate:** Attachment is to stones, logs, leaves, and other underwater objects such as discarded tires, bottles, pipes, etc. Organisms in this group are primarily composed of the primary producer plants and animals, such as green algae, diatoms, aquatic mosses, caddisfly larvae, and freshwater sponges.

2. **The Use of Hooks and Suckers:** These organisms have the unusual ability to remain attached and withstand even the strongest rapids. Two Diptera larvae, *Simulium* and *Blepharocera*, are examples.

3. **A Sticky Undersurface:** Snails and flatworms are examples of organisms that are able to use their sticky undersurfaces to adhere to underwater surfaces.

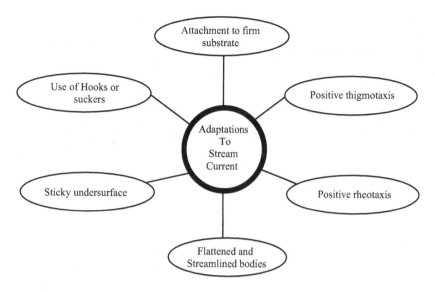

FIGURE 13.15 Adaptations to stream current.

4. **Flattened and Streamlined Bodies:** All macro-consumers have streamlined bodies, i.e., the body is broad in front and tapers posteriorly to offer minimum resistance to the current. All nektons such as fish, amphibians, and insect larvae exhibit this adaptation. Some organisms have flattened bodies, which enable them to stay under rocks and in narrow places. Examples are water penny, a beetle larva, mayfly, and stonefly nymphs.

5. **Positive Rheotaxis** (*rheo*: current; *taxis*: arrangement): An inherent behavioral trait of stream animals (especially those capable of swimming) is to orient themselves upstream and swim against the current.

6. **Positive Thigmotaxis** (*thigmo:* touch, contact): Another inherent behavior pattern for many stream animals is to cling close to a surface or keep the body in close contact with the surface. This is the reason that stonefly nymphs (when removed from one environment and placed into another) will attempt to cling to just about anything, including each other.

It would take an entire text to describe the great number of adaptations made by aquatic organisms to their surroundings in streams. For our purposes, instead, we will cover those special adaptations that are germane to this discussion. The important thing to remember is that an aquatic organism can adapt to its environment in several basic ways.

Types of Adaptive Changes

Adaptive changes are classified as genotypic, phenotypic, behavioral, or ontogenetic:

1. **Genotypic Changes:** Tend to be great enough to separate closely related animals into species, such as mutations or recombination of genes. A salmonid is an example that has evolved a subterminal mouth (i.e., below the snout) in order to eat from the benthos.

2. **Phenotypic Changes:** Are the changes that an organism might make during its lifetime to better utilize its environment (e.g., a fish that changes sex from female to male because of an absence of males).

3. **Behavioral Changes:** Have little to do with body structure or type; a fish might spend more time under an overhang to hide from predators.

4. **Ontogenetic Changes:** Take place as an organism grows and matures (e.g., a coho salmon that inhabits streams when young and migrates to the sea when older, changing its body chemistry to allow it to tolerate saltwater).

Specific Adaptations

Specific adaptations observed in aquatic organisms include mouths, shape, color, aestivation, and schooling.

1. **Mouths:** Aquatic organisms such as fish change mouth shape (morphology) depending on the food they eat. The arrangement of the jawbones and even other head bones, the length and width of gill rakers, the number, shape, and location of teeth, and barbels all change to allow fish to eat just about anything found in a stream.

2. **Shape:** Changes to allow fish to do different things in the water. Some organisms have body shapes that push them down in the water against the substrate and will enable them to hold their place against even strong currents (e.g., chubs, catfish, dace, and sculpins). Other organisms, especially predators, have evolved an arrangement and shape of fins that allow them to lurk without moving; they lunge suddenly to catch their prey (e.g., bass, perch, pike, trout, and sunfish).

3. **Color:** May change within hours to camouflage, or within days, or may be genetically predetermined. Fish tend to turn dark in clear water and pale in muddy water.

4. **Aestivation:** Helps fish survive in arid desert climates, where streams may dry up from time to time. *Aestivation* refers to the ability of some fish to burrow into the mud and wait out the dry period.

5. **Schooling:** Serves as protection for many fish, particularly those that are subject to predation.

BENTHIC LIFE

The benthic habitat is found in the streambed or benthos. As mentioned, the streambed is comprised of various physical and organic materials where erosion and/or deposition are continuous characteristics. Erosion and deposition may occur simultaneously and alternately at different locations in the same streambed. Where channels are exceptionally deep and taper slowly to meet the relatively flattened streambed, habitats may form on the slopes of the channel. These habitats are referred to as littoral habitats. Shallow channels may dry up periodically in accordance with weather changes. The streambed is then exposed to open air and may take on the characteristics of a wetland.

Silt and organic materials settle and accumulate in the streambed of slowly flowing streams. These materials decay and become the primary food resource for the invertebrates inhabiting the streambed. Productivity in this habitat depends upon the breakdown of these organic materials by herbivores. Bottom-dwelling organisms do not use all-organic materials; a substantial amount becomes part of the streambed in the form of peat.

In faster-moving streams, organic materials do not accumulate so easily. Primary production occurs in a different type of habitat found in the riffle regions where there are shoals and rocky regions for organisms to adhere to. Therefore, plants that can root themselves in the streambed dominate these regions. By plants, we are referring mostly to forms of algae, often microscopic and filamentous, that can cover rocks and debris that settle into the streambed during the summer months.

Note: If you have ever stepped into a stream, the green, slippery slime on the rocks in the streambed is representative of this type of algae.

Although the filamentous algae seem well-anchored, strong currents can easily lift them from the streambed and carry them downstream where they become a food resource for low-level consumers. One factor that greatly influences the productivity of a stream is the width of the channel; a direct relationship exists between stream width and the richness of bottom organisms. Bottom-dwelling organisms are very important to the ecosystem as they provide food for larger benthic organisms by consuming detritus.

BENTHIC PLANTS AND ANIMALS

Vegetation is not common in the streambed of slow-moving streams; however, some may anchor themselves along the banks. Algae (mainly green and blue-green) as well as common types of water moss attach themselves to rocks in fast-moving streams. Mosses and liverworts often climb up the sides of the channel onto the banks as well. Some plants, similar to the reeds of wetlands with long stems and narrow leaves, are able to maintain roots and withstand the current. Aquatic insects and invertebrates dominate slow-moving streams. Most aquatic insects are in their larval and nymph forms, such as the blackfly, caddisfly, and stonefly. Adult water beetles and water bugs are also abundant. Insect larvae and nymphs provide the primary food source for many fish species, including American eel and brown bullhead catfish. Representatives of crustaceans, rotifers, and nematodes (flatworms) are sometimes present. The abundance of leeches, worms, and mollusks (especially freshwater mussels) varies with stream conditions but generally favors low phosphate conditions. Larger animals found in slow-moving streams and rivers include newts, tadpoles, and frogs. As mentioned, the important characteristic of all life in streams is adaptability to withstand currents.

BENTHIC MACROINVERTEBRATES

The emphasis on aquatic insect studies, which has expanded exponentially in the last three decades, has been largely ecological. With regard to benthic macroinvertebrates, they are small organisms without backbones that are visible to the naked eye and large enough to be easily collected. Freshwater macroinvertebrates are ubiquitous; even polluted waters contain some representative of this diverse and ecologically important group of organisms. Benthic macroinvertebrates are aquatic organisms that spend at least part of their life cycle on the stream bottom. Examples include aquatic insects—such as stoneflies, mayflies, caddisflies, midges, and beetles—as well as crayfish, worms, clams, and snails. Most hatch from eggs and mature from larvae to adults. The majority of the insects spend their larval phase on the river bottom and, after a few weeks to several years, emerge as winged adults. Aquatic beetles, true bugs, and other groups remain in the water as adults. Macroinvertebrates typically collected from the stream substrate are either aquatic larvae or adults.

In practice, stream ecologists observe indicator organisms and their responses to determine the quality of the stream environment. There are a number of methods for determining water quality based on biological characteristics. A wide variety of indicator organisms (biotic groups) are used for biomonitoring. The most often used include algae, bacteria, fish, and macroinvertebrates.

Notwithstanding their popularity, in this text, we use benthic macroinvertebrates for a number of other reasons. Simply, they offer a number of advantages:

1. They are ubiquitous, so they are affected by per-
   turbations in many different habitats.
2. They are species-rich, so the large number of species produces a range of responses.
3. They are sedentary, so they stay put, which
   allows determination of the spatial extent of a
   perturbation.
4. They are long-lived, which allows temporal
   changes in abundance and age structure to be
   followed.
5. They integrate conditions temporally, so like any
   biotic group, they provide evidence of conditions
   over long periods.

In addition, benthic macroinvertebrates are preferred as
bioindicators because they are easily collected and handled
by samplers; they require no special culture protocols.
They are visible to the naked eye and samplers can easily distinguish their characteristics. They have a variety
of fascinating adaptations to stream life. Certain benthic
macroinvertebrates have very specific tolerances and thus
excellent indicators of water quality. Useful benthic macroinvertebrate data are easy to collect without expensive
equipment. The data obtained from macroinvertebrate sampling can indicate the need for additional data collection,
possibly including water analysis and fish sampling.

In short, we focus on benthic macroinvertebrates (with
regard to water quality in streams and lakes) because some
cannot survive in polluted water while others can survive
or even thrive in polluted water. In a healthy stream, the
benthic community includes a variety of pollution-sensitive
macroinvertebrates. In an unhealthy stream or lake, there
may be only a few types of non-sensitive macroinvertebrates
present. Thus, the presence or absence of certain benthic
macroinvertebrates is an excellent indicator of water quality.

Moreover, it may be difficult to identify stream or lake
pollution with water analysis, which can only provide information for the time of sampling (a snapshot in time). Even
the presence of fish may not provide information about a
polluted stream because fish can move away to avoid polluted water and then return when conditions improve.
However, most benthic macroinvertebrates cannot move to
avoid pollution. Thus, a macroinvertebrate sample may provide information about pollution that is not present during
sample collection.

Obviously, before we can use benthic macroinvertebrates to gauge water quality in a stream (or for any other
reason), we must be familiar with the macroinvertebrates
that are commonly used as bioindicators. Samplers need to
be aware of basic insect structures before they can classify
the macroinvertebrates they collect. Structures that need
to be stressed include head, eyes (compound and simple),
antennae, mouth (no emphasis on parts), segments, thorax,
legs and leg parts, gills, abdomen, etc. Samplers also need
to be familiar with insect metamorphosis—both complete
and incomplete—as most of the macroinvertebrates collected are in larval or nymph stages.

Note: Information on basic insect structures is beyond
the scope of this text. Thus, we highly recommend "the"
standard guide to aquatic insects of North America: Merritt
and Cummins (1996).

IDENTIFICATION OF BENTHIC MACROINVERTEBRATES

Before identifying and describing the key benthic macroinvertebrates significant to water and wastewater operators, it
is important first to provide foundational information. We
characterize benthic macroinvertebrates using two important descriptive classifications: trophic groups and mode of
existence. In addition, we discuss their relationship in the
food web; that is, what or whom they eat.

1. *Trophic groups*: Of the trophic groups (i.e., feeding groups) that Merritt and Cummins identified
   for aquatic insects, only five are likely to be found
   in a stream using typical collection and sorting
   methods (Merritt and Cummins 1996):

   Shredders: These have strong, sharp mouthparts that allow them to shred and chew coarse
   organic material such as leaves, algae, and rooted
   aquatic plants. These organisms play an important
   role in breaking down leaves or larger pieces of
   organic material to a size that other macroinvertebrates can use. Shredders include certain stonefly
   and caddisfly larvae, sowbugs, scuds, and others.

   Collectors: These gather the finest suspended
   matter in the water. To do this, they often sieve
   the water through rows of tiny hairs. These sieves
   of hairs may be displayed in fans on their heads
   (blackfly larvae) or on their forelegs (some mayflies). Some caddisflies and midges spin nets
   and catch their food in them as the water flows
   through.

   Scrapers: These scrape algae and diatoms off
   surfaces of rocks and debris using their mouthparts. Many of these organisms are flattened to
   hold onto surfaces while feeding. Scrapers include
   water pennies, limpets, snails, net-winged midge
   larvae, certain mayfly larvae, and others.

   Piercers: These herbivores pierce plant tissues
   or cells and suck the fluids out. Some caddisflies
   do this.

   Predators: Predators eat other living creatures.
   Some of these are *engulfers*, meaning they eat their
   prey completely or in parts. This is very common
   in stoneflies, dragonflies, and caddisflies. Others
   are *piercers*, similar to the herbivorous piercers
   except they eat live animal tissues.

2. *Mode of Existence* (Habit, Locomotion,
   Attachment, Concealment):

   Skaters: Adapted for "skating" on the surface where they feed as scavengers on organisms trapped in the surface film (example: water
   striders).

Planktonic: Inhabiting the open-water limnetic zone of standing waters (lentic: lakes, bogs, ponds). Representatives may float and swim in the open water but usually exhibit a diurnal vertical migration pattern (example: phantom midges) or float at the surface to obtain oxygen and food, diving when alarmed (example: mosquitoes).

Divers: Adapted for swimming by "rowing" with the hind legs in lentic habitats and lotic pools. Representatives come to the surface to obtain oxygen, dive and swim when feeding or alarmed, and may cling to or crawl on submerged objects such as vascular plants (examples: water boatmen, predaceous diving beetle).

Swimmers: Adapted for "fishlike" swimming in lotic or lentic habitats. Individuals usually cling to submerged objects, such as rocks (lotic riffles) or vascular plants (lentic), between short bursts of swimming (for example: mayflies).

Clingers: Representatives have behavioral (e.g., fixed retreat construction) and morphological (e.g., long, curved tarsal claws, dorsoventral flattening, and ventral gills arranged as a sucker) adaptations for attachment to surfaces in stream riffles and wave-swept rocky littoral zones of lakes (e.g., mayflies and caddisflies).

Sprawlers: Inhabiting the surface of floating leaves of vascular hydrophytes or fine sediments, usually with modifications for staying on top of the substrate and maintaining the respiratory surfaces free of silt (e.g., mayflies, dobsonflies, and damselflies).

Climbers: Adapted for living on vascular hydrophytes or detrital debris (e.g., overhanging branches, roots, and vegetation along streams, and submerged brush in lakes) with modifications for moving vertically on stem-type surfaces (e.g., dragonflies and damselflies).

Burrowers: Inhabiting the fine sediments of streams (pools) and lakes. Some construct discrete burrows, which may have sand grain tubes extending above the surface of the substrate or individuals may ingest their way through the sediments (e.g., mayflies and midges).

MACROINVERTEBRATES AND THE FOOD WEB

In a stream or lake, the two possible sources of primary energy are (1) photosynthesis by algae, mosses, and higher aquatic plants and (2) imported organic matter from streamside/lakeside vegetation (e.g., leaves and other parts of vegetation). Simply put, a significant portion of the food that is eaten grows right in the stream or lake, such as algae, diatoms, nymphs and larvae, and fish. Food that originates from within the stream is called *autochthonous*. Most food in a stream, however, comes from outside the stream—especially the case in small, heavily wooded streams,

where there is normally insufficient light to support substantial in-stream photosynthesis; thus, energy pathways are supported largely by imported energy. Leaves provide a large portion of this imported energy. Worms drown in floods and are washed in. Leafhoppers and caterpillars fall from trees. Adult mayflies and other insects mate above the stream, lay their eggs in it, and then die in it. All of this food from outside the stream is called *allochthonous*.

Macroinvertebrates, like all other organisms, are classified and named. Macroinvertebrates are classified and named using a *taxonomic hierarchy*. The taxonomic hierarchy for the caddisfly (a macroinvertebrate insect commonly found in streams) is shown below.

Kingdom: Animalia (animals)
Phylum: Arthropoda ("jointed legs")
Class: Insecta (insect)
Order: Trichoptera (caddisfly)
Family: Hydropsychidae
                      (net-spinning caddis)
Genus species: *Hydropsyche morosa*

BENTHIC MACROINVERTEBRATES IN RUNNING WASTERS

As mentioned, macroinvertebrates are the best-studied and most diverse animals in streams. We therefore devote our discussion to the various macroinvertebrate groups. While it is true that non-insect macroinvertebrates, such as Oligochaeta (worms), Hirudinea (leeches), and Acari (water mites), are frequently encountered groups in lotic environments, insects are among the most conspicuous inhabitants of streams. In most cases, it is the larval stages of these insects that are aquatic, whereas the adults are terrestrial. Typically, the larval stage is much extended, while the adult lifespan is short. Lotic insects are found among many different orders, and brief accounts of their biology are presented in the following sections. First, however, we present a brief glossary that includes select terminology used below and also includes other important terms mentioned earlier that are often associated with the description of aquatic macroinvertebrates. The source of most of these definitions is F.R. Spellman's (1996) *Stream Ecology and Self Purification* (Boca Raton, FL: CRC Press; Merritt and Cummins 1996)

MACROINVERTEBRATE GLOSSARY

Abdomen: The third main division of the body; behind the head and thorax.

Anterior: In front (before).

Apical: Near or pertaining to the end of any structure; part of the structure that is farthest from the body.

Basal: Pertaining to the end of any structure that is nearest to the body.

Burrower: An animal that uses a variety of structures designed for moving and burrowing into sand and silt or building tubes within loose substrate.

Carapace: The hardened part of some arthropods that spreads like a shield over several segments of the head and thorax.

Caudal Filament: A threadlike projection at the end of the abdomen, like a tail.

Clinger: An animal that uses claws or hooks to cling to surfaces such as rocks, plants, or other hard surfaces and often moves slowly along these surfaces.

Concentric: A growth pattern on the opercula of some gastropods, marked by a series of circles that lie entirely within each other; compare multi-spiral and pauci-spiral.

Crawler: An animal whose main means of locomotion is moving slowly along the bottom, usually has some type of hooks, claws, or specially designed feet to help hold it to surfaces.

Detritus: Disintegrated or broken-up mineral or organic material.

Dextral: The curvature of a gastropod shell where the opening is visible on the right when the spire is pointing up.

Distal: Near or toward the free end of any appendage; that part farthest from the body.

Dorsal: Pertaining to or situated on the back or top, especially of the thorax and abdomen.

Elytra: Hardened shell-like mesothoracic wings of adult beetles (Coleoptera).

Femur: The leg section between the tibia and coxa of Arthropoda, comparable to an upper arm or thigh.

Flagellum: A small, finger-like or whip-like projection.

Gill: Any structure especially adapted for the exchange of dissolved gases between an animal and a surrounding liquid.

Glossae: A lobe or lobes located front and center on the labium; in Plecoptera, the lobes are between the paraglossae.

Hemimetabolism: Incomplete metamorphosis.

Holometabolism: Complete metamorphosis.

Labium: Lower mouthpart of an arthropod, like a jaw or lip.

Labrum: Upper mouthpart of an arthropod consisting of a single usually hinged plate above the mandibles.

Lateral: Feature or marking located on the side of a body or other structure.

Ligula: Forming the ventral wall of an arthropod's oral cavity.

Lobe: A rounded projection or protuberance.

Mandibles: The first pair of jaws in insects.

Maxillae: The second pair of jaws in insects.

Multi Spiral: The growth pattern on the opercula of some gastropods marked by several turns from the center to the edge.

Operculum: A lid or covering structure, like a door to an opening.

Palpal Lobes: The grasping pincers at the end of the lower jaw in Odonata.

Pauci-Spiral: A growth pattern on the opercula of some gastropods marked by a few turns from the center to the edge.

Periphyton: Algae and associated organisms that live attached to underwater surfaces.

Posterior: Behind; opposite of anterior.

Proleg: Any projection or appendage that serves to support locomotion or attachment.

Prothorax: The first thoracic segment closest to the head.

Rostrum: A beak or beak-like mouthpart.

Sclerite: A hardened area of an insect's body wall, usually surrounded by softer membranes.

Seta – (pl. Setae): Hair-like projection.

Sinistral: The curvature of a gastropod shell where the opening is seen on the left when the spire is pointing up.

INSECT MACROINVERTEBRATES

The most important insect groups in streams are Ephemeroptera (mayflies), Plecoptera (stoneflies), Trichoptera (caddisflies), Diptera (true flies), Coleoptera (beetles), Hemiptera (bugs), Megaloptera (alderflies and dobsonflies), and Odonata (dragonflies and damselflies). The identification of these different orders is usually easy, and there are many keys and specialized references (e.g., Merritt and Cummins 1996) available to help in identifying species. In contrast, for some genera and species, specialist taxonomists can often only diagnose, particularly the Diptera. As mentioned, insect macroinvertebrates are ubiquitous in streams and are often represented by many species. Although the numbers refer to aquatic species, a majority are to be found in streams.

Mayflies (Order: Ephemeroptera)

Description: Wing pads may be present on the thorax; three pairs of segmented legs attach to the thorax; one claw occurs at the end of the segmented legs; gills occur on the abdominal segments and are attached mainly to the sides of the abdomen, but sometimes extend over the top and bottom of the abdomen; gills consist of either flat plates or filaments; their long, thin caudal (tails filaments) usually occur at the end of the abdomen, but there may only be two in some kinds.

Streams and rivers are generally inhabited by many species of mayflies; in fact, most species are restricted to streams. For the experienced freshwater ecologist who looks upon a mayfly nymph, recognition is obtained through trained observation: abdomen with leaf-like or feather-like gills, legs with a single tarsal claw, generally (but not always) with three cerci (three 'tails'; two cerci, and between them usually a terminal filament; see Figure 13.16). The experienced ecologist knows that mayflies are hemimetabolous insects (i.e., where larvae or nymphs resemble wingless adults) that go through many postembryonic molts, often in the range of 20 to 30. For some species, body length increases about 15% for each instar.

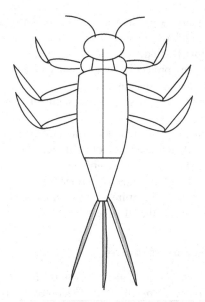

FIGURE 13.16   Mayfly (Ephemeroptera order).

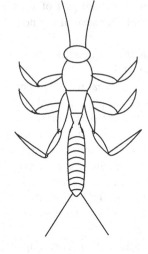

FIGURE 13.17   Stonefly (Plecoptera order).

Mayfly nymphs are mainly grazers or collector-gatherers feeding on algae and fine detritus, although a few genera are predatory. Some members filter particles from the water using hair-fringed legs or maxillary palps. Shredders are rare among mayflies. In general, mayfly nymphs tend to live mostly in unpolluted streams, where densities can reach up to 10,000/m², contributing substantially to secondary producers.

Adult mayflies resemble nymphs, but usually possess two pairs of long, lacy wings folded upright; adults usually have only two cerci. The adult lifespan is short, ranging from a few hours to a few days, rarely up to two weeks, and the adults do not feed. Mayflies are unique among insects in having two winged stages, the subimago and the imago. The emergence of adults tends to be synchronous, thus ensuring the survival of enough adults to continue the species.

Stoneflies (Order: Plecoptera)

Description: Long, thin antennae project in front of the head; wing pads are usually present on the thorax but may only be visible in older larvae; three pairs of segmented legs attach to the thorax; two claws are located at the end of the segmented legs; gills occur in the thorax region, usually on the legs or bottom of the thorax, or there may be no visible gills (usually there are none or very few gills on the abdomen); gills are either single or branched filaments; two long, thin tails project from the rear of the abdomen. Stoneflies have very low tolerance to many insults; however, several families are tolerant of slightly acidic conditions.

Although many freshwater ecologists would maintain that the stonefly is a well-studied group of insects, this is not exactly the case. Despite their importance, less than 5%–10% of stonefly species are well known with respect to

life history, trophic interactions, growth, development, spatial distribution, and nymphal behavior. Notwithstanding our lack of extensive knowledge regarding stoneflies, enough is known to provide an accurate characterization of these aquatic insects. We know, for example, that stonefly larvae are characteristic inhabitants of cool, clean streams (i.e., most nymphs occur under stones in well-aerated streams). While they are sensitive to organic pollution, or more precisely to low oxygen concentrations accompanying organic breakdown processes, stoneflies seem rather tolerant of acidic conditions. Lack of extensive gills at least partly explains their relative intolerance of low oxygen levels.

Stoneflies are drab-colored, small- to medium-sized 1/6 to 2¼ inches (4–60 mm), rather flattened insects. Stoneflies have long, slender, many-segmented antennae and two long narrow antenna-like structures (cerci) on the tip of the abdomen (see Figure 13.17). The cerci may be long or short. At rest, the wings are held flat over the abdomen, giving a "square-shouldered" look compared to the roof-like position of most caddisflies and the vertical position of the mayflies. Stoneflies have two pairs of wings. The hindwings are slightly shorter than the forewings and much wider, having a large anal lobe that is folded fanwise when the wings are at rest. This fanlike folding of the wings gives the order its name: "*pleco*" (folded or plaited) and "*-ptera*" (wings). The aquatic nymphs are generally very similar to mayfly nymphs except that they have only two cerci at the tip of the abdomen. Stoneflies have chewing mouthparts. They may be found anywhere in a non-polluted stream where food is available. Many adults, however, do not feed and have reduced or vestigial mouthparts.

Stoneflies have a specific niche in high-quality streams, where they are very important as a fish food source at specific times of the year (winter to spring, especially) and of the day. They complement other important food sources, such as caddisflies, mayflies, and midges.

Caddisflies (Order: Trichoptera)

Description: The head has a thick, hardened skin; the antennae are very short, usually not visible; no wing pads occur on the thorax; the top of the first thorax always has a hardened plate, and in several families, the second and third sections of the thorax have a hardened plate; three pairs of segmented legs attach to the thorax; the abdomen has thin, soft skin; there is a singe of branched gills on the abdomen in many families, but some have no visible gills; a pair of prolegs with one claw on each, is situated at the end of the abdomen; most families construct various kinds of retreats consisting of a wide variety of materials collected from the streambed.

Trichoptera (Greek: *trichos*, a hair; *ptera*, wing) is one of the most diverse insect orders living in the stream environment, and caddisflies have nearly a worldwide distribution (the exception: Antarctica). Caddisflies may be categorized broadly into free-living (roving and net-spinning) and case-building species.

Caddisflies are described as medium-sized insects with bristle-like and often long antennae. They have membranous, hairy wings (which explains the Latin name "Trichos"), which are held tent-like over the body when at rest; most are weak fliers. They have greatly reduced mouthparts and five tarsi. The larvae are mostly caterpillar-like and have a strongly sclerotized (hardened) head with very short antennae and biting mouthparts. They have well-developed legs with a single tarsus. The abdomen is usually ten-segmented; in case-bearing species the first segment bears three papillae, one dorsally and the other two laterally, which help hold the insect centrally in its case, allowing a good flow of water passed the cuticle and gills. The last or anal segment bears a pair of grappling hooks.

In addition to being aquatic insects, caddisflies are superb architects. Most caddisfly larvae (see Figure 13.18) live in self-designed, self-built houses called *cases*. They spin out silk and either live in silk nets or use the silk to stick together bits of whatever is lying on the stream bottom. These houses are so specialized that you can usually identify a caddisfly larva to genus if you can see its house (case). With nearly 1,400 species of caddisflies in North America (north of Mexico), this is a good thing!

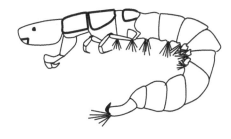

FIGURE 13.18   Caddis (Hydropsyche) larvae.

Caddisflies are closely related to butterflies and moths (Order: Lepidoptera). They live in most stream habitats and that is why they are so diverse (have so many species). Each species has special adaptations that allow it to live in the environment in which it is found.

Mostly herbivorous, most caddisflies feed on decaying plant tissue and algae. Their favorite algae are diatoms, which they scrape off rocks. Some of them, though, are predaceous.

Caddisfly larvae can take a year or two to change into adults. They then change into *pupae* (the inactive stage in the metamorphosis of many insects, following the larval stage and preceding the adult form) while still inside their cases for their metamorphosis. It is interesting to note that caddisflies, unlike stoneflies and mayflies, go through a "complete" metamorphosis.

Caddisflies remain as pupae for 2–3 weeks and then emerge as adults. When they leave their pupae, splitting their case, they must swim to the surface of the water to escape it. The winged adults fly in the evening and at night, and some are known to feed on plant nectar. Most of them will live less than a month; like many other winged stream insects, their adult lives are brief compared to the time they spend in the water as larvae.

Caddisflies are sometimes grouped by the kinds of cases they make into five main groups: free-living forms that do not make cases, saddle-case makers, purse-case makers, net-spinners and retreat-makers, and tube-case makers. Caddisflies demonstrate their architectural talents in the cases they design and make. For example, a caddisfly might make a perfect four-sided box case of bits of leaves and bark or tiny bits of twigs. It may make a clumsy dome of large pebbles. Others make rounded tubes out of twigs or very small pebbles. In our experience in gathering caddisflies, we have come to appreciate not only their architectural ability but also their flair in the selection of construction materials. For example, we have found many caddisfly cases constructed of silk emitted through an opening at the tip of the labium, used together with bits of ordinary rock mixed with sparkling quartz and red garnet, green peridot, and bright fool's gold.

In addition to the protection their cases provide them, the cases provide another advantage. The cases actually help caddisflies breathe. They move their bodies up and down, back and forth inside their cases, and this makes a current that brings them fresh oxygen. The less oxygen there is in the water, the faster they have to move. It has been seen that caddisflies inside their cases get more oxygen than those that are outside of their cases—and this is why stream ecologists think that caddisflies can often be found even in still waters, where dissolved oxygen is low, in contrast to stoneflies and mayflies.

True Flies (Order: Diptera)

Description: The head may be a capsule-like structure with thick, hard skin; the head may be partially reduced so that it appears to be part of the thorax, or it may be greatly reduced with only the mouthparts visible. No wing pads occur on the thorax. False-legs (pseudo-legs) may extend from various sections of the thorax and abdomen, composed, entirely of soft skin, but some families have hardened plates scattered on various body features. The larval stages do not have segmented leg features.

True or two- (*Di*-) winged (*ptera*) flies not only include the flies that we are most familiar with, like fruit flies and houseflies, but they also include midges (see Figure 13.19), mosquitoes, craneflies (see Figure 13.20), and others. Houseflies and fruit flies live only on land, and we do not concern ourselves with them. Some, however, spend nearly their whole lives in water; they contribute to the ecology of streams.

True flies are in the order Diptera and are one of the most diverse orders of the class Insecta, with about 120,000 species worldwide. Dipteran larvae occur almost everywhere except Antarctica and deserts where there is no running water. They may live in a variety of places within a stream: buried in sediments, attached to rocks, beneath stones, in saturated wood or moss, or in silken tubes, attached to the stream bottom. Some even live below the stream bottom.

True fly larvae may eat almost anything, depending on their species. Those with brushes on their heads use them to strain food out of the water that passes through. Others may eat algae, detritus, plants, and even other fly larvae.

The longest part of the true fly's life cycle, like that of mayflies, stoneflies, and caddisflies, is the larval stage. It may remain an underwater larva anywhere from a few hours to 5 years. The colder the environment, the longer it takes to mature. It pupates and emerges, becoming a winged adult. The adult may live for 4 months—or it may only live for a few days. While reproducing, it will often eat plant nectar for the energy it needs to produce its eggs. Mating sometimes takes place in aerial swarms. The eggs are deposited back in the stream; some females will crawl along the stream bottom, losing their wings, to search for the perfect place to lay their eggs. Once they lay them, they die.

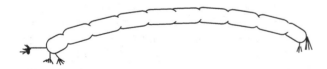

FIGURE 13.19 Midge larvae.

Diptera serve an important role in cleaning water and breaking down decaying material, and they are a vital food source (i.e., they play pivotal roles in the processing of food energy) for many of the animals living in and around streams. However, the true flies most familiar to us are the midges, mosquitoes, and craneflies because they are pests. Some midge flies and mosquitoes bite; the cranefly, however, does not bite but looks like a giant mosquito.

Like mayflies, stoneflies and caddisflies, true flies are mostly in larval form. Like caddisflies, you can also find their pupae, as they are holometabolous insects (going through complete metamorphosis). Most of them are free-living; that is, they can travel around. Although none of the true fly larvae have the six jointed legs we see on the other insects in the stream, they sometimes have strange little almost-legs (prolegs) to move around with. Others may move somewhat like worms do, and some—the ones that live in waterfalls and rapids—have a row of six suction discs that they use to move much like a caterpillar does. Many use silk pads and hooks at the ends of their abdomens to hold themselves fast to smooth rock surfaces.

Beetles (Order: Coleoptera)

Description: The head has thick hardened skin; the thorax and abdomen of most adult families have moderately hardened skin, while several larvae have a soft-skinned abdomen. No wing pads on the thorax in most larvae are seen, but wing pads are usually visible in adults. Three pairs of segmented legs attach to the thorax. No structures or projections extend from the sides of the abdomen in most adult families, but some larval stages have flat plates or filaments. There are no prolegs or long tapering filaments at the end of the abdomen. Beetles are one of the most diverse insect groups but are not as common in aquatic environments.

Of the more than one million described species of insect, at least one-third are beetles, making the Coleoptera not only the largest order of insects but also the most diverse order of living organisms. Even though it is the most special order of terrestrial insects, surprisingly their diversity is not so apparent in running waters. Coleoptera belongs to the infraclass Neoptera, division Endopterygota. Members of this order have an anterior pair of wings (the *elytra*) that are hard and leathery and not used in flight; the membranous hindwings, which are used for flight, are concealed under the elytra when the organisms are at rest. Only 10% of the 350,000 described species of beetles are aquatic.

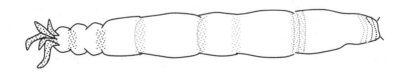

FIGURE 13.20 Cranefly larvae.

Beetles are holometabolous. Eggs of aquatic coleopterans hatch in 1 or 2 weeks, with diapause occurring rarely. Larvae undergo three to eight molts. The pupal phase of all coleopterans is technically terrestrial, making this life stage of beetles the only one that has not successfully invaded the aquatic habitat. A few species have diapausing prepupae, but most complete their transformation to adults in 2–3 weeks. Terrestrial adults of aquatic beetles are typically short-lived and sometimes nonfeeding, like those of the other orders of aquatic insects. The larvae of Coleoptera are morphologically and behaviorally different from the adults, and their diversity is high.

Aquatic species occur in two major suborders, the Adephaga and the Polyphaga. Both larvae and adults of six beetle families are aquatic: Dytiscidae (predaceous diving beetles), Elmidae (riffle beetles), Gyrinidae (whirligig beetles), Halipidae (crawling water beetles), Hydrophilidae (water scavenger beetles), and Noteridae (burrowing water beetles). Five families, Chrysomelidae (leaf beetles), Limnichidae (marsh-loving beetles), Psephenidae (water pennies), Ptilodactylidae (toe-winged beetles), and Scirtidae (marsh beetles), have aquatic larvae and terrestrial adults, as do most of the other orders of aquatic insects. Adult Limnichids, however, readily submerge when disturbed. Three families have species that are terrestrial as larvae and aquatic as adults: Curculionidae (weevils), Dryopidae (long-toed water beetles), and Hydraenidae (moss beetles), a highly unusual combination among insects. Because they provide a greater understanding of a freshwater body's condition (i.e., they are useful indicators of water quality), we focus our discussion on the riffle beetle, water penny, and whirligig beetle.

Riffle beetle larvae (most commonly found in running waters, hence the name Riffle Beetle) are up to ¾ inches long (see Figure 13.21). Their body is not only long but also hard, stiff, and segmented. They have six long segmented legs on the upper middle section of the body; the back end has two tiny hooks and short hairs. Larvae may take 3 years to mature before they leave the water to form a pupa; adults return to the stream. Riffle beetle adults are considered better indicators of water quality than larvae because they have been subjected to water quality conditions over a longer period. They walk very slowly under the water (on the stream bottom), and do not swim on the surface. They have small oval-shaped bodies (see Figure 13.22) and are typically about ¼ inches in length. Both adults and larvae of most species feed on fine detritus with associated microorganisms that are scraped from the substrate, although others may be xylophagous, that is, wood-eating (e.g., *Lara*, Elmidae). Predators do not seem to include riffle beetles in their diet, except perhaps for eggs, which are sometimes attacked by flatworms.

The adult *water penny* is inconspicuous and is often found clinging tightly in a sucker-like fashion to the undersides of submerged rocks, where it feeds on attached algae. The body is broad, slightly oval, and flat in shape, ranging from 4 to 6 mm (1/4 inches) in length. The body is covered

FIGURE 13.21   Riffle beetle larvae.

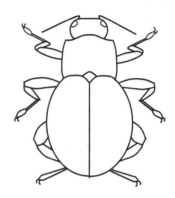

FIGURE 13.22   Riffle beetle adult.

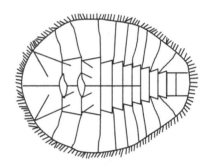

FIGURE 13.23   Water penny larvae.

with segmented plates and looks like a tiny round leaf (see Figure 13.23). It has six tiny, jointed legs (underneath). The color ranges from light brown to almost black. There are 14 water penny species in the United States. They live predominantly in clean, fast-moving streams. Aquatic larvae live 1 year or more (they are aquatic); adults (they are terrestrial) live on land for only a few days. They scrape algae and plants from surfaces.

Whirligig beetles are common inhabitants of streams and are normally found on the surface of quiet pools. Their bodies have pincher-like mouthparts, and they have six segmented legs in the middle of the body, which end in tiny claws. Many filaments extend from the sides of the abdomen. They have four hooks at the end of the body and no tail (see Figure 13.24).

Note: When disturbed, whirligig beetles swim erratically or dive while emitting defensive secretions.

As larvae, they are benthic predators, whereas the adults live on the water surface, attacking dead and living organisms trapped in the surface film. They occur on the surface in aggregations of up to thousands of individuals. Unlike the mating swarms of mayflies, these aggregations serve primarily to confuse predators. Whirligig beetles

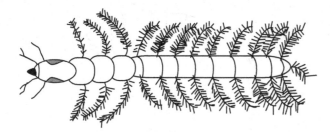

FIGURE 13.24 Whirligig beetle larva.

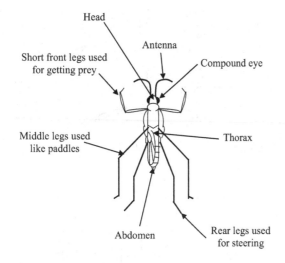

FIGURE 13.25 Water strider.

have other interesting defensive adaptations. For example, Johnston's organ at the base of the antennae enables them to echolocate using surface wave signals. Their compound eyes are divided into two pairs, one above and one below the water surface, enabling them to detect both aerial and aquatic predators; and they produce noxious chemicals that are highly effective at deterring predatory fish.

Water Strider ("Jesus bugs") (Order: Hemiptera)

Description: The most distinguishing characteristic of the order is the mouthparts that are modified into an elongated, sucking beak. Most adults have hemelytra, which are modified leathery forewings. Some adults and all larvae lack wings; however, most mature larvae possess wing pads. Both adults and larvae have three pairs of segmented legs with two tarsal claws at the end of each leg. Many families are also able to utilize atmospheric oxygen. This order is generally not used for the biological assessment of flowing waters due to their ability to use atmospheric oxygen.

It is fascinating to sit on a log at the edge of a stream pool and watch the drama that unfolds among the small water animals. Among the star performers in small streams are the water bugs. These are aquatic members of that large group of insects called the "true bugs," most of which live on land. Moreover, unlike many other types of water insects, they do not have gills but get their oxygen directly from the air. The most conspicuous and commonly known are the water striders or water skaters. These ride the top of the water, with only their feet making dimples in the surface film. Like all insects, the water striders have a three-part body (head, thorax, and abdomen), six jointed legs, and two antennae. They have a long, dark, narrow body (see Figure 13.25). The underside of the body is covered with water-repellent hair. Some water striders have wings, while others do not. Most water striders are over 5 mm (0.2 inches) long. Water striders eat small insects that fall on the water's surface and larvae. Water striders are very sensitive to motion and vibrations on the water's surface. They use this ability to locate prey. They push their mouth into their prey, paralyze it, and suck the insect dry. Predators of the water strider, like birds, fish, water beetles, backswimmers, dragonflies, and spiders, take advantage of the fact that water striders cannot detect motion above or below the surface of the water.

Alderflies and Dobsonflies (Order: Megaloptera)

Description: The head and thorax have thick, hardened skin, while the abdomen has thin, soft skin. The prominent chewing mouthparts project in front of the head. There are no wing pads on the thorax. Three pairs of segmented legs attach to the thorax. Seven or eight pairs of stout tapering filaments extend from the abdomen. The end of the abdomen has either a pair of prolegs with two claws on each proleg or a single long tapering filament with no prolegs.

Larvae of all species of Megaloptera ("large wing") are aquatic and attain the largest size of all aquatic insects. Megaloptera is a medium-sized order with fewer than 5,000 species worldwide. Most species are terrestrial; in North America, 64 aquatic species occur. In running waters, alderflies (Family: Sialidae) and dobsonflies (Family: Corydalidae; sometimes called hellgrammites or toe biters) are particularly important, as they are voracious predators, having large mandibles with sharp teeth.

Alderfly brownish-colored larvae possess a single tail filament with distinct hairs. The body is thick-skinned, with six to eight filaments on each side of the abdomen; gills are located near the base of each filament. Mature body size: 0.5–1.25 inches (see Figure 13.26). Larvae are aggressive predators, feeding on other adult aquatic macroinvertebrates (they swallow their prey without chewing); as secondary consumers, they are eaten by larger predators. Female alderflies deposit eggs on vegetation that overhangs water. Larvae hatch and fall directly into the water (i.e., into quiet but moving water). Adult alderflies are dark, with long wings folded back over the body; they only live a few days.

Dobsonfly larvae are extremely ugly (thus, they are rather easy to identify) and can be quite large, ranging anywhere from 25 to 90 mm (1–3.5 inches) in length. The body is stout, with eight pairs of appendages on the abdomen.

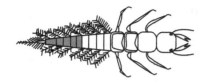

FIGURE 13.26 Alderfly larva.

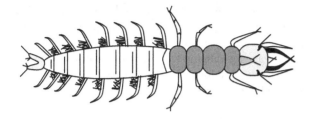

FIGURE 13.27 Dobsonfly larva.

Brush-like gills at the base of each appendage look like "hairy armpits" (see Figure 13.27). The elongated body has spiracles (spines) and three pairs of walking legs near the upper body, with one pair of hooked legs at the rear. The head bears four segmented antennae, small compound eyes, and strong mouthparts (large chewing pinchers). Coloration varies from yellowish, brown, gray, and black, often mottled. Dobsonfly larvae, commonly known as hellgrammites, are customarily found along stream banks under and between stones. As indicated by the mouthparts, they are predators and feed on all kinds of aquatic organisms.

Dragonflies and Damselflies (Order: Odonata)

Description: Dragonflies: The lower lip (labium) is long and elbowed, folding back against the head when not feeding, thus concealing other mouthparts. Wing pads are present on the thorax and three pairs of segmented legs attach to the thorax. There are no gills on the sides of the abdomen. Dragonflies have three pointed structures that may occur at the end of the abdomen, forming a pyramid-shaped opening. Their bodies are long and stout or somewhat oval. Damselflies have three flat gills at the end of the abdomen, forming a tail-like structure and their bodies are long and slender.

The Odonata (dragonflies, suborder Anisoptera; and damselflies, suborder Zygoptera) is a small order of conspicuous, hemimetabolous insects (lacking a pupal stage) with about 5,000 named species and 23 families worldwide. Odonata is a Greek word meaning toothed one. It refers to the serrated teeth located on the insect's chewing mouthparts (mandibles). Characteristics of dragonfly and damselfly larvae include:

- Large eyes.
- Three pairs of long segmented legs on the upper middle section (thorax) of the body.
- Large scoop-like lower lip that covers the bottom of the mouth.
- No gills on the sides or underneath the abdomen.

Note: Dragonflies and damselflies are unable to fold their four elongated wings back over the abdomen when at rest.

Dragonflies and damselflies are medium to large insects, with two pairs of long, equal-sized wings. The body is long and slender, with short antennae. Immature stages are aquatic, and the development occurs in three stages (egg, nymph, and adult).

Dragonflies are also known as darning needles. Myths about dragonflies warned children to keep quiet or less the dragonfly's "darning needles" would sew the child's mouth shut. The nymphal stage of dragonflies is grotesque creatures, robust, and stoutly elongated. They do not have long "tails" (see Figure 13.28). They are commonly gray, greenish, or brown to black in color. They are medium to large aquatic insects, ranging in size from 15 to 45 mm. The legs are short and used for perching. They are often found on submerged vegetation and at the bottom of streams in the shallows. They are rarely found in polluted waters. Food consists of other aquatic insects, annelids, small crustacea, and mollusks. Transformation occurs when the nymph crawls out of the water, usually onto vegetation. It splits its skin and emerges prepared for flight. The adult dragonfly is a strong flier, capable of great speed (>60 mph) and maneuverability (fly backward, stop on a dime, zip 20 feet straight up, and slip sideways in the blink of an eye!). When at rest the wings remain open and out to the sides of the body. A dragonfly's freely movable head has large, hemispherical eyes (nearly 30,000 facets each), which the insects use to locate prey with their excellent vision. Dragonflies eat small insects, mainly mosquitoes (large numbers of mosquitoes), while in flight. Depending on the species, dragonflies lay hundreds of eggs by dropping them into the water and leaving them to hatch or by inserting eggs singly into a slit in the stem of a submerged plant. The incomplete metamorphosis (egg, nymph, mature nymph, and adult) can take 2–3 years. Nymphs are often covered by algal growth.

Note: Adult dragonflies are sometimes called "mosquito hawks" because they eat a large number of mosquitoes, which they catch while they are flying.

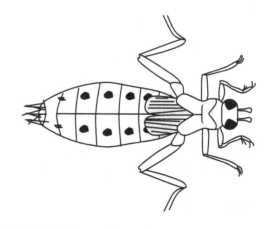

FIGURE 13.28 Dragonfly nymph.

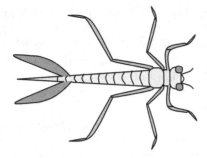

FIGURE 13.29   Damselfly nymph.

Damselflies are smaller and more slender than dragon-flies. They have three long, oar-shaped feathery tails, which are actually gills, and long, slender legs (see Figure 13.29). They are gray, greenish, or brown to black in color. Their habits are similar to those of dragonfly nymphs, and they emerge from the water as adults in the same manner. The adult damselflies are slow and seem uncertain in flight. Their wings are commonly black or clear, and their bodies are often brilliantly colored. When at rest, they perch on vegetation with their wings closed upright. Damselflies mature in 1–4 years and adults live for a few weeks or months. Unlike dragonflies, adult damselflies rest with their wings held vertically over their backs. They mostly feed on live insect larvae.

Note: Relatives of the dragonflies and damselflies are some of the most ancient flying insects. Fossils have been found of giant dragonflies with wingspans of up to 720 mm (28.4 inches) that lived long before the dinosaurs!

NON-INSECT MACROINVERTEBRATES

Non-insect macroinvertebrates are important to our discussion of stream and freshwater ecology because many of them are used as bioindicators of stream quality. Three frequently encountered groups in running water systems are Oligochaeta (worms), Hirudinea (leeches), and Gastropoda (lung-breathing snails). They are by no means restricted to running water conditions; the great majority of them occupy slow-flowing marginal habitats where the sedimentation of fine organic materials takes place.

Oligochaeta (Family Tuificidae, Genus: Tubifex)

Tubifex worms (commonly known as sludge worms) are unique in that they build tubes. Sometimes, there are as many as 8,000 individuals per square meter. They attach themselves within the tube and wave their posterior end in the water to circulate it and make more oxygen available to their body surface. These worms are commonly red since their blood contains hemoglobin. *Tubifex* worms may be very abundant in situations where other macroinvertebrates are absent; they can survive in very low oxygen levels and can live without oxygen at all for short periods. They are commonly found in polluted streams and feed on sewage or detritus.

Hirudinea (Leeches)

Despite the many different families of leeches, they all share common characteristics. They are soft-bodied worm-like creatures that are flattened when extended. Their bodies are dull in color, ranging from black to brown and reddish to yellow, often with a brilliant pattern of stripes or diamonds on the upper body. Their size varies within species but generally ranges from 5 mm to 45 cm when extended. Leeches are very good swimmers, but they typically move in an inchworm fashion. They are carnivorous and feed on other organisms ranging from snails to warm-blooded animals. Leeches are found in warm protected shallows under rocks and other debris.

Gastropoda (Lung-Breathing Snail)

Lung-breathing snails (pulmonates) may be found in streams that are clean. However, their dominance may indicate that dissolved oxygen levels are low. These snails are different from *right-handed snails* because they do not breathe underwater by use of gills but instead have a lung-like sac called a pulmonary cavity, which they fill with air at the surface of the water. When the snail takes in air from the surface, it makes a clicking sound. The air taken in can enable the snail to breathe underwater for long periods, sometimes hours.

Lung-breathing snails have two characteristics that help us identify them. First, they have no operculum or hard cover over the opening to their body cavity. Second, snails are either "right-handed" or "left-handed," and the lung-breathing snails are "left-handed." We can tell the difference by holding the shell so that its tip is upward and the opening is toward us. If the opening is to the *left* of the axis of the shell, the snail is termed sinistral—that is, it is left-handed. If the opening is to the *right* of the axis of the shell, the snail is termed dextral—that is, it is right-handed, and it breathes with gills. Snails are animals of the substrate and are often found creeping along on all types of submerged surfaces in water from 10 cm to 2 m deep.

Before the Industrial Revolution of the 1800s, metropolitan areas were small and sparsely populated. Thus, river and stream systems within or next to early communities received insignificant quantities of discarded waste. Early on, these river and stream systems were able to compensate for the small amount of waste they received; when wounded (polluted), nature has a way of fighting back. In the case of rivers and streams, nature provides their flowing waters with the ability to restore themselves through their own self-purification process. It was only when humans gathered in great numbers to form great cities that the stream systems were not always able to recover from having received great quantities of refuse and other waste. What exactly is it that man does to rivers and streams? What man does to rivers and streams is upset the delicate balance between pollution and the purification process. That is, we tend to unbalance the aquarium.

SUMMARY OF KEY TERMS

- **Abiotic Factor:** The nonliving part of the environment composed of sunlight, soil, mineral elements, moisture, temperature, topography, minerals, humidity, tide, wave action, wind, and elevation.

Important Note: Every community is influenced by a particular set of abiotic factors. While it is true that the abiotic factors affect the community members, it is also true that the living (biotic) factors may influence the abiotic factors. For example, the amount of water lost through the leaves of plants may add to the moisture content of the air. Also, the foliage of a forest reduces the amount of sunlight that penetrates the lower regions of the forest. The air temperature is therefore much lower than in non-shaded areas (Tomera 1989).

- **Autotroph:** (green plants) fixes the energy of the sun and manufactures food from simple, inorganic substances.
- **Biogeochemical Cycles:** are cyclic mechanisms in all ecosystems by which biotic and abiotic materials are constantly exchanged.
- **Biotic Factor (Community):** the living part of the environment composed of organisms that share the same area, are mutually sustaining, interdependent, and constantly fixing, utilizing, and dissipating energy.
- **Community:** in an ecological sense, the community includes all the populations occupying a given area.
- **Consumers and Decomposers:** dissipate energy fixed by the producers through food chains or webs. The available energy decreases by 80%–90% during transfer from one trophic level to another.
- **Ecology:** is the study of the interrelationship of an organism or a group of organisms and their environment.
- **Ecosystem:** is the community and the non-living environment functioning together as an ecological system.
- **Environment:** is everything that is important to an organism in its surroundings.
- **Heterotrophs:** (animals) use food stored by the autotroph, rearrange it, and finally decompose complex materials into simple inorganic compounds. Heterotrophs may be carnivorous (meat-eaters), herbivorous (plant-eaters), or omnivorous (plant- and meat-eaters).
- **Homeostasis:** is a natural occurrence during which an individual population or an entire ecosystem regulates itself against negative factors and maintains an overall stable condition.

- **Niche:** is the role that an organism plays in its natural ecosystem, including its activities, resource use, and interaction with other organisms.
- **Pollution:** is an adverse alteration to the environment by a pollutant.

THE BOTTOM LINE

Streamflow, geology, elevation, temperature, DO, seasonal life cycle patterns, substrate, and riparian habitat are factors that influence the abundance and diversity of macroinvertebrates in a waterbody. The truth be told, these natural factors can have an impact on macroinvertebrates, but the impact pales in comparison to the effect brought about by human activities. Consider, for example, that human-caused turbidity and sedimentation in a waterbody from reduced riparian vegetation or other causes of erosion can eliminate food sources and habitat for macroinvertebrates. In addition, excess nutrients promote algal blooms. The eventual death and the decomposition of the excessive algae deplete dissolved oxygen, reducing macroinvertebrate survival. Toxic pollutants such as pesticides, metals, and other contaminants can shift the abundances of macroinvertebrates toward favoring more pollution-tolerant species. Increasing or decreasing the pH of a waterbody affects macroinvertebrate survival by weakening shells or reducing alkaline-intolerant species.

CHAPTER REVIEW QUESTIONS

13.1 The major ecological unit is _____.

13.2 Those organisms residing within or on the bottom sediment are _____.

13.3 Organisms attached to plants or rocks are referred to as _____.

13.4 Small plants and animals that move about with the current are _____.

13.5 Free-swimming organisms belong to which group of aquatic organisms?

13.6 Organisms that live on the surface of the water are _____.

13.7 The movement of new individuals into a natural area is referred to as _____.

13.8 Fixes energy of the sun and makes food from simple inorganic substances _____.

13.9 The freshwater habitat that is characterized by normally calm water is _____.

13.10 The amount of oxygen dissolved in water and available for organisms is the _____.

REFERENCES

ASTM. 1969. *Manual on Water*. Philadelphia, PA: American Society for Testing and Materials.

Barlocher, R., and Kendrick, L., 1975. Leaf conditioning by microorganisms. *Oecologia*; 20:359–362.

Benfield, E.F., 1996. Leaf Breakdown in Streams Ecosystems, in *Methods in Stream Ecology*, Hauer, F.R., and Lambertic, G.A., (eds.). San Diego: Academic Press, pp. 579–590.

Benfield, E.F., Jones, D.R., and Patterson, M.F., 1977. Leaf pack processing in a pastureland stream. *Oikos*; 29:99–103.

Carson, R., 1962. *Silent Spring*. Boston: Houghton Mifflin.

Clements, E.S., 1960. *Adventures in Ecology*. New York: Pageant Press.

Crossley, D.A., Jr., House, G.J., Snider, R.M., Snider, R.J., and Stinner, B.R., 1984. The Positive Interactions in Agroecosystems, in *Agricultural Ecosystems*, Lowrance, R., Stinner, B.R., and House, G.J., (eds.). New York: John Wiley & Sons.

Cummins, K.W., 1974. Structure and function of stream ecosystems. *Bioscience*; 24:631–641.

Cummins, K.W., and Klug, M.J., 1979. Feeding ecology of stream invertebrates. *Annual Review of Ecology and Systematics*; 10:631–641.

Darwin, C., 1998. *The Origin of Species*, Suriano, G., (ed.). New York: Grammercy.

Dolloff, C.A., and Webster, J.R., 2000. Particulate Organic Contributions from Forests to Streams: Debris Isn't So Bad, in *Riparian Management in Forests of the Continental Eastern United States*, Verry, E.S., Hornbeck, J.W., and Dolloff, C.A., (eds.). Boca Raton, FL: Lewis Publishers.

Ecology. 2007. *Ecology*. Accessed 02/19/23 @ https://www.newworldencyclopedia.org/preview/.

Ecosystem Topics. 2000. *Ecosystem Services Research Trends*. Accessed 12/12/23 @https://www.mdpi.com.

Marshall, P., 1950. *Mr. Jones, Meet the Master*. Fleming H. Revel Company, NY.

McCafferty, P.W., 1981. *Aquatic Entomology*. Boston: Jones and Bartlett Publishers, Inc.

Merrit, R.W., and Cummins, K.W., 1996. *An Introduction to the Aquatic Insects of North America*, 3rd ed. Dubuque, IA: Kendall/Hunt Publishing.

Odum, E.P., 1952. *Fundamentals of Ecology*, 1st ed. Philadelphia: W.B. Saunders Co.

Odum, E.P., 1971. *Fundamentals of Ecology*, 3rd ed. Philadelphia: Saunders.

Odum, E.P., 1983. *Basic Ecology*. Philadelphia: Saunders College Publishing.

Odum, E.P., 1984. Properties of agroecosystems, in *Agricultural Ecosystems*, Lowrance, R., Stinner, B.R., and House, G.J., (eds.). New York: John Wiley & Sons.

Paul, R.W., Jr., Benfield, E.F., and Cairns, J., Jr., 1978. Effects of thermal discharge on leaf decomposition in a river ecosystem. *Verhandlugen der Internationalen Vereinigung fur Thoeretsche and Angewandte Limnologie*; 20:1759–1766.

Peterson, R.C., and Cummins, K.W., 1974. Leaf processing in woodland streams. *Freshwater Biology*; 4:345–368.

Price, P.W., 1984. *Insect Ecology*. New York: John Wiley & Sons.

Ramalay, F., 1940. *The Growth of a Science*. Boulder, CO: University of Colorado Boulder; 26:3–14.

Smith, R.L., 1996. *Ecology and Field biology*. New York: HarperCollins College Publishers.

Smith, C.H., 2007. *Karl Ludwig Willdenow*. Accessed 02/09/24 @ https://www.wku.edu/~smithch/chronob/WILL1765.htm.

Smith, T.M., and Smith, R.L., 2006. *Elements of Ecology*, 6th ed. San Francisco: Pearson, Benjamin Cummings.

Spellman, F.R., 1996. *Stream Ecology and Self-Purification*. Lancaster, PA: Technomic Publishing Company.

Suberkoop, K., Godshalk, G.L., & Klug, M.J., 1976. Changes in the chemical composition of leaves during processing in a woodland stream. *Ecology*; 57:720–727.

Tansley, A.G., 1935. The use and abuse of vegetational concepts and terms. *Ecology*; 16:284–307.

Tomera, A.N., 1989. *Understanding Basic Ecological Concepts*. Portland, ME: J. Weston Walch, Publisher.

USDA. 1982. *Agricultural Statistics 1982*. Washington, DC: U.S. Government Printing Office.

USDA. 1999. *Autumn Colors—How Leaves Change Color*. Accessed 02/08/23 @ https://www.na.fs.fed.us/spfo/pubs/misc/autumn/autumn_colors.htm.

USDA. 2007. *Agricultural Ecosystems and Agricultural Ecology*. Accessed 02/11/23 @ https://nrcs.usda.gov/technical/ECS/agecol/ecosystem.html.

14 Water Quality

INTRODUCTION

The quality of water, whether it is used for drinking, irrigation, or recreational purposes, is significant for health in both developing and developed countries worldwide. The first problem with water is rather obvious: a source of water must be found. Second, when accessible water is found, it must be suitable for human consumption. Meeting the water needs of those who populate Earth is an ongoing challenge. New approaches to meeting these water needs will not be easy to implement, as economic and institutional structures still encourage the wasting of water and the destruction of ecosystems (Gleick 2001). Again, finding a water source is the first problem. Finding a source of water that is safe to drink is the other problem.

Water quality is important; it can have a major impact on health both through outbreaks of waterborne diseases and by contributing to background disease rates. Accordingly, water quality standards are essential to protect public health.

In this text, *water quality* refers to the characteristics or range of characteristics that make water appealing and useful. Keep in mind that *useful* also means nonharmful or nondisruptive to either ecology or the human condition within the very broad spectrum of possible water uses. For example, the absence of odor, turbidity, or color are desirable immediate qualities. However, there are imperceptible qualities that are also important, such as chemical qualities. The presence of materials like toxic metals (e.g., mercury and lead), excessive nitrogen and phosphorus, or dissolved organic material may not be readily perceived by the senses but can exert substantial negative impacts on the health of a stream and/or on human health. The ultimate impact of these imperceptible water qualities (chemicals) on the user may be nothing more than a loss of aesthetic values. On the other hand, water containing chemicals could also lead to a reduction in biological health or an outright degradation of human health.

Simply stated, the importance of water quality cannot be overstated.

With regard to water and wastewater treatment operations, water quality management begins with a basic understanding of how water moves through the environment, is exposed to pollutants, and transports and deposits pollutants. The Hydrologic (Water) Cycle, depicted in Figure 14.1, illustrates the general links between the atmosphere, soil, surface waters, groundwater, and plants.

THE WATER CYCLE

The water cycle describes how water moves through the environment and identifies the links between groundwater, surface water, and the atmosphere (see Figure 14.1). It involves several key processes. As illustrated, water is taken from the Earth's surface to the atmosphere by evaporation from the surface of lakes, rivers, streams, and oceans. This evaporation occurs when the sun heats the water. The sun's heat energizes surface molecules, allowing them to break free from the attractive forces binding them together, and then evaporate and rise as invisible vapor into the atmosphere. Water vapor is also emitted from plant leaves through a process called *transpiration*. Every day, an actively growing plant transpires five to ten times as much water as it can hold at once. As water vapor rises, it cools and eventually condenses, usually on tiny particles of dust in the air. When it condenses, it becomes a liquid again or turns directly into a solid (ice, hail, or snow). These water particles then collect and form clouds. The atmospheric water formed in clouds eventually falls to Earth as precipitation, which can contain contaminants from air pollution. The precipitation may fall directly onto surface waters, be intercepted by plants or structures, or fall onto the ground. Most precipitation falls in coastal areas or at high elevations. Some of the water that falls at high elevations becomes runoff, which is water that runs over the ground (sometimes collecting nutrients from the soil) to lower elevations, forming streams, lakes, and fertile valleys. The water we see is known as *surface water*. Surface water can be categorized into five types: oceans, lakes, rivers and streams, estuaries, and wetlands.

Because the amount of rain and snow remains almost constant, while population and usage per person are both increasing rapidly, water is in short supply. In the United States alone, water usage is four times greater today than it was in 1900. In homes, this increased use is directly related to the rise in the number of bathrooms, garbage disposals, home laundries, and lawn sprinklers. In industry, usage has increased 13 times since 1900.

Approximately 170,000 small-scale suppliers provide drinking water to over 200+ million Americans through 60,000 community water supply systems and to nonresidential locations such as schools, factories, and campgrounds. The rest of Americans are served by private wells. Most of the drinking water used in the United States comes from groundwater. Untreated water drawn from groundwater and surface waters, used as a drinking water supply, can contain contaminants that pose a threat to human health.

Note: The USEPA (2002) reported that American households use approximately 146,000 gallons of freshwater annually and that Americans drink 1 billion glasses of tap water per day.

Obviously, with a limited amount of drinking water available the water that is accessible must be reused, or we

DOI: 10.1201/9781003581901-17

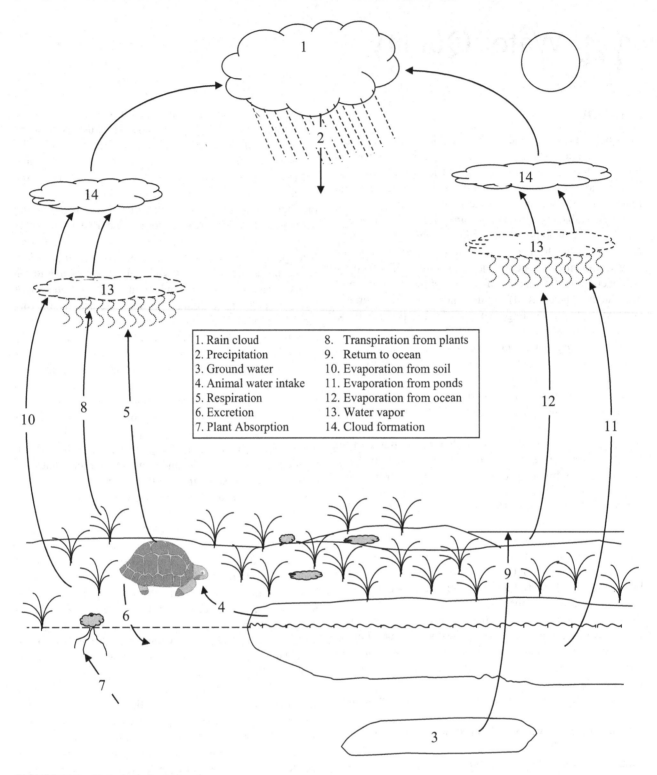

FIGURE 14.1   Hydrologic (water) cycle.

will face an inadequate supply to meet the needs of all users. Water use and reuse are complicated by pollution. Pollution is relative and difficult to define. For example, floods and animals (dead or alive) are polluters, but their effects are local and tend to be temporary. Today, water is polluted from many sources and in many forms. It may appear as excess aquatic weeds, oil slicks, a decline in sport fishing, or an increase in carp, sludge worms, and other organisms that readily tolerate pollution. Maintaining water quality is crucial because water pollution is detrimental not only to health but also to recreation, commercial fishing, aesthetics, and private, industrial, and municipal water supplies.

At this point, the reader might be asking: With all the recent publicity about pollution and the enactment of new environmental regulations, hasn't water quality in the United States improved recently? The answer:

Given the recent progress toward achieving fishable-swimmable waters under the Clean Water Act (CWA), one might think so. However, the 1994 *National Water Quality Inventory Report to Congress* indicated that 63% of the nation's lakes, rivers, and estuaries meet designated uses—only a slight increase over what was reported in 1992.

The main culprit is *nonpoint source pollution* (NPS) (to be discussed in detail later). NPS is the leading cause of impairment for rivers, lakes, and estuaries. Impaired sources are those that do not fully support designated uses, such as fish consumption, drinking water supply, groundwater recharge, aquatic life support, or recreation. According to Fortner and Schechter (1996), the five leading sources of water quality impairment in rivers are agriculture, municipal wastewater treatment plants, habitat and hydrologic modification, resource extraction, and urban runoff and storm sewers.

The health of rivers and streams is directly linked to the integrity of habitat along the river corridor and in adjacent wetlands. Stream quality deteriorates if activities damage vegetation along riverbanks and in nearby wetlands. Trees, shrubs, and grasses filter pollutants from runoff and reduce soil erosion. The removal of vegetation also eliminates shade that moderates stream temperature, which, in turn, affects the availability of dissolved oxygen in the water column for fish and other aquatic organisms. Lakes, reservoirs, and ponds may receive water carrying pollutants from rivers and streams, melting snow, runoff, or groundwater. Lakes may also receive pollution directly from the air.

Thus, in attempting to answer the original question—hasn't water quality in the United States improved recently?—the best answer is likely: We are holding our own in controlling water pollution, but more progress is needed. This, however, understates an important point: when it comes to water quality, continuous progress is essential.

WATER QUALITY STANDARDS

The effort to regulate drinking water and wastewater effluent has increased since the early 1900s. Beginning with an effort to control the discharge of wastewater into the environment, preliminary regulatory efforts focused on protecting public health. The goal of this early wastewater treatment program was to remove suspended and floatable material, treat biodegradable organics, and eliminate pathogenic organisms. Thus, regulatory efforts were directed toward constructing wastewater treatment plants in an effort to alleviate the problem. Then a problem soon developed: progress. Progress in the sense that time marched on, and with it came the proliferation of city growth in the United States where it became increasingly difficult to find land required for wastewater treatment and disposal. Wastewater professionals soon recognized the need to develop methods of treatment that would accelerate "nature's way" (the natural purification of water) under controlled conditions in treatment facilities of comparatively smaller size.

Regulatory influence on water-quality improvements in both wastewater and drinking water took a giant step forward in the 1970s. The Water Pollution Control Act Amendments of 1972 (Clean Water Act) established national water pollution control goals. At about the same time, the Safe Drinking Water Act (SDWA), passed by Congress (1974), started a new era in the field of drinking water supply to the public.

CLEAN WATER ACT

In 1972, Congress adopted the Clean Water Act (CWA), which established a framework for achieving its national objective "to restore and maintain the chemical, physical, and biological integrity of the nation's waters." Congress decreed that, where attainable, water quality "provides for the protection and propagation of fish, shellfish, and wildlife and provides for recreation in and on the water." These goals are referred to as the "fishable and swimmable" goals of the Act.

Before the CWA, there were no *specific* national water pollution control goals or objectives. Current standards require that municipal wastewater be given secondary treatment (to be discussed in detail later) and that most effluents meet the conditions shown in Table 14.1. The goal, via secondary treatment (i.e., the biological treatment component of a municipal treatment plant), was set so that the principal components of municipal wastewater—suspended solids, biodegradable material, and pathogens—could be reduced to acceptable levels. Industrial dischargers are required to treat their wastewater to the level obtainable by the *best available technology* (BAT) for wastewater treatment in that particular type of industry.

Moreover, a National Pollutant Discharge Elimination System (NPDES) program was established based on uniform technological minimums with which each point source discharger must comply. Under NPDES, each municipality and industry discharging effluent into streams is assigned discharge permits. These permits reflect the secondary treatment and best available technology standards. Water quality standards are the benchmark against which monitoring data are compared to assess the health of waters and to develop Total Maximum Daily Loads in impaired waters. They are also used to calculate water-quality-based discharge limits in permits issued under the National Pollutant Discharge Elimination System (NPDES).

DID YOU KNOW?

In 1775, British engineer Alexander Cumming, a Scottish watchmaker and instrument inventor, patented a toilet with a siphon trap—a pocket of water in a lazy-S pipe—that kept the smell of the sewer (cesspool) from backing up into the bathroom.

TABLE 14.1

Minimum National Standards for Secondary Treatment

Characteristic of Discharge	Unit of Measure Concentration	Average 30-day Concentration	Average 7-day
BOD$_5$	mg/L	30	45
Suspended solids	mg/L	30	45
Concentration	pH units	Within the range of 6.0–9.0	

Source: Federal Register, Secondary Treatment Regulations, 40 CFR Part 133, 1988.

SAFE DRINKING WATER ACT

The Safe Drinking Water Act of 1974 mandated the USEPA to establish drinking water standards for all public water systems serving 25 or more people or having 15 or more connections. Pursuant to this mandate, the EPA has established maximum contaminant levels for drinking water delivered through public water distribution systems. The maximum contaminant levels (MCLs) for inorganics, organic chemicals, turbidity, and microbiological contaminants are shown in Table 14.2. The EPA's primary regulations are mandatory and must be complied with by all public water systems to which they apply. If analysis of the water produced by a water system indicates that an MCL for a contaminant is being exceeded, the system must take steps to stop providing the water to the public or initiate treatment to reduce the contaminant concentration to below the MCL.

The USEPA has also issued guidelines to the states regarding secondary drinking water standards. These appear in Table 14.3. These guidelines apply to drinking water contaminants that may adversely affect the aesthetic qualities of the water (i.e., those qualities that make water appealing and useful), such as odor and appearance. These qualities have no known adverse health effects, and thus secondary regulations are not mandatory. However, most drinking water systems comply with the limits; they have learned through experience that the odor and appearance of drinking water are not a problem until customers complain. One thing is certain: they will complain.

Implementing the Safe Drinking Water Act[1]

On December 3, 1998, at the oceanfront of Fort Adams State Park in Newport, Rhode Island, President Clinton

TABLE 14.2

EPA Primary Drinking Water Standards

1. Inorganic Contaminant Levels

Contaminants	Level (mg/L)
Arsenic	0.05
Barium	1.00
Cadmium	0.010
Chromium	0.05
Lead	0.05
Mercury	0.002
Nitrate	10.00
Selenium	0.01
Silver	0.05

2. Organic Contaminant Levels

Chemical	Maximum Contaminant Level (MCL) mg/L
Chlorinated hydrocarbons	
Endrin	0.0002
Lindane	0.004
Methoxychlor	0.1
Toxaphene	0.005
Chlorophenyls	
2,4-D	0.1
2,4,5-TP silvex	0.01

(Continued)

TABLE 14.2 (*Continued*)
EPA Primary Drinking Water Standards

3. Maximum Levels of Turbidity

Reading Basis	Maximum Contaminant Level (MCL) Turbidity Units
Turbidity reading (monthly average)	1 TU or up to 5 TUs if the water supplier can demonstrate to the state that the higher turbidity does not interfere with disinfection maintenance of an effective disinfection agent throughout the distribution system, or microbiological determinants.
Turbidity reading (based on average of two consecutive days)	5 Tus

4. Microbiological Contaminants

Test Method Used	Monthly Basis	Individual Sample Basis	
		Fewer than 20 Samples/Month	More than 20 Samples/Month
Membrane filter technique	1/100 mL average daily	Number of coliform bacteria not to exceed:	
		4/100 mL in more than one sample	4/100 mL in more than 5% of samples
Fermentation		Coliform bacteria shall not be present in:	
10 mL standard portions	More than 10% of the portions	Three or more portions in more than one sample	Three or more portions in more than 5% of samples
100-mL standard portions	More than 60% of the portions	Five portions in more than one sample	Five portions in more than 20% of the samples

Source: Adapted from USEPA. 1975. *National Interim Primary Drinking Water Regulations*, Federal Register, Part IV.

TABLE 14.3
Secondary Maximum Contaminant Levels

Contaminant	Level	Adverse Effect
Chloride	250 mg/L	Causes taste
Color	15 cu	Appearance problems
Copper	1 mg/L	Tastes and odors
Corrosivity	Noncorrosive	Tastes and odors
Fluoride	2 mg/L	Dental fluorosis
Foaming agents	0.5 mg/L	Appearance problems
Iron	0.3 mg/L	Appearance problems
Manganese	0.05 mg/L	Discolors laundry
Odor	3 TON	Unappealing to drink
pH	6.5–8.5	Corrosion or scaling
Sulfate	250 mg/L	Has laxative effect
Total dissolved solids	500 mg/L	Taste, corrosive
Zinc	5 mg/L	Taste, appearance

Cu, color unit; TON, threshold odor number.
Source: Adapted from McGhee (1991, p. 161).

made remarks to the community of Newport, announcing a significant part of the 1996 SDWA and amendments. The expectation was that the new requirements would protect most of the nation from dangerous contaminants while adding only about $2 to many monthly water bills.

The rules require approximately 13,000 municipal water suppliers to use better filtering systems to screen out Cryptosporidium and other microbes, ensuring that U.S. community water supplies are safe from microbial contamination.

Consumer Confidence Report Rule

The requirement for community water systems to provide **Annual Drinking Water Quality Reports** (consumer right-to-know information) to customers is important. While water systems are free to enhance their reports in any useful way, each report must provide consumers with the following fundamental information about their drinking water:

- The lake, river, aquifer, or other source of the drinking water. Consider the following information provided by the city of XXXXX:

  The City of XXXXX receives its raw (untreated) water from eight reservoirs, two rivers and four deep wells. From these sources, raw water is pumped to one of the Department of Utilities' two water treatment plants, where it is filtered and disinfected. Once tested for top quality, XXXXX drinking water is pumped on demand to your tap.

- A brief summary of the susceptibility to contamination of the local drinking water source, based on the source water assessments by the states. Consider the following information provided by the City of XXXXX.

  Contaminants that may be present in source (raw) water include:

  - **Microbial Contaminants**: such as viruses and bacteria, which may come from sewage treatment plants, septic systems, agricultural livestock operations and wildlife
  - **Inorganic Contaminants**: such as salts and metals, which can be naturally occurring or result from urban stormwater runoff, industrial or domestic wastewater discharges, oil and gas production, mining, or farming.
  - **Pesticides and Herbicides**: which may come from a variety of sources such as agriculture, urban stormwater runoff, and residential uses.

- **Radioactive Contaminants:** which can be naturally occurring or be the result of oil and gas production and mining activities.
- How to get a copy of the water system's complete source water assessment. Consider the following information taken from the City XXXXX's Consumer Confidence Report (annual drinking water quality report):

  For a copy of the water system's complete source water assessment and questions regarding this report, contact XXXXX's Water Quality Lab at xxx-xxx-xxxx. For more information about decisions affecting your drinking water quality, you may attend XXXXX City Council meetings. For times and agendas, call the XXXXX Clerk's office at xxx-xxx-xxxx.

- The level (or range of levels) of any contaminant found in local drinking water, as well as the EPA's health-based standard (maximum contaminant level) for comparison.
- The likely source of that contaminant in the local drinking water supply.
- The potential health effects of any contaminant detected in violation of an EPA health standard, along with an accounting of the system's actions to restore safe drinking water.
- The water system's compliance with other drinking water-related rules.
- An educational statement for vulnerable populations about avoiding *Cryptosporidium*.
- Educational information on nitrate, arsenic, or lead in areas where these contaminants may be a concern. In regard to the levels of contaminants, their possible sources, and the levels detected in local drinking water supplies, consider the information provided by City XXXXX to its ratepayers in the annual drinking water quality report for 2011.
- Phone numbers of additional sources of information, including the water system and the EPA's Safe Drinking Water Hotline (800-426-4791).

This information supplements the public notification that water systems must provide to their customers upon discovering any violation of a contaminant standard. This annual report should not be the primary notification of potential health risks posed by drinking water; instead, it provides customers with water quality information from the previous year.

WATER QUALITY CHARACTERISTICS OF WATER AND WASTEWATER

In this section, individual pollutants and stressors that affect water quality are described.

Knowledge of the parameters and characteristics most commonly associated with water and wastewater treatment processes is essential for water and wastewater operators.

Water and wastewater practitioners are encouraged to use a holistic approach to managing water quality problems. It is important to note that when this text refers to water quality, the definition used is predicated on the intended use of the water. Many parameters have evolved that qualitatively reflect the impact that various contaminants (impurities) have on selected water uses; the following sections provide a brief discussion of these parameters.

Physical Characteristics of Water/Wastewater

The physical characteristics of water and wastewater that we are interested in are more germane to the discussion at hand; namely, a category of parameters and characteristics that can be used to describe water quality. One such category is the physical characteristics of water, which are apparent to the senses of smell, taste, sight, and touch. Solids, turbidity, color, taste and odor, and temperature also fall into this category.

Solids

Other than gases, all contaminants of water contribute to the solids content. Classified by their size and state, chemical characteristics, and size distribution, solids can be dispersed in water in both suspended and dissolved forms. In regard to size, solids in water and wastewater can be classified as suspended, settleable, colloidal, or dissolved. Solids are also characterized as being *volatile* or *nonvolatile*. The distribution of solids is determined by computing the percentage of filterable solids by size range. Solids typically include inorganic solids such as silt, sand, gravel, and clay from riverbanks, as well as organic matter such as plant fibers and microorganisms from natural or human-made sources. We use the term "siltation" to describe the suspension and deposition of small sediment particles in water bodies. In flowing water, many of these contaminants result from the erosive action of water flowing over surfaces.

Sedimentation and siltation can severely alter aquatic communities. Sedimentation may clog and abrade fish gills, suffocate eggs and aquatic insect larvae on the bottom, and fill in the pore space between bottom cobbles where fish lay their eggs. Suspended silt and sediment interfere with recreational activities and aesthetic enjoyment at streams and lakes by reducing water clarity and filling in lakes. Sediment may also carry other pollutants into surface waters. Nutrients and toxic chemicals may attach to sediment particles on land and ride the particles into surface waters, where the pollutants may settle with the sediment or detach and become soluble in the water column.

Suspended solids are a measure of the weight of relatively insoluble materials in the ambient water. These materials enter the water column as soil particles from land surfaces or as sand, silt, and clay from stream bank erosion or channel scour. Suspended solids can include both organic (detritus and biosolids) and inorganic (sand or finer colloids) constituents.

In water, suspended material is objectionable because it provides adsorption sites for biological and chemical agents. These adsorption sites create a protective barrier for attached microorganisms against the chemical action of chlorine. In addition, suspended solids in water may be biologically degraded, resulting in objectionable by-products. Thus, the removal of these solids is of great concern in the production of clean, safe drinking water and wastewater effluent.

In water treatment, the most effective means of removing solids from water is filtration. It should be pointed out, however, that not all solids, such as colloids and other dissolved solids, can be removed by filtration. In wastewater treatment, suspended solids are an important water-quality parameter and are used to measure the quality of the wastewater influent, monitor the performance of several processes, and assess the quality of effluent. Wastewater is normally 99.9% water and 0.1% solids. If a wastewater sample is evaporated, the remaining solids are called *total solids*. As shown in Table 12.1, the USEPA has set a maximum suspended solids standard of 30 mg/L for most treated wastewater discharges.

Turbidity

One of the first things that is noticed about water is its clarity. The clarity of water is usually measured by its *turbidity*. Turbidity is a measure of the extent to which light is either absorbed or scattered by suspended material in water. Both the size and surface characteristics of the suspended material influence absorption and scattering. Although algal blooms can make waters turbid, in surface water, most turbidity is related to the smaller inorganic components of the suspended solids burden, primarily clay particles. Microorganisms and vegetable material may also contribute to turbidity. Wastewaters from industry and households usually contain a wide variety of turbidity-producing materials. Detergents, soaps, and various emulsifying agents contribute to turbidity.

In water treatment, turbidity is useful in defining drinking-water quality. In wastewater treatment, turbidity measurements are particularly important whenever ultraviolet radiation (UV) is used in the disinfection process. For UV to be effective in disinfecting wastewater effluent, UV light must be able to penetrate the stream flow. Obviously, turbid stream flow reduces the effectiveness of irradiation (penetration of light).

The colloidal material associated with turbidity provides adsorption sites for microorganisms and chemicals that may be harmful or cause undesirable tastes and odors. Moreover, the adsorptive characteristics of many colloids provide protective sites for microorganisms against disinfection processes. Turbidity in running waters interferes with light penetration and photosynthetic reactions.

Turbidity in water bodies is caused by both natural and human-induced factors. Natural factors include (USEPA 2021):

- Runoff caused by precipitation and severe weather.
- Spring snowmelt and precipitation.

- Dead organic matter in the water column.
- Wildfire debris (i.e., wood ash) washing into surface water.
- Summer algal growth in still waters and slower-moving running waters.
- Bottom-feeding animals moving sediments around.
- Small floating organisms suspended in the water column (plankton, algae, cyanobacteria).
- Turbulence from windstorms and rain events that disturb bottom sediment (resuspension).

Human-caused factors that increase turbidity include:

- Stream bank erosion adding soil to water.
- Changes in land use (farming, forestry, construction, and urban development) that cause soil to be carried in runoff to surface water.
- Untreated wastewater discharges.
- Disturbance and resuspension of bottom sediments during dredging or boating activities.
- Algal growth due to fertilizer use and resulting increases in nutrients in the water, especially in lakes or slower-moving rivers.
- Urban runoff carrying particles from impervious surfaces to surface water.

Color

Color is another physical characteristic by which the quality of water can be judged. Pure water is colorless. Water takes on color when foreign substances, such as organic matter from soils, vegetation, minerals, and aquatic organisms, are present. Color can also be contributed to water by municipal and industrial wastes. Color in water is classified as either *true color* or *apparent color*. Water whose color is partly due to dissolved solids that remain after the removal of suspended matter is known as true color. Color contributed by suspended matter is said to have apparent color. In water treatment, true color is the most difficult to remove.

Note: Water has an intrinsic color, and this color has a unique origin. Intrinsic color is easy to discern, as can be seen in Crater Lake, Oregon, which is known for its intense blue color. The appearance of the lake varies from turquoise to deep navy blue, depending on whether the sky is hazy or clear. Pure water and ice have a pale blue color.

The obvious problem with colored water is that it is not acceptable to the public. Given a choice, the public prefers clear, uncolored water. Another problem with colored water is its effect on laundering, papermaking, manufacturing, textiles, and food processing. The color of water has a profound impact on its marketability for both domestic and industrial use.

In water treatment, color is not usually considered unsafe or unsanitary, but it is a treatment problem in regard to exerting a chlorine demand, which reduces the effectiveness of chlorine as a disinfectant. In wastewater treatment, color is not necessarily a problem but rather an indicator of the *condition* of the wastewater. Condition refers to the age of the wastewater, which, along with odor, provides a qualitative indication of its age. Early in the flow, wastewater is a light brownish-gray color. The color of wastewater containing dissolved oxygen (DO) is normally gray. Black-colored wastewater, usually accompanied by foul odors and containing little or no DO, is said to be septic. Table 14.4 provides wastewater color information. As the travel time in the collection system increases (flow becomes increasingly more septic) and more anaerobic conditions develop, the color of the wastewater changes from gray to dark gray and ultimately to black.

Taste and Odor

Taste and odor are used jointly in the vernacular of water science. The term "odor" is used in wastewater; taste, obviously, is not a consideration. Domestic sewage should have a musty odor. Bubbling gas and/or foul odors may indicate industrial wastes, anaerobic (septic) conditions, and operational problems. Refer to Table 14.5 for typical wastewater odors, possible problems, and solutions.

In wastewater, odors are a major concern, especially for those who reside in close proximity to a wastewater treatment plant. These odors are generated by gases produced by the decomposition of organic matter or by substances added to the wastewater. Because these substances are volatile, they are readily released into the atmosphere at any

point where the waste stream is exposed, particularly if there is turbulence at the surface.

Most people would argue that all wastewater is the same; it has a disagreeable odor. It is hard to argue against the disagreeable odor. However, one wastewater operator told us that wastewater "smelled great—smells just like money to me—money in the bank," she said.

This was one operator's view. We also received another opinion on odor problems resulting from wastewater operations. This particular opinion, given by an odor control manager, was quite different. His statement was "that odor control is a never-ending problem." She also pointed out that to combat this difficult problem, odors must be contained. In most urban plants, it has become necessary to physically cover all source areas such as treatment basins, clarifiers, aeration basins, and contact tanks to prevent odors from leaving the processes. These contained spaces must then be positively vented to wet-chemical scrubbers to prevent the buildup of a toxic concentration of gas.

As mentioned, in drinking water, taste and odor are not normally a problem until the consumer complains. The issue is that most consumers find taste and odor in water aesthetically displeasing. Although taste and odor do not directly present a health hazard, they can cause customers to seek water that tastes and smells good but may not be safe to drink. Most consumers consider water tasteless and odorless. Thus, when consumers find that their drinking water has a taste or odor, or both, they automatically associate the drinking water with contamination.

Water contaminants are attributable to contact with nature or human use. Taste and odor in water are caused by a variety of substances such as minerals, metals, and salts from the soil, constituents of wastewater, and end products produced in biological reactions. When water has a taste but no accompanying odor, the cause is usually inorganic contamination. Water that tastes bitter is typically alkaline, while salty water is commonly the result of metallic salts. However, when water has both taste and odor, the likely cause is organic materials. The list of possible organic contaminants is too long to record here; however petroleum-based

TABLE 14.4
Significance of Color in Wastewater

Unit Process	Color	Problem Indicated
Influent of Plant	Gray	None
	Red	Blood or other industrial wastes
	Green, Yellow	Other Industrial wastes not pretreated (paints, etc.)
	Red or other soil	Surface runoff into influent
	Black	Septic conditions or industrial flows

TABLE 14.5
Odors in Wastewater Treatment Plant

Odor	Location	Problem	Possible Solution
Earthy, musty	Primary and secondary units	No problem (normal)	None required
Hydrogen sulfide	Influent	Septic	Aerate, chlorinate, (rotten egg odor)ozonize
	Primary clarifier	Septic sludge	Remove sludge
	Activated sludge		
	Trickling filters	Septic conditions	More air/less BOD
	Secondary clarifiers	"	Remove sludge
	Chlorine contact	"	Remove sludge
	General plant	"	Good house-keeping
Chlorine like	Chlorine contact tank	Improper chlorine dosage	Adjust chlorine dosage controls
Industrial odors	General plant	Inadequate pretreatment	Enforce sewer use regulation

products lead the list of offenders. Taste- and odor-producing liquids and gases in water are produced by the biological decomposition of organics. A prime example of one of these is hydrogen sulfide, known best for its characteristic "rotten-egg" taste and odor. Certain species of algae also secrete an oily substance that may produce both taste and odor. When certain substances combine (such as organics and chlorine), the synergistic effect produces taste and odor.

In water treatment, one of the common methods used to remove taste and odor is to oxidize the materials that cause the problem. Oxidants, such as potassium permanganate and chlorine, are used. Another common treatment method is to feed powdered activated carbon before the filter. The activated carbon has numerous small openings that absorb the components that cause the odor and taste. These contained spaces must then be positively vented to wet-chemical scrubbers to prevent the buildup of toxic concentrations of gas.

Temperature

Heat is added to surface and groundwater in many ways, some of which are natural and others artificial. For example, heat is added by natural means to Yellowstone Lake, Wyoming. The lake, one of the world's largest freshwater lakes, resides in a caldera situated at more than 7,700 feet (the largest high-altitude lake in North America). When one attempts to swim in Yellowstone Lake (without a wetsuit), the bitter cold of the water literally takes one's breath away. However, if it were not for the hydrothermal discharges that occur in Yellowstone, the water would be even colder. Regarding human-heated water, this most commonly occurs whenever a raw water source is used for cooling water in industrial operations. The influent to industrial facilities is at normal ambient temperature. However, when it is used to cool machinery and industrial processes and then discharged back to the receiving body, it is often heated. The problem with heat or temperature increases in surface waters is that it affects the solubility of oxygen in water, the rate of bacterial activity, and the rate at which gases are transferred to and from the water.

Note: It is important to point out that in the examination of water or wastewater, temperature is not normally used to evaluate either. However, temperature is one of the most important parameters in natural surface-water systems, as surface waters are subject to great temperature variations.

Water temperature does determine, in part, how efficiently certain water treatment processes operate. For example, temperature affects the rate at which chemicals dissolve and react. When water is cold, more chemicals are required for efficient coagulation and flocculation to take place. When water temperature is high, the result may be a higher chlorine demand due to increased reactivity and there is often an increased level of algae and other organic matter in raw water. Temperature also has a pronounced effect on the solubility of gases in water.

Ambient temperature (the temperature of the surrounding atmosphere) has the most profound and universal effect on the temperature of shallow natural water systems. When water is used by industry to dissipate process waste heat, the discharge locations into surface waters may experience localized temperature changes that can be quite dramatic. Other sources of increased temperatures in running water systems result from clear-cutting practices in forests (where protective canopies are removed) and from irrigation flows returned to a body of running water.

In wastewater treatment, the temperature of wastewater varies greatly, depending on the type of operations being conducted at a particular installation. However, wastewater is generally warmer than the water supply because of the addition of warm water from industrial activities and households. Wide variation in the wastewater temperature indicates heated or cooled discharges, often of substantial volume. They have any number of sources. For example, decreased temperatures after a snowmelt or rain event may indicate serious infiltration. In the treatment process itself, temperature not only influences the metabolic activities of the microbial population but also has a profound effect on factors such as gas-transfer rates and the settling characteristics of the biological solids.

CHEMICAL CHARACTERISTICS OF WATER

Another category used to define or describe water quality is its chemical characteristics. The most important chemical characteristics are Total Dissolved Solids (TDS), Alkalinity, Hardness, Fluoride, Metals, Organics, and Nutrients. Chemical impurities can be either natural, human-made (industrial), or introduced into raw water sources by enemy forces. Some chemical impurities cause water to behave as either an acid or a base. Because either condition has an important bearing on the water treatment process, the pH value must be determined. Generally, the pH influences the corrosiveness of the water, the chemical dosages necessary for proper disinfection, and the ability to detect contaminants. The principal contaminants found in water are shown in Table 14.6. These chemical constituents are important because each one affects water use in some manner; each one either restricts or enhances specific uses. The pH of water is important. As pH rises, for example, the equilibrium (between bicarbonate and carbonate) increasingly favors the formation of carbonate, which often results in the precipitation of carbonate salts. If you have ever had flow in a pipe system interrupted or a heat-transfer problem in your water heater system, then carbonate salts that formed a hard-to-dissolve scale within the system are most likely the cause. It should be pointed out that not all carbonate salts have a negative effect on their surroundings. Consider, for example, the case of blue marl lakes; they owe their unusually clear, attractive appearance to carbonate salts.

Water has been called the *universal solvent*. This is, of course, a fitting description. The solvent capabilities of water are directly related to its chemical characteristics or parameters. As mentioned, in water-quality management, total dissolved solids (TDS), alkalinity, hardness, fluorides,

TABLE 14.6
Chemical Constituents Commonly Found in Water

Constituents

Calcium
Fluorine
Magnesium
Nitrate
Sodium
Silica
Potassium
Silica
Potassium
TDS
Iron
Hardness
Manganese
Color
Bicarbonate
pH
Carbonate
Turbidity
Sulfate
Temperature
Chloride

metals, organics, and nutrients are the major chemical parameters of concern.

Total Dissolved Solids (TDS)

Because of water's solvent properties, minerals dissolved from rocks and soil as water passes over and through them produce Total Dissolved Solids (TDS; comprised of any minerals, salts, metals, cations, or anions dissolved in water). TDS constitutes a part of total solids (TS) in water; it is the material remaining in water after filtration. Dissolved solids may be organic or inorganic. Water may be exposed to these substances within the soil, on surfaces, and in the atmosphere. The organic dissolved constituents of water come from the decay products of vegetation, from organic chemicals, and from organic gases.

Dissolved solids can be removed from water by distillation, electro-dialysis, reverse osmosis, or ion exchange. It is desirable to remove these dissolved minerals, gases, and organic constituents because they may cause psychological effects and produce aesthetically displeasing color, taste, and odors. While it is desirable to remove many of these dissolved substances from water, it is not prudent to remove them all. This is the case, for example, because pure, distilled water has a flat taste. Further, water has an equilibrium state with respect to dissolved constituents. Thus, if water is out of equilibrium or undersaturated, it will aggressively dissolve materials it comes into contact with. Because of this problem, substances that are readily dissolvable are sometimes added to pure water to reduce its tendency to dissolve plumbing.

Alkalinity

Another important characteristic of water is its *alkalinity,* which is a measure of water's ability to neutralize acid, or really, an expression of buffering capacity. The major chemical constituents of alkalinity in natural water supplies are the bicarbonate, carbonate, and hydroxyl ions. These compounds are mostly the carbonates and bicarbonates of sodium, potassium, magnesium, and calcium. These constituents originate from carbon dioxide (from the atmosphere and as a by-product of microbial decomposition of organic material) and from their mineral origin (primarily from chemical compounds dissolved from rocks and soil). Highly alkaline waters are unpalatable; however, this condition has little known significance for human health. The principal problem with alkaline water is the reactions that occur between alkalinity and certain substances in the water. Alkalinity is important for fish and aquatic life because it protects or buffers against rapid pH changes. Moreover, the resultant precipitate can foul water system appurtenances. In addition, alkalinity levels affect the efficiency of certain water treatment processes, especially the coagulation process.

Hardness

Hardness is due to the presence of multivalent metal ions, which come from minerals dissolved in water. Hardness is based on the ability of these ions to react with soap to form a precipitate or soap scum. In freshwater, the primary ions are calcium and magnesium; however, iron and manganese may also contribute. Hardness is classified as *carbonate* hardness or *noncarbonate* hardness. Carbonate hardness is equal to alkalinity but a non-carbonate fraction may include nitrates and chlorides. Hardness is either temporary or permanent. Carbonate hardness (temporary hardness) can be removed by boiling. Noncarbonate hardness cannot be removed by boiling and is classified as permanent.

Hardness values are expressed as an equivalent amount or equivalent weight of calcium carbonate (*the equivalent weight* of a substance is its atomic or molecular weight divided by n). Water with a hardness of less than 50 ppm is considered soft. Above 200 ppm, domestic supplies are usually blended to reduce the hardness value. Table 14.7 The U.S. Geological Survey uses the following classification:

The impact of hardness can be measured in economic terms. Soap consumption points this out; that is, soap

TABLE 14.7
Classification of Hardness

Range of Hardness [mg/L (ppm) as CaCO₃]	Descriptive Classification
1–50	Soft
51–150	Moderately hard
151–300	Hard
Above 300	Very hard

consumption represents an economic loss to the water user. When washing with a bar of soap, there is a need to use more soap to "get a lather" whenever washing in hard water. There is another problem with soap and hardness. When using a bar of soap in hard water, when lather is finally built up, the water has been "softened" by the soap. The precipitate formed by the hardness and soap (soap curd) adheres to just about anything (tubs, sinks, dishwashers) and may stain clothing, dishes, and other items. There is also a personal problem: the residues of the hardness-soap precipitate may precipitate into the pores, causing skin to feel rough and uncomfortable. Today these problems have been largely reduced by the development of synthetic soaps and detergents that do not react with hardness. However, hardness still leads to other problems, such as scaling and a laxative effect. Scaling occurs when carbonate hard water is heated, and calcium carbonate and magnesium hydroxide are precipitated out of solution, forming a rock-hard scale that clogs hot water pipes and reduces the efficiency of boilers, water heaters, and heat exchangers. Hardness, especially with the presence of magnesium sulfates, can lead to the development of a laxative effect on new consumers.

The use of hard water does offer some advantages, though, in that: (1) hard water aids in the growth of teeth and bones; (2) hard water reduces toxicity to many by preventing poisoning from lead oxide in lead pipelines; and (3) soft waters are suspected to be associated with cardiovascular diseases (Rowe and Abdel-Magid 1995).

Fluoride

We purposely fluoridate a range of everyday products, notably toothpaste and drinking water, because for decades we have believed that fluoride in small doses has no adverse effects on health to offset its proven benefits in preventing dental decay. The jury is still out, however, on the real benefits of fluoride, even in small amounts. Fluoride is seldom found in appreciable quantities in surface waters and appears in groundwater in only a few geographical regions. However, fluoride is sometimes found in a few types of igneous or sedimentary rocks. Fluoride is toxic to humans in large quantities and is also toxic to some animals. For example, certain plants used for fodder have the ability to store and concentrate fluoride. When animals consume this forage, they ingest an enormous overdose of fluoride. Animals' teeth become mottled, they lose weight, give less milk, grow spurs on their bones, and become so crippled that they must be destroyed (Koren 1991).

Fluoride used in small concentrations (about 1.0 mg/L in drinking water) can be beneficial. Experience has shown that drinking water containing a proper amount of fluoride can reduce tooth decay by 65% in children between the ages of 12 and 15. When large concentrations are used (>2.0 mg/L), discoloration of teeth may result. Adult teeth are not affected by fluoride. The EPA sets the upper limits for fluoride based on ambient temperatures because people drink more water in warmer climates; therefore, fluoride concentrations should be lower in these areas.

Note: How does the fluoridization of a drinking water supply actually work to reduce tooth decay? Fluoride combines chemically with tooth enamel when permanent teeth are forming. The result is teeth that are harder, stronger, and more resistant to decay.

Metals

Although iron and manganese are most commonly found in groundwaters, surface waters may also contain significant amounts at times. Metal ions are dissolved in groundwater and surface water when the water is exposed to rock or soil containing the metals, usually in the form of metal salts. Metals can also enter with discharges from sewage treatment plants, industrial plants, and other sources. The metals most often found in the highest concentrations in natural waters are calcium and magnesium. These are usually associated with a carbonate anion and come from the dissolution of limestone rock. As mentioned in the discussion of hardness, the higher the concentration of these metal ions, the harder the water; however, in some waters other metals can contribute to hardness. Calcium and magnesium are non-toxic and are normally absorbed by living organisms more readily than the other metals. Therefore, if the water is hard, the toxicity of a given concentration of a toxic metal is reduced. Conversely, in soft, acidic water the same concentrations of metals may be more toxic. In natural water systems, other nontoxic metals are generally found in very small quantities. Most of these metals cause taste problems well before they reach toxic levels. Fortunately, toxic metals are present in only minute quantities in most natural water systems. However, even in small quantities, toxic metals in drinking water are harmful to humans and other organisms. Arsenic, barium, cadmium, chromium, lead, mercury, and silver are toxic metals that may be dissolved in water. Arsenic, cadmium, lead, and mercury, all cumulative toxins, are particularly hazardous. These metals are concentrated by the food chain, thereby posing the greatest danger to organisms near the top of the chain.

Organics

Organic chemicals in water primarily emanate from synthetic compounds that contain carbon, such as PCBs, dioxin, and DDT (all toxic organic chemicals). These synthesized compounds often persist and accumulate in the environment because they do not readily breakdown in natural ecosystems. Many of these compounds can cause cancer in people and birth defects in other predators near the top of the food chain, such as birds and fish. The presence of organic matter in water is troublesome for the following reasons: "(1) color formation, (2) taste and odor problems, (3) oxygen depletion in streams, (4) interference with water treatment processes, and (5) the formation of halogenated compounds when chlorine is added to disinfect water" (Tchobanglous and Schroeder 2003).

Generally, the source of organic matter in water is from decaying leaves, weeds, and trees; the amount of these materials present in natural waters is usually low. The general category of "organics" in natural waters includes organic matter whose origins could be from both natural sources and human activities. It is important to distinguish natural organic compounds from organic compounds that are solely man-made (anthropogenic), such as pesticides and other synthetic organic compounds.

Many organic compounds are soluble in water, and surface waters are more prone to contamination by natural organic compounds than groundwaters. In water, dissolved organics are usually divided into two categories: *biodegradable* and *nonbiodegradable*. Biodegradable (break down) material consists of organics that can be utilized for nutrients (food) by naturally occurring microorganisms within a reasonable length of time. These materials usually consist of alcohols, acids, starches, fats, proteins, esters, and aldehydes. They may result from domestic or industrial wastewater discharges, or they may be end products of the initial microbial decomposition of plant or animal tissue. The principal problem associated with biodegradable organics is the effect resulting from the action of microorganisms. Moreover, some biodegradable organics can cause color, taste, and odor problems.

Oxidation and *reduction* play an important accompanying role in microbial utilization of dissolved organics. In oxidation, oxygen is added, or hydrogen is deleted from elements of the organic molecule. Reduction occurs when hydrogen is added to or oxygen is deleted from elements of the organic molecule. The oxidation process is by far more efficient and is predominant when oxygen is available. In *oxygen-present (aerobic)* environments, the end products of microbial decomposition of organics are stable and acceptable compounds. On the other hand, *oxygen-absent (anaerobic)* decomposition results in unstable and objectionable end products.

The quantity of oxygen-consuming organics in water is usually determined by measuring the *biochemical oxygen demand (BOD)*: the amount of dissolved oxygen needed by aerobic decomposers to break down the organic materials in a given volume of water over a 5-day incubation period at 20°C (68°F).

Nonbiodegradable organics are resistant to biological degradation. For example, constituents of woody plants such as tannins and lignin acids, phenols, and cellulose are found in natural water systems and are considered refractory (resistant to biodegradation). In addition, some polysaccharides with exceptionally strong bonds and benzene with its ringed structure are essentially nonbiodegradable. An example is benzene associated with the refining of petroleum. Some organics are toxic to organisms and thus are nonbiodegradable. These include organic pesticides and compounds that have combined with chlorine. Pesticides and herbicides have found widespread use in agriculture, forestry (silviculture), and mosquito control. Surface streams are contaminated via runoff and wash-off by

rainfall. These toxic substances are harmful to some fish, shellfish, predatory birds, and mammals. Some compounds are toxic to humans.

Nutrients

Nutrients (biostimulants) are essential building blocks for healthy aquatic communities, but excess nutrients (especially nitrogen and phosphorus compounds) over stimulate the growth of aquatic weeds and algae. Excessive growth of these organisms, in turn, can clog navigable waters, interfere with swimming and boating, outcompete native submerged aquatic vegetation, and, with excessive decomposition, lead to oxygen depletion. Oxygen concentrations can fluctuate daily during algal blooms, rising during the day as algae perform photosynthesis and falling at night as algae continue to respire, which consumes oxygen. Beneficial bacteria also consume oxygen as they decompose the abundant organic food supply in dying algae cells.

Plants require large amounts of the nutrients carbon, nitrogen, and phosphorus; otherwise growth will be *limited*.

Carbon is readily available from a number of natural sources including alkalinity, decaying products of organic matter, and dissolved carbon dioxide from the atmosphere. Since carbon is readily available, it is seldom the *limiting nutrient*. This is an important point because it suggests that identifying and reducing the supply of a particular nutrient can control algal growth. In most cases, nitrogen and phosphorus are essential growth factors and are the limiting factors in aquatic plant growth. Freshwater systems are most often limited by phosphorus.

Nitrogen gas (N_2), which is extremely stable, is the primary component of the earth's atmosphere. Major sources of nitrogen include runoff from animal feedlots, fertilizer runoff from agricultural fields, from municipal wastewater discharges, and from certain bacteria and blue-green algae that can obtain nitrogen directly from the atmosphere. In addition, certain forms of acid rain can also contribute nitrogen to surface waters.

Nitrogen in water is commonly found in the form of nitrate (NO_3). Nitrate in drinking water can lead to a serious problem. Specifically, nitrate poisoning in infants, including animals, can cause serious problems and even death. Bacteria commonly found in the intestinal tract of infants can convert nitrate to highly toxic nitrites (NO_2). Nitrite can replace oxygen in the bloodstream and result in oxygen starvation, which causes a bluish discoloration of the infant ("blue baby" syndrome).

In aquatic environments, phosphorus is found in the form of phosphate. Major sources of phosphorus include phosphates in detergents, fertilizer and feedlot runoff, and municipal wastewater discharges.

DID YOU KNOW?

Trying to find the typhoid bacillus in a sample of water is slim, but if you find a high count of *E. coli*, it is proof enough that the water is potentially infectious.

CHEMICAL CHARACTERISTICS OF WASTEWATER

The chemical characteristics of wastewater consist of three parts: (1) organic matter, (2) inorganic matter, and (3) gases. Metcalf and Eddy (2003) point out that in "wastewater of medium strength, about 75% of the suspended solids and 40% of the filterable solids are organic in nature." The organic substances of interest in this discussion include proteins, oil and grease, carbohydrates, and detergents (surfactants).

Organic Substances

Proteins are nitrogenous organic substances of high molecular weight found in the animal kingdom and, to a lesser extent, in the plant kingdom. The amount present varies from a small percentage found in tomatoes and other watery fruits and in the fatty tissues of meat to a high percentage in lean meats and beans. All raw foodstuffs, both plant and animal, contain proteins. Proteins consist wholly or partially of very large numbers of amino acids. They also contain carbon, hydrogen, oxygen, sulfur, phosphorus, and a fairly high and constant proportion of nitrogen. The molecular weight of proteins is quite high.

Proteinaceous materials constitute a large part of the wastewater biosolids; biosolids particles that do not consist of pure protein will be covered with a layer of protein that will govern their chemical and physical behavior (Coakley 1975). Moreover, the protein content ranges between 15% and 30% of the organic matter present for digested biosolids, and 28%–50% in the case of activated biosolids. Proteins and urea are the chief sources of nitrogen in wastewater. When proteins are present in large quantities, microorganisms decompose them producing end products that have objectionable foul odors. During this decomposition process, proteins are hydrolyzed to amino acids and then further degraded to ammonia, hydrogen sulfide, and simple organic compounds.

Oils and *grease* are another major component of foodstuffs. They are also usually related to spills or other releases of petroleum products. Minor oil and grease problems can result from wet weather runoff from highways or the improper disposal in storm drains of motor oil. They are insoluble in water but dissolve in organic solvents such as petroleum, chloroform, and ether. Fats, oils, waxes, and other related constituents found in wastewater are commonly grouped under the term "grease." Fats and oils are contributed to domestic wastewater in butter, lard, margarine, and vegetable fats and oils. Fats, which are compounds of alcohol and glycerol, are among the more stable of organic compounds and are not easily decomposed by bacteria. However, they can be broken down by mineral acids, resulting in the formation of fatty acids and glycerin. When these glycerides of fatty acids are liquid at ordinary temperature, they are called oils, and those that are solid are called fats.

The grease content of wastewater can cause many problems in wastewater treatment unit processes. For example, high grease content can cause clogging of filters, nozzles, and sand beds (Gilcreast et al. 1953). Moreover, grease can coat the walls of sedimentation tanks and decompose, increasing the amount of scum. Additionally, if grease is not removed before the discharge of the effluent, it can interfere with the biological processes in surface waters and create unsightly floating matter and films (Rowe and Abdel-Magid 1995). In the treatment process, grease can coat trickling filters and interfere with the activated sludge process, which, in turn, interferes with the transfer of oxygen from the liquid to the interior of living cells (Sawyer et al. 1994).

Carbohydrates, which are widely distributed in nature and found in wastewater, are organic substances that include starch, cellulose, sugars, and wood fibers; they contain carbon, hydrogen, and oxygen. Sugars are soluble, while starches are insoluble in water. The primary function of carbohydrates in higher animals is to serve as a source of energy. In lower organisms, e.g., bacteria, carbohydrates are utilized to synthesize fats and proteins as well as energy. In the absence of oxygen, the end products of the decomposition of carbohydrates are organic acids, alcohols, and gases such as carbon dioxide and hydrogen sulfide. The formation of large quantities of organic acids can affect the treatment process by overtaxing the buffering capacity of the wastewater, resulting in a drop in pH and a cessation of biological activity.

Detergents (surfactants) are large organic molecules that are slightly soluble in water and cause foaming in wastewater treatment plants and in the surface waters into which the effluent is discharged. Probably the most serious effect detergents can have on wastewater treatment processes is their tendency to reduce the oxygen uptake in biological processes. According to Rowe and Abdel-Magid (1995), "Detergents affect wastewater treatment processes by (1) lowering the surface or interfacial tension of water and increasing its ability to wet surfaces with which they come in contact; (2) emulsifying grease and oil, deflocculating colloids; (3) inducing flotation of solids and giving rise to foams; and (4) possibly killing useful bacteria and other living organisms." Since the development and increasing use of synthetic detergents, many of these problems have been reduced or eliminated.

Inorganic Substances

Several inorganic components are common to both wastewater and natural waters and are important in establishing and controlling water quality. The inorganic load in water is the result of discharges of treated and untreated wastewater, various geologic formations, and inorganic substances left in the water after evaporation. Natural waters dissolve rocks and minerals with which they come in contact. As mentioned, many of the inorganic constituents found in natural waters are also found in wastewater. Many of these constituents are added via human use. These inorganic constituents include pH, chlorides, alkalinity, nitrogen, phosphorus, sulfur, toxic inorganic compounds, and heavy metals.

When the *pH* of water or wastewater is considered, we are simply referring to the hydrogen ion concentration. Acidity, or the concentration of hydrogen ions, drives many chemical reactions in living organisms. A pH value of 7 represents a neutral condition. A low pH value (less than 5) indicates acidic conditions, while a high pH (greater than 9) indicates alkaline conditions. Many biological processes, such as reproduction, cannot function in acidic or alkaline waters. Acidic conditions also aggravate toxic contamination problems because sediments release toxicants in acidic waters.

Many of the important properties of wastewater are due to the presence of weak acids, bases, and their salts. The wastewater treatment process is made up of several different unit processes (these are discussed later). It can be safely stated that one of the most important unit processes in the overall wastewater treatment process is disinfection. pH has an effect on disinfection, particularly in regard to disinfection using chlorine. For example, with increases in pH, the amount of contact time needed for disinfection using chlorine increases. Common sources of acidity include mine drainage, runoff from mine tailings, and atmospheric deposition.

In the form of the Cl^- ion, *chloride* is one of the major inorganic constituents in water and wastewater. Sources of chlorides in natural waters are (1) leaching of chloride from rocks and soils; (2) in coastal areas, saltwater intrusion; (3) from agricultural, industrial, domestic, and human wastewater; and (4) from the infiltration of groundwater into sewers adjacent to saltwater. The salty taste produced by chloride concentration in potable water is variable and depends on the chemical composition of the water. In wastewater, the chloride concentration is higher than in raw water because sodium chloride (salt) is a common part of the diet and passes unchanged through the digestive system. Because conventional methods of waste treatment do not remove chloride to any significant extent, higher than usual chloride concentrations can be taken as an indication that the body of water is being used for waste disposal (Metcalf & Eddy 2003).

As mentioned earlier, *alkalinity* is a measure of the buffering capacity of water and, in wastewater, helps to resist changes in pH caused by the addition of acids. Alkalinity is caused by chemical compounds dissolved from soil and geologic formations and is mainly due to the presence of hydroxyl and bicarbonate ions. These compounds are mostly the carbonates and bicarbonates of calcium, potassium, magnesium, and sodium. Wastewater is usually alkaline. Alkalinity is important in wastewater treatment because anaerobic digestion requires sufficient alkalinity to ensure that the pH will not drop below 6.2; if alkalinity does drop below this level, the methane bacteria cannot function. For the digestion process to operate successfully, the alkalinity must range from about 1,000 to 5,000 mg/L as calcium carbonate. Alkalinity in wastewater is also important when chemical treatment is used, in biological nutrient removal, and whenever ammonia is removed by air stripping.

In domestic wastewater, "nitrogen compounds result from the biological decomposition of proteins and form urea discharged in body waste" (Peavy et al. 1987). In wastewater treatment, biological treatment cannot proceed unless *nitrogen*, in some form, is present. Nitrogen must be present in the form of organic nitrogen (N), ammonia (NH_3), nitrite (NO_2), or nitrate (NO_3). Organic nitrogen includes such natural constituents as peptides, proteins, urea, nucleic acids, and numerous synthetic organic materials. Ammonia is present naturally in wastewaters. It is produced primarily by the de-aeration of organic nitrogen-containing compounds and by the hydrolysis of urea. Nitrite, an intermediate oxidation state of nitrogen, can enter a water system through use as a corrosion inhibitor in industrial applications. Nitrate is derived from the oxidation of ammonia.

Nitrogen data is essential in evaluating the treatability of wastewater by biological processes. If nitrogen is not present in sufficient amounts, it may be necessary to add it to the waste to make it treatable. When the treatment process is complete, it is important to determine how much nitrogen is in the effluent. This is important because the discharge of nitrogen into receiving waters may stimulate algal and aquatic plant growth. These, of course, exert a high oxygen demand at nighttime, which adversely affects aquatic life and has a negative impact on the beneficial use of water resources.

Phosphorus (P) is a macronutrient that is necessary for all living cells and is a ubiquitous constituent of wastewater. It is primarily present in the form of phosphates—the salts of phosphoric acid. Municipal wastewaters may contain 10–20 mg/L phosphorus as P, much of which comes from phosphate builders in detergents. Because of noxious algal blooms that occur in surface waters, there is much interest in controlling the amount of phosphorus compounds that enter surface waters in domestic and industrial waste discharges and natural runoff. This is particularly the case in the United States because approximately 15% of the population contributes wastewater effluents to lakes, resulting in *eutrophication* (or cultural enrichment) of these water bodies. Eutrophication occurs when the nutrient levels exceed the affected waterbody's ability to assimilate them. Eutrophication leads to significant changes in water quality. Reducing phosphorus inputs to receiving waters can control this problem.

Sulfur (S) is required for the synthesis of proteins and is released in their degradation. The sulfate ion occurs naturally in most water supplies and is present in wastewater as well. Sulfate is reduced biologically to sulfide, which in turn can combine with hydrogen to form hydrogen sulfide (H_2S). H_2S is toxic to animals and plants. H_2S in interceptor systems can cause severe corrosion to pipes and appurtenances. Moreover, in certain concentrations, H_2S is a deadly toxin.

Toxic inorganic compounds such as copper, lead, silver, arsenic, boron, and chromium are classified as priority pollutants and are toxic to microorganisms. Thus, they must be taken into consideration in the design and operation of a biological treatment process. When introduced into a treatment process, these contaminants can kill off the microorganisms needed for treatment and thus stop the treatment process.

Heavy metals are major toxicants found in industrial wastewaters; they may adversely affect the biological treatment of wastewater. Mercury, lead, cadmium, zinc, chromium, and plutonium are among the so-called heavy metals—those with a high atomic mass. (It should be noted that the term heavy metals is rather loose and is taken by some to include arsenic, beryllium, and selenium, which are not really metals and are better termed toxic metals.) The presence of any of these metals in excessive quantities will interfere with many beneficial uses of water because of their toxicity. Urban runoff is a major source of lead and zinc in many water bodies.

Note: Lead is a toxic metal that is harmful to human health; there is *no* safe level for lead exposure. It is estimated that up to 20% of the total lead exposure in children can be attributed to a waterborne route, i.e., consuming contaminated water. The lead comes from the exhaust of automobiles using leaded gasoline, while zinc comes from tire wear.

BIOLOGICAL CHARACTERISTICS OF WATER AND WASTEWATER

Specialists or practitioners who work in the water or wastewater treatment field must not only have a general understanding of the microbiological principles presented in Chapter 11, but they must also have some knowledge of the biological characteristics of water and wastewater. This knowledge begins with an understanding that water may serve as a medium in which thousands of biological species spend part, if not all, of their life cycles. It is important to understand that, to some extent, all members of the biological community are water-quality parameters, because their presence or absence may indicate, in general terms the characteristics of a given body of water.

The presence or absence of certain biological organisms is of primary importance to the water/wastewater specialist. These are, of course, the *pathogens*. Pathogens are organisms that are capable of infecting or transmitting diseases in humans and animals. It should be pointed out that these organisms are not native to aquatic systems and usually require an animal host for growth and reproduction. They can, however, be transported by natural water systems. These waterborne pathogens include species of bacteria, viruses, protozoa, and parasitic worms (helminths). In the following sections, a brief review of each of these species is provided.

Bacteria

The word *bacteria* (singular: bacterium) comes from the Greek word meaning "rod" or "staff," a shape characteristic of many bacteria. Recall that bacteria are single-celled microscopic organisms that multiply by splitting in two (binary fission). In order to multiply they need carbon dioxide if they are autotrophs, or organic compounds (dead vegetation, meat, sewage) if they are heterotrophs. Their energy comes either from sunlight if they are photosynthetic or from chemical reactions if they are chemosynthetic. Bacteria are present in air, water, earth, rotting vegetation, and the intestines of animals. Human and animal wastes are the primary sources of bacteria in water. These sources of bacterial contamination include runoff from feedlots, pastures, dog runs, and other land areas where animal wastes are deposited. Additional sources include seepage or discharge from septic tanks and sewage treatment facilities. Bacteria from these sources can enter wells that are either open at the land surface or do not have watertight casings or caps. Gastrointestinal disorders are common symptoms of most diseases transmitted by waterborne pathogenic bacteria. In wastewater treatment processes, bacteria are fundamental, especially in the degradation of organic matter, which takes place in trickling filters, activated biosolids processes, and biosolids digestion.

Viruses

A *virus* is an entity that carries the information needed for its replication but does not possess the machinery for such replication (Sterritt and Lester 1988). Thus, it is an obligate parasite that requires a host in which to live. Viruses are the smallest biological structures known, so they can only be seen with the aid of an electron microscope. Waterborne viral infections are usually indicated by disorders of the nervous system rather than the gastrointestinal tract. Viruses that are excreted by human beings may become a major health hazard to public health. Waterborne viral pathogens are known to cause poliomyelitis and infectious hepatitis. Testing for viruses in water is difficult because: (1) they are small; (2) they are present in low concentrations in natural waters; (3) there are numerous varieties; (4) they are unstable; and (5) there are limited identification methods available. Because of these testing problems and the uncertainty of viral disinfection, the direct recycling of wastewater and the practice of land application of wastewater is a cause of concern (Peavy et al. 1987).

Protozoa

Protozoa (singular: protozoan) are mobile, single-celled, complete, self-contained organisms that can be free-living or parasitic, pathogenic or nonpathogenic, and microscopic or macroscopic. Protozoa range in size from two to several hundred microns in length. They are highly adaptable and widely distributed in natural waters, although only a few are parasitic. Most protozoa are harmless, only a few cause illness in humans—*Entamoeba histolytica* (amebiasis) being one of the exceptions. Because aquatic protozoa form cysts during adverse environmental conditions, they are difficult to deactivate by disinfection and must undergo filtration to be removed.

Worms (Helminths)

Worms are normal inhabitants of organic mud and organic slime. They have aerobic requirements but can metabolize solid organic matter that is not readily degraded by other microorganisms. Water contamination may result from human and animal waste that contains worms. Worms pose hazards primarily to those individuals who come into direct contact with untreated water. Thus, swimmers in surface water polluted by sewage or stormwater runoff from cattle feedlots, as well as sewage plant operators, are at particular risk.

THE BOTTOM LINE

This book is slanted toward chemical, physical, and biological water parameters.

CHAPTER REVIEW QUESTIONS

14.1 Those characteristics or range of characteristics that make water appealing and useful: _____.

14.2 Process by which water vapor is emitted by leaves: _____

14.3 Water we see: _____

14.4 Leading cause of impairment for rivers, lakes, and estuaries: _____

14.5 All contaminants of water contribute to the _____.

14.6 The clarity of water is usually measured by its: _____

14.7 Water has been called the: _____

14.8 A measure of water's ability to neutralize acid: _____

14.9 A pH value of 7 represents a: _____

14.10 There is no safe level for _____ exposure.

NOTE

1 Information from this section is from Spellman & Drinan (2012) *The Drinking Water Handbook*, 2nd ed. Boca Raton, FL: CRC Press.

REFERENCES

Coakley, P., 1975. Developments in our Knowledge of Sludge Dewatering Behavior. *8th Public Health Engineering Conference held in the Department of Civil Engineering*, Loughborough, University of Technology.

Fortner, B., and Schechter, D., 1996. U.S. water quality shows little improvement over 1992 inventory. *Water Environ Technol*; 8(2):10–14.

Gilcreast, F.W., Sanderson, W.W., and Elmer, R.P., 1953. Two methods for the determination of grease in sewage. *Sewage Ind Wastes*; 25:1379.

Gleick, P.H., 2001. Freshwater forum. *U.S. Water News*; 18(6):22–27.

Koren, H., 1991. *Handbook of Environmental Health and Safety: Principles and Practices*. Chelsea, MI: Lewis Publishers.

McGhee, T.J., 1991. *Water Supply and Sewerage*, 6th ed. New York: McGraw-Hill.

Metcalf & Eddy, Inc., 2003. *Wastewater Engineering: Treatment, Disposal, Reuse*, 4th ed. New York: McGraw-Hill.

Peavy, H.S., Rowe, D.R., and Tchobanglous, G., 1987. *Environmental Engineering*. New York: McGraw-Hill, Inc.

Rowe, D.R., and Abdel-Magid, I.M., 1995. *Handbook of Wastewater Reclamation and Reuse*. Boca Raton, FL: Lewis Publishers.

Sawyer, C.N., McCarty, A.L., and Parking, G.F., 1994. *Chemistry for Environmental Engineering*. New York: McGraw-Hill.

Sterritt, R.M., and Lester, J.M. 1988. *Microbiology for Environmental and Public Health Engineers*. London: E. and F.N. Spoon.

Tchobanglous, G., and Schroeder, E.D., 2003. *Water Quality*. Reading, MA: Addison-Wesley Publishing Company.

USEPA. 2002. *National Recommended Water Quality*. Washington, DC: US Environmental Protection Agency.

USEPA. 2021. *Turbidity. EPA 841F2100&D*. Washington, DC: United States Environmental Protection Agency.

15 Biomonitoring, Monitoring, Sampling, and Testing

INTRODUCTION

In January, we take our nets to a no-name stream in the foothills of the Blue Ridge Mountains in Virginia to conduct a special kind of macroinvertebrate monitoring—looking for "winter stoneflies." Winter stoneflies have an unusual life cycle. Soon after hatching in early spring, the larvae bury themselves in the streambed. They spend the summer lying dormant in the mud, thereby avoiding problems like overheated streams, low oxygen concentrations, fluctuating flows, and heavy predation. In late November, they emerge, grow quickly for a couple of months, and then lay their eggs in January.

January monitoring of winter stoneflies helps in interpreting the results of spring and fall macroinvertebrate surveys. In spring and fall, a thorough benthic survey is conducted based on *Protocol II* of the *USEPA's Rapid Bioassessment Protocols for Use in Streams and Rivers*. Some sites in various rural streams have poor diversity and sensitive families. Is the lack of macroinvertebrate diversity because of specific warm-weather conditions, high water temperature, low oxygen, or fluctuating flows, or is some toxic contamination present? In the January screening, if winter stoneflies are plentiful, seasonal conditions were probably to blame for the earlier results; if winter stoneflies are absent, the site probably suffers from toxic contamination (based on our rural location, probably emanating from non-point sources), which is present year-round. Though different genera of winter stoneflies are found in our region (southwestern Virginia), Allocapnia is sought because it is present even in the smallest streams.

WHAT IS BIOMONITORING?

The life in, and physical characteristics of, a stream ecosystem provide insight into the historical and current status of its quality. The assessment of a water body ecosystem based on organisms living in it is called *biomonitoring*. The assessment of the system based on its physical characteristics is called a habitat assessment. Biomonitoring and habitat assessments are two tools that stream ecologists use to assess the water quality of a stream.

Biological monitoring involves the use of organisms (called assemblages)—periphytons, fish, and macroinvertebrates—to assess environmental conditions. Biological observation is more representative, as it reveals cumulative effects as opposed to chemical observation, which is representative only at the actual time of sampling.

Again, the presence of different assemblages of organisms is used in conducting biological assessments and/

or biosurveys. In selecting the appropriate for a particular biomonitoring situation, the advantages of using each assemblage must be considered along with the objectives of the program. Some of the advantages of using periphytons (algae), benthic macroinvertebrates, and fish in a biomonitoring program are presented in this section.

Important Note: Periphytons are a complex matrix of benthic-attached algae, cyanobacteria, heterotrophic microbes, and detritus that is attached to submerged surfaces in most aquatic ecosystems.

ADVANTAGES OF USING PERIPHYTON

1. Algae generally have rapid reproduction rates and very short life cycles, making them valuable indicators of short-term impacts.
2. As primary producers, algae are most directly affected by physical and chemical factors.
3. Sampling is easy, inexpensive, requires few people, and creates minimal impact on resident biota.
4. Relatively standard methods exist for evaluation of functional and non-taxonomic structural (biomass, chlorophyll measurements) characteristics of algal communities.
5. Algal assemblages are sensitive to some pollutants that may not visibly affect other aquatic assemblages or may only affect other organisms at higher concentrations (i.e., herbicides) (Carins and Dickson, 1971; APHA, 1971; Patrick, 1973; Rodgers et al., 1979; Weitzel, 1979; Karr, 1981; USEPA, 1983).

ADVANTAGES OF USING FISH

1. Fish are good indicators of long-term (several years) effects and broad habitat conditions because they are relatively long-lived and mobile (Karr et al., 1986)
2. Fish assemblages include a range of species that represent a variety of trophic levels (omnivores, herbivores, insectivores, planktivores, piscivores). They tend to integrate the effects of lower trophic levels; thus, fish assemblage structure is reflective of integrated environmental health.
3. Fish are at the top of the aquatic food web and are consumed by humans, making them important for assessing contamination.
4. Fish are relatively easy to collect and identify at the species level. Most specimens can be sorted and

DOI: 10.1201/9781003581901-18

identified in the field by experienced fisheries professionals, and subsequently released unharmed.

5. Environmental requirements of most fish are comparatively well known. Life history information is extensive for many species, and information on fish distributions is commonly available.

6. Aquatic life uses (water quality standards) are typically characterized in terms of fisheries (cold water, cool water, warm water, sport, forage). Monitoring fish provides a direct evaluation of "fishability" and fish "propagation," which emphasizes the importance of fish to anglers and commercial fishermen.

7. Fish account for nearly half of the endangered vertebrate species and subspecies in the United States (Warren and Burr, 1994).

ADVANTAGES OF USING MACROINVERTEBRATES

As discussed earlier, benthic macroinvertebrates are the larger organisms such as aquatic insects, insect larvae, and crustaceans that live in the bottom portions of a waterway for part of their life cycle. They are ideal for use in biomonitoring, as they are ubiquitous, relatively sedentary, and long-lived. They provide a cross-section of the situation, as some species are extremely sensitive to pollution, while others are more tolerant. However, like toxicity testing, biomonitoring does not tell you why animals are present or absent. As mentioned, benthic macroinvertebrates are excellent indicators for several reasons:

1. Biological communities reflect overall ecological integrity (i.e., chemical, physical, and biological integrity). Therefore, biosurvey results directly assess the status of a water body relative to the primary goal of the Clean Water Act (CWA).

2. Biological communities integrate the effects of different stressors and thus provide a broad measure of their aggregate impact.

3. Because they are ubiquitous, communities integrate the stressors over time and provide an ecological measure of fluctuating environmental conditions.

4. Routine monitoring of biological communities can be relatively inexpensive because they are easy to collect and identify.

5. The status of biological communities is of direct interest to the public as a measure of a particular environment.

6. Where criteria for specific ambient impacts do not exist (e.g., non-point sources that degrade habitats), biological communities may be the only practical means of evaluation.

Benthic macroinvertebrates act as continuous monitors of the water they live in. Unlike chemical monitoring, which provides information about water quality at the time of measurement (a snapshot), biological monitoring can provide information about past and/or episodic pollution (a videotape). This concept is analogous to miners who took canaries into deep mines with them to test for air quality. If the canary died, the miners knew the air was bad and they had to leave the mine. Biomonitoring a water body ecosystem uses the same theoretical approach. Aquatic macroinvertebrates are subject to pollutants in the water body. Consequently, the health of the organisms reflects the quality of the water they live in. If the pollution levels reach a critical concentration, certain organisms will migrate away, fail to reproduce, or die, eventually leading to the disappearance of those species at the polluted site. Normally, these organisms will return if conditions improve in the system (Bly and Smith, 1994).

Biomonitoring (and the related term, bioassessment) surveys are conducted before and after an anticipated impact to determine the effect of the activity on the water body habitat. Moreover, surveys are performed periodically to monitor water body habitats and watch for unanticipated impacts. Finally, biomonitoring surveys are designed to reference conditions or to set biocriteria (serve as monitoring thresholds to signal future impacts, regulatory actions, etc.) for determining whether an impact has occurred (Camann, 1996).

Note: The primary justification for bioassessment and monitoring is that the degradation of water body habitats affects the biota using those habitats. Therefore, the living organisms themselves provide the most direct means of assessing real environmental impacts. Although the focus of this text is on macroinvertebrate protocols, the periphyton and fish protocols are discussed briefly before an in-depth discussion of the macroinvertebrate protocols.

PERIPHYTON PROTOCOLS

Benthic algae (periphyton or phytobenthos) are primary producers and an important foundation of many stream food webs (Bahls, 1993; Stevenson, 1996). These organisms also stabilize substrates and serve as habitats for many other organisms. Because benthic algal assemblages are attached to substrates, their characteristics are affected by physical, chemical, and biological disturbances that occur in the stream reach during the time in which the assemblage develop.

Diatoms, in particular, are useful ecological indicators because they are found in abundance in most lotic ecosystems. Diatoms and many other algae can be identified as species by experienced algologists. The great number of species provides multiple, sensitive indicators of environmental change and the specific conditions of their habitat. Diatom species are differentially adapted to a wide range of ecological conditions.

Periphyton indices of biotic integrity have been developed and tested in several regions (Kentucky Department of Environmental Protection, 1993; Hill, 1997). Because the ecological tolerances for many species are known, changes in community composition can be used to diagnose the environmental stressors affecting ecological health, as well as to assess biotic integrity (Stevenson, 1998; Stevenson and Pan, 1999). Periphyton protocols may be used by themselves, but they are most effective when used with one or more of the other assemblages and protocols. They should be used with habitat and benthic macroinvertebrate assessments, particularly because of the close relation between periphyton and these elements of stream ecosystems.

Presently, few states have developed protocols for periphyton assessment. Montana, Kentucky, and Oklahoma have developed periphyton bioassessment programs. Other states are exploring the possibility of developing periphyton programs. Algae have been widely used to monitor water quality in rivers in Europe, where many different approaches have been used for sampling and data analysis.

FISH PROTOCOLS

Monitoring of the fish assemblage is an integral component of many water quality management programs, and its importance is reflected in the aquatic life use-support designations of many states. Assessments of the fish assemblage must measure the overall structure and function of the ichthyofaunal community to adequately evaluate biological integrity and protect surface water resource quality. Fish bioassessment data quality and comparability are assured through the utilization of qualified fisheries professionals and consistent methods.

In the fish protocol, the principal evaluation mechanism utilizes the technical framework of the Index of Biotic Integrity (IBI)—a fish assemblage assessment approach developed by Karr (1981). The IBI incorporates the zoogeographic, ecosystem, community, and population aspects of the fish assemblage into a single ecologically-based index. Calculation and interpretation of the IBI involve a sequence of activities, including fish sample collection, data tabulation, and regional modification and calibration of metrics and expectation values.

The fish protocol involves careful, standardized field collection, species identification and enumeration, and analyses using aggregated biological attributes or quantification of the numbers of key species. The role of experienced fisheries scientists in the adaptation and application of the bioassessment protocol and the taxonomic identification of fish cannot be overemphasized. The fish bioassessment protocols survey yields an objective, discrete measure of the condition of the fish assemblage. Although the fish survey can usually be completed in the field by qualified fish biologists, difficult species identifications will require laboratory confirmation. Data provided by the fish bioassessment protocols can serve to assess use attainment, develop biological criteria, prioritize sites for further evaluation, provide a reproducible impact assessment, and evaluate the status and trends of the fish assemblage.

Fish collection procedures must focus on a multi-habitat approach—sampling habitats in relative proportion to their local representation (as determined during site reconnaissance). Each sample reach should contain riffle, run, and pool habitats when available. Whenever possible, the reach should be sampled sufficiently upstream of any bridge or road crossing to minimize the hydrological effects on overall habitat quality. Wadeability and accessibility may ultimately govern the exact placement of the sample reach. A habitat assessment is performed, and physical/chemical parameters are measured concurrently with fish sampling to document and characterize available habitat specifics within the sample reach.

MACROINVERTEBRATE PROTOCOLS

Benthic macroinvertebrates, by indicating the extent of oxygenation in a stream, can be regarded as indicators of the intensity of pollution from organic waste. The responses of aquatic organisms in water bodies to large quantities of organic waste are well-documented and occur in a predictable cyclical manner. For example, upstream from the discharge point, a stream can support a wide variety of algae, fish, and other organisms. However, in sections of the water body where oxygen levels drop below 5 ppm, only a few types of worms are able to survive.

DETERMINING INCREMENTAL CHANGE IN AQUATIC ECOSYSTEMS

The truth be told (based on experience and current practice), most states (and many water biomonitoring personnel) are using biological assessment information to support their water quality management programs. The standard biological assessment and criteria toolbox includes biological indices, the biological condition gradient (BCG), models, statistical approaches and guidance.

DID YOU KNOW?

As streamflow courses downstream, oxygen levels recover, and species that can tolerate low oxygen levels (such as gar, catfish, and carp) begin to appear. Eventually, at some point further downstream, a clean water zone re-establishes itself, and a more diverse and desirable community of organisms returns. Due to this characteristic pattern of alternating levels of dissolved oxygen (in response to the dumping of large amounts of biodegradable organic material), a stream, as stated above, goes through a cycle called an *oxygen sag curve*. Its state can be determined using the biotic index as an indicator of oxygen content.

The biotic index and biological condition gradient (BCG) are explained in the following sections, but first, to ensure understanding, key terms and definitions are provided.

Key Terms and Definitions (USEPA, 2002)

Assemblage: An association of interacting populations of organisms in a stream or other habitat. Examples of assemblages used for biological assessments include algae, amphibians, birds, fish, macroinvertebrates (e.g., insects, crayfish, clams, snails), and vascular plants.

Attribute: A measurable component of a biological assemblage. In the context of biological assessments, attributes include the ecological processes or characteristics of an individual or assemblage of species that are expected, but not empirically shown, to respond to a gradient of human disturbance.

Benthos: The bottom fauna of bodies of water.

Biological Assessment: Using biomonitoring data of samples of living organisms to evaluate the condition or health of a stream or other habitat.

Biological Integrity: "The ability of an aquatic ecosystem to support and maintain a balanced, adaptive community of organisms having a species composition, diversity, and functional organization comparable to that of natural habitats within a region" (Karr and Dudley, 1981).

Biological Magnification: The increase in the concentration of some material in organisms compared with its concentration in the environment.

Biological Monitoring: Sampling the biota of a place (e.g., a stream or other habitat).

Biota: The plants and animals living in a habitat.

Community: All the groups of organisms living together in the same area, usually interacting or depending on each other for existence.

Competition: Utilization by different species of limited resources such as food or nutrients, refugia, space, ovi position sites, or other resources necessary for reproduction, growth, and survival.

Composition: The taxonomic grouping of organisms, such as fish, algae, or macroinvertebrates, relating primarily to the kinds and numbers of organisms in the group.

Continuum: A gradient of change.

Disturbance: "Any discrete event in time that disrupts ecosystems, communities, or population structure and changes resources, substrate availability, or the physical environment" (Pickett and White, 1985). Natural disturbances include fire, drought, and floods. Human-caused disturbances are referred to as "human influence" and tend to be more persistent over time, e.g., plowing, clear-cutting of forests, and conducting urban stormwater into wetlands (aka constructed wetlands).

Diversity: A combination of the number of taxa and their relative abundance. A variety of diversity indexes have been developed to calculate diversity.

Dominance: The relative increase in the abundance of one or more species in relation to the abundance of other species in samples from a habitat.

Ecosystem: The community plus its habitat; the connotation is of an interacting system.

Ecoregion: A region defined by the similarity of climate, landform, soil, potential natural vegetation, hydrology, and other ecologically relevant variables.

Functional Feeding Groups (FFGs): Groupings of different invertebrates based on the mode of food acquisition rather than the category of food eaten. The groupings relate to the morphological structures, behaviors, and life history attributes that determine the mode of feeding by invertebrates. Examples of invertebrate FFGs are shredders, which chew live plant tissue or plant litter, and scrapers, which scape periphyton and associated matter from substrates (Merritt and Cummins, 1996).

Functional Groups: A means of dividing organisms into groups, often based on their method of feeding (e.g., shredder, scraper, filterer, predator) type of food (e.g., fruit, seeds, nectar, insects), or habits (e.g., burrower, climber, clinger).

Gradient of Human Influence: The relative ranking of sample sites within a regional wetland class based on the estimation of the degree of human disturbance (e.g., pollution and physical alteration of habitats).

Habitat: The sum of the physical, chemical, and biological environment occupied by individuals of a particular species, population, or community.

Hydrology: The science of dealing with the properties, distribution, and circulation of water both on the surface and under ground.

Impact: A change in the chemical, physical (including habitat), or biological quality or condition of a body of water caused by external forces.

Impairment: Adverse changes occurring to an ecosystem or habitat. An impaired wetland has some degree of human influence affecting it.

Index of Biologic Integrity (IBI): An investigative expression of biological condition that is composed of multiple metrics. It is similar to economic indexes used for expressing the condition of the economy. The IBI was first developed by Dr. James Karr to help resource managers sample, evaluate, and describe the condition of small warm water streams in central Illinois and Indiana (Karr, 1981). The phrase "biological integrity" comes from the 1972 Clean Water Act, which established "restoration and maintenance of the chemical, physical, and biological integrity of the nation's waters." "Integrity" implies an unimpaired condition or quality or state of being complete. "Biologic integrity" is based on the premise that the status of living organism provides the most direct and effective measure of the "integrity of water."

Intolerant Taxa: Taxa that tend to decrease in wetlands or other habitats that have higher levels of human disturbances, such as chemical pollution or siltation.

Least Impaired Site: Sample sites within a regional wetland class that exhibit the least degree of detrimental effect. Such sites help anchor gradients of human disturbance and are commonly referred to as reference sites.

Macroinvertebrates: Animals without backbones that are caught with a 500–800-μm mesh net. Macroinvertebrates do not include zooplankton or ostracods (aka seed shrimp), which are generally smaller than 200 μm in size.

Metric: An attribute with empirical change in value along a gradient of human disturbance.

Most Impaired Site: Sample sites that help anchor gradients of human disturbance and serve as important references, although they are commonly referred to as reference sites.

Munsell Color: The color of soil based on its chroma and hue as determined by a chart in the book *Munsell Soil Color Charts* (Munsell Color, 1975).

Mottles: Spots or blotches of different colors or shades of color interspersed within the dominant color in a soil layer, usually resulting from the presence of periodic reducing soil conditions.

Omnivores: Organisms that consume both plant and animal material.

Population: A set of organisms belonging to the same species and occupying a particular area at the same time.

Predator: An animal that feeds on other animals.

Reference Site: As used with an Index of Biological Integrity, a minimally impaired site that is representative of the expected ecological conditions and integrity of other sites of the same type and region.

Taxa: A grouping of organisms given a forma taxonomic name, such as species, genus, or family. The singular form is taxon.

Taxa Richness: The number of distinct species or taxa that are found in an assemblage, community, or sample.

Tolerance: The biological ability of different species or populations to survive successfully within a certain range of environmental conditions.

Tolerance (Range of): The set of conditions within which an organism, taxon, or population can survive.

Tolerant Taxa: This means taxa that tend to increase in wetlands or other habitats that have higher levels of human disturbances, such as chemical pollution or siltation.

Trophic: Pertaining to feeding and energy transfers.

THE BIOTIC INDEX

The biotic index is a systematic survey of macroinvertebrate organisms. Since the diversity of species in a stream is often a good indicator of the presence of pollution, the biotic index can be used to correlate with stream quality. Observation of the types of species present or missing is used as an indicator of stream pollution. The biotic index, used in the determination of the types, species, and numbers of biological organisms present in a stream, is commonly used as an auxiliary to BOD determination in assessing stream pollution. The biotic index is based on two principles:

1. A large dumping of organic waste into a stream tends to restrict the variety of organisms at a certain point in the stream.
2. As the degree of pollution in a stream increases key organisms tend to disappear in a predictable order. The disappearance of particular organisms tends to indicate the water quality of the stream.

Several different forms of the biotic index exist. In Great Britain, for example, the Trent Biotic Index (TBI), the Chandler score, the Biological Monitoring Working Party (BMWP) score, and the Lincoln Quality Index (LQI) are widely used. Most forms use a biotic index that ranges from 0 to 10. The most polluted stream, which contains the smallest variety of organisms, is at the lowest end of the scale (0); the cleanest streams are at the highest end (10). A stream with a biotic index greater than 5 will support game fish; on the other hand, a stream with a biotic index of less than 4 will not support game fish.

As mentioned, because they are easy to sample, macroinvertebrates have predominated in biological monitoring. Macroinvertebrates are a diverse group and demonstrate tolerances that vary between species. Thus, discrete differences tend to appear, containing both tolerant and sensitive indicators. Macroinvertebrates can be easily identified using identification keys that are portable and easily used in field settings. Present knowledge of macroinvertebrate tolerances and responses to stream pollution is well documented. In the United States, for example, the Environmental Protection Agency (EPA) required states into incorporate narrative biological criteria into their water quality standards by 1993. The National Park Service (NPS) has collected macroinvertebrate samples from American streams since 1984. Through their sampling efforts, the NPS has been able to derive quantitative biological standards (Huff, 1993).

The biotic index provides a valuable measure of pollution, especially for species that are very sensitive to a lack of oxygen. An example of an organism commonly used in biological monitoring is the stonefly. Stonefly larvae live underwater and survive best in well-aerated, unpolluted waters with clean gravel bottoms. When stream water quality deteriorates due to organic pollution, stonefly larvae cannot survive. The degradation of stonefly larvae has an exponential effect on other insects and fish that feed on the larvae; when the stonefly larvae disappear, many insects and fish disappear as well (O'Toole, 1986).

Table 15.1 shows a modified version of the BMWP biotic index. Considering that the BMWP biotic index indicates ideal stream conditions, it accounts for the sensitivities of different macroinvertebrate species, which are represented by diverse populations and are excellent indicators of

TABLE 15.1

BMWP Score System (Modified for Illustrative Purposes)

Families	Common-Name Examples	Score
Heptageniidae	Mayflies	10
Leuctridae	Stoneflies	9–10
Aeshnidae	Dragonflies	8
Polycentropidae	Caddisflies	7
Hydrometridae	Water Strider	6–7
Gyrinidae	Whirligig beetle	5
Chironomidae	Mosquitoes	2
Oligochaeta	Worms	1

pollution. These aquatic macroinvertebrates are organisms large enough to be seen by the unaided eye. Moreover, most aquatic macroinvertebrate species live for at least a year, and they are sensitive to stream water quality both in the short term and over the long term. For example, mayflies, stoneflies, and caddisflies are aquatic macroinvertebrates that are considered clean-water organisms; they are generally the first to disappear from a stream if water quality declines and are, therefore, given a high score. On the other hand, tubificid worms (which are tolerant to pollution) are given a low score.

In Table 15.1, a score from 1 to 10 is given for each family present. A site score is calculated by adding the individual family scores. The site score or total score is then divided by the number of families recorded to derive the Average Score Per Taxon (ASPT). High ASPT scores result from taxa such as stoneflies, mayflies, and caddisflies being present in the stream. A low ASPT score is obtained from streams that are heavily polluted and dominated by tubificid worms and other pollution-tolerant organisms. From Table 15.1, it can be seen that organisms with high scores, especially mayflies and stoneflies, are the most sensitive, while others, such as dragonflies and caddisflies, are also very sensitive to any pollution (deoxygenation) of their aquatic environment.

As noted earlier, the benthic macroinvertebrate biotic index employs certain benthic macroinvertebrates to gauge the water quality (or relative health) of a water body (stream or river). Benthic macroinvertebrates are classified into three groups based on their sensitivity to pollution. The number of taxa in each of these groups is tallied and assigned a score. The scores are then summed to yield a score that can be used as an estimate of the quality of the water body.

METRICS WITHIN THE BENTHIC MACROINVERTEBRATES

Table 15.2 provides a sample index of macroinvertebrates and their sensitivity to pollution. The three groups, based on their sensitivity to pollution, are:

Group One: Indicators of Poor Water Quality
Group Two: Indicators of Moderate Water Quality
Group Three: Indicators of Good Water Quality

TABLE 15.2
Sample Index of Macroinvertebrates

Group One (Sensitive)	Group Two (Somewhat Sensitive)	Group Three (Tolerant)
Stonefly larva	Alderfly larva	Aquatic worm
Caddisfly larva	Damselfly larva	Midgefly larva
Water penny larva	Cranefly larva	Blackfly larva
Riffle beetle adult	Beetle adult	Leech
Mayfly larva	Dragonfly larva	Snails
Gilled snail	Sowbugs	

BIOLOGICAL CONDITION GRADIENT

The Biological Condition Gradient (BCG) is a conceptual, scientific framework for interpreting biological responses to increasing effects of stressors on aquatic ecosystems. The framework was developed based on common patterns of biological responses to stressors observed empirically by aquatic biologists and ecologists from different geographical areas of the United States.

Note that what the BCG was designed to accomplish is to contribute to the EPA biological assessment and criteria toolbox, which includes biological indices, models, statistical approaches, and guidance. It was built upon and complemented by other approaches to provide a more refined and detailed measure of biological condition and will help states to:

- Identify and protect high-quality waters
- Develop biological criteria
- More precisely define and measure biological conditions for specific waters
- Evaluate the potential for improvement in degraded waters and track improvements
- Clearly communicate the likely impact of water quality management decisions to the public

These applications support Clean Water Act (CWA) programs such as 305(b) assessments and reports, 303(d) listing of impaired waters, and TMDL program implementation.

So, what does the Biological Condition Gradient (BCG) tell us about biological responses to increasing levels of stress? There are two components of the BCG:

- **Attributes** are measurable components of a BCG (Karr and Chu, 1999).
- Levels are discrete levels of biological condition across a stressor-response curve.
    Example: Level 1=natural; level 6=severely altered from natural.

Note: To date, BCG models for streams, rivers, and coral reefs have primarily utilized attributes I–VI, which characterize changes in taxa in response to stress. An in-depth explanation of the BCG is provided by USEPA 2016-EPA/822F-16/002.

In summary, it can be said that unpolluted streams normally support a wide variety of macroinvertebrates and other aquatic organisms with relatively few of any one kind. Any significant change in the normal population usually indicates pollution. With regard to BCG, its framework can be considered analogous to a field-based dose-response curve where the dose (x-axis) represents increasing levels of anthropogenic stress, and the response (y-axis) represents biological condition.

BIOLOGICAL SAMPLING IN STREAMS

A few years ago, we were preparing to perform benthic macroinvertebrate sampling protocols in a wadable section of one of the countless reaches of the Yellowstone River, WY. It was autumn, windy, and cold. Before we stepped into the slow-moving frigid waters, we stood for a moment at the bank and took in the surroundings. The palette of autumn is austere in Yellowstone. The coniferous forests west of the Mississippi lack the bronzes, the coppers, the peach-tinted yellows, and the livid scarlets that set the mixed stands of the East aflame. All we could see in that line was the quaking aspen and its gold. This autumnal gold, which provides the closest thing to eastern autumn in the West, is mined from the narrow, rounded crowns of *Populus tremuloides*. The aspen trunks stand stark white and antithetical against the darkness of the firs and pines, the shiny pale gold leaves sensitive to the slightest rumor of wind. Agitated by the slightest hint of breeze, the gleaming upper surfaces bounced the sun into our eyes. Each tree scintillated like a show of gold coins in free fall. The aspens' bright, metallic flash seemed, in all their glittering motion, to make a valiant dying attempt to fill the spectrum of fall.

Because they were bright and glorious, we did not care that they could not approach the colors of an eastern autumn. While nothing is comparable to experiencing leaf-fall in autumn along the Appalachian Trail, the fact that this autumn was not the same simply did not matter. This spirited display of gold against dark green lightened our hearts and eased the task that was before us, warming the thought of the bone-chilling water and all. With the aspens gleaming gold against the pines and firs, it simply did not seem to matter. Notwithstanding the glories of nature alluded to above, one should not be deceived. Conducting biological sampling in a water body is not only the nuts and bolts of biological sampling, but it is also very hard and important work.

BIOLOGICAL SAMPLING: PLANNING

When planning a biological sampling outing, it is important to determine the precise objectives. One key consideration is whether sampling will be conducted at a single point or multiple, isolated points. Additionally, the frequency of sampling must be decided. Will sampling take place hourly, daily, weekly, monthly, or at even longer intervals? Whatever sampling frequency is chosen, the entire process will likely continue over a protracted period (i.e., preparing for biological sampling in the field might take several months from the initial planning stages to the time when actual sampling occurs). An experienced freshwater ecologist should be centrally involved in all aspects of planning.

In *Monitoring Water Quality: Intensive Stream Bioassay* (08/18/2000), the USEPA recommends that the following issues be considered when planning the sampling program:

Availability of reference conditions for the chosen area
Appropriate dates to sample in each season
Appropriate sampling gear
Availability of laboratory facilities

Sample storage
Data management
Appropriate taxonomic keys, metrics, or measurement for macroinvertebrate analysis
Habitat assessment consistency
A USGS topographical map
Familiarity with safety procedures

When the initial objectives (issues) have been determined and the plan devised, the sampler can then move to other important aspects of the sampling procedure. Along with the items just mentioned, it is imperative that the sampler understands what biological sampling is all about.

Sampling is one of the most basic and important aspects of water quality management (Tchobanoglous and Schroeder, 1985). Biological sampling allows for rapid and general water quality classification. Rapid classification is possible because quick and easy cross-checking between stream biota and a standard stream biotic index can be performed. Biological sampling is typically used for general water quality classification in the field because sophisticated laboratory apparatus is usually not available. Additionally, stream communities often show a great deal of variation in basic water quality parameters such as DO, BOD, suspended solids, and coliform bacteria. This variation can be observed in eutrophic lakes that may experience fluctuations in oxygen saturation from less than 0.5 mg/L in a single day, and the concentration of suspended solids may double immediately after heavy rain. Moreover, the sampling method chosen must consider the differences in the habits and habitats of the aquatic organisms.

The first step toward the accurate measurement of a stream's water quality is to ensure that the sampling targets those organisms (i.e., macroinvertebrates) that are most likely to provide the information being sought. Second, it is essential that representative samples be collected. Laboratory analysis is meaningless if the sample collected is not representative of the aquatic environment being analyzed. As a rule, samples should be taken at many locations as often as possible. If, for example, you are studying the effects of sewage discharge into a stream, you should first take at least six samples upstream of the discharge, six samples at the discharge, and at least six samples at several points below the discharge for 2–3 days (the six–six sampling rule). If these samples show wide variability, then the number of samples should be increased. On the other hand, if the initial samples exhibit little variation, then a reduction in the number of samples may be appropriate (Kittrell, 1969).

When planning the biological sampling protocol (using biotic indices as the standards), remember that when sampling is conducted in a stream, findings are based on the presence or absence of certain organisms. Thus, the absence of these organisms must be a function of pollution and not of some other ecological problem. The preferred (favored in this text) aquatic group for biological monitoring in streams is the macroinvertebrates, which are usually retained by 30-mesh sieves (pond nets).

SAMPLING STATIONS

After determining the number of samples to be taken, sampling stations (locations) must be identified. Several factors determine where the sampling stations should be set up. These factors include stream habitat types, the position of the wastewater effluent outfalls, stream characteristics, stream developments (dams, bridges, navigation locks, and other man-made structures), the self-purification characteristics of the stream, and the nature of the objectives of the study (Velz, 1970). The stream habitat types used in this discussion are those that are associated with macroinvertebrate assemblages in stream ecosystems. Some combination of these habitats would be sampled in a multi-habitat approach to benthic sampling (Barbour et al., 1997):

1. **Cobble (Hard Substrate):** Cobble is prevalent in the riffles (and runs), which are a common feature throughout most mountain and piedmont streams. In many high-gradient streams, this habitat type will be dominant. However, riffles are not a common feature of most coastal or other low-gradient streams. Sample shallow areas with coarse substrates (mixed gravel, cobble, or larger) by holding the bottom of the dip net against the substrate and dislodging organisms by kicking (this is where the "designated kicker," your sampling partner, comes into play) the substrate for 0.5 m upstream of the net.
2. **Snags:** Snags and other woody debris that have been submerged for a relatively long period (not recent deadfall) provide excellent colonization habitat. Sample submerged woody debris by jabbing into medium-sized snag material (sticks and branches). The snag habitat may be kicked first to help dislodge organisms, but only after placing the net downstream of the snag. Accumulated woody material in pool areas is considered snag habitat. Large logs should be avoided because they are generally difficult to sample adequately.
3. **Vegetated Banks:** When lower banks are submerged and have roots and emergent plants associated with them, they are sampled in a fashion similar to snags. Submerged areas of undercut banks are good habitats to sample. Sample banks with protruding roots and plants by jabbing into the habitat. Bank habitat can be kicked first to help dislodge organisms, but only after placing the net downstream.
4. **Submerged Macrophytes:** Submerged macrophytes are seasonal in their occurrence and may not be a common feature of many streams, particularly those that are high gradient. Sample aquatic plants that are rooted on the bottom of the stream in deep water by drawing the net through the vegetation from the bottom to the surface of the water (maximum of 0.5 m each jab). In shallow water, sample by bumping or jabbing the net along the bottom in the rooted area, avoiding sediments where possible.
5. **Sand (and Other Fine Sediment):** Usually the least productive macroinvertebrate habitat in streams, this habitat may be the most prevalent in some streams. Sample banks of unvegetated or soft soil by bumping the net along the surface of the substrate rather than dragging the net through the soft substrate; this reduces the amount of debris in the sample.

It is usually impossible to go out and count each and every macroinvertebrate present in a waterway. This would be comparable to counting different sizes of grains of sand on the beach. Thus, in a biological sampling program (i.e., based on our experience), the most common sampling methods are the *transect* and the *grid*. Transect sampling involves taking samples along a straight line either at uniform or at random intervals (see Figure 15.1). The transect involves the cross-section of a lake or stream or the longitudinal section of a river or stream. The transect sampling

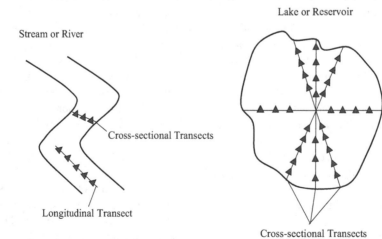

FIGURE 15.1 Transect sampling.

Stream or
River

Lake or Reservoir

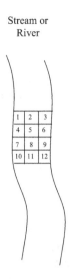

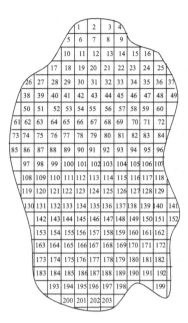

FIGURE 15.2 Grid sampling.

method allows for a more complete analysis by including variations in habitat.

In grid sampling, an imaginary grid system is placed over the study area. The grids may be numbered, and random numbers are generated to determine which grids should be sampled (see Figure 15.2). This type of sampling method allows for quantitative analysis because the grids are all of a certain size. For example, to sample a stream for benthic macroinvertebrates, grids that are $0.25\,m^2$ may be used. Then, the weight or number of benthic macroinvertebrates per square meter can be determined.

Random sampling requires that each possible sampling location have an equal chance of being selected. Numbering all sampling locations, and then using a computer, calculator, or a random numbers table to collect a series of random numbers can achieve this. An illustration of how to put the random numbers to work is provided in the following example. Given a pond that has 300 grid units, find eight random sampling locations using the following sequence of random numbers taken from a standard random numbers table: 101, 209, 007, 018, 099, 100, 017, 069, 096, 033, 041, 011. The first eight numbers of the sequence could be selected, and only those grids would be sampled to obtain a random sample.

SAMPLING FREQUENCY AND NOTES

(The sampling procedures that follow have been suggested by USEPA (2000).) After establishing the sampling methodology and the sampling locations, the frequency of sampling must be determined. The more samples collected, the more reliable the data will be. A frequency of once a week or once a month will be adequate for most aquatic studies.

Usually, the sampling period covers an entire year so that yearly variations may be included. The details of sample collection will depend on the type of problem being solved and will vary with each study. When a sample is collected, it must be carefully identified with the following information:

1. **Location:** name of water body and place of study; longitude and latitude
2. Date and time
3. **Site:** point of sampling (sampling location)
4. Name of collector
5. **Weather:** temperature, precipitation, humidity, wind, etc.
6. **Miscellaneous:** any other important information, such as observations
7. **Field Notebook:** on each sampling day, notes on field conditions should be written. For example, miscellaneous notes and weather conditions can be entered. Additionally, notes that describe the condition of the water are also helpful (color, turbidity, odor, algae, etc.). All unusual findings and conditions should also be entered.

MACROINVERTEBRATE SAMPLING EQUIPMENT

In addition to the appropriate sampling equipment described in the Sample Devices section, assemble the following equipment:

1. Jars (two, at least quart size), plastic, wide-mouth with a tight cap; one should be empty and the other filled about two-thirds with 70% ethyl alcohol
2. Hand lens, magnifying glass, or field microscope

Most professional biological monitoring programs employ sieve buckets as a holding container for composited samples. These buckets have a mesh bottom that allows water to drain out while the organisms and debris remain. This material can then be easily transferred to the alcohol-filled jars. However, sieve buckets can be expensive. Many volunteer programs employ alternative equipment, such as the two regular buckets described in this section. Regardless of the equipment, the process for compositing and transferring the sample is basically the same. The decision is one of cost and convenience.

FIGURE 15.3   Sieve bucket.

3. Fine-point forceps
4. Heavy-duty rubber gloves
5. Plastic sugar scoop or ice-cream scoop
6. Kink net (for rocky-bottom streams) or dip net (for muddy-bottom streams)
7. Buckets (two; see Figure 15.3)
8. String or twine (50 yards); tape measure
9. Stakes (four)
10. Orange (a stick, an apple, or a fish float may also be used in place of an orange) to measure velocity
11. Reference maps indicating general information pertinent to the sampling area, including the surrounding roadways, as well as a hand-drawn station map
12. Station ID tags
13. Spray water bottle
14 Pencils (at least two)

MACROINVERTEBRATE SAMPLING IN ROCKY-BOTTOM STREAMS

Rocky-bottom streams are defined as those with bottoms made up of gravel, cobbles, and boulders in any combination. They usually have definite riffle areas. As mentioned, riffle areas are fairly well-oxygenated and, therefore, are prime habitats for benthic macroinvertebrates. In these streams, we use the *rocky-bottom sampling method*. This method of macroinvertebrate sampling is used in streams that have riffles and gravel/cobble substrates. Three samples are to be collected at each site, and a composite sample is obtained (i.e., one large total sample).

Step 1: *A site should have already been located on a map, with its latitude and longitude indicated.*
A. Samples will be taken in three different spots within a 100-yard stream site. These spots may be three separate riffles, one large riffle with different current velocities, or, if no riffles are present, three-run areas with gravel or cobble substrate. Combinations are also possible (if, for example, your site has only one small

riffle and several run areas). Mark off the 100-yard stream site. If possible, it should begin at least 50 yards upstream of any human-made modification of the channel, such as a bridge, dam, or pipeline crossing. Avoid walking in the stream because this might dislodge macroinvertebrates and later sampling results.
B. Sketch the 100-yard sampling area. Indicate the location of the three sampling spots on the sketch. Mark the downstream site as Site 1, the middle site as Site 2, and the upstream site as Site 3.

Step 2: *Get into place.*
A. Always approach sampling locations from the downstream end and sample the site farthest downstream first (Site 1). This prevents biasing of the second and third collections with dislodged sediment of macroinvertebrates. Always use a clean kick-seine, relatively free of mud and debris from previous uses. Fill a bucket about one-third full of stream water and fill your spray bottle.
B. Select a 3-ft by 3-ft riffle area for sampling at Site 1. One member of the team, the net holder, should position the net at the downstream end of this sampling area. Hold the net handles at a 45° angle to the water's surface. Be sure that the bottom of the net fits tightly against the streambed so that no macroinvertebrates escape under the net. You may use rocks from the sampling area to anchor the net against the stream bottom. Do not allow any water to flow over the net.

Step 3: *Dislodge the macroinvertebrates.*
A. Pick up any large rocks in the 3-ft by 3-ft sampling area and rub them thoroughly over the partially filled bucket so that any macroinvertebrates clinging to the rocks will be dislodged into the bucket. Then place each cleaned rock outside of the sampling area. After sampling is completed, rocks can be returned to the stretch of stream they came from.

B. The member of the team designated as the "kicker" should thoroughly stir up the sampling area with their feet, starting at the upstream edge of the 3-ft by 3-ft sampling area and working downstream toward the net. All dislodged organisms will be carried by the stream flow into the net. Be sure to disturb the first few inches of stream sediment to dislodge burrowing organisms. As a guide, disturb the sampling area for about 3 min, or until the area is thoroughly worked over.

C. Any large rocks used to anchor the net should be thoroughly rubbed into the bucket as above.

Step 4: *Remove the net.*

A. Next, remove the net without allowing any of the organisms it contains to wash away. While the net holder grabs the top of the net handles, the kicker grabs the bottom of the net handles and the net's bottom edge. Remove the net from the stream with a forward scooping motion.

B. Roll the kick net into a cylinder shape and place it vertically in the partially filled bucket. Pour or spray water down the net to flush its contents into the bucket. If necessary, pick debris and organisms from the net by hand. Release back into the stream any fish, amphibians, or reptiles caught in the net.

Step 5: *Collect the second and third samples.*

A. Once all of the organisms have been removed from the net, repeat the steps above at Sites 2 and 3. Put the samples from all three sites into the same bucket. Combining the debris and organisms from all three sites into the same bucket is called compositing.

Note: If your bucket is nearly full of water after you have washed the net clean, let the debris and organisms settle to the bottom. Then, cup the net over the bucket and pour the water through the net into a second bucket. Inspect the water in the second bucket to be sure no organisms came through.

Step 6: *Preserve the sample.*

A. After collecting and compositing all three samples, it is time to preserve the sample. All team members should leave the stream and return to a relatively flat section of the stream bank with their equipment. The next step will be to remove large pieces of debris (leaves, twigs, and rocks) from the sample. Carefully remove the debris one piece at a time. While holding the material over the bucket, use the forceps, spray bottle, and your hands to pick, rub, and rinse the leaves, twigs, and rocks to remove any attached organisms. Use a magnifying lens and forceps to find and remove small organisms clinging to the debris. When

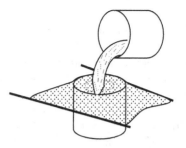

FIGURE 15.4 Pouring sample water through the net.

satisfied that the material is clean, discard it back into the stream.

B. The water will have to be drained before transferring material to the jar. This process will require two team members. Place the kick net over the second bucket, which has not yet been used and should be completely empty. One team member should push the center of the net into bucket #2, creating a small indentation or depression. Then, hold the sides of the net closely over the mouth of the bucket. The second person can now carefully pour the remaining contents of bucket #1 onto a small area of the net to drain the water and concentrate the organisms. Use care when pouring so that organisms are not lost over the side of the net (see Figure 15.4). Use the spray bottle, forceps, sugar scoop, and gloved hands to remove all material from bucket #1 onto the net. When you are satisfied that bucket #1 is empty, use your hands and the sugar scoop to transfer the material from the net into the empty jar. Bucket #2 captured the water and any organisms that might have fallen through the netting during pouring. As a final check, repeat the process above, but this time, pour bucket #2 over the net into bucket #1. Transfer any organisms on the net into the jar.

C. Now, fill the jar (so that all material is submerged) with the alcohol from the second jar. Put the lid tightly back onto the jar and gently turn the jar upside down two or three times to distribute the alcohol and remove air bubbles.

D. Complete the sampling station ID tag. Be sure to use a pencil, not a pen, because the ink will run in the alcohol! The tag includes your station number, the stream, and location (e.g., upstream from a road crossing), date, time, and the names of the members of the collecting team. Place the ID tag into the sample container, writing side facing out, so that identification can be seen clearly.

Rocky-Bottom Habitat Assessment

The habitat assessment (including measuring general characteristics and local land use) for a rocky-bottom stream is conducted in a 100-yard section of the stream that includes the riffles from which organisms were collected.

Step 1: *Delineate the habitat assessment boundaries.*
   A. Begin by identifying the most downstream riffle that was sampled for macroinvertebrates. Using a tape measure or twine, mark off a 100-yard section extending 25 yards below the downstream riffle and about 75 yards upstream.
   B. Complete the identifying information on the field data sheet for the habitat assessment site. On the stream sketch, be as detailed as possible, and be sure to note which riffles were sampled.

Step 2: *Describe the general characteristics and local land use on the field sheet.*
   1. For safety reasons, as well as to protect the stream habitat, it is best to estimate the following characteristics rather than actually wading into the stream to measure them.
      A. Water appearance can be a physical indicator of water pollution.
         1. **Clear:** colorless, transparent
         2. **Milky:** cloudy-white or gray, not transparent; might be natural or due to pollution
         3. **Foamy:** might be natural or due to pollution, generally detergents or nutrients (foam that is several inches high and does not brush apart easily is generally due to pollution)
         4. **Turbid:** cloudy brown due to suspended silt or organic material
         5. **Dark Brown:** might indicate that acids are being released into the stream due to decaying plants
         6. **Oily Sheen:** multicolored reflection might indicate oil floating in the stream, although some sheens are natural
         7. **Orange:** might indicate acid drainage
         8. **Green:** might indicate that excess nutrients are being released into the stream
      B. Water odor can be a physical indicator of water pollution
         1. None or natural smell
         2. **Sewage:** might indicate the release of human waste material
         3. **Chlorine:** might indicate that a sewage treatment plant is over-chlorinating its effluent
         4. **Fishy:** might indicate the presence of excessive algal growth or dead fish
         5. **Rotten Eggs:** might indicate sewage pollution (the presence of a natural gas)

   C. Water temperature can be particularly important for determining whether the stream is suitable as habitat for some species of fish and macroinvertebrates that have distinct temperature requirements. Temperature also has a direct effect on the amount of dissolved oxygen available to aquatic organisms. Measure the temperature by submerging a thermometer for at least 2 min in a typical stream run. Repeat once and average the results.
   D. The width of the stream channel can be determined by estimating the width of the streambed that is covered by water from bank to bank. If it varies widely along the stream, estimate an average width.
   E. Local land use refers to the part of the watershed within one-quarter mile upstream of and adjacent to the site. Note which land uses are present, as well as which ones seem to be having a negative impact on the stream. Base observations on what can be seen, what was passed on the way to the stream, and, if possible, what is noticed when leaving the stream.

Step 3: *Conduct the habitat assessment.* The following information describes the parameters that will be evaluated for rocky-bottom habitats. Use these definitions when completing the habitat assessment field data sheet. The first two parameters should be assessed directly at the riffle(s) or run(s) that were used for the macroinvertebrate sampling. The last eight parameters should be assessed in the entire 100-yard section of the stream.
   A. *Attachment sites* for macroinvertebrates are essentially the amount of living space or hard substrates (rocks, snags) available for adequate insects and snails. Many insects begin their life underwater in streams and need to attach themselves to rocks, logs, branches, or other submerged substrates. The greater the variety and number of available living spaces or attachment sites, the greater the variety of insects in the stream. Optimally, cobble should predominate, and boulders and gravel should be common. The availability of suitable living spaces for macroinvertebrates decreases as cobble becomes less abundant and boulders, gravel, or bedrock become more prevalent.
   B. *Embeddedness* refers to the extent to which rocks (gravel, cobble, and boulders) are surrounded by, covered, or sunken into the silt, sand, or mud of the stream bottom. Generally, as rocks become embedded, fewer living spaces are available to macroinvertebrates and fish for shelter, spawning, and egg incubation.

Note: To estimate the percent of embeddedness, observe the amount of silt or finer sediments overlaying and surrounding the rocks. If kicking does not dislodge the rocks or cobbles, they might be greatly embedded.

C. *Shelter* for fish includes the relative quantity and variety of natural structures in the stream, such as fallen trees, logs, and branches; cobble and large rocks; and undercut banks that are available to fish for hiding, sleeping, or feeding. A wide variety of submerged structures in the stream provide fish with many living spaces; the more living spaces in a stream, the more types of fish the stream can support.

D. *Channel alteration* is a measure of large-scale changes in the shape of the stream channel. Many streams in urban and agricultural areas have been straightened, deepened (e.g., dredged), or diverted into concrete channels, often for flood control purposes. Such streams have far fewer natural habitats for fish, macroinvertebrates, and plants than do naturally meandering streams. Channel alteration is present when the stream runs through a concrete channel, when artificial embankments, riprap, and other forms of artificial bank stabilization or structures are present; when the stream is very straight for significant distances; when dams, bridges, and flow-altering structures such as combined sewer overflow (CSO) are present; when the stream is of uniform depth due to dredging; and when other such changes have occurred. Signs that indicate the occurrence of dredging include straightened, deepened, and otherwise uniform stream channels, as well as the removal of streamside vegetation to provide dredging equipment access to the stream.

E. *Sediment deposition* is a measure of the amount of sediment that has been deposited in the stream channel and the changes to the stream bottom that have occurred as a result of the deposition. High levels of sediment deposition create an unstable and continually changing environment that is unsuitable for many aquatic organisms. Sediments are naturally deposited in areas where the stream flow is reduced, such as in pools and bends, or where flow is obstructed. These deposits can lead to the formation of islands, shoals, or point bars (sediments that build up in the stream, usually at the beginning of a meander) or can result in the complete filling of pools. To determine whether these sediment deposits are new, look for vegetation growing on them: new sediments will not yet have been colonized by vegetation.

F. *Stream velocity and depth* combinations are important to the maintenance of healthy aquatic communities. Fast water increases the amount of dissolved oxygen in the water, keeps pools from being filled with sediment, and helps food items like leaves, twigs, and algae move more quickly through the aquatic system. Slow water provides spawning areas for fish and shelters macroinvertebrates that might be washed downstream in higher stream velocities. Similarly, shallow water tends to be more easily aerated (i.e., it holds more oxygen), but deeper water stays cooler longer. Thus, the best stream habitat includes all of the following velocity/depth combinations and can maintain a wide variety of organisms.

slow (<1 ft/sec), shallow (<1.5 ft)

slow, deep

fast, deep

fast, shallow

Measure stream velocity by marking off a 10-ft section of stream run and measuring the time it takes an orange stick or other floating biodegradable object to float the 10 ft. Repeat five times in the same 10-ft section and determine the average time. Divide the distance (10 ft) by the average time (seconds) to determine the velocity in feet per second.

Measure the stream depth by using a stick of known length and taking readings at various points within your stream site, including riffles, runs, and pools. Compare velocity and depth at various points within the 100-yard site to see how many of the combinations are present.

G. *Channel flow* status is the percent of the existing channel that is filled with water. The flow status changes as the channel enlarges or as flow decreases because of dams and other obstructions, diversions for irrigation, or drought. When water does not cover much of the streambed, the living area for aquatic organisms is limited.

Note: For the following parameters, evaluate the conditions of the left and right stream banks separately. Define the "left" and "right" banks by standing at the downstream end of the study stretch and looking upstream. Each bank is evaluated on a scale of 0–10.

H. *Bank vegetation protection* measures the amount of the stream bank that is covered by natural (i.e., growing wild and not obviously planted) vegetation. The root system of plants growing on stream banks helps hold soil in place, reducing erosion. Vegetation on banks provides shade for fish and macroinvertebrates and serves as a food source by dropping leaves and other organic matter into the stream. Ideally, a variety of vegetation should be present, including trees, shrubs, and grasses. Vegetation disruption can occur when the grasses and plants on the stream banks are mowed or grazed, or when the trees and shrubs are cut back or cleared.

I. *Condition of banks* measures erosion potential and whether the stream banks are eroded. Steep banks are more likely to collapse and suffer from erosion than gently sloping banks and are, therefore, considered to have erosion potential. Signs of erosion include crumbling, unvegetated banks, exposed tree roots, and exposed soil.

J. The *riparian vegetative zone* is defined as the width of natural vegetation from the edge of the stream bank. The riparian vegetative zone is a buffer zone to pollutants entering a stream from runoff. It also controls erosion and provides stream habitat and nutrient input into the stream.

Note: A wide, relatively undisturbed riparian vegetative zone reflects a healthy stream system; narrow, far less useful riparian zones occur when roads, parking lots, fields, lawns, and other artificially cultivated areas, bare soil, rock, or buildings are near the stream bank. The presence of "old fields" (i.e., previously developed agricultural fields allowed to revert to natural conditions) should rate higher than fields in continuous or periodic use. In arid areas, the riparian vegetative zone can be measured by observing the width of the area dominated by riparian or water-loving plants, such as willows, marsh grasses, and cottonwood trees.

MACROINVERTEBRATE SAMPLING IN MUDDY-BOTTOM STREAMS

In muddy-bottom streams, as in rocky-bottom streams, the goal is to sample the most productive habitat available and look for the widest variety of organisms. The most productive habitat is the one that harbors a diverse population of pollution-sensitive macroinvertebrates. Samples should be taken using a D-frame net (see Figure 15.5) to jab at the habitat and scoop up the organisms that are dislodged. The idea is to collect a total sample that consists of 20 jabs taken from a variety of habitats. Use the following method of macroinvertebrate sampling in streams that have muddy-bottom substrates.

Step 1: *Determine which habitats are present.*

1. Muddy-bottom streams usually have four habitats: vegetated bank margins, snags and logs, aquatic vegetation beds and decaying organic matter, and silt/sand/gravel substrate. It is generally best to concentrate sampling efforts

on the most productive habitat available, yet to sample other principal habitats if they are present. This ensures that you will secure as wide a variety of organisms as possible. Not all habitats are present in all streams or are present in significant amounts. If the sampling areas have not been preselected, determine which of the following habitats are present.

Note: Avoid standing in the stream while making habitat determinations.

A. Vegetated bank margins consist of overhanging bank vegetation and submerged root mats attached to banks. The bank margins may also contain submerged, decomposing leaf packs trapped in root wads or lining the streambanks. This is generally a highly productive habitat in a muddy stream, and it is often the most abundant type of habitat.

B. Snags and logs consist of submerged wood, primarily dead trees, logs, branches, roots, cypress knees, and leaf packs lodged between rocks or logs. This is also a very productive muddy-bottom stream habitat.

C. Aquatic vegetation beds and decaying organic matter consist of beds of submerged, green/leafy plants that are attached to the stream bottom. This habitat can be as productive as vegetated bank margins, snags, and logs.

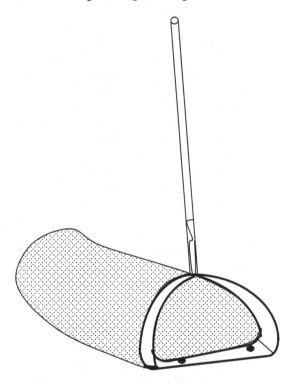

FIGURE 15.5 D-frame aquatic net.

D. Silt/sand/gravel substrate includes sandy, silty, or muddy stream bottoms; rocks along the stream bottom; and/or wetted gravel bars. This habitat may also contain algae-covered rocks (*Aufwuchs*). This is the least productive of the four muddy-bottom stream habitats, and it is always present in one form or another (e.g., silt, sand, mud, or gravel might predominate).

Step 2: *Determine how many times to jab in each habitat type.* The sampler's goal is to jab 20 times. The D-frame net (see Figure 15.5) is 1 ft wide, and a jab should be approximately 1 ft in length. Thus, 20 jabs equal 20 ft^2 of combined habitat.

A. If all four habitats are present in plentiful amounts, jab the vegetated banks ten times and divide the remaining ten jabs amounting to the remaining three habitats.

B. If three habitats are present in plentiful amounts, and one is absent, jab the silt/sand/gravel substrate, the least productive habitat, five times and divide the remaining 15 jabs between the other two more productive habitats.

C. If only two habitats are present in plentiful amounts, the silt/sand/gravel substrate will most likely be one of those habitats. Jab the silt/sand/gravel substrate five times and the more productive habitat 15 times.

D. If some habitats are plentiful and others are sparse, sample the sparse habitats to the extent possible, even if you can take only one or two jabs. Take the remaining jabs from the plentiful habitat(s). This rule also applies if you cannot reach a habitat because of unsafe stream conditions. Jab 20 times.

Note: Because the sampler might need to make an educated guess to decide how many jabs to take in each habitat type, it is critical that the sampler note on the field data sheet, how many jabs were taken in each habitat. This information can be used to help characterize the findings.

Step 3: *Get into place.*
1. Outside and downstream of the first sampling location (first habitat), rinse the dip net and check to make sure it does not contain any macroinvertebrates or debris from the last time it was used. Fill a bucket approximately one-third full with clean stream water. Also, fill the spray bottle with clean stream water. This bottle will be used to wash the net between jabs and after sampling is completed.

Note: This method of sampling requires only one person to disturb the stream habitats. While one person is sampling, a second person should stand outside the sampling area, holding the bucket and spray bottle. After every few jabs, the sampler should hand the net to the second person, who can then rinse the contents of the net into the bucket.

Step 4: *Dislodge the macroinvertebrates.*
1. Approach the first sample site from downstream, and sample while walking upstream. Sample in the four habitat types as follows:
A. Sample vegetated bank margins by jabbing vigorously, with an upward motion, brushing the net against vegetation and roots along the bank. The entire jab motion should occur underwater.
B. To sample snags and logs, hold the net with one hand under the section of submerged wood being sampled. With the other hand (which should be gloved), rub about 1 ft^2 of area on the snag or log. Scoop organisms, bark, twigs, or other organic matter dislodged into the net. Each combination of log rubbing and net scooping is one jab.
C. To sample aquatic vegetation beds, jab vigorously, with an upward motion, against or through the plant bed. The entire jab motion should occur underwater.
D. To sample a silt/sand/gravel substrate, place the net with one edge against the stream bottom and push it forward about a foot (in an upstream direction) to dislodge the first few inches of silt, sand, gravel, or rocks. To avoid gathering a net full of mud, periodically sweep the mesh bottom of the net back and forth in the water, making sure that water does not run over the top of the net. This will allow fine silt to rinse out of the net. When 20 jabs have been completed, rinse the net thoroughly in the bucket. If necessary, pick any clinging organisms from the net by hand, and put them in the bucket.

Step 5: *Preserve the sample.*
1. Look through the material in the bucket and immediately return any fish, amphibians, or reptiles to the stream. Carefully remove large pieces of debris (leaves, twigs, and rocks) from the sample. While holding the material over the bucket, use the forceps, spray bottle, and your hands to pick, rub, and rinse the leaves, twigs, and rocks to remove any attached organisms. Use the magnifying lens and forceps to find and remove small organisms clinging to the debris. When satisfied that the material is clean, discard it back into the stream.
2. Drain the water before transferring material to the jar. This process will require two people. One person should place the net into the second bucket, like a sieve (this bucket, which

has not yet been used, should be completely empty) and hold it securely. The second person can now carefully pour the remaining contents of bucket #1 onto the center of the net to drain the water and concentrate the organisms. Use care when pouring so that organisms are not lost over the side of the net. Use the spray bottle, forceps, sugar scoop, and gloved hands to remove all the material from bucket #1 onto the net. When satisfied that bucket #1 is empty, use your hands and the sugar scoop to transfer all the material from the net into the empty jar. The contents of the net can also be emptied directly into the jar by turning the net inside out into the jar. Bucket #2 captured the water and any organisms that might have fallen through the netting. As a final check, repeat the process above, but this time, pour bucket #2 over the net, into bucket #1. Transfer any organisms on the net into the jar.

3. Fill the jar (so that all material is submerged) with alcohol. Put the lid tightly back onto the jar and gently turn the jar upside down two or three times to distribute the alcohol and remove air bubbles.

4. Complete the sampling station ID tag (see Figure 15.6). Be sure to use a pencil, not a pen, because the ink will run in the alcohol. The tag should include your station number, the stream, and location (e.g., upstream from a road crossing), date, time, and the names of the members of the collecting crew. Place the ID tag into the sample container, writing side facing out, so that identification can be seen clearly.

Note: To prevent samples from being mixed up, samplers should place the ID tag inside the sample jar.

MUDDY-BOTTOM STREAM HABITAT ASSESSMENT

The muddy-bottom stream habitat assessment (which includes measuring general characteristics and local land use) is conducted in a 100-yard section of the stream that includes the habitat areas from which organisms were collected.

Note: Reference made previously, and in the following sections, about a field data sheet (habitat assessment field data sheet) assumes that the sampling team is using either the standard forms provided by the USEAP, the USGS, State Water Control Authorities, or generic forms put together by the sampling team. The source of the form and the exact type of form are not important. Some type of data recording field sheet should be employed to record pertinent data.

Step 1: *Delineate the habitat assessment boundaries.*

1. Begin by identifying the downstream point that was sampled for macroinvertebrates. Using your tape measure or twine, mark off a 100-yard section extending 25 yards below the downstream sampling point and about 75 yards upstream.

2. Complete the identifying information on the field data sheet for the habitat assessment site. On the stream sketch, be as detailed as possible, and be sure to note which habitats were sampled.

Step 2: *Record General Characteristics and Local Land Use on the data field sheet.*

1. For safety reasons as well as to protect the stream habitat, it is best to estimate these characteristics rather than to actually wade into the stream to measure them. For instructions on completing these sections of the field data sheet, see the rocky-bottom habitat assessment instructions.

Step 3: *Conduct the habitat assessment.*

1. The following information describes the parameters to be evaluated for muddy-bottom

---

Station ID Tag

Station # _____

Stream _____

Location _____

Date/Time _____

Team Members: _____

_____

FIGURE 15.6   Station ID tag.

habitats. Use these definitions when completing the habitat assessment field data sheet.

A. *Shelter* for fish and attachment sites for macroinvertebrates are essentially the amount of living space and shelter (rocks, snags, and undercut banks) available for fish, insects, and snails. Many insects attach themselves to rocks, logs, branches, or other submerged substrates. Fish can hide or feed in these areas. The greater the variety and number of available shelter sites or attachment sites, the greater the variety of fish and insects in the stream.

Note: Many of the attachment sites result from debris falling into the stream from the surrounding vegetation. When debris first falls into the water, it is termed a new fall, and it has not yet been "broken down" by microbes (conditioned) for macroinvertebrate colonization. Leaf material or debris that is conditioned is called old fall. Leaves that have been in the stream for some time lose their color, turn brown or dull yellow, become soft and supple with age, and might be slimy to the touch. Woody debris becomes blackened or dark in color; smooth bark becomes coarse and partially disintegrated, creating holes and crevices. It might also be slimy to the touch.

B. *Poor substrate characterization* evaluates the type and condition of bottom substrates found in pools. Pools with firmer sediment types (e.g., gravel, sand) and rooted aquatic plants support a wider variety of organisms than do pools with substrates dominated by mud or bedrock and no plants. In addition, a pool with one uniform substrate type will support far fewer types of organisms than will a pool with a wide variety of substrate types.

C. *Pool variability* rates the overall mixture of pool types found in the stream according to size and depth. The four basic types of pools are large-shallow, large-deep, small-shallow, and small-deep. A stream with many pool types will support a wide variety of aquatic species. Rivers with low sinuosity (few bends) and monotonous pool characteristics do not have sufficient quantities and types of habitats to support a diverse aquatic community.

D. *Channel alteration*

E. *Sediment deposition*

F. *Channel sinuosity* evaluates the sinuosity or meandering of the stream. Streams that meander provide a variety of habitats (such as pools and runs) and stream velocities and reduce the energy from current surges during storm events. Straight stream segments are characterized by even stream depth and unvarying velocity, and they are prone to flooding. To evaluate this parameter, imagine how much longer the stream would be if it were straightened out.

G. *Channel flow status*

H. *Bank vegetative protection*

I. *Condition of banks*

J. *Riparian vegetative zone width*

Note: Whenever stream sampling is to be conducted, it is a good idea to have a reference collection on hand. A reference collection is a sample of locally found macroinvertebrates that have been identified, labeled, and preserved in alcohol. The program advisor, along with a professional biologist/entomologist, should assemble the reference collection, properly identify all samples, preserve them in vials, and label them. This collection may then be used as a training tool and, in the field, as an aid in macroinvertebrate identification.

POST-SAMPLING ROUTINE

After completing the stream characterization and habitat assessment, make sure that all of the field data sheets have been completed properly and that the information is legible. Be sure to include the site's identifying name and the sampling date on each sheet. This information will function as a quality control element. Before leaving the stream location, make sure that all sampling equipment/devices have been collected and rinsed properly. Double-check to see that sample jars are tightly closed and properly identified. All samples, field sheets, and equipment should be returned to the team leader at this point. Keep a copy of the field data sheet(s) for comparison with future monitoring trips and for personal records. The next step is to prepare for macroinvertebrate laboratory work. This step includes all the work needed to set up a laboratory for processing samples into subsamples and identifying macroinvertebrates to the family level. A professional biologist, entomologist, or freshwater ecologist, or the professional advisor, should supervise the identification procedure. (The actual laboratory procedures after the sampling and collecting phase are beyond the scope of this text.)

SAMPLING DEVICES

In addition to the sampling equipment mentioned previously, it may be desirable to employ, depending on stream conditions, the use of other sampling devices. Additional sampling devices commonly used, and discussed in the following sections, include dissolved oxygen and temperature monitors, sampling nets (including the D-frame aquatic net), sediment samplers (dredges), plankton samplers, and Secchi disks. The methods described below are approved by the USEPA. More detailed coverage is available in APHA (1998).

Dissolved Oxygen and Temperature Monitor

As mentioned, the dissolved oxygen (DO) content of a stream sample can provide the investigator with vital information, as DO content reflects the stream's ability to maintain aquatic life.

The Winkler DO with Azide Modification Method

The Winkler DO with azide modification method is commonly used to measure DO content. The Winkler Method is best suited for clean waters. The Winkler Method can be used in the field but is better suited for laboratory work where better accuracy may be achieved. The Winkler Method adds a divalent manganese solution followed by a strong alkali to a 300 mL BOD bottle of the stream water sample. Any DO rapidly oxidizes an equivalent amount of divalent manganese to basic hydroxides of higher balance states. When the solution is acidified in the presence of iodide, oxidized manganese again reverts to the divalent state, and iodine, equivalent to the original DO content of the sample, is liberated. The amount of iodine is then determined by titration with a standard, usually thiosulfate, solution.

Fortunately for the field biologist, this is the age of miniaturized electronic circuit components and devices; thus, it is not too difficult to obtain portable electronic measuring devices for DO and temperature that are of quality construction and have better than moderate accuracy. These modern electronic devices are usually suitable for laboratory and field use. The device may be subjected to severe abuse in the field; therefore, the instrument must be durable, accurate, and easy to use. Several quality DO monitors are available commercially.

When using a DO monitor, it is important to calibrate (standardize) the meter prior to use. Calibration procedures can be found in *Standard Methods* (latest edition) or in the manufacturer's instructions for the meter to be used. Determining the air temperature, the DO at saturation for that temperature, and then adjusting the meter so that it reads the saturation value usually accomplishes meter calibration. After calibration, the monitor is ready for use. As mentioned, all recorded measurements, including water temperatures and DO readings, should be entered in a field notebook.

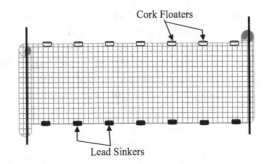

FIGURE 15.7 Two-person seine net.

Sampling Nets

A variety of sampling nets are available for use in the field. The two-person seine net shown in Figure 15.7 is 20 ft long × 4 ft deep with 1/8-in mesh and is utilized to collect a variety of organisms. Two people, each holding one end, walk upstream, and small organisms are gathered in the net. Dip nets are used to collect organisms in shallow streams. The Surber sampler (which collects macroinvertebrates stirred up from the bottom; see Figure 15.8) can be used to obtain a quantitative sample (number of organisms/square foot). It is designed for sampling riffle areas in streams and rivers up to a depth of about 450 mm (18 in). It consists of two folding stainless steel frames set at right angles to each other. The frame is placed on the bottom, with the net extending downstream. Using your hand or a rake, all sediment enclosed by the frame is dislodged. All organisms are caught in the net and transferred to another vessel for counting.

The D-frame aquatic dip net is ideal for sweeping over vegetation or for use in shallow streams.

Sediment Samplers (Dredges)

A sediment sampler or dredge is designed to obtain a sample of the bottom material in a slow-moving stream and the organisms in it. The simple homemade dredge shown in Figure 15.9 works well in water too deep to sample effectively with handheld tools. The homemade dredge is fashioned from a #3 coffee can and a smaller can with a tight-fitting plastic lid (peanut cans work well). To use the homemade dredge, first invert it underwater so it can fill with water and no air is trapped. Then, lower the dredge as quickly as possible with the "down" line. The idea is to bury the open end

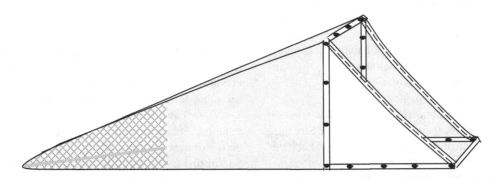

FIGURE 15.8 Surber sampler.

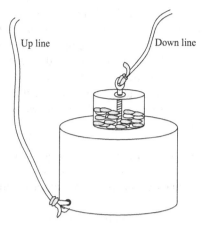

FIGURE 15.9 Home-made dredge.

of the coffee can in the bottom. Then, quickly pull the "up" line to bring the can to the surface with a minimum loss of material. Dump the contents into a sieve or observation pan to sort. It works best in bottoms composed of sediment, mud, sand, and small gravel. The bottom sampling dredge can be used for a number of different analyses. Because the bottom sediments represent a good area in which to find macroinvertebrates and benthic algae, the communities of organisms living on or in the bottom can be easily studied quantitatively and qualitatively. A chemical analysis of the bottom sediment can be conducted to determine what chemicals are available to organisms living in the bottom habitat.

Plankton Sampler

(More detailed information on plankton sampling can be found in AWRI (2000).) *Plankton* (meaning to drift) are distributed throughout the stream and, in particular, in pool areas. They are found at all depths and are comprised of plant (phytoplankton) and animal (zooplankton) forms. Plankton shows a distribution pattern that can be associated with the time of day and the seasons. The three fundamental sizes of plankton are *nanoplankton, microplankton,* and *macroplankton.* The smallest are nanoplankton which range in size from 5 to 60 μm (millionths of a meter). Because of their small size, most nanoplankton will pass through the pores of a standard sampling net. Special fine mesh nets can be used to capture the larger nanoplankton. Most planktonic organisms fall into the microplankton or net plankton category. Their sizes range from the largest nanoplankton to about 2 mm (thousandths of a meter). Nets of various sizes and shapes are used to collect microplankton. The nets collect the organisms by filtering water through fine-meshed cloth. The plankton nets on the vessels are used to collect microplankton. The third group of plankton, as associated with size, is called macroplankton. They are visible to the naked eye. The largest can be several meters long.

The plankton net or sampler (see Figure 15.10) is a device that makes it possible to collect phytoplankton and zooplankton samples. For quantitative comparisons of different samples, some nets have a flow meter used to determine the amount of water passing through the collecting net.

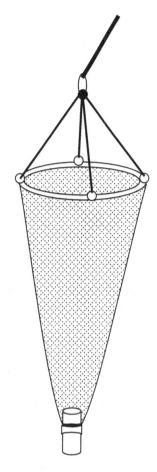

FIGURE 15.10 Plankton net.

The plankton net or sampler provides a means of obtaining samples of plankton from various depths so that distribution patterns can be studied. Considering the depth of the water column that is sampled can make quantitative determinations. The net can be towed to sample plankton at a single depth (horizontal tow) or lowered into the water to sample the water column (vertical tow). Another possibility is oblique tows, where the net is lowered to a predetermined depth and raised at a constant rate as the vessel moves forward.

After towing and removal from the stream, the sides of the net are rinsed to dislodge the collected plankton. If a quantitative sample is desired, a certain quantity of water is collected. If the plankton density is low, then the sample may be concentrated using a low-speed centrifuge or some other filtering device. A definite volume of the sample is studied under a compound microscope for counting and identification of plankton.

Secchi Disk

For determining water turbidity or the degree of visibility in a stream, a Secchi disk is often used (Figure 15.11). The Secchi disk originated with Father Pietro Secchi, an astrophysicist and scientific advisor to the Pope, who was requested to measure transparency in the Mediterranean

FIGURE 15.11   Secchi disk.

Sea by the head of the Papal Navy. Secchi used some white disks to measure the clarity of water in the Mediterranean in April of 1865. Various sizes of disks have been used since that time, but the most frequently used disk is an 8-in diameter metal disk painted in alternating black and white quadrants. The disk shown in Figure 15.11 is 20 cm in diameter; it is lowered into the stream using a calibrated line. To use the Secchi disk properly, it should be lowered into the stream water until it is no longer visible. At the point where it is no longer visible, a measurement of the depth is taken. This depth is called the *Secchi disk transparency light extinction coefficient*. The best results are usually obtained in the early morning and before late afternoon.

Miscellaneous Sampling Equipment

Several other sampling tools and devices are available for use in sampling a stream. For example, consider the standard sand-mud sieve. Generally made of heavy-duty galvanized 1/8 in. mesh screen supported by a water-sealed 24×15×3-in wood frame; this device is useful for collecting burrowing organisms found in soft bottom sediments. Moreover, no stream sampling kit would be complete without a collecting tray, collecting jars of assorted sizes, heavy-duty plastic bags, small pipets, large 2-oz pipets, a fine mesh straining net, and a black china marking pencil. In addition, depending upon the quantity of material to be sampled, it is prudent to include several 3- and 5-gal collection buckets in the stream sampling.

THE BOTTOM LINE ON BIOLOGICAL SAMPLING

This discussion has stressed the practice of biological monitoring, employing the use of biotic indices as key measuring tools. We emphasized biotic indices not only for their simplicity of use but also for the relative accuracy they provide, although their development and use can sometimes be derailed. The failure of a monitoring protocol to assess environmental conditions accurately or to protect running waters usually stems from conceptual, sampling, or analytical pitfalls. Biotic indices can be combined with other

tools for measuring the condition of ecological systems in ways that enhance or hinder their effectiveness. The point is, like any other tool, they can be misused. However, that biotic indices can be, and are, misused does not mean that the indices' approach itself is useless. Thus, to ensure that the biotic indices approach is not useless, it is important for the practicing freshwater ecologist and water sampler to remember a few key guidelines:

1. Sampling everything is not the goal. As Botkin (1990) note, biological systems are complex and unstable in space and time, and samplers often feel compelled to study all components of this variation. Complex sampling programs proliferate. However, every study need not explore everything. Freshwater samplers and monitors should avoid the temptation to sample all the unique habitats and phenomena that make freshwater monitoring so interesting. Concentration should be placed on the central components of a clearly defined research agenda (a sampling/monitoring protocol)—detecting and measuring the influences of human activities on the water body's ecological system.
2. In regard to the influence of human activities on the water body's ecological system, we must see protecting biological conditions as a central responsibility of water resource management. One thing is certain: until biological monitoring is seen as essential to track the attainment of that goal and biological criteria as enforceable standards mandated by the Clean Water Act, life in the nation's freshwater systems will continue to decline.

Biomonitoring is only one of several tools available to the water practitioner. No matter the tool employed, all results depend upon proper biomonitoring techniques. Biological monitoring must be designed to obtain accurate results—present approaches need to be strengthened. In addition, "the way it's always been done" must be reexamined, and efforts must be undertaken to do what works to keep freshwater systems alive. We can afford nothing less.

DID YOU KNOW?

There is increasing, and for some parameters mandatory, use of automatic online sampling and monitoring. It has the advantage of yielding constant and immediate results. Time is saved, and problems due to carriage to the laboratory are eliminated. Meters are installed directly into the process flow, and the water is monitored as it flows by. Typical tests that are performed this way include pH, DO, temperature, turbidity, and chlorine residue. The signal may be read on-site or transmitted to the lab or control room for remote readout. Frequent cleaning and calibration of the monitoring instruments are essential (Hauser, 1995).

DRINKING WATER QUALITY MONITORING

When we speak of water quality monitoring, we refer to a monitoring practice based on three criteria: (1) to ensure, to the extent possible, that the water is not a danger to public health; (2) to ensure that the water provided at the tap is as aesthetically pleasing as possible; and (3) to ensure compliance with applicable regulations. To meet these goals, all public systems must monitor water quality to some extent. The degree of monitoring employed depends on local needs and requirements, and on the type of water system; small water systems using good-quality water from deep wells may only need to provide occasional monitoring, but systems using surface water sources must test water quality frequently (AWWA, 1995).

Drinking water must be monitored to provide adequate control of the entire water drawing/treatment/conveyance system. *Adequate control* is defined as monitoring employed to assess the present level of water quality so that action can be taken to maintain the required level (whatever that might be). We define *water quality monitoring* as the sampling and analysis of water constituents and conditions. When we monitor, we collect data. As a monitoring program is developed, deciding the reasons for collecting the information is important. The reasons are defined by establishing a set of objectives, which includes a description of who will collect the information.

It may come as a surprise to know that today the majority of people collecting data are not water and wastewater operators; instead, many are volunteers. These volunteers have a stake in their local stream, lake, or other water body, and in many cases are proving they can successfully carry out a water quality-monitoring program.

IS THE WATER GOOD OR BAD?

(Much of the information presented in the following sections is based on our personal experience.) To answer the question of whether the water is good or bad, we must consider two factors. First, we return to the basic principles of water quality monitoring—sampling and analyzing water constituents and conditions. These constituents include:

1. Introduced pollutants, such as pesticides, metals, and oil.
2. Constituents found naturally in water that can nevertheless be affected by human sources, such as dissolved oxygen, bacteria, and nutrients.

The magnitude of their effects is influenced by properties such as pH and temperature. For example, temperature influences the quantity of dissolved oxygen that water can contain, and pH affects the toxicity of ammonia, for example.

The second factor to be considered is that the only valid way to answer this question is to conduct tests that must be compared to some form of water quality standards. If simply assigning a "good" and "bad" value to each test factor were possible, the meters and measuring devices in water quality test kits would be much easier to make. Instead of fine graduations, they could simply have a "good" and a "bad" zone.

Water quality—the difference between "good" and "bad" water—must be interpreted according to the intended use of the water. For example, the perfect balance of water chemistry that assures a sparkling clear, sanitary swimming pool would not be acceptable as drinking water and would be a deadly environment for many biota (Table 15.3). In another example, widely different levels of fecal coliform bacteria are considered acceptable, depending on the intended use of the water.

State and local water quality practitioners as well as volunteers have been monitoring water quality conditions for many years. In fact, until the past decade or so (until biological monitoring protocols were developed and began to take hold), water quality monitoring was generally considered the primary way of identifying water pollution problems. Today, professional water quality practitioners and volunteer program coordinators alike are moving toward approaches that combine chemical, physical, and biological monitoring methods to achieve the best picture of water quality conditions. Water quality monitoring can be used for many purposes:

1. **To Identify Whether Waters Are Meeting Designated Uses:** All states have established specific criteria (limits on pollutants) identifying what concentrations of chemical pollutants are allowable in their waters. When chemical pollutants exceed maximum or minimum allowable concentrations, waters may no longer be able to support the beneficial uses—such as fishing, swimming, and drinking—for which they have been designated (see Table 15.4). Designated or intended uses and the specific criteria that protect them (along with antidegradation statements that say waters should not be allowed to deteriorate below existing or anticipated uses) together form water quality standards. State water quality professionals assess water quality by comparing the concentrations of

TABLE 15.3
Total Residual Chlorine (TRC) mg/L

0.06	Toxic to striped bass larvae
0.31	Toxic to white perch larvae
0.5–1.0	Typical drinking water residual
1.0–3.0	Recommended for swimming pools

TABLE 15.4
Fecal Coliform Bacteria per 100 mL of Water

Desirable	Permissible	Type of Water Use
0	0	Potable and well water (for drinking)
<200	<1,000	Primary contact water (for swimming)
<1,000	<5,000	Secondary contact water (boating & fishing)

chemical pollutants found in streams to the criteria in the state's standards and judge whether streams are meeting their designated uses. Water quality monitoring, however, might be inadequate for determining whether aquatic life needs are being met in a stream. While some constituents (such as dissolved oxygen and temperature) are important for maintaining healthy fish and aquatic insect populations, other factors (such as the physical structure of the stream and the condition of the habitat) play an equal or greater role. Biological monitoring methods are generally better suited to determining whether aquatic life is supported.

2. **To Identify Specific Pollutants and Sources of Pollution:** Water quality monitoring helps link sources of pollution to water body quality problems because it identifies specific problem pollutants. Since certain activities tend to generate certain pollutants (bacteria and nutrients are more likely to come from an animal feedlot than an automotive repair shop), a tentative link can be formed that would warrant further investigation or monitoring.
3. **To Determine Trends:** Chemical constituents that are properly monitored (i.e., using consistent time of day and on a regular basis using consistent methods) can be analyzed for trends over time.
4. **To Screen for Impairment:** Finding excessive levels of one or more chemical constituents can serve as an early warning "screen" for potential pollution problems.

State Water Quality Standards Programs

Each state has a program to set standards for the protection of each body of water within its boundaries. Standards for each body of water are developed that:

1. Depend on the water's designated use
2. Are based on USEPA national water quality criteria and other scientific research into the effects of specific pollutants on different types of aquatic life and on human health
3. May include limits based on the biological diversity of the body of water (the presence of food and prey species)

State water quality standards set limits on pollutants and establish water quality levels that must be maintained for each type of water body based on its designated use. Resources for this type of information include:

1. USEPA Water Quality Criteria Program
2. U.S. Fish and Wildlife Service Habitat Suitability Index Models (for specific species of local interest)

Monitoring test results can be plotted against these standards to provide a focused, relevant, required assessment of water quality.

TABLE 15.5
Water Quality Problems and Pollution Sources

Source	Common Associated Chemical Pollutants
Cropland	Turbidity, phosphorus, nitrates, temperature, and total solids
Forestry harvest	Turbidity, temp., total solids
Grazing land	Fecal bacteria, turbidity, phosphorus
Industrial discharge	Temperature, conductivity, total solids, toxics, pH
Mining	pH, alkalinity, total dissolved solids
Septic systems	Fecal bacteria, (i.e., *Escherichia coli*, *enterococcus*), nitrates, dissolved oxygen/biochemical oxygen demand, conductivity, temperature
Sewage treatment	Dissolved oxygen and BOD, turbidity, conductivity, phosphorus, nitrates, fecal bacteria, temperature, total solids, pH
Construction	Turbidity, temperature, dissolved oxygen and BDO, total solids, toxics
Urban runoff	Turbidity, phosphorus, nitrates, temperature, conductivity, dissolved oxygen, BOD

Designing a Water Quality Monitoring Program

The first step in designing a water quality-monitoring program is to determine the purpose of the monitoring. This aids in the selection of parameters to monitor. This decision should be based on factors, including:

1. Types of water quality problems and pollution sources that will likely be encountered (see Table 15.5)
2. Cost of available monitoring equipment
3. Precision and accuracy of available monitoring equipment
4. Capabilities of monitors

Note: We discuss the parameters most commonly monitored by drinking water practitioners in streams (i.e., we assume, for illustration and discussion purposes, that our water source is a surface water stream) in detail in this section. They include dissolved oxygen, biochemical oxygen demand, temperature, pH, turbidity, total orthophosphate, nitrates, total solids, conductivity, total alkalinity, fecal bacteria, apparent color, odor, and hardness. When monitoring water supplies under the Safe Drinking Water Act (SDWA) or the National Pollutant Discharge Elimination System (NPDES), utilities must follow test procedures approved by the USEPA for these purposes. Additional testing requirements under these and other federal programs are published as amendments in the Federal Register.

Except when monitoring discharges for specific compliance purposes, a large number of approximate measurements can provide more useful information than one or two accurate analyses. Because water quality and chemistry continually change, making periodic, representative measurements and observations that indicate the range of water quality is necessary rather than testing the quality at any single moment.

The more complex a water system is, the more time is required to observe, understand, and draw conclusions regarding the cause and effect of changes in the particular system.

GENERAL PREPARATION AND SAMPLING CONSIDERATIONS

The sections that follow detail specific equipment considerations and analytical procedures for each of the most common water quality parameters. Sampling devices should be corrosion-resistant, easily cleaned, and capable of collecting desired samples safely and in accordance with test requirements. Whenever possible, assign a sampling device to each sampling point. Sampling equipment must be cleaned on a regular schedule to avoid contamination.

Note: Some tests require special equipment to ensure the sample is representative. Dissolved oxygen and fecal bacteria sampling require special equipment and/or procedures to prevent the collection of nonrepresentative samples.

CLEANING PROCEDURES

Reused sample containers and glassware must be cleaned and rinsed before the first sampling run and after each run by following Method A or Method B described below. The most suitable method depends on the parameter being measured.

METHOD A: GENERAL PREPARATION OF SAMPLING CONTAINERS

Use the following method when preparing all sample containers and glassware for monitoring conductivity, total solids, turbidity, pH, and total alkalinity. Wearing latex gloves:

1. Wash each sample bottle or piece of glassware with a brush and phosphate-free detergent.
2. Rinse three times with cold tap water.
3. Rinse three times with distilled or deionized water.

METHOD B: ACID WASH PROCEDURES

Use this method when preparing all sample containers and glassware for monitoring nitrates and phosphorus. Wearing latex gloves:

1. Wash each sample bottle or piece of glassware with a brush and phosphate-free detergent.
2. Rinse three times with cold tap water.
3. Rinse with 10% hydrochloric acid.
4. Rinse three times with deionized water.

SAMPLE TYPES

Two types of samples are commonly used for water quality monitoring: grab samples and composite samples. The type of sample used depends on the specific test, the reason the sample is being collected, and the applicable regulatory requirements.

Grab samples are taken all at once, at a specific time and place. They are representative only of the conditions at the time of collection. Grab samples must be used to determine pH, total residual chlorine (TRC), dissolved oxygen (DO), and fecal coliform concentrations. Grab samples may also be used for any test that does not specifically prohibit their use.

Note: Before collecting samples for any test procedure, it is best to review the sampling requirements of the test.

Composite samples consist of a series of individual grab samples collected over a specified period in proportion to flow. The individual grab samples are mixed together in proportion to the flow rate at the time the sample was collected to form the composite sample. This type of sample is taken to determine average conditions in a large volume of water whose properties vary significantly over the course of a day.

COLLECTING SAMPLES FROM A STREAM

In general, sample away from the stream bank in the main current. Never sample stagnant water. The outside curve of the stream is often a good place to sample because the main current tends to hug this bank. In shallow stretches, carefully wade into the center current to collect the sample. A boat is required for deep sites. Try to maneuver the boat into the center of the main current to collect the water sample. When collecting a water sample for analysis in the field or at the lab, follow the steps below.

Whirl-pak® Bags

To collect water samples using Whirl-Pak bags, use the following procedures:

1. Label the bag with the site number, date, and time.
2. Tear off the top of the bag along the perforation above the wire tab just before sampling. Avoid touching the inside of the bag. If you accidentally touch the inside of the bag, use another one.
3. **Wading:** Try to disturb as little bottom sediment as possible. In any case, be careful not to collect water that contains bottom sediment. Stand facing upstream. Collect the water samples in front of you.

   Boat: Carefully reach over the side and collect the water sample on the upstream side of the boat.
4. Hold the two white pull-tabs in each hand and lower the bag into the water on your upstream side with the opening facing upstream. Open the bag midway, between the surface and the bottom by pulling the white pull-tabs. The bag should begin to fill with water. You may need to "scoop" water into the bag by drawing it through the water upstream and away from you. Fill the bag no more than ¾ full!
5. Lift the bag out of the water. Pour out excess water. Pull on the wire tabs to close the bag. Continue holding the wire tabs and flip the bag over at least four to five times quickly to seal the bag. Do not try to squeeze the air out of the top of the bag. Fold the ends of the bag, being careful not to puncture it. Twist them together, forming a loop.

6. Fill in the bag number and/or site number on the appropriate field data sheet. *This is important.* It is the only way the lab specialist will know which bag goes with which site.
7. If samples are to be analyzed in a lab, place the sample in the cooler with ice or cold packs. Take all samples to the lab.

Screw-Cap Bottles

To collect water samples using screw-cap bottles, use the following procedures (see Figure 15.12).

1. Label the bottle with the site number, date, and time.
2. Remove the cap from the bottle just before sampling. Avoid touching the inside of the bottle or the cap. If you accidentally touch the inside of the bottle, use another one.
3. **Wading:** Try to disturb as little bottom sediment as possible. In any case, be careful not to collect water that has sediment from bottom disturbance.

Fill bottle to shoulder

FIGURE 15.12 Sampling bottle; filling to shoulder ensures collecting enough sample—do not overfill.

Stand facing upstream. Collect the water sample on your upstream side, in front of you. You may also tape your bottle to an extension pole to sample from deeper water.

    Boat: Carefully reach over the side and collect the water sample on the upstream side of the boat.

4. Hold the bottle near its base and plunge it (opening downward) below the water surface. If you are using an extension pole, remove the cap, turn the bottle upside down, and plunge it into the water, facing upstream. Collect a water sample 8–12 in beneath the surface, or midway between the surface and the bottom if the stream reach is shallow.
5. Turn your bottle underwater into the current and away from you. In slow-moving stream reaches, push the bottle underneath the surface and away from you in the upstream direction.
6. Leave a 1-in air space (except for DO and BOD samples). Do not fill the bottle completely (so that the sample can be shaken just before analysis). Recap the bottle carefully, remembering not to touch the inside.
7. Fill in the bottle number and/or site number on the appropriate field data sheet. This is important because it tells the lab specialist which bottle goes with which site.
8. If the samples are to be analyzed in the lab, place them in the cooler for transport to the lab.

SAMPLE PRESERVATION AND STORAGE

Samples can change very rapidly. However, no single preservation method will serve for all samples and constituents. If analysis must be delayed, follow the instructions for sample preservation and storage listed in *Standard Methods* or those specified by the laboratory that will eventually process the samples (see Table 15.6). In general,

TABLE 15.6

Recommended Sample Storage and Preservation Techniques

Test Factor	Container Type	Preservation	Max. Storage Time Recommended/ Regulatory
Alkalinity	P, G	Refrigerate	24 h/14 days
BOD	P, G	Refrigerate	6 h/48 h
Conductivity	P, G	Refrigerate	28 days/28 days
Hardness	P, G	Lower pH to <2	6 months/6 months
Nitrate	P, G	Analyze ASAP	48 h/48 h
Nitrite	P, G	Analyze ASAP	none/48 h
Odor	G	Analyze ASAP	6 h/NR
Oxygen, dissolved			
Electrode	G	Immediately analyze	0.5 h/stat
Winkler	G	"Fix" immediately	8-h/8 h
pH	P, G	Immediately analyze	2 h/stat
Phosphate	G(A)	Filter immediately, refrigerate	48 h/NR
Salinity	G, wax seal or use wax seal	Immediately analyze	6 months/NR
Temperature	P, G	Immediately analyze	Stat/stat
Turbidity	P, G	Analyze the same day	24 h/48 h store in dark up to 24 h, refrigerate

handle the sample in a way that prevents changes from biological activity, physical alterations, or chemical reactions. Cool the sample to reduce biological and chemical reactions. Store it in darkness to suspend photosynthesis. Fill the sample container completely to prevent the loss of dissolved gases. Metal cations such as iron and lead, as well as suspended particles may adsorb onto container surfaces during storage.

STANDARDIZATION OF METHODS

References used for sampling and testing must correspond to those listed in the most current federal regulations. For the majority of tests, comparing the results of either different water quality monitors or the same monitors over the course of time requires some form of standardization of the methods. The American Public Health Association (APHA) recognized this requirement when, in 1899, the Association appointed a committee to draw up standard procedures for the analysis of water. The report (published in 1905) constituted the first edition of what is now known as *Standard Methods for the Examination of Water and Wastewater* or *Standard Methods*. This book is now in its 20th edition and serves as the primary reference for water testing methods and as the basis for most USEPA-approved methods.

TEST METHODS FOR DRINKING WATER & WASTEWATER

The material presented in this section is based on personal experience and adaptations from *Standard Methods, the Federal Register,* and *The Monitor's Handbook,* LaMotte Company, Chestertown, Maryland, 1992. Descriptions of general methods to help you understand how each works in specific test kits follow. Always use the specific instructions included with the equipment and individual test kits. Most water analyses are conducted either by titrimetric analyses or colorimetric analyses. Both methods are easy to use and provide accurate results.

TITRIMETRIC METHODS

Titrimetric analyses are based on adding a solution of known strength (the titrant, which must have an exact known concentration) to a specific volume of a treated sample in the presence of an indicator. The indicator produces a color change indicating that the reaction is complete. Titrants are generally added by a titrator (microburette) or a precise glass pipette.

COLORIMETRIC METHODS

Colorimetric standards are prepared as a series of solutions with increasing known concentrations of the constituent to be analyzed. Two basic types of colorimetric tests are commonly used:

1. The pH is a measure of the concentration of hydrogen ions (the acidity of a solution), determined by the reaction of an indicator that varies in color, depending on the hydrogen ion levels in the water.
2. Tests that determine the concentration of an element or compound are based on Beer's Law. Simply put, this law states that the higher the concentration of a substance, the darker the color produced in the test reaction, and therefore the more light absorbed. Assuming a constant path length, the absorption increases exponentially with concentration.

VISUAL METHODS

The Octet Comparator uses standards that are mounted in a plastic comparator block. It employs eight permanent translucent color standards and built-in filters to eliminate optical distortion. The sample is compared using either of two viewing windows. Two devices that can be used with the comparator are the B-color Reader, which neutralizes color or turbidity in water samples, and a view path, which intensifies faint colors of low concentrations for easy distinction.

ELECTRONIC METHODS

Although the human eye is capable of differentiating color intensity, interpretation is quite subjective. Electronic colorimeters consist of a light source that passes through a sample and is measured on a photodetector with an analog or digital readout. Besides electronic colorimeters, specific electronic instruments are manufactured for lab and field determination of many water quality factors, including pH, total dissolved solids (TDS)/conductivity, dissolved oxygen, temperature, and turbidity.

DISSOLVED OXYGEN TESTING

In this section and the sections that follow, we discuss several water quality factors that are routinely monitored in drinking water operations. We do not discuss the actual test procedures to analyze each water quality factor; we refer you to the latest edition of *Standard Methods* for the correct procedure to use in conducting these tests.

A stream system used as a source of water produces and consumes oxygen. It gains oxygen from the atmosphere and from plants because of photosynthesis. Because of running water's churning, it dissolves more oxygen than does still water, such as in a reservoir behind a dam. Respiration by aquatic animals, decomposition, and various chemical reactions consume oxygen.

Oxygen is actually poorly soluble in water. Its solubility is related to pressure and temperature. In water supply systems, *dissolved oxygen (DO)* in raw water is considered the necessary element to support the life of many aquatic organisms. From the drinking water practitioner's point of view, DO is an important indicator of the water treatment process and an important factor in corrosivity.

Wastewater from sewage treatment plants often contains organic materials that are decomposed by microorganisms, which use oxygen in the process. (The amount of oxygen consumed by these organisms in breaking down the waste is known as the biochemical oxygen demand (BOD). We include a discussion of BOD and how to monitor it later.) Other sources of oxygen-consuming waste include stormwater runoff from farmland or urban streets, feedlots, and failing septic systems.

Oxygen is measured in its dissolved form as dissolved oxygen (DO). If more oxygen is consumed than produced, dissolved oxygen levels decline and some sensitive animals may move away, weaken, or die. DO levels fluctuate over a 24-h period and seasonally. They vary with water temperature and altitude. Cold water holds more oxygen than warm water (see Table 15.7), and water holds less oxygen at higher altitudes. Thermal discharges (such as water used to cool machinery in a manufacturing plant or a power plant) raise the temperature of water and lower its oxygen content. Aquatic animals are most vulnerable to lowered DO levels in the early morning on hot summer days when stream flows are low, water temperatures are high, and aquatic plants have not been producing oxygen since sunset.

Sampling and Equipment Considerations

In contrast to lakes, where DO levels are most likely to vary vertically in the water column, changes in DO in rivers and streams move horizontally along the course of the waterway. This is especially true in smaller, shallow streams. In larger, deeper rivers, some vertical stratification of dissolved oxygen might occur. The DO levels in and below riffle areas, waterfalls, or dam spillways are typically higher than those in pools and slower-moving stretches. If you wanted to measure the effect of a dam, sampling for DO behind the dam, immediately below the spillway, and upstream of the dam would be important. Because DO levels are critical to fish, a good place to sample is in the pools that fish tend to favor or in the spawning areas they use.

An hourly time profile of DO levels at a sampling site is a valuable set of data because it shows the change in DO levels from the low point (just before sunrise) to the high point (sometime near midday). However, this might not be practical for a volunteer monitoring program. Note the time of your DO sampling to help judge when in the daily cycle the data were collected.

Dissolved oxygen is measured either in milligrams per liter (mg/L) or "percent saturation." Milligrams per liter are the amount of oxygen in a liter of water. Percent saturation is the amount of oxygen in a liter of water relative to the total amount of oxygen that the water can hold at that temperature. DO samples are collected using a special BOD bottle: a glass bottle with a "turtleneck" and a ground stopper. You can fill the bottle directly in the stream if the stream is wadeable or boatable, or you can use a sampler dropped from a bridge or boat into water deep enough to submerge it. Samplers can be made or purchased.

Winkler Method (Azide Modification)

The Winkler Method (azide modification) involves filling a sample bottle completely with water (no air is left to bias the test). The dissolved oxygen is then "fixed" using a series of reagents that form a titrated acid compound. Titration involves the drop-by-drop addition of a reagent that neutralizes the acid compound, causing a change in the color of the solution. The point at which the color changes is the "endpoint" and is equivalent to the amount of oxygen dissolved in the sample. The sample is usually fixed and titrated in the field at the sample site. Preparing the sample in the field and delivering it to a lab for titration is possible. The azide modification method is best suited for relatively clean waters; otherwise, substances such as color, organics, suspended solids, sulfide, chlorine, and ferrous and ferric iron can interfere with test results. If fresh azide is used, nitrite will not interfere with the test.

In testing, iodine is released in proportion to the amount of DO present in the sample. By using sodium thiosulfate with starch as the indicator, the sample can be titrated to determine the amount of DO present. The chemicals used include:

1. Manganese sulfate solution
2. Alkaline azide-iodide solution
3. Sulfuric acid—concentrated
4. Starch indicator

TABLE 15.7
Maximum DO Concentrations vs. Temperature Variations

Temperature °C	DO (mg/L)	Temperature °C	DO (mg/L)
0	14.60	23	8.56
1	14.19	24	8.40
2	13.81	25	8.24
3	13.44	26	8.09
4	13.09	27	7.95
5	12.75	28	7.81
6	12.43	29	7.67
7	12.12	30	7.54
8	11.83	31	7.41
9	11.55	32	7.28
10	11.27	33	7.16
11	11.01	34	7.05
12	10.76	35	6.93
13	10.52	36	6.82
14	10.29	37	6.71
15	10.07	38	6.61
16	9.85	39	6.51
17	9.65	40	6.41
18	9.45	41	6.31
19	9.26	42	6.22
20	9.07	43	6.13
21	8.90	44	6.04
22	8.72	45	5.95

5. Sodium thiosulfate solution 0.025 N, or phenylarsine solution 0.025 N, or potassium biniodate solution 0.025 N
6. Distilled or deionized water

The equipment used includes:

1. Burette, graduated to 0.1 mL
2. Burette stand
3. 300 mL BOD bottles
4. 500 mL Erlenmeyer flasks
5. 1.0 mL pipets with elongated tips
6. Pipet bulb
7. 250 mL graduated cylinder
8. Laboratory-grade water rinse bottle
9. Magnetic stirrer and stir bars (optional)

Procedure

The procedure for the Winkler method is:

1. Collect the sample in a 300 mL BOD bottle.
2. Add 1 mL of manganous sulfate solution to the surface of the liquid.
3. Add 1 mL of alkaline-iodide-azide solution to the surface of the liquid.
4. Stopper the bottle and mix by inverting it.
5. Allow the floc to settle halfway in the bottle, remix, and allow it to settle again.
6. Add 1 mL of concentrated sulfuric acid to the surface of the liquid.
7. Re-stopper the bottle, rinse the top with laboratory-grade water, and mix until the precipitate is dissolved.
8. The liquid in the bottle should appear clear and have an amber color.
9. Measure 201 mL from the BOD bottle into an Erlenmeyer flask.
10. Titrate with 0.025 N PAO or thiosulfate to a pale yellow color and note the amount of titrant.
11. Add 1 mL of starch indicator solution.
12. Titrate until the blue color first disappears.
13. Record the total amount of titrant.

Calculation

To calculate the DO concentration when the modified Winkler titration method is used:

$$DO, mg/L = \frac{(\text{Burette}_{Final}, mL - \text{Burette}_{Start}, mL) \times N \times 8,000}{\text{Sample Volume, mL}}$$

(15.1)

Note: Using a 200-mL sample and a 0.025 N (N = Normality of the solution used to titrate the sample) titrant reduces this calculation to:

$$DO, mg/L = mL \text{ Titrant Used}$$

Example 15.1

Problem: The operator titrates a 200-mL DO sample. The burette reading at the start of the titration was 0.0 mL. At the end of the titration, the burette read 7.1 mL. The concentration of the titrating solution was 0.025 N. What is the DO concentration in mg/L?

Solution:

$$DO, mg/L = \frac{(7.1 \text{ mL} - 0.0 \text{ mL}) \times 0.025 \times 8,000}{200 \text{ mL}} = 7.1 \text{ mL}$$

Dissolved oxygen field kits using the Winkler method are relatively inexpensive, especially compared to a meter and probe. Field kits run between $35 and $200, and each kit comes with enough reagents to run 50–100 DO tests. Replacement reagents are inexpensive, and you can buy them already measured out for each test in plastic pillows. You can also purchase the reagents in larger quantities in bottles and measure them out with a volumetric scoop. The advantage of the pillows is that they have a longer shelf life and are much less prone to contamination or spillage. Buying larger quantities in bottles has the advantage of considerably lower cost per test.

The major factor in the expense for the kits is the method of titration used—eyedropper or syringe-type titrator. Eyedropper and syringe-type titration are less precise than digital titration because a larger drop of titrant is allowed to pass through the dropper opening, and on a micro-scale, the drop size (and thus the volume of titrant) can vary from drip to drip. A digital titrator or a burette (a long glass tube with a tapered tip like a pipette) permits much more precision and uniformity for the titrant it allows to pass.

If a high degree of accuracy and precision in DO results is required, a digital titrator should be used. A kit that uses an eyedropper-type or syringe-type titrator is suitable for most other purposes. The lower cost of this type of DO field kit might be attractive if several teams of samplers and testers are relied on at multiple sites at the same time.

Meter and Probe

A *dissolved oxygen meter* is an electronic device that converts signals from a probe placed in the water into units of DO in milligrams per liter. Most meters and probes also measure temperature. The probe is filled with a salt solution and has a selectively permeable membrane that allows DO to pass from the stream water into the salt solution. The DO that has diffused into the salt solution changes the electric potential of the salt solution, and this change is sent by electric cable to the meter, which converts the signal to milligrams per liter on a scale that the user can read.

Methodology

If samples are to be collected for analysis in the laboratory, a special APHA sampler or the equivalent must be used. This is the case because, if the sample is exposed to

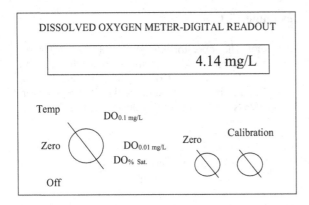

FIGURE 15.13 Dissolved oxygen meter.

or mixed with air during collection, test results can change dramatically. Therefore, the sampling device must allow the collection of a sample that is not mixed with atmospheric air and allows for at least 3X bottle overflow (see Figure 15.13). Again, because the DO level in a sample can change quickly, only grab samples should be used for dissolved oxygen testing. Samples must be tested immediately (within 15 min) after collection.

Note: Samples collected for analysis using the modified Winkler titration method may be preserved for up to 8 h by adding 0.7 mL of concentrated sulfuric acid or by adding all the chemicals required by the procedure. Samples collected from the aeration tank of the activated sludge process must be preserved using a solution of copper sulfate-sulfamic acid to inhibit biological activity.

The advantage of using the DO meter method is that the meter can be used to determine DO concentration directly (see Figure 15.13). In the field, a direct reading can be obtained using a probe (see Figure 15.14) or by collecting samples for testing in the laboratory using a laboratory probe (Figure 15.15).

Note: The field probe can be used for laboratory work by placing a stirrer in the bottom of the sample bottle, but the laboratory probe should never be used in any situation where the entire probe might be submerged.

The probe used in the determination of DO consists of two electrodes, a membrane, and a membrane-filling solution. Oxygen passes through the membrane into the filling solution and causes a change in the electrical current passing between the two electrodes. The change is measured and displayed as the concentration of DO. To be accurate, the probe membrane must be in proper operating condition, and the meter must be calibrated before use. The only chemical used in the DO meter method during normal operation is the electrode-filling solution. However, in the Winkler DO method, chemicals are required for meter calibration.

Calibration prior to use is important. Both the meter and the probe must be calibrated to ensure accurate results. The frequency of calibration is dependent on the frequency of use. For example, if the meter is used once a day, then calibration should be performed before use. There are three methods available for calibration: saturated water,

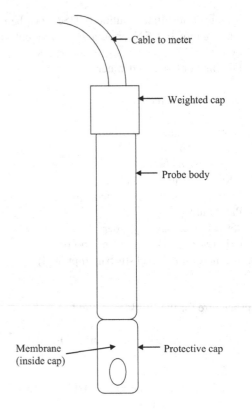

FIGURE 15.14 Dissolved oxygen-field probe.

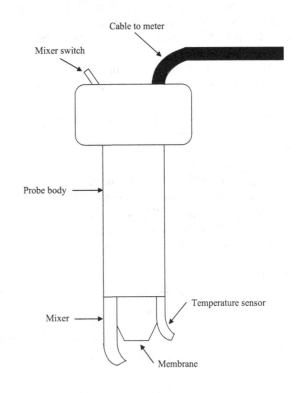

FIGURE 15.15 Dissolved oxygen-lab probe.

saturated air, and the Winkler method. It is important to note that if the Winkler method is not used as a routine calibration method, periodic checks using this method are recommended.

Procedure

It is important to keep in mind that the meter and probe supplier's operating procedures should always be followed. Normally, the manufacturer's recommended procedure will include the following generalized steps:

1. Turn the DO meter on and allow 15 min for it to warm up.
2. Turn the meter switch to zero and adjust as needed.
3. Calibrate the meter using the saturated air, saturated water, or Winkler azide procedure for calibration.
4. Collect the sample in 300 mL bottle, or place the field electrode directly in the stream.
5. Place the laboratory electrode in the BOD bottle without trapping air against the membrane, and turn on the stirrer.
6. Turn the meter switch to the temperature setting, and measure the temperature.
7. Turn the meter switch to DO mode and allow 10 sec for the meter reading to stabilize.
8. Read DO mg/L from the meter and record the results.

No calculation is necessary using this method because the results are read directly from the meter.

Dissolved oxygen meters are expensive compared to field kits that use the titration method. Meter/probe combinations run between $500 and $1,200, including a long cable to connect the probe to the meter. The advantage of a meter/probe is that DO and temperature can be quickly read at any point where the probe is inserted into the stream. DO levels can be measured continuously at a certain point. The results are read directly as milligrams per liter, unlike the titration methods, in which the final titration result might have to be converted by an equation to milligrams per liter. However, DO meters are more fragile than field kits, and repairs to a damaged meter can be costly. The meter/probe must be carefully maintained, and it must be calibrated before each sample run, and if many tests are done, between samplings. Because of the expense, a small water/wastewater facility might only have one meter/probe, which means that only one team of samplers can sample DO, and they must test all the sites. With field kits, on the other hand, several teams can sample simultaneously.

BIOCHEMICAL OXYGEN DEMAND TESTING

Biochemical oxygen demand (BOD) measures the amount of oxygen consumed by microorganisms in decomposing organic matter in stream water. BOD also measures the chemical oxidation of inorganic matter (the extraction of oxygen from water via chemical reaction). A test is used to measure the amount of oxygen consumed by these organisms during a specified period of time (usually 5 days at 20°C). The rate of oxygen consumption in a stream is affected by several variables: temperature, pH, the presence of certain kinds of microorganisms, and the type of organic and inorganic material in the water. BOD directly affects the amount of dissolved oxygen in water bodies. The greater the BOD, the more rapidly oxygen is depleted in the water body, leaving less oxygen available to higher forms of aquatic life. The consequences of high BOD are the same as those for low dissolved oxygen: aquatic organisms become stressed, suffocate, and die. Most river waters used as water supplies have a BOD of less than 7 mg/L; therefore, dilution is not necessary.

Sources of BOD include leaves and wood debris; dead plants and animals; animal manure; effluents from pulp and paper mills, wastewater treatment plants, feedlots, and food-processing plants; failing septic systems; and urban stormwater runoff.

Note: To evaluate raw water's potential for use as a drinking water supply, it is usually sampled, analyzed, and tested for biochemical oxygen demand when turbid, polluted water is the only source available.

Sampling Considerations

Biochemical oxygen demand is affected by the same factors that affect dissolved oxygen. Aeration of stream water—by rapids and waterfalls, for example—will accelerate the decomposition of organic and inorganic material. Therefore, BOD levels at a sampling site with slower, deeper waters might be higher for a given column of organic and inorganic material than the levels at similar site in highly aerated waters. Chlorine can also affect BOD measurement by inhibiting or killing the microorganisms that decompose the organic and inorganic matter in a sample. If sampling in chlorinated waters (such as those below the effluent from a sewage treatment plant), neutralizing the chlorine with sodium thiosulfate is necessary (APHA, 1998).

Biochemical oxygen demand measurement requires taking two samples at each site. One is tested immediately for dissolved oxygen, and the second is incubated in the dark at 20°C for 5 days, and then tested for dissolved oxygen remaining. The difference in oxygen levels between the first test and the second test [in milligrams per liter (mg/L)] is the amount of BOD. This represents the amount of oxygen consumed by microorganisms and used to break down the organic matter present in the sample bottle during the incubation period. Because of the 5-day incubation, the tests are conducted in a laboratory.

Sometimes, by the end of the 5-day incubation period, the dissolved oxygen level is zero. This is especially true for rivers and streams with a lot of organic pollution. Since it is not possible to know when the zero point was reached, determining the BOD level is also impossible. In this case, diluting the original sample by a factor that results in a final dissolved oxygen level of at least 2 mg/L is necessary. Special dilution water should be used for the dilutions (APHA, 1998).

Some experimentation is needed to determine the appropriate dilution factor for a particular sampling site. The result is the difference in dissolved oxygen between the first measurement and the second, after multiplying the second result by the dilution factor. *Standard Methods* prescribes all phases of procedures and calculations for BOD determination. A BOD test is not required for monitoring water supplies.

BOD Sampling, Analysis, and Testing

The approved biochemical oxygen demand sampling and analysis procedure measures the DO depletion (biological oxidation of organic matter in the sample) over a 5-day period under controlled conditions (20°C in the dark). The test is performed using a specified incubation time and temperature. Test results are used to determine plant loadings, plant efficiency, and compliance with NPDES effluent limitations. The duration of the test (5 days) makes it difficult to use the data effectively for process control.

The standard BOD test does not differentiate between oxygen used to oxidize organic matter and that used to oxidize organic and ammonia nitrogen to more stable forms. Because many biological treatment plants now control treatment processes to achieve oxidation of the nitrogen compounds, there is a possibility that BOD test results for plant effluent and some process samples may produce BOD test results based on both carbon and nitrogen oxidation. To avoid this situation, a nitrification inhibitor can be added. When this is done, the test results are known as *carbonaceous BOD* (CBOD). A second uninhibited BOD test should also be run whenever CBOD is determined.

When taking a BOD sample, no special sampling container is required. Either a grab or composite sample can be used. BOD_5 samples can be preserved by refrigeration at or below 4°C (not frozen)—composite samples must be refrigerated during collection. The maximum holding time for preserved samples is 48 h.

Using the incubation of dissolved oxygen approved test method, a sample is mixed with dilution water in several different concentrations (dilutions). The dilution water contains nutrients and materials to provide an optimum environment. Chemicals used include dissolved oxygen, ferric chloride, magnesium sulfate, calcium chloride, phosphate buffer, and ammonium chloride.

Note: Remember all chemicals can be dangerous if not used properly and in accordance with the recommended procedures. Review appropriate sections of the individual chemical materials safety data sheet (MSDS) to determine proper methods for handling and for safety precautions that should be taken.

Sometimes it is necessary to add (seed) healthy organisms to the sample. The DO of the dilution and the dilution water is determined. If seed material is used, a series of dilutions of seed material must also be prepared. The dilutions and dilution blanks are incubated in the dark for 5 days at 20°C ± 1°C. At the end of 5 days, the DO of each dilution and the dilution blanks are determined. For the test results to be valid, certain criteria must be met:

1. Dilution water blank DO change must be ≤0.2 mg/L.
2. Initial DO must be >7.0 mg/L but ≤9.0 mg/L (or saturation at 20°C and test elevation).
3. Sample dilution DO depletion must be ≥2.0 mg/L.
4. Sample dilution residual DO must be ≥1.0 mg/L.

TABLE 15.8
BOD_5 Test Procedure

(1) Fill two bottles with BOD dilution water; insert stoppers.
(2) Place the sample in two BOD bottles; fill them with dilution water; and insert stoppers.
(3) Test for dissolved oxygen (DO).
(4) Incubate for 5 days.
(5) Test for DO.
(6) Add 1 mL $MnSO_4$ below the surface.
(7) Add 1 mL alkaline Kl below the surface.
(8) Add 1 mL H_2SO_4.
(9) Transfer 203 mL to the flask.
(10) Titrate with PAO or thiosulfate.

5. Sample dilution initial DO must be ≥7.0 mg/L.
6. Seed correction should be ≥0.6 but ≤1.0 mg/L.

The BOD_5 test procedure consists of ten steps (for unchlorinated water) as shown in Table 15.8.

Note: BOD_5 is calculated individually for all sample dilutions that meet the criteria. The reported result is the average of the BOD_5 of each valid sample dilution.

BOD_5 Calculation

Unlike the direct reading instrument used in DO analysis, BOD results require calculation. Several criteria are used in selecting which BOD_5 dilutions should be used for calculating test results. Consult a laboratory testing reference manual such as *Standard Methods* (APHA, 1998), for this information. Currently, there are two basic calculations for BOD_5. The first is used for samples that have not been seeded, while the second must be used whenever BOD_5 samples are seeded. In this section, we illustrate the calculation procedure for unseeded samples.

$$BOD_5 (Unseeded) = \frac{(DO_{start}, mg/L - DO_{final}, mg/L) \times 300 mL}{Sample Volume, mL}$$

(15.2)

Example 15.2

Problem: The BOD_5 test is completed. Bottle 1 of the test had a DO of 7.1 mg/L at the start of the test. After 5 days, Bottle 1 had a DO of 2.9 mg/L. Bottle 1 contained 120 mg/L of sample.

Solution:

$$BOD_5 (Unseeded) = \frac{(7.1 mg/L - 2.9 mg/L) \times 300 mL}{120 mL}$$

$$= 0.5 mg/L$$

If the BOD_5 sample has been exposed to conditions that could reduce the number of healthy, active organisms, the sample must be seeded with organisms. Seeding requires the use of a correction factor to remove the BOD_5 contribution of the seed material:

$$Seed\ Correction = \frac{Seed\ Material\ BOD_5 \times Seed\ in\ Dilution,\ mL}{300\ mL}$$

(15.3)

BOD_5 (Seeded)

$$= \frac{\left[(DO_{start},\ mg/L - DO_{final},\ mg/L) - Seed\ Corr.\right] \times 300}{Sample\ Volume,\ mL}$$

(15.4)

TEMPERATURE MEASUREMENT

As mentioned, an ideal water supply should have at all times, an almost constant temperature or one with minimal variation. Knowing the temperature of the water supply is important because the rates of biological and chemical processes depend on it. Temperature affects the oxygen content of the water (oxygen levels decrease as temperature increases), the rate of photosynthesis by aquatic plants, the metabolic rates of aquatic organisms, and the sensitivity of organisms to toxic wastes, parasites, and diseases. Causes of temperature changes include weather, removal of shading streambank vegetation, impoundments (a body of water confined by a barrier, such as a dam), and the discharge of cooling water, urban stormwater, and groundwater inflows to the stream.

Sampling and Equipment Considerations

Temperature—for example, in a stream—varies with width and depth, and the temperature of well-sunned portions of a stream can be significantly higher than the shaded portions on a sunny day. In a small stream, the temperature will be relatively constant as long as the stream is uniformly in sun or shade. In a large stream, temperature can vary considerably with width and depth, regardless of shade. If safe to do so, temperature measurements should be collected at varying depths and across the surface of the stream to obtain vertical and horizontal temperature profiles. This can be done at each site at least once to determine the necessity of collecting a profile during each sampling visit. Temperature should be measured at the same place every time.

Temperature is measured in the stream with a thermometer or a meter. Alcohol-filled thermometers are preferred over mercury-filled ones because they are less hazardous if broken. Armored thermometers for field use can withstand more abuse than unprotected glass thermometers and are worth the additional expense. Meters for other tests, such as pH (acidity) or dissolved oxygen, also measure temperature and can be used instead of a thermometer.

HARDNESS MEASUREMENT

Hardness refers primarily to the amount of calcium and magnesium in the water. Calcium and magnesium enter water mainly through the leaching of rocks. Calcium is an important component of aquatic plant cell walls and the shells and bones of many aquatic organisms. Magnesium is an essential nutrient for plants and is a component of the chlorophyll molecule. Hardness test kits express results in ppm of $CaCO_3$, but these results can be converted directly to calcium or magnesium concentrations:

$$Calcium\ Hardness\ as\ ppm\ CaCO_3 \times 0.40 = ppm\ Ca$$

(15.5)

$$Magnesium\ Hardness\ as\ ppm\ CaCO_3 \times 0.24 = ppm\ Mg$$

(15.6)

Note: Because of less contact with soil minerals and more contact with rain, surface raw water is usually softer than groundwater.

As a rule, when hardness is greater than 150 mg/L, softening treatment may be required for public water systems. Hardness determination through testing is required to ensure treatment efficiency.

Note: Keep in mind that when measuring calcium hardness, the concentration of calcium is routinely measured separately from total hardness. Its concentration in water can range from 0 to several thousand mg/L as $CaCO_3$. Likewise, when measuring magnesium hardness, magnesium is routinely determined by subtracting calcium hardness from total hardness. There is usually less magnesium than calcium in natural water. Lime dosage for water softening operations is partly based on the concentration of magnesium hardness in the water.

In the hardness test, the sample must be carefully measured, and then a buffer is added to the sample to adjust the pH for the test and an indicator to signal the titration endpoint. The indicator reagent is normally blue in a sample of pure water, but if calcium or magnesium ions are present, the indicator combines with them to form a red-colored complex. The titrant in this test is EDTA (ethylenediaminetetraacetic acid, used with its salts in the titration method), a "chelant" that "pulls" the calcium and magnesium ions away from the red-colored complex. The EDTA is added dropwise to the sample until all the calcium and magnesium ions have been "chelated" away from the complex and the indicator returns to its normal blue color. The amount of EDTA required to cause the color change is a direct indication of the amount of calcium and magnesium ions in the sample.

Some hardness kits include an additional indicator that is specific for calcium. This type of kit will provide three readings: total hardness, calcium hardness, and magnesium hardness. For information on interference, precision, and accuracy, consult the latest edition of *Standard Methods*.

pH MEASUREMENT

pH is defined as the negative log of the hydrogen ion concentration of a solution. This is a measure of the ionized hydrogen in solution. Simply put, it is the relative acidity or basicity of the solution. The chemical and physical properties, and the reactivity of almost every component in water are dependent upon pH. It relates to corrosivity, contaminant solubility, and the water's conductance, and has a secondary MCL range set at 6.5–8.5.

Analytical and Equipment Considerations

The pH can be analyzed in the field or the lab. If analyzed in the lab, it must be measured within 2 h of sample collection because the pH will change due to carbon dioxide from the air as it dissolves in the water, bringing the pH toward 7. If your program requires a high degree of accuracy and precision in pH results, the pH should be measured with a laboratory-quality pH meter and electrode. Meters of this quality range in cost from around $250 to $1,000. Color comparators and pH "pocket pals" are suitable for most other purposes, and the cost of either of these is in the $50 range. The lower cost of the alternatives might be attractive if multiple samplers are used to sample several sites at the same time.

pH Meters

A pH meter measures the electric potential (millivolts) across an electrode when immersed in water. This electric potential is a function of the hydrogen ion activity in the sample; therefore, pH meters can display results in either millivolts (mV) or pH units. A pH meter consists of a *potentiometer*, which measures electric potential where it meets the water sample; a reference electrode, which provides a constant electric potential; and a *temperature-compensating device*, which adjusts the readings according to the temperature of the sample (since pH varies with temperature). The reference and glass electrodes are frequently combined into a single probe called a *combination electrode*. A wide variety of meters are available, but the most important part of the pH meter is the electrode. Thus, purchasing a good, reliable electrode and following the manufacturer's instructions for proper maintenance is important. Infrequently used or improperly maintained electrodes are subject to corrosion, which makes them highly inaccurate.

pH "Pocket Pals" and Color Comparators

pH "pocket pals" are electronic handheld "pens" that are dipped in water, providing a digital readout of the pH. They can be calibrated to only one pH buffer. (Lab meters, on the other hand, can be calibrated to two or more buffer solutions and thus are more accurate over a wide range of pH measurements.) Color comparators involve adding a reagent to the sample that colors the water. The intensity of the color is proportional to the pH of the sample and is then matched against a standard color chart. The color chart equates particular colors to associated pH values, which can be determined by matching the colors from the chart to the color of the sample. For instructions on how to collect and analyze samples, refer to *Standard Methods* (APHA, 1998).

TURBIDITY MEASUREMENT

Turbidity is a measure of water clarity—how much the material suspended in water decreases the passage of light through it. Turbidity consists of suspended particles in the water and may be caused by a number of materials, both organic and inorganic. These particles are typically in the size range of 0.004 mm (clay) to 1.0 mm (sand). The occurrence of turbid source waters may be permanent or temporary and it can affect the color of the water. Higher turbidity increases water temperatures because suspended particles absorb more heat. This, in turn, reduces the concentration of dissolved oxygen (DO) because warm water holds less DO than cold. Higher turbidity also reduces the amount of light penetrating the water, which decreases photosynthesis and the production of DO. Suspended materials can clog fish gills, reducing resistance to disease in fish, lowering growth rates, and affecting egg and larval development. As the particles settle, they can blanket the stream bottom (especially in slower waters) and smother fish eggs and benthic macroinvertebrates.

Turbidity also affects treatment plant operations. For example, turbidity hinders disinfection by shielding microbes, some of which are pathogens, from the disinfectant. Obviously, this is the most significant reason for turbidity monitoring; the test for it is an indication of the effectiveness of filtration in water supplies. It is important to note that turbidity removal is the principal reason for chemical addition, settling, coagulation, and filtration in potable water treatment. Sources of turbidity include:

1. Soil erosion
2. Waste discharge
3. Urban runoff
4. Eroding stream banks
5. Large numbers of bottom feeders (such as carp), which stir up bottom sediments
6. Excessive algal growth

Sampling and Equipment Considerations

Turbidity can be useful as an indicator of the effects of runoff from construction, agricultural practices, logging activity, discharges, and other sources. Turbidity often increases sharply during rainfall, especially in developed watersheds, which typically have relatively high proportions of impervious surfaces. The flow of stormwater runoff from impervious surfaces rapidly increases stream velocity, which raises the erosion rates of streambanks and channels. Turbidity can also rise sharply during dry weather if earth-disturbing activities occur in or near a stream without erosion control practices in place.

Regular monitoring of turbidity can help detect trends that might indicate increasing erosion in developing watersheds. However, turbidity is closely related to stream flow and velocity and should be correlated with these factors. Comparisons of changes in turbidity over time should therefore be made at the same point at the same flow.

Keep in mind that turbidity is not a measurement of the amount of suspended solids present or the rate of sedimentation of a stream because it measures only the amount of light that is scattered by suspended particles. Measurement of total solids is a more direct measurement of the amount of material suspended and dissolved in water.

Turbidity is generally measured using a turbidity meter or *turbidimeter*. The turbidimeter is a modern nephelometer, originally a box containing a light bulb that directed light at a sample. The amount of light scattered at right angles by the turbidity particles was measured as an indication of the turbidity in the sample and registered as nephelometric turbidity units (NTU). The turbidimeter uses a photoelectric cell to register the scattered light on an analog or digital scale, and the instrument is calibrated with permanent turbidity standards composed of the colloidal substance formazan. Meters can measure turbidity over a wide range—from 0 to 1,000 NTUs. A clear mountain stream might have a turbidity of around 1 NTU, whereas a large river like the Mississippi might have a dry-weather turbidity of 10 NTUs. Because these values can jump into the hundreds of NTUs during runoff events, the turbidity meter used should be reliable over the range in which you will be working. Meters of this quality cost about $800. Many meters in this price range are designed for field or lab use (USEPA, 2000).

An operator may also take samples to a lab for analysis. Another approach discussed previously, is to measure transparency (an integrated measure of light scattering and absorption) instead of turbidity. Water clarity/transparency can be measured using a *Secchi disk* (see Figure 15.11) or a transparency tube. The Secchi disk can only be used in deep, slow-moving rivers, while the transparency tube (a comparatively new development) is gaining acceptance in and around the country but is not yet in widespread use.

Using a Secchi Disk

A Secchi disk is a black and white disk that is lowered by hand into the water to the depth at which it vanishes from sight (see Figure 15.11). The distance to the vanishing point is then recorded—the clearer the water, the greater the distance. Secchi disks are simple to use and inexpensive. For river monitoring, they have limited use, however, because in most cases, the river bottom will be visible, and the disk will not reach a vanishing point. Deeper, slower-moving rivers are the most appropriate places for Secchi disk measurement, although the current might require that the disk be extra-weighted so it does not sway and make measurement difficult. Secchi disks cost about $50 but can be homemade.

The line attached to the Secchi disk must be marked in waterproof ink according to the units designated by the sampling program. Many programs require samplers to measure to the nearest 1/10 m. Meter intervals can be tagged (e.g., with duct tape) for ease of use. To measure water clarity with a Secchi disk:

1. Check to make sure that the Secchi disk is securely attached to the measured line.
2. Lean over the side of the boat and lower the Secchi disk into the water, keeping your back to the sun to block glare.

3. Lower the disk until it disappears from view. Lower it one-third of a meter and then slowly raise the disk until it just reappears. Move the disk up and down until you find the exact vanishing point.
4. Attach a clothespin to the line at the point where the line enters the water. Record the measurement on your data sheet. Repeating the measurement provides you with a quality control check.

The key to consistent results is to train samplers to follow standard sampling procedures, and if possible, have the same individual take the reading at the same site throughout the season.

Transparency Tube

Pioneered by Australia's Department of Conservation, the *transparency tube* is a clear, narrow plastic tube marked in units with a dark pattern painted on the bottom. Water is poured into the tube until the pattern disappears. Some U.S. volunteer monitoring programs (e.g., the Tennessee Valley Authority (TVA) Clean Water Initiative and the Minnesota Pollution Control Agency (MPCA)) are testing the transparency tube in streams and rivers. The MPCA uses tubes marked in centimeters and has found tube readings to relate fairly well to lab measurements of turbidity and total suspended solids, although it does not recommend the transparency tube for applications where precise and accurate measurement is required or in highly colored waters. The TVA and MPCA recommend the following sampling considerations:

1. Collect the sample in a bottle or bucket in mid-stream and at mid-depth if possible. Avoid stagnant water and sample as far from the shoreline as is safe. Avoid collecting sediment from the bottom of the stream.
2. Face upstream as you fill the bottle or bucket.
3. Take readings in open but shaded conditions. Avoid direct sunlight by turning your back to the sun.
4. Carefully stir or swish the water in the bucket or bottle until it is homogeneous, taking care not to produce air bubbles (these scatter the light and affect the measurement). Then pour the water slowly into the tube while looking down the tube. Measure the depth of the water column in the tube at the point where the symbol just disappears.

ORTHOPHOSPHATE MEASUREMENT

Earlier we discussed the nutrients phosphorus and nitrogen. Both phosphorus and nitrogen are essential nutrients for the plants and animals that make up the aquatic food web. Because phosphorus is the nutrient in short supply in most freshwater systems, even a modest increase in phosphorus can (under the right conditions) set off a whole chain of undesirable events in a stream, including accelerated plant growth, algae blooms, low dissolved oxygen, and the death of certain fish, invertebrates, and other aquatic animals. Phosphorus comes from many sources, both natural and human. These include soil and rocks, wastewater treatment plants, runoff from fertilized lawns and cropland, failing septic systems, runoff from animal manure storage areas, disturbed land areas, drained wetlands, water treatment, and commercial cleaning preparations.

Forms of Phosphorus

Phosphorus has a complicated story. Pure, elemental phosphorus (P) is rare. In nature, phosphorus usually exists as part of a phosphate molecule (PO_4). Phosphorus in aquatic systems occurs as organic phosphate and inorganic phosphate. Organic phosphate consists of a phosphate molecule associated with a carbon-based molecule, as in plant or animal tissue. Phosphate that is not associated with organic material is inorganic, the form required by plants. Animals can use either organic or inorganic phosphate. Both organic and inorganic phosphate can either be dissolved in the water or suspended (attached to particles in the water column).

The Phosphorus Cycle

Phosphorus cycles through the environment, changing form as it does so. Aquatic plants take in dissolved inorganic phosphorus, as it becomes part of their tissues. Animals obtain the organic phosphorus they need by eating either aquatic plants, other animals, or decomposing plant and animal material. In water bodies, as plants and animals excrete wastes or die, the organic phosphorus they contain sinks to the bottom, where bacterial decomposition converts it back to inorganic phosphorus, both dissolved and attached to particles. This inorganic phosphorus returns to the water column when animals, human activity, interactions, or water currents stir up the bottom. Then plants take it up, and the cycle begins again. In a stream system, the phosphorus cycle tends to move phosphorus downstream as the current carries decomposing plant and animal tissue and dissolved phosphorus. It becomes stationary only when it is taken up by plants or bound to particles that settle to the bottom of ponds.

In the field of water quality chemistry, phosphorus is described by several terms. Some of these terms are chemistry-based (referring to chemically based compounds), and others are methods-based (describing what is measured by a particular method). The term *orthophosphate* is a chemistry-based term that refers to the phosphate molecule by itself. More specifically, orthophosphate is simple phosphate, or reactive phosphate, i.e., Na_3PO_4 (sodium phosphate, tribasic) and NaH_2PO_4 (sodium phosphate, monobasic). Orthophosphate is the only form of phosphate that can be directly tested for in the laboratory and is the form that bacteria use directly for metabolic processes. Reactive phosphorus is a corresponding method-based term that describes what is actually being measured when the test for orthophosphate is performed. Because the lab procedure isn't quite perfect, mostly orthophosphate is obtained along with a small fraction of some other forms. More complex inorganic phosphate compounds are referred to as *condensed phosphates* or *polyphosphates*. The method-based term for these forms is *acid hydrolyzable*.

Testing Phosphorus

Testing phosphorus is challenging because it involves measuring very low concentrations, down to 0.01 milligrams per liter (mg/L) or even lower. Even such very low concentrations of phosphorus can have a dramatic impact on streams. Less sensitive methods should be used only to identify serious problem areas. While many tests for phosphorus exist, only four are likely to be performed by most samplers. The total orthophosphate test is largely a measure of orthophosphate. Because the sample is not filtered, the procedure measures both dissolved and suspended orthophosphate. The USEPA-approved method for measuring this is known as the ascorbic acid method. Briefly, a reagent (either liquid or powder) containing ascorbic acid and ammonium molybdate reacts with orthophosphate in the sample to form a blue compound. The intensity of the blue color is directly proportional to the amount of orthophosphate in the water.

The total phosphate test measures all the forms of phosphorus in the sample (orthophosphate, condensed phosphate, and organic phosphate) by first "digesting" (heating and acidifying) the sample to convert all the other forms to orthophosphate; then the orthophosphate is measured by the ascorbic acid method. Because the sample is not filtered, the procedure measures both dissolved and suspended orthophosphate. The dissolved phosphorus test measures the fraction of the total phosphorus that is in solution in the water (as opposed to being attached to suspended particles). It is determined by first filtering the sample and then analyzing the filtered sample for total phosphorus. Insoluble phosphorus is calculated by subtracting the dissolved phosphorus result from the total phosphorus result.

All these tests have one thing in common—they all depend on measuring orthophosphate. The total orthophosphate test measures the orthophosphate that is already present in the sample. The others measure what is already present and what is formed when the other forms of phosphorus are converted to orthophosphate by digestion. Monitoring phosphorus involves two basic steps:

1. Collecting a water sample
2. Analyzing it in the field or lab for one of the types of phosphorus described above

Sampling and Equipment Considerations

Sample containers made of either some form of plastic or Pyrex® glass are acceptable to USEPA. Because phosphorus molecules tend to "absorb" (attach) to the inside surface of sample containers, if containers are to be reused, they must be acid-washed to remove absorbed phosphorus. The container must be able to withstand repeated contact with hydrochloric acid. Plastic containers, either high-density polyethylene or polypropylene, might be preferable to glass from a practical standpoint because they are better able to withstand breakage.

Some programs use disposable, sterile, plastic Whirl-Pak® bags. The size of the container depends on the sample amount needed for the phosphorus analysis method chosen, and the amount needed for other analyses to be performed.

All containers that will hold water samples or come into contact with reagents used in the orthophosphate test must be dedicated. They should not be used for other tests to eliminate the possibility that reagents containing phosphorus will contaminate the labware. All labware should be acid-washed.

The only form of phosphorus this text recommends for field analysis is total orthophosphate, which uses the ascorbic acid method on an untreated sample. Analysis of any of the other forms requires adding potentially hazardous reagents, heating the sample to boiling, and using too much time and equipment to be practical. In addition, analysis for other forms of phosphorus is prone to errors and inaccuracies in field situations. Pretreatment and analysis for these other forms should be handled in a laboratory.

Ascorbic Acid Method for Determining Orthophosphate

In the ascorbic acid method, a combined liquid or prepackaged powder reagent consisting of sulfuric acid, potassium antimonyl tartrate, ammonium molybdate, and ascorbic acid (or comparable compounds) is added to either 50 or 25 mL of the water sample. This colors the sample blue in direct proportion to the amount of orthophosphate in the sample. Absorbance or transmittance is then measured after 10 min but before 30 min, using a color comparator with a scale in milligrams per liter that increases with the increase in color hue, or an electronic meter that measures the amount of light absorbed or transmitted at a wavelength of 700–880 nm (again, depending on the manufacturer's directions).

A color comparator may be useful for identifying heavily polluted sites with high concentrations (greater than 0.1 mg/L). However, matching the color of a treated sample to a comparator can be very subjective, especially at low concentrations, and lead to variable results.

A field spectrophotometer or colorimeter with a 2.5-cm light path and an infrared photocell (set for a wavelength of 700–880 nm) is recommended for the accurate determination of low concentrations (between 0.2 and 0.02 mg/L). Use of a meter requires that a prepared known standard concentration be analyzed ahead of time to convert the absorbance readings of a stream sample to milligrams per liter, or that the meter reads directly in milligrams per liter.

For information on how to prepare standard concentrations and how to collect and analyze samples, refer to *Standard Methods* and USEPA's *Methods for Chemical Analysis of Water and Wastes* (2nd ed., 1991, Method 365.2).

NITRATES MEASUREMENT

As mentioned, *nitrates* are a form of nitrogen found in several different forms in terrestrial and aquatic ecosystems. These forms of nitrogen include ammonia (NH_3), nitrates (NO_3), and nitrites (NO_2). Nitrates are essential plant nutrients, but excess amounts can cause significant water quality problems. Together with phosphorus, excess nitrates can accelerate eutrophication, causing dramatic increases in aquatic plant growth and changes in the types of plants and animals that live in the stream. This, in turn, affects dissolved oxygen, temperature, and other indicators. Excess nitrates can cause hypoxia (low levels of dissolved oxygen) and can become toxic to warm-blooded animals at higher concentrations (10 mg/L or higher) under certain conditions. The natural level of ammonia or nitrate in surface water is typically low (less than 1 mg/L); in the effluent of wastewater treatment plants, it can range up to 30 mg/L. Conventional potable water treatment plants cannot remove nitrates. High concentrations must be prevented by controlling the input at the source. Sources of nitrates include wastewater treatment plants, runoff from fertilized lawns and cropland, failing on-site septic systems, runoff from animal manure storage areas, and industrial discharges that contain corrosion inhibitors.

Sampling and Equipment Considerations

Nitrates from land sources end up in rivers and streams more quickly than other nutrients like phosphorus because they dissolve in water more readily than phosphorus, which has an attraction for soil particles. As a result, nitrates serve as a better indicator of the possibility of sewage or manure pollution during dry weather. Water that is polluted with nitrogen-rich organic matter might show low nitrates. The decomposition of organic matter lowers the dissolved oxygen level, which in turn slows the rate at which ammonia is oxidized to nitrite (NO_2) and then to nitrate (NO_3). Under such circumstances, monitoring for nitrites or ammonia (which are considerably more toxic to aquatic life than nitrate) might also be necessary. (See *Standard Methods* sections 4500-NH3 and 4500-NH2 for appropriate nitrite methods.) Water samples to be tested for nitrate should be collected in glass or polyethylene containers that have been prepared using Method B (described previously). Two methods are typically used for nitrate testing: the cadmium reduction method and the nitrate electrode. The more commonly used cadmium reduction method produces a color reaction measured either by comparison to a color wheel or by using a spectrophotometer. A few programs also use a nitrate electrode, which can measure in the range of 0–100-mg/L nitrate. A newer colorimetric immunoassay technique for nitrate screening is also now available.

Cadmium Reduction Method

In the *cadmium reduction method*, nitrate is reduced to nitrite by passing the sample through a column packed with activated cadmium. The sample is then quantitatively measured for nitrite.

More specifically, the cadmium reduction method is a colorimetric method that involves contact between the nitrate in the sample and cadmium particles, which causes nitrates to be converted to nitrites. The nitrites then react with another reagent to form a red color, with intensity proportional to the original amount of nitrate. The color is measured either by comparison to a color wheel with a scale in milligrams per liter that increases with the intensity of color hue or by using an electronic spectrophotometer that measures the amount of light absorbed by the treated sample at a 543-nm wavelength. The absorbance value is converted to the equivalent concentration of nitrate against a standard curve. Methods for making standard solutions and standard curves are presented in *Standard Methods*.

Before each sampling run, the sampling/monitoring supervisor should create this curve. The curve is developed by making a set of standard concentrations of nitrate, reacting them, and developing the corresponding color, then plotting the absorbance value for each concentration against concentration. A standard curve could also be generated for the color wheel. The use of the color wheel is appropriate only if nitrate concentrations are greater than 1 mg/L. For concentrations below 1 mg/L, a spectrophotometer should be used. Matching the color of a treated sample at low concentrations to a color wheel (or cubes) can be very subjective and can lead to variable results. Color comparators can, however, be effectively used to identify sites with high nitrates.

This method requires that the samples being treated are clear. If a sample is turbid, it should be filtered through a 0.45-μm filter. Be sure to test to make sure the filter is nitrate-free. If copper, iron, or other metals are present in concentrations above several mg/L, the reaction with the cadmium will slow down and the reaction time must be increased.

The reagents used for this method are often prepackaged for different ranges, depending on the expected concentration of nitrate in the stream. Manufacturers, for example, provide reagents for the following ranges: low (0–0.40 mg/L), medium (0–15 mg/L), and high (0–30 mg/L). Determining the appropriate range for the stream being monitored is important.

Nitrate Electrode Method

A *nitrate electrode* (used with a meter) is similar in function to a dissolved oxygen meter. It consists of a probe with a sensor that measures nitrate activity in the water; this activity affects the electric potential of a solution in the probe. This change is then transmitted to the meter, which converts the electric signal to a scale that is read in millivolts; then the millivolts are converted to mg/L of nitrate by plotting them against a standard curve. The accuracy of the electrode can be affected by high concentrations of chloride or bicarbonate ions in the sample water. Fluctuating pH levels can also affect the meter reading.

Nitrate electrodes and meters are expensive compared to field kits that employ the cadmium reduction method. (The expense is comparable, however, if a spectrophotometer is used rather than a color wheel.) Meter/probe combinations run between $700 and $1,200, including a long cable to connect the probe to the meter. If the program has

a pH meter that displays readings in millivolts, it can be used with a nitrate probe, and no separate nitrate meter is needed. Results are read directly as milligrams per liter.

Although nitrate electrodes and spectrophotometers can be used in the field, they have certain disadvantages. These devices are more fragile than the color comparators and are therefore more at risk of breaking in the field. They must be carefully maintained and calibrated before each sample run, and if many tests are being run, calibration should occur between samplings. This means that samples are best tested in the lab. Note that samples to be tested with a nitrate electrode should be at room temperature, whereas color comparators can be used in the field with samples at any temperature.

SOLIDS MEASUREMENT

Solids in water are defined as any matter that remains as residue upon evaporation and drying at 103°C. They are separated into two classes: suspended solids and dissolved solids.

$$\underset{\text{(nonfilterable residue)}}{\text{Total Solids}} = \underset{}{\text{Suspended Solids}} + \underset{\text{(filterable residue)}}{\text{Dissolved Solids}}$$

As shown above, *total solids* are the sum of dissolved solids plus suspended and settleable solids in water. In natural freshwater bodies, dissolved solids consist of calcium, chlorides, nitrates, phosphorus, iron, sulfur, and other ions—particles that will pass through a filter with pores of around 2 μm (0.002 cm) in size. Suspended solids include silt and clay particles, plankton, algae, fine organic debris, and other particulate matter. These are particles that will not pass through a 2-μm filter.

The concentration of total dissolved solids affects the water balance in the cells of aquatic organisms. An organism placed in water with a very low level of solids (distilled water, for example) swells because water tends to move into its cells, which have a higher concentration of solids. An organism placed in water with a high concentration of solids shrinks somewhat because the water in its cells tends to move out. This, in turn, affects the organism's ability to maintain the proper cell density, making it difficult to keep its position in the water column. It might float up or sink to a depth to which it is not adapted, and it might not survive.

Higher concentrations of suspended solids can serve as carriers of toxins, which readily cling to suspended particles. This is particularly a concern where pesticides are being used on irrigated crops. Where solids are high, pesticide concentrations may increase well beyond those of the original application as the irrigation water travels down irrigation ditches. Higher levels of solids can also clog irrigation devices and might become so high that irrigated plant roots will lose water rather than gain it.

A high concentration of total solids will make drinking water unpalatable and might have an adverse effect on people who are not used to drinking such water. Levels of total solids that are too high or too low can also reduce the efficiency of wastewater treatment plants, as well as the operation of industrial processes that use raw water.

Total solids affect water clarity. Higher solids decrease the passage of light through water, thereby slowing photosynthesis by aquatic plants. The water heats up more rapidly and holds more heat; this, in turn, might adversely affect aquatic life adapted to a lower temperature regime.

Sources of total solids include industrial discharges, sewage, fertilizers, road runoff, and soil erosion. Total solids are measured in milligrams per liter (mg/L).

Solids Sampling and Equipment Considerations

When conducting solids testing, many factors affect the accuracy of the test or result in wide variations in results for a single sample, including:

1. Drying temperature
2. Length of drying time
3. Condition of desiccator and desiccant
4. Non-representative samples' lack of consistency in the test procedure
5. Failure to achieve constant weight prior to calculating results

Several precautions can help increase the reliability of test results:

1. Use extreme care when measuring samples, weighing materials, and drying or cooling samples.
2. Check and regulate oven and furnace temperatures frequently to maintain the desired range.
3. Use an indicator drying agent in the desiccator that changes color when it is no longer effective—change or regenerate the desiccant when necessary.
4. Keep the desiccator cover greased with the appropriate type of grease—this will seal the desiccator and prevent moisture from entering as the test glassware cools.
5. Check ceramic glassware for cracks and glass fiber filters for possible holes. A hole in a glass filter will cause solids to pass through and yield inaccurate results.
6. Follow the manufacturer's recommendations for the care and operation of analytical balances.

Total solids are important to measure in areas where discharges from sewage treatment plants, industrial plants, or extensive crop irrigation may occur. In particular, streams and rivers in arid regions where water is scarce and evaporation is high tend to have higher concentrations of solids and are more readily affected by human introduction of solids from land use activities.

Total solids measurements can be useful as an indicator of the effects of runoff from construction, agricultural practices, logging activities, sewage treatment plant discharges,

and other sources. As with turbidity, concentrations often increase sharply during rainfall, especially in developed watersheds. They can also rise sharply during dry weather if earth-disturbing activities occur in or near the stream without erosion control practices in place. Regular monitoring of total solids can help detect trends that might indicate increasing erosion in developing watersheds. Total solids are closely related to stream flow and velocity and should be correlated with these factors. Any change in total solids over time should be measured at the same site at the same flow.

Total solids are measured by weighing the amount of solids present in a known volume of sample; this is accomplished by weighing a beaker, filling it with a known volume, evaporating the water in an oven and completely drying the residue, then weighing the beaker with the residue. The total solids concentration is equal to the difference between the weight of the beaker with the residue and the weight of the beaker without it. Since the residue is so light, the lab needs a balance that is sensitive to weights in the range of 0.0001 g. Balances of this type are called analytical or Mettler balances, and they are expensive (around $3,000). The technique requires that the beakers be kept in a desiccator, a sealed glass container that contains material that absorbs moisture and ensures that the weighing is not biased by water condensing on the beaker. Some desiccants change color to indicate moisture content. Measurement of total solids cannot be done in the field. Samples must be collected using clean glass or plastic bottles or Whirl-Pak® bags and taken to a laboratory where the test can be run.

Total Suspended Solids

As mentioned, the term "solids" refers to any material suspended or dissolved in water and wastewater. Although normal domestic wastewater contains a very small amount of solids (usually less than 0.1%), most treatment processes are designed specifically to remove or convert solids to a form that can be removed or discharged without causing environmental harm.

In sampling for total suspended solids (TSS), samples may be either grab or composite and can be collected in either glass or plastic containers. TSS samples can be preserved by refrigeration at or below 4°C (not frozen). However, composite samples must be refrigerated during collection. The maximum holding time for preserved samples is 7 days.

Test Procedure

To conduct a TSS test procedure, a well-mixed measured sample is poured into a filtration apparatus and, with the aid of a vacuum pump or aspirator, is drawn through a pre-weighed glass fiber filter. After filtration, the glass filter is dried at 103°C–105°C, cooled, and reweighed. The increase in weight of the filter and solids compared to the filter alone represents the total suspended solids. An example of the specific test procedure used for total suspended solids is given below:

1. Select a sample volume that will yield between 10 and 200 mg of residue with a filtration time of 10 min or less.

Note: If the filtration time exceeds 10 min, increase the filter area or decrease the volume to reduce filtration time.

Note: For non-homogeneous samples or samples with very high solids concentrations (i.e., raw wastewater or mixed liquor), use a larger filter to ensure a representative sample volume can be filtered.

2. Place the pre-weighed glass fiber filter on the filtration assembly in a filter flask.
3. Mix the sample well and measure the selected volume of the sample.
4. Apply suction to the filter flask, and wet the filter with a small amount of laboratory-grade water to seal it.
5. Pour the selected sample volume into the filtration apparatus.
6. Draw the sample through the filter.
7. Rinse the measuring device into the filtration apparatus with three successive 10 mL portions of laboratory-grade water. Allow complete drainage between rinsing.
8. Continue suction for 3 min after the filtration of the final rinse is completed.
9. Remove the glass filter from the filtration assembly (membrane filter funnel or clean Gooch crucible). If using the large disks and membrane filter assembly, transfer the glass filter to a support (aluminum pan or evaporating dish) for drying.
10. Place the glass filter with solids and support (pan, dish, or crucible) in a drying oven.
11. Dry the filter and solids to constant weight at 103°C–105°C (for at least 1 h).
12. Cool to room temperature in a desiccator.
13. Weigh the filter and support, and record the constant weight in the test record.

TSS Calculations

To determine the total suspended solids concentration in mg/L, we use the following equations:

1. To determine the weight of dry solids in grams

Dry Solids, g

= Wt. of Dry Solids and Filter, g − Wt. of Dry Filter, g

$$(15.7)$$

To determine the weight of dry solids in milligrams (mg)

Dry Solids, mg

= Wt. of Solids and Filter, g − Wt. of Dry Filter, g

$$(15.8)$$

3. To determine the TSS concentration in mg/L

$$\text{TSS, mg/L} = \frac{\text{Dry Solids, mg} \times 1{,}000 \text{ mL}}{\text{mL sample}} \qquad (15.9)$$

Example 15.3

Problem: Using the data provided below, calculate total suspended solids (TSS):

Sample Volume, mL	250 mL
Weight of Dry Solids and Filter, g	2.305 g
Weight of Dry Filter, g	2.297 g

Solution:

Dry Solids, g = 2.305 g − 2.297 g = 0.008 g

Dry Solids, mg = 0.008 g × 1,000 mg/g = 8 mg

$$\text{TSS, mg/L} = \frac{8.0 \times 1{,}000 \text{ mL/L}}{250 \text{ mL}} = 32.0 \text{ mg/L}$$

Volatile Suspended Solids Testing

When the total suspended solids are ignited at 550°C ± 50°C, the volatile (organic) suspended solids of the sample are converted to water vapor and carbon dioxide and are released into the atmosphere. The solids that remain after ignition (ash) are the inorganic or fixed solids. In addition to the equipment and supplies required for the total suspended solids test, you need the following:

1. Muffle furnace (550°C ± 50°C)
2. Ceramic dishes
3. Furnace tongs
4. Insulated gloves

Test Procedure

An example of the test procedure used for volatile suspended solids is as follow:

Place the weighed filter with solids and support from the total suspended solids test in the muffle furnace. Ignite the filter, solids, and support at 550°C ± 50°C for 15–20 min. Remove the ignited solids, filter, and support from the furnace, and partially air cool. Cool to room temperature in a desiccator. Weigh the ignited solids, filter, and support on an analytical balance, and record the weight of the ignited solids, filter, and support.

Total Volatile Suspended Solids Calculations

To calculate total volatile suspended solids (TVSS), the following information is required:

1. Weights of dry solids, filter, and support in grams
2. Weight of ignited solids, filter, and support in grams

Tot. Vol. Suspended. Solids, mg/L

$$= \frac{(A-C) \times 1,000 \text{ mg/g} \times 1,000 \text{ mL/L}}{\text{Sample Vol., mL}} \quad (15.10)$$

where

A = Weight of Dried Solids, Filter, and Support
Weight of Ignited Solids, Filter, and Support

Example 15.4

Problem: Using the data provided below, calculate the total volatile suspended solids:

Weight of dried solids, filter, and support = 1.6530 g
Weight of ignited solids, filter, and support = 1.6330 g
Sample volume = 100 mL

Solution:

$$\text{TVSS.} = \frac{(1.6530 \text{ g} - 1.6330 \text{ g}) \times 1,000 \text{ mg/g} \times 1,000 \text{ mL}}{100 \text{ mL}}$$

$$= \frac{0.02 \times 1,000,000 \text{ mg/L}}{100}$$

$$= 200 \text{ mg/L}$$

Note: *Total fixed suspended solids* (TFSS) is the difference between the total volatile suspended solids (TVSS) and the total suspended solids (TSS) concentrations.

$$\text{TFSS (mg/L)} = \text{TTS} - \text{TVSS} \quad (15.11)$$

Example 15.5

Problem: Using the data provided below, calculate the total fixed suspended solids:

Total Fixed Suspended Solids = 202 mg/L

Total Volatile Suspended Solids = 200 mg/L

Solution:

Total Fixed Suspended Solids, mg/L
= 202 mg/L − 200 mg/L = 2 mg/L

CONDUCTIVITY TESTING

Conductivity is a measure of the capacity of water to pass an electrical current. Conductivity in water is affected by the presence of inorganic dissolved solids such as chloride, nitrate, sulfate, and phosphate anions (ions that carry a negative charge) or sodium, magnesium, calcium, iron, and aluminum cations (ions that carry a positive charge). Organic compounds like oil, phenol, alcohol, and sugar do not conduct electrical current very well and therefore have low conductivity when in water. Conductivity is also affected by temperature: the warmer the water, the higher the conductivity.

Conductivity in streams and rivers is primarily affected by the geology of the area through which the water flows. Streams that run through areas with granite bedrock tend to have lower conductivity because granite is composed of more inert materials that do not ionize (dissolve into ionic components) when washed into the water. On the other hand, streams that run through areas with clay soils tend to have higher conductivity, because of the presence of materials that ionize when washed into the water. Groundwater inflows can have the same effects, depending on the bedrock they flow through.

Discharges to streams can change the conductivity depending on their makeup. A failing sewage system would raise the conductivity because of the presence of chloride, phosphate, and nitrate, while an oil spill would lower conductivity.

The basic unit of measurement of conductivity is the mho or siemens. Conductivity is measured in micromhos per centimeter (μmhos/cm) or microsiemens per centimeter (μs/cm). Distilled water has conductivity in the range of 0.5–3 μmhos/cm. The conductivity of rivers in the United States generally ranges from 50 to 1,500 μmhos/cm. Studies of inland freshwaters indicate that streams supporting good mixed fisheries have a range between 150 and 500 μmhos/cm. Conductivity outside this range could indicate that the water is not suitable for certain species of fish or macroinvertebrates. Industrial waters can range as high as 10,000 μmhos/cm.

Sampling, Testing, and Equipment Considerations

Conductivity is useful as a general measure of source water quality. Each stream tends to have a relatively constant range of conductivity that, once established, can be used as a baseline for comparison with regular conductivity measurements. Significant changes in conductivity could indicate that a discharge or some other source of pollution has entered a stream. The conductivity test is not routine in potable water treatment, but when performed on source water, it is a good indicator of contamination. Conductivity readings can also be used to indicate wastewater contamination or saltwater intrusion.

Note: Distilled water used for potable water analyses at public water supply facilities must have a conductivity of no more than 1 μmhos/cm.

Conductivity is measured with a probe and a meter. Voltage is applied between two electrodes in a probe immersed in the sample water. The drop in voltage caused by the resistance of the water is used to calculate the conductivity per centimeter. The meter converts the probe measurement to micromhos per centimeter (μmhos/cm) and displays the result for the user.

Note: Some conductivity meters can also be used to test for total dissolved solids and salinity. The total dissolved solids concentration in milligrams per liter (mg/L) can also be calculated by multiplying the conductivity result by a factor between 0.55 and 0.9, which is empirically determined, see *Standard Methods,* Methods #2510 (APHA, 1998).

Suitable conductivity meters cost about \$350. Meters in this price range should also measure temperature and automatically compensate for temperature in the conductivity reading. Conductivity can be measured in the field or the lab. In most cases, collecting samples in the field and taking them to a lab for testing is probably better. In this way, several teams can collect samples simultaneously. If testing in the field is important, meters designed for field use can be obtained for around the same cost mentioned above. If samples will be collected in the field for later measurement, the sample bottle should be a glass or polyethylene bottle that has been washed in phosphate-free detergent and rinsed thoroughly with both tap and distilled water. Factory-prepared Whirl-pak® bags may be used.

TOTAL ALKALINITY

Alkalinity is defined as the ability of water to resist a change in pH when acid is added; it relates to the pH buffering capacity of the water. Almost all natural waters have some alkalinity. These alkaline compounds in the water, such as bicarbonates (baking soda is one type), carbonates, and hydroxides, remove H^+ ions and lower the acidity of the water (which means increased pH). They usually do this by combining with the H^+ ions to make new compounds. Without this acid-neutralizing capacity, any acid added to a stream would cause an immediate change in the pH. Measuring alkalinity is important in determining a stream's ability to neutralize acidic pollution from rainfall or wastewater—one of the best measures of the sensitivity of the stream to acid inputs. Alkalinity in streams is influenced by rocks and soils, salts, certain plant activities, and certain industrial wastewater discharges.

Total alkalinity is determined by measuring the amount of acid (e.g., sulfuric acid) needed to bring the sample to a pH of 4.2. At this pH, all the alkaline compounds in the sample are "used up." The result is reported as milligrams per liter of calcium carbonate (mg/L $CaCO_3$).

Testing for alkalinity in potable water treatment is most important with regard to its relation to coagulant addition; that is, it is important that there exists enough natural alkalinity in the water to buffer chemical acid addition so that floc formation will be optimal and turbidity removal can proceed. In water softening, proper chemical dosage will depend on the type and amount of alkalinity in the water. For corrosion control, the presence of adequate alkalinity in a water supply neutralizes any acid tendencies and prevents it from becoming corrosive.

Analytical and Equipment Considerations

For total alkalinity, a double endpoint titration using a pH meter (or pH "pocket pal") and a digital titrator or burette is recommended. This can be done in the field or the lab. If alkalinity must be analyzed in the field, a digital titrator should be used instead of a burette because burettes are fragile and more difficult to set up. The alkalinity method described below was developed by the Acid Rain Monitoring

Project of the University of Massachusetts Water Resources Research Center (River Watch Network, 1992).

Burettes, Titrators, and Digital Titrators for Measuring Alkalinity

The total alkalinity analysis involves titration. In this test, titration is the addition of small, precise quantities of sulfuric acid (the reagent) to the sample until the sample reaches a certain pH (known as an endpoint). The amount of acid used corresponds to the total alkalinity of the sample. Alkalinity can be measured using a burette, titrator, or digital titrator (described below).

1. A burette is a long, graduated glass tube with a tapered tip like a pipette and a valve that opens to allow the reagent to drop out of the tube. The amount of reagent used is calculated by subtracting the original volume in the burette from the column left after the endpoint has been reached. Alkalinity is calculated based on the amount used.
2. Titrators forcefully expel the reagent by using a manual or mechanical plunger. The amount of reagent used is calculated by subtracting the original volume in the titrator from the volume left after the endpoint has been reached. Alkalinity is then calculated based on the amount used or is read directly from the titrator.
3. Digital titrators have counters that display numbers. A plunger is forced into a cartridge containing the reagent by turning a knob on the titrator. As the knob turns, the counter changes in proportion to the amount of reagent used. Alkalinity is then calculated based on the amount used. Digital titrators cost approximately \$100.

Digital titrators and burettes allow for much more precision and uniformity in the amount of titrant that is used.

FECAL COLIFORM BACTERIA TESTING

Much of the information in this section is from USEPA (1985). Fecal coliform bacteria are non-disease-causing organisms that are found in the intestinal tract of all warm-blooded animals. Each discharge of body wastes contains large amounts of these organisms. The presence of fecal coliform bacteria in a stream or lake indicates the presence of human or animal wastes. The number of fecal coliform bacteria present is a good indicator of the amount of pollution in the water. EPA's 2001 Total Coliform Rule 816-F-01-035:

1. Is intended to improve public health protection by reducing fecal pathogens to minimal levels through the control of total coliform bacteria, including fecal coliforms and *Escherichia coli (E. coli)*.
2. Establishes a maximum contaminant level (MCL) based on the presence or absence of total coliforms, modifies monitoring requirements including

testing for fecal coliforms or *E. coli,* requires the use of a sample siting plan, and also requires sanitary surveys for systems collecting fewer than five samples per month.

3. Applies to all public water systems.
4. Has resulted in a reduction in the risk of illness from disease-causing organisms associated with sewage or animal wastes. Disease symptoms may include diarrhea, cramps, nausea, and possibly jaundice, along with associated headaches and fatigue.

Fecal coliforms are used as indicators of possible sewage contamination because they are commonly found in human and animal feces. Although they are not generally harmful themselves, they indicate the possible presence of pathogenic (disease-causing) bacteria and protozoans that also live in human and animal digestive systems. Their presence in streams suggests that pathogenic microorganisms might also be present and that swimming in and/or eating shellfish from the waters might present a health risk. Since testing directly for the presence of a large variety of pathogens is difficult, time-consuming, and expensive, water is usually tested for coliforms and fecal streptococci instead. Sources of fecal contamination to surface waters include wastewater treatment plants, on-site septic systems, domestic and wild animal manure, and storm runoff. In addition to the possible health risks associated with the presence of elevated levels of fecal bacteria, they can also cause cloudy water, unpleasant odors, and an increased oxygen demand.

Note: In addition to the most commonly tested fecal bacteria indicators, total coliforms, fecal coliforms, and E. coli—fecal streptococci and enterococci are also commonly used as bacteria indicators. The focus of this presentation is on total coliforms and fecal coliforms.

Fecal coliforms are widespread in nature. All members of the total coliform group can occur in human feces, but some can also be present in animal manure, soil, submerged wood, and other places outside the human body. The usefulness of total coliforms as an indicator of fecal contamination depends on the extent to which the bacterial species found are fecal and human in origin. For recreational waters, total coliforms are no longer recommended as an indicator. For drinking water, total coliforms are still the standard test because their presence indicates contamination of a water supply by an outside source.

Fecal coliforms, a subset of total coliform bacteria, are more fecal-specific in origin. However, even this group contains a genus, *Klebsiella,* with species that are not necessarily fecal in origin. *Klebsiella* are commonly associated with textile and pulp and paper mill wastes. If these sources discharge to a local stream, consideration should be given to monitoring more fecal and human-specific bacteria. For recreational waters, this group was the primary bacteria indicator until relatively recently, when USEPA began recommending *E. coli* and enterococci as better indicators of health risk from water contact. Fecal coliforms are still being used in many states as indicator bacteria.

USEPA's Total Coliform Rule

Under EPA's Total Coliform Rule, sampling requirements are specified as follows:

Routine Sampling Requirements

1. Total coliform samples must be collected at sites that are representative of water quality throughout the distribution system according to a written sample siting plan subject to state review and revision.
2. Samples must be collected at regular time intervals throughout the month, except groundwater systems serving 4,900 persons or fewer may collect them on the same day.
3. Monthly sampling requirements are based on the population served (see Table 15.9 for the minimum sampling frequency).

TABLE 15.9

Public Water System ROUTINE Monitoring Frequencies

Population	Minimum Samples/Month
25–1,000[a]	1
1,001–2,500	2
2,501–3,300	3
3,301–4,100	4
4,101–4,900	5
4,901–5,800	6
5,801–6,700	7
6,701–7,600	8
7,601–8,500	9
8,501–12,900	10
12,901–17,200	15
17,201–21,500	20
21,501–25,000	25
25,001–33,000	30
33,001–41,000	40
41,001–50,000	50
50,001–59,000	60
59,001–70,000	70
70,000–83,000	80
83,001–96,000	90
96,001–130,000	100
130,000–220,000	120
220,001–320,000	150
320,001–450,000	180
450,001–600,000	210
600,001–780,000	240
780,001–970,000	270
970,001–1,230,000	330
1,520,001–1,850,000	360
1,850,001–2,270,000	390
2,270,001–3,020,000	420
3,020,001–3,960,000	450
≥3,960,001	480

[a] Includes PWSs, which have at least 15 service connections, but serve <25 people.

4. A reduced monitoring frequency may be available for systems serving 1,000 persons or fewer and using only groundwater if a sanitary survey conducted within the past 5 years shows the system is free of sanitary defects (the frequency may be no less than 1 sample/quarter for community systems and 1 sample/year for non-community systems).
5. Each total coliform-positive routine sample must be tested for the presence of fecal coliforms or E. coli.

Repeat Sampling Requirements

1. Within 24 h of learning of a total coliform-positive ROUTINE sample result, at least three REPEAT samples must be collected and analyzed for total coliforms.
2. One REPEAT sample must be collected from the same tap as the original sample.
3. One REPEAT sample must be collected within five service connections upstream.
4. One REPEAT sample must be collected within five service connections downstream.
5. Systems that collect one ROUTINE sample per month or fewer must collect a fourth REPEAT sample.
6. If any REPEAT sample is total coliform-positive:
7. The system must analyze the total coliform-positive culture for fecal coliforms or E. coli.
8. The system must collect another set of REPEAT samples, as before, unless the MCL has been violated and the system has notified the state.

Additional Routine Sample Requirements

A positive *routine* or *repeat* total coliform result requires a minimum of five *routine* samples to be collected the following month that the system provides water to the public unless waived by the state.

Other Total Coliform Rule Provisions

1. Systems collecting fewer than five ROUTINE samples per month must have a sanitary survey every 5 years (or every 10 years if it is a non-community water system using protected and disinfected groundwater).
2. Systems using surface water or groundwater under the direct influence of surface water (GWUDI) and meeting filtration avoidance criteria must collect and analyze one coliform sample each day when the turbidity of the source water exceeds 1 NTU. This sample must be collected from a tap near the first service connection.

Compliance

Compliance is based on the presence or absence of total coliforms. Moreover, compliance is determined each calendar month the system serves water to the public (or each calendar month that sampling occurs for systems on reduced monitoring). The results of routine and REPEAT samples are used to calculate compliance.

Regarding violations, a monthly MCL violation is triggered if a system collecting fewer than 40 samples per month has more than one ROUTINE/REPEAT sample per month that is total coliform-positive. In addition, a system collecting at least 40 samples per month has more than 5.0% of the ROUTINE/REPEAT samples in a month that are total coliform-positive, which is technically in violation of the Total Coliform Rule. An acute MCL violation is triggered if any public water system has any fecal coliform- or E. coli-positive REPEAT sample or has a fecal coliform- or E. coli-positive ROUTINE sample followed by a total coliform-positive REPEAT sample.

The Total Coliform Rule also has requirements for public notification and reporting. For example, for a monthly MCL violation, the violation must be reported to the state no later than the end of the next business day after the system learns of the violation. The public must be notified within 14 days. For an acute MCL violation, the violation must be reported to the state no later than the end of the next business day after the system learns of the violation. The public must be notified within 72 h. Systems with ROUTINE or REPEAT samples that are fecal coliform- or E. coli-positive must notify the state by the end of the day they are notified of the result or by the end of the next business day if the state office is already closed.

Sampling and Equipment Considerations

Bacteria can be difficult to sample and analyze for many reasons. Natural bacteria levels in streams can vary significantly; bacterial conditions are strongly correlated with rainfall, making the comparison of wet and dry weather bacteria data problematic. Many analytical methods have a low level of precision yet can be quite complex to accomplish, and absolutely sterile conditions are essential to maintain while collecting and handling samples. The primary equipment decision to make when sampling for bacteria is what type and size of sample container you will use. Once you have made that decision, the same straightforward collection procedure is used, regardless of the type of bacteria being monitored.

When monitoring bacteria, it is critical that all containers and surfaces with which the sample will come into contact be sterile. Containers made of either some form of plastic or Pyrex glass are acceptable to the USEPA. However, if the containers are to be reused, they must be sturdy enough to survive sterilization using heat and pressure. The containers can be sterilized by using an autoclave, a machine that sterilizes with pressurized steam. If using an autoclave, the container material must be able to withstand high temperatures and pressure. Plastic containers—either high-density polyethylene or polypropylene—might be preferable to glass from a practical standpoint because they will better withstand breakage. In any case, be sure to check the manufacturer's specifications to see whether the container can withstand 15 min in an autoclave at a temperature of 121°C without melting (extreme caution is advised when working with an autoclave). Disposable, sterile plastic Whirl-Pak® bags are used by several programs. The size of the container depends on the sample amount needed for the bacteria analysis method you choose and the amount needed for other analyses. The two basic methods for analyzing water samples for bacteria in common use are membrane filtration and multiple tube fermentation methods (described later).

Given the complexity of the analysis procedures and the equipment required, field analysis of bacteria is not recommended. Bacteria can either be analyzed by the volunteer at a well-equipped lab or sent to a state-certified lab for analysis. If you send a bacteria sample to a private lab, make sure that the lab is certified by the state for bacteria analysis. Consider state water quality labs, university and college labs, private labs, wastewater treatment plant labs, and hospitals. You might need to pay these labs for analysis. On the other hand, if you have a modern lab with the proper equipment and properly trained technicians, the fecal coliform testing procedures described in the following section will be helpful. A note of caution: if you decide to analyze your samples in your own lab, be sure to carry out a quality assurance/quality control program.

Fecal Coliform Testing

Federal regulations cite two approved methods for the determination of fecal coliform in water: (1) the multiple tube fermentation or most probable number (MPN) procedure; and (2) the membrane filter (MF) procedure.

Note: Because the MF procedure can yield low or highly variable results for chlorinated wastewater, the USEPA requires verification of results using the MPN procedure to resolve any controversies. However, do not attempt to perform the fecal coliform test using the summary information provided in this handbook. Instead, refer to the appropriate reference cited in the Federal Regulations for a complete discussion of these procedures.

Basic Equipment and Techniques

Whenever microbiological testing of water samples is performed, certain general considerations and techniques will be required. Because these are basically the same for each test procedure, they are reviewed here prior to the discussion of the two methods.

1. **Reagents and Media:** All reagents and media utilized in performing microbiological tests on water samples must meet the standards specified in the references cited in the Federal Regulations.
2. **Reagent Grade Water:** Deionized water that is tested annually and found to be free of dissolved metals and bactericidal or inhibitory compounds is preferred for use in preparing culture media and test reagents, although distilled water may be used.
3. **Chemicals:** All chemicals used in fecal coliform monitoring must be ACS reagent grade or equivalent.

4. **Media:** To ensure uniformity in the test procedures, the use of dehydrated media is recommended. Sterilized, prepared media in sealed test tubes, ampoules, or dehydrated media pads are also acceptable for use in this test.
5. **Glassware and Disposable Supplies:** All glassware, equipment, and supplies used in microbiological testing should meet the standards specified in the references cited in the Federal Regulations.

Sterilization

All glassware used for bacteriological testing must be thoroughly cleaned using a suitable detergent and hot water. The glassware should be rinsed with hot water to remove all traces of residual detergent and, finally, rinsed with distilled water. Laboratories should use a detergent certified to meet bacteriological standards or, at a minimum, rinse all glassware after washing with two tap water rinses followed by five distilled water rinses. For the sterilization of equipment, either a hot air sterilizer or an autoclave can be used. When using the hot air sterilizer, all equipment should be wrapped in high-quality (Kraft) paper or placed in containers prior to hot air sterilization. All glassware, except those in metal containers, should be sterilized for a minimum of 60 min at 170°C. Sterilization of glassware in metal containers should require a minimum of 2 h. Hot air sterilization cannot be used for liquids. An autoclave can be used to sterilize sample bottles, dilution water, culture media, and glassware by autoclaving at 121°C for 15 min.

Sterile Dilution Water Preparation

The dilution water used for making sample serial dilutions is prepared by adding 1.25 mL of stock buffer solution and 5.0 mL of magnesium chloride solution to 1,000 mL of distilled or deionized water. The stock solutions of each chemical should be prepared as outlined in the reference cited by the Federal Regulations. The dilution water is then dispensed in sufficient quantities to produce 9 or 99 mL in each dilution bottle following sterilization. If the membrane filter procedure is used, additional 60- to 100-mL portions of dilution water should be prepared and sterilized to provide the rinse water required by the procedure.

Serial Dilution Procedure

At times, the density of the organisms in a sample makes it difficult to accurately determine the actual number of organisms in the sample. When this occurs, the sample size may need to be reduced to as small as one millionth of a milliliter. In order to obtain such small volumes, a technique known as serial dilutions has been developed.

Bacteriological Sampling

To obtain valid test results that can be utilized in evaluating the process efficiency of water quality, proper technique, equipment, and sample preservation are critical. These factors are especially important in bacteriological sampling:

1. **Sample Dechlorination:** When samples of chlorinated effluents are to be collected and tested, the sample must be dechlorinated. Prior to sterilization, place enough sodium thiosulfate solution (10%) in a clean sample container to produce a concentration of 100 mg/L in the sample (for a 120-mL sample bottle, 0.1 mL is usually sufficient). Sterilize the sample container as previously described.

2. **Sample Procedure:**
   A. Keep the sample bottle unopened after sterilization until the sample is to be collected.
   B. Remove the bottle stopper and hood or cap as one unit. Do not touch or contaminate the cap or the neck of the bottle.
   C. Submerge the sample bottle in the water to be sampled.
   D. Fill the sample bottle approximately ¾ full, but not less than 100 mL.
   E. Aseptically replace the stopper or cap on the bottle.
   F. Record the date, time, and location of sampling, as well as the sampler's name and any other descriptive information pertaining to the sample.

3. **Sample Preservation and Storage:** Examination of bacteriological water samples should be performed immediately after collection. If testing cannot be started within 1 h of sampling, the sample should be iced or refrigerated at 4°C or less. The maximum recommended holding time for fecal coliform samples from wastewater is 6 h.

The storage temperature and holding time should be recorded as part of the test data.

Multiple-Tube Fermentation Technique

The multiple fermentation technique for fecal coliform testing is useful in determining the fecal coliform density in most water, solid, or semisolid samples. Wastewater testing typically requires the use of presumptive and confirming test procedures. This method is recognized as the preferred choice for any samples that may be controversial (enforcement-related). The technique is based on the most probable number of bacteria present in a sample that produces gas in a series of fermentation tubes with various volumes of diluted samples. The MPN is obtained from charts based on statistical studies of known concentrations of bacteria.

The technique utilizes a two-step incubation procedure (see Figure 15.16). The sample dilutions are first incubated in lauryl (sulfonate) tryptose broth for 24–48 h (presumptive test). Positive samples are then transferred to EC broth and incubated for an additional 24 h (confirming test). Positive samples from this second incubation are used to statistically determine the MPN from the appropriate reference chart. A single-media, 24-h procedure is also acceptable. In this procedure, sample dilutions are inoculated in A-1 media and incubated for 3 h at 35°C, followed by incubation for the remaining 20 h at 44.5°C. Positive samples from these inoculations are then used to statistically determine the MPN value from the appropriate chart.

Fecal Coliform MPN Presumptive Test Procedure

The procedure for the fecal coliform MPN presumptive test is described below:

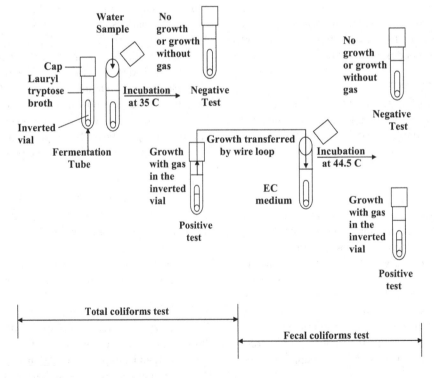

FIGURE 15.16 Multiple-tube fermentation technique.

1. Prepare dilutions and inoculate five fermentation tubes for each dilution.
2. Cap all tubes and transfer them to the incubator.
3. Incubate for 24+2 h at 35°C±0.5°C.
4. Examine the tubes for gas:
   A. Gas present=Positive test – transfer
   B. No gas=Continue incubation
5. Incubate for a total of 48±3 h at 35°C±0.5°C
6. Examine the tubes for gas:
   A. Gas present=Positive test – transfer
   B. No gas=Negative test

Note: Keep in mind that the fecal coliform MPN confirming procedure using A-1 broth is used to determine the MPN/100 mL. The MPN procedure for fecal coliform determinations requires a minimum of three dilutions with five tubes per dilution.

Calculation of Most Probable Number (MPN)/100 mL
Calculation of the MPN test results requires the selection of a valid series of three consecutive dilutions. The number of positive tubes in each of the three selected dilution inoculations is used to determine the MPN/100 mL. In selecting the dilution inoculations for the calculation, each dilution is expressed as a ratio of positive tubes to tubes inoculated in the dilution, i.e., three positive out of five inoculated (3/5). There are several rules to follow in determining the most valid series of dilutions. In the following examples, four dilutions were used for the test:

1. Using the confirming test data, select the highest dilution showing all positive results (with no lower dilution showing less than all positive) and the next two higher dilutions.
2. If a series shows all negative values with the exception of one dilution, select the series that places the only positive dilution in the middle of the selected series.
3. If a series shows a positive result in a dilution higher than the selected series (using rule #1), it should be incorporated into the highest dilution of the selected series.

After selecting the valid series, the MPN/1,000 mL is determined by locating the selected series on the MPN reference chart. If the selected dilution series matches the dilution series on the reference chart, the MPN value from the chart is the reported value for the test. If the dilution series used for the test does not match the dilution series on the chart, the test result must be calculated.

$$MPN/10 \text{ mL} = MPN_{chart} \times \frac{\text{sample vol. in 1st dilution}_{chart}}{\text{sample vol. In 1st dilution}_{sample}}$$

(15.12)

Membrane Filtration Technique
The membrane filtration technique can be useful for determining the fecal coliform density in wastewater effluents, except for primary treated wastewater that has not been chlorinated or wastewater containing toxic metals or phenols. Chlorinated secondary or tertiary effluents may be tested using this method, but results are subject to verification by the MPN technique. The membrane filter technique utilizes a specially designed filter pad with uniformly sized pores (openings) that are small enough to prevent bacteria from entering the filter (see Figure 15.17). Another unique characteristic of the filter allows liquids, such as the media, placed under the filter to pass upward through the filter to provide the nourishment required for bacterial growth.

Note: In the membrane filter method, the number of colonies grown estimates the number of coliforms.

Membrane Filter Procedure
The procedure for the membrane filter method is described below:

1. Sample Filtration
   A. Select a filter and aseptically separate it from the sterile package.
   B. Place the filter on the support plate with the grid side up.
   C. Place the funnel assembly on the support; secure it as needed.

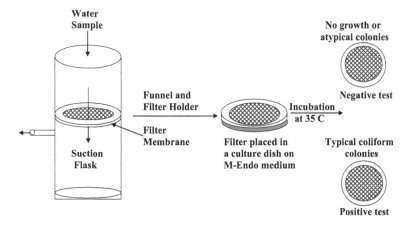

FIGURE 15.17 Membrane filter technique.

D. Pour 100 mL of sample or serial dilution onto the filter, then apply vacuum.

Note: The sample size and/or necessary serial dilution should produce a growth of 20–60 fecal coliform colonies on at least one filter. The selected dilutions must also be capable of showing permit excursions.

E. Allow all of the liquid to pass through the filter.
F. Rinse the funnel and filter with three portions (20–30 mL) of sterile, buffered dilution water. (Allow each portion to pass through the filter before adding the next.)

Note: Filtration units should be sterile at the start of each filtration series and should be sterilized again if the series is interrupted for 30 min or more. A rapid interim sterilization can be accomplished by exposing the units to ultraviolet (UV) light, flowing steam, or boiling water for 2 minutes.

2. Incubation
   A. Place an absorbent pad into the culture dish using sterile forceps.
   B. Add 1.8–2.0 mL of M-FC media to the absorbent pad.
   C. Discard any media not absorbed by the pad.
   D. Filter the sample through the sterile filter.
   E. Remove the filter from the assembly, and place it on the absorbent pad (grid side up).
   F. Cover the culture dish.
   G. Seal the culture dishes in a weighted plastic bag.
   H. Incubate the filters in a water bath for 24 h at 44.5°C ± 0.2°C.

Colony Counting

Upon completion of the incubation period, the surface of the filter will display growths of both fecal coliform and non-fecal coliform bacterial colonies. Fecal coliform colonies will appear blue in color, while non-fecal coliform colonies will appear gray or cream-colored. When counting the colonies, the entire surface of the filter should be scanned using a 10× to 15× binocular, wide-field dissecting microscope. The desired range for valid fecal coliform determination is 20–60 colonies per filter. If multiple sample dilutions are used for the test, counts for each filter should be recorded on the laboratory data sheet.

1. **Too Many Colonies:** Filters that show growth over the entire surface, without individually identifiable colonies, should be recorded as "confluent growth." Filters with a very high number of colonies (greater than 200) should be recorded as TNTC (too numerous to count).
2. **Not Enough Colonies:** If no single filter meets the desired minimum colony count (20 colonies), the

sum of the individual filter counts and the respective sample volumes can be used to calculate the colonies/100 mL.

Note: In each case, adjustments in sample dilution volumes should be made to ensure future tests meet the criteria for obtaining a valid test result.

Calculations

The fecal coliform density can be calculated using the following formula:

$$\text{Colonies}/100\text{ mL} = \frac{\text{Colonies counted}}{\text{Sample volume, mL}} \times 100 \text{ mL} \quad (15.13)$$

Note: The USEPA criterion for fecal coliform bacteria in bathing waters is a logarithmic mean of 200/100 mL, based on a minimum of five samples taken over a 30-day period, with no more than 10% of the total samples exceeding 400/100 mL. Because shellfish may be eaten without being cooked, the strictest coliform criterion applies to shellfish cultivation and harvesting. The USEPA criterion states that the mean fecal coliform concentration should not exceed 14/100 mL, with no more than 10% of the samples exceeding 43/100 mL.

Interferences

Large amounts of turbidity, algae, or suspended solids may interfere with this technique by blocking the filtration of the sample through the membrane filter. Dilution of these samples to prevent this problem may make the test inappropriate for samples with low fecal coliform densities because the sample volumes after dilution may be too small to provide representative results. The presence of large amounts of non-coliform bacteria in the samples may also prohibit the use of this method.

Note: Many NPDES discharge permits require fecal coliform testing. Results for fecal coliform testing must be reported as a geometric mean (average) of all the test results obtained during a reporting period. A geometric mean, unlike an arithmetic mean or average, dampens the effect of very high or low values that might otherwise cause a non-representative result.

APPARENT COLOR TESTING/ANALYSIS

Color in water often originates from organic sources: decomposition of leaves and other forest debris such as bark, pine needles, etc. Tannin and lignin, organic compounds, dissolve in water. Some organics bond with iron to produce soluble color compounds. Biodegrading algae from recent blooms may cause significant color. Though less likely a source of color in water, possible inorganic sources of color include salts of iron, copper, and potassium permanganate added in excess at the treatment plant.

Note: Noticeable color is an objectionable characteristic that makes the water psychologically unacceptable to the consumer.

Recall that *true color* is dissolved. It is measured colorimetrically and compared against an EPA color standard. Apparent color may be caused by suspended material (turbidity) in the water. It is important to point out that even though it may also be objectionable in the water supply, it is not meant to be measured in the color analysis or test. Probably the most common cause of apparent color is particulate oxidized iron.

By using established color standards, people in different areas can compare test results. Over the years, several attempts have been made to standardize the method of describing the "apparent" color of water using comparisons to color standards. *Standard Methods* (APHA, 1998) recognize the visual comparison method as a reliable method of analyzing water from the distribution system. One of the visual comparison methods is the Forel-Ule Color Scale, consisting of a dozen shades ranging from deep blue to khaki green, typical of offshore and coastal bay waters. By using established color standards, people in different areas can compare test results. Another visual comparison method is the Borger Color System, which provides an inexpensive, portable color reference for shades typically found in natural waters; it can also be used for its original purpose—describing the colors of insects and larvae found in streams or lakes. The Borger Color System also allows the recording of the color of algae and bacteria on streambeds. To ensure reliable and accurate descriptions of apparent color, use a system of color comparison that is reproducible and comparable to the systems used by other groups.

Note: Do not leave color standard charts and comparators in direct sunlight.

Measured levels of color in water can serve as indicators for a number of conditions. For example, transparent water with a low accumulation of dissolved minerals and particulate matter usually appears blue and indicates low productivity. Yellow to brown color normally indicates that the water contains dissolved organic materials, humic substances from soil, peat, or decaying plant material. Deeper yellow to reddish colors indicate some algae and dinoflagellates. A variety of yellows, reds, browns, and grays are indicative of soil runoff.

Note: Color by itself has no health significance in drinking water. A secondary MCL is set at 15 color units, and it is recommended that community supplies provide water that has less color.

When treating color in water, alum and ferric coagulation are often effective. They remove apparent color and often much of the true color. Oxidation of color-causing compounds to a non-colored version is sometimes effective. Activated carbon treatment may adsorb some of the organics causing color. For apparent color problems, filtration is usually effective in trapping the colored particles.

Odor Analysis of Water

Odor is expected in wastewater—the fact is, any water-containing waste, especially human waste, has a detectable

TABLE 15.10
Descriptions of Odors

Nature of Odor	Description	Examples
Aromatic	Spicy	Camphor, cloves, lavender
Balsamic	Flowery	Geranium, violet, vanilla
Chemical	Industrial wastes or treatments chlorinous	Chlorine
	Hydrocarbon	Oil refinery wastes
	Medicinal	Phenol and iodine
	Sulfur	Hydrogen sulfide
Disagreeable	Fishy	Dead algae
	Pigpen	Algae
	Septic	Stale sewage
Earthy	Damp earth	
	Peaty	Peat
Grassy	Crushed grass	Musty
	Decomposing straw	Moldy
Vegetable	Root vegetables	Damp cellar

Source: Adapted from APHA (1998).

(expected) odor associated with it. *Odor* in a raw water source (for potable water) is caused by several constituents. For example, chemicals that may come from municipal and industrial waste discharges, or natural sources such as decomposing vegetable matter or microbial activity, may cause odor problems. Odor affects the acceptability of drinking water, the aesthetics of recreational water, and the taste of aquatic foodstuffs.

The human nose can accurately detect a wide variety of smells, making it the best odor-detection and testing device presently available. To measure odor, collect a sample in a large-mouthed jar. After waving off the air above the water sample with your hand, smell the sample. Use the list of odors provided in Table 15.10—a system of qualitative description that helps monitors describe and record detected odors. Record all observations (see *Standard Methods*).

When treating for odor in water, removal depends upon the source of the odor. Some organic substances that cause odor can be removed with powdered activated carbon. If the odor is of gaseous origin, scrubbing (aeration) may remove it. Some odor-causing chemicals can be oxidized to odorless chemicals with chlorine, potassium permanganate, or other oxidizers. Settling may remove some material which, when later dissolved in the water, may have potential odor-causing capacity. Unfortunately, the test for odor in water is subjective—there is no scientific means of measurement, and it is not very accurate.

To test odor in water intended for potable use, a sample is generally heated to 60°C. Odor is observed and recorded. A threshold odor number (TON) is assigned. TON is found using the following equation:

$$\text{TON} = \frac{\text{Total Volume of Water Sample}}{\text{Lowest Sample Volume with Odor}} \quad (15.14)$$

CHLORINE RESIDUAL TESTING/ANALYSIS

Chlorination is the most widely used means of disinfecting water in the U.S. When chlorine gas is dissolved in (pure) water, it forms hypochlorous acid, hypochlorite ion, and hydrogen chloride (hydrochloric acid). The total concentration of HOCl and OCl ions is known as *free chlorine residual*. Currently, federal regulations cite six approved methods for the determination of total residual chlorine (TRC).

1. **DPD:** spectrophotometric
2. **Titrimetric:** amperometric direct
3. **Titrimetric:** iodometric direct
4. **Titrimetric:** iodometric back
   A. **Starch Iodine Endpoint:** iodine titrant
   B. **Starch Iodine Endpoint:** iodate titrant
5. Amperometric endpoint
6. DPD-FAS titration
7. Chlorine electrode

All of these test procedures are approved methods and, unless prohibited by the plant's NPDES discharge permit, can be used for effluent testing. Based on the most popular method usage in the United States, discussion is limited to

1. DPD-spectrophotometric
2. DPD-FAS titration
3. Titrimetric—amperometric direct

Note: Treatment facilities required to meet "non-detectable" total residual chlorine limitations must use one of the test methods specified in the plant's NPDES discharge permit.

For information on any of the other approved methods, refer to the appropriate reference cited in the federal regulations.

DPD-Spectrophotometry

Diethyl-*p*-phenylenediamine (DPD) reacts with chlorine to form a red color. The intensity of the color is directly proportional to the amount of chlorine present. This color intensity is measured using a colorimeter or spectrophotometer. The meter reading can be converted to a chlorine concentration using a graph developed by measuring the color intensity produced by solutions with precisely known concentrations of chlorine. In some cases, spectrophotometers or colorimeters are equipped with scales that display chlorine concentration directly. In these cases, there is no requirement to prepare a standard reference curve. If the direct reading colorimeter is not used, the chemicals required include:

1. Potassium dichromate solution 0.100 N
2. Potassium iodide crystals
3. Standard ferrous ammonium sulfate solution 0.00282 N
4. Concentrated phosphoric acid
5. Sulfuric acid solution (1+5)
6. Barium diphenylamine sulfonate 0.1%

If an indicator is not used, a DPD indicator and phosphate buffer (DPD prepared indicator and buffer+indicator together) are required.

To conduct the test, a direct readout colorimeter designed to meet the test specifications, or a spectrophotometer (wavelength of 515 nm and light path of at least 1 cm), or a filter photometer with a filter having maximum transmission in the wavelength range of 490–520 nm and a light path of at least 1 cm are required. In addition, for direct readout colorimeter procedures, a sample test vial is required. When the direct readout colorimeter procedure is not used, the equipment required includes:

1. 250 mL Erlenmeyer flask
2. 10 mL measuring pipettes
3. 15 mL test tubes
4. 1 mL pipettes (graduated to 0.1 mL)
5. Sample cuvettes with 1 cm light path

Note: A cuvette is a small, often tubular laboratory vessel, often made of glass.

Procedure

Note: For direct readout colorimeters, follow the procedure supplied by the manufacturer.

1. Prepare a standard curve for TRC concentrations from 0.05 to 4.0 mg/L—chlorine versus percent transmittance.

Note: Instructions on how to prepare the TRC concentration curve or a standard curve are normally included in the spectrophotometer manufacturer's operating instructions.

2. Calibrate the colorimeter in accordance with the manufacturer's instructions using a laboratory-grade water blank.
3. Add one prepared indicator packet (or tablet) of the appropriate size to match the sample volume to a clean test tube or cuvette; or
   A. Pipette 0.5 mL phosphate buffer solution.
   B. Pipette 0.5 mL DPD indicator solution.
   C. Add 0.1-g Kl (potassium iodide) crystals to a clean tube or cuvette.
4. Add 10 mL of the sample to the cuvette.
5. Stopper the cuvette, and swirl to mix the contents well.
6. Let it stand for 2 min.
7. Verify the wavelength of the spectrophotometer or colorimeter, and check and set the 0% T using the laboratory-grade water blank.
8. Place the cuvette in the instrument, read %T, and record the reading.
9. Determine mg/L TRC from the standard curve.

Note: Calculations are not required in this test because TRC, mg/L, is read directly from the meter or the graph.

DPD-FAS Titration

The amount of ferrous ammonium sulfate solution required to just remove the red color from a total residual chlorine sample that has been treated with a DPD indicator can be used to determine the concentration of chlorine in the sample. This is known as a titrimetric test procedure. The chemicals used in the test procedure include the following:

1. DPD prepared indicator (buffer and indicator together)
2. Potassium dichromate solution 0.100 N
3. Potassium iodide crystals
4. Standard ferrous ammonium sulfate solution 0.00282 N
5. Concentrated phosphoric acid
6. Sulfuric acid solution (1 + 5)
7. Barium diphenylamine sulfonate 0.1%

Note: DPD indicator and/or phosphate buffer is not required if the prepared indicator is used.

The equipment required for this text procedure includes the following:

1. 250 mL graduated cylinder
2. 5 mL measuring pipettes
3. 500 mL Erlenmeyer flask
4. 50 mL burette (graduate to 0.1 mL)
5. Magnetic stirrer and stir bars

Procedure

1. Add the contents of a prepared indicator packet (or tablet) to the Erlenmeyer flask, or
   A. Pipette 5 mL phosphate buffer solution into the Erlenmeyer flask
   A. Pipette 5 mL DPD indicator solution into the flask
   A. Add 1-g Kl crystals to the flask
2. Add 100 mL of sample to the flask.
3. Swirl the flask to mix the contents.
4. Let the flask stand for 2 min.
5. Titrate with ferrous ammonium sulfate (FAS) until the red color first disappears.
6. Record the amount of titrant.

The calculation required in this procedure is

$$\text{TRC, mg/L} = \text{mL of FAS used} \qquad (15.15)$$

In this test procedure, phenylarsine oxide is added to a treated sample to determine when the test reaction has been completed. The volume of phenylarsine oxide (PA) used can then be used to calculate the TRC. The chemicals used include:

1. Phenylarsine oxide solution 0.00564 N
2. Potassium dichromate solution 0.00564 N
3. Potassium iodide solution 5%

4. Acetate buffer solution (pH 4.0)
5. Standard arsenite solution 0.1 N

Equipment used includes

1. 250 mL graduated cylinder
2. 5 mL measuring pipettes
3. Amperometric titrator

Procedure

1. Prepare the amperometric titrator according to the manufacturer.
2. Add a 200-mL sample.
3. Place the container on the titrator stand and turn on the mixer.
4. Add 1-g Kl crystals or 1 mL Kl solution.
5. Pipette 1 mL of pH 4 (acetate) buffer into the container.
6. Titrate with 0.0056 N PAO.

When conducting the test procedure, as the downscale endpoint is neared, slow the titrant addition to 0.1-mL increments and note the titrant volume used after each increment. When no needle movement is noted, the endpoint has been reached. Subtract the final increment from the burette reading to determine the final titrant volume. For this procedure, the only calculation normally required is

$$\text{TRC, mg/L} = \text{mL PAO used} \qquad (15.16)$$

FLUORIDES

It has long been accepted that a moderate amount of fluoride ions (F^-) in drinking water contributes to good dental health—it has been added to many community water supplies throughout the U.S. to prevent dental caries in children's teeth. Fluoride is seldom found in appreciable quantities in surface waters and appears in groundwater in only a few geographical regions. Fluorides are used to make ceramics and glass. Fluoride is toxic to humans in large quantities and to some animals. The chemicals added to potable water in treatment plants are:

NaF: Sodium fluoride, solid

Na_2SiF_6: Sodium silicofluoride, solid

H_2SiF_6: Hexafluorosilicic acid; most widely used

Analysis of the fluoride content of water can be performed using the colorimetric method. In this test, fluoride ion reacts with zirconium ion and produces zirconium fluoride, which bleaches an organic red dye in direct proportion to its concentration. This can be compared to standards and read calorimetrically.

THE BOTTOM LINE

By understanding the role of biomonitoring, monitoring, sampling, and testing, water and wastewater operators can better plan what individual capabilities they should use and when.

CHAPTER REVIEW QUESTIONS

15.1 Explain (in simple terms) the methods involved in analyzing total coliforms in an effluent sample.

15.2 How soon after the sample is collected must the pH be tested?

15.3 What is a grab sample?

15.4 When is it necessary to use a grab sample?

15.5 What is a composite sample?

15.6 List three sample rules for sample collection.

15.7 What is the acceptable preservation method for suspended solids samples?

15.8 Most solids test methods are based upon weight changes. What can cause changes in weight during the testing procedure?

REFERENCES

APHA. 1971. *Standard Methods for the Examination of Water and Wastewater*, 17th ed. Washington, DC: APHA.

APHA. 1998. *Standard Methods for the Examination of Water and Wastewater*, 20th ed. Washington, DC: American Water Works Association.

AWRI. 2000. *Plankton Sampling*. Allendale, MI: Robert Annis Water Resource Institute, Grand Valley state University.

AWWA. 1995. *Water Treatment*, 2nd ed. Denver: American Water Works Association.

Bahls, L.L., 1993. *Periphyton Bioassessment Methods for Montana Streams*. Helena, MT: Montana Water Quality Bureau, Department of Health and Environmental Science.

Barbour, M.T., Gerritsen, J., Snyder, B.D., and Stibling, J.B., 1997. *Revision to Rapid Bioassessment Protocols for Use in Streams and Rivers, Periphytons, Benthic Macroinvertebrates, and Fish*. Washington, DC: United States Environmental Protection Agency.

Bly, T.D., and Smith, G.F., 1994. *Biomonitoring Our Streams: What's It All About?* Nashville, TN: U.S. Geological Survey.

Botkin, D.B., 1990. *Discordant Harmonies*. New York: Oxford University Press.

Camann, M., 1996. Freshwater Aquatic Invertebrates: Biomonitoring. Accessed 11/10/23 @ https://www.humboldt.edu.

Carins, J., Jr., and Dickson, K.L., 1971. A simple method for the biological assessment of the effects of waste discharges on aquatic bottom-dwelling organisms. *J Water Pollut Control Fed*; 43:755–772.

Hauser, B.A., 1995. *Practical Hydraulics Handbook*, 2nd ed. Boca Raton, FL: Lewis Publishers.

Hill, B.H., 1997. The use of periphyton assemblage data in an index of biotic integrity. *Bull N Am Benthol Soc*; 14:158.

Huff, W.R., 1993. Biological indices define water quality standard. *Water Environ Technol*; 5:21–22.

Karr, J.R., 1981. Assessment of biotic integrity using fish communities. *Fisheries*; 6(6):21–27.

Karr, J.R., and Chu, E.W., (1999). *Restoring Life In Running Waters: Better Biological Monitoring*. Washington, DC: Island Press.

Karr, J.R., Fausch, K.D., Hagermeier, P.L., Yant, P.R., and Schlosser, I.J., 1986. *Assessing Biological Integrity in Running Waters: A Method and Its Rationale*. Special Publications 5. Champaign, IL: Illinois Natural History Survey.

Kentucky Department of Environmental Protection. 1993. *Methods for Assessing Biological Integrity of Surface Waters*. Frankfort, KY: Kentucky Department of Environmental Protection.

Kittrell, F.W., 1969. *A Practical Guide to Water Quality Studies of Streams*. Washington, DC: U.S. Department of Interior.

Merritt, R., and K. Cummins (1996). *Introduction to Aquatic Insects of North America*. Dubuque, IA: Kendall Hunt Publishing.

Munsell Color (1975). *Munsell Color System*. Washington, DC: US Department of Agriculture.

O'Toole, C., Ed., 1986. *The Encyclopedia of Insects*. New York: Facts on File, Inc.

Patrick, R., 1973. Use of Algae, Especially Diatoms, in the Assessment of Water Quality, in *Biological Methods for the Assessment of Water Quality*, Carins, J., and Dickson, K.L., (eds.). Special Technical Publications 528. Philadelphia: American Society of Testing and Materials.

Pickett, S.T., and White, P.S., 1985. *The Ecology of Natural Disturbance and Patch Dynamics*. Orlando, FL: Academia Press.

River Watch Network (1992). *Colorado River Watch*. Accessed 12/12/23 @ https://crwn.lcra.org

Rodgers, J.H., Jr., Dickson, K.L., and Cairns, J., Jr., 1979. A Review and Analysis of Some Methods Used to Measure Functional Aspects of Periphyton, in *Methods and Measurements of Periphyton Communities: A Review*, Weitzel, R.L., (ed.). Special Technical Publications 690. Philadelphia, PA: American Society for Testing and Materials.

Stevenson, R.J., 1996. An Introduction to Algal Ecology in Freshwater Benthic Habitats, in *Algal Ecology: Freshwater Benthic Ecosystems*, Stevenson, R.J., Bothwell, M., and Lowe, R.L., (eds.). Sand Diego, CA: Academic Press, pp. 3–30.

Stevenson, R.J., 1998. Diatom indicators of stream and wetland stressors in a risk management framework. *Environ Monit Assess*; 51:107–108.

Stevenson, R.J., and Pan, Y., 1999. Assessing Ecological Conditions in Rivers and Streams with Diatoms, in *The Diatoms: Application to the Environmental and Earth Sciences*, Stoermer, E.F., and Smol, J.P., (eds.). Cambridge, UK: Cambridge University Press, pp. 11–40.

Tchobanoglous, G., and Schroeder, E.D., 1985. *Water Quality*. Reading, MA: Addison-Wesley.

USEPA. 1983. *Technical Support Manual: Waterbody Survey and Assessments for Conducting Use Attainability Analyses*. Washington, DC: Environmental Protection Agency.

USEPA. 1985. *Fecal Coliform Testing*. Accessed 12/12/23 @ .https://archive.epa.gov/water/archive/web/html/

USEPA. 2000. *Monitoring Water Quality: Intensive Stream Bioassay*. Washington, DC: U.S. Environmental Protection Agency.

USEPA. 2002. 6Developing Metrics and Indexes of Biological Integrity. EPA822-R-02-016. Accessed 04/21/24 @ https://www.epa.gov/sites/default/files/documents/wetlands_gmetrics.pdf.

Velz, C.J., 1970. *Applied Stream Sanitation*. New York: Wiley Inter-Science.

Warren, M.L., Jr., and Burr, B.M., 1994. Status of freshwater fishes of the US: overview of an imperiled fauna. *Fisheries*; 19(1):6–18.

Weitzel, R.L., 1979. Periphyton Measurements and Applications, in *Methods and Measurements of Periphyton Communities: A Review*, Weitzel, R.L., (ed.). Special Technical Publications 690. Philadelphia, PA: America Society for Testing and Material.

Part IV

Water and Water Treatment

16 Potable Water Source

INTRODUCTION

Note that because of the huge volume and flow conditions, the quality of natural water cannot be modified significantly within the body of water. Accordingly, humans must augment Nature's natural processes with physical, chemical, and biological treatment procedures. Essentially, this quality control approach is directed at the water withdrawn, which is treated from a source for a specific use.

Before presenting a discussion of potential potable water supplies available to us at the current time, we must define potable water:

> Potable water is water fit for human consumption and domestic use, which is sanitary and normally free of minerals, organic substances, and toxic agents in excess or reasonable amounts for domestic usage in the area served, and normally adequate in quantity for the minimum health requirements of the persons served.

With regard to a potential potable water supply, the keywords are "quality and quantity." If we have a water supply unfit for human consumption, we have a quality problem. If we do not have an adequate supply of quality water, we have a quantity problem. In this chapter, we discuss surface water and groundwater hydrology and the mechanical components associated with the collection and conveyance of water from its source to the public water supply system for treatment. We also discuss the development of well supplies. To better comprehend the material presented in this chapter, we have provided the following list of key terms and their definitions.

KEY TERMS AND DEFINITIONS

Annular Space: The space between the casing and the wall of the hole.

Aquifer: A porous, water-bearing geologic formation.

Caisson: A large pipe placed in a vertical position.

Cone of Depression: As the water in a well is drawn down, the water near the well drains or flows into it. The water will drain further back from the top of the water table into the well as the drawdown increases.

Confined Aquifer: An aquifer that is surrounded by formations of less permeable or impermeable material.

Contamination: The introduction of toxic materials, bacteria, or other deleterious agents into water that make it unfit for its intended use.

Drainage Basin: An area from which surface runoff or groundwater recharge is carried into a single drainage system. It is also called a catchment area, watershed, and drainage area.

Drawdown: The distance or difference between the static level and the pumping level. When the drawdown for any particular capacity well and rate pump bowls is determined, the pumping level is known for that capacity. The pump bowls are located below the pumping level so that they will always be underwater. When the drawdown is fixed or remains steady, the well is then furnishing the same amount of water as is being pumped.

Groundwater: Subsurface water occupying a saturated geological formation from which wells and springs are fed.

Hydrology: The applied science pertaining to the properties, distribution, and behavior of water.

Impermeable: A material or substance water will not pass through.

Overland Flow: The movement of water on and just under the earth's surface.

Permeable: A material or substance that water can pass through.

Porosity: The ratio of pore space to total volume; that portion of a cubic foot of soil that is air space and could therefore contain moisture.

Precipitation: The process by which atmospheric moisture is discharged onto the earth's crust. Precipitation takes the form of rain, snow, hail, and sleet.

Pumping Level: The level at which the water stands when the pump is operating.

Radius of Influence: The distance from the well to the edge of the cone of depression, the radius of a circle around the well from which water flows into the well.

Raw Water: The untreated water to be used after treatment for drinking water.

Recharge Area: An area from which precipitation flows into underground water sources.

Specific Yield: The geologist's method for determining the capacity of a given well and the production of a given water-bearing formation; it is expressed as gallons per minute per foot of drawdown.

Spring: A surface feature, where without the help of man, water issues from rock or soil onto the land or into a body of water, the place of issuance being relatively restricted in size.

Static Level: The height to which the water will rise in the well when the pump is not operating.

Surface Runoff: The amount of rainfall that passes over the surface of the earth.

Surface Water: The water on the earth's surface as distinguished from water underground (groundwater).

Unconfined Aquifer: An aquifer that sits on an impervious layer but is open on the top to local infiltration. The recharge for an unconfined aquifer is local. It is also called a water table aquifer.

DOI: 10.1201/9781003581901-20

Water Rights: The rights acquired under the law to use the water accruing in surface or groundwater for a specified purpose in a given manner and usually within the limits of a given time period.

Water Table: The average depth or elevation of the groundwater over a selected area; the upper surface of the zone of saturation, except where that surface is formed by an impermeable body.

Watershed: A drainage basin from which surface water is obtained.

HYDROLOGIC CYCLE

Although we have discussed the hydrologic cycle (water cycle) earlier in the text, to gain a better understanding of the hydrologic cycle (water cycle), it is important to review it again (see Figure 16.1). The hydrologic cycle is a cycle without a beginning or end. As stated earlier, it transports the earth's water from one location to another. As can be seen in Figure 16.1, it consists of precipitation, surface runoff, infiltration, percolation, and evapotranspiration. In the hydrologic cycle, water from streams, lakes, and oceans evaporated by the sun, together with evaporation from the earth and transpiration from plants, furnishes the atmosphere with moisture. Masses of warm air laden with moisture are either forced to cooler upper regions or encounter cool air masses, where the masses condense and form clouds. This condensed moisture falls to earth in the form of rain, snow, and sleet. Part of the precipitation runs off to streams and lakes. Part enters the earth to supply vegetation and rises through the plants to transpire from the leaves, and part seeps or percolates deeply into the ground to supply wells, springs, and the baseflow (dry weather flow) of streams. The cycle constantly repeats itself; a cycle without end.

Note: How long water that falls from the clouds takes to return to the atmosphere varies tremendously. After a short summer shower, most of the rainfall on land can evaporate into the atmosphere in only a matter of minutes. A drop of rain falling on the ocean may take as long as 37,000 years before it returns to the atmosphere, and some water has been in the ground or caught in glaciers for millions of years.

SOURCES OF WATER

Approximately 40 million cubic miles of water cover or reside within the earth. The oceans contain about 97% of

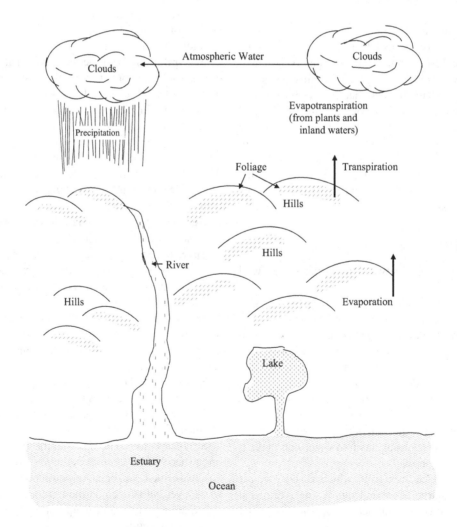

FIGURE 16.1 Natural water cycle.

all water on earth. The other 3% is freshwater: (1) snow and ice on the surface of the earth contains about 2.25% of the water; (2) usable groundwater is approximately 0.3%; and (3) surface freshwater is less than 0.5%. In the United States, for example, the average rainfall is approximately 2.6 feet (a volume of 5,900 km³). Of this amount, approximately 71% evaporates (about 4,200 cm³), and 29% goes to stream flow (about 1,700 km³).

Beneficial freshwater uses include manufacturing, food production, domestic and public needs, recreation, hydro-electric power production, and flood control. Stream flow withdrawn annually is about 7.5% (440 km³). Irrigation and industry use almost half of this amount (3.4% or 200 km³/year). Municipalities use only about 0.6% (35 km³/year) of this amount. Historically, in the United States, water usage is increasing (as might be expected). For example, in 1990, 40 billion gallons of freshwater were used. In 1975, the total increased to 455 billion gallons. Projected use in 2002 is about 725 billion gallons.

The primary sources of freshwater include:

1. Captured and stored rainfall in cisterns and water jars
2. Groundwater from springs, artesian wells, and drilled or dug wells
3. Surface water from lakes, rivers, and streams
4. Desalinized seawater or brackish groundwater
5. Reclaimed wastewater

Current federal drinking water regulations actually define three distinct and separate sources of freshwater. They are surface water, groundwater, and groundwater under the direct influence of surface water (GUDISW). This last classification is the result of the *Surface Water Treatment Rule (SWTR)*. The definition of what conditions constitute GUDISW, while specific, is not obvious. This classification is discussed in detail later.

SURFACE WATER

Surface waters are not uniformly distributed over the Earth's surface. In the United States, for example, only about 4 percent of the landmass is covered by rivers, lakes, and streams. The volumes of these freshwater sources depend on geographic, landscape, and temporal variations, and the impact of human activities. *Surface water* is water that is open to the atmosphere and results from *overland flow* (i.e., *runoff* that has not yet reached a definite stream channel). Put a different way, surface water is the result of *surface runoff.* For the most part, however, surface (as used in the context of this text) refers to water flowing in streams and rivers, as well as water stored in natural or artificial lakes, man-made impoundments such as lakes made by damming a stream or river; springs that are affected by a change in level or quantity; shallow wells that are affected by precipitation; wells drilled next to or in a stream or river; rain catchments; and/or muskeg and tundra ponds.

ADVANTAGES AND DISADVANTAGES OF SURFACE WATER

The biggest advantage of using a surface water supply as a water source is that these sources are readily located; finding surface water sources does not demand sophisticated training or equipment. Many surface water sources have been used for decades and even centuries (in the United States., for example), and considerable data are available on the quantity and quality of the existing water supply. Surface water is also generally softer (not mineral-laden), which makes its treatment much simpler.

The most significant disadvantage of using surface water as a water source is *pollution*. Surface waters are easily contaminated (polluted) with microorganisms that cause waterborne diseases and chemicals that enter the river or stream from surface runoff and upstream discharges. Another problem with many surface water sources is *turbidity,* which fluctuates with the amount of precipitation. Increases in turbidity increase treatment costs and operator time. Surface water temperatures can be a problem because they fluctuate with ambient temperature, making consistent water quality production at a waterworks plant difficult. Drawing water from a surface water supply might also present problems; intake structures may clog or become damaged from winter ice, or the source may be so shallow that it completely freezes in the winter. *Water rights* cause problems too—removing surface water from a stream, lake, or spring requires a legal right. The lingering, seemingly unanswerable question is, who owns the water?

Using surface water as a source means that the purveyor is obligated to meet the requirements of the Surface Water Treatment Rule (SWTR) and the Interim Enhanced Surface Water Treatment Rule (IESWTR). [Note: This rule only applies to large public water systems (PWS) that serve more than 10,000 people. It tightened controls on DBPs and turbidity and regulates *Cryptosporidium*.]

SURFACE WATER HYDROLOGY

To properly manage and operate water systems, a basic understanding of the movement of water and the factors that affect water quality and quantity is important—in other words, *hydrology*. A discipline of applied science, hydrology includes several components, including the physical configuration of the watershed, geology, soils, vegetation, nutrients, energy, wildlife, and the water itself. The area from which surface water flows is called a *drainage basin* or catchment area. With a surface water source, this drainage basin is most often called, in nontechnical terms, a *watershed* (when dealing with groundwater, we call this area a *recharge area).*

Key Point: The area that directly influences the quantity and quality of surface water is called the drainage basin or watershed.

When you trace on a map the course of a major river from its meager beginnings to its seaward path, its flow becomes larger and larger. While every tributary brings a

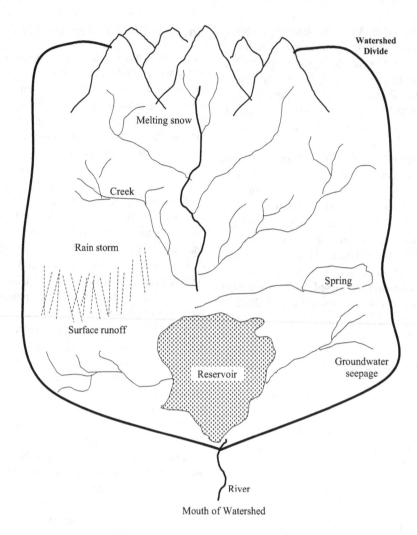

FIGURE 16.2 Watershed.

sudden increase, between tributaries, the river grows gradually from overland flow entering it directly (see Figure 16.2). Not only does the river grow, but its whole watershed or drainage basin, basically the land it drains into, grows too, in the sense that it embraces an ever-larger area. The area of the watershed is commonly measured in square miles, sections, or acres. When taking water from a surface water source, knowing the size of the watershed is desirable.

RAW WATER STORAGE

Raw water (i.e., water that has not been treated) is stored for single or multiple uses, such as navigation, flood control, hydroelectric power, agriculture, water supply, pollution abatement, recreation, and flow augmentation. The primary reason for storing water is to meet peak demands and/or to store water to meet demands when the flow of the source is below the demand. Raw water is stored in natural storage sites (such as lakes, muskeg, and tundra ponds) or man-made storage areas such as dams. Man-made dams are either masonry or embankment dams. If embankment dams are used, they are typically constructed of local materials with an impermeable clay core.

SURFACE WATER INTAKES

Withdrawing water from a river, lake, or reservoir so that it may be conveyed to the first unit process of treatment requires an intake structure. Intakes have no standard design and range from a simple pump suction pipe sticking out into the lake or stream to expensive structures costing several thousand dollars. Typical intakes include submerged intakes, floating intakes, infiltration galleries, spring boxes, and roof catchments. Their primary functions are to supply the highest quality water from the source and protect piping and pumps from clogging as a result of wave action, ice formation, flooding, and submerged debris. A poorly conceived or constructed intake can cause many problems. Failure of the intake could result in water system failure.

For a small stream, the most common intake structures used are small gravity dams placed across the stream or a submerged intake. In the gravity dam type, a gravity line or pumps can remove water from behind the dam. In the submerged intake type, water is collected in a diversion and carried away by gravity or pumped from a caisson. Another common intake used on small and large streams is an end-suction centrifugal pump or submersible pump placed on a float. The float is secured to the bank, and the water is pumped to a storage area.

Often, the intake structure placed in a stream is an infiltration gallery. The most common infiltration galleries are built by placing well screens or perforated pipes into the streambed, which are covered with clean, graded gravel. When water passes through the gravel, coarse filtration removes some turbidity and organic material. The water collected by the perforated pipe flows to a caisson placed next to the stream and is removed from the caisson by gravity or pumping. Intakes used in springs are normally implanted into the water-bearing strata, then covered with clean, washed rock and sealed, usually with clay. The outlet is piped into a spring box. In some locations, a primary source of water is rainwater, which is collected from the roofs of buildings with a device called a roof catchment.

After determining that a water source provides a suitable quality and quantity of raw water, choosing an intake location includes determining the following

1. Best quality water location
2. Dangerous currents
3. Sandbar formation
4. Wave action
5. Ice storm factors
6. Flood factors
7. Navigation channel avoidance
8. Intake accessibility
9. Power availability
10. Floating or moving object damage factors
11. Distance from pumping station
12. Upstream uses that may affect water quality

SURFACE WATER SCREENS

Generally, screening devices are installed to protect intake pumps, valves, and piping. A coarse screen of vertical steel bars with openings of 1–3 inches, placed in a near-vertical position, excludes large objects. It may be equipped with a trash truck rack rake to remove accumulated debris. A finer screen with 3/8-inch openings removes leaves, twigs, small fish, and other material passing through the bar rack. Traveling screens consist of wire mesh trays that retain solids as water passes through them. Drive chains and sprockets raise the trays into a head enclosure, where debris is removed by water sprays. The screen travel pattern is intermittent and controlled by the amount of accumulated material.

Note: When considering what type of screen to be employ, the most important consideration is ensuring that they can be easily maintained.

SURFACE WATER QUALITY

Surface waters should be of adequate quality to support aquatic life and be aesthetically pleasing, and waters used as sources of supply should be treatable by conventional processes to provide potable supplies that meet drinking water standards. Many lakes, reservoirs, and rivers are maintained at a quality suitable for swimming, water skiing, and boating, as well as for drinking water. Whether the surface water supply is taken from a river, stream, lake, spring, impoundment, reservoir, or dam, surface water quality varies widely, especially in rivers, streams, and small lakes. These water bodies are not only susceptible to waste discharge contamination but also to "flash" contamination (which can occur almost immediately and not necessarily over time). Lakes are subject to summer/winter stratification (turnover) and algal blooms. Pollution sources range from runoff (agricultural, residential, and urban) to spills, municipal and industrial wastewater discharges, recreational users, and natural occurrences. Surface water supplies are difficult to protect from contamination and must always be treated.

GROUNDWATER

As mentioned, part of the precipitation that falls on land infiltrates the surface, percolates downward through the soil under the force of gravity, and becomes *groundwater*. Groundwater, like surface water, is extremely important to the hydrologic cycle and our water supplies. Almost half of the people in the U.S. drink public water from groundwater supplies. Overall, more water exists as groundwater than surface water in the United States, including the water in the Great Lakes. However, sometimes pumping it to the surface is not economical, and in recent years, pollution of groundwater supplies from improper disposal has become a significant problem.

We find groundwater in saturated layers called *aquifers* under the earth's surface. Three types of aquifers exist: unconfined, confined, and springs. Aquifers are made up of a combination of solid materials, such as rock and gravel, and open spaces called pores. Regardless of the type of aquifer, the groundwater within is in a constant state of motion, caused by gravity or by pumping.

The actual amount of water in an aquifer depends on the amount of space available between the various grains of material that make up the aquifer. The amount of space available is called *porosity*. The ease of movement through an aquifer depends on how well the pores are connected. For example, clay can hold a lot of water and has high porosity, but the pores are not connected, making it difficult for water to move through the clay. The ability of an aquifer to allow water to infiltrate is called *permeability*.

The aquifer that lies just under the earth's surface is called the *zone of saturation,* or unconfined aquifer (see Figure 16.3). The top of the zone of saturation is the *water table*. An unconfined aquifer is only contained at the bottom and depends on local precipitation for recharge. This type of aquifer is often called a water table aquifer.

Unconfined aquifers are a primary source of shallow well water (see Figure 16.3). These wells are shallow (and not desirable as a public drinking water source). They are subject to local contamination from hazardous and toxic materials—such as fuel and oil, as well as septic tanks

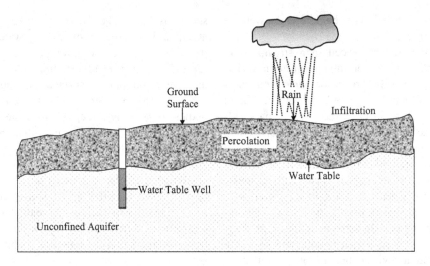

FIGURE 16.3   Unconfined aquifer.

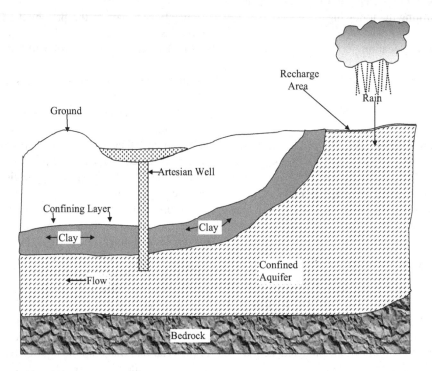

FIGURE 16.4   Confined aquifer.

and agricultural runoff—which provide increased levels of nitrates and microorganisms. These wells may be classified as groundwater under the direct influence of surface water (GUDISW) and therefore require treatment to control microorganisms.

A *confined aquifer* is sandwiched between two impermeable layers that block the flow of water. The water in a confined aquifer is under hydrostatic pressure and does not have a free water table (see Figure 16.4). Confined aquifers are called *artesian* aquifers. Wells drilled into artesian aquifers are called artesian wells and commonly yield large quantities of high-quality water. An artesian well is any well where the water in the well casing rises above the saturated strata. Wells in confined aquifers are normally referred to as deep wells and are not generally affected by local hydrological events.

A confined aquifer is recharged by rain or snow in the mountains where the aquifer lies close to the surface of the Earth. Because the recharge area is some distance from areas of possible contamination, the possibility of contamination is usually very low. However, once contaminated, confined aquifers may take centuries to recover.

Groundwater naturally exits the earth's crust in areas called *springs*. The water in a spring can originate from a water table aquifer or a confined aquifer. Only water from a confined spring is considered desirable for a public water system.

GROUNDWATER QUALITY

Generally, groundwater possesses high chemical, bacteriological, and physical quality. When pumped from an aquifer composed of a mixture of sand and gravel, and if not directly influenced by surface water, groundwater is often used without filtration. It can also be used without disinfection if it has a low coliform count. However, as mentioned, groundwater can become contaminated. When septic systems fail, saltwater intrudes, improper disposal of wastes occurs, improperly stockpiled chemicals leach, underground storage tanks leak, hazardous materials spill, fertilizers and pesticides are misplaced, and mines are improperly abandoned, groundwater can become contaminated. To understand how an underground aquifer becomes contaminated, you must understand what occurs when pumping takes place within the well.

When groundwater is removed from its underground source (i.e., from the water-bearing stratum) via a well, water flows toward the center of the well. In a water table aquifer, this movement causes the water table to sag toward the well. This sag is called the *cone of depression.* The shape and size of the cone depend on the relationship between the pumping rate and the rate at which water can move toward the well. If the rate is high, the cone is shallow, and its growth stabilizes. The area included in the cone of depression is called the *cone of influence,* and any contamination in this zone will be drawn into the well.

GROUNDWATER UNDER THE DIRECT INFLUENCE OF SURFACE WATER

Water under the direct influence of surface water (GUDISW) is not classified as a groundwater supply. A supply designated as GUDISW must be treated under the state's surface water rules rather than the groundwater rules. The *Surface Water Treatment Rule* of the *Safe Drinking Water Act* requires each site to determine which groundwater supplies are influenced by surface water (i.e., when surface water can infiltrate a groundwater supply and contaminate it with *Giardia,* viruses, turbidity, and organic material from the surface water source). To determine whether a groundwater supply is under the direct influence of surface water, the USEPA has developed procedures that focus on significant and relatively rapid shifts in water quality characteristics, including turbidity, temperature, and pH. When these shifts can be closely correlated with rainfall or other surface water conditions, or when certain indicator organisms associated with surface water are found, the source is said to be under the direct influence of surface water.

SURFACE WATER QUALITY AND TREATMENT REQUIREMENTS

Public water systems (PWS) must comply with applicable federal and state regulations and must provide quantity and quality water supplies including proper treatment (where/when required) and competent, qualified waterworks operators. The USEPA's regulatory requirements insist that all public water systems using any surface or groundwater under the direct influence of surface water must disinfect and may be required by the state to filter unless the water source meets certain requirements and site-specific conditions. Treatment technique requirements are established in lieu of Maximum Contaminant Levels (MCLs) for *Giardia,* viruses, heterotrophic plate count bacteria, *Legionella,* and turbidity. Treatment must achieve at least 99.9% removal (3-log removal) and/or inactivation of *Giardia lamblia* cysts and 99.9% removal and/or inactivation of viruses.

Qualified operators (as determined by the state) must operate all systems. To avoid filtration, waterworks must satisfy the following criteria:

1. Fecal coliform concentration must not exceed 20/100 mL, or the total coliform concentration must not exceed 100/100 mL before disinfection in more than 10% of the measurements for the previous 6 months, calculated each month.
2. Turbidity levels must be measured every 4 hours by grab samples or continuous monitoring. The turbidity level may not exceed 5 Nephelometric Turbidity Unit (NTU). If the turbidity exceeds 5 NTU, the water supply system must install filtration unless the state determines that the event is unusual or unpredictable and that it does not occur more than twice in any 1 year or five times in any consecutive 10 years.

It is important, when considering the choice of water source, that the source presents minimal risks of contamination from wastewater and contains a minimum of impurities that may be hazardous to health. Acute (immediate) health effects, such as those presented by exposure to *Giardia lamblia,* and chronic effects (those that take longer to affect health) must be guarded against. Maximum contaminant levels (MCLs) must be monitored to ensure that the maximum permissible level of contaminants in water is not exceeded.

Note: Primary MCL is based on health considerations, while secondary MCL is based on aesthetic considerations (taste, odor, and appearance).

The Public Water System must also provide water free of pathogens (disease-causing microorganisms; bacteria, protozoa, spores, viruses, etc.). Chemical quality must also be monitored to ensure the prevention of inorganic and organic contamination.

In 1996, the USEPA finalized the Stage 1 Disinfectants/Disinfection By-products (D/DBP) and Interim Enhanced Surface Water Treatment rules and implemented them in 1998. These amendments tightened controls on DBPs and turbidity and regulated *Cryptosporidium.* Highlights of these changes include the following:

Stage 1 D/DBP Rule

a. Tightened the total trihalomethane standard to 0.080 mg/L.
b. Sets new DBP standards for five haloacetic acids (0.060 mg/L), chlorite (1.0 mg/L), and bromate (0.010 mg/L).
c. Established new standards for disinfectant residuals (4.0 mg/L for chlorine, 4.0 mg/L for chloramines, and 0.8 mg/L for chlorine dioxide).
d. Requires systems using surface water or groundwater directly influenced by surface water to implement enhanced coagulation or softening to remove DBP precursors unless systems meet alternative criteria.
e. Applies to all community and nontransient-noncommunity systems that disinfect, including those serving fewer than 10,000 people.
  i. MCLGs
     For maximum contaminant level goals (MCLGs), however, the USEPA opted to retain the chloroform MCLG at zero instead of loosening it to 0.3 mg/L as outlined in the Spring 1998 Notice of Data Availability. The EPA also loosened the chlorite MCLG from 0.08 to 0.8 mg/L, loosened the maximum residual disinfectant level goal for chlorine dioxide from 0.3 to 0.8 mg/L, and set no MCLG for the DBP chloral hydrate (control of which will be covered by the other requirements).
  ii. Chloroform
     In dropping its plan to loosen the chloroform MCLG, the USEPA has backed away (for now) from its first attempt to set a level higher than zero MCLG for a carcinogenic contaminant, opting for more time to allow the issue to be discussed by stakeholders and the Science Advisory Board, which is slated to produce a chloroform report by November 1999. The USEPA outlined what it termed a "compelling" case for recognizing a safe (or threshold) exposure level for chloroform, one of the regulated trihalomethanes.

Interim Enhanced Surface Water Treatment (IESWT) Rule

This treatment optimization rule, which only applies to large public water systems (those serving more than 10,000 people) that use surface water or groundwater directly influenced by surface water, is the first to directly regulate *Cryptosporidium* (crypto):

a. The rule sets a crypto MCLG of zero.
b. Requires systems that filter to remove 99% (2 log) of crypto oocysts.

c. Adds crypto control to watershed protection requirements for systems operating under filtration waivers.
d. Is particular to the genus *Cryptosporidium,* not to the *Cryptosporidium parvum* species.
  i. Turbidity
     The rule requires continuous turbidity monitoring of individual filters and tightens allowable turbidity limits for combined filter effluent, cutting the maximum from 5 to 1 NTU and the average monthly limit from 0.5 to 0.3 NTU.
  ii. Benchmarking
     Systems must determine within 15 months of promulgation whether they must establish a disinfection benchmark to ensure maintenance of microbial protection as systems comply with new DBP standards. Because the determination is based on whether the PWS exceeds annual average levels of THMs or haloacetic acids, systems that lack such data must begin collecting it within 3 months of promulgation to have a year's worth by the 15-month deadline.
     The rule also requires states to conduct periodic sanitary surveys of all surface water systems, regardless of size, and covers all new treated-water reservoirs.

Regulatory Deadlines

Large surface water systems (those serving over 10,000) were required to comply with the Stage 1 D/DBP and IESWT rules by December 2001. Smaller surface water systems and all groundwater systems were required to comply with the Stage 1 D/DBP Rule by December 2003.

PUBLIC WATER SYSTEM QUALITY REQUIREMENTS

Many factors affect the use of water, including climate, economic conditions, type of community (i.e., residential, commercial, industrial), integrity of the distribution system (waste pressure/leaks in the system), and water cost. In the United States, the typical per capita usage is approximately 150 gallons/day (gpd) per person. Each residential connection requires approximately 400 gpd per connection. Keep in mind that fire-fighting requirements at a standard fire flow of 500 gpm will be used in 1 minute which a family of five normally uses in 24 hours. Water pressure delivered to each service connection should (at a minimum) reach 20 psi under all flow conditions.

WELL SYSTEMS

The most common method for withdrawing groundwater is to penetrate the aquifer with a vertical well and then pump the water up to the surface. In the past, when some-one wanted a well, they simply dug (or hired someone to dig) and hoped (gambled) that they would find water in a quantity suitable for their needs. Today, in most locations in the United States, for example, developing a well supply usually involves a more complicated step-by-step process. Local, state, and federal requirements specify the actual requirements for the development of a well supply in the United States. The standard sequence for developing a well supply generally involves a seven-step process. This process includes:

Step 1: *Application*—Depending on location, filling out and applying (to the applicable authorities) to develop a well supply is standard procedure.

Step 2: *Well Site Approval*—Once the application has been made, local authorities check various local geological and other records to ensure that the siting of the proposed well coincides with mandated guidelines for approval.

Step 3: *Well drilling*—The well is then drilled.

Step 4: *Preliminary engineering report*—After the well is drilled and the results documented, a preliminary engineering report is made on the suitability of the site to serve as a water source. This procedure involves performing a pump test to determine if the well can supply the required amount of water. The well is generally pumped for at least 6 hours at a rate equal to or greater than the desired yield. A stabilized drawdown should be obtained at that rate and the original static level should be recovered within 24 hours after pumping stops. During this test period, samples are taken and tested for bacteriological and chemical quality.

Step 5: *Submission of documents for review and approval*—The application and test results are submitted to an authorized reviewing authority that determines if the well site meets approval criteria.

Step 6: *Construction permit*—If the site is approved, a construction permit is issued.

Step 7: *Operation permit*—When the well is ready for use, an operation permit is issued.

Well Site Requirements

To protect the groundwater source and provide high-quality, safe water, the waterworks industry has developed standards and specifications for wells. The following listing includes industry standards and practices, as well as those items included in example State Department of Environmental Compliance regulations.

Note: Check with your local regulatory authorities to determine well site requirements.

1. Minimum well lot requirements
   a. 50 feet from well to all property lines

   b. All-weather access road provided
   c. Lot graded to divert surface runoff
   d. Recorded well plat and dedication document
2. Minimum well location requirements
   a. At least 50 feet horizontal distance from any actual or potential sources of contamination involving sewage
   b. At least 50 feet horizontal distance from any petroleum or chemical storage tank, pipeline, or similar source of contamination; except where plastic-type well casing is used, the separation distance must be at least 100 feet
3. Vulnerability assessment
   a. Wellhead area = 1,000 ft radius from the well
   b. What is the general land use of the area (residential, industrial, livestock, crops, undeveloped, other)?
   c. What are the geologic conditions (sinkholes, surface, subsurface)?

Type of Wells

Water supply wells may be characterized as shallow or deep. In addition, wells are classified as follows:

1. **Class I:** cased and grouted to 100 feet
2. **Class II A:** cased to a minimum of 100 feet and grouted to 20 feet
3. **Class II B:** cased and grouted to 50 feet

Note: During the well development process, mud and silt forced into the aquifer during the drilling process are removed, allowing the well to produce the best-quality water at the highest rate from the aquifer.

Shallow Wells

Shallow wells are those that are less than 100 feet deep. Such wells are not particularly desirable for municipal supplies since the aquifers they tap are likely to fluctuate considerably in depth, making the yield somewhat uncertain. Municipal wells in such aquifers cause a reduction in the water table (or phreatic surface) that affects nearby private wells, which are more likely to utilize shallow strata. Such interference with private wells may result in damage suits against the community. Shallow wells may be dug, bored, or driven.

1. **Dug Wells:** Dug wells are the oldest type of well and date back many centuries; they are dug by hand or by a variety of unspecialized equipment. They range in size from approximately 4 to 15 feet in diameter and are usually about 20–40 feet deep. Such wells are usually lined or cased with concrete or brick. Dug wells are prone to failure from drought or heavy pumpage. They are vulnerable to contamination and are not acceptable as a public water supply in many locations.

2. **Driven Wells:** Driven wells consist of a pipe casing terminating in a point slightly greater in diameter than the casing. The pointed well screen and the lengths of pipe attached to it are pounded down or driven in the same manner as a pile, usually with a drop hammer, to the water-bearing strata. Driven wells are usually 2–3 inches in diameter and are used only in unconsolidated materials. This type of shallow well is not acceptable as a public water supply.

3. **Bored Wells:** Bored wells range from 1 to 36 inches in diameter and are constructed in unconsolidated materials. The boring is accomplished with augers (either hand or machine-driven) that fill with soil and are then drawn to the surface to be emptied. The casing may be placed after the well is completed (in relatively cohesive materials) but must advance with the well in noncohesive strata. Bored wells are not acceptable as a public water supply.

Deep Wells

Deep wells are the usual source of groundwater for municipalities. Deep wells tap thick and extensive aquifers that are not subject to rapid fluctuations in water (piezometric surface—the height to which water will rise in a tube penetrating a confined aquifer) level and that provide a large and uniform yield. Deep wells typically yield water of more constant quality than shallow wells, although the quality is not necessarily better. Deep wells are constructed by a variety of techniques; we discuss two of these techniques (jetting and drilling) below.

1. **Jetted Wells:** Jetted well construction commonly employs a jetting pipe with a cutting tool. This type of well cannot be constructed in clay, hardpan, or where boulders are present. Jetted wells are not acceptable as a public water supply.

2. **Drilled Wells:** Drilled wells are usually the only type of well allowed for use in most public water supply systems. Several different methods of drilling are available, all of which are capable of drilling wells of extreme depth and diameter. Drilled wells are constructed using a drilling rig that creates a hole into which the casing is placed. Screens are installed at one or more levels when water-bearing formations are encountered.

WELL COMPONENTS

The components that make up a well system include the well itself, the building, the pump, and the related piping system. In this section, we focus on the components that make up the well itself. Many of these components are shown in Figure 16.5.

Well Casing

A well is a hole in the ground called the borehole. The hole is protected from collapse by placing a casing inside the borehole. The well casing prevents the walls of the hole from collapsing and prevents contaminants (either surface or subsurface) from entering the water source. The casing also provides a column of stored water and housing for the pump mechanisms and pipes. Well casings constructed of steel or plastic material are acceptable. The well casing must extend a minimum of 12 inches above grade.

Grout

To protect the aquifer from contamination, the casing is sealed to the borehole near the surface and near the bottom, where it passes into the impermeable layer, with grout. This sealing process keeps the well from being polluted by surface water and seals out water from water-bearing strata that have undesirable water quality. Sealing also protects the casing from external corrosion and restrains unstable soil and rock formations. Grout consists of near cement that is pumped into the annular space (it is completed within 48 hours of well construction); it is pumped under continuous pressure starting at the bottom and progressing upward in one continuous operation.

Well Pad

The well pad provides a ground seal around the casing. The pad is constructed of reinforced concrete, 6 feet × 6 feet (6-inch thick), with the well head located in the middle. The well pad prevents contaminants from collecting around the well and seeping down into the ground along the casing.

Sanitary Seal

To prevent contamination of the well, a sanitary seal is placed at the top of the casing. The type of seal varies depending on the type of pump used. The sanitary seal contains openings for power and control wires, pump support cables, a drawdown gauge, discharge piping, pump shaft, and air vent while providing a tight seal around them.

Well Screen

Screens can be installed at the intake point(s) on the end of a well casing or the end of the inner casing of gravel-packed well. These screens perform two functions: (1) supporting the borehole and (2) reducing the amount of sand that enters the casing and the pump. They are sized to allow the maximum amount of water while preventing the passage of sand, sediment, and gravel.

Casing Vent

The well casing must have a vent to allow air into the casing as the water level drops. The vent terminates 18 inches above the floor with a return bend pointing downward. The opening of the vent must be screened with #24 mesh stainless steel to prevent the entry of vermin and dust.

Drop Pipe

The drop pipe or riser is the line leading from the pump to the wellhead. It ensures adequate support so that an aboveground pump does not move and that a submersible pump is

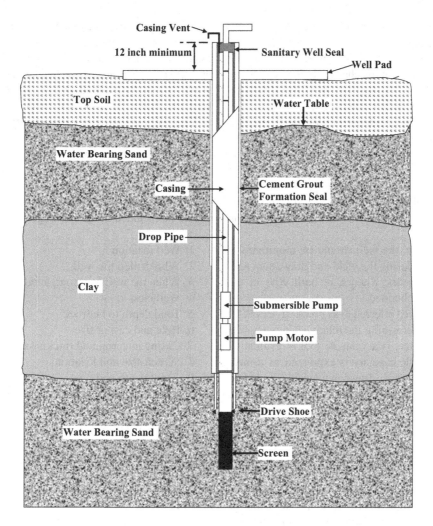

FIGURE 16.5 Components of a well.

not lost down the well. This pipe is either steel or PVC, with steel being the most desirable.

Miscellaneous Well Components

Miscellaneous well components include:

Gauge and Airline: measures the water level of the well.

Check Valve: located immediately after the well, it prevents system water from returning to the well. It must be located above ground and be protected from freezing.

Flowmeter: required to monitor the total amount of water withdrawn from the well, including any water blown off.

Control Switches: controls for well pump operation.

Blowoff: valved and located between the well and storage tank; used to flush the well of sediment or turbid or super-chlorinated water.

Sample Taps: (a) Raw water sample tap—located before any storage or treatment to permit sampling of the water directly from the well. (b) Entry point sample tap—located after treatment.

Control Valves: isolates the well for testing or maintenance or is used to control water flow.

WELL EVALUATION

After a well is developed, conducting a pump test determines if it can supply the required amount of water. The well is generally pumped for at least 6 hours (many states require a 48-hour yield and drawdown test) at a rate equal to or greater than the desired yield. *Yield* is the volume or quantity of water per unit of time discharged from a well (GPM, cubic feet/sec). Regulations usually require that a well produce a minimum of 0.5 gallons/minute per residential connection. Drawdown is the difference between the static water level (the level of the water in the well when it has not been used for some time and has stabilized) and the pumping water level in a well. Drawdown is measured by using an airline and pressure gauge to monitor the water level during the 48 hours of pumping.

The procedure calls for the airline to be suspended inside the casing down into the water. At the other end are the pressure gauge and a small pump. Air is pumped into the line (displacing the water) until the pressure stops increasing.

The gauge's highest pressure reading is recorded. During the 48 hours of pumping, the yield and drawdown are monitored more frequently during the beginning of the testing period because the most dramatic changes in flow and water level usually occur then. The original static level should be recovered within 24 hours after pumping stops.

Testing is accomplished on a bacteriological sample for analysis by the MPN method every half hour during the last 10 hours of testing. The results are used to determine if chlorination is required or if chlorination alone will be sufficient to treat the water. Chemical, physical, and radiological samples are collected for analyses at the end of the test period to determine if treatment other than chlorination may be required.

Note: Recovery from the well should be monitored at the same frequency as during the yield and drawdown testing and for at least the first 8 hours, or until 90% of the observed drawdown is obtained.

Specific capacity (often referred to as *productivity index*) is a test method for determining the relative adequacy of a well and, over some time, is a valuable tool in evaluating well production. Specific capacity is expressed as a measure of well yield per unit of drawdown (yield divided by drawdown). When conducting this test, if possible, always run the pump for the same length of time and at the same pump rate.

WELL PUMPS

Pumps are used to move the water out of the well and deliver it to the storage tank/distribution system. The type of pump chosen for use should provide optimum performance based on location and operating conditions, required capacity, and total head. Two types of pumps commonly installed in groundwater systems are *line shaft turbines* and *submersible turbines*. Whichever type of pump is used, it is rated based on pumping capacity expressed in gpm (e.g., 40 gpm), not on horsepower.

ROUTINE OPERATION AND RECORDKEEPING REQUIREMENTS

Ensuring the proper operation of a well requires close monitoring; wells should be visited regularly. During routine monitoring visits, check for any unusual sounds in the pump, line, or valves, and any leaks. In addition, routinely cycle valves to ensure good working conditions. Check motors to make sure they are not overheating. Check the well pump to guard against short cycling. Collect a water sample for a visual check for sediment. Also, check chlorine residual and treatment equipment. Measure gallons on the installed meter for 1 minute to obtain the pump rate in gpm (look for gradual trends or big changes). Check the water level in the well at least monthly (possibly more often in summer or during periods of low rainfall). Finally, from recorded meter readings, determine gallons used and compare with water consumed to identify possible distribution system leaks. Along with meter readings, other records must be accurately

and consistently maintained for water supply wells. This recordkeeping is absolutely imperative. The records (an important resource for troubleshooting) can be useful when problems develop or helpful in identifying potential problems. A properly operated and managed waterworks facility keeps the following records of well operation.

The *well log* provides documentation of the materials found in the borehole and at what depths. The well log includes the depths at which water was found, the casing length and type, the depth at which different types of soils were found, the testing procedure, well development techniques, and well production. In general, the following items should be included in the well log:

1. Well location
2. Who drilled the well
3. When the well was completed
4. Well class
5. Total depth to bedrock
6. Hole and casing size
7. Casing material and thickness
8. Screen size and locations
9. Grout depth and type
10. Yield and drawdown (test results)
11. Pump information (type, HP, capacity, intake depth, and model #)
12. Geology of the hole
13. A record of yield and drawdown data

Pump data should be collected and maintained. This data should include:

- Pump brand and model number
- Rate capacity
- Date of installation
- Maintenance performed
- Date replaced
- Pressure reading or water level when the pump will cut on and off
- Pumping time (hours per day the pump is running)
- Output in gallons per minute

A record of water quality should be kept and maintained, including bacteriological, chemical and physical (inorganic, metals, nitrate/nitrite, VOC), and radiological reports.

System-specific Monthly Operation Reports should be kept and maintained. These reports should contain information and data from meter readings/total gallons per day/month, chlorine residuals, amount and type of chemicals used, turbidity readings, physical parameters (pH, temperature), pumping rate, total population served, and total number of connections.

A record of water level (static and dynamic levels) should be kept and maintained, in addition to a record of any changes in conditions (such as heavy rainfall, high consumption, leaks, and earthquakes) and a record of specific capacity.

Well Maintenance

Wells do not have an infinite life, and their output is likely to reduce over time due to hydrological or mechanical factors. Protecting the well from possible contamination is an important consideration. If proper well location (based on knowledge of local geological conditions and a vulnerability assessment of the area) is chosen, potential problems can be minimized.

During the initial assessment, ensuring that the well is not located in a sinkhole area is important. A determination must be made regarding where unconsolidated or bedrock aquifers may be subject to contamination. Several other important determinations must also be made: Is the well located on a floodplain? Is it located next to a drainfield for septic systems or near a landfill? Are petroleum/gasoline storage tanks nearby? Is pesticide/plastic manufacturing conducted near the well site?

Along with proper well location, proper well design and construction are essential to prevent wells from acting as conduits for the vertical migration of contaminants into the groundwater. Basically, the pollution potential of a well depends on how well it was constructed. Contamination can occur during the drilling process, and an unsealed or unfinished well provides an avenue for contamination. Any opening in the sanitary seal or break in the casing may cause contamination, as can the reversal of water flow. In routine well maintenance, corroded casings or screens are sometimes withdrawn and replaced, but this is difficult and not always successful. Simply constructing a new well may be less expensive.

Troubleshooting Well Problems

During operation, various problems may develop. For example, the well may pump sand or mud. When this occurs, the well screen may have collapsed or corroded, causing the screen's slot openings to become enlarged (allowing debris, sand, and mud to enter). If the well screen is not the problem, the pumping rate should be checked, as it may be too high. Below are a few other well problems, their probable causes, and the required remediation:

- If the water is white, the pump might be sucking air; reduce the pump rate.
- If water rushes backward when the pump shuts off, check the valve, as it may be leaking.
- If the well yield has decreased, check the static water level. A downward trend in static water levels suggests that the aquifer is becoming depleted, which could be due to the following:
  - Local overdraft (well spacing too close)
  - General overdraft—pumpage exceeds recharge
  - Temporary decrease in recharge—dry cycles
  - Permanent decrease in recharge—less flow in rivers

- Decreased specific capacity (if it has dropped 10%–15%, take steps to determine the cause; it may be a result of incrustation)

Note: *Incrustation* occurs when clogging, cementation, or stoppage of a well screen and water-bearing formation happens. Incrustations on screens and adjacent aquifer materials result from chemical or biological reactions at the air–water interface in the well. The chief encrusting agent is calcium carbonate, which cements the gravel and sand grains together. Incrustation can also result from carbonates of magnesium, clays and silts, or iron bacteria. Treatment involves pulling the screen and removing the incrusted material, replacing the screen, or treating the screen and water-bearing formation with acids. If severe, treatment may involve rehabilitating the well.

- The pump rate is dropping, but the water level is not—probable cause is pump impairment.
- Impellers might be worn.
- There is a change in the hydraulic head against which the pump is working (the head may change as a result of corrosion in the pipelines, a higher pressure setting, or possibly a newly elevated tank).

Well Abandonment

In the past, the common practice was simply to walk away and forget about a well when it ran dry. Today, while dry or failing wells are still abandoned, we know that they must be abandoned with care (and not completely forgotten). An abandoned well can become a convenient (and dangerous) receptacle for waste, thus contaminating the aquifer. An improperly abandoned well could also become a haven for vermin or, worse, a hazard for children. A temporarily abandoned well must be sealed with a watertight cap or wellhead seal. The well must be maintained to ensure that it does not become a source or channel of contamination during temporary abandonment.

When a well is permanently abandoned, all casing and screen materials may be salvaged. The well should be checked from top to bottom to assure that no obstructions interfere with plugging and sealing operations. Prior to plugging, the well should be thoroughly chlorinated. Bored wells should be completely filled with cement grout. If the well was constructed in an unconsolidated formation, it should be completely filled with cement grout or clay slurry introduced through a pipe that initially extends to the bottom of the well. As the pipe is raised, it should remain submerged in the top layers of grout while the well is filled.

Wells constructed in consolidated rock or that penetrate zones of consolidated rock can be filled with sand or gravel opposite the zones of consolidated rock. The sand or gravel fill is terminated 5 feet below the top of the consolidated rock. The remainder of the well is filled with sand-cement grout.

THE BOTTOM LINE

Source water refers to bodies of water such as rivers, streams, lakes, reservoirs, springs, groundwater, and recycled water (a.k.a. reused water). that provide water to public drinking-water supplies and wells.

CHAPTER REVIEW QUESTIONS

16.1 When water is withdrawn from a well, a _____ will develop.

16.2 How far should the well casing extend above the ground or well-house floor?

16.3 A well casing should be grouted for at least 10 feet, with the first 20 feet grouted with _____.

16.4 List three sources of drinking water.

16.5 Explain GUDISW.

16.6 What are two advantages of surface water sources?

16.7 Define hydrology:

16.8 The area inside the cone of depression is called the _____.

16.9 A spring is an example of what type of water source?

16.10 Describe the function of the bar screen at a surface water intake.

17 Water Treatment Operations

INTRODUCTION

Municipal water treatment operations and associated treatment unit processes are designed to provide reliable, high-quality water service for customers, and to preserve and protect the environment for future generations.

Water management officials and treatment plant operators are tasked with exercising responsible financial management, ensuring fair rates and charges, providing responsive customer service, providing a consistent supply of safe potable water for consumption by users, and promoting environmental responsibility.

While studying this chapter on water and the next one on wastewater treatment plant operation, keep in mind the major point in both chapters: Water and wastewater treatment plant design can be taught in school, but when operating the plants, experience, attention to detail, and operator common sense are most important.

In the past, water quality was described as "wholesome and delightful" and "sparkling to the eye." Today, we describe water quality as "safe and healthy to drink."

This chapter focuses on water treatment operations and the various unit processes currently used to treat raw source water before it is distributed to users. In addition, it discusses the reasons for water treatment and the basic theories associated with individual treatment unit processes. Water treatment systems are installed to remove materials that cause disease and/or create nuisances. At its simplest level, the basic goal of water treatment operations is to protect public health, with a broader goal to provide potable and palatable water. The bottom line: The water treatment process functions to provide water that is safe to drink and pleasant in appearance, taste, and odor.

In this text, water treatment is defined as any unit process that changes/alters the chemical, physical, and/or bacteriological quality of water to make it safe for human consumption and/or appeal to the customer. Treatment also is used to protect the water distribution system components from corrosion.

Many water treatment unit processes are commonly used today. The processes used depend on the evaluation of the nature and quality of the particular water to be treated and the desired quality of the finished water. In the water treatment unit, processes are employed to treat raw water. One thing is certain: as new USEPA regulations take effect, many more processes will come into use in an attempt to produce water that complies with all current regulations, despite the conditions of source water.

Small water systems tend to use a smaller number of the wide array of unit treatment processes available, in part because they usually rely on groundwater as the source, and in part because small size makes many sophisticated processes impractical (i.e., too expensive to install, too expensive to operate, too sophisticated for limited operating staff). This chapter concentrates on those individual treatment unit processes usually found in conventional water treatment systems, corrosion control methods, and fluoridation. A summary of basic water treatment processes (many of which are discussed in this chapter) is presented in Table 17.1.

WATERWORKS OPERATOR (A.K.A. THE FLUID MECHANIC)

The operation of a water treatment system, regardless of its size or complexity, requires operators. To perform their functions at the highest knowledge and experience level possible, operators must understand the basic principles and theories behind many complex water treatment concepts and systems. Under new regulations, waterworks operators must be certified or licensed. Actual water treatment protocols and procedures are important; however, without proper implementation, they are nothing more than hollow words occupying space on reams of paper. This is where the waterworks operator comes in. Successfully treating water requires skill, dedication, and vigilance. The waterworks operator must not only be highly trained and skilled but also conscientious—the ultimate user demands nothing less. The role of the waterworks operator can be succinctly stated:

- Waterworks operators provide water that complies with state Waterworks Regulations, water that is safe to drink and ample in quantity and pressure without interruption.
- Waterworks operators must know their facilities.
- Waterworks operators must be familiar with bacteriology, chemistry, and hydraulics.
- Waterworks operators must stay abreast of technological changes and keep current with water supply information.
- When operating a waterworks facility, waterworks operator duties include:
  - Maintaining the distribution system
  - Collecting or analyzing water samples
  - Operating chemical feed equipment
  - Keeping records
  - Operating treatment unit processes
  - Performing sanitary surveys of the water supply watershed
  - Operating a cross-connection control program

DOI: 10.1201/9781003581901-21

TABLE 17.1
Basic Water Treatment Processes

Process/Step	Purpose
Screening	Removes large debris (leaves, sticks, fish) that can foul or damage plant equipment
Chemical pretreatment	Conditions the water for removal of algae and other aquatic nuisances
Presedimentation	Removes gravel, sand, silt, and other gritty materials
Microstraining	Removes algae, aquatic plants, and small debris
Chemical feed and rapid mix	Adds chemicals (coagulants, pH, adjusters, etc.) to water
Coagulation/flocculation	Converts nonsettleable or settable particles
Sedimentation	Removes settleable particles
Softening	Removes hardness-causing chemicals from water
Filtration	Removes particles of solid matter which can include biological contamination and turbidity
Disinfection	Kills disease-causing organisms
Adsorption using granular activated	Removes radon and many organic chemicals such as pesticides, solvents, and trihalomethanes
Aeration	Removes volatile organic chemicals (VOCs), radon H_2S, and other dissolved gases; oxidizes iron and manganese
Corrosion control	Prevents scaling and corrosion
Reverse osmosis, electrodialysis	Removes nearly all inorganic contaminants
Ion exchange	Removes some inorganic contaminants, including hardness-causing chemicals
Activated alumina	Removes some inorganic contamination
Oxidation filtration	Removes some inorganic contaminants (e.g., iron, manganese, radium)

Source: Adapted from AWWA (2010b).

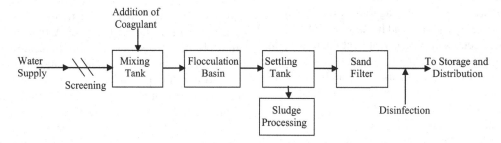

FIGURE 17.1 Conventional water treatment model.

PURPOSE OF WATER TREATMENT

As mentioned, the purpose of water treatment is to condition, modify and/or remove undesirable impurities to provide water that is safe, palatable, and acceptable to users. While this is the obvious and expected purpose of treating water, various regulations also require water treatment. Some regulations state that if the contaminants listed under various regulations are found in excess of maximum contaminant levels (MCLs), the water must be treated to reduce those levels. If a well or spring source is surface-influenced, treatment is required, regardless of the actual presence of contamination. Some impurities affect the aesthetic qualities (taste, odor, color, and hardness) of the water; if they exceed secondary MCLs established by the USEPA and the state, the water may need to be treated.

If we assume that the water source used to feed a typical water supply system is groundwater (usually the case in the United States), several common groundwater problems may require water treatment. Keep in mind that water that must

be treated for any one of these problems may also exhibit several other problems. Among these problems are:

- Bacteriological contamination
- Hydrogen sulfide odors
- Hard water
- Corrosive water
- Iron and manganese

STAGES OF WATER TREATMENT

Earlier, it was stated that the focus of our discussion in this text is on the conventional water treatment model. Figure 17.1 presents the basic conventional model, clearly illustrating that water treatment involves various stages, unit processes, or a train of processes combined to form one treatment system. Note that a given waterworks may contain all the unit processes discussed in the following sections or any combination of them. One or more of these stages may be used to treat any one or more of the source water problems listed above. Also, note that the model shown in Figure 17.1 does not necessarily apply to very small water systems. In some small systems, water treatment may consist of nothing more than the removal of water via pumping from a groundwater source to storage to distribution. In some small water supply operations, disinfection may be added because it is required. Although it is likely that the basic model shown in Figure 17.1 does not reflect the type of treatment process used in most small systems, we use it in this handbook for illustrative and instructive purposes because higher-level licensure requires operators, at a minimum, to learn these processes.

PRETREATMENT

Simply stated, water pretreatment (also called preliminary treatment) is any physical, chemical, or mechanical process used before the main water treatment processes. It can include screening, presedimentation, and chemical addition (see Figure 17.1). Pretreatment in water treatment operations usually consists of oxidation or other treatments for the removal of tastes and odors, iron and manganese, trihalomethane precursors, or entrapped gases (like hydrogen sulfide). Unit processes may include chlorine, potassium permanganate, or ozone oxidation, activated carbon addition, aeration, and presedimentation. Pretreatment of surface water supplies accomplishes the removal of certain constituents and materials that interfere with or place an unnecessary burden on conventional water treatment facilities.

Based on our experience and according to the Texas Water Utilities Association's *Manual of Water Utility Operations*, 8th edition, Hutto, Texas, typical pretreatment processes include the following:

1. Removal of debris from water sourced from rivers and reservoirs that would clog pumping equipment.
2. Destratification of reservoirs to prevent anaerobic decomposition, which could result in reducing iron and manganese from the soil to a soluble state. This can cause subsequent removal problems in the treatment plant. The production of hydrogen sulfide and other taste- and odor-producing compounds also results from stratification.
3. Chemical treatment of reservoirs to control the growth of algae and other aquatic organisms that could result in taste and odor problems.
4. Presedimentation to remove excessively heavy silt loads prior to the treatment processes.
5. Aeration to remove dissolved odor-causing gases, such as hydrogen sulfide and other dissolved gases or volatile constituents, and to aid in the oxidation of iron and manganese, although manganese or high concentrations of iron are not removed in the detention provided in conventional aeration units.

6. Chemical oxidation of iron and manganese, sulfides, taste- and odor-producing compounds, and organic precursors that may produce trihalomethanes upon the addition of chlorine.
7. Adsorption for the removal of tastes and odors.

Note: An important point to keep in mind is that in small systems using groundwater as a source, pretreatment may be the only treatment process used.

Note: Pretreatment may be incorporated as part of the total treatment process or may be located adjacent to the source before the water is sent to the treatment facility.

AERATION

Aeration is commonly used to treat water that contains trapped gases (such as hydrogen sulfide) that can impart an unpleasant taste and odor to the water. Just allowing the water to rest in a vented tank will (sometimes) drive off much of the gas, but usually some form of forced aeration is needed. Aeration works well (about 85% of the sulfides may be removed) whenever the pH of the water is less than 6.5. Aeration may also be useful in oxidizing iron and manganese, oxidizing humic substances that might form trihalomethanes when chlorinated, eliminating other sources of taste and odor, or imparting oxygen to oxygen-deficient water.

Note: Iron is a naturally occurring mineral found in many water supplies. When the concentration of iron exceeds 0.3 mg/L, red stains will occur on fixtures and clothing. This increases customer costs for cleaning and replacement of damaged fixtures and clothing. Manganese, like iron, is a naturally occurring mineral found in many water supplies. When the concentration of manganese exceeds 0.05 mg/L, black stains occur on fixtures and clothing. As with iron, this increases customer costs for cleaning and replacement of damaged fixtures and clothing. Iron and manganese are commonly found together in the same water supply. We discuss iron and manganese later.

SCREENING

Screening is usually the first major step in the water pretreatment process (see Figure 17.1). It is defined as the process whereby relatively large and suspended debris is removed or retained from the water before it enters the plant. River water, for example, typically contains suspended and floating debris varying in size from small rocks to logs. Removing these solids is important, not only because these items have no place in potable water, but also because this river trash may cause damage to downstream equipment (clogging and damaging pumps, etc.), increase chemical requirements, impede hydraulic flow in open channels or pipes, or hinder the treatment process. The most important criteria used in the selection of a particular screening system for water treatment technology are the screen opening size and flow rate; they range in size from microscreens to trash racks. Other important criteria include costs related to operation and equipment, plant hydraulics, debris handling requirements, and operator qualifications and availability. Large surface water treatment plants may employ a variety of screening devices including trash screens (or trash rakes), traveling water screens, drum screens, bar screens, or passive screens.

CHEMICAL ADDITION

Much of the procedural information presented in this section applies to both water and wastewater operations. Two of the major chemical pretreatment processes used in treating water for potable use are iron and manganese and hardness removal. Another chemical treatment process that is not necessarily part of the pretreatment process, but is also discussed in this section, is corrosion control. Corrosion prevention is affected by chemical treatment— not only in the treatment process but also in the distribution process. Before discussing each of these treatment methods in detail, however, it is important to describe chemical addition, chemical feeders, and chemical feeder calibration.

When chemicals are used in the pretreatment process, they must be the proper ones, fed in the proper concentration, and introduced to the water at the proper locations. Determining the proper amount of chemical to use is accomplished by testing. The operator must test the raw water periodically to determine if the chemical dosage should be adjusted. For surface supplies, checking must be done more frequently than for groundwater (remember, surface water supplies are subject to change on short notice, while groundwater generally remains stable). The operator must be aware of the potential for interactions between various chemicals and how to determine the optimum dosage (e.g., adding both chlorine and activated carbon at the same point will minimize the effectiveness of both processes, as the adsorptive power of the carbon will be used to remove the chlorine from the water).

Note: Sometimes, using too many chemicals can be worse than not using enough.

Prechlorination (distinguished from chlorination used in disinfection at the end of treatment) is often used as an oxidant to help with the removal of iron and manganese. However, currently, concern for systems that prechlorinate is prevalent because of the potential for the formation of total trihalomethanes (TTHMs), which form as a by-product of the reaction between chlorine and naturally occurring compounds in raw water. The USEPA's TTHM standard does not apply to water systems that serve less than 10,000 people, but operators should be aware of the impact and causes of TTHMs. Chlorine dosage or application points may be changed to reduce problems with TTHMs.

Note: TTHMs such as chloroform are known or suspected to be carcinogenic and are limited by water and state regulations.

Note: To be effective, pretreatment chemicals must be thoroughly mixed with the water. Short-circuiting or plug flows of chemicals that do not come in contact with most of the water will not result in proper treatment.

All chemicals intended for use in drinking water must meet certain standards. Thus, when ordering water treatment chemicals, the operator must be assured that they meet all appropriate standards for drinking water use.

Chemicals are normally fed with dry chemical feeders or solution (metering) pumps. Operators must be familiar with all of the adjustments needed to control the rate at which the chemical is fed to the water (or wastewater). Some feeders are manually controlled and must be adjusted by the operator when the raw water quality or the flow rate changes; other feeders are paced by a flow meter to adjust the chemical feed so it matches the water flow rate. Operators must also be familiar with chemical solutions and feeder calibration.

As mentioned, a significant part of a waterworks operator's important daily operational functions includes measuring quantities of chemicals and applying them to water at preset rates. Normally accomplished semi-automatically by the use of electro-mechanical-chemical feed devices, waterworks operators must still know what chemicals to add, how much to add to the water (or wastewater), and the purpose of the chemical addition.

Chemical Solutions

A *water solution* is a homogeneous liquid made of the *solvent* (the substance that dissolves another substance) and the *solute* (the substance that dissolves in the solvent). Water is the solvent (see Figure 17.2). The solute (whatever it may be) may dissolve up to a certain limit. This is called its *solubility*—that is, the solubility of the solute in a particular solvent (water) at a particular temperature and pressure.

Note: Temperature and pressure influence the stability of solutions, but not through filtration, because only suspended material can be eliminated by filtration or sedimentation.

Remember, in chemical solutions, the substance being dissolved is called the solute, and the liquid present in the greatest amount in a solution (that does the dissolving) is called the solvent. The operator should also be familiar with another term, *concentration*—the amount of solute

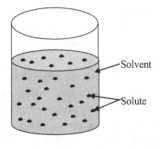

FIGURE 17.2 Solution with two components: solvent and solute.

dissolved in a given amount of solvent. Concentration is measured as:

$$\% \text{ Strength} = \frac{\text{Wt. of solute}}{\text{Wt. of solution}} \times 100$$

$$= \frac{\text{Wt. of solute}}{\text{Wt. of solute + solvent}} \times 100 \qquad (17.1)$$

Example 17.1

Problem: If 30 lb of chemical is added to 400 lb of water, what is the percent strength (by weight) of the solution?

Solution:

$$\% \text{ Strength} = \frac{30 \text{ lb solute}}{400 \text{ lb water}} \times 100$$

$$= \frac{30 \text{ lb solute}}{30 \text{ lb solute} + 400 \text{ lb water}} \times 100$$

$$= \frac{30 \text{ lb solute}}{430 \text{ lb solute/water}}$$

$$\% \text{ Strength} = 7.0 \text{ (rounded)}$$

Important to the process of making accurate computations of chemical strength is a complete understanding of the dimensional units involved. For example, operators should understand exactly what *milligrams per liter (mg/L)* signify.

$$\text{Milligrams per Liter (mg/L)} = \frac{\text{Milligrams of Solute}}{\text{Liters of Solution}}$$
$$(17.2)$$

Another important dimensional unit commonly used when dealing with chemical solutions is *parts per million (ppm)*.

$$\text{Parts per Million (ppm)} = \frac{\text{Parts of Solute}}{\text{Million Parts of Solution}}$$
$$(17.3)$$

Note: "Parts" is usually a weight measurement.
   An example is:

$$9 \text{ ppm} = \frac{9 \text{ lb solids}}{1,000,000 \text{ lb solution}}$$

Or

$$9 \text{ ppm} = \frac{9 \text{ mg solids}}{1,000,000 \text{ mg solution}}$$

This leads us to two important parameters that operators should commit to memory.

Concentrations—Units and Conversions

$$1 - \text{mg/L} = 1 \text{ ppm}$$

$$1\% = 10,000 \text{ mg/L}$$

When working with chemical solutions, you should also be familiar with two chemical properties we briefly described earlier: density and specific gravity. *Density* is defined as the weight of a substance per unit of its volume; for example, pounds per cubic foot or pounds per gallon. *Specific gravity* is defined as the ratio of the density of a substance to a standard density.

$$\text{Density} = \frac{\text{Mass of substance}}{\text{Volume of substance}} \qquad (17.4)$$

Here are a few key facts about density (of water):

- measured in units of lb/cf, lb/gal, or mg/L
- density of water = 62.5 lbs/cf = 8.34 lb/gal
- other densities: concrete = 130 lb/cf, alum (liquid, @ 60°F) = 1.33, and hydrogen peroxide (35%) = 1.132

$$\text{Specific Gravity} = \frac{\text{Density of substance}}{\text{Density of water}} \qquad (17.5)$$

Here are a few facts about specific gravity:

- Specific gravity has no units.
- Specific gravity of water = 1.0
- Specific gravity of concrete = 2.08, alum (liquid, @ 60°F) = 1.33, and hydrogen peroxide (355) = 1.132

CHEMICAL FEEDERS

Simply put, a chemical feeder is a mechanical device for measuring a quantity of chemical and applying it to water at a preset rate.

Types of Chemical Feeders

Two types of chemical feeders are commonly used: solution (or liquid) feeders and dry feeders. Liquid feeders apply chemicals in solutions or suspensions, while dry feeders apply chemicals in granular or powdered forms. In a solution feeder, the chemical enters the feeder and leaves the feeder in a liquid state; in a dry feeder, the chemical enters and leaves the feeder in a dry state.

Solution Feeders

Solution feeders are small, positive displacement metering pumps of three types: (1) reciprocating (piston-plunger or diaphragm types); (2) vacuum type (e.g., gas chlorinator); or (3) gravity feed rotameter (e.g., drip feeder). Positive displacement pumps are used in high-pressure, low-flow applications; they deliver a specific volume of liquid for each stroke of a piston or rotation of an impeller.

Dry Feeders

Two types of dry feeders are volumetric and gravimetric, depending on whether the chemical is measured by volume (volumetric type) or weight (gravimetric type). Simpler and less expensive than gravimetric pumps, volumetric dry feeders are also less accurate. Gravimetric dry feeders are extremely accurate, deliver high feed rates, and are more expensive than volumetric feeders.

Chemical Feeder Calibration

Chemical feeder calibration ensures effective control of the treatment process. Obviously, chemical feed without some type of metering and accounting of chemicals used adversely affects the water treatment process. Chemical feeder calibration also optimizes the economy of operation; it ensures the optimum use of expensive chemicals. Finally, operators must have accurate knowledge of each feeder's capabilities in specific settings. When a certain dose must be administered, the operator must rely on the feeder to deliver the correct amount of chemical. Proper calibration ensures that chemical dosages can be set with confidence. At a minimum, chemical feeders must be calibrated on an annual basis. During operation, when the operator changes chemical strength or chemical purity or makes any adjustment to the feeder, or when the treated water flow changes, the chemical feeder should be calibrated. Ideally, any time maintenance is performed on chemical feed equipment, calibration should be conducted.

What factors affect chemical feeder calibration (i.e., feed rate)? For solution feeders, calibration is affected any time the solution strength changes, any time a mechanical change is introduced in the pump (change in stroke length or stroke frequency), and/or whenever the flow rate changes. In the dry chemical feeder, calibration is affected any time chemical purity changes, mechanical damage occurs (e.g., belt change), and/or whenever the flow rate changes. In the calibration process, calibration charts are usually used or created to fit the calibration equipment. The calibration chart is also affected by certain factors, including changes in chemical, changes in the flow rate of water being treated, and/or a mechanical change in the feeder.

Calibration Procedures

When calibrating a positive displacement pump (liquid feeder), the operator should always refer to the manufacturer's technical manual. Keeping in mind the need to refer to the manufacturer's specific guidelines, we provide general examples here of calibration procedures for simple positive-displacement pumps and dry feeder calibration procedures.

Calibration Procedure: Positive Displacement Pump

The following equipment is required:

- Graduated cylinder (1,000 mL or less)
- Stopwatch
- Calculator
- Graph paper
- Plain paper
- Straight edge

The procedure is as follows:

1. Fill the graduated cylinder with solution.
2. Insert the pump suction line into the graduated cylinder.
3. Run the pump 5 min at the highest setting (100%).
4. Divide the mL of liquid withdrawn by 5 min to determine the pumping rate (mL/min) and record it on plain paper.
5. Repeat steps 3 and 4 at 100% setting.
6. Repeat steps 3 and 4 for 20%, 50%, and 70% settings twice.
7. Average the mL/min pumped for each setting.
8. Calculate the weight of the chemical pumped for each setting.
9. Calculate the dosage for each setting.
10. Graph dosage versus setting.

Calibration Procedure: Dry Feeder

The following equipment is required:

- Weighing pan
- Balance
- Stopwatch
- Plain paper
- Graph paper
- Straight edge
- Calculator

The procedure is as follows:

1. Weigh the pan and record it.
2. Set the feeder to 100% setting.
3. Collect samples for 5 min.
4. Calculate the weight of the sample and record it in the table.
5. Repeat steps 3 and 4 twice.
6. Repeat steps 3 and 4 for settings of 25%, 50%, and 75%.
7. Calculate the average sample weight per minute for each setting and record it in the table.
8. Calculate the weight fed per day for each setting.
9. Plot the weight fed per day versus the setting on graph paper.

Note: Pounds per day (lbs/day) is not normally useful information for setting the feed rate on a feeder. This is because process control usually determines a dosage in ppm, mg/L, or grains/gal. A separate chart may be necessary for another conversion based on the individual treatment facility's flow rate.

Example 17.2

To demonstrate that performing a chemical feed procedure is not as simple as opening a bag of chemicals and dumping the contents into the feed system, we provide a real-world example below.

Problem: Consider the chlorination dosage rates below.

Setting		Dosage
100%	111/121	0.93 mg/L
70%	78/121	0.66 mg/L
50%	54/121	0.45 mg/L
20%	20/121	0.16 mg/L

Solution:

This is not a good dosage setup for a chlorination system. Maintenance of a chlorine residual at the ends of the distribution system should be within 0.5–1.0 ppm. At 0.9 ppm, the dosage will probably result in this range—depending on the chlorine demand of the raw water and detention time in the system. However, the pump is set at its highest setting. We have room to decrease the dosage, but cannot increase the dosage without changing the solution strength in the solution tank. In this example, doubling the solution strength to 1% provides the ideal solution, resulting in the following chart changes.

Setting		Dosage
100%	222/121	1.86 mg/L
70%	154/121	1.32 mg/L
50%	108/121	0.90 mg/L
20%	40/121	0.32 mg/L

This is ideal because the dosage we want to feed is at the 50% setting for our chlorinator. We can now easily increase or decrease the dosage whereas the previous setup only allowed the dosage to decrease.

IRON AND MANGANESE REMOVAL

Iron and manganese are frequently found in groundwater and some surface waters. They do not cause health-related problems but are objectionable because they may cause aesthetic problems. Severe aesthetic problems may cause consumers to avoid an otherwise safe water supply in favor of one of unknown or questionable quality or may cause them to incur unnecessary expenses for bottled water. Aesthetic problems associated with iron and manganese include the discoloration of water (iron=reddish water, manganese=brown or black water), staining of plumbing fixtures, imparting a bitter taste to the water, and stimulating the growth of microorganisms.

As mentioned, there are no direct health concerns associated with iron and manganese, although the growth of iron bacteria in slimes may cause indirect health problems. Economic problems include damage to textiles, dye, paper, and food. The iron residue (or tuberculation) in pipes increases the pumping head, decreases the carrying capacity, may clog pipes, and may corrode through pipes.

Note: Iron and manganese are secondary contaminants. Their secondary maximum contaminant levels (SMCLs) are iron=0.3 mg/L and manganese=0.05 mg/L.

Iron and manganese are most likely found in groundwater supplies, industrial waste, and acid mine drainage, and as by-products of pipeline corrosion. They may accumulate in lake and reservoir sediments, causing possible problems during lake/reservoir turnover. They are not usually found in running waters (streams, rivers, etc.).

Iron and Manganese Removal Techniques

Chemical precipitation treatments for iron and manganese removal are called *deferrization* and *demanganization*. The usual process is *aeration;* dissolved oxygen in the chemical causing precipitation; chlorine or potassium permanganate may be required.

Precipitation

Precipitation (or pH adjustment) of iron and manganese from water in their solid forms can be affected in treatment plants by adjusting the pH of the water by adding lime or other chemicals. Some of the precipitates will settle out with time, while the rest is easily removed by sand filters. This process requires the pH of the water to be in the range of 10–11.

Note: Although the precipitation or pH adjustment technique for treating water containing iron and manganese is effective, the pH level must be adjusted higher (10–11 range) to cause the precipitation, which means that the pH level must then also be lowered (to the 8.5 range or a bit lower) to use the water for consumption.

Oxidation

One of the most common methods of removing iron and manganese is through the process of oxidation (another chemical process), usually followed by settling and filtration. Air, chlorine, or potassium permanganate can oxidize these minerals. Each oxidant has advantages and disadvantages, and each operates slightly differently:

1. **Air:** To be effective as an oxidant, the air must come in contact with as much of the water as possible. Aeration is often accomplished by bubbling diffused air through the water, by spraying the water up into the air, or by trickling the water over rocks, boards, or plastic packing materials in an aeration tower. The more finely divided the drops of water, the more oxygen comes in contact with the water and the dissolved iron and manganese.
2. **Chlorine:** This is one of the most popular oxidants for iron and manganese control because it is also widely used as a disinfectant. Iron and manganese control by prechlorination can be as simple as adding a new chlorine feed point in a facility already feeding chlorine. It also provides a pre-disinfecting step that can help control bacterial growth through the rest of the treatment system.

The downside to chlorine use; however, is that when chlorine reacts with the organic materials found in surface water and some groundwaters, it forms TTHMs. This process also requires that the pH of the water be in the range of 6.5–7; because many groundwaters are more acidic than this, pH adjustment with lime, soda ash, or caustic soda may be necessary when oxidizing with chlorine.

3. **Potassium Permanganate:** This is the best oxidizing chemical to use for manganese removal. An extremely strong oxidant, it has the additional benefit of producing manganese dioxide during the oxidation reaction. Manganese dioxide acts as an adsorbent for soluble manganese ions, providing removal to extremely low levels.

The oxidized compounds form precipitates that are removed by a filter. Note that sufficient time should be allowed from the addition of the oxidant to the filtration step. Otherwise, the oxidation process will be completed after filtration, creating insoluble iron and manganese precipitates in the distribution system.

Ion Exchange

The ion exchange process is used primarily to soften hard waters, but it also removes soluble iron and manganese. The water passes through a bed of resin that adsorbs undesirable ions from the water, replacing them with less troublesome ions. When the resin has given up all its donor ions, it is regenerated with strong salt brine (sodium chloride); the sodium ions from the brine replace the adsorbed ions and restore the ion exchange capabilities.

Sequestering

Sequestering or stabilization may be used when the water contains mainly a low concentration of iron, and the volumes needed are relatively small. This process does not actually remove the iron or manganese from the water but complexes (binds it chemically) it with other ions in a soluble form that is not likely to come out of solution (i.e., not likely oxidized).

Aeration

The primary physical process uses air to oxidize the iron and manganese. The water is either pumped into the air or allowed to fall over an aeration device. The air oxidizes the iron and manganese, which is then removed using a filter. Adding lime to raise the pH is often part of the process. While this is called a physical process, removal is accomplished by chemical oxidation.

Potassium Permanganate Oxidation and Manganese Greensand

The continuous regeneration potassium greensand filter process is another commonly used filtration technique for iron and manganese control. Manganese greensand is a mineral (glauconite) treated with alternating solutions of manganous chloride and potassium permanganate. The result is a sand-like (zeolite) material coated with a layer of manganese dioxide—an adsorbent for soluble iron and manganese. Manganese greensand has the ability to capture (adsorb) soluble iron and manganese that may have escaped oxidation, as well as the capability of physically filtering out the particles of oxidized iron and manganese. Manganese greensand filters are generally set up as pressure filters, totally enclosed tanks containing the greensand. The process of adsorbing soluble iron and manganese "uses up" the greensand by converting the manganese dioxide coating to manganic oxide, which does not have the adsorption property. The greensand can be regenerated in much the same way as ion exchange resins, by washing the sand with potassium permanganate.

Hardness Treatment

Hardness in water is caused by the presence of certain positively charged metallic irons in solution in the water. The most common hardness-causing ions are calcium and magnesium; others include iron, strontium, and barium. As a general rule, groundwaters are harder than surface waters, thus, hardness is frequently of concern to the small water system operator. This hardness is derived from contact with soil and rock formations such as limestone. Although rainwater itself will not dissolve many solids, the natural carbon dioxide in the soil enters the water and forms carbonic acid (HCO), capable of dissolving minerals. Where soil is thick (contributing more carbon dioxide to the water) and limestone is present, hardness is likely to be a problem. The total amount of hardness in water is expressed as the sum of its calcium carbonate ($CaCO_3$) and its magnesium hardness. However, for practical purposes, hardness is expressed as calcium carbonate. This means that regardless of the amount of the various components that make up hardness, they can be related to a specific amount of calcium carbonate (e.g., hardness is expressed as mg/L as $CaCO_3$—milligrams per liter as calcium carbonate).

Note: The two types of water hardness are temporary hardness and permanent hardness. Temporary hardness, also known as carbonate hardness, can be removed by boiling, while permanent hardness, also known as noncarbonate hardness, cannot be removed by boiling.

Hardness is a concern in domestic water consumption because hard water increases soap consumption, leaves a soapy scum in the sink or tub, can cause water heater electrodes to burn out quickly, can cause discoloration of plumbing fixtures and utensils, and is perceived as less desirable water. In industrial water use, hardness is a concern because it can cause boiler scale and damage to industrial equipment.

The objection of customers to hardness often depends on the amount of hardness they are used to. People familiar with water with a hardness of 20 mg/L might think that a hardness of 100 mg/L is too much. On the other hand, a person who has been using water with a hardness of 200 mg/L

TABLE 17.2
Classification of Hardness

Classification	mg/L CaCO₃
Soft	0–75
Moderately hard	75–150
Hard	150–300
Very hard	Over 300

might think that 100 mg/L is very soft. Table 17.2 lists the classifications of hardness.

Hardness Calculation

Recall that hardness is expressed as mg/L as $CaCO_3$. The mg/L of Ca and Mg must be converted to mg/L as $CaCO_3$ before they can be added. The hardness (in mg/L as $CaCO_3$) for any given metallic ion is calculated using the formula:

$$Hardness\,(mg/L\ as\ CaCO_3) = M\,(mg/L) \times \frac{50}{Eq.\ Wt.\ of\ M}$$

(17.6)

where:

M = metal ion concentration (mg/L)
Eq. Wt. = equivalent weight = gram molecular weight ÷ valence

Treatment Methods

Two common methods are used to reduce hardness:

• **Ion Exchange Process:** The ion exchange process is the most frequently used method for softening water. Accomplished by charging a resin with sodium ions, the resin exchanges the sodium ions for calcium and/or magnesium ions. Naturally occurring and synthetic cation exchange resins are available. Natural exchange resins include such substances as aluminum silicate, zeolite clays (zeolites are hydrous silicates found naturally in the cavities of lavas, such as greensand; glauconite zeolites; or synthetic, porous zeolites), humus, and certain types of sediments. These resins are placed in a pressure vessel, and salt brine is flushed through the resins. The sodium ions in the salt brine attach to the resin, which is now said to be charged. Once charged, water is passed through the resin, and the resin exchanges the sodium ions attached to it for calcium and magnesium ions, thus removing them from the water. The zeolite clays are most common because they are quite durable, can tolerate extreme ranges in pH, and are chemically stable. They have relatively limited exchange capacities, however, so they should be used only for water with moderate total hardness. One of the results is that the water may be more corrosive than before.

Another concern is that the addition of sodium ions to the water may increase the health risk for those with high blood pressure.

• **Cation Exchange Process:** The cation exchange process takes place with little or no intervention from the treatment plant operator. Water containing hardness-causing cations (Ca^{++}, Mg^{++}, Fe^{+3}) is passed through a bed of cation exchange resin. The water coming through the bed contains hardness near zero, although it will have elevated sodium content. (The sodium content is not likely to be high enough to be noticeable, but it could be high enough to pose problems for people on highly restricted salt-free diets.) The total lack of hardness in the finished water is likely to make it very corrosive, so normal practice bypasses a portion of the water around the softening process. The treated and untreated waters are blended to produce an effluent with a total hardness of around 50–75 mg/L as $CaCO_3$.

CORROSION CONTROL

Water operators add chemicals (e.g., lime or sodium hydroxide) to water at the source or at the waterworks to control corrosion. Using chemicals to achieve a slightly alkaline chemical balance prevents the water from corroding distribution pipes and consumers' plumbing. This keeps substances like lead from leaching out of plumbing and into the drinking water.

For our purposes, we define *corrosion* as the conversion of a metal to a salt or oxide with a loss of desirable properties such as mechanical strength. Corrosion may occur over an entire exposed surface or may be localized at micro- or macroscopic discontinuities in the metal. In all types of corrosion, a gradual decomposition of the material occurs, often due to an electrochemical reaction. Corrosion may be caused by (1) stray current electrolysis, (2) galvanic corrosion caused by dissimilar metals, or (3) differential concentration cells. Corrosion starts at the surface of a material and moves inward.

The adverse effects of corrosion can be categorized according to health, aesthetics, economic effects, and/or other effects. The corrosion of toxic metal pipes made from lead creates a serious *health hazard*. Lead tends to accumulate in the bones of humans and animals. Signs of lead intoxication include gastrointestinal disturbances, fatigue, anemia, and muscular paralysis. Lead is not a natural contaminant in either surface waters or groundwaters, and the MCL of 0.005 mg/L in source waters is rarely exceeded. It is a corrosion by-product from high-lead solder joints in copper and lead piping. Small doses of lead can lead to developmental problems in children. The USEPA's Lead and Copper Rule addresses the matter of lead in drinking water exceeding specified action levels.

Note: The USEPA's Lead and Copper Rule requires that a treatment facility achieve optimum corrosion control.

Since lead and copper contamination generally occurs after water has left the public water system, the best way for the water system operator to find out if customer water is contaminated is to test water that has come from a household faucet.

With regard to the Lead and Copper Rule, it is important to note that the EPA made minor changes in 1999 to the original rule. These minor revisions (also known as the Lead and Copper Rule Minor Revisions or LCRMR) streamline requirements, promote consistent national implementation, and, in many cases, reduce the burden for water systems. The LCRMR does not change the action levels of 0.015 mg/L for lead and 1.3 mg/L for copper, or the Maximum Contaminant Level Goals established by the 1991 Lead and Copper Rule ("the rule"), which are 0 mg/L for lead and 1.3 mg/L for copper. They also do not affect the rule's basic requirements to optimize corrosion control and, if appropriate, treat source water, deliver public education, and replace lead service lines (USEPA, 1990a).

Cadmium is the only other toxic metal found in samples from plumbing systems. Cadmium is a contaminant found in zinc. Its adverse health effects are best known for being associated with severe bone and kidney syndrome in Japan. The PMCL for cadmium is 0.01 mg/L.

Note: Water systems should try to supply water free of lead and have no more than 1.3 mg of copper (mg/L). This is a non-enforceable health goal.

Aesthetic effects that result from the corrosion of iron are characterized by "pitting" and are a consequence of the deposition of ferric hydroxide and other products and the solution of iron—*tuberculation*. Tuberculation reduces the hydraulic capacity of the pipe. Corrosion of iron can cause customer complaints of reddish or reddish-brown staining of plumbing fixtures and laundry. Corrosion of copper lines can cause customer complaints of bluish or blue-green stains on plumbing fixtures. Sulfide corrosion of copper and iron lines can cause a blackish color in the water. The by-products of microbial activity (especially iron bacteria) can cause foul tastes and/or odors in the water.

The economic effects of corrosion may include the need for water main replacement, especially when tuberculation reduces the flow capacity of the main. Tuberculation increases pipe roughness, causing an increase in pumping costs and reducing distribution system pressure. Tuberculation and corrosion can cause leaks in distribution mains and household plumbing. Corrosion of household plumbing may require extensive treatment, public education, and other actions under the Lead and Copper Rule.

Other effects of corrosion include the short service life of household plumbing caused by pitting. The buildup of mineral deposits in the hot water system may eventually restrict hot water flow. Also, the structural integrity of steel water storage tanks may deteriorate, causing structural failures. Steel ladders in clear wells or water storage tanks may corrode, introducing iron into the finished water. Steel parts in flocculation tanks, sedimentation basins, clarifiers, and filters may also corrode.

Types of Corrosion

Three types of corrosion occur in water mains: galvanic, tuberculation, and pitting:

- *Galvanic* occurs when two dissimilar metals are in contact and are exposed to a conductive environment—a potential exists between them, and current flows. This type of corrosion is the result of an electrochemical reaction in which the flow of electric current itself is an essential part of the reaction.
- *Tuberculation* refers to the formation of localized corrosion products scattered over the surface in the form of knob-like mounds. These mounds increase the roughness of the inside of the pipe, increasing resistance to water flow and decreasing the C-factor of the pipe.
- *Pitting* is localized corrosion that is generally classified as pitting when the diameter of the cavity at the metal surface is the same as or less than the depth.

Factors Affecting Corrosion

The primary factors affecting corrosion are pH, alkalinity, hardness (calcium), dissolved oxygen, and total dissolved solids. Secondary factors include temperature, velocity of water in pipes, and carbon dioxide (CO_2).

Determination of Corrosion problems

To determine if corrosion is taking place in water mains, materials removed from the distribution system should be examined for signs of corrosion damage. A primary indicator of corrosion damage is pitting. (Note: measure the depth of pits to gauge the extent of damage.) Another common method used to determine if corrosion or scaling is taking place in distribution lines is by inserting special steel specimens of known weight (called *coupons*) in the pipe and examining them for corrosion after a period of time. Detecting evidence of leaks, conducting flow tests, and performing chemical tests for dissolved oxygen and toxic metals, as well as customer complaints (red or black water and/or laundry and fixture stains), are also used to indicate corrosion problems.

Formulas can also be used to determine corrosion (to an extent). The *Langelier Saturation Index* (L.I.) and *Aggressive Index* (A.I.) are two commonly used indices. The L.I. is a method used to determine if water is corrosive. A.I. refers to waters that have low natural pH, are high in dissolved oxygen, are low in total dissolved solids, and have low alkalinity and low hardness. These waters are very aggressive and can be corrosive. Both the Langelier Saturation and Aggressive Indices are typically used as starting points in determining the adjustments required to produce a film:

L.I. approximately 0.5
A.I. value of 12 or higher

Note: L.I. and A.I. are based on the dissolving and precipitation of calcium carbonate; therefore, the respective indices may not actually reflect the corrosive nature of the particular water for a specific pipe material. However, they can be useful tools in selecting materials or treatment options for corrosion control.

As mentioned, one method used to reduce the corrosive nature of water is *chemical addition*. The selection of chemicals depends on the characteristics of the water, where the chemicals can be applied, how they can be applied and mixed with water, and the cost of the chemicals.

If the product of the calcium hardness times the alkalinity of the water is less than 100 treatments may be required. Both lime and CO_2 may be required for proper treatment of the water. If the calcium hardness and alkalinity levels are between 100 and 500, either lime or soda ash (Na_2CO_3) will be satisfactory. The decision regarding which chemical to use depends on the cost of the equipment and chemicals. If the product of the calcium hardness times the alkalinity is greater than 500, either lime or caustic (NaOH) may be used. Soda ash will be ruled out because of the expense.

The chemicals chosen for the treatment of public drinking water supplies modify the water characteristics, making the water less corrosive to the pipe. Modification of water quality can increase the pH of the water, reducing the hydrogen ions available for galvanic corrosion, as well as reducing the solubility of copper, zinc, iron, lead, and calcium, and increasing the possibility of forming carbonate protective films.

Calcium carbonate stability is the most effective means of controlling corrosion. Lime, caustic soda, or soda ash is added until the pH and alkalinity indicate the water is saturated with calcium carbonate. Saturation does *not* always assure non-corrosiveness. Utilities should also exercise caution when applying sodium compounds since high sodium content in water can be a health concern for some customers. By increasing the alkalinity of the water, the bicarbonate and carbonate available to form a protective carbonate film increase. By decreasing the dissolved oxygen of the water, the rate of galvanic corrosion is reduced, along with the possibility of iron tuberculation.

Inorganic phosphates used include:

a. Zinc phosphates, which can cause algae blooms in open reservoirs.
b. Sodium silicate, which is used by individual customers, such as apartments, houses, and office buildings.
c. Sodium polyphosphates (tetrasodium pyrophosphate or sodium hexametaphosphate), which control scale formation in supersaturated waters and are known as *sequestering agents*.
d. Silicates (SiO_2), which form a film; an initial dosage of 12–16 mg/L for about 30 days will adequately coat the pipes, and a 1.0 mg/L concentration should be maintained thereafter.

Caution: Great care and caution must be exercised anytime feeding corrosion control chemicals into a public drinking water system!

Another corrosion control method is *aeration*. Aeration works to remove carbon dioxide (CO_2); it can be reduced to about 5 mg/L. *Cathodic protection,* often employed to control corrosion, is achieved by applying an outside electric current to the metal to reverse the electromechanical corrosion process. The application of direct current (D-C) prevents normal electron flow. Cathodic protection uses a sacrificial metal electrode (a magnesium anode) that corrodes instead of the pipe or tank. *Linings, coatings,* and *paints* can also be used in corrosion control. Slip-lining with plastic liner, cement mortar, zinc or magnesium, polyethylene, epoxy, and coal tar enamels are some of the materials that can be used.

Caution: Before using any protective coatings, consult the district engineer first!

Several *corrosive-resistant pipe materials* are used to prevent corrosion, including:

1. PVC plastic pipe
2. Aluminum
3. Nickel
4. Silicon
5. Brass
6. Bronze
7. Stainless steel
8. Reinforced concrete

In addition to internal corrosion problems, waterworks operators must also be concerned with external corrosion problems. The primary culprit involved in the external corrosion of distribution system pipe is soil. The measure of corrosivity of soil is *soil resistivity*. If the soil resistivity is greater than 5,000 Ω/cm, serious corrosion is unlikely, and steel pipe may be used under these conditions. If soil resistivity is less than 500 Ω/cm, plastic PVC pipe should be used. For intermediate ranges of soil resistivity (500–5,000 Ω/cm), use ductile iron pipe, lining, and coating.

Common operating problems associated with corrosion control include:

1. **$CaCO_3$ Not Depositing a Film**: Usually a result of poor pH control (out of the normal range of 6.5–8.5). This may also cause excessive film deposition.
2. **Persistence of Red Water Problems**: Most probably a result of poor flow patterns, insufficient velocity, tuberculation of the pipe surface, and the presence of iron bacteria.
   a. *Velocity*—Chemicals need to contact the pipe surface. Dead ends and low-flow areas should have a flushing program; dead ends should be looped.
   b. *Tuberculation*—The best approach is to clean with a "pig." In extreme cases, clean the pipe

with metal scrapers and install a cement-mortar lining.

c. *Iron bacteria*—Slime prevents film contact with the pipe surface. Slime will grow and lose coating. Pipe cleaning and disinfection programs are needed.

COAGULATION

The primary purpose of surface-water treatment is chemical clarification by coagulation and mixing, flocculation, sedimentation, and filtration. These unit processes, along with disinfection, work to remove particles, naturally occurring organic matter (NOM—i.e., bacteria, algae, zooplankton, and organic compounds), and microbes from water and to produce non-corrosive water. Specifically, coagulation/flocculation works to destabilize particles and agglomerate dissolved and particulate matter. Sedimentation removes solids and provides ½ log Giardia and 1 log virus removal. Filtration removes solids and provides 2 log Giardia and 1 log virus removal. Finally, disinfection provides microbial inactivation and ½ Giardia and 2 log Virus removal.

From Figure 17.3, it can be observed that following the screening and other pretreatment processes, the next unit process in a conventional water treatment system is a mixer where chemicals are added in what is known as coagulation. The exception to this unit process configuration occurs in small systems using groundwater when chlorine or other taste and odor control measures are introduced at the intake and are the extent of treatment.

Materials present in raw water may vary in size, concentration, and type. Dispersed substances in the water may be classified as suspended, colloidal, or solution. Suspended particles may vary in mass and size and depend on the water flow. High flows and velocities can carry larger material. As velocities decrease, the suspended particles settle according to size and mass.

Other material may be in solution; for example, salt dissolves in water. Matter in the colloidal state does not dissolve, but the particles are so small that they will not settle out of the water. Color (as in tea-colored swamp water) is mainly due to colloids or extremely fine particles of matter in suspension. Colloidal and solute particles in water are electrically charged. Because most of the charges are alike (negative) and repel each other, the particles stay dispersed and remain in the colloidal or soluble state.

Suspended matter will settle without treatment if the water is still enough to allow it to settle. The rate of settling of particles can be determined, as this settling follows certain laws of physics. However, much of the suspended matter may be so slow in settling that the normal settling processes become impractical, and if colloidal particles are present, settling will not occur. Moreover, water drawn from a raw water source often contains many small unstable (unsticky) particles; therefore, sedimentation alone is usually an impractical way to obtain clear water in most locations, and another method of increasing the settling rate must be used: coagulation, which is designed to convert stable (unstick) particles to unstable (sticky) particles.

The term *coagulation* refers to the series of chemical and mechanical operations by which coagulants are applied and made effective. These operations are comprised of two distinct phases: (1) rapid mixing to disperse coagulant chemicals violently into the water being treated, and (2) flocculation to agglomerate small particles into well-defined floc by gentle agitation for a much longer time.

Coagulation results from adding salts of iron or aluminum to the water. The coagulant must be added to the raw water and perfectly distributed into the liquid; such uniformity of chemical treatment is reached through rapid agitation or mixing. Common coagulants (salts) include:

- Alum—aluminum sulfate
- Sodium aluminate
- Ferric sulfate
- Ferrous sulfate
- Ferric chloride
- Polymers

Coagulation is the reaction between one of these salts and water. The simplest coagulation process occurs between alum and water. *Alum*, or aluminum sulfate, is produced by a chemical reaction of bauxite ore and sulfuric acid. The normal strength of liquid alum is adjusted to 8.3%, while the strength of dry alum is 17%.

When alum is placed in water, a chemical reaction occurs that produces positively charged aluminum ions. The overall result is the reduction of electrical charges and the formation of a sticky substance—the formation of *floc,* which when properly formed, will settle. These two destabilizing factors are the major contributions that coagulation makes to the removal of turbidity, color, and microorganisms.

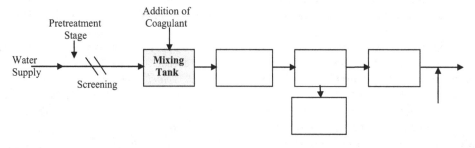

FIGURE 17.3   Coagulation.

Liquid alum is preferred in water treatment because it has several advantages over other coagulants, including the following:

1. Ease of handling
2. Lower costs
3. Less labor required to unload, store, and convey
4. Elimination of dissolving operations
5. Less storage space required
6. Greater accuracy in measurement and control
7. Elimination of the nuisance and unpleasantness of handling dry alum
8. Easier maintenance

The formation of floc is the first step of coagulation; for the greatest efficiency, rapid, intimate mixing of the raw water and the coagulant must occur. After mixing, the water should be slowly stirred so that the very small, newly formed particles can attract and enmesh colloidal particles, holding them together to form larger floc. This slow mixing is the second stage of the process (flocculation), covered later.

Several factors influence the coagulation process—pH, turbidity, temperature, alkalinity, and the use of polymers. The degree to which these factors influence coagulation depends on the coagulant used. The raw water conditions, optimum pH for coagulation, and other factors must be considered before deciding which chemical is to be fed and at what levels.

To determine the correct chemical dosage, a Jar Test or Coagulation Test is performed. *Jar tests* (widely used for many years by the water treatment industry) simulate full-scale coagulation and flocculation processes to determine optimum chemical dosages. It is important to note that jar testing is only an attempt to achieve a ballpark approximation of the correct chemical dosage for the treatment process. The test conditions are intended to reflect the normal operation of a chemical treatment facility. The test can be used to:

- Select the most effective chemical.
- Select the optimum dosage.
- Determine the value of a flocculant aid and the proper dose.

The testing procedure requires a series of samples to be placed in testing jars (see Figure 17.4) and mixed at 100 ppm. Varying amounts of the process chemical or specified amounts of several flocculants are added (one volume/sample container). The mix is continued for 1 min. Next, the mixing is slowed to 30 rpm to provide gentle agitation, and then the floc is allowed to settle. The flocculation period and settling process are observed carefully to determine the floc strength, settleability, and clarity of the supernatant liquor (defined: as the water that remains above the settled floc). Additionally, the supernatant can be tested to determine the efficiency of the chemical addition for the removal of TSS, BOD_5, and phosphorus.

The equipment required for the jar test includes a six-position, variable-speed paddle mixer (see Figure 17.4), six two-quart wide-mouthed jars, an interval timer, assorted glassware, pipettes, graduates, and so forth. The jar testing procedure follows:

1. Place an appropriate volume of water sample in each of the jars (250–1,000 mL samples may be used, depending on the size of the equipment being used). Start the mixers and set them to 100 rpm.
2. Add the previously selected amounts of the chemical being evaluated. (Initial tests may use wide variations in chemical volumes to determine the approximate range. This is then narrowed in subsequent tests.)
3. Continue mixing for 1 min.

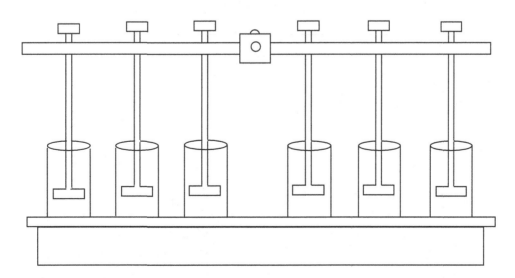

FIGURE 17.4 Variable-speed paddle mixer used in jar testing procedure.

4. Reduce the mixer speed to a gentle agitation (30 rpm) and continue mixing for 20 min. Again, time and mixer speed may be varied to reflect the facility.

Note: During this time, observe the floc formation—how well the floc holds together during the agitation (floc strength).

5. Turn off the mixer and allow solids to settle for 20–30 min. Observe the settling characteristics, the clarity of the supernatant, the settleability of the solids, the flocculation of the solids, and the compatibility of the solids.
6. Perform phosphate tests to determine removals.
7. Select the dose that provided the best treatment based on the observations made during the analysis.

Note: After initial ranges and/or chemical selections are completed, repeat the test using a smaller range of dosages to optimize performance.

MIXING AND FLOCCULATION

As we see in Figure 17.5, flocculation follows coagulation in the conventional water treatment process. *Flocculation* is the physical process of slowly mixing the coagulated water to increase the probability of particle collision—unstable particles collide and stick together to form fewer larger flocs. Through experience, we find that effective mixing reduces the required amount of chemicals and greatly improves the sedimentation process, which results in longer filter runs and higher quality finished water.

The goal of flocculation is to form a uniform, feather-like material similar to snowflakes—a dense, tenacious floc that entraps fine, suspended, and colloidal particles and carries them down rapidly in the settling basin. Proper flocculation requires 15 to 45 min. The time is based on water chemistry, water temperature, and mixing intensity. Temperature is the key component in determining the amount of time required for floc formation. To increase the speed of floc formation and the strength and weight of the floc, polymers are often added.

SEDIMENTATION

After raw water and chemicals have been mixed and the floc formed, the water containing the floc (which has a higher specific gravity than water) flows to the sedimentation or settling basin (see Figure 17.6). *Sedimentation* is also called clarification. Sedimentation removes settleable solids by gravity. Water moves slowly through the sedimentation tank/basin with a minimum of turbulence at entry and exit points with minimum short-circuiting. Sludge accumulates at the bottom of the tank/basin. Typical tanks or basins used in sedimentation include conventional rectangular basins, conventional center-feed basins, peripheral-feed basins, and spiral-flow basins.

In conventional treatment plants, the amount of detention time required for settling can vary from 2 to 6 h. Detention time should be based on the total filter capacity when the filters are passing 2 gpm/ft^2 of superficial sand area. For plants with higher filter rates, the detention time is based on a filter rate of 3–4 gpm/ft^2 of sand area. The time requirement is dependent on the weight of the floc, the temperature of the water, and how quiescent (still) the basin is.

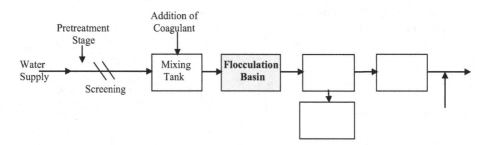

FIGURE 17.5   Mixing and flocculation.

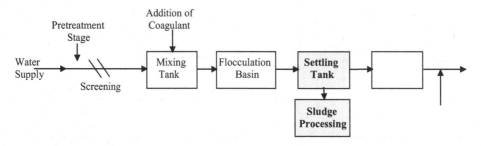

FIGURE 17.6   Sedimentation.

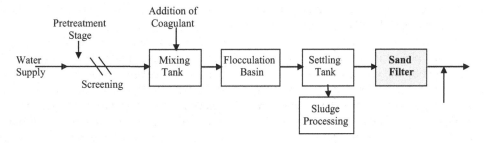

FIGURE 17.7 Filtration.

A number of conditions affect sedimentation: (1) uniformity of flow of water through the basin; (2) stratification of water due to differences in temperature between water entering and water already in the basin; (3) release of gases that may collect in small bubbles on suspended solids, causing them to rise and float as scum rather than settle as sludge; (4) disintegration of previously formed floc; and (5) size and density of the floc.

FILTRATION

In the conventional water treatment process, *filtration* usually follows coagulation, flocculation, and sedimentation (see Figure 17.7). At present, filtration is not always used in small water systems. However, recent regulatory requirements under USEPA's Interim Enhanced Surface Water Treatment rules may make water filtering necessary for most water supply systems. Water filtration is a physical process of separating suspended and colloidal particles from water by passing it through a granular material. The process of filtration involves straining, settling, and adsorption. As floc passes into the filter, the spaces between the filter grains become clogged, reducing these openings and increasing removal. Some material is removed merely because it settles on a media grain. One of the most important processes is the adsorption of the floc onto the surface of individual filter grains. This helps collect the floc and reduces the size of the openings between the filter media grains. In addition to removing silt and sediment, floc, algae, insect larvae, and any other large elements, filtration also contributes to the removal of bacteria and protozoans such as *Giardia lamblia* and *Cryptosporidium*. Some filtration processes are also used for iron and manganese removal.

TYPES OF FILTER TECHNOLOGIES

The Surface Water Treatment Rule (SWTR) specifies four filtration technologies, although the SWTR also allows the use of alternate filtration technologies (e.g., cartridge filters). The specified technologies are (1) slow sand filtration/rapid sand filtration, (2) pressure filtration, (3) diatomaceous earth filtration, and (4) direct filtration. Of these, all but rapid sand filtration are commonly employed in small water systems that use filtration. Each type of filtration system has advantages and disadvantages. Regardless of the type of filter, however, filtration involves the processes of *straining*

(where particles are captured in the small spaces between filter media grains), *sedimentation* (where the particles land on top of the grains and stay there), and *adsorption* (where a chemical attraction occurs between the particles and the surfaces of the media grains).

Slow Sand Filters

The first slow sand filter was installed in London in 1829 and was used widely throughout Europe, though not in the United States. By 1900, rapid sand filtration began taking over as the dominant filtration technology, and a few slow sand filters are in operation today. However, with the advent of the Safe Drinking Water Act and its regulations (especially the Surface Water Treatment Rule) and the recognition of the problems associated with *Giardia lamblia* and *Cryptosporidium* in surface water, the water industry is reexamining slow sand filters. However, low technology requirements may prevent many state water systems from using this type of equipment.

On the plus side, slow sand filtration is well-suited for small water systems. It is a proven, effective filtration process with relatively low construction costs and low operating costs (it does not require constant operator attention). It is quite effective for water systems as large as 5,000 people; beyond that, surface area requirements and the manual labor required to recondition the filters make rapid sand filters more effective. The filtration rate is generally in the range of 45–150 gal/day/ ft². Components making up a slow sand filter include the following:

- A covered structure to hold the filter media
- An underdrain system
- Graded rock that is placed around and just above the underdrain
- The filter media, consisting of 30–55 in of sand with a grain size of 0.25–0.35 mm
- Inlet and outlet piping to convey the water to and from the filter, and the means to drain filtered water to waste

The area above the top of the sand layer is flooded with water to a depth of 3–5 ft, and the water is allowed to trickle down through the sand. An overflow device prevents excessive water depth. The filter must have provisions for filling it from the bottom up, and it must be equipped with a loss-of-head gauge, a rate-of-flow control device (such as an orifice or butterfly valve), a weir or effluent pipe that assures that the water level cannot drop below the sand surface, and filtered waste sample taps.

When the filter is first placed in service, the head loss through the media caused by the resistance of the sand is about 0.2 ft (i.e., a layer of water 0.2 ft deep on top of the filter will provide enough pressure to push the water downward through the filter). As the filter operates, the media becomes clogged with the material being filtered out of the water, and the head loss increases. When it reaches about 4–5 ft, the filter needs to be cleaned.

For the efficient operation of a slow sand filter, the water being filtered should have an average turbidity of less than 5 TU, with a maximum of 30 TU. Slow sand filters are not backwashed the way conventional filtration units are. The 1–2 in of material must be removed periodically to keep the filter operating.

Rapid Sand Filters

The rapid sand filter, which is similar in some ways to the slow sand filter, is one of the most widely used filtration units. The major difference is in the principle of operation; that is, in the speed or rate at which water passes through the media. In operation, water passes downward through a sand bed that removes the suspended particles. The suspended particles consist of the coagulated matter remaining in the water after sedimentation, as well as a small amount of uncoagulated suspended matter.

Some significant differences exist in construction, control, and operation between slow sand filters and rapid sand filters. Because of the construction and operation of the rapid sand filtration with its higher filtration, the land area needed to filter the same quantity of water is reduced. Components of a rapid sand filter include:

- Structure to house media
- Filter media
- Gravel media support layer
- Underdrain system
- Valves and piping system
- Filter backwash system
- Waste disposal system

Usually 2–3 ft deep, the filter media is supported by approximately 1 ft of gravel. The media may be fine sand or a combination of sand, anthracite coal, and coal (dual-multimedia filter). Water is applied to a rapid sand filter at a rate of 1.5 gallons per minute per square foot of filter media surface. When the rate is between 4 and 6 gpm/ft², the filter is referred to as a high-rate filter; at a rate over 6 gpm/ft², the filter is called ultra-high-rate. These rates compare to the slow sand filtration rate of 45–150 gal/day/ ft². High-rate and ultra-high-rate filters must meet additional conditions to ensure proper operation.

Generally, raw water turbidity is not that high. However, even if raw water turbidity values exceed 1,000 TU, properly operated rapid sand filters can produce filtered water with a turbidity of well under 0.5 TU. The time the filter is in operation between cleanings (filter runs) usually lasts from 12 to 72 h, depending on the quality of the raw water; the end of the run is indicated by the head loss approaching 6–8 ft. Filter *breakthrough* (when filtered material is pulled through the filter into the effluent) can occur if the head loss becomes too great. Operation with head loss that is too high can also cause *air binding* (which blocks part of the filter with air bubbles), increasing the flow rate through the remaining filter area.

Rapid sand filters have the advantage of lower land requirements, and they have other advantages, too. For example, rapid sand filters cost less, are less labor-intensive to clean, and offer higher efficiency with highly turbid waters. On the downside, the operation and maintenance costs of rapid sand filters are much higher because of the increased complexity of the filter controls and backwashing system.

When *backwashing* a rapid sand filter, the filter is cleaned by passing treated water backward (upwards) through the filter media and agitating the top of the media. The need for backwashing is determined by a combination of filter run time (i.e., the length of time since the last backwashing), effluent turbidity, and head loss through the filter. Depending on the raw water quality, the run time varies from one filtration plant to another (and may even vary from one filter to another in the same plant).

Note: Backwashing usually requires 3–7% of the water produced by the plant.

Pressure Filter Systems

When raw water is pumped or piped from the source to a gravity filter, head (pressure) is lost as the water enters the floc basin. When this occurs, pumping the water from the plant clear well to the reservoir is usually necessary. One way to reduce pumping is to place the plant components into pressure vessels, thus maintaining the head. This type of arrangement is called a pressure filter system. Pressure filters are also quite popular for iron and manganese removal and for the filtration of water from wells. They may be placed directly in the pipeline from the well or pump with little head loss. Most pressure filters operate at a rate of about 3 gpm/ft².

Operationally the same and consisting of components similar to those of a rapid sand filter, the main difference between a rapid sand filtration system and a pressure filtration system is that the entire pressure filter is contained within a pressure vessel. These units are often highly automated and are usually purchased as self-contained units with all necessary piping, controls, and equipment contained in a single unit. They are backwashed in much the same manner as the rapid sand filter.

The major advantage of the pressure filter is its low initial cost. They are usually prefabricated, with standardized designs. A major disadvantage is that the operator is unable to observe the filter in the pressure filter and so is unable to determine the condition of the media. Unless the unit has an automatic shutdown feature for high effluent turbidity, driving filtered material through the filter is possible.

Diatomaceous Earth Filters

Diatomaceous earth is a white material made from the skeletal remains of diatoms. The skeletons are microscopic and, in most cases, porous. There are different grades of diatomaceous earth, and the grade is selected based on filtration requirements. These diatoms are mixed in a water slurry and fed onto a fine screen called a *septum,* usually made of stainless steel, nylon, or plastic. The slurry is fed at a rate of 0.2 lb/ft^2 of filter area. The diatoms collect in a pre-coat over the septum, forming an extremely fine screen. Diatoms are fed continuously with the raw water, causing the buildup of a filter cake approximately 1/8–1/5 in thick. The openings are so small that the fine particles that cause turbidity are trapped on the screen. Coating the septum with diatoms gives it the ability to filter out very small, microscopic material. The fine screen and the buildup of filtered particles cause a high head loss through the filter. When the head loss reaches a maximum level (30 psi on a pressure-type filter or 15 in of mercury on a vacuum-type filter), the filter cake must be removed by backwashing.

A slurry of diatoms is fed with raw water during filtration in a process called *body feed.* The body feed prevents premature clogging of the septum cake. These diatoms are caught on the septum, increasing the head loss and preventing the cake from clogging too rapidly with the particles being filtered. While the body feed increases head loss, the increases are more gradual than if body feed were not used.

Diatomaceous earth filters are relatively low in cost to construct, have high operating costs, and can cause frequent operating problems if not properly operated and maintained. They can be used to filter raw surface waters or surface-influenced groundwaters, with low turbidity (<5 NTU), and low coliform concentrations (no more than 50 coliforms/100 mL) and may also be used for iron and manganese removal following oxidation. Filtration rates are between 1.0 and 1.5 gpm/ft^2.

Direct Filtration

Direct filtration is a treatment scheme that omits the flocculation and sedimentation steps prior to filtration. Coagulant chemicals are added, and the water is passed directly onto the filter. All solids removal takes place on the filter, which can lead to much shorter filter runs, more frequent backwashing, and a greater percentage of finished water used for backwashing. The lack of a flocculation process and sedimentation basin reduces construction costs but increases the requirement for skilled operators and high-quality instrumentation. Direct filtration should only be used where the water flow rate and raw water quality are fairly consistent and where the incoming turbidity is low.

Alternative Filters

A *cartridge filter system* can be employed as an alternative filtering system to reduce turbidity and remove *Giardia.*

A cartridge filter is made of synthetic media contained in a plastic or metal housing. These systems are normally installed in a series of three or four filters. Each filter contains media that is successively smaller than the previous filter. The media sizes typically range from 50 to 5μ or less. The filter arrangement depends on the quality of the water, the capability of the filter, and the quantity of water needed. The USEPA and state agencies have established criteria for the selection and use of cartridge filters. Generally, cartridge filter systems are regulated in the same manner as other filtration systems.

Because of new regulatory requirements and the need to provide more efficient removal of pathogenic protozoans (e.g., *Giardia and Cryptosporidium*) from water supplies, *membrane filtration systems* are finding increased application in water treatment systems. A *membrane* is a thin film separating two different phases of a material, acting as a selective barrier to the transport of matter operated by some driving force. Simply put, a membrane can be regarded as a sieve with very small pores. Membrane filtration processes are typically pressure, electrically, vacuum, or thermally driven. The types of drinking water membrane filtration systems include microfiltration, ultrafiltration, nanofiltration, and reverse osmosis. In a typical membrane filtration process, there is one input and two outputs. Membrane performance is largely a function of the properties of the materials to be separated and can vary during operation.

COMMON FILTER PROBLEMS

Two common types of filter problems occur: those caused by filter runs that are too long (infrequent backwash), and those caused by inefficient backwash (cleaning). A filter that is run too long can cause *breakthrough* (the pushing of debris removed from the water through the media and into the effluent) and *air binding* (the trapping of air and other dissolved gases in the filter media). Air binding occurs when the rate at which water exits the bottom of the filter exceeds the rate at which the water penetrates the top of the filter. When this happens, a void and partial vacuum occur inside the filter media. The vacuum causes gases to escape from the water and fill the void. When the filter is backwashed, the release of these gases may cause a violent upheaval in the media and destroy the layering of the media bed, gravel, or underdrain. Two solutions to these problems are as follows: (1) check the filtration rates to ensure they are within the design specifications and (2) remove the top 1 inch of media and replace it with new media. This keeps the top of the media from collecting the floc and sealing the entrance into the filter media.

Another common filtration problem is associated with poor backwashing practices: the formation of *mud balls* that get trapped in the filter media. In severe cases, mud balls can completely clog a filter. Poor agitation of the surface of the filter can form a crust on top of the filter; the crust later cracks under the water pressure, causing uneven distribution of water through the filter media. Filter cracking can be corrected by removing the top 1-in of the filter media, increasing the backwash rate, or checking the effectiveness of the surface wash (if installed). Backwashing at too high a rate can cause the filter media to wash out of the filter over the effluent troughs and may damage the filter underdrain system. Two possible solutions are as follows: (1) check the backwash rate to ensure that it meets the design criteria, and (2) check the surface wash (if installed) for proper operation.

FILTRATION AND COMPLIANCE WITH TURBIDITY REQUIREMENTS

Under the1996 Safe Drinking Water Act (SDWA) Amendments, the USEPA must supplement the existing 1989 Surface Water Treatment Rule (SWTR) with the Interim Enhanced Surface Water Treatment Rule (IESWTR) to improve protection against waterborne pathogens. Key provisions established in the IESWTR include (USEPA, 1998):

- A maximum contaminant level goal (MCLG) of zero for *Cryptosporidium;* 2-log (99%) *Cryptosporidium* removal requirement for systems that filter.
- Strengthened combined filter effluent turbidity performance standards.
- Individual filter turbidity monitoring provisions.
- Disinfection benchmark provisions to ensure continued levels of microbial protection while facilities take the necessary steps to comply with new disinfection byproduct standards.
- Inclusion of *Cryptosporidium* in the definition of groundwater under the direct influence of surface water (GWUDI) and in the watershed control requirements for unfiltered public water systems.
- Requirements for covers on new finished water reservoirs.
- Sanitary surveys for all surface water systems regardless of size.

This section outlines the regulatory, reporting, and recordkeeping requirements that all waterworks operators should be familiar with, along with additional compliance aspects of the IESWTR related to turbidity.

IESWTR Regulatory Requirements

The Interim Enhanced Surface Water Treatment Rule contains several key provisions including strengthened combined filter effluent turbidity performance standards and individual filter turbidity monitoring.

Applicability

Entities potentially regulated by the IESWTR are public water systems that use surface water or groundwater under the direct influence of surface water and serve at least 10,000 people (including industries, state, local, tribal, or federal governments). To determine whether your facility may be

regulated by this action, you should carefully examine the applicability criteria of Subpart H (systems subject to the Surface Water Treatment Rule) and Subpart P (Subpart H systems that serve 10,000 or more people) of the final rule.

Note: Systems subject to the turbidity provisions of the IESWTR are a subset of systems regulated by the IESWTR, which utilize rapid granular filtration (i.e., conventional filtration treatment and direct filtration) or other filtration processes (excluding slow sand and diatomaceous earth filtration).

Combined Filter Effluent Monitoring

Under the SWTR, a Subpart H system that provides filtration treatment, must monitor turbidity in the combined filter effluent. Turbidity measurements must be performed on representative samples of the system's filtered water every 4 h (or more frequently) that the system serves water to the public. A public water system may substitute continuous turbidity monitoring for grab sample monitoring if it validates the continuous measurement for accuracy regularly using a protocol approved by the State. The turbidity performance requirements of the IESWTR require that all surface water systems that use conventional treatment or direct filtration and serve a population of 10,000 people must meet two distinct filter effluent limits: a maximum limit and a 95% limit. These limits, set forth in the IESWTR, are outlined below for the different types of treatment employed by systems.

1. **Conventional Treatment or Direct Filtration:** For conventional and direct filtration systems (including those systems utilizing in-line filtration), the turbidity level of representative samples of a system's filtered water (measured every 4 h) must be less than or equal to **0.3 NTU** in at least 95% of the measurements taken each month. The turbidity level of representative samples of a system's filtered water must not exceed **1 NTU** at any time. Conventional filtration is defined as a series of processes, including coagulation, flocculation, sedimentation, and filtration, resulting in substantial particulate removal. Direct filtration is defined as a series of processes including coagulation and filtration but excluding sedimentation, resulting in substantial particle removal.
2. **Other Treatment Technologies (Alternative Filtration):** For other filtration technologies (those technologies other than conventional, direct, slow sand, or diatomaceous earth filtration), a system may demonstrate to the State, using pilot plant studies or other means, that the alternative filtration technology, in combination with disinfection treatment, consistently achieves 99.9% removal and/or inactivation of *Giardia lamblia* cysts, 99.99% removal and/or inactivation of viruses, and 99% removal of *Cryptosporidium* oocysts. For a system that makes this demonstration, then representative samples of a system's filtered

water must be less than or equal to a value determined by the State which is indicative of 2-log Cryptosporidium removal, 3-log Giardia removal, and 4-log virus removal in at least 95% of the measurements taken each month. The turbidity level of representative samples of a system's filtered water must not exceed a maximum turbidity value determined by the State. Examples of such technologies include bag or cartridge filtration, microfiltration, and reverse osmosis. The USEPA recommends a protocol similar to the "Protocol for Equipment Verification Testing for Physical Removal of Microbiological and Particulate Contaminants" prepared by NSF International with support from the USEPA.

3. **Slow Sand & Diatomaceous Earth Filtration:** The IESWTR does not contain new turbidity provisions for slow sand or diatomaceous earth (DE) filtration systems. Utilities utilizing either of these filtration processes must continue to meet the requirements for their respective treatment as outlined in the SWTR (1 NTU 95%, 5 NTU max).

4. **Systems that Utilize Lime Softening:** Systems that practice lime softening, may experience difficulty in meeting the turbidity performance requirements due to residual lime floc carryover inherent in the process. The USEPA allows such systems to acidify turbidity samples prior to measurement using a protocol approved by the states. The chemistry supporting this decision is well-documented in environmental chemistry texts.

   The USEPA recommends that acidification protocols lower the pH of samples to <8.3 to ensure an adequate reduction in carbonate ions and a corresponding increase in bicarbonate ions. Acid should consist of either hydrochloric acid or sulfuric acid of Standard Lab Grade. Care should be taken when adding acid to samples. Operators should always follow the sampling guidelines as directed by their supervisors and standard protocols. If systems choose to use acidification, the USEPA recommends that systems maintain documentation regarding the turbidity with and without acidification, as well as pH values and the quantity of acid added to the sample.

Individual Filter Monitoring

In addition to the combined filter effluent monitoring discussed above, those systems that use **conventional treatment or direct filtration** (including in-line filtration) must conduct continuous monitoring of turbidity for each individual filter using an approved method in §141.74. (a) and must calibrate turbidimeters using the procedure specified by the manufacturer. Systems must record the results of individual filter monitoring every 15 min. If the individual filter does not provide water that contributes to the combined filter effluent (i.e., it is not operating, is filtering to waste, or is recycled), the system does not need to record or monitor the turbidity for that specific filter.

Note: Systems that utilize filtration other than conventional or direct filtration are not required to conduct individual filter monitoring, although the USEPA recommends that such systems consider individual filter monitoring.

If a failure occurs in the continuous turbidity monitoring equipment, the system must conduct grab sampling every 4 h in lieu of continuous monitoring but must return to 15-min monitoring no more than five working days following the failure of the equipment.

Reporting and Recordkeeping

Distinct reporting and recordkeeping requirements for the turbidity provisions of the IESWTR for both systems and states include the following:

System Reporting Requirements

Under the IESWTR, systems are tasked with specific reporting requirements associated with combined filter effluent monitoring and individual filter effluent monitoring.

- **Combined Filter Effluent Reporting:** Turbidity measurements, as required by §141.173, must be reported within 10 days after each month the system serves water to the public. Information that must be reported includes:
  1. The total number of filtered water turbidity measurements taken during the month.
  2. The number and percentage of filtered water turbidity measurements taken during the month that are less than or equal to the turbidity limits specified in §141.173. (0.3 NTU for conventional and direct filtration and the turbidity limit established by the state for other filtration technologies)
  3. The date and value of any turbidity measurements taken during the month that exceed 1 NTU for systems using conventional filtration treatment or direct filtration and the maximum limit established by the state for other filtration technologies. This reporting requirement is similar to the reporting requirement currently found under the SWTR.

- **Individual Filter Requirements:** Systems utilizing conventional and direct filtration must report that they have conducted individual filter monitoring in accordance with the requirements of the IESWTR within 10 days after the end of each month the system serves water to the public. Additionally, systems must report individual filter turbidity measurements within 10 days after the end of each month the system serves water to the public **only if** measurements demonstrate one of the following:

- Any individual filter has a measured turbidity level greater than 1.0 NTU in two consecutive measurements taken 15 min apart. The system must report the filter number, the turbidity measurement, and the date(s) on which the exceedance occurred. In addition, the system must either produce a filter profile for the filter within 7 days of the exceedance (if the system is not able to identify an obvious reason for the abnormal filter performance) and report that the profile has been produced or report the obvious reason for the exceedance.

- Any individual filter that has a measured turbidity level greater than 0.5 NTU in two consecutive measurements taken 15 min apart at the end of the first 4 h of continuous filter operation after the filter has been backwashed or otherwise taken offline. The system must report the filter number, the turbidity, and the date(s) on which the exceedance occurred. In addition, the system must either produce a filter profile for the filter within 7 days of the exceedance (if the system is not able to identify an obvious reason for the abnormal filter performance) and report that the profile has been produced or report the obvious reason for the exceedance.

- Any individual filter with a measured turbidity level of greater than 1.0 NTU in two consecutive measurements taken 15 minutes apart at any time in three consecutive months. The system must report the filter number, the turbidity measurement, and the date(s) on which the exceedance occurred. In addition, the system shall conduct a self-assessment of the filter.

- Any individual filter with a measured turbidity level of greater than 2.0 NTU in two consecutive measurements taken 15 minutes apart at any time in each of two consecutive taken 15 minutes apart at any time in two consecutive months. The system must report the filter number, the turbidity measurement, and the date(s) on which the exceedance occurred. In addition, the system shall contact the state or a third party approved by the state to conduct a comprehensive performance evaluation.

State Reporting Requirements

Under §142.15, each state that has primary enforcement responsibility, is required to submit quarterly reports to the Administrator of the EPA on a schedule and in a format prescribed by the Administrator, which includes:

1. New violations by public water systems in the State during the previous quarter concerning state regulations adopted to incorporate the requirements of national primary drinking water regulations.

2. New enforcement actions taken by the state during the previous quarter against public water systems concerning state regulations adopted to incorporate the requirements of national primary drinking water standards.

Any violations or enforcement actions related to turbidity would be included in the quarterly report noted above. The USEPA has developed a State Implementation Guidance Manual which includes additional information on state reporting requirements.

System Recordkeeping Requirements

Systems must maintain the results of individual filter monitoring taken under §141.174 for at least 3 years. These records must be readily available for state representatives to review during Sanitary Surveys on other visits.

State Recordkeeping Requirements

Records of turbidity measurements must be kept for not less than 1-year. The information retained must be set forth in a form which makes possible comparison with limits specified in §§141.71, 141.73, 141.173, and 141.175. Records of decisions made on a system-by-system and case-by-case basis under provisions of part 141, subpart H, or subpart P, must be made in writing and kept by the state (this includes records regarding alternative filtration determinations). The USEPA has developed a State Implementation Guidance Manual that includes additional information on state recordkeeping requirements.

Additional Compliance Issues include additional compliance issues associated with the IESWTR.

Schedule

The IESWTR was published on December 16, 1998, and became effective on February 16, 1999. The SDWA requires, within 24 months following the promulgation of a rule, that the Primacy Agencies adopt any state regulations necessary to implement the rule. Under Sec. 14.13, these rules must be at least as stringent as those required by the USEPA. Thus, primary agencies must promulgate regulations that are at least as stringent as the IESWTR by December 17, 2000. Beginning December 17, 2001, systems serving at least 10,000 people must meet the turbidity requirements in §141.173.

Individual Filter Follow-up Action

Based on the monitoring results obtained through continuous filter monitoring, a system may have to conduct one of the following follow-up actions due to persistently high turbidity levels at an individual filter:

- Filter profile
- Individual filter self-assessment
- Comprehensive performance evaluation

These specific requirements are found in section 141.175(b)(1)-(4).

Abnormal Filter Operations—Filter Profile

A filter profile must be produced if no obvious reason for abnormal filter performance can be identified. A filter profile is a graphical representation of individual filter performance based on continuous turbidity measurements or total particle counts versus time for an entire filter run, from startup to backwash, inclusively that includes assessment of filter performance while another filter is being backwashed. The run length during this assessment should be representative of typical plant filter runs. The profile should include an explanation of the cause of any filter performance spikes during the run. Examples of possible abnormal filter operations that may be obvious to operators include the following:

- Outages or maintenance activities at processes within the treatment train
- Coagulant feed pump or equipment failure
- Filters being run at significantly higher loading rates than approved

It is important to note that while the reasons for abnormal filter operation may appear obvious, they could be masking other reasons that are more difficult to identify. These may include situations such as:

- Distribution in filter media
- Excessive or insufficient coagulant dosage
- Hydraulic surges due to pump changes or other filters being brought on/offline.

Systems need to use the best professional judgment and discretion when determining whether to develop a filter profile. Attention at this stage will help systems avoid the other forms of follow-up action described below.

Individual Filter Self-Assessment

A system must conduct an individual filter self-assessment for any individual filter that has a measured turbidity level greater than 1.0 NTU in two consecutive measurements taken 15 min apart in each of three consecutive months. The system must report the filter number, the turbidity measurement, and the dates on which the exceedance occurred.

Comprehensive Performance Evaluation

A system must conduct a comprehensive performance evaluation (CPE) if any individual filter has a measured turbidity level greater than 2.0 NTU in two consecutive measurements taken 15 min apart in two consecutive months. The system must report the filter number, the turbidity measurement, and the date(s) on which the exceedance occurred. The system shall contact the state or a third party approved by the state to conduct a comprehensive performance evaluation.

Note: The USEPA has developed a guidance document called, *Handbook: Optimizing Water Treatment Plant Performance Using the Composite Correction Program* [EPA/625/6–91/027 (1998)].

Notification

The IESWTR contains two distinct types of notification: state and public. It is important to understand the differences between each and the respective requirements.

- **State Notification:** Systems are required to notify states under §141.31. Systems must report to the state within 48 h if they fail to comply with any national primary drinking water regulation. Within 10 days of completing each public notification required pursuant to §141.32, the system must submit to the state a representative copy of each type of notice distributed, published, posted, and/or made available to persons served by the system and/or the media. The water supply system must also submit to the State (within the time stated in the request made by the State) copies of any records required to be maintained under §141.33 or copies of any documents then in existence which the State or the Administrator is entitled to inspect pursuant to the authority of Section 1445 of the Safe Drinking Water Act or the equivalent provisions of the State Law.

Notification

The IESWTR specifies that the public notification requirements of the Safe Drinking Water Act (SDWA) and the implementation regulations of 40 CFR §141.32 must be followed. These regulations divide public notification requirements into two tiers. These tiers are defined as follows:

- *TIER 1*
  Failure to comply with MCL
  Failure to comply with the prescribed treatment technique
  Failure to comply with a variance or exemption schedule
- *TIER 2*
  Failure to comply with monitoring requirements
  Failure to comply with a testing procedure prescribed by an NPDWR
  Operating under a variance/exemption. This is not considered a violation, but public notification is required.

Certain general requirements must be met by all public notices. All notices must provide a clear and readily understandable explanation of the violation, any potential adverse health effects, the population at risk, the steps the system is taking to correct the violation, the necessity of seeking alternate water supplies (if any), and any preventive measures the consumer should take. The notice must be conspicuous and must not contain any unduly technical language, small print, or similar problems. The notice must include the telephone number of the owner, operator, or designee of the public water system as a source of additional information concerning the violation where appropriate. The notice must be bi- or multilingual if appropriate.

Tier 1 Violations

In addition, the public notification rule requires that when providing notification on potential adverse health effects in Tier 1 public notices and in notices regarding the granting and continued existence of a variance or exemption, the owner or operator of a public water system must include certain mandatory health effects language. For violations of treatment technique requirements for filtration and disinfection, the mandatory health effects language is:

> The USEPA sets drinking water standards and has determined that the presence of microbiological contaminants is a health concern at certain levels of exposure. If water is inadequately treated, microbiological contaminants in that water cause disease. Disease symptoms may include diarrhea, cramps, nausea, and possibly jaundice, and any associated headaches and fatigue. These symptoms, however, are not just associated with disease-causing organisms in drinking water, but also may be caused by a number of factors other than your drinking water. USEPA has set enforceable requirements for treating drinking water to reduce the risk of these adverse health effects. Treatment such as filtering and disinfection the water removes or destroys microbiological contaminants. Drinking water which is treated to meet UEPA requirements is associated with little to none of this risk and should be considered safe.

Furthermore, the owner or operator of a community water system must give a copy of the most recent notice for any Tier 1 violations to all new billing units or hookups prior to or at the time service begins.

The medium for performing public notification and the time period in which notification must be sent vary depending on the type of violation and are specified in §141.32. For Tier 1 violations, the owner or operator of a public water system must give notice:

> By publication in a local daily newspaper as soon as possible but in no case later than 14 days after the violation or failure. If the area does not have a daily newspaper, then notice shall be given by publication in a weekly newspaper of general circulation in the area, and
>
> By either direct mail delivery or hand delivery of the notice, either by itself or with the water bill no later than 45 days after the violation or failure. The Primacy Agency may waive the requirement if it determines that the owner or operator has corrected the violation with 45 days.

Although the IESWTR does not specify any acute violations, the Primacy Agency may specify some Tier 1 violations as posing an acute risk to human health; examples might include:

- A waterborne outbreak in an unfiltered supply
- The turbidity of filtered water exceeds 1.0 NTU at any time
- Failure to maintain a disinfectant residual of at least 0.2 mg/L in the water being delivered to the distribution system.

For these violations or any others defined by the Primacy Agency as "acute" violations, the system must furnish a copy of the notice to the radio and television stations serving the area as soon as possible, but in no case later than 72 h after the violation. Depending on the circumstances particular to the system, as determined by the Primacy Agency, the notice may instruct that all water be boiled prior to consumption.

Following the initial notice, the owner or operator must give notice at least once every 3 months by mail delivery (either with the water bill or separately) or by hand delivery for as long as the violation or failures exist.

There are two variations on these requirements. First, the owner or operator of a community water system in an area not served by a daily or weekly newspaper must give notice within 14 days after the violation by hand delivery or continuous posting of a notice of the violation. The notice must continue for as long as the violation exists. Notice by hand delivery must be repeated at least every 3 months for the duration of the violation. Secondly, the owner or operator of a noncommunity water system (i.e., one serving a transitory population) may give notice by hand delivery or by continuously posting the notice in conspicuous places in the area served by the system. Notice must be given within 14 days after the violation. If notice is given by posting, it must continue as long as the violations exist. Notice given by hand delivery must be repeated at least every 3 months for as long as the violation exists.

Tier 2 Violations

For Tier 2 violations (i.e., violations of 40 CFR §§141.74 and 141.174) notice must be given within 3 months after the violation by publication in a daily newspaper of general circulation, or if there is no daily newspaper, then in a weekly newspaper. In addition, the owner or operator shall give notice by mail (either by itself or with the water bill) or by hand delivery at least once every 3 months for as long as the violation exists. Notice of a variance or exemption must be given every 3 months from the date it is granted for as long as it remains in effect.

If a daily or weekly newspaper does not serve the area, the owner or operator of a community water system must give notice by continuous posting in conspicuous places in the area served by the system. This must continue as long as the violation exists or the variance or exemption remains in effect. Notice by hand delivery must be repeated at least every 3 months for the duration of the violation or the variance or exemption.

For noncommunity water systems, the owner or operator may give notice by hand delivery or continuous posting in conspicuous places; beginning within 3 months of the violation or the variance or exemption. Posting must continue for the duration of the violation, variance, or exemption, and notice by hand delivery must be repeated at least every 3 months during this period.

The Primacy Agency may allow for owner or operator to provide less frequent notice for minor monitoring violations (as defined by the Primacy Agency if the EPA has approved the Primacy Agency's substitute requirements contained in a program revision application).

Variances and Exemptions

As with the SWTR, no variances from the requirements in §141 are permitted for subpart H systems. Under Section 1416(a), the USEPA or a State may exempt a public water system from any requirements related to an MCL or treatment technique of a National Primary Drinking Water Regulation (NPDWR) if it finds that (1) due to compelling factors (which may include economic factors such as qualifications of the Public Water System (PWS)as serving a disadvantaged community), the PWS is unable to comply with the requirement or implement measures to develop an alternative source of water supply; (2) the exemption will not result in an unreasonable health risk; and (3) the PWS was in operation on the effective date of the NPDWR, or for a system that was not in operation by that date, only if no reasonable alternative source of drinking water is available to the new system; and (4) management or restructuring changes (or both) cannot reasonably result in compliance with the Act or improve the quality of drinking water.

DISINFECTION

Disinfection is a unit process used in both water and wastewater treatment. Many of the terms, practices, and applications discussed in this section apply to both types of treatment. However, there are also some differences—mainly in the types of disinfectants used and their applications—between disinfection in water and wastewater treatment. Thus, in this section, we will discuss disinfection as it applies to water treatment, and later we will cover disinfection as it applies to wastewater treatment. Much of the information presented in this section is based on personal experience and USEPA (1999b).

To comply with the SDWA regulations, the majority of PWSs use some form of water treatment. The 1995 Community Water System Survey reports that in the United States, 99% of surface water systems provide some treatment to their water, with 99% of these treatment systems using disinfection/oxidation as part of the treatment process. Although 45% of groundwater systems provide no treatment, 92% of those groundwater plants that do provide some form of treatment include disinfection/oxidation as part of the treatment process (USEPA, 1997). Regarding groundwater supplies, what is the public health concern? According to USEPA's Bruce Maclear [in What is the Ground Water Disinfection Rule, @ www.groc.org/winter96/gwdr.htm.],

> There are legitimate concerns for public health from microbial contamination of groundwater systems. Micro-organisms and other evidence of fecal contamination have been detected in a large number of wells tested, even those wells that had been previously judged not vulnerable to

such contamination. The scientific community believes that microbial contamination of groundwater is real and widespread. Public health impact from this contamination, while not well quantified, appear to be large. Disease outbreaks have occurred in many groundwater systems. Risk estimates suggest several million illnesses each year. Additional research is underway to better characterize the nature and magnitude of the public health problem.

The most commonly used disinfectants/oxidants (in no particular order) are chlorine, chlorine dioxide, chloramines, ozone, and potassium permanganate.

As mentioned, the process used to control waterborne pathogenic organisms and prevent waterborne disease is called *disinfection*. The goal of proper disinfection in a water system is to destroy all disease-causing organisms. Disinfection should not be confused with sterilization. *Sterilization* is the complete killing of all living organisms. Waterworks operators disinfect by destroying organisms that might be dangerous; they do not attempt to sterilize water.

Disinfectants are also used to achieve other specific objectives in drinking water treatment. These other objectives include nuisance control (e.g., for zebra mussels and Asiatic clams), oxidation of specific compounds (i.e., taste- and odor-causing compounds, iron, and manganese), and use as a coagulant and filtration aid. The goals of this section are to:

- Provide a brief overview of the need for disinfection in water treatment.
- Provide basic information that is common to all disinfectants.
- Discuss other uses for disinfectant chemicals (i.e., as oxidants).
- Describe trends in DBP formation and the health effects of DBPs found in water treatment.
- Discuss micro-organisms of concern in water systems, their associated health impacts, and the inactivation mechanisms and efficiencies of various disinfectants.
- Summarize current disinfection practices in the United States, including the use of chlorine as a disinfectant and an oxidant.

In water treatment, disinfection is almost always accomplished by adding chlorine or chlorine compounds after all other treatment steps (see Figure 17.8), although in the United States ultraviolet (UV) light and potassium permanganate and ozone processes may also be encountered.

The effectiveness of disinfection in a drinking water system is measured by testing for the presence or absence of coliform bacteria. *Coliform bacteria* found in water are generally not pathogenic, though they are good indicators of contamination. Their presence indicates the possibility of contamination, and their absence indicates the possibility that the water is potable—if the source is adequate, the waterworks history is good, and an acceptable chlorine residual is present.

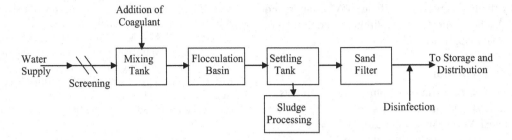

FIGURE 17.8 Disinfection.

Desired characteristics of a disinfectant include the following:

- It must be able to deactivate or destroy any type or number of disease-causing microorganisms that may be in a water supply within a reasonable time, within expected temperature ranges, and despite changes in the character of the water (pH, for example).
- It must be nontoxic.
- It must not add unpleasant taste or odor to the water.
- It must be readily available at a reasonable cost and be safe and easy to handle, transport, store, and apply.
- It must be quick and easy to determine the concentration of the disinfectant in the treated water.
- It should persist within the disinfected water at a high enough concentration to provide residual protection throughout the distribution.

NEED FOR DISINFECTION IN WATER TREATMENT

SIDE BAR 17.1   A SHERLOCK HOLMES-TYPE AT THE PUMP[1]

He wandered the foggy, filthy, garbage-strewn, corpse-ridden streets of 1854 London searching, making notes, always looking … seeking a murderous villain (no; not the Ripper, but a killer just as insidious and unfeeling)—and find the miscreant, he did. He acted; he removed the handle from a water pump. And, fortunately for untold thousands of lives, his was the correct action—the lifesaving action.

He was a detective—of sorts. No, not the real Sherlock Holmes—but absolutely as clever, as skillful, as knowledgeable, as intuitive—and definitely as driven. His real name: Dr. John Snow. His middle name? Common Sense. Snow's master criminal, his target? A mindless, conscienceless, brutal killer: cholera.

Let's take a closer look at this medical super sleuth and at his quarry, the deadly cholera—and at Doctor Snow's actions to contain the spread of cholera. More to the point, let's look at Dr. Snow's subsequent impact on the water treatment (disinfection) of raw water used for potable and other purposes.

Dr. John Snow—an unassuming—and creative—London obstetrician, Dr. John Snow (1813–1858) achieved prominence in the mid-nineteenth century for proving his theory (in his *On the Mode of Communication of Cholera*) that cholera is a contagious disease caused by a "poison" that reproduces in the human body and is found in the vomitus and stools of cholera patients. He theorized that the main (though not the only) means of transmission was water contaminated with this poison. His theory was not held in high regard at first because a commonly held and popular counter-theory stated that diseases are transmitted by inhalation of vapors. Many theories of cholera's cause were expounded. In the beginning, Snow's argument did not cause a great stir; it was only one of many hopeful theories proposed during a time when cholera was causing great distress. Eventually, Snow was able to prove his theory. We will describe how Snow accomplished this later, but for now, let's take a look again at Snow's target: cholera.

Cholera

According to the U.S. Centers for Disease Control (CDC) and as mentioned, cholera is an acute, diarrheal illness caused by infection of the intestine with the bacterium *Vibrio cholera*. The infection is often mild or without symptoms, but sometimes can be quite severe. Approximately 1 in 20 infected persons have severe disease symptoms such as profuse watery diarrhea, vomiting, and leg cramps. In these persons, rapid loss of body fluids leads to dehydration and shock. Without treatment, death can occur within hours.

> **DID YOU KNOW?**
>
> You don't need to be a rocket scientist to figure out just how deadly cholera was during the London cholera outbreak of 1854. Comparing the state of "medicine" at that time to ours is like comparing the speed potential of a horse and buggy to a state-of-the-art NASCAR race car today. Simply stated: cholera was the classic epidemic disease of the nineteenth century, as the plague had been for the fourteenth. Its defeat reflected both common sense and progress in medical knowledge—and the enduring changes in European and American social thought.

How does a person contract cholera? Good question. Again, we refer to the CDC for our answer. A person may contract cholera (even today) by drinking water or eating food contaminated with the cholera bacterium. In an epidemic, the source of the contamination is usually feces of an infected person. The disease can spread rapidly in areas with inadequate treatment of sewage and drinking water. Disaster areas often pose special risks. For example, the aftermath of Hurricane Katrina in New Orleans raised concerns for a potential cholera outbreak.

Cholera bacterium also lives in brackish river and coastal waters. Raw shellfish have been a source of cholera, with a few people in the United States having contracted it from eating shellfish from the Gulf of Mexico. The disease is not likely to spread directly from one person to another; therefore, casual contact with an infected person is not a risk for transmission of the disease.

Flashback to 1854 London

The information provided in the preceding section was updated and provided by the Centers for Disease Control (CDC) in 1996. Basically, for our purposes, the CDC confirms the fact that cholera is a waterborne disease. Today, we know quite a lot about cholera and its transmission, as well as how to prevent infection and how to treat it. But what did they know about cholera in the 1850s? Not much. However, one thing is certain: they knew cholera was a deadly killer. And that was just about all they knew—until Dr. Snow proved his theory. He believed that cholera is a contagious disease caused by a poison that reproduces in the human body and is found in the vomitus and stools of cholera victims. He also believed that the main means of transmission was contaminated water.

Dr. Snow's theory was correct, of course, as we know it today. The question is, how did he prove his theory correct 20 years before the development of the germ theory? The answer to this provides us with an account of one of the all-time legendary quests for answers in epidemiological research—and an interesting story!

Dr. Snow proved his theory in 1854, during yet another severe cholera epidemic in London. Though ignorant of the concept of bacteria carried in water (germ theory), Snow traced an outbreak of cholera to a water pump located at the intersection of Cambridge and Broad Street (London). How did he isolate this source to this particular pump? He began his investigation by determining in which area in London persons with cholera lived and worked. He then used this information to map the distribution of cases on what epidemiologists call a "spot map." His map indicated that the majority of the deaths occurred within 250 yards of that communal water pump. The water pump was used regularly by most of the area residents. Those who did not use the pump remained healthy. Suspecting the

Broad Street pump as the plague's source, Snow had the water pump handle removed and thus ended the cholera epidemic.

Sounds like a rather simple solution, doesn't it? For us, it is simple, but remember that in that era, aspirin had not yet been formulated—to say nothing of other medical miracles we now take for granted—antibiotics, for example. Dr. John Snow, by the methodical process of elimination and linkage (Sherlock Holmes would have been impressed—and he was), proved his point and his theory. Specifically, he painstakingly documented the cholera cases and correlated the comparative incidence of cholera among subscribers to the city's two water companies. He learned that one company drew water from the lower Thames River, while the other company obtained water from the upper Thames. Snow discovered that cholera was much more prevalent among customers of the water company that drew its water from the lower Thames, where the river had become contaminated with London sewage. Snow tracked and pinpointed the Broad Street pump's water source. You guessed it: the contaminated lower Thames, of course.

Dr. Snow the obstetrician became the first effective practitioner of scientific epidemiology. His creative use of logic, common sense (removing the handle from the pump) and scientific information enabled him to solve a major medical mystery—to discern the means by which cholera was transmitted—and earned him the title "the father of field epidemiology." Today, Dr. John Snow is known as the father of modern epidemiology.

Pump Handle Removal—To Water Treatment (Disinfection)—Dr. John Snow's major contribution to the medical profession, to society, and humanity, in general, can be summarized rather succinctly: he determined and proved that the deadly disease cholera is a waterborne disease (Dr. John Snow's second medical accomplishment was that he was the first person to administer anesthesia during childbirth).

What does all of this have to do with water treatment (disinfection)? Actually, Dr. Snow's discovery—his stripping of a mystery to its barest bones—has quite a lot to do with water treatment. Combating any disease is rather difficult without a determination of how the disease is transmitted—how it travels from vector or carrier to receiver. Dr. Snow established this connection, and from his work, and the work of others, progress was made in understanding and combating many different waterborne diseases.

Today, sanitation problems in developed countries (those with the luxury of adequate financial and technical resources) deal more with the consequences that arise from inadequate commercial food preparation and the results of bacteria becoming resistant to disinfection techniques and antibiotics. We simply flush our toilets to rid ourselves of unwanted wastes and turn on our taps to take in high-quality drinking water supplies, from which we've all but eliminated cholera and epidemic diarrheal diseases. This is generally the case in most developed countries today—but it certainly wasn't true in Dr. Snow's time.

The progress in water treatment from that notable day in 1854 [when Snow made the "connection" (actually the "disconnection" of the handle from the pump) between deadly cholera and its means of transmission, its "communication"] to the present reads like a chronology of discovery leading to our modern water treatment practices. This makes sense, of course, because over time, pivotal events and discoveries occur—events that have a profound effect on how we live today. Let's take a look at a few elements of the important chronological progression that evolved from the simple removal of a pump handle to the advanced water treatment (disinfection) methods we employ today to treat our water supplies.

After Snow's discovery (that cholera is a waterborne disease emanating primarily from human waste), events began to drive the water/wastewater treatment process. In 1859, 4 years after Snow's discovery, the British Parliament was suspended during the summer because the stench coming from the Thames was unbearable. According to one account, the river began to "seethe and ferment under a burning sun." As was the case in many cities at this time, storm sewers carried a combination of storm water, sewage, street debris and other wastes to the nearest body of water. In the 1890s, Hamburg, Germany, suffered a cholera epidemic. Detailed studies by Koch tied the outbreak to the contaminated water supply. In response to the epidemic, Hamburg was among the first cities to use chlorine as part of a wastewater treatment regimen. About the same time, the town of Brewster, New York, became the first U.S. city to disinfect its treated wastewater. Chlorination of drinking water was used temporarily in 1896, and its first known continuous use for water supply disinfection occurred in Lincoln, England, and Chicago in 1905. Jersey City, New Jersey, became one of the first routine users of chlorine in 1908.

Time marched on and with it came an increased realization of the need to treat and disinfect both water supplies and wastewater. Between 1910 and 1915, technological improvements in gaseous and then solution feed of elemental chlorine (Cl_2) made the process more practical and efficient. Disinfection of water supplies and chlorination of treated wastewater for odor control increased over the next several decades. In the United States, disinfection, in one form or another, is now being used by more than 15,000 out of approximately 16,000 Publicly Owned Treatment Works (POTWs). The significance of this number becomes apparent when you consider that fewer than 25 of the 600-plus POTWs in the United States in 1910 were using disinfectants.

Although the epidemiological relation between water and disease had been suggested as early as the 1850s, it was not until the establishment of the germ theory of disease by Pasteur in the mid-1880s that water as a carrier of disease-producing organisms was understood. As related in Sidebar 17.1, in the 1850s, while London experienced the "Broad Street Well" cholera epidemic, Dr. John Snow conducted his now-famous epidemiological study. Dr. Snow concluded that the well had become contaminated by a visitor with the disease who had arrived in the vicinity. Cholera was one of the first diseases to be recognized as capable of being waterborne. Also, this incident was probably the first reported disease epidemic attributed to the direct recycling of non-disinfected water. Now, over 100 years later, the list of potential waterborne diseases due to pathogens is considerably larger and includes bacterial, viral, and parasitic microorganisms, as shown in Tables 17.3–17.5, respectively.

A major cause of the number of disease outbreaks in potable water is contamination of the distribution system from cross-connections and back siphonage with non-potable

TABLE 17.3
Waterborne Diseases from Bacteria

Causative Agent	Disease
Salmonella typhosa	Typhoid fever
S. paratyphi	Paratyphoid fever
S. schottinulleri	
S. hirschfeldi C.	
Shigella flexneri	Bacillary dysentery
Sh. dysenteriae	
Sh. sonnel	
Sh. paradysinteriae	
Vibrio comma	Cholera
V. cholerae	
Pasteurella tularensis	Tularemia
Brucella melitensis	Brucellosis
Leptospira icterohaemorrhagica	Leptospirosis
Enteropathogenic E. coli	Gastroenteritis

TABLE 17.4
Waterborne Diseases from Human Enteric Viruses

Group	Subgroup
Enterovirus	Poliovirus
	Echovirus
	Coxsackie virus
	A
	B
Reovirus	
Adenovirus	
Hepatitis	

water. However, outbreaks resulting from distribution system contamination are usually quickly contained and result in relatively few illnesses compared to contamination of the source water or a breakdown in the treatment system, which typically produces many cases of illness per incident. When considering the number of cases, the major causes of disease outbreaks are source water contamination and treatment deficiencies (White, 1992). Historically, about 46% of the outbreaks in public water systems have been found to be related to deficiencies in source water and treatment systems, with 92% of the causes of illness due to these two particular problems.

All natural waters support biological communities. Because some microorganisms can be responsible for public health problems, the biological characteristics of the source water are one of the most important parameters in water treatment. In addition to public health problems, microbiology can also affect the physical and chemical water quality and treatment plant operation.

PATHOGENS OF PRIMARY CONCERN

Table 17.6 shows the attributes of three groups of pathogens of concern in water treatment, namely bacteria, viruses, and protozoa.

1. **Bacteria:** Recall that bacteria are single-celled organisms typically ranging in size from 0.1 to 10 μm. Shape, components, size, and the manner

in which they grow can characterize the physical structure of the bacterial cell. Most bacteria can be grouped by shape into four general categories: spheroid, rod, curved rod or spiral, and filamentous. Cocci, or spherical bacteria, are approximately 1–3 μm in diameter. Bacilli (rod-shaped bacteria) are variable in size and range from 0.3 to 1.5 μm in width (or diameter) and from 1.0 to 10.0 μm in length. Vibrio, or curved rod-shaped bacteria, typically vary in size from 0.6 to 1.0 μm in width (or diameter) and from 2 to 6 μm in length. Spirilla (spiral bacteria) can be found in lengths of up to 50 μm, whereas filamentous bacteria can occur in lengths more than 100 μm.

2. **Viruses:** Viruses are microorganisms composed of the genetic material deoxyribonucleic acid (DNA) or ribonucleic acid (RNA) and a protective protein coat (either single, double, or partially double-stranded). All viruses are obligate parasites, unable to carry out any form of metabolism, and are completely dependent upon host cells for replication. Viruses are typically 0.01–0.1 μm in size and are very species-specific concerning infection, typically attacking only one type of host. Although the principal modes of transmission for the hepatitis B virus and poliovirus are through food, personal contact, or exchange of body fluids, these viruses can be transmitted through potable water. Some viruses, such as retroviruses (including the HIV group), appear to be too fragile for water transmission to be a significant danger to public health (Spellman, 2007).

3. **Protozoa:** Protozoans are single-celled eukaryotic microorganisms without cell walls that utilize bacteria and other organisms for food. Most protozoa are free-living in nature and can be encountered in water; however, several species are parasitic and live on or in host organisms. Host organisms can vary from primitive organisms such as algae to highly complex organisms such as human beings. Several species of protozoa known to utilize human beings as hosts are shown in Table 17.7.

TABLE 17.5
Waterborne Diseases from Parasites

Causative Agent	Symptoms
Ascaris lumbricoides (round worm)	Ascariasis
Cryptosporidium muris and *Cryptosporidium parvum*	Cryptosporidiosis
Entamoeba histolytica	Amebiasis
Giardia lamblia	Giardiasis
Naegleria gruberi	Amoebic meningoencephalitis
Schistosoma mansoni	Schistosomiasis
Taenia saginata (beef tapeworm)	Taeniasis

TABLE 17.6
Attributes of the Three Waterborne Pathogens in Water Treatment

Organism	Size (μm)	Mobility	Point(s) of Origin	Resistance to Disinfection
Bacteria	0.1–10	Motile, nonmotile	Humans and animals, water, and contaminated food	Type specific-bacterial spores typically have the highest resistance, whereas vegetative bacteria have the lowest resistance
Viruses	0.01–0.1	Nonmotile	Humans and animals, polluted water; contaminated food	Generally, more resistant than vegetative bacteria
Protozoa	1–20	Motile, nonmotile	Humans and animals, sewage decaying vegetation, and water	More resistant than viruses or vegetative bacteria

TABLE 17.7

Human Parasitic Protozoans

Protozoan	Host(s)	Disease	Transmission
Acanthamoeba castellanii	Fresh water, sewage, humans, soil	Amoebic meningoencephalitis	Gains entry through abrasions, ulcers, and as a secondary invader during other infections
Balantidium coli	Pigs, humans	Balantidiasis (dysentery)	Contaminated water
Cryptosporidium parvum	Animals, humans	Cryptosporidiosis	Person-to-person or animal-to-person contact, ingestion of fecally contaminated water or food, or contact with fecally contaminated environmental surfaces.
Entamoeba histolytica	Humans	Amoebic dysentery	Contaminated water
Giardia lamblia (gastroenteritis)	Animals, humans	Giardiasis	Contaminated water
Naegleria fowleri	Soil, water, humans, and decaying vegetation	Primary amoebic meningoencephalitis	Nasal inhalation with subsequent penetration of nasopharynx; exposure from swimming in fresh-water lakes

RECENT WATERBORNE DISEASE OUTBREAKS

Within the past 50 years, several pathogenic agents never before associated with documented waterborne outbreaks have appeared in the United States. Enteropathogenic *E. coli* and *Giardia lamblia* were first identified as the etiological agents responsible for waterborne outbreaks in the 1960s. The first recorded *Cryptosporidium* infection in humans occurred in the mid-1970s. Also, during that time, there was the first recorded outbreak of pneumonia caused by *Legionella pneumophila*. Recently, there have been numerous documented waterborne disease outbreaks caused by *E. coli*, *G. lamblia*, *Cryptosporidium*, and *L. pneumophila*.

Escherichia coli

The first documented case of waterborne disease outbreaks in the United States associated with enteropathogenic *E.= coli* occurred in the 1960s. Various serotypes of *E. coli* have been implicated as the etiological agents responsible for disease in newborn infants, usually the result of cross-contamination in nurseries. Now, there have been several well-documented outbreaks of *E. coli* associated with adult waterborne disease. In 1975, the etiologic agent of a large outbreak at Crater Lake National Park was *E. coli* serotype 06:H16 (Craun, 1981).

Giardia lamblia

Similar to *Escherichia coli*, *Giardia lamblia* was first identified in the 1960s as being associated with waterborne outbreaks in the United States. Recall that *G. lamblia* is a flagellated protozoan that is responsible for giardiasis, a disease that can range from being mildly to extremely debilitating. *Giardia* is currently one of the most commonly identified pathogens responsible for waterborne disease outbreaks. The life cycle of *Giardia* includes a cyst stage when the organism remains dormant and is extremely resilient (i.e., the cyst can survive some extreme environmental conditions). Once ingested by a warm-blooded animal, the life cycle of *Giardia* continues with excystation.

The cysts are relatively large (8–14 μm) and can be removed effectively by filtration using diatomaceous earth, granular media, or membranes. Giardiasis can be acquired by ingesting viable cysts from food or water or by direct contact with fecal material. In addition to humans, wild and domestic animals have been implicated as hosts. Between 1972 and 1981, 50 waterborne outbreaks of giardiasis occurred, with about 20,000 reported cases (Craun and Jakubowski, 1996). Currently, no simple and reliable method exists to assay *Giardia* cysts in water samples. Microscopic methods for detection and enumeration are tedious and require examiner skill and patience. *Giardia* cysts are relatively resistant to chlorine, especially at higher pH and low temperatures.

Cryptosporidium

Cryptosporidium is a protozoan similar to *Giardia*. It forms resilient oocysts as part of its life cycle. The oocysts are smaller than *Giardia* cysts, typically about 4–6 μm in diameter. These oocysts can survive under adverse conditions until ingested by a warm-blooded animal and then continue with excystation. Due to the increase in the number of outbreaks of Cryptosporidiosis, a tremendous amount of research has focused on *Cryptosporidium* within the last 10 years. Medical interest has increased because of its occurrence as a life-threatening infection in individuals with depressed immune systems. As previously mentioned, in 1993, the largest documented waterborne disease outbreak in the United States occurred in Milwaukee and was determined to be caused by *Cryptosporidium*. An estimated 403,000 people became ill, 4,400 people were hospitalized, and 100 people died. The outbreak was associated with a deterioration in raw water quality and a simultaneous decrease in the effectiveness of the coagulation-filtration process, which led to an increase in the turbidity of treated water and inadequate removal of *Cryptosporidium* oocysts.

Legionella pneumophila

An outbreak of pneumonia occurred in 1976 at the annual convention of the Pennsylvania American Legion. A total of 221 people were affected by the outbreak, and 35 of those afflicted died. The cause of pneumonia was not determined immediately despite an intense investigation by the Centers for Disease Control. Six months after the incident, microbiologists were able to isolate a bacterium from the autopsy lung tissue of one of the Legionnaires. The bacterium responsible for the outbreak was found to be distinct from other known bacteria and was named *Legionella pneumophila* (Witherell et al., 1988). Following the discovery of this organism, other *Legionella*-like organisms were discovered. Legionnaires' disease does not appear to be transmitted person-to-person. Epidemiological studies have shown that the disease enters the body through the respiratory system. *Legionella* can be inhaled in water particles less than 5 μm in size from facilities such as cooling towers, hospital hot water systems, and recreational whirlpools.

MECHANISM OF PATHOGEN INACTIVATION

The three primary mechanisms of pathogen inactivation are:

- Destroying or impairing cellular structural organization by attacking major cell constituents, such as destroying the cell wall or impairing the functions of semi-permeable membranes.
- Interfering with energy-yielding metabolism through enzyme substrates in combination with prosthetic groups of enzymes, thus rendering them non-functional.
- Interfering with biosynthesis and growth by preventing the synthesis of normal proteins, nucleic acids, coenzymes, or the cell wall.

Depending on the disinfectant and microorganism type, combinations of these mechanisms can also be responsible for pathogen inactivation. In water treatment, it is believed that the primary factors controlling disinfection efficiency are: (1) the ability of the disinfectant to oxidize or rupture the cell wall; and (2) the ability of the disinfectant to diffuse into the cell and interfere with cellular activity (Montgomery, 1985). In addition, it is important to point out that disinfection is effective in reducing waterborne diseases because most pathogenic organisms are more sensitive to disinfection than are nonpathogens. However, disinfection is only as effective as the care used in controlling the process and assuring that all of the water supply is continually treated with the amount of disinfectant required producing safe water.

OTHER USES OF DISINFECTANTS IN WATER TREATMENT

Disinfectants are used for more than just disinfection in drinking water treatment. While the inactivation of pathogenic organisms is a primary function, disinfectants are also used as oxidants in drinking water treatment for several other functions:

- Minimization of DBP formation
- Control of nuisance Asiatic clams and zebra mussels
- Oxidation of iron and manganese
- Prevention of regrowth in the distribution system and maintenance of biological stability
- Removal of taste and odors through chemical oxidation
- Improvement of coagulation and filtration efficiency
- Prevention of algal growth in sedimentation basins and filters
- Removal of color

A brief discussion of these additional oxidant uses follows.

Minimization of DBP Formation

Strong oxidants may play a role in disinfection and DBP control strategies in water treatment. Several strong oxidants, including potassium permanganate and ozone, may be used to control DBP precursors.

Note: Potassium permanganate can be used to oxidize organic precursors at the head of the treatment plant, thus minimizing the formation of byproducts at the downstream disinfection stage of the plant. The use of ozone for oxidation of DBP precursors is currently being studied. Early work has shown that the effects of ozonation, prior to chlorination, were highly site-specific and unpredictable. The key variables that seem to determine the effect of ozone are dose, pH, alkalinity, and the nature of the organic material. Ozone has been shown to be effective for DBP precursor reduction at low pHs. However, at higher pH (i.e., above 7.5), ozone may actually increase the amount of chlorination byproduct precursors.

Control of Nuisance Asiatic Clams and Zebra Mussels

The Asiatic clam (*Corbicula fluminea*) was introduced to the United States from Southeast Asia in 1938 and now inhabits almost every major river system south of 40° latitude. Asiatic clams have been found in the Trinity River, TX; the Ohio River at Evansville, IN; New River at Narrows and Glen Lyn, VA; and the Catawba River in Rock Hill, SC. This animal has invaded many water utilities, clogging source water transmission systems, valves, screens, and meters; damaging centrifugal pumps; and causing taste and odor problems (Britton and Morton, 1982; Sinclair, 1964; Cameron et al., 1989).

Cameron et al. (1989) investigated the effectiveness of several oxidants to control the Asiatic clam in both the juvenile and adult phases. As expected, the adult clam was found to be much more resistant to oxidants than the juvenile form. In many cases, the traditional method of control, free chlorination, cannot be used because of the formation of excessive amounts of THMs. As shown in Table 17.8, Cameron et al. (1989) compared the effectiveness of four oxidants for controlling the juvenile Asiatic clam in terms of the LT50 (time required for 50% mortality). Monochloramine was

TABLE 17.8

The Effects of Various Oxidants on Mortality of the Asiatic Clam

Chemical	Residual (mg/L)	Temperature (°C)	pH	Life Stage	LT50 (Days)
Free chlorine	0.5	23	8.0	Adult	8.7
	4.8	21	7.9	Adult	5.9
	4.7	16	7.8	Juvenile	4.8
Potassium	1.1	17	7.6	Juvenile	7.9
Permanganate	4.8	17	7.6	Juvenile	8.6
Chlorine	1.2	24	6.9	Juvenile	0.7
dioxide	4.7	22	6.6	Juvenile	0.6

Source: Adaptation from Cameron et al. (1989).

found to be the best for controlling the juvenile clams without forming THMs. The effectiveness of monochloramine increased greatly as the temperature increased. Clams can tolerate temperatures between 2°C and 35°C.

In a similar study, Belanger et al. (1991) studied the biocidal potential of total residual chlorine, monochloramine, monochloramine plus excess ammonia, bromine, and copper for controlling the Asiatic clam. Belanger et al. (1991) showed that monochloramine with excess ammonia was the most effective for controlling the clams at 30°C. Chlorination at 0.25–0.40 mg/L total residual chlorine at 20°C–25°C controlled clams of all sizes but had minimal effect at 12°C–15°C (as low as zero mortality). As in other studies, the toxicity of all the biocides was highly dependent on temperature and clam size.

The zebra mussel (*Dreissena polymorpha*) is a recent addition to the fauna of the Great Lakes. It was first found in Lake St. Clair in 1988, though it is believed that this native of the Black and Caspian seas was brought over from Europe in ballast water around 1985. The zebra mussel population in the Great Lakes has expanded very rapidly, both in size and geographical distribution (Herbert et al., 1989; Roberts, 1990). Lang (1994) reported that zebra mussels have been found in the Ohio River, Cumberland River, Arkansas River, Tennessee River, and the Mississippi River south to New Orleans.

Klerks and Fraleigh (1991) evaluated the effectiveness of hypochlorite, permanganate, and hydrogen peroxide with iron for their effectiveness in controlling adult zebra mussels. Both continuous and intermittent 28-day static renewal tests were conducted to determine the impact of intermittent dosing. Intermittent treatment proved to be much less effective than continuous dosing. The hydrogen peroxide-iron combination (1–5 mg/L with 25% iron) was less effective in controlling the zebra mussel than either permanganate or hypochlorite. MnO^{-4} (0.5–2.5 mg $KnnO_4$/L) was usually less effective than hypochlorite (0.5–10 mg Cl_2/L).

Van Benschoten et al. (1995) developed a kinetic model to predict the rate of mortality of the zebra mussel in response to chlorine. The model shows the relationship between chlorine residual and temperature on the exposure time required to achieve 50% and 95% mortality. Data were collected for chlorine residuals between 0.5 and 3.0 mg Cl_2/L and temperatures from 0.3°C to 24°C. The results show a strong dependence on temperature and required contact times ranging from 2 days to more than a month, depending on environmental factors and the level of mortality required.

Brady et al. (1996) compared the efficiency of chlorine in controlling the growth of zebra mussels and quagga mussel (*Dreissena bugensis*). The quagga mussel is a newly identified mollusk within the Great Lakes that is similar in appearance to the zebra mussel. Full-scale chlorination treatment found a significantly higher mortality for the quagga mussel. The required contact time for 100% mortality for quagga and zebra mussels was 23 and 37 days, respectively, suggesting that chlorination programs designed to control zebra mussels should also be effective for controlling populations of quagga mussels.

Matisoff et al. (1996) evaluated chlorine dioxide (CIO_2) to control adult zebra mussels using simple, intermittent, and continuous exposures. A single 30-min exposure to 20 mg/L chlorine dioxide or higher concentration induced at least 50% mortality, while sodium hypochlorite produced only 26% mortality, and permanganate and hydrogen peroxide were totally ineffective when dosed at 30 mg/L for 30 min under the same conditions. These high dosages, even though only used for a short period, may not allow application directly in water for certain applications due to byproducts that remain in the water. Continuous exposure to chlorine dioxide for 4 days was effective at concentrations above 0.5 mg/L (LC50=0.35 mg/L), and 100% mortality was achieved at chlorine dioxide concentrations above 1 mg/L.

These studies all show that the dose required to induce mortality in these nuisance organisms is extremely high, both in terms of chemical dose and contact time. The potential impact on DBPs is significant, especially when the water is high in organic content and has a high propensity to form THMs and other DBPs.

Oxidation of Iron and Manganese

Iron and manganese occur frequently in groundwater but are less problematic in surface waters. Although not harmful to human health at the low concentrations typically found in water, these compounds can cause staining and taste problems. These compounds are readily treated by oxidation to produce a precipitate that is removed in subsequent sedimentation and filtration processes. Almost all the common oxidants except chloramines, will convert ferrous (2+) iron to the ferric (3+) state and manganese (2+) to the (4+) state, which will precipitate as ferric hydroxide and manganese dioxide, respectively (AWWA, 2010). The precise chemical composition of the precipitate will depend on the nature of the water, temperature, and pH.

TABLE 17.9
Oxidant Doses Required for Oxidation of Iron and Manganese

Oxidant	Iron (II) (mg/mg Fe)	Manganese (II) (mg/mg Mn)
Chlorine	0.62	0.77
Chlorine dioxide	1.21	2.45
Ozone	0.43	0.85
Oxygen	0.14	0.29
Potassium permanganate	0.94	1.92

Source: Adapted from Culp and Culp (1974b).

Table 17.9 shows that oxidant doses for iron and manganese control are relatively low. In addition, the reactions are relatively rapid, on the order of seconds, while DBP formation occurs over hours. Therefore, with proper dosing, residual chlorine during iron and manganese oxidation is relatively low and short-lived. These factors reduce the potential for DBP formation as a result of oxidation for iron and manganese removal.

Prevention of Re-growth in Distribution System & Maintenance of Biological Stability

Biodegradable organic compounds and ammonia in treated water can cause microbial growth in the distribution system. *Biological stability* refers to a condition wherein the water quality does not enhance biological growth in the distribution system. Biological stability can be accomplished in several ways:

- Removing nutrients from the water prior to distribution.
- Maintaining a disinfectant residual in the treated water.
- Combining nutrient removal and disinfectant residual maintenance.

To maintain biological stability in the distribution system, the Total Coliform Rule (TCR) requires that treated water have a residual disinfectant of 0.2 mg/L when entering the distribution system. A measurable disinfectant residual must be maintained in the distribution system, or the utility must show through monitoring that the heterotrophic plate count (HPC) remains less than 500/100 mL. A system remains in compliance as long as 95% of samples meet these criteria. Chlorine, monochloramine, and chlorine dioxide are typically used to maintain a disinfectant residual in the distribution system. Filtration can also be used to enhance biological stability by reducing the nutrients in the treated water.

The level of secondary disinfectant residual maintained is low, typically in the range of 0.1–0.3 mg/L, depending on the distribution system and water quality. However, because the contact times in the system are quite long, it is possible

to generate significant amounts of DBPs in the distribution system, even at low disinfectant doses. Distribution system problems associated with the use of combined chlorine residual (chloramines), or no residual, have been documented in several instances. The use of combined chlorine is characterized by an initial satisfactory phase in which chloramine residuals are easily maintained throughout the system and bacterial counts are very low. However, problems may develop over a period of years, including increased bacterial counts, reduced combined chlorine residual, increased taste and odor complaints, and reduced transmission main carrying capacity. Conversion of the system to free-chlorine residual produces an initial increase in consumer complaints of taste and odors resulting from the oxidation of accumulated organic material. Also, it is difficult to maintain a free-chlorine concentration at the ends of the distribution system (AWWA, 2010).

Removal of Taste and Odors through Chemical Oxidation

Taste and odors in drinking water are caused by several sources, including microorganisms, decaying vegetation, hydrogen sulfide, and specific compounds of municipal, industrial, or agricultural origin. Disinfectants themselves can also create taste and odor problems. In addition to a specific taste- and odor-causing compound, the sanitary impact is often accentuated by a combination of compounds. More recently, significant attention has been given to tastes and odors from specific compounds such as geosmin, 2-methylisoborneol (MIB), and chlorinated inorganic and organic compounds (AWWARF, 1987).

Oxidation is commonly used to remove taste- and odor-causing compounds. Because many of these compounds are very resistant to oxidation, advanced oxidation processes (ozone/hydrogen peroxide, ozone/UV, etc.) and ozone by itself are often used to address taste and odor problems. The effectiveness of various chemicals to control taste and odors can be site-specific. Suffet et al. (1986) found that ozone is generally the most effective oxidant for use in taste and odor treatment. They found ozone doses of 2.5–2.7 mg/L and 10 min of contact time (residual 0.2 mg/L) significantly reduce levels of taste and odors. Lalezary et al. (1986) used chlorine, chlorine dioxide, ozone, and permanganate to treat earthy-musty-smelling compounds. In that study, chlorine dioxide was found most effective, although none of the oxidants were able to remove geosmin and MIB by more than 40%–60%. Potassium permanganate has been used in doses of 0.25–20 mg/L.

Prior experiences with taste and odor treatment indicate that oxidant doses are dependent on the source or the water and causative compounds. In general, small doses can be effective for many taste and odor compounds, but some of the difficult-to-treat compounds require strong oxidants such as ozone and/or advanced oxidation processes or alternative technologies such as granular activated carbon (GAC) adsorption.

Improvement of Coagulation and Filtration Efficiency

Oxidants, specifically ozone, have been reported to improve coagulation and filtration efficiency. Others, however, have found no improvement in effluent turbidity from oxidation. Prendiville (1986) collected data from a large treatment plant showing that preozonation was more effective than prechlorination in reducing filter effluent turbidities. The cause of improved coagulation is not clear, but several possibilities have been offered (Gurol and Pidatella, 1983; Reckhow et al., 1986), including:

- Oxidation of organics into more polar forms
- Oxidation of metal ions to yield insoluble complexes such as ferric iron complexes
- Change in the structure and size of suspended particles.

Prevention of Algal Growth in Sedimentation Basins and Filters

Prechlorination is often used to minimize operational problems associated with biological growth in water treatment plants (AWWA, 2010). Prechlorination will prevent slime formation on filters, pipes, and tanks and reduce potential taste and odor problems associated with such slimes. Many sedimentation and filtration facilities operate with a small chlorine residual to prevent the growth of algae and bacteria in laundries (in like equipment or machinery) and on the filter surfaces. This practice has increased in recent years as utilities take advantage of additional contact time in the treatment units to meet disinfection requirements under the SWTR.

Removal of Color

Free chlorine is used for color removal. A low pH is favored. Humic compounds which have a high potential for DBP formation, cause color. The chlorine dosage and kinetics for color removal are best determined through bench studies.

TYPES OF DISINFECTION BYPRODUCTS AND DISINFECTION RESIDUALS

Table 17.10 provides a list compiled by the USEPA of DBPs and disinfection residuals that may be of health concern. The table includes both the disinfectant residuals and the specific byproducts produced by the disinfectants of interest in drinking water treatment. These contaminants of concern are grouped into four distinct categories: disinfectant residuals, inorganic byproducts, organic oxidation byproducts, and halogenated organic byproducts.

The production of DBPs depends on the type of disinfectant, the presence of organic material (e.g., TOC), bromide ion, and other environmental factors as discussed in this section. By removing DBP precursors, the formation of DBPs can be reduced. The health effects of DBPs and disinfectants are generally evaluated through epidemiological studies and/or toxicological studies using laboratory

TABLE 17.10

List of Disinfection Byproducts and Disinfection Residuals

Disinfectant Residuals	Halogenated Organic Byproducts
Free chlorine	Trihalomethanes
Hypochlorous acid	Chloroform
Hypochlorite ion	Bromodichloromethane
Chloramines	Dibromochloromethane
Monochloramine	Bromoform
Chlorine dioxide	Haloacetic acids
Inorganic byproducts	Monochloroacetic acid
Chlorate ion	Dichloroacetic acid
Chlorite ion	Trichloroacetic acid
Bromate ion	Monobromoacetic acid
Iodate ion	Dibromoacetic acid
Hydrogen peroxide	Haloacetonitriles
Ammonia	Dichloroacetronitrile
Organic Oxidation	Bromochloroacetonitrile
Byproducts	
Aldehydes	Dibromoacetonitrile
Formaldehyde	Trichloroacetonitrile
Acetaldehyde	Haloketones
Glyoxal	1,1-Dichloropropanone
Hexanal	1,1,1-Trichloropropanone
Heptanal	Chlorophenols
Carboxylic acids	2-Chlorophenol
Hexanoic acid	2,4-Dichlorophenol
Heptanoic acid	2,4,6-Trichlorophenol
Oxalic acid	Chloropicrin
Assimilable organic carbon	Chloral hydrate cyanogen chloride
	N-Organochloramines

animals. Table 17.11 indicates the cancer classifications of both disinfectants and DBPs as of January 1999. The classification scheme used by the USEPA is shown at the bottom of Table 17.11. The USEPA classification scheme for carcinogenicity weighs both animal studies and epidemiologic studies, but places greater weight on evidence of carcinogenicity in humans.

DISINFECTION BYPRODUCT FORMATION

Halogenated organic byproducts are formed when natural organic matter (NOM) reacts with free chlorine or free bromine. Free chlorine can be introduced to water directly as a primary or secondary disinfectant, with chlorine dioxide, or with chloramines. Free bromine results from the oxidation of the bromide ion in source water. Factors affecting the formation of halogenated DBPs include the type and concentration of natural organic matter, oxidant type and dose, time, bromide ion concentration, pH, organic nitrogen concentration, and temperature. Organic nitrogen significantly influences the formation of nitrogen-containing DBPs such as haloacetonitriles, halopicrins, and cyanogen halides. The parameter TOX represents the concentration of total

TABLE 17.11
Status of Health Information for Disinfectants and DBPs

Contaminant	Cancer Classification
Chloroform	Probable human carcinogen
Bromodichloromethane	Probable human carcinogen
Dibromochloromethane	Possible human carcinogen
Bromoform	Probable human carcinogen
Monochloroacetic acid	------------------------------------
Dichloroacetic acid	Probable human carcinogen
Trichloroacetic acid	Possible human carcinogen
Dichloroacetonitrile	Possible human carcinogen
Bromochloroacetonitrile	------------------------------------
Dibromoacetonitrile	Possible human carcinogen
Trichloroacetonitrile	------------------------------------
1,1-Dichloropropanone	------------------------------------
1,1,1-Trichloropropanone	------------------------------------
2-Chlorophenol	Not classifiable
2,4-Dichlorphenol	Not classifiable
2,4,6-Trichlorophenol	Probable human carcinogen
Chloropicrin	------------------------------------
Chloral hydrate	Possible human carcinogen
Cyanogen chloride	------------------------------------
Formaldehyde	Probable human carcinogen
Chlorate	------------------------------------
Chlorite	Not classifiable
Bromate	Probable human carcinogen
Ammonia	Not classifiable
Hypochlorous acid	------------------------------------
Hypochlorite	------------------------------------
Monochloramine	------------------------------------
Chlorine dioxide	Not classifiable

Source: USEPA (1996).

DBP Precursors

Numerous researchers have documented that NOM is the principal precursor of organic DBP formation. Chlorine reacts with NOM to produce a variety of DBPs, including THMs, haloacetic acids (HAAs), and others. Ozone reacts with NOM to produce aldehydes, organic acids, and aldo- and keto-acids; many of these are produced by chlorine as well (Stevens, 1976; Singer and Harrington., 1993). Natural waters contain mixtures of both humic and non-humic organic substances. NOM can be subdivided into a hydrophobic fraction composed primarily of humic material and a hydrophilic fraction composed primarily of fulvic material. The type and concentration of NOM are often assessed using surrogate measures. Although surrogate parameters have limitations, they are used because they may be measured more easily, rapidly, and inexpensively than the parameter of interest, often allowing on-line monitoring of the operation and performance of water treatment plants. Surrogates used to assess NOM include:

- Total and dissolved organic carbon (TOC and DOC)
- Specific ultraviolet light absorbance (SUVA), which is the absorbance at 254 nm wavelength (UV-254) divided by DOC (SUVA=(UV-254/DOC)100 in L/mg-m)
- THM formation potential (THMFP)—a test measuring the quantity of THMs formed with a high dosage of free chlorine and a long reaction time
- TTHM Stimulated Distribution System (SDS)—a test to predict the TTHM concentration at some selected point in a given distribution system, where the conditions of the chlorination test simulate the distribution system at the point desired.

organic halides in a water sample (calculated as chloride). In general, less than 50% of the TOX content has been identified, despite evidence that several of these unknown halogenated byproducts of water chlorination may be harmful to humans (Reckhow et al., 1990; Singer and Chang, 1989).

Nonhalogenated DBPs are also formed when strong oxidants react with organic compounds found in water. Ozone and peroxone oxidation of organics lead to the production of aldehydes, aldo- and keto-acids, organic acids, and, when bromide ion is present, brominated organics. Many of the oxidation byproducts are biodegradable and appear as biodegradable dissolved organic carbon (BDOC) and assimilable organic carbon (AOC) in treated water.

Bromide ion plays a key role in DBP formation. Ozone or free chlorine oxidizes bromide ion to hypobromate ion/hypobromous acid, which subsequently forms brominated DBPs. Brominated organic byproducts include compounds such as bromoform, brominated acetic acids and acetonitrile, bromopicrin, and cyanogen bromide. Only about one third of the bromide ions incorporated into byproducts have been identified.

On average, about 90% of the TOC is dissolved. DOC is defined as the TOC able to pass through a 0.45 μm filter. UV absorbance is a good technique for assessing the presence of DOC through a 0.45 μm filter. UV absorbance is a good technique for assessing the presence of DOC because DOC primarily consists of humic substances, which contain aromatic structures that absorb light in the UV spectrum. Oxidation of DOC reduces the UV absorbance of the water due to the oxidation of some of the organic bonds that absorb UV absorbance. Complete mineralization of organic compounds to carbon dioxide usually does not occur underwater treatment conditions; therefore, the overall TOC concentration usually remains constant.

Concentrations of DBPs vary seasonally and are typically greatest in the summer and early fall for several reasons:

- The rate of DBP formation increases with increasing temperature.
- The nature of organic DBP precursors varies with the season.

- Due to warmer temperatures, chlorine demand may be greater during the summer months requiring higher dosages to maintain disinfection.

If the bromide ion is present in source waters, it can be oxidized to hypobromous acid, which can react with NOM to form brominated DBPs, such as bromoform. Furthermore, under certain conditions, ozone may react with the hypobromite ion to form bromate ion (Singer, 1992).

The ratio of bromide ion to the chlorine dose affects THM formation and bromine substitution of chlorine. Increasing the bromide ion to chlorine dose ratio shifts the speciation of THMs to produce more brominated forms. In the Krasner (1989) study, the chlorine dose was roughly proportional to TOC concentration. As TOC was removed through the treatment train, the chlorine doses decrease and TTHM formation declined. However, at the same time, the bromide ion to chlorine dose increased, thereby shifting TTHM concentrations to the more brominated THMs. Therefore, improving the removal of NOM prior to chlorination can shift the speciation of halogenated byproducts toward more brominated forms.

Chloropicrin is produced by the chlorination of humic materials in the presence of nitrate ion. Thibaud et al. (1988) chlorinated humic compounds in the presence of bromide ion to demonstrate the formation of brominated analogs to chloropicrin.

Impacts of pH on DBP Formation

The pH of water being chlorinated has an impact on the formation of halogenated byproducts. THM formation increases with increasing pH. Trichloroacetic acid, dichloroacetonitrile, and trichloropropanone formation decrease with increased pH. Overall TOX formation decreases with increasing pH. Based on chlorination studies of humic material in model systems, high pH tends to favor chloroform formation over the formation of trichloroacetic acid and other organic halides. Accordingly, water treatment plants practicing precipitative softening at pH values greater than 9.5–10 are likely to have a higher fraction of TOX attributable to THMs than plants treating surface waters by conventional treatment in pH ranges of 6–8 (Reckhow and Singer, 1985).

Because the application of chlorine dioxide and chloramines may introduce free chlorine into water, chlorination byproducts that may be formed would be influenced by pH, as discussed above. Ozone application to bromide ion-containing waters at high pH favors the formation of bromate ion, while application at low pH favors the formation of brominated organic byproducts.

The pH also impacts enhanced coagulation (i.e., for ESWTR compliance) and Lead and Copper Rule compliance. These issues are addressed in USEPA's *Microbial and Disinfection Byproduct Simultaneous Compliance Guidance Manual.*

Organic Oxidation Byproducts

Organic oxidation byproducts are formed by reactions between NOM and all oxidizing agents added during drinking water treatment. Some of these byproducts are halogenated, as discussed in the previous section, while others are not. The types and concentrations of organic oxidation byproducts produced depend on the type and dosage of the oxidant being used, the chemical characteristics and concentration of the NOM being oxidized, and other factors such as pH and temperature.

Inorganic Byproducts and Disinfectants

Table 17.12 shows some of the inorganic DBPs that are produced or remain as residuals during disinfection. As discussed earlier, bromide ion reacts with strong oxidants to form bromate ions and other organic DBPs. Chlorine dioxide and chloramines leave residuals that are of concern for health considerations, as well as for taste and odor.

DBP Control Strategies

In 1983, the USEPA identified technologies, treatment techniques, and plant modifications that community water systems could use to comply with the maximum contaminant level for TTHMs. The principal treatment modifications involved moving the point of chlorination downstream in the water treatment plant, improving the coagulation process to enhance the removal of DBP precursors, and using chloramines to supplement or replace the use of free chlorine (Singer and Harrington, 1993). Moving the point of chlorination downstream in the treatment train is often very effective in reducing DBP formation because it allows the NOM precursor concentration to be reduced during treatment prior to chlorine addition. Replacing prechlorination with pre-oxidation using an alternate disinfectant that produces fewer DBPs is another option for reducing formation of chlorinated byproducts. Other options to control the formation of DBPs include source water quality control, DBP precursor removal, and disinfection strategy selection. An overview of each is provided below.

Source Water Quality Control

Source water control strategies involve managing the source water to lower the concentrations of NOM and bromide

---

TABLE 17.12

Inorganic DBPs Produced During Disinfection

Disinfectant	Inorganic Byproducts or Disinfectant Residual Discussed
Chlorine dioxide	Chlorine dioxide, chlorite ion, chlorate ion, bromate ion
Ozone	Bromate ion, hydrogen peroxide
Chloramination	Monochloramine, dichloramine, trichloramine, ammonia

ions. Research has shown that algal growth leads to the production of DBP precursors (Oliver and Shindler, 1980). Therefore, nutrient and algal management is one method of controlling the DBP formation potential of source waters. Control of bromide ions in source waters may be accomplished by preventing brine or saltwater intrusion into the water source.

Precursor Removal

Raw water can include DBP precursors in both dissolved and particulate forms. For the dissolved precursors to be removed in conventional treatment, they must be converted to particulate form for subsequent removal during settling and filtering. The THM formation potential generally decreases by about 50% through conventional coagulation and settling, indicating the importance of moving the point of chlorine application after coagulation and settling (and even filtration) to control TOX as well as TTHM formation (Singer and Harrington, 1993). Conventional systems can lower the DBP formation potential of water prior to disinfection by further removing precursors with enhanced coagulation, GAC adsorption, or membrane filtration prior to disinfection. Precursor removal efficiencies are site-specific and vary with different source waters and treatment techniques.

Aluminum (alum) and iron (ferric) salts can remove variable amounts of NOM. For alum, the optimal pH for NOM removal is in the range of 5.5–6.0. The addition of alum decreases pH and may allow the optimal pH range to be reached without acid addition. However, waters with very low or very high alkalinities may require the addition of base or acid to reach the optimal NOM coagulation pH (Singer, 1992).

Granular activated carbon adsorption can be used following filtration to remove additional NOM. For most applications, empty bed contact times in excess of 20 min are required, with regeneration frequencies on the order of 2–3 months. These long control times and frequent regeneration requirements make GAC an expensive treatment option. In cases where prechlorination is practiced, the chlorine rapidly degrades GAC. The addition of a disinfectant to the GAC bed can result in specific reactions in which previously absorbed compounds leach into the treated water.

Membrane filtration has been shown to be effective in removing DBP precursors in some instances. In pilot studies, ultrafiltration (UF) with a molecular weight cutoff (MWCO) of 100,000 Da was ineffective for controlling DBP formation. However, when little or no bromide ion was present in source water, nanofiltration (NF) membranes with MWCOs of 400–800 Da effectively controlled DBP formation (Laine, 1993).

In waters containing bromide ion, higher bromoform concentrations were observed after chlorination of membrane permeate (compared with raw water). This occurs as a result of filtration removing NOM while concentrating bromide ions in the permeate, thus providing a higher ratio of bromide ions to NOM than in raw water. This reduction in chlorine demand increases the ratio of bromide to chlorine, resulting in higher bromoform concentrations after chlorination of NF membrane permeate (compared with the raw water). TTHMs were lower in chlorinated permeate than in chlorinated raw water. However, due to the shift in the speciation of THMs toward more brominated forms, bromoform concentrations were actually greater in chlorinated treated water than in chlorinated raw water. The use of spiral-wound NF membranes (200–300 Da) more effectively controlled the formation of brominated THMs, but pretreatment of the water was necessary. Significant limitations in the use of membranes are the disposal of the waste brine generated, fouling of membranes, the cost of membrane replacement, and increasing energy costs.

Disinfection byproduct regulations require enhanced coagulation as an initial step for the removal of DBP precursors. In addition to meeting MCLs and MRDLs, some water suppliers must also meet treatment requirements to control the organic material (DBP precursors) in the raw water that combines with disinfectant residuals to form DBPs. Systems using conventional treatment are required to control precursors (measured as TOC) by using enhanced coagulation or enhanced softening. A system must remove a specified percentage of TOC (based on raw water quality) prior to the point of continuous disinfection (Table 17.13).

Systems using ozone followed by biologically active filtration or chlorine dioxide that meet specific criteria would be required to meet the TOC removal requirements prior to the addition of a residual disinfectant. Systems able to reduce TOC by a specified percentage level have met the DBP treatment technique requirements. If the system does not meet the percent reduction, it must determine its alternative minimum TOC removal level. The primacy agency approves the alternative minimum TOC removal possible for the system based on the relationship between coagulant dose and TOC in the system, based on the results of bench or pilot-scale testing. Enhanced coagulation is determined in part by the coagulant dose, where an incremental addition of 10 mg/L of alum (or an equivalent amount of ferric salt) results in a TOC removal below 0.3 mg/L.

TABLE 17.13

Required Removal of TOC by Enhanced Coagulation for Surface Water Systems Using Convention Treatment (Percent Reduction)

Source Water TOC (mg/L)	Source Water Alkalinity (mg/L as CaCO₃)		
	0–60	>60–120	>120
>2.0–4.0	35.0	25.0	15.0
>4.0–8.0	45.0	35.0	25.0
>8.0	50.0	40.0	30.0

DISINFECTION STRATEGY SELECTION

In addition to improving the raw or pre-disinfectant water quality, alternative disinfection strategies can be used to control DBPs. These strategies include the following:

- Use an alternative or supplemental disinfectant or oxidant such as chloramines or chlorine dioxide that will produce fewer DBPs.
- Move the point of chlorination to reduce TTHM formation and, where necessary, substitute chloramines, chlorine dioxide, or potassium permanganate for chlorine as a pre-oxidant.
- Use two different disinfectants or oxidants at various points in the treatment plant to avoid DBP formation at locations where precursors are still present in high quantities
- Use powdered activated carbon for THM precursor or TTHM reduction seasonally or intermittently.
- Maximize precursor removal.

CT FACTOR

One of the most important factors for determining or predicting the germicidal efficiency of any disinfectant is the CT factor, a version of the Chick-Watson law (Watson, 1908). The CT factor is defined as the product of the residual disinfectant concentration, C, in mg/L, and the contact time, T, in minutes, that the residual disinfectant is in contact with the water. The USEPA developed CT values for the inactivation of *Giardia* and viruses under the SWTR. Table 17.14 compares the CT values for virus inactivation using chlorine, chlorine dioxide, ozone, chloramine, and ultraviolet light disinfection under specified conditions. Table 17.15 shows the CT values for the inactivation of *Giardia* cysts using chlorine, chloramine, chlorine dioxide, and ozone under specified conditions. The CT values shown in Tables 17.14 and 17.15 are based on water temperatures of 10°C and pH values in the range of 6–9. CT values for chlorine disinfection are based on a free chlorine residual. Note that chlorine is less effective as pH increases from 6 to 9. In addition, for a given CT value, a low C and a high T

TABLE 17.14

CT Values for Inactivation of Viruses

Disinfectant (@ 10°C)	Units	Inactivation	
		2-log	3-log
Chlorine	mg·min/L	4	4
Chloramine	mg·min/L	643	1,067
Chlorine dioxide	mg·min/L	4.2	12.8
Ozone	mg·min/L	0.5	0.8
UV	mW·s/cm²	21	36

Source: Modified from data obtained from AWWA (1991).

TABLE 17.15

CT Values for Inactivation of *Giardia* Cysts

Disinfectant (@ 10°C)	Inactivation (mg·min/L)		
	1-log	2-log	3-log
Chlorine	35	69	104
Chloramine	615	1,240	1,850
Chlorine dioxide	7.7	15	23
Ozone	0.48	0.95	1.43

Source: Modified from data obtained from AWWA (1991).

are more effective than the reverse (i.e., a high C and a low T). For all disinfectants, as temperature increases, effectiveness increases.

DISINFECTANT RESIDUAL REGULATORY REQUIREMENTS

One of the most important factors for evaluating the merits of alternative disinfectants is their ability to maintain the microbial quality in the water distribution system. Disinfectant residuals may serve to protect the distribution system against regrowth. The Surface Water Treatment Rule (SWTR) requires that filtration and disinfection be provided to ensure that the total treatment of the system achieves at least a 3-log removal/inactivation of *Giardia* cysts and a 4-log removal/inactivation of viruses. In addition, the disinfection process must demonstrate through continuous monitoring and recording that the disinfection residual in the water entering the distribution system is never less than 0.2 mg/L for more than 4h (Snead et al., 1980).

Several of the alternative disinfectants examined in the handbook cannot be used to meet the residual requirements stated in the SWTR. For example, if either ozone or ultraviolet light disinfection is used as the primary disinfectant, a secondary disinfectant such as chlorine or chloramines should be utilized to obtain a residual in the distribution system.

Disinfectant byproduct formation continues in the distribution system due to reactions between the residual disinfectant and organics in the water. Koch et al. (1991) found that with a chlorine dose of 3–4 mg/L, THM and HAA concentrations increase rapidly during the first 24h in the distribution system. After the initial 48h, the subsequent increase in THMs is very small. Chloral hydrate concentrations continued to increase after the initial 24h, but at a reduced rate. Haloketones actually decreased in the distribution system.

Nieminski et al. (1993) evaluated DBP formation in the simulated distribution systems of treatment plants in Utah. Finished water chlorine residuals ranged from 0.4 to 2.8 mg/L. Generally, THM values in the distribution system studies increased by 50%–100% (range of 30%–200%) of the plant effluent value after a 24-h contact time. The 24-h

THM concentration was essentially the same as the 7-day THM formation potential. HAA concentrations in the simulated distribution system were about 100% (range of 30%–200%) of the HAA in the plant effluent. The 7-day HAA formation potential was sometimes higher than or below the distribution system values. If chlorine is used as a secondary disinfectant, one should therefore anticipate a 100% increase in the plant effluent THMs or plan to reach the 7-day THM formation level in the distribution system.

SUMMARY OF CURRENT NATIONAL DISINFECTION PRACTICES

Most water treatment plants disinfect water prior to distribution. The 1995 Community Water Systems Survey reports that 81% of all community water systems provide some form of treatment on all or a portion of their water sources. The survey also found that virtually all surface water systems provide some treatment of their water. Of those systems reporting no treatment, 80% rely on groundwater as their only water source.

The most commonly used disinfectants/oxidants are chlorine, chlorine dioxide, chloramines, ozone, and potassium permanganate. Chlorine is predominantly used in surface and groundwater disinfection treatment systems; more than 60% of the treatment systems use chlorine as a disinfectant/oxidant. Potassium permanganate, on the other hand, is used by many systems, but its application is primarily for oxidation rather than for disinfection.

Permanganate will have a beneficial impact on disinfection since it is a strong oxidant that will reduce the chemical demand for the ultimate disinfection chemical. Chloramine is used by some systems and is more frequently used as a post-treatment disinfectant.

The International Ozone Association (1997) conducted a survey of ozone facilities in the United States. The survey documented the types of ozone facilities, size, objective of ozone application, and year of operation. The International Ozone Association (1997) summarizes the findings of its survey of water treatment plants in the United States in Table 17.16. The most common use of ozone is for oxidation of iron and manganese, and for taste and odor control. Twenty-four of the 158 ozone facilities used GAC following ozonation. In addition to the 158 operating ozone facilities, the survey identified 19 facilities under construction and

another 30 under design. The capacity of the systems ranges from less than 25 gpm to exceeding 500 mg. Nearly half of the operating facilities have a capacity exceeding 1 mg. Rice et al. (1998) found that as of May 1998, 264 drinking water plants in the United States were using ozone.

SUMMARY OF METHODS OF DISINFECTION

The methods of disinfection include:

Heat: possibly the first method of disinfection. Disinfection is accomplished by boiling water for 5–10 min. This method is good, obviously, only for household quantities of water when bacteriological quality is questionable.

Ultraviolet (UV) Light: while a practical method of treating large quantities, adsorption of UV light is very rapid, so the use of this method is limited to non-turbid waters close to the light source.

Metal Ions: silver, copper, mercury.

pH Adjustment: to under 3.0 or over 11.0

Oxidizing Agents: bromine, ozone, potassium permanganate, and chlorine

The vast majority of drinking water systems in the United States use chlorine for disinfection (Spellman, 2007). Along with meeting the desired characteristics listed above, chlorine has the added advantage of a long history of use—it is fairly well understood. Although some small water systems may use other disinfectants, we focus on chlorine in this handbook and provide only a brief overview of other disinfectant alternatives.

Note: One of the recent developments in chlorine disinfection is the use of multiple and interactive disinfectants. In these applications, chlorine is combined with a second disinfectant to achieve improved disinfection efficiency and/or effective DBP control.

Note: As described earlier, the 1995 Community Water System Survey indicated that all surface water and groundwater systems in the United States use chlorine for disinfection (Spellman, 2007).

CHLORINATION

The addition of chlorine or chlorine compounds to water is called *chlorination*. Chlorination is considered to be the single most important process for preventing the spread of waterborne disease. Chlorine has many attractive features that contribute to its wide use in industry. Five of the key attributes of chlorine are:

- It damages the cell wall.
- It alters the permeability of the cell (the ability to pass water in and out through the cell wall).
- It alters the cell protoplasm.
- It inhibits the enzyme activity of the cell, so it is unable to use its food to produce energy.
- It inhibits cell reproduction.

TABLE 17.16
Ozone Application in Water Treatment Plants in the U.S.

Ozone Objective	Number of Plants	% Plants
THM control	50	32
Disinfection	63	40
Iron/manganese, taste and odor control	92	58
Total	158	—

Some concerns regarding chlorine usage that may impact its uses include:

- Chlorine reacts with many naturally occurring organic and inorganic compounds in water to produce undesirable DBPs.
- Hazards associated with using chlorine, specifically chlorine gas, require special treatment and response programs.
- High chlorine doses can cause taste and odor problems.

Chlorine is used in water treatment facilities primarily for disinfection. Because of chlorine's oxidizing powers, it has been found to serve other useful purposes in water treatment, such as (White, 1992):

- Taste and odor control
- Prevention of algal growth
- Maintenance of clear filter media
- Removal of iron and manganese
- Destruction of hydrogen sulfide
- Bleaching of certain organic colors
- Maintenance of distribution system water quality by controlling slime growth
- Restoration and preservation of pipeline capacity
- Restoration of well capacity and water main sterilization
- Improved coagulation

Chlorine is available in a number of different forms: (1) as pure elemental gaseous chlorine (a greenish-yellow gas possessing a pungent and irritating odor that is heavier than air, nonflammable, and nonexplosive). When released to the atmosphere, this form is toxic and corrosive; (2) as solid calcium hypochlorite (in tablets or granules); or (3) as a liquid sodium hypochlorite solution (in various strengths).

The selection of one form of chlorine over the others for a given water system depends on the amount of water to be treated, the configuration of the water system, the local availability of the chemicals, and the skill of the operator.

One of the major advantages of using chlorine is the effective residual it produces. A residual indicates that disinfection is complete, and the system has acceptable bacteriological quality. Maintaining a residual in the distribution system provides an additional line of defense against pathogenic organisms that could enter the system and helps prevent the regrowth of microorganisms that were injured but not killed during the initial disinfection stage.

Chlorine Terminology

Often, it is difficult for new waterworks operators to understand the terms used to describe the various reactions and processes used in chlorination. Common chlorination terms include the following:

- **Chlorine Reaction:** Regardless of the form of chlorine used for disinfection, the reaction in water is the same. The same amount of disinfection can be expected, provided the same amount of available chlorine is added to the water. The standard term for the concentration of chlorine in water is milligrams per liter (mg/L) or parts per million (ppm); these terms indicate the same quantity.
- **Chlorine Dose:** The amount of chlorine added to the system. It can be determined by adding the desired residual for the finished water to the chlorine demand of the untreated water. Dosage can be measured in either milligrams per liter (mg/L) or pounds per day, with mg/L being the most common.
- **Chlorine Demand:** The amount of chlorine used by iron, manganese, turbidity, algae, and microorganisms in the water. Because the reaction between chlorine and microorganisms is not instantaneous, demand is relative to time. For instance, the demand 5 min after applying chlorine will be less than the demand after 20 min. Demand, like dosage, is expressed in mg/L. The chlorine demand is as follows:

$$Cl_2 \text{ demand} = Cl_2 \text{ dose} - Cl_2 \text{ residual} \quad (17.7)$$

- **Chlorine Residual:** The amount of chlorine (determined by testing) that remains after the demand is satisfied. Residual, like demand, is based on time. The longer the time after dosage, the lower the residual will be until all of the demand has been satisfied. Residual, like dosage and demand, is expressed in mg/L. The presence of a *free residual* of at least 0.2–0.4 ppm usually provides a high degree of assurance that the disinfection of the water is complete. *Combined residual* is the result of combining free chlorine with nitrogen compounds. Combined residuals are also called chloramines. *Total chlorine residual* is the mathematical combination of free and combined residuals. The total residual can be determined directly with standard chlorine residual test kits.
- **Chorine Contact Time:** One of the key items in predicting the effectiveness of chlorine on microorganisms. It is the interval (usually only a few minutes) between the time when chlorine is added to the water and the time the water passes by the sampling point, contact time is the "T" in CT. CT is calculated based on the free chlorine residual prior to the first customer, multiplied by the contact time in minutes.

$$CT = \text{Concentration} \times \text{Contact Time} = \text{mg/L} \times \text{minutes} \quad (17.8)$$

A certain minimum time period is required for the disinfecting action to be completed. The contact time is usually a fixed condition determined by the rate of flow of the water and the distance from the chlorination point to the first consumer connection. Ideally, the contact time should not be less than 30 min, but more time is needed at lower chlorine doses, in cold weather, or under other conditions.

Pilot studies have shown that specific CT values are necessary for the inactivation of viruses and *Giardia*. The required CT value will vary depending on pH, temperature, and the organisms to be killed. Charts and formulas are available to make this determination. The USEPA has set a CT value of three-log ($CT_{99.9}$) inactivation to ensure the water is free of *Giardia*. State drinking water regulations include charts giving this value for different pH and temperature combinations. Filtration, in combination with disinfection, must provide a 3-log removal/inactivation of *Giardia*. Charts in the USEPA Surface Water Treatment Rule Guidance Manual list the required CT values for various filter systems. Under the 1996 Interim Enhanced Surface Water Treatment (IESWT) rules, the USEPA requires systems that filter to remove 99% (2 log) of *Cryptosporidium* oocysts. To ensure that the water is free of viruses, a combination of filtration and disinfection to provide a 4-log removal of viruses has been judged the best for drinking water safety—99.99% removal. Viruses are inactivated (killed) more easily than cysts or oocysts.

Chlorine Chemistry

The reactions of chlorine with water and the impurities that might be in the water are quite complex, but a basic understanding of these reactions can aid the operator in keeping the disinfection process operating at its highest efficiency. When dissolved in pure water, chlorine reacts with the H^+ ions and the OH^- radicals in the water. Two of the products of this reaction (the actual disinfecting agents) are *hypochlorous acid*, HOCl, and the *hypochlorite radical*, OCl^-. If microorganisms are present in the water, the HOCl and the OCl^- penetrate the microbe cells and react with certain enzymes. This reaction disrupts the organisms' metabolism and kills them. The chemical equation for hypochlorous acid is as follows:

$$\frac{Cl_2}{(chlorine)} + \frac{H_2O}{(water)} \leftrightarrow \frac{HOCl}{(hypochlorous\ acid)}$$
$$+ \frac{HCl}{(hydrochloric\ acid)} \quad (17.9)$$

Note: The symbol $\leftrightarrow$ indicates that the reactions are reversible.

Hypochlorous acid (HOCl) is a weak acid—meaning it dissociates slightly into hydrogen and hypochlorite ions—but a strong oxidizing and germicidal agent. Hydrochloric acid (HCl) in the above equation is a strong acid and retains more of the properties of chlorine. HCl tends to lower the pH of the water, especially in swimming pools where the water is recirculated and continually chlorinated. The total hypochlorous acid and hypochlorite ions in water constitute the *free available chlorine*. Hypochlorites act in a manner similar to HCl when added to water because hydrochloric acid is formed.

When chlorine is first added to water containing some impurities, the chlorine immediately reacts with the dissolved inorganic or organic substances and is then unavailable for disinfection. The amount of chlorine used in this initial reaction is the *chlorine demand* of the water. If dissolved ammonia (NH_3) is present in the water, the chlorine will react with it to form compounds called *chloramines*. Only after the chlorine demand is satisfied and the reaction with all the dissolved ammonia is complete is the chlorine actually available in the form of HOCl and OCl^-. The equation for the reaction of hypochlorous acid (HOCl) and ammonia (NH_3) is as follows:

$$\frac{HOCl}{(hypochlorous\ acid)} + \frac{NH_3}{(ammonia)}$$
$$\leftrightarrow \frac{NH_2Cl}{(monochloramine)} + \frac{H_2O}{(water)} \quad (17.10)$$

Note: The chlorine, as hypochlorous acid and hypochlorite ions remaining in the water after the above reactions are complete, is known as *free available chlorine,* and it is a very active disinfectant.

Breakpoint Chlorination

To produce a free chlorine residual, enough chlorine must be added to the water to reach what is referred to as *breakpoint chlorination,* which is the point at which nearly complete oxidation of nitrogen compounds is reached; any residual beyond the breakpoint is mostly free chlorine (see Figure 17.9). When chlorine is added to natural waters, it begins combining with and oxidizing the chemicals in the water before it begins disinfecting. Although residual chlorine will be detectable in the water, it will be in a combined form with weak disinfecting power. As shown in Figure 17.9, adding more chlorine at this point actually decreases the chlorine residual as the additional chlorine destroys the combined chlorine compounds. At this stage, the water may have a strong swimming pool or medicinal taste and odor. Free chlorine has the highest disinfecting power. The point at which most of the combined chlorine compounds have been destroyed and free chlorine starts to form is the *breakpoint.*

The chlorine breakpoint of water can only be determined through experimentation. This simple experiment requires 20 1,000-mL beakers and a solution of chlorine. Place the raw water in the beakers and dose them with progressively larger amounts of chlorine. For instance, you might start with zero in the first beaker, then 0.5 and

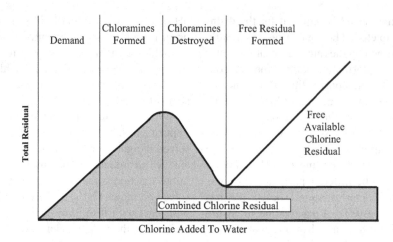

FIGURE 17.9   Breakpoint chlorination curve.

1.0 mg/L, and so on. After a period of time, say 20 min, test each beaker for total chlorine residual and plot the results.

Breakpoint Chlorination Curve

Refer to Figure 17.9 for the following explanation. Where the curve starts, no residual exists, even though there was a dosage. This is called the initial demand and is due to micro-organisms and interfering agents using the chlorine. After the initial demand, the curve slopes upward. Chlorine combining to form chloramines produces this part of the curve. All of the residual measured on this part of the curve is combined residual. At some point, the curve begins to drop back toward zero. This portion of the curve results from a reduction in combined residual, which occurs because enough chlorine has been added to destroy (oxidize) the nitrogen compounds used to form combined residuals. The breakpoint is the point where the downward slope of the curve breaks upward. At this point, all of the nitrogen compounds that could be destroyed have been destroyed. After the breakpoint, the curve starts upward again, usually at a 45° angle. Only on this part of the curve can free residuals be found. The distance that the breakpoint is above zero is a measure of the remaining combined residual in the water. This combined residual exists because some of the nitrogen compounds will not have been oxidized by chlorine. If the irreducible combined residual is more than 15% of the total residual, chlorine odor and taste complaints will be high.

Gas Chlorination

Gas chlorine is provided in 100-lb or 1-ton containers. Chlorine is placed in the container as a liquid. The liquid boils at room temperature, reducing to a gas and building pressure in the cylinder. At a room temperature of 70°F, a chlorine cylinder will have a pressure of 85 psi; 100/150-lb cylinders should be maintained in an upright position and chained to the wall. To prevent a chlorine cylinder from rupturing in a fire, the cylinder valves are

equipped with special fusible plugs that melt between 158°F and 164°F.

Chlorine gas is 99.9% chlorine. A gas chlorinator meters the gas flow and mixes it with water, which is then injected as a water solution of pure chlorine. As the compressed liquid chlorine is withdrawn from the cylinder, it expands as a gas, withdrawing heat from the cylinder. Care must be taken not to withdraw the chlorine at too fast a rate: if the operator attempts to withdraw more than about 40 lbs of chlorine per day from a 150-lb cylinder, it will freeze up.

Note: All chlorine gas feed equipment sold today is vacuum operated. This safety feature ensures that if a break occurs in one of the components in the chlorinator, the vacuum will be lost, and the chlorinator will shut down without allowing gas to escape.

Chlorine gas is a highly toxic lung irritant, and special facilities are required for storing and housing it. Chlorine gas will expand to 500 times its original compressed liquid volume at room temperature (1 gal of liquid chlorine will expand to about 67 ft³). Its advantage as a drinking water disinfectant is the convenience afforded by a relatively large quantity of chlorine available for continuous operation for several days or weeks without the need for mixing chemicals. Where water flow rates are highly variable, the chlorination rate can be synchronized with the flow.

Chlorine gas has a very strong, characteristic odor that can be detected by most people at concentrations as low as 3.5 ppm. Highly corrosive in moist air, it is extremely toxic and irritating in concentrated form. Its toxicity ranges from throat irritation at 15 ppm to rapid death at 1,000 ppm. Although chlorine does not burn, it supports combustion, so open flames should never be used around chlorination equipment.

When changing chlorine cylinders, an accidental release of chlorine may occasionally occur. To handle this type of release, an approved (NIOSH-approved) Self-Contained Breathing Apparatus (SCBA) must be worn. Special emergency repair kits are available from the Chlorine Institute for use by emergency response teams to deal with chlorine leaks. Because chlorine gas is 2.5 times heavier than air, exhaust, and inlet air ducts should be installed at floor level. A leak of chlorine gas can be found by using the fumes from a strong ammonia mist solution. A white cloud develops when ammonia mist and chlorine combine.

HYPOCHLORINATION

Combining chlorine with calcium or sodium produces hypochlorites. Calcium hypochlorites are sold in powder or tablet forms and can contain chlorine concentrations of up to 67%. Sodium hypochlorite is a liquid (bleach, for example) and is found in concentrations up to 16%. Chlorine concentrations of household bleach range from 4.75% to 5.25%. Most small system operators find using these liquid or dry chlorine compounds more convenient and safer than chlorine gas.

The compounds are mixed with water and fed into the water with inexpensive solution feed pumps. These pumps are designed to operate against high system pressures but can also be used to inject chlorine solutions into tanks, though injecting chlorine into the suction side of a pump is not recommended as the chlorine may corrode the pump impeller.

Calcium hypochlorite can be purchased as tablets or granules, with approximately 65% available chlorine (10 lbs of calcium hypochlorite granules contain only 6.5 lbs of chlorine). Normally, 6.5 lbs of calcium hypochlorite will produce a concentration of 50 mg/L chlorine in 10,000 gal of water. Calcium hypochlorite can burn (at 350°F) if combined with oil or grease. When mixing calcium hypochlorite, operators must wear chemical safety goggles, a cartridge breathing apparatus, and rubberized gloves. Always place the powder in the water. Placing the water into the dry powder could cause an explosion.

Sodium hypochlorite is supplied as a clear, greenish-yellow liquid in strengths from 5.25% to 16% of available chlorine. Often referred to as "bleach," it is, in fact, used for bleaching. As we stated earlier, common household bleach is a solution of sodium hypochlorite containing 5.25% available chlorine. The amount of sodium hypochlorite needed to produce a 50 mg/L-chlorine concentration in 10,000 gal of water can be calculated using the solutions equation:

$$C_1 V_1 = C_2 V_2 \qquad (17.11)$$

where:

C = the solution concentration in mg/L or %
V = the solution volume in the liters, gal., qt., etc.
1.0% = 10,000 mg/L

In this example, C_1 and V_1 are associated with the sodium hypochlorite, and C_2 and V_2 are associated with the 10,000 gal of water with a 50 mg/L chlorine concentration. Therefore,

$$C_1 = 5.25\%$$

$$C_1 = \frac{(5.25\%)(10,000 \text{ mg/L})}{1.0\%} = 52,500 \text{ mg/L}$$

V_1 = unknown volume of sodium hypochlorite
C_2 = 50 mg/L
V_2 = 10,000 gal

$$(52,500 \text{ mg/L})(V_1) = (50 \text{ mg/L})(10,000 \text{ gal})$$

$$V_1 = \frac{(50 \text{ mg/L})(10,000 \text{ gal})}{52,500 \text{ mg/L}}$$

$$V_1 = 9.52 \text{ gal of sodium hypochlorite}$$

Sodium hypochlorite solutions are introduced to the water in the same manner as calcium hypochlorite solutions. The purchased stock "bleach" is usually diluted with water to produce a feed solution pumped into the water system.

Hypochlorites are stored properly to maintain their strengths. Calcium hypochlorite must be stored in airtight containers in cool, dry, dark locations. Sodium hypochlorite degrades relatively quickly even when properly stored; it can lose more than half of its strength in 3–6 months. Operators should purchase hypochlorite in small quantities to ensure they are used while still strong. Old chemicals should be discarded safely.

The pumping rate of a chemical metering pump is usually manually adjusted by varying the length of the piston or diaphragm stroke. Once the stroke is set, the hypochlorinator feeds accurately at that rate. However, chlorine measurements must be made occasionally at the beginning and end of the well pump cycle to ensure correct dosage. A metering device may be used to vary the hypochlorinator feed rate, synchronized with the water flow rate. Where a well pump is used, the hypochlorinator is connected electrically with the pump's on-off controls to ensure that chlorine solution is not fed into the pipe when the well is not pumping.

Determining Chlorine Dosage

Proper disinfection requires the calculation of the amount of chlorine that must be added to the water to produce the required dosage. The type of calculation used depends on the form of chlorine being used. The basic chlorination calculation used is the same one used for all chemical addition calculations—the *pounds formula*. The pounds formula is

$$\text{Pounds} = \text{mg/L} \times 8.34 \times \text{MG} \qquad (17.12)$$

where

Pounds = pounds of available chlorine required
mg/L = desired concentration in milligrams per liter
8.24 = conversion factor
MG = millions of gallons of water to be treated

Example 17.3

Problem: Calculate the number of pounds of gaseous chlorine needed to treat 250,000 gal of water with 1.2 mg/L of chlorine.

Solution:

$$Pounds = 1.2\,mg/L \times 8.34 \times 0.25\,MG$$
$$= 2.5\,lb$$

Note: Hypochlorites contain less than 100% available chlorine. Thus, we must use more hypochlorite to get the same number of pounds of chlorine into the water.

If we substitute calcium hypochlorite with 65% available chlorine in our example, 2.5 lb of available chlorine is still needed, but more than 2.5 lbs of calcium hypochlorite is needed to provide that much chlorine. To determine how much of the chemical is needed, divide the pounds of chlorine needed by the decimal form of the percent available chlorine. Since 65% is the same as 0.65, we need to add:

$$\frac{2.5\,lb}{0.65\ available\ chlorine} = 3.85\,lb\,Ca(OCl) \qquad (17.13)$$

to get that much chlorine.

In practice, because most hypochlorite is fed as solutions, we often need to know how much chlorine solution we need to feed. In addition, the practical problems faced in day-to-day operations are never so clearly stated as the practice problems we work on. For example, small water systems do not usually deal with water flow in million gallons per day. Real-world problems usually require a lot of intermediate calculations to get everything ready to plug into the pounds formula.

Example 17.4

Problem: We have raw water with a chlorine demand of 2.2 mg/L. We need a final residual of 1.0 mg/L at the entrance to the distribution system. We can use sodium hypochlorite or calcium hypochlorite granules as the source of chlorine. If the well output is 65 gallons per minute (gpm) the chemical feed pump can inject 100 milliliters per minute (mL/min) at the 50% setting. What is the required strength of the chlorine solution we will feed? What volume of 5.20% sodium hypochlorite will be needed to produce 1 gal of the chlorine feed solution? How many pounds of 65% calcium hypochlorite will be needed to mix each gallon of solution?

Solution:

- **Step 1:** Determine the amount of chlorine to be added to the water (the chlorine dose). The dose is defined as the chlorine demand of the water plus the desired residual, or in this case:

$$Q_1 = 65\ gal/min \qquad Q_t = Q_1 + Q_2$$
$$CL = 0.0\ mg/L \qquad C_t = 3.2\ mg/L$$

Well pump →→→ Treated Water

↑

Chlorination Metering Pump

$$Q_2 = 100\ mL/min$$

$$C_2 = unknown$$

$$Dose = demand + residual$$
$$= 2.2\,mg/L + 1.0\,mg/L$$
$$= 3.2\,mg/L$$

To obtain a 1.0 mg/L residual when the water enters the distribution system, we must add 3.2 mg/L of chlorine to the water.
- **Step 2:** Determine the strength of the chlorine feed solution that would add 3.2 mg/L of chlorine to 65 gal/min of water when fed at a rate of 100 mL/ min. To make this calculation, consider the diagram of this chlorination system in Figure 17.10. The well pump below is producing water at a flow rate (Q_1) of 65 gpm, with a chlorine concentration (C_1) of 0.0 mg/L. The metering pump will add 100 mL/min (Q_2) of chlorine solution, but we do not know its concentration (C_2) yet. The finished water will have been dosed with a chlorine concentration of 3.2 mg/L (C_1) and will enter the distribution system at a rate (Q_t) of 65 gal + 100 mL/min, with a free residual chlorine concentration of 1.0 mg/L after the chlorine contact time.
- **Step 3:** Convert the metering pump flow (100 mL/min) to gpm so we can calculate Q_1. To do this, we use standard conversion factors:

$$\frac{100\ mL}{min} \times \frac{1\ L}{1{,}000\ mL} \times \frac{1\ gal}{3.785\ L} = 0.026\ gal/min$$

So, now we know that Q_2 is 0.026 gpm and that Q_t is 65 gpm + 0.026 gpm = 65,026 gpm.

$$Q1 = 65\ gal/min \qquad Qt = Q1 + Q2$$
$$CL = 0.0\ mg/l \qquad Ct = 3.2\ mg/l$$

Well pump →→→Treated Water

↑

Chlorination Metering Pump

$$Q2 = 100\ mL/min$$

$$C2 = unknown$$

FIGURE 17.10 Chlorination system problem.

- **Step 4:** We must now use what is known as the *standard mass balance equation*. The mass balance equation says that the flow rate times the concentration of the output is equal to the flow rate times the concentration of each of the inputs added together, or in equation form,

$$\left(Q_1 \times C_1\right) + \left(Q_2 \times C_2\right) = \left(Q_1 \times C_1\right) \quad (17.14)$$

Substituting the numbers given and those we have calculated so far gives us:

$$(65\,\text{gpm} \times 0.0\,\text{mg/L}) + (0.026\,\text{gpm} \times \text{mg/L}) = [(65 + 0.026)\,\text{gpm}$$

$$\times 3.2\,\text{mg/L}]$$

$$0 + 0.026x = 65.026 \times 3.2$$

$$0.026x = 208.1$$

$$x = \frac{208.1}{0.026}$$

$$x = 8{,}004\,\text{mg/L}$$

This is the answer to the first part of the question, the required strength of the chlorine feed solution. Since a 1% solution is equal to 10,000 mg/L, a solution strength of 8,004 mg/L is approximately a 0.80% solution.

To determine the required volume of bleach per gallon to produce this 0.80% solution, we go back to the pounds formula:

$$\begin{aligned}\text{Pounds} &= \text{mg/L} \times 8.34 \times \text{MG}\\ &= 8{,}004\,\text{mg/L} \times 8.34 \times 0.000001\,\text{MG}\\ &= 0.067\,\text{lb chlorine/gal solution}\end{aligned}$$

Note: Remember to convert gallons to MG.

Recalling that the bleach, like water, weighs 8.34 lb/gal and contains approximately 5.20% (0.0520) available chlorine,

$$1\,\text{gal bleach} = 8.34\,\text{lb/gal} \times 0.0520$$

$$= 0.43\,\text{lb available chlorine per gallon of bleach}$$

If 1 gal of bleach contains 0.43 lb of available chlorine, how many gallons of bleach do we need to provide the 0.067 lb of chlorine we need for each gallon of chlorine feed solution? To determine the gallons of bleach needed, we use the simple ratio equation:

$$\frac{1\,\text{gal}}{0.443\,\text{lb Cl}_2} = \frac{x\,\text{gal}}{0.067\,\text{lb Cl}_2}$$

$$x = \frac{1\,\text{gal} \times 0.067\,\text{lb}}{0.43\,\text{lb}}$$

$$= 0.16\text{-gal bleach per gal solution}$$

Or, to determine the required volume of bleach per gallon to produce this 0.80% solution, we use the solutions equation and calculate it this way:

$$C_1 V_1 = C_2 V_2$$
$$C_1 = 0.80\%$$
$$V_1 = 1.0\,\text{gal}$$
$$C_2 = 5.20\%$$
$$V_2 = \text{unknown}$$

$$0.80\%\ \text{solution} \times 1.0\,\text{gal} = 5.20\%\ \text{solution} \times \text{gal}$$

$$x = \frac{0.80\% \times 1.0\,\text{gal}}{5.20\%}$$

$$= 0.15\text{-gal bleach per gal solution}$$

To summarize, 1 gal of household bleach:

- Contains 0.85-gal water
- Contains 5.20% or 52,000 mg/L of available chlorine
- Contains 0.15 gal of available chlorine
- Contains 0.43 lb of available chlorine
- Weighs 8.34 lb

The third part of the problem requires determining the pounds of calcium hypochlorite needed for each gallon of feed solution. We know that we need 0.066 lb of chlorine for each gallon of solution and that HTH contains 65% of available chlorine; that is 1.0 lb of HTH contains 0.65 lb of available chlorine. Using the ratio equation,

$$\frac{1\,\text{lb HTH}}{0.65\,\text{lb Cl}_2} = \frac{x\,\text{lb HTH}}{0.067\,\text{lb Cl}_2}$$

$$x = \frac{1\,\text{lb} \times 0.067\,\text{lb}}{0.65\,\text{lb}}$$

Chlorine Generation

Onsite generation of chlorine has recently become practical. These generation systems, using only salt and electric power, can be designed to meet disinfection and residual standards and to operate unattended at remote sites. Considerations for chlorine generation include cost, concentration of the brine produced, and availability of the process (AWWA, 2010).

Chlorine

Chlorine gas can be generated by several processes including the electrolysis of alkaline brine or hydrochloric acid, the reaction between sodium chloride and nitric acid, or hydrochloric acid oxidation—about 70% of the chlorine produced in the United States is manufactured from the electrolysis of salt brine and caustic solutions in a diaphragm cell (White, 1992). Because chlorine is a stable compound, it is typically produced off-site by a chemical manufacturer. Once produced, chlorine is packaged as a liquefied gas under pressure for delivery to the site in railcars, tanker trucks, or cylinders.

Sodium Hypochlorite

Dilute sodium hypochlorite solutions (less than 1%) can be generated electrochemically on-site from salt brine solution.

TABLE 17.17

Chlorine Uses and Doses

Application	Typical Dose	Optimal pH	Reaction Time	Effectiveness	Other Considerations
Iron	0.62 mg/mg Fe	7.0	Less than 1 h	Good	
Manganese	0.77 mg/mg Mn	7–8	1–3-h	Slow kinetics	Reaction time
		9.5	Minutes		Increases at lower pH
Biological growth	1–2 mg/L	6–8	NA	Good	DBP formation
Taste/odor	Varies	6–8	Varies	Varies	Effectiveness depends on compound
Color removal	Varies	4.0–6.8	Minutes	Good	DBP formation
Zebra	2–5 mg/L		Shock level	Good	DBP formation
Mussels	0.2–0.5 mg/L		Maintenance level		
Asiatic clams	0.3–0.5 mg/L		Continuous	Good	DBP formation

Sources: Adapted in part from White (1992), Connell (1996), and Williams and Culp (1986).

Typically, sodium hypochlorite solutions are referred to as *liquid bleach* or *Javelle water*. Generally, the commercial or industrial-grade solutions produced have hypochlorite strengths of 10%–16%. The stability of sodium hypochlorite solution depends on the hypochlorite concentration, the storage temperature, the length of storage (time), the impurities in the solution, and exposure to light. The decomposition of hypochlorite over time can affect the feed rate and dosage, as well as produce undesirable byproducts such as chlorite ions or chlorate (Gordon et al., 1995). Because of the storage problems, many systems are investigating the onsite generation of hypochlorite in lieu of its purchase from a manufacturer or vendor.

Calcium Hypochlorite

To produce calcium hypochlorite, hypochlorous acid is made by adding chlorine monoxide to water and then neutralizing it with lime slurry to create a solution of calcium hypochlorite. Generally, the final product contains up to 70% available chlorine and 4%–6% lime. Storage of calcium hypochlorite is a major safety consideration. It should never be stored where it is subject to heat or allowed to contact any organic material of an easily oxidized nature (Spellman, 2007).

Primary Uses and Points of Application of Chlorine

Uses

The main usage of chlorine in drinking water treatment is for disinfection. However, chlorine has also found application for a variety of other water treatment objectives, such as the control of nuisance organisms, oxidation of taste and odor compounds, oxidation of iron and manganese, color removal, and as a general treatment aid to filtration and sedimentation processes (White, 1992). Table 17.17 presents a summary of chlorine uses and doses.

Points of Application

At conventional surface water treatment plants, chlorine is typically added for prechlorination at either the raw water

TABLE 17.18

Typical Chlorine Points of Application

Point of Application

Raw water intake

Flash mixer (prior to sedimentation)

Filter influent

Filter clearwell

Distribution system

Sources: Adapted in part from Connell (1996), White (1992) and AWWA (1990).

TABLE 17.19

Typical Chlorine Dosages at Water Treatment Plants

Chlorine Compound	Range of Doses (mg/L)
Calcium hypochlorite	0.6–5
Sodium hypochlorite	0.2–2
Chlorine gas	1–16L

Source: Science Applications International Corporation (SAIC), 1998, San Diego, CA. as adapted from USEPA's review of public water systems' initial Sampling Plans which were required by USEPA's Information Collection Rule (ICR).

intake or flash mixer, for intermediate chlorination ahead of the filters, for post-chlorination at the filter clear well, or rechlorination of the distribution system (Connell, 1996). Table 17.18 summarizes typical points of application.

Typical Doses

Table 17.19 shows the typical dosages for the various forms of chlorine. The wide range of chlorine gas dosages most likely represents its use as both an oxidant and a disinfectant. While sodium hypochlorite and calcium hypochlorite can also serve as both an oxidant and a disinfectant, their higher costs may limit their use.

Factors Affecting Chlorination

Disinfection by chlorination is a rather straightforward process, but several factors (interferences) can affect the ability of chlorine to perform its main function: disinfection. *Turbidity* is one such interference. As we described earlier, turbidity is a general term that describes particles suspended in the water. Water with high turbidity appears cloudy. Turbidity interferes with disinfection when microorganisms "hide" from the chlorine within the particles causing turbidity. This problem is magnified when turbidity comes from organic particles, such as those from sewage effluent. To overcome this, the length of time the water is exposed to the chlorine or the chlorine dose must be increased, although highly turbid waters may still shield some microorganisms from the disinfectant. The USEPA took this type of problem into consideration in its Surface Water Treatment Rule and the 1996 Amendments (the Interim Enhanced Surface Water Treatment (IESWT) rule) that tightens controls on DBPs and turbidity and regulates *Cryptosporidium*. The rule requires continuous turbidity monitoring of individual filters and tightens allowable turbidity limits for combined filter effluent, cutting the maximum from 5 to 1 NTU and the average monthly limit from 0.5 to 0.3 NTU.

Note: Recall that the IESWT rule applies to large public water systems (those serving more than 10,000 people) that use surface water or groundwater directly influenced by surface water and are the first to directly regulate *Cryptosporidium*.

Temperature affects the solubility of chlorine, the rate at which disinfecting ions are produced, and the proportion of highly effective forms of chlorine that will be present in the water. More importantly, however, temperature affects the rate at which chlorine reacts with the microorganisms themselves. As water temperature decreases, the rate at which the chlorine can pass through the microorganisms' cell wall decreases, making the chlorine less effective as a disinfectant. Along with the presence of turbidity-causing agents such as suspended solids and organic matter, chemical compounds in the water may influence chlorination: high alkalinity, nitrates, manganese, iron, and hydrogen sulfide.

Measuring Chlorine Residual

During normal operations, waterworks operators perform many operating checks and tests on unit processes throughout the plant. One of the most important and frequent operating checks is the *chlorine residual test*. This test must be performed whenever a distribution sample is collected for microbiological analysis and should be done frequently where the treatment facility discharges to the water distribution system to ensure that the disinfection system is working properly.

To test for residual chlorine, several methods are available. The most common and convenient is the *DPD Color Comparator Method*. This method uses a small portable test kit with prepared chemicals that produce a color

reaction indicating the presence of chlorine. By comparing the color produced by the reaction with a standard, we can determine the approximate chlorine residual concentration of the sample. DPD color comparator chlorine residual test kits are available from the manufacturers of chlorination equipment.

Note: The color comparator method is acceptable for most groundwater systems and for chlorine residual measurements in the distribution system. However, keep in mind that the methods used to take chlorine residual measurements for controlling the disinfection process in surface water systems and in some groundwater systems where adequate disinfection is essential must be approved by standard methods.

Pathogen Inactivation and Disinfection Efficacy
Inactivation Mechanisms

Research has shown that chlorine is capable of producing lethal events at or near the cell membrane as well as affecting DNA. In bacteria, chlorine was found to adversely affect cell respiration, transport, and possibly DNA activity. Chlorination was found to cause an immediate decrease in oxygen utilization in both *Escherichia coli* and *Candida parapsilosis* studies. The results also found that chlorine damages the cell wall membrane, promotes leakage through the cell membrane, and produces lower levels of DNA synthesis for *Escherichia coli*, *Candida parapsilosis*, and *Mycobacterium fortuitum* bacteria. This study also showed that chlorine inactivation is rapid and does not require bacterial reproduction (Hass and Englebrecht, 1980). These observations rule out mutation or lesions as the principal inactivation mechanisms since these mechanisms require at least one generation of replication for inactivation to occur.

Environmental Effects

Several environmental factors influence the inactivation efficiency of chlorine, including water temperature, pH, contact time, mixing, turbidity, interfering substances, and the concentration of available chlorine. In general, the highest levels of pathogen inactivation are achieved with high chlorine residuals, long contact times, high water temperature, and good mixing, combined with low pH, low turbidity, and the absence of interfering substances. Of the environmental factors, pH and temperature have the most impact on pathogen inactivation by chlorine. The effects of pH and temperature on pathogen inactivation are discussed below.

- **pH:** The germicidal efficiency of hypochlorous acid (HOCl) is much higher than that of the hypochlorite ion (OCl⁻). The distribution of chlorine species between HOCl and OCl⁻ is determined by pH, as discussed above. Because HOCl dominates at low pH, chlorination provides more effective disinfection at low pH. At high pH, OCl⁻ dominates, which causes a decrease in disinfection efficiency. The inactivation efficiency of gaseous chlorine and hypochlorite is the same at the same pH after chlorine addition. Note, however, that the addition of gaseous chlorine will decrease the pH while the addition of hypochlorite will increase the pH of the water. Therefore, without pH adjustment to maintain the same treated water pH, gaseous chlorine will have greater disinfection efficiency than hypochlorite.

- The impact of pH on chlorine disinfection has been demonstrated in the field. For example, virus inactivation studies have shown that 50% more contact time is required at pH 7.0 than at pH 6.0 to achieve comparable levels of inactivation. These studies also demonstrated that a rise in pH from 7.0 to 8.8 or 9.0 requires six times the contact time to achieve the same level of virus inactivation (Culp and Culp, 1974). Although these studies found a decrease in inactivation with increasing pH, some studies have shown the opposite effect. A 1972 study reported that viruses were more sensitive to free chlorine at high pH than at low pH (Scarpino et al., 1972).
- **Temperature:** For typical drinking water treatment temperatures, pathogen inactivation increases with temperature. Virus studies indicated that the contact time should be increased by two to three times to achieve comparable inactivation levels when the water temperature is lowered by 10°C (Clarke et al., 1962).
- **Disinfection Efficacy:** Since its introduction, numerous investigations have been made to determine the germicidal efficiency of chlorine. Although there are widespread differences in the susceptibility of various pathogens, the general order of increasing chlorine disinfection difficulty is bacteria, viruses, and then protozoa.
- **Bacteria Inactivation:** Chlorine is an extremely effective disinfectant for inactivating bacteria. A study conducted during the 1940s investigated the inactivation levels as a function of time for *E. coli*, *Pseudomonas aeruginosa*, *Salmonella typhi*, and *Shigella dysenteriae* (Butterfield et al., 1943). Study results indicated that HOCl is more effective than OCl⁻ for the inactivation of these bacteria. These results have been confirmed by several researchers who concluded that HOCl is 70–80 times more effective than OCl⁻ for inactivating bacteria (Williams and Culp, 1986).
- **Virus Inactivation:** Chlorine has been shown to be a highly effective viricide. One of the most comprehensive virus studies was conducted in 1971 using treated Potomac estuary water (Liu et al., 1971). The tests were performed to determine the resistance of 20 different enteric viruses to free chlorine under constant conditions of 0.5 mg/L free chlorine and a pH and temperature

of 7.8°C and 2°C, respectively. In this study, the least resistant virus was found to be reovirus, which required 2.7 min to achieve 99.99% inactivation (4 log removal). The most resistant virus was found to be a poliovirus, which required more than 60 min for 99.99% inactivation. The corresponding CT range required to achieve 99.99% inactivation for all 20 viruses was between 1.4 and over 30 mg·min/L. Virus survival studies have also been conducted on a variety of both laboratory and field strains (AWWA, 1991). All of the virus inactivation tests in this study were performed at a free chlorine residual of 0.4 mg/L, a pH of 7.0, a temperature of 5°C, and contact times of either 10, 100, or 1,000 min. Test results showed that of the 20 cultures tested only two poliovirus strains reached 99.99% inactivation after 10 min (CT=4 mg·min/L), six poliovirus strains reached 99.99% inactivation after 100 min (CT=40 mg·min/L), and 11 of the 12 polioviruses plus one *Coxsackievirus* strain (12 out of a total of 20 viruses) reached 99.99% inactivation after 1,000 min (CT=400 mg·min/L).

- **Protozoa Inactivation:** Chlorine has been shown to have limited success in inactivating protozoa. Data obtained during a 1984 study indicated that the resistance of *Giardia* cysts is two orders of magnitude higher than that of enteroviruses and more than three orders of magnitude higher than enteric bacteria (Hoff et al., 1984). CT requirements for *Giardia* cyst inactivation when using chlorine as a disinfectant have been determined for various pH and temperature conditions. The CT values increase at low temperatures and high pH. Chlorine has little impact on the viability of *Cryptosporidium* oocysts when used at the relatively low doses encountered in water treatment (e.g., 5 mg/L). Approximately 40% removals (0.2 log) of *Cryptosporidium* were achieved at CT values of both 30 and 3,600 mg·min/L (Finch et al., 1994). Another study determined that "no practical inactivation was observed" when oocysts were exposed to free chlorine concentrations ranging from 5 to 80 mg/L at pH 8, a temperature of 22°C, and contact times of 48–245 min (Gyurek et al., 1996). CT values ranging from 3,000 to 4,000 mg·min/L were required to achieve 1-log of *Cryptosporidium* inactivation at pH 6.0 and a temperature of 22°C. During this study, one trial in which oocysts were exposed to 80 mg/L of free chlorine for 120 min was found to produce greater than 3-logs of inactivation.

Disinfection By-Products (DBPs)

Halogenated organics are formed when natural organic matter (NOM) reacts with free chlorine or free bromine. Free chlorine is normally introduced into water directly as a primary or secondary disinfectant. Free bromine results from the oxidation of the bromide ion in the source water by chlorine. Factors affecting the formation of these halogenated DBPs include the type and concentration of NOM, chlorine form and dose, time, bromide ion concentration, pH, organic nitrogen concentration, and temperature. Organic nitrogen significantly influences the formation of nitrogen-containing DBPs, including haloacetonitriles (Reckhow et al., 1990), halopicrins, and cyanogen halides. Because most water treatment systems have been required to monitor total trihalomethanes (TTHM) in the past, most water treatment operators are probably familiar with some of the requirements involved in the D/DBP Rule. The key points of the DBP Rule and some of the key changes water supply systems are required to comply with are summarized below:

- **Chemical Limits and Testing:** Testing requirements will include total trihalomethanes (TTHM) and five haloacetic acids (HAA5). The Maximum Contaminant Level (MCL) for TTHM is 0.080 mg/L for surface water systems. In addition, a new MCL of 0.060 mg/L has been established for haloacetic acids (HAA5). New MCLs have been established for bromate (0.010 mg/L) and chlorite (1.0 mg/L). Bromate monitoring is required for systems that use ozone. Chlorite monitoring will only be required for systems that use chlorine dioxide (i.e., sodium and calcium hypochlorite are not included). Maximum Residual Disinfectant Levels (MRDLs) will be established for total chlorine (4.0 mg/L) and chlorine dioxide (0.8 mg/L).
- **Operational Requirements:** Analytical requirements for measuring chlorine residual have been changed to require digital equipment (i.e., no color wheels or analog test kits). The test kit must have a detection limit of at least 0.1 mg/L.
- **Monitoring and Reporting:** Individual state requirements will differ, but at a minimum, the following requirements are listed.
  - Surface water system monitoring requirements include four quarterly samples per treatment plant (Source Treatment Unit or STU) for TTHM and HAA5. One of these quarterly samples, or 25% of the total samples, must be collected at the maximum residence time location. The remaining samples must be collected at representative locations throughout the entire distribution system. Compliance is based on a running annual average computed quarterly.
  - For surface water systems using conventional filtration or lime softening, a D/DBP monthly operating report for Total Organic Carbon (TOC) removal will be required to be completed and filed with the state EPA. This report

will include TOC, alkalinity, and Specific Ultraviolet Absorption (SUVA) parameters. There will be an additional monthly operating report for bromate, chlorite, chlorine dioxide, and chlorine residual. More information will be provided closer to the compliance date.
- TTHM monitoring results may indicate the possible need for additional treatment, including the best available technology for reducing DBPs. This may include granular activated carbon, enhanced coagulation (for surface water systems using conventional filtration), or enhanced softening (for systems using lime softening).
- Operators are required to develop and implement a sample monitoring plan for disinfectant residuals and disinfection byproducts. The plan will need to be submitted to and approved by the state EPA. Disinfection residual monitoring compliance for total chlorine, including chloramines, will be based on a running annual average, computed quarterly, from the monthly average of all samples collected under this rule. Disinfectant residual monitoring compliance for chlorine dioxide will be based on consecutive daily samples. Disinfectant residual monitoring will be required at the same distribution point and time as total coliform monitoring. In addition, if the operator feeds ozone or chlorine dioxide, a sample monitoring plan for bromate or chlorite, respectively, must be submitted to and approved by the state EPA.

Application Methods

Different application methods are used depending on the form of chlorine. The following paragraphs describe the typical application methods for chlorine, sodium hypochlorite, and calcium hypochlorite:

Chlorine: Liquefied chlorine gas is typically evaporated into gaseous chlorine prior to metering. The heat required for evaporation can be provided by either a liquid chlorine evaporator or ambient heat from the storage container. Once the compressed liquid chlorine is evaporated, the chlorine gas is typically fed under vacuum conditions. Either an injector or a vacuum induction mixer usually creates the required vacuum. The injector uses water flowing through a venturi to draw the chlorine gas into aside stream of carrier water to form a concentrated chlorine solution. This solution is then introduced into the process water through a diffuser or mixed with a mechanical mixer. A vacuum induction mixer uses the motive force of the mixer to create a vacuum and draws the chlorine gas directly into the process water at the mixer.

Sodium Hypochlorite: Sodium hypochlorite solutions degrade over time. For example, a 12.5% hypochlorite solution will degrade to 10% in 30 days under "best case" conditions. Increased temperature, exposure to light, and contact with metals increase the rate of sodium hydroxide degradation (White, 1992). Sodium hypochlorite solution is typically fed directly into the process water using metering pump. Similar to chlorine solution, sodium hypochlorite is mixed with the process water with either a mechanical mixer or an induction mixer. Sodium hypochlorite solution is typically not diluted prior to mixing to reduce scaling problems.

Calcium Hypochlorite: Commercial high-level calcium hypochlorite contains at least 70% available chlorine. Under normal storage conditions, calcium hypochlorite loses 3%–5% of its available chlorine in a year. Calcium hypochlorite comes in powder, granular, and compressed tablet forms. Typically, calcium hypochlorite solution is prepared by mixing powdered or granular calcium hypochlorite with a small flow of water. The high-chlorinated solution is then flow-paced into the drinking water supply (USEPA, 1991).

Safety and Handling Considerations

Chlorine gas is a strong oxidizer. The U.S. Department of Transportation classifies chlorine as a poisonous gas (Connell, 1996). Fire codes typically regulate the storage and use of chlorine. In addition, facilities storing more than 2,500 lbs of chlorine are subject to the following two safety programs:

- Process Safety Management (PSM) standard regulated by the Occupational Safety and Health Administration (OSHA) under 29 CFR 1910.119.
- The Risk Management Program (RMP) Rule administered by the EPA under Section 112(r) of the Clean Air Act (CAA).

These regulations (as well as local and state codes and regulations) must be considered during the design and operation of chlorination facilities at a water treatment plant.

Sodium hypochlorite solution is a corrosive liquid with an approximate pH of 12.

Therefore, typical precautions for handling corrosive materials, such as avoiding contact with metals, including stainless steel, should be used (AWWA, 2010). Sodium hypochlorite solutions may contain chlorate. Chlorate is formed during both the manufacturing and storage of sodium hypochlorite (due to the degradation of the product). Chlorate formation can be minimized by reducing the degradation of sodium hypochlorite by limiting storage time, avoiding high temperatures, and reducing light exposure (Gordon et al., 1995). Spill contaminant must be provided for sodium hypochlorite storage tanks. Typical spill containment structures include containment for the entire contents of the largest tank (plus freeboard for rainfall or fire sprinklers), no uncontrolled floor drains, and separate containment areas for each incompatible chemical.

Calcium hypochlorite is an oxidant and, as such, should be stored separately from organic materials that can be readily oxidized. It should also be stored away from sources of heat. Improperly stored calcium hypochlorite has caused spontaneous combustion fires.

Advantages and Disadvantages of Chlorine Use

The following list presents selected advantages and disadvantages of using chlorine as a disinfection method for drinking water. Because of the wide variation in system size, water quality, and dosages applied, some of these advantages and disadvantages may not apply to a particular system (Masschelein, 1992).

Advantages

- Chlorine oxidizes soluble iron, manganese, and sulfides.
- Chlorine enhances color removal.
- Chlorine enhances taste and odor.
- Chlorine may enhance coagulation and filtration of particulate contaminants.
- Chlorine is an effective biocide.
- Chlorine is easiest disinfectant method, regardless of system size.
- Chlorine is the most widely used disinfection method and, therefore, the best known.
- Chlorine is available as calcium and sodium hypochlorite (the use of these solutions is more advantageous for smaller systems than chlorine gas because they are easier to use, safer, and need less equipment compared to chlorine gas).
- Chlorine provides a residual.

Disadvantages

- Chlorine may cause a deterioration in the coagulation/filtration of dissolved organic substances.
- Chlorine forms halogen-substituted byproducts.
- Chlorine-finished water could have taste and odor problems, depending on the water quality and dosage.
- Chlorine gas is a hazardous corrosive gas.
- Special leak containment and scrubber facilities may be required for chlorine gas.
- Typically, sodium and calcium hypochlorite are more expensive than chlorine gas.
- Sodium hypochlorite degrades over time and with exposure to light.
- Sodium hypochlorite is a corrosive chemical.
- Calcium hypochlorite must be stored in a cool, dry place because of its reaction with moisture and heat.

TABLE 17.20
Summary of Chlorine Disinfection

Consideration	Description
Generation	Chlorination may be performed using chlorine gas or other chlorinated compounds that may be in liquid or solid form. Chlorine gas can be generated by several processes including the electrolysis of alkaline brine or hydrochloric acid, the reaction between sodium chloride and nitric acid, or hydrochloric acidoxidation. Since chlorine is a stable compound, chlorine gas, sodium hypochlorite, and calcium hypochlorite are typically produced off-site by a chemical manufacturer.
Primary uses	The primary use of chlorination is disinfection. Chlorine also serves as an oxidizing agent for taste and odor control, preventing algal growths, maintaining clear filter media, removing iron and manganese, destroying hydrogen sulfide, color removal, maintaining the water quality at the distribution systems, and improving coagulation.
Inactivation efficiency	The general order of increasing chlorine disinfection difficulty is bacteria, viruses, and then protozoa. Chlorine is an extremely effective disinfectant for inactivating bacteria and a highly effective viricide. However, chlorine is less effective against *Giardia* cysts. *Cryptosporidium* oocysts are highly resistant to chlorine.
Byproduct formation	When added to the water, free chlorine reacts with NOM and bromide to form DBPs, primarily THMs, some haloacetic acids (HAAs), and others.
Point of application	Raw water storage, pre-coagulation/post-raw water storage, pre-sedimentation/post-coagulation, post-sedimentation/prefiltration, post-filtration (disinfection), or in the distribution system.
Special considerations	Because chlorine is such a strong oxidant and extremely corrosive, special storage and handling should be considered in planning a water treatment plant. Additionally, health concerns associated with the handling and use of chlorine are an important consideration.

- A precipitate may form in a calcium hypochlorite solution because of impurities; therefore, an antiscalant chemical may be needed.
- Higher concentrations of hypochlorite solutions are unstable and will produce chlorate as a byproduct.
- Chlorine is less effective at high pH.
- Chlorine forms oxygenated byproducts that are biodegradable and can enhance subsequent biological growth if a chlorine residual is not maintained.
- Chlorine can cause the release of constituents bound in the distribution system (e.g., arsenic) by changing the redox state.

Chlorine Summary Table

Table 17.20 presents a summary of the considerations for the use of chlorine as a disinfectant.

ARSENIC AND OTHER REMOVAL FROM DRINKING WATER

(Much of the following information is based on USEPA, 2000.) Operators may be familiar with the controversy created when newly elected President George W. Bush placed the pending Arsenic Standard on temporary hold. The President prevented the implementation of the Arsenic Standard to give scientists time to review the Standard, take a closer look at the possible detrimental effects on the health and well-being of consumers in certain geographical areas of the United States, and to give economists time to determine the actual cost of implementation.

President Bush's decision caused quite a stir, especially among environmentalists, the media, and others who felt that the Arsenic Standard should be enacted immediately to protect affected consumers. The President, who understood the emotional and political implications of shelving the Arsenic Standard, also understood the staggering economic implications involved in implementing the new, tougher standard. Many view the President's decision as the wrong one. Others view his decision as the right one; they base their opinion on the old adage, "It is best to make scientific judgments based on good science instead of on "feel good" science." Whether the reader shares the latter view or not, the point is that arsenic levels in potable water supplies are required to be reduced to a set level and in the future will have to be reduced to an even lower level. Accordingly, water treatment plants affected by the existing arsenic requirements and the pending tougher arsenic requirements should be familiar with the technologies for the removal of arsenic from water supplies. In this section, we describe a number of these technologies.

ARSENIC EXPOSURE

Arsenic (As) is a naturally occurring element present in food, water, and air. Known for centuries to be an effective poison, some animal studies suggest that arsenic may be an essential nutrient at low concentrations. Non-malignant skin alterations, such as keratosis and hypo- and hyper-pigmentation, have been linked to arsenic ingestion, and skin cancers have developed in some patients. Additional studies indicate that arsenic ingestion may result in internal malignancies, including cancers of the kidney, bladder, liver, lung, and other organs. Vascular system effects have also been observed, including peripheral vascular disease, which, in its most severe form, results in gangrene or Blackfoot Disease. Other potential effects include neurologic impairment.

The primary route of exposure to arsenic for humans is ingestion. Exposure via inhalation is considered minimal, though there are regions where elevated levels of airborne arsenic occur periodically (Hering and Chiu, 1998). Arsenic occurs in two primary forms: organic and inorganic. Organic species of arsenic are predominantly found in foodstuffs, such as shellfish, and include forms such as monomethyl arsenic acid (MMAA), dimethyl arsenic acid (DMAA), and arseno-sugars. Inorganic arsenic occurs in two valence states: arsenite and arsenate. In natural surface waters, arsenate is the dominant species.

ARSENIC REMOVAL TECHNOLOGIES

Some of the arsenic removal technologies discussed in this section are traditional treatment processes that have been tailored to improve the removal of arsenic from drinking water. Several treatment techniques discussed here are at the experimental stage with regard to arsenic removal, and some have not been demonstrated at full scale. Although some of these processes may be technically feasible, their cost may be prohibitive. Technologies discussed in this section are grouped into four broad categories: prescriptive processes, adsorption processes, ion exchange processes, and separation (membrane) processes. Each category is discussed here, with at least one treatment technology described in each category.

Prescriptive Processes

Coagulation/Filtration

Coagulation/flocculation (C/F) is a treatment process by which the physical or chemical properties of dissolved colloidal or suspended matter are altered such that agglomeration is enhanced to an extent that the resulting particles will settle out of solution by gravity or will be removed by filtration. Coagulants change the surface charge properties of solids to allow for agglomeration and/or enmeshment of particles into a flocculated precipitate. In either case, the final products are larger particles, or floc, which more readily filter or settle under the influence of gravity.

The coagulation/filtration process has traditionally been used to remove solids from drinking water supplies. However, the process is not restricted to the removal of particles. Coagulants render some dissolved species [e.g., natural organic matter (NOM), inorganics, and hydrophobic synthetic organic compounds (SOCs)] insoluble and the metal hydroxide particles produced by the addition of metal salt coagulants (typically aluminum sulfate, ferric chloride, or ferric sulfate) can adsorb other dissolved species. Major components of a basic coagulation/filtration facility include chemical feed systems, mixing equipment, basins for rapid mixing, flocculation, settling, filter media, sludge handling equipment, and filter backwash facilities. Settling may not be necessary in situations where the influent particle concentration is very low. Treatment plants without settling are known as direct filtration plants.

Iron/Manganese Oxidation

Iron/Manganese (Fe/Mn) oxidation is commonly used by facilities treating groundwater. The oxidation process used to remove iron and manganese leads to the formation of hydroxides that remove soluble arsenic by precipitation or adsorption reactions. Arsenic removal during iron precipitation is fairly efficient. Removal of 2 mg/L of iron achieved a 92.5% removal of arsenic from a 10 μg/L arsenate initial concentration by adsorption alone. Even the removal of 1 mg/L of iron resulted in the removal of 83% of influent arsenic from a source with 22 μg/L arsenate. Indeed, field studies of iron removal plants have indicated that this treatment can feasibly reach 3 g/L.

The removal efficiencies achieved by iron removal are not as high or as consistent as those realized by activated alumina or ion exchange (Edwards, 1994). Note, however, that arsenic removal during manganese precipitation is relatively ineffective when compared to iron, even when removal by both adsorption and coprecipitation are considered. For instance, precipitation of 3 mg/L manganese removed only 69% of arsenate of a 12.5 μg/L arsenate influent concentration.

Oxidation filtration technologies may be effective arsenic removal technologies. Research on oxidation filtration technologies has primarily focused on greensand filtration. As a result, the following discussion focuses on the effectiveness of greensand filtration as an arsenic removal technology. Substantial arsenic removal has been observed using greensand filtration (Subramanian et al., 1997). The active material in "greensand" is glauconite, a green, iron-rich, clay-like mineral that has ion exchange properties. Glauconite often occurs in nature as small pellets mixed with other sand particles, giving a green color to the sand. The glauconite sand is treated with $KMnO_4$ until the sand grains are coated with a layer of manganese oxides, particularly manganese dioxide. The principle behind this arsenic removal treatment is multifaceted and includes oxidation, ion exchange, and adsorption. Arsenic compounds displace species from the manganese oxide (presumably OH^- and H_2O), becoming bound to the greensand surface—in effect, an exchange of ions. The oxidative nature of the manganese surface converts arsenite to arsenate, which is then adsorbed to the surface. As a result of the transfer of electrons and adsorption of arsenate, reduced manganese (Mn) is released from the surface.

The effectiveness of greensand filtration for arsenic removal depends on the influent water quality. Subramanian et al. (1997) showed a strong correlation between influent Fe concentration and arsenic percent removal. Removal increased from 41% to more than 80% as the Fe/As ratio increased from 0 to 20 when treating tap water with a spiked arsenite concentration of 200 mg/L. The tap water contained 366 mg/L sulfate and 321 mg/L TDS; neither constituent seemed to affect arsenic removal. The authors also point out that the influent manganese concentration may play an important role. Divalent ions, such as calcium, can also compete with arsenic for adsorption sites. Water quality would need to be carefully evaluated for applicability for treatment using greensand. Other researchers have also reported substantial arsenic removal using this technology, including arsenic removals of greater than 90% for the treatment of groundwater (Subramanian et al., 1997).

As with other treatment media, greensand must be regenerated when its oxidative and adsorptive capacity has been exhausted. Greensand filters are regenerated using a solution of excess potassium permanganate ($KMnO_4$). Like other treatment media, the regeneration frequency will depend on the influent water quality in terms of constituents that will degrade the filter capacity. Regenerant disposal for greensand filtration has not been addressed in previous research.

Coagulation-Assisted Microfiltration

Arsenic is removed effectively by the coagulation process. Microfiltration is used as a membrane separation process to remove particulates, turbidity, and microorganisms. In coagulation-assisted microfiltration technology, microfiltration is used in a manner similar to a conventional gravity filter. The advantages of microfiltration over conventional filtration are outlined below (Muilenberg, 1997):

- More effective microorganism barrier during the coagulation process upsets
- Smaller floc sizes can be removed (smaller amounts of coagulants are required)
- Increased total plant capacity

Vickers et al. (1997) reported that microfiltration exhibited excellent arsenic removal capability. This report is corroborated by pilot studies conducted by Clifford, which found that coagulation-assisted microfiltration could reduce arsenic levels below 2 ug/L in waters with a pH of between 6 and 7, even when the influent concentration of Fe is approximately 2.5 mg/L (Clifford et al., 1997). These studies also found that the same level of arsenic removal could be achieved by this treatment process even if source water sulfate and silica levels were high. Further, coagulation-assisted microfiltration can reduce arsenic levels to an even greater extent at a slightly lower pH (approximately 5.5).

The addition of a coagulant did not significantly affect the membrane-cleaning interval, although the solids level in the membrane system increased substantially. With an iron and manganese removal system, it is critical that all of the iron and manganese be fully oxidized before they reach the membrane to prevent fouling (Muilenberg, 1997).

Enhanced Coagulation

The Disinfectant/Disinfection By-Product (D/DBP) Rule requires the use of enhanced coagulation treatment for the reduction of disinfection byproduct (DBP) precursors for surface water systems that have sedimentation capabilities. The enhanced process involves modifications to the existing coagulation process such as increasing the coagulant dosage, reducing the pH, or both. Cheng et al. (1994) conducted bench, pilot, and demonstration-scale studies to examine arsenate removals during enhanced coagulation.

The enhanced coagulation conditions in these studies included an increase in alum and ferric chloride coagulant dosage from 10 to 30 mg/L, a decrease in pH from 7 to 5.5, or both. Results from these studies indicated the following:

- Greater than 90% arsenate removal can be achieved under enhanced coagulation conditions. Arsenate removals greater than 90% were easily attained under all conditions when ferric chloride was used.
- Enhanced coagulation using ferric salts is more effective for arsenic removal than enhanced coagulation using alum. With an influent arsenic concentration of 5 μg/L, ferric chloride achieved 96% arsenate removal with a dosage of 10 mg/L and no acid addition. When alum was used, 90% arsenate removal could not be achieved without reducing the pH.
- Lowering pH during enhanced coagulation improved arsenic removal by alum coagulation. With ferric coagulation, pH does not have a significant effect between 5.5 and 7.0.

Note: Post-treatment pH adjustment may be required for corrosion control when the process is operated at a low pH.

Lime Softening

Recall that hardness is predominantly caused by calcium and magnesium compounds in solution. Lime softening removes this hardness by creating a shift in the carbonate equilibrium. The addition of lime to water raises the pH. Bicarbonate is converted to carbonate as the pH increases, and as a result, calcium is precipitated as calcium carbonate. Soda ash (sodium carbonate) is added if insufficient bicarbonate is present in the water to remove hardness to the desired level. Softening for calcium removal is typically accomplished at a pH range of 9–9.5. For magnesium removal, excess lime is added beyond the point of calcium carbonate precipitation. Magnesium hydroxide precipitates at pH levels greater than 10.5. Neutralization is required if the pH of the softened water is excessively high (above 9.5) for potable use. The most common form of pH adjustment in softening plants is recarbonation with carbon dioxide.

Lime softening has been widely used in the United States for reducing hardness in large water treatment systems. Lime softening, excess lime treatment, split lime treatment, and lime-soda softening are all common in municipal water systems. All of these treatment methods are effective in reducing arsenic. Arsenite or arsenate removal is pH-dependent and oxidation of arsenite is the predominant form. Considerable amounts of sludge are produced in a lime softening system and its disposal is expensive. Large-capacity systems may find it economically feasible to install recalcination equipment to recover and reuse the lime sludge and reduce disposal problems. The construction of a new lime softening plant for the removal of arsenic would not generally be recommended unless hardness must also be reduced.

Adsorptive Processes

Activated Alumina

Activated alumina is a physical/chemical process by which ions in the feed water are sorbed to the oxidized activated alumina surface. Activated alumina is considered an adsorption process, although the chemical reactions involved are actually an exchange of ions. Activated alumina is prepared through the dehydration of $Al(OH)_3$ at high temperatures and consists of amorphous and gamma alumina oxide (AWWA, 2010; Clifford and Lin, 1985). Activated alumina is used in packed beds to remove contaminants such as fluoride, arsenic, selenium, silica, and NOM. Feed water is continuously passed through the bed to remove contaminants. The contaminant ions are exchanged with the surface hydroxides on the alumina. When adsorption sites on the activated alumina surface become filled, the bed must be regenerated. Regeneration is accomplished through a sequence of rinsing with regenerant, flushing with water, and neutralizing with acid. The regenerant is a strong base, typically sodium hydroxide; the neutralizer is a strong acid, typically sulfuric acid. Many studies have shown that activated alumina is an effective treatment technique for arsenic removal. Factors such as pH, arsenic oxidation state, competing ions, empty bed contact time, and regeneration have significant effects on the removals achieved with activated alumina. Other factors include spent regenerant disposal, alumina disposal, and secondary water quality.

Ion Exchange

Ion exchange is a physical/chemical process by which an ion in the solid phase is exchanged for an ion in the feed water. This solid phase is typically a synthetic resin that has been chosen to preferentially adsorb the particular contaminant of concern. To accomplish this exchange of ions, feed water is continuously passed through a bed of ion exchange resin beads in a downflow or upflow mode until the resin is exhausted. Exhaustion occurs when all sites on the resin beads have been filled by contaminant ions. At this point, the bed is regenerated by rinsing the ion exchange column with a regenerant—a concentrated solution of ions initially exchanged from the resin. The number of bed volumes that can be treated before exhaustion varies with resin type and influent water quality. Typically, from 300 to 60,000 bed volumes (BV) can be treated before regeneration is required. In most cases, regeneration of the bed can be accomplished with only 1–5 BV of regenerant followed by 2–20 BV of rinse water. Important considerations in the applicability of the ion exchange process for the removal of a contaminant include water quality parameters such as pH, competing ions, resin type, alkalinity, and influent arsenic concentration. Other factors include the affinity of the resin for the contaminant, spent regenerant and resin disposal requirements, secondary water quality effects, and design operating parameters.

Membrane Processes

Membranes are a selective barrier, allowing some constituents to pass while blocking the passage of others. The movement

TABLE 17.21
Typical Pressure Ranges for Membrane Processes

Membrane Process	Pressure Range (psi)
MF	5–45
UF	7–100
NF	50–150
RO	100–150

of constituents across a membrane requires a driving force (i.e., a potential difference between the two sides of the membrane). Membrane processes are often classified by the type of driving force, including pressure, concentration, electrical potential, and temperature. The processes discussed here include only pressure-driven and electrical potential-driven types. Pressure-driven membrane processes are often classified by pore size into four categories: microfiltration (MF), ultrafiltration (UF), nanofiltration (NF), reverse osmosis (RO), and high-pressure processes (i.e., NF and UF). Typical pressure ranges for these processes are given in Table 17.21. NF and RO primarily remove constituents through chemical diffusion. MF and UF primarily remove constituents through physical sieving. An advantage of high-pressure processes is that they tend to remove a broader range of constituents than low-pressure processes. However, the drawback to broader removal is the increase in energy required for high-pressure processes (Aptel and Buckley, 1996). Electrical potential-driven membrane processes can also be used for arsenic removal. These processes include, for this document, only electrodialysis reversal (EDR). In terms of achievable contaminant removal, EDR is comparable to RO. The separation process used in EDR, however, is ion exchange.

Alternative Technologies

Iron Oxide Coated Sand

Iron-oxide-coated sand is a rare process that has shown some tendency for arsenic removal. Iron oxide-coated sand consists of sand grains coated with ferric hydroxide which are used in fixed bed reactors to remove various dissolved metal species. The metal ions are exchanged with the surface hydroxides on the iron oxide-coated sand. Iron oxide-coated sand exhibits selectivity in the adsorption and exchange of ions present in the water. Like other processes, when the bed is exhausted it must be regenerated by a sequence of operations consisting of rinsing with regenerant, flushing with water, and neutralizing with strong acid. Sodium hydroxide is the most common regenerant, and sulfuric acid is the most common neutralizer. Several studies have shown that iron oxide-coated sand is effective for arsenic removal. Factors such as pH, arsenic oxidation state, competing ions, and regeneration have significant effects on the removal achieved with iron oxide-coated sand.

Sulfur-Modified Iron

A patented Sulfur-Modified Iron (SMI) process for arsenic removal has recently been developed (Hydrometrics, 1998).

The process consists of three components: (1) finely divided metallic iron; (2) powdered elemental sulfur or other sulfur compounds; and (3) an oxidizing agent. The powdered iron, powdered sulfur, and the oxidizing agent (H_2O_2 in preliminary tests) are thoroughly mixed and then added to the water to be treated. The oxidizing agent serves to convert arsenite to arsenate. The solution is then mixed and settled. Using the sulfur-modified iron process on several water types, high adsorptive capacities were obtained with a final arsenic concentration of 0.050 mg/L. Arsenic removal was influenced by pH. Approximately 20 mg/L of arsenic per gram of iron was removed at pH 8, and 50 mg of arsenic per gram of iron was removed at pH 7. Arsenic removal seems to be very dependent on the iron-to-arsenic ratio.

Packed-bed column tests demonstrated significant arsenic removal at residence times of 5–15 min. Significant removal of both arsenate and arsenite was measured. The highest adsorption capacity measured was 11 mg of arsenic removed per gram of iron. Flow distribution problems were evident, as several columns became partially plugged, and better arsenic removal was observed with reduced flow rates.

Spent media from the column tests were classified as non-hazardous waste. Projected operating costs for sulfur-modified iron, when the process is operated below a pH of 8, are much lower than alternative arsenic removal technologies such as ferric chloride addition, reverse osmosis, and activated alumina. Cost savings would increase proportionally with increased flow rates and increased arsenic concentrations.

Possible treatment systems using sulfur-modified iron include continuous stirred tank reactors, packed bed reactors, fluidized bed reactors, and passive in situ reactors. Packed bed and fluidized bed reactors appear to be the most promising for successful arsenic removal in pilot-scale and full-scale treatment systems based on present knowledge of the sulfur-modified iron process.

Granular Ferric Hydroxide

A new removal technique for arsenate, which has recently been developed at the Technical University of Berlin (Germany), Department of Water Quality Control, is adsorption on granular ferric hydroxide in fixed bed reactors. This technique combines the advantages of the coagulation-filtration process, efficiency, and small residual mass, with the fixed bed adsorption on activated alumina, and sample processing. Demers and Renner (1992) reported that the application of granular ferric hydroxide in test adsorbers showed a high treatment capacity of 30,000–40,000 bed volumes, with an effluent arsenate concentration never exceeding 10 μ/L. The typical residual mass was in the range of 5–25 g/m³ of treated water. The residue was a solid with an arsenate content of 1–10 g/kg. Table 17.22 summarizes data from the adsorption tests.

The competition of sulfate on arsenate adsorption was not very strong. Phosphate, however, competed strongly with arsenate, which reduced arsenate removal with GFH.

TABLE 17.22
Adsorption Tests on GFH

	Units	Test 1	Test 2	Test 3	Test 4
		Raw Water Parameters			
pH		7.8	7.8	8.2	7.6
Arsenate concentration	μg/L	100–800	21	16	15–20
Phosphate conc.	μg/L	0.70	0.22	0.15	0.30
Conductivity	μS/cm	780	480	200	460
Adsorption capacity for arsenate	g/kg	8.5	4.5	3.2	N/D
		Adsorber			
Bed height	M	0.24	0.16	0.15	0.82
Filter rate 15	m/h	6–10	7.6		5.7
Treatment cap.	BV	34,000	37,000	32,000	85,000
Max. effluent conc.	μg/L	10	10	10	7
Arsenate content of GFH	g/kg	8.5	1.4	0.8	1.7
Mass of spent GFH (dry weight)	g/m³	20.5	12	18	8.6

N/D, not determined.

Arsenate adsorption decreases with pH, which is typical for anion adsorption. At high pH values, GFH outperforms alumina. Below a pH of 7.6, the performance is comparable.

A field study reported by Simms et al. (2000) confirms the efficacy of GFH for arsenic removal. Throughout this study, a 5.3 mgd GFH located in the United Kingdom was found to reduce average influent arsenic concentrations from 20 g/L to less than 10 g/L reliably and consistently for 200,000 BV (over a year of operation) at an empty bed contact time (EBCT) of 3 min. Despite the insignificant head loss, routine backwashing was conducted on a monthly basis to maintain media conditions and to reduce the possibility of bacterial growth. The backwash was not hazardous and could be recycled or disposed of by a sanitary sewer. At the time of replacement, arsenic loading on the media was 2.3%. Leachate tests conducted on the spent media found that arsenic did not leach from the media.

The most significant weakness of this technology appears to be its cost. Currently, GFH media costs approximately $4,000/ton. However, if a GFH bed can be used several times longer than an alumina bed, for example, it may prove to be the more cost-effective technology. Indeed, the system profiled in the field study presented above tested activated alumina as well as GFH and found that GFH was sufficiently more efficient so that smaller adsorption vessels and less media could be used to achieve the same level of arsenic removal (reducing costs). In addition, unlike activated alumina, GFH does not require pre-oxidation.

A treatment for leaching arsenic from the media to enable the regeneration of GFH seems feasible, but it results in the generation of an alkaline solution with high levels of arsenate, which requires further treatment to obtain solid waste. Thus, direct disposal of spent GFH should be favored.

Iron Filings

Iron filings and sand may be used to reduce inorganic arsenic species to iron co-precipitates, mixed precipitates, and, in conjunction with sulfates, arsenopyrites. This type of process is essentially a filter technology, much like greensand filtration, wherein the source water is filtered through a bed of sand and iron filings. Unlike some technologies, such as ion exchange, sulfate is introduced in this process to encourage arsenopyrite precipitation. This arsenic removal method was originally developed as a batch arsenic remediation technology. It appears to be quite effective in this use. Bench-scale tests indicate an average removal efficiency of 81%, with much higher removals at lower influent concentrations. This method was tested at arsenic levels of 20,000 ppb, and at 2,000 ppb, it consistently reduced arsenic levels to less than 50 ppb (the current MCL). While it is quite effective in this capacity, its use as a drinking water treatment technology appears to be limited. In batch tests, a residence time of approximately 7 days was required to reach the desired arsenic removal. In flowing conditions, even though removals averaged 81% and reached greater than 95% at 2,000 ppb arsenic, there is no indication that this technology can reduce arsenic levels below approximately 25 ppb. Additionally, there are no data to indicate how the technology performs at normal source water arsenic levels. This technology needs to be further evaluated before it can be recommended as an approved arsenic removal technology for drinking water.

Photooxidation

Researchers at the Australian Nuclear Science and Technology Organization (ANSTO) have found that, in the presence of light and naturally occurring light-absorbing materials, the oxidation rate of arsenite by oxygen can be increased ten-thousandfold. The oxidized arsenic, now arsenate, can then be effectively removed by co-precipitation. ANSTO evaluated both UV lamp reactors and sunlight-assisted photo-oxidation using acidic, metal-bearing water from an abandoned gold, silver, and lead mine. Air sparging was required for sunlight-assisted oxidation due to the high initial arsenate concentration (12 mg/L). Tests demonstrated that near-complete oxidation of arsenite could be achieved using the photochemical process. Analysis of process waters showed that 97% of the arsenic in the process stream was present as arsenate. Researchers also concluded that arsenite was preferentially oxidized in the presence of excess dissolved Fe (22:1 iron to arsenic mole ratio). This contrasts with conventional plants, where dissolved Fe represents an extra chemical oxidant demand that has to be satisfied during the oxidation of arsenite.

Photooxidation of the mine water, followed by co-precipitation was able to reduce arsenic concentrations to as low as 17 µg/L, which meets the current MCL for arsenic. Initial total arsenic concentrations were unknown, though the arsenite concentration was given as approximately 12 mg/L, which is considerably higher than typical raw water arsenic concentrations. ANSTO reported that the residuals from this process are environmentally stable and passed the Toxicity Characteristic Leaching Procedure (TCLP) test necessary to declare waste non-hazardous and suitable for landfill disposal. Based on the removals achieved and residual characteristics, it is expected that photo-oxidation followed by co-precipitation would be an effective arsenic removal technology. However, this technology is still largely experimental and should be further evaluated before being recommended as an approved arsenic removal technology for drinking water.

SIDEBAR 17.2 PLASTIC SOUP—MICROPLASTICS

The term *microplastics* (or MPs) was introduced by Thompson and colleagues in 2004 to denote plastic particles less than 5 mm in diameter (Thompson et al., 2004). There are two types of microplastics: primary and secondary. Primary microplastics are defined as being less than 5 mm in size. Secondary microplastics are slightly larger and are formed from larger pieces of plastic.

Microplastics come in many forms including beads, resin pellets, fibers, and fragments. Plastic microfibers are fibers, such as nylon and polyester, that are used to make clothing, furnishings, and even fishing nets and lines. Resin pellets are melted and used to create larger plastic items. In addition, secondary microplastics are generated when larger pieces of plastic are fragmented by weathering actions, including the effects of ultraviolet rays, wind, and wave action. Current information on the use of tiny plastic abrasives (commonly called microbeads or nanobeads), especially in pharmaceuticals and personal care products (PPCPs) such as toothpaste, face washes, cosmetics, and home cleaning products, along with synthetic fabrics shedding during laundering, has shown that micro- and nanoparticle-sized plastics are pervasive in water bodies (Eriksen et al., 2013). Many of the microplastics are generated as stated above and manifest from daily activities and products that we use without giving them a second, third, or fourth thought. These household activities and products include:

- Body wash
- Tires
- Wet wipes
- Skincare products
- Dishwasher/laundry detergent pods

- Toothpaste
- Synthetic clothing (gives the stretch factor to the clothing)
- Tea bags

From this list, it is obvious that many of the contributors to microplastics in oceans (and other biomes) are the result of the use of pharmaceuticals and personal care products (PPCPs). Again, we contribute to microplastic pollution simply because we do not think about it; ignorance of the ingredients of our everyday personal care products and their fate is common. That is, we just think about using such products to keep us clean and healthy without ever giving a thought to their ultimate fate and impact on the environment. We often do not think about our environment until it offends us—until it is visible or in one way or another reaches out and grabs us.

'Until it is visible' is a key phrase here. Why? Well, in water and wastewater treatment, when raw influent is examined, the materials (in this case, plastics) that are relatively large (i.e., large enough to see with the naked eye and/or with low magnification) indicate what needs to be accomplished via filtration or other unit processes to remove the visible plastics. Removal can be verified by sampling and observing the treated effluent before outfalling the effluent to the receiving water body (or routing it to additional purification to drinking water quality, potentially for potable water usage).

Note that the jury is still out on determining the best way to remove microplastics from water, but as we continue experimentation and analyses of results, we will develop or enhance present technology to remove microplastics and maybe even remove some contaminants not yet identified but potentially found in drinking water.

THE BOTTOM LINE

The Safe Drinking Water Act gives the USEPA the responsibility for setting national drinking water standards that protect the health of the 250 million people who get their water from public water systems. Other people get their water from private wells, which are not subject to federal regulations. Since 1974, the USEPA has set national standards for over 80 contaminants that may occur in drinking water.

Although the USEPA and state governments set and enforce standards, local governments and private water suppliers have direct responsibility for the quality of the water that flows to the customer's tap. Water systems test and treat their water, maintain the distribution systems that deliver water to consumers, and report on their water quality to the state. States and the USEPA provide technical assistance to water suppliers and can take legal action against systems that fail to provide water that meets state and USEPA standards. As mentioned, the USEPA has set standards for more than 180 contaminants that may occur in drinking water and pose a risk to human health. The USEPA sets these standards to protect the health of everybody, including vulnerable groups like children. The contaminants fall into two groups according to the health effects that they cause. Local water suppliers normally alert customers through the local media, direct mail, or other means if there is a potential acute or chronic health effect from compounds in the drinking water. Customers may want to contact them for additional information specific to their area.

Acute effects occur within hours or days of the time that a person consumes a contaminant. People can suffer acute health effects from almost any contaminant if they are exposed to extraordinarily high levels (as in the case of a spill). In drinking water, microbes, such as bacteria and viruses, are the contaminants with the greatest chance of reaching levels high enough to cause acute health effects. Most people's bodies can fight off these microbial contaminants the way they fight off germs, and these acute contaminants typically don't have permanent effects. Nonetheless, when high enough levels occur, they can make people ill and can be dangerous or deadly for a person whose immune system is already weak due to HIV/AIDS, chemotherapy, steroid use, and another reason.

Chronic effects occur after people consume a contaminant at levels over the USEPA's safety standards for many years. The drinking water contaminants that can have chronic effects are chemicals (such as disinfection by-products, solvents, and pesticides), radionuclides (such as radium), and minerals (such as arsenic). Examples of these chronic effects include cancer, liver or kidney problems, or reproductive difficulties.

The *real bottom line* is that *we are all* **Ultimately Responsible for Drinking Water Quality.**

CHAPTER REVIEW QUESTIONS

17.1 What two minerals are primarily responsible for causing "hard water?"

17.2 The power of a substance to resist pH changes is referred to as a (n):

17.3 What chemical is used as a titrant when analyzing a water sample for carbon dioxide (CO_2)?

17.4 How many pounds of chlorine a day will be used if the dosage is 1.2 mg/L for a flow of 1,600,000 gal/day?

17.5 The specific capacity of a well having a yield of 60 gal/min with a drawdown of 25 ft is:

17.6 What is the chlorine demand in mg/L if the chlorine dosage is 1.0 mg/L and the residual chlorine is 0.5 mg/L?

17.7 If the chlorine dosage is 6 mg/L, what must the chlorine residual be if the chlorine demand is 3.3 mg/L?

17.8 A water treatment plant has a daily flow of 3.1 MGD. If the chlorinator setting is 220 lbs/day, and the chlorine demand is 6.9 mg/L, what is the chlorine residual?

17.9 A well log is best described as

17.10 The type of well construction not normally permitted for a public water supply is

17.11 Name three water quality tests routinely performed on water samples collected from storage tanks:

17.12 Paint for the interior of a drinking water storage tank must be approved by:

17.13 Potable water is water that _____.

17.14. A test of the effluent in the clear well shows that the required dosage of chlorine is 0.6 mg/L. The average daily flow at the treatment plant is 1 MGD. If we are using a hypochlorite solution with 68% available chlorine, how many lbs/day of hypochlorite will be required?

17.15 A waterworks conveys piped water to the _____.

17.16 The hydrologic cycle describes:

17.17 Is *Giardia lamblia* a chronic or acute health threat?

17.18 Disinfection can be defined as:

17.19 As disinfectants, they are as common as household bleach:

17.20 Effective disinfectants must:

17.21 How many pounds of chlorine will be used if the dosage is 0.4 mg/L for a flow of 5,300,000 gpd?

17.22 If the chlorine dosage is 10 mg/L, what must the chlorine residual be if the demand is 2.6 mg/L?

17.23 There are a number of possible interferences with chlorine disinfection; name one.

17.24 Given: Dose of soda ash = 0.8 mg/L
Flow = 2.6 MGD
Find: Feed rate of soda ash in lbs/day

17.25 A circular clarifier handles a flow of 0.75 MGD. The clarifier has a 20-ft radius, and a depth of 10-ft. Find the detention time.

17.26 A water treatment plant operates at a rate of 2 MGD. The dosage of alum is 35 ppm (or mg/L); how many pounds of alum are used per day?

17.27 Is the following statement true? By weight, more pounds (65%) of hypochlorite material (HTH) than pounds of chlorine are needed to get the same number of pounds of available chlorine into the water.

17.28 _____ is the amount of chlorine present in water after a specified time period.

17.29 For a potable water system to be contaminated by water from a non-potable system through a cross-connection, two conditions must exist simultaneously. What are they?

17.30 A piping arrangement that could allow a non-toxic substance (such as milk, beer, or orange juice) to contaminate a potable water system would be classified as a _____ hazard situation.

17.31 When firefighting, main breaks or heavy water usage withdraw more water from a potable water system than is being supplied to the system, _____ pressure may develop in the potable system.

17.32 What type of pump uses rollers and tubing?

17.33 List three watershed management practices:

17.34 A filter plant has three filters, each measuring 10 ft long × 7 ft wide. One filter is out of service, and the other two together are capable of filtering 280 gpm. How many gallons per square feet per minute will each filter have?

17.35 A filter having an area of 300 ft^2 is ready to be backwashed. Assuming a rate of 15 gal/ft^2/min and 8 min of backwash is required, what is the amount of water in gallons required for each backwash?

17.36 If water travels 600 ft in 5 min, the velocity is:

17.37 Where should you look to find information about the hazards associated with the various chemicals you come into contact with at your treatment plant?

17.38 What two water unit treatment processes have done the most to eradicate or reduce the level of waterborne disease in the United States?

17.39 If the discharge pressure is lower than the pump's rating, the pump will:

17.40 The most effective chlorine compound for killing or inactivating pathogens is:

17.41 The group of microorganisms that form cysts and thus become resistant to disinfection is:

17.42 The Surface Water Treatment Rule contains operational and monitoring requirements to ensure that:

17.43 The Aggressive Index is an indicator of:

17.44 List three factors that influence coagulant dose:

17.45 Flow is through a 2.5-ft wide rectangular channel that is 1.4 ft deep and measures 11.2 cfs. What is the average velocity?

17.46 A cylindrical tank is 100 ft high and 20 ft in diameter. How many gallons of water will it contain?

17.47 What is the correct sequence for running a jar test?

17.48 The purpose of coagulation and flocculation is to accomplish:

17.49 The goal of chemical precipitation is to:

17.50 To achieve optimum removal of hardness, how much lime and soda ash should be added?

17.51 What percentage of positive samples cannot be exceeded for bacteriological compliance monitoring of a distribution system?

17.52 What is the velocity in ft/min if water travels 1,500 ft in 4 min? What is the velocity in feet/second?

17.53 A type of valve in a water distribution system used to isolate a damaged line would be:

17.54 The Lead and Copper Rule requires that a treatment facility:

17.55 Fluoridation at 0.2 mg/L below optimum cuts effectiveness by:

17.56 The chemical normally used in a fluoride saturator is:

17.57 Long-term consumption of water with a fluoride concentration of 3.0 mg/L or more may cause:

17.58 If a raw water source has a fluoride ion concentration of 0.15 mg/L and the optimum concentration for the fluoride ion is 0.9 mg/L, then the desired fluoride dose is:

17.59 The purpose of the Jar Test in water treatment is to determine the:

17.60 Inorganic phosphate addition normally functions to:

17.61 The flow in a 6-in pipe is 350 gpm. What is the average velocity?

17.62 What is used to oxidize iron and/or manganese?

17.63 List three factors that affect corrosion:

17.64 Activated carbon removes taste and odor-producing substances by what method?

17.65 To achieve good coagulation in low alkalinity waters, an additional source of alkalinity is most effective when added:

17.66 To prevent media loss, supplemental backwash air flow and surface sweeps should be turned off:

17.67 The most common complaint concerning taste and odor primarily involves:

17.68 The concentration of volatile organic compounds (VOCs) is usually greater in groundwater. True or False?

17.69 VOCs are suspected of being potential carcinogens. True or False?

17.70 A round tank 30 ft in diameter is filled with water to a depth of 15 ft. How many gallons of water are in the tank?

17.71 What chemical substance would you use to reduce trihalomethane formation?

17.72 The cause of brownish-black water is:

17.73 Waters that cause bluish-green stains on household fixtures often contain:

17.74 Water hardness is caused by:

17.75 When Fe^{+2} is chemically changed to Fe^{+3}, the iron atom:

17.76 Alkalinity is caused by:

17.77 If the head loss, in feet, at any level in the filter bed exceeds the depth of the water above the same point (static head), a vacuum can result. This situation is referred to as:

17.78 Sedimentation is the removal of settleable solids by:

17.79 In sedimentation, the _____ zone decreases the velocity of the incoming water and distributes the flow evenly across the basin.

17.80 As water enters a sedimentation basin, _____ flow distribution is important to achieve proper velocity throughout the basin.

17.81 List the three purposes of enhanced coagulation:

17.82 For a water plant that performs bacteriological analysis, what is the maximum hold time from when the sample is collected until analysis is started in the lab?

17.83 When performing a chlorine residual test with an amperometric titrator, what reagent is used as the titrant?

17.84 How many gallons of water fell into a 20-acre reservoir if the water level of the pond rose 2 in after a storm event?

17.85 To maintain a 0.5 mg/L chlorine residual throughout the distribution system, a chlorine dosage of 1.2 mg/L is required at the clear well. If the average daily flow reaches 2.0 MGD, how many lb/day of chlorine must be added?

17.86 A cylindrical water storage tank is 70 ft tall and 30 ft in diameter. The tank is 40% full. What is the water pressure in PSI at the base of the tank?

17.87 The concentration of volatile organic compounds (VOCs) is usually greater in _____.

17.88 List three methods to remove VOCs:

17.89 The addition of _____ _____ is a method to reduce trihalomethane formation.

17.90 The ability of soil to allow water to pass through it is called its _____.

17.91 The top of an aquifer is called the _____.

17.92 Diseases that are carried by water are referred to as _____ diseases.

17.93 A chemical that combines with suspended particles in water is called a _____.

17.94 When water freezes, its volume becomes _____.

17.95 The most commonly used algaecide is _____.

17.96 Carbonate hardness can be removed by adding _____ to the water.

17.97 To destroy coliforms and pathogens, water must be _____.

17.98 BOD helps the plant operator determine how much _____ will be needed to stabilize the organic matter.

17.99 Bacteria are produced by dividing in half, which is called _____.

NOTE

1 This section is adapted for Spellman (2006).

REFERENCES

Aptel, P., and Buckley, C.A., 1996. Categories of Membrane Operations, in *Water Treatment Membrane Processes*. New York: McGraw-Hill.

AWWA. 1991. *Guidance Manual for Compliance with the Filtration and Disinfection Requirements for Public Works Systems Using* Surface Water Sources. Denver, CO: American Water Works Association.

AWWA. 2010a. *A Handbook on Drinking Water.* New York: McGraw Hill.

AWWA. 2010b. *Introduction to Water Treatment,* AWWA (American Water Works Association), Vol. 2, 1984. Denver, CO.

AWWARF. 1987. *Identification and Treatment of Tastes and Odors in Drinking Water.* Denver, CO: The Water Research Foundation.

Belanger, S., Cherry, D.S., Farris, J.L., Sappington, K.G., and Cairns, J., 1991. Sensitivity of the Asiatic clam to various biocidal control agents. *J Am Water Works Assoc;* 83(10):79.

Brady, T.J., Van Benschoten, J.E., and Jensen, J.N., 1996. Chlorination effectiveness for zebra and quagga mussels. *J Am Water Works Assoc;* 88(1):107–110.

Britton, J.C., and Morton, B.A., 1982. Dissection guide, field and laboratory manual for the introduced bivalve Corbicula fluminea. *Malacol Rev;* 3(1):145.

Butterfield, C.T., Wattie, E., Megregian, S., and Chambers, C.W., 1943. Influence of pH and temperature on the survival of coliforms and enteric pathogens when exposed to chlorine. *Public Health Rep;* 58:1837–1866.

Cameron, G.N., Symons, J.M., Spencer, S.R., and Ja, J.Y., 1989. Minimizing THM formation during control of the Asiatic clam: a comparison of biocides. *J Am Water Works Assoc;* 81(10):53–62.

Cheng, R.C., Liang, S., Wang, H.-C., and Beuhler, M.D., 1994. Enhanced coagulation for arsenic removal. *J Am Water Works Assoc;* 9:79–90.

Clarke, N.A., Berg, G., Kabler, P.W., Chang, S.L., and Taft, R.A., 1962. Human Enteric Viruses in Water, Source, Survival, and Removability. *International Conference on Water Pollution Research,* New York.

Clifford, D.A., and Lin, C.C., 1985. *Arsenic (Arsenite) and Arsenic (Arsenate) Removal from Drinking Water in San Ysidro, New Mexico.* Texas: University of Houston.

Clifford, D.A., et al. 1997. *Final Report: Phases 1 & 2 City of Albuquerque Arsenic Study Field Studies on Arsenic Removal in Albuquerque, New Mexico Using the University of Houston/EPA Mobile Drinking Water Treatment Research Facility,* Houston Texas: University of Houston.

Connell, G.F., 1996. *The Chlorination/Chloramination Handbook.* Denver, CO: American Water Works Association.

Craun, G.F., 1981. Outbreaks of waterborne disease in the United States. *J Am Water Works Assoc;* 73(7):360.

Craun, G.F., and Jakubowski, W., 1996. Status of Waterborne Giardiasis Outbreaks and Monitoring Methods. *American Water Resources Association, Water Related Health Issue Symposium,* Atlanta, GA.

Culp G.L., and Culp, R.L., 1974b. *New Concepts in Water Purification.* New York: Van Nostrand Reinhold Company.

Culp, G.L., and Culp, R.L., 1974a. Outbreaks of waterborne disease in the United States. *J Am Water Works Assoc;* 73(7):360.

Demers, L.D., and Renner, R.C., 1992. *Alternative Disinfection Technologies for Small Drinking Water Systems.* Denver, CO: AWWA and AWWART.

Edwards, M.A., 1994. Chemistry of arsenic removal during coagulation and Fe-Mn oxidation. *J Am Water Works Assoc;* 86:64–77.

Eriksen, M., Maximento, N., Theil, M., Cummins, A., Latin, G., Wilson, S., Hafner, J., Zellers, A., and Rifman, S., 2013. Plastic marine pollution in the South Pacific subtropical gyre. *Marine Pollut Bull;* 68:71–76.

Finch, G.R., et al. 1994. Ozone *and Chlorine Inactivation of Cryptosporidium. Conference Proceedings, Water Quality Technology Conference, Part II,* San Francisco, CA.

Gordon, G., Adam, L., and Bubnis, B., 1995. *Minimizing Chlorate Ion Formation in Drinking Water When Hypochlorite Ion is the Chlorinating Agent.* Denver, CO: AWWA-AWWARF.

Gurol, M.D., and Pidatella, M.A., 1983. Study on Ozone-Induced Coagulation. *Conference Proceedings, ASCE Environmental Engineering Division Specialty Conference,* Medicine, A., and Anderson, M., (eds.), Boulder, CO.

Gyurek, L.L., et al., 1996. Disinfection of *Cryptosporidium Parvum* Using Single and Sequential Application of Ozone and Chlorine Species. *Conference Proceedings, AWWA Water Quality Technology Conference,* Boston, MA.

Hass, C.N., and Englebrecht, R.S., 1980. Physiological alterations of vegetative microorganisms resulting from aqueous chlorination. *J Water Pollut Control Fed;* 52(7):1976–1989.

Herbert, P.D.N., Muncaster, B.W., and Mackie, G.L., 1989. Ecological and genetic studies on *Dresissmena polymorpha* (Pallas): a new mollusc in the great Lakes. *Can J Fisheries Aquatic Sci;* 46:187.

Hering, J.G., and Chiu, V.Q., 1998. The Chemistry of Arsenic: Treatment and Implications of Arsenic Speciation and Occurrence. *AWWA Inorganic Contaminants Workshop,* San Antonio, TX.

Hoff, J.C., et al., 1984. Disinfection and the Control of Waterborne Giardiasis. *Conference Proceedings, ASCE Specialty Conference.*

Hydrometrics. 1998. *Second Interim Report on the Sulfur-Modified Iron (SMI) Process for Arsenic Removal.* Helena, MT: Hydrometrics, Inc.

IOA. 1997. *Survey of Water Treatment Plants.* Stanford, CT: International Ozone Association.

Klerks, P.L., and Fraleigh, P.C., 1991. Controlling adult zebra mussels with oxidants. *J Am Water Works Assoc;* 83(12):92–100.

Koch, B., Krasner, S.W., Sclimenti, M.J., and Schimpff, W.K., 1991. Predicting the formation of DBPs by the simulate distribution system. *J Am Water Works Assoc;* 83(10):62–70.

Krasner, S.W., 1989. The occurrence of disinfection byproducts in US drinking water. *J Am Water Works Assoc;* 81(8):41–53.

Laine, J.M., 1993. Influence of bromide on low-pressure membrane filtration for controlling DBPs in surface waters. *J Am Water Works Assoc;* 85(6):87–99.

Lalezary, S., Pirbazari, M., and McGuire, M.J., 1986. Oxidation of five earthy-musty taste and odor compounds. *J Am Water Works Assoc;* 78(3):62.

Lang, C.L., 1994. The Impact of the Freshwater Macrofouling Zebra Mussel (*Dretssena Polymorpha*) on Drinking Water Supplies. *Conference Proceedings AWWA Water Quality Technology Conference Part II,* San Francisco, CA.

Liu, O.C., Seraichekas, H.R., Akin, E.W., Brashear, D.A., Katz, E.L., and Hill, W.J., Jr., 1971. Relative resistance of Twenty Human Enteric Viruses to free Chlorine. Virus and Water A Quality: Occurrence and Control. *Conference Proceedings, Thirteenth Water Quality Conference,* Urban-Champaign, Illinois: University of Illinois.

Masschelein, W.J., 1992. *Unit Processes in Drinking Water Treatment.* New York, Brussels, Hong Kong: Marcel Decker D.D.

Matisoff, G., Brooks, G., and Bourland, B.I., 1996. Toxicity of chlorine dioxide to adult zebra mussels. *J Am Water Works Assoc;* 88(8):93–106.

Montgomery, J.M., 1985. *Water Treatment Principles and Design.* New York: John Wiley & Sons.

Muilenberg, T., 1997. Microfiltration Basics: Theory and Practice. *Proceedings Membrane Technology Conference,* New Orleans, LA.

Nieminski, E.C., Chaudhuri, S., and Lamoreaux, T., 1993. The occurrence of DBPs in Utah drinking waters. *J Am Water Works Assoc*; 85(9):98–105.

Oliver, B.G., and Shindler, D.B., 1980. Trihalomethanes for chlorination of aquatic algae. *Environ Sci Technol*; 14(12):1502.

Prendiville, P.W., 1986. Ozonation at the 900 cfs Los Angeles water purification plant. *Ozone Sci Eng*; 8:77.

Reckhow, D.A., and Singer, P.C., 1985. Mechanisms of Organic Halide formation during Fulvic Acid Chlorination and Implications with Respect to Prezonation, in *Water Chlorination: Chemistry, Environmental Impact and Health Effects*, Vol 5, Jolley, R.L., et al. (eds.). Chelsea, MI: Lewis Publishers.

Reckhow, D.A., et al. 1986. Ozone as a Coagulant Aid. *Seminar Proceedings, Ozonation, Recent Advances and Research Needs. AWWA Annual Conference*, Denver, CO.

Reckhow, D.A., Singer, P.C., and Malcolm, R.L., 1990. Chlorination of humic materials: byproduct formation and chemical interpretations. *Environ Sci Technol*; 24(11):1655.

Rice, R.G., et al. 1998. Ozone Treatment for Small Water Systems. *Presented at First International Symposium on Safe Drinking Water in Small Systems, NSF International*, Arlington, VA.

Roberts, R., 1990. Zebra mussel invasion threatens US waters. *Science*; 249:1370.

Scarpino, P.V., Berg, G., Chang, S.L., Dahling, D., and Lucas, M., 1972. A comparative study of the inactivation of viruses in water by chlorine. *Water Res*; 6:959.

Simms, J., Upton, J., and Barnes, J., 2000. Arsenic Removal Studies and the Design of a 20,000 m³ Per Day Plant in U.K. *AWWA Inorganic Contaminants Workshop*, Albuquerque, N.M.

Sinclair, R.M., 1964. Clam pests in Tennessee water supplies. *J Am Water Works Assoc*; 56(5):592.

Singer, P.C., 1992. Formation and Characterization of Disinfection by Products. *Presented at the First International Conference on the Safety of Water Disinfection: Balancing Chemical and Microbial Risks,* Chicago, IL.

Singer, P.C., and Chang, S.D., 1989. Correlations between trihalomethanes and total organic halides formed during water treatment. *J Am Water Works Assoc*; 81(8):61–65.

Singer, P.C., and Harrington, G.W., 1993. Coagulation of DBP Precursors: Theoretical and Practical Considerations. *Conference Proceedings. AWWA Water Quality Technology Conference*, Miami, FL.

Snead, M.C., Olivieri, V.P., Kruse, C.W., and Kawata, K., 1980. *Benefits of Maintaining a Chlorine Residual in Water Supply Systems.* EPA600/2–80-010. Cincinnati, OH: U.S. Environmental Protection Agency

Spellman, F.R., 2006. *Environmental Science and Technology*, 2nd ed. Rockville, MD: Government Institutes Press.

Spellman, F.R., 2007. *The Science of Water*, 2nd ed. Boca Raton, FL: CRC Press.

Stevens, A.A., 1976. Chlorination of organics in drinking water. *J Am Water Works Assoc*; 8(11):615.

Subramanian, K.D., Viraraghavan, T., Phommavong, T., and Tanjore, S., 1997. Manganese greensand for removal of arsenic in drinking water. *Water Qual Res J Can*; 32(3):551561.

Suffet, I.H., et al., 1986. Removal of Tastes and Odors by Ozonation. *Conference Proceedings, AWWA Seminar on Ozonation: Recent Advances and Research Needs*, Denver, CO.

Thibaud, H., De Laat, J., and Doré, M., 1988. Effects of bromide concentration on the production of chloropicrin during chlorination of surface waters: formation of brominated trihalonitromethanes. *Water Res*; 22(3):381.

Thompson, R.C., Jon, A.W.G., McGonigle, D., and Russell, A.E., 2004. Lost at sea: where is all the plastic? *Science*; 304(5672):838.

USEPA. 1991. *Manual of Individual and Non-Public Works Supply Systems.* Office of Water, EPA5709–91–004. Washington, DC: U.S. Environmental Protection Agency.

USEPA. 1996. *Drinking Water Regulations and Health Advisories.* EPA 822-B-96-002. Washington, DC. Washington, DC: U.S. Environmental Protection Agency.

USEPA. 1997. *Community Water System Survey—Volumes I and II; Overview.* EPA 815-R-97-001a. Washington, DC: U.S. Environmental Protection Agency.

USEPA. 1998. *National Primary Drinking Water Regulations: Interim Enhanced Surface Water Treatment Final Rule.* 63 FR 69477. Washington, DC: U.S. Environmental Protection Agency.

USEPA. 1999a. *Guidance Manual: Alternative Disinfectants and Oxidants, Chapter 1 and 2.* Washington, DC: U.S. Environmental Protection Agency.

USEPA. 1999b. *Lead and Copper Rule Minor Revisions: Fact Sheet.* EPA 815-F-99-010. Washington, DC: U.S. Environmental Protection Agency.

USEPA. 2000. *Technologies and Costs for the Removal of Arsenic from Drinking Water.* EPA-815-R-00–028. Washington, DC: US Environmental Protection Agency.

Van Benschoten, J.E., Jensen, J.N., Harrington, D., DeGirolamo, D.J., 1995. Zebra mussel mortality with chorine. *J Am Water Works Assoc*; 87(5):101–108.

Vickers, J.C., Braghetta, A., and Hawkins, R.A., 1997. Bench Scale Evaluation of Microfiltration for Removal of Particles and Natural Organic Matter. *Proceedings Membrane Technology Conference*, New Orleans, LA.

Watson, H.E., 1908. A note on the variation of the rate of disinfection with change in the concentration of the disinfectant. *J Hygiene*; 8:538.

White, G.C., 1992. *Handbook of Chlorination and Alternative Disinfectants.* New York: Van Nostrand Reinhold.

Williams, R.B., and Culp, G.L., 1986. *Handbook of Public Water Systems.* New York: Van Nostrand Reinhold.

Witherell, L.E., Duncan, R.W., Stone, K.M., Stratton, L.J., Orciari, L., Kappel, S., and Jillson, D.A., 1988. Investigation of Legionella pneumophila in drinking water. *J Am Water Works Assoc*; 80(2):88–93.

Part V

Wastewater and Wastewater Treatment

18 Wastewater Treatment Operations

INTRODUCTION

The Code of Federal Regulations (CFR) 40 CFR Part 403 regulations were established in the late 1970s and early 1980s to help Publicly Owned Treatment Works (POTW) control industrial discharges to sewers. These regulations were designed to prevent pass-through and interference at the treatment plants and interference in the collection and transmission systems.

Pass-through occurs when pollutants literally "pass through" a POTW without being properly treated and cause the POTW to have an effluent violation or increase the magnitude or duration of a violation.

Interference occurs when a pollutant discharge causes a POTW to violate its permit by inhibiting or disrupting treatment processes, treatment operations, or processes related to sludge use or disposal.

Unit operations (unit processes), which are the components linked together to form a process train (as shown in Figure 18.1; keep in mind the caboose attached to this train is treated and cleaned wastewater, which, when out-falled, is usually cleaner than the water in the receiving body), are commonly divided based on the fundamental mechanisms acting within them (i.e., physical, chemical, and biochemical). Physical operations are those, such as sedimentation, that are governed by the laws of physics (gravity). Chemical operations are those in which strict chemical reactions occur, such as precipitation. Biochemical operations are those that use living microorganisms to destroy or transform pollutants through enzymatically catalyzed chemical reactions (Grady et al., 2011).

WASTEWATER OPERATORS (AKA FLUID MECHANICS)

Like waterworks operators, wastewater operators are highly trained and artful practitioners and technicians of their trade. Moreover, wastewater operators, like waterworks operators are required by the states to be licensed or certified to operate a wastewater treatment plant. When learning wastewater operator skills, there are a number of excellent texts available to aid in the training process. Many of these texts are listed in Table 18.1.

WASTEWATER TREATMENT PROCESS: THE MODEL

Figure 18.1 shows a basic schematic of an example wastewater treatment process providing primary and secondary treatment using the *Activated sludge process*. This is the model, the prototype, the paradigm used in the book.

Though it is true that in secondary treatment (which provides BOD removal beyond what is achievable by simple sedimentation), there are actually three commonly used approaches–trickling filter, activated sludge, and oxidation ponds—we focus, for instructive and illustrative purposes, on the activated sludge process throughout this handbook. The purpose of Figure 18.1 is to allow the reader to follow the treatment process step-by-step as it is presented (and as it is actually configured in the real world) and to assist in understanding how all the various unit processes sequentially follow and tie into each other. Therefore, we begin certain sections (which discuss unit processes) with frequent reference to Figure 18.1. It is important to begin these sections in this manner because wastewater treatment is a series of individual steps (unit processes) that treat the waste stream as it makes its way through the entire process. Thus, it logically follows that a pictorial presentation, along with pertinent written information, enhances the learning process. It should also be pointed out, however, that even though the model shown in Figure 18.1 does not include all unit processes currently used in wastewater treatment, we do not ignore the other major processes: trickling filters, rotating biological contactors (RBCs), and oxidation ponds.

WASTEWATER TERMINOLOGY AND DEFINITIONS

Wastewater treatment technology, like many other technical fields, has its own unique terms with their own meanings. Though some of the terms are unique, many are common to other professions. Remember that the science of wastewater treatment is a combination of engineering, biology, mathematics, hydrology, chemistry, physics, and other disciplines. Therefore, many of the terms used in engineering, biology, mathematics, hydrology, chemistry, physics, and others are also used in wastewater treatment. Those terms not listed or defined in the following section will be defined as they appear in the text.

Activated Sludge: the solids formed when microorganisms are used to treat wastewater using the activated sludge treatment process. It includes organisms, accumulated food materials, and waste products from the aerobic decomposition process.

Advanced Waste Treatment: treatment technology to produce an extremely high-quality discharge.

Aerobic: conditions in which free, elemental oxygen is present. Also used to describe organisms, biological activity, or treatment processes that require free oxygen.

Anaerobic: conditions in which no oxygen (free or combined) is available. Also used to describe organisms,

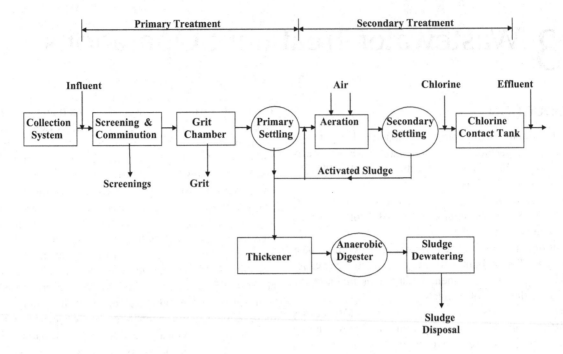

FIGURE 18.1   Schematic of an example wastewater treatment process providing primary and secondary treatment using the activated sludge process.

TABLE 18.1
Recommended Reference/Study Materials

1.   *Advanced Waste Treatment, A Field Study Program*, 2nd ed., Kerri, K., et al. California State University, Sacramento, CA.

2.   *Aerobic Biological Wastewater Treatment Facilities*, Environmental Protection Agency, EPA 430/9-77-006, Washington, D.C., 1977.

3.   *Anaerobic Sludge Digestion*, Environmental Protection Agency, EPA-430/9-76-001, Washington, D.C., 1977.

4.   *Annual Book of ASTM Standards,* Section 11, *"Water and Environmental Technology,"* American Society for Testing Materials (ASTM), Philadelphia, P.A.

5.   *Guidelines Establishing Test Procedures for the Analysis of Pollutants.* Federal Register (40 CFR 136), April 4, 1995 Volume 60, No. 64, Page 17160.

6.   *Handbook of Water Analysis*, 2nd ed., HACH Chemical Company, P.O. Box 389, Loveland, CO, 1992.

7.   *Industrial Waste Treatment, A Field Study Program, Volume I*, Kerri, K. et al. California State University, Sacramento, CA.

8.   *Industrial Waste Treatment, A Field Study Program, Volume 2*, Kerri, K. et al. California State University, Sacramento, CA.

9.   *Methods for Chemical Analysis of Water and Wastes*, U.S. Environmental Protection Agency, Environmental Monitoring Systems Laboratory-Cincinnati (EMSL-CL), EPA-6000/4-79-020, Revised March 1983 and 1979 (where applicable).

10. *O & M of Trickling Filters, RBC and Related Processes, Manual of Practice OM-10*, Water Pollution Control Federation (now called Water Environment Federation), Alexandria, VA, 1988.

11. *Operation of Wastewater Treatment Plants, A Field Study Program, Volume I*, 4th ed., Kerri, K., et al. California State University, Sacramento, CA.

12. *Operation of Wastewater Treatment Plants, A Field Study Program, Volume II*, 4th ed., Kerri, K., et al. California State University, Sacramento, CA.

13. *Standard Methods for the Examination of Water and Wastewater*, 18th ed., American Public Health Association, American Water Works Association-Water Environment Federation, Washington, D.C., 1992.

14. *Treatment of Metal Waste streams*, K. Kerri, et. al., California State University, Sacramento, CA.

15. *Basic Math Concepts: For Water and Wastewater Plant Operators.* Joanne K. Price, Lancaster, PA: Technomic Publishing Company, 1991.

16. *The Science of Water,* 4th ed. F.R. Spellman. Boca Raton, FL: CRC Press, 2020.

17. *Water and Wastewater Infrastructure: Energy Efficiency and Sustainability.* F.R. Spellman. Boca Raton, CRC Press, 2013.

biological activity, or treatment processes that function in the absence of oxygen.

Anoxic: conditions in which no free, elemental oxygen is present; the only source of oxygen is combined oxygen, such as that found in nitrate compounds. Also used to describe biological activity or treatment processes that function only in the presence of combined oxygen.

Average Monthly Discharge Limitation: the highest allowable discharge over a calendar month.

Average Weekly Discharge Limitation: the highest allowable discharge over a calendar week.

Biochemical Oxygen Demand (BOD$_5$): the amount of organic matter that can be biologically oxidized under controlled conditions (5 days @ 20°C in the dark).

Biosolids: From *Merriam-Webster's Collegiate Dictionary, Tenth Edition* (1998): biosolid *n* (1977) Springfield, Massachusetts– solid organic matter recovered from a sewage treatment process and used especially as fertilizer—usually used in the plural.

NOTE: In this text, biosolids are used in many places (activated sludge being the exception) to replace the standard term sludge. The author (along with others in the field) views the term sludge as an ugly four-letter word that is inappropriate for in describing biosolids. Biosolids are a product that can be reused; they have some value. Because biosolids have some value, they should not be classified as a "waste" product, and when biosolids for beneficial reuse are addressed, they are not considered waste.

Buffer: A substance or solution that resists changes in pH.

Carbonaceous Biochemical Oxygen Demand, CBOD$_5$: the amount of biochemical oxygen demand that can be attributed to carbonaceous material.

Chemical Oxygen Demand (COD): the amount of chemically oxidizable materials present in the wastewater.

Clarifier: a device designed to permit solids to settle or rise and be separated from the flow. Also known as a settling tank or sedimentation basin.

Coliform: a type of bacteria used to indicate possible human or animal contamination of water.

Combined Sewer: a collection system that carries both wastewater and stormwater flows.

Comminution: a process to shred solids into smaller, less harmful particles.

Composite Sample: a combination of individual samples taken in proportion to flow.

Daily Discharge: the discharge of a pollutant measured during a calendar day or any 24-h period that reasonably represents a calendar day for sampling. Limitations expressed as weight is the total mass (weight) discharged over the day.

Limitations expressed in other units are average measurements for the day.

Daily Maximum Discharge: the highest allowable values for a daily discharge.

Detention Time: the theoretical time that water remains in a tank at a given flow rate.

Dewatering: the removal or separation of a portion of water present in a sludge or slurry.

Discharge Monitoring Report (DMR): the monthly report required by the treatment plant's NPDES discharge permit.

Dissolved Oxygen (DO): free or elemental oxygen that is dissolved in water.

Effluent: the flow leaving a tank, channel, or treatment process.

Effluent Limitation: any restriction imposed by the regulatory agency on quantities, discharge rates, or concentrations of pollutants that are discharged from point sources into state waters.

Facultative: organisms that can survive and function in the presence or absence of free, elemental oxygen.

Fecal Coliform: a type of bacteria found in the bodily discharges of warm-blooded animals. Used as an indicator organism.

Floc: solids that join together to form larger particles that will settle better.

Flume: a flow rate measurement device.

Food-to-Microorganism Ratio (F/M): an activated sludge process control calculation based on the amount of food (BOD$_5$ or COD) available per pound of mixed liquor volatile suspended solids.

Grab Sample: an individual sample collected at a randomly selected time.

Grit: heavy inorganic solids such as sand, gravel, eggshells, or metal filings.

Industrial Wastewater: wastes associated with industrial manufacturing processes.

Infiltration/Inflow: extraneous flows in sewers, defined by Metcalf & Eddy in *Wastewater Engineering: Treatment, Disposal, Reuse*, 3rd. Ed., New York: McGraw-Hill, Inc., pp. 29–31, 1991 as follows:

- *Infiltration*: water entering the collection system through cracks, joints, or breaks.
- *Steady Inflow*: water discharged from cellar and foundation drains, cooling water discharges, and drains from springs and swampy areas. This type of inflow is steady, identified, and measured along with infiltration.
- *Direct Flow*: those types of inflow that have a direct stormwater runoff connection to the sanitary sewer and cause an almost immediate increase in wastewater flows. Possible sources are roof leaders, yard and area drains, manhole covers, cross connections from storm drains and catch basins, and combined sewers.
- *Total Inflow*: the sum of the direct inflow at any point in the system plus any flow discharged from the system upstream through overflows, pumping station bypasses, and the like.
- *Delayed Inflow*: stormwater that may require several days or more to drain through the sewer system. This category can include the discharge of

sump pumps from cellar drainage as well as the slowed entry of surface water through manholes in ponded areas.

Influent: the wastewater entering a tank, channel, or treatment process.

Inorganic: mineral materials such as salt, ferric chloride, iron, sand, gravel, etc.

License: a certificate issued by the State Board of Waterworks/Wastewater Works Operators authorizing the holder to perform the duties of a wastewater treatment plant operator.

Mean Cell Residence Time (MRCT): the average length of time a mixed liquor suspended solids particle remains in the activated sludge process. May also be known as sludge retention time.

Mixed Liquor: the combination of return-activated sludge and wastewater in the aeration tank.

Mixed Liquor Suspended Solids (MLSS): the suspended solids concentration of the mixed liquor.

Mixed Liquor Volatile Suspended Solids (MLVSS): the concentration of organic matter in the mixed liquor suspended solids.

Milligrams/Liter (mg/L): a measure of concentration. It is equivalent to parts per million (ppm).

Nitrogenous Oxygen Demand (NOD): a measure of the amount of oxygen required to biologically oxidize nitrogen compounds under specified conditions of time and temperature.

NPDES Permit: National Pollutant Discharge Elimination System permit that authorizes the discharge of treated wastes and specifies the condition, which must be met for discharge.

Nutrients: substances required to support living organisms. Usually refers to nitrogen, phosphorus, iron, and other trace metals.

Organic: materials that consist of carbon, hydrogen, oxygen, sulfur, and nitrogen. Many organics are biologically degradable. All organic compounds can be converted to carbon dioxide and water when subjected to high temperatures.

Pathogenic: disease-causing. A pathogenic organism is capable of causing illness.

Point Source: any discernible, defined, and discrete conveyance from which pollutants are or may be discharged.

Part per Million: an alternative (but numerically equivalent) unit used in chemistry is milligrams per liter (mg/L). As an analogy, think of a ppm as being equivalent to a full shot glass in a swimming pool.

Return Activated Sludge Solids (RASS): the concentration of suspended solids in the sludge flow being returned from the settling tank to the head of the aeration tank.

Sanitary Wastewater: wastes discharged from residences and from commercial, institutional, and similar facilities, which include both sewage and industrial wastes.

Scum: the mixture of floatable solids and water, which is removed from the surface of the settling tank.

Septic: a wastewater which has no dissolved oxygen present. Generally characterized by black color and rotten egg (hydrogen sulfide) odors.

Settleability: a process control test used to evaluate the settling characteristics of the activated sludge. Readings taken at 30–60 min are used to calculate the settled sludge volume (SSV) and the sludge volume index (SVI).

Settled Sludge Volume: the volume in percent occupied by an activated sludge sample after 30–60 min of settling. Normally written as SSV with a subscript to indicate the time of the reading used for calculation (SSV_{60}) or (SSV_{30}).

Sewage: wastewater containing human wastes.

Sludge: the mixture of settleable solids and water, which is removed from the bottom of the settling tank.

Sludge Retention Time (SRT): See Mean Cell Residence Time.

Sludge Volume Index (SVI): a process control calculation used to evaluate the settling quality of the activated sludge. Requires the SSV_{30} and mixed liquor suspended solids test results to calculate.

Storm Sewer: a collection system designed to carry only stormwater runoff.

Storm Water: runoff resulting from rainfall and snowmelt.

Supernatant: in a digester, it is the amber-colored liquid above the sludge.

Wastewater: the water supply of the community after it has been soiled by use.

Waste Activated Sludge Solids (WASS): the concentration of suspended solids in the sludge, which is being removed from the activated sludge process.

Weir: a device used to measure wastewater flow.

Zoogleal Slime: the biological slime that forms on fixed film treatment devices. It contains a wide variety of organisms essential to the treatment process.

MEASURING PLANT PERFORMANCE

To evaluate how well a plant or treatment unit process is operating, *performance efficiency* or *percent (%) removal* is used. The results can be compared with those listed in the plant's operation and maintenance manual (O & M) to determine if the facility performs as expected. In this chapter, sample calculations often used to measure plant performance/efficiency are presented.

PLANT PERFORMANCE/EFFICIENCY

The calculation used for determining the performance (percent removal) for a digester differs from that used for the performance (percent removal) for other processes. Care must be taken to select the right formula.

$$\% \text{ Removal} = \frac{\left[\begin{array}{l} \text{Influent Concentration} - \\ \text{Effluent Concentration} \end{array} \right] \times 100}{\text{Influent Concentration}} \quad (18.1)$$

Example 18.1

Problem: The influent BOD_5 is 247 mg/L, and the plant effluent BOD is 17 mg/L. What is the percent removal?

Solution:

$$\% \text{ Removal} = \frac{(247 \text{ mg/L} - 17 \text{ mg/L}) \times 100}{247 \text{ mg/L}} = 93\%$$

UNIT PROCESS PERFORMANCE AND EFFICIENCY

Equation (18.1) is used again to determine unit process efficiency. The concentration entering the unit and the concentration leaving the unit (i.e., primary, secondary, etc.) are used to determine the unit performance.

$$\% \text{ Removal} = \frac{\left[\begin{array}{l} \text{Influent Concentration} - \\ \text{Effluent Concentration} \end{array}\right] \times 100}{\text{Influent Concentration}}$$

Example 18.2

Problem: The primary influent BOD is 235 mg/L, and the primary effluent BOD is 169 mg/L. What is the percent removal?

$$\% \text{ Removal} = \frac{(235 \text{ mg/L} - 169 \text{ mg/L}) \times 100}{235 \text{ mg/L}}$$

PERCENT VOLATILE MATTER REDUCTION IN SLUDGE

The calculation used to determine *percent volatile matter reduction* is more complicated because of the changes occurring during sludge digestion.

$$\% \text{ V.M. Reduction} = \frac{\left(\% \text{ V.M.}_{\cdot\text{in}} - \% \text{ V.M.}_{\cdot\text{out}}\right) \times 100}{\left[\% \text{ V.M.}_{\cdot\text{in}} - \left(\% \text{ V.M.}_{\cdot\text{in}} \times \% \text{ V.M.}_{\cdot\text{out}}\right)\right]} \quad (18.2)$$

V.M. = Volatile Matter

Example 18.3

Problem: Using the digester data provided below, determine the % Volatile Matter Reduction for the digester.
Data:
Raw Sludge Volatile Matter = 74%
Digested Sludge Volatile Matter = 54%

$$\% \text{ Volatile Matter Reduction} = \frac{(0.74 - 0.54) \times 100}{[0.74 - (0.74 \times 0.54)]} = 59\%$$

HYDRAULIC DETENTION TIME

The term *detention time* or *hydraulic detention time (HDT)* refers to the average length of time (theoretical time) that a drop of water, wastewater, or suspended particles remains in a tank or channel. It is calculated by dividing the volume of water/wastewater in the tank by the flow rate through the tank. The units of flow rate used in the calculation depend on whether the detention time is to be calculated in seconds, minutes, hours, or days. Detention time is used in conjunction with various treatment processes, including sedimentation and coagulation-flocculation. Generally, in practice, detention time refers to associated with the amount of time required for a tank to empty. The range of detention time varies with the process. For example, in a tank used for sedimentation, detention time is commonly measured in minutes. The calculation methods used to determine detention time are illustrated in the following sections.

Detention Time in Days

$$\text{HDT, Days} = \frac{\text{Tank Volume, ft}^3 \times 7.48 \text{ gal/ft}^3}{\text{Flow, gal/day}} \quad (18.3)$$

Example 18.4

Problem: An anaerobic digester has a volume of 2,400,000 gal. What is the detention time in days when the influent flow rate is 0.07 MGD?

Solution:

$$\text{D.T., Days} = \frac{2,400,000 \text{ gal}}{0.07 \text{ MGD} \times 1,000,000 \text{ gal/MG}}$$

$$\text{D.T., Days} = 34 \text{ days}$$

Detention Time in Hours

$$\text{HDT, h} = \frac{\text{Tank Volume, ft}^3 \times 7.48 \text{ gal/ft}^3 \times s24 \text{ h/day}}{\text{Flow, gal/day}}$$

$$(18.4)$$

Example 18.5

Problem: A settling tank has a volume of 44,000 ft³. What is the detention time in hours when the flow is 4.15 MGD?

$$\text{D.T., h} = \frac{44,000 \text{ ft}^3 \times 7.48 \text{ gal/ft}^3 \times 24 \text{ h/day}}{4.15 \text{ MGD} \times 1,000,000 \text{ gal/MG}}$$

$$\text{D.T., h} = 1.9 \text{ h}$$

Detention Time in Minutes

$$\text{HDT, min} = \frac{\text{Tank Vol., ft}^3 \times 7.48 \text{ gal/ft}^3 \times 1,440 \text{ min/day}}{\text{Flow, gal/day}}$$

$$(18.5)$$

Example 18.6

Problem: A grit channel has a volume of 1,340 ft³. What is the detention time in minutes when the flow rate is 4.3 MGD?

Solution:

$$\text{D.T., Min} = \frac{1{,}340 \text{ ft}^3 \times 7.48 \text{ gal/ft}^3 \times 1{,}440 \text{ min/day}}{4{,}300{,}000 \text{ gal/day}}$$

$$= 3.36 \text{ min}$$

Note: The tank volume and the flow rate must be in the same dimensions before calculating the hydraulic detention time.

WASTEWATER SOURCES AND CHARACTERISTICS

Wastewater treatment is designed to use natural purification processes (self-purification processes of streams and rivers) to the maximum level possible. It is also designed to complete these processes in a controlled environment rather than over many miles of stream or river. Moreover, the treatment plant is designed to remove other contaminants that are not normally subjected to natural processes, as well as treat the solids generated through the treatment unit steps. The typical wastewater treatment plant is designed to achieve many different purposes:

- Protect public health
- Protect public water supplies
- Protect aquatic life
- Preserve the best uses of the waters
- Protect adjacent lands

Wastewater treatment is a series of steps. Each step can be accomplished using one or more treatment processes or types of equipment. The major categories of treatment steps are:

- **Preliminary Treatment**: Removes materials that could damage plant equipment or would occupy treatment capacity without being treated.
- **Primary Treatment:** Removes settleable and floatable solids (may not be present in all treatment plants).
- **Secondary Treatment:** Removes BOD_5 and dissolved and colloidal suspended organic matter by biological action; organics are converted to stable solids, carbon dioxide, and more organisms.
- **Advanced Waste Treatment:** Uses physical, chemical, and biological processes to remove additional BOD_5, solids, and nutrients (not present in all treatment plants).

- **Disinfection:** Removes microorganisms to eliminate or reduce the possibility of disease when the flow is discharged.
- **Sludge Treatment:** Stabilizes the solids removed from wastewater during treatment, inactivates pathogenic organisms, and/or reduces the volume of the sludge by removing water.

The various treatment processes described above are discussed in detail later.

WASTEWATER SOURCES

The principal sources of domestic wastewater in a community are residential areas and commercial districts. Other important sources include institutional and recreational facilities, stormwater runoff, and groundwater infiltration. Each source produces wastewater with specific characteristics. In this section, wastewater sources and the specific characteristics of wastewater are described.

Wastewater is generated by five major sources: human and animal wastes, household wastes, industrial wastes, stormwater runoff, and groundwater infiltration.

1. **Human and Animal Wastes:** Contain the solid and liquid discharges of humans and animals and are considered by many to be the most dangerous from a human health viewpoint. The primary health hazard is presented by the millions of bacteria, viruses, and other microorganisms (some of which may be pathogenic) present in the waste stream.
2. **Household Wastes:** Wastes, other than human and animal wastes, discharged from the home. Household wastes usually contain paper, household cleaners, detergents, trash, garbage, and other substances discharged into the sewer system.
3. **Industrial Wastes:** Includes industry-specific materials, which can be discharged from industrial processes into the collection system. Typically contains chemicals, dyes, acids, alkalis, grit, detergents, and highly toxic materials.
4. **Storm Water Runoff:** Many collection systems are designed to carry both the wastes of the community and stormwater runoff. When a storm event occurs, the waste stream can contain large amounts of sand, gravel, and other grit, as well as excessive amounts of water.
5. **Groundwater Infiltration:** Groundwater enters older, improperly sealed collection systems through cracks or unsealed pipe joints. This can this add large amounts of water to wastewater flows, along with additional grit.

Note that wastewater can be classified according to the sources of flows: domestic, sanitary, industrial, combined, and stormwater:

1. **Domestic (Sewage) Wastewater:** Mainly contains human and animal wastes, household wastes, small amounts of groundwater infiltration, and small amounts of industrial wastes.
2. **Sanitary Wastewater:** Consists of domestic wastes and significant amounts of industrial wastes. In many cases, the industrial wastes can be treated without special precautions. However, in some cases, industrial wastes will require special precautions or a pretreatment program to ensure they do not cause compliance problems for the wastewater treatment plant.
3. **Industrial Wastewater:** Consists solely industrial wastes. Often, the industry will determine that it is safer and more economical to treat its waste independently of domestic waste.
4. **Combined Wastewater:** Is the combination of sanitary wastewater and stormwater runoff. All the wastewater and stormwater of the community is transported through one system to the treatment plant.
5. **Storm Water:** A separate collection system (no sanitary waste) that carries stormwater runoff, including street debris, road salt, and grit.

WASTEWATER CHARACTERISTICS

Wastewater contains many different substances, which can be used to characterize it. The specific substances and amounts or concentrations of each will vary, depending on the source. Thus, it is difficult to "precisely" characterize wastewater. Instead, wastewater characterization is usually based on and applied to average domestic wastewater. Wastewater is characterized in terms of its physical, chemical, and biological characteristics.

Note: Keep in mind that other sources and types of wastewater can dramatically change the characteristics.

Physical Characteristics

The *physical characteristics* of wastewater are based on color, odor, temperature, and flow.

- **Color:** Fresh wastewater is usually a light brownish-gray color. However, typical wastewater is gray and has a cloudy appearance. The color of the wastewater will change significantly if allowed to go septic (if travel time in the collection system increases). Typical septic wastewater will have a black color.
- **Odor:** Odors in domestic wastewater are usually caused by gases produced by the decomposition of organic matter or by other substances added to the wastewater. Fresh domestic wastewater has a musty odor. If the wastewater is allowed to go septic, this odor will change significantly to a rotten egg odor associated with the production of hydrogen sulfide (H_2S).

- **Temperature:** The temperature of wastewater is commonly higher than that of the water supply because of the addition of warm water from households and industrial plants. However, significant amounts of infiltration or stormwater flow can cause major temperature fluctuations.
- **Flow:** The actual volume of wastewater is commonly used as a physical characterization of wastewater and is normally expressed in terms of gallons per person per day. Most treatment plants are designed using an expected flow of 100–200 gal/person/day. This figure may have to be revised to reflect the degree of infiltration or storm flow the plant receives. Flow rates will vary throughout the day. This variation, which can be as much as 50%–200% of the average daily flow, is known as *diurnal flow variation*.

Note: *Diurnal* means occurring in a day or each day; daily.

Chemical Characteristics

When describing the chemical characteristics of wastewater, the discussion generally includes topics such as organic matter, the measurement of organic matter, inorganic matter, and gases. For the sake of simplicity, this handbook specifically describes chemical characteristics in terms of alkalinity, biochemical oxygen demand (BOD), chemical oxygen demand (COD), dissolved gases, nitrogen compounds, pH, phosphorus, solids (organic, inorganic, suspended, and dissolved solids), and water.

- **Alkalinity:** A measure of the wastewater's capability to neutralize acids. It is measured in terms of bicarbonate, carbonate, and hydroxide alkalinity. Alkalinity is essential to buffer (hold the neutral pH) of the wastewater during the biological treatment processes.
- **Biochemical Oxygen Demand (BOD):** A measure of the amount of biodegradable matter in the wastewater. It is normally measured by a 5-day test conducted at 20°C. The BOD_5 of domestic waste is normally in the range of 100–300 mg/L.
- **Chemical Oxygen Demand (COD):** A measure of the amount of oxidizable matter present in the sample. The COD is normally in the range of 200–500 mg/L. The presence of industrial wastes can increase this significantly.
- **Dissolved Gases:** Gases that are dissolved in wastewater. The specific gases and normal concentrations are based on the composition of the wastewater. Typical domestic wastewater contains oxygen in relatively low concentrations, carbon dioxide, and hydrogen sulfide (if septic conditions exist).
- **Nitrogen Compounds:** The type and amount of nitrogen present will vary from the raw

wastewater to the treated effluent. Nitrogen follows a cycle of oxidation and reduction. Most of the nitrogen in untreated wastewater will be in the forms of organic nitrogen and ammonia nitrogen. Laboratory tests exist for the determination of both of these forms. The sum of these two forms of nitrogen is also measured and is known as *Total Kjeldahl Nitrogen (TKN)*. Wastewater will normally contain between 20 and 85 mg/L of nitrogen. Organic nitrogen will normally be in the range of 8–35 mg/L, and ammonia nitrogen will be in the range of 12–50 mg/L.

- **pH:** A method of expressing the acid condition of the wastewater. pH is expressed on a scale of 1–14. For proper treatment, wastewater pH should normally be in the range of 6.5–9.0 (ideal range is 6.5–8.0).
- **Phosphorus:** Essential to biological activity and must be present in at least minimum quantities; otherwise, secondary treatment processes will not perform adequately. Excessive amounts can cause stream damage and excessive algal growth. Phosphorus will normally be in the range of 6–20 mg/L. The removal of phosphate compounds from detergents has had a significant impact on the amounts of phosphorus in wastewater.
- **Solids:** Most pollutants found in wastewater can be classified as solids. Wastewater treatment is generally designed to remove solids or to convert solids to a form that is more stable or can be removed. Solids can be classified by their chemical composition (organic or inorganic) or by their physical characteristics (settleable, floatable, and colloidal). The concentration of total solids in wastewater is normally in the range of 350–1,200 mg/L.

  Organic Solids: consist of carbon, hydrogen, oxygen, and nitrogen and can be converted to carbon dioxide and water by ignition at 550°C. Also known as fixed solids or loss on ignition.

  Inorganic Solids: Mineral solids that are unaffected by ignition. They are also known as fixed solids or ash.

  Suspended Solids: Will not pass through a glass fiber filter pad. They can be further classified as Total Suspended Solids (TSS), Volatile Suspended Solids, and/or Fixed Suspended Solids. They can also be separated into three components based on settling characteristics: settleable solids, floatable solids, and colloidal solids. Total suspended solids in wastewater are normally in the range of 100–350 mg/L.

  Dissolved Solids: will pass through a glass fiber filter pad. They can also be classified as Total Dissolved Solids (TDS), volatile dissolved solids, and fixed dissolved solids. Total dissolved solids are normally in the range of 250–850 mg/L.

- **Water:** Always the major constituent of wastewater. In most cases, water makes up 99.5%–99.9% of the wastewater. Even in the strongest wastewater, the total amount of contamination present is less than 0.5% of the total, and in average-strength wastes, it is usually less than 0.1%.

Biological Characteristics and Processes

(Note: The biological characteristics of water were discussed in detail earlier in this text). After undergoing the physical aspects of treatment (i.e., screening, grit removal, and sedimentation) in preliminary and primary treatment, wastewater still contains some suspended solids and other solids that are dissolved in the water. In a natural stream, such substances are a source of food for protozoa, fungi, algae, and several varieties of bacteria. In secondary wastewater treatment, these same microscopic organisms (which are one of the main reasons for treating wastewater) are allowed to work as fast as they can to biologically convert the dissolved solids to suspended solids which will physically settle out at the end of secondary treatment.

Raw wastewater influent typically contains millions of organisms. The majority of these organisms are non-pathogenic; however, several pathogenic organisms may also be present (these may include the organisms responsible for diseases such as typhoid, tetanus, hepatitis, dysentery, gastroenteritis, and others). Many of the organisms found in wastewater are microscopic (microorganisms); they include algae, bacteria, protozoans (such as amoeba, flagellates, free-swimming ciliates, and stalked ciliates), rotifers, and viruses. Table 18.2 is a summary of typical domestic wastewater characteristics.

WASTEWATER COLLECTION SYSTEMS

Wastewater collection systems collect and convey wastewater to the treatment plant. The complexity of the system depends on the size of the community and the type of system selected. Methods of collection and conveyance of wastewater include gravity systems, force main systems, vacuum systems, and combinations of all three types of systems.

TABLE 18.2
Typical Domestic Wastewater Characteristics

Characteristic	Typical Characteristic
Color	Gray
Odor	Musty
Dissolved oxygen	>1.0 mg/L
pH	6.5–9.0
TSS	100–350 mg/L
BOD$_5$	100–300 mg/L
COD	200–500 mg/L
Flow	100–200 gal/person/day
Total nitrogen	20–85 mg/L
Total phosphorus	6–20 mg/L
Fecal coliform	500,000–3,000,000 MPN/100 mL

GRAVITY COLLECTION SYSTEM

In a *gravity collection system*, the collection lines are sloped to permit the flow to move through the system with as little pumping as possible. The slope of the lines must keep the wastewater moving at a velocity (speed) of 2–4 fps. Otherwise, at lower velocities, solids will settle out causing clogged lines, overflows, and offensive odors. To keep collection system lines at a reasonable depth, wastewater must be lifted (pumped) periodically so that it can continue flowing downhill to the treatment plant. Pump stations are installed at selected points within the system for this purpose.

FORCE MAIN COLLECTION SYSTEM

In a typical *force main collection system*, wastewater is collected at central points and pumped under pressure to the treatment plant. The system is normally used for conveying wastewater over long distances. The use of the force main system allows the wastewater to flow to the treatment plant at the desired velocity without using sloped lines. It should be noted that the pump station discharge lines in a gravity system are considered to be force mains since the content of the lines is under pressure.

Note: Extra care must be taken when performing maintenance on force main systems since the content of the collection system is under pressure.

VACUUM SYSTEM

In a *vacuum collection system*, wastewaters are collected at central points and then drawn toward the treatment plant under vacuum. The system consists of a large amount of mechanical equipment and requires significant maintenance to function properly. Generally, vacuum-type collection systems are not economically feasible.

PUMPING STATIONS

Pumping stations provide the motive force (energy) to keep the wastewater moving at the desired velocity. They are used in both force main and gravity systems. They are designed in several different configurations and may use different sources of energy to move the wastewater (i.e., pumps, air pressure, or vacuum). One of the more commonly used types of pumping station designs is the wet well/dry well design.

Wet-Well/Dry-Well Pumping Stations

The Wet Well/Dry Well pumping station consists of two separate spaces or sections separated by a common wall. Wastewater is collected in one section (the wet well section), while the pumping equipment and, in many cases, the motors and controllers are located in a second section known as the dry well. There are many different designs for this type of system, but in most cases, the pumps selected

for this system are of a centrifugal design. There are a couple of major considerations in selecting centrifugal designs: (1) it allows for the separation of mechanical equipment (pumps, motors, controllers, wiring, etc.) from the potentially corrosive atmosphere (sulfides) of the wastewater; and (2) this type of design is usually safer for workers because they can monitor, maintain, operate, and repair equipment without entering the pumping station wet well.

Note: Most pumping station wet wells are confined spaces. To ensure safe entry into such spaces, compliance with OSHA's 29 CFR 1910.146 (Confined Space Entry Standard) is required.

WET-WELL PUMPING STATIONS

Another type of pumping station design is the *wet well* type. This type consists of a single compartment, which collects the wastewater flow. The pump is submerged in the wastewater with motor controls located in the space or has a weatherproof motor housing located above the wet well. In this type of station, a submersible centrifugal pump is normally used.

Pneumatic Pumping Stations

The *pneumatic pumping station* consists of a well and a control system, which controls the inlet and outlet value operations and provides pressurized air to force or "push" the wastewater through the system. The exact method of operation depends on the system design. When operating, wastewater in the wet well reaches a predetermined level and activates an automatic valve, which closes the influent line. The tank (wet well) is then pressurized to a predetermined level. When the pressure reaches the predetermined level, the effluent line valve is opened and the pressure pushes the waste stream out of the discharge line.

PUMPING STATION WET WELL CALCULATIONS

Calculations normally associated with pumping station wet well design (determining design lift or pumping capacity, etc.) are usually left up to design and mechanical engineers. However, on occasions, wastewater operators or interceptor's technicians may be called upon to make certain basic calculations. Usually, these calculations determine either pump capacity without influent (e.g., to check the pumping rate of the station's constant speed pump) or pump capacity with influent (e.g., to check how many gallons per minute the pump is discharging). In this section, we use examples to describe instances of how and where these two calculations are made.

Example 18.7 Determining Pump Capacity without Influent

Problem: A pumping station wet well is 10 ft by 9 ft. The operator needs to check the pumping rate of the station's constant speed pump. To do this, the influent valve to the wet well is closed for a 5-min test, and the level in the well drops 2.2 ft. What is the pumping rate in gallons per minute?

Solution:
Using the length and width of the well, we can find the area of the water surface:

$$10 \text{ ft} \times 9 \text{ ft} = 90 \text{ ft}^2$$

The water level dropped by 2.2 ft. From this, we can find the volume of water removed by the pump during the test:

$$\text{Area} \times \text{Depth} = \text{Volume} \qquad (18.6)$$

$$90 \text{ ft}^2 \times 2.2 \text{ ft} = 198 \text{ ft}^3$$

One cubic foot of water holds 7.48 gal. We can convert this volume in cubic feet to gallons.

$$198 \text{ ft}^3 \times \frac{7.48 \text{ gal}}{1 \text{ ft}^3} = 1,481 \text{ gal}$$

The test was done for 5 min. From this information, a pumping rate can be calculated.

$$\frac{1,481 \text{ gal}}{5 \text{ min}} = \frac{296.2}{1 \text{ min}} = 296.2 \text{ gpm}$$

Example 18.8 Determining Pump Capacity with Influent

Problem: A wet well is 8.2 ft by 9.6 ft. The influent flow to the well, measured upstream, is 365 gpm. If the wet well rises 2.2 in. in 5 min, how many gallons per minute is the pump discharging?

Solution:

$$\text{Influent} = \text{Discharge} + \text{Accumulation} \qquad (18.7)$$

$$\frac{365 \text{ gal}}{1 \text{ min}} = \text{Discharge} + \text{Accumulation}$$

We want to calculate the discharge. The influent is known, and we have enough information to calculate the accumulation.

$$\text{Volume accumulated} = 8.2 \text{ ft} \times 9.6 \text{ ft} \times 2.2 \text{ in} \times \frac{1 \text{ ft}}{12 \text{ in}}$$

$$\times \frac{7.48 \text{ gal}}{1 \text{ ft}^3} = 108\text{-gal}$$

$$\text{Accumulation} = \frac{108 \text{ gal}}{5 \text{ min}} = \frac{21.6 \text{ gal}}{1 \text{ min}} = 21.6 \text{ gpm}$$

Using Equation (18.7):

$$\text{Influent} = \text{Discharge} + \text{Accumulation}$$

$$365 \text{ gpm} = \text{Discharge} + 21.6$$

Subtracting from both sides:

$$365 \text{ gpm} - 21.6 \text{ gpm} = \text{Discharge} + 21.6 \text{ gpm} - 21.6 \text{ gpm}$$

$$343.4 \text{ gpm} = \text{Discharge}$$

The wet well pump is discharging 343.4 gal each minute.

PRELIMINARY TREATMENT

The initial stage in the wastewater treatment process (following collection and influent pumping) is *preliminary treatment*. Raw influent entering the treatment plant may contain many kinds of materials (trash). The purpose of preliminary treatment is to protect plant equipment by removing these materials that could cause clogs, jams, or excessive wear to plant machinery. In addition, the removal of various materials at the beginning of the treatment process saves valuable space within the treatment plant.

Preliminary treatment may include many different processes, each designed to remove a specific type of material that is a potential problem for the treatment process. Processes include wastewater collection, such as influent pumping, screening, shredding, grit removal, flow measurement, preaeration, chemical addition, and flow equalization; the major processes are shown in Figure 18.1. In this section, we describe and discuss each process and its importance in the treatment process.

Note: As mentioned, not all treatment plants will include all of the processes shown in Figure 18.1. Specific processes have been included to facilitate discussion of major potential problems with each process and its operation; this information may be important to the wastewater operator.

SCREENING

The purpose of *screening* is to remove large solids such as rags, cans, rocks, branches, leaves, roots, etc. from the flow before the flow moves on to downstream processes.

Note: Typically, a treatment plant will remove anywhere from 0.5 to 12 ft³ of screenings for each million gallons of influent received.

A *bar screen* traps debris as wastewater influent passes through. Typically, a bar screen consists of a series of parallel, evenly spaced bars or a perforated screen placed in a channel (see Figure 18.2). The waste stream passes through the screen and the large solids (*screenings*) are trapped on the bars for removal.

Note: The screenings must be removed frequently enough to prevent accumulation that will block the screen and cause the water level in front of the screen to build up.

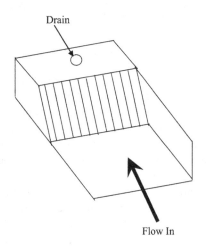

FIGURE 18.2 Basic bar screen.

The bar screen may be coarse (2–4-in openings) or fine (0.75–2.0-in openings). The bar screen may be manually cleaned (bars or screens are placed at an angle of 30° for easier solids removal—see Figure 18.2) or mechanically cleaned (bars are placed at a 45°–60° angle to improve mechanical cleaner operation).

The screening method employed depends on the design of the plant, the amount of solids expected, and whether the screen is for constant or emergency use only.

Manually Cleaned Screens

Manually cleaned screens are cleaned at least once per shift (or often enough to prevent buildup that may cause reduced flow into the plant) using a long-toothed rake. Solids are manually pulled to the drain platform and allowed to drain before storage in a covered container. The area around the screen should be cleaned frequently to prevent a buildup of grease or other materials, which can cause odors, slippery conditions, and insect and rodent problems. Because screenings may contain organic matter as well as large amounts of grease, they should be stored in a covered container. Screenings can be disposed of by burial in approved landfills or by incineration. Some treatment facilities grind the screenings into small particles, which are then returned to the wastewater flow for further processing and removal later in the process.

Operational Problems

Manually cleaned screens require a certain amount of operator attention to maintain optimum operation. Failure to clean the screen frequently can lead to septic wastes entering the primary; surge flows after cleaning, and/or low flows before cleaning. On occasion, when such operational problems occur, it becomes necessary to increase the frequency of the cleaning cycle. Another operational problem is excessive grit in the bar screen channel. Improper design or construction or insufficient cleaning may cause this problem. The corrective action required is either to correct the design problem or increase cleaning frequency and flush the channel regularly. Another common problem

with manually cleaned bar screens is their tendency to clog frequently. This may be caused by excessive debris in the wastewater or the screen being too fine for its current application. The operator should locate the source of the excessive debris and eliminate it. If the screen is the problem, a coarser screen may need to be installed. If the bar screen area is filled with obnoxious odors, flies, and other insects, it may be necessary to dispose of screenings more frequently.

Mechanically Cleaned Screens

Mechanically cleaned screens use a mechanized rake assembly to collect the solids and move them (carry them) out of the wastewater flow for discharge to a storage hopper. The screen may be continuously cleaned or cleaned on a time- or flow-controlled cycle. As with the manually cleaned screen, the area surrounding the mechanically operated screen must be cleaned frequently to prevent the buildup of materials that can cause unsafe conditions. As with all mechanical equipment, operator vigilance is required to ensure proper operation and that proper maintenance is performed. Maintenance includes lubricating equipment and maintaining it in accordance with the manufacturer's recommendations or the plant's O&M Manual (Operations & Maintenance Manual). Screenings from mechanically operated bar screens are disposed of in the same manner as screenings from manually operated screens: landfill disposal, incineration, or grinding into smaller particles for return to the wastewater flow.

Operational Problems

Many of the operational problems associated with mechanically cleaned bar screens are the same as those for manual screens: septic wastes entering the primary; surge flows after cleaning; excessive grit in the bar screen channel; and/or the screen clogging frequently. The same corrective actions employed for manually operated screens would be applied to these problems in mechanically operated screens. In addition to these problems, however, mechanically operated screens also have other problems, including the cleaner not operating at all and the rake not operating while the motor does. These are mechanical problems that could be caused by a jammed cleaning mechanism, broken chain, a broken cable, or a broken shear pin. Authorized and fully trained maintenance operators should be called in to handle these types of problems.

Screening Safety

The screening area is the first location where the operator is exposed to the wastewater flow. Any toxic, flammable, or explosive gases present in the wastewater can be released at this point. Operators who frequent enclosed bar screen areas should be equipped with personal air monitors. Adequate ventilation must be provided. It is also important to remember that due to the grease attached to the screenings, this area of the plant can be extremely slippery. Routine cleaning is required to minimize this problem.

Note: Never override safety devices on mechanical equipment. Overrides can result in dangerous conditions, injuries, and major mechanical failure.

Screenings Removal Computations

Operators responsible for screening disposal are typically required to keep a record of the amount of screenings removed from the wastewater flow. To maintain accurate screenings records, the volume of screenings withdrawn must be determined. Two methods are commonly used to calculate the volume of screenings withdrawn.

$$\text{Screenings Removed, ft}^3\text{/day} = \frac{\text{Screenings, ft}^3}{\text{days}} \quad (18.8)$$

$$\text{Screenings Removed, ft}^3\text{/MG} = \frac{\text{Screenings, ft}^3}{\text{Flow, MG}} \quad (18.9)$$

Example 18.9

Problem: A total of 65 gal of screenings are removed from the wastewater flow during a 24-h period. What are the removal of the screening reported as ft³/day?

Solution:
First, convert gallons screenings to ft³:

$$\frac{65 \text{ gal}}{7.48 \text{ gal/ft}^3} = 8.7 \text{ ft}^3 \text{ screenings}$$

Next, calculate screenings removed as ft³/day:

$$\text{Screenings Removed} \left(\text{ft}^3\text{/day}\right) = \frac{8.7 \text{ ft}^3}{1 \text{ day}} = 8.7 \text{ ft}^3\text{/day}$$

Example 18.10

Problem: For 1 week, a total of 310 gal of screenings were removed from the wastewater screens. What is the average screening removal in ft³/day?

Solution:
First, gallons of screenings must be converted to cubic feet of screenings:

$$\frac{310 \text{ gal}}{7.48 \text{ gal/ft}^3} = 41.4 \text{ ft}^3 \text{ screenings}$$

$$\text{Screenings Removed, ft}^3\text{/day} = \frac{41.4 \text{ ft}^3}{7} = 5.9 \text{ ft}^3\text{/day}$$

SHREDDING

As an alternative to screening, *shredding* can be used to reduce solids to a size that can enter the plant without causing mechanical problems or clogging. Shredding processes include comminution (comminute means to cut up) and barminution devices.

Comminution

The *comminutor* is the most common shredding device used in wastewater treatment. In this device, all the wastewater flow passes through the grinder assembly. The grinder consists of a screen or slotted basket, a rotating or oscillating cutter, and a stationary cutter. Solids pass through the screen and are chopped or shredded between the two cutters. The comminutor will not remove solids that are too large to fit through the slots, and it will not remove floating objects. These materials must be removed manually. Maintenance requirements for comminutors include aligning, sharpening, and replacing cutters, as well as corrective and preventive maintenance performed in accordance with the plant's O&M Manual.

Operational Problems

Common operational problems associated with comminutors include output containing coarse solids. When this occurs, it is usually a sign that the cutters are dull or misaligned. If the system does not operate at all, the unit is either clogged, jammed, has a broken shear pin or coupling, or the electrical power is shut off. If the unit stalls or jams frequently, this usually indicates cutter misalignment, excessive debris in the influent, or dull cutters.

Note: Only qualified maintenance operators should perform maintenance on shredding equipment.

Barminution

In barminution, the *barminutor* uses a bar screen to collect solids, which are then shredded and passed through the bar screen for removal in a later process. In operation, each device's cutter alignment and sharpness are critical factors in effective operation. Cutters must be sharpened or replaced, and alignment must be checked in accordance with the manufacturer's recommendations. Solids that are not shredded, must be removed daily, stored in closed containers, and disposed of by burial or incineration. Barminutor operational problems are similar to those listed above for comminutors. Preventive and corrective maintenance, as well as lubrication must be performed by qualified personnel and in accordance with the plant's O&M Manual. Because of higher maintenance requirements, the barminutor is less frequently used.

GRIT REMOVAL

The purpose of *grit removal* is to remove the heavy inorganic solids that could cause excessive mechanical wear. Grit is heavier than inorganic solids and includes sand, gravel, clay, egg shells, coffee grounds, metal filings, seeds, and other similar materials. Several processes or devices are used for grit removal. All of the processes are based on the fact that grit is heavier than organic solids, which should be kept in suspension for treatment in the following processes. Grit removal may be accomplished in grit chambers or by the centrifugal separation of sludge. These processes use gravity/velocity, aeration, or centrifugal force to separate the solids from the wastewater.

Gravity/Velocity-Controlled Grit Removal

Gravity/velocity-controlled grit removal is normally accomplished in a channel or tank where the speed or velocity of the wastewater is controlled to about 1 fps (ideal), allowing grit to settle while the organic matter remains suspended. As long as the velocity is controlled in the range of 0.7–1.4 fps, grit removal will remain effective. Velocity is controlled by the amount of water flowing through the channel, the depth of the water in the channel, the width of the channel, or the cumulative width of channels in service.

Process Control Calculations

The velocity of the flow in a channel can be determined either by the float and stopwatch method or by channel dimensions.

Example 18.11 Velocity by Float and Stopwatch

$$\text{Velocity, fps} = \frac{\text{Distance Traveled, (ft)}}{\text{Time Required, (sec)}} \quad (18.10)$$

Problem: A float takes 25 sec to travel 34 ft in a grit channel. What is the velocity of the flow in the channel?

Solution:

$$\text{Velocity, fps} = \frac{34 \text{ ft}}{25 \text{ sec}} = 1.4 \text{ fps}$$

Example 18.12 Velocity by Flow and Channel Dimensions

This calculation can be used for a single channel or tank or for multiple channels or tanks with the same dimensions and equal flow. If the flows through each unit of the same dimensions are unequal, the velocity for each channel or tank must be computed individually.

$$\text{Velocity, fps} = \frac{\text{Flow, MGD} \times 1.55 \text{ cfs/MGD}}{\text{\#Chan. in Ser.} \times \text{Chan Width, ft} \times \text{Water D, ft}} \quad (18.11)$$

Problem: The plant is currently using two grit channels. Each channel is 3 ft wide and has a water depth of 1.2 ft. What is the velocity when the influent flow rate is 3.0 MGD?

Solution:

$$\text{Velocity, fps} = \frac{3.0 \text{ MGD} \times 1.55 \text{ cfs/MGD}}{2 \text{ Channels} \times 3 \text{ ft} \times 1.2 \text{ ft}}$$

$$\text{Velocity, fps} = \frac{4.65 \text{ cfs}}{7.2 \text{ ft}^2} = 0.65 \text{ fps}$$

Note: The channel dimensions must always be in feet. Convert inches to feet by dividing by 12 in/ft.

Example 18.13 Required Settling Time

This calculation can be used to determine the time required for a particle to travel from the surface of the liquid to the bottom at a given settling velocity. To compute the settling time, the settling velocity in fps must be provided or determined experimentally in a laboratory.

$$\text{Settling Time (sec)} = \frac{\text{Liquid Depth in (ft)}}{\text{Settling, Velocity, fps}} \quad (18.12)$$

Problem: The plant's grit channel is designed to remove sand, which has a settling velocity of 0.085 fps. The channel is currently operating at a depth of 2.2 ft. How many seconds will it take for a sand particle to reach the channel bottom?

Solution:

$$\text{Settling Time, sec} = \frac{2.2 \text{ ft}}{0.085 \text{ fps}} = 25.9 \text{ sec}$$

Example 18.14 Required Channel Length

This calculation can be used to determine the length of the channel required to remove an object with a specified settling velocity.

$$\text{Required Channel Length} = \frac{\text{Channel Depth, ft} \times \text{Flow Velocity, fps}}{\text{Settling Velocity, fps}} \quad (18.13)$$

Problem: The plant's grit channel is designed to remove sand, which has a settling velocity of 0.070 fps. The channel is currently operating at a depth of 3 ft. The calculated velocity of flow through the channel is 0.80 fps. The channel is 35 ft long. Is the channel long enough to remove the desired sand particle size?

Solution:

$$\text{Required Channel Length, ft} = \frac{3 \text{ ft} \times 0.80 \text{ fps}}{0.070 \text{ fps}} = 34.3 \text{ ft}$$

Yes, the channel is long enough to ensure all of the sand will be removed.

Cleaning

Gravity-type systems may be manually or mechanically cleaned. Manual cleaning normally requires that the channel be taken out of service, drained, and manually cleaned. Mechanical cleaning systems operate continuously or on a time cycle. Removal should be frequent enough to prevent grit carryover into the rest of the plant.

Note: Before and during cleaning activities, always ventilate the area thoroughly.

Operational Observations/Problems/Troubleshooting

Gravity/velocity-controlled grit removal normally occurs in a channel or tank where the speed or velocity of the wastewater is controlled to about 1 fps (ideal), allowing grit to settle while organic matter remains suspended. As long as the velocity is controlled in the range of 0.7–1.4 fps, grit removal remains effective. Velocity is controlled by the amount of water flowing through the channel, the depth of the water in the channel, the width of the channel, or the cumulative width of channels in service. During operation, the operator must pay particular attention to grit characteristics, evidence of organic solids in the channel, signs of grit carryover into the plant, mechanical problems, and grit storage and disposal (housekeeping).

Aeration

Aerated grit removal systems use aeration to keep lighter organic solids in suspension while allowing heavier grit particles to settle out. Aerated grit removal may be manually or mechanically cleaned; however, the majority of systems are mechanically cleaned. During normal operation, adjusting the aeration rate produces the desired separation. This requires observation of mixing and aeration, as well as a sampling of fixed suspended solids. Actual grit removal is controlled by the rate of aeration. If the rate is too high, all solids remain in suspension. If the rate is too low, both grit and organics will settle out. The operator observes the same types of conditions as those listed for the gravity/velocity-controlled system but must also pay close attention to the air distribution system to ensure proper operation.

Centrifugal Force

The *cyclone degritter* uses a rapid spinning motion (centrifugal force) to separate the heavy inorganic solids or grit from the light organic solids. This unit process is normally used on primary sludge rather than the entire wastewater flow. The critical control factor for the process is the inlet pressure. If the pressure exceeds the recommendations of the manufacturer, the unit will flood, and grit will carry through with the flow. Grit is separated from the flow, washed, and discharged directly to a storage container. Grit removal performance is determined by calculating the percent removal for inorganic (fixed) suspended solids. The operator observes the same kinds of conditions listed for the gravity/velocity-controlled and aerated grit removal systems, except the air distribution system. Typical problems associated with grit removal include mechanical malfunctions and rotten egg odor in the grit chamber (hydrogen sulfide formation), which can lead to metal and concrete corrosion problems. A low recovery rate of grit is another typical problem. Bottom scour, over-aeration, or not enough detention time normally causes this. When these problems occur, the operator must make the required adjustments or repairs to correct the problem.

Grit Removal Calculations

Wastewater systems typically average 1–15 ft^3 of grit per million gallons of flow (sanitary systems: 1–4 ft^3/MG; combined wastewater systems average from 4 to 15 ft^3/MG of flow), with higher ranges during storm events. Generally, grit is disposed of in sanitary landfills. Because of this practice, for planning purposes, operators must keep accurate records of grit removal. Most often, the data is reported as cubic feet of grit removed per million gallons of flow:

$$\text{Cubic Removed, ft}^3/\text{MG} = \frac{\text{Grit Volume, ft}^3}{\text{Flow, MG}} \quad (18.14)$$

Over a given period, the average grit removal rate at a plant (at least a seasonal average) can be determined and used for planning purposes. Typically, grit removal is calculated in cubic yards because excavation is normally expressed in terms of cubic yards.

$$\text{Grit}\left(\text{yd}^3\right) = \frac{\text{Total Grit}\left(\text{ft}^3\right)}{27\ \text{ft}^3/\text{yd}^3} \quad (18.15)$$

Example 18.15

Problem: A treatment plant removes 10 ft^3 of grit in 1 day. How many cubic feet of grit are removed per million gallons if the plant flow is 9 MGD?

Solution:

$$\text{Grit Removed, ft}^3/\text{MG} = \frac{\text{Grit Volume, ft}^3}{\text{Flow, MG}}$$

$$= \frac{10\ \text{ft}^3}{9\ \text{MG}} = 1.1\ \text{ft}^3/\text{MG}$$

Example 18.16

Problem: The total daily grit removed for a plant is 250 gal. If the plant flow is 12.2 MGD, how many cubic feet of grit are removed per MG flow?

Solution:

First, convert the gallon grit removed to ft^3:

$$\frac{250\ \text{gal}}{7.48\ \text{gal/ft}^3} = 33\ \text{ft}^3$$

Next, complete the calculation of ft^3/MG:

$$\text{Grit Removal, ft}^3/\text{MG} = \frac{\text{Grit Vol., ft}^3}{\text{Flow, MG}}$$

$$= \frac{33\ \text{ft}^3}{12.2\ \text{MGD}} = 2.7\ \text{ft}^3/\text{MGD}$$

Example 18.17

Problem: The monthly average grit removal is 2.5 ft^3/MG. If the monthly average flow is 2,500,000 gpd, how many cubic yards must be available for the grit disposal pit to have a 90-day capacity?

Solution:
First, calculate the grit generated each day:

$$\frac{(2.5\ \text{ft}^3)}{\text{MG}}(2.5\ \text{MGD}) = 6.25\ \text{ft}^3\text{each day}$$

The ft³ grit generated for 90 days would be

$$\frac{(6.25\ \text{ft}^3)}{\text{day}}(90\ \text{days}) = 562.5\ \text{ft}$$

Convert ft³ grit to yd³ grit:

$$\frac{562.5\ \text{ft}^3}{27\ \text{ft}^3/\text{yd}^3} = 21\ \text{yd}^3$$

PREAERATION

In the *preaeration process* (diffused or mechanical), we aerate wastewater to achieve and maintain an aerobic state (to freshen septic wastes), strip off hydrogen sulfide (to reduce odors and corrosion), agitate solids (to release trapped gases and improve solids separation and settling), and reduce BOD_5. All of this can be accomplished by aerating the wastewater for 10–30 min. To reduce BOD_5, preaeration must be conducted for 45 to 60 min.

Operational Observations/Problems/Troubleshooting

In preaeration grit removal systems, the operator is concerned with maintaining proper operation and must be alert to any possible mechanical problems. In addition, the operator monitors dissolved oxygen levels and the impact of preaeration on the influent.

CHEMICAL ADDITION

Chemical addition to the waste stream is done (either via dry chemical metering or solution feed metering) to improve settling, reduce odors, neutralize acids or bases, reduce corrosion, reduce BOD_5, improve solids and grease removal, reduce loading on the plant, add or remove nutrients, add organisms, and/or aid subsequent downstream processes. The particular chemical and amount used depend on the desired result. Chemicals must be added at a point where sufficient mixing will occur to obtain maximum benefit. Chemicals typically used in wastewater treatment include chlorine, peroxide, acids and bases, mineral salts (ferric chloride, alum, etc.), and bioadditives and enzymes.

Operational Observations/Problems/Troubleshooting

When adding chemicals to the waste stream to remove grit, the operator monitors the process for evidence of mechanical problems and takes proper corrective actions when necessary. The operator also monitors the current chemical feed rate and dosage. The operator ensures that mixing at the point of addition is accomplished in accordance with Standard Operating Procedures and monitors the impact of chemical addition on the influent.

EQUALIZATION

The purpose of *flow equalization* (whether by surge, diurnal, or complete methods) is to reduce or remove the wide swings in flow rates normally associated with wastewater treatment plant loading; it minimizes the impact of storm flows. The process can be designed to prevent flows above maximum plant design hydraulic capacity; reduce the magnitude of diurnal flow variations; and eliminate flow variations. Flow equalization is accomplished using mixing or aeration equipment, pumps, and flow measurement. Normal operation depends on the purpose and requirements of the flow equalization system. Equalized flows allow the plant to perform at optimum levels by providing stable hydraulic and organic loading. The downside to flow equalization is in additional costs associated with the construction and operation of the flow equalization facilities.

During normal operations, the operator must monitor all mechanical systems involved with flow equalization, watch for mechanical problems, and take the appropriate corrective action. The operator also monitors dissolved oxygen levels, the impact of equalization on the influent, and water levels in equalization basins, making necessary adjustments.

AERATED SYSTEMS

Aerated grit removal systems use aeration to keep the lighter organic solids in suspension while allowing the heavier grit particles to settle out. Aerated grit removal may be manually or mechanically cleaned; however, the majority of the systems are mechanically cleaned. In normal operation, the aeration rate is adjusted to produce the desired separation, which requires observation of mixing, aeration, and sampling of fixed suspended solids. Actual grit removal is controlled by the rate of aeration. If the rate is too high, all of the solids remain in suspension. If the rate is too low, both the grit and the organics will settle out.

CYCLONE DEGRITTER

The *cyclone degritter* uses a rapid spinning motion (centrifugal force) to separate the heavy inorganic solids or grit from the light organic solids. This unit process is normally used on primary sludge rather than the entire wastewater flow. The critical control factor for the process is the inlet pressure. If the pressure exceeds the recommendations of the manufacturer, the unit will flood and grit will carry through with the flow. Grit is separated from the flow and discharged directly to a storage container. Grit removal performance is determined by calculating the percent removal of inorganic (fixed) suspended solids.

PRELIMINARY TREATMENT SAMPLING AND TESTING

During the normal operation of grit removal systems (with the exception of the screening and shredding processes), the plant operator is responsible for sampling and testing as shown in Table 18.3.

TABLE 18.3

Sampling and Testing Grit Removal Systems

Process	Location	Test	Frequency
Grit removal (velocity)	Influent	Suspended solids (fixed)	Variable
	Channel	Depth of grit	Variable
	Grit	Total solids—fixed	Variable
	Effluent	Suspended solids (fixed)	Variable
Grit removal (aerated)	Influent	Suspended solids (fixed)	Variable
	Channel	Dissolved oxygen	Variable
	Grit	Total solids—fixed	Variable
	Effluent	Suspended solids (fixed)	Variable
Chemical addition	Influent	Jar test	Variable
Preaeration	Influent	Dissolved oxygen	Variable
	Effluent	Dissolved oxygen	Variable
Equalization	Effluent	Dissolved oxygen	Variable

OTHER PRELIMINARY TREATMENT PROCESS CONTROL CALCULATIONS

The desired velocity in sewers is approximately 2 fps at peak flow because this velocity normally prevents solids from settling from the lines. However, when the flow reaches the grit channel, the velocity should decrease to about 1 fps to permit the heavy inorganic solids to settle. In the example calculations that follow, we describe how the velocity of the flow in a channel can be determined by the float and stopwatch method and channel dimensions.

Example 18.18 Velocity by Float and Stopwatch

$$\text{Velocity, fps} = \frac{\text{Distance Traveled (ft)}}{\text{Time required (sec)}} \quad (18.16)$$

Problem: A float takes 30 sec to travel 37 ft in a grit channel. What is the velocity of the flow in the channel?

Solution:

$$\text{Velocity, fps} = \frac{37\,\text{ft}}{30\,\text{sec}} = 1.2\,\text{fps}$$

Example 18.19 Velocity by Flow and Channel Dimensions

This calculation can be used for a single channel or tank or multiple channels or tanks with the same dimensions and equal flow. If the flow through each of the unit dimensions is unequal, the velocity for each channel or tank must be computed individually.

$$\text{Velocity, fps} = \frac{\text{Flow, MGD} \times 1.55\,\text{cfs/MGD}}{\substack{\text{\# Chan in Ser} \times \text{Chan Width,} \\ \text{ft} \times \text{Water Depth, ft}}} \quad (18.17)$$

Problem: The plant is currently using two grit channels. Each channel is 3 ft wide and has a water depth of 1.3 ft. What is the velocity when the influent flow rate is 4.0 MGD?

Solution:

$$\text{Velocity, fps} = \frac{4.0\,\text{MGD} \times 1.55\,\text{cfs/MGD}}{2\,\text{Channels} \times 3\,\text{ft} \times 1.3\,\text{ft}}$$

$$\text{Velocity, fps} = \frac{6.2\,\text{cfs}}{7.8\,\text{ft}^2} = 0.79\,\text{fps}$$

Note: Because 0.79 is within the 0.7–1.4 level, the operator of this unit would not make any adjustments.

Note: The channel dimensions must always be in feet. Convert inches to feet by dividing by 12 in/ft.

Example 18.20 Required Settling Time

This calculation can be used to determine the time required for a particle to travel from the surface of the liquid to the bottom at a given settling velocity. To compute the settling time, the settling velocity in fps must be provided or determined experimentally in a laboratory.

$$\text{Settling Time, sec} = \frac{\text{Liquid Depth, ft}}{\text{Settling, Velocity, fps}} \quad (18.18)$$

Problem: The plant's grit channel is designed to remove sand, which has a settling velocity of 0.080 fps. The channel is currently operating at a depth of 2.3 ft. How many seconds will it take for a sand particle to reach the channel bottom?

Solution:

$$\text{Settling Time, sec} = \frac{2.3 \text{ ft}}{0.080 \text{ fps}} = 28.7 \text{ sec}$$

Example 18.21 Required Channel Length

This calculation can be used to determine the length of the channel required to remove an object with a specified settling velocity.

$$\text{Required Channel Length} = \frac{\begin{array}{c}\text{Channel Depth, ft} \\ \times \text{Flow Velocity, fps}\end{array}}{0.080 \text{ fps}}$$

$$(18.19)$$

Problem: The plant's grit channel is designed to remove sand, which has a settling velocity of 0.080 fps. The channel is currently operating at a depth of 3 ft. The calculated velocity of flow through the channel is 0.85 fps. The channel is 36 ft long. Is the channel long enough to remove the desired sand particle size?

Solution:

$$\text{Required Channel Length} = \frac{3 \text{ ft} \times 0.85 \text{ fps}}{0.080 \text{ fps}} = 31.9 \text{ ft}$$

Yes, the channel is long enough to ensure all the sand will be removed.

Caution: Before and during cleaning activities, always ventilate thoroughly.

PRIMARY TREATMENT (SEDIMENTATION)

The purpose of primary treatment (primary sedimentation or primary clarification) is to remove settleable organic and floatable solids. Normally, each primary clarification unit can be expected to remove 90%–95% settleable solids, 40%–60% total suspended solids, and 25%–35% BOD_5.

Note: Performance expectations for settling devices used in other areas of plant operation are normally expressed as overall unit performance rather than settling unit performance.

Sedimentation may be used throughout the plant to remove settleable and floatable solids. It is used in primary, and secondary, as well as advanced wastewater treatment processes. In this section, we focus on primary treatment or primary clarification, which uses large basins in which primary settling is achieved under relatively quiescent conditions (see Figure 18.1). Within these basins, mechanical scrapers collect the primary settled solids into a hopper, from which they are pumped to a sludge-processing area. Oil, grease, and other floating materials (scum) are skimmed from the surface. The effluent is discharged over weirs into a collection trough.

PROCESS DESCRIPTION

In primary sedimentation, wastewater enters a settling tank or basin. Velocity is reduced to approximately 1 ft/min.

Note: Notice that the velocity is based on minutes instead of seconds, as was the case in the grit channels. A grit channel velocity of 1 ft/sec would be 60 ft/min.

Solids that are heavier than water settle to the bottom, while solids that are lighter than water float to the top. Settled solids are removed as sludge, and floating solids are removed as scum. Wastewater leaves the sedimentation tank over an effluent weir and proceeds to the next step in treatment. Detention time, temperature, tank design, and the condition of the equipment control the efficiency of the process.

Overview of Primary Treatment

- Primary treatment reduces the organic loading on downstream treatment processes by removing a large amount of settleable, suspended, and floatable materials.
- Primary treatment reduces the velocity of the wastewater through a clarifier to approximately 1–2 ft/min, so settling and floatation can occur. Slowing the flow enhances the removal of suspended solids in wastewater.
- Primary settling tanks remove floated grease and scum, remove the settled sludge solids, and collect them for pumping to disposal or further treatment.
- Clarifiers may be rectangular or circular. In rectangular clarifiers, wastewater flows from one end to the other, and settled sludge is moved to a hopper at one end, either by flights set on parallel chains or by a single bottom scraper set on a traveling bridge. Floating material (mostly grease and oil) is collected by a surface skimmer.
- In circular tanks, the wastewater usually enters the middle and flows outward. Settled sludge is pushed to a hopper in the middle of the tank bottom, and a surface skimmer removes floating material.
- Factors affecting primary clarifier performance include:
  - Rate of flow through the clarifier
  - Wastewater characteristics (strength; temperature; amount and type of industrial waste; and the density, size, and shapes of particles)
  - Performance of pretreatment processes
  - Nature and amount of any wastes recycled to the primary clarifier

Key factors in primary clarifier operation include the following concepts:

$$\text{Retention time (h)} = \frac{(\text{Vol., gal})(24 \text{ h/day})}{\text{Flow, gal/day}}$$

$$\text{Surface loading rate } \left(\text{gal/day/ft}^2\right) = \frac{Q \text{ (gal/day)}}{\text{Surface area } \left(\text{ft}^2\right)}$$

$$\text{Solids Loading Rate } \left(\text{lb/day/ft}^2\right) = \frac{\text{Solids into Clarifier (lb/day)}}{\text{Surface Area } \left(\text{ft}^2\right)}$$

$$\text{Weir overflow rate } \left(\text{gal/day/lineal ft}\right) = \frac{Q \text{ (gal/day)}}{\text{Weir length (lineal ft)}}$$

TYPES OF SEDIMENTATION TANKS

Sedimentation equipment includes septic tanks, two-story tanks, and plain settling tanks or clarifiers. All three devices may be used for primary treatment while plain settling tanks are normally used for secondary or advanced wastewater treatment processes.

Septic Tanks

Septic tanks are prefabricated tanks that serve as a combined settling and skimming tank and an unheated-unmixed anaerobic digester. They provide long settling times (6–8 h or more) but do not separate decomposing solids from the wastewater flow. When the tank becomes full, solids will be discharged with the flow. The process is suitable for small facilities (i.e., schools, motels, homes, etc.) but, due to the long detention times and lack of control, it is unsuitable for larger applications.

Two Story (Imhoff) Tank

The *two-story or Imhoff tank*, named after German engineer Karl Imhoff (1876–1965), is similar to a septic tank in the removal of settleable solids and the anaerobic digestion of solids. The difference is that the two-story tank consists of a settling compartment where sedimentation is accomplished, a lower compartment where settled solids and digestion take place, and gas vents. Solids are removed from the wastewater by settling a pass from the settling compartment into the digestion compartment through a slot at the bottom of the settling compartment. The design of the slot prevents solids from returning to the settling compartment. Solids decompose anaerobically in the digestion section. Gases produced as a result of the solids decomposition are released through the gas vents running along each side of the settling compartment.

Plain Settling Tanks (Clarifiers)

The plain settling tank or clarifier optimizes the settling process. Sludge is removed from the tank for processing in other downstream treatment units. Flow enters the tank, is slowed and distributed evenly across the width and depth of the unit, passes through the unit, and leaves over the effluent weir. Detention time within the primary settling tank is from 1 to 3 h (2-h average). Sludge removal is accomplished frequently on either a continuous or intermittent basis. Continuous removal requires additional sludge treatment processes to remove the excess water resulting from the removal of sludge which contains less than 2%–3% solids. Intermittent sludge removal requires the sludge to be pumped from the tank on a schedule frequent enough to prevent large clumps of solids from rising to the surface but infrequent to obtain 4%–8% solids in the sludge withdrawn.

Scum must be removed from the surface of the settling tank frequently. This is normally a mechanical process but may require manual start-up. The system should be operated frequently enough to prevent excessive buildup and scum carryover but not so frequently as to cause hydraulic overloading of the scum removal system. Settling tanks require housekeeping and maintenance. Baffles (prevent floatable solids, and scum, from leaving the tank), scum troughs, scum collectors, effluent troughs and effluent weirs require frequent cleaning to prevent heavy biological growths and solids accumulations. Mechanical equipment must be lubricated and maintained as specified in the manufacturer's recommendations or in accordance with procedures listed in the plant O&O&M manual.

Process control sampling and testing are used to evaluate the performance of the settling process. Settleable solids, dissolved oxygen, pH, temperature, total suspended solids, and BOD_5, as well as sludge solids and volatile matter testing, are routinely carried out.

OPERATOR OBSERVATIONS, PROCESS PROBLEMS & TROUBLESHOOTING

Before identifying a primary treatment problem and proceeding with appropriate troubleshooting efforts, the operator must be cognizant of what constitutes "normal" operation (i.e., is there a problem, or is the system operating as per design?). Several important items of normal operation can have a strong impact on performance. The following section discusses the important operational parameters and normal observations.

Primary Clarification: Normal Operation

In primary clarification, wastewater enters a settling tank or basin. Velocity reduces to approximately 1 ft/min.

Note: Notice that the velocity is based on minutes instead of seconds, as was the case in the grit channels. A grit channel velocity of 1 ft/sec would be 60 ft/min.

Solids that are heavier than water settle to the bottom, while solids that are lighter than water float to the top. Settled solids are removed as sludge, and floating solids are removed as scum. Wastewater leaves the sedimentation tank over an effluent weir and proceeds to the next step in treatment. Detention time, temperature, tank design, and the condition of the equipment control the efficiency of the process.

Primary Clarification: Operational Parameters (Normal Observations)

- **Flow Distribution:** Normal flow distribution is indicated by flow to each in-service unit being equal and uniform. There is no indication of short-circuiting. The surface-loading rate is within design specifications.
- **Weir Condition:** Weirs are level; the flow over the weir is uniform; and the weir overflow rate is within design specifications.
- **Scum Removal:** The surface is free of scum accumulations; the scum removal does not operate continuously.
- **Sludge Removal:** No large clumps of sludge appear on the surface; the system operates as designed; the pumping rate is controlled to prevent coning or buildup; and the sludge blanket depth is within desired levels.
- **Performance:** The unit is removing expected levels of BOD_5, TSS, and settleable solids.
- **Unit Maintenance:** Mechanical equipment is maintained in accordance with planned schedules; equipment is available for service as required.

To assist the operator in judging primary treatment operation, several process control tests can be used for process evaluation and control. These tests include the following:

- pH—normal range: 6.5–9.0
- Dissolved oxygen—normal range: <1.0 mg/L
- Temperature—varies with climate and season
- Settleable solids, mL/L—influent: 5–15 mL/L

    –effluent: 0.3–5 mL/L

- BOD_5, mg/L—influent: 150–400 mg/L

    –effluent: 50–150 mg/L

- % solids—4%–8%
- % volatile matter—40%–70%
- Heavy metals—as required
- Jar tests—as required

Note: Testing frequency should be determined based on the process influent, effluent variability, and the available resources. All should be performed periodically to provide reference information for evaluation of performance.

Process Control Calculations

As with many other wastewater treatment plant unit processes, process control calculations aid in determining the performance of the sedimentation process. Process control calculations are used in the sedimentation process to determine

- Percent removal
- Hydraulic detention time
- Surface loading rate (surface settling rate)
- Weir overflow rate (weir loading rate)
- Sludge pumping
- Percent total solids (% TS)

In the following sections, we take a closer look at a few of these process control calculations and example problems.

Note: The calculations presented in the following sections allow you to determine values for each function performed. Keep in mind that an optimally operated primary clarifier should have values in an expected range.

Percent Removal

The expected range of % removal for a primary clarifier is

- Settleable solids   90%–95%
- Suspended solids   40%–60%
- BOD_5 25%–35%

Detention Time

The primary purpose of primary settling is to remove settleable solids. This is accomplished by slowing the flow down to approximately 1 ft/min. At this velocity, the flow will stay in the primary tank for 1.5 to 2.5 h. The length of time the water stays in the tank is called the hydraulic detention time.

Surface Loading Rate (Surface Settling Rate/Surface Overflow Rate)

Surface loading rate is the number of gallons of wastewater passing over 1 ft² of tank per day. This can be used to compare actual conditions with design. Plant designs generally use a surface-loading rate of 300–1,200 gal/day/ft². Other terms used synonymously with surface loading rate include *surface overflow rate* and *surface settling rate*.

$$\text{Surface Settling Rate, gpd/ft}^2 = \frac{\text{Flow, gal/day}}{\text{Settling Tank Area, ft}^2}$$

(18.20)

Example 18.22

Problem: The settling tank is 120 ft in diameter, and the flow to the unit is 4.5 MGD. What is the surface loading rate in gal/day/ft²?

Solution:

$$\text{Surface Loading Rate} = \frac{4.5\,\text{MGD} \times 1,000,000\,\text{gal/MGD}}{0.785 \times 120\,\text{ft} \times 120\,\text{ft}}$$

$$= 398\,\text{gpd/ft}^2$$

Example 18.23

Problem: A circular clarifier has a diameter of 50 ft. If the primary effluent flow is 2,150,000 gpd, what is the surface overflow rate in gpd/ft²?

Solution:

$$\text{Area} = (0.785)(50 \text{ ft})(50 \text{ ft})$$

$$\text{Surface Overflow Rate} = \frac{\text{Flow, gpd}}{\text{Area }\left(\text{ft}^2\right)}$$

$$= \frac{2,150,000}{(0.785)(50 \text{ ft})(50 \text{ ft})}$$

$$= 1,096 \text{ gpd/ft}^2$$

Weir Overflow Rate (Weir Loading Rate)

The weir overflow rate or weir loading rate is the amount of water leaving the settling tank per linear foot of weir. The result of this calculation can be compared with design. Normally, weir overflow rates of 10,000–20,000 gal/day/ft are used in the design of a settling tank.

$$\text{Weir Overflow Rate, gpd/ft} = \frac{\text{Flow, gal/day}}{\text{Weir Length, ft}} \quad (18.21)$$

Example 18.24

Problem: The circular settling tank is 90 ft in diameter and has a weir along its circumference. The effluent flow rate is 2.55 MGD. What is the weir overflow rate (gpd/ft)?

Solution:

$$\text{Weir Overflow, gpd/ft} = \frac{2.55 \text{ MGD} \times 1,000,000 \text{ gal/MG}}{3.14 \times 90 \text{ ft}}$$

$$= 9,023 \text{ gpd/ft}$$

Sludge Pumping

Determination of sludge pumping (the quantity of solids and volatile solids removed from the sedimentation tank) provides accurate information needed for process control of the sedimentation process.

(1) Sol. Pumped, lb/day = Pump Rate ×

Pump Time × 8.34 lbs/gal × % Solids   (18.22)

(2) Vol. Solids/lb/day = Pump Rate ×

Pump Time × 8.34 × % Solids × % Vol.   (18.23)

Example 18.25

Problem: The sludge pump operates 20 min/h. The pump delivers 20 gal/min of sludge. Laboratory tests

indicate that the sludge is 5.2% solids and 66% volatile matter. How many pounds of volatile matter are transferred from the settling tank to the digester?

Solution:

$$\text{Pump Time} = 20 \text{ min/h}$$

$$\text{Pump Rate} = 20 \text{ gpm}$$

$$\% \text{ Solids} = 5.2\%$$

$$\% \text{ V.M.} = 66\%$$

$$\text{Vol. Solids, lb/day} = 20 \text{ gpm} \times (20 \text{ min/h} \times 24 \text{ h/day})$$

$$\times 8.34 \text{ lbs/gal} \times 0.052 \times 0.66$$

$$= 2,748 \text{ lb/day}$$

Percent Total Solids (% TS)
Example 18.26

Problem: A settling tank sludge sample is tested for solids. The sample and dish weight 74.69 g. The dish alone weighs 21.2 g. After drying, the dish with the dry solids weighs 22.3 g.

What is the percent total solids (% TS) of the sample?

Solution:

Sample + Dish	74.69 g	Dish + Dry Solids	22.3 g
Dish alone	− 21.2 g	Dish Alone	− 21.2 g
Sample Weight	53.49-g	Dry Solids Weight	1.1 g

$$\frac{1.1 \text{ g}}{53.49 \text{ g}} \times 100\% = 2\%$$

BOD and SS Removal

To calculate the pounds of BOD or suspended solids removed each day, you need to know the mg/L BOD or SS removed and the plant flow. Then, you can use the mg/L to lb/day equation.

$$\text{SS Removed} = \text{mg/L} \times \text{MGD} \times 8.34 \text{ lb/gal} \quad (18.24)$$

Example 18.27

Problem: If 120 mg/L suspended solids are removed by a primary clarifier, how many lb/day suspended solids are removed when the flow is 6,230,000 gpd?

Solution:

$$\text{SS Removed} = 120 \text{ mg/L} \times 6.25 \text{ MGD} \times 8.34 \text{ lb/gal}$$

$$= 6,255 \text{ lb/day}$$

Example 18.28

Problem: The flow to a secondary clarifier is 1.6 MGD. If the influent BOD concentration is 200 mg/L, and the effluent BOD concentration is 70 mg/L, how many pounds of BOD are removed daily?

$$\text{lb/day BOD removed} = 200 \text{ mg/L} - 70 \text{ mg/L} = 130 \text{ mg/L}$$

After calculating mg/L BOD removed, calculate lb/day BOD removed:

$$\text{BOD removed, lb/day} = (130 \text{ mg/L})(1.6 \text{ MGD})(8.34 \text{ lb/gal})$$

$$= 1{,}735 \text{ lb/day}$$

PROBLEM ANALYSIS

In primary treatment (as is also clear in the operation of other unit processes), the primary function of the operator is to identify the causes of process malfunctions; develop solutions; and prevent recurrence. In other words, the operator's goal is to perform problem analysis or troubleshooting on unit processes when required and restore the unit processes to optimal operating conditions. The immediate goal in problem analysis is to solve the immediate problem. The long-term goal is to ensure that the problem does not pop up again, causing poor performance in the future. This section covers a few indicators or observations of operational problems with the primary treatment process. The observations presented are not all-inclusive but highlight the most frequently confronted problems.

Causal Factors for Poor Suspended Solids Removal (Primary Clarifier)

- Hydraulic overload
- Sludge buildup in tanks and decreased volume allow solids to scour out tanks
- Strong recycle flows
- Industrial waste concentrations
- Wind currents
- Temperature currents

Causal Factors for Floating Sludge

- Sludge becoming septic in the tank
- Damaged or worn collection equipment
- Recycled waste sludge
- Primary sludge pumps malfunction
- The sludge withdrawal line plugged
- Return of well-nitrified waste-activated sludge
- Too few tanks in service
- Damaged or missing baffles

Causal Factors for Septic Wastewater or Sludge

- Damaged or worn collection equipment
- Infrequent sludge removal
- Insufficient industrial pretreatment

- Septic sewage from the collection system
- Strong recycle flows
- Primary sludge pump malfunction
- The sludge withdrawal line plugged
- Sludge collectors do not run often enough
- Septage dumpers

Causal Factors for Too Low Primary Sludge Solids Concentrations

- Hydraulic overload
- Overpumping of sludge
- Collection system problems
- Decreased influent solids loading

Casual Factors for Too High Primary Sludge Solids Concentrations

- Excessive grit and compacted material
- Primary sludge pump malfunction
- The sludge withdrawal line plugged
- Sludge retention time is too long
- Increased influent loadings

EFFLUENT FROM SETTLING TANKS

Upon completion of screening, degritting, and settling in sedimentation basins, large debris, grit, and many settleable materials have been removed from the waste stream. What is left is referred to as *primary effluent*. Usually cloudy and frequently gray, primary effluent still contains large amounts of dissolved food and other chemicals (nutrients). These nutrients are treated in the next step of the treatment process (secondary treatment), which is discussed in the next section.

Note: Two of the most important nutrients left to remove are phosphorus and ammonia. While we want to remove these two nutrients from the waste stream, we do not want to remove too much. Carbonaceous microorganisms in secondary treatment (biological treatment) need both phosphorus and ammonia.

SECONDARY TREATMENT

The main purpose of *secondary treatment* (sometimes referred to as biological treatment) is to provide biochemical oxygen demand (BOD) removal beyond what is achievable by primary treatment. There are three commonly used approaches, all of which take advantage of the ability of microorganisms to convert organic wastes (via biological treatment), into stabilized, low-energy compounds. Two of these approaches, the *trickling filter* [and/or its variation, the *rotating biological contactor (RBC)*] and the *activated sludge* process, sequentially follow normal primary treatment. The third, *ponds* (oxidation ponds or lagoons), however, can provide equivalent results without preliminary treatment. In this section, we present a brief overview of the secondary treatment process followed by a detailed discussion of wastewater treatment ponds (used primarily in smaller treatment plants), trickling filters, and RBCs. We then shift focus to the activated sludge process—the secondary treatment process, which is used primarily in large installations and is the main focus of the handbook.

Secondary treatment refers to those treatment processes that use biological processes to convert dissolved, suspended, and colloidal organic wastes to more stable solids that can either be removed by settling or discharged to the environment without causing harm. Exactly what is the secondary treatment? As defined by the Clean Water Act (CWA), secondary treatment produces an effluent with no more than 30 mg/L BOD_5 and 30 mg/L total suspended solids.

Note: The CWA also states that ponds and trickling filters will be included in the definition of secondary treatment even if they do not meet the effluent quality requirements continuously.

Most secondary treatment processes decompose solids aerobically producing carbon dioxide, stable solids, and more organisms. Since solids are produced, all of the biological processes must include some form of solids removal (settling tank, filter, etc.). Secondary treatment processes can be separated into two large categories: fixed film systems and suspended growth systems.

Fixed film systems are processes that use a biological growth (biomass or slime) MENTslime. When the wastewater and slime are in contact, the organisms remove and oxidize the organic solids. The media may be stone, redwood, synthetic materials, or any other substance that is durable (capable of withstanding weather conditions for many years), provides a large area for slime growth while providing open space for ventilation, and is not toxic to the organisms in the biomass. Fixed film devices include trickling filters and rotating biological contactors (RBCs). *Suspended growth systems* are processes, which use a biological growth, which is mixed with the wastewater. Typical suspended growth systems consist of various modifications of the activated sludge process.

TREATMENT PONDS

Wastewater treatment can be accomplished using *ponds* (aka, *lagoons*). Ponds are relatively easy to build and manage, accommodate large fluctuations in the flow, and can also provide treatment that approaches conventional systems (producing a highly purified effluent) at a much lower cost. It is the cost (the economics) that drives many managers to decide on the pond option. The actual degree of treatment provided depends on the type and number of ponds used. Ponds can be used as the sole type of treatment, or they can be used in conjunction with other forms of wastewater treatment; that is, other treatment processes followed by a pond or a pond followed by other treatment processes.

Stabilization ponds (also known as treatment ponds) have been used for the treatment of wastewater for over 3,000 years. The first recorded construction of a pond system in the United States was at San Antonio, Texas, in 1901. Today, over 8,000 wastewater treatment ponds are in place, involving more than 50% of the wastewater treatment facilities in the United States (CWNS, 2000). Facultative ponds account for 62%, aerated ponds 25%, anaerobic 0.04%, and total containment 12% of the pond treatment systems. They treat a variety of wastewater from domestic wastewater to complex industrial wastes, and they function under a wide range of weather conditions, from tropical to arctic. Ponds can be used alone or in combination with other wastewater treatment processes. As our understanding of pond operating mechanisms has increased, different types of ponds have been developed for application in specific types of wastewater under local environmental conditions. This handbook focuses on municipal wastewater treatment pond systems.

> **DID YOU KNOW?**
>
> A pond can be judged as an "Attractive Nuisance." The term "Attractive Nuisance" is a legal expression that implies the pond could be attractive to potential users, such as duck hunters, fishermen, or playing children. Since ponds have fairly steep slopes, the potential for someone falling in and drowning is a significant legal problem that must be a concern. It is important that adequate fencing and signing be provided.

While the tendency in the United States has been for smaller communities to build ponds, in other parts of the world, including Australia, New Zealand, Mexico, Latin America, Asia, and Africa, treatment ponds have been built for large cities. As a result, our understanding of the biological, biochemical, physical, and climatic factors that interact to transform the organic compounds, nutrients and pathogenic organisms found in sewage into less harmful chemicals and unviable organisms (i.e., dead or sterile) has grown since 1983. A wealth of experience has been built up as civil, sanitary, or environmental engineers, operators, public works managers and public health and environmental agencies have gained more experience with these systems. While some of this information makes its way into technical journals and textbooks, there is a need for a less formal presentation of the subject for those working in the field every day (USEPA, 2011).

Ponds are designed to enhance the growth of natural ecosystems that are either anaerobic (providing conditions for bacteria that grow in the absence of oxygen (O_2) environments), aerobic (promoting the growth of O_2 producing and/or requiring organs, such as algae and bacteria), or facultative, which is a combination of the two. Ponds are managed to reduce concentrations of biochemical oxygen demand (BOD), TSS, and coliform numbers (fecal or total) to meet water quality requirements.

Types of Ponds

Ponds can be classified based on their location in the system, by the type of waste they receive, and the main biological process occurring in the pond. First, we look at the types of ponds according to their location and the type of waste they receive: Raw Sewage Stabilization ponds (see Figure 18.3), oxidation ponds, and polishing ponds. Then, in the following section, we look at ponds classified by the type of processes occurring within the pond: *Aerobic Ponds, Anaerobic Ponds, Facultative Ponds*, and *Aerated Ponds*.

Ponds Based on Location and Types of Waste They Receive

Raw Sewage Stabilization Pond

The raw sewage stabilization pond is the most common type of pond (see Figure 18.3). With the exception of screening and shredding, this type of pond receives no prior treatment. Generally, raw sewage stabilization ponds are designed to provide a minimum of 45 days detention time and to receive no more than 30 lbs of BOD_5 per day per acre. The quality of the discharge is dependent on the time of the year. Summer months produce high BOD_5 removal but excellent suspended solids removal. The pond consists of an influent structure, pond berm or walls, and an effluent structure designed to permit the selection of the best quality effluent. Normal operating depth of the pond is 3–5 ft. The process occurring in the pond involves bacteria decomposing the organics in the wastewater (aerobically and anaerobically) and algae using the products of the bacterial action to produce oxygen (photosynthesis). Because this type of pond is the most commonly used in wastewater treatment, the process that occurs within the pond is described in greater detail in the following text.

When wastewater enters the stabilization pond several processes begin to occur. These include settling, aerobic decomposition, anaerobic decomposition, and photosynthesis (see Figure 18.3). Solids in the wastewater will settle to the bottom of the pond. In addition to the solids in the wastewater entering the pond, solids, which are produced by the biological activity, will also settle to the bottom. Eventually, this will reduce the detention time and the performance of the pond. When this occurs (20–30 years normal) the pond will have to be replaced or cleaned.

Bacteria and other microorganisms use organic matter as a food source. They use oxygen (aerobic decomposition), organic matter, and nutrients to produce carbon dioxide, water, and stable solids, which may settle out more organisms. The carbon dioxide is an essential component of the photosynthesis process occurring near the surface of the pond. Organisms also use the solids that settle out as food material; however, the oxygen levels at the bottom of the pond are extremely low, so the process used is anaerobic decomposition. The organisms use the organic matter to produce gases (hydrogen sulfide, methane, etc.), which are dissolved in the water, stable solids, and more organisms. Near the surface of the pond, a population of green algae will develop, using the carbon dioxide produced by the bacterial population, along with nutrients and sunlight, to produce more algae and oxygen, which dissolves into the water. The dissolved oxygen is then used by organisms in the aerobic decomposition process.

When compared with other wastewater treatment systems involving biological treatment, a stabilization pond treatment system is the simplest to operate and maintain. Operation and maintenance activities include collecting and testing samples for dissolved oxygen (D.O.), and pH,

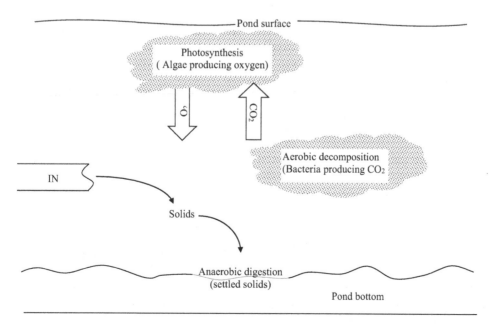

FIGURE 18.3 Stabilization pond processes.

removing weeds and other debris (scum) from the pond, mowing the berms, repairing erosion, and removing burrowing animals.

Note: Dissolved oxygen and pH levels in the pond will vary throughout the day. Normal operation will result in very high D.O. and pH levels due to the natural processes occurring.

Note: When operating properly the stabilization pond will exhibit a wide variation in both dissolved oxygen and pH. This is due to the photosynthesis occurring in the system.

Oxidation Pond

An oxidation pond, which is normally designed using the same criteria as a stabilization pond, receives flows that have passed through a stabilization pond or primary settling tank. This type of pond provides biological treatment, additional settling, and some reduction in the number of fecal coliforms present.

Polishing Pond

A polishing pond, which uses the same equipment as a stabilization pond, receives flow from an oxidation pond or other secondary treatment systems. Polishing ponds remove additional BOD_5, solids, fecal coliforms, and some nutrients. They are designed to provide 1–3 days of detention time and normally operate at a depth of 5–10 ft. Excessive detention time or too shallow a depth can result in algae growth, increasing influent suspended solids concentrations.

Ponds Based on the Type of Processes Occurring Within

The type of processes occurring within the pond may also classify ponds. These include the aerobic, anaerobic, facultative, and aerated processes.

Aerobic Ponds

In aerobic ponds, also known as oxidation ponds or high-rate aerobic ponds, which are not widely used, oxygen is present throughout the pond, and all biological activity involves aerobic decomposition. They are usually 30–45 cm deep, allowing light to penetrate the entire pond. Mixing is often provided to keep algae at the surface, maintaining maximum rates of photosynthesis and O_2 production and preventing algae from settling and producing an anaerobic bottom layer. The rate of photosynthetic production of O_2 may be enhanced by surface re-aeration. O_2 and aerobic bacteria biochemically stabilize the waste. Detention time is typically 2–6 days.

These ponds are appropriate for treatment in warm, sunny climates. They are used where a high degree of BOD_5 removal is desired, but land area is limited. The chief advantage of these ponds is that they produce a stable effluent during short detention times with low land and energy requirements. However, their operation is somewhat more complex than that of facultative ponds and, unless the algae are removed, the effluent will contain high TSS. While the

shallow depths allow penetration of ultraviolet (UV) light that may reduce pathogens, shorter detention times work against effective coliform and parasite die-off. Since they are shallow, bottom paving or veering is usually necessary to prevent aquatic plants from colonizing the ponds. The Advanced Integrated Wastewater Pond System® (AIWPS®) uses the high-rate pond to maximize the growth of microalgae using a low-energy paddle wheel (USEPA, 2011).

Anaerobic Ponds

Anaerobic ponds are normally used to treat high-strength industrial wastes; that is, they receive heavy organic loading, so much so that there is no aerobic zone—no oxygen is present and all biological activity is anaerobic decomposition. They are usually 2.5–4.5 m in depth and have detention times of 5–50 days. The predominant biological treatment reactions are bacterial acid formation and methane fermentation.

Anaerobic ponds are usually used for the treatment of strong industrial and agricultural (food processing) wastes, as a pretreatment step in municipal systems, or where an industry is a significant contributor to a municipal system. The biochemical reactions in an anaerobic pond produce hydrogen sulfide (H_2S) and other odorous compounds. To reduce odors, the common practice is to recirculate water from a downstream facultative or aerated pond. This provides a thin aerobic layer at the surface of the anaerobic pond, which prevents odors from escaping into the air. A cover may also be used to contain odors. The effluent from anaerobic ponds usually requires further treatment prior to discharge (USEPA, 2011).

Facultative Pond

The facultative pond, which may also be called an oxidation or photosynthetic pond, is the most common type of pond (based on the processes occurring). Oxygen is present in the upper portions of the pond and aerobic processes are occurring. No oxygen is present in the lower levels of the pond, where the processes occurring are anoxic and anaerobic. Facultative ponds are usually 0.9–2.4 m deep or deeper, with an aerobic layer overlying an anaerobic layer. Recommended detention times vary from 5 to 50 days in warm climates and 90–180 days in colder climates (NEIWPCC, 1988). Aerobic treatment processes in the upper layer provide odor control, and nutrient and BOD removal. Anaerobic fermentation processes, such as sludge digestion, denitrification, and some BOD removal, occur in the lower layer. The key to the successful operation of this type of pond is O_2 production by photosynthetic algae and/or re-aeration at the surface.

Facultative ponds are used to treat raw municipal wastewater in small communities and for primary or secondary effluent treatment in small or large cities. They are also used in industrial applications, usually in the process line after aerated or anaerobic ponds, to provide additional treatment prior to discharge. Commonly achieved effluent

BOD values, as measured in the BOD_5 test, range from 20 to 60 mg/L, and TSS levels may range from 30 to 150 mg/L. The size of the pond needed to treat BOD loadings depends on specific conditions and regulatory requirements.

Aerated Pond

Facultative ponds overloaded due to unplanned additional sewage volume or higher strength influent from a new industrial connection may be modified by the addition of mechanical aeration. Ponds originally designed for mechanical aeration are generally 2–6 m deep with detention times of 3–10 days. For colder climates, 20–40 days is recommended. Mechanically aerated ponds require less land area but have greater energy requirements. When aeration is used, the depth of the pond and/or the acceptable loading levels may increase. Mechanical or diffused aeration is often used to supplement natural oxygen production or to replace it.

DID YOU KNOW?

A pond should be drawn down in the fall after the first frost, when the algae concentration drops off, the BOD is still low, and the receiving stream temperature is low with accompanying high dissolved oxygen.

A pond should be drawn down in the spring before the algae concentration increases, when the BOD level is acceptable, and when the receiving stream flows are high (low temperature with high dissolved oxygen helps). During the actual discharge, the effluent must be sampled for BOD, suspended solids, and pH at a frequency specified in the discharge permit.

Elements of Pond Processes

The Organisms

Although our understanding of wastewater pond ecology is far from complete, general observations about the interactions of macro- and microorganisms in these biologically driven systems support our ability to design, operate, and maintain them.

Bacteria

In this section, we discuss other types of bacteria found in the pond; these organisms help to decompose complex organic constituents in the influent into simple, non-toxic compounds. Certain pathogenic bacteria and other microbial organisms (viruses, protozoa) associated with human waste enter the system with the influent; the wastewater treatment process is designed so that neither numbers will be reduced adequately to meet public health standards.

- *Aerobic bacteria* are found in the aerobic zone of a wastewater pond and are primarily the same type as those found in an activated sludge process or the Zoogleal mass of a trickling filter. The most frequently isolated bacteria include *Beggiatoa alba*, *Sphaerotilus natans*, *Achromobacter*, *Alcaligenes*, *Flavobacterium*, *Pseudomonas*, and *Zoogloea* spp. (Spellman, 2000; Lynch and Poole, 1979; Pearson, 2005). These organisms decompose the organic materials present in the aerobic zone into oxidized end products.

- *Anaerobic bacteria* are hydrolytic bacteria that convert complex organic material into simple alcohols and acids, primarily amino acids, glucose, fatty acids, and glycerols (Spellman, 2000; Brockett, 1976; Pearson, 2005; Paterson and Curtis, 2005). Acidogenic bacteria convert the sugars and amino acids into acetate, ammonia (NH_3), hydrogen (H), and carbon dioxide (CO_2). Methanogenic bacteria break down these products further into methane (CH_4) and CO_2 (Gallert and Winter, 2005).

- *Cyanobacteria*, formerly classified as blue-green algae, are autotrophic organisms that are able to synthesize organic compounds using CO_2 as the major carbon source. Cyanobacteria produce O_2 as a by-product of photosynthesis, providing an O_2 source for other organisms in the ponds. They are found in very large numbers as blooms when environmental conditions are suitable (Gaudy and Gaudy, 1980). Commonly encountered cyanobacteria include *Oscillatoria*, *Arthrospira*, *Spirulina*, and *Microcystis* (Vasconcelos and Pereira, 2001; Spellman, 2000).

- *Purple sulfur bacteria* (Chromatiaceae) may grow in any aquatic environment to which light of the required wavelength penetrates, provided that CO_2, nitrogen (N), and a reduced form of sulfur (S) or H are available. Purple sulfur bacteria occupy the anaerobic layer below the algae, cyanobacteria, and other aerobic bacteria in a pond. They are commonly found at a specific depth in a thin layer where light and nutrient conditions are at an optimum (Gaudy and Gaudy, 1980; Pearson, 2005). Their biochemical conversion of odorous sulfide compounds to elemental S or sulfate (SO_4) helps to control odor in facultative and anaerobic ponds.

Algae

Algae constitute a group of aquatic organisms that may be unicellular or multicellular, motile or immotile, and, depending on the phylogenetic family, have different combinations of photosynthetic pigments. As autotrophs, algae need only inorganic nutrients, such as N, phosphorus (P), and a suite of microelements, to fix CO_2 and grow in the presence of sunlight. Algae do not fix atmospheric N; they require an external source of inorganic N in the form of nitrate (NO_3) or NH_3. Some algal species can use amino acids and other organic N compounds. Oxygen is a by-product of these reactions.

Algae are generally divided into three major groups, based on the color reflected from the cells by the chlorophyll and other pigments involved in photosynthesis. Green and brown algae are common in wastewater ponds; red algae occur infrequently. The algal species that is dominant at any particular time is thought to be primarily a function of temperature, although the effects of predation, nutrient availability, and toxins are also important.

Green algae (Chlorophyta) include unicellular, filamentous, and colonial forms. Some green algal genera commonly found in facultative and aerobic ponds are *Euglena, Phacus, Chlamydomonas, Ankistrodesmus, Chlorella, Micractinium, Scenedesmus, Selenastrum, Dictyosphaerium,* and *Volvox.*

Chrysophytes, or brown algae, are unicellular and may be flagellated, and they include the diatoms. Certain brown algae are responsible for toxic red blooms. Brown algae found in wastewater ponds include the diatoms *Navicula* and *Cyclotella.*

Red algae (Rhodophyta) include a few unicellular forms but are primarily filamentous (Gaudy and Gaudy, 1980; Pearson, 2005).

Importance of Interactions between Bacteria and Algae

It is generally accepted that the presence of both algae and bacteria is essential for the proper functioning of a treatment pond. Bacteria break down the complex organic waste components found in anaerobic and aerobic pond environments into simple compounds, which are then available for uptake by the algae. Algae, in turn, produce the O_2 necessary for the survival of aerobic bacteria.

In the process of pond reactions involving the biodegradation and mineralization of waste material by bacteria and the synthesis of new organic compounds in the form of algal cells, a pond effluent might contain a higher-than-acceptable TSS. Although this form of TSS does not contain the same constituents as the influent TSS, it does contribute to turbidity and needs to be removed before the effluent is discharged. Once concentrated and removed, depending on regulatory requirements, algal TSS may be used as a nutrient for use in agriculture or as a feed supplement (Grolund, 2002).

Invertebrates

Although bacteria and algae are the primary organisms, waste stabilization is accomplished, and predator life forms play a role in wastewater pond ecology. It has been suggested that the planktonic invertebrate *Cladocera* spp. and the benthic invertebrate family Chironomidae are the most significant fauna in the pond community in terms of stabilizing organic material. The cladocerans feed on the algae, promoting flocculation and settling of particulate matter. This results in better light penetration and algal growth at greater depths. The settled matter is further broken down and stabilized by the benthic-feeding

Chironomidae. Predators, such as rotifers, often control the population levels of certain of the smaller life forms in the pond, thereby influencing the succession of species throughout the seasons.

Mosquitoes can present a problem in some ponds. Aside from their nuisance characteristics, certain mosquitoes are also vectors for diseases such as encephalitis, malaria, and yellow fever, and they constitute a hazard to public health that must be controlled. *Gambusia*, commonly called mosquito fish, have been introduced to eliminate mosquito problems in some ponds in warm climates (Ullrich, 1967; Pipes, 1961; Pearson, 2005), but their introduction has been problematic as they can out-compete native fish that also feed on mosquito larvae. There are also biochemical controls, such as the larvicides *Bacillus thuringiensis* israelensis (Bti), and Abate®, which may be effective if the product is applied directly to the area containing mosquito larvae. The most effective means of controlling mosquitoes in ponds is the control of emergent vegetation (USEPA, 2011).

Biochemistry in a Pond

Photosynthesis

Photosynthesis is the process whereby organisms use solar energy to fix CO_2 and obtain the reducing power to convert it to organic compounds. In wastewater ponds, the dominant photosynthetic organisms include algae, cyanobacteria, and purple sulfur bacteria (Pipes, 1961; Pearson, 2005).

Photosynthesis may be classified as oxygenic or anoxygenic, depending on the source of reducing power a particular organism uses. In oxygenic photosynthesis, water serves as the source of reducing power, with O_2 as a by-product. The equation representing oxygenic photosynthesis is:

$$H_2O + sunlight \rightarrow 1/2O_2 + 2H^+ + 2e^- \qquad (18.25)$$

Oxygenic photosynthetic algae and cyanobacteria convert CO_2 to organic compounds, which serve as the major source of chemical energy for other aerobic organisms. Aerobic bacteria need the O_2 produced to function in their role as primary consumers in degrading complex organic waste material.

Anoxygenic photosynthesis does not produce O_2 and, in fact, occurs in the complete absence of O_2. The bacteria involved in anoxygenic photosynthesis are largely strict anaerobes, unable to function in the presence of O_2. They obtain energy by reducing inorganic compounds. Many photosynthetic bacteria utilize reduced S compounds or elemental S in anoxygenic photosynthesis according to the following equation:

$$H_2S \rightarrow S^o + 2H^+ + 2e^- \qquad (18.26)$$

Respiration

Respiration is a physiological process by which organic compounds are oxidized into CO_2 and water. Respiration is also an indicator of cell material synthesis. It is a complex process that consists of many interrelated biochemical reactions (Pearson, 2005). Aerobic respiration, common to species of bacteria, algae, protozoa, invertebrates, and higher plants and animals, may be represented by the following equation:

$$C_2H_{12}O_6 + 6O_2 + enzymes \rightarrow 6CO_2 + 6H_2O + new\ cells \tag{18.27}$$

The bacteria involved in aerobic respiration are primarily responsible for the degradation of waste products.

In the presence of light, respiration and photosynthesis can occur simultaneously in algae. However, the respiration rate is low compared to the photosynthesis rate, which results in a net consumption of CO_2 and production of O_2. In the absence of light, on the other hand, algal respiration continues while photosynthesis stops, resulting in a net consumption of O_2 and the production of CO_2 (USEPA, 2011).

Nitrogen Cycle

The N cycle occurring in a wastewater treatment pond consists of several biochemical reactions mediated by bacteria. Organic N and NH_3 enter with the influent wastewater. Organic N in fecal matter and other organic materials undergo conversion to NH_3 and ammonium ion NH_4^+ by microbial activity. The NH_3 may volatilize into the atmosphere. The rate of gaseous NH_3 losses to the atmosphere is primarily a function of pH, surface-to-volume ratio, temperature, and the mixing conditions. An alkaline pH shifts the equilibrium of NH_3 gas and NH_4^+ towards gaseous NH_3 production, while the mixing conditions affect the magnitude of the mass transfer coefficient.

Ammonium is nitrified to nitrite (NO_2^-) by the bacterium *Nitrosomonas* and then to NO_3^- by *Nitrobacter*. The overall nitrification reaction is:

$$NH_4^+ + 2O_2 \rightarrow NO_3^- + 2H^+ + H_2O \tag{18.28}$$

The NO_3^- produced in the nitrification process, as well as a portion of the NH_4^- produced from ammonification, can be assimilated by organisms to produce cell protein and other N-containing compounds. The NO_3^- may also be denitrified to form $NO_2^=$ and then N gas. Several species of bacteria may be involved in the denitrification process, including *Pseudomonas, Micrococcus, Achromobacter,* and *Bacillus.* The overall denitrification reaction is

$$6NO_3^- + 5CH_3OH \rightarrow 3N_2 + 5CO_2 + 7H_2O + 6OH^- \tag{18.29}$$

Nitrogen gas may be fixed by certain species of cyanobacteria when N is limited. This may occur in *N*-poor industrial ponds, but rarely in municipal or agricultural ponds (USEPA, 1975, 1993).

Nitrogen removal in facultative wastewater ponds can occur through any of the following processes: (1) gaseous NH_3 stripping to the atmosphere, (2) NH_4^- assimilation in algal biomass, (3) NO_3^- uptake by floating vascular plants and algae, and (4) biological nitrification-denitrification. Whether NH_4^- is assimilated into algal biomass depends on the biological activity in the system and is affected by several factors such as temperature, organic load, detention time, and wastewater characteristics.

Dissolved Oxygen (DO)

Oxygen is a partially soluble gas. Its solubility varies in direct proportion to the atmospheric pressure at any given temperature. DO concentrations of approximately 8 mg/L are generally considered to be the maximum available under local ambient conditions. In mechanically aerated ponds, the limited solubility of O_2 determines its absorption rate (Sawyer et al., 1994).

The natural sources of DO in ponds are photosynthetic oxygenation and surface re-aeration. In areas of low wind activity, surface re-aeration may be relatively unimportant, depending on the water depth. Where surface turbulence is created by excessive wind activity, surface re-aeration can be significant. Experiments have shown that DO in wastewater ponds varies almost directly with the level of photosynthetic activity, which is low at night and early morning and rises during daylight hours to a peak in the early afternoon. At increased depth, the effects of photosynthetic oxygenation and surface re-aeration decrease, as the distance from the water-atmosphere interface increases and light penetration decreases. This can result in the establishment of a vertical gradient. The microorganisms in the pond will segregate along the gradient.

pH and Alkalinity

In wastewater ponds, the H ion concentration, expressed as pH, is controlled through the carbonate buffering system represented by the following equations:
where:

$$CO_2 + H_2O \leftrightarrow H_2CO_3 \leftrightarrow HCO_3^- + H^+ \tag{18.30}$$

$$HCO_3^- \leftrightarrow CO_3^{-2} + H^+ \tag{18.31}$$

$$CO_3^{-2} + H_2O \leftrightarrow HCO_3^- + OH^- \tag{18.32}$$

$$OH^- + H^+ \leftrightarrow H_2O \tag{18.33}$$

The equilibrium of this system is affected by the rate of algal photosynthesis. In photosynthetic metabolism, CO_2 is removed from the dissolved phase, forcing the equilibrium of the first expression (18.30) to the left. This tends to decrease the hydrogen ion (H^+) concentration and the bicarbonate (HCO_3^-) alkalinity. The effect of the decrease in HCO_3^- concentration is to force the third Equation (18.32) to the left and the fourth (18.33) to the right, both of which decrease total alkalinity. The decreased alkalinity associated with photosynthesis will simultaneously reduce the carbonate hardness present in the waste. Because of the close correlation between pH and photosynthetic activity, there is a diurnal fluctuation in pH when respiration is the dominant metabolic activity.

Physical Factors

Light

The intensity and spectral composition of light penetrating a pond surface significantly affect all resident microbial activity. In general, activity increases with increasing light intensity until the photosynthetic system becomes light-saturated. The rate at which photosynthesis increases in proportion to an increase in light intensity, as well as the level at which an organism's photosynthetic system becomes light-saturated, depends upon the particular biochemistry of the species (Lynch and Poole, 1979; Pearson, 2005). In ponds, photosynthetic O_2 production has been shown to be relatively constant within the range of 5,380–53,8000 lumens/m² light intensity with a reduction occurring at higher and lower intensities (Pipes, 1961; Paterson and Curtis, 2005).

The spectral composition of available light is also crucial in determining photosynthetic activity. The ability of photosynthetic organisms to utilize available light energy depends primarily on their ability to absorb the available wavelengths. This absorption ability is determined by the specific photosynthetic pigment of the organism. The main photosynthetic pigments are chlorophylls and phycobilins. Bacterial chlorophyll differs from algal chlorophyll in both chemical structure and absorption capacity. These differences allow the photosynthetic bacteria to live below dense algal layers where they can utilize light not absorbed by the algae (Lynch and Poole, 1979; Pearson, 2005).

The quality and quantity of light penetrating the pond surface to any depth depend on the presence of dissolved and particulate matter, as well as the water absorption characteristics. The organisms themselves contribute to water turbidity, further limiting the depth of light penetration. Given the light penetration interferences, photosynthesis is significant only in the upper pond layers. This region of net photosynthetic activity is called the euphotic zone (Lynch and Poole, 1979; Pearson, 2005).

Light intensity from solar radiation varies with the time of day and the difference in latitudes. In cold climates, light penetration can be reduced during the winter by ice and snow cover. Supplementing the treatment ponds with mechanical aeration may be necessary in these regions during that time of year.

Temperature

Temperature at or near the surface of the aerobic environment of a pond determines the succession of predominant species of algae, bacteria, and other aquatic organisms. Algae can survive at temperatures of 5°C–40°C. Green algae show the most efficient growth and activity at temperatures of 30°C–35°C. Aerobic bacteria are viable within a temperature range of 10°C–40°C; 35°C–40°C is optimum for cyanobacteria (Anderson and Zweig, 1962; Gloyna, 1976; Paterson and Curtis, 2005; Crites et al., 2006).

As the major source of heat for these systems is solar radiation, a temperature gradient can develop in a pond with depth. This will influence the rate of anaerobic decomposition of solids that have settled at the bottom of the pond. The bacteria responsible for anaerobic degradation are active in temperatures from 15°C to 65°C. When they are exposed to lower temperatures, their activity is reduced.

The other major source of heat is the influent water. In sewerage systems with no major inflow or infiltration problems, the influent temperature is higher than that of the pond contents. Cooling influences are exerted by evaporation, contact with cooler groundwater, and wind action.

The overall effect of temperature in combination with light intensity is reflected in the fact that nearly all investigators report improved performance during summer and autumn months when both temperature and light are at their maximum. The maximum practical temperature of wastewater ponds is likely less than 30°C, indicating that most ponds operate at less than optimum temperature for anaerobic activity (Oswald, 1990, 1996; Paterson and Curtis, 2005; Crites et al., 2006; USEPA, 2011).

During certain times of the year, cooler, denser water remains at depth, while the warmer water stays at the surface. Water temperature differences may cause ponds to stratify throughout their depth. As the temperature decreases during the fall and the surface water cools, stratification decreases and the deeper water mixes with the cooling surface water. This phenomenon is called *mixis*, or pond or lake overturn. As the density of water decreases and the temperature falls below 4°C, winter stratification can develop. When the ice cover breaks up and the water warms, a spring overturn can also occur (Spellman, 1996).

Pond overturn, which releases odorous compounds into the atmosphere, can generate complaints from property owners living downwind of the pond. The potential for pond overturn during certain times of the year is the reason why regulations may specify that ponds be located downwind, based on prevailing winds during overturn periods, and away from dwellings.

Wind

Prevailing and storm-generated wind should be factored into pond design and siting as they influence performance and maintenance in several significant ways:

- **Oxygen Transfer and Dispersal:** By producing circulatory flows, winds provide the mixing needed for O_2 transfer and diffusion below the surface of facultative ponds. This mixing action also helps disperse microorganisms and augments the movement of algae, particularly green algae.

- **Prevention of Short-Circuiting and Reduction of Odor Events:** Care must be taken during design to position the pond inlet/outlet axis perpendicular to the direction of prevailing winds to reduce short-circuiting, which is the most common cause of poor performance. Consideration must also be made for the transport and fate of odors generated by treatment by-products in anaerobic and facultative ponds.
- **Disturbance of Pond Integrity:** Waves generated by strong prevailing or storm winds are capable of eroding or overtopping embankments. Some protective material should extend 1 or more ft above and below the water level to stabilize earthen berms.
- Various studies indicate that wind effects can reduce hydraulic retention time.

Pond Nutritional Requirements

To function as designed, the wastewater pond must provide sufficient macro- and micronutrients for microorganisms to grow and populate the system adequately. It should be understood that a treatment pond system should be neither overloaded nor underloaded with wastewater nutrients.

DID YOU KNOW?

The variation in pH in a facultative pond normally occurs in the upper aerobic zone, while the anaerobic and facultative zones will be relatively constant. This variation happens due to the changes that occur in the concentration of dissolved carbon dioxide. When carbon dioxide is dissolved in water it forms a weak carbonic acid which tends to lower pH. The relationship between algae and bacteria affects the carbon dioxide levels. During intense photosynthesis, algae use carbon dioxide and produce oxygen to be used by bacteria to assimilate organic wastes. The algae use much of the carbon dioxide and the pH can rise significantly (pH in the 11–12 range is not uncommon). During the night or cloudy weather, the algae respire, and active photosynthesis do not occur. The bacteria continue to use up oxygen and produce carbon dioxide. This can cause a significant drop in the pond's pH, especially if the influent wastewater has low alkalinity. This same pH swing can occur in natural ponds, lakes, and stream impoundments. During peak summer algae activity, the dissolved oxygen of stream impoundments has varied from dawn levels of less than 1 mg/L to later afternoon values of 13–15 mu/L (supersaturation).

Nitrogen can be a limiting nutrient for primary productivity in a pond. The conversion of organic N to various other N forms results in a total net loss (Assenzo and Reid, 1966;

Pano and Middlebrooks, 1982; Middlebrooks et al. 1982; Middlebrooks and Pano, 1983; Craggs, 2005). This N loss may be due to algal uptake or bacterial action. Both mechanisms likely contribute to the overall total N reduction. Another factor contributing to the reduction of total N is the removal of gaseous NH_3 under favorable environmental conditions. Regardless of the specific removal mechanism involved, NH_3 removal in facultative wastewater ponds has been observed at levels greater than 90%, with the major removal occurring in the primary cell of a multicell pond system (Middlebrooks et al., 1982; Shilton, 2005; Crites et al., 2006; USEPA, 2011).

Phosphorus

Phosphorus (P) is most often the growth-limiting nutrient in aquatic environments. Municipal wastewater in the United States is normally enriched in P, even though restrictions on P-containing compounds in laundry detergents in some states have resulted in reduced concentrations since the 1970s. As of 1999, 27 states and the District of Columbia had passed laws prohibiting the manufacture and use of laundry detergents containing P. However, phosphate (PO_4^{-3}) content limits in automatic dishwashing detergents and other household cleaning agents containing P remain unchanged in most states. With a contribution of approximately 15%, the concentration of P from wastewater treatment plants is still adequate to promote growth in aquatic organisms.

In aquatic environments, P occurs in three forms: (1) particulate P, (2) soluble organic P, and (3) inorganic P. Inorganic P, primarily in the form of orthophosphate ($OP(OR)_3$), is readily utilized by aquatic organisms. Some organisms may store excess P as polyphosphate. At the same time, some PO_4^{-3} is continuously lost to sediments, where it is locked up in insoluble precipitates (Lynch and Poole, 1979; Craggs, 2005; Crites et al., 2006).

Phosphorus removal in ponds occurs via physical mechanisms such as adsorption, coagulation, and precipitation. The uptake of P by organisms for metabolic function as well as for storage can also contribute to its removal. Removal in wastewater ponds has been reported to range from 30% to 95% (Assenzo and Reid, 1966; Pearson, 2005; Crites et al., 2006).

Algae discharged in the final effluent may introduce organic P to receiving waters. Excessive algal "afterblooms" observed in waters receiving effluents have, in some cases, been attributed to N and P compounds remaining in the treated wastewater.

Sulfur

Sulfur (S) is a required nutrient for microorganisms, and it is usually present in sufficient concentration in natural waters. Because S is rarely limiting, its removal from wastewater is usually not considered necessary. Ecologically, S compounds such as hydrogen sulfide (H_2S and sulfuric acid (H_2SO_4) are toxic, while the oxidation of certain S compounds is an important energy source for some aquatic bacteria (Lynch and Poole, 1979; Pearson, 2005).

Carbon

The decomposable organic C content of a waste is traditionally measured in terms of its BOD_5, or the amount of O_2 required under standardized conditions for the aerobic biological stabilization of the organic matter over a certain period of time. Since complete treatment by biological oxidation can take several weeks, depending on the organic material and the organisms present, standard practice is to use the BOD_5 as an index of the organic carbon content or organic strength of a waste. The removal of BOD_5 is a primary criterion by which treatment efficiency is evaluated.

BOD_5 reduction in wastewater ponds ranging from 50% to 95% has been reported in the literature. Various factors affect the rate of reduction of BOD_5. A very rapid reduction occurs in a wastewater pond during the first 5–7 days. Subsequent reductions take place at a sharply reduced rate. BOD_5 removals are generally much lower during winter and early spring than in summer and early fall. Many regulatory agencies recommend that pond operations do not include discharge during cold periods.

Process Control Calculations for Stabilization Ponds

Process control calculations are an important part of wastewater treatment operations, including pond operations. More significantly, process control calculations are an important part of state wastewater licensing examinations—you simply cannot master the licensing examinations without being able to perform the required calculations. Thus, as with previous sections (and with sections to follow), whenever possible, example process control problems are provided to enhance your knowledge and skill.

Determining Pond Area in Acres

$$\text{Area, acres} = \frac{\text{Area, ft}^2}{43,560 \text{ ft}^2/\text{acre}} \quad (18.34)$$

Determining Pond Volume in Acre-Ft

$$\text{Volume, acre-feet} = \frac{\text{Volume, ft}^3}{43,560 \text{ ft}^3/\text{acre-ft}} \quad (18.35)$$

Determining Flow Rate in Acre-Ft/Day

$$\text{Flow, acre-ft/day} = \text{Flow, MGD} \times 3.069 \text{ acre-ft/MG} \quad (18.36)$$

Note: Acre-feet is a unit that can cause confusion, especially for those not familiar with pond or lagoon operations. 1 acre-ft is the volume of a box with a 1-acre top and 1 ft of depth—but the top doesn't have to be an even number of acres in size to use acre-feet. 1 acre-ft = 325,851 gal.

Determining Flow Rate in Acre-in/Day

$$\text{Flow, acre-in/day} = \text{Flow, MGD} \times 36.8 \text{ acre-in/MG} \quad (18.37)$$

Hydraulic Detention Time in Days

$$\text{Hydraulic Detention Time, Days} = \frac{\text{Pond Volume, acre-ft}}{\text{Influent Flow, acre-ft/Day}} \quad (18.38)$$

Note: Normally, hydraulic detention time ranges from 30 to 120 days for stabilization ponds.

Example 18.29

Problem: A stabilization pond has a volume of 53.5 acre-ft. What is the detention time on days when the flow is 0.30 MGD?

Solution:

$$\text{Flow, acre-ft/day} = 0.30 \text{ MGD} \times 3.069$$
$$= 0.92 \text{ acre-ft/day}$$

$$\text{D.T. Days} = \frac{53.5 \text{ acre}}{0.92 \text{ acre-ft/day}} = 58.2 \text{ days}$$

Hydraulic Loading, Inches/Day (Overflow Rate)

$$\text{Hydraulic Loading, in/day} = \frac{\text{Influent Flow, acre-in/day}}{\text{Pond Area, acre}} \quad (18.39)$$

$$\text{Pop. Loading, People/acre/day} = \frac{\text{Pop. Served by System, People}}{\text{Pond Area, Acres}} \quad (18.40)$$

Note: Population loading normally ranges from 50 to 500 people per acre.

Organic Loading

Organic loading can be expressed as pounds of BOD_5 per acre per day (most common), pounds BOD_5 per acre-foot per day or people per acre per day.

$$\text{Organic L, lb BOD}_5/\text{acre/day} = \frac{BOD_5, \text{mg/L} \times \text{Infl. flow,} \text{MGD} \times 8.34}{\text{Pond Area, Acres}} \quad (18.41)$$

Note: Normal range is 10–50 lbs BOD_5 per day per acre.

Example 18.30

Problem: A wastewater treatment pond has an average width of 380 ft and an average length of 725 ft. The influent flow rate to the pond is 0.12 MGD with a BOD concentration of 160 mg/L. What is the organic loading rate to the pond in pounds per day per acre (lbs/day/acre)?

Solution:

$$725 \text{ ft} \times 380 \text{ ft} \times \frac{1 \text{ acre}}{43,560 \text{ ft}^2} = 6.32 \text{ acre}$$

$$0.12 \text{ MGD} \times 160 \text{ mg/L} \times 8.34 \text{ lb/gal} = 160.1 \text{ lb/day}$$

$$\frac{160.1 \text{ lb/day}}{6.32 \text{ acre}} = 25.3 \text{ lb/day/acre}$$

DID YOU KNOW?

Common maintenance problems associated with pond systems include:
- **Weed Control:** cattails and other rooted aquatic plants.
- **Algae Control:** blue-green and associated floating algae mats
- **Burrowing Animals:** muskrats and turtles.
- Duckweed control and removal.
- Floating sludge mats.
- **Dike Vegetation:** mowing and removing woody plants.
- **Dike Erosion:** riprap and proper vegetation.
- Fence maintenance to restrict access.
- **Mechanical Equipment:** pumps, blowers, etc.

TROUBLESHOOTING WASTEWATER PONDS

Common problems in wastewater treatment pond operation are shown in Table 18.4.

DID YOU KNOW?

Duckweed must be physically removed with a rake, push board, or broom. With sufficient wind, the duckweed will be pushed to one side or corner of a pond. This is an ideal time to rake it out. It is important that duckweed not be allowed to become too abundant, as it reduces oxygen transfer at the water surface, reduces light penetration and photosynthesis, and, upon decomposing, can cause both odor and BOD problems.

TRICKLING FILTERS

Trickling filters have been used to treat wastewater since the 1890s. It was found that if settled wastewater was passed over rock surfaces, slime grew on the rocks and the water became cleaner. Today, we still use this principle; however, in many installations, instead of rocks, we use plastic media. In most wastewater treatment systems, the *trickling filter* follows primary treatment and includes a secondary settling tank or clarifier, as shown in Figure 18.4. Trickling filters are widely used for the treatment of domestic and industrial wastes. The process is a fixed-film biological treatment method designed to remove BOD_5 and suspended solids.

A trickling filter consists of a rotating distribution arm that sprays and evenly distributes liquid wastewater over a circular bed of fist-sized rocks, other coarse materials, or synthetic media (see Figure 18.5). The spaces between the media allow air to circulate easily so that aerobic conditions can be maintained. The spaces also allow wastewater to trickle down through, around and over the media. A layer of biological slime that absorbs and consumes the wastes trickling through the bed covers the media material. The organisms aerobically decompose the solids producing more organisms and stable wastes, which either become part of the slime or are discharged back into the wastewater flowing over the media. This slime consists mainly of bacteria, but it may also include algae, protozoa, worms, snails, fungi, and insect larvae. The accumulating slime occasionally sloughs off (*sloughings*) individual media materials (see Figure 18.6) and is collected at the bottom of the filter, along with the treated wastewater, and passed on to the secondary settling tank, where it is removed. The overall performance of the trickling filter is dependent on hydraulic and organic loading, temperature, and recirculation.

Trickling Filter Definitions

To clearly understand the correct operation of the trickling filter, the operator must be familiar with certain terms.

Note: The following list of terms applies to the trickling filter process. We assume that other terms related to other units within the treatment system (plant) are already familiar to operators.

- **Biological Towers:** a type of trickling filter that is very deep (10–20 ft). Filled with lightweight synthetic media, these towers are also known as oxidation or roughing towers, or (because of their extremely high hydraulic loading) super-rate trickling filters.
- **Biomass:** the total mass of organisms attached to the media. Similar to solids inventory in the activated sludge process, it is sometimes referred to as *zoogleal slime*.
- **Distribution Arm:** the device most widely used to apply wastewater evenly over the entire surface of the media. In most cases, the force of the wastewater being sprayed through the orifices moves the arm.
- **Filter Underdrain:** the open space provided under the media to collect the liquid (wastewater and sloughings) and to allow air to enter the filter. It has a sloped floor to collect the flow to a central channel for removal.
- **Hydraulic Loading:** the amount of wastewater flow applied to the surface of the trickling filter media. It can be expressed in several ways: flow per square foot of surface per day (gpd/ft²); flow per acre per day (MGAD); or flow per acre-foot per day (MGAFD). The hydraulic loading includes all flow entering the filter.

TABLE 18.4

Common Problems in Wastewater Treatment Pond Operation

Problem	Possible Causes	Possible Solutions
Odors	Organic overload	Increase aeration capacity.
	Poor aeration or mixing	After the aerator run time, change or supplement the type of aerator to eliminate ice cover.
	Previous ice-covered ponds	Increase aerator run time, and change the type of aerator to eliminate ice over
	Duckweed growth	Increase aerator run time, chemical treatment (Diquat), physical removal (harvest), and add Ducks or geese.
	Excess weed growth on pond banks harboring flies, and mosquitoes, organics.	Physical removal by pulling, mowing, burning, or chemical treatment (Diquat). In winter, lower the pond level and allow ice to freeze trapping grease around the weed stem. Increase water level.
Poor BOD removal	Organic overloading	Increase aeration capacity.
	Short-circuiting	Improve inlet-outlet conditions, add bubbles, add recirculation to improve mixing, and add or improve aeration of ponds.
	Ice-covered ponds	Change or add an aerator.
	The recent reduction in pond temperature	Increase hydraulic detention time.
	Algal bloom	Increase mechanical mixing; add physical shade (Aquashade, Photoblue), floating cover such as swimming pool cover. Styrofoam sheets or balls, duckweed cover. Addition of algal predators such as daphnia. Add chemicals (copper sulfate). Addition of constructed wetlands to polish effluent.
High effluent TSS	Algal bloom	See algal bloom solution above.
	Excess pond mixing or short-circuiting	See short-circuiting solution above.
	Spring or fall turnover	Add different types of aeration to eliminate stratification or add supplemental aeration, and recirculation.
	Excessive solids buildup in the bottoms of ponds	Physically remove solids by pump or sludge barge; proper sludge disposal in conformance with State and Federal regulators
Poor fecal coliform removal	Chlorine residual too low or poor chlorine contact chamber design	Increase chlorine feed rate, provide 40:1 1:w ratio, and provide a minimum 30 min detention time at peak flow.
	Increase in chlorine demanding substance in effluent (H_2S, NO_2)	Remove solids from the chlorine contact chamber and increase the chlorine feed rate.
	Waterfowl contamination	Increase chlorine feed rate.
High pH	Algal stripping of carbon dioxide and bicarbonate alkalinity	See the algal bloom solution above.
Low pH	Accumulation of organic sludge stuck in the "acid phase"	Physically remove sludge by pumping or sludge dredging.
	Extensive nitrification	Increase aerator run time and add recirculation.
	Organic overloading	Increase aeration capacity and add recirculation.
	Excessive daphnia growth	Increase aeration capacity or run time.
Water weeds	Poor circulation, maintenance	Pull weeds by hand if new growth.
	Insufficient water depth.	Mow weeds with a sickle or bar mover.
		Lower the water level to expose weeds, then burn with a gas burner.
		Allow the surface to freeze at a low water level and the floating ice will pull the weeds as it rises.
		Increase water depth to above the tops of weeds.
		Use riprap. Caution: If weeds get started in the riprap, they will be difficult to remove but can be sprayed with EPA-approved herbicides.
		To control duckweed, use rakes or push with a boat, then physically remove duckweed from the pond.

(Continued)

TABLE 18.4 (Continued)
Common Problems in Wastewater Treatment Pond Operation

Problem	Possible Causes	Possible Solutions
Burrowing animals	Bank conditions that attract animals. High population in the area adjacent to ponds.	Remove food supplies such as cattails and burr reed from ponds and adjacent areas. Muskrats prefer a partially submerged tunnel, completely. They may be discouraged by raising and lowering the level 6–8 in over several weeks. If the problem persists, check with the local game commission officer for approved methods of removal, such as live trapping, etc.
Dike vegetation	Poor maintenance.	Periodic mowing is the best method. Sow dikes with a mixture of fescue and blue grasses on the shore and short native grasses elsewhere. It is desirable to select a grass that will form a good sod and drive out tall weeds by binding the soil and "out compete" undesirable growth. Spray with approved weed control chemicals. Some small animals, such as sheep, have been used. This may increase fecal coliform, especially in the discharge cell. Not recommended with pond systems utilizing synthetic pond liners. Practice "rotation grazing" to prevent destroying individual species of grasses. An example schedule for rotation grazing is a 3-pond system: Graze each pond area for 2 months over a 6-month grazing season.
Scum	The pond bottom is turning over with sludge floating to the surface. Poor circulation and wind action. High amounts of grease and oil in influent will also cause scum.	Use rakes, a portable pump to get a water jet, or motor boats to break up scum formations. Broken scum usually sinks. Any remaining scum should be skimmed and disposed of by burial or hauled to a landfill with the approval of a regulatory agency.
Insects	Poor circulation and maintenance.	Keep the pond clear of weeds and allow wave action on the bank to prevent mosquitoes from hanging out. Keep the pond free from scum. As for stocking fish, note that Gambusia does not eat mosquito larvae any faster than other fish species. Spray with EPA-approved larvicide as a last resort. Check with state regulatory officials for approved chemicals.
Blue-Green algae	Blue-green algae is an indicator of incomplete treatment, overloading, and/or poor nutrient balance.	Refer to common causes of pond effluent non-compliance. WARNING! Prior to using copper sulfate, see the explanation below. Apply three applications of a solution of copper sulfate. If the total alkalinity is above 50 mg/L apply $1,200 \text{kg/m}^3$ (10 lbs/MG) of copper sulfate per million gallons in the cell. If alkalinity is below 50 mg/L, reduce the amount of copper sulfate to 600kg/m^3 (5 lbs/MG). **Note:** Some states do not approve the use of copper sulfate since in concentrations greater than 1 mg/L, it is toxic to certain organisms and fish. Break up algal blooms by motorboat or a portable pump and hose. Motorboat motors should be air-cooled as algae may plug up water-cooled motors. **Important:** In the past copper sulfate has been used to control algae. It is recommended that the operator check with the regulatory agency for the discharge permit. It should also be noted that prolonged use of copper sulfate may cause a buildup of copper in the benthic sludges, making it difficult to dispose of the sludges when pond cleaning becomes necessary.
Lightly loaded ponds	Overdesign, low seasonal flow	Correct by increasing the loading by reducing the number of cells in use. Use series operation.

(Continued)

TABLE 18.4 (Continued)
Common Problems in Wastewater Treatment Pond Operation

Problem	Possible Causes	Possible Solutions
Overloading	Short-circuiting, industrial wastes, poor design, infiltration, new construction (service area expansion), inadequate treatment, and weather conditions.	Bypass the cell and let it rest. Use parallel operation. Apply recirculation of pond effluent. Look at possible short-circuiting. Install supplementary aeration equipment.
Decreasing trend in pH	A decreasing pH is followed by a drop in DO as the green algae die off. This is most often caused by overloading, long periods of adverse weather, or higher animals, such as daphnia, feeding on the algae.	Bypass the cell and let it rest. Use parallel operation. Apply recirculation of pond effluent. Check for possible short-circuiting. Install supplementary aeration equipment if the problem is persistent and due to overloading.
Low DO	Poor light penetration, low detention time, high BOD loading, or toxic industrial wastes. (Daytime DO should drop below 1.0 mg/L during warm months.)	Increase aerator running time. Remove weeds such as duckweed if covering greater than 40% of the pond. Reduce organic loading to primary cell(s) by going to parallel operation. Add supplemental aeration (surface aerators, diffusers, and/or daily operation of a motor boat.) Add recirculation by using a portable pump to return final effluent to the head works. Apply sodium nitrate. Determine if the overload is due to an industrial source and remove it.
Short-circuiting	Poor wind action due to trees or poor arrangement of inlet and outlet locations. It may also be due to the shape of the pond, weed growth, or irregular bottom.	Cut trees and grow at least 150 m (500 ft) away from the pond if in the direction of the prevailing wind. Install baffling around the inlet location to improve distribution. Add recirculation to improve mixing. Provide new inlet-outlet locations, including multiple inlets and manifolds. Clean out weeds. Fill in irregular bottoms. Add directional surface mixers or aerators to mix and retard flow.
Anaerobic conditions	Overloading short-circuiting, poor operation, or toxic discharge.	Change from a series to parallel operation to divide the load. Helpful if conditions exist at a certain time each year and are not persistent. Add supplemental aeration if the pond is continuously overloaded. Change inlets and outlets to eliminate short-circuiting. See how to correct short-circuiting. Add recirculation (temporary use portable pumps) to provide oxygen and mixing. In some cases, temporary help can be obtained by adding sodium nitrate. Eliminate sources of toxic discharges.

Sources: USEPA (1977) and Richard and Bowman (1991).

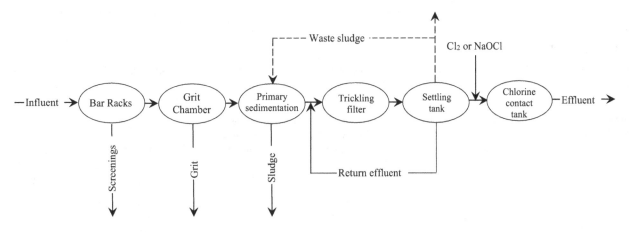

FIGURE 18.4 Simplified flow diagram of trickling filter used for wastewater treatment.

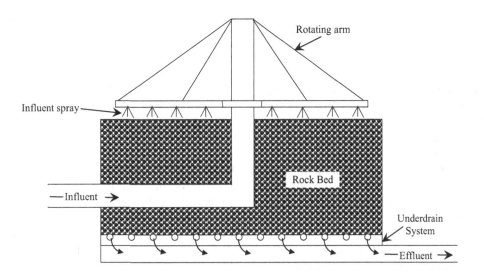

FIGURE 18.5 Schematic of a cross-section of a trickling filter.

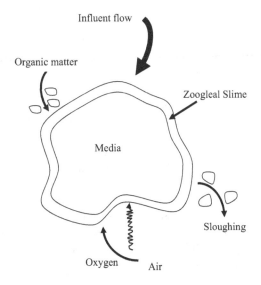

FIGURE 18.6 Filter media showing biological activities that take place on the surface area.

TABLE 18.5

Trickling Filter Classification

Filter Class	Standard	Intermediate	High-Rate	Super High-Rate	Roughing
Hydraulic loading gpd/ft^2	25–90	90–230	230–900	350–2,100	>900
Organic loading BOD/1,000 ft^3	5–25	15–30	25–300	Up to 300	>300
Sloughing frequency	Seasonal	Varies	Continuous	Continuous	Continuous
Distribution	Rotary	Rotary fixed	Rotary fixed	Rotary	Rotary fixed
Recirculation	No	Usually	Always	Usually	Not usually
Media depth, ft	6–8	6–8	3–8	Up to 40	3–20
Media type	Rock	Rock	Rock	Plastic	Rock
	Plastic	Plastic	Plastic	Plastic	
	Wood	Wood	Wood	Wood	
Nitrification	Yes	Some	Some	Limited	None
Filter flies	Yes	Variable	Variable	Very few	Not usually
BOD removal	80%–85%	50%–70%	65%–80%	65%–85%	40%–65%
TSS removal	80%–85%	50%–70%	65%–80%	65%–85%	40%–65%

- **High-Rate Trickling Filters:** a classification (see Table 18.5) in which the organic loading is in the range of 25–100 lbs of BOD$_5$ per 1,000 ft^3 of media per day. The standard rate filter may also produce a highly nitrified effluent.
- **Media:** an inert substance placed in the filter to provide a surface for microorganisms to grow on. The media can be filed stone, crushed stone, slag, plastic, or redwood slats.
- **Organic loading:** the amount of BOD$_5$ or chemical oxygen demand (COD) applied to a given volume of filter media. It does not include the BOD$_5$ or COD contributed to any recirculated flow and is commonly expressed as pounds of BOD$_5$ or COD/1,000 ft^3 of media.
- **Recirculation:** the return of filter effluent back to the head of the trickling filter. It can help level flow variations and assist in resolving operational problems, such as ponding, filter flies, and odors.
- **Roughing Filters:** a classification of trickling filters (see Table 18.5) in which the organic is in excess of 200 lbs of BOD$_5$/1,000 ft^3 of media per day. A roughing filter is used to reduce the loading on other biological treatment processes, producing an industrial discharge that can be safely treated in a municipal treatment facility.
- **Sloughing:** the process in which excess growths break away from the media and wash through the filter to the underdrains with the wastewater. These sloughings must be removed from the flow by settling.
- **Staging:** the practice of operating two or more trickling filters in series. The effluent of one filter is used as the influent of the next. This practice can produce a higher quality effluent by removing additional BOD$_5$ or COD.

Trickling Filter Equipment

The trickling filter distribution system is designed to spread wastewater evenly over the surface of the entire media. The most common system is the rotary distributor which moves above the surface of the media and sprays the wastewater on the surface. The force of the water leaving the orifices drives the rotary system. The distributor's arms usually have small plates below each orifice to spread the wastewater into a fan-shaped distribution system. The second type of distributor is the fixed nozzle system. In this system, the nozzles are fixed in place above the media and are designed to spray the wastewater over a fixed portion of the media. This system is frequently used with deep bed synthetic media filters.

Note: Trickling filters that use ordinary rock are normally only about 3 m in depth because of structural problems caused by the weight of the rocks—which also requires the construction of beds that are quite wide, in many applications, up to 60 ft in diameter. When synthetic media is used, the bed can be much deeper.

No matter which type of media is selected, the primary consideration is that it must be capable of providing the desired film location for the development of the biomass. Depending on the type of media used and the filter classification, the media may be 3–20 ft or more in depth.

The underdrains are designed to support the media, collect the wastewater and sloughings, carry them out of the filter, and provide ventilation to the filter.

Note: In order to ensure sufficient airflow to the filter the underdrains should never be allowed to flow more than 50% full of wastewater.

The effluent channel is designed to carry the flow from the trickling filter to the secondary settling tank. The secondary settling tank provides 2–4 h of detention time to separate the sloughing materials from the treated wastewater. Design, construction, and operation are similar to those of the primary settling tank. Longer detention times are provided because the sloughing materials are lighter and settle more slowly.

Recirculation pumps and piping are designed to recirculate (and thus improve the performance of the trickling filter or settling tank) a portion of the effluent back to be mixed with the filter influent. When recirculation is used, obviously, pumps and metering devices must be provided.

Filter Classifications

Trickling filters are classified by hydraulic and organic loading. Moreover, the expected performance and the construction of the trickling filter are determined by the filter classification. Filter classifications include standard rate, intermediate rate, high rate, super high rate (plastic media), and roughing rate types. Standard rate, high rate, and roughing rate are the filter types most commonly used. The *standard rate filter* has a hydraulic loading (gpd/ft^3) of 25 to 90; a seasonal sloughing frequency; does not employ recirculation; and typically has an 80%–85% BOD$_5$ removal rate and 80%–85% TSS removal rate. The *high-rate filter* has a hydraulic loading (gpd/ft^3) of 230–900; a continuous sloughing frequency; always employs recirculation; and typically has a 65%–80% BOD$_5$ removal rate and 6%–80% TSS removal rate. The *roughing filter* has a hydraulic loading (gpd/ft^3) of >900; a continuous sloughing frequency; does not normally include recirculation; and typically has a 40%–65% removal rate and 40%–65% TSS removal rate.

Standard Operating Procedures

Standard operating procedures for trickling filters include sampling and testing, observation, recirculation, maintenance, and expectations of performance. Collection of influent and process effluent samples to determine performance and monitor the process condition of trickling filters is required. Dissolved oxygen, pH, and settleable solids testing should be collected daily. BOD$_5$ and suspended solids testing should be done as often as practical to determine the percent removal.

The operation and condition of the filter should be observed daily. Items to observe include the distributor movement, uniformity of distribution, evidence of operation or mechanical problems, and the presence of objectionable odors. In addition to the items above the normal observation for a settling tank should also be performed.

Recirculation is used to reduce organic loading, improve sloughing, reduce odors, and reduce or eliminate filter fly or ponding problems. The amount of recirculation is dependent on the design of the treatment plant and the operational requirements of the process. Recirculation flow may be expressed as a specific flow rate (i.e., 2.0-MGD). In most cases, it is expressed as a ratio (3:1, 0.5:1.0, etc.). The recirculation is always listed as the first number and the influent flow is listed as the second number. Because the second number in the ratio is always 1.0, the ratio is sometimes written as a single number (dropping the: 1.0)

Flows can be recirculated from various points following the filter to various points before the filter. The most common form of recirculation removes flow from the filter effluent or settling tank and returns it to the influent of the trickling filter as shown in Figure 18.7.

Maintenance requirements include lubrication of mechanical equipment, removal of debris from the surface and orifices, as well as adjustment of flow patterns and maintenance associated with the settling tank.

General Process Description

The trickling filter process involves spraying wastewater over a solid media such as rock, plastic, or redwood slats (or laths). As the wastewater trickles over the surface of the media, the growth of microorganisms (bacteria, protozoa, fungi, algae, helminths or worms, and larvae) develops. This growth is visible as a shiny slime very similar to the slime found on rocks in a stream. As the wastewater passes over this slime, the slime adsorbs the organic (food) matter. This organic matter is used for food by the microorganisms. At the same time, air moving through the open spaces in the filter transfers oxygen to the wastewater. This oxygen is then transferred to the slime to keep the outer layer aerobic. As the microorganisms use the food and oxygen, they produce more organisms, carbon dioxide, sulfates, nitrates, and other stable by-products; these materials are then discarded from the slime back into the wastewater flow and are carried out of the filter:

$$\text{Organics} + \text{Organisms} + O_2 = \text{More Organisms}$$

$$+ CO_2 + \text{Solid Wastes}$$

(18.42)

The growth of the microorganisms and the buildup of solid wastes in the slime make it thicker and heavier. When this slime becomes too thick, the wastewater flow breaks off parts of the slime. These must be removed in the final settling tank. In some trickling filters, a portion of the filter effluent is returned to the head of the trickling filter to level out variations in flow and improve operations (recirculation).

Overview and Brief Summary of the Trickling Filter Process

A trickling filter consists of a bed of coarse media, usually rocks or plastic, covered with microorganisms.

Note: Trickling filters that use ordinary rock are normally only about 10 ft in depth because of structural problems caused by the weight of the rocks, which also requires the construction of beds that are quite wide—in many

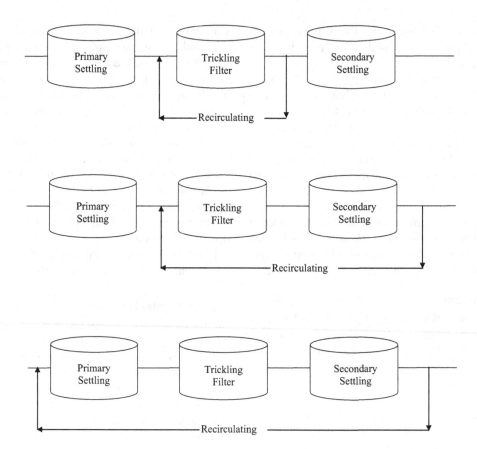

FIGURE 18.7   Common forms of recirculation.

applications, up to 60 ft in diameter. When synthetic media are used, the bed can be much deeper.

- The wastewater is applied to the media at a controlled rate, using a rotating distributor arm or fixed nozzles. Organic material is removed by contact with the microorganisms as the wastewater trickles down through the media openings. The treated wastewater is collected by an underdrain system.

Note: To ensure sufficient airflow to the filter, the underdrains should never be allowed to flow more than 50% full of wastewater.

- The trickling filter is usually built into a tank that contains the media. The filter may be square, rectangular, or circular.
- The trickling filter does not provide any actual filtration. The filter media provides a large amount of surface area that the microorganisms can cling to and grow in slime that forms on the media as they feed on the organic material in the wastewater.
- The slime growth on the trickling filter media periodically sloughs off and is settled and removed in a secondary clarifier that follows the filter.
- Key factors in trickling filter operation include the following concepts:

Hydraulic Loading Rate

$$\frac{\text{gal/day}}{\text{ft}^2} = \frac{\text{Flow, gal/day (including recirculation)}}{\text{Media top surface, ft}^2}$$

Organic Loading Rate

$$\frac{\text{lb/day}}{1{,}000\ \text{ft}^3} = \frac{\text{BOD in filter, lb/day}}{\text{Media Vol, }1{,}000\ \text{ft}^3}$$

Recirculation

$$\text{Ratio} = \frac{\text{Recirculation flow, MGD}}{\text{Ave. influent flow, MGD}}$$

Operator Observations

Trickling filter operation requires routine observation, meter readings, process control sampling and testing, and process control calculations. Comparison of daily results with expected "normal" ranges is the key to identifying problems and implementing appropriate corrective actions.

1. **Slime:** The operator checks the thickness of the slime to ensure that it is thin and uniform (normal) or thick and heavy (indicating organic overload). The operator is also concerned with ensuring

that excessive recirculation is not taking place and checks slime toxicity (if any). The operator is also concerned about the color of the slime: green slime is normal; dark green/black slime indicates organic overload; other colors may indicate industrial waste or chemical additive contamination. The operator should check the subsurface growth of the slime to ensure that it is normal (thin and translucent). If the growth is thick and dark, organic overload conditions are indicated. Distribution arm operation is a system function important to slime formation; it must be checked regularly for proper operation. For example, the distribution of slime should be even and uniform. Striped conditions indicate clogged orifices or nozzles.

2. **Flow:** Flow distribution must be checked to ensure uniformity. If it is non-uniform, the arms are not level, or the orifices are plugged. Flow drainage is also important. Drainage should be uniform and rapid. If not, ponding may occur from media breakdown or debris on the surface.

3. **Distributor:** The movement of the distributor is critical to the proper operation of the trickling filter. Movement should be uniform and smooth. Chattering or noisy operation may indicate bearing failure. The distributor seal must be checked to ensure there is no leakage.

4. **Recirculation:** The operator must check the rate of recirculation to ensure that it is within design specifications. Rates above design specifications indicate hydraulic overloading, while rates below design specifications indicate hydraulic underloading.

Note: *Recirculation* reduces the organic loading, improves sloughing, reduces odors, and reduces or eliminates filter fly or ponding problems. The amount of recirculation needed depends on the design of the treatment plant and the operational requirements of the process. Recirculation flow may be expressed as a specific flow rate (i.e., 2.0-MGD). In most cases, it is expressed as a ratio (3:1, 0.5:1.0, etc.). The recirculation is always listed as the first number, and the influent flow is listed as the second number.

Note: Because the second number in the ratio is always 1.0, the ratio is sometimes written as a single number (dropping the: 1.0).

5. **Media:** The operator should check to ensure that the medium is uniform.

Process Control Sampling and Testing

To ensure proper operation of the trickling filter, sampling and scheduling are important. However, for samples and the tests derived from the samples to be beneficial,

operators must perform a variety of daily or variable tests. Individual tests and sampling may be needed daily, weekly, or monthly, depending on seasonal changes. The frequency may be lower during normal operations and higher during abnormal conditions. The information gathered through the collection and analysis of samples from various points in the trickling filter process helps determine the current status of the process, as well as identifying and correcting operational problems. The following routine sampling points and types of tests will permit the operator to identify normal and abnormal operating conditions.

Filter influent tests
Dissolved oxygen
pH
Temperature
Settleable solids
BOD5
Suspended solids
Metals

Recirculated flow
Dissolved oxygen
pH
Flow rate
Temperature

Filter effluent
Dissolved oxygen
pH
Jar tests

Process effluent
Dissolved oxygen
pH
Settleable solids
BOD5
Suspended solids

Troubleshooting Operational Problems

(Much of the information in this section is based on USEPA's *Field Manual Performance Evaluation and Troubleshooting at Municipal Wastewater Treatment Facilities*, Washington, DC, current editions. The following sections are not all-inclusive; they do not cover all of the operational problems associated with the trickling filter process. They do provide, however, information on the most common operational problems.

Ponding

Symptoms:

- Small pools or puddles of water on the surface of the media
- Decreased performance in the removal of BOD and TSS
- Possible odors due to anaerobic conditions in the media
- Poor airflow through the media

Causal Factors:

- Inadequate hydraulic loading to keep the media voids flushed clear
- Application of high-strength wastes without sufficient recirculation to provide dilution
- Non-uniform media
- Degradation of the media due to aging or weathering
- Medium is uniform but is too small
- Debris (moss, leaves, sticks) or living organisms (snails) that clog the void spaces

Corrective Actions (listed in increasing impact on the quality of the plant effluent):

- Remove all leaves, sticks, and other debris from the media.
- Increase recirculation of dilute, high-strength wastes to improve sloughing and keep voids open.
- Use a high-pressure stream of water to agitate and flush the ponded area.
- Rake or fork the ponded area.
- Dose the filter with chlorine solution for 2–4 h. The specific dose of chlorine required will depend on the severity of the ponding problem. When using elemental chlorine, the dose must be sufficient to provide a residual at the orifices of 1–50 mg/L. If the filter is severely clogged, higher residuals may be needed to unload the majority of the biomass. If the filter cannot be dosed with elemental chlorine, chlorinated lime or HTH powder may be used. Dosing should be in the range of 8–10 lbs of chlorine per 1,000 ft² of media.
- If the filter design permits, the filter media can be flooded for a period of 4 h. Remember, if the filter is flooded, care must be taken to prevent hydraulic overloads of the final settling tank. The trickling filter should be drained slowly during low flow periods.
- Dry the media. By stopping the flow to the filter, the slime will dry and loosen. When the flow is restarted, the loosened slime will flow out of the filter. The amount of drying time will be dependent on the thickness of the slime and the amount of removal desired. Time may range from a few hours to several days.

Note: Portions of the media can be dried without taking the filter out of service by plugging the orifices that normally service the area.

Note: If these corrective actions do not provide the desired improvement, the media must be carefully inspected. Remove a sample of the media from the affected area. Carefully clean it, inspect for its solidity, and determine its size uniformity (3–5 in). If it is acceptable, the media must

be carefully replaced. If the media appear to be decomposing or are not uniform, then they should be replaced.

Odors

Frequent offensive odors usually indicate an operational problem. These foul odors occur within the filter periodically and are normally associated with anaerobic conditions. Under normal circumstances, a slight anaerobic slime layer forms due to the inability of oxygen to penetrate all the way to the media. However, under normal operation, the outer slime layers will remain aerobic, and no offensive odors will be produced.

Causal Factors:

- Excessive organic loading due to poor filter effluent quality (recirculation); poor primary treatment operation; poor control of the sludge treatment process that results in high BOD_5 recycle flows
- Poor ventilation because of submerged or obstructed underdrains, clogged vent pipes, or clogged void spaces
- The filter is overloaded hydraulically and/or organically
- Poor housekeeping

Corrective Actions:

- Evaluate the operation of the primary treatment process. Eliminate any short-circuiting. Determine any other actions that can be taken to improve the performance of the primary process.
- Evaluate and adjust the control of sludge treatment processes to reduce the BOD_5 or recycle flows.
- Increase the recirculation rate to add additional dissolved oxygen (DO) to the filter influent. Do not increase the recirculation rate if the flow rate through the underdrains would cause less than 50% open space.
- Maintain aerobic conditions in filter influent.
- Remove debris from the media surface.
- Flush underdrains and vent pipes.
- Add one of the commercially available masking agents to reduce odors and prevent complaints.
- Add chlorine at a 1–2 mg/L residual for several hours at low flow. This will reduce activity and cut down on the oxygen demand. Chlorination only treats symptoms; a permanent solution must be determined and instituted.

High Clarifier Effluent Suspended Solids and/or BOD
Symptom:

- The effluent from the trickling filter process settling unit contains a high concentration of suspended solids.

Causal Factors:

- Recirculated flows are too high, causing hydraulic overloading of the settling tank. In multiple unit operations, the flow is not evenly distributed.
- The settling tank baffles or skirts have corroded or broken.
- Sludge collection mechanism is broken or malfunctioning.
- Effluent weirs are not level.
- Short-circuiting occurs because of temperature variations.
- Improper sludge withdrawal rate or frequency.
- Excessive solids loading from excessive sloughing.

Corrective Actions:

- Check hydraulic loading and adjust recirculated flow if hydraulic loading is too high.
- Adjust flow to ensure equal distribution.
- Inspect sludge removal equipment and repair broken parts.
- Monitor sludge blanket depth and sludge solids concentration; adjust withdrawal rate and/or frequency to maintain aerobic conditions in the settling tank.
- Adjust the effluent weir to obtain equal flow over all parts of the weir length.
- Determine the temperature in the clarifier at various points and depths throughout the clarifier. If depth temperatures are consistently 1°F–2°F lower than surface readings, a temperature problem exists. Baffles may be installed to help break up these currents.
- High sloughing rates because of the biological activity or temperature changes may create excessive solids loading. The addition of 1–2 mg/L of the cationic polymer may help improve solids capture. Remember, if polymer addition is used, solids withdrawal must be increased.
- High sloughings because of organic overloading, toxic wastes, or wide variations in influent flow are best controlled at their source.

Filter Flies
Symptoms:

- The trickling filter and surrounding area become populated with large numbers of very small flying insects (psychoid moths; also known as sewer gnats, drain flies, filter flies, or sewer flies).

Causal Factors:

- Poor housekeeping
- Insufficient recirculation
- Intermittent wet and dry conditions
- Warm weather

Corrective Actions (note that corrective actions for filter fly problems revolve around the need to disrupt the fly's life cycle, which is 7–10 days in warm weather):

- Increase the recirculation rate to obtain a hydraulic loading of at least 200 gpd/ft^2. At this rate, filter fly larvae are normally flushed out of the filter.
- Clean filter walls and remove weeds, brush, and shrubbery around the filter. This helps reduce the area for fly breeding.
- Dose the filter periodically with low chlorine concentrations (less than 1 mg/L). This normally destroys larvae.
- Dry the filter media for several hours.
- Flood the filter for 24 h.
- Spray the area around the filter with insecticide. Do not use insecticide directly on the media, because of the chance of carryover and unknown effects on the slime populations.

Freezing
Symptoms:

- Decreased air temperature results in visible ice formation and decreased performance.
- Distributed wastes are in a thin film or spray, which is more likely to cause ice formation.

Causal Factors:

- Recirculation causes increased temperature drops and losses.
- Strong prevailing winds cause heat losses.
- Intermittent dosing allows water to stand too long, causing freezing.

Corrective Actions (aimed to reducing heat loss as the wastes move through the filter):

- Reduce recirculation as much as possible to minimize cooling effects.
- Operate two-stage filters in parallel to reduce heat loss.
- Adjust splash plates and orifices to obtain a coarse spray.
- Construct a windbreak or plant evergreens or shrubs in the direction of the prevailing wind.
- If intermittent dosing is used, leave dump gates open.
- Cover pump wet wells and dose tanks to reduce heat losses.
- Cover filter media to reduce heat loss.
- Remove ice before it becomes large enough to cause a stoppage of arms.

Note: During periods of cold weather, the filter will show decreased performance. However, the filter should not be shut off for extended periods. Freezing of the moisture trapped within the media causes expansion and may lead to structural damage.

Process Calculations

Several calculations are useful for the operation of a trickling filter, including total flow, hydraulic loading, and organic loading.

Total Flow

If the recirculated flow rate is given, the total flow is

Total Flow, MGD = Influent Flow, MGD +

Recirculation Flow, MGD   (18.43)

Total Flow, gpd = Total Flow, MGD × 1,000,000 gal/MG

Note: The total flow to the tricking filter includes the influent flow and the recirculated flow. This can be determined using the recirculation ratio.

Total Flow, MGD = influent Flow × (Recirculation Rate + 1.0)

Example 18.31

Problem: The trickling filter is currently operating with a recirculation rate of 1.5. What is the total flow applied to the filter when the influent flow rate is 3.65 MGD?

Solution:

$$\text{Total Flow, MGD} = 3.65 \text{ MGD} \times (1.5 + 1.0)$$
$$= 9.13 \text{ MGD}$$

Hydraulic Loading

Calculating the hydraulic loading rate is important in accounting for both the primary effluent and the recirculated trickling filter effluent. Both of these are combined before being applied to the surface of the filter. The hydraulic loading rate is calculated based on the surface area of the filter.

Example 18.32

Problem: A trickling filter 90 ft in diameter is operated with a primary effluent of 0.488 MGD and a recirculated effluent flow rate of 0.566 MGD. Calculate the hydraulic loading rate on the filter in units of gpd/ft².

Solution: The primary effluent and recirculated trickling filter effluent are applied together across the surface of the filter; therefore,

$$0.488 \text{ MGD} + 0.566 \text{ MGD} = 1.054 \text{ MGD} = 1,054,000 \text{ gpd}$$

$$\text{Circular surface area} = 0.785 \times (\text{diameter})^2$$
$$= 0.785 \times (90 \text{ ft})^2$$
$$= 6,359 \text{ ft}^2$$

$$\frac{1,054,000 \text{ gpd}}{6,359 \text{ ft}^2} = 165.7 \text{ gpd/ft}^2$$

Organic Loading Rate

As mentioned earlier, trickling filters are sometimes classified by the organic loading rate applied. The organic loading rate is expressed as a certain amount of BOD applied to a certain volume of media.

Example 18.33

Problem: A trickling filter 50 ft in diameter receives a primary effluent flow rate of 0.445 MGD. Calculate the organic loading rate in units of pounds of BOD applied per day per 900 ft³ of media volume. The primary effluent BOD concentration is 85 mg/L, and the media depth is 9 ft.

Solution:

$$0.445 \text{ MGD} \times 85 \text{ mg/L} \times 8.34 \text{ lb/gal} = 315.5 \text{ BOD applied/day}$$

$$\text{Surface Area} = 0.785 \times (50)^2$$
$$= 1962.5 \text{ ft}^2$$

$$\text{Area} \times \text{Depth} = \text{Volume}$$

$$1,962.5 \text{ ft}^2 \times 9 \text{ ft} = 17,662.5 \text{ (TF Volume)}$$

Note: To determine the pounds of BOD per 1,000 ft³ in a volume of thousands of cubic feet, we must set up the equation as shown below.

$$\frac{315.5 \text{ lb BOD/day}}{17,662.5} \times \frac{1,000}{1,000}$$

Regrouping the numbers and the units together:

$$= \frac{315.5 \text{ lb} \times 1,000}{17,662.5} \times \frac{\text{lb BOD/day}}{1,000 \text{ ft}^3}$$

$$= 17.9 \frac{\text{lb BOD/day}}{1,000 \text{ ft}^3}$$

Settling Tank

In the operation of settling tanks that follow trickling filters, various calculations are routinely made to determine detention time, surface settling rate, hydraulic loading, and sludge pumping.

SIDEBAR 18.1 DEPLOYING BARLEY STRAW TO REDUCE ALGAL GROWTH IN WASTEWATER TREATMENT POND SYSTEM

(Note: The following guidance document was developed by the Illinois Environmental Protection Agency, Author: Charles E. Corley (1983). Springfield, Ill.)

Question: If you are not Anheuser Busch, what does one do with barley?

Answer: The days of filling a feedbag of a horse-drawn milk wagon are long gone; the sensible use is to float it, the straw portion at least, on your pond, ditch, impoundment, or reservoir. Why? Well, as some people have found from German Valley to Manchester, Illinois, and England respectively, your water will be clearer, cleaner, and lower in suspended solids due to algae.

Studied in the United Kingdom, the decomposition of barley straw and the observed effects on algae have been recorded since the late 1980s by multiple researchers, including Dr. Jonathon Newman, University of Bristol, Department of Agricultural Sciences, Reading, U.K. Over 10 years' worth of observations has distinguished "rotting barley straw" as an effective inhibitor of the color and suspended solids attributed to various types of algae. The research was done in "impoundments," slow-moving "canals," and many other bodies of water; and has been confirmed by laboratory studies. This has led researchers to propound: "Decomposing barely straw inhibits the growth of both filamentous and blue-green algae species in all types of water bodies so far assessed" (Newman and Barrett, 1993).

What causes barley to be so effective is not truly identified. Again, researchers have analyzed many chemical constituents produced by rotting barley straw (Newman, 1999). No one chemical is predominant and the combined effect appears to be the controlling factor. Not the presence of the straw, but the decomposition products appear to provide the effect. Other straws and plant material have been tested and dismissed in preference to barley (Newman, 1999). For example, green plant materials like alfalfa and hay impart an organic load on the system while wheat straw, corn, and lavender stalks, to quite common Illinois plant

materials the less so, seem to have poorer effects and longevity. Despite the uncertainty of the exact mechanism or product that produces the benefit, it is a benefit. One easily measured and observed, at that.

Transferring the technology, if one can refer to rotting barley straw as technology, to wastewater systems at best would seem a stretch. Newman's own studies indicate that algae growth continues with sufficient nutrient concentrations (Newman, 1999). Further, algae and fungi appear to be affected while all other aquatic animal and plant life are not. Nor is dissolved oxygen. Therefore, no detrimental conditions would be expected, and if any benefit accrues in wastewater systems, all would be positive steps.

The application of this truly natural and beneficial product to water bodies of all types is fundamentally simple. Bundle it, float it, and watch it rot! There is no need to search for the "right" type of barley straw or the vintage years, if there is one. Contact the nearest, cheapest, and most readily available source and have at it. A slight oversimplification, perhaps, but the years of observation have demonstrated these few basics—all confirmed by trials in Illinois communities and industries.

A few basics have been displayed in use in Illinois, including these: First, the straw must be floating throughout the application period. When allowed to sink, it is thought that it becomes a detrimental organic load. Second, since the original uses were in surface water ponds and impoundments and not wastewater systems, repeated applications are necessary from spring through warm weather. Warm weather and wind action on the surface are two necessary ingredients. Also, keeping the straw loosely packed inside a long open-web material, such as a common snow fence, is ideal and preferred to the more open-weave Christmas wrap where straws can escape.

Success can be found in all corners of our state. From Gardner to Ohio and Sorrento to Hudsonville, barley straw decomposition abounds. Measurable and observable benefits without any detrimental environmental effects abound. First used in Gardner at the wastewater pond system, it reduced the use of copper sulfate while improving the effluent suspended solids for weeks in the hottest part of the summer of 2000. The operator at Sorrento experienced similar benefits during the summer of 2002 at the water plant where lower turbidity was demonstrated and few applications of copper sulfate were needed. These two have a side benefit of reduced applications and reduced costs of an admittedly useful but hazardous material: copper sulfate. Other wastewater applications include the ash pond treatment at the Ameren Hudsonville Generating Station. Barley straw here reduced the algae count in the effluent, along with the suspended solids while positively affecting the pH of the discharge to the

Wabash River. The use of the straw in Ohio was done late in the summer of 2001. Not expecting a huge margin of success as a result of sludge pockets in all pond cells, the floating barley straw booms were effective in keeping the effluent suspended solids from exceeding the permitted limits for weeks.

These and other stories could be repeated throughout Illinois with willing participants and experimentation-minded communities (**Note**: No conflict with the EPA Federal Insecticide Fungicide and Rodenticide Act (FIFRA) would be exempted from use in the privacy of one's own non-public water body or pond, which the above were. Barley straw has been promoted without apparent conflict in the landscape pond industry for decades.) Who knows, the result might be cleaner, clearer ponds with fewer green discharges to Illinois' surface waters. Better water quality. What a concept!

Rotating Biological Contractors (RBCs)

The *rotating biological contactor (RBC)* is a biological treatment system (see Figure 16.9) and is a variation of the attached growth idea provided by the trickling filter. Still relying on microorganisms that grow on the surface of a medium, the RBC is instead a **fixed film** biological treatment device; the basic biological process, however, is similar to that occurring in the trickling filter. An RBC consists of a series of closely spaced (mounted side by side), circular, plastic (synthetic) disks that are typically about 3.5 m in diameter and attached to a rotating horizontal shaft (see Figure 18.8). Approximately 40% of each disk is submerged in a tank containing the wastewater to be treated. As the RBC rotates, the attached biomass film (zoogleal slime) that grows on the surface of the disk moves into and out of the wastewater. While submerged in the wastewater, the microorganisms absorb organics; while they are rotated out of the wastewater, they are supplied with needed oxygen for aerobic decomposition. As the zoogleal slime reenters the wastewater, excess solids and waste products are stripped off the media as sloughings. These sloughings are transported with the wastewater flow to a settling tank for removal.

Modular RBC units are placed in series (see Figure 18.9) simply because a single contactor is not sufficient to achieve the desired level of treatment; the resulting treatment achieved exceeds conventional secondary treatment. Each individual contactor is called a stage, and the group is known as a train. Most RBC systems consist of two or more trains with three or more stages in each. The key advantage in using RBCs instead of trickling filters is that RBCs are easier to operate under varying load conditions since it is easier to keep the solid medium wet at all times. Moreover, the level of nitrification that can be achieved by an RBC system is significant—especially when multiple stages are employed.

RBC Equipment

The equipment that makes up an RBC includes the rotating biological contactor (the media: either standard or high density), a center shaft, a drive system, a tank, baffles, housing or a cover, and a settling tank. The *rotating biological contactor* consists of circular sheets of synthetic material (usually plastic) that are mounted side by side on a shaft. The sheets (media) contain large amounts of surface area for the growth of the biomass. The *center shaft* provides support for the disks of media and must be strong enough to support the weight of the media and the biomass.

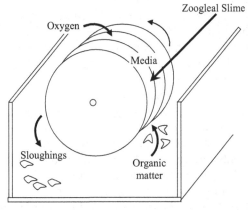

FIGURE 18.8 Rotating biological contactor (RBC) cross-section and treatment system.

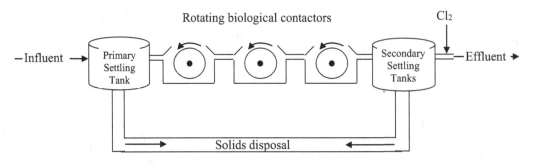

FIGURE 18.9 Rotating biological contactor (RBC) treatment system.

Experience has indicated that a major problem has been the collapse of the support shaft. The *drive system* provides the motive force to rotate the disks and shaft. The drive system may be mechanical, air-driven, or a combination of each. When the drive system does not provide uniform movement of the RBC, major operational problems can arise.

The *tank* holds the wastewater that the RBC rotates in. It should be large enough to permit variation in the liquid depth and detention time. *Baffles* are required to permit proper adjustment of the loading applied to each stage of the RBC process. Adjustments can be made to increase or decrease the submergence of the RBC. RBC stages are normally enclosed in some type of protective structure (*cover*) to prevent the loss of biomass due to severe weather changes (snow, rain, temperature, wind, sunlight, etc.). In many instances, this housing greatly restricts access to the RBC. The *settling tank* is provided to remove the sloughing material created by the biological activity and is similar in design to the primary settling tank. The settling tank provides 2–4-h detention times to permit the settling of lighter biological solids.

RBC Operation

During normal operation, operator vigilance is required to observe the RBC movement, slime color, and appearance. However, if the unit is covered, observations may be limited to that portion of the media that can be viewed through the access door. Slime color and appearance can indicate process conditions; for example:

- **Gray, Shaggy Slime Growth**: indicates normal operation
- **Reddish Brown, Golden Shaggy Growth:** indicates nitrification
- **White Chalky Appearance:** indicates high sulfur concentrations
- **No Slime:** indicates severe temperature or pH changes

Sampling and testing should be conducted daily for dissolved oxygen content and pH. BOD_5 and suspended solids testing should also be accomplished to aid in assessing performance.

RBC Expected Performance

The RBC normally produces a high-quality effluent with BOD_5 at 85%–95% and suspended solids removal at 85%–95%. The RBC treatment process may also significantly reduce (if designed for this purpose) the levels of organic nitrogen and ammonia nitrogen.

Operator Observations

Rotating biological filter operation requires routine observation, process control sampling and testing, troubleshooting, and process control calculations. Comparison of daily results with expected "normal" ranges is the key to identifying problems and appropriate corrective actions.

Note: If the RBC is covered, observations may be limited to the portion of the media that can be viewed through the access door.

1. **Rotation:** The operator routinely checks the operation of the RBC to ensure that smooth, uniform rotation is occurring (normal operation). Erratic, non-uniform rotation indicates a mechanical problem or uneven slime growth. If no movement is observed, mechanical problems or extreme excess of slime growth are indicated.
2. **Slime Color/Appearance:** Gray, shaggy slime growth on the RBC indicates normal operation. Reddish brown or golden brown shaggy growth indicates normal during nitrification. A very dark brown, shaggy growth (with worms present) indicates very old slime. White chalky growth indicates high influent sulfur/sulfide levels. No visible slime growth on the RBC indicates a severe pH or temperature change.

RBC Process Control Sampling and Testing

For process control, the RBC process does not require large amounts of sampling and testing to provide the information required. The frequency of performing suggested testing depends on available resources and the variability of the process. The frequency may be lower during normal operation and higher during abnormal conditions. The following routine sampling points and types of tests will permit the operator to identify normal and abnormal operating conditions:

1. *RBC train influent tests*
   Dissolved oxygen
   pH
   Temperature
   Settleable solids
   BOD_5
   Suspended solids
   Metals
2. *RBC test*
   Speed of rotation
3. *RBC train effluent tests*
   Dissolved oxygen
   pH
   Jar tests
4. *Process effluent tests*
   Dissolved oxygen
   pH
   Settleable solids
   BOD_5
   Suspended solids

Troubleshooting Operational Problems

(Much of the information in this section is based on material provided by USEPA in *Performance Evaluation and Troubleshooting at Municipal Wastewater Treatment*

Facilities, Washington, DC, current edition.) The following sections are not all-inclusive; they do not cover all of the operational problems associated with the rotating biological contactor process. They do, however, provide information on the most common operational problems.

1. White Slime
    Symptoms:
    • white slime on most of the disk area
    Causal Factors:
    • high hydrogen sulfide in influent
    • septic influent
    • first stage overloaded
    Corrective Actions:
    • aerate RBC or plant influent
    • add sodium nitrate or hydrogen peroxide to influent
    • adjust baffles between stages 1 and 2 to increase the fraction of total surface area in the first stage
2. Excessive Sloughing
    Symptoms:
    • loss of slime
    Causal Factors:
    • excessive pH variance
    • toxic influent
    Corrective Actions:
    • implement/enforce a pretreatment program
    • install pH control equipment
    • equalize flow to acclimate organisms
3. RBC Rotation
    Symptoms:
    • RBC rotation is uneven
    Causal Factors:
    • mechanical growth
    • uneven growth
    Corrective Actions:
    • repair mechanical problem
    • increase rotational speed
    • adjust baffles to decrease loading
    • increase sloughing
4. Solids
    Symptoms:
    • solids accumulating in reactors
    Causal Factors:
    • inadequate pretreatment
    Corrective Actions:
    • identify and correct the grit removal problem
    • identify and correct the primary settling problem
5. Shaft Bearings
    Symptoms:
    • shaft bearings running hot or failing
    Causal Factor:
    • inadequate maintenance
    Corrective Action:
    • follow the manufacturer's recommendations

6. Drive Motor
    Symptoms:
    • drive motor running hot
    Causal Factors:
    • inadequate maintenance
    • improper chain drive alignment
    Corrective Actions:
    • follow the manufacturer's recommendations
    • adjust alignment

RBC: Process Control Calculations

Several process control calculations may be useful in the operation of an RBC. These include soluble BOD, total media area, organic loading rate, and hydraulic loading rate. Settling tank calculations and sludge pumping calculations may be helpful for the evaluation and control of the settling tank following the RBC.

RBC Soluble BOD

The soluble BOD_5 concentration of the RBC influent can be determined experimentally in the laboratory, or it can be estimated using the suspended solids concentration and the "K" factor. The "K" factor is used to approximate the BOD_5 (particulate BOD) contributed by the suspended matter. The K factor must be provided or determined experimentally in the laboratory. The K factor for domestic wastes is normally in the range of 0.5–0.7.

$$\text{Soluble } BOD_5 = \text{Total } BOD_5$$
$$- (\text{K Factor} \times \text{Total Suspended Solids})$$
$$(18.44)$$

Example 18.34

Problem: The suspended solids concentration of wastewater is 250 mg/L. If the normal K-value at the plant is 0.6, what is the estimated particulate biochemical oxygen demand (BOD) concentration of the wastewater?

Solution:

Note: The K-value of 0.6 indicates that about 60% of the suspended solids are organic suspended solids (particulate BOD).

$$(250 \text{ mg/L})(0.6) = 150 \text{ mg/L (Particulate BOD)}$$

Example 18.35

Problem: A rotating biological contactor receives a flow of 2.2 MGD with a BOD content of 170 mg/L and a suspended solids (SS) concentration of 140 mg/L. If the K-value is 0.7, how many pounds of soluble BOD enter the RBC daily?

Solution:

$$\text{Total BOD} = \text{Particulate BOD} + \text{Soluble BOD}$$

$$170 \text{ mg/L} = (140 \text{ mg/L})(0.7) + x \text{ mg/L}$$

$$170 \text{ mg/L} = 98 \text{ mg/L} + x \text{ mg/L}$$

$$170 \text{ mg/L} - 98 \text{ mg/L} = x$$

$$x = 72 \text{ mg/L Soluble BOD}$$

Now, we can determine lb/day soluble BOD:

$$(\text{mg/L Soluble BOD})(\text{MGD Flow})(8.34 \text{ lb/gal}) = \text{lb/day}$$

$$(72 \text{ mg/L})(2.2 \text{ MGD})(8.34 \text{ lb/gal}) = 1{,}321 \text{ lb/day soluble BOD}$$

RBC Total Media Area

Several process control calculations for the RBC use the total surface area of all the stages within the train. As was the case with the soluble BOD calculation, plant design information or information supplied by the unit manufacturer must provide the individual stage areas (or the total train area), because physical determination of this would be extremely difficult.

$$\text{Total Area} = \text{1st Stage Area} + \text{2nd Stage Area} + \cdots + \text{nth Stage Area} \tag{18.45}$$

RBC Organic Loading Rate

If the soluble BOD concentration is known, the organic loading on an RBC can be determined. Organic loading on an RBC based on soluble BOD concentration can range from 3 to 4 lb/day/1,000 ft^2.

Example 18.36

Problem: An RBC has a total media surface area of 102,500 ft^2 and receives a primary effluent flow rate of 0.269 MGD. If the soluble BOD concentration of the RBC influent is 159 mg/L, what is the organic loading rate in lbs/1,000 ft^2?

Solution:

$$0.269 \text{ MGD} \times 159 \text{ mg/L} \times \frac{8.34 \text{ lb}}{1 \text{ gal}} = 356.7 \text{ lb/day}$$

$$\frac{356.7 \text{ lb/day}}{102{,}500 \text{ ft}^2} \times \frac{1{,}000 \text{ (number)}}{1{,}000 \text{ (unit)}} = 3.48 \text{ lb/day/1,000 ft}^2$$

RBC Hydraulic Loading Rate

The manufacturer normally specifies the RBC media surface area and the hydraulic loading rate is based on the media surface area, usually in square feet (ft^2). Hydraulic loading on an RBC can range from 1 to 3 gpd/ft^2.

Example 18.37

Problem: An RBC treats a primary effluent flow rate of 0.233 MGD. What is the hydraulic loading rate in gpd/ft^2 if the media surface area is 96,600 ft^2?

Solution:

$$\frac{233{,}000 \text{ gpd}}{96{,}600 \text{ ft}^2} = 2.41 \text{ gpd/ft}^2$$

ACTIVATED SLUDGE

The biological treatment systems discussed to this point [ponds, trickling filters, and rotating biological contactors (RBCs)] have been around for years. The trickling filter, for example, has been around and successfully used since the late 1800s. The problem with ponds, trickling filters, and RBCs is that they are temperature-sensitive, and remove less BOD, and, in the case of trickling filters, cost more to build than the activated sludge systems that were later developed.

Note: Although trickling filters and other systems cost more to build than activated sludge systems, it is important to point out that activated sludge systems cost more to operate due to the need for energy to run pumps and blowers.

As shown in Figure 18.1, the activated sludge process follows primary settling. The basic components of an activated sludge sewage treatment system include an aeration tank and a secondary basin, settling basin, or clarifier (see Figure 18.10). Primary effluent is mixed with settled solids recycled from the secondary clarifier and is then introduced into the aeration tank. Compressed air is injected continuously into the mixture through porous diffusers located at the bottom of the tank, usually along one side.

Wastewater is fed continuously into an aerated tank, where the microorganisms metabolize and biologically flocculate the organics. Microorganisms (activated sludge) are settled from the aerated mixed liquor under quiescent conditions in the final clarifier and are returned to the aeration tank. Left uncontrolled, the number of organisms would eventually become too great; therefore, some must periodically be removed (wasted). A portion of the concentrated solids from the bottom of the settling tank must be removed from the process (waste-activated sludge or WAS). The clear supernatant from the final settling tank is the plant effluent.

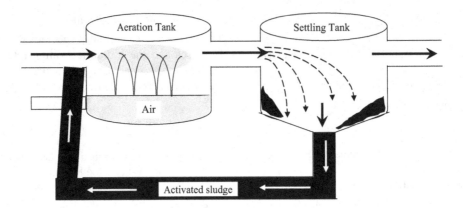

FIGURE 18.10 The activated sludge process.

ACTIVATED SLUDGE TERMINOLOGY

To better understand the discussion of the activated sludge process presented in the following sections, you must understand the terms associated with the process. Some of these terms have been used and defined earlier in the text, but we list them here again to refresh your memory. Review these terms and remember them. They are used throughout the discussion.

- **Adsorption:** taking in or reception of one substance into the body of another by molecular or chemical actions and distribution throughout the absorber.
- **Activated:** to speed up a reaction. When applied to sludge, it means that many aerobic bacteria and other microorganisms are in the sludge particles.
- **Activated Sludge:** a floc or solid formed by microorganisms. It includes organisms, accumulated food materials, and waste products from the aerobic decomposition process.
- **Activated Sludge Process:** a biological wastewater treatment process in which a mixture of influent and activated sludge is agitated and aerated. The activated sludge is subsequently separated from the treated mixed liquor by sedimentation and is returned to the process as needed. The treated wastewater overflows the weir of the settling tank in which separation from the sludge takes place.
- **Adsorption:** the adherence of dissolved, colloidal, or finely divided solids to the surface of solid bodies when they come into contact.
- **Aeration:** mixing air and a liquid by one of the following methods: spraying the liquid in the air; diffusing air into the liquid; or agitating the liquid to promote the surface adsorption of air.
- **Aerobic:** a condition in which "free" or dissolved oxygen is present in the aquatic environment. Aerobic organisms must be in the presence of dissolved oxygen to be active.
- **Bacteria:** single-celled organisms that play a vital role in the stabilization of organic waste.

- **Biochemical Oxygen Demand (BOD):** a measure of the amount of food available to the microorganisms in a particular waste. It is measured by the amount of dissolved oxygen used up during a specific time period (usually 5 days, expressed as BOD_5).
- **Biodegradable:** from "degrade" (to wear away or break down chemically) and "bio" (by living organisms). Put it all together, and you have a "substance, usually organic, which can be decomposed by biological action."
- **Bulking:** a problem in activated sludge plants that results in poor settleability of sludge particles.
- **Coning:** a condition that may be established in a sludge hopper during sludge withdrawal, when part of the sludge moves toward the outlet while the remainder tends to stay in place. This leads to the development of a cone or channel of moving liquids surrounded by relatively stationary sludge.
- **Decomposition:** generally, in waste treatment, decomposition refers to the process of changing waste matter into simpler, more stable forms that will not harm the receiving stream.
- **Diffuser:** a porous plate or tube through which air is forced and divided into tiny bubbles for distribution in liquids. Commonly made of carborundum, aluminum, or silica sand.
- **Diffused Air Aeration:** a diffused air-activated sludge plant takes air, compresses it, and then discharges the air below the water surface into the aerator through some type of air diffusion device.
- **Dissolved Oxygen:** atmospheric oxygen dissolved in water or wastewater, usually abbreviated as DO.

Note: The typical required DO for a well-operated activated sludge plant is between 2.0 and 2.5 mg/L.

- **Facultative:** facultative bacteria can use either molecular (dissolved) oxygen or oxygen obtained from food materials. In other words, facultative bacteria can live under aerobic or anaerobic conditions.

- **Filamentous Bacteria:** organisms that grow in thread or filamentous form.
- **Food-to-Microorganisms Ratio:** a process control calculation used to evaluate the amount of food (BOD or COD) available per pound of mixed liquor volatile suspended solids. This may be written as F/M ratio.

$$\frac{\text{Food}}{\text{Microorganism}} = \frac{\text{BOD, lb/day}}{\text{MLVSS, lb}}$$

$$= \frac{\text{Flow, MGD} \times \text{BOD, mg/L} \times 8.34 \text{ lb/gal}}{\text{Vol., MG} \times \text{MLVSS, mg/L} \times 8.34 \text{ lb/gal}}$$

- **Fungi:** multicellular aerobic organisms.
- **Gould Sludge Age:** a process control calculation used to evaluate the amount of influent suspended solids available per pound of mixed liquor suspended solids.
- **Mean Cell Residence Time (MCRT):** the average length of time mixed liquor suspended solids particle remains in the activated sludge process. This is usually written as MCRT and may also be referred to as *sludge retention rate* (STR).

$$\text{MCRT, days} = \frac{\text{Solids in Activated Sludge Process, lbs}}{\text{Solids Removed from Process, lb/day}}$$

- **Mixed Liquor:** the contribution of return activated sludge and wastewater (either influent or primary effluent) that flows into the aeration tank.
- **Mixed Liquor Suspended Solids (MLSS):** the suspended solids concentration of the mixed liquor. Many references use this concentration to represent the amount of organisms in the activated sludge process. This is usually written as MLSS.
- **Mixed Liquor Volatile Suspended Solids (MLVSS):** the organic matter in the mixed liquor suspended solids. This can also be used to represent the amount of organisms in the process. This is normally written as MLVSS.
- **Nematodes:** microscopic worms that may appear in biological waste treatment systems.
- **Nutrients:** substances required to support plant organisms. Major nutrients are carbon, hydrogen, oxygen, sulfur, nitrogen, and phosphorus.
- **Protozoa:** single-celled animals that are easily observed under the microscope at a magnification of 100×. Bacteria and algae are prime sources of food for advanced forms of protozoa.
- **Return Activated Sludge:** the solids returned from the settling tank to the head of the aeration tank. This is normally written as RAS.
- **Rising Sludge:** rising sludge occurs in the secondary clarifiers or activated sludge plant when the sludge settles to the bottom of the clarifier, is

compacted, and then rises to the surface in a relatively short time.
- **Rotifers:** multicellular animals with flexible bodies and cilia near their mouths used to attract food. Bacteria and algae are their major sources of food.
- **Secondary Treatment:** a wastewater treatment process used to convert dissolved or suspended materials into a form that can be removed.
- **Settleability:** a process control test used to evaluate the settling characteristics of the activated sludge. Readings taken at 30–60 min are used to calculate the settled sludge volume (SSV) and the sludge volume index (SVI).
- **Settled Sludge Volume:** the volume of mL/L (or percent) occupied by an activated sludge sample after 30 or 60 min of settling. Normally written as SSV with a subscript to indicate the time of the reading used for calculation (SSV_{30} or SSV_{60}).
- **Shock Load:** the arrival at a plant of a waste toxic to organisms, in sufficient quantity or strength to cause operating problems, such as odor or sloughing off of the growth of slime on the trickling filter media. Organic overloads can also cause a shock load.
- **Sludge Volume Index:** a process control calculation used to evaluate the settling quality of the activated sludge. It requires the SSV_{30} and mixed liquor suspended solids test results to calculate.

$$\text{Sludge Vol. Index (SVI), mL/g} = \frac{(30 \text{ min settled vol., mL/L}) (1,000 \text{ mg/g})}{\text{Mixed Liquor Suspended Solids, mg/L}}$$

- **Solids:** material in the solid state.
  Dissolved: solids present in solution. Solids that will pass through a glass fiber filter.
  Fixed: also known as inorganic solids. The solids that are left after a sample is ignited at 550°C (centigrade) for 15 min.
  Floatable Solids: solids that will float to the surface of still water, sewage, or other liquids. Usually composed of grease particles, oils, light plastic materials, etc. Also called *scum*.
  Non-settleable: finely divided suspended solids that will not sink to the bottom in still water, sewage, or other liquids in a reasonable period, usually 2 h. Non-settleable solids are also known as colloidal solids.
  Suspended: the solids that will not pass through a glass fiber filter.
  Total: the solids in water, sewage, or other liquids; it includes the suspended solids and dissolved solids.

Volatile: the organic solids. Measured as the solids that are lost on ignition of the dry solids at 550°C.

- **Waste Activated Sludge:** the solids being removed from the activated sludge process. This is normally written as WAS.

Activated Sludge Process: Equipment

The equipment requirements for the activated sludge process are more complex than those of other processes discussed. Equipment includes an *aeration tank, aeration, system-settling tank, return sludge,* and *waste sludge system.* These are discussed in the following sections.

Aeration Tank

The *aeration tank* is designed to provide the required detention time (depending on the specific modification) and ensure that the activated sludge and the influent wastewater are thoroughly mixed. Tank design normally attempts to ensure that no dead spots are created.

Aeration

Aeration can be mechanical or diffused. Mechanical aeration systems use agitators or mixers to mix air and mixed liquor. Some systems use a sparge ring to release air directly into the mixer. Diffused aeration systems use pressurized air released through diffusers near the bottom of the tank. Efficiency is directly related to the size of the air bubbles produced. Fine bubble systems have a higher efficiency. The diffused air system has a blower to produce large volumes of low-pressure air (5–10 psi), airlines to carry the air to the aeration tank, and headers to distribute the air to the diffusers, which release the air into the wastewater.

Settling Tank

Activated sludge systems are equipped with plain *settling tanks* designed to provide 2–4 h of hydraulic detention time.

Return Sludge

The return sludge system includes pumps, a timer or variable speed drive to regulate pump delivery, and a flow measurement device to determine actual flow rates.

Waste Sludge

In some cases, the *waste-activated sludge* withdrawal is accomplished by adjusting valves on the return system. When a separate system is used it includes pump(s), a timer or variable speed drive, and a flow measurement device.

Overview of Activated Sludge Process

The activated sludge process is a treatment technique in which wastewater and reused biological sludge full of living microorganisms are mixed and aerated. The biological solids are then separated from the treated wastewater in a clarifier and returned to the aeration process or wasted. The microorganisms are mixed thoroughly with the incoming organic material, and they grow and reproduce by using the organic material as food. As they grow and are mixed with air, the individual organisms cling together (flocculate). Once flocculated, they more readily settle in the secondary clarifiers.

The wastewater being treated flows continuously into an aeration tank where air is injected to mix the wastewater with the returned activated sludge and to supply the oxygen needed by the microbes to live and feed on the organics. Aeration can be supplied by injection through air diffusers at the bottom of the tank or by mechanical aerators located at the surface. The mixture of activated sludge and wastewater in the aeration tank is called the "mixed liquor." The mixed liquor flows to a secondary clarifier where the activated sludge is allowed to settle.

The activated sludge is constantly growing, and more is produced than can be returned for use in the aeration basin. Some of this sludge must, therefore, be wasted in a sludge handling system for treatment and disposal. The volume of sludge returned to the aeration basins is normally 40%–60% of the wastewater flow. The rest is wasted.

Factors Affecting Operation of the Activated Sludge Process

Several factors affect the performance of an activated sludge system. These include the following:

- Temperature
- Return rates
- Amount of oxygen available
- Amount of organic matter available
- pH
- Waste rates
- Aeration time
- Wastewater toxicity

To obtain the desired level of performance in an activated sludge system, a proper balance must be maintained between the amount of food (organic matter), organisms (activated sludge), and oxygen (dissolved oxygen, DO). The majority of problems with the activated sludge process result from an imbalance among these three items.

To fully appreciate and understand the biological processes taking place in a normally functioning activated sludge process, the operator must have knowledge of the key players in the process: the organisms. This makes sense when you consider that the heart of the activated sludge process is the mass of settleable solids formed by aerating wastewater containing biologically degradable compounds in the presence of microorganisms. Activated sludge consists of organic solids plus bacteria, fungi, protozoa, rotifers, and nematodes.

Growth Curve

To understand the microbiological population and its function in an activated sludge process, the operator must be familiar with the microorganism *growth curve*. In the presence of excess organic matter, the microorganisms multiply at a fast rate, and the demand for food and oxygen is at its peak. Most of this is used for the production of new cells. This condition is known as the *log growth phase*. As time continues, the amount of food available for the organisms declines. Floc begins to form while the growth rate of bacteria and protozoa begins to decline. This is referred to as the *declining growth phase*. The *endogenous respiration* phase occurs as the food available becomes extremely limited, and the organism mass begins to decline. Some of the microorganisms may die and break apart, thus releasing organic matter that can be consumed by the remaining population.

The actual operation of an activated-sludge system is regulated by three factors: (1) the quantity of air supplied to the aeration tank, (2) the rate of activated-sludge recirculation, and (3) the amount of excess sludge withdrawn from the system. Sludge wasting is an important operational practice because it allows the operator to establish the desired concentration of MLSS, the food/microorganism ratio, and the sludge age.

Note: Air requirements in an activated sludge basin are governed by (1) biological oxygen demand (BOD) loading and the desired removal effluent; (2) volatile suspended solids concentration in the aerator; and (3) suspended solids concentration of the primary effluent.

Activated Sludge Formation

The formation of activated sludge is dependent on three steps. The first step is the transfer of food from wastewater to organisms. The second step is the conversion of waste to a usable form. The third is the flocculation step.

1. **Transfer:** Organic matter (food) is transferred from the water to the organisms. Soluble material is absorbed directly through the cell wall. Particulate and colloidal matter is adsorbed to the cell wall, where it is broken down into simpler soluble forms and then absorbed through the cell wall.
2. **Conversion:** Food matter is converted to cell matter by synthesis and oxidation into end products such as CO_2, H_2O, NH_3, stable organic waste, and new cells.
3. **Flocculation:** Flocculation is the gathering of fine particles into larger particles. This process begins in the aeration tank and is the basic mechanism for the removal of suspended matter in the final clarifier. The concentrated *bio-floc* that settles and forms the sludge blanket in the secondary clarifier is known as activated sludge.

Activated Sludge: Performance-Controlling Factors

To maintain the working organisms in the activated sludge process, the operator must ensure that a suitable environment is maintained by being aware of the many factors influencing the process and by monitoring them repeatedly "Control" is defined as maintaining the proper solids (floc mass) concentration in the aerator for the incoming water (food) flow by adjusting the return and waste sludge pumping rates and regulating the oxygen supply to maintain a satisfactory level of dissolved oxygen in the process.

Aeration

The activated sludge process must receive sufficient aeration to keep the activated sludge in suspension and to satisfy the organisms' oxygen requirements. Insufficient mixing results in dead spots, septic conditions, and/or loss of activated sludge.

Alkalinity

The activated sludge process requires sufficient alkalinity to ensure that pH remains in the acceptable range of 6.5–9.0. If organic nitrogen and ammonia are being converted to nitrate (nitrification), sufficient alkalinity must be available to support this process as well.

Nutrients

The microorganisms of the activated sludge process require nutrients (nitrogen, phosphorus, iron, and other trace metals) to function. If sufficient nutrients are not available, the process will not perform as expected. The accepted minimum ratio of carbon to nitrogen, phosphorus, and iron is 100 parts carbon to five parts nitrogen, one part phosphorus, and 0.5 parts iron.

pH

The pH of the mixed liquor should be maintained within the range of 6.5–9.0 (6.0–8.0 is ideal). Gradual fluctuations within this range will normally not upset the process. Rapid fluctuations or fluctuations outside this range can reduce organism activity.

Temperature

As temperature decreases, the activity of the organisms will also decrease. Cold temperatures also require longer recovery times for systems that have been upset. Warm temperatures tend to favor denitrification and filamentous growth.

Note: The activity level of bacteria within the activated sludge process increases with the temperature rise.

Toxicity

Sufficient concentrations of elements or compounds that enter a treatment plant and can kill the microorganisms (the activated sludge) are known as toxic waste (shock level). Common to this group are cyanides and heavy metals.

Note: A typical example of a toxic substance added by operators is the uninhabited use of chlorine for odor control or the control of filamentous organisms (prechlorination). Chlorination is for disinfection. Chlorine is a toxicant and should not be allowed to enter the activated sludge process; it is not selective with respect to the type of organisms damaged or killed. It may kill the organisms that should be retained in the process as workers. Chlorine is very effective in disinfecting the plant effluent after treatment by the activated sludge process, however.

Hydraulic Loading

Hydraulic loading is the amount of flow entering the treatment process. When compared with the design capacity of the system, it can be used to determine if the process is hydraulically overloaded or underloaded. If more flow is entering the system than it was designed to handle, the system is hydraulically overloaded. If less flow is entering the system than it was designed for, the system is hydraulically underloaded. Generally, the system is more affected by overloading than by underloading. Overloading can be caused by stormwater, infiltration of groundwater, excessive return rates, or many other causes. Underloading normally occurs during periods of drought or in the period following initial startup when the plant has not reached its design capacity. Excess hydraulic flow rates through the treatment plant will reduce the efficiency of the clarifier by allowing activated sludge solids to rise in the clarifier and pass over the effluent weir. This loss of solids in the effluent degrades effluent quality and reduces the amount of activated sludge in the system, which in turn reduces process performance.

Organic Loading

Organic loading is the amount of organic matter entering the treatment plant. It is usually measured as biochemical oxygen demand (BOD). An organic overload occurs when the amount of BOD entering the system exceeds the design capacity of the system. An organic underload occurs when the amount of BOD entering the system is significantly less than the design capacity of the plant. Organic overloading may occur when the system receives more waste than it was designed to handle. It can also occur when an industry or other contributor discharges more waste into the system than originally planned. Wastewater treatment plant processes can also cause organic overloads by returning high-strength wastes from the sludge treatment processes.

Regardless of the source, organic overloading of the plant results in increased demand for oxygen. This demand may exceed the air supply available from the blowers. When this occurs, the activated sludge process may become septic. Excessive wasting can also result in a type of organic overload. The food available exceeds the number of activated sludge organisms, resulting in increased oxygen demand and very rapid growth.

Organic underloading may occur when a new treatment plant is initially put into service. The facility may not receive enough waste to allow the plant to operate at its design level. Underloading can also occur when excessive amounts of activated sludge are allowed to remain in the system. When this occurs, the plant will have difficulty developing and maintaining good activated sludge.

Activated Sludge Modifications

First developed in 1913, the original activated sludge process has been modified over the years to provide better performance for specific operating conditions or different influent waste characteristics.

1. **Conventional Activated Sludge**
   - Employing the conventional activated sludge modification requires primary treatment.
   - Conventional activated sludge provides excellent treatment; however, large aeration tank capacity is required, and construction costs are high.
   - In operation, the initial oxygen demand is high. The process is also very sensitive to operational problems (e.g., bulking).
2. **Step Aeration**
   - Step aeration requires primary treatment.
   - It provides excellent treatment.
   - Operational characteristics are similar to conventional.
   - It distributes organic loading by splitting influent flow.
   - It reduces oxygen demand at the head of the system.
   - It reduces solids loading on the settling tank.
3. **Complete Mix**
   - It may or may not include primary treatment.
   - It distributes waste, return, and oxygen evenly throughout the tank.
   - Aeration may be more efficient.
   - It maximizes tank use.
   - It permits a higher organic loading.

Note: During the complete mix, activated sludge process organisms are in a declining phase on the growth curve.

4. **Pure Oxygen**
   - It requires primary treatment.
   - It permits higher organic loading.
   - It uses higher solids levels.
   - It operates at higher F: M ratios.
   - It uses covered tanks.
   - It involves potential safety hazards (pure oxygen).
   - Oxygen production is expensive.
5. **Contact Stabilization**
   - Contact stabilization does not require primary treatment.
   - During operation, organisms collect organic matter (during contact).

- Solids and activated sludge are separated from the flow via settling.
- Activated sludge and solids are aerated for 3–6 h (stabilization).

Note: Return sludge is aerated before it is mixed with influent flow.

- The activated sludge oxidizes available organic matter.
- While the process is complicated to control, it requires less tank volume than other modifications and can be prefabricated as a *package* unit for flows of 0.05–1.0 MGD.
- A disadvantage is that common process control calculations do not provide usable information.

Extended Aeration

- It does not require primary treatment.
- It is used frequently for small flows such as schools and subdivisions.
- It uses 24-h aeration.
- It produces low BOD effluent.
- It produces the least amount of waste-activated sludge.
- The process is capable of achieving 95% or more removals of BOD.
- it can produce effluent low in organic and ammonia nitrogen.

Oxidation Ditch

- It does not require primary treatment.
- The oxidation ditch process is similar to the extended aeration process.

Table 18.6 lists the process parameters for each of the four most commonly used activated sludge modifications.

Extended Aeration: Package Plants

One of the most common types of modified active sludge processes that provide biological treatment for the removal of biodegradable organic waste under aerobic conditions is the extended aeration process called the package plant. Package plants are pre-manufactured treatment facilities used to treat wastewater in small communities or on individual properties. According to manufacturers, package plants can be designed to treat flows as low as 0.002 MGD or as high as 0.5 MGD, although they more commonly treat flows between 0.01 and 0.25 MGD (Metcalf and Eddy, 1991).

In operation, air may be supplied to the extended aeration package plant by mechanical or diffused aeration to provide the oxygen required to sustain the aerobic biological process. Mixing must be provided by aeration or mechanical means to maintain the microbial organisms in contact with the dissolved organics. In addition, the pH must be controlled to optimize the biological process and essential nutrients must be present to facilitate biological growth and the continuation of biological degradation. Wastewater enters the treatment system and is typically screened immediately to remove large suspended, settleable, or floating solids that could interfere with or damage equipment downstream in the process. Wastewater may then pass through a grinder to reduce large particles that are not captured in the screening process. If the plant requires the flow to be regulated, the effluent will then flow into equalization basins which regulate peak wastewater flow rates. Wastewater then enters the aeration chamber, where it is mixed and oxygen is provided to the microorganisms. The mixed liquor then flows to a clarifier or settling chamber where most microorganisms settle to the bottom of the clarifier and a portion is pumped back to the incoming wastewater at the beginning of the plant. This returned material is the return-activated sludge (RAS). The material that is not returned, the waste-activated sludge (WAS), is removed for treatment and disposal.

TABLE 18.6
Activated Sludge Modifications

Parameter	Conventional	Contact Stabilization	Extended Aeration	Oxidation Ditch
Aeration time, h	4–8	0.5–1.5 (contact) 3–6 (reaeration)	24	24
Settling time, h	2–4	2–4	2–4	2–4
Return rate, % of influent flow	25–100	25–100	25–100	25–100
MLSS, mg/L	1,500–4,000	1,000–3,000 3,000–8,000	2,000–6,000	2,000–6,000
D.O., mg/L	1–3	1–3	1–3	1–3
SSV_{30}, mL/L	400–700	400–700 (contact)	400–700	400–700
Food: mass ratio lbs BOD_5/lb MLVSS	0.2–0.5	0.2–0.6 (contact)	0.05–0.15	0.05–0.15
MCRT (whole system [d])	5–15	N/A	20–30	20–30
% Removal BOD_5	85%–95%	85%–95%	85%–95%	85%–95%
% Removal TSS	85%–95%	85%–95%	85%–95%	85%–95%
Primary treatment	Yes	No	No	No

The clarified wastewater then flows over a weir and into a collection channel before being diverted to the disinfection system (USEPA, 2000).

Typically located in small municipalities, suburban subdivisions, apartment complexes, highway rest areas, trailer parks, small institutions, and other sites where flow rates are below 0.1 MGD, extended aeration package plants consist of a steel tank that is compartmentalized into flow equalization, aeration, clarification, disinfection, and aerated sludge holding/digestion segments. Extended aeration systems are typically manufactured to treat wastewater flow rates between 0.002 and 0.1 MGD. Using concrete tanks may be preferable for larger sizes (Sloan Equipment, 1999).

Extended aeration plants are usually started up using "seed sludge" from another sewage plant. It may take as many as 2–4 weeks from the time it is seeded for the plant to stabilize (Sloan Equipment, 1999). These systems are also useful for areas requiring nitrification.

Key internal components of extended aeration treatment package plants consist of the following: transfer pumps to move wastewater between the equalization and aeration zones; a bar screen and/or grinder to decrease the size of large solids; an aeration system consisting of blowers and diffusers for the equalization, aeration, and sludge holding zones; transfer pumps to move wastewater between the equalization and aeration zones; an airlift pump for returning sludge; a skimmer and effluent weir for the clarifier; and UV, liquid hypochlorite, or tablet modules used in the disinfection zone. Blowers and the control panel containing switches, lights, and motor starters are typically attached to either the top or one side of the package plant (Sloan Equipment, 1999).

Advantages and Disadvantages

Advantages:

- Plants are easy to operate, as many are manned for a maximum of 2 or 3 h/day.
- Extended aeration processes are often better at handling organic loading and flow fluctuations, as there is a greater detention time for the nutrients to be assimilated by microbes.
- Systems are easy to install, as they are shipped in one or two pieces and then mounted on an onsite concrete pad, above or below grade.
- Systems are odor-free, can be installed in most locations, have a relatively small footprint, and can be landscaped to match the surrounding area.
- Extended aeration systems have a relatively low sludge yield due to long sludge ages, can be designed to provide nitrification, and do not require a primary clarifier.

Disadvantages:

- Extended aeration plants do not achieve denitrification or phosphorus removal without additional unit processes.
- Flexibility is limited to adapting to changing effluent requirements resulting from regulatory changes.
- A longer aeration period requires more energy.
- Systems require a larger amount of space and tankage than other "higher rate" processes, which have shorter aeration detention times.

Oxidation Ditches

An oxidation ditch is a modified extended aeration-activated sludge biological treatment process that utilizes long solids retention times (SRTs) to remove biodegradable organics. Oxidation ditches are typically complete mix systems, but they can be modified to approach plug flow conditions. (**Note:** as conditions approach plug flow, diffused air must be used to provide enough mixing. The system will also no longer operate as an oxidation ditch.) Typical oxidation ditch treatment systems consist of a single or multi-channel configuration within a ring, oval, or horseshoe-shaped basin. As a result, oxidation ditches are called "racetrack type" reactors. Horizontally or vertically mounted aerators provide circulation, oxygen transfer, and aeration in the ditch.

Preliminary treatment, such as bar screens and grit removal, normally precedes the oxidation ditch. Primary settling prior to an oxidation ditch is sometimes practiced but is not typical in this design. Tertiary filters may be required after clarification, depending on the effluent requirements. Disinfection is required, and reaeration may be necessary prior to final discharge. Flow to the oxidation ditch is aerated and mixed with return sludge from a secondary clarifier. Surface aerators, such as brush rotors, disc aerators, draft tube aerators, or fine bubble diffusers are used to circulate the mixed liquor. The mixing process entrains oxygen into the mixed liquor to foster microbial growth, and the motive velocity ensures contact of microorganisms with the incoming wastewater. The aeration sharply increases the dissolved oxygen (DO) concentration but decreases as biomass uptakes oxygen as the mixed liquor travels through the ditch. Solids are maintained in suspension as the mixed liquor travels through the ditch. If design SRTs are selected for nitrification, a high degree of nitrification will occur. Oxidation ditch effluent is usually settled in a separate secondary clarifier. An anaerobic tank may be added prior to the ditch to enhance biological phosphorus removal.

An oxidation ditch may also be operated to achieve partial denitrification. One of the most common design modifications for enhanced nitrogen removal is known as the Modified Ludzack-Ettinger (MLE) process. In this process, an anoxic tank is added upstream of the ditch, along with mixed liquor recirculation from the aerobic zone to the tank, to achieve higher levels of denitrification. In the aerobic basin, autotrophic bacteria (nitrifiers) convert ammonia-nitrogen to nitrite-nitrogen and then to nitrate-nitrogen. In the anoxic zone, heterotrophic bacteria convert nitrate-nitrogen to nitrogen gas which is released into the atmosphere. Some mixed liquor from the aerobic basin is recirculated to the anoxic zone to provide mixed liquor with a high concentration of nitrate-nitrogen to the anoxic zone.

Several manufacturers have developed modifications to the oxidation ditch design to remove nutrients in conditions cycled or phased between the anoxic and aerobic states. While the mechanics of operation differ by manufacturer, in general, the process consists of two separate aeration basins: the first anoxic and the second aerobic. Wastewater and RAS are introduced into the first reactor which operates under anoxic conditions. Mixed liquor then flows into the second reactor operating under aerobic conditions. The process is then reversed, and the second reactor begins to operate under anoxic conditions.

With regard to applicability, the oxidation ditch process is a fully demonstrated secondary wastewater treatment technology, applicable in any situation where activated sludge treatment (conventional or extended aeration) is appropriate. Oxidation ditches are applicable in plants that require nitrification because the basins can be sized using an appropriate SRT to achieve nitrification at the mixed liquor minimum temperature. This technology is very effective in small installations, small communities, and isolated institutions because it requires more land than conventional treatment plants (USEPA, 2000).

There are currently more than 9,000 municipal oxidation ditch installations in the United States (Spellman, 2007). Nitrification to less than 1 mg/L ammonia nitrogen consistently occurs when ditches are designed and operated for nitrogen removal. An excellent example of an upgrade to the MLE process is provided in the following case. Keep in mind that the motivation for this upgrade was twofold: to increase optimal plant operation (DO optimization) and to conserve energy.

Advantages and Disadvantages

Advantages:

The main advantage of the oxidation ditch is the ability to achieve removal performance objectives with low operational requirements and operation and maintenance costs. Some specific advantages of oxidation ditches include:

- An added measure of reliability and performance over other biological processes owing to a constant water level and continuous discharge, which lowers the weir overflow rate and eliminates the periodic effluent surge common to other biological processes, such as SBRS.
- Long hydraulic retention time and complete mixing minimize the impact of a shock load or hydraulic surge.
- Produces less sludge than other biological treatment processes, owing to extended biological activity during the activated sludge process.
- Energy-efficient operations result in reduced energy costs compared with other biological treatment processes.

Disadvantages:

- Effluent suspended solids concentrations are relatively high compared to other modifications of the activated sludge process.
- Requires a larger land area than other activated sludge treatment options. This can prove costly, limiting the feasibility of oxidation ditches in urban, suburban, or other areas where land acquisition costs are relatively high (USEPA, 2000).

ACTIVATED SLUDGE PROCESS CONTROL PARAMETERS

When operating an activated sludge process, the operator must be familiar with the many important process control parameters that must be monitored frequently and adjusted occasionally to maintain optimal performance.

Alkalinity

Monitoring alkalinity in the aeration tank is essential to the control of the process. Insufficient alkalinity will reduce organism activity and may result in low effluent pH and, in some cases, extremely high chlorine demand in the disinfection process.

Dissolved Oxygen (DO)

The activated sludge process is an aerobic process that requires some dissolved oxygen to be present at all times. The amount of oxygen required is dependent on the influent food (BOD), the activity of the activated sludge, and the degree of treatment desired.

pH

Activated sludge microorganisms can be injured or destroyed by wide variations in pH. The pH of the aeration basin will normally be in the range of 6.5–9.0. Gradual variations within this range will not cause any major problems; however, rapid changes of one or more pH units can have a significant impact on performance. Industrial waste discharges, septic wastes, or significant amounts of stormwater flows may produce wide variations in pH. pH should be monitored as part of the routine process control testing schedule. Sudden changes or abnormal pH values may indicate an industrial discharge of strongly acidic or alkaline wastes. Because these wastes can upset the environmental balance of the activated sludge, the presence of wide pH variations can result in poor performance. Processes undergoing nitrification may show a significant decrease in effluent pH.

Mixed Liquor Suspended Solids, Mixed Liquor Volatile Suspended Solids and Mixed

Liquor Total Suspended Solids

The mixed liquor suspended solids (MLSS) or mixed liquor volatile suspended solids (MLVSS) can be used to represent the activated sludge or microorganisms present in the process. Process control calculations, such as sludge age and sludge volume index, cannot be calculated unless the MLSS is determined. Adjust the MLSS and MLVSS by increasing or decreasing the waste sludge rates. The mixed liquor total suspended solids (MLTSS) is an important activated sludge control parameter. To increase the MLTSS, for example, the operator must decrease the waste rate and/or increase the MCRT. The MCRT must be decreased to prevent the MLTSS from changing when the number of aeration tanks in service is reduced.

Note: In performing the Gould Sludge Age Test, assume that the source of the MLTSS in the aeration tank is influent solids.

Return Activated Sludge Rate and Concentration

The sludge rate is a critical control variable. The operator must maintain a continuous return of activated sludge to the aeration tank; otherwise, the process will show a drastic decrease in performance. If the rate is too low, solids remain in the settling tank, resulting in solids loss and a septic return. If the rate is too high, the aeration tank can become hydraulically overloaded, causing reduced aeration time and poor performance. The return concentration is also important because it may be used to determine the return rate required to maintain the desired MLSS.

Waste-Activated Sludge Flow Rate

Because the activated sludge contains living organisms that grow, reproduce, and produce waste matter, the amount of activated sludge is continuously increasing. If the activated sludge is allowed to remain in the system too long, the performance of the process will decrease. If too much activated sludge is removed from the system, the solids become very light and will not settle quickly enough to be removed in the secondary clarifier.

Temperature

Because temperature directly affects the activity of the microorganisms, accurate monitoring of temperature can help identify the causes of significant changes in organization populations or process performance.

Sludge Blanket Depth

The separation of solids and liquids in the secondary clarifier results in a blanket of solids. If solids are not removed from the clarifier at the same rate they enter, the blanket will increase in depth. If this occurs, the solids may carry over into the process effluent. The sludge blanket depth may be affected by other conditions, such as temperature variation, toxic wastes, or sludge bulking. The best sludge blanket depth is dependent upon factors such as hydraulic load, clarifier design, sludge characteristics, and many more. The best blanket depth must be determined on an individual basis by experimentation.

Note: In measuring sludge blanket depth, it is general practice to use a 15–20 ft long clear plastic pipe marked at 6-in intervals, the pipe is equipped with a ball valve at the bottom.

ACTIVATED SLUDGE OPERATIONAL CONTROL LEVELS

(Much of the information in this section is based on *Activated Sludge Process Control, Part II*, 2nd ed. Richmond, VA: Virginia Water Control Board, 1990.) The operator has two methods available to operate an activated sludge system. The operator can wait until the process performance deteriorates and make drastic changes, or the operator can establish *normal* operational levels and make minor adjustments to keep the process within the established operational levels.

Note: Control levels can be defined as the upper and lower values for a process control variable that can be expected to produce the desired effluent quality.

Although the first method will guarantee that plant performance is always maintained within effluent limitations, the second method has a much higher probability of achieving this objective. This section discusses methods used to establish *normal* control levels for the activated sludge process. Several major factors should be considered when establishing control levels for the activated sludge system. These include the following:

- Influent characteristics
- Industrial contributions
- Process side streams
- Seasonal variations
- Required effluent quality

Influent Characteristics

Influent characteristics were discussed earlier. However, a major area to consider when evaluating influent characteristics is the nature and volume of industrial contributions to the system. Waste characteristics (BOD, solids, pH, metals, toxicity, and temperature), volume, and discharge pattern (continuous, slug, daily, weekly, etc.) should be evaluated when determining if a waste will require pretreatment by the industry or adjustments to operational control levels.

Industrial Contributions

One or more industrial contributors produce a significant portion of the plant loading (in many systems). Identifying and characterizing all industrial contributors is important. Remember that the volume of waste generated may not be as important as the characteristics of the waste. Extremely high-strength wastes can result in organic overloading and/or poor performance because of insufficient nutrient availability. A second consideration is the presence of materials that, even in small quantities, are toxic to the process microorganisms or that create a toxic condition in the plant effluent or plant sludge. Industrial contributions to a biological treatment system should be thoroughly characterized prior to acceptance, monitored frequently, and controlled by either local ordinances or by the implementation of a pretreatment program.

Process Side Streams

Process side streams are flows produced in other treatment processes that must be returned to the wastewater system for treatment prior to disposal. Examples of process side streams include the following:

- Thickener supernatant
- Aerobic and anaerobic digester supernatant
- Liquids removed by sludge dewatering processes (filtrate, centrate, and subnate)
- Supernatant from heat treatment and chlorine oxidation sludge treatment processes

Testing these flows periodically to determine both their quantity and strength is important. In many treatment systems, a significant part of the organic and/or hydraulic loading for the plant is generated by side stream flows. The contribution of the plant side stream flows can significantly change the operational control levels of the activated sludge system.

Seasonal Variations

Seasonal variations in temperature, oxygen solubility, organism activity, and waste characteristics may require several *normal* control levels for the activated sludge process. For example, during the cold months of the year, aeration tank solids levels may have to be maintained at significantly higher levels than are required during warm weather. Likewise, the aeration rate may be controlled by the mixing requirements of the system during the colder months and by the oxygen demand of the system during the warm months.

Control Levels at Startup

Control levels for an activated sludge system during startup are usually based on design engineer recommendations or information available from recognized reference sources.

Although these levels provide a starting point, you should recognize that both the process control parameter sensitivity and control levels should be established on a plant-by-plant basis.

During the first 12 months of operation, you should evaluate all potential process control options to determine the following:

- Sensitivity to effluent quality changes
- Seasonal variability
- Potential problems

VISUAL INDICATORS FOR INFLUENT OR AERATION TANK

Wastewater operators are required to monitor or make certain observations of treatment unit processes to ensure optimum performance and to adjust when required. In monitoring the operation of an aeration tank, the operator should look for three physical parameters (turbulence, surface foam

and scum, and sludge color and odor), that aid in determining how the process is operating and indicate if any operational adjustments should be made. This information should be recorded each time operational tests are performed. We summarize aeration tank and secondary settling tank observations in the following sections. Remember that many of these observations are very subjective and must be based on experience. Plant personnel must be properly trained on the importance of ensuring that recorded information is consistent throughout the operating period.

Turbulence

Normal operation of an aeration basin includes a certain amount of turbulence. This turbulent action is, of course, required to ensure a consistent mixing pattern. However, whenever excessive, deficient, or non-uniform mixing occurs, adjustments may be necessary to airflow, or diffusers may need cleaning or replacement.

Surface Foam and Scum

The type, color, and amount of foam or scum present may indicate the required wasting strategy to be employed. Types of foam include the following:

- **Fresh, Crisp, White Foam:** moderate amounts of crisp white foam are usually associated with activated sludge processes producing an excellent final effluent. *Adjustment:* None, normal operation.
- **Thick, Greasy, Dark Tan Foam:** a thick greasy dark tan, or brown foam or scum normally indicates old sludge that is over-oxidized, has high mixed liquor concentration, and a waste rate is too high. *Adjustment:* Indicates old sludge; more wasting is required.
- **White Billowing Foam:** large amounts of a white, soap suds-like foam indicate very young, under-oxidized sludge. *Adjustment:* Young sludge; less wasting is required.

Sludge Color and Odor

Though not as reliable an indicator of process operations as foam, sludge colors and odors are also useful indicators. Colors and odors that are important include the following:

1. *Chocolate brown/earthy odor* indicates normal operation.
2. *Light tan or brown/no odor* indicates sand and clay from infiltration/inflow.
   Adjustment: Decrease wasting of extremely young sludge.
3. *Dark brown/earthy odor* indicates old sludge with high solids.
   Adjustment: Increase wasting.
4. *Black color/rotten egg odor* indicates septic conditions; low dissolved oxygen concentration; and airflow rate that is too low.
   Adjustment: Increase aeration.

Mixed Liquor Color

A light chocolate brown mixed liquor color indicates a well-operated activated sludge process.

FINAL SETTLING TANK (CLARIFIER) OBSERVATION

Settling tank observations include flow pattern (normally uniform distribution), settling, the amount and type of solids leaving with the process effluent (normally very low), and the clarity or turbidity of the process effluent (normally very clear). Observations should include the following conditions:

1. **Sludge Bulking:** occurs when solids are evenly distributed throughout the tank and leave over the weir in large quantities.
2. **Sludge Solids Washout:** sludge blanket is down but solids are flowing over the effluent weir in large quantities. Control tests indicate good-quality sludge.
3. **Clumping:** large "clumps" or masses of sludge (several inches or more) rise to the top of the settling tank.
4. **Ashing:** fine particles of gray to white material flowing over the effluent weir in large quantities.
5. **Straggler Floc:** small, almost transparent, very fluffy, buoyant solid particles (1/8–1/4 in diameter rising to the surface). This is usually accompanied by very clean effluent. New growth is most noted in the early morning hours. Sludge age is slightly below optimum.
6. **Pin Floc:** very fine solid particles, usually less than 1/32 in diameter, suspended throughout lightly turbid liquid. This is usually the result of over-oxidized sludge.

PROCESS CONTROL TESTING AND SAMPLING

The activated sludge process generally requires more sampling and testing to maintain adequate process control than any of the other unit processes in the wastewater treatment system. During periods of operational problems, both the parameters tested and the frequency of testing may increase substantially. Process control testing may include settleability testing to determine the settled sludge volume; suspended solids testing to determine influent and mixed liquor suspended solids; return activated sludge solids and waste activated sludge concentrations; determination of the volatile content of the mixed liquor suspended solids; dissolved oxygen and pH of the aeration tank; BOD_5 and/or chemical oxygen demand (COD) of the aeration tank influent and process effluent; and microscopic evaluation of the activated sludge to determine the predominant organisms. The following sections describe most of the common process control tests.

Aeration Influent Sampling

pH is tested daily with a sample taken from the aeration tank influent and process effluent. pH is normally close to 7.0 (normal) with the best pH range being from 6.5 to 8.5 (however, a pH range of 6.5–9.0 is satisfactory). A pH of >9.0 may indicate toxicity from an industrial waste contributor. A pH of <6.5 may indicate a loss of flocculating organisms; potential toxicity; industrial waste contributor; or acid storm flow. Keep in mind that the effluent pH may be lower because of nitrification.

Temperature

Temperature is important because as the:

Temperature Increases	Organism Activity Increases
	Aeration Efficiency Decreases
	Oxygen Solubility Decreases
Temperature Decreases	Organism Activity Decreases
	Aeration Efficiency Increases
	Oxygen Solubility Increases

Dissolved Oxygen

The content of dissolved oxygen (DO) in the aeration process is critical to performance. DO should be tested at least daily (peak demand). Optimum is determined for individual plants, but normal is from 1 to 3 mg/L. If the system contains too little D.O., the process will become septic. If it contains too much D.O., energy and money are wasted.

Settled Sludge Volume (Settleability)

Settled Sludge Volume or SSV is determined at specified times during sample testing; 30 and 60-min observations are used for control. Subscript (SSV_{30} and SSV_{60}) indicates settling time. The test is performed on an aeration tank effluent sample.

$$SSV = \frac{\text{Milliliters of Settled Sludge 1,000 mL/L}}{\text{Milliliters of Sample}} \quad (18.46)$$

$$\%SSV = \frac{\text{Milliliters of Settled Sludge} \times 100}{\text{Milliliters of Sample}} \quad (18.47)$$

Under normal conditions, sludge settles as a mass, producing a clear supernatant with SSV_{60} in the range of 400–700 mL/L. When higher values are indicated, this may suggest excessive solids (old sludge) and/or bulking conditions. Rising solids (if sludge is well oxidized) may appear after two or more hours. However, rising solids in less than 1 h indicates a problem.

Note: Running the settleability test with a diluted sample can assist in determining if the activated sludge is old (too many solids) or bulking (not settling). Old sludge will settle to a more compact level when diluted.

Centrifuge Testing

The centrifuge test provides a quick, relatively easy control test for the solids level in the aerator but does not usually correlate with MLSS results. The results are directly affected by variations in sludge quality.

Alkalinity

Alkalinity is essential for biological activity. Nitrification needs 7.3-mg/L of alkalinity per mg/L of total Kjeldahl nitrogen.

BOD₅

BOD₅

Testing that shows an increase in BOD₅ indicates increased organic loading, while a decrease in BOD₅ indicates decreased organic loading.

Total Suspended Solids (TSS)

An increase in total suspended solids indicates an increase in organic loading, while a decrease in TSS indicates a decrease in organic loading.

Total Kjeldahl Nitrogen

Total Kjeldahl nitrogen determination is required to monitor the process nitrification status and to determine alkalinity requirements.

Ammonia Nitrogen

Determination of ammonia nitrogen is required to monitor the process nitrification status.

Metals

Metal contents are measured to determine toxicity levels.

Aeration Tank

pH

The normal pH range in the aeration tank is 6.5–9.0. Decrease in pH indicate process side streams or insufficient alkalinity available.

Dissolved Oxygen (DO)

The normal range of dissolved oxygen in an aeration tank is 1–3 mg/L; however, keep in mind that the typical required DO for a well-operated activated sludge plant is between 2.0 and 2.5 mg/L. Decreases in dissolved oxygen levels may indicate increased activity, increased temperature, increased organic loading, or decreased MLSS/MLVSS. An increase in dissolved oxygen could be indicative of decreased activity, decreased temperature, decreased organic loading, increased MLSS/MLVSS, or influent toxicity.

Dissolved Oxygen Profile

All dissolved oxygen profile readings should be >0.5 mg/L. Readings of <0.5 mg/L indicate inadequate aeration or poor mixing.

Mixed Liquor Suspended Solids (MLSS)

The range of mixed liquor volatile suspended solids is determined by the process modification used. When MLSS levels increase, more solids, organisms, and older, more oxidized sludge are typical.

Microscopic Examination

The activated sludge process cannot operate as designed without the presence of microorganisms. Thus, a microscopic examination of an aeration basin sample to determine the presence and type of microorganisms is important. Different species prefer different conditions; therefore, the presence of different species can indicate process conditions.

Note: It is important to point out that during microscopic examination, identifying all organisms present is not required, but identification of the predominant species is required.

Table 18.7 lists process conditions indicated by the presence and population of certain microorganisms.

AERATION TANK COLOR CONDITION INDICATOR

- The black aeration tank color indicates low dissolved oxygen concentration.
- Dark greasy aeration greasy aeration tank foam indicates high mixed liquor concentration.
- The red aeration tank color indicates a ferric chloride spill.
- White foam on the aeration tank indicates waste rate is too high.
- Dark greasy foam on the aeration tank indicates the waste rate is too low.
- The black color of mixed liquor indicates that the air flow rate is too low.

Interpretation

Routine process control identification can be limited to the general category of organisms present. For troubleshooting more difficult problems, a more detailed study of organism distribution may be required (the knowledge required to perform this type of detailed study is beyond the scope of this text). The major categories of organisms found in the activated sludge are:

- Protozoa
- Rotifers
- Filamentous organisms

Note: Bacteria are the most important microorganisms in the activated sludge. They perform most of the stabilization

TABLE 18.7
Process Condition versus Organisms Present/Population

Process Condition	Organism Population
Poor BOD$_5$ & TSS removal	Predominance of amoeba and flagellates
No floc formation	Mainly dispersed bacteria
Very cloudy effluent	A few ciliates present
Poor quality effluent	Predominance of amoeba and flagellates
Dispersed bacteria	Some free-swimming ciliates
Some free-swimming ciliates	
Some floc formation	
Cloudy effluent	
Satisfactory effluent	Predominance of free-swimming
Good floc formation	ciliates
Good settleability	Few amoeba and flagellates
Good clarity	
High-quality effluent	Predominance of stalked ciliates
Excellent floc formation	Some free-swimming ciliates
Excellent settleability	A few rotifers
High effluent clarity	A few flagellates
Effluent high TSS and low	Predominance of rotifers
BOD$_5$	Large numbers of stalked ciliates
High-settled sludge volume	A few free-swimming ciliates
Cloudy effluent	No flagellates

or oxidation of organic matter and are normally present in extremely large numbers. However, they are not normally visible with a conventional microscope operating at the recommended magnification and are not included in Table 16.7, which lists of indicator organisms.

Note: The presence of free-swimming and stalked ciliates, some flagellates, and rotifers in mixed liquor indicates a balanced, properly settling environment.

Protozoa

Protozoa are secondary feeders in the activated sludge process (secondary as feeders, but nonetheless definitely important to the activated sludge process). Their principal function is to remove (eat or crop) dispersed bacteria and help produce a clear process effluent. To help gain an appreciation for the role of protozoa in the activated sludge process, consider the following explanation.

The activated sludge process is typified by the successive development of protozoa and mature floc particles. This succession can be indicated by the type of dominant protozoa present. At the start of the activated process (or recovery from an upset condition), amoebas dominate.

Note: Amoebas have very flexible cell walls and move by shifting fluids within the cell wall. Amoeba predominates during process startup or during recovery from severe plant upsets.

As the process continues uninterrupted or without upset, small populations of bacteria begin to grow in a logarithmic fashion, which, as the population increases, develop into mixed liquor. When this occurs, the flagellates dominate.

Note: Flagellated protozoa typically have single hair-like flagella or a "tail" that they use for movement. The flagellate predominates when the MLSS and bacterial populations are low and the organic load is high. As the activated sludge gets older and denser, the flagellates decrease until they are seldom used.

When the sludge attains an age of about 3 days, lightly dispersed floc particles begin to form (flocculation "grows" fine solids into larger, more settleable solids), and bacteria increase. At this point, free-swimming ciliates dominate.

Note: The free-swimming ciliated protozoa have hair-like projections (cilia) that cover all or part of the cell. The cilia are used for motion and create currents that carry food to the organism. The free-swimming ciliates are sometimes divided into two sub-categories: free swimmers and crawlers. The free swimmers are usually seen moving through the fluid portion of the activated sludge, while the crawlers appear to be "walking" or "grazing" on the activated sludge solids. The free-swimming ciliated protozoa usually predominate when a large number of dispersed bacteria are present that can be used as food. Their predominance indicates a process nearing optimum conditions and effluent quality.

The process continues with floc particles beginning to stabilize, taking on irregular shapes and starting to show filamentous growth. At this stage, the crawling ciliates dominate. Eventually, mature floc particles develop and increase in size, and large numbers of crawling and stalked ciliates are present. When this occurs, the succession process has reached its terminal point. The succession of protozoan and mature floc particle development just described in detail the occurrence of phases of development in a step-by-step progression. Protozoan succession is also based on other factors, including dissolved oxygen and food availability.

Probably the best way to understand protozoan succession based on dissolved oxygen and food availability is to view the wastewater treatment plant's aeration basin as a "stream within a container." Using the *saprobity system* to classify the various phases of the activated sludge process in relation to the self-purification process that takes place in a stream, you can see a clear relationship between the two processes based on available dissolved oxygen and food supply. Any change in the relative numbers of bacteria in the activated sludge process has a corresponding change in the microorganism population. Decreases in bacteria increase competition between protozoa and result in the secession of dominant groups of protozoa.

The success or failure of protozoa to capture bacteria depends on several factors. Those with more advanced locomotion capabilities are able to capture more bacteria. Individual protozoan feeding mechanisms are also important in the competition for bacteria. At the beginning of the activated sludge process, amoebas and flagellates are the first protozoan groups to appear in large numbers. They can survive on smaller quantities of bacteria because their energy requirements are lower than those of other protozoan types. Because few bacteria are present, competition for dissolved substrates is low. However, as the bacteria population increases, these protozoa are not able to compete for available food. This is when the next group of protozoa (the free-swimming protozoa) enters the scene.

The free-swimming protozoa take advantage of the large populations of bacteria because they are better equipped with food-gathering mechanisms than the amoebas and flagellates. The free swimmers are important for their insatiable appetites for bacteria and also in floc formation. Secreting polysaccharides and mucoproteins that are absorbed by bacteria—which make the bacteria "sticky" through biological agglutination (biological gluing together)—allows them to stick together and, more importantly, to stick to floc. Thus, large quantities of floc are prepared for removal from secondary effluent and are either returned to aeration basins or wasted. The crawlers and stalked ciliates succeed the free swimmers.

Note: Stalked ciliated protozoa are attached directly to the activated sludge solids by a stalk. In some cases, the stalk is rigid and fixed in place, while in others, the organism can move (contract or expand the stalk) to change its position. The stalked ciliated protozoa normally have several cilia that are used to create currents, which carry bacteria and organic matter to them. The stalked ciliated protozoa predominate when the dispersed bacteria population decreases and does not provide sufficient food for the free swimmers. Their predominance indicates a stable process operating at optimum conditions.

The free swimmers are replaced in part because the increasing level of mature floc retards their movement. Additionally, the type of environment provided by the presence of mature floc is more suited to the needs of the crawlers and stalked ciliates. The crawlers and stalked ciliates also aid in floc formation by adding weight to floc particles, thus enabling removal.

Rotifers

Rotifers are a higher life form normally associated with clean, unpolluted waters. Significantly larger than most of the other organisms observed in activated sludge, rotifers can use other organisms as well as organic matter as their food source. Rotifers are usually the predominant organism; the effluent will usually be cloudy (pin of ash floc) and will have very low BOD_5.

Filamentous Organisms

Filamentous organisms (bacteria, fungi, etc.) occur whenever the environment of the activated sludge favors their predominance. They are normally present in small amounts and provide the basic framework for floc formation. When the environmental conditions (i.e., pH, nutrient levels, DO, etc.) favor their development, they become the predominant organisms. When this occurs, they restrict settling, and the condition known as "bulking" occurs.

Note: Microorganisms examination of activated sludge is a useful control tool. In attempting to identify the microscopic contents of a sample, the operator should try to identify the predominant groups of organisms.

Note: During the microscopic examination of the activated sludge, a predominance of amoebas indicates that the activated sludge is very young.

Settling Tank Influent

Dissolved Oxygen

The dissolved oxygen level of the activated sludge-settling tank should be 1–3 mg/L; lower levels may result in rising sludge.

pH

The normal pH range in an activated sludge-settling tank should be maintained between 6.5 and 9.0. Decreases in pH may indicate alkalinity deficiency.

Alkalinity

A lack of alkalinity in an activated sludge-settling tank will prevent nitrification.

Total Suspended Solids

MLSS sampling and testing are required for determining solids loading, mass balance, and return rates.

Settled Sludge Volume (Settleability)

Settled sludge volume (SSV) is determined at specified times during sample testing. Thirty- and 60-min observations

- **Normal Operation:** When the process is operating properly, the solids will settle as a "blanket" (a mass), with a crisp or sharp edge between the solids and the liquor above. The liquid over the solids will be clear, with little or no visible solids remaining in suspension. The settled sludge volume at the end of 30–60 min will be in the range of 400–700 mL.
- **Old or Over-Oxidized Activated Sludge:** When the activated sludge is over-oxidized, the solids will settle as discrete particles. The edge between the solids and liquid will be fuzzy, with a large number of visible solids (pin floc, ash floc, etc.) in the liquid. The settled sludge volume at the end of 30 or 60 min will be greater than 700 mL.
- **Young or Under-Oxidized Activated Sludge:** When the activated sludge is under-oxidized, the solids settle as discrete particles, and the boundary between the solids and the liquid is poorly defined. Large amounts of small visible solids are suspended in the liquid. The settled sludge volume after 30–60 min will usually be less than 400 mL.
- **Bulking Activated Sludge:** When the activated sludge is experiencing a bulking condition, very little or no settling is observed.

572
Handbook of Water and Wastewater Treatment Plant Operations

$$SSV = \frac{\text{Milliliters of Settled Sludge } 1{,}000 \text{ mL}}{\text{Milliliters of Sample}} \quad (18.48)$$

$$\%SSV = \frac{\text{Milliliters of Settled Sludge} \times 100}{\text{Milliliters of Sample}} \quad (18.49)$$

Note: Running the settleability test with a diluted sample can assist in determining if the activated sludge is old (too many solids) or bulking (not settling). Old sludge will settle to a more compact level when diluted.

Flow

Monitoring flow in settling tank influent is important for determining mass balance.

Jar Tests

Jar tests are performed as required on settling tank influent and are beneficial in determining the best flocculant aid and appropriate doses to improve solids capture during periods of poor settling.

Settling Tank

Sludge Blanket Depth

As mentioned, sludge blanket depth refers to the distance from the surface of the liquid to the solids-liquid interface or the thickness of the sludge blanket as measured from the bottom of the tank to the solids-liquid interface. Part of the operator's sampling routine, this measurement is taken directly in the final clarifier. Sludge blanket depth depends on hydraulic load, return rate, clarifier design, waste rate, sludge characteristics, and temperature. If all other factors remain constant, the blanket depth will vary with the amount of solids in the system and the return rate; thus, it will vary throughout the day.

Note: The depth of the sludge blanket indicates sludge quality; it is used as a trend indicator. Many factors affect the test results.

Suspended Solids and Volatile Suspended Solids

Suspended solids and volatile suspended solids concentrations of the mixed liquor (mixed liquor suspended solids), RAS, and WAS are routinely sampled and tested because they are critical for process control.

Settling Tank Effluent

BOD₅ and Total Suspended Solids

BOD_5 and total suspended solids testing is conducted variably (daily, weekly, and monthly). Increases indicate that treatment performance is decreasing, while decreases indicate that treatment performance is increasing.

Total Kjeldahl Nitrogen

Total Kjeldahl nitrogen sampling and testing is variable. An increase in TKN indicates that nitrification is decreasing, while a decrease in TKN indicates that nitrification is increasing.

Nitrate Nitrogen

Nitrate nitrogen sampling and testing are variable. Increases in nitrate nitrogen indicate nitrification is increasing or the industrial contribution of nitrates. A decrease indicates reduced nitrification.

Flow

Settling tank effluent flow is sampled and tested daily. Results are required for several process control calculations.

Return-Activated Sludge and Waste-Activated Sludge

Total Suspended Solids and Volatile Suspended Solids

Total suspended solids and total volatile suspended solids concentrations of the mixed liquor (mixed liquor suspended solids—MLSS), the RAS, and WAS are routinely sampled (using either grab or composite samples) and tested because they are critical to process control. The results of the suspended and volatile suspended tests can be used directly or to calculate process control figures such as mean cell residence time (MCRT) or food or mass ratio (F:M). In most situations, increasing the MLSS produces an older, denser sludge, while decreasing the MLSS produces a younger, less dense sludge.

Note: Control of the sludge wasting rate by constant MLVSS concentration involves maintaining a certain concentration of volatile suspended solids in the aeration tank.

Note: The activated sludge aeration tank should be observed daily. Included in this daily observation should be a determination of the type and amount of foam, mixing uniformity, and color.

Flow

The flow of return-activated sludge is tested daily. Test results are required to determine mass balance and for control of sludge blanket, MLSS, and MLVSS. For waste-activated sludge, flow is sampled and tested whenever sludge is wasted. Results are required to determine mass balance and to control solids level in the process.

Process Control Adjustments

In the routine performance of their duties, wastewater operators make process control adjustments to various unit processes, including the activated sludge process. The following section provides a summary of the process controls available for the activated sludge process and the result that will occur from the adjustment of each.

Return Rate

Condition: Return rate too high

> *Result:*
> - Hydraulic overloading of aeration and settling tanks
> - Reduced aeration time
> - Reduced settling time
> - Loss of solids over time

Condition: Return rate too low

> *Result:*
> - Septic return
> - Solids buildup in the settling tank
> - Reduced MLSS in aeration tank
> - Loss of solids over the weir

Waste Rate

Condition: Waste rate too high

> *Result:*
> - Reduced MLSS
> - Decreased sludge density
> - Increased SVI
> - Decreased MCRT
> - Increased F: M Ratio

Condition: Waste rate too low

> *Result:*
> - Increased MLSS
> - Increased sludge density
> - Decreased SVI
> - Increased MCRT
> - Decreased F: M ratio

Aeration Rate

Condition: Aeration rate too high

> *Result:*
> - Wasted energy
> - Increased operating cost
> - Rising solids
> - Breakup of activated sludge

Condition: Aeration rate too low

> *Result:*
> - Septic aeration tank
> - Poor performance
> - Loss of nitrification

Troubleshooting Operational Problems

Without a doubt, the most important dual function performed by the wastewater operator is identifying process control problems and implementing the appropriate actions to correct the problem(s). In this section, typical aeration system operational problems are listed with their symptoms, causes, and the appropriate corrective actions required to restore the unit process to a normal or optimal performance level.

SYMPTOM 1

The solids blanket is flowing over the effluent weir (classic bulking). Settleability test shows no settling.

> *Cause:*
> 1. Organic overloading
> *Corrective Action*: Reduce organic loading
> *Cause:*
> 2. Low pH
> *Corrective Action*: Add alkalinity
> *Cause:*
> 3. Filamentous growth
> *Corrective Action*: Add nutrients; add chlorine or peroxide to return
> *Cause:*
> 4. Nutrient deficiency
> *Corrective Action*: Add nutrients
> *Cause:*
> 5. Toxicity
> *Corrective Action*: Identify source; implement pretreatment
> *Cause:*
> 6. Over aeration
> *Corrective Action:* Reduce aeration during low flow periods

SYMPTOM 2

Solids settled properly in settleability test but large amounts of solids were lost over effluent weir.

> *Cause:*
> 1. Billowing solids due to short-circuiting.
> *Corrective Action*: Identify short-circuiting cause and eliminate if possible.

SYMPTOM 3

Large amounts of small pinhead-sized solids leave the settling tank.

> *Cause:*
> 1. Old sludge.
> *Corrective Action:* Reduce sludge age (gradual change is best); increase waste rate.
> *Cause:*
> 2. Excessive turbulence
> *Corrective Action:* Decrease turbulence (adjust aeration during low flows).

SYMPTOM 4

A large amount of light floc (low BOD_5 and high solids) leaving the settling tank.

> *Cause:*
>   1.   Extremely old sludge
> *Corrective Action:* Reduce age; increase waste.

SYMPTOM 5

Large amounts of small translucent particles (1/16–1/8 in) are leaving the settling tank.

> *Cause:*
>   1.   Rapid solids growth
> *Corrective Action:* Increase sludge age.
> *Cause:*
>   2.   Slightly young, activated sludge
> *Corrective Action*: Decrease waste.

SYMPTOM 6

Solids settle properly but rise to the surface within a short time. Many small (1/4 in) to large (several feet) clumps of solids on the surface of the settling tank.

> *Cause:*
>   1.   Denitrification
> *Corrective Action:* Increase rate of return; adjust sludge age to eliminate nitrification.
> *Cause:*
>   2.   over aeration
> *Corrective Action:* Reduce aeration.

SYMPTOM 7

Return-activated sludge has a rotten egg odor.

> *Cause:*
>   1.   Return is septic
> *Corrective Action:* Increase aeration rate
> *Cause:*
>   2.   The return rate is too low
> *Corrective Action:* Increase the rate of return

SYMPTOM 8

Activated sludge organisms die for a short time.

> *Cause:*
>   1.   Influent contained toxic material
> *Corrective Action:* Isolate activated sludge (if possible); return all available solids; stop wasting; increase return rate; implement pretreatment program.

SYMPTOM 9

The surface of the aeration tank is covered with thick, greasy foam.

> *Cause:*
>   1.   Extremely old, activated sludge
> *Corrective Action:* Reduce activated sludge age; increase wasting; use foam control sprays.
> *Cause:*
>   2.   Excessive grease and oil in the system
> *Corrective Action:* Improve grease removal; use foam control sprays; implement a pretreatment program.
> *Cause:*
>   3.   Froth forming bacteria
> *Corrective Action:* Remove froth-forming bacteria.

SYMPTOM 10

Large "clouds" of billowing white foam on the surface of the aeration tank.

> *Cause:*
>   1.   Young activated sludge
> *Corrective Action:* Increase sludge age; decrease wasting; use foam control sprays.
> *Cause:*
>   2.   Low solids in aeration tank
> *Corrective Action:* Increase sludge age; decrease wasting; use foam control sprays.
> *Cause:*
>   3.   Surfactants (detergents)
> *Corrective Action:* Eliminate surfactants; use foam control sprays; add antifoam.

Process Control Calculations

As with other wastewater treatment unit processes, process control calculations are important tools used by the operator to optimize and control process operations. In this section, we review the most frequently used activated sludge calculations.

Settled Sludge Volume

Settled sludge volume (SSV) is the volume that settled activated sludge occupies after a specified time. The settling time may be shown as a subscript (i.e., SSV_{60} indicates that the reported value was determined at 60 min). The settled sludge volume can be determined for any time interval; however, the most common values are the 30 min reading (SSV_{30}) and 60 min reading (SSV_{60}). The settled sludge volume can be reported as milliliters of sludge per liter of sample (Ml/L) or as a percent settled sludge volume.

$$\text{Settled Sludge Volume (mL/L)} = \frac{\text{Settled Sludge Vol (mL)}}{\text{Sample Volume (L)}}$$

$$(18.50)$$

Note: 1,000 mL = 1 L

$$\text{Sample Vol. (L)} = \frac{\text{Sample Volume, mL}}{1,000 \text{ mL/L}} = 720 \text{ mL/L}$$

$$(18.51)$$

$$\text{\% Settled Sludge Volume} = \frac{\text{Settled Sludge Volume, mL}}{\text{Sample Volume, mL}}$$

$$(18.52)$$

Example 18.38

Problem: Using the information provided in the table, calculate the SSV 30 and the % SSV 60.

Time	Milliliters
Start	2,500
15 min	2,250
30 min	1,800
45 min	1,700
60 min	1,600

Solution:

$$\text{Settled Sludge Vol. } (SSV_{30}) = \frac{1,800 \text{ mL}}{2.5 \text{ L}} = 720 \text{ mL/L}$$

$$\text{\% Settled Sludge Vol., } (SSV_{60}) = \frac{1,600 \text{ mL} \times 100}{2,500 \text{ mL}} = 64\%$$

Estimated Return Rate

Many different methods are available for the estimation of the proper return sludge rate. A simple method described in the *Operation of Wastewater Treatment Plants, Field Study Program* (1986)—developed by the California State University, Sacramento uses the 60-min percent settled sludge volume. The $\%SSV_{60}$ can provide an approximation of the appropriate return-activated sludge rate. The results of this calculation can then be adjusted based on sampling and visual observations to develop the optimum return sludge rate.

Note: The $\%SSV_{60}$ must be converted to a decimal percent and the total flow rate (wastewater flow and current return rate in million gallons per day must be used).

$$\text{Est. Return Rate, MGD} = \left(\begin{array}{c} \text{Infl. Flow, MGD} + \\ \text{Current Return Flow, MGD} \end{array} \right) \times \%SSV_{60}$$

- Assumes $\%SSV_{60}$ is representative
- Assumes return rate, in percent equals $\%SSV_{60}$

The actual return rate is normally set slightly higher to ensure organisms are returned to the aeration tank as quickly as possible. The rate of return must be adequately controlled to prevent the following:

- Aeration and settling hydraulic overloads
- Low MLSS levels in the aerator
- Organic overloading of aeration
- Solids loss due to excessive sludge blanket depth

Example 18.39

Problem: The influent flow rate is 4.2 MGD, and the current return-activated sludge flow rate is 1.5 MGD. The SSV_{60} is 38%. Based on this information, what should be the return sludge rate in million gallons per day (MGD)?

Solution:

$$\text{Return, MGD} = (4.2 \text{ MGD} + 1.5 \text{ MGD}) \times 0.38 = 2.2 \text{ MGD}$$

Sludge Volume Index

The sludge volume index (SVI) is a measure of the settling quality (a quality indicator) of the activated sludge. As the SVI increases, the sludge settles slower, does not compact as well, and is likely to increase effluent suspended solids. As the SVI decreases the sludge becomes denser, settling is more rapid, and the sludge is becoming older. SVI is the volume in milliliters occupied by 1 g of activated sludge. The Settled Sludge Volume, mL/L, and the mixed liquor suspended solids (MLSS), mg/L are required for this calculation.

$$\text{Sludge Volume Index } (SVI) = \frac{\text{SSV, mL/L} \times 1,000}{\text{MLSS mg/L}}$$

$$(18.53)$$

Example 18.40

Problem: The SSV_{30} is 365 mL/L, and the MLSS is 2,365 mg/L. What is the SVI?

Solution:

$$\text{Sludge Volume Index} = \frac{365 \text{ mL/L} \times 1,000}{2,365 \text{ mg/L}} = 154.3$$

SVI equals 154.3—what does this mean? It means that the system is operating normally with good settling and low effluent turbidity. How do we know this? Another good question. We know this because we compare the 154.3 results with the parameters listed below to obtain the expected condition (the result).

SVI Value	Expected Condition (Indicates)
Less than 100	Old sludge—possible pin floc Effluent turbidity increasing
100–200	Normal operation—good settling Low effluent turbidity
Greater than 250	Bulking sludge—poor settling High effluent turbidity

The SVI is best used as a trend indicator to evaluate what is occurring compared to previous SVI values. Based on this evaluation, the operator may determine if the SVI trend is increasing or decreasing (refer to the following chart).

SVI Value	Result	Adjustment
Increasing	Sludge is becoming less dense.	Decrease waste
	Sludge is either younger or bulking.	Increase return rate
	Sludge will settle more slowly.	
	Sludge will compact less.	
Decreasing	Sludge is becoming denser.	Increase waste rate
	Sludge is becoming older.	
	Sludge will settle more rapidly.	Decrease return rate
	Sludge will compact more with no other process changes.	
Holding constant	No changes indicated	
	Sludge should continue to have its current characteristics.	

Waste Activated Sludge

The quantity of solids removed from the process as *waste-activated sludge* is an important process control parameter that operators need to be familiar with—and, more importantly, know how to calculate.

$$\text{Waste, lb/day} = \text{WAS Conc., mg/L} \times \text{WAS Flow,}$$

$$\text{MGD} \times 8.34 \text{ lb/MG/mg/L} \qquad (18.54)$$

Example 18.41

Problem: The operator wastes 0.44 MGD of activated sludge. The waste-activated sludge has a solids concentration of 5,540 mg/L. How many pounds of waste-activated sludge are removed from the process?

Solution:

Waste, lb/day = 5,540 mg/L × 0.44 MGD × 8.34 lb/MG/mg/L

= 20,329.6 lbs/day

Food to Microorganism Ratio (F/M Ratio)

The food-to-microorganism ratio (F/M ratio) is a process control calculation used in many activated sludge facilities to control the balance between available food materials (BOD or COD) and available organisms (mixed liquor volatile suspended solids, MLVSS). The chemical oxygen demand (COD) test is sometimes used because the results are available in a relatively short period of time. To calculate the F/M ratio, the following information is required:

- Aeration tank influent flow rate, MGD
- Aeration tank influent BOD or COD, mg/L
- Aeration tank MLVSS, mg/L
- Aeration tank volume, MG

$$\text{F/M Ratio} = \frac{\begin{array}{c}\text{Prim. Eff. COD/BOD mg/L} \times \\ \text{Flow MGD} \times .34 \text{ lb/mg/L/MG}\end{array}}{\begin{array}{c}\text{MLVSS mg/L} \times \text{Aerator Volume,} \\ \text{MG} \times 8.34 \text{ lb/mg/L/MG}\end{array}}$$

$$(18.55)$$

Typical F/M Ratio for activated sludge processes is shown in the following:

Process	lb BOD$_5$ / lb MLVSS	lb COD / lb MLVSS
Conventional	0.2–0.4	0.5–1.0
Contact stabilization	0.2–0.6	0.5–1.0
Extended aeration	0.05–0.15	0.2–0.5
Oxidation ditch	0.05–0.15	0.2–0.5
Pure oxygen	0.25–1.0	0.5–2.0

Example 18.42

Problem: Given the following data, what is the F/M ratio?

Primary effluent flow	2.5 MGD	Aeration volume	0.65 MG
Primary effluent BOD	145 mg/L	Settling volume	0.30 MG
Primary effluent TSS	165 mg/L	MLSS	3,650 mg/L
Effluent flow	2.2 MGD	MLVSS	2,550 mg/L
Effluent BOD	22 mg/L	% Waste volatile	71%
Effluent TSS	16 mg/L	Desired F/M	0.3

Solution:

$$\text{F/M Ratio} = \frac{145 \text{ mg/L} \times 2.2 \text{ MGD} \times 8.34 \text{ lbs/mg/L/MG}}{2,550 \text{ mg/L} \times 0.65 \text{ MG} \times 8.34 \text{ lbs/mg/L/MG}}$$

$$= 0.19 \text{ lbs BOD/lb MLVSS}$$

Note: If the MLVSS concentration is not available, it can be calculated if the % volatile matter (% V.M.) of the Mixed Liquor Suspended Solids (MLSS) is known (see Equation 18.56).

$$\text{MLVSS} = \text{MLSS} \times \% \text{ (decimal) Volatile Matter (V.M)}$$

$$(18.56)$$

Note: The "F" value in the F/M ratio for computing loading to an activated sludge process can be either BOD or COD. Remember that the reason for sludge production in the activated sludge process is to convert BOD to bacteria. One advantage of using COD over BOD for the analysis of organic load is that COD is more accurate. However, an overall advantage of the BOD test over the COD test is that the BOD test shows the actual impact of the wastewater.

Example 18.43

Problem: The aeration tank contains 2,985 mg/L of MLSS. Laboratory tests indicate that the MLSS is 66%

volatile matter. What is the MLVSS concentration in the aeration tank?

Solution:

$$\text{MLVSS, mg/L} = 2{,}985 \text{ mg/L} \times 0.66 = 1{,}970 \text{ mg/L}$$

F/M Ratio Control

Maintaining the F/M ratio within a specified range can be an excellent control method. Although the F/M ratio is affected by adjustments to the return rates, the most practical method for adjusting the ratio is through waste rate adjustments.

Increasing the rate will:

1. Decrease the MLVSS
2. Increase the F/M ratio

Decreasing the waste rate will:

1. Increase the MLVSS
2. Decrease the F/M ratio

The desired F/M ratio must be established on a plant-by-plant basis. Comparison of F/M ratios with plant effluent quality is the primary means to identify the most effective range for individual plants, when the range of F/M values that produce the desired effluent quality is established.

Required MLVSS Quantity (Pounds)

The pounds of MLVSS required in the aeration tank to achieve the optimum F/M ratio can be determined from the average influent food (BOD or COD) and the desired F/M ratio.

$$\text{MLVSS, lb} = \frac{\text{Primary Effluent BOD or COD} \times \text{Flow, MGD} \times 8.34}{\text{Desired F/M Ratio}}$$

$$(18.57)$$

The required pounds of MLVSS determined by this calculation can then be converted to a concentration value by:

$$\text{MLVSS, mg/L} = \frac{\text{Desired MLVSS, lbs}}{\left[\text{Aeration Volume, MG} \times 8.34\right]} \quad (18.58)$$

Example 18.44

Problem: The aeration tank influent flow is 4.0 MGD, and the influent COD is 145 mg/L. The aeration tank volume is 0.65 MG. The desired F/M ratio is 0.3 lb COD/lb MLVSS.

1. How many pounds of MLVSS must be maintained in the aeration tank to achieve the desired F/M ratio?

2. What is the required concentration of MLVSS in the aeration tank?

Solution:

$$\text{MLVSS, lb} = \frac{145 \text{ mg/L} \times 4.0 \text{ MGD} \times 8.34 \text{ lb/gal}}{0.3 \text{ lb COD/lb MLVSS}}$$

$$= 16{,}124 \text{ lb MLVSS}$$

$$\text{MLVSS, mg/L} = \frac{16{,}124 \text{ MLVSS}}{\left[0.65 \text{ MG} \times 8.34\right]}$$

$$= 2{,}974 \text{ mg/L MLVSS}$$

Calculating Waste Rates Using F/M Ratio

Maintaining the desired F/M ratio is accomplished by controlling the MLVSS level in the aeration tank. This may be accomplished by adjustment of return rates; however, the most practical method is proper control of the waste rate.

$$\text{Waste Vol. Solids, lb/day} = \text{Actual MLVSS, lb}$$

$$- \text{Desired MLVSS, lb} \quad (18.59)$$

If the desired MLVSS is greater than the actual MLVSS, wasting is stopped until the desired level is achieved.

Practical considerations require that the required waste quantity be converted to a required volume of waste per day. This is accomplished by converting the waste pounds to a flow rate in million gallons per day or gallons per minute.

$$\text{Waste, MGD} = \frac{\text{Waste Volatile, lb/day}}{\left[\text{Waste Volatile Concentration, mg/L} \times 8.34\right]}$$

$$(18.60)$$

Note: When F/M ratio is used for process control, the volatile content of the waste activated sludge should be determined.

Example 18.45

Problem: Given the following information, determine the required waste rate in gallons per minute to maintain an F/M ratio of 0.17-lb COD/lb MLVSS:

Primary Effluent COD	140 mg/L
Primary Effluent Flow	2.2 MGD
MLVSS, mg/L	3,549 mg/L
Aeration Tank Volume	0.75 MG

Waste Volatile Concentration 4,440 mg/L (Volatile Solids)

Solution:

$$\text{Actual MLVSS, lb} = 3.549 \text{ mg/L} \times 0.75 \text{ MG} \times 8.34 = 22{,}199 \text{ lb}$$

$$\text{Required MLVSS, lb} = \frac{140 \text{ mg/L} \times 2.2 \text{ MGD} \times 8.34}{0.17 \text{ lb COD/lb MLVSS}}$$

$$= 15{,}110 \text{ lb MLVSS}$$

$$\text{Waste, lb/day} = 22{,}199 \text{ lb} - 15{,}110 \text{ lb} = 7{,}089 \text{ lb}$$

$$\text{Waste, MGD} = \frac{7{,}089 \text{ lb/day}}{[4{,}440 \text{ mg/L} \times 8.34]} = 0.19 \text{ MGD}$$

$$\text{Waste, gpm} = \frac{0.19 \text{ MGD} \times 1{,}000{,}000 \text{ gpd/MGD}}{1{,}440 \text{ min/day}} = 132 \text{ gpm}$$

Mean Cell Residence Time (MCRT)

Mean Cell Residence Time (MCRT), sometimes referred to as *sludge retention time,* is a process control calculation used for activated sludge systems. The MCRT calculation illustrated in Example 18.46 uses the entire volume of the activated sludge system (aeration and settling).

$$\text{MRCT, day} = \frac{\begin{bmatrix} \text{MLSS mg/L} \times \\ (\text{Aeration Vol., MG} + \text{Clarifier Vol., MG} \times 8.34] \end{bmatrix}}{\begin{bmatrix} (\text{WAS, mg/L} \times \text{WAS flow, MGD} \times 8.34) + \\ (\text{TSS out, mg/L} \times \text{Flow} \times 8.34) \end{bmatrix}}$$

$$(18.61)$$

Note: Due to the length of the MCRT equation, the units for the conversion factor 8.34 have not been included. The dimensions for the 8.34 conversion factor are lb/mg/L/MG.

Note: MCRT can be calculated using only the aeration tank solids inventory. When comparing plant operational levels to reference materials, it is important to determine which calculation the reference manual uses to obtain its example values. Other methods are available to determine the clarifier solids concentration. However, the simplest method assumes that the average suspended solids concentration is equal to the aeration tank's solids concentration.

Example 18.46

Problem: Given the following data, what is the MCRT?

Influent flow	4.2 MGD	Aeration volume	1.20 MG
Influent BOD	135 mg/L	Settling volume	0.60 MG
Influent TSS	150 mg/L	MLSS	3,350 mg/L
Effluent flow	4.2 MGD	Waste rate	0.080 MGD
Effluent BOD	22 mg/L	Waste conc.	6,100 mg/L
Effluent TSS	10 mg/L	Desired MCRT	8.5 days

$$\text{MRCT} = \frac{[3{,}350 \text{ mg/L} \times (1.2 \text{ MG} + 0.6 \text{ MG}) \times 8.34]}{\begin{array}{c} [6{,}100 \text{ mg/L} \times 0.08 \text{ MGD} \times 8.34) \\ + (10 \text{ mg/L} \times 4.2 \text{ MGD} \times 8.34)] \end{array}}$$

$$\text{MRCT} = 11.4 \text{ days}$$

Mean Cell Residence Time Control

Because it provides an accurate evaluation of process conditions and takes all aspects of the solids inventory into account, the MCRT is an excellent process control tool. Increases in the waste rate will decrease the MCRT, as will large losses of solids over the effluent weir. Reductions in the waste rate will result in increased MCRT values.

Note: You should remember these important process control parameters.

Process Parameters and Impact on MCRT/MCRT Impact on Parameters:

- To increase F/M, decrease MCRT.
- To increase MCRT, decrease the waste rate.
- As MCRT increases, MLTSS and 30-min settling increases.
- The return sludge rate has no impact on MCRT.
- MCRT has no impact on F/M change when the number of aeration tanks in service is reduced.

Typical MCRT Values

The following chart lists the various aeration process modifications and associated MCRT values.

Process	MCRT, Days
Conventional	5–15
Step aeration	5–15
Contact stabilization (contact)	5–15
Extended aeration	20–30
Oxidation ditch	20–30
Pure oxygen	8–20

Control Values for MCRT

Control values for the MCRT are normally established based on effluent quality. Once the MCRT range required to produce the desired effluent quality is established, it can be used to determine the waste rate required to maintain it.

Waste Quantities/Requirements

MCRT for process control requires the determination of the optimum range for MCRT values. This is accomplished by comparing the effluent quality with MCRT values. When the optimum MCRT is established, the quantity of solids to be removed (wasted) is determined by:

$$\text{Waste, lb/day} = \left[\frac{\text{MLSS} \times (\text{Aer., MG} + \text{Clar., MG}) \times 8.34}{\text{Desired MCRT}} \right]$$

$$- [\text{TSS}_{\text{out,}} \times \text{Flow} \times 8.34]$$

$$(18.62)$$

Example 18.47

$$\frac{3{,}400 \text{ mg/L} \times (1.4 \text{ MG} + 0.50 \text{ MG}) \times 8.34}{8.6 \text{ days}}$$

$$- [10 \text{ mg/L} \times 5.0 \text{ MGD} \times 8.34]$$

$$\text{Waste Quality, lb/day} = 5,848 \text{ lb}$$

Waste Rate in Million Gallons/Day

When the quantity of solids to be removed from the system is known, the desired waste rate in million gallons per day can be determined. The unit used to express the rate (MGD, gpd, or gpm) is a function of the volume of waste to be removed and the design of the equipment.

$$\text{Waste, MGD} = \frac{\text{Waste, lbs/day}}{\text{WAS Concentrations, mg/L} \times 8.34} \quad (18.63)$$

$$\text{Waste, gpm} = \frac{\text{Waste MGD} \times 1,000,000 \text{ gpd/MGD}}{1,440 \text{ min/day}} \quad (18.64)$$

Example 18.48

Problem: Given the following data, determine the required waste rate to maintain an MCRT of 8.8 days.

MLSS, mg/L	2,500 mg/L
Aeration volume	1.20 MG
Clarifier volume	0.20 MG
Effluent TSS	11 mg/L
Effluent flow	5.0 MGD
Waste concentrations	6,000 mg/L

Solution:

$$\text{Waste, lb/day} = \frac{2,500 \text{ mg/L} \times (1.20 + 0.20) \times 8.34}{8.8 \text{ days}}$$

$$-\left[11 \text{ mg/L} \times 5.0 \text{ MGD} \times 8.34 \right]$$

$$= 3,317 \text{ lb/day} - 459 \text{ lb/day}$$

$$= 2,858 \text{ lb/day}$$

$$\text{Waste, MGD} = \frac{2,858 \text{ lb/day}}{\left[6,000 \text{ mg/L} \times 8.34 \right]} = 0.057 \text{ MGD}$$

$$\text{Waste, gpm} = \frac{0.057 \text{ MGD} \times 1,000,000 \text{ gpd/MGD}}{1,440 \text{ min/day}} = 40 \text{ gpm}$$

Mass Balance

Mass balance is based on the fact that solids and BOD are not "lost" in the treatment system. In simple terms, the mass balance concept states that "what comes in must equal waste that goes out." The concept can be used to verify operational control levels and to determine if potential problems exist within the plant's process control monitoring program.

Note: If influent and effluent values do not correlate within 10%–15%, it usually indicates either a sampling or testing error or a process control discrepancy.

Mass balance procedures for evaluating the operation of a settling tank and a biological process are described in this section. Operators should recognize that, although the procedures are discussed in reference to the activated sludge process, the concepts can be applied to any settling or biological process.

Mass Balance: Settling Tank Suspended Solids

The settling tank mass balance calculation assumes that no suspended solids are produced in the settling tank. Any settling tank operation can be evaluated by comparing the solids entering the unit with the solids leaving the tank as effluent suspended solids or as sludge solids. If sampling and testing are accurate and representative, and process control and operation are appropriate, the quantity of suspended solids entering the settling tank should equal (±10%) the quantity of suspended solids leaving the settling tank as sludge, scum, and effluent total suspended solids.

Note: In most instances, the amount of suspended solids leaving the process as scum is so small that it is ignored in the calculation.

Mass Balance Calculation

$$\text{Total Suspended Solids in, lb} = \text{TSS}_{in} \times \text{Flow, MGD} \times 8.34$$
$$(18.65)$$

$$\text{Total Suspended Solids out, lb} = \text{TSS}_{out} \times \text{Flow, MGD} \times 8.34$$

$$\text{Sludge Solids} = \text{Sludge Pumped, gal} \times \%\text{Solids} \times 8.34$$

$$\%\text{ Mass Balance} = \frac{\left[\text{TSS}_{in}, \text{lb} - (\text{TSS}_{out}, \text{lb} + \text{Sludge Sol, lb}) \right] \times 100}{\text{TSS}_{in}, \text{lb}}$$

Explanation of Results

1. **If the Mass Balance Is ±15% or Less:** The process is considered to be in balance. Sludge removal should be adequate, with the sludge blanket depth remaining stable. Sampling is considered to be producing representative samples that are being tested accurately.
2. **If the Mass Balance Is Greater Than ±15%:** This indicates that more solids are entering the settling tank than are being removed. The sludge blanket depth should be increasing, effluent solids may also be increasing, and effluent quality is decreasing.

   If changes described are not occurring, the mass balance may indicate that sample type, location, times, or procedures and/or testing procedures are not producing representative results.

3. **If the Mass Balance Is Greater Than—15%:**
This indicates that fewer solids are entering the settling tank than are being removed. The sludge blanket depth should be decreasing, and the sludge solids concentration may also be decreasing. This could adversely impact sludge treatment processes.

If the changes described are not occurring, the mass balance may indicate that sample type, location, times, or procedures and/or testing procedures are not producing representative results.

Example 18.49

Problem: Given the following data, determine the solids mass balance for the settling tank.

Influent	Flow	2.6 MGD
	TSS	2,445 mg/L
Effluent	Flow	2.6 MGD
	TSS	17 mg/L
Return	Flow	0.5 MGD
	TSS	8,470 mg/L

Solution:

Solids in, lb/day $= 2{,}445$ mg/L $\times 2.6$ MGD $\times 8.34$

$$= 53{,}017 \text{ lb/day}$$

Solids out, lb/day $= 17$ mg/L $\times 2.6$ MGD $\times 8.34 = 369$ lb/day

Sludge Solids out, lb/day $= 8{,}470$ mg/L $\times 0.5$ MGD $\times 8.34$

$$= 35{,}320 \text{ lb/day}$$

$$\text{Mass Balance} = \frac{\left[\begin{array}{c} 53{,}017 \text{ lb/day} - \\ (369 \text{ lb/day} + 35{,}320 \text{ lb/day}) \end{array}\right] \times 100}{53{,}017 \text{ lb/day}}$$

$$= 32.7\%$$

The value indicates that:

1. The sampling point/collection procedure, or laboratory procedure is producing inaccurate data upon which to make process control decisions.
   - Or, more solids are entering the settling tank each day than are being removed. This should result in either (1) a solids buildup in the settling tank or (2) a loss of solids over the effluent weir.

   Investigate further to determine the specific cause of the imbalance.

Mass Balance: Biological Process

Solids are produced whenever biological processes are used to remove organic matter from wastewater. The mass balance for an aerobic biological process must consider both the solids removed by physical settling processes and the solids produced by the biological conversion of soluble organic matter to insoluble suspended matter or organisms. Research has shown that the amount of solids produced per pound of BOD_5 removed can be predicted based on the type of process being used. Although the exact amount of solids produced can vary from plant to plant, research has developed a series of K factors that can be used to estimate the solids production for plants using a particular treatment process. These average factors provide a simple method to evaluate the effectiveness of a facility's process control program. The mass balance also provides an excellent mechanism to evaluate the validity of process control and effluent monitoring data generated. Table 18.8 lists the average K factors in pounds of solids produced per pound of BOD removed for selected processes.

Conversion Factor

Conversion factors depend on the activated sludge modification involved. Factors generally range from 0.5 to 1.0 lb of solids per pound of BOD removed (see Table 18.8).

Mass Balance Calculation

$$BOD_5\text{in, lb} = BOD, \text{mg/L} \times Flow, MGD \times 8.34 \qquad (18.66)$$

$$BOD_5\text{out, lb} = BOD, \text{mg/L} \times Flow, MGD \times 8.34$$

$$\text{Solids Produced, lb/day} = [BOD \text{ in, lb} - BOD \text{ out, lb}] \times K$$

$$\text{TSS out, lb/day} = TSS_{out}, \text{mg/L} \times Flow, MGD \times 8.34$$

TABLE 18.8
Conversion Factors K

Process	lb Solids / lb BOD$_5$ Removed
Primary	1.7
Activated sludge with primary	0.7
Activated sludge without primary	
Conventional	0.85
Step feed	0.85
Extended aeration	0.65
Oxidation ditch	0.65
Contact stabilization	1.00
Trickling filter	1.00
Rotating biological contactor	1.00

Waste, lb/day = waste, mg/L × Flow, MGD × 8.34

Solids Removed, lb/day = TSS_{out}, lb/day + Waste, lb/day

$$\% \text{ Mass Balance} = \frac{(\text{Solids Produced} - \text{Solids Removed}) \times 100}{\text{Solids Produced}}$$

Explanation of Results

If the mass balance is ±15% or less, the process sampling, testing, and process control are within acceptable levels. If the balance is greater than ±15%, investigate further to determine if the discrepancy represents a process control problem or is the result of non-representative sampling and inaccurate testing.

Sludge Waste Based upon Mass Balance

The mass balance calculation predicts the amount of sludge that will be produced by a treatment process. This information can then be used to determine what the waste rate must be, under current operating conditions, to maintain the current solids level.

$$\text{Waste Rate, MGD} = \frac{\text{Solids Produced, lb/day}}{(\text{Waste Concentration} \times 8.34)} \quad (18.67)$$

Example 18.50

Problem: Given the following data, determine the mass balance of the biological process and the appropriate waste rate to maintain current operating conditions:

Influent	Flow	1.1 MGD
	BOD_5	220 mg/L
	TSS	240 mg/L
Effluent	Flow	1.5 MGD
	BOD_5	18 mg/L
	TSS	22 mg/L
Waste	Flow	24,000 gpd
	TSS	8,710 mg/L

BOD_5 in = 220 mg/L × 1.1 MGD × 8.34 = 2,018 lb/day

BOD_5 out = 18 mg/L × 1.1 MGD × 8.34 = 165 lb/day

BOD_5 Removed = 2,018 lb/day − 165 lb/day = 1,853 lb/day

Solids Produced = 1,853 lb/day × 0.65 lb/lb BOD_5

= 1,204 lb solids/day

Solids Out, lb/day = 22 mg/L × 1.1 MGD × 8.34 = 202 lb/day

Sludge Out, lb/day = 8,710 mg/L × 0.024 MGD × 8.34

= 1,743 lb/day

Solids Removed, lb/day = (292 lb/day + 1,743 lb/day)

= 1,945 lb/day

$$\text{Mass Balance} = \frac{(1,204 \text{ lb Solids/day} - 1,945 \text{ lb/day}) \times 100}{1,204 \text{ lb/day}}$$

= 62%

The mass balance indicates that:

1. The sampling point(s), collection methods, and/or laboratory testing procedures are producing non-representative results.
2. The process is removing significantly more solids than is required. Additional testing should be performed to isolate the specific cause of the imbalance.

To assist in the evaluation, the waste rate based on the mass balance information can be calculated.

$$\text{Waste, GPD} = \frac{\text{Solids Produced, lb/day}}{(\text{Waste TSS, mg/L} \times 8.34)} \quad (18.68)$$

$$\text{Waste, GPD} = \frac{1,204 \text{ lb/day} \times 1,000,000}{8,710 \text{ mg/L} \times 8.34} = 16,575 \text{ gpd}$$

SOLIDS CONCENTRATION: SECONDARY CLARIFIER

The solids concentration in the secondary clarifier can be assumed to be equal to the solids concentration in the aeration tank effluent. It may also be determined in the laboratory using a core sample taken from the secondary clarifier. The secondary clarifier solids concentration can be calculated as an average of the secondary effluent suspended solids and the return activated sludge suspended solids concentration.

ACTIVATED SLUDGE PROCESS RECORDKEEPING REQUIREMENTS

Wastewater operators soon learn that recordkeeping is a major requirement and responsibility of their jobs. Records are important (essential) for process control, providing information on the cause of problems, making seasonal changes, and ensuring compliance with regulatory agencies. Records should include sampling and testing data, process control calculations, meter readings, process adjustments, operational problems and corrective actions taken, and process observations.

DISINFECTION OF WASTEWATER

Like drinking water, liquid wastewater effluent is disinfected. Unlike drinking water, wastewater effluent is disinfected not directly protect a drinking water supply (direct end-of-pipe connection) but instead is treated to protect public health in general. This is particularly important when the secondary effluent is discharged into a body of water used for swimming or as a water supply for a downstream community. In the treatment of water for human consumption, treated water is typically chlorinated (although ozonation is also currently being applied in many cases). Chlorination is the preferred disinfection method in potable water supplies because of chlorine's unique ability to provide a residual. This chlorine residual is important because when treated water leaves the waterworks facility and enters the distribution system, the possibility of contamination is increased. The residual works to continuously disinfect water right up to the consumer's tap.

In this section, we discuss basic chlorination and dechlorination. In addition, we describe ultraviolet (UV) irradiation, ozonation, bromine, chlorine, and no disinfection. Keep in mind that much of the chlorination material presented in the following is similar to the chlorination information presented earlier.

CHLORINE DISINFECTION

Chlorination for disinfection, as shown in Figure 18.1, follows all other steps in conventional wastewater treatment. The purpose of chlorination is to reduce the population of organisms in the wastewater to levels low enough to ensure that pathogenic organisms will not be present in sufficient quantities to cause disease when discharged.

Note: Chlorine gas is heavier than air (vapor density of 2.5). Therefore, exhaust from a chlorinator room should be taken from the floor level.

Note: The safest action to take in the event of a major chlorine container leak is to call the fire department.

Note: You might wonder why chlorination of critical waters, such as natural trout streams, is not normal practice. This practice is strictly prohibited because chlorine and its by-products (i.e., chloramines) are extremely toxic to aquatic organisms.

Chlorination Terminology

Remember that there are several terms used in the discussion of disinfection by chlorination. Because it is important for the operator to be familiar with these terms, we repeat key terms again.

- **Chlorine:** a strong oxidizing agent that has strong disinfecting capability. A yellow-green gas that is extremely corrosive and toxic to humans in extremely low concentrations in air.
- **Contact Time:** the length of time the disinfecting agent and the wastewater remain in contact.

- **Demand:** the chemical reactions that must be satisfied before a residual or excess chemical will appear.
- **Disinfection:** refers to the selective destruction of disease-causing organisms. Not all the organisms are not destroyed during the process. This differentiates disinfection from sterilization, which is the destruction of all organisms.
- **Dose:** the amount of chemical being added in milligrams/liter.
- **Feed Rate:** the amount of chemical being added in pounds per day.
- **Residual:** the amount of disinfecting chemical remaining after the demand has been satisfied.
- **Sterilization:** the removal of all living organisms.

Wastewater Chlorination: Facts and Process Description

Chlorine Facts

- Elemental chlorine (Cl_2—gaseous) is a yellow-green gas, 2.5 times heavier than air.
- The most common use of chlorine in wastewater treatment is for disinfection. Other uses include odor control and activated sludge bulking control. Chlorination takes place prior to the discharge of the final effluent to the receiving waters (see Figure 18.1).
- Chlorine may also be used for nitrogen removal through a process called *breakpoint chlorination*. For nitrogen removal, enough chlorine is added to the wastewater to convert all the ammonium nitrogen to gas. To do this, approximately 10 mg/L of chlorine must be added for every 1 mg/L of ammonium nitrogen in the wastewater.
- For disinfection, chlorine is fed manually or automatically into a chlorine contact tank or basin, where it contacts flowing wastewater for at least 30 min to destroy disease-causing microorganisms (pathogens) found in treated wastewater.
- Chlorine may be applied as a gas, a solid, or in liquid hypochlorite form.
- Chlorine is a very reactive substance. It has the potential to react with many different chemicals (including ammonia) as well as with organic matter. When chlorine is added to wastewater, several reactions occur:
  1. Chlorine will react with any reducing agent (i.e., sulfide, nitrite, iron, and thiosulfate) present in wastewater. These reactions are known as *chlorine demand*. The chlorine used for these reactions is not available for disinfection.
  2. Chlorine also reacts with organic compounds and ammonia compounds to form chlor-organics and chloramines. Chloramines are part of the group of chlorine compounds that have disinfecting properties and show up as part of the chlorine residual test.

3. After all of the chlorine demands are met, the addition of more chlorine will produce free residual chlorine. Producing free residual chlorine in wastewater requires very large additions of chlorine.

Hypochlorite Facts

Hypochlorite, though there are some minor hazards associated with its use (skin irritation, nose irritation, and burning eyes), is relatively safe to work with. It is normally available in dry form as a white powder, pellet, or tablet, or in liquid form. It can be added directly using a dry chemical feeder or dissolved and fed as a solution.

Note: In most wastewater treatment systems, disinfection is accomplished by means of a combined residual.

Wastewater Chlorination Process Description

Chlorine is a very reactive substance. Chlorine is added to wastewater to satisfy all chemical demands, that is, to react with certain chemicals (such as sulfide, sulfite, ferrous iron, etc.). When these initial chemical demands have been satisfied, chlorine will react with substances such as ammonia to produce chloramines and other substances which, although not as effective as chlorine, have disinfecting capability. This produces a combined residual, which can be measured using residual chlorine test methods. If additional chlorine is added, free residual chlorine can be produced. Due to the chemicals normally found in wastewater, chlorine residuals are normally combined rather than free residuals. Control of the disinfection process is normally based on maintaining total residual chlorine of at least 1.0 mg/L for a contact time of at least 30 min at design flow.

Note: Residual level, contact time, and effluent quality affect disinfection. Failure to maintain the desired residual levels for the required contact time will result in lower efficiency and an increased probability that disease organisms will be discharged.

Based on water quality standards, **total residual limitations** on chlorine are:

- **Fresh Water:** Less than 11 ppb total residual chlorine.
- **Estuaries:** Less than 7.5 ppb for halogen-produced oxidants.
- **Endangered Species:** The use of chlorine is prohibited.

Chlorination Equipment

Hypochlorite Systems

Depending on the form of hypochlorite selected for use, special equipment that will control the addition of hypochlorite to the wastewater is required. Liquid forms require the use of metering pumps, which can deliver varying flows of hypochlorite solution. Dry chemicals require the use of a feed system designed to provide variable doses of the form used. The tablet form of hypochlorite requires the use of a tablet chlorinator designed specifically to provide the desired dose of chlorine. The hypochlorite solution or dry feed system dispenses the hypochlorite, which is then mixed with the flow. The treated wastewater then enters the contact tank to provide the required contact time.

Chlorine Systems

Because of the potential hazards associated with the use of chlorine, the equipment requirements are significantly greater than those associated with hypochlorite use. The system most widely used is a **solution feed system**. In this system, chlorine is removed from the container at a flow rate controlled by a variable orifice. Water moving through the chlorine injector creates a vacuum, which draws the chlorine gas to the injector and mixes it with the water. The chlorine gas reacts with the water to form hypochlorous and hydrochloric acid. The solution is then piped to the chlorine contact tank and dispersed into the wastewater through a diffuser. Larger facilities may withdraw the liquid form of chlorine and use evaporators (heaters) to convert it to gas form. Small facilities will normally draw the gas form of chlorine from the cylinder. As gas is withdrawn, the liquid will be converted to gas form. This requires heat energy and may result in chlorine line freeze-up if the withdrawal rate exceeds the available energy levels.

Chlorination: Operation

In either type of system, normal operation requires adjustment of feed rates to ensure the required residual levels are maintained. This normally requires chlorine residual testing and adjustment based on the results of the test. Other activities include the removal of accumulated solids from the contact tank, collection of bacteriological samples to evaluate process performance, and maintenance of safety equipment (respirator-air pack, safety lines, etc.). Hypochlorite operation may also include make-up solution (solution feed systems), adding powder or pellets to the dry chemical feeder, or tablets to the tablet chlorinator.

Chlorine operations include adjustment of chlorinator feed rates, inspection of mechanical equipment, testing for leaks using ammonia swabs (white smoke means leaks), changing containers (which requires more than one person for safety), and adjusting the injector water feed rate when required. Chlorination requires routine testing of plant effluent for total residual chlorine and may also require collection and analysis of samples to determine the fecal coliform concentration in the effluent.

Troubleshooting Operation Problems

On occasion, operational problems with the plant's disinfection process develop. The wastewater operator must not only be able to recognize these problems but also correct them. For proper operation, the chlorination process requires routine observation, meter readings, process control and testing, and various process control calculations. Comparison of daily results with expected "normal" ranges is the key to identifying problems during the troubleshooting process and taking the appropriate corrective action (if required). In this section, we review normal operational and performance factors and point out various problems that can occur with the plant's disinfection process, the causes, and the corrective action(s) that should be taken.

Operator Observations

The operator should consider the following items:

1. **Flow Distribution:** the operator monitors the flow to ensure that it is evenly distributed between all units in service, and that the flow through each individual unit is uniform, with no indication of short-circuiting.
2. **Contact Tank:** the contact tanks or basins must be checked to ensure that there is no excessive accumulation of scum on the surface; that there is no indication of solids accumulation on the bottom; and that mixing appears to be adequate.
3. **Chlorinator:** the operator should check to ensure there is no evidence of leakage; operating pressure/vacuum is within specified levels; current chlorine feed settling is within expected levels; in-line cylinders have sufficient chlorine to ensure continuous feed; and the exhaust system is operating as designed.

Factors Affecting Performance

Operators must be familiar with those factors that affect chlorination performance. Any item that interferes with the chlorine reactions or increases the demand for chlorine can affect performance and, in turn, may produce non-disinfectant products. Factors affecting chlorination performance include:

1. **Effluent Quality:** Poor quality effluents have higher chlorine demands; high concentrations of solids prevent chlorine-organism contact; and incomplete nitrification can cause extremely high chlorine demand.
2. **Mixing:** To be effective, chlorine must be in contact with the organisms. Poor mixing results in poor chlorine distribution. The installation of baffles and the use of a high length-to-width ratio will improve mixing and contact.
3. **Contact Time:** The chlorine disinfection process is time-dependent. As the contact time decreases, process effectiveness decreases. A minimum of 30 min of contact must be available at the design flow.
4. **Residual Levels:** The chlorine disinfection process is total residual chlorine (TRC) dependent. The concentration of residual must be sufficient to ensure the desired reactions occur. At the design contact time, the required minimum total residual chlorine concentration is 1.0 mg/L.

Process Control Sampling and Testing

To ensure proper operation of the chlorination process, the operator must perform process control testing for the chlorination process. [Note: The process performance evaluation is based on the bacterial content (fecal coliform) of the final effluent.] Process control testing consists of performing a *total chlorine residual test* on chlorine contact effluent. The frequency of the testing is specified in the plant permit. The normal expected range of results is also specified in the plant permit.

Troubleshooting

The following sections present common operational problems, symptoms, causal factors, and corrective actions associated with chlorination system use in wastewater treatment.

SYMPTOM 1

The coliform count fails to meet the required standards for disinfection.

1. **Cause:** Inadequate chlorination equipment capacity
   Corrective Action: Replace equipment as necessary to provide treatment based on maximum flow through the pipe.
2. **Cause:** Inadequate chlorine residual control
   Corrective Action: Use a chlorine residual analyzer to monitor and control chlorine dosage automatically.
3. **Cause:** Short-circuiting in the chlorine contact chamber
   Corrective Action: Install baffling in the chlorine contact chamber; install the mixing device in the chlorine contact chamber.
4. **Cause:** Solids buildup in the contact chamber
   Corrective Action: Clean the contact chamber.
5. **Cause:** Chlorine residual is too low
   Corrective Action: Increase contact time and/ or increase the chlorine feed rate.

SYMPTOM 2

Low chlorine gas pressure at the chlorinator.

1. **Cause:** Insufficient number of cylinders connected to the system
   Corrective Action: Connect enough cylinders to the system so that the feed rate does not exceed the recommended withdrawal rate for cylinders.
2. **Cause:** Stoppage or restriction of flow between cylinders and chlorinator
   Corrective Action: Disassemble the chlorine header system at the point where cooling begins, locate stoppage, and clean with solvent.

SYMPTOM 3

No chlorine gas pressure is available at the chlorinator.

1. **Cause:** Chlorine cylinders empty or not connected to the system
   Corrective Action: Connect cylinders or replace empty cylinders.
2. **Cause:** Plugged or damaged pressure-reducing valve
   Corrective Action: Repair the reducing valve after shutting cylinder valves and decreasing gas in the header system.

SYMPTOM 4

The chlorinator will not feed any chlorine.

1. **Cause:** Pressure reducing valve in the chlorinator is dirty
   Corrective Action: Disassemble the chlorinator and clean the valve stem and seat; precede the valve with filter/sediment trap.
2. **Cause:** The chlorine cylinder is hotter than chlorine control apparatus (chlorinator)
   Corrective Action: Reduce the temperature in the cylinder area; do not connect a new cylinder, which has been sitting in the sun.

SYMPTOM 5

Chlorine gas escaping from the chlorine pressure-reducing valve (CPRV).

1. **Cause:** The main diaphragm of CPRV ruptured
   Corrective Action: Disassemble the valve and diaphragm; inspect the chlorine supply system for moisture intrusion.

SYMPTOM 6

Inability to maintain chlorine feed rate without icing of chlorine system.

1. **Cause:** Insufficient evaporator capacity
   Corrective Action: Reduce feed rate to 75% of evaporator capacity. If this eliminates the problem then the main diaphragm of CPRV is ruptured.
2. **Cause:** External CPRV cartridge is clogged
   Corrective Action: Flush and clean cartridge.

SYMPTOM 7

The chlorinator system is unable to maintain sufficient water bath temperature to keep external CPRV open.

1. **Cause:** Heating element malfunction
   Corrective Action: Remove and replace the heating element.

SYMPTOM 8

Inability to obtain maximum feed rate from chlorinator.

1. **Cause:** Inadequate chlorine gas pressure
   Corrective Action: Increase pressure, and replace empty or low cylinders.
2. **Cause:** Water pump injector clogged with deposits
   Corrective Action: Clean injector parts using muriatic acid. Rinse parts with fresh water and place back in service.
3. **Cause:** Leak in vacuum relief valve
   Corrective Action: Disassemble the vacuum relief valve and replace all springs.
4. **Cause:** Vacuum Leak in joints, gaskets, tubing, etc. in the chlorinator system
   Corrective Action: Repair all vacuum leaks by tightening joints, replacing gaskets, and replacing tubing and/or compression nuts.

SYMPTOM 9

Inability to maintain adequate chlorine feed rate.

1. **Cause:** Malfunction or deterioration of the chlorine water supply pump
   Corrective Action: Overhaul the pump (if a turbine pump is used, try closing the valve to maintain proper discharge pressure).

SYMPTOM 10

Chlorine residual is too high in plant effluent to meet requirements.

1. **Cause:** Chlorine residual too high
   Corrective Action: Install dechlorination facilities.

SYMPTOM 11

Wide variation in chlorine residual produced in the effluent.

1. **Cause:** Chlorine flow proportion meter capacity inadequate to meet plant flow rates
   Corrective Action: Replace with higher capacity chlorinator meter.
2. **Cause:** Malfunctioning controls
   Corrective Action: Call the manufacturer's technical representative.
3. **Cause:** Solids settled in the chlorine contact chamber

Corrective Action: Clean the chlorine contact tank.

4. **Cause:** Flow proportioning control device not zeroed or spanned correctly

    Corrective Action: Re-zero and span the device in accordance with the manufacturer's instructions.

SYMPTOM 12

Unable to obtain chlorine residual.

1. **Cause:** High chemical demand

    Corrective Action: Locate and correct the source of the high demand.
2. **Cause:** Test interference

    Corrective Action: Add sulfuric acid to samples to reduce interference.

SYMPTOM 13

The chlorine residual analyzer, recorder, and controller do not control chlorine residual properly.

1. **Cause:** Electrodes fouled

    Corrective Action: Clean electrodes.
2. **Cause:** Loop time is too long

    Corrective Action: Reduce control loop time by:

    1. Moving the injector closer to the point of application.
    2. Increasing the velocity in the sample line to the analyzer.
    3. Moving the cell closer to the sample point.
    4. Moving the sample point closer to the point of application.
3. **Cause:** Insufficient potassium iodide being added for the amount of residual being measured

    Corrective Action: Adjust potassium iodide feed to correspond with the chlorine residual being measured.
4. **Cause:** Buffer additive system is malfunctioning

    Corrective Action: Repair buffer additive system.
5. **Cause:** Malfunctioning of analyzer cell

    Corrective Action: Call authorized service personnel to repair electrical components.
6. **Cause:** Poor mixing of chlorine at the point of application

    Corrective Action: Install a mixing device to cause turbulence at the point of application.
7. **Cause:** Rotameter tube range is improperly set

    Corrective Action: Replace the rotameter with a proper range of feed rate.

Dechlorination

The purpose of *dechlorination* is to remove chlorine and reaction products (chloramines) before the treated waste stream is discharged into its receiving waters. Dechlorination follows chlorination—usually at the end of the contact tank to the final effluent. Sulfur dioxide gas, sodium sulfate, sodium metabisulfate, or sodium bisulfates are the chemicals used to dechlorinate. No matter which chemical is used to dechlorinate, its reaction with chlorine is instantaneous.

Chlorination Environmental Hazards and Safety

Chlorine is an extremely toxic substance that can when released into the environment, cause severe damage. For this reason, most state regulatory agencies have established a chlorine water quality standard (e.g., in Virginia, 0.011 mg/L in fresh waters for total residual chlorine and 0.0075 mg/L for chlorine-produced oxidants in saline waters). Studies have indicated that above these levels chlorine can reduce shellfish growth and destroy sensitive aquatic organisms. This standard has resulted in many treatment facilities being required to add process to remove the chlorine before discharge. As mentioned, the process, known as dechlorination, uses chemicals that react quickly with chlorine to convert it to a less harmful form. Elemental chlorine is a chemical with potentially fatal hazards associated with it. For this reason, many different state and federal agencies regulate the transport, storage, and use of chlorine. All operators required to work with chlorine should be trained in proper handling techniques. They should also be trained to ensure that all procedures for the storage transport, handling, and use of chlorine are following appropriate state and federal regulations.

Chlorine: Safe Work Practice

Because of the inherent dangers involved with handling chlorine, each facility using chlorine (for any reason) should ensure that a written safe work practice is in place and is followed by plant operators. A sample safe work practice for handling chlorine is provided in the below:

WORK: CHEMICAL HANDLING—CHLORINE

PRACTICE

1. Plant personnel *must* be trained and instructed on the use and handling of chlorine, chlorine equipment, chlorine emergency repair kits, and other chlorine emergency procedures.
2. Use extreme care and caution when handling chlorine.
3. Lift chlorine cylinders only with an approved and load-tested device.
4. Secure chlorine cylinders into position immediately. *Never* leave a cylinder suspended.
5. Avoid dropping chlorine cylinders.

6. Avoid banging chlorine cylinders into other objects.

7. Store chlorine 1-ton cylinders in a cool, dry place away from direct sunlight or heating units. Railroad tank cars are direct sunlight compensated.

8. Store chlorine 1-ton cylinders on their sides only (horizontally).

9. Do not stack unused or used chlorine cylinders.

10. Provide positive ventilation to the chlorine storage area and chlorinator room.

11. *Always* keep chlorine cylinders at ambient temperature. *Never* apply direct flame to a chlorine cylinder.

12. Use the oldest chlorine cylinder in stock first.

13. Always keep valve protection hoods in place until the chlorine cylinders are ready for connection.

14. Except to repair a leak, do not tamper with the fusible plugs on chlorine cylinders.

15. Wear SCBA whenever changing a chlorine cylinder and have at least one other person with a standby SCBA unit outside the immediate area.

16. Inspect all threads and surfaces of the chlorine cylinder and have at least one other person with a standby SCBA unit outside the immediate area.

17. Use new lead gaskets each time a chlorine cylinder connection is made.

18. Use only the specified wrench to operate chlorine cylinder valves.

19. Open chlorine cylinder valves slowly; no more than one full turn.

20. Do not hammer, bang, or force chlorine cylinder valves under any circumstances.

21. Check for chlorine leaks as soon as the chlorine cylinder connection is made. Leaks are checked for by gently expelling ammonia mist from a plastic squeeze bottle filled with approximately 2 oz of liquid ammonia solution. Do not put liquid ammonia on valves or equipment.

22. Correct all minor chlorine leaks at the chlorine cylinder connection immediately.

23. Except for automatic systems, draw chlorine from only one manifold chlorine cylinder at a time. *Never* simultaneously open two or more chlorine cylinders connected to a common manifold pulling liquid chlorine. Two or more cylinders connected to a common manifold pulling gaseous chlorine are acceptable.

24. Wear SCBA and chemical protective clothing covering your face, arms, and hands before entering an enclosed chlorine area to investigate a chlorine odor or chlorine leak—a two-person rule is required.

25. Provide positive ventilation to a contaminated chlorine atmosphere before entering whenever possible.

26. Have at least two personnel present before entering a chlorine atmosphere: One person to enter the chlorine atmosphere, the other to observe in the event of an emergency. *Never* enter a chlorine atmosphere unattended. Remember: OSHA mandates that only fully qualified Level III HAZMAT responders are authorized to aggressively attack a hazardous materials leak such as chlorine.

27. Use supplied air-breathing equipment when entering a chlorine atmosphere. *Never* use canister-type gas masks when entering a chlorine atmosphere.

28. Ensure that supplied air-breathing apparatus has been properly maintained in accordance with the plant's Self-Contained Breathing Apparatus Inspection Guidelines as specified in the plant's Respiratory Protection Program.

29. Stay upwind from all chlorine leak danger areas unless involved with making repairs. Look to plant windsocks for wind direction.

30. Contact trained plant personnel to repair chlorine leaks.

31. Roll uncontrollable leaking chlorine cylinders so that the chlorine escapes as a gas, not as a liquid.

32. Stop leaking chlorine cylinders or leaking chlorine equipment (by closing off valve(s) if possible) prior to attempting repair.

33. Connect uncontrollable leaking chlorine cylinders to the chlorination equipment and feed the maximum chlorine feed rate possible.

34. Keep leaking chlorine cylinders at the plant site. Chlorine cylinders received at the plant site must be inspected for leaks prior to taking delivery from the shipper. *Never* ship a leaking chlorine cylinder back to the supplier after it has been accepted (bill of lading has been signed by plant personnel) from the shipper; instead, repair or stop the leak first.

35. Keep moisture away from a chlorine leak. *Never* put water onto a chlorine leak.

36. Call the fire department or rescue squad if a person is incapacitated by chlorine.

37. Administer CPR (use a barrier mask if possible) immediately to a person who has been incapacitated by chlorine.

38. Breathe shallow rather than deep if exposed to chlorine without the appropriate respiratory protection.

39. Place a person who does not have difficulty breathing and is heavily contaminated with chlorine into a deluge shower. Remove their clothing under the water and flush all body parts that were exposed to chlorine.

40. Flush eyes contaminated with chlorine with copious quantities of lukewarm running water for at least 15 min.

41. Drink milk if your throat is irritated by chlorine.

42. *Never* store other materials in chlorine cylinder storage areas; substances like acetylene and propane are not compatible with chlorine.

Chlorination Process Calculations

Several calculations may be useful in operating a chlorination system. Many of these calculations are discussed and illustrated in this section.

Chlorine Demand

Chlorine demand is the amount of chlorine in milligrams per liter that must be added to the wastewater to complete all of the chemical reactions that must occur before producing a residual.

$$\text{Chlorine Demand} = \text{Chlorine Dose, mg/L}$$
$$- \text{Chlorine Residual, mg/L} \quad (18.69)$$

Example 18.51

Problem: The plant effluent currently requires a chlorine dose of 7.1 mg/L to produce the required 1.0 mg/L chlorine residual in the chlorine contact tank. What is the chlorine demand in milligrams per liter?

Solution:

$$\text{Chlorine Demand mg/L} = 7.1 \text{ mg/L} - 1.0 \text{ mg/L} = 6.1 \text{ mg/L}$$

Chlorine Feed Rate

The chlorine feed rate is the amount of chlorine added to the wastewater in pounds per day.

$$\text{Chlorine Feed Rate} = \text{Dose, mg/L} \times \text{Flow,}$$
$$\text{MGD} \times 8.34 \text{ lb/mg/L/MG} \quad (18.70)$$

Example 18.52

Problem: The current chlorine dose is 5.55 mg/L. What is the feed rate in pounds per day if the flow is 22.89 MGD?

Solution:

$$\text{Feed, lbs/day} = 5.55 \text{ mg/L} \times 22.89 \text{ MGD} \times 8.34 \text{ lb/mg/L/MG}$$
$$= 1,060 \text{ lb/day}$$

Chlorine Dose

The chlorine dose is the concentration of chlorine being added to the wastewater, expressed in milligrams per liter.

$$\text{Dose, mg/L} = \frac{\text{Chlorine Feed Rate, lbs/day}}{\text{Flow in Million gal/day} \times 8.34 \text{ lbs/mg/L/MG}} \quad (18.71)$$

Example 18.53

Problem: Three hundred twenty (320) pounds of chlorine are added per day to a wastewater flow of 5.60 MGD. What is the chlorine dose in milligrams per liter?

$$\text{Dose, mg/L} = \frac{320 \text{ lb/day}}{5.60 \text{ MGD} \times 8.34 \text{ lb/mg/L/MG}} = 6.9 \text{ mg/L}$$

Available Chlorine

When hypochlorite forms of chlorine are used, the available chlorine is listed on the label. In these cases, the amount of chemical added must be converted to the actual amount of chlorine using the following calculation.

$$\text{Available Chlorine} = \text{Amount of Hypochlorite}$$
$$\times \% \text{ Available Chlorine} \quad (18.72)$$

Example 18.54

Problem: The calcium hypochlorite used for chlorination contains 62.5% available chlorine. How many pounds of chlorine are added to the plant effluent if the current feed rate is 30 lbs of calcium hypochlorite per day?

Solution:

$$\text{Quantity of Chlorine} = 30 \text{ lbs} \times 0.625 = 18.75 \text{ lb Chlorine}$$

Required Quantity of Dry Hypochlorite

This calculation is used to determine the amount of hypochlorite needed to achieve the desired dose of chlorine:

$$\text{Hypochlorite Quantity, lb/day} = \frac{\substack{\text{Required Chlorine Dose, mg/L} \times \text{Flow,} \\ \text{MGD} \times 8.34 \text{ lb/mg/L/MG}}}{\% \text{ Available Chlorine}} \quad (18.73)$$

Example 18.55

Problem: The laboratory reports that the chlorine dose required to maintain the desired residual level is 8.5 mg/L. Today's flow rate is 3.25 MGD. The hypochlorite powder used for disinfection contains 70% available chlorine. How many pounds of hypochlorite must be used?

Solution:

$$\text{Hypochlorite Quantity} = \frac{8.5 \text{ mg/L} \times 3.25 \text{ MGD} \times 8.34 \text{ lb/mg/L/MG}}{0.70}$$
$$= 329 \text{ lb/day}$$

Required Quantity of Liquid Hypochlorite

$$\text{Hypochlorite needed, gal/day} = \frac{\begin{array}{c}\text{Required Chlorine Dose, mg/L}\\ \times \text{Flow, MGD} \times 8.34 \text{ lb/mg/L/MG}\end{array}}{\% \text{ Available Chlorine} \times 8.34} \\ \times \text{Hypochlorite solution spec. gravity}$$

$$(18.74)$$

Example 18.56

Problem: The chlorine dose is 8.8 mg/L, and the flow rate is 3.28 MGD. The hypochlorite solution contains 71% available chlorine and has a specific gravity of 1.25. How many pounds of hypochlorite must be used?

Solution:

$$\text{Hypochlorite Quantity} = \frac{\begin{array}{c}8.8 \text{ mg/L} \times 3.28 \text{ MGD}\\ \times 8.34 \text{ lb/mg/L/MG}\end{array}}{0.71 \times 8.34 \text{ lb/gal} \times 1.25} \\ = 32.5 \text{ gal/day}$$

Ordering Chlorine

Because disinfection must be continuous, the supply of chlorine must never be allowed to run out. The following calculation provides a simple method for determining when additional supplies must be ordered. The process consists of three steps:

Step 1: Adjust the flow and use variations if projected changes are provided.
Step 2: If an increase in flow and/or required dosage is projected, adjust the current flow rate and/or dose to reflect the projected change.
Step 3: Calculate the projected flow and dose:

$$\text{Projected Flow} = \text{Current Flow, MGD} \times (1.0 + \% \text{ Change})$$

$$(18.75)$$

$$\text{Projected Dose} = \text{Current Dose, mg/L} \times (1.0 + \% \text{ Change})$$

Example 18.57

Problem: Based on the available information for the past 12 months, the operator projects that the effluent flow rate will increase by 7.5% during the next year. If the average daily flow has been 4.5 MGD, what will be the projected flow for the next 12 months?

Solution:

$$\text{Projected Flow, MGD} = 4.5 \text{ MGD} \times (1.0 + 0.075) = 4.84 \text{ MGD}$$

To determine the amount of chlorine required for a given period:

$$\text{Chlorine Required} = \text{Feed Rate, lb/day} \times \text{no. of days required}$$

Example 18.58

Problem: The plant currently uses 90 lb of chlorine per day. The town wishes to order enough chlorine to supply the plant for 4 months (assuming 31 days/month). How many pounds of chlorine should be ordered to provide the needed supply?

Solution:

$$\text{Chlorine required} = 90 \text{ lb/day} \times 124 \text{ days} = 11,160 \text{ lb}$$

Note: In some instances, projections for flow or dose changes are not available, but the plant operator wishes to include an extra amount of chlorine as a safety factor. This safety factor can be stated as a specific quantity or as a percentage of the projected usage. The safety factor as a specific quantity can be expressed as:

$$\text{Total Required Cl}_2 = \text{Chlorine Required, lbs} + \text{Safety Factor}$$

Note: Because chlorine is only shipped in full containers unless asked specifically for the amount of chlorine required or used during a specified period, all decimal parts of a cylinder are rounded up to the next highest number of full cylinders.

ULTRAVIOLET IRRADIATION

Although ultraviolet (UV) disinfection was recognized as a method for achieving disinfection in the late nineteenth century, its application virtually disappeared with the evolution of chlorination technologies. However, in recent years, there has been a resurgence in its use in the wastewater field, largely as a consequence of concern for the discharge of toxic chlorine residuals. Even more recently, UV has gained more attention due to the tough new regulations on chlorine use imposed by both OSHA and USEPA. Because of this relatively recent increased regulatory pressure, many facilities are actively engaged in substituting chlorine with other disinfection alternatives. Moreover, UV technology itself has made many improvements, making UV an attractive disinfection alternative. Ultraviolet light has very good germicidal qualities and is very effective in destroying microorganisms. It is used in hospitals, biological testing facilities, and many other similar locations. In wastewater treatment, the plant effluent is exposed to ultraviolet light of a specified wavelength and intensity for a specified contact period. The effectiveness of the process depends on:

- UV light intensity
- Contact time
- Wastewater quality (turbidity)
- For any one treatment plant, disinfection success is directly related to the concentration of colloidal and particulate constituents in the wastewater.

The Achilles' heel of UV disinfection for wastewater is turbidity. If the wastewater quality is poor, the ultraviolet light will be unable to penetrate the solids and the effectiveness of the process decreases dramatically. For this reason, many states limit the use of UV disinfection to facilities that can reasonably be expected to produce an effluent containing ≤30 mg/L or less of BOD_5 and total suspended solids.

The main components of a UV disinfection system are mercury arc lamps, a reactor, and ballasts. The source of UV radiation is either the low-pressure or medium-pressure mercury arc lamp with low or high intensities. Note that in the operation of UV systems, UV lamps must be readily available when replacements are required. The best lamps are those with a stated operating life of at least 7,500 h and that do not produce significant amounts of ozone or hydrogen peroxide. The lamps must also meet technical specifications for intensity, output, and arc length. If the UV light tubes are submerged in the waste stream, they must be protected inside quartz tubes, which not only protect the lights but also make cleaning and replacement easier.

Contact tanks must be used with UV disinfection. They must be designed with the banks of UV lights in a horizontal position, either parallel or perpendicular to the flow, or with banks of lights placed in a vertical position perpendicular to the flow.

Note: The contact tank must provide, at a minimum, a 10-sec exposure time.

It was stated earlier that turbidity has been the problem with using UV in wastewater treatment—and this is the case. However, if turbidity is its Achilles' heel, then the need for increased maintenance (compared to other disinfection alternatives) is the toe of the same foot. UV maintenance requires that the tubes be cleaned on a regular basis or as needed. In addition, periodic acid washing is required to remove chemical buildup.

Routine monitoring is required. Monitoring to check for bulb burnout, buildup of solids on quartz tubes, and UV light intensity is necessary.

Note: UV light is extremely hazardous to the eyes. Never enter an area where UV lights are in operation without proper eye protection. Never look directly into the ultraviolet light.

Advantages and Disadvantages

Advantages:

- UV disinfection is effective at inactivating most viruses, spores, and cysts.
- UV disinfection is a physical process rather than a chemical disinfectant, which eliminates the need to generate, handle, transport, or store toxic/hazardous, or corrosive chemicals.
- There is no residual effect that can be harmful to humans or aquatic life.
- UV disinfection is user-friendly for operators.
- UV disinfection has a shorter contact time compared to other disinfectants (approximately 20–30 sec with low-pressure lamps).
- UV disinfection equipment requires less space than other methods.

Disadvantages:

- Low dosages may not effectively inactivate some viruses, spores, and cysts.
- Organisms can sometimes repair and reverse the destructive effects of UV through a "repair mechanism," known as photo reactivation, or in the absence of light known as "dark repairs."
- A preventive maintenance program is necessary to control fouling of tubes.
- Turbidity and total suspended solids (TSS) in the wastewater can render UV disinfection ineffective. UV disinfection with low-pressure lamps is not as effective for secondary effluent with TSS levels above 30 mg/L.
- UV disinfection is not as cost-competitive when chlorination-dechlorination is used and fire codes are met (USEPA, 1999a).

MICROBIAL REPAIR

Many microorganisms have enzyme systems that repair damage caused by UV light. Repair mechanisms are classified as photo repair and dark repair (Knudson, 1985). Microbial repair can increase the UV dose needed to achieve a given degree of inactivation of a pathogen, but the process does not prevent activation. Even though microbial repair can occur, neither photo repair nor dark repair is anticipated to affect the performance of drinking water UV disinfection.

PHOTO REPAIR

In photo repair (or photoreactivation), enzymes energized by exposure to light between 310 and 490 nm (near and in the visible range) break the covalent bonds that form the pyrimidine dimmers. Photoreceptor requires reactivating light and repair only pyrimidine dimers (Jagger, 1967). Knudson (1985) found that bacteria have the enzymes necessary for photo repair. Unlike bacteria, viruses lack the necessary enzymes for repair but can repair using the enzymes of a host cell (Rauth, 1965). Linden et al. (2002) did not observe photo repair of *Giardia* at UV doses typical of UV disinfection applications (16 and 40 mJ/cm^2). However, unpublished data from the same study show *Giardia* reactivation in light conditions at very low UV doses (0.5 mJ/cm^2, USEPA, 2006). Shin et al. (2001) reported that Cryptosporidium does not regain infectivity after inactivation by UV light. One study showed that Cryptosporidium can undergo some DNA photo repair (Oguma et al., 2001). Even though the DNA is repaired, however, infectivity is not restored.

DARK REPAIR

Dark repair is defined as any repair process that does not require the presence of light. The term is somewhat misleading because dark repair can also occur in the presence of light. Excision repair, a form of dark repair, is an enzyme-mediated process in which the damaged section of DNA is removed and regenerated using the existing complementary strand of DNA. As such, excision repair can occur only with double-stranded DNA and RNA. The extent of dark repair varies with the microorganism. With bacteria and protozoa, dark repair enzymes start to act immediately following exposure to UV light; therefore, reported dose-response data are assumed to account for dark repair.

Knudson (1985) found that bacteria can undergo dark repair, but some lack the enzymes needed for dark repair. Viruses also lack the necessary enzymes for repair but can repair using the enzymes of a host cell (Rauth, 1965). Oguma et al. (2001) used an assay that measures the number of dimers formed in nucleic acid to show that dark repair occurs in *Cryptosporidium*, even though the microorganism did not regain infectivity. Linden et al. (2002) did not observe dark repair of Giardia at UV doses typical for UV disinfection applications (16 and 40 mJ/cm^2). Shin et al. (2001) reported that *Cryptosporidium* does not regain infectivity after inactivation by UV light.

Applicability

When choosing a UV disinfection system, there are three critical areas to consider. The first is primarily determined by the manufacturer, the second by design and Operation and Maintenance (O&M); and the third must be controlled at the treatment facility. Choosing a UV disinfection system depends on three critical factors listed below:

- **Hydraulic Properties of the Reactor:** Ideally, a UV disinfection system should have a uniform flow with enough axial motion (radial mixing) to maximize exposure to UV radiation. The path that an organism takes in the reactor determines the amount of UV radiation it will be exposed to before inactivation. A reactor must be designed to eliminate short-circuiting and/or dead zones, which can result in inefficient use of power and reduced contact time.
- **Intensity of the UV Radiation:** Factors affecting the intensity include the age of the lamps, lamp fouling, and the configuration and placement of lamps in the reactor.
- **Wastewater Characteristics:** These include the flow rate, suspended and colloidal solids, initial bacterial density, and other physical and chemical parameters. Both the concentration of TSS and the concentration of particle-associated microorganisms determine how much UV radiation ultimately reaches the target organism. The higher these concentrations, the lower the UV radiation absorbed by the organisms. UV disinfection can be used in plants of various sizes that provide secondary or advanced levels of treatment.

Operation and Maintenance

The proper O&M of a UV disinfection system ensures that sufficient UV radiation is transmitted to the organisms to render them sterile. All surfaces between the UV radiation and the target organism must be clean, and the ballasts, lamps, and reactors must be functioning at peak efficiency. Inadequate cleaning is one of the most common causes of a UV system's ineffectiveness. The quartz sleeves or Teflon tubes need to be cleaned regularly using mechanical wipes, ultrasonics, or chemicals. The cleaning frequency is very site-specific; some systems need to be cleaned more often than others.

Chemical cleaning is most commonly done with citric acid. Other cleaning agents include mild vinegar solutions and sodium hydrosulfite. A combination of cleaning agents should be tested to find the agent most suitable for the wastewater characteristics without producing harmful or toxic by-products. Non-contact reactor systems are most effectively cleaned using sodium hydrosulfite.

Any UV disinfection should be pilot-tested prior to the full-scale platform to ensure that it will meet discharge permit requirements for a particular site.

The average lamp life ranges from 8,760 to 14,000 working hours, and the lamps are usually replaced after 12,000h of use. Operating procedures should be set to reduce the on/off cycles of the lamps since efficacy is reduced with repeated cycles.

The ballast must be compatible with the lamps and should be ventilated to protect it from excessive heating, which may shorten its life or even result in fires. Although the life cycle of ballasts is approximately 10–15years, they are usually replaced every 10years. Quartz sleeves will last about 5–8years but are generally replaced every 5years (USEPA, 1999a).

OZONATION

Ozone is a strong oxidizing gas that reacts with most organic and many inorganic molecules. It is produced when oxygen molecules separate, collide with other oxygen atoms, and form a molecule consisting of three oxygen atoms. For high-quality effluents, ozone is a very effective disinfectant. Current regulations for domestic treatment systems limit the use of ozonation to filtered effluents unless the system's effectiveness can be demonstrated prior to installation.

Note: Effluent quality is the key performance factor for ozonation.

For the ozonation of wastewater, the facility must have the capability to generate pure oxygen along with an ozone generator. A contact tank with a ≥10-min contact time at design average daily flow is required. Off-gas monitoring for process control is also required. In addition, safety equipment capable of monitoring ozone in the atmosphere and a ventilation system to prevent ozone levels from exceeding 0.1 ppm is required.

The actual operation of the ozonation process consists of monitoring and adjusting the ozone generator and monitoring the control system to maintain the required ozone concentration in the off-gas. The process must also be evaluated periodically using biological testing to assess its effectiveness.

Note: Ozone is an extremely toxic substance. Concentrations in the air should not exceed 0.1 ppm. It also has the potential to create an explosive atmosphere. Sufficient ventilation and purging capabilities should be provided.

Note: Ozone has certain advantages over chlorine for the disinfection of wastewater: (1) Ozone increases DO in the effluent; (2) ozone has a briefer contact time; (3) ozone has no undesirable effects on marine organisms; and (4) ozone decreases turbidity and odor.

Advantages and Disadvantages

Advantages:

- Ozone is more effective than chlorine in destroying viruses and bacteria.
- The ozonation process utilizes a short contact time (approximately 10–30min).

- There are no harmful residuals that need to be removed after ozonation because ozone decomposes rapidly.
- After ozonation, there is no regrowth of microorganisms, except for those protected by the particulates in the wastewater stream.
- Ozone is generated onsite, and thus, there are fewer safety problems associated with shipping and handling.
- Ozonation elevates the dissolved oxygen (DO) concentration of the effluent. The increase in DO can eliminate the need for reaeration and also raise the level of DO in the receiving stream.

Disadvantages:

- Low dosage may not effectively inactivate some viruses, spores, and cysts.
- Ozonation is a more complex technology than chlorine or UV disinfection, requiring complicated equipment and efficient contacting systems.
- Ozone is very reactive and corrosive, thus requiring corrosion-resistant materials such as stainless steel.
- Ozonation is not economical for wastewater with high levels of suspended solids (SS), biochemical oxygen demand (BOD), chemical oxygen demand, or total organic carbon.
- Ozone is extremely irritating and possibly toxic, so off-gases from the contactor must be destroyed to prevent worker exposure.
- The cost of treatment can be relatively high in capital and power intensity.

Applicability

Ozone disinfection is generally used at medium to large-sized plants after at least secondary treatment. In addition to disinfection, another common use for ozone in wastewater treatment is odor control.

Ozone disinfection is the least used method in the U.S. although this technology has been widely accepted in Europe for decades. Ozone treatment can achieve higher levels of disinfection than either chlorine or UV; however, the capital costs as well as maintenance expenditures are not competitive with available alternatives. Ozone is therefore used only sparingly, primarily in special cases where alternatives are not effective (USEPA, 1999b).

Operation and Maintenance

Ozone generation uses a significant amount of electrical power. Thus, constant attention must be given to the system to ensure that power is optimized for controlled disinfection performance.

There must be no leaking connections in or surrounding the ozone generator. The operator must regularly monitor the appropriate subunits to ensure that they are not overheated. Therefore, the operator must check for leaks routinely since a very small leak can cause unacceptable ambient ozone concentrations. The ozone monitoring equipment must be tested and calibrated as recommended by the equipment manufacturer.

Like oxygen, ozone has limited solubility and decomposes more rapidly in water than in air. This factor, along with ozone reactivity, requires that the ozone contactor be well covered and that the ozone diffuses into the wastewater as effectively as possible.

Ozone in gaseous form is explosive once it reaches a concentration of 240 g/m³. Since most ozonation systems never exceed a gaseous ozone concentration of 50–200 g/m³, this is generally not a problem. However, ozone in gaseous form will remain hazardous for a significant amount of time; thus, extreme caution is needed when operating the ozone gas systems.

It is important that the ozone generator, distribution, contactor, off-gas, and ozone destructor inlet piping be purged before opening the various systems or subsystems. When entering the ozone contactor, personnel must recognize the potential for oxygen deficiencies or trapped ozone gas, despite best efforts to purge the system. The operator should be aware of all emergency operating procedures required if a problem occurs. All safety equipment should be available for operators to use in case of an emergency. Key O&M parameters include:

- Clean feed gas with a dew point of −60°C (−76°F), or lower, must be delivered to the ozone generator. If the supply gas is moist, the reaction of ozone and moisture will yield a very corrosive condensate on the inside of the ozonator. The output of the generator could be lowered by the formation of nitrogen oxides (such as nitric acid).
- Maintain the required flow of generator coolant (air, water, or other liquids).
- Lubricate the compressor or blower in accordance with the manufacturer's specifications. Ensure that all compressor sealing gaskets are in good condition.
- Operate the ozone generator within its design parameters. Regularly inspect and clean the ozonator, air supply, and dielectric assemblies, and monitor the temperature of the ozone generator.
- Monitor the ozone gas feed and distribution system to ensure that the necessary volume comes into sufficient contact with the wastewater.
- Maintain ambient levels of ozone below the limits of applicable safety regulations.

BROMINE CHLORIDE

Bromine chloride is a mixture of bromine and chlorine that forms hydrocarbons and hydrochloric acid when mixed with water. It is an excellent disinfectant that reacts quickly and does not produce long-term residuals.

Note: Bromine chloride is an extremely corrosive compound in the presence of low moisture concentrations.

The reactions that occur when bromine chloride is added to the wastewater are similar to those occurring when chlorine is added. The major difference is the production of bromine compounds rather than chloramines. The bromine compounds are excellent disinfectants but are less stable and dissipate quickly. In most cases, the bromamines rapidly decay into other, less toxic compounds and are undetectable in the plant effluent. The factors that affect performance are similar to those affecting the performance of the chlorine disinfection process. Effluent quality, contact time, etc. have a direct impact on the performance of the process.

NO DISINFECTION

In a very limited number of cases, treated wastewater discharges without disinfection are permitted. These are approved on a case-by-case basis. Each request must be evaluated based on the point of discharge, the quality of the discharge, the potential for human contact, and many other factors.

ADVANCED WASTEWATER TREATMENT

Advanced wastewater treatment is defined as the method(s) and/or process(es) that remove more contaminants (suspended and dissolved substances) from wastewater than are taken out by conventional biological treatment. Put another way, advanced wastewater treatment is the application of a process or system that follows secondary treatment or that includes phosphorus removal or nitrification in conventional secondary treatment.

Advanced wastewater treatment is used to augment conventional secondary treatment because secondary treatment typically removes only between 85% and 95% of the biochemical oxygen demand (BOD) and total suspended solids (TSS) in raw sanitary sewage. Generally, this leaves 30 mg/L or less of BOD and TSS in the secondary effluent. To meet stringent water-quality standards, this level of BOD and TSS in secondary effluent may not prevent a violation of water-quality standards—the plant may not meet a permit. Thus, advanced wastewater treatment is often used to remove additional pollutants from treated wastewater.

In addition to meeting or exceeding the requirements of water-quality standards, treatment facilities use advanced wastewater treatment for other reasons as well. For example, sometimes conventional secondary wastewater treatment is not sufficient to protect the aquatic environment. In a stream, for example, when periodic flow events occur, the stream may not provide the amount of dilution of effluent needed to maintain the necessary dissolved oxygen (DO) levels for aquatic organism survival.

Secondary treatment has other limitations. It does not significantly reduce the effluent concentration of nitrogen and phosphorus (important plant nutrients) in sewage. An overabundance of these nutrients can over-stimulate plant and algae growth, creating water quality problems. For example, if discharged into lakes, these nutrients contribute to algal blooms and accelerated eutrophication (lake aging). Also, the nitrogen in the sewage effluent may be present mostly in the form of ammonia compounds. If present in high enough concentrations, ammonia compounds are toxic to aquatic organisms. Yet another problem with these compounds is that they exert a *nitrogenous* oxygen demand in the receiving water as they convert to nitrates. This process is called nitrification.

Note: The term *tertiary treatment* is commonly used as a synonym for advanced wastewater treatment. However, these two terms do not have precisely the same meaning. Tertiary suggests a third step that is applied after primary and secondary treatment.

Advanced wastewater treatment can remove more than 99% of the pollutants from raw sewage and can produce effluent of almost potable (drinking) water quality. However, advanced treatment is not cost-free. The cost of advanced treatment, for operation and maintenance as well as for retrofit of present conventional processes, is very high (sometimes doubling the cost of secondary treatment). Therefore, a plan to install advanced treatment technology calls for careful study—the benefit-to-cost ratio is not always large enough to justify the additional expense.

Even considering the expense, the application of some form of advanced treatment is not uncommon. These treatment processes can be physical, chemical, or biological. The specific process used is based on the purpose of the treatment and the quality of the effluent desired.

CHEMICAL TREATMENT

The purpose of chemical treatment is to remove:

- Biochemical oxygen demand (BOD)
- Total suspended solids (TSS)
- Phosphorus
- Heavy metals
- Other substances that can be chemically converted to a settleable solid

Chemical treatment is often accomplished as an "add-on" to existing treatment systems or by means of separate facilities specifically designed for chemical addition. In each case, the basic process necessary to achieve the desired results remains the same:

- Chemicals are thoroughly mixed with the wastewater.
- The chemical reactions that occur form solids (coagulation).
- The solids are mixed to increase particle size (flocculation).
- Settling and/or filtration (separation) then remove the solids.

The specific chemical used depends on the pollutant to be removed and the characteristics of the wastewater. Chemicals may include the following:

- Lime
- Alum (aluminum sulfate)
- Aluminum salts
- Ferric or ferrous salts
- Polymers
- Bioadditives

OPERATION, OBSERVATION, AND TROUBLESHOOTING PROCEDURES

Operation and observation of the performance of chemical treatment processes depend on the pollutant being removed and on process design. Operational problems associated with chemical treatment processes used in advanced treatment usually revolve around problems with floc formation, settling characteristics, removal in the settling tank, and sludge (in the settling tank) turning anaerobic. To correct these problems, the operator must be able to recognize the applicable problem indicators through proper observation. In the following sections, we list common indicators/observations of operational problems, along with the applicable causal factors and corrective actions.

1. **Poor Floc Formation and Settling Characteristics**
   Causal Factors:
   - insufficient chemical dispersal during the rapid mix
   - excessive detention time in the rapid mix
   - improper coagulant dosage
   - excessive flocculation speed
   Corrective Actions (Where Applicable)
   - increase the speed of the rapid mixer
   - reduce detention time to15–60 sec
   - correct dosage (determined by jar testing)
   - reduce flocculation speed
2. **Good Floc Formation, Poor Removal in Settling Tank**
   Causal Factors:
   - excessive velocity between flocculation and settling
   - settling tank operational problem
   Corrective Action:
   - reduce velocity to an acceptable range
3. **Settling Tank Sludge is Turning Anaerobic**
   Causal Factors:
   - a sludge blanket has developed in the settling tank
   - excessive organic carryover from secondary treatment

- increase sludge withdrawal to eliminate the blanket
- correct secondary treatment operational problems

MICROSCREENING

Microscreening (also called *microstraining*) is an advanced treatment process used to reduce suspended solids. The microscreens are composed of specially woven steel wire fabric mounted around the perimeter of a large revolving drum. The steel wire cloth acts as a fine screen, with openings as small as 20 μm (or millionths of a meter)—small enough to remove microscopic organisms and debris. The rotating drum is partially submerged in the secondary effluent, which must flow into the drum and then outward through the microscreen. As the drum rotates, captured solids are carried to the top where a high-velocity water spray flushes them into a hopper or backwash tray mounted on the hollow axle of the drum. Backwash solids are recycled to plant influent for treatment. These units have found the greatest application in the treatment of industrial waters and final polishing filtration of wastewater effluents. Expected performance for suspended solids removal is 95%–99%, but the typical suspended solids removal achieved with these units is about 55%. The normal range is from 10% to 80%.

According to Metcalf & Eddy (2003), the functional design of the microscreen unit involves the following considerations: (1) The characterization of the suspended solids with respect to the concentration and degree of flocculation; (2) the selection of unit design parameter values that will not only ensure capacity to meet maximum hydraulic loadings with critical solids characteristics but also provide desired design performance over the expected range of hydraulic and solids loadings; (3) the provision of backwash and cleaning facilities to maintain the capacity of the screen.

Operation, Observation, and Troubleshooting Procedures

Microscreen operators typically perform sampling and testing on influent and effluent TSS and monitor screen operation to ensure proper operation. Operational problems generally consist of a gradual decrease in throughput rate, leakage at the ends of the drum, reduced screen capacity, hot or noisy drive systems, erratic drum rotation, and sudden increases in effluent solids.

1. **Decrease in Throughput Rate (from Slime Growth)**
   Causal Factors:
   - inadequate cleaning
   - spray nozzles plugged
   Corrective Actions (Where Applicable):
   - increase backwash pressure (60–120 psi)
   - add hypochlorite upstream of the unit
   - unclog nozzles

2. **Decreased Performance from Leakage at Ends of the Drum**
   Causal Factor:
   - defective/leaking units
   Corrective Actions:
   - tighten the tension on sealing bands
   - replace sealing bands if excessive tension is required

3. **Screen Capacity Reduced after Shutdown Period**
   Causal Factor:
   - The screen is fouled.
   Corrective Actions:
   - Clean screen prior to shutdown.
   - Clean screen with hypochlorite.

4. **Drive System is Running Hot or Noisy**
   Causal Factor:
   - Inadequate lubrication
   Corrective Action:
   - Fill to specified level with recommended oil

5. **Erratic Drum Rotation**
   Causal Factors:
   - Improper drive belt adjustment
   - Drive belts worn out
   Corrective Actions:
   - Adjust tension to a specified level
   - Replace drive belts

6. **Sudden Increase in Effluent Solids**
   Causal Factors:
   - Hole in screen fabric
   - Screws that secure fabric are loose
   - The solids collection trough is overflowing
   Corrective Actions (Where Applicable):
   - Repair fabric
   - Tighten screws
   - Reduce microscreen influent flow rate

7. **Decreased Screen Capacity after High-Pressure Washing**
   Causal Factor:
   - Iron or manganese oxide film on fabric
   Corrective Action:
   - Clean screen with inhibited acid cleaner. Follow the manufacturer's instructions.

FILTRATION

The purpose of *filtration* processes used in advanced treatment is to remove suspended solids. The specific operations associated with a filtration system depend on the equipment used. A general description of the process follows.

Filtration Process Description

Wastewater flows to a filter (gravity or pressurized). The filter contains single, dual, or multimedia. Wastewater flows through the media, which removes solids. The solids remain in the filter. Backwashing the filter as needed

Handbook of Water and Wastewater Treatment Plant Operations

removes trapped solids. Backwash solids are returned to the plant for treatment. The processes typically remove 95%–99% of the suspended matter.

Operation, Observation, and Troubleshooting Procedures

Operators routinely monitor filter operation to ensure optimal performance and to detect operational problems based on indications or observations of equipment malfunctions or process suboptimal performance. Operational problems typically encountered in filter operations are discussed in the sections that follow.

1. **High Effluent Turbidity**
   Causal Factors:
   - The filter requires backwashing
   - Prior chemical treatment is inadequate
   Corrective Actions (Where Applicable):
   - Backwash the unit as soon as possible.
   - Adjust/control chemical dosage properly.
2. **High Head Loss through the Filter**
   Causal Factor:
   - The filter requires backwashing
   Corrective Action:
   - Backwash the unit as soon as possible.
3. **High Head Loss through the Unit Right after Backwashing**
   Causal Factors:
   - The backwash cycle was insufficient
   - Surface scour/wash arm inoperative
   Corrective Actions (Where Applicable):
   - Increase the backwash time.
   - Repair air scour or surface scrubbing arm.
4. **Backwash Water Requirement Exceeds 5%**
   Causal Factors:
   - excessive solids in the filter influent
   - excessive filter aid dosage
   - surface washing/air scour not operating
   - surface washing/air scour not operated long enough during the backwash cycle
   - excessive backwash cycle used
   Corrective Actions (Where Applicable):
   - Improve treatment prior to filtration.
   - Reduce control/filter aid dose rates.
   - Repair mechanical problems.
   - Increase surface wash cycle time.
   - Adjust backward cycle length.
5. **Filter Surface Clogging**
   Causal Factors:
   - Inadequate prior treatment (single media filters)
   - Excessive filter aid dosage (dual or mixed media filters)
   - Inadequate surface wash cycle
   - Inadequate backwash cycle

Corrective Actions (Where Applicable):
- Improve prior treatment.
- Replace single media with dual/mixed media.
- Reduce or eliminate filter aid.
- Provide adequate surface wash cycle.
- Provide adequate backwash cycle.

6. **Short Filter Runs**
   Causal Factor:
   - High head loss
   Corrective Action:
   - See corrective actions, Section 5.
7. **Filter Effluent Turbidity Increases Rapidly**
   Causal Factors:
   - Inadequate filter aid dosage
   - Filter aid system mechanical failure
   - Filter aid requirements have changed
   Corrective Actions (where applicable)
   - Increase chemical dosage.
   - Repair feed system.
   - Adjust filter aid dose rate (do jar test).
8. **Mud Ball Formation**
   Causal Factors:
   - Inadequate backwash flow rate
   - Inadequate surface wash
   Corrective Actions (where applicable):
   - Increase backwash flow to specified levels.
   - Increase surface wash cycle.
9. **Gravel Displacement**
   Causal Factor:
   - Air enters the underdrains during the backwash cycle
   Corrective Actions:
   - Control backwash volume.
   - Control backwash water head.
   - Replace media (severe displacement).
10. **Medium is Lost during the Backwash Cycle**
    Causal Factors:
    - Excessive backwash flows
    - Excessive auxiliary scour
    - Air attached to filter media, causing it to float
    Corrective Actions (where applicable):
    - Reduce backwash flow rate.
    - Stop auxiliary scour several minutes before the end of the backwash cycle.
    - Increase backwash frequency to prevent bubble displacement and/or maintain maximum operating water depth above the filter surface.
11. **Filter Backwash Cycle not Effective during Warm Weather**
    Causal Factor:
    - Decreased water viscosity due to higher temperatures
    Corrective Action:
    - Increase backwash rate until required bed expansion is achieved.

12. **Air Binding Causes Premature Headloss Increase**
    Causal Factors:
    - Air bubbles are produced by exposing an influent containing high dissolved oxygen levels to less than atmospheric pressure
    - Pressure drops occur during the changeover to the backwash cycle
    Corrective Actions (Where Applicable):
    - Increase backwash frequency.
    - Maintain maximum operating water depth.

Membrane Bioreactors (MBRs)

The most commonly used technologies for performing secondary treatment of municipal wastewater rely on microorganisms suspended in wastewater to treat it. These technologies work well in many situations, but they have several drawbacks, including the difficulty of growing the right types of microorganisms and the physical requirement of a large plant site. In the past 20 years, the use of microfiltration membrane bioreactors (MBRs) has overcome many of the limitations of conventional systems. The advantage of MBRs is the combination of suspended growth biological with solids removal via filtration. The membranes can be designed for and operated in small spaces and with high removal efficiency of contaminants such as nitrogen, phosphorus, bacteria, biological oxygen demand, and total suspended solids. In effect, the membranes can be designed to replace secondary clarifiers and sand filters in a typical activated sludge treatment system.

The bottom line: membrane filtration allows a higher biomass concentration to be maintained, thereby allowing smaller bioreactors to be used (USEPA, 2007a).

Biological Nitrification

Biological nitrification is a microbial process in which reduced nitrogen compounds, primarily ammonia, are sequentially oxidized to nitrite and nitrate. Nitrification occurs naturally in drinking water or during secondary disinfection when ammonia is added to form chloramines.

Biological Denitrification

Biological denitrification removes nitrogen from wastewater. When bacteria come into contact with a nitrified element in the absence of oxygen, they reduce the nitrates to nitrogen gas, which escapes the wastewater. The denitrification process can be performed in either an anoxic-activated sludge system (suspended growth) or a column system (fixed growth). The denitrification process can remove up to 85% or more of nitrogen. After effective biological treatment, little oxygen-demanding material is left in the wastewater when it reaches the denitrification stage. The denitrification reaction will only occur if an oxygen demand source exists when no dissolved oxygen is present in the wastewater. An oxygen demand source is usually added to reduce the nitrates quickly. The most common demand source added is soluble BOD or methanol. Approximately 3 mg/L of methanol is added for every 1 mg/L of nitrate-nitrogen. Suspended growth denitrification reactors are mixed mechanically, but only enough to keep the biomass from settling without adding unwanted oxygen. Submerged filters of different types of media may also be used to provide denitrification. A fine media downflow filter is sometimes used to provide both denitrification and effluent filtration. A fluidized sand bed, where wastewater flows upward through a media of sand or activated carbon at a rate to fluidize the bed may also be used. Denitrification bacteria grow on the media.

Observational Operations, Problems, and Troubleshooting

Operators monitor performance by observing various parameters. Parameters or other indicators/observations that demonstrate process malfunction or suboptimal performance indicate the need for various corrective actions. We discuss several of these indicators and observations of poor process performance, their causal factors, and corrective actions in the sections that follow.

1. **Process Effluent: Sudden Increase in BOD$_5$**
   Causal Factor:
   - Excessive methanol or other organic matter present
   Corrective Actions (as Required):
   - Reduce methanol addition.
   - Install an automated methanol control system.
   - Install an aerated stabilization unit for the removal of excess methanol.
2. **Sudden Increase in Effluent Nitrate Concentration**
   Causal Factors:
   - Inadequate methanol control
   - Denitrification pH is outside the 7.0–7.5 range required for the process
   - Loss of solids from the denitrification process due to pump failure
   - Excessive mixing introduces dissolved oxygen
   Corrective Actions (Where Applicable):
   - Identify/correct control problem.
   - Correct pH problem in the nitrification process.
   - Adjust pH at process influent.
   - Correct denitrification sludge return.
   - Increase denitrification sludge waste rate.
   - Decrease denitrification sludge waste rate.
   - Transfer sludge from carbonaceous units to denitrification units.
   - Reduce mixer speed.
   - Remove some mixers from the service.

3. **High Head Loss (Packed Bed Nitrification)**
   Causal Factors:
   - Excessive solids in the unit
   - Nitrogen gas accumulating in the unit
   Corrective Action:
   - Backwash the unit 1–2 min, then return to service.

4. **Out of Service Packed Bed Unit Binds on Startup**
   Causal Factor:
   - Solids have floated to the top during shutdown
   Corrective Action:
   - Backwash units before removing them from service and immediately before placing them in servicing.

Note: The following case study demonstrates how retrofitting an existing aeration system for biological phosphorus removal and nitrogen removal (with a developed denitrification sludge blanket) can make a significant innovative contribution to BNR methodology. For those seeking a more in-depth treatment of this case study it can be obtained from USEPA (2008).

CARBON ADSORPTION

The main purpose of *carbon adsorption* used in advanced treatment processes is the removal of refractory organic compounds (non-BOD$_5$) and soluble organic material that are difficult to eliminate through biological or physical/chemical treatment. In the carbon adsorption process, wastewater passes through a container filled either with carbon powder or carbon slurry. Organics adsorb onto the carbon (i.e., organic molecules are attracted to the activated carbon surface and are held there) with sufficient contact time. A carbon system usually has several columns or basins used as contactors. Most contact chambers are either open concrete gravity-type systems or steel pressure containers applicable to either upflow or downflow operation. With use, carbon loses its adsorptive capacity. The carbon must then be regenerated or replaced with fresh carbon. As head loss develops in carbon contactors, they are backwashed with clean effluent in much the same way that effluent filters are backwashed. Carbon used for adsorption may be in a granular form or in powdered form.

Note: Powdered carbon is too fine for use in columns and is usually added to the wastewater, then later removed by coagulation and settling.

Operational, Observations, and Troubleshooting

With regard to the carbon adsorption system for advanced wastewater treatment, operators are primarily interested in monitoring the system to prevent the excessive head loss, reduce levels of hydrogen sulfide in the carbon contactor, ensure that the carbon is not fouled, and ensure corrosion of metal parts and damage to concrete in contactors is minimal.

1. **Excessive Head Loss**
   Causal Factors:
   - Highly turbid influent
   - Growth and accumulation of biological solids in unit
   - Excessive carbon fines due to deterioration during handling
   - Inlet or outlet screens plugged
   Corrective Actions (Where Applicable):
   - Backwash the unit vigorously.
   - Correct problem in prior treatment steps.
   - Operate as an expanded upflow bed to remove solids continuously.
   - Increase frequency of backwashing for downflow beds.
   - Improve soluble BOD$_5$ removal in prior treatment steps.
   - Remove carbon from the unit and wash out fines.
   - Replace carbon with harder carbon.
   - Backflush screens.

2. **Hydrogen Sulfide in Carbon Contactor**
   Causal Factors:
   - Low/no dissolved oxygen and/or nitrate in contactor influent
   - High influent BOD$_5$ concentrations
   - Excessive detention time in carbon contactor
   Corrective Actions (Where Applicable):
   - Add air, oxygen, or sodium nitrate to the unit influent.
   - Improve soluble BOD$_5$ removal in prior treatment steps.
   - Precipitate sulfides already formed with iron on chlorine.
   - Reduce detention time by removing contactor(s) from service.
   - Backwash units more frequently and more violently, using air scour or surface wash.

3. **Large Decrease in COD Removed/lb of Carbon Regenerated**
   Causal Factor:
   - Carbon is fouled and losing efficiency
   Corrective Action:
   - Improve regeneration process performance.

4. **Corrosion of Metal Parts/Damage to Concrete in Contactors**
   Causal Factors:
   - Hydrogen sulfide in carbon contactors
   - Holes in protective coatings exposed to dewatered carbon
   Corrective Actions:
   - See corrective actions in number 2 above.
   - Repair protective coatings.

LAND APPLICATION

The application of secondary effluent onto a land surface can provide an effective alternative to the expensive and complicated advanced treatment methods discussed previously and the biological nutrient removal (BNR) system discussed later. A high-quality polished effluent (i.e., effluent with high levels of TSS, BOD, phosphorus, nitrogen compounds, as well as reduced refractory organics) can be obtained through the natural processes that occur as the effluent flows over the vegetated ground surface and percolates through the soil. Limitations are involved with the land application of wastewater effluent. For example, the process needs large land areas. Soil type and climate are also critical factors in controlling the design and feasibility of a land treatment process.

Type and Modes of Land Application

Three basic types or modes of land application or treatment are commonly used: irrigation (slow rate), overland flow, and infiltration-percolation (rapid rate). The basic objectives of these types of land applications and the conditions under which they can function vary. In *irrigation* (also called slow rate), wastewater is sprayed or applied (usually by ridge-and-furrow surface spreading or by sprinkler systems) to the surface of the land. Wastewater enters the soil, where crops growing on the irrigation area utilize available nutrients. Soil organisms stabilize the organic content of the flow. Water returns to the hydrologic (water) cycle through evaporation or by entering the surface water or groundwater.

The irrigation land application method provides the best results (compared with the other two types of land application systems) with respect to advanced treatment levels of pollutant removal. Not only are suspended solids and BOD significantly reduced by filtration of the wastewater, but also biological oxidation of the organics in the top few inches of soil occurs. Nitrogen is removed primarily by crop uptake, and phosphorus is removed by adsorption within the soil. Expected performance levels for irrigation include:

- BOD_5—98%
- Suspended solids—98%
- Nitrogen—85%
- Phosphorus—95%
- Metals—95%

The overland flow application method utilizes physical, chemical, and biological processes as the wastewater flows in a thin film down a relatively impermeable surface. In this process, wastewater sprayed over sloped terraces flows slowly over the surface. Soil and vegetation remove suspended solids, nutrients, and organics. A small portion of the wastewater evaporates, while the remainder flows to collection channels. The collected effluent is then discharged into surface waters. Expected performance levels for overflow include:

- BOD_5—92%
- Suspended solids—92%
- Nitrogen—70%–90%
- Phosphorus—40%–80%
- Metals—50%

In the infiltration-percolation (rapid rate) land application process, wastewater is sprayed/pumped to spreading basins (a.k.a. recharge basins or large ponds). Some wastewater evaporates, while the remainder percolates/infiltrates into the soil. Solids are removed by filtration. Water recharges the groundwater system. Most of the effluent percolates to the groundwater; very little of it is absorbed by vegetation. The filtering and adsorption action of the soil removes most of the BOD, TSS, and phosphorus from the effluent; however, nitrogen removal is relatively poor. Expected performance levels for infiltration-percolation include:

- BOD_5—85%–99%
- Suspended solids—98%
- Nitrogen—0%–50%
- Phosphorus—60%–95%
- Metals—50%–95%

Operational Observations, Problems, and Troubleshooting

Performance levels are dependent on the land application process used. To be effective, operators must monitor the operation of the land application process employed. Experience has shown that these processes can be very effective, but problems arise when the flow contains potentially toxic materials that may become concentrated in the crops being grown on the land. Along with this problem, other problems are common, including ponding, deterioration of distribution piping systems, malfunctioning sprinkler heads, waste runoff, irrigated crop die-off, poor crop growth, and too much flow rate.

1. **In Irrigated Areas, Water is Ponding**
   Causal Factors:
   - Excessive application rate
   - Inadequate drainage because of groundwater levels
   - Damaged drainage wells
   - Inadequate good withdrawal rates
   - Damaged drain tiles
   - Broken pipe in the distribution system
   Corrective Actions (Where Applicable):
   - Reduce application rate to an acceptable level.
   - Irrigate in portions of the site where groundwater is not a problem.
   - Store wastewater until the condition is corrected.
   - Repair drainage wells.
   - Increase drainage well pumping rates.
   - Repair damaged drain tiles.
   - Repair pipe.

2. **Deterioration of Distribution Piping**

Causal Factors:

- Effluent remains in the pipe for long periods
- Different metals used in the same line

  Corrective Actions (Where Applicable):

- Drain pipe after each use.
- Coat steel valves.
- Install cathodic/anodic protection.

3. **No Flow from Source Sprinkler Nozzles**

Causal Factors:

- Nozzles clogged

  Corrective Action:

- Repair/replace screen on irrigation pump inlet.

4. **Wastes Running Off Irrigation Area**

Causal Factors:

- The high sodium adsorption ratio has caused clay soil to become impermeable
- Solids seal the soil surface
- The application rate is greater than the soil infiltration rate
- Break in distribution piping
- Soil permeability has decreased because of the continuous application of wastewater
- Rain has saturated the soil

  Corrective Actions (Where Applicable):

- Feed calcium and magnesium to maintain sodium adsorption ratio (SAR) to less than 9.
- Strip crop area.
- Reduces application rate to an acceptable level.
- Repair system.
- Allow a 2–3-day rest period between each application.
- Store wastewater until the soil has drained.

5. **Irrigated Crop is Dead**

Causal Factors:

- Too much or not enough water has been applied
- Wastewater contains toxic materials in toxic concentrations.
- Excessive insecticide or herbicide applied.
- Inadequate drainage has flooded the root zone of the crop.

  Corrective Actions (Where Applicable):

- Adjust the application rate to the appropriate level.
- Eliminate the source of toxicity.
- Apply only as permitted/directed.

6. **Poor Crop Growth**

Causal Factors:

- Too little nitrogen (N) or phosphorus (P)
- The timing of nutrient applications does not coincide with plant nutrient needs

  Corrective Actions (Where Applicable):

- Increase application rate to supply N and P.

- Augment N and P of wastewater with commercial fertilizer applications.
- Adjust the application schedule to match crop needs.

7. **Irrigation Pump: Normal psi but above Average Flow Rate**

Causal Factors:

- Broken main, riser, or lateral.
- Leaking gasket.
- The sprinkler head or nozzle is missing.
- Too many distribution laterals are in service at one time.

  Corrective Actions (Where Applicable):

- Locate and repair problems.
- Locate and replace defective gasket.
- Correct valving to adjust a number of laterals in service.

8. **Irrigation Pump: Above Average psi, Below Average Flow**

Causal Factor:

- Blockage in system

  Corrective Action:

- Locate and correct blockage.

9. **Irrigation Pump: Below Average psi and Flow Rate**

Causal Factors:

- Worn impeller
- Partially clogged pump inlet screen

  Corrective Actions (Where Applicable):

- Replace impeller.
- Clean screen.

10. **Excessive Erosion Occurring**

Causal Factors:

- Excessive application rates
- Inadequate crop coverage

  Coverage Actions (Where Applicable):

- Reduce application rate.

11. **Odor Complaints**

Causal Factors:

- Wastes are turning septic during transport to the treatment/irrigation site
- Storage reservoirs are septic

  Corrective Actions (Where Applicable):

- Aerate or chemically treat wastes during transport.
- Install the cover over the discharge point; collect and treat gases before release.
- Improve pretreatment.
- Aerate storage reservoirs.

12. **Center Pivot Irrigation Rigs Stuck in Mud**

Causal Factors:

- Excessive application rates
- Improper rig or tires
- Poor drainage

Corrective Actions:
- Reduce application rate.
- Install tire with higher flotation capabilities.

13. **Nitrate in Groundwater near Irrigation Site Increasing**

   Causal Factors:
   - Nitrogen application rate does not balance with crop needs.
   - Applications occur during dormant periods.
   - The crop is not being properly harvested and removed.

   Corrective Actions (Where Applicable):
   - Change to crop with higher nitrogen requirement.
   - Adjust the schedule to apply only during active growth periods.
   - Harvest and remove crops as required.

BIOLOGICAL NUTRIENT REMOVAL (BNR)

Nitrogen and phosphorus are the primary causes of cultural eutrophication (i.e., nutrient enrichment due to human activities) in surface waters. The most recognizable manifestations of this eutrophication are algal blooms that occur during the summer. Chronic symptoms of over-enrichment include low dissolved oxygen, fish kills, murky water, and depletion of desirable flora and fauna. In addition, the increase in algae and turbidity increases the need to chlorinate drinking water, which, in turn, leads to higher levels of disinfection by-products that have been shown to increase the risk of cancer (USEPA, 2007c). Excessive amounts of nutrients can also stimulate the activity of microbes, such as *Pfiesteria*, which may be harmful to human health (USEPA, 2007d).

Approximately 25% of all water body impairments are due to nutrient-related causes (e.g., nutrients, oxygen depletion, algal growth, ammonia, harmful algal blooms, biological integrity, and turbidity) (USEPA, 2007d). In efforts to reduce the number of nutrient impairments, many point source discharges have received more stringent effluent limits for nitrogen and phosphorus. To achieve these new, lower effluent limits, facilities have begun to look beyond traditional treatment technologies.

Recent experience has reinforced the concept that biological nutrient removal (BNR) systems are reliable and effective in removing nitrogen and phosphorus. The process is based on the principle that, under specific conditions, microorganisms will remove more phosphorus and nitrogen than is required for biological activity; thus, treatment can be accomplished without the use of chemicals. Not having to use, and therefore purchase, chemicals to remove nitrogen and phosphorus potentially has numerous cost-benefit implications. In addition, because chemicals are not required, chemical waste products are not produced, reducing the need to handle and dispose of waste. Several patented processes are available for this purpose. Performance depends on the biological activity and the process employed.

Description

As mentioned, biological nutrient removal (BNR) removes total nitrogen (TN) and total phosphorus (TP) from wastewater through the use of microorganism under different environmental conditions in the treatment process (Metcalf & Eddy, 2003).

Nitrogen Removal

Total effluent nitrogen comprises ammonia, nitrate, particulate organic nitrogen, and soluble organic nitrogen. The biological processes that primarily remove nitrogen are nitrification and denitrification (USEPA, 2007b). During nitrification, ammonia is oxidized to nitrate by one group of autotrophic bacteria, most commonly *Nitrosomonas* (Metcalf & Eddy, 2003). Nitrite is then oxidized to nitrate by another autotrophic bacteria group, the most common being *Nitrobacter*.

Denitrification involves the biological reduction of nitrite to nitric oxide, nitrous oxide, and nitrogen gas (Metcalf & Eddy, 2003). Both heterotrophic and autotrophic bacteria are capable of denitrification. The most common and widely distributed denitrifying bacteria are Pseudomonas species, which can use hydrogen, methanol, carbohydrates, organic acids, alcohols, benzoates, and other aromatic compounds for denitrification (Metcalf & Eddy, 2003).

In BNR systems, nitrification is the controlling reaction because ammonia-oxidizing bacteria lack functional diversity, have stringent growth requirements, and are sensitive to environmental conditions (USEPA, 2007b). Note that nitrification by itself does not actually remove nitrogen from wastewater. Rather, denitrification is needed to convert the oxidized form of nitrogen (nitrate) to nitrogen gas. Nitrification occurs in the presence of oxygen under aerobic conditions, and denitrification occurs in the absence of oxygen under anoxic conditions (USEPA, 2007c).

Table 18.9 summarizes the removal mechanisms applicable to each form of nitrogen.

TABLE 18.9
Mechanisms Involved in the Removal of Total Nitrogen

Form of Nitrogen	Common Removal Mechanism	Technology Limit (mg/L)
Ammonia-N	Nitrification	<0.5
Nitrate-N	Denitrification	1–2
Particulate organic-N	Solids separation	<1.0
Soluble organic-N	None	00.5–1.5

Source: USEPA (2007c).

Note that organic nitrogen is not removed biologically; rather, only the particulate fraction can be removed through solids separation via sedimentation or filtration.

Phosphorus Removal

Total effluent phosphorus comprises soluble and particulate phosphorus. Particulate phosphorus can be removed from wastewater through solids removal. To achieve low effluent concentrations, the soluble fraction of phosphorus must also be targeted. Table 18.10 shows the removal mechanisms for phosphorus.

Biological phosphorus removal relies on phosphorus uptake by anaerobic heterotrophs capable of storing orthophosphate in excess of their biological growth requirements. The treatment process can be designed to promote the growth of these organisms, known as phosphate-accumulating organisms (PAOs), in mixed liquor (Phosphate-accumulating organisms ASCE/EWRI, 2006). Under anaerobic conditions, PAOs convert readily available organic matter [e.g., volatile fatty acids (VFAs)] to carbon compounds called polyhydroxyalkanoates (PHAs). PAOs use energy generated during the breakdown of polyphosphate molecules to create PHAs. This breakdown results in the release of phosphorus (WEF, 1998).

Under subsequent aerobic conditions in the treatment process, PAOs use the stored PHAs as energy to take up the phosphorus that was released in the anaerobic zone, as well as any additional phosphate present in the wastewater. In addition to reducing the phosphate concentration, the process renews the polyphosphate pool in the return sludge so that the process can be repeated (USEPA, 2007b).

Some PAOs use nitrate instead of free oxygen to oxidize stored PHAs and take up phosphorus. These denitrifying PAOs remove phosphorus in the anoxic zone rather than the aerobic zone (USEPA, 2007b).

As shown in Table 18.10, phosphorus can also be removed from wastewater through chemical precipitation. Chemical precipitation primarily uses aluminum and iron coagulants or lime to form chemical flocs with phosphorus. These flocs are then settled out to remove phosphorus from the wastewater (Spellman, 2007). However, compared to the biological removal of phosphorus, chemical processes have higher operating costs, produce more sludge, and result in the added chemicals in the sludge (Metcalf &

TABLE 18.10

Mechanisms Involved in the Removal of Total Phosphorus

Form of Phosphorus	Common Removal Mechanism	Technology Limit (mg/L)
Soluble phosphorus	Microbial uptake	
	Chemical precipitation	0.1
Particulate phosphorus	Solids removal	<0.05

Source: USEPA (2007c).

Eddy, 2003). When TP levels close to 0.1 mg/L are needed, a combination of biological and chemical processes may be less costly than either process by itself.

Process

Several BNR process configurations are available. Some BNR systems are designed to remove only TN or TP, while others remove both. The configuration most appropriate for any particular system depends on the target effluent quality, operator experience, influent quality, and existing treatment processes if retrofitting an existing facility. BNR configuration varies based on the sequencing of environmental conditions (i.e., aerobic, anaerobic, and anoxic) and timing (USEPA, 2007b, 2007c).

Note: Anoxic is a condition in which oxygen is available only in the combined form (e.g., NO_2^- or NO_3^-. However, anaerobic is a condition in which neither free nor combined oxygen is available (WEF, 1998).

Common BNR system configurations include:

- **Modified Ludzack-Ettinger (MLE) Process:** continuous-flow suspended-growth process with an initial anoxic stage followed by an aerobic stage; used to remove
- **A²/O Process:** MLE process preceded by an initial anaerobic stage; used to remove both TN and TP
- **Step Feed Process:** alternating anoxic and aerobic stages; however, influent flow is split to several feed locations and the recycle sludge stream is sent to the beginning of the process; used to remove TN
- **Bardenpho Process (Four-Stage):** continuous-flow suspended-growth process with alternating anoxic/aerobic/anoxic/aerobic stages; used to remove TN
- **Modified Bardenpho Process:** Bardenpho process with the addition of an initial anaerobic zone; used to remove both TN and TP
- **Sequencing Batch Reactor (SBR) Process:** suspended-growth batch process sequenced to simulate the four-stage process; used to remove TN (TP removal is inconsistent)
- **Modified University of Cape Town (UCT) Process:** A²/O Process with a second anoxic stage where the internal nitrate recycle is returned; used to remove both TN and TP
- **Rotating Biological Contactor (RBC) Process:** continuous-flow process using RBCs with sequential anoxic/aerobic stages; used to remove TN
- **Oxidation Ditch: a** continuous-flow process using looped channels to create time-sequenced anoxic, aerobic, and anaerobic zones; used to remove both TN and TP.

Although the exact configurations of each system differ, BNR systems designed to remove TN must have an aerobic zone for nitrification and an anoxic zone for denitrification, and BNR systems designed to remove TP must have an anaerobic zone free of dissolved oxygen and nitrate. Often, sand or other media filtration is used as a polishing step to remove particulate matter when low TN and TP effluent concentrations are required. Sand filtration can also be combined with attached growth denitrification filters to further reduce soluble nitrates and effluent TN levels (WEF, 1998).

Choosing which system is most appropriate for a particular facility primarily depends on the target effluent concentrations, and whether the facility will be constructed as new or retrofit with BNR to achieve more stringent effluent limits. New plants have more flexibility and options when deciding which BNR configuration to implement because they are not constrained by existing treatment units and sludge handling procedures.

Retrofitting an existing plant with BNR capabilities should involve consideration of the following factors (Park, 2012):

- Aeration basin size and configuration
- Clarifier capacity
- Type of aeration system
- Sludge processing units
- Operator skills

The aeration basin size and configuration dictate which BNR configurations are the most economical and feasible. Available excess capacity reduces the need for additional basins and may allow for a more configuration (e.g., 5-stage Bardenpho versus 4-stage Bardenpho configuration). The need for additional basins can result in the need for more land if the space needed is not available. If land is not available, another BNR process configuration may have to be considered.

Clarifier capacity influences the return activated sludge (RAS) rate and effluent suspended solids, which in turn, affects effluent TN and TP levels. If the existing facility configuration does not allow for a preanoxic zone so that nitrates can be removed prior to the anaerobic zone, then the clarifier should be modified to have a sludge blanket just deep enough to prevent the release of phosphorus to the liquid. However, if a preanoxic zone is feasible, a sludge blanket in the clarifier may not be necessary (WEF, 1998). The existing clarifiers also remove suspended solids including particulate nitrogen and phosphorus, and thus reduce total nitrogen and phosphorus concentrations.

The aeration system will most likely need to be modified to accommodate an anaerobic zone and to reduce the DO concentration in the return sludge. Such modifications could be as simple as removing aeration equipment from the zone designated for anaerobic conditions or changing the type of pump used for the recycled sludge stream (to avoid the introduction of oxygen).

The manner in which sludge is processed at a facility is important in designing nutrient removal systems. Sludge is recycled within the process to provide the organisms necessary for the TN and TP removal mechanisms to occur. The content and volume of sludge recycled directly impact the system's performance. Thus, sludge handling processes may be modified to achieve optimal TN and TP removal efficiencies. For example, some polymers in sludge dewatering could inhibit nitrification when recycled. Also, because aerobic digestion of sludge processes nitrates, denitrification, and phosphorus uptake rates may be lowered when the sludge is recycled (WEF, 1998).

Operators should be able to adjust the process to compensate for constantly varying conditions. BNR processes are very sensitive to influent conditions, which are influenced by weather events, sludge processing, and other treatment processes (e.g., recycling after filter backwashing). Therefore, operator skills and training are essential for achieving target TN and TP effluent concentrations (USEPA, 2007c).

Performance

Table 18.11 provides a comparison of the TN and TP removal capacities of common BNR configurations. Note that site-specific conditions dictate the performance of each process and that the table is only meant to provide a general comparison of treatment performance among the various BNR configurations.

The limit of technology (LOT), at least for larger treatment plants, is 3 mg/L for TN and 0.1 mg/L for TP (USEPA, 2007b). However, some facilities may be able to achieve concentrations lower than these levels due to site-specific conditions. Table 18.12 provides TN and TP effluent concentrations for various facilities using BNR processes.

LOT levels (i.e., TN less than 3 mg/L and TP less than 0.1 mg/L) have not been demonstrated at treatment plants with capacities of less than 0.1 mgd (USEPA, 2007b). BNR for TN removal may be feasible and cost-effective. However, BNR for TP removal is often not cost-effective at small treatment plants (USEPA, 2007b). Therefore, performance data for TP removal at small treatment plants is limited. Table 18.13 summarizes the TN levels achievable with various BNR configurations.

TABLE 18.11
Comparison of Common BNR Configurations

Process	Nitrogen Removal	Phosphorus Removal
MLE	Good	None
A²/O	Good	Good
Step feed	Moderate	None
Four-stage Bardenpho	Excellent	None
Modified Bardenpho	Excellent	Good
SBR	Moderate	Inconsistent
Modified UCT	Good	Excellent
Oxidation ditch	Excellent	Good

Source: USEPA (2007a).

TABLE 18.12

Treatment Performance of Various BNR Process Configurations

Treatment Plant (State)	Treatment Process Description	Flow (mgd)	Average Effluent Concentration (mg/L— Represents 2003–2006 Average)	
			TN	TP
Annapolis (MD)	Bardenpho (4-Stage)	13	7.1	0.66
Back Rick (MD)	MLE	180	7.6	0.19
Bowie (MD)	Oxidation Ditch	3.3	6.6	0.20
Cambridge (MD)	MLE	8.1	3.2	0.34
Cape Coral (FL)	Modified Bardenpho	8.5	1.0	0.2
Cox Creek (MD)	MLE	15	9.7	0.89
Cumberland (MD)	Step feed	15	7.0	1.0
Frederick (MD)	A²O	7	7.2	1.0
Freedom District (MD)	MLE	3.5	7.8	0.51
Largo (FL)	A²/O	15	2.3	No data
Medford Lakes (NJ)	Bardenpho (5-stage)	0.37	2.6	0.09
Palmetto (FL)	Bardenpho (4-stage)	1.4	3.2	0.82
Piscataway (MD)	Step feed	30	2.7	0.09
Seneca (MD)	MLE	20	6.4	0.08
Sod Run (MD)	Modified A²/O	20	9.2	0.86
Westminster (MD)	MLE-A²/O	5	5.3	0.79

Sources: USEPA (2006), Gannett Fleming (2012), and Park (2012).

TABLE 18.13

BNR Performance of Small Systems (Less than 0.1 mgd)

BNR Process	Achievable TN Effluent Quality (mg/L)
MLE	10
Four-stage Bardenpho	6
Three-stage Bardenpho	6
SBR	8
RBC	12

Source: USEPA (2007c).

Operation and Maintenance

For BNR systems to result in low TN and TP effluent concentrations, proper operation and control of the systems are essential. Operators should be trained to understand how temperature, dissolved oxygen (DO) levels, pH, filamentous growth, and recycle loads affect system performance.

Biological nitrogen removal reaction rates are temperature-dependent. Nitrification and denitrification rates increase as temperature increases (until a maximum temperature is reached). In general, nitrification rates double for every 8°C–10°C rise in temperature (WEF, 1998). The effect of temperature on biological phosphorus removal is not completely understood (WEF, 1998), although rates usually slow at temperatures above 30°C (USEPA, 2007b).

DO must be present in the aerobic zone for phosphorus uptake to occur. However, it is important not to over-aerate. DO concentrations around 1 mg/L are sufficient. Over-aeration can lead to secondary release of phosphorus due to cell lysis, high DO levels in the internal mixed liquor recycle (which could reduce TP and TN removal rates), and increased operation and maintenance (O&M) costs (USEPA, 2007b).

There is evidence that both nitrification and phosphorus removal rates decrease when pH levels drop below 6.9. Nitrification results in the consumption of alkalinity. As alkalinity is consumed, pH decreases. Thus, treatment plants with low influent alkalinity may have reduced nitrification rates (WEF, 1998). Glycogen-accumulating organisms may also compete with PAOs at pH values less than 7.

Filamentous growth can cause poor settling of particulate nitrogen and phosphorus in final clarifiers. However, many conditions necessary to achieve good BNR rates, such as low DO, longer solids retention times, and good mixing, also promote filament growth (USEPA, 2007b). Therefore, operators may need to identify the dominant filaments present in the system so that they can design strategies to target their removal (e.g., chlorinating recycle streams and chemical addition as a polishing step) while still maintaining nutrient removal rates.

Nitrogen and phosphorus removal efficiencies are a function of the percentage and content of the mixed liquor recycle rate to the anoxic zone and the RAS recycle rate to the anaerobic zone (WEF, 1998). The mixed liquor recycle stream supplies active biomass that enables nitrification and denitrification. Optimizing the percentage and content of this recycle stream results in optimal TN removal. The RAS recycle rate should be kept as low as possible to reduce the amount of nitrates introduced to the anaerobic zone because nitrates interfere with TP removal. In addition, the type of pump used to recycle the activated sludge is important to avoid aeration and increased DO concentrations in the anaerobic zone (Spellman, 2007).

Note: The following case study illustrates how Kalispell, Montana, developed ways to minimize recycle loads from its sludge-handling processes while producing the lowest phosphorus concentration achieved entirely by a biological process. This facility was selected as a case study because of good biological phosphorus removal and nitrification using a modified University of Cape Town (UCT) process with fermenter technology in a cold region. For those seeking a more in depth treatment of this case study, it can be obtained from USEPA (2008).

Enhanced Biological Nutrient Removal (EBNR)

Removing phosphorus from wastewater in secondary treatment processes has evolved into innovative *enhanced biological nutrient removal* (EBNR) technologies. An EBNR treatment process promotes the production of phosphorus-accumulating organisms that utilize more phosphorus in their metabolic processes than a conventional secondary biological treatment process (USEPA, 2007b). The average total phosphorus concentration in raw domestic wastewater is usually between 6 and 8 mg/L, and the total phosphorus concentration in municipal wastewater after conventional secondary treatment is routinely reduced to 3 or 4 mg/L. Whereas, EBNR incorporated into the secondary treatment system can often reduce total phosphorus concentrations to 0.3 mg/L or less. Facilities using EBNR significantly reduce the amount of phosphorus to be removed through the subsequent chemical addition and tertiary filtration processes. This improves the efficiency of the tertiary process and significantly reduces the costs of chemicals used to remove phosphorus. Facilities using EBNR reported that their chemical dosing was cut in half after EBNR was installed to remove phosphorus (USEPA, 2007b).

Treatment provided by these WWTPs also removes other pollutants that commonly affect water quality to very low levels (USEPA, 2007b). Biochemical oxygen demand (BOD) and total suspended solids are routinely less than 2 mg/L and fecal coliform bacteria are less than 10 fcu/100 mL. Turbidity of the final effluent is very low which allows for effective disinfection using ultraviolet light rather than chlorination. Recent studies report that wastewater treatment plants using EBNR also significantly reduce the amount of pharmaceuticals and personal healthcare products from municipal wastewater, compared to the removal accomplished by conventional secondary treatment. The following section describes some of the EBNR treatment technologies presently being used in various U.S. locations.

0.5 MGD Capacity Plant

- **Advanced Phosphorus Treatment Technology:** Chemical addition, two-stage filtration
- **Treatment Process Description (Liquid Only):** Treatment consists of grit removal and screening; extended aeration and secondary clarification (in a combined aeration basin/clarifier); chemical addition for flocculation using polyaluminum silicate sulfate (PASS); and filtration through two-stage DynaSand® filters.

1.5-MGD Capacity Plant

- **Advanced Phosphorus Treatment Technology:** BNR, chemical addition, tertiary settlers, and filtration
- **Treatment Process Description (Liquid Side Only):** Accomplished by screening and grit removal in the headworks; activated sludge biological treatment; biological aerated filter (IDI BioFor™ for nitrification); chemical coagulation using alum; flocculation and clarification using tube settler (IDI Densadeg™); filtration (Single Stage Parkson DynaSand® filters); disinfection and dechlorination. The DynaSand filter reject rate is reported to be about 15%–20%. The DynaSand filters are configured in four, two-cell units for a total of eight filter beds which are each 8 ft deep. Influent concentrations of total phosphorus are typically measured at about 6 mg/L (a very typical value for untreated domestic wastewater). The aeration basins are operated with an anoxic zone to provide for the biological removal of phosphorus. About 60% of the influent phosphorus is removed through the biological treatment process. Sodium sulfate is added to maintain alkalinity through the treatment process for phosphorus removal. Approximately 100–120 mg/L of sodium sulfate is applied to the wastewater just upstream of where alum is added. Alum is used to precipitate phosphorus. The alum dose is typically 135 mg/L and is used with 0.5–1.0 mg/L cationic polymer.

Note: A DynaSand® filter is a continuous backwash, upflow, deep bed, granular media filter.

1.55-MGD Capacity Plant

- **Advanced Phosphorus Treatment Technology:** Chemical addition, two-stage filtration
- **Treatment Process Description (Liquid Only):** Treatment consists of grit removal and screening; extended aeration and secondary clarification; chemical addition for flocculation using aluminum chloride (added to the wastewater at both the secondary clarifiers and the distribution header

for the DynaSand filters); and filtration through two-stage DynaSand filters; disinfection with chlorine and dechlorination with sulfur dioxide. Chlorine is added to the filter influent to control biological growth in the filters.

2 MGD CAPACITY PLANT

- **Advanced Phosphorus Treatment Technology:** BNR, chemical addition, two-stage filtration
- **Treatment Process (Liquid Side Only):** Screening and grit removal; BNR activated sludge (Bardenpho™ 5 Stage [Anaerobic Basin, Anoxic Basin, Oxidation Ditch Aeration Basin, Anoxic Basin, Reaeration Basin]); clarifiers (two parallel rectangular); chemical addition using alum and polymer; effluent polishing and filtration [using four US Filter Memcor™ filter modules]; and UV disinfection. The US Filter units utilize two-stage filtration in which the first stage is upflow through a plastic media with air scour. The second stage of filtration is through a downflow, mixed media with backwash cleaning. The concentration of alum used for coagulation was reported to be 95 mg/L.

2.6-MGD CAPACITY PLANT

- **Advanced Phosphorus Treatment Technology:** BNR, chemical addition, tertiary settlers, and filtration
- **Treatment Process Description (Liquid Side Only):** Treatment includes screening and grit removal; aeration basins; secondary clarification; chemical coagulation and flocculation using alum and polymer; tertiary clarification (rectangular convention with plate settlers); mixed media bed filters (5 ft deep); and disinfection (the filtration process removes enough fecal coliform so that conventional disinfection is not normally required). The average alum dose is 70 mg/L in the wastewater and varies from 50 to 180 mg/L. A greater dose of alum is applied during the winter period. The polymer dose concentration is about 0.1 mg/L.

3-MGD CAPACITY PLANT

- **Advanced Phosphorus Treatment Technology:** BNR, Chemical Addition, tertiary settlers, and filtration
- **Treatment Process Description (Liquid Side Only):** Treatment consists of screening and grit removal; biological nutrient removal; chemical coagulation and flocculation using polymer and alum; clarification via tube settlers; filtration through mixed media bed filters; and disinfection with chlorine and dechlorination (using sodium bisulfate).

4.8-MGD CAPACITY PLANT

- **Advanced Phosphorus Treatment Technology:** Multi-point chemical addition, tertiary settling, and filtration
- **Treatment Process Description (Liquid Only):** Treatment consists of screening and grit removal; primary clarification; trickling filters; intermediate clarification (with polymer addition to aid settling); rotating biological contactors; secondary clarification; chemical addition using poly-aluminum chloride; filtration through mixed media traveling bed filters; and ultraviolet disinfection. The final effluent is discharged down a cascading outfall to achieve reaeration prior to mixing in the receiving water. Approximately 1 mgd/day of final effluent is utilized by the local power company for cooling water.

5-MGD CAPACITY PLANT

- **Advanced Phosphorus Treatment Technology:** BNR, filtration
- **Treatment Process Description (Liquid Side Only):** Involves screening and grinding; primary clarification; biological nutrient removal (BNR) in the contact basins; secondary clarification; filtration through single-pass Dynasand filters (four cells with four filters per cell); and UV disinfection.

24-MGD CAPACITY PLANT

- **Advanced Phosphorus Treatment Technology:** BNR, chemical addition, filtration
- **Treatment Process Description (Liquid Only):** Treatment consists of screening and grit removal; primary clarification; biological treatment with enhanced biological nutrient removal; secondary clarification; chemical addition of alum and polymer for phosphorus removal; tertiary clarification; filtration through dual media gravity bed filters and disinfection. Lime is added to the biological process to maintain pH and alkalinity. A two-stage fermenter is operated to produce volatile fatty acids which are added to the biological contact basin. The enhanced biological nutrient removal process at times reduces total phosphorus to levels that are less than the 0.11 mg/L permit limitation. However, this performance is not achieved during the entire period when the seasonal phosphorus limitations are in effect. The tertiary treatment with chemical addition and filtration assures that the final effluent is of consistently good quality. Some of the treated effluent are reclaimed for irrigation.

39-MGD CAPACITY PLANT

- **Advanced Phosphorus Treatment Technology:** Chemical Addition, Filtration
- **Treatment Process Description (Liquid Only):** Treatment consists of screening and grit removal; alum addition; primary clarification; extended aeration; secondary clarification; flocculation using alum and polymer; tertiary clarification; filtration; disinfection (with chlorine) and dechlorination. Wastewater is treated in two separate trains. Four 60 ft diameter ClariCone® tertiary clarifiers are used in one treatment train to provide contact with six monomedia anthracite gravity flow bed filters. The other treatment train uses conventional clarifiers for tertiary settling, followed by filtration through four dual media gravity flow bed filters. Phosphorus is removed in four locations within this system: alum-enhanced removal in the primary clarifiers; biological removal in the aeration basins; chemical flocculation and removal in the tertiary clarifiers; and removal through filtration.

42-MGD CAPACITY PLANT

- **Advanced Phosphorus Treatment Technology:** Chemical (high lime) and tertiary filtration
- **Treatment Process Description (Liquid Only):** Treatment consists of conventional methods that remove 90% of most incoming pollutants; screening; grit removal; primary clarification; aerobic biological selectors; activated sludge aeration basins with nitrification/denitrification processes; and secondary clarification. A chemical advanced treatment—high lime process—is used to reduce phosphorus to below 0.10 mg/L, capture organics from secondary treatment, precipitate heavy metals, and serve as a barrier to viruses: lime slurry is added to rapid mix basins (to achieve pH of 11); anionic polymer added in flocculation basins; chemical clarification occurs; the first stage recarbonation lowers pH to 10; recarbonation clarifiers collect precipitated calcium carbonate; the second stage recarbonation lowers pH to 7; and storage occurs in ballast ponds. Physical advanced treatment meets stringent limits for TSS (1 mg/L), and COD (10 mg/L) including alum and/or polymer addition, multimedia filters, and activated carbon contactors. Disinfection is achieved by a chlorination/dechlorination process.

54-MGD CAPACITY PLANT

- **Advanced Phosphorus Treatment Technology:** BNR, multi-point chemical addition, tertiary settling, and filtration

- **Treatment Process Description (Liquid Only):** Treatment consists of screening; grit removal; primary settling with the possible addition of ferric chloride and polymer; methanol or volatile fatty acid added to biological reactor basins to aid BNR; ferric chloride and polymer added prior to secondary settling; alum addition and mixing; tertiary clarification with inclined plate settlers; dual media gravity bed filtration; UV disinfection; and post aeration.

67-MGD CAPACITY PLANT

- **Advanced Phosphorus Treatment Technology:** BNR, chemical addition, tertiary clarification, and filtration
- **Treatment Process Description (Liquid Only):** Treatment consists of screening; primary clarification; biological treatment with enhanced biological nutrient removal (BNR); polymer addition as needed; secondary clarification; equalization and storage in retention ponds; tertiary clarification with ferric chloride addition to removing phosphorus; disinfection with sodium hypochlorite; and filtration through dual/mono media gravity bed filters.

SOLIDS (SLUDGE/BIOSOLIDS) HANDLING

The wastewater treatment unit processes described to this point remove solids and BOD from the waste stream before the liquid effluent is discharged to its receiving waters. What remains to be disposed of is a mixture of solids and wastes, called *process residuals*—more commonly referred to as *sludge* or *biosolids*.

Note: *Sludge* is the commonly accepted name for wastewater solids. However, if wastewater sludge is used for beneficial reuse (e.g., as a soil amendment or fertilizer), it is commonly called *biosolids*.

The most costly and complex aspect of wastewater treatment can be the collection, processing, and disposal of sludge. This is the case because the quantity of sludge produced may be as high as 2% of the original volume of wastewater, depending somewhat on the treatment process being used.

Because sludge can be as much as 97% water content, and because the cost of disposal will be related to the volume of sludge being processed, one of the primary purposes or goals (along with stabilizing it so it is no longer objectionable or environmentally damaging) of sludge treatment is to separate as much of the water from the solids as possible. Sludge treatment methods may be designed to accomplish both of these purposes.

Note: Sludge treatment methods are generally divided into three major categories: thickening, stabilization, and dewatering. Many of these processes include complex sludge treatment methods (i.e., heat treatment, vacuum filtration, incineration, and others).

SLUDGE: BACKGROUND INFORMATION

When we speak of *sludge* or *biosolids*, we are referring to the same substance or material; each is defined as the suspended solids removed from wastewater during sedimentation, and then concentrated for further treatment and disposal or reuse. The difference between the terms *sludge* and *biosolids* is determined by the way they are managed.

Note: The task of disposing of, treating, or reusing wastewater solids is called *sludge* or *biosolids management*.)

Sludge is typically seen as wastewater solids that are "disposed" of. Biosolids are the same substance managed for reuse—commonly called beneficial reuse (e.g., for land application as a soil amendment, such as biosolids compost). Note that even as wastewater treatment standards have become more stringent because of increasing environmental regulations, the volume of wastewater sludge has also increased. Also note that before sludge can be disposed of or reused, it requires some form of treatment to reduce its volume, stabilize it, and inactivate pathogenic organisms.

Sludge forms initially as a 3%–7% suspension of solids, and with each person typically generating about 4 gal of sludge per week, the total quantity generated each day, week, month, and year is significant. Because of the volume and nature of the material, sludge management is a major factor in the design and operation of all water pollution control plants.

Note: Wastewater solids treatment, handling, and disposal account for more than half of the total costs in a typical secondary treatment plant.

SOURCES OF SLUDGE

Wastewater sludge is generated in primary, secondary, and chemical treatment processes. In primary treatment, the solids that float or settle are removed. The floatable material makes up a portion of the solid waste known as scum. Scum is not normally considered sludge; however, it should be disposed of in an environmentally sound way. The settleable material that collects on the bottom of the clarifier is known as *primary sludge*. Primary sludge can also be referred to as raw sludge because it has not undergone decomposition. Raw primary sludge from a typical domestic facility is quite objectionable and has a high percentage of water, two characteristics that make handling difficult.

Solids not removed in the primary clarifier are carried out of the primary unit. These solids are known as *colloidal suspended solids*. The secondary treatment system (i.e., trickling filter, activated sludge, etc.) is designed to change those colloidal solids into settleable solids that can be removed. Once in the settleable form, these solids are removed in the secondary clarifier. The sludge at the bottom of the secondary clarifier is called *secondary sludge*. Secondary sludges are light and fluffy and more difficult to process than primary sludges—in short, secondary sludges do not dewater well.

The addition of chemicals and various organic and inorganic substances prior to sedimentation and clarification may increase the solids capture and reduce the amount of solids lost in the effluent. This *chemical addition* results in the formation of heavier solids, which trap the colloidal solids or convert dissolved solids to settleable solids. The resultant solids are known as *chemical sludges*. As chemical usage increases, so does the quantity of sludge that must be handled and disposed of. Chemical sludges can be very difficult to process; they do not dewater well and contain lower percentages of solids.

SLUDGE CHARACTERISTICS

The composition and characteristics of sewage sludge vary widely and can change considerably with time. Notwithstanding these facts, the basic components of wastewater sludge remain the same. The only variations occur in the quantity of the various components as the type of sludge and the process from which it originated changes. The main component of all sludges is *water*. Prior to treatment, most sludge contains 95%–99+% water (see Table 18.14). This high water content makes sludge handling and processing extremely costly in terms of both money and time. Sludge handling may represent up to 40% of the capital cost and 50% of the operation cost of a treatment plant. As a result, the importance of optimum design for handling and disposal of sludge cannot be overemphasized. The water content of the sludge is present in several different forms. Some forms can be removed by several sludge treatment processes, thus allowing the same flexibility in choosing the optimum sludge treatment and disposal method.

The various forms of water and their approximate percentages for a typical activated sludge are shown in Table 18.15. The forms of water associated with sludges include:

- **Free Water:** water that is not attached to sludge solids in any way. This can be removed by simple gravitational settling.
- **Floc Water:** water that is trapped within the floc and travels with it. Its removal is possible by mechanical de-watering.

TABLE 18.14
Typical Water Content of Sludges

Water Treatment Process	% Moisture of Sludge	Generated lb Water/ lb Sludge Solids
Primary sedimentation	95	19
Trickling filter		
Humus—low rate	93	13.3
Humus—high rate	97	32.3
Activated sludge	99	99

Source: USEPA (1978).

TABLE 18.15
Distribution of Water in an
Activated Sludge

Water Type	% Volume
Free water	75
Floc water	20
Capillary water	2
Particle water	2.5
Solids	0.5
Total	**100**

Source: USEPA (1978).

- **Capillary Water:** water that adheres to the individual particles and can be squeezed out of shape and compacted.
- **Particle Water:** water that is chemically bound to the individual particles and can't be removed without inclination.

From a public health point of view, the second and probably more important component of sludge is the *solid matter.* Representing 1% to 8% of the total mixture, these solids are extremely unstable. Wastewater solids can be classified into two categories based on their origin—organic and inorganic. *Organic solids* in wastewater, simply put, are materials that are or were at one time alive and that will burn or volatilize at 550°C after 15 min in a muffle furnace. The percent organic material within sludge will determine how unstable it is.

The inorganic material within sludge will determine how stable it is. The *inorganic solids* are those solids that were never alive and will not burn or volatilize at 550°C after 15 min in a muffle furnace. Inorganic solids are generally not subject to breakdown by biological action and are considered stable. Certain inorganic solids, however, can create problems when related to the environment; for example, heavy metals such as copper, lead, zinc, mercury, and others. These can be extremely harmful if discharged.

Organic solids may be subject to biological decomposition in either an aerobic or anaerobic environment. The decomposition of organic matter (with its production of objectionable by-products) and the possibility of toxic organic solids within the sludge compound the problems of sludge disposal.

The pathogens in domestic sewage are primarily associated with insoluble solids. Primary wastewater treatment processes concentrate these solids into sewage sludge, so untreated or raw primary sewage sludges have higher quantities of pathogens than the incoming wastewater. Biological wastewater treatment processes such as lagoons, trickling filters, and activated sludge treatment may substantially reduce the number of pathogens in wastewater (USEPA, 1989). These processes may also reduce the number of pathogens in sewage sludge by creating adverse conditions for pathogen survival.

Nevertheless, the resulting biological sewage sludges may still contain sufficient levels of pathogens to pose a public health and environmental concern. Moreover, insects, birds, rodents, and domestic animals may transport sewage sludge, and pathogens from sewage sludge to humans and animals. Vectors are attracted to sewage sludge as a food source, and the reduction of the attraction of vectors to sewage sludge to prevent the spread of pathogens is a focus of current regulations. Sludge-borne pathogens and vector attraction are discussed in the following section.

Sludge Pathogens and Vector Attraction

As discussed earlier, a pathogen is an organism capable of causing disease. Pathogens infect humans through several different pathways including ingestion, inhalation, and dermal contact. The infective dose, or the number of pathogenic organisms to which a human must be exposed to become infected, varies depending on the organism and the health status of the exposed individual.

Pathogens that propagate in the enteric or urinary system of humans and are discharged in feces or urine pose the greatest risk to public health regarding the use and disposal of sewage sludge. Pathogens are also found in the urinary and enteric systems of other animals and may propagate in non-enteric settings.

As mentioned earlier, the four major types of human pathogenic (disease-causing) organisms (bacteria, viruses, protozoa, and helminths) may all be present in domestic sewage. The actual species and quantity of pathogens present in the domestic sewage from a particular municipality (and the sewage sludge produced when treating the domestic sewage) depend on the health status of the local community and may vary substantially at different times. The level of pathogens present in treated sewage sludge (biosolids) also depends on the reductions achieved by the wastewater and sewage sludge treatment processes.

If improperly treated sewage sludge were illegally applied to land or placed on a surface disposal site, humans and animals could be exposed to pathogens directly by coming into contact with sewage sludge, or indirectly by consuming drinking water or food contaminated by sewage sludge pathogens. Insects, birds, rodents, and even farm workers could contribute to these exposure routes by transporting sewage sludge and sewage sludge pathogens away from the site. Potential routes of exposure include:

Direct Contact
- Touching the sewage sludge.
- Walking through an area—such as a field, forest, or reclamation area—shortly after sewage sludge application.
- Handling soil from fields where sewage sludge has been applied.
- Inhaling microbes that become airborne (via aerosols, dust, etc.) during sewage sludge spreading or by strong winds, plowing, or cultivating the soils after application.

Indirect Contact

- Consumption of pathogen-contaminated crops grown on sewage sludge-amended soil or other food products that have been contaminated by contact with these crops or field workers, etc.
- Consumption of pathogen-contaminated milk or other food products from animals contaminated by grazing in pastures or fed crops grown on sewage sludge-amended fields.
- Ingestion of drinking water or recreational waters contaminated by runoff from nearby land application sites or by organisms from sewage sludge migrating into groundwater aquifers.
- Consumption of inadequately cooked or uncooked pathogen-contaminated fish from water contaminated by runoff from a nearby sewage sludge application site.
- Contact with sewage sludge or pathogens transported away from the land application or surface disposal site by rodents, insects, or other vectors, including grazing animals or pets.

DID YOU KNOW?

The purpose of USEPA's Part 503 regulation is to place barriers in the pathway of exposure either by reducing the number of pathogens in the treated sewage sludge (biosolids) to below detectable limits, in the case of Class A treatment, or, in the case of Class B treatment, by preventing direct or indirect contact with any pathogens possibly present in the biosolids. Each potential pathway has been studied to determine how the potential for public health risk can be alleviated.

One of the lesser impacts on public health can be from inhalation of airborne pathogens. Pathogens may become airborne via the spray of liquid biosolids from a splash plate or high-pressure hose or in fine particulate dissemination as dewatered biosolids are applied or incorporated. While high-pressure spray applications may result in some aerosolization of pathogens, this type of equipment is generally used on large, remote sites such as forests, where the impact on the public is minimal. Fine particulates created by the application of dewatered biosolids or the incorporation of biosolids into the soil may cause very localized fine particulate/dusty conditions, but particles in dewatered biosolids are too large to travel far, and the fine particulates do not spread beyond the immediate area. The activity of applying and incorporating biosolids may create dusty conditions. However, the biosolids are moist materials and do not add to the dusty condition, and by the time biosolids have dried sufficiently to create fine particulates, the pathogens have been reduced (Yeager and Ward, 1981).

With regard to vector attraction reduction, it can be accomplished in two ways: by treating the sewage sludge to the point at which vectors will no longer be attracted to the sewage sludge and by placing a barrier between the sewage sludge and vectors.

Note: Before moving on to a discussion of the fundamentals of sludge treatment methods, it is important to begin by covering sludge pumping calculations. It is important to point out that it is difficult (if not impossible) to treat the sludge unless it is pumped into the specific sludge treatment process.

SLUDGE PUMPING CALCULATIONS

Wastewater operators are often called upon to make various process control calculations. An important calculation involves sludge pumping. The sludge pumping calculations that the operator may be required to make during plant operations (and should be known for licensure examinations) are covered in this section.

Estimating Daily Sludge Production

The calculation for *estimation of the required sludge-pumping rate* provides a method to establish an initial pumping rate or to evaluate the adequacy of the current withdrawal rate:

$$\text{Est. Pump Rate} = \frac{\begin{array}{c}(\text{Influent TSS Conc.} - \text{Effluent TSS Conc.}) \times \\ \text{Flow} \times 8.34\end{array}}{\begin{array}{c}\% \text{ Solids in Sludge} \times 8.34 \times \\ 1{,}440 \text{ min/Day}\end{array}}$$

(18.76)

Example 18.59

Problem: The sludge withdrawn from the primary settling tank contains 1.4-% solids. The unit influent contains 285 mg/L TSS and the effluent contains 140 mg/L TSS. If the influent flow rate is 5.55 MGD, what is the estimated sludge withdrawal rate in gallons per minute (assuming the pump operates continuously)?

Solution:

$$\text{Sludge Rate, gpm} = \frac{(285 \text{ mg/L} - 140 \text{ mg/L}) \times 5.55 \times 8.34}{0.014 \times 8.34 \times 1{,}440 \text{ min/day}}$$

$$= 40 \text{ gpm}$$

Note: The following information is used for examples 18.60–18.65

Operating time	15 min/cycle
Frequency	24 times/day
Pump rate	120 gpm
Solids	3.70%
Volatile matter	66%

Sludge Pumping Time

The *Sludge Pumping Time* is the total time the pump operates during a 24-hour period in minutes.

Pump Op. Time = Time/Cycle, min × Frequency, cycles/day

$$(18.77)$$

Example 18.60

Problem: What is the pump operating time?

Solution:

Pump operating time = 15 min/h × 24 (cycles)/day = 360 min/day

Gallons of Sludge Pumped per Day

Sludge, gpd = Operating Time, min/day × Pump Rate, gpm

$$(18.78)$$

Example 18.61

Problem: What is the sludge pumped/Day in gallons?

Solution:

Sludge, gpd = 360 min/day × 120 gpm = 43,200 gpd

Pounds Sludge Pumped per Day

Sludge, lb/day = Gallons of Sludge Pumped × 8.34 lb/gal

$$(18.79)$$

Example 18.62

Problem: What is the sludge pumped per day in gallons?

Solution:

Sludge, lb/day = 43,200 gal/day × 8.34 lb/gal = 360,300 lb/day

Pounds Solids Pumped per Day

Solids Pumped, lbs/day = Sludge Pumped, gpd × % Solids

$$(18.80)$$

Example 18.63

Problem: What are the solids pumped per day?

Solution:

Solids Pumped lb/day = 360,300 lb/day × 0.0370 = 13,331 lb/day

Pounds of Volatile Matter (VM) Pumped per Day

Vol. Matter (lb/day) = Solids Pumped, lbs/day × % Volatile Matter

$$(18.81)$$

Example 18.64

Problem: What is the volatile matter in pounds per day?

Solution:

Volatile Matter, lb/day = 13,331 lb/day × 0.66 = 8,798 lb/day

Note: If we wish to calculate the pounds of solids or the pounds of volatile solids removed per day, the individual equations demonstrated above can be combined into a single calculation.

Solids, lb/day = Pump Time, min/cycle × Frequency,

cycles/day × Rate, gpm × 8.34 lb/gal × solids Vol.

Matter, lb/day = Time, min/cyc. × Freq. Cycles/day ×

Rate, gpm × 8.34 × % Solids × % V.M.

$$(18.82)$$

Example 18.65

Solids, lb/day = 15 min/cycle × 24 cycle/day × 120 gpm ×

8.34 × 0.0370 = 13,331 lb/day

V.M., lb/day = 15 min/cycle × 24 cycle/day × 120 gpm

×8.34 × 0.0370 × 0.66

= 8,798 lb/day

Sludge Production in Pounds/Million Gallons

A common method of expressing sludge production is in pounds of sludge per million gallons of wastewater treated.

$$\text{Sludge, lb/MG} = \frac{\text{Total Sludge Production, lb}}{\text{Total Wastewater Flow, MG}} \quad (18.83)$$

Problem: Records show that the plant has produced 85,000 gal of sludge during the past 30 days. The average daily flow for this period was 1.2 MGD. What was the plant's sludge production in pounds per million gallons?

Solution:

$$\text{Sludge, lb/MG} = \frac{85,000 \text{ gal} \times 8.34 \text{ lb/gal}}{1.2 \text{ MGD} \times 30 \text{ days}} = 19,692 \text{ lb/MG}$$

Sludge Production in Wet Tons/Year

Sludge production can also be expressed in terms of the amount of sludge (water and solids) produced per year. This is normally expressed in wet tons per year.

$$\text{Sludge, Wet Tons/year} = \frac{\begin{array}{c}\text{Sludge Prod., lb/MG} \times \\ \text{Ave. Daily Flow, MGD} \times 365 \text{ days/year}\end{array}}{2{,}000 \text{ lb/ton}} \quad (18.84)$$

Example 18.66

Problem: The plant is currently producing sludge at the rate of 16,500 lb/MG. The current average daily wastewater flow rate is 1.5 MGD. What will be the total amount of sludge produced per year in wet tons per year?

Solution:

$$\text{Sludge, Wet Tons/year} = \frac{\begin{array}{c}16{,}500 \text{ lb/MG} \times 1.5 \text{ MGD} \times \\ 365 \text{ days/year}\end{array}}{2{,}000 \text{ lb/ton}}$$

$$= 4{,}517 \text{ Wet Tons/year}$$

Important Note: The release of wastewater solids without proper treatment could result in severe damage to the environment. Obviously, we must have a system to treat the volume of material removed as sludge throughout the system. Release without treatment would defeat the purpose of environmental protection. A design engineer can choose from many processes when developing sludge treatment systems. No matter what the system or combination of systems chosen, the ultimate purpose will be the same: the conversion of wastewater sludges into a form that can be handled economically and disposed of without damaging the environment or creating nuisance conditions. Leaving either condition unmet will require further treatment. The degree of treatment will generally depend on the proposed method of disposal. Sludge treatment processes can be classified into several major categories. In this handbook, we discuss the processes of thickening, digestion (or stabilization), dewatering, incineration, and land application. Each of these categories has then been further subdivided according to the specific processes that are used to accomplish sludge treatment. As mentioned, the importance of adequate, efficient sludge treatment cannot be overlooked when designing wastewater treatment facilities. The inadequacies of a sludge treatment system can severely affect a plant's overall performance capabilities. The inability to remove and process solids as fast as they accumulate in the process can lead to the discharge of large quantities of solids to receiving waters. Even with proper design and capabilities in place, no system can be effective unless it is properly operated. Proper operation requires proper operator performance. Proper operator performance begins and ends with proper training.

SLUDGE THICKENING

The solids content of primary, activated, trickling-filter, or even mixed sludge (i.e., primary plus activated sludge) varies considerably, depending on the characteristics of the sludge. Note that the sludge removal and pumping facilities and the method of operation also affect the solids content. *Sludge thickening* (or *concentration*) is a unit process used to increase the solids content of the sludge by removing a portion of the liquid fraction. By increasing the solids content, more economical treatment of the sludge can be achieved. Sludge thickening processes include:

- Gravity Thickeners
- Flotation Thickeners
- Solids Concentrators

Gravity Thickening

Gravity thickening is most effective on primary sludge. In operation, solids are withdrawn from primary treatment (and sometimes secondary treatment) and pumped to the thickener. The solids buildup in the thickener forms a solids blanket on the bottom. The weight of the blanket compresses the solids on the bottom and "squeezes" the water out. By adjusting the blanket thickness the percent solids in the underflow (solids withdrawn from the bottom of the thickener) can be increased or decreased. The supernatant (clear water), which rises to the surface, is returned to the wastewater flow for treatment. Daily operations of the thickening process include pumping, observation, sampling and testing, process control calculations, maintenance, and housekeeping.

Note: The equipment employed in thickening depends on the specific thickening processes used.

Equipment used for gravity thickening consists of a thickening tank, which is similar in design to the settling tank used in primary treatment. Generally, the tank is circular and provides equipment for continuous solids collection. The collector mechanism uses heavier construction than that in a settling tank because the solids being moved are more concentrated. The gravity thickener pumping facilities (i.e., pump and flow measurement) are used for the withdrawal of thickened solids.

Solids concentrations achieved by gravity thickening are typically 8%–10% solids from primary underflow, 2%–4% solids from waste-activated sludge, 7%–9% solids from trickling filter residuals, and 4%–9% from combined primary and secondary residuals. The performance of gravity thickening processes depends on various factors, including:

- Type of sludge
- Condition of influent sludge
- Temperature
- Blanket depth

- Solids loading
- Hydraulic loading
- Solids retention time
- Hydraulic detention time

Flotation Thickening

Flotation thickening is used most efficiently for waste sludges from suspended-growth biological treatment processes, such as the activated sludge process. In operation, recycled water from the flotation thickener is aerated under pressure. During this time, the water absorbs more air than it would under normal pressure. The recycled flow, together with chemical additives (if used), is mixed with the flow. When the mixture enters the flotation thickener, the excess air is released in the form of fine bubbles. These bubbles become attached to the solids and lift them toward the surface. The accumulation of solids on the surface is called the **float cake**. As more solids are added to the bottom of the float cake, it becomes thicker, and water drains from the upper levels of the cake. The solids are then moved up an inclined plane by a scraper and discharged. The supernatant leaves the tank below the surface of the float solids and is recycled or returned to the waste stream for treatment. Typically, flotation thickener performance is 3%–5% solids for waste-activated sludge with polymer addition and 2%–4% solids without polymer addition.

The flotation thickening process requires pressurized air, a vessel for mixing the air with all or part of the process residual flow, a tank for the flotation process to occur, solids collector mechanisms to remove the float cake (solids) from the top of the tank and accumulated heavy solids from the bottom of the tank. Since the process normally requires chemicals to be added to improve separation, chemical mixing equipment, storage tanks, and metering equipment to dispense the chemicals at the desired dose are required. The performance of the dissolved air-thickening process depends on various factors:

- Bubble size
- Solids loading
- Sludge characteristics
- Chemical selection
- Chemical dose

Solids Concentrators

Solids concentrators (belt thickeners) usually consist of a mixing tank, chemical storage and metering equipment, and a moving porous belt. In operation, the process residual flow is chemically treated and then spread evenly over the surface of the moving porous belt. As the flow is carried down the belt (similar to a conveyor belt), the solids are mechanically turned or agitated, and water drains through the belt. This process is primarily used in facilities where space is limited.

OPERATIONAL OBSERVATIONS, PROBLEMS, AND TROUBLESHOOTING PROCEDURES

As with other unit treatment processes, proper operation of sludge thickeners depends on operator observation. The operator must make routine adjustments of sludge addition and withdrawal rates to achieve the desired blanket thickness. Sampling and analysis of influent sludge, supernatant, and thickened sludge are also required. Sludge addition and withdrawal should be continuous, if possible, to achieve optimum performance. Mechanical maintenance is also required. Expected performance ranges for gravity and dissolved air flotation thickeners are:

- Primary sludge—8%–19% Solids
- Waste-activated sludge—2%–4% Solids
- Trickling filter sludge—7%–9% Solids
- Combined sludges—4%–9% Solids

Typical operational problems with sludge thickeners include odors, rising sludge, thickened sludge below the desired solids concentration, dissolved air concentration too low, effluent flow containing excessive solids, and torque alarm conditions.

INDICATORS OF POOR PROCESS PERFORMANCE

1. **Odors and Rising Sludge**
   Causal Factors:
   - The sludge withdrawal rate is too low.
   - The overflow rate is too low.
   - Septicity in the thickener.
   Corrective Actions (Where Applicable):
   - Increase sludge withdrawal rate.
   - Increase influent flow rate.
   - Add chlorine, permanganate, or peroxide to the influent.
2. **Thickened Sludge below Desired Solids Concentration**
   Causal Factors:
   - The overflow rate is too high.
   - The sludge withdrawal rate is too high.
   - Short-circuiting.
   Corrective Actions (Where Applicable):
   - Decrease influent sludge flow rate.
   - Decrease pump rate for sludge withdrawal.
   - Identify cause and correct.
3. **Torque Alarm Activated**
   Causal Factors:
   - Heavy sludge accumulation
   - Collector mechanism jammed
   Corrective Actions (Where Applicable):
   - Agitate sludge blanket to decrease density.
   - Increase sludge withdrawal rate.
   - Attempt to locate and remove obstacles.
   - Dewater the tank and remove obstacles.

Dissolved Air Flotation Thickener

1. **Float Solids Concentration Too Low**

   Causal Factors:
   - The skimmer speed is too high.
   - The unit is overloaded.
   - Insufficient polymer dose.
   - Excessive air-to-solids ratio.
   - Low dissolved air levels.

   Corrective Actions (Where Applicable):
   - Adjust the skimmer speed to permit concentration to occur.
   - Stop sludge flow through the unit/purge with recycles flow.
   - Determine the proper chemical dose and adjust.
   - Reduce airflow to the pressurization tank.
   - Identify malfunction and correct it.

2. **Dissolved Air Concentration Too Low**

   Causal Factor:
   - mechanical malfunction

   Corrective Action:
   - Identify the cause and correct

3. **Effluent (Subnatant) Flow Contains Excessive Solids**

   Causal Factors:
   - The unit is overloaded.
   - The chemical dose is too low.
   - The skimmer is not operating.
   - Low solids: air ratio.
   - Solids buildup in thickener.

   Corrective Actions (Where Applicable):
   - Turn off sludge flow.
   - Purge unit with recycling.
   - Determine the proper chemical dose and blow.
   - Turn the skimmer on.
   - Adjust skimmer speed.
   - Increase airflow to the pressurization system.
   - Remove sludge from the tank.

Process Calculations (Gravity/Dissolved Air Flotation)

Sludge thickening calculations are based on the concept that the solids in the primary or secondary sludge are equal to the solids in the thickened sludge. Assuming a negligible amount of solids is lost in the thickener overflow, the solids are the same. Note that water is removed to thicken the sludge, resulting in a higher percent solids.

Estimating Daily Sludge Production

Equation (18.85) provides a method to establish an initial pumping rate or evaluate the adequacy of the current pump rate.

$$\text{Est. Pump Rate} = \frac{(\text{Infl. TSS Conc.} - \text{Eff. TSS Conc.}) \times \text{Flow} \times .34}{\%\text{Solids in Sludge} \times 8.34 \times 1,440 \text{ min/day}}$$

$$(18.85)$$

Example 18.67

Problem: The sludge withdrawn from the primary settling tank contains 1.5% solids. The unit influent contains 280 mg/L TSS, and the effluent contains 141 mg/L. If the influent flow rate is 5.55 MGD, what is the estimated sludge withdrawal rate in gallons per minute (assuming the pump operates continuously)?

Solution:

$$\text{Sludge Rate, gpm} = \frac{(280 \text{ mg/L} - 141 \text{ mg/L}) \times 5.55 \text{ MGD} \times 8.34}{0.015 \times 8.34 \times 1,440 \text{ min/day}}$$

$$= 36 \text{ gpm}$$

Surface Loading Rate (gpd/ft²)

The surface loading rate (surface settling rate) is hydraulic loading—the amount of sludge applied per square foot of gravity thickener:

$$\text{Surface. Loading, gal/day/ft}^2 = \frac{\text{Sludge Applied to the Thickener, gpd}}{\text{Thickener Area, ft}^2}$$

$$(18.86)$$

Example 18.68

Problem: The 70-ft-diameter gravity thickener receives 32,000 gpd of sludge. What is the surface loading in gallons per square foot per day?

Solution:

$$\text{Surface Loading} = \frac{32,000 \text{ gpd}}{0.785 \times 70 \text{ ft} \times 70 \text{ ft}} = 8.32 \text{ gpd/ft}^2$$

Solids Loading Rate, lb/day/ft²

The solids loading rate is the pounds of solids per day being applied to 1 ft² of tank surface area. The calculation uses the surface area of the bottom of the tank. It assumes that the floor of the tank is flat and has the same dimensions as the surface.

$$\text{Solids Ldg Rate, lb/day/ft} = \frac{\begin{array}{c}\%\text{ Sludge Solids} \times \text{Sludge Flow,}\\ \text{gpd} \times 8.34 \text{ lb/gal}\end{array}}{\text{Thickener Area, ft}^2}$$

$$(18.87)$$

Example 18.69

Problem: The thickener influent contains 1.6-% solids. The influent flow rate is 39,000 gpd. The thickener is 50 ft in diameter and 10 ft deep. What is the solid loading in pounds per day?

Solution:

$$\text{Solids Ldg Rate, lb/day/ft}^2 = \frac{0.016 \times 39,000 \text{ gpd} \times 8.34 \text{ lb/gal}}{0.785 \times 50 \text{ ft} \times 50 \text{ ft}}$$

$$= 2.7 \text{ /b/ft}^2$$

Concentration Factor

The concentration factor (CF) represents the increase in concentration resulting from the thickener:

$$CF = \frac{\text{Thickened Sludge Concentration, \%}}{\text{Influent Sludge Concentration, \%}} \quad (18.88)$$

Example 18.70

Problem: The influent sludge contains 3.5% solids. The thickened sludge solids concentration is 7.7%. What is the concentration factor?

Solution:

$$CF = \frac{7.7\%}{3.5\%} = 2.2$$

Air-to-Solids Ratio

The air-to-solids ratio is the ratio of air being applied to the pounds of solids entering the thickener:

$$\text{Air: Solids Ratio} = \frac{\text{Air Flow ft}^3/\text{min} \times 0.075 \text{ lb/ft}^3}{\text{Sludge Flow, gpm} \times \% \text{ Solids} \times 8.34 \text{ lb/gal}}$$
$$(18.89)$$

Example 18.71

Problem: The sludge pumped to the thickener is 0.85% solids. The airflow is 13 cfm. What is the air-to-solids ratio if the current sludge flow rate entering the unit is 50 gpm?

Solution:

$$\text{Air: Solids Ratio} = \frac{13 \text{ cfm} \times 0.075 \text{ lb/ft}}{50 \text{ gpm} \times 0.0085 \times 8.34 \text{ lb/gal}}$$
$$= 0.28$$

Recycle Flow in Percent

The amount of recycle flow expressed as a percent:

$$\text{Recycle, \%} = \frac{\text{Recycle Flow Rate, gpm} \times 100}{\text{Sludge Flow, gpm}} = 175\%$$
$$(18.90)$$

Example 18.72

Problem: The sludge flow to the thickener is 80 gpm. The recycle flow rate is 140 gpm. What is the percent recycled?

Solution:

$$\% \text{ Recycle} = \frac{140 \text{ gpm} \times 100}{80 \text{ gpm}} = 175\%$$

SLUDGE STABILIZATION

The purpose of sludge stabilization is to reduce volume, stabilize the organic matter, and eliminate pathogenic organisms to permit reuse or disposal. The equipment required for stabilization depends on the specific process used. Sludge stabilization processes include:

- Aerobic Digestion
- Anaerobic Digestion
- Composting
- Lime Stabilization
- Wet Air Oxidation (Heat Treatment)
- Chemical Oxidation (Chlorine Oxidation)
- Incineration

Aerobic Digestion

Equipment used for *aerobic digestion* consists of an aeration tank (digester), which is similar in design to the aeration tank used for the activated sludge process. Either diffused or mechanical aeration equipment is necessary to maintain aerobic conditions in the tank. Solids and supernatant removal equipment are also required. In operation, process residuals (sludge) are added to the digester and aerated to maintain a dissolved oxygen (D.O.) concentration of 1.0 mg/L. Aeration also ensures that the tank contents are well mixed. Generally, aeration continues for approximately 20 days of retention time. Periodically, aeration is stopped, and the solids are allowed to settle. Sludge and the clear liquid supernatant are withdrawn as needed to provide more room in the digester. When no additional volume is available, mixing is stopped for 12–24 h before solids are withdrawn for disposal. Process control testing should include alkalinity, pH, % solids, % volatile solids for influent sludge, supernatant, digested sludge, and digester contents. Normal operating levels for an aerobic digester are listed in Table 18.16.

A typical operational problem associated with an aerobic digester is pH control. When the pH drops, for example, it may indicate normal biological activity or low influent alkalinity. This problem can be corrected by adding alkalinity (lime, bicarbonate, etc.).

TABLE 18.16
Aerobic Digester Normal Operating Levels

Parameter	Normal Levels
Detention time, days	10–20
Volatile solids loading, lbs/ft³/day	0.1–0.3
D.O., mg/L	1.0
pH	5.9–7.7
Volatile solids reduction	40%–50%

Process Control Calculations for the Aerobic Digester

Wastewater operators who operate aerobic digesters are required to make certain process control calculations. Moreover, licensing examinations typically include aerobic digester problems for determining volatile solids loading, digestion time, digester efficiency, and pH adjustment. These process control calculations are explained in the following sections.

Volatile Solids Loading

Volatile solids loading for the aerobic digester is expressed in pounds of volatile solids entering the digester per day per cubic foot of digester capacity:

$$\text{Volatile Solids Loading} = \frac{\text{Volatile Solids Added, lbs/day}}{\text{Digester Volume, ft}^3} \quad (18.91)$$

Example 18.73

Problem: The aerobic digester is 25 ft in diameter and has an operating depth of 24 ft. The sludge added to the digester daily contains 1,350 lb of volatile solids. What is the volatile solids loading in pounds per day per cubic foot?

Solution:

$$\text{Volatile Solids Loading} = \frac{1,350 \text{ lb/day}}{0.785 \times 25 \text{ ft} \times 25 \text{ ft} \times 24 \text{ ft}}$$

$$= 0.11 \text{ lb/day/ft}^3$$

Digestion Time, Days

Digestion time is the theoretical time the sludge remains in the aerobic digester:

$$\text{Digestion Time, Days} = \frac{\text{Digester Volume, gal}}{\text{Sludge Added, gpd}} \quad (18.92)$$

Example 18.74

Problem: Digester volume is 240,000 gal. Sludge is being added to the digester at the rate of 13,500 gpd. What is the digestion time in days?

Solution:

$$\text{Digestion Time, Days} = \frac{240,000 \text{ gal}}{13,500 \text{ gpd}} = 17.8 \text{ days}$$

Digester Efficiency (% Reduction)

To determine digester efficiency or the percentage of reduction, a two-step procedure is required. First, the percentage of Volatile Matter Reduction must be calculated, followed by the percentage of moisture reduction.

STEP 1: CALCULATE VOLATILE MATTER

Due to the changes occurring during sludge digestion, the calculation used to determine the percentage of volatile matter reduction is more complicated.

$$\% \text{ Reduction} = \frac{(\% \text{ Volatile Matter}_{in} - \% \text{ Volatile Matter}_{out}) \times 100}{[\% \text{ Vol. Matter}_{in} - (\% \text{ Vol. Matter}_{in} \times \% \text{ Vol. Matter}_{out}]} \quad (18.93)$$

Example 18.75

Problem: Using the digester data provided below, determine the percentage of volatile matter reduction for the digester.
Raw Sludge Volatile Matter 71%
Digested Sludge Volatile Matter 53%

Solution:

$$\% \text{ Vol. Matter Reduction} = \frac{(0.71 - 0.53) \times 100}{[0.71 - (0.71 \times 0.53)]} = 53.9 \text{ or } 54\%$$

STEP 2: CALCULATE MOISTURE REDUCTION

$$\% \text{ Moisture Reduction} = \frac{(\% \text{ Moisture}_{in} - \% \text{ Moisture}_{out}) \times 100}{[\% \text{ Moisture}_{in} - (\% \text{ Moisture}_{in} \times \% \text{ Moisture}_{out})]} \quad (18.94)$$

Example 18.76

Problem: Using the digester data provided below, determine the percentage of moisture reduction for the digester.

Note: Percent moisture = 100% – percent solids

Solution:

Raw sludge	% Solids	6%
	% Moisture	94% (100% – 6%)
Digested sludge	% Solids	15%
	% Moisture	85% (100% – 15%)

$$\% \text{ Reduction} = \frac{(0.94 - 0.85) \times 100}{\left[0.94 - (0.94 \times 0.85)\right]} = 64\%$$

pH Adjustment

Occasionally, the pH of the aerobic digester will fall below the levels required for good biological activity. When this occurs, the operator must perform a laboratory test to determine the amount of alkalinity required to raise the pH to the desired level. The results of the lab test must then be converted to the actual quantity of chemical (usually lime) required by the digester.

$$\text{Chem. Required, lb} = \frac{\text{Chemical Used in Lab Test, mg}}{\text{Sample Volume, L}}$$

$$\times \text{Dig. Vol, MG} \times 8.34$$

(18.95)

Example 18.77

Problem: The lab reports that it took 225 mg of lime to increase the pH of a 1-L sample of the aerobic digester contents to pH 7.2. The digester volume is 240,000 gal. How many pounds of lime will be required to increase the digester pH to 7.2?

Solution:

$$\text{Chemical Required, lb} = \frac{225 \text{ mg} \times 240,000 \text{ gal} \times 3.785 \text{ L/gal}}{1 \text{ L} \times 454 \text{ g/lb} \times 1,000 \text{ mg/g}}$$

$$= 450 \text{ lb}$$

Anaerobic Digestion

Anaerobic digestion is the traditional method of sludge stabilization. It involves using bacteria that thrive in the absence of oxygen and is slower than aerobic digestion but has the advantage that only a small percentage of the wastes are converted into new bacterial cells. Instead, most of the organics are converted into carbon dioxide and methane gas.

Note: In an anaerobic digester, the entrance of air should be prevented because of the potential for air to mix with the gas produced in the digester, which could create an explosive mixture.

Equipment used in anaerobic digestion includes a sealed digestion tank with either a fixed or a floating cover, heating and mixing equipment, gas storage tanks, solids and supernatant withdrawal equipment, and safety equipment (e.g., vacuum relief, pressure relief, flame traps, explosion-proof electrical equipment).

In operation, process residual (thickened or unthickened sludge) is pumped into the sealed digester. The organic matter digests anaerobically by a two-stage process. Sugars, starches, and carbohydrates are converted to volatile acids, carbon dioxide, and hydrogen sulfide. The volatile acids are then converted to methane gas. This operation can occur in a single tank (single stage) or in two tanks (two stages). In a single-stage system, supernatant and/or digested solids must be removed whenever flow is added. In a two-stage operation, solids and liquids from the first stage flow into the second stage each time fresh solids are added. The supernatant is withdrawn from the second stage to provide additional treatment space. Periodically, solids are withdrawn for dewatering or disposal. The methane gas produced in the process may be used for many plant activities.

Note: The primary purpose of a secondary digester is to allow for solids separation.

Various performance factors affect the operation of the anaerobic digester. For example, the % Volatile Matter in raw sludge, digester temperature, mixing, volatile acids/alkalinity ratio, feed rate, % solids in raw sludge, and pH are all important operational parameters that the operator must monitor.

Along with being able to recognize normal/abnormal anaerobic digester performance parameters, wastewater operators must also know and understand normal operating procedures. Normal operating procedures include sludge additions, supernatant withdrawal, sludge withdrawal, pH control, temperature control, mixing, and safety requirements. Important performance parameters are listed in Table 18.17.

Sludge Additions

Sludge must be pumped (in small amounts) several times each day to achieve the desired organic loading and optimum performance.

Note: Keep in mind that, in fixed cover operations, additions must be balanced by withdrawals. If not, structural damage occurs.

Supernatant Withdrawal

Supernatant withdrawal must be controlled for maximum sludge retention time. When sampling, sample all draw-off points and select the level with the best quality.

Sludge Withdrawal

Digested sludge is withdrawn only when necessary—always leave at least 25% seed.

TABLE 18.17
Anaerobic Digester—Sludge Parameters

Raw Sludge Solids	Impact
<4% Solids	Loss of alkalinity
	Decreased sludge retention time
	Increased heating requirements
	Decreased volatile acid: alk ratio
4%–8% Solids	Normal operation
>8% Solids	Poor mixing
	Organic overloading
	Decreased volatile acid: alk ratio

pH Control

pH should be adjusted to maintain 6.8 to 7.2 pH by adjusting feed rate, sludge withdrawal, or alkalinity additions.

Note: The buffer capacity of an anaerobic digester is indicated by the volatile acid/alkalinity relationship. Decreases in alkalinity cause a corresponding increase in ratio.

Temperature Control

If the digester is heated, the temperature must be controlled to a normal temperature range of 90°F–95°F. Never adjust the temperature by more than 1°F/day.

Mixing

If the digester is equipped with mixers, mixing should be accomplished to ensure organisms are exposed to food materials.

Safety

Anaerobic digesters are inherently dangerous—several catastrophic failures have been recorded. To prevent such failures, safety equipment such as pressure relief and vacuum relief valves, flame traps, condensate traps, and gas collection safety devices are installed. These critical safety devices must be checked and maintained for proper operation.

Note: Because of the inherent danger involved with working inside anaerobic digesters, they are automatically classified as permit-required confined spaces. Therefore, all operations involving internal entry must be made in accordance with OSHA's confined space entry standard.

Process Control Monitoring/Testing/Troubleshooting

During operation, anaerobic digesters must be monitored and tested to ensure proper operation. Testing should be accomplished to determine supernatant pH, volatile acids, alkalinity, BOD or COD, total solids, and temperature. Sludge (in & out) should be routinely tested for % solids and % volatile matter. Normal operating parameters are listed in Table 18.18.

As with all other unit processes, the wastewater operator is expected to recognize problematic symptoms with anaerobic digesters and effect the appropriate corrective action(s). Symptoms, causes, and corrective actions are discussed below.

SYMPTOM 1

Digester gas production is reduced; pH drops below 6.8; and/or volatile acids/alkalinity ratio increases.

Cause:
  Digester souring
  Organic overloading
  Inadequate mixing

TABLE 18.18
Anaerobic Digester: Normal Operating Ranges

Parameter	Normal Range
Sludge retention time	
Heated	30–60 days
Unheated	180+ days
Volatile solids loading	0.04–0.1 lbs V.M/day/ft^3
Operating temperature	
Heated	90°F–95°F
Unheated	Varies with season
Mixing	
Heated - primary	Yes
Unheated - secondary	No
% Methane in gas	60%–72%
% Carbon dioxide in gas	28%–40%
pH	6.8–7.2
Volatile acids: alkalinity ratio	≤0.1
Volatile solids reduction	40%–60%
Moisture reduction	40%–60%

  Low alkalinity
  Hydraulic overloading
  Toxicity
  Loss of digestion capacity

Corrective Actions:
  Add alkalinity (digested sludge, lime, etc.); improve temperature control; improve mixing; eliminate toxicity; clean digester.

SYMPTOM 2

Gray foam oozing form digester.

Cause:
  Rapid gasification
  Foam-producing organisms present
  Foam-producing chemicals present
Corrective Actions:
  Reduce mixing; reduce feed rate; mix slowly by hand; clean all contaminated equipment.

Anaerobic Digester: Process Control Calculations

Process control calculations involved with anaerobic digester operation include determining the required seed volume, volatile acid-to-alkalinity ratio, sludge retention time, estimated gas production, volatile matter reduction, and percent moisture reduction in digester sludge. Examples of how to make these calculations are provided in the following sections.

Required Seed Volume in Gallons

$$\text{Seed Volume (gal)} = \text{Digester Volume} \times \% \text{ Seed} \quad (18.96)$$

Example 18.78

Problem: The new digester requires a 25% seed to achieve normal operation within the allotted time. If the digester volume is 266,000 gal, how many gallons of seed material will be required?

Solution:

$$\text{Seed Volume} = 266,000 \times 0.25 = 66,500 \text{ gal}$$

Volatile Acids to Alkalinity Ratio

The volatile acids to alkalinity ratio can be used to control the operation of an anaerobic digester.

$$\text{Ratio} = \frac{\text{Volatile Acids Concentration}}{\text{Alkalinity Concentration}} \quad (18.97)$$

Example 18.79

Problem: The digester contains 240 mg/L of volatile acids and 1,860-mg/L of alkalinity. What is the volatile acids-to-alkalinity ratio?

$$\text{Ratio} = \frac{240 \text{ mg/L}}{1,860 \text{ mg/L}} = 0.13$$

Note: Increases in the ratio normally indicate a potential change in the operation condition of the digester as shown in Table 18.19.

Sludge Retention Time

Sludge retention time (SRT) is the length of time the sludge remains in the digester.

$$\text{SRT, days} = \frac{\text{Digester Vol} \ (\text{gal})}{\text{Sludge Vol. Added per Day} \ (\text{gpd})} \quad (18.98)$$

Example 18.80

Problem: Sludge is added to a 525,000-gal digester at the rate of 12,250 gal/day.

Solution:

$$\text{SRT} = \frac{525,000 \text{ gal}}{12,250 \text{ gpd}} = 42.9 \text{ days}$$

TABLE 18.19
Digester Condition

Operating Condition	V.A./Alkalinity Ratio
Optimum	≤0.1
Acceptable range	0.1–0.3
% Carbon dioxide in gas increases	≥0.5
pH decreases	≥0.8

Estimated Gas Production in Cubic Feet/Day

The rate of gas production is normally expressed as the volume of gas (ft^3) produced per pound of volatile matter destroyed. The total cubic feet of gas a digester will produce per day can be calculated by:

$$\text{Gas Prod.} \left(\text{ft}^3\right) = \text{Vol. Matter In, lb/day}$$
$$\times \% \text{ Vol. Matter Red.} \times \text{Prod. Rate ft}^3\text{/lb} \quad (18.99)$$

Example 18.81

Problem: The digester receives 11,450 lb of volatile matter per day. Currently the volatile matter reduction achieved by the digester is 52%. The rate of gas production is 11.2 ft^3 of gas per pound of volatile matter destroyed.

Solution:

$$\text{Gas Prod.} = 11,450 \text{ lb/day} \times 0.52 \times 11.2 \text{ ft}^3\text{/lb} = 66,685 \text{ ft}^3\text{/day}$$

Percent Volatile Matter Reduction

Because of the changes occurring during sludge digestion, the calculation used to determine the percent volatile matter reduction is more complicated.

$$\% \text{ VM Reduction} = \frac{\left(\% \text{ VM}_{\text{in}} - \% \text{ VM}_{\text{out}}\right) \times 100}{\left[\% \text{ VM}_{\text{in}} - \left(\% \text{ VM}_{\text{in}} \times \% \text{ VM}_{\text{out}}\right)\right]} \quad (18.100)$$

Example 18.82

Problem: Using the data provided below, determine the percent Volatile Matter Reduction for the digester.
   Raw Sludge Volatile Matter   74%
   Digested Sludge Volatile Matter   55%

$$\% \text{ Volatile Matter Reduction} = \frac{\left(0.74 - 0.55\right) \times 100}{\left[0.74 - \left(0.74 \times 0.55\right)\right]} = 57\%$$

Percent Moisture Reduction in Digested Sludge

$$\% \text{ Moisture Reduction} = \frac{\left(\% \text{ Moisture}_{\text{in}} - \% \text{ Moisture}_{\text{out}}\right) \times 100}{\left[\begin{array}{c}\% \text{ Moisture}_{\text{in}} \\ - \left(\% \text{ Moisture}_{\text{in}} \times \% \text{ Moisture}_{\text{out}}\right)\end{array}\right]}$$
$$(18.101)$$

Example 18.83

Problem: Using the digester data provide below, determine the percent Moisture Reduction and percent Volatile Matter Reduction for the digester.

Solution:

> Raw Sludge % Solids 6%
> Digested Sludge % Solids   14%

Note: Percent Moisture = 100% − Percent Solids

$$\% \text{ Moisture Reduction} = \frac{(0.94 - 0.86) \times 100}{[0.94 - (0.94 \times 0.86)]} = 61\%$$

Other Sludge Stabilization Processes

In addition to aerobic and anaerobic digestion, other sludge stabilization processes include composting, lime stabilization, wet air oxidation, and chemical (chlorine) oxidation. These stabilization processes are briefly described in this section.

Composting

The purpose of composting sludge is to stabilize the organic matter, reduce volume, and eliminate pathogenic organisms. In a *composting operation,* dewatered solids are usually mixed with a bulking agent (i.e., hardwood chips) and stored until biological stabilization occurs. The composting mixture is ventilated during storage to provide sufficient oxygen for oxidation and to prevent odors. After the solids are stabilized, they are separated from the bulking agent. The composted solids are then stored for curing and applied to farmlands or other beneficial uses. The expected performance of the composting operation for both percent volatile matter reduction and percent moisture reduction ranges from 40% to 60%.

Definitions of Key Terms

Aerated Static Pile: A composting system using controlled aeration from a series of perforated pipes running underneath each pile and connected to a pump that draws or blows air through the piles.

Aeration (for Composting): Bringing about contact between air and composted solid organic matter by means of turning or ventilating to allow microbial aerobic metabolism (bio-oxidation).

Aerobic: A composting environment characterized by bacteria active in the presence of oxygen (aerobes); generates more heat and is a faster process than anaerobic composting.

Anaerobic: A composting environment characterized by bacteria active in the absence of oxygen (anaerobes).

Bagged Biosolids: Biosolids that are sold or given away in a bag or other container (i.e., either an open or closed vessel containing 1 metric ton or less of biosolids).

Bioaerosols: Biological aerosols that can pose potential health risks during the composting and handling of organic materials. Bioaerosols are suspensions of particles in the air consisting partially or wholly of microorganisms. The bioaerosols of concern during composting include actinomycetes, bacteria, viruses, molds, and fungi.

Biosolids Composting: This is the process involving the aerobic biological degradation or bacterial conversion of dewatered biosolids, which works to produce compost that can be used as a soil amendment or conditioner.

Biosolids Quality Parameters: The EPA determined that three main parameters of concern should be used in gauging biosolids quality: (1) the relevant presence or absence of pathogenic organisms, (2) pollutants, and (3) the degree of attractiveness of the biosolids to vectors. There can be several possible biosolids qualities. To express or describe those biosolids meeting the highest quality for all three of these biosolids quality parameters, the term Exceptional Quality or EQ has come into common use.

Bulking Agents: Materials, usually carbonaceous such as sawdust or woodchips, added to a compost system to maintain airflow by preventing settlement and compaction of the compost.

Bulk Biosolids: Biosolids that are not sold or given away in a bag or other container for application to the land.

Carbon to Nitrogen Ration (C:N Ratio): A ratio representing the quantity of carbon (C) concerning the quantity of nitrogen (N) in soil or organic material; it determines the composting potential of a material and serves to indicate product quality.

Compost: The end product (innocuous humus) remaining after the composting process is completed.

Curing: The late stage of composting, after much of the readily metabolized material has been decomposed, which provides additional stabilization and allows further decomposition of cellulose and lignin (found in woody-like substances).

Curing Air: Curing piles are aerated primarily for moisture removal to meet final product moisture requirements and to keep odors from building up in the compost pile as biological activity dissipates. Final product moisture requirements and summer ambient conditions are used to determine air requirements for moisture removal during the curing process.

Endotoxins: Toxins produced within a microorganism and released upon the destruction of the cell in which they are produced. Endotoxins can be carried by airborne dust particles at composting facilities.

EPA's 503 Regulation: To ensure that sewage sludge (biosolids) is used or disposed of in a way that protects both human health and the environment, under the authority of the Clean Water Act as amended, the U.S. Environmental Protection Agency (EPA) promulgated, at 40 CFR Part 503, Phase I of the risk-based regulation that governs the final use or disposal of sewage sludge (biosolids).

Exceptional Quality (EQ) Sludge (Biosolids): Although this term is not used in 40 CFR Part 503, it has become a shorthand term for biosolids that meet the pollutant concentrations in Table 3 of Part 503.13(b)(3); one of the six class A pathogen reduction alternatives in 503.32(a); and one of the vector attraction reduction options in 503.33(b)(1)-(8) (Spellman, 1997).

Feedstock: Decomposable organic material used for the manufacture of compost.

Heat Removal and Temperature Control: The biological oxidation process for composting biosolids is an exothermic reaction. The heat given off by the composting process can raise the temperature of the compost pile high enough to destroy the organisms responsible for biodegradation. Therefore, the compost pile is aerated to control the temperatures of the compost process by removing excess heat to maintain optimum temperature for organic solids degradation and pathogen reduction. Optimum temperatures are typically between 50°C and 60°C (122°F and 140°F). Using summer ambient air conditions, aeration requirements for heat removal can be calculated.

Metric Ton: One (1) metric ton, or 1,000 kg, equals about 2,205 lbs, which is larger than the short ton (2,000 lb) usually referred to in the British system of units. The metric ton unit is used throughout this text.

Moisture Removal: When temperature increases, the quantity of moisture in saturated air increases. Air is required for the composting process to remove water that is present in the mix and produced by the oxidation of organic solids. The quantity of air required for moisture removal is calculated based on the desired moisture content for the compost product and the psychometric properties of the ambient air supply. Air requirements for moisture removal are calculated from summer ambient air conditions and required final compost characteristics.

Oxidation Air: The composting process requires oxygen to support aerobic biological oxidation of degradable organics in the biosolids and wood chips. Stoichiometric requirements for oxygen are related to the extent of organic solids degradation expected during the composting cycle time.

Pathogen Organisms: Specifically, Salmonella and E. coli bacteria, enteric viruses, or visible helminth ova.

Peaking Air: The rate of organic oxidation, and therefore the heat release, can vary greatly during the composting process. If sufficient aeration capacity is not provided to meet peak requirements for heat or moisture removal, temperature limits for the process may be exceeded. Peaking air rates are typically 1.9 times the average aeration rate for heat removal.

Pollutant: An organic substance, an inorganic substance, a combination of organic and inorganic substances, or a pathogenic organism that, after discharge and upon exposure, ingestion, inhalation, or assimilation into an organism—either directly from the environment or indirectly by ingestion through the food chain—could, based on information available to the EPA, cause death, disease, behavioral abnormalities, cancer, genetic mutations, physiological malfunctions, or physical deformations in either organisms or their offspring.

Stability: The state or condition in which the composted material can be stored without giving rise to nuisances or can be applied to the soil without causing problems; the desired degree of stability for finished compost is one in which the readily decomposed compounds are broken down and only the decomposition of the more resistant biologically decomposable compounds remains to be accomplished.

Vectors: Refers to the degree of attractiveness of biosolids to flies, rats, and mosquitoes that could come into contact with pathogenic organisms and spread disease.

Aerated Static Pile (ASP)

Three methods of composting wastewater biosolids are common. Each method involves mixing dewatered wastewater solids with a bulking agent to provide carbon and increase porosity. The resulting mixture is piled or placed in a vessel where microbial activity causes the temperatures of the mixture to rise during the "active composing" period. The specific temperatures that must be achieved and maintained for successful composting vary based on the method and the use of the end product. After active composting, the material is cured and distributed. Again, there are three commonly employed composting methods, but we only describe the aerated static pile (ASP) method because it is commonly used. For an in-depth treatment of the other two methods, windrow, and in-vessel, we refer you to Spellman (1997).

The ASP Model Composting Facility uses a homogenized mixture of bulking agent (coarse hardwood chips) and dewatered biosolids that is piled by front-end loaders onto a large concrete composting pad, where it is mechanically aerated via PVC plastic pipe embedded within the concrete slab. This ventilation procedure is part of the 26-day period of "active" composting when adequate air and oxygen are necessary to

DID YOU KNOW?

Aeration is an important process control parameter in the aerated static pile composting system. Air is required to supply oxygen for the biological degradation of organic solids in the biosolids and wood chips. Aeration is also needed for the removal of heat generated by the biological activity in the compost pile and excess moisture from the compost mix. Fans are used to ensure that sufficient quantities of air are supplied to meet composting process requirements and to provide the process control flexibility necessary for optimizing operations.

support aerobic biological activity in the compost mass and to reduce the heat and moisture content of the compost mixture. Keep in mind that a compost pile without a properly sized air distribution system can lead to the onset of anaerobic conditions and the appearance of putrefactive odors.

For illustration and discussion purposes, we assume a typical overall composting pad area is approximately 200 ft by 240 ft, consisting of 11 blowers and 24 pipe troughs (troffs). Three blowers are 20 hp, 2,400 cfm, variable speed drive units capable of operating in either the positive or negative aeration mode. Blowers A, B, and C are each connected to two piping troughs that run the full length of the pad.

The two troughs are connected at the opposite end of the composting pad to create an "aeration pipe loop." The other eight blowers are rated at 3 hp, 1,200 cfm and are arranged one blower per six troughs at half-length, feeding 200 cfm per trough. These blowers can be operated in the positive or negative aeration mode. Aeration piping within the six pipe troughs is perforated PVC plastic pipe, 6 in inside diameter and 1/4 in wall thickness. Perforation holes/orifices vary in size from 7/32 to 1/2 in, increasing in diameter as the distance from the blower increases.

The variable speed motor drives installed with blowers A, B, and C are controlled by five thermal probes mounted at various depths in the compost pile, and various parameters are fed back to the recorder, whereas the other eight blowers are constant speed, controlled by a timer that cycles them on and off. To ensure optimum composting operations, it is important to verify that these thermal probes are calibrated regularly. In the constant speed system, thermal probes are installed but all readings are taken and recorded manually.

For water and leachate drainage purposes, all aeration piping within the troughs slopes downward, with the highest point at the center of the composting pad. Drain caps located at each end of the pipe length are manually removed regularly so that any build-up of debris or moisture will not interfere with the airflow.

The actual construction process involved in building the compost pile will be covered in detail later, but for now, a few key points should be made. For example, prior to the piling of the mixture on the composting pad, an 18-in layer of wood chips is used as a base material. The primary purpose of the wood chips base is to keep the composting mixture clear of the aeration pipes, which reduces clogging of the air distribution openings in the pipes and allows free air circulation. A secondary benefit is that the wood chips insulate the composting mixture from the pad. The compost pad is like a heat sink and this insulating barrier improves the uniformity of heat distribution within the composting mixture.

Construction of Composting Pile

The ASP Model uses extended pile construction. The following discussion details the procedure used in their formation and operation.

Compost Pile Formation Procedure

1. Check to ensure the non-perforated section of the aeration pipe will extend at least 8–10 ft under the slope at each end of the compost pile. This practice is necessary to prevent the short-circuiting of air which could result in "cold spots" with inadequate pathogen destruction.
2. Check to ensure the aeration pipe is not damaged. Replace the pipe if necessary.
3. Fill the trough area around the aeration pile with a suitable bulking agent such as woodchips. Replace it when the bulking agent or composted material becomes compacted in the troughs.
4. Place a 3–18 in base layer of bulking agent in an area approximately 4 ft on either side of the aeration troughs. The purpose of installing a bulking agent base is to improve air distribution, absorb moisture, and prevent clogging of the trenches.
5. Construct over the bulking agent base with a front-end loader an initial pile with a triangular cross-section to a convenient height of 7–10 ft. Be careful not to compact the compost mix: a compacted mix will not compost properly. If the compost pile is constructed outdoors, care should be taken while forming piles to ensure that tops of the piles form peaks or rounded dome-like tops to allow for rainwater runoff; otherwise, unwanted moisture may pool on the pile tops, retarding the composting process.
6. Blanket both ends and the exterior side of the first pile with either 8–12 in of cured screened compost or 16–20 in of unscreened compost or bulking agent. The purpose of this "blanket" is to provide insulation and prevent the escape of odorous gases.
7. Form subsequent piles parallel to the first. Extend the base following procedures 1–4 outlined previously. Then loosely place the compost mixture next to the previous pile to form an extended pile with a trapezoidal cross section. As the piles are made, blanket the tops and ends with 8–12 in of cured screened compost or 16–20 in of unscreened compost or bulking agent.
8. Hand rake the top and sides of the piles smoothly to prevent water pockets from forming. Sweep excess woodchips from around the base of the piles.
9. At the end of each day, dust the uncovered side of the compost pile with approximately 1–3 in of screened or unscreened compost or bulking agent for overnight control.
10. Place an appropriate sign on the pile to indicate the date started and the date to be taken down (26 days).
11. Close aeration dampers in troughs that have no compost piles.

Compost Pile Operation Procedure

1. Set blowers to operate intermittently. Aeration rates may vary from 200 to 1,200 cf/h/dry ton. Generally, the blowers operate for 3–25 min each 1/2 h cycle. The blower operating cycle should be adjusted depending on interior oxygen levels. As a guideline: If the oxygen level is <5%, the blower on time should be increased; if the oxygen level is >15%, the blower on time should be decreased.
2. Check all blowers each day to ensure they are operating correctly.

3. Adjust aeration dampers as necessary to ensure even distribution of air.
4. Check all drains each day to ensure they are operating properly. Drain water from 8-in fiberglass headers as necessary.
5. Compost is to be sampled and tested as directed.
6. Temperatures should increase to 50°C within a few days.
7. If temperatures of 55°C have been maintained at all pile monitoring points for a minimum of 3 days, the compost pile can be removed after 21–26 days.

Note: Piles are to be removed with a front-end loader. Care should be taken to break into the pile on the upwind side.

8. Compost that has not maintained pile temperatures of 55°C for at least 3 days must be recycled back through the composting process.

Advantages and Disadvantages

Biosolids composting has grown in popularity for the following reasons (WEF, 1995; Spellman, 1997):

- Lack of availability of landfill space for solids disposal.
- Composting economics are more favorable when landfill tipping fees escalate.
- Emphasis on beneficial reuse at federal, state, and local levels.
- Ease of storage, handling, and use of the composted product.
- The addition of biosolids compost to soil increases the soil's phosphorus, potassium, nitrogen, and organic carbon content.

Composted biosolids can also be used in various land applications. Compost mixed with appropriate additives creates a material useful in wetland and mineland restoration. The high organic matter content and low nitrogen content common in compost provide a strong organic substrate that mimics wetland soils, prevents the overloading of nitrogen, and absorbs ammonium to prevent transport to adjacent surface waters (Peot, 1998). Compost-amended strip-mine spoils produce a sustainable cover of appropriate grasses, in contrast to inorganic-only amendments which seldom provide such a good or sustainable cover (Sopper, 1993).

Compost-enriched soil can also help suppress diseases and ward off pests. These beneficial uses of compost can help growers save money, reduce the use of pesticides, and conserve natural resources. Compost also plays a role in the bioremediation of hazardous sites and pollution prevention. Compost has proven effective in degrading or altering many types of contaminants, such as wood preservatives, solvents, heavy metals, pesticides, petroleum products, and explosives. Some municipalities are using compost to filter stormwater runoff before it is discharged to remove hazardous

chemicals picked up when stormwater flows over surfaces such as roads, parking lots, and lawns. Additional uses for compost include soil mulch for erosion control, silviculture crop establishment, and production media (USEPA, 1997). Limitations of biosolids composting may include:

- Odor production at the composting site.
- Survival and presence of primary pathogens in the product.
- Dispersion of secondary pathogens such as *Aspergillus fumigatus*, particulate matter, and other airborne allergens.
- Lack of consistency in product quality with reference to metals, stability, and maturity.

SIDEBAR 18.2 ASPERGILLUS FUMIGATUS

Compost facility workers and the public who live near a composting facility can be impacted by *Aspergillus fumigatus* exposure. More specifically, *Aspergillus fumigatus* can cause *aspergillosis* in man. Aspergillosis is a disease usually caused by the inhalation of airborne spores but can also occur through ingestion or via wounds (USEPA, 2007b). Man is not the only species affected by aspergillosis; it can affect a wide range of animals including horses, cattle, sheep, and poultry. The disease caused by the *Aspergillus* organism is an acute or chronic inflammatory infection primarily of the respiratory tract and the ear (Burnett and Schuster, 1973). *Aspergillus fumigatus* has been found to grow on wood, grass, compost, rubber, and green leaves (Epstein and Epstein, 1989).

It is interesting to note that *Aspergillus f.* is not considered a hazard to healthy individuals. It is the susceptible individuals (for example, those with chronic pulmonary function problems) who can be infected and that must be protected. To aid in preventing infection, decision-makers involved with managing biosolids-derived composting facilities should incorporate a medical screening process into their hiring procedures. Pre-screening compost workers prior to their assignment at the compost site and requiring subsequent annual physical examinations, including annual pulmonary function testing is prudent practice.

With regard to protecting the public from infection by *Aspergillus fumigatus*, decision-makers at the composting site should take certain measures to control generation and exposure. For example, when siting new facilities, critical evaluations should be made on their proximity to private residences and public facilities. Because there is a potential risk of increased exposure to *Aspergillus* spores in areas located downwind from a composting site, meteorological conditions that could transport bioaerosols off-site should be evaluated (Toomey, 1994).

TABLE 18.20
Levels of Aspergillus f. at a Biosolids Composting Facility

Location	Concentration (Colony-Forming Units/m²)
Mixing area	110–120
Near tear down pile	8–24
Compost pile	12–15
Front-end loader operation	11–79
Periphery of compost site	2

Source: Adapted from Epstein and Epstein (1989).

Additional preventive measures are available. For example, the inclusion of a sufficient buffer area around the compost site should be incorporated. However, if an adequate buffer area is not feasible because of a lack of available land, enclosing the process, increasing the mechanization of the process, and implementing good management practices can help to mitigate the dispersion of bioaerosols (Millner, 1995).

Table 18.20 shows locations where *Aspergillus f.* is usually present and the typical levels that can be expected at a biosolids composting facility. From Table SB.2, it should be apparent that there is a direct relationship between *Aspergillus f.* levels and compost site activities.

DID YOU KNOW?

More than 7.3 billion chickens, ducks, and turkeys are raised for commercial sale in the United States each year, according to the U.S. Department of Agriculture's National Agricultural Service. About 37 million birds (18%–25%) die from disease or other natural causes before they are marketable. As more poultry is consumed, these numbers are expected to climb.

Composting is a viable and cost-effective option for disposing of poultry mortalities as compared to incineration or burial. Pathogens in poultry carcasses are destroyed during composting by the high temperatures (130°F–155°F) inherent in the process.

During composting, various odor control techniques can be used. As a result, this type of compost is not only safe for crop application but it can also be safely sold by farmers. In fact, selling excess compost could even be a source of additional income for farmers. Markets for high-quality compost include professional growers (such as horticultural greenhouses and nurseries), homeowners, turf growers, and crop farmers (such as corn and wheat farmers). Professional growers alone purchase $250 million per year in compost products.

Composting Odor Problems

Odors from a composting operation can be a nuisance and a potential irritant. Consider the following:

> When a backyard cookout is canceled because of a local malodor, or when the homeowner feels he must close his windows and install air purifiers, or when he operates his air-conditioning system when outside temperatures do not require air cooling, these behaviors may be translated into dollar costs. In fact, the courts often recognize such actions as evidence that odorous emissions are damaging and that compensation should be made by the offender.
>
> —*(Cheremisinoff and Young, 1981)*

In the preceding statement, Cheremisinoff and Young point out (with regard to the interface between the public and industry) a well-known fact: When impairment of use and enjoyment of property results from the operation of an industrial process, such as composting, serious complaints from political leaders, community administrators, and the public are almost guaranteed to occur.

It should be pointed out that impairment of use and enjoyment of property is only one of the effects associated with the industrial production of nuisance conditions, such as malodors. As an example, consider the effects of malodors with respect to dollar costs: (1) When malodors pervade a community or industrial setting, there can be a decrease in the values of existing property. (2) A dollar cost is involved with the loss of use of surrounding open land; that is, some areas surrounding industrial complexes are designated for recreational use, but how many people are going to seek recreation in an area where the odor is offensive? (3) The direct personal effect of offensive odor production is another factor that must be considered. For example, it is not unusual for people to complain that malodors cause them to experience a loss of sleep, loss of appetite, and nausea. When the public begins to complain, and when their complaints turn into lawsuits, obviously, the industrial complex responsible for the production of the offensive odors can be put in a serious financial bind.

Since the major operational problem associated with composting is reportedly the production of odors, composting process odor control is a matter that must be given serious consideration (McGhee, 1991). Those planners, designers, decision-makers, and engineers who are tasked with the responsibility of planning, designing, engineering, and funding composting facilities generally consider the need to "site" the proposed composting facility in a location that is suitable for its intended operation. This suitability of location is important. For example, when a compost site is in the planning stages, consideration must be given to accessibility. That is, can the truck loads of biosolids cake be delivered economically to the proposed site? Are the roads compatible with small private vehicles as well as large Ram-E-Jec-type trucks and trailers? Does the proposed site afford enough room for the entire composting operation?

Can bulking agents be easily obtained, transported, and stored on-site? Is there enough surrounding land to form an adequate buffer zone around the site?

When potential odor problems become a planning, designing, and engineering concern (as they should be in the biosolids composting process), conventional practice calls for the inclusion of an adequate plant site buffer zone. The question becomes: What exactly is an adequate buffer zone?

In years past, when cities were smaller than they are today, finding a composting site with a large enough natural buffer zone was not a difficult undertaking. Today, however, the situation is different. This difference can be seen in present-day cities in the United States where suburban areas have expanded in population and in size. In the past, it was not unusual to build a composting facility in an area where few people lived and where hundreds of undeveloped acres of forest and wide open spaces were the norm. Over time, however, the population and suburbia grew in all directions. What might have been an isolated composting facility with an extensive buffer zone 20 years ago today is surrounded by urban sprawl.

The point is that in the past when composting facilities were built, in most cases, little thought was given to future growth near or around the composting site. Thus, when originally built, the composting facility had little to worry about in relation to odor complaints from the public. Today, this is no longer the case.

In addressing odor control problems with biosolids-derived composting facilities, two main scenarios are addressed in this text: new construction and established facilities.

In new construction, planners, designers, and engineers have the luxury of basing their plans, designs, and processes on data (lessons learned) derived from other composting operations. This is a huge advantage in that, for the past 25 years, several biosolids-derived composting operations have been in operation throughout the United States. Data obtained from these composting operations can be used to select the proper process, design the proper facility, and properly train managers in site operations—all of which may result in an operation free of malodors (Corbitt, 1990).

However, new construction does face one major hurdle that is not normally experienced by older operations: siting. As pointed out earlier, the potential problems of finding a suitable site with an accompanying buffer zone are very real and troublesome. The never-ending encroachment of population and industry into what originally was "virgin" land space is a reality. The Not-In-My-Backyard (NIMBY) syndrome is real. This is especially the case when there is the perception that the new site will not be free of malodors. The reality is that if a community perceives the siting of a compost facility in its "backyard" as a potential nuisance that will pollute the environment, decrease land values, and affect the quality of life, then the planners, designers, and engineers may be up against overwhelming opposition (Outwater, 1994). Even when planners, designers, and engineers go to the citizens and openly and honestly present their plans, designs, and processes, the struggle they face to bring the public on board and get them to buy in on having a composting facility as a neighbor is not to be underestimated.

The second scenario deals with established facilities. There is an old saying: odors are not a problem until the neighbors complain. Long-established composting facilities may have enjoyed, at earlier times, a lack of neighbors and thus a lack of public awareness of their facility and the odors generated by their composting process. In most cases, this is no longer the case. As stated earlier, massive expansion of urban development has encroached upon what used to be remote areas around most composting facilities. It is probably safe to say that if the local community has expanded into and has become a close neighbor of an existing compost facility and if the composting process is not carefully managed (controlled), the production of odors can become a definite problem (Tchobanoglous et al., 1993); that is, the neighbors will complain.

Another issue arises when attempting to get people to agree on the desirability or undesirability of an odor. The perception of odor and how well it might be received by people has a lot to do with the association of the odors with their sources (Vesilind, 1980). As a case in point, consider the following example.

When a person, for the first time—not knowing the source, detects the heavy, earthy smell of biosolids-derived compost, he or she may like it or dislike it. The point is that each person perceives odor differently. For those who are not offended by the odor of composting biosolids, they may not give it a second thought. However, a few days or weeks later, if this same person is driving to work with a neighbor who is familiar with the source of that heavy, earthy compost odor and the informed passenger passes this information on to the unknowing person, then the unknowing person's perception has a good chance of changing. This phenomenon should not come as a surprise, considering that most individuals do not associate flushing their toilets with wastewater treatment and its ancillary processes (composting). However, when one finds out, like the unknowing person in the car, people may perceive the heavy, earthy biosolids-derived compost odor in a very different way.

Along with knowledge of the odor's source having a bearing on a person's sensitivity to odor, odor has another characteristic that may impact people: the individual's sensitivity. As far as sensitivity is concerned, the key point to remember is that odor is subjective; that is, what is offensive to one individual may not be to another (Outwater, 1994).

When attempting to address the factors that completely characterize an odor, it may be wise to refer to the four independent factors described by Metcalf & Eddy (1991) and listed here as follows: "intensity, character, hedonics, and detectability" (p. 58).

Intensity refers to the perceived strength of the odor as measured by an olfactometer. The character of an odor refers to any mental associations made by the subject sensing the odor. Hedonics refers to the relative degree of pleasantness or unpleasantness as perceived by the subject, which is usually determined "by using a scale estimating the magnitudes of the aesthetic qualities found in odors" (Lue-Hing et al., 1992, p. 197). Odor detectability refers to the number of dilutions that are needed to reduce an odor to its minimum detection point.

Measuring Odors

In measuring odors, panel or dilution methods are usually used. The dilution method is used in the water treatment process to detect odors in water and will not be discussed here. The panel method involves using ten or more people who make a judgment about the odor. These individual judgments are recorded and then analyzed. According to Vesilind (1980), the results of the panel method can be used to determine an "average opinion of the strength and nuisance value" of certain odors (p. 42).

When the panel method is used to measure odor, the parameter normally used for detecting odor is expressed as the number of effective dilutions-50 (ED-50). ED-50 is the number of fresh air dilutions required to reduce the odor level of a sample so that only 50% of the panel can smell it. Odor standards are based on odor control parameters such as ED-50, ED-10, ED-5, and others.

Note: In setting the odor control parameter for a compost facility that is to be constructed or for an existing facility, a set point of ED-50 is not practical. The point is that at an odor control set point of ED-50, the odor level would be reduced (diluted) for only 50% of those who live near or come in close proximity to the compost facility.

Malodorous Compounds in Biosolids-Derived Compost

To characterize composting as a smelly process is to correctly state the case. Whether or not this smelly process is offensive to the subject is another issue; it depends almost entirely on individual sensitivity.

It is interesting to note that ingredients important to the composting process itself all smell. These smelly but important ingredients include the following: amines, aromatics, terpenes, organic and inorganic sulfur, and fatty acids.

Generally associated with fats and oil-based industrial operations, amines are more commonly known for their distinct fishy odor. In composting, amines are a by-product of microbial decomposition and generally form during anaerobic fermentation. Aromatics are usually present in biosolids and are volatilized during aeration. When woodchips are used as the bulking agent in the biosolids mix, aromatics are produced during aerobic composting as the lignin (in woodchips) breaks down. Likewise, terpenes (which are products of wood) are also present in compost piles that use woodchips as the bulking agent.

Probably, most wastewater specialists have not been exposed to hydrogen sulfide and its characteristic rotten egg odor. Under normal circumstances, when biosolids are received at the composting site, any hydrogen sulfide emissions are quickly reduced when the biosolids and bulking

agent are mixed and formed into aerobic piles. However, there can be a problem with hydrogen sulfide emissions if the mix is incorrect or if the biosolids are too wet. When the biosolids are wet, they tend to form into clumps. These clumps can become anaerobic and will form and release hydrogen sulfide.

Whether described as "stinking like a skunk" or smelling like "decayed cabbage," organic sulfurs are generally present in all biosolids-derived composting piles. Of the various organic sulfur compounds found in compost piles probably the best known is methyl mercaptans (which smells like decayed cabbage). Fatty acids are generally produced under anaerobic conditions and do not add to odor generation problems unless the pile is allowed to go anaerobic.

Bottom Line on Composting Odor Control

At a biosolids composting facility, any sensible odor control management plan must consider all the areas and components of the composting process that might cause odors to be generated. While it is true that most odor problems are generated in the composting and curing process air systems, it is also true that, in enclosed composting operations, odors generated from ancillary processes within the enclosure must be considered. Moreover, enclosed systems must have a way to control or scrub airflow within the structure prior to its release to the outside environment. Usually, not all process areas at a composting facility are enclosed. Keep in mind that these open areas, e.g., biosolids handling and mixing areas, can also cause odor control problems.

Lime Stabilization

Lime or alkaline stabilization can achieve the minimum requirements for both Class A (no detectable pathogens) and Class B (a reduced level of pathogens) biosolids with respect to pathogens, depending on the amount of alkaline material added and other processes employed. Generally, alkaline stabilization meets the Class B requirements when the pH of the mixture of wastewater solids and alkaline material is at 12 or above after 2 h of contact.

Class A requirements can be achieved when the pH of the mixture is maintained at or above 12 for at least 72 h, with a temperature of 52°C maintained for at least 12 h during this time. In one process, the mixture is air-dried to over 50% solids after the 72-h period of elevated pH. Alternatively, the process may be manipulated to maintain temperatures at or above 70°F for 30 or more minutes while maintaining the pH requirement of 12. This higher temperature can be achieved by overdosing with lime (that is, adding more than is needed to reach a pH of 12), by using a supplemental heat source, or by using a combination of the two. Monitoring for fecal coliforms or *Salmonella* sp. is required prior to release by the generator for use.

Materials that may be used for alkaline stabilization include hydrated lime, quicklime (calcium oxide), fly ash, lime and cement kiln dust, and carbide lime. Quicklime is commonly used because it has a high heat of hydrolysis (491 British thermal units) and can significantly enhance pathogen destruction. Fly ash, lime kiln dust, or cement kiln dust are often used for alkaline stabilization because of their availability and relatively low cost.

The alkaline-stabilized product is suitable for application in many situations, such as landscaping, agriculture, and mine reclamation. The product serves as a lime substitute, a source of organic matter, and a specialty fertilizer. The addition of alkaline stabilized biosolids results in more favorable conditions for vegetative growth by improving soil properties such as pH, texture, and water-holding capacity. Appropriate applications depend on the needs of the soil and crops that will be grown, as well as the pathogen classification. For example, a Class B material would not be suitable for blending in a topsoil mix intended for use in home landscaping but is suitable for agriculture, mine reclamation, and landfill cover where the potential for contact with the pulse is lower and access can be restricted. Class A alkaline-stabilized biosolids are useful in agriculture and as a topsoil blend ingredient. Alkaline-stabilized biosolids provide pH adjustment, nutrients, and organic matter, reducing reliance on other fertilizers.

Alkaline-stabilized biosolids are also useful as daily landfill cover. They satisfy the federal requirement that landfills must be covered with soil or soil-like material at the end of each day (40 CFR 258). In most cases, lime-stabilized biosolids are blended with other soil to achieve the proper consistency for daily cover.

As previously mentioned, alkaline-stabilized biosolids are excellent for land reclamation in degraded areas, including acid mine spills or mine tailings. Soil conditions at such sites are very unfavorable for vegetative growth, often due to acid content, lack of nutrients, elevated levels of heavy metals, and poor soil texture. Alkaline-stabilized biosolids help to remedy these problems, making conditions more favorable for plant growth and reducing erosion potential. In addition, once a vegetative cover is established, the quality of mine drainage improves.

Advantages and Disadvantages

Alkaline stabilization offers several advantages, including:

- Consistency with the EPA's national beneficial reuse policy. It results in a product suitable for a variety of uses and is usually able to be sold.
- Simple technology requiring few special skills for reliable operation.
- Easy to construct with readily available parts.
- Small land area required.
- Flexible operation that is easily started and stopped.

Several possible disadvantages should be considered in evaluating this technology:

- The resulting product is not suitable for use on all soil. For example, alkaline soils common in southwestern states will not benefit from the addition of a high pH material.
- The volume of material to be managed and moved off-site is increased by approximately 15%–50% in comparison with other stabilization techniques, such as digestion. This increased volume results in higher transportation costs when material is moved off-site.
- There is potential for odor generation at both the processing and end-use sites.
- There is a potential for dust production.
- There is a potential for pathogen regrowth if the pH drops below 9.5 while the material is stored prior to use.
- The nitrogen content in the final product is lower than that in several other biosolids products. During processing, nitrogen is converted to ammonia, which is lost to the atmosphere through volatilization. In addition, plant-available phosphorus can be reduced through the formation of calcium phosphate.
- There are fees associated with proprietary processes (Class A stabilization).

Thermal Treatment

Thermal treatment (or wet air oxidation) subjects sludge to high temperature and pressure in a closed reactor vessel. The high temperature and pressure rupture the cell walls of any microorganisms present in the solids and cause chemical oxidation of the organic matter. This process substantially improves dewatering and reduces the volume of material for disposal. It also produces a very high-strength waste, which must be returned to the wastewater treatment system for further treatment.

Chlorine Oxidation

Chlorine oxidation also occurs in a closed vessel. In this process chlorine (100–1,000 mg/L) is mixed with a recycled solids flow. The recycled flow and process residual flow are mixed in the reactor. The solids and water are separated after leaving the reactor vessel. The water is returned to the wastewater treatment system, and the treated solids are dewatered for disposal. The main advantage of chlorine oxidation is that it can be operated intermittently. The main disadvantage is the production of extremely low pH and high chlorine content in the supernatant.

Stabilization Operation

Depending on the stabilization process employed, the operational components vary. In general, operations include pumping, observations, sampling and testing, process control calculations, maintenance, and housekeeping. The performance of the stabilization process will also vary with the type of process used. Generally, stabilization processes can produce a 40%–60% reduction of both volatile matter (organic content) and moisture.

Sludge Dewatering

Digested sludge removed from the digester is still mostly liquid. The primary objective of dewatering biosolids is to reduce moisture and, consequently, volume to a degree that will allow for economical disposal or reuse. Epstein and Alpert (1984) make the point that if the biosolids cake is higher in solids content, it reduces the need for space, fuel, labor, equipment, and the size of the receiving facility, e.g., a composting facility.

Probably one of the best summarizations of the various reasons why it is important to dewater biosolids is given by Metcalf & Eddy (1991) in the following: (1) The costs of transporting biosolids to the ultimate disposal site are greatly reduced when biosolids volume is reduced; (2) dewatered biosolids allow for easier handling; (3) dewatering biosolids (reduction in moisture content) allows for more efficient incineration; (4) if composting is the beneficial reuse choice, dewatered biosolids decrease the amount and, therefore, the cost of bulking agents; (5) with the USEPA's new 503 rule, dewatering biosolids may be required to render the biosolids less offensive; and (6) when landfilling is the ultimate disposal option, dewatering biosolids is required to reduce leachate production.

Again, the point being made here is that the importance of adequately dewatering biosolids for proper disposal/reuse can't be overstated.

The unit processes that are most often used for dewatering biosolids are (1) vacuum filtration, (2) pressure filtration, (3) centrifugation, and (4) drying beds. The solids content achievable by various dewatering techniques is shown in Table 18.21. The biosolids cake produced by common dewatering processes has a consistency similar to dry, crumbly bread pudding (Spellman, 1996). This non-fluid, dewatered, dry, crumbly cake product is easily handled, non-offensive, and can be land applied manually and by conventional agricultural spreaders (Outwater, 1994).

Dewatering processes are usually divided into natural air drying and mechanical methods (USEPA, 1997). Natural dewatering methods include those methods in which moisture is removed by evaporation and gravity or induced drainage, such as sand beds, biosolids lagoons, paved beds, Phragmites reed beds, vacuum-assisted beds,

Wedgewater beds, and dewatering via freezing. These natural dewatering methods are less controllable than mechanical dewatering methods but are typically less expensive. Moreover, these natural dewatering methods require less power because they rely on solar energy, gravity, and biological processes as the source of energy for dewatering. Mechanical dewatering processes include pressure filters, vacuum filters, belt filters, and centrifuges. The aforementioned air-drying and mechanical dewatering processes will be discussed in greater detail later in this text.

Biosolids Characteristics Affecting Dewatering

Vesilind (1980) makes the following point: "Sludge [biosolids] is composed of diverse solid particles suspended in an impure water continuum. Attempts to characterize sludge [biosolids] 'particles' have been hampered by the fact that they are dynamic—dispersing and reforming, depending on biological, chemical, and physical conditions" (p.43). Now that Vesilind's characterization of biosolids particles has been given its proper emphasis as one of the main factors affecting dewatering, it is the water portion of biosolids that will begin the following discussion of biosolids characteristics, to be followed by the other factors that affect dewatering.

Water contained in the biosolids particle exists in four phases: free water, colloidal water, intercellular water, and capillary water (Outwater, 1994). According to Vesilind (1980), free, or bulk, water is not associated with and not influenced by suspended solids; it can be easily separated and removed from biosolids by gravity. To remove colloidal and capillary water, biosolids must be chemically conditioned first, and then mechanical methods, such as centrifugation or belt presses, are used to remove the water. **Intercellular** water is much more difficult to remove. For intercellular water to be separated from the biosolids particle, the cell structure must first be broken. This is usually accomplished by thermal treatment. Basically, thermal treatment is accomplished with either direct or indirect drying.

As related earlier, the degree to which biosolids may be dewatered depends on several factors. For example, factors such as the source of biosolids and their prior treatment can change biosolids characteristics prior to dewatering and thus change its ability to be dewatered. Several characteristics can be used to define the ability of biosolids to be dewatered. It should be mentioned, however, that due to the sophisticated equipment and specialized training required, some of these characteristics are difficult to measure in most wastewater treatment plants, while others are readily measured with standard equipment available at most plants.

Characteristics Affecting Dewatering

It is important to note that the characteristics listed and described in the following sections relate to the difficulty in forcing biosolids particles (solids) closer together or to the difficulty of forcing water through the voids between the biosolids particles. Keep in mind that to facilitate the

TABLE 18.21
Solids Content of Dewatered Biosolids

Dewatering Method	Approximate Solids Content, %
Lagoons/Ponds	30
Drying beds	40
Filter press	35–45
Vacuum filtration	25
Standard centrifuge	20–25
High G/High solids centrifuge	25–40

dewatering process, biosolids are usually conditioned first by a variety of means. These biosolids conditioning processes basically involve the treatment of biosolids with various chemicals, thermal treatment processes, or blending processes. The biosolids characteristics that most significantly affect dewatering are as follows:

- pH
- volatile solids to fixed solids ratio
- septicity
- temperature
- compressibility
- particle size
- surface charge and hydration

These characteristics and their interrelationships are discussed in the following sections.

pH: pH affects the surface charge on biosolids particles. This characteristic is important because it determines the type of polymer to be used for conditioning. As a general rule of thumb, if the biosolids are conditioned with lime and have a high pH, **anionic** (having a negative charge) polymers are recommended for use. On the other hand, if the pH level is slightly above or below neutral, then **cationic** (having a positive charge) polymers are recommended.

Volatile Solids to Fixed Solids Ratio: As the percentage of fixed solids increases, assuming other factors are equal, biosolids are easier to dewater.

Septicity: When biosolids go septic, dewatering becomes more difficult and requires the use of more chemicals.

Temperature: The temperature of biosolids is indirectly related to the viscosity of the water present in the biosolids mass. That is, as the biosolids temperature increases, the viscosity of the water in the biosolids decreases. In centrifugation, in particular, viscosity is very important. This is the case because the terminal settling velocity during centrifugal acceleration varies according to an inverse linear relationship with the viscosity of the water. More will be said on this subject later.

Compressibility: Biosolids particles are compressible. This characteristic tends to deform and reduce the void space between particles. When this void space deformation and reduction occurs, the movement of water is inhibited which, in turn, reduces the rate of dewaterability.

Particle Size: The most important factor influencing the dewaterability of biosolids is particle size. The surface area for a given biosolids mass increases as the average particle size decreases. The effects of increasing the surface area include:

- Increased frictional resistance to the movement of water.
- Increased attraction of water to the particle surface due to more adsorption sites.
- Greater electrical repulsion between biosolids particles due to a larger area of negatively charged surface.

Particle size is directly influenced by prior treatment and the biosolids source.

Particle Surface Charge and Hydration: Biosolids particles have a negative surface charge and repel each other as they are forced together (remember the rule: like charges repel, while unlike charges attract). As the biosolids particles are forced more closely together the repulsive force increases dramatically. Water molecules are attracted to the surface of the biosolids particles, making dewatering more difficult. Along with the particle surface charge, the water of hydration must be considered. Water of hydration is chemically bound to the solids and is difficult to remove; great expenditures of thermal energy are required to remove it.

Sand Drying Beds

Sand beds have been used successfully for years to dewater sludge. Composed of a sand bed (consisting of a gravel base, underdrains, and 8–12 in of filter-grade sand), drying beds include an inlet pipe, splash pad, containment walls, and a system to return filtrate (water) for treatment. In some cases, the sand beds are covered to protect drying solids from the elements.

In operation, solids are pumped to the sand bed and allowed to dry by first draining off excess water through the sand and then by evaporation. This is the simplest and cheapest method for dewatering sludge. Moreover, no special training or expertise is required. However, there are sometimes downsides; namely, drying beds require a great deal of manpower to clean, can create odor and insect problems, and may cause sludge buildup during inclement weather.

According to Metcalf & Eddy (1991), four types of drying are commonly used in dewatering biosolids: (1) sand, (2) paved, (3) artificial media, and (4) vacuum-assisted. In addition to these commonly used dewatering methods, a few innovative methods of natural dewatering will also be discussed in this section. The innovative natural dewatering methods include experimental work on biosolids dewatering via freezing. Moreover, dewatering biosolids with aquatic plants, which have been tested and installed in several sites throughout the United States, will also be discussed.

Drying beds are generally used for dewatering well-digested biosolids. Attempting to air-dry raw biosolids is generally unsuccessful and may result in odor and vector control problems. Biosolids drying beds consist of a perforated or open-joint drainage system in a support medium, usually gravel, covered with a filter medium, usually sand, but it can consist of extruded plastic or wire mesh. Drying beds are usually separated into workable sections by wood, concrete, or other materials. Drying beds may be enclosed or open to the weather. They may rely entirely on natural drainage and evaporation processes or may use a vacuum to assist the operation (both types are discussed in the following sections).

Traditional Sand Drying Beds

This is the oldest biosolids dewatering technique and consists of 6–12 in of coarse sand underlain by layers of graded gravel ranging from 1/8 to 1/4 in at the top and 3/4–1-1/2 in at the bottom. The total gravel thickness is typically about 1 ft. Graded natural earth (4–6 in) usually makes up the bottom, with a web of drain tile placed on 20–30-ft centers. Sidewalls and partitions between bed sections are usually made of wooden planks or concrete and extend about 14 in above the sand surface (McGhee, 1991).

Large open areas of land are required for sand drying biosolids. For example, it is not unusual to have drying beds that are up to 125+ ft long and from 20 to 35 ft in width. Even at the smallest wastewater treatment plants, it is normal practice to provide at least two drying beds.

The actual dewatering process occurs as a result of two different physical processes: evaporation and drainage. The liquor that drains off the biosolids goes to a central sump, which pumps it back to the treatment process to undergo further treatment. The operation is very much affected by climate. In wet climates, it may be necessary to cover the beds with a translucent material that will allow at least 85% of the sun's ultraviolet radiation to pass through.

Typical loading rates for primary biosolids in dry climates range up to 200 kg/(m²×year) and from 60 to 125 kg/(m²×year) for mixtures of primary and waste activated biosolids.

When a drying bed is put into operation, it is generally filled with digested biosolids to a depth ranging from 8 to 12 in. The actual drying time is climate-sensitive; that is, drying can take from a few weeks to a few months, depending on the climate and the season. After dewatering, the biosolids solids content will range from about 20% to 35%, and, more importantly, the volume will have been reduced by up to 85%. Upon completion of the drying process, the dried biosolids are generally removed from the bed with handheld forks or front-end loaders. It is important to note that in the dried biosolids removal process, a small amount of sand is lost, and the bed must be refilled and graded periodically.

The bottom line: dried solids removed from a biosolids drying bed can be either incinerated or land-filled.

Drying Bed Operational Capacity

One of the primary considerations that must be considered when designing a biosolids drying bed is the determination of how many pounds of solids can be dried for every square foot of drying bed each year. To make this determination for the biosolids drying bed model described here, it is necessary to list certain parameters that will be needed to calculate the result.

The following example lists these vital parameters and shows the calculation that is necessary to determine the answer.

SIDEBAR 18.3 MODEL SAND DRYING BED

In this particular case, the biosolids drying bed is 40 ft long and 32 ft wide. The treated biosolids typically have a total solids concentration of 5% and fill the bed to a depth of 1 ft. If it takes an average of 20 days for the biosolids to dry and 1 day to remove the dried solids, the number of pounds of solids that can be dried for every square foot of drying bed area each year can be determined by performing the following calculation:

Calculation:

First, calculate the gallons of biosolids added to the drying bed.

$$40 \text{ ft} \times 32 \text{ ft} \times 1 \text{ ft} \times \frac{7.48 \text{ gal}}{1 \text{ ft}^3} = 9{,}574.4 \text{ gal}$$

Next, calculate the pounds of solids added to the drying bed.

$$9{,}574.4 \text{ gal} \times \frac{5\%}{100\%} \times \frac{8.34 \text{ lb}}{1 \text{ gal}} = 3{,}992.5 \text{ lb}$$

For the 21-day cycle, calculate pounds dried each day.

$$\frac{3{,}992.5 \text{ lb}}{21 \text{ day}} = 190.1 \text{ lb/day}$$

For an entire year, calculate the total pounds generated.

$$\frac{190.1 \text{ lb}}{1 \text{ day}} \times \frac{365 \text{ day}}{1 \text{ year}} = 69{,}386.5 \text{ lb/year}$$

Determine what this is for each square foot of bed space.

$$\frac{69{,}386.5 \text{ lb}}{1 \text{ year}} \times \frac{1}{(40 \text{ ft} \times 32 \text{ ft})} = 54.2 \text{ lb/year/ft}^2$$

The treatment plant model used in this example is equipped with biosolids drying beds for the drying of digested biosolids withdrawn from the digesters. When it is necessary to waste some of the digested biosolids in these beds, it is necessary to open the valves for discharge to the particular bed that is to be used. The digested biosolids will be forced from the digester through the 4-in biosolids draw-off line by gravity.

Before the digested biosolids are distributed to a biosolids drying bed, it is necessary to have the sand bed prepared for the best drying results. This preparation of the bed should include the following:

1. Remove all old biosolids as soon as convenient after they have dewatered sufficiently. Well-dewatered biosolids have a cracked, sponge appearance when squeezed and can be readily forked. Its moisture content should be no more than about 70%.
2. Digested or undigested biosolids should never be added to a bed already covered with biosolids.

3. Remove all weeds and other vegetation that might be present. Herbicides or other types of weed killers may be used if growths are extensive. Hand-weeding is also satisfactory.

4. The sand bed should be leveled and scarified using rakes, spikes, or a spring-tooth harrow. Re-leveling after scarification is recommended just prior to adding the digested biosolids. This will reduce the surface compaction of the sand and improve its filtering ability. When necessary, clean, coarse sand should be added to maintain approximately the same depth at all times.

After the biosolids drying bed has been properly prepared, the digested biosolids from the digester can be applied. This procedure of withdrawal should begin by opening the withdrawal valves wide open to the bed to be used. This valve should be left wide open for only a short period to allow the biosolids pipeline from the digester to the bed to be cleared of any grit or compacted material. Well-compacted material, including grit, at the inlet end of the line, may make rodding or back-flushing necessary. Once full flow is established, the valve should be closed just enough to maintain a constant flow. This valve adjustment should prevent the lighter, watery biosolids from coming to the withdrawal pipe inlet. In addition, it will prevent scouring of the bed surface.

Biosolids should be deposited on the bed to a depth not greater than 8–10 in; however, experience will usually establish the most effective depth of biosolids application consistent with the biosolids characteristics and local weather conditions. Normally, with good drying conditions, well-digested biosolids at this depth should dewater sufficiently to be ready for removal in 1 to 2 weeks. If the biosolids are very high in solids, it may require up to 3 weeks or longer unless they are deposited at a lesser depth.

Another operating feature that is necessary to control when withdrawing biosolids from the digester is the mixing of the digester contents. It will be necessary usually to discontinue mixing prior to withdrawal. The length of time should be neither too long nor too short before withdrawal. The best time will be determined from experience.

After the biosolids have been applied to the drying bed surface, the best time for removal depends on:

1. Suitability of the subsequent disposal method;
2. The need to remove the next batch of biosolids from the digesters; and
3. The moisture content of the biosolids on the beds.

Biosolids cake can be removed from the beds by shovel or fork at a moisture content of 60%–70%. However, if it is allowed to dry until the moisture content is reduced to about 40%, it will weigh only about half to two-thirds as much and will handle better.

When the decision has been made to remove the dried biosolids, the best method should be used to reduce manual labor requirements. When it is necessary to employ manual labor for removal, one of the best tools to use is a shovel-like fork with several tines an inch or so apart, such as an ensilage, coal, or stone fork. Using a fork allows biosolids to be removed with much less loss of sand than with a shovel. Nevertheless, some sand will cling to the bottom of the biosolids cake, and eventually, this sand will need to be replaced as mentioned previously.

After the operator forks or shovels the dried cake into a wheelbarrow and hauls the biosolids from the bed, it can either be hauled away to land disposal immediately or stockpiled for a year or two before land disposal. The advantage of stockpiling is that the biosolids will be more easily pulverized and therefore better for disposal on agricultural lands.

Paved Drying Beds

The main reason for using paved drying beds is that they alleviate the problem of mechanical biosolids removal equipment damaging the underlying piping networks. The beds are paved with concrete or asphalt and are generally sloped toward the center where a sump-like area with underlying pipes is arranged. These dewatering beds, like biosolids lagoons, depend on evaporation for the dewatering of the applied solids. Paved drying beds are usually rectangular with a center drainage strip and can be heated via buried pipes in the paved section. They are generally covered to prevent rain incursion.

In this type of natural dewatering, solids contents of 45%–50% can be achieved within 35 days in dry climates under normal conditions (McGhee, 1991). The operation of paved drying beds involves applying the biosolids to a depth of about 12 in. The settled surface area is routinely mixed by a special vehicle-mounted machine that is driven through the bed. Mixing is important because it breaks up the crust and exposes wet surfaces to the environment. The supernatant is decanted like biosolids lagoons. Biosolids loadings in relatively dry climates range from about 120 to 260 kg/(m²/year). High capital costs and larger land requirements than for sand beds are the two major disadvantages of paved drying beds.

In attempting to determine the bottom area dimensions of a paved drying bed, Metcalf & Eddy (1991) recommend computation by trial using the following equation:

$$A = \frac{1.04\, S\left[(1-S_d)/S_d - (1-S_e)/S_e\right] + (62.4)(P)(A)}{(62.4)(K_e)(E_p)}$$

where

A = bottom area of paved bed, ft².
S = annual biosolids production, dry solids, lb
S_d = percent dry solids in the biosolids after decanting, as a decimal
S_e = percent dry solids required from final disposal, as a decimal
P = annual precipitation, ft

K_e = reduction factor for evaporation from biosolids versus a free water surface.

   Use 0.6 for a preliminary estimate; pilot test to determine factor for final design

E_p = free water pan evaporation rate for the area, ft/year

Although the construction and operation methodologies for biosolids drying beds are well-known and widely accepted, this is not to say that the wastewater industry has not attempted to incorporate further advances into their construction and operation. For example, in an attempt to reduce the amount of dewatered biosolids that must be manually removed from drying beds, attempts have been made to construct drying beds in a specific manner whereby they can be planted with reeds; namely, the *Phragmites communis* variety (to be covered in greater detail later). The intent of augmenting the biosolids drying bed with reeds is to effect further desiccation. Moreover, tests have shown that the plants extend their root systems into the biosolids mass. This extended root system has the added benefit of helping to establish a rich microflora, which eventually feeds on the organic content of the biosolids. It is interesting to note that normal plant activity works to keep the system aerobic.

Artificial Media Drying Beds

The first artificial media drying beds, developed in England in 1970, used a stainless steel medium called Wedgewire. Later, as the technology advanced, the stainless steel fine wire screen mesh (Wedgewire) was replaced with a high-density polyurethane medium (Wedgewater). Polyurethane is less expensive than the stainless steel medium but has a shorter life expectancy.

Wedgewire beds are similar in concept to vacuum-assisted drying beds (to be described later). The medium used in Wedgewire beds consists of a septum with wedge-shaped slots about 0.01 in. wide. Initially, the bed is filled with water to a level above the wire screen. Chemically conditioned biosolids are then added and, after a brief holding period, are allowed to drain through the screen. Since excess water cannot return to the biosolids through capillary action, the biosolids dewater faster with this process (McGhee, 1991; Corbitt, 1990).

Advantages

1. Clogging is reduced;
2. Drainage is constant and rapid;
3. Throughput is greater than sand beds; and
4. This type of drying bed is relatively easy to maintain.

Disadvantages

The principal disadvantage is the capital costs, which are higher than sand or paved drying beds.

The Wedgewater method can normally dewater between 0.5 and 1.0 lb/ft^2) of dry matter per charge with the loading rate depending on the initial solids concentration of the waste biosolids applied (WEF, 1995). Another disadvantage of this method is that the biosolids to be removed is still relatively wet (about 10% dry solids) which may create disposal problems.

Wedgewater beds work on the same principle as sand drying beds but, instead of sand, they use interlocked high-density polyurethane 12 in^2 tiles that are formed for installation on a prefabricated sloped bed. Each polyurethane panel creates a capillary action that drains the water quickly.

The Wedgewater system, when compared to conventional drying beds, has several advantages. For example, with a loading capacity of about 2 lbs of dry solids per square foot, the Wedgewater method exceeds the capacity of sand drying beds by a 2–1 margin. Moreover, under favorable conditions, the biosolids content reaches approximately 18% in about 3 days, compared to more than 3 weeks on a sand drying bed (Outwater, 1994). In addition, it is interesting to note that Wedgewater beds require less surface area (about 1/16 as much) than conventional sand drying beds. Metcalf & Eddy (1991) cite additional advantages for this method of dewatering. For example, these units can be easily cleaned and their filtrate normally contains low suspended solids.

To receive optimum operation from Wedgewater beds, it is important to initially flood the bed to a level just above the polyurethane media surface. Then, the chemically conditioned biosolids are applied. In the initial stages of operation, the operator carefully controls the drainage rate. The purpose of this initial operating step is to ensure the establishment of a hydraulic continuum from the top of the biosolids bed to the bottom of the bed. Providing such a saturated profile will quicken water flow through the bed more effectively than when biosolids are applied to an unsaturated (dry) surface.

Vacuum-Assisted Drying Beds

For small plants that process small quantities of biosolids and have limited land area, vacuum-assisted drying beds may be the preferred method of dewatering biosolids. Vacuum-assisted drying beds normally employ a small vacuum to accelerate the dewatering of biosolids applied to a porous medium plate. This porous medium is set above an aggregate-filled support underdrain which, as the name implies, drains to a sump. The small vacuum is applied to this underdrain, working to extract free water from the biosolids. With biosolids loadings of less than 10 kg/m^2/cycle, the time required to dewater conditioned biosolids is about 1 day (McGhee, 1991). Using this method of dewatering, it is possible to achieve a solids content of >30%, although 20% solids is a more normal expectation.

Removal of dewatered biosolids is usually accomplished with mechanized machinery such as front-end loaders. Once the solids have been removed, it is important to wash the surface of the bed with high-pressure hoses to ensure residuals are removed. The main advantage of this dewatering method is the reduced time needed for dewatering, which reduces the effects of inclement climatic conditions on biosolids drying. The main disadvantage of this type of dewatering may be its dependence on adequate chemical conditioning for successful operation (Metcalf & Eddy, 1991).

Natural Methods of Dewatering Biosolids

Two of the innovative methods of natural biosolids dewatering are discussed in this section: (1) dewatering via freezing and (2) dewatering using aquatic plants.

Note: Dewatering via freezing is primarily in the pilot study stage. Experimental work in this area has led to pilot plants and has yielded various mathematical models, but no full-scale operations (Outwater, 1994).

Freeze-Assisted Drying Beds: In freeze-assisted drying, low winter temperatures accelerate the dewatering process. Freezing biosolids labors to part the water from the solids. The free water drains quickly when the granular mass is thawed. It is not unusual to attain a solids concentration greater than 25% when the mass thaws and drains.

Determining the feasibility of freezing biosolids in a particular area depends on the depth of frost penetration. To determine the maximum depth of frost penetration for an area consult published sources or local records.

In attempting to calculate the depth of frost penetration, it may be helpful to use the following equation for 3 in (75-mm) layers (McGhee, 1991):

$$Y = 1.76\,F_p - 101$$

where

Y = total depth of biosolids, cm
F_p = the maximum depth of frost penetration, cm

In this example, the biosolids are applied in 75-mm layers. As soon as the first layer is frozen, another layer is applied. The goal is to fill the bed with biosolids by the end of winter. It must be pointed out that in using this layered method of freeze-dewatering biosolids, it is important to ensure that each layer is frozen before the next is applied. Moreover, any snow or debris that falls should be removed from the surface; if not, it will serve to insulate the biosolids. Outwater (1994) points out that "to ensure successful performance at all times, the design should be based on the warmest winter in the past 20 years and on a layer thickness which will freeze in a reasonable amount of time if freeze-thaw cycles occur during the winter" (p. 86). Another rule of thumb to use in deciding whether or not biosolids freezing is a viable dewatering option is that biosolids freezing is unlikely to be a practical concept unless frost penetration is assured.

Dewatering Using Aquatic Plants: Using aquatic plants to dewater biosolids was developed in Germany in the 1960s. In this first project, reed beds were constructed in submerged wetlands. To date, although considered to be an innovative dewatering technology without specific design criteria, several hundred systems are operating in various sectors of the globe. In the reed bed system, a typical sand drying bed for biosolids is modified. Instead of removing the dewatered biosolids from the beds after each application, *Phragmites communis* (reeds) are planted in the sand. For the next several years (5–10 years), biosolids are added and then the beds are emptied.

In the reed bed operation, biosolids are spread on the surface of the bed via troughs or gravity-fed pipes. When the bed is filled to capacity, about 4 in of standing liquid will remain on the surface until it evaporates or drains down through the bed, where tile drains return it to the treatment process.

In an aquatic reed bed, the reeds perform the important function of developing (as described earlier) a rich microflora near the root zone, which feeds on the organic material in biosolids. **Phragmites** reeds are particularly suitable for this application because they are resistant to biosolid contaminants. Although the roots penetrate the finer gravel and sand, they do not penetrate through lower areas where the larger stones or fragments are located. This is important because root penetration to the lower bed levels could interfere with free drainage.

Another advantage of using *Phragmites* reeds in the aquatic plant drying bed is their growing pattern. *Phragmites* roots grow and extend themselves through rhizomes (defined as an underground horizontal stem, often thickened and tuber-shaped, and possessing buds, nodes, and scale-like leaves). From each rhizome, several plants branch off and grow vertically through the biosolids (Chambers et al., 1999). This vertical growth aids in dewatering by providing channels through which the water drains. The reeds also absorb some of this water, which is then given off to the atmosphere through evapotranspiration, resulting in the biosolids being reduced to about 97% solids. In addition to desiccating the biosolids deposits, the reeds function to cause broad mineralization.

Phragmites reed beds are operated year-round. In the fall, the reeds are harvested, leaving their root systems intact. Depending on contaminant concentrations, the harvested reeds, can be incinerated, land-filled, or composted.

It takes about 8 years to fill an average reed bed. When this occurs, it must be taken out of service and allowed to stand fallow for 6 months to a year. This fallow period allows for the stabilization of the top surface layer. The resulting biosolids product is dry and crumbles in the hands (it is friable). If contamination levels are within acceptable limits as per the EPA's 503 Rule, the dewatered biosolids product can be land-applied.

Operation of a Phragmites reed bed has limitations. For example, to ensure a successful dewatering operation, it is prudent to hire the services of an agronomist who is familiar with plant growth, care, and control of plant pests such as aphids. Moreover, this dewatering system is designed for regions that experience four distinct seasons, i.e., northern exposures. This is the case because Phragmites require a dormancy or wintering over period for proper root growth.

Additionally, reed beds are not suitable for large-scale operations for operational reasons and may also be cost-prohibitive due to the cost of land.

The jury is still out on how effective reed bed dewatering systems are. This is the case because the technology is relatively new, and none of the beds has been emptied yet; thus, it is difficult to predict the quality of the end product.

Stabilization Performance Factors

In sludge drying beds, various factors affect the length of time required to achieve the desired solids concentrations. The major factors and their impact on drying bed performance include the following:

- **Climate:** Drying beds in cold or moist climates will require significantly longer drying times to achieve an adequate level of percent solids concentration in the dewatered sludge.
- **Depth of Applied Sludge:** The depth of the sludge drawn onto the bed has a major impact on the required drying time. Deeper sludge layers require longer drying times. Under ideal conditions, a well-digested sludge drawn to a depth of approximately 8 in will require approximately 3 weeks to reach the desired 40%–60% solids.
- **Type of Sludge Applied:** The quality and solids concentration of the drying media will affect the time requirements.
- **Bed Cover:** Covered drying beds prevent rewetting of the sludge during storm events. In most cases, this reduces the average drying time required to reach the desired solids levels.

Operational Observations, Problems, and Troubleshooting Procedures

Although drying beds involve two natural processes—drainage and evaporation—that normally work well enough on their own, a certain amount of preparation and operator attention is still required to maintain optimum drying performance. For example, in the preparation stage, all debris is removed from the raked and leveled media surface. Then, all openings to the bed are sealed. After the bed is properly prepared, the sludge lines are opened, and sludge is allowed to flow slowly onto the media. The bed is filled to the desired operating level (8–12 in). The sludge line is closed and flushed, and the bed drain is opened. Water begins to drain. The sludge remains on the media until the desired % solids (40%–60%) is achieved. Later, the sludge is removed. In most operations, manual removal is required to prevent damage to the underdrain system. The sludge is disposed of in an approved landfill or by land application as a soil conditioner.

Operational Problems

In the operation of a sludge drying bed, the operator observes the operations, looks for various indicators of operational problems, and makes process adjustments as required.

1. **Sludge Takes a Long Time to De-Water**
   Causal Factors:
   - The applied sludge is too deep.
   - Sludge was applied to a dirty bed.
   - The drain system is plugged or broken.
   - Insufficient design capacity.
   - Inclement weather/poor drying conditions.

   Corrective Actions (Where Applicable):
   - Allow the bed to dry to minimum acceptable % solids and remove.
   - Use the described procedure below to determine the appropriate sludge depth.
     1. Clean the bed and apply smaller depth of sludge (i.e., 6–8 in).
     2. Measure the decrease in depth (drawdown) at the end of 3 days of drying.
     3. Use a sludge depth equal to twice the 3-day drawdown depth for future applications.
   - After the sludge has dried, remove the sludge and 0.5–1.0 in of sand. Add clean sand.
   - Allow sludge to dry to minimum allowable % solids and remove.
   - Use an external water source (with backflow prevention) to slowly flush underdrains.
   - Repair/replace underdrains as required.
   - Prevent damage to underdrains by draining during freezing weather.
   - Use polymer to increase bed performance.
   - Cover or enclose the beds.

2. **Influent Sludge is Very Thin**
   Causal Factor:
   - "Coning" is occurring in the digester.

   Corrective Action:
   - Reduce the rate of sludge withdrawal.

3. **Sludge Feed Lines Plug Frequently**
   Causal Factor:
   - Solids and/or grit are accumulating in time.

   Corrective Actions:
   - Open lines fully at the start of each withdrawal cycle.
   - Flush lines at the end of each withdrawal cycle.

4. **Flies Breeding in the Drying Sludge**
   Causal Factors:
   - Inadequately digested sludge
   - Natural insect reproduction

   Corrective Actions:
   - Break the sludge crust and apply a larvicide (borax).
   - Use insecticide (if approved) to remove adult insects.
   - Remove sludge as soon as possible.

5. **Objectionable Odors When Sludge is Applied to the Bed**
   Causal Factor:
   - Raw or partially digested sludge is being applied to the bed.

Corrective Actions:
- Add lime to the sludge to control odors and potential insect and rodent problems.
- Remove the sludge as quickly as possible.
- Identify and correct the digester problem.

ROTARY VACUUM FILTRATION

Rotary vacuum filters have also been used for many years to dewater sludge. The vacuum filter includes filter media (belt, cloth, or metal coils), media support (drum), vacuum system, chemical feed equipment, and conveyor belt(s) to transport the dewatered solids. In operation, chemically treated solids are pumped to a vat or tank in which a rotating drum is submerged. As the drum rotates, a vacuum is applied to the drum. Solids collect on the media and are held there by the vacuum as the drum rotates out of the tank. The vacuum removes additional water from the captured solids. When solids reach the discharge zone, the vacuum is released and the dewatered solids are discharged onto a conveyor belt for disposal. The media is then washed prior to returning to the start of the cycle.

Types of Rotary Vacuum Filters

The three principal types of rotary vacuum filters are rotary drum, coil, and belt. The *rotary drum* filter consists of a cylindrical drum rotating partially submerged in a vat or pan of conditioned sludge. The drum is divided length-wise into several sections that are connected through internal piping to ports in the valve body (plant) at the hub. This plate rotates in contact with a fixed valve plate with similar parts, which are connected to a vacuum supply, a compressed air supply, and an atmosphere vent. As the drum rotates, each section is thus connected to the appropriate service.

The *coil-type* vacuum filter uses two layers of stainless-steel coils arranged in a corduroy fashion around the drum. After a dewatering cycle, the two layers of springs leave the drum bed and are separated from each other so that the cake is lifted off the lower layer and discharged from the upper layer. The coils are then washed and reapplied to the drum. The coil filter is used successfully for all types of sludges; however, sludges with extremely fine particles or ones that are resistant to flocculation dewater poorly with this system.

The media on a *belt filter* leaves the drum surface at the end of the drying zone and passes over a small-diameter discharge roll to aid in cake discharge. Washing of the media occurs next. Then, the media are returned to the drum and the vat for another cycle. This type of filter normally has a small-diameter curved bar between the point where the belt leaves the drum and the discharge roll. This bar primarily aids in maintaining the belt's dimensional stability.

Filter Media

Drum and belt vacuum filters use natural or synthetic fiber materials. On the drum filter, the cloth is stretched and secured to the surface of the drum. In the belt filter, the cloth is stretched over the drum and through the pulley system. The installation of a blanket requires several days. The cloth will (with proper care) last several hundred to several thousand hours. The life of the blanket depends on the cloth selected, the conditioning chemicals, backwash frequency, and cleaning (i.e., acid bath) frequency.

Filter Drum

The filter drum is a maze of pipework running from a metal screen and wooden skeleton and connecting to a rotating valve port at each end of the drum. The drum is equipped with a variable speed drive to turn the drum from 1/8 to 1 rpm. Normally, solids pickup is indirectly related to the drum speed. The drum is partially submerged in a vat containing the conditioned sludge. Normally, submergence is limited to 1/5 or less of the filter surface at a time.

Chemical Conditioning

Sludge dewatered using vacuum filtration is normally chemically conditioned just prior to filtration. Sludge conditioning increases the percentage of solids captured by the filter and improves the dewatering characteristics of the sludge. However, conditional sludge must be filtered as quickly as possible after chemical addition to obtain these desirable results.

Operational Observation, Problems, and Troubleshooting Procedures

In operation, the rotating drum picks up chemically treated sludge. A vacuum is applied to the inside of the drum to draw the sludge onto the outside of the drum cover. This porous outercover, or filter medium, allows the filtrate or liquid to pass through into the drum while retaining the filter cake (dewatered sludge) on the medium. In the cake release/discharge mode, slight air pressure is applied to the drum's interior. Dewatered solids are lifted from the medium and scraped off by a scraper blade. The solids drop onto a conveyor to transport for further treatment or disposal. The filtrate water is returned to the plant for treatment.

While in operation, the operator observes drum speed, sludge pickup, filter cake thickness/appearance, chemical feed rates, sludge depth in the vat, and overall equipment operation. Sampling and testing are routinely performed on influent sludge solids concentration, filtrate BOD and solids, and sludge cake solids concentration. We cover the indicators/observations of vacuum filter operational problems and causal factors, along with recommended corrective actions in the following sections.

1. **High Solids in Filtrate**
   Causal Factors:
   - Improper coagulant dosage
   - Filter media binding
   Corrective Actions (Where Applicable):
   - Adjust coagulant dosage.
   - Recalibrate the coagulant feeder.
   - Clean synthetic cloth with steam and detergent.

- Clean steel coil with an acid bath.
- Clean cloth with water or replace cloth.

2. **Thin Filter Cake and Poor Dewatering**
   Causal Factors:
   - Filter media binding
   - Improper chemical dosage
   - Inadequate vacuum
   - Drum speed too high
   - Drum submergence too low

   Corrective Actions (Where Applicable):
   - Clean synthetic cloth with steam and detergent.
   - Clean steel cloth with an acid bath.
   - Clean cloth with water or replace cloth.
   - Adjust coagulant dosage.
   - Recalibrate the coagulant feeder.
   - Repair vacuum system.
   - Reduce drum speed.
   - Increase drum submergence.

3. **Vacuum Pump Stops**
   Causal Factors:
   - The power to drive the motor is off.
   - Lack of seal water.
   - Broken drive belt.

   Corrective Action (Where Applicable)
   - Reset heater, breaker, etc.; restart
   - Starts seal water flow.
   - Replace the drive belt.

4. **Drum Stops Rotating**
   Causal Factor:
   - The power to drive the motor is off.
     Corrective Action:
   - Reset heater, breaker, etc.; restart

5. **Receiver Vibrating**
   Causal Factors:
   - The filtrate pump is clogged.
   - Loose bolts and gasket around inspection plate.
   - Worn ball check valve in filtrate pump.
   - Air leaks in the suction line.
   - Dirty drum face.
   - Seal strips are missing.

   Corrective Actions (Where Applicable):
   - Clear pump.
   - Tighten bolts and gasket.
   - Replace ball check.
   - Seal leaks.
   - Clean the face with a pressure hose.
   - Replace missing seal strips.

6. **High Vat Level**
   Causal Factors:
   - Improper chemical conditioning.
   - The feed rate is too high.
   - The drum speed is too slow.
   - The filtrate pump is off or clogged.
   - The drain line is plugged.
   - Vacuum pump stopped.
   - Seal strips are missing.

Corrective Actions:
- Change coagulant dosage.
- Reduce the feed rate.
- Increase the drum speed.
- Turn on or clean the pump.
- Clean the drain line.
- Replace seal strips.

7. **Low Vat Level**
   Causal Factors:
   - The feed rate is too low.
   - The vat drain valve is open.

   Corrective Actions (Where Applicable):
   - Increase the feed rate.
   - Close the vat drain valve.

8. **Vacuum Pump Drawing High Amperage**
   Causal Factors:
   - The filtrate pump is clogged.
   - Improper chemical conditioning.
   - High vat level.
   - The cooling water flow to the vacuum pump is too high.

   Corrective Actions (where applicable):
   - Clear pump clog.
   - Adjust coagulant dosage.
   - Decrease cooling water flow rate.

9. **Scale Buildup on Vacuum Pump Seals**
   Causal Factor:
   - Hard, unstable water
   Corrective Action:
   - Add sequestering agent.

Process Control Calculations

Filter Yield (lb/h/ft²): Vacuum Filter

Probably the most frequent calculation vacuum filter operators have to make is determining filter yield. Example 18.84 illustrates how this calculation is made.

Example 18.84

Problem: Thickened, thermal condition sludge is pumped to a vacuum filter at a rate of 50 gpm. The vacuum area of the filter is 12 ft wide with a drum diameter of 9.8 ft. If the sludge concentration is 12%, what is the filter yield in lb/h/ft²? Assume the sludge weighs 8.34 lb/gal.

Solution: First, calculate the filter surface area.

$$\text{Area of a cylinder side} = 3.14 \times \text{Diameter} \times \text{Length}$$

$$= 3.14 \times 9.8 \text{ ft} \times 12 \text{ ft} = 369.3 \text{ ft}^2$$

Next, calculate the pounds of solids per hour:

$$\frac{50 \text{ gpm}}{1 \text{ min}} \times \frac{60 \text{ min}}{1 \text{ h}} \times \frac{8.34 \text{ lb}}{1 \text{ gal}} \times \frac{12\%}{100\%} = 3{,}002.4 \text{ lb/h}$$

Dividing the two:

$$\frac{3{,}002.4 \text{ lb/h}}{369.3 \text{ ft}^2} = 8.13 \text{ lb/h/ft}^2$$

Pressure Filtration

Pressure filtration differs from vacuum filtration in that the liquid is forced through the filter media by positive pressure instead of a vacuum. Several types of presses are available, but the most commonly used types are plate-and-frame presses and belt presses. *Filter presses* include the belt or plate-and-frame types. The belt filter includes two or more porous belts, rollers, and related handling systems for chemical makeup and feed, as well as supernatant and solids collection and transport.

The plate-and-frame filter consists of a support frame, filter plates covered with a porous material, a hydraulic or mechanical mechanism for pressing the plates together, and related handling systems for chemical makeup and feed, as well as supernatant and solids collection and transport. In the plate-and-frame filter, solids are pumped (sandwiched) between the plates. Pressure (200–250 psi) is applied to the plates and water is "squeezed" from the solids. At the end of the cycle, the pressure is released, and as the plates separate the solids drop out onto a conveyor belt for transport to storage or disposal. Performance factors for plate-and-frame presses include feed sludge characteristics, the type and amount of chemical conditioning, operating pressures, and the type and amount of precoat.

The belt filter uses a coagulant (polymer) mixed with the influent solids. The chemically treated solids are discharged between two moving belts. First, water drains from the solids by gravity. Then, as the two belts move between a series of rollers, pressure "squeezes" additional water out of the solids. The solids are then discharged onto a conveyor belt for transport to storage/disposal. Performance factors for the belt press include sludge feed rate, belt speed, belt tension, belt permeability, chemical dosage, and chemical selection.

Filter presses have lower operation and maintenance costs than vacuum filters or centrifuges. They typically produce a good quality cake and can be batch-operated. However, construction and installation costs are high. Moreover, chemical addition is required, and the presses must be operated by skilled personnel.

Operational Observations, Problems, and Troubleshooting

Most plate and filter press operations are partially/fully automated. The operation consists of observation, maintenance, sampling, and testing. Operation of belt filter presses consists of preparation of conditioning chemicals, chemical feed rate adjustments, sludge feed rate adjustments, belt alignment, belt speed, belt tension adjustments, sampling and testing, and maintenance. We include common operational problems, causal factors, and recommended corrective actions for the plate press and belt filter press in the following sections.

1. **Plate Press: Plates Fail to Seal**
   Causal Factors:
   - Poor alignment
   - Inadequate shimming

Corrective Actions (where applicable):
- Realign parts
- Adjust shimming of stay bosses.

2. **Plate Press: Cake Discharge Difficult**
   Causal Factors:
   - Inadequate precoat
   - Improper conditioning
   Corrective Actions (where applicable):
   - Increase precoat, feed at 25–40 psig.
   - Change conditioner type or dosage (use filter leaf test to determine).

3. **Plate Press: Filter Cycle Times Excessive**
   Causal Factors:
   - Improper conditioning
   - Feed solids are low
   Corrective Actions (Where Applicable):
   - Change chemical dosage.
   - Improve thickening operation.

4. **Plate Press: Filter Cake Sticks to Conveyors**
   Causal Factor:
   - Improper conditioning of chemical/dosage
   Corrective Action:
   - Increase inorganic conditioner dose.

5. **Plate Press: Precoat Pressures Gradually Increase**
   Causal Factors:
   - Improper sludge conditioning
   - Improper precoat feed
   - Filter media plugged
   - Calcium buildup in media
   Corrective Actions (where applicable):
   - Change chemical dosage
   - Decrease feed for a few cycles, then optimize.
   - Wash filter media.
   - Wash media with inhibited hydrochloric acid.

6. **Plate Press: Frequent Media Binding**
   Causal Factors:
   - Inadequate precoat
   - The initial feed rate is too high (no precoat)
   Corrective Actions (Where Applicable):
   - Increase precoat.
   - Reduce feed rate/develop initial cake slowly.

7. **Plate Press: Excessive Moisture in Cake**
   Causal Factors:
   - Improper conditioning
   - Filter cycle too short
   Corrective Actions (Where Applicable):
   - Change chemical dosage.
   - Lengthen the filter cycle.

8. **Plate Press: Sludge Blowing Out of Press**
   Causal Factor:
   - Obstruction between plates.
   Corrective Action:
   - Shut down the feed pump, hit press closure drive, restart the feed pump, and clean after cycle.

9. **Plate Press: Leaks around the Lower Faces of Plates**

Causal Factor:
- Wet cake soiling media on lower faces

Corrective Action:
- See Item 7 above.

10. **Belt Press: Filter Cake Discharge is Difficult**

Causal Factors:
- Wrong conditioning chemical selected
- Improper chemical dosage
- Changing sludge characteristics
- Wrong application point

Corrective Actions (Where Applicable):
- Change conditioning chemical.
- Adjust chemical dosage.
- Change chemical and/or sludge
- Adjust application point.

11. **Belt Press: Sludge Leaking from Belt Edges**

Causal Factors:
- Excessive belt tension.
- The belt speed is too low.
- Excessive sludge feed rate.

  Corrective Actions (Where Applicable):
- Reduce belt tension.
- Increase belt speed.
- Reduce sludge feed rate.

12. **Belt Press: Excessive Moisture in Filter Cake**

Causal Factors:
- Improper belt speed/drainage time
- Wrong conditioning chemical
- Improper chemical dosage
- Inadequate belt washing
- Wrong belt weave/material

Corrective Actions (Where Applicable):
- Adjust belt speed.
- Change conditioning chemical.
- Adjust chemical dosage.
- Clear spray nozzles/adjust sprays.
- Replace belt.

13. **Belt Press: Excessive Belt Wear along Edges**

Causal Factors:
- Roller misalignment
- Improper belt tension
- Tension/alignment control system

Corrective Actions (where applicable):
- Correct roller alignment.
- Correct tension.
- Repair tracking and alignment system controls.

14. **Belt Press: Belt Shifts or Seizes**

Causal Factors:
- Uneven sludge distribution
- Inadequate/uneven belt washing

Corrective Actions:
- Adjust feed for uniform sludge distribution.
- Clean and adjust belt-washing sprays.

Filter Press Process Control Calculations

As part of the operating routine for filter presses, operators are called upon to make certain process control calculations. The process control calculation most commonly used in operating the belt filter press determines the hydraulic loading rate on the unit. The most commonly used process control calculation in the operation of plate and filter presses determines the pounds of solids pressed per hour. Both of these calculations are demonstrated below.

Hydraulic Loading Rate: Belt Filter Press

Example 18.85

Problem: A belt filter press receives a daily sludge flow of 0.30 gal. If the belt is 60 in. wide, what is the hydraulic loading rate on the unit in gallons per minute for each foot of belt width (gpm/ft)?

Solution:

$$\frac{0.30 \text{ MG}}{1d} \times \frac{1,000,000 \text{ gal}}{1 \text{ MG}} \times \frac{1d}{1,440 \text{ min}} = \frac{208.3 \text{ gal}}{1 \text{ min}}$$

$$60 \text{ in} \times \frac{1 \text{ ft}}{12 \text{ in}} = 5 \text{ ft}$$

$$\frac{208.3 \text{ gal}}{5 \text{ ft}} = 41.7 \text{ gpm/ft}$$

Pounds of Solids Pressed Per Hour for Plate and Frame Press

Example 18.86

Problem: A plate and frame filter press can process 850 gal of sludge during its 120-min operating cycle. If the sludge concentration is 3.7% and the plate surface area is 140 ft², how many pounds of solids are pressed per hour for each square foot of plate surface area?

Solution:

$$850 \text{ gal} \times \frac{3.7\%}{100\%} \times \frac{8.34 \text{ lb}}{1 \text{ gal}} = 262.3 \text{ lb}$$

$$\frac{262.3 \text{ lb}}{120 \text{ min}} \times \frac{60 \text{ min}}{1 \text{ h}} = 131.2 \text{ lb/h}$$

$$\frac{131.2 \text{ lb/h}}{140 \text{ ft}^2} = 0.94 \text{ lb/h/ft}^2$$

CENTRIFUGATION

Centrifuges of various types have been used in dewatering operations for at least 30 years and appear to be gaining in popularity. Depending on the type of centrifuge used, in addition to centrifuge pumping equipment for solids feed and centrate removal, chemical makeup and feed equipment, and support systems for the removal of dewatered solids are required.

Operational Observations, Problems, and Troubleshooting

Generally, in operation, the centrifuge spins at a very high speed. The centrifugal force it creates "throws" the solids out of the water. Chemically conditioned solids are pumped into the centrifuge. The spinning action "throws" the solids to the outer wall of the centrifuge. The centrate (water) flows inside the unit to a discharge point. The solids held against the outer wall are scraped to a discharge point by an internal scroll moving slightly faster or slower than the centrifuge's speed of rotation. In the operation of the continuous feed solids bowl conveyor-type centrifuge (the most common type currently used), and other commonly used centrifuges, solid/liquid separation occurs as a result of rotating the liquid at high speeds to cause separation by gravity.

In the solid bowl type, the solid bowl has a rotating unit with a bowl and a conveyor. The unit has a conical section at one end that acts as a drainage device. The conveyor screw pushes the sludge solids to outlet ports and the cake to a discharge hopper. The sludge slurry enters the rotating bowl through a feed pipe leading into the hollow shaft of the rotating screw conveyor. The sludge is distributed through ports into a pool inside the rotating bowl. As the liquid sludge flows through the hollow shaft toward the overflow device, the fine solids settle into the wall of the rotating bowl. The screw conveyor pushes the solids to the conical section, where the solids are forced out of the water and the water drains back into the pool.

Expected % solids for centrifuge de-watered sludge are in the range of 10%–15%. The expected performance depends on the type of sludge being de-watered, as shown in Table 18.22.

Centrifuge operation depends on various performance factors.

TABLE 18.22
Expected % of Solids for Centrifuge Dewatered Sludges

Type of Sludge	% Solids
Raw sludge	25%–35%
Anaerobic digestion	15%–30%
Activated sludge	8%–10%
Heat treated	30%–50%

- Bowl design: length/diameter ratio; flow pattern
- Bowl speed
- Pool volume
- Conveyor design
- Relative conveyor speed
- Type/condition of sludge
- Type and amount of chemical conditioning
- Operating pool depth
- Relative conveyor speed (if adjustable)

Centrifuge operators often find that the operation of centrifuges can be simple, clean, and effluent. In most cases, chemical conditioning is required to achieve optimum concentrations. Operators soon discover that centrifuges are noisemakers; units run at very high speeds and produce high-level noise, which can cause loss of hearing with prolonged exposure. Therefore, when working in an area where a centrifuge is in operation, special care must be taken to provide hearing protection.

The actual operation of a centrifugation unit requires the operator to control and adjust chemical feed rates; observe the unit operation and performance; control and monitor centrate returned to the treatment system; and perform required maintenance as outlined in the manufacturer's technical manual.

The centrifuge operator must be trained to observe and recognize (as with other unit processes) operational problems that may occur with centrifuge operation. We cover several typical indicators and/or observations of centrifuge problems, along with causal factors and suggested corrective actions (troubleshooting procedures), in the following sections.

1. **Poor Centrate Clarity**
   Causal Factors:
   - The feed rate is too high
   - Wrong plate dam position
   - Worn conveyor flights
   - Speed too high
   - High-feed sludge solids concentration
   - Improper chemical conditioning
   Corrective Actions (Where Applicable):
   - Adjust sludge feed rate.
   - Increase pool depth.
   - Repair/replace conveyor.
   - Change the pulley setting to obtain a lower speed.
   - Dilute sludge feed.
   - Adjust chemical dosage.
2. **Solids Cake Not Dry Enough**
   Causal Factors:
   - The feed rate is too high
   - Wrong plate dam position
   - Speed too low
   - Excessive chemical conditioning
   - Influent too warm

Corrective Actions (If Applicable):
- Reduce sludge feed rate.
- Decrease pool depth to increase dryness.
- Change the pulley setting to obtain a higher speed.
- Adjust chemical dosage.
- Reduce influent temperature

3. **Torque Control Keeps Tripping**

Causal Factors:
- The feed rate is too high.
- Feed solids concentration is too high.
- Foreign material (i.e., tramp iron) in machine.
- The gear unit is misaligned.
- The gear unit has a mechanical problem.

Corrective Actions (Where Applicable):
- Reduce flows.
- Dilute flows.
- Remove conveyor/clear foreign materials.
- Correct gear unit alignment.
- Repair gear unit.

4. **Excess Vibration**

Causal Factors:
- Improper lubrication.
- Improper adjustment of vibration isolators.
- Discharge funnels contact centrifuges.
- The portion of conveyor flights may be plugged (causing an imbalance).
- Gearbox improperly aligned.
- Pillow box bearings are damaged.
- The bowl is out of balance.
- Parts not tightly assembled.
- Uneven wear on the conveyor.

Corrective Actions (Where Applicable):
- Lubricate according to the manufacturer's instructions.
- Adjust isolators.
- Reposition slip joints at funnels.
- Flush centrifuge.
- Align gearbox.
- Replace bearings.
- Return rotating parts to the factory for re-balancing
- Tighten parts.
- Resurface and re-balance.

5. **Sudden Increase in Power Consumption**

Causal Factors:
- Contact between bowl exterior and accumulated solids in case
- Effluent pipe plugged

Corrective Actions (Where Applicable):
- Apply hard surfacing to areas with wear.
- Clear solids discharge.

6. **Gradual Increase in Power Consumption**

Causal Factor:
- Conveyor blade wear

Corrective Action:
- Replace blades.

7. **Spasmodic Surging of Solids Discharge**

Causal Factors:
- The pool depth is too low.
- The conveyor helix is rough.
- Feed pipe too near drainage deck.
- Excessive vibration.

Corrective Actions (Where Applicable):
- Increase pool depth.
- Refinish the conveyor blade area.
- Move the feed pipe to the effluent end (if applicable).

8. **Centrifuge Shuts Down on Will Not Start**

Causal Factors:
- Blown fuses.
- The overload relay is tripped.
- Motor overheated/thermal protectors tripped.
- Torque control is tripped.
- The vibration switch is tripped.

Corrective Actions (Where Applicable):
- Replace fuses.
- Flush centrifuge and reset relay.
- Flush centrifuge, and reset thermal protectors.

SLUDGE INCINERATION

Not surprisingly, incinerators produce the maximum solids and moisture reductions. The equipment required depends on whether the unit is a multiple hearth or fluid-bed incinerator. Generally, the system will require a source of heat to reach ignition temperature, a solids feed system, and ash handling equipment. It is important to note that the system must also include all required equipment (e.g., scrubbers) to achieve compliance with air pollution control requirements. Solids are pumped to the incinerator. The solids are dried and then ignited (burned). As they burn, the organic matter is converted to carbon dioxide and water vapor, and the inorganic matter is left behind as ash or "fixed" solids. The ash is then collected for reuse or disposal.

Process Description

The incineration process first dries and then burns the sludge. The process involves the following steps:

1. The temperature of the sludge feed is raised to 212°F.
2. Water evaporates from the sludge.
3. The temperature of the water vapor and air mixture increases.
4. The temperature of the dried sludge volatile solids rises to the ignition point.

Note: Incineration will achieve maximum reductions if sufficient fuel, air, time, temperature, and turbulence are provided.

Incineration Processes

Multiple Hearth Furnace

The *multiple-hearth furnaces* consist of a circular steel shell surrounding several hearths. Scrappers (rabble arms) are connected to a central rotating shaft. Units range from 4.5 to 21.5 ft in diameter and have from 4 to 11 hearths. In operation, de-watered sludge solids are placed on the outer edge of the top hearth. The rotating rabble arms move them slowly to the center of the hearth. At the center of the hearth, the solids fall through ports to the second level. The process is repeated in the opposite direction. Hot gases are generated by burning on lower hearths dry solids. The dry solids pass to the lower hearths. The high temperature on the lower hearths ignites the solids. Burning continues to completion. Ash materials are discharged to lower cooling hearths where they are discharged for disposal. Air flow inside the center column and rabble arms continuously cool internal equipment.

Fluidized Bed Furnace

The *fluidized bed* incinerator consists of a vertical circular steel shell (reactor) with a grid to support a sand bed and an air system to provide warm air to the bottom of the sand bed. The evaporation and incineration processes take place within the superheated sand bed layer. In operation, air is pumped to the bottom of the unit. The airflow expands (fluidizes) the sand bed inside. The fluidized bed is heated to its operating temperature (1,200°F–1,500°F). Auxiliary fuel is added when needed to maintain the operating temperature. The sludge solids are injected into the heated sand bed. Moisture immediately evaporates. Organic matter ignites and reduces to ash. Residues are ground to fine ash by the sand movement. Fine ash particles flow up and out of the unit with exhaust gases. Ash particles are removed using common air pollution control processes. Oxygen analyzers in the exhaust gas stack control the airflow rate.

Note: Because these systems retain a high amount of heat in the sand, the system can be operated for as little as 4 h/day with little or no reheating.

Operational Observations, Problems, and Troubleshooting

The operator of an incinerator monitors various performance factors to ensure optimal operation. These performance factors include feed sludge volatile content, feed sludge moisture content, operating temperature, sludge feed rate, fuel feed rate, and air feed rate.

Note: To ensure that the volatile material is ignited, the sludge must be heated between 1,400°F and 1,700°F.

To be sure that operating parameters are in the correct range, the operator monitors and adjusts the sludge feed rate, airflow, and auxiliary fuel feed rate. All maintenance conducted on an incinerator should be in accordance with the manufacturer's recommendations.

OPERATIONAL PROBLEMS

The operator of a multiple hearth or fluidized bed incinerator must be able to recognize operational problems using various indicators or through observations. We discuss these indicators/observations, causal factors, and recommended corrective actions in the following sections.

1. **Multiple Hearth: Incinerator Temperature Too High**
   Causal Factors:
   - Excessive fuel feed rate
   - Greasy solids
   - Thermocouple burned out

   Corrective Actions (Where Applicable):
   - Decrease fuel feed rate.
   - Reduce sludge feed rate.
   - Increase air feed rate.
   - Replace thermocouple.

2. **Multiple Hearth: Furnace Temperature Too Low**
   Causal Factors:
   - The moisture content of the sludge has increased.
   - Fuel system malfunction.
   - Excessive air feed rate.
   - Flame out.

   Corrective Actions (Where Applicable):
   - Increase fuel feed rate until the dewatering operation improves.
   - Establish proper fuel feed rate.
   - Decrease air feed rate.
   - Increase sludge feed rate.
   - Relight furnace.

3. **Multiple Hearths: Oxygen Content of Stack Gas Too High**
   Causal Factors:
   - The sludge feed rate is too low.
   - Sludge feed system blockage.
   - The air feed rate is too high.

   Corrective Actions (Where Applicable).
   - Increase sludge feed rate.
   - Clear any feed system blockages.
   - Decrease air feed rate.

4. **Multiple Hearths: Oxygen Content of Stack Gas Too Low**
   Causal Factors:
   - The volatile or grease content of the sludge has increased.
   - The air feed rate is too low.

   Corrective Actions (Where Applicable):
   - Increase air feed rate.
   - Decrease sludge feed rate.
   - Increase air feed rate.

5. **Multiple Hearths: Furnace Refractories Deteriorated**
   Causal Factor:
   - Rapid startup/shutdown of furnace
   Corrective Actions:
   - Repair furnace refractories.
   - Follow specified startup/shutdown procedures.

6. **Multiple Hearths: Unusually High Cooling Effect**
   Causal Factor:
   - Air leak
   Corrective Action:
   - Locate and repair the leak.

7. **Multiple Hearths: Short Hearth Life**
   Causal Factor:
   - Uneven firing
   Corrective Action:
   - Fire hearths are equal on both sides.

8. **Multiple Hearths: Center Shaft Shear Pin Failure**
   Causal Factors:
   - Rabble's arm is dragging on the hearth.
   - Debris is caught under the arm.
   Corrective Actions (Where Applicable):
   - Adjust the rabble arm to eliminate rubbing.
   - Remove debris.

9. **Multiple Hearth: Scrubber Temperature Too High**
   Causal Factor:
   - Low water flow to scrubber
   Corrective Action:
   - Adjust water flow to the proper level.

10. **Multiple Hearth: Stack Gas Temperatures Too Low**
    Causal Factors:
    - Inadequate fuel feed supply
    - Excessive sludge feed rate
    Corrective Actions (Where Applicable):
    - Increase fuel feed rate.
    - Decrease sludge feed rate.

11. **Multiple Hearth: Stack Gas Temperatures Too High**
    Causal Factors:
    - Sludge has a higher volatile content (heat value).
    - Excessive fuel feed rate.
    Corrective Actions (Where Applicable):
    - Increase air feed rate.
    - Decrease sludge feed rate.
    - Decrease fuel feed rate.

12. **Multiple Hearth: Furnace Burners Slagging Up**
    Causal Factor:
    - Burner design
    Corrective Action:
    - Replace burners with newer designs that reduce slagging.

13. **Multiple Hearths: Rabble Arms Dropping**
    Causal Factors:
    - Excessive hearth temperatures
    - Loss of cooling air
    Corrective Actions (Where Applicable):
    - Maintain temperatures within the proper range.
    - Discontinue injection of scum into the hearth.
    - Repair the cooling air system immediately.

14. **Multiple Hearths: Excessive Air Pollutants in Stack Gas**
    Causal Factors:
    - Incomplete combustion—insufficient air
    - Air pollution control malfunction
    Corrective Actions (Where Applicable):
    - Raise air-to-fuel ratio.
    - Repair/replace broken equipment.

15. **Multiple Hearths: Flashing or Explosions**
    Causal Factor:
    - Scum or grease additions
    Corrective Action:
    - Remove scum/grease before incineration.

16. **Fluidized Bed: Bed Temperature Falling**
    Causal Factors:
    - Inadequate fuel supply
    - Excessive sludge feed rate
    - Excessive sludge moisture levels
    - Excessive airflow
    Corrective Actions (Where Applicable):
    - Increase fuel supply.
    - Repair fuel system malfunction.
    - Decrease sludge feed rate.
    - Correct sludge dewatering process problem.
    - Decrease airflow rate.

17. **Fluidized Bed: Low (<3%) Oxygen in Exhaust Gas**
    Causal Factors:
    - Low air flow rate
    - The fuel feed rate is too high
    Corrective Actions (Where Applicable):
    - Increase blower air feed rate.
    - Reduce fuel feed rate.

18. **Fluidized Bed Excessive (>6%) Oxygen in Exhaust Gas**
    Causal Factor:
    - The sludge feed rate is too low.
    Corrective Actions (Where Applicable):
    - Increase sludge feed rate.
    - Adjust the fuel feed rate to maintain a steady bed temperature.

19. **Fluidized Bed: Erratic Bed Depth on Control Panel**
    Causal Factor:
    - Bed pressure taps plugged with solids
    Corrective Actions (Where Applicable):
    - Tap a metal rod into the pressure tap pipe when the unit is not in operation.

- Apply compressed air to the pressure tap while the unit is in operation (follow the manufacturer's safety guidelines).

20. **Fluidized Bed: Preheat Burner Fails and Alarm Sounds**
Causal Factors:
- The pilot flame is not receiving fuel.
- The pilot flame is not receiving a spark.
- Defective pressure regulator(s).
- The pilot flame ignites but the flame scanner malfunctions.

Corrective Actions (Where Applicable):
- Correct fuel system problem.
- Replace defective parts.
- Replace defective regulator(s).
- Clear scanner sight glass.
- Replace defective scanner.

21. **Fluidized Bed: Bed Temperature Too High**
Causal Factors:
- The bed gun fuel feed rate is too high.
- Grease or high organic content in sludge (high heat value).

Corrective Actions (Where Applicable):
- Reduce bed gun fuel feed rate.
- Increase airflow rate.
- Decrease sludge fuel rate.

22. **Fluidized Bed: Bed Temperature Reads off the Scale**
Causal Factor:
- Thermocouple burned out

Corrective Action:
- Replace thermocouple

23. **Fluidized Bed: Scrubber Inlet Shows High Temperature**
Causal Factors:
- No water flowing in the scrubber.
- Spray nozzles are plugged.
- Ash water is not recirculating.

Corrective Actions (Where Applicable):
- Open valves to provide water.
- Correct system malfunction to provide required pressure.
- Clear nozzles and strainers.
- Repair/replace the recirculation pump.
- Unclog the scrubber discharge line.

24. **Fluidized Bed: Poor Bed Fluidization**
Causal Factor:
- Sand leakage through support plate during shutdown.

Corrective Actions (Where Applicable):
- Clear wind box.
- Clean the wind box at least once per month.

LAND APPLICATION OF BIOSOLIDS

The purpose of land application of biosolids is to dispose of the treated biosolids in an environmentally sound manner by recycling nutrients and soil conditioners. To be land-applied, wastewater biosolids must comply with state and federal biosolids management/disposal regulations. Biosolids must not contain materials that are dangerous to human health (i.e., toxicity, pathogenic organisms, etc.) or dangerous to the environment (i.e., toxicity, pesticides, heavy metals, etc.). Treated biosolids are land applied by either direct injection or application and plowing in (incorporation).

PROCESS CONTROL: SAMPLING AND TESTING

Land application of biosolids requires precise control to avoid problems. The quantity and quality of biosolids applied must be accurately determined. For this reason, the operator's process control activities include biosolids sampling/testing functions. Biosolids sampling and testing include the determination of % solids, heavy metals, organic pesticides and herbicides, alkalinity, total organic carbon (TOC), organic nitrogen, and ammonia nitrogen.

PROCESS CONTROL CALCULATIONS

As previously mentioned, process control calculations include determining disposal costs, plant-available nitrogen (PAN), application rates (dry tons and wet tons/acre), metals loading rates, maximum allowable applications based on metals loading, and site life based on metals loading.

DISPOSAL COST

The cost of disposal of biosolids can be determined by

$$\text{Cost} = \text{Wet tons/year} \times \% \text{ solids} \times \text{cost/dry ton} \quad (18.102)$$

Example 18.87

Problem: The treatment system produces 1,925 wet tons of biosolids for disposal each year. The biosolids are 16% solids. A contractor disposes of the biosolids for $28.00 per dry ton. What is the annual cost of sludge disposal?

Solution:

$$\text{Cost} = 1,925 \text{ wet tons/year} \times 0.16 \times \$28.00/\text{dry ton} = \$8,624$$

PLANT AVAILABLE NITROGEN (PAN)

One factor considered when land applying biosolids is the amount of nitrogen in the biosolids available to the plants grown on the site. This includes ammonia nitrogen and organic nitrogen. The organic nitrogen must be mineralized for plant consumption. Only a portion of the organic nitrogen is mineralized per year. The mineralization factor (f_1) is assumed to be 0.20. The amount of ammonia nitrogen available is directly related to the time elapsed between applying the biosolids and incorporating (plowing) the sludge into the soil. We provide volatilization rates based on the example below.

$$\text{Pan, lb/dry ton} = \left[\begin{pmatrix} \text{Org. Nit., mg/kg} \times f_1) + \\ (\text{Amm.Nit., mg/kg} \times V_1) \end{pmatrix} \right] \times 0.002 \text{ lb/dry ton}$$

(18.103)

where

> f_1 = Mineral rate for organic nitrogen (assume 0.20)
> V_1 = Volatilization rate ammonia nitrogen
> V_1 = 1.00 if biosolids are injected
> V_1 = 0.85 if biosolids are plowed in within 24 h
> V_1 = 0.70 if biosolids are plowed in within 7 days

Example 18.88

Problem: The biosolids contain 21,000 mg/kg of organic nitrogen and 10,000 mg/kg of ammonia nitrogen. The biosolids are incorporated into the soil within 24 h after application. What is the plant-available nitrogen (PAN) per dry ton of solids?

Solution:

$$\text{PAN, lb/dry ton} = \left[\begin{matrix} (21{,}000 \text{ mg/kg} \times 0.20) + \\ (10{,}000 \times 0.85) \end{matrix} \right] \times 0.002$$

$$= 25.4 \text{ lb PAN/dry ton}$$

APPLICATION RATE BASED ON CROP NITROGEN REQUIREMENT

In most cases, the application rate of domestic biosolids to crop lands will be controlled by the amount of nitrogen the crop requires. The biosolids application rate based upon the nitrogen requirement is determined by the following:

1. Using an agriculture handbook to determine the nitrogen requirement of the crop to be grown
2. Determining the amount of sludge in dry tons required to provide this much nitrogen

$$\text{Dry tons/acre} = \frac{\text{Plant Nitrogen Requirement, lb/acre}}{\text{Plant Available Nitrogen, lb/dry ton}}$$

(18.104)

Example 18.89

Problem: The crop to be planted on the land application site requires 150 lb of nitrogen per acre. What is the required biosolids application rate if the PAN of the biosolids is 25 lb/dry ton?

Solution:

$$\text{Dry tons/acre} = \frac{150 \text{ lb nitrogen/acre}}{26 \text{ lb/dry ton}} = 6 \text{ dry tons/acre}$$

METALS LOADING

When biosolids are land-applied, metal concentrations are closely monitored, and their loading on land application sites is calculated.

$$\text{Loading, lb/acre} = \text{Metal Conc., mg/kg} \times 0.002 \text{ lb/dry}$$

$$\text{ton} \times \text{Appl. Rate, dry tons/acre}$$

(18.105)

Example 18.90

Problem: The biosolids contain 14 mg/kg of lead. Biosolids are currently being applied to the site at a rate of 10 dry tons/acre. What is the metals loading rate for lead in pounds per acre?

Solution:

$$\text{Loading Rate, lb/acre} = 14 \text{ mg/kg} \times 0.002 \text{ lb/dry}$$

$$\text{ton} \times 10 \text{ dry tons} = 0.28 \text{ lb/acre}$$

MAXIMUM ALLOWABLE APPLICATIONS BASED UPON METALS LOADING

If metals are present, they may limit the total number of applications a site can receive. Metals loading is normally expressed in terms of the maximum total amount of metal that can be applied to a site during its use.

$$\text{Applications} = \frac{\begin{array}{c} \text{Max. Allowable Cumulative Load} \\ \text{for the metal, lb/acre} \end{array}}{\text{Metal Loading, lb/acre/application}}$$

(18.106)

Example 18.91

Problem: The maximum allowable cumulative lead loading is 48.0 lb/acre. Based on the current loading of 0.30 lb/acre, how many applications of biosolids can be made to this site?

Solution:

$$\text{Applications} = \frac{48.0 \text{ lb/acre}}{0.30 \text{ lb/acre}} = 160 \text{ applications}$$

SITE LIFE BASED ON METALS LOADING

The maximum number of applications based on metals loading and the number of applications per year can be used to determine the maximum site life.

$$\text{Site Life, years} = \frac{\text{Maximum Allowable Applications}}{\text{Number of Applications Planned Per Year}}$$

(18.107)

Example 18.92

Problem: Biosolids are currently applied to a site twice annually. Based on the lead content of the biosolids, the maximum number of applications is determined to be 120 applications. Based on the lead loading and the application rate, how many years can this site be used?

Solution:

$$\text{Site Life} = \frac{120 \text{ applications}}{2 \text{ applications/year}} = 60 \text{ years}$$

Note: When more than one metal is present, the calculations must be performed for each metal. The site life would then be the lowest value generated by these calculations.

Note: The following case study illustrates how the Clark County Water Reclamation Facility in Las Vegas, Nevada, developed a *Process Today's Sludge Today* policy. For those seeking a more in-depth treatment of this case study, it can be obtained from USEPA (2008).

PERMITS, RECORDS, AND REPORTS

Permits, records, and reports play a significant role in wastewater treatment operations. In fact, in regard to the "permit," one of the first things any new operator quickly learns is the importance of "making permit" each month. In this chapter, we briefly cover NPDES Permits and other pertinent records and reports with which the wastewater operator must be familiar.

Note: The discussion that follows is general in nature; it does not necessarily apply to any state in particular but instead provides an overview of permits, records, and reports that are an important part of wastewater treatment plant operations. For specific guidance on requirements for your locality, refer to your state water control board or other authorized state agency for information. In this handbook, the term "Board" signifies the state-reporting agency.

DEFINITIONS

There are several definitions that should be discussed prior to discussing the permit requirements for records and reporting. These definitions are listed below.

- **Average Monthly Limitation:** the highest allowable average over a calendar month, calculated by adding all of the daily values measured during the month and dividing the sum by number of daily values measured during the month.
- **Average Weekly Limitation:** the highest allowable average over a calendar week, calculated by adding all the daily values measured during the calendar week and dividing the sum by the number of daily values determined during the week.
- **Average Daily Limitation:** the highest allowable average over a 24-h period, calculated by adding all the values measured during the period and dividing the sum by the number of values determined during the period.
- **Average Hourly Limitation:** the highest allowable average for a 60-min period, calculated by adding all the values measured during the period and dividing the sum by the number of values determined during the period.
- **Daily Discharge:** means the discharge of a pollutant measured during a calendar day or any 24 h period that reasonably represents the calendar for the purpose of sampling. For pollutants with limitations expressed in units of weight, the daily discharge is calculated as the total mass of the pollutant discharged over the day. For pollutants with limitations expressed in other units, the daily discharge is calculated as the average measurement of the pollutant over the day.
- **Maximum Daily Discharge:** the highest allowable value for a daily discharge.
- **Effluent Limitation:** any restriction by the State Board on quantities, discharge rates, or concentrations of pollutants that are discharged from point sources into State waters.
- **Maximum Discharge:** the highest allowable value for any single measurement.
- **Minimum Discharge:** the lowest allowable value for any single measurement.
- **Point Source:** any discernible, defined, and discrete conveyance, including but not limited to any pipe, ditch, channel, tunnel, conduit, well, discrete fissure, container, rolling stock, vessel, or other floating craft, from which pollutants are or may be discharged. This definition does not include return flows from irrigated agricultural land.
- **Discharge Monitoring Report:** forms for reporting self-monitoring results of the permittee.
- **Discharge Permit:** State Pollutant Discharge Elimination System permit, which specifies the terms and conditions under which a point source discharge to State waters is permitted.

NPDES Permits

In the United States, all treatment facilities that discharge to State waters must have a discharge permit issued by the State Water Control Board or other appropriate State agency. This permit is known on the national level as the National Pollutant Discharge Elimination System (NPDES) permit and the state level as the (State) Pollutant Discharge Elimination System (state-PDES) permit. The permit states the specific conditions that must be met to legally discharge treated wastewater to State waters. The permit contains general requirements (applying to every discharger) and specific requirements (applying only to the point source specified in the permit). A general permit is a discharge permit that covers a specified class of dischargers. It is developed to allow dischargers within the specified category to discharge under specified conditions. All discharge permits contain general conditions. These conditions are standard for all dischargers and cover a broad series of requirements. Read the general conditions of the treatment facility's permit carefully. Permittees must retain certain records.

Monitoring

- Date, time, and exact place of sampling or measurements
- Name(s) of the individual(s) performing sampling or measurement
- Date(s) and time(s) analyses were performed
- Name(s) of the individuals who performed the analyses
- Analytical techniques or methods used
- Observations, readings, calculations, bench data, and results
- Instrument calibration and maintenance
- Original strip chart recordings for continuous monitoring
- Information used to develop reports required by the permit
- Data used to complete the permit application

Note: All records must be kept for at least 3 years (longer at the request of the State Board).

Reporting

Generally, reporting must be made under the following conditions/situations (requirements may vary depending on the state regulatory body with reporting authority):

- **Unusual or Extraordinary Discharge Reports:** must notify the Board by telephone within 24 h of the occurrence and submit a written report within 5 days. The report must include:
  1. Description of the non-compliance and its cause.
  2. Non-compliance date(s), time(s), and duration.
  3. Steps planned/taken to reduce/eliminate.
  4. Steps planned/taken to prevent reoccurrence.

- **Anticipated Non-Compliance:** Must notify the Board at least 10 days in advance of any changes to the facility or activity that may result in non-compliance.
- **Compliance Schedules:** Must report compliance or non-compliance with any requirements contained in compliance schedules no later than 14 days following the scheduled date for completion of the requirement.
- **24-H Reporting:** Any non-compliance which may adversely affect State waters or may endanger public health must be reported orally within 24 h of the time the permittee becomes aware of the condition. A written report must be submitted within 5 days.
- **Discharge Monitoring Reports (DMRs):** Reports self-monitoring data generated during a specified period (normally 1 month). When completing the DMR, remember:
  - More frequent monitoring must be reported
  - All results must be used to complete reported values
  - Pollutants monitored by an approved method but not required by the permit must be reported
  - No empty blocks on the form should be left blank
  - Averages are arithmetic unless noted otherwise
  - Appropriate significant figures should be used
  - All bypasses and overflows must be reported
  - The licensed operator must sign the report
  - A responsible official must sign the report
  - The department must receive the report by the 10th of next month

Sampling and Testing

The general requirements of the permit specify minimum sampling and testing that must be performed on the plant discharge. Moreover, the permit will specify the frequency of sampling, sample type, and length of time for composite samples. Unless a specific method is required by the permit, all sample preservation and analysis must follow the requirements outlined in the Federal Regulations *Guidelines Establishing Test Procedures for the Analysis of Pollutants Under the Clean Water Act* (40 CFR 136).

Note: All samples and measurements must be representative of the nature and quantity of the discharge.

Effluent Limitations

The permit sets numerical limitations on specific parameters contained in the plant discharge. Limits may be expressed as

- Average monthly quantity (kg/day)
- Average monthly concentration (mg/L)
- Average weekly quantity (kg/day)
- Average weekly concentration (mg/L)

- Daily quantity (kg/day)
- Daily concentration (mg/L)
- Hourly average concentration (mg/L)
- Instantaneous minimum concentration (mg/L)
- Instantaneous maximum concentration (mg/L)

Compliance Schedules

If the facility requires additional construction or other modifications to fully comply with the final effluent limitations, the permit will contain a schedule of events to be completed to achieve full compliance.

Special Conditions

Any special requirements or conditions set for the approval of the discharge will be contained in this section. Special conditions may include:

- Monitoring required to determine effluent toxicity
- Pretreatment program requirements

Licensed Operator Requirements

The permit will specify, based on the treatment system's complexity and the volume of flow treated, the minimum license classification required to be the designated responsible charge operator.

Chlorination/Dechlorination Reporting

Several reporting systems apply to chlorination or chlorination followed by dechlorination. It is best to review this section of the specific permit for guidance. If confused contact the appropriate State Regulatory Agency.

REPORTING CALCULATIONS

Failure to accurately calculate report data will result in violations of the permit. The basic calculations associated with completing the DMR are covered below.

Average Monthly Concentration

The average monthly concentration is the average of the results of all tests performed during the month.

$$AMC, mg/L = \frac{\sum Test_1 + Test_2 + Test_3 + \cdots + Test_n}{N \,(Tests \; during \; month)} \quad (18.108)$$

Average Weekly Concentration (AWC)

The average weekly concentration (AWC) is the result of all the tests performed during a calendar week. A calendar week must start on Sunday and end on Saturday and be completely within the reporting month. A weekly average is not computed for any week that does not meet these criteria.

$$AWC, mg/L = \frac{\sum Test_1 + Test_2 + Test_3 + \cdots + Test_n}{N \,(tests \; during \; calendar \; week)} \quad (18.109)$$

Average Hourly Concentration

The average hourly concentration is the average of all the test results collected during a 60-min period.

$$AHC, mg/L = \frac{\sum Test_1 + Test_2 + Test_3 + \cdots + Test_n}{N \,(tests \; during \; a \; 60\text{-minute} \; period)} \quad (18.110)$$

Daily Quantity (kg/day)

Daily quantity is the quantity of a pollutant in kilograms per day discharged during a 24-h period.

$$kg/day = Concentration, mg/L \times Flow, MGD \times 3.785 \; kg/MG/mg/L \quad (18.111)$$

Average Monthly Quantity (AMQ)

Average monthly quantity (AMQ) is the average of all the individual daily quantities determined during the month.

$$AMQ, kg/day = \frac{\sum DQ_1 + DQ_2 + DQ_3 + \cdots + DQ_n}{N \,(Tests \; during \; month)} \quad (18.112)$$

Average Weekly Quantity

The average weekly quantity is the average of all the daily quantities determined during a calendar week. A calendar week must start on Sunday and end on Saturday and be completely within the reporting month. A weekly average is not computed for any week that does not meet these criteria.

$$AWQ, kg/day = \frac{\sum DQ_1 + DQ_2 + DQ_3 + \cdots + DQ_n}{N \,(tests \; during \; calendar \; week)} \quad (18.113)$$

Minimum Concentration

The minimum concentration is the lowest instantaneous value recorded during the reporting period.

Maximum Concentration

Maximum concentration is the highest instantaneous value recorded during the reporting period.

Bacteriological Reporting

Bacteriological reporting is used for reporting fecal coliform test results. To make this calculation the geometric mean calculation is used and all monthly geometric means are computed using all the test values. Note that weekly geometric means are computed using the same selection criteria discussed for average weekly concentration and quantity calculations. The easiest method for making this calculation requires a calculator that can perform logarithmic (log) or Nth root functions.

$$\text{Geometric Mean} = \text{Antilog}\left[\frac{\text{Log } X_1 + \log_2 + \log X_3 + \cdots + \log X_n}{\text{N, Number of Tests}}\right]$$

(18.114)

or

$$\text{Geometric Mean} = \sqrt[n]{X_1} \times X_2 \times \cdots \times X_n$$

CHAPTER REVIEW QUESTIONS

18.1 Who must sign the DMR?
18.2 What does the COD test measure?
18.3 Give three reasons for treating wastewater.
18.4 Name two types of solids based on physical characteristics.
18.5 Define organic and inorganic.
18.6 Name four types of microorganisms, which may be present in wastewater.
18.7 When organic matter is decomposed aerobically, what materials are produced?
18.8 Name three materials or pollutants, which are not removed by the natural purification process.
18.9 What are the used water and solids from a community that flows to a treatment plant called?
18.10 Where do disease-causing bacteria in wastewater come from?
18.11 What does the term pathogenic mean?
18.12 What is wastewater called that comes from the household?
18.13 What is wastewater called that comes from industrial complexes?
18.14 The lab test indicates that a 500-g sample of sludge contains 22 g of solids. What are the percent solids in the sludge sample?
18.15 The depth of water in the grit channel is 28 in. What is the depth in feet?
18.16 The operator withdraws 5,250 gal of solids from the digester. How many pounds of solids have been removed?
18.17 Sludge added to the digester causes a 1,920 ft³ change in the volume of sludge in the digester. How many gallons of sludge have been added?
18.18 The plant effluent contains 30 mg/L solids. The effluent flow rate is 3.40 MGD. How many pounds per day of solids are discharged?
18.19 The plant effluent contains 25 mg/L BOD₅. The effluent flow rate is 7.25 MGD. How many kilograms per day of BOD₅ are being discharged?
18.20 The operator wishes to remove 3,280 lbs/day of solids from the activated sludge process. The waste-activated sludge concentration is 3,250 mg/L. What is the required flow rate in million gallons per day?
18.21 The plant influent includes an industrial flow, which contains 240 mg/L BOD. The industrial

flow is 0.72 MGD. What is the population equivalent for the industrial contribution in people per day?
18.22 The label of hypochlorite solution states that the specific gravity of the solution is 1.1288. What is the weight of 1 gal of the hypochlorite solution?
18.23 What must be done to the cutters in a comminutor to ensure proper operation?
18.24 What is grit? Give three examples of material, which is considered to be grit.
18.25 The plant has three channels in service. Each channel is 2 ft wide and has a water depth of 3 ft. What is the velocity in the channel when the flow rate is 8.0 MGD?
18.26 The grit from the aerated grit channel has a strong hydrogen sulfide odor upon standing in a storage container. What does this indicate and what action should be taken to correct the problem?
18.27 What is the purpose of primary treatment?
18.28 What is the purpose of the settling tank in the secondary or biological treatment process?
18.29 The circular settling tank is 90 ft in diameter and has a depth of 12 ft. The effluent weir extends around the circumference of the tank. The flow rate is 2.25 MGD. What is the detention time in hours, surface loading rate in gallons/day/ft² and weir overflow rate in gallons/day/foot?
18.30 Give three classifications of ponds based on their location in the treatment system.
18.31 Describe the processes occurring in a raw sewage stabilization pond (facultative).
18.32 How do changes in the season affect the quality of the discharge from a stabilization pond?
18.33 What is the advantage of using mechanical or diffused aeration equipment to provide oxygen?
18.34 Name three classifications of trickling filters and identify the classification that produces the highest quality effluent.
18.35 Microscopic examination reveals a predominance of rotifers. What process adjustment does this indicate is required?
18.36 Increasing the wasting rate will _____ the MLSS, _____ the return concentration, _____ the MCRT, _____ the F/M ratio, and _____ the SVI.
18.37 The plant currently uses 45.8 lbs of chlorine per day. Assuming the chlorine usage will increase by 10% during the next year, how many 2,000-lb cylinders of chlorine will be needed for the year (365 days)?
18.38 The plant has six 2,000-lb cylinders on hand. The current dose of chlorine being used to disinfect the effluent is 6.2 mg/L. The average effluent flow rate is 2.25 MGD. Allowing 15 days for ordering and shipment, when should the next order for chlorine be made?

18.39 The plant feeds 38 lbs of chlorine per day and uses 150-lb cylinders. Chlorine use is expected to increase by 11% next year. The chlorine supplier has stated that the current price of chlorine ($0.170/lb) will increase by 7.5% next year. How much money should the Town budget for chlorine purchases for the next year (365 days)?

18.40 The sludge pump operates 30 min every 3 h. The pump delivers 70 gpm. If the sludge is 5.1% solids and has a volatile matter content of 66%, how many pounds of volatile solids are removed from the settling tank each day?

18.41 The aerobic digester has a volume of 63,000 gal. The laboratory test indicates that 41 mg of lime were required to increase the pH of a 1-L sample of digesting sludge from 6.0 to the desired 7.1. How many pounds of lime must be added to the digester to increase the pH of the unit to 7.4?

18.42 The digester has a volume of 73,500 gal. Sludge is added to the digester at the rate of 2,750 gal/day. What is the sludge retention time in days?

18.43 The raw sludge pumped to the digester contains 72% volatile matter. The digested sludge removed from the digester contains 48% volatile matter. What is the % Volatile Matter Reduction?

18.44 What does NPDES stand for?

18.45 How can primary sludge be freshened going into a gravity thickener?

18.46 A neutral solution has what pH value?

18.47 Why is the seeded BOD test required for some samples?

18.48 What is the foremost advantage of the COD over the BOD?

18.49 High mixed liquor concentration is indicated by a _____ aeration tank foam.

18.50 What typically happens to the activity level of bacteria when the temperature is increased?

18.51 List three factors other than food that affect the growth characteristics of activated sludge.

18.52 What are the characteristics of facultative organisms?

18.53 BOD measures the amount of _____ material in wastewater.

18.54 The activated sludge process requires _____ _____ in the aeration tank to be successful.

18.55 The activated sludge process cannot be successfully operated with a _____ clarifier.

18.56 The activated biosolids process can successfully remove _____ BOD.

18.57 Successful operation of a complete mix reactor in the endogenous growth phase is _____.

18.58 The bacteria in the activated biosolids process are either _____ or _____.

18.59 Step feed activated biosolids processes have _____ mixed liquor concentrations in different parts of the tank.

18.60 An advantage of contact stabilization compared to complete mix is _____ aeration tank volume.

18.61 Increasing the _____ of wastewater increases the BOD in the activated biosolids process.

18.62 Bacteria need phosphorus to successfully remove _____ in the activated biosolids process.

18.63 The growth rate of microorganisms is controlled by the _____.

18.64 Adding chlorine just before the _____ _____ can control alga growth.

18.65 The purpose of the secondary clarifier in an activated biosolids process is:

18.66 The _____ growth phase should occur in a complete mix-activated biosolids process.

18.67 The typical DO value for activated biosolids plants is between _____ and _____ mg/l.

18.68 In the activated biosolids process, what change would an operator normally expect to make when the temperature decreases from 25°C to 15 °C?

18.69 In the activated biosolids process, what change must be made to increase the MLVSS?

18.70 In the activated biosolids process, what change must be made to increase the F/M?

18.71 What does the "Gould Sludge Age" assume to be the source of the MLVSS in the aeration tank?

18.72 What is one advantage of complete mix over plug flow?

18.73 The grit in the primary sludge is causing excessive wear on primary treatment sludge pumps. The plant uses an aerated grit channel. What action should be taken to correct this problem?

18.74 When the Mean Cell Residence Time (MCRT) increases, the mixed liquor-suspended solids (MLSS) concentration in the aeration tank:

18.75 Exhaust air from a chlorine room should be taken from where?

18.76 If chlorine costs $0.21/lb, what is the daily cost to chlorinate a 5-mgd flow rate at a chlorine feed rate of 2.6 mg/L?

18.77 What is the term that describes a normal aerobic system from which the oxygen has temporarily been depleted?

18.78 The ratio that describes the minimum amount of nutrients theoretically required for an activated sludge system is 100:5:1. What are the elements that fit this ratio?

18.79 A flotation thickener is best used for what type of sludge?

18.80 Drying beds are/are not (circle correct choice) an example of a sludge stabilization process.

18.81 The minimum flow velocity in collection systems should be:

18.82 What effect will the addition of chlorine, acid, alum, carbon dioxide, or sulfuric acid have on the pH of wastewater?

18.83 An amperometric titrator is used to measure:

18.84 The normal design detention time for the primary clarifier is:

18.85 The volatile Acids: Alkalinity ratio in an anaerobic digester should be approximately:

18.86 The surface loading rate in a final clarifier should be approximately:

18.87 In a conventional effluent chlorination system the chlorine residual measured is mostly in the form of:

18.88 For a conventional activated biosolids process, the Food: Microorganism (F/M) ratio should be in the range of:

18.89 Denitrification in a final clarifier can cause clumps of sludge to rise to the surface. The sludge flocs attach to small sticky bubbles of _____ gas.

18.90 An anaerobic digester is covered and kept under positive pressure to:

18.91 During the summer months, the major source of oxygen added to a stabilization pond is:

18.92 Which solids cannot be removed by vacuum filtration?

18.93 The odor recognition threshold for H_2S is reported to be as low as:

18.94 Explain the reasons for using ponds to treat wastewater.

18.95 Discuss the advantages and disadvantages of pond systems compared to bio-mechanical systems for wastewater treatment.

18.96 Describe the following types of ponds.

   A. Aerobic
   B. Anaerobic
   C. Aerated
   D. Facultative

18.97 List ways most ponds gain dissolved oxygen.

18.98 Explain why dissolved oxygen concentrations vary with pond depth.

18.99 List ways to measure the dissolved oxygen level of a pond.

18.100 Given data, calculate pond surface area in acres.

Given: Pond Length = 400 ft
Pond Width = 300 ft
Formula: (1 acre = 43,500 ft²)

$$\text{Area of Pond} = \text{Length (ft)} \times \text{Width (ft)} = \text{area in ft}^2$$

$$\text{Area of Pond (in acres)} = \text{area of pond} \div \text{area of 1 acre}$$

REFERENCES

Anderson, J.B., and Zwieg, H.P., 1962. Biology of waste stabilization ponds. *Southwest Water Works J*; 44(2):15–18.

Assenzo, J.R., and Reid, G.W., 1966. Removing nitrogen and phosphorus by bio-oxidation ponds in central Oklahoma. *Water Sewage Works*; 13(8):294–299.

Brockett, O.D., 1976. Microbial reactions in facultative ponds-1. The anaerobic nature of oxidation pond sediments. *Water Res*; 10(1):45–49.

Burnett, G.W., and Schuster, G.S., 1973. *Pathogenic Microbiology*. St Louis: C.V. Mosby Company.

Chambers, R.M., Meyerson, L.A., and Saltonstall, K., 1999. Expansion of *Phragmites australis* into tidal wetlands of North America. *Aquatic Bot*; 64:261–273.

Cheremisinoff, P.N., and Young, R.A., 1981. *Pollution Engineering Practice Handbook*. Ann Arbor, MI: Ann Arbor Science Publishers, Inc.

Corbitt, R.A., 1990. *Standard Handbook of Environmental Engineering*. New York: McGraw-Hill, Inc.

Craggs, R., 2005. Nutrients, in Pond Treatment Technology, Hilton, A., (ed.). London, UK: IWA Publishing.

Crites, R.W., Middlebrooks, E.J., and Reed, S.C., 2006. *Natural Wastewater Treatment Systems*. Boca Raton, FL: CRC, Taylor and Francis Group.

CWNS (Clean Watersheds Needs Survey). 2000. *Report to Congress*, EPA-832-R-10–002. Washington, DC.

Epstein, E., and Alpert, J.E., 1984. Sludge dewatering and compost economics. *BioCycle*; 25(10):31–34.

Epstein, E., and Epstein, J., 1989. Public health issues and composting. *BioCycle*; 30(8):50–53.

Gallert, C., and Winter, J., 2005. Bacterial Metabolism in Wastewater Treatment Systems, in *Environmental Biotechnology*, Jordening, H.J., and Winter, J., (eds.), 211–221, Wiley VCH, NY.

Gannett, F., 2012. *Refinement of Nitrogen Removal from Municipal Wastewater Treatment Plants. Prepared for the Maryland Department of the Environment*. https://www.mde.state.md.us/assets/documetns/BRE%20GAnnet%20Flemin-GMB%200 presentation.pdf. Accessed 12/26/23.

Gaudy, A.F., Jr., and Gaudy, E.T., 1980. *Microbiology for Environmental Scientists and Engineers*. New York, NY: McGraw Hill.

Gloyna, E.F., 1976. Facultative Waste Stabilization Pond Design, in *Ponds as a Waste Treatment Alternative*, Gloyna, E.F., Malina, J.F., Jr., and Davis, E.M., (eds.). Water Resources Symposium No. 9. Austin, TX: University of Texas Press.

Grady, C.P.L, Jr., Daigger, G.T., Lover, N.G., and Filipe, C.D.M., 2011. *Biological Wastewater Treatment*, 3rd ed. Boca Raton, FL: CRC Press.

Grolund, E., 2002. Microalgae at Wastewater Treatment in Cold Climates. Department of Environmental Engineering. SE 971 87 LULEA Sweden, Lic Thesis 2002:35.

Jagger, J., 1967. *Introduction to Research in Ultraviolet Photobiology*. Englewood Cliffs, NJ: Prentice-Hall, Inc.

Knudson, G.B., 1985. Photoreactivation of UV-irradiated Legionella pneumoplila and other Legionella species. *Appl Environ Microbiol*; 49:975–980.

Linden, K.G., Shin, G.A., Faubert, G., Cairns, W., and Sobsey, M.D., 2002. UV disinfection of *Giardia Lamblia* cysts in water. *Environ Sci Technol*; 36:2519–2522.

Lue-Hing, C., Zenz, D.R., and Kuchenrither, R., 1992. *Municipal Sewage Sludge Management: Processing, Utilization, and Disposal*. Lancaster, PA: Technomic Publishing Company, Inc.

Lynch, J.M., and Poole, N.J., 1979. *Microbial Ecology, A Conceptual Approach*. New York, NY: John Wiley & Sons.

McGhee, T.J., 1991. *Water Supply and Sewerage*. New York, NY: McGraw-Hill, Inc.

Metcalf & Eddy, Inc. 1991. *Wastewater Engineering: Treatment, Disposal, Reuse*, 3rd ed. New York, NY: McGraw-Hill.

Metcalf & Eddy, Inc. 2003. *Wastewater Engineering: Treatment, Disposal*, Reuse, 4th ed. New York, NY: McGraw-Hill.

Middlebrooks, E.J., and Pano, A., 1983. Nitrogen removal in aerated lagoons. *Water Res*; 17(10):1369–1378.

Middlebrooks, E.J., Middlebrooks, C.H., Reynolds, J.H., Watters, G.Z., Reed, S.C., and George, D. B., 1982. *Wastewater Stabilization Lagoon Design, Performance and Upgrading*. New York: Macmillan Publishing Co. Inc.

Millner, P., (ed.), (1995). Bioaerosols and composting. *BioCycle*; 36(1):48–54.

NEIWPCC. 1988. *Guides for the Design of Wastewater Treatment Works TR-16*. Wilmington, MA: New England Interstate Water Pollution Control Commission.

Newman, J., 1999. Control *of Algae with Barley Straw*. Sonning, UK: Information sheet 3 IACR CTR for Aquatic Plant Management.

Newman, J., and Barrett, P.R.F., 1993. Control of Microcystis aeruginosa by decomposing barley straw. *J Aqut Plant Management*; 31:203–206.

Oguma, K., Katayama, H., Mitani, H., Morita, S., Hirata, T., and Ohgaki, S., 2001. Determination of pyrimidine dimmers in Escherichia coli and Cryptosporidium parvum during UV light inactivation, photoreactivation, and dark repair. *Appl Environ* Microbiol; 67:4630–4637.

Oswald, W.J., 1990. Advanced Integrated Wastewater Pond Systems: Supplying Water and Saving the Environment for Six Billion People. *Proceedings of the ASCE Convention, Environmental Engineering Division*, San Francisco, CA, November 5–8.

Oswald, W.J., 1996. *A Syllabus on Advanced Integrated Pond Systems®*. Berkeley, CA: University of California.

Outwater, A.B., 1994. *Reuse of Sludge and Minor Wastewater Residuals*. Boca Raton, FL: Lewis Publishers.

Pano, A., and Middlebrooks, E.J., 1982. Ammonia nitrogen removal in facultative waste water stabilization ponds. *J Water Pollution Cont Fed*; 54(4):2148.

Park, J., 2012. *Biological Nutrient Removal Theories and Design*. https://www.dnr.state.wi,us/org/water/wm/ww/biophos/bnr_remvoal.htm. Accessed 12/10/23.

Paterson, C., and Curtis, T., 2005. Physical and Chemical Environments, in *Pond Treatment Technology*, Shilton, A., (ed.). London, UK: IWA Publishing.

Pearson, H., 2005. Microbiology of Waste Stabilization Ponds, in *Pond Treatment Technology*, Shilton, A., (ed.). London, UK: IWA Publishing.

Peot, C., 1998. Compost Use in Wetland Restoration. Design for Success. *Published in Proceedings of the 12th Annual Residual and Biosolids Management Conference*, Water Environment Federation, Alexandra, Virginia.

Phosphate-accumulating organisms ASCE/EWRI, 2006. *Biological nutrient removal*. Accessed 12/12/23. @http://iaspub.epa.gov/water/national_rept.contol.

Pipes, W.O., Jr., 1961. Basic biology of stabilization ponds. *Water Sewage Works*; 108(4):131–136.

Rauth, A.M., 1965. The physical state of viral nucleic acid and the sensitivity of viruses to ultraviolet light. *Biophys J*; 5:257–273.

Richard, M., and Bowman, D., 1991. Troubleshooting the Aerated and Facultative Waste Treatment lagoon. *Presented at the USEPA's Natural/Constructed Wetlands Treatment System Workshop*, Denver, CO.

Sawyer, C.N., McCarty, P.L., and Parkin, G.F., 1994. *Chemistry for Environmental Engineering*. New York, NY: McGraw Hill.

Shilton, A., (ed.), 2005. *Pond Treatment Technology*. London: IWA Publishing.

Shin, G.A., Linden, K.G., Arrowood, M.J., Faubert, G., and Sosbey, M.D., 2001. DNA Repair of UV-Irradiated *Cryptosporidium parvum* Oocysts and *Giardia lamblia* Cysts. *Proceedings of the First International Ultraviolet Association Congress*, Washington, DC, June 14–16.

Sloan Equipment. 1999. *Aeration Products*. Maryland: Owings Mills.

Sopper, W.E., 1993. *Municipal Sludge Use in Land Reclamation*. Boca Raton, FL: Lewis Publishers.

Spellman, F.R., 1996. *Stream Ecology and Self-Purification*. Boca Raton, FL: CRC Press.

Spellman, F.R., 1997. *Wastewater Biosolids to Compost*. Boca Raton, FL: CRC Press.

Spellman, F.R., 2000. *Microbiology for Water and Wastewater Operators*. Boca Raton, FL: CRC Press.

Spellman, F.R., 2007. *The Science of Water*, 2nd ed. Boca Raton, FL: CRC Press.

Tchobanoglous, G., Theisen, H., and Vigil, S.A., 1993. *Integrated Solid Waste Management*. New York: McGraw-Hill, Inc.

Toomey, W.A., 1994. Meeting the challenge of yard trimmings diversion. *BioCycle*; 35(5):55–58.

Ullrich, A.H., 1967. Use of wastewater stabilization ponds in two different systems. *J Water Pollution Cont Fed*; 39(6):965–977.

USEPA. 1975. *Process Design Manual for Nitrogen Control*, EPA-625/1–75–007. Cincinnati, OH: Center for Environmental Research Information.

USEPA. 1977. *Operations Manual for Stabilization Ponds*. EPA-430/9–77-012, NTIS No. PB-279443. Washington, DC: Office of Water Program Operations.

USEPA. 1978. *Operations Manual, Sludge Handling and Conditioning*. EPA-430/9-78-002. Washington, DC: United States Environmental Protection Agency.

USEPA. 1989. *Technical Support Document for Pathogen Reducing in Sewage Sludge*. NTIS No. PB89–136618. Springfield, VA: National Technical Information Service.

USEPA. 1993. *Manual: Nitrogen Control*. EPA-625/R-93/010. Cincinnati, OH: United States Environmental Protection Agency.

USEPA. 1997. *Innovative Uses of Compost: Disease Control for Plants and Animals*. Washington, DC: United States Environmental Protection Agency.

USEPA. 1999a. *Wastewater Technology Fact Sheet: Ultraviolet Disinfection*. Washington, DC: United States Environmental Protection Agency.

USEPA. 1999b. *Wastewater Technology Fact Sheet: Ozone Disinfection*. Washington, DC: United States Environmental Protection Agency.

USEPA. 2000. *Wastewater Technology fact Sheet Package Plants*. Washington, DC: United States Environmental Protection Agency.

USEPA. 2006. *UV Disinfection Guidance Manual*. Washington, DC: United States Environmental Protection Agency.

USEPA. 2007a. *Wastewater Management Fact Sheet: Membrane Bioreactors*. Washington, DC: United States Environmental Protection Agency.

USEPA. 2007b. *Advanced Wastewater Treatment to Achieve Low Concentration of Phosphorus*. Washington, DC: Environmental Protection Agency.

USEPA. 2007c. *Biological Nutrient Removal Processes and Costs*. Washington, DC: United States Environmental Protection Agency.

USEPA. 2007d. *National* Section 303*(d) List Fact Sheet*. Accessed 07/22/24 @ https://iaspub.epa.gov/waters/national_rept.control.

USEPA. 2008. *Municipal Nutrient Removal Technologies Reference Document Volume* 2—Appendices. Washington, DC: United States Environmental Protection Agency.

USEPA. 2011. *Principles of Design and Operations of Wastewater treatment Pond Systems for Plant operators, Engineers, and Managers*. Washington, DC: U.S. Environmental Protection Agency.

Vasconcelos, V.M., and Pereira, E., 2001. Cyanobacteria diversity and toxicity in a wastewater treatment plant (Portugal). *Water Res*; 35(5):1354–1357.

Vesilind, P.A., 1980. *Treatment and Disposal of Wastewater Sludges*, 2nd ed. Ann Arbor, MI: Ann Arbor Science Publishers, Inc.

WEF (Water Environment Federation). 1995. *Wastewater Residuals Stabilization. Manual of Practice FD-9*. Alexandria, VA: WEF.

WEF (Water Environment Federation). 1998. *Design of Municipal Wastewater Treatment Plants, Manual of Practice No. 8*, 4th ed. Vol. 2. Alexandria, VA: WEF.

Yeager, J.G., and Ward, R.I., 1981. Effects of moisture content on long-term survival and regrowth of bacteria in wastewater sludge. *Appl Environ Microbiol*; 41(5):1117–1122.

19 Practice Examination

Note: Answers to Chapter Review Questions and Practice Exam are provided in Appendix A.

EXAM

19.1 What is one of the first questions that should be answered before planning entry into a permit-required confined space?
- A. Who is going to enter?
- B. Can this task be accomplished without entering the permit space?
- C. Why is permit entry required?
- D. Who is the attendant?

19.2 What are raw water intakes designed to remove?
- A. pH
- B. THCM
- C. Debris and fish
- D. Hardness

19.3 Which of the following is more likely to be found in groundwater?
- A. Fish
- B. Microorganisms
- C. Branches and other woody debris
- D. Minerals

19.4 To knock down algae growth in reservoirs, what is commonly used?
- A. Soap
- B. Hydrogen peroxide
- C. Copper sulfate
- D. Ferric oxide

19.5 OSHA addresses confined space hazards in two specific, *comprehensive* standards. One of the standards covers General Industry, and the other covers:
- A. Agriculture
- B. Long shoring
- C. Shipyards
- D. Space travel

19.6 How are moving parts in a centrifugal pump cooled?
- A. Bearing grease
- B. Air flow
- C. Leakage within the pump
- D. Installed cooling package

19.7 A single sample of water collected at a particular time and place is called a:
- A. Composite sample
- B. Grab sample
- C. Total sample
- D. Glass of water

19.8 Which of the following will increase the pH of water?
- A. More water
- B. Sodium hypochlorite
- C. Gravel
- D. Food coloring

19.9 The Groundwater Rule is designed to:
- A. Protect the groundwater from microorganisms
- B. Set rules on chlorine addition
- C. Decrease organic contamination
- D. Decrease inorganic contamination

19.10 What type of pump has a piston within the casing?
- A. Venturi
- B. Centrifugal
- C. Turbine
- D. Reciprocating

19.11 When a new chemical is introduced into the workplace, what must accompany it?
- A. Chemical dipper
- B. Safety Data Sheet
- C. Container
- D. First aid kit

19.12 OSHA's definition of confined spaces in general industry includes:
- A. The space being more than 4 ft deep
- B. Limited or restricted means for entry and exit
- C. The space being designed for short-term occupancy
- D. Having only natural ventilation

19.13 The depression of a water table is known as:
- A. Cone of depression
- B. Radius of influence
- C. Static water level during a drought
- D. Well depression

19.14 Which of the following would *not* constitute a hazardous atmosphere under the permit-required confined space standard?
- A. Less than 19.5% oxygen
- B. More than the IDLH of hydrogen sulfide
- C. Enough combustible dust that obscures vision at a distance of 5 ft
- D. 5% of LEL

19.15 OSHA's review of accident data indicates that most confined space deaths and injuries are caused by the following three hazards:
- A. Electrical, Falls, Toxics
- B. Asphyxiates, Flammables, Toxics
- C. Drowning, Flammables, Entrapment
- D. Asphyxiates, Explosions, Engulfment

19.16 What parameter is the best way to monitor filter performance?
   A. Turbidity
   B. Dissolved solids
   C. Contact time
   D. Hardness

19.17 What causes water to look dirty or cloudy?
   A. Suspended material
   B. Bicarbonates
   C. Hardness
   D. Dissolved solids

19.18 Ammonia is sprayed near a chlorine gas leak to produce what?
   A. Neutral gas
   B. White smoke
   C. Black smoke
   D. Green smoke

19.19 A ball valve or plug valve is a type of:
   A. Foot valve
   B. Control valve
   C. Corporation stop
   D. Needle valve

19.20 Toxic gases in confined spaces can result from:
   A. Products stored in the space and the manufacturing processes
   B. Work being performed inside the space or in adjacent areas
   C. Desorption from porous walls and decomposing organic matter
   D. All of the above

19.21 Oxygen deficiency in confined spaces does *not* occur by:
   A. Consumption by chemical reactions and combustion
   B. Absorption by porous surfaces, such as activated charcoal
   C. Leakage around valves, fittings, couplings, and hoses of oxy-fuel gas welding equipment
   D. Displacement by other gases

19.22 What is installed in a water pipeline to minimize the number of customers affected by a line break?
   A. An air gap
   B. Backflow preventers
   C. A sufficient number of valves
   D. A control valve

19.23 The best way to measure your plant's performance against the best system in operation is to:
   A. Work at both plants
   B. Read the literature
   C. Benchmark
   D. Ask for opinions

19.23 Which of the following actions has had little effect on the natural water cycle?
   A. Dredging
   B. Damming
   C. Damaged ecological niches
   D. Desertification

19.25 Which water system supplies water to the same population year-round?
   A. Non-community
   B. Transient system
   C. Community
   D. Non-transient

19.26 What reading (in $\%O_2$) would you expect to see on an oxygen meter after an influx of 10% nitrogen into a permit space?
   A. 5.0%
   B. 11.1%
   C. 18.9%
   D. 90.0%

19.27 What type of water system is typically used in campgrounds?
   A. Transient water system
   B. Community water system
   C. Nontransient water system
   D. County water system

19.28 An attendant is which of the following?
   A. A person who makes a food run to the local 7–11 store for refreshments for the crew inside the confined space.
   B. A person who often enters a confined space while other personnel are within the same space.
   C. A person who watches over a confined space while other employees are in it and only leaves if he or she must use the restroom.
   D. A person with no other duties assigned, other than to remain immediately outside the entrance to the confined space, and who may render assistance as needed to personnel inside the space. The attendant never enters the confined space and never leaves the space unattended while personnel are within the space.

19.29 Water quality is defined in terms of:
   A. Location
   B. Source
   C. Size of water supply
   D. Water characteristics

19.30 Treatment plant safety depends principally on:
   A. Upper management support
   B. Number of hazards present
   C. Attitude of outside contractors
   D. Luck of the draw

19.31 Potable water must be free of:
   A. Rocks
   B. Fish
   C. Non-sanitary properties
   D. Vegetation

19.32 USEPA classifies water systems; which of the following systems is not included in their classifications?
   A. Nontransient
   B. Non-community
   C. Private commune supplies
   D. Community

19.33 Per 1910.146, an atmosphere that contains a substance at a concentration exceeding a permissible exposure limit, intended solely to prevent long-term (chronic) adverse health effects, is *not* considered to be a hazardous atmosphere on that basis alone.
- A. True
- B. False

19.34 Besides regulations and cleanup efforts, our main waterways are still polluted.
- A. True
- B. False

19.35 Which of the following is not a source of drinking water?
- A. Freshwater lakes
- B. Rivers
- C. Glaciers
- D. Moon rocks

19.36 A drainage basin is naturally shunted toward:
- A. Streams
- B. Creeks
- C. Brooks
- D. All of the above

19.37 With regard to water supply factor Q, Q refers to:
- A. Quick and quality
- B. Quality and quick
- C. Quantity and quality
- D. None of the above

19.38 Water that is unaccounted for in a distribution system is attributable to:
- A. Theft
- B. Leaks
- C. Transpiration
- D. Inaccurate meter readings

19.39 Of the following chemical substances, which one is a simple asphyxiate *and* flammable?
- A. Carbon monoxide (CO)
- B. Methane (CH_4)
- C. Hydrogen sulfide (H_2S)
- D. Carbon dioxide (CO_2)

19.40 Glaciers and ice caps hold the largest amount of freshwater.
- A. True
- B. False

19.41 Reengineering is:
- A. Tearing down the old and rebuilding with the new
- B. Engineering out a problem
- C. Systematic transformation of an existing system to a new form
- D. Unengineering the engineering

19.42 Entry into a permit-required confined space is considered to have occurred:
- A. When an entrant reaches into a space too small to enter
- B. As soon as any part of the body breaks the plane of an opening into the space

C. Only when there is clear intent to fully enter the space (therefore, reaching into a permit space would not be considered entry)
D. When the entrant says, "I'm going in now."

19.43 Do human activities slow or speed up water flow in canals and ditches?
- A. Increase the rate of water flow.
- B. Have no effect
- C. Slow water flow
- D. Stop the flow of water

19.44 If the LEL of a flammable vapor is 1% by volume, how many parts per million is 10% of the LEL?
- A. 10 ppm
- B. 100 ppm
- C. 1,000 ppm
- D. 10,000 ppm

19.45 Groundwater, in relationship to surface water, has the following advantages:
- A. Is usually clean
- B. Is easy to get to
- C. Flows naturally
- D. Is available everywhere

19.46 The principal operation of most combustible gas meters, used for permit entry testing, is:
- A. Electric arc
- B. Double displacement
- C. Electrochemical
- D. Catalytic combustion

19.47 Groundwater sources have some disadvantages when compared to surface water sources. Which of the following are disadvantages?
- A. Contamination is usually hidden from view
- B. Operating costs are usually higher
- C. Groundwater may be subject to saltwater intrusion
- D. All of the above

19.48 Water can't be destroyed or lost for practical use.
- A. True
- B. False

19.49 The proper testing sequence for confined spaces is the following:
- A. Toxics, Flammables, Oxygen
- B. Oxygen, Flammables, Toxics
- C. Oxygen, Toxics, Flammables
- D. Flammables, Toxics, Oxygen

19.50 The main reason public utility officials take a hard look at privatizing includes:
- A. To save money
- B. To keep OSHA from knocking on the door
- C. To avoid unionization
- D. It's a way of raising rates without ratepayers knowing.

19.51 Circle the following true statement(s).
- A. Employers must document that they have evaluated their workplace to determine if any spaces are permit-required confined spaces.

    B. If employers decide that their employees will enter permit spaces, they shall develop and implement a written permit space program.

    C. Employers do not have to comply with any of 1910.146, if they have identified the permit spaces, and have told their employees not to enter those spaces.

    D. The employer must identify permit-confined spaces by posting signage.

19.52 ROAD Gangers are those who:
    A. Work on road gangs
    B. Lay pipe in roadways
    C. Are retired on active duty
    D. Paint stripes on roadways

19.53 The benchmarking includes the following steps:
    A. Start, stop, adapt
    B. Research, observe, adapt
    C. Adapt, start, stop
    D. Observe, stop, start

19.54 Is it true that benchmarking is an objective-setting process?
    A. True
    B. False

19.55 Is it true that the public has no clue as to water's true economic value?
    A. True
    B. False

19.56 Circle the following true statement(s).
    A. Under paragraph (c)(5) (i.e., alternate procedures), continuous monitoring can be used in lieu of continuous forced air ventilation if no hazardous atmosphere is detected.
    B. Continuous forced air ventilation eliminates atmospheric hazards.
    C. Continuous atmospheric monitoring is required if employees are entering permit spaces using alternate procedures under paragraph (c)(5).
    D. Periodic atmospheric monitoring is required when making entries using alternate procedures under paragraph (c)(5).

19.57 Sustainable development can be defined as that which meets the needs of future generations without compromising the ability of present generations to meet their needs.
    A. True
    B. False

19.58 OSHA's position allows employers the option of making a space *eligible* for the application of alternate procedures for entering permit spaces under paragraph (c)(5) by first temporarily "eliminating" all non-atmospheric hazards, then controlling atmospheric hazards by continuous forced air ventilation.
    A. True
    B. False

19.59 Water contracts (gets smaller) when it freezes.
    A. True
    B. False

19.60 Respirators allowed for entry into and escape from immediately dangerous to life or health (IDLH) atmospheres are _____.
    A. Airline
    B. Self-contained breathing apparatus (SCBA)
    C. Gas mask
    D. Air purifying
    E. A and B

19.61 Which of the following is a measure of water's ability to neutralize acids?
    A. pH
    B. Hardness
    C. Fluoride level
    D. Alkalinity

19.62 What two minerals are primarily responsible for causing "hard water?"
    A. Salt and borax
    B. Hydrogen sulfide and sulfur dioxide
    C. Alum and chlorine
    D. Calcium (Ca) and magnesium

19.63 Water has a high surface tension.
    A. True
    B. False

19.64 The power of a substance to resist pH changes is referred to as a(n):
    A. Antidote
    B. Fix
    C. Buffer
    D. Neutral substance

19.65 Which of the following complicates the use or reuse of water?
    A. Sunlight
    B. Rainfall
    C. Recreational use
    D. Pollution

19.66 Condensation is water coming out of the air.
    A. True
    B. False

19.67 Which of the following is a leading cause of impairment to freshwater bodies?
    A. Rivers
    B. Nonpoint pollution
    C. Air pollution
    D. Snowfall

19.68 Which of the following is not a feature of water?
    A. Appearance
    B. Wetness
    C. Smell
    D. Odor

19.69 Circle the following *false* statement(s):
    A. If all hazards within a permit space are eliminated without entry into the space, the permit space may be reclassified as a non-permit confined space under paragraph (c)(7).

B. Minimizing the amounts of regulation that apply to spaces whose hazards have been eliminated encourages employers to remove all hazards from permit spaces.

C. A certification containing only the date, location of the space, and the signature of the person determining that all hazards have been eliminated shall be made available to each employee entering a space that has been reclassified under paragraph (c)(7).

D. An example of eliminating an engulfment hazard is requiring an entrant to wear a full-body harness attached directly to a retrieval system.

19.70 If water is turbid, it is
   A. Smelly
   B. Wet
   C. Cloudy
   D. Clear

19.71 In order for an employee to be approved for respirator use on the job, what is required?
   A. Training
   B. Medical approval
   C. Fit-testing
   D. All of the above

19.72 Solids are classified as
   A. Settable
   B. Dissolved
   C. Colloidal
   D. All of the above

19.73 Colloidal material is composed of
   A. Leaves
   B. Branches
   C. Silt
   D. Decayed fish

19.74 Taste and odor in drinking water may be
   A. No hazard
   B. A tooth destroyer
   C. A contributor to kidney stones
   D. A laxative

17.75 More things can be dissolved in sulfuric acid than in water.
   A. True
   B. False

19.76 Rainwater is the purest form of water.
   A. True
   B. False

19.77 Circle the following *false* statement(s):
   A. Compliance with OSHA's Lockout/Tagout Standard is considered to eliminate electro-mechanical hazards.
   B. Compliance with the requirements of the Lockout/Tagout Standard, is not considered to eliminate hazards created by flowable materials, such as steam, natural gas, and other substances that can cause hazardous atmospheres or engulfment hazards in a confined space.

C. Techniques used in isolation are blanking, blinding, misaligning, or removing sections of line soil pipes, and a double and bleed system.

D. Water is considered to be an atmospheric hazard.

19.78 It takes more energy to heat water at room temperature to 212°F than it does to change 212°F water to steam.
   A. True
   B. False

19.79 Circle the following *false* statement(s):
   A. Compliance with OSHA's Lockout/Tagout Standard is considered to eliminate electro-mechanical hazards.
   B. Compliance with the requirements of the Lockout/Tagout Standard is not considered to eliminate hazards created by flowable materials, such as steam, natural gas, and other substances that can cause hazardous atmospheres or engulfment hazards in a confined space.
   C. Techniques used in isolation are blanking, blinding, misaligning, or removing sections of line soil pipes, and a double and bleed system.

19.80 If you evaporate an 8-in glass full of water from the Great Salt Lake (with a salinity of about 20% by weight), you will end up with about 1 in of salt.
   A. True
   B. False

19.81 Pretreatment of water is used to oxidize which of the following:
   A. Iron and manganese
   B. Entrapped gases
   C. Removal of tastes and odors
   D. All of the above

19.82 Sea water is slightly more basic (the pH value is higher) than most natural water.
   A. True
   B. False

19.83 Circle the following true statement(s):
   A. "Alarm only" devices that do not provide numerical readings, are considered acceptable direct reading instruments for initial (pre-entry) or periodic (assurance) testing.
   B. Continuous atmospheric testing must be conducted during permit space entry.
   C. Under alternate procedures, OSHA will accept a minimal "safe for entry" level as 50% of the level of flammable or toxic substances that would constitute a hazardous atmosphere.
   D. The results of air sampling required by 1910.146, which show the composition of an atmosphere to which an employee is exposed, are *not* exposure records under 1910.1020.

19.84 Example(s) of simple asphyxiates are:
   A. Nitrogen (N_2)
   B. Carbon monoxide (CO)
   C. Carbon dioxide (CO_2)
   D. A and C

19.85 Which statement(s) is/are true about combustible gas meters (CGMs)?
   A. CGMs can measure all types of gases.
   B. The percent of oxygen will affect the operation of CGMs.
   C. Most CGMs can measure only pure gases.
   D. CGMs will indicate the lower explosive limit for explosive dusts.

19.86 Solids in water may be classified by:
   A. Where they come from
   B. Size and state
   C. Whether they dissolve or not
   D. None of the above

19.87 Colloidal material is never beneficial in water.
   A. True
   B. False

19.88 Why are the adsorption sites on suspended material present in water objectionable?
   A. They can provide protective barriers against the chemical treatment of microorganisms.
   B. Enhance bad smell problems
   C. Make treatment by screening impossible
   D. There is no problem

19.89 What can cause water to turn the color of tea?
   A. Iron contamination
   B. Air contamination
   C. Decayed vegetation
   D. Saltwater contamination

19.90 Taste and odor of water are usually not an issue until:
   A. The water boils on its own
   B. The water is viscous
   C. The customer complains
   D. None of the above

19.91 Circle the following true statement(s):
   A. An off-site rescue service should have a permit space program before performing confined space rescues.
   B. The only respirator that a rescuer can wear into an IDLH atmosphere is a self-contained breathing apparatus.
   C. Only members of in-house rescue teams shall practice making permit space rescues at least once every 12 months.
   D. Each member of the rescue team shall be trained in basic first aid and CPR.
   E. To facilitate non-entry rescue, with no exceptions, retrieval systems shall be used whenever an authorized entrant enters a permit space.

19.92 Hydrogen sulfide gives off a characteristically odor of:
   A. Rotten egg odor
   B. Sweet odor
   C. Flowery odor
   D. No odor

19.93 The chemical that gives off a rotten cabbage odor is:
   A. Hydrogen sulfide
   B. Calcium oxide

   C. Organic sulfides
   D. Alum

19.94 The chemical composition of groundwater is changed by:
   A. Atmospheric pollution
   B. The rocks it comes into contact with
   C. Pollution
   D. None of the above

19.95 The Permit-Required Confined Space standard requires the employer to initially:
   A. Train employees to recognize confined spaces
   B. Measure the levels of air contaminants in all confined spaces
   C. Evaluate the workplace to determine if there are any confined spaces
   D. Develop an effective confined space program

19.96 Which of the following would increase alkalinity in the water?
   A. Dirt
   B. Sulfur dioxide
   C. Manganese
   D. $CaCO_3$

19.97 What should be done with a water sample that can't be analyzed right away?
   A. Throw it away
   B. Freeze it
   C. Add chemicals
   D. Refrigerate it

19.98 The water table is:
   A. Aquifer water
   B. Atmospheric pressure
   C. Water under the table
   D. None of the above

19.99 In a pump, what is the purpose of the shaft sleeve?
   A. Makes for a tighter connection of the impeller
   B. Acts as a water slinger ring
   C. Protects the shaft from wear
   D. Prevents contamination inside the pump

19.100 If an employer decides that he/she will contract out all confined space work, then the employer:
   A. Has no further requirement under the standard
   B. Must label all spaces with a keep-out sign
   C. Must train workers on how to rescue people from confined spaces
   D. Must effectively prevent all employees from entering confined spaces

19.101 Which of the following is a primary reason for ensuring the safety of drinking water?
   A. Reducing costs to customers
   B. Ridding an area of excess water
   C. Ensuring public health
   D. Political considerations

19.102 99.999% of anything is equivalent to:
   A. 5-log
   B. 6-log
   C. No log
   D. Excessive water quality

19.103 Which of the following are water quality factors?
A. Fecal bacteria
B. Odor
C. Hardness
D. All of the above

19.104 In chemical addition to water, what is most important?
A. Using the correct additives
B. Using the correct amount of additive
C. Metering additives
D. Using the best brand of chemical

19.105 Water pollutants include:
A. Snow
B. Rain
C. Oil
D. Grass seed

19.106 Not required on a permit for confined space entry is:
A. Names of all entrants
B. Name(s) of entry supervisors(s)
C. The date of entry
D. The ventilation requirements of the space

19.107 Which of the following is more likely to be found in groundwater?
A. Minerals
B. Salt
C. Giardia
D. Organic matter

19.108 Which of the following is the definition of a grab sample?
A. Represents flow
B. A single sample of water collected at a particular time and place which represents the composition of the water at that time and place.
C. Represents dosage
D. A composite sample

19.109 Beer's Law has to do with what?
A. A sample's color based on its concentration
B. Taste
C. Odor
D. pH

19.110 What causes water turbidity?
A. Bicarbonate
B. Hardness
C. Suspended material
D. Dissolved solids

19.111 Which water sampling parameter must be taken in the field?
A. Salinity
B. Fecal coliform
C. Color
D. pH & temperature

19.112 Circle the following training requirement that is identical for the entrant, attendant, and entry supervisor.

A. Know the hazards that may be faced during entry
B. The means of summoning rescue personnel
C. The schematic of the space, to ensure all can get around in the space
D. The proper procedure for putting on, and using, a self-contained breathing apparatus

19.113 Which of the following describes the trickle of water from the land surface toward an aquifer?
A. Conveyance
B. Suction
C. Evapotranspiration
D. Percolation

19.114 Which of the following must always be done before repairing a machine?
A. Notify people affected
B. Lockout/Tagout equipment
C. Notify the supervisor
D. Wipe down equipment

19.115 What reading must be taken in the field when obtaining a sample?
A. pH
B. Color
C. Chlorine residual
D. Hardness

19.116 Attendants can:
A. Perform other activities when the entrant is on break inside the confined space.
B. Summon rescue services, as long as he/she does not exceed a 200-ft radius around the confined space.
C. Enter the space to rescue a worker, but only when wearing an SCBA and connected to a lifeline.
D. Order evacuation if a prohibited condition occurs.

19.117 An oxygen-enriched atmosphere is considered by 1910.146 to be:
A. Greater than 22% oxygen
B. Greater than 23.5% oxygen
C. Greater than 20.9% oxygen
D. Greater than 25% oxygen when the nitrogen concentration is greater than 75%.

19.118 The following confined space, which would be permit-required, is:
A. A grain silo with inward-sloping walls
B. A 10-gal methylene chlorine reactor vessel
C. An overhead crane cab which moves over a steel blast furnace
D. All of the above

19.119 Performing laboratory work safely requires:
A. Adding acid to water
B. Mixing acid with a pH of 3 with a base of pH 12.5
C. Adding water to acid
D. Titrating water to acid

19.120 What are colloidal particles?
- A. Sand
- B. Large particles
- C. Total solids
- D. Very small particles that do not dissolve

19.121 A written permit space program requires:
- A. That the employer purchases SCBAs and lifelines, while employees purchase safety shoes and corrective lens safety glasses.
- B. That the employer test all permit-required confined spaces at least once per year, or before entry, whichever is more stringent.
- C. That the employer provides one attendant for each entrant up to five, and one for each two entrants, when there are more than five
- D. That the employer develops a system to prepare, issue, and cancel entry permits.

19.122 Bernoulli's principle states that the total energy of a hydraulic fluid is
- A. Flexible
- B Adjustable
- C. Flat
- D. Constant

19.123 The difference, or the drop, between static water level and the pumping water head is known as:
- A. Drawdown
- B. Draw up
- C. Slope
- D. Potential energy

19.124 What causes water motion?
- A. Inertia
- B. Energy
- C. Force
- D. Momentum

19.125 Of the following, which is *not* a duty of the confined space entrant?
- A. Properly use all assigned equipment
- B. Communicate with the attendant
- C. Exit when told to
- D. Continually test the level of toxic chemicals in the space

19.126 This is present when water is in motion:
- A. Potential energy
- B. Force
- C. Kinetic energy
- D. Inertia

19.127 Friction created as water encounters the surface of a pipe causes:
- A. Major head loss
- B. No loss
- C. Minor head loss
- D. Heat

19.128 Iron and manganese are most likely to reside in which of the following water sources?
- A. River
- B. Groundwater
- C. Stream
- D. Lake

19.129 Alkalinity is:
- A. pH of 7
- B. The capacity of water to neutralize acid
- C. pH of 5
- D. The property of neutralizing a base

19.130 Force per unit area is:
- A. Force
- B. Head
- C. Hydrostatic charge
- D. Pressure

19.131 Of the following, which is *not* a duty of the entry supervisor?
- A. Summon rescue services
- B. Terminate entry
- C. Remove unauthorized persons
- D. Endorse the entry permit

19.132 What is head loss measured in feet commonly called?
- A. Slope
- B. Head gain
- C. Head
- D. Static system

19.133 Chlorine is added to the water distribution system to ensure residual chlorine presence within the system to ensure continuous disinfection.
- A. True
- B. False

19.134 Why do we have secondary MCLs?
- A. For health reasons
- B. Just for the heck of it
- C. For aesthetic reasons
- D. Because of carcinogens

19.135 An aquifer is typically composed of:
- A. Rocks
- B. Melted rocks
- C. Sand and gravel
- D. Silt

19.136 What is the force as it rounds a bend in a pipe called?
- A. Inertia
- B. Dynamic load
- C. Thrust
- D. Cavitation

19.137 When designing ventilation systems for permit space entry:
- A. The air should be blowing into space
- B. The air should always be exhausting out of space
- C. The configuration, contents, and tasks determine the type of ventilation methods used
- D. Larger ducts and bigger blowers are better

19.138 What is the action called when there is a sudden change in water pressure and direction in a pipe?
- A. Pressure loss
- B. Water hammer
- C. Pressure increase
- D. Dynamic load

19.139 Which one of the following is used to lower the pH in water?
   A. Sand
   B. Carbon dioxide
   C. Ammonia
   D. Sulfur

19.140 What does a gallon of water weigh in pounds?
   A. 50 lbs
   B. 62.14 lbs
   C. 8.34 lbs
   D. It is weightless

19.141 Of the following, which is *not* a duty of the attendant
   A. Know accurately how many entrants are in the space
   B. Communicate with entrants
   C. Summon rescue services when necessary
   D. Continually test the level of toxic chemicals in the space

19.142 Circle the following true statement(s):
   A. Carbon monoxide gas should be ventilated from the bottom
   B. The mass of air going into a space equals the amount leaving
   C. Methane gas should be ventilated from the bottom
   D. Gases flow by the inverse law of proportion

19.143 Turbidity is described as:
   A. Cloudiness or haziness of water
   B. Dissolved solids
   C. Total solids
   D. Mud

19.144 A disease-causing microorganism is called:
   A. Algae
   B. Fungi
   C. A pathogen
   D. PCP

19.145 Which of the following is used to mix coagulants?
   A. Diffuser
   B. Mechanical mixing
   C. Hydraulic mixing
   D. All of the above

19.146 Hot work is going to be performed in a solvent reactor vessel that is 10 ft high and 6 ft in diameter. Which of the following is the *preferred* way to do this?
   A. Use submerged arc-welding equipment
   B. Inert the vessel with nitrogen and provide a combination airline with an auxiliary SCBA respirator, for the welder
   C. Fill the tank with water and use underwater welding procedures
   D. Pump all the solvent out, ventilate for 24 h, and use non-sparking welding sticks
   E Clean the reactor vessel, then weld per 1910.252

19.147 Your pump is making a pinging noise; what is the most likely cause?
   A. Water hammer
   B. Downstream valves closed
   C. Frequent cycling
   D. Cavitation

19.148 The efficiency of removing suspended solids in the sedimentation process is controlled by:
   A. Flowrate
   B. Amount of sediment
   C. Chemicals added
   D. Temperature

19.149 Bacteria is colloidal.
   A. True
   B. False

19.150 Chlorine is not considered free residual chlorine.
   A. True
   B. False

19.151 The certification of training required for attendants, entrants, and entry supervisors must contain (circle all that apply):
   A. The title of each person trained
   B. The signature or initials of each person trained
   C. The signature or initials of the trainer
   D. The topics covered by the training

19.152 What is a static head?
   A. Horizontal distance between a reference point to the water surface
   B. The output of a centrifugal pump
   C. Friction loss
   D. The vertical distance between a reference point to the water surface when water is not moving.

19.153 What is the purpose of packing and a mechanical seal?
   A. Increases head
   B. Supports the shaft assembly
   C. Prevents pump water leakage
   D. Prevents bearing grease leakage

19.154 Coliform is used as an indicator organism because:
   A. Saves time, money, and cost of analysis
   B. Why not?
   C. Easily detected
   D. Provides a wealth of information

19.155 A rest stop is an example of a community water system.
   A. True
   B. False

19.156 Hardness is least likely to impact the formation of flocs during coagulation and flocculation.
   A. True
   B. False

19.157 If excessive water leaks from the stuffing box
   A. Impeller has crumbled
   B. Broken check valve

C. Packing gland needs to be tightened

D. Bearing failure

19.158 _____ is installed on the suction pipe of a pump to prevent water from draining out of the pump.

A. Foot valve

B. Slinger ring

C. Pressure valve

D. Emergency valve

19.159 What is used to restrict leakage from the impeller discharge?

A. Slinger ring

B. Wear rings

C. Lantern ring

D. Shaft sleeve

19.160 A thermocline in a reservoir is a thin layer of water in which temperature changes more rapidly with depth than it does in the layers above or below.

A. True

B. False

19.161 What agency is responsible for protecting worker safety and health?

A. EPA

B. NRA

C. AARP

D. OSHA

19.162 Which of the following is a zone of a clarifier?

A. Top

B. Sludge

C. Bottom

D. Aerator

19.163 Which of the following is a positive displacement pump?

A. Feeder pump

B. Sub pump

C. Piston pump

D. Well pump

19.164 What is the main purpose of sealing water in pumps?

A. Prevent water hammer

B. Protect bearings

C. Cool packing

D. Protect the slinger ring

19.165 Which of the following impacts the color of water?

A. Salt

B. Sulfur dioxide

C. Vegetation

D. Carbon dioxide

19.166 In which of the following would it be likely to find coliform?

A. Sand

B. Tap water

C. Lake

D. Deep well

19.167 What is an example of natural organic matter?

A. Vegetation

B. Salt

C. Dirt

D. Iron

19.168 Federal Paragraph 1910.146(g) requires that training of all employees whose work is regulated by the permit-required confined space standard shall be provided:

A. On an annual basis

B. When the employer believes there are inadequacies in the employee's knowledge of the company's confined space procedures

C. When the union demands it

D. All of the above

19.169 What is the function of a velocity pump?

A. Piston

B. Propeller meter

C. Motor controller

D. Relief valve

19.170 Short-circuiting in a clarifier is a shorter settling time.

A. True

B. False

19.171 Adsorption is substances sticking to the surface of a media.

A. True

B. False

19.172 What do you think is the most common water quality complaint from customers?

A. The well runs dry

B. The water tastes bad

C. The water is rust-filled

D. Taste and odor issues

19.173 Which of the following causes hardness in water?

A. Iron

B. Calcium

C. Dirt

D. Leaves

19.174 Which of the following ranges of pH is considered to be basic?

A. 3–5

B. 7

C. 12–14

D. 0–6

19.175 Blackish water may be due to precipitated manganese.

A. True

B. False

19.176 What is the drawback of using UV for disinfection?

A. Not expensive

B. Electrocution

C. Longer contact time

D. High turbidity water

19.177 The _____ pump has a piston inside its casing.

A. Centrifugal
B. Venturi
C. Reciprocating
D. Peristaltic

19.178 Scaling is a common problem with hard water:
A. True
B. False

19.179 The problem with soft water is that it causes:
A. Scaling
B. Blockage
C. Water hammer
D. Corrosion

19.180 Chlorine gas is brown in color.
A. True
B. False

19.181 The _____ is responsible for enforcing drinking water standards.
A. EPA
B. State
C. FDA
D. USDA

19.182 _____ control fluid flow through piping systems.
A. Robots
B. Valves
C. Accumulators
D. Receivers

19.183 Which is the smallest pathogen?
A. Cyst
B. Protozoa
C. Virus
D. Bacteria

19.184 What does a high C factor for a pipe indicate?
A. No corrosion protection
B. Easy to bend
C. Smooth pipe interior
D. Large pipe

19.185 Decaying vegetation in water is an example of:
A. Suspended solids
B. Dirt
C. Soluble solids
D. Colloids

19.186 When air pockets form at the high points in a pipeline, the cause is usually _____.
A. Flow rate too high
B. Excessive air bubbles
C. Oil contamination
D. Pump impeller cracked

19.187 _____ happens during the course of a filter run time.
A. Headloss increases
B. No head loss
C. Maximum head loss
D. Headloss decreases

19.188 Surge tanks are used to control _____.
A. Leaks
B. Air bubbles
C. Water hammer
D. Corrosion

19.189 A coupling (a Dresser) is used to:
A. Repair clamps
B. Saddle a pipe
C. Install a volute
D. Connect a new water main to an existing water main.

19.190 A metallic flat surface used to minimize leakage around the pump shaft is known as a:
A. Mechanical seal
B. Air gap
C. Bearing ring
D. Slinger ring

19.191 When a new pipe suffers tuberculation, what does this indicate?
A. Scaling
B. Low pH
C. Water is corrosive
D. Neutral pH

19.192 The valve type commonly used to control pressure and flow is:
A. Emergency relief valve
B. Check valve
C. Foot valve
D. Globe valve

19.193 Deposition of calcium carbonate in a water main is known as _____.
A. Overloading
B. Scaling
C. Pigging
D. Corrosion

19.194 What is a regulator station used for?
A. Feed chlorine
B. Prevent corrosion
C. Maintain an acceptable water pressure within a piping system
D. Release air within the pipe

19.195 A turbine pump is best described as a:
A. Piston pump
B. Air pump
C. Centrifugal pump
D. Jet pump

19.196 A chain of flight collectors is commonly used to:
A. Remove soluble materials from water
B. Remove sludge from the clarifier bottom
C. Has no function
D. Protect water from contamination

19.197 When scheduling preventive maintenance, how often should it be accomplished?
A. Refer to the manufacturer's recommendation
B. Never
C. Annually
D. Every day

19.198 The purpose of a check valve is to:
A. Throttle flow
B. Stop flow

C. Speed up the flow

D. Allow water to flow in one direction only

19.199 If you add lime to water, what happens to pH?

A. Goes acidic

B. Increases pH

C. Decreases turbidity

D. Increases turbidity

19.200 What type of valve does a corporation stop?

A. Relief valve

B. Check valve

C. Foot valve

D. Ball valve or plug valve

19.201 What is the best procedure for ensuring proper equipment operation?

A. All new equipment

B. Preventive maintenance

C. 24 h operator surveillance

D. Good luck

19.202 A _____ describes groundwater that flows naturally from the ground.

A. Stream

B. Well

C. Spring

D. Waterfall

19.203 What is the most critical problem that may be encountered when entering a confined space?

A. Black widow spiders

B. Lack of oxygen

C. Trip hazards

D. Low temperature

19.204 Which of the following are commonly used coagulants?

A. Ferric sulfate

B. Alum

C. Polymer

D. All of the above

19.205 What causes water to look cloudy?

A. High pH

B. Wastewater

C. Hardness

D. Suspended solids

19.206 If you are exposed to high noise levels for only an hour a day it will not lessen your hearing acuity.

A. True

B. False

19.207 What is a plant vulnerability study used for?

A. Ensure security

B. Evaluate potential threats

C. Find out who is vulnerable

D. Prepare for OSHA inspection

19.208 Why is a chlorine residual important?

A. Oxidizes iron and manganese

B. Provides chlorine residual to save on costs

C. Oxidizes nitrate

D. Continues treatment after treatment

19.209 One of the most prominent industrial wastes found in surface water is:

A. Milk

B. Caustic

C. Organic solvents

D. None of the above

19.210 Who is responsible for the proper operation and maintenance of plant equipment?

A. Top manager

B. CEO

C. Custodian

D. All plant employees

19.211 When the flow rate in the sedimentation process is increased efficiency is lowered.

A. True

B. False

19.212 Which of the following mechanisms occurs in filter media?

A. Sedimentation

B. Adsorption

C. Straining

D. All of the above

19.213 What is the main disadvantage of performing preventive maintenance on plant equipment?

A. Okay equipment could be damaged

B. Too expensive

C. Someone might be injured

D. Might shut down plant operations

19.214 Why does the water industry have a high incidence of injury?

A. Almost all types of hazards are present

B. Workplace violence

C. Too many off-hours

D. Very liberal on-the-job-injury reporting system

19.215 Turbidity measurement is used to evaluate filtration efficiency.

A. True

B. False

19.216 Absence of plant safety rules can lead to:

A. Increased injury rate

B. Lawsuits

C. Abuse of workers' compensation

D. All of the above

19.217 The _____ process does not remove pathogens.

A. Filtration

B. Disinfection

C. Screening

D. Sedimentation

19.218 Rupture in pipe(s) is the most common source of waterborne disease.

A. True

B. False

19.219 Humans can't detect chlorine gas contamination at 3 ppm.

A. True

B. False

19.220 Safety starts at the top. What is the main factor that brings this statement to reality?

A. Unlimited funds in the safety budget

B. Great insurance coverage

C. Top management supports safety

D. Employees buy into safety programs

17.221 Energy is expensive. What is the best way to save on energy costs?

A. Turn out the lights when not in use

B. Turn off equipment not needed

C. Purchase only energy-efficient equipment

D. All of the above

19.222 _____ has a rotten egg odor.

A. Dirt

B. Sulfur dioxide

C. Hydrogen sulfide

D. Alum

19.223 Which of the following is a recommended way to protect and conserve plant infrastructure?

A. Purchase a good insurance policy

B. Proper operating procedures

C. If it runs, leave it alone mantra

D. Increase funding

19.224 Why is/are public water utilities "invisible" for many people?

A. Piping is buried underground and treatment plants are located in remote locations

B. Many people have no idea where their drinking water comes from

C. Many people do not care where their water comes from

D. Water is not often thought about until the tap is dry or they are dying of thirst

19.225 When are Safety Data Sheets required to be available to workers?

A. Once per year

B. Whenever the supervisor provides them

C. 24/7

D. When asked for

19.226 Confined spaces can be killers due to:

A. Wild animals within

B. Lack of oxygen

C. Piping galleries

D. One way in and out

19.227 What does a pH of 3 indicate?

A. Strong acid

B. Weak acid

C. Neutral

D. Very strong base

19.228 Total dissolved solids in water give the water a bad taste.

A. True

B. False

19.229 Who is authorized to provide first aid in the workplace?

A. Anyone

B. Outside responders only

C. Trained & certified first aid personnel

D. Human resources specialists

19.230 A pH of 7.5 is ideal for disinfecting bacteria.

A. True

B. False

19.231 Who is authorized to remove a lockout/tagout device?

A. Supervisor

B. Anyone

C. General manager

D. None of the above

19.232 The centrifugal pump is not a positive displacement pump.

A. True

B. False

19.233 Which of the following imparts velocity to water?

A. Pump housing

B. Slinger ring

C. Impeller

D. Bearings

19.234 Algal blooms have an increasing pH of water in a reservoir.

A. True

B. False

19.235 Coagulants have a positive charge.

A. True

B. False

19.236 The Hazard Communication Standard requires employers to:

A. Label chemical hazards

B. Train employees

C. Provide safety data sheets

D. All of the above

19.237 High-velocity backwashing of filter systems can result in _____.

A. Breakdown of the filter

B. Excessive spent media

C. No problem

D. Failure to remove contaminants

19.238 _____ neutralizes the negative charges on colloids during water treatment.

A. pH

B. Coagulation and flocculation

C. Ferric chloride

D. Alum

19.239 _____ inactivates many of the pathogens in water.

A. Alum

B. Sodium hydroxide

C. Disinfection

D. Filtering

19.240 What is a dial indicator used for?

A. Feeding chemicals

B. Changing bearings

C. Checking pump vibration

D. Measuring alignment

19.241 _____ is formed when chlorine is added to water.
A. Hypochlorous acid
B. Sludge
C. Alum
D. Hypochlorite

19.242 During hot work operations, why is a fire watch required?
A. To prevent fire
B. To comply with OSHA
C. To create extra jobs
D. To satisfy insurance requirements

19.243 After a filter is backwashed, turbidity momentarily _____ in the filter effluent.
A. Remains the same
B. Causes head loss to decrease
C. Increases
D. Decreases

19.244 Five log removal is what?
A. 10^{15}
B. 10%
C. 99.999%
D. 5%

19.245 Torch cutting is dangerous because:
A. Can cause fire
B. Can burn operators
C. Can seep into cracks, out of sight of the operator
D. All of the above

19.246 Heterotrophic uses sunlight for energy.
A. True
B. False

19.247 Taste and odor problems in raw surface water are usually caused by _____.
A. High pH
B. Algae
C. Dirt
D. Sand

19.248 _____ is used to remove turbidity.
A. Alum
B. Carbon
C. Sand filter
D. Hypochlorite

19.249 Those who are required to wear full-face respirators in the performance of their daily duties should be clean-shaven.
A. True
B. False

19.250 The typical source of lead in drinking water is _____.
A. Water supply
B. Groundwater supply
C. Artesian water
D. Corrosion of pipe

19.251 A grab sample is
A. A composite sample
B. A sample that is taken to represent the flow
C. A single sample collected at a particular time and place that represents the composition of the water only at that time and place
D. A sample that is not adequate for making periodic measurements

19.252 In gravity filtration, the common filtration rates are
A. 12–5.0 GPM/ft².
B. 21–30 GPM/ft².
C. 2–10 GPM/ft².
D. 23–28 GPM/ft².

19.253 What type of pump has a piston in its casing?
A. Volume pump
B. Stirling pump
C. Turbine pump
D. Reciprocating pump

19.254 What chemical do you spray near a chlorine gas leak to produce white smoke?
A. Ajax
B. Hydrogen peroxide
C. Ammonia
D. Baking soda

19.255 The following is an indicator organism for pathogens:
A. Giardia
B. Coliform
C. TB
D. Fungi

Appendix A
Answers to Chapter Review Questions/Problems

CHAPTER 3

3.1 A pattern or point of view that determines what is seen as reality.

3.2 A change in the way things are understood and done.

3.3 1. assessing and protecting drinking water sources
   2. optimizing treatment processes
   3. ensuring the integrity of distribution systems
   4. effecting correct cross-connection control procedures
   5. continuous monitoring and testing of the water before it reaches the tap

3.4 Water/wastewater operations are usually low-profile activities and much of the water/wastewater infrastructure is buried underground.

3.5 Secondary

3.6 Privatization means allowing private enterprises to compete with the government in providing public services, such as water and wastewater operations. Re-engineering is the systematic transformation of an existing system into a new form to realize quality improvements in operation, systems capability, functionality, and performance at lower cost, schedule, or risk to the customer.

3.7 A process for rigorously measuring your performance vs "best-in-class" operations and using the analysis to meet and exceed the best in class.

3.8 Planning, research, observation, analysis, adaptation

CHAPTER 4 ANSWERS

4.1 Operators are exposed to the full range of hazards and work under all weather conditions.

4.2 Plants are upgrading to computerized operations.

4.3 Computerized maintenance management system

4.4 HAZMAT emergency response technician 24-h certification

4.5 Safe Drinking Water Act

CHAPTER 5 ANSWERS

5.1 Answers will vary

5.2 Answers will vary

CHAPTER 6

MATCHING ANSWERS

6.1 o
6.2 c
6.3 t
6.4 j
6.5 s
6.6 p
6.7 d
6.8 i
6.9 e
6.10 Q
6.11 U
6.12 K
6.13 A
6.14 V
6.15 R
6.16 W
6.17 L
6.18 X
6.19 M
6.20 Y
6.21 F
6.22 B
6.23 Z
6.24 H
6.25 N
6.26 G

CHAPTER 7

7.1 $(0.785)(70\,\text{ft})(70\,\text{ft})(25\,\text{ft})\ 7.48\ \text{gal/ft}^3 = 719{,}295.5\ \text{gal}$

7.2 $(60\,\text{ft})(20\,\text{ft})(10\,\text{ft}) = 12{,}000\,\text{ft}^3$

7.3 $(20\,\text{ft})(60\,\text{ft})(12\,\text{ft})(7.48\ \text{gal/ft}^3) = 107{,}712\ \text{gal}$

7.4 $(20\,\text{ft})(40\,\text{ft})(12\,\text{ft})(7.48\ \text{ft}^3) = 71{,}808\ \text{gal}$

7.5 $(0.785)(60\,\text{ft})(60\,\text{ft})(12\,\text{ft})(7.48\ \text{gal/ft}^3) = 253{,}662\ \text{gal}$

7.6 $(20\,\text{ft})(50\,\text{ft})(16\,\text{ft})(7.48\ \text{gal/ft}^3) = 119{,}680\ \text{gal}$

7.7 $(4\,\text{ft})(6\,\text{ft})(340\,\text{ft}) = 8{,}160\,\text{ft}^3$

7.8 $(0.785)(0.83\,\text{ft})(0.83\,\text{ft})(1{,}600\,\text{ft})(7.48\ \text{gal/ft}^3) = 6{,}472\ \text{gal}$

   $5\,\text{ft} + 10\,\text{ft} / 2(4\,\text{ft})(800\,\text{ft})(7.48\,\text{gal/ft}^3)$

7.9 $= (7.5\,\text{ft})(4\,\text{ft})(800\,\text{ft})(7.48\,\text{gal/ft}^3)$

   $= 179{,}520\,\text{gal}$

7.10 $(0.785)(0.66)(0.66)(2,250\,\text{ft})(7.48\,\text{gal/ft}^3)=5,755\,\text{gal}$

7.11 $(5\,\text{ft})(4\,\text{ft})(1,200\,\text{ft})(7.48\,\text{gal/ft}^3)=179,520\,\text{gal}$

7.12 $\dfrac{(4\,\text{ft})(4\,\text{ft})(1,200\,\text{ft})}{27\,\text{ft}^3/\text{yd}^3}=711\,\text{yd}^3$

7.13 $(500\,\text{yds})(1\,\text{yd})(1.33\,\text{yd})=665\,\text{yd}^3$

7.14 $(900\,\text{ft})(3\,\text{ft})(3\,\text{ft})=8,100\,\text{ft}^3$

7.15 $(700\,\text{ft})(6.5\,\text{ft})(3.5\,\text{ft})=15,925\,\text{ft}^3$

7.16 $(0.785)(90\,\text{ft})(90\,\text{ft})(25\,\text{ft})(7.48\,\text{ft}^3)=1,189,040\,\text{gal}$

7.17 $(80\,\text{ft})(16\,\text{ft})(20\,\text{ft})=25,600\,\text{ft}^3$

7.18 $(0.785)(0.67\,\text{ft})(0.67\,\text{ft})(4,000\,\text{ft})(7.48\,\text{ft}^3)=10,543\,\text{gal}$

7.19 $(1,200\,\text{ft})(3\,\text{ft})(3\,\text{ft})=10,800\,\text{ft}^3$

7.20 $\dfrac{(3\,\text{ft})(4\,\text{ft})(1,200\,\text{ft})}{27\,\text{ft}^3/\text{yd}^3}=533\,\text{yd}^3$

7.21 $(30\,\text{ft})(80\,\text{ft})(12\,\text{ft})(7.48\,\text{gal/ft}^3)=215,424\,\text{gal}$

7.22 $(8\,\text{ft})(3.5\,\text{ft})(3,000\,\text{ft})(7.48\,\text{gal/ft}^3)=628,320\,\text{gal}$

7.23 $(0.785)(70\,\text{ft})(70\,\text{ft})(19\,\text{ft})(7.48\,\text{gal/ft}^3)=546,665\,\text{gal}$

7.24 $(0.785)(25\,\text{ft})(25\,\text{ft})(30\,\text{ft})(7.48\,\text{gal/ft}^3)=110,096\,\text{gal}$

7.25 $(2.4\,\text{ft})(3.7\,\text{ft})(2.5\,\text{fps})(60\,\text{sec/min})=1,332\,\text{cfm}$

7.26 $(20\,\text{ft})(12\,\text{ft})(0.8\,\text{fpm})(7.48\,\text{gal/ft}^3)=1,436\,\text{gpm}$

7.27 $\dfrac{(4\,\text{ft}+6\,\text{ft})}{2}(3.3\,\text{ft})(130\,\text{fpm})=5(3.3\,\text{ft})(130\,\text{fpm})$
$$=2,145\,\text{cfm}$$

7.28 $(0.785)(0.66)(0.66)(2.4\,\text{fps})(7.48\,\text{gal/ft}^3)(60\,\text{sec/min})=368\,\text{gpm}$

7.29 $(0.785)(3\,\text{ft})(3\,\text{ft})(4.7\,\text{fpm})(7.48\,\text{gal/ft}^3)=248\,\text{gpm}$

7.30 $(0.785)(0.83\,\text{ft})(0.83\,\text{ft})(3.1\,\text{fps})(7.48\,\text{gal/ft}^3)(60\,\text{sec/min})(0.5)=376\,\text{gpm}$

7.31 $(6\,\text{ft})(2.6\,\text{ft})(x\,\text{fps})(60\,\text{sec/min})(7.48\,\text{gal/ft}^3)=14,200\,\text{gpm}$
$x=2.03\,\text{ft}$

7.32 $(0.785)(0.67)(0.67)(x\,\text{fps})(7.48\,\text{gal/ft}^3)(60\,\text{sec/min})=584\,\text{gpm}$
$x=3.7\,\text{fps}$

7.33 $550\,\text{ft}/208\,\text{sec}=2.6\,\text{fps}$

7.34 $(0.785)(0.83\,\text{ft})(0.83\,\text{ft})(2.4\,\text{fps})=(0.785)(0.67\,\text{ft})(0.67\,\text{ft})(x\,\text{fps})$
$x=3.7\,\text{fps}$

7.35 $500\,\text{ft}/92\,\text{sec}=5.4\,\text{fps}$

7.36 $(0.785)(0.67)(0.67)(3.2\,\text{fps})=(0.785)(0.83\,\text{ft})(0.83\,\text{ft})(x\,\text{fps})$
$x=2.1\,\text{fps}$

7.37 $35.3\,\text{MGD}/7=5\,\text{MGD}$

7.38 $121.4\,\text{MG}/30\,\text{days}=4.0\,\text{MGD}$

7.39 $1,000,000\times0.165=165,000\,\text{gpd}$

7.40 $3,335,000\,\text{gal}/1,440\,\text{min}=2,316\,\text{gpm}$

7.41 $(8\,\text{cfs})(7.48\,\text{gal/ft}^3)(60\,\text{sec/min})=3,590\,\text{gpm}$

7.42 $(35\,\text{gps})(60\,\text{sec/min})(1,440\,\text{min/day})=3,024,000\,\text{gpd}$

7.43 $\dfrac{4,570,000\,\text{gpd}}{(1,440\,\text{min/day})(7.48\,\text{gal/ft}^3)}=424\,\text{cfm}$

7.44 $(6.6\,\text{MGD})(1.55\,\text{cfs/MGD})=10.2\,\text{cfs}$

7.45 $\dfrac{(445,875\,\text{cfd})(7.48\,\text{gal/ft}^3)}{1,440\,\text{min/day}}=2,316\,\text{gpm}$

7.46 $(2450\,\text{gpm})(1,440\,\text{min/day})=3,528,000\,\text{gpd}$

7.47 $(6\,\text{ft})(2.5\,\text{ft})(x\,\text{fps})(7.48\,\text{gal/ft}^3)(60\,\text{sec/min})=14,800\,\text{gpm}$
$x=2.2\,\text{fps}$

7.48 $(4.6\,\text{ft})(3.4\,\text{ft})(3.6\,\text{fps})(60\,\text{sec/min})=3,378\,\text{cfm}$

7.49 $373.6/92\,\text{days}=4.1\,\text{MGD}$

7.50 $(12\,\text{ft})(12\,\text{ft})(0.67\,\text{fpm})(7.48\,\text{gal/ft}^3)=722\,\text{gpm}$

7.51 $(0.785)(0.67)(0.67\,\text{ft})(x\,\text{fps})(7.48\,\text{gal/ft}^3)(60\,\text{sec/min})=510\,\text{gpm}$
$x=3.2\,\text{fps}$

7.52 $(10\,\text{cfs})(7.48\,\text{gal/ft}^3)(60\,\text{sec/min})=4,488\,\text{gpm}$

7.53 $134.6/31\,\text{days}=4.3\,\text{MGD}$

7.54 $(5.2\,\text{MGD})(1.55\,\text{cfs/MGD})=8.1\,\text{cfs}$

7.55 $(0.785)(2\,\text{ft})(2\,\text{ft})(3.3\,\text{fpm})(7.48\,\text{gal/ft}^3)=77.5\,\text{gpm}$

7.56 $\dfrac{(1,825,000\,\text{gpd})}{(1,440\,\text{min/day})(7.48\,\text{gal/ft}^3)}=169\,\text{cfm}$

7.57 $(0.785)(0.5\,\text{ft})(0.5\,\text{ft})(2.9\,\text{fps})(7.48\,\text{gal/ft}^3)(60\,\text{sec/min})=255\,\text{gpm}$

7.58 $(0.785)(0.83\,\text{ft})(0.83\,\text{ft})(2.6\,\text{fps})=(0.785)(0.67\,\text{ft})(0.67\,\text{ft})(x\,\text{fps})$
$x=4.0\,\text{fps}$

7.59 $(2,225\,\text{gpm})(1,440\,\text{min/day})=3,204,000\,\text{gpd}$

7.60 $5,350,000\,\text{gal}/1,440\,\text{min/day}=3,715\,\text{gpm}$

7.61 $(2.5\,\text{mg/L})(5.5\,\text{MGD})(8.34\,\text{lb/gal})=115\,\text{lbs/day}$

7.62 $(7.1\,\text{mg/L})(4.2\,\text{MGD})(8.34\,\text{lb/gal})=249\,\text{lbs/day}$

7.63 $(11.8\,\text{mg/L})(4.8\,\text{MGD})(8.34\,\text{lb/gal})=472\,\text{lbs/day}$

7.64 $\dfrac{(10\,\text{mg/L})(1.8\,\text{MGD})(8.34\,\text{lb/gal})}{0.65}=231\,\text{lbs/day}$

7.65 $(60\,\text{mg/L})(0.086\,\text{MGD})(8.34\,\text{lb/gal})=43\,\text{lbs}$

7.66 $(2,220\,\text{mg/L})(0.225)(8.34\,\text{lb/gal})=4,166\,\text{lbs}$

7.67 $\dfrac{(8\,\text{mg/L})(0.83\,\text{MGD})(8.34\,\text{lb/gal})}{0.65}=85\,\text{lb/day}$

7.68 $(450\,\text{mg/L})(1.84\,\text{MGD})(8.34\,\text{lb/gal})=6,906\,\text{lb/day}$

7.69 $(25\,\text{mg/L})(2.90\,\text{MGD})(8.34\,\text{lb/gal})=605\,\text{lb/day}$

7.70 $(260\,\text{mg/L})(5.45\,\text{MGD})(8.34\,\text{lb/gal})=11,818\,\text{lb/day}$

7.71 $(144\,\text{mg/L})(3.66\,\text{MGD})(8.34\,\text{lb/gal})=4,396\,\text{lb/day}$

7.72 $(290\,\text{mg/L})(3.31\,\text{MGD})(8.34\,\text{lb/gal})=8,006\,\text{lb/day}$

7.73 $(152\,\text{mg/L})(5.7\,\text{MGD})(8.34\,\text{lb/gal})=7,226\,\text{lb/day}$

7.74 $(188\,\text{mg/L})(1.92\,\text{MGD})(8.34\,\text{lb/gal})=3,010\,\text{lb/day SS}$

7.75 $(184\,\text{mg/L})(1.88\,\text{MGD})(8.34\,\text{lb/day})=2,885\,\text{lb/day SS}$

7.76 $(150\,\text{mg/L})(4.88\,\text{MGD})(8.34\,\text{lbs/gal})=6,105\,\text{lbs/day BOD}$

7.77 $(205\,\text{mg/L})(2.13)(8.34\,\text{lb/gal})=3,642\,\text{solids}$

7.78 $(115\,\text{mg/L})(4.20\,\text{MGD})(8.34\,\text{lb/gal})=4,028\,\text{lb/day}$

7.79 $(2,230\,\text{mg/L})(0.40\,\text{MG})(8.34\,\text{lb/gal})=7,439\,\text{lb SS}$

7.80 $(1,890\,\text{mg/L})(0.41\,\text{MG})(8.34\,\text{lb/gal})=6,463\,\text{lb MLVSS}$

7.81 $(3,125\,\text{mg/L})(0.18\,\text{MG})(8.34\,\text{lb/gal})=4,691\,\text{lb MLVSS}$

7.82 $(2,250\,\text{mg/L})(0.53\,\text{MG})(8.34\,\text{lb/gal})=9,945\,\text{lbs MLSS}$

7.83 $(2,910\,\text{mg/L})(0.63\,\text{MG})(8.34\,\text{lb/gal})=15,290\,\text{lbs MLSS}$

7.84   (6,150 mg/L)(x MGD)(8.34 lb/gal) = 5,200 lbs/day
         x = 0.10 MGD

7.85   (6,200 mg/L)(x MGD)(8.34 lb/gal) = 4,500 lb/day
           (a) x = 0.09
           (b) 90,000 gpd/1,440 min/day = 62.5 gpm

7.86   (6,600 lb/day)(x MGD) (8.34 lb/gal) = 6,070 lb/day
         x = 0.11 MGD
           = 110,000 gpd ÷ 1,440 min/day = 76 gpm

7.87   (6,350 mg/L)(x MGD)(8.34 lb/day) = 7,350 lb/day
         x = 0.14 MGD
           = 140,000 gpd ÷ 1,440 min/day
           = 97 gpm

7.88   (7240 mg/L)(x MGD)(8.34 lbs/gal) = 5,750 lbs/day
         x = 0.10 MGD
           = 100,000 gpd ÷ 1,440 min/day = 69 gpm

7.89   (2.5 mg)(3.65 MGD)(8.34 lbs/gal) = 76.1 lb/day

7.90   (17 mg/L)(2.10 MGD)(8.34 lb/gal) = 298 lb/day BOD

7.91   (190 mg/L)(4.8 MGD)(8.34 lb/gal) = 7,606 lb/day SS Rem.

7.92   (9.7 mg/L)(5.5 MGD)(8.34 lb/gal) = 445 lb/day

7.93   (305 mg/L)(3.5 MGD)(8.34 lb/gal) = 8,903 lb/day

7.94 $\dfrac{(10 \text{ mg/L})(3.1 \text{ MGD})(8.34 \text{ lb/gal})}{0.65}$

         = 398 lb/day Hypochlorite

7.95   (210 mg/L)(3.44 MGD)(8.34 lb/gal) = 6,025 lb/day Solids

7.96   (60 mg/L)(0.09 MG)(8.34 lb/gal) = 45 lb Chlorine

7.97   (2,720 mg/L)(0.52 MG)(8.34 lb/gal) = 11,796 lb MLSS

7.98   (5,870 mg/L)(x MGD)(8.34 lb/gal) = 5,480 lb/day
           x = 0.11 MGD

7.99   (120 mg/L)(3.312 MGD)(8.34 lb/gal) = 3,315 lb/day BOD

7.100 (240 mg/L)(3.18 MGD)(8.34 lb/gal) = 6,365 lb BOD

7.101 (196 mg/L)(1.7 MGD)(8.34 lb/gal) = 2,779 lb/day BOD Removed

7.102 (x mg/L)(5.3 MGD)(8.34 lb/day) = 330 lb/day
           x = 7.5 mg/L

7.103 (5,810 mg/L)(x MGD)(8.34 lb/gal) = 5,810 mg/L
           x = 0.12 MGD
             = 120,000 gpd ÷ 1,440 min/day
             = 83 gpm

7.104 $\dfrac{3,400,000 \text{ gpd}}{(0.785)(100 \text{ ft})(100 \text{ ft})} = 433 \text{ gpd/ft}^2$

7.105 $\dfrac{4,525,000 \text{ gpd}}{(0.785)(90 \text{ ft})(90 \text{ ft})} = 712 \text{ gpd/ft}^2$

7.106 $\dfrac{3,800,000 \text{ gpd}}{870,000 \text{ ft}^2} = 4.4 \text{ gpd/ft}^2$

7.107 $\dfrac{280,749 \text{ ft}^3\text{day}}{696,960 \text{ ft}^2} = 0.4 \text{ ft/day}$

         (0.4 ft/day)(12 in/ft) = 4.8 in/day

7.108 $\dfrac{5,280,000 \text{ gpd}}{(0.785)(90 \text{ ft})(90 \text{ ft})} = 830 \text{ gpd/ft}^2$

7.109 $\dfrac{4.4 \text{ acre-ft/day}}{20 \text{ acre}} = 0.22 \text{ ft/day}$

                 = 3 in/day

7.110 $\dfrac{2,050,000 \text{ gpd}}{(70 \text{ ft})(25 \text{ ft})} = 1,171 \text{ gpd/ft}^2$

7.111 $\dfrac{2,440,000 \text{ gpd}}{(0.785)(60 \text{ ft})(60 \text{ ft})} = 863 \text{ gpd/ft}^2$

7.112 $\dfrac{3,450,000 \text{ gpd}}{(110 \text{ ft})(50 \text{ ft})} = 627 \text{ gpd/ft}^2$

7.113 $\dfrac{1,660,000 \text{ gpd}}{(25 \text{ ft})(70 \text{ ft})} = 949 \text{ gpd/ft}^2$

7.114 $\dfrac{2,660,000 \text{ gpd}}{(0.785)(70 \text{ ft})(70 \text{ ft})} = 691 \text{ gpd/ft}^2$

7.115 $\dfrac{2,230 \text{ gpm}}{(40 \text{ ft})(20 \text{ ft})} = 2.8 \text{ gpm/ft}^2$

7.116 $\dfrac{3,100 \text{ gpm}}{(40 \text{ ft})(25 \text{ ft})} = 3.1 \text{ gpm/ft}^2$

7.117 $\dfrac{2,500 \text{ gpm}}{(26 \text{ ft})(60 \text{ ft})} = 1.6 \text{ gpm/ft}^2$

7.118 $\dfrac{1,528 \text{ gpm}}{(40 \text{ ft})(20 \text{ ft})} = 1.9 \text{ gpm/ft}^2$

7.119 $\dfrac{2,850 \text{ gpm}}{880 \text{ ft}^2} = 3.2 \text{ gpm/ft}^2$

7.120 $\dfrac{4,750 \text{ gpm}}{(14 \text{ ft})(14 \text{ ft})} = 24 \text{ gpm/ft}^2$

7.121 $\dfrac{4,900 \text{ gpm}}{(20 \text{ ft})(20 \text{ ft})} = 12 \text{ gpm/ft}^2$

7.122 $\dfrac{3,400 \text{ gpm}}{(25 \text{ ft})(15 \text{ ft})} = 9 \text{ gpm/ft}^2$

7.123 $\dfrac{3,300 \text{ gpm}}{(75 \text{ ft})(30 \text{ ft})} = 4.4 \text{ gpm/ft}^2$

7.124 $\dfrac{3,800 \text{ gpm}}{(15 \text{ ft})(20 \text{ ft})} = 12.7 \text{ gpm/ft}^2$

7.125 $\dfrac{3,770,000 \text{ gal}}{(15 \text{ ft})(30 \text{ ft})} = 8,378 \text{ gal/ft}^2$

7.126 $\dfrac{1,860,000 \text{ gal}}{(20 \text{ ft})(15 \text{ ft})} = 6,200 \text{ gal/ft}^2$

7.127 $\dfrac{3,880,000 \text{ gal}}{(25 \text{ ft})(20 \text{ ft})} = 7,760 \text{ gal/ft}^2$

7.128 $\dfrac{1,410,200 \text{ gal}}{(20 \text{ ft})(14 \text{ ft})} = 5,036 \text{ gal/ft}^2$

7.129 $\dfrac{5,425,000 \text{ gal}}{(30 \text{ ft})(20 \text{ ft})} = 9,042 \text{ gal/ft}^2$

7.130 $\dfrac{1,410,000 \text{ gpd}}{163 \text{ ft}} = 8,650 \text{ gpd/ft}$

7.131 $\dfrac{2,120,000 \text{ gpd}}{(3.14)(60 \text{ ft})} = 11,253 \text{ gpd/ft}$

7.132 $\dfrac{2,700,00 \text{ gpd}}{240 \text{ ft}} = 11,250 \text{ gpd/ft}$

7.133 $\dfrac{(1,400 \text{ gpm})(1,440 \text{ min/day})}{(3.14)(80 \text{ ft})} = 8,025 \text{ gpd/ft}$

7.134 $\dfrac{2,785 \text{ gpm}}{189 \text{ ft}} = 14.7 \text{ gpm/ft}$

7.135 $\dfrac{(210 \text{ mg/L})(2.45 \text{ MGD})(8.34 \text{ lb/gal})}{25.1 \quad 1,000\text{-ft}^3}$

$= 171 \text{ lbs BOD/day/1,000 ft}^3$

7.136 $\dfrac{(170 \text{ mg/L})(0.120 \text{ MGD})(8.34 \text{ lb/gal})}{3.5 \text{ acre}}$

$= 49 \text{ lbs BOD/day/acre}$

7.137 $\dfrac{(120 \text{ mg/L})(2.85 \text{ MGD})(8.34 \text{ lb/gal})}{34 \quad 1,000\text{-ft}^3}$

$= 84 \text{ lb BOD/day/1,000 ft}^3$

7.138 $\dfrac{(140 \text{ mg/L})(2.20 \text{ MGD})(8.34 \text{ lb/gal})}{900 \quad 1,000\text{-ft}^3}$

$= 2.9 \text{ lb BOD/day/1,000 ft}^3$

$(0.785)(90 \text{ ft})(90 \text{ ft})(4 \text{ ft}) = 25,434$

7.139 $\dfrac{(150 \text{ mg/L})(3.5 \text{ MGD})(8.34 \text{ lb/gal})}{25.4 \quad 1,000\text{-ft}^3}$

$= 172 \text{ lb BOD/day/1,000 ft}^3$

7.140 $\dfrac{(200 \text{ mg/L})(3.42 \text{ MGD})(8.34 \text{ lb/gal})}{(1,875 \text{ mg/L})(0.42 \text{ MG})(8.34 \text{ lb/gal})} = 0.9$

7.141 $\dfrac{(190 \text{ mg/L})(3.24 \text{ MGD})(8.34 \text{ lb/gal})}{(1,710 \text{ mg/L})(0.28 \text{ MG})(8.34 \text{ lb/gal})} = 1.3$

7.142 $\dfrac{(151 \text{ mg/L})(2.25 \text{ MGD})(8.34 \text{ lb/gal})}{x \text{ lb MLVSS}} = 0.9$

$x = 3,148 \text{ lb MLVSS}$

7.143 $\dfrac{(160 \text{ mg/L})(2.10 \text{ MGD})(8.34 \text{ lb/gal})}{(1,900 \text{ mg/L})(0.255 \text{ MG})(8.34 \text{ lb/gal})} = 0.7$

7.144 $\dfrac{(180 \text{ mg/L})(3.11 \text{ MGD})(8.34 \text{ lb/gal})}{(x \text{ mg/L})(0.88 \text{ MG})(8.34 \text{ lb/gal})} = 0.5$

$x = 1,262 \text{ mg/L MLVSS}$

7.145 $\dfrac{(2,650 \text{ mg/L})(3.60 \text{ MGD})(8.34 \text{ lb/gal})}{(0.785)(70 \text{ ft})(70 \text{ ft})}$

$= 20.7 \text{ lbs MLSS/day/ft}^2$

7.146 $\dfrac{(2,825 \text{ mg/L})(4.25 \text{ MGD})(8.34 \text{ lb/gal})}{(0.785)(80 \text{ ft})(80 \text{ ft})}$

$= 19.9 \text{ lbs MLSS/day/ft}^2$

7.147 $\dfrac{(x \text{ mg/L})(3.61 \text{ MGD})(8.34 \text{ lb/gal})}{(0.785)(60 \text{ ft})(60 \text{ ft})} = 26 \text{ lbs/day/ft}^2$

$x = 2,441 \text{ mg/L MLSS}$

7.148 $\dfrac{(2,210 \text{ mg/L})(3.3 \text{ MGD})(8.34 \text{ lb/gal})}{(0.785)(60 \text{ ft})(60 \text{ ft})}$

$= 21.5 \text{ lb MLSS/day/ft}^2$

7.149 $\dfrac{(x \text{ mg/L})(3.11 \text{ MGD})(8.34 \text{ lb/gal})}{(0.785)(60 \text{ ft})(60 \text{ ft})} = 20 \text{ lbs MLSS/day/ft}^2$

$x = 2,174 \text{ mg/L MLSS}$

7.150 $\dfrac{12,110 \text{ lb VS/day}}{33,100 \text{ ft}^3} = 0.37 \text{ lbs VS/day/ft}^3$

7.151 $\dfrac{(124,000 \text{ lb/day})(0.065)(0.70)}{(0.785)(60 \text{ ft})60 \text{ ft})(25 \text{ ft})} = 0.08 \text{ lbs VS/day/ft}^2$

7.152 $\dfrac{(141,000 \text{ lb/day})(0.06)(0.71)}{(0.785)(50 \text{ ft})(50 \text{ ft})(20 \text{ ft})} = 0.15 \text{ lbs VS/day/ft}^3$

7.153 $\dfrac{(21,200 \text{ gpd})(8.34 \text{ lb/gal})(0.055)(0.69)}{(0.785)(40 \text{ ft})(40 \text{ ft})(16 \text{ ft})}$

$= 0.33 \text{ VS/day/ft}^3$

7.154 $\dfrac{(22,000 \text{ gpd})(8.6 \text{ lb/gal})(0.052)(0.70)}{(0.785)(50 \text{ ft})(50 \text{ ft})(20 \text{ ft})}$

$= 0.18 \text{ lbs VS/day/ft}^3$

7.155 $\dfrac{2,050 \text{ lb VS Added/day}}{32,400 \text{ lb VS}} = 0.06$

7.156 $\dfrac{620 \text{ lb VS Added/day}}{(174,600 \text{ lb})(0.061)(0.65)} = 0.09$

7.157 $\dfrac{(63,200 \text{ lb/day})(0.055)(0.73)}{(115,000 \text{ gal})(8.34 \text{ lb/gal})(0.066)(0.59)} = 0.07$

7.158 $\dfrac{x \text{ lb VS Added/day}}{(110,000 \text{ gal})(8.34 \text{ lb/gal})(0.059)(0.58)} = 0.08$

$x = 2,511 \text{ lb/day VS}$

7.159 $\dfrac{(7,900 \text{ gpd})(8.34 \text{ lb/gal})(0.048)(0.73)}{x \text{ lb VS}} = 0.06$

$x = 38,477 \text{ lb VS}$

7.160 $1,733 \text{ people}/5.3 \text{ acre} = 327 \text{ people/acre}$

7.161 $4,112 \text{ people}/10 \text{ acre} = 411 \text{ people/acre}$

7.162 $\dfrac{(1,765 \text{ mg/L})(0.381 \text{ MGD})(8.34 \text{ lb/gal})}{0.2 \text{ lb/day}}$

$= 28,040 \text{ people}$

7.163 $\dfrac{6,000 \text{ people}}{x \text{ acre}} = 420 \text{ people/acre}$

$x = 14.3 \text{ acre}$

7.164 $\dfrac{(2,210 \text{ mg/L})(0.100 \text{ MGD})(8.34 \text{ lb/gal})}{0.2 \text{ lb/day}}$

$= 9,216 \text{ people}$

7.165 $\dfrac{2,250,000 \text{ gpd}}{(0.785)(80 \text{ ft})(80 \text{ ft})} = 448 \text{ gpd/ft}^2$

7.166 $\dfrac{2,960 \text{ gpm}}{190 \text{ ft}^2} = 15.6 \text{ gpm/ft}^2$

7.167 $\dfrac{2,100,000 \text{ gpd}}{(3.14)(80 \text{ ft})} = 8,360 \text{ gpd/ft}$

7.168 $\dfrac{3,300,000 \text{ gpd}}{(0.785)(90 \text{ ft})(90 \text{ ft})} = 519 \text{ gpd/ft}^2$

7.169 $\dfrac{(161 \text{ mg/L})(2.1 \text{ MGD})(8.34 \text{ lb/gal})}{x \text{ lb MLVSS}} = 0.7$

$x = 4,028 \text{ lb MLVSS}$

7.170 $\dfrac{500 \text{ lb/day VS Added/day}}{(182,000 \text{ lb})(0.064)(0.67)} = 0.06$

7.171 $\dfrac{(2,760 \text{ mg/L})(3.58 \text{ MGD})(8.34 \text{ lb/gal})}{(0.785)(80 \text{ ft})(80 \text{ ft})}$

$= 16 \text{ lbs/day/ft}^2$

7.172
7.173 $\dfrac{(115,000 \text{ lb/day})(0.071)(0.70)}{(0.785)(70 \text{ ft})(70 \text{ ft})(21 \text{ ft})} = 0.09$

$\dfrac{4.15 \text{ acre-ft/day}}{25 \text{ acre}} = 0.17 \text{ ft/day}$

$= (0.17 \text{ ft/day})(12 \text{ in/ft}) = 2.0 \text{ in/day}$

7.174 $\dfrac{(174 \text{ mg/L})(3.335 \text{ MGD})(8.3 \text{ lb/gal})}{(x \text{ mg/L})(0.287 \text{ MG})(8.34 \text{ lb/gal})} = 0.5$

$x = 4,033 \text{ mg/L MLVSS}$

7.175 $\dfrac{2,000,000 \text{ gpd}}{(80 \text{ ft})(25 \text{ ft})} = 1,000 \text{ gpd/ft}^2$

7.176 $\dfrac{1,785,000 \text{ gal}}{(25 \text{ ft})(20 \text{ ft})} = 3,570 \text{ gal/ft}^2$

7.177 $\dfrac{(150 \text{ mg/L})(2.69 \text{ MGD})(8.34 \text{ lb/gal})}{(1,920 \text{ mg/L})(0.31 \text{ MG})(8.34 \text{ lb/gal})} = 0.68$

7.178 $\dfrac{x \text{ lb VS added/day}}{(24,500 \text{ gal})(8.34 \text{ lb/gal})(0.055)(0.56)} = 0.09$

$x = 566 \text{ lb/day}$

7.179 $\dfrac{3,083 \text{ gpm}}{(40 \text{ ft})(30 \text{ ft})} = 2.6 \text{ gpm/ft}^2$

7.180 $\dfrac{(115 \text{ mg/L})(3.3 \text{ MGD})(8.34 \text{ lb/gal})}{20.1 \quad 1,000\text{-ft}^3}$

$= 157 \text{ lb BOD/day/1,000 ft}^3$

7.181 $\dfrac{2,560,000 \text{ gpd}}{(3.14)(80 \text{ ft})} = 10,191 \text{ gpd/ft}$

7.182 $1,900 \text{ people/5.5 acre} = 345 \text{ people/acre}$

7.183 $\dfrac{(140 \text{ mg/L})(2.44 \text{ MGD})(8.34 \text{ lb/gal})}{750 \quad 1,000\text{-ft}^2}$

$= 3.8 \text{ lb BOD/day/1,000 ft}^2$

7.184 $\dfrac{2,882 \text{ gpm}}{(40 \text{ ft})(30 \text{ ft})} = 2.4 \text{ gpm/ft}^2$

7.185 $\dfrac{(30 \text{ ft})(16 \text{ ft})(8 \text{ ft})(7.48 \text{ gal/ft}^2)}{1,007 \text{ gpm}} = 29 \text{ min}$

7.186 $\dfrac{(80 \text{ ft})(20 \text{ ft})(12 \text{ ft})(7.48 \text{ gal/ft}^3)}{75,000 \text{ gph}} = 1.9 \text{ h}$

7.187 $\dfrac{(3 \text{ ft})(4 \text{ ft})(3 \text{ ft})(7.48 \text{ gal/ft}^3)}{(6 \text{ gpm})(60 \text{ min/h})} = 0.75 \text{ h}$

7.188 $\dfrac{(0.785)(80 \text{ ft})(80 \text{ ft})(10 \text{ ft})(7.48 \text{ gal/ft}^3)}{216,667 \text{ gpd}} = 1.7 \text{ h}$

7.189 $\dfrac{(500 \text{ ft})(600 \text{ ft})(6 \text{ ft})(7.48 \text{ gal/ft}^3)}{222,500 \text{ gpd}} = 60.5 \text{ days}$

7.190 $\dfrac{12,300 \text{ lb MLSS}}{2,750 \text{ lb/day}} = 4.5 \text{ days}$

7.191 $\dfrac{(2,820 \text{ mg/L MLSS})(0.49 \text{ MG})(8.34 \text{ lb/gal})}{(132 \text{ mg/L})(0.988 \text{ MGD})(8.34 \text{ lb/gal})}$

$= 10.6 \text{ days}$

7.192 $\dfrac{(2,850 \text{ mg/L MLSS})(0.20 \text{ MG})(8.34 \text{ lb/gal})}{(84 \text{ mg/L})(1.52 \text{ MGD})(8.34 \text{ lb/gal})}$

$= 4.5 \text{ days}$

7.193 $\dfrac{(x \text{ mg/L MLSS})(0.205 \text{ MG})(8.34 \text{ lb/gal})}{(80 \text{ mg/L})(2.10 \text{ MGD})(8.34 \text{ lb/gal})} = 6 \text{ days}$

$x = 4,917 \text{ mg/L MLSS}$

7.194 $\dfrac{x \text{ lb MLSS}}{1,610 \text{ lb/day SS}} = 5.5 \text{ days}$

$x = 8,855 \text{ lb MLSS}$

7.195 $\dfrac{(3,300 \text{ mg/L})(0.50 \text{ MG})(8.34 \text{ lb/gal})}{1,610 \text{ lb/day Wasted} + 340 \text{ lb/day in SE}} = 7.1 \text{ days}$

$(2,750 \text{ mg/L MLSS})$

7.196 $\dfrac{(0.360 \text{ MG})(8.35 \text{ lb/gal})}{(5,410 \text{ mg/L})(0.0192 \text{ MG})(8.34 \text{ lb/gal})}$

$+ (16 \text{ mg/L SS})(2.35 \text{ MGD})(8.34 \text{ lb/gal})$

$= \dfrac{8,257 \text{ lb}}{866 \text{ lb/day} + 314 \text{ lb/day}} = 7.0 \text{ days}$

$(2,550 \text{ mg/L MLSS})$

7.197 $\dfrac{(1.8 \text{ MG})(8.34 \text{ lb/gal})}{(6,240 \text{ mg/L SS})(0.085 \text{ MGD})(8.34 \text{ lb/gal})}$

$+ (20 \text{ mg/L})(2.8 \text{ MGD})(8.34 \text{ lb/gal})$

$\dfrac{38,281 \text{ lb MLSS}}{4,424 \text{ lb/day} + 467 \text{ lbs/day}} = 8 \text{ days}$

$$\frac{\dfrac{(x\ mg/L)(0.970\ MG)}{(8.34\ lb/gal)}}{(6{,}340\ mg/L)(0.032\ MGD)(8.34\ lb/gal)} = 8\ days$$

7.198

$$\frac{(x\ mg/L)(0.970\ MG)(8.34\ lb/gal)}{+(20\ mg/L)(2.6\ MGD)(8.34\ lb/gal)} = 8\ days$$

$$\frac{(x\ mg/L)(0.970\ MG)(8.34\ lb/gal)}{1{,}692\ lb/day + 434\ lb/day} = 8\ days$$

$$\frac{(x\ mg/L)(0.970\ MG)(8.34\ lb/gal)}{2{,}126} = 8\ days$$

$$x = 2{,}100\ mg/L\ MLSS$$

7.199 $\dfrac{(75\ ft)(30\ ft)(14\ ft)(7.48\ gal/ft^3)}{68{,}333\ gph} = 3.5\ h$

7.200 $\dfrac{12{,}600\ lb\ MLSS}{2{,}820\ lb/day} = 4.5\ days$

7.201 $\dfrac{(3{,}120\ mg/L\ MLSS)(0.48\ MG)(8.34\ lb/gal)}{1{,}640\ lb/day\ wasted + 320\ lb/day}$
$= 6.4\ days$

7.202 $\dfrac{(40\ ft)(20\ ft)(10\ ft)(7.48\ gal/ft^3)}{1{,}264\ gpm} = 47\ min$

7.203 $\dfrac{(2{,}810\ mg/L\ MLSS)(0.325\ MG)(8.34\ lb/gal)}{(6{,}100\ mg/L)(0.0189\ MGD)(8.34\ lb/gal)+}$
$(18\ mg/L)(2.4\ MGD)(8.34\ lb/gal)$
$= \dfrac{7{,}617\ lb\ MLSS}{962\ lb/day + 360\ lb/day} = 5.8\ days$

7.204 $\dfrac{(3{,}250\ mg/L)(0.33\ MG)(8.34\ lb/gal)}{(100\ mg/L)(2.35\ MGD)(8.34\ lb/gal)} = 4.6\ days$

7.205 $\dfrac{(2{,}408\ mg/L)(1.9\ MG)(8.34\ lb/gal)}{(6{,}320\ mg/L)(0.0712\ MGD)(8.34\ lb/gal)+}$
$(25\ mg/L)(2.85\ MGD)(8.34\ lb/gal)$
$= \dfrac{38{,}157\ lb}{3{,}753\ lb/day + 594\ lb/day} = 9.8\ days$

7.206 $\dfrac{(2{,}610\ mg/L)(0.15\ MG)(8.34\ lb/gal)}{(140\ mg/L)(0.92\ MGD)(8.34\ lb/gal)} = 3\ days$

7.207 $\dfrac{(0.785)(6\ ft)(6\ ft)(4\ ft)(7.48\ gal/ft^3)}{12\ gpm} = 70\ min$

7.208 $\dfrac{x\ lbs\ MLSS}{(140\ mg/L)(2.14\ MGD)(8.34\ lb/gal)} = 6\ days$
$x = 14{,}992\ lb\ MLSS$

7.209 $\dfrac{(400\ ft)(440\ ft)(6\ ft)(7.48\ gal/ft^3)}{200{,}000\ gpd} = 39.5$

$$\frac{\dfrac{(x\ mg/L\ MLSS)(0.64\ MG)}{(8.34\ lb/gal)}}{(6{,}310\ mg/L)(0.034\ MGD)(8.34\ lb/gal)+} = 8\ days$$

7.210

$$\frac{(x\ mg/L)(0.64\ MG)(8.34\ lb/gal)}{(12\ mg/L)(2.92\ MGD)(8.34\ lb/gal)} = 8\ days$$

$$\frac{(x\ mg/L)(0.64\ MG)(8.34\ lb/gal)}{1{,}789\ lb/day + 292\ lb/day} = 8\ days$$

$$\frac{(x\ mg/L)(0.64\ MG)(8.34\ lb/gal)}{2{,}081\ lb/day} = 8\ days$$

$$x = 3{,}141\ mg/L\ MLSS$$

7.211 $\dfrac{89\ mg/L\ Rem.}{110\ mg/L} \times 100 = 81\%$

7.212 $\dfrac{216\ mg/L\ Rem.}{230\ mg/L} \times 100 = 94\%$

7.213 $\dfrac{200\ mg/L\ Rem.}{260\ mg/L} \times 100 = 77\%$

7.214 $\dfrac{175\ mg/L\ Rem.}{310\ mg/L} \times 100 = 56\%$

7.215 $4.9 = \dfrac{x\ lb/day\ Solids}{(3{,}700\ gal)(8.34\ lb/gal)} \times 100 = 56\%$
$x = 1{,}512\ lb/day\ Solids$

7.216 $\dfrac{0.87\ g\ Sludge}{12.87\ g\ Sludge} \times 100 = 6.8\%$

7.217 $\dfrac{1{,}450\ lb/day\ Solids}{x\ lb/day\ Sludge} \times 100 = 3.3\%$
$x = 43{,}939\ lb/day$

7.218 $4.4 = \dfrac{258\ lb/day}{(x\ gpd)(8.34\ lb/gal)} \times 100$
$x = 703\ gpd$

7.219 $3.6 = \dfrac{x\ lb/day\ Solids}{291{,}000\ lb/day\ Sludge} \times 100$
$x = 10{,}476\ lb/day\ Solids$

7.220 $\dfrac{\dfrac{(3{,}100\ gpd)(8.34\ lb/gal)(4.4)}{100}+}{\dfrac{(4{,}100\ gp)(8.34\ lb/gal)(3.6)}{100}} \times 100$
$\dfrac{(3{,}100\ gpd)(8.34\ lb/gal)+}{(4{,}100\ gpd)(8.34\ lb/gal)}$

$\dfrac{1{,}138\ lb/day\ Solids + 1{,}231\ lb/day\ Solids}{25{,}854\ lb/day\ Sludge + 34{,}194\ lb/day\ Sludge} \times 100$

$\dfrac{2{,}369\ lb/day\ Solids}{60{,}048\ lb/day\ Sludge} \times 100$
$= 3.9\%$

$$\frac{(8,100\ \text{gpd})(8.34\ \text{lb/gal})(5.1)}{100}+$$

$$\frac{\dfrac{(7,000\ \text{gpd})(8.34\ \text{lb/gal})(4.1)}{100}}{(8,100\ \text{gpd})(8.34\ \text{lb/gal})+}\times 100$$

7.221 $(7,000\ \text{gpd})(8.34\ \text{lb/gal})$

3,445 lb/day Solids +

$$=\frac{2,394\ \text{lb/day Solids}}{67,554\ \text{lb/day Sludge}+}\times 100$$

58,380 lb/day Sludge

$$=\frac{5,839\ \text{lb/day Solids}}{125,934\ \text{lb/day Sludge}}\times 100$$

= 4.6%

$$\frac{(4,750\ \text{gpd})(8.34\ \text{lb/gal})(4.7)}{100}+$$

$$\frac{\dfrac{(5250\ \text{gpd})(8.34\ \text{lb/gal})(3.5)}{100}}{(4,750\ \text{gpd})(8.34\ \text{lb/gal})+}\times 100$$

7.222 $(5,250\ \text{gpd})(8.34\ \text{lb/gal})$

$$=\frac{1,862\ \text{lb/day}+1,532\ \text{lb/day}}{39,615+43,785}\times 100$$

$$=\frac{3,394\ \text{lb/day Solids}}{83,400\ \text{lb/day Sludge}}\times 100$$

= 4.1%

$$\frac{(8,925\ \text{gpd})(8.34\ \text{lb/gal})(4.0)}{100}+$$

$$\frac{\dfrac{(11,340\ \text{gpd})(8.34\ \text{lb/gal})(6.6)}{100}}{(8,925\ \text{gpd})(8.34\ \text{lb/gal})+}\times 100$$

7.223 $(11,340\ \text{gpd})(8.34\ \text{lb/gal})$

$$=\frac{2,977\ \text{lb/day}+6,242\ \text{lb/day}}{74,435\ \text{lb/day}+94,576\ \text{lb/day}}\times 100$$

$$=\frac{9,219\ \text{lb/day}}{169,011\ \text{lb/day}}\times 100$$

= 5.5%

7.224 (3,250 lb/day Solids)(0.65) = 2,113 lb/day VS

7.225   (4,120 gpd)(8.34 lb/gal)(0.07)(0.70)

= 1,684 lb/day VS

7.226 98 ft − 91 ft = 7 ft drawdown

7.227 125 ft − 110 ft = 15 ft drawdown

7.228 161 ft − 144 ft = 17 ft drawdown

(3.7 psi)(2.31 ft/psi)

= 8.5 ft water depth in sounding line

= 112 ft − 8.5 ft

7.229

= 103.5 ft

= 103.5 ft − 86 ft

= 17.5 ft

(4.6 psi)(2.31 ft/psi)

= 10.6 water depth in sounding line

= 150 ft − 10.6 ft

7.230

= 139.4 ft

= 171 ft − 139.4 ft

= 31.4 ft drawdown

7.231 300/20 = 15 gpm/ft of drawdown

7.232 420 gal/5 min = 84 gpm

7.233 810 gal/5 min = 162 gpm

856 gal/5 min = 171 gpm

7.234

(171 gpm)(60 min/h) = 10,260 gph

(0.785)(1 ft)(1 ft)(12 ft)

7.235 $\dfrac{(7.48\ \text{gal/ft}^3)(12\ \text{round trips})}{5\ \text{min}}=169\ \text{gpm}$

750 gal/5 min = 150 gpm

7.236 (150 gpm)(60 min/h)=9,000 gph

(9,000 gph)(10 h/day) = 90,000 gal/day

7.237 200 gpm/28 ft = 7.1 gpm/ft

7.238 620 gpm/21 ft = 29.5 gpm/ft

7.239 1,100 gpm/41.3 ft = 26.6 gpm/ft

$$\frac{x\ \text{gpm}}{42.8\ \text{ft}}=33.4\ \text{fpm/ft}$$

7.240   x = (33.4)(42.8)

x = 1,430 gpm

(0.785)(0.5 ft)(0.5 ft)(140 ft)$(7.48\ \text{gal/ft}^3)$

= 206 gal

7.241

(40 mg/L)(0.000206 MG)(8.34 lb/gal)

= 0.07 lb Chlorine

7.242 (0.785)(1 ft)(1 ft)(109 ft)$(7.48\ \text{gal/ft}^3)$ = 640 gal

(40 mg/L)(0.000640 MG)(8.34 lb/gal) = 0.21 lb Chlorine

$$(0.785)(1 \text{ ft})(1 \text{ ft})(109 \text{ ft})\left(7.48 \text{ gal/ft}^3\right) = 633 \text{ gal}$$

$$(0.785)(0.67 \text{ ft})(0.67 \text{ ft})(40 \text{ ft})\left(7.48 \text{ gal/ft}^3\right) = 105 \text{ gal}$$

7.243 $633 + 105 \text{ gal} = 738 \text{ gal}$

$$(110 \text{ mg/L})(0.000738 \text{ gal})(8.34 \text{ lb/gal})$$

$$= 0.68 \text{ lb Chlorine}$$

$$(x \text{ mg/L})(0.000540 \text{ gal})(8.34 \text{ lb/gal}) = 0.48 \text{ lb}$$

7.244 $x = \dfrac{0.48}{(0.000540)(8.34)}$

$\qquad x = 107 \text{ mg/L}$

$\dfrac{0.09 \text{ lb chlorine}}{5.25/100} = 1.5 \text{ lb}$

7.245 $\dfrac{1.5 \text{ lb}}{8.34 \text{ lb/gal}} = 0.18 \text{ gal}$

$(0.18 \text{ gal})(128 \text{ fluid oz./gal}) = 23 \text{ fl oz}$

$(0.785)(0.5 \text{ ft})(0.5 \text{ ft})(120 \text{ ft})\left(7.48 \text{ ga/ft}^3\right) = 176 \text{ gal}$

7.246 $\dfrac{\left(50 \text{ mg/L chlorine}\right)(0.000176 \text{ MG})(8.34 \text{ lb/gal})}{65/100}$

$= 0.1 \text{ calcium hypochlorite}$

$(0.1 \text{ lb})(16 \text{ oz/lb}) = 1.6 \text{ oz calcium hypochlorite}$

$(0.785)(1.5 \text{ ft})(1.5 \text{ ft})(105 \text{ ft})\left(748 \text{ gal/ft}^3\right)$

$= 1,387 \text{ gal}$

7.247 $\dfrac{(100 \text{ mg/L})(0.001387 \text{ MG})(8.34 \text{ lb/gal})}{25/100}$

$= 4.6 \text{ lb chloride of lime}$

$\dfrac{\left(60 \text{ mg/L}\right)(0.000240 \text{ MG})(8.34 \text{ lb/gal})}{5.25/100} = 2.3 \text{ lb}$

7.248 $\dfrac{2.3 \text{ lb}}{8.34 \text{ lb/gal}} = 0.3 \text{ gal}$

$(0.3 \text{ gal})(128 \text{ fl oz/gal})$

$= 38.4 \text{ fl oz sodium hypochlorite}$

7.249 $(4.0 \text{ psi})(2.31 \text{ ft/psi}) = 9.2 \text{ ft}$

7.250 $(94 \text{ ft} + 24 \text{ ft}) + (3.6 \text{ psi})(2.31 \text{ ft/psi}) = 118 \text{ ft} + 8.3 \text{ ft}$

$\qquad\qquad\qquad\qquad\qquad = 126.3 \text{ ft}$

7.251 $(400 \text{ ft})(110 \text{ ft})(14 \text{ ft})\left(7.48 \text{ gal/ft}^3\right) = 4,607,680 \text{ gal}$

$(400 \text{ ft})(110 \text{ ft})(30 \text{ ft} \times 0.4 \text{ average depth})$

7.252 $\left(7.48 \text{ gal/ft}^3\right)$

$= 3,949,440 \text{ gal}$

7.253 $\dfrac{(200 \text{ ft})(80 \text{ ft})(12 \text{ ft})}{43,560 \text{ ft}^3/\text{acre-ft}} = 4.4 \text{ acre-ft}$

7.254 $\dfrac{(320 \text{ ft})(170 \text{ ft})(16 \text{ ft})(0.4)}{43,560 \text{ ft}^3/\text{acre-ft}} = 8.0 \text{ acre-ft}$

7.255 $\dfrac{\left(0.5 \text{ mg/L chlorine}\right)(20 \text{ MG})(8.34 \text{ lb/gal})}{25/100}$

$= 334 \text{ lb copper sulfate}$

7.256 $131.9 \text{ ft} - 93.5 \text{ ft} = 38.4 \text{ ft}$

7.257 $\dfrac{707 \text{ gal}}{5 \text{ min}} = 141 \text{ gpm}$

$(141 \text{ gpm})(60 \text{ min/h}) = 8,460 \text{ gph}$

7.258 $\dfrac{(0.785)(1 \text{ ft})(1 \text{ ft})(12 \text{ ft})\left(7.48 \text{ gal/ft}^3\right)(8 \text{ round trips})}{5 \text{ gpm}}$

$= 141 \text{ gpm}$

7.259 $(3.5 \text{ psi})(2.31 \text{ ft/psi})$

$= 8.1 \text{ water depth in sounding line}$

$= 167 \text{ ft} - 8.1 \text{ ft}$

$= 158.9 \text{ ft pumping water level, ft}$

$\text{Drawdown, ft} = 158.9 \text{ ft} - 141 \text{ ft}$

$\qquad\qquad\qquad = 17.9 \text{ ft drawdown}$

7.260 $\dfrac{610 \text{ gpm}}{28 \text{ ft drawdown}} = 21.8 \text{ gpm/ft}$

$(0.785)(0.5 \text{ ft})(0.5 \text{ ft})(150 \text{ ft})\left(7.48 \text{ gal/ft}^3\right) = 220 \text{ gal}$

7.261 lb chlorine required: (55 mg/L)

$(0.000220 \text{ MG})(8.34 \text{ lb/gal}) = 0.10 \text{ lb chlorine}$

7.262 $780 \text{ gal/5 min} = 156 \text{ gpm}$

$(156 \text{ gal/min})(60 \text{ min/h})(8 \text{ h/day}) = 74,880 \text{ gal/day}$

$(x \text{ mg/L})(0.000610 \text{ MG})(8.34 \text{ lb/gal}) = 0.47 \text{ lb}$

7.263 $x = \dfrac{0.47}{(0.000610)(8.34)}$

$= 92.3 \text{ mg/L}$

$(0.785)(1 \text{ ft})(1 \text{ ft})(89)\left(7.48 \text{ gal/ft}^3\right) = 523 \text{ gal}$

$(0.785)(0.67)(0.67)(45 \text{ ft})\left(7.48 \text{ gal/ft}^3\right) = 119 \text{ gal}$

7.264 $523 \text{ gal} + 119 \text{ gal} = 642 \text{ gal}$

$(100 \text{ mg/L})(0.000642 \text{ MG})(8.34 \text{ lb/gal})$

$= 0.54 \text{ lb chlorine}$

$$\frac{0.3 \text{ lb chlorine}}{5.25/100} = 5.7 \text{ lbs}$$

7.265 $\dfrac{5.7 \text{ lb}}{8.34 \text{ lb/gal}} = 0.68 \text{ gal}$

(0.68 gal)(128 fl oz/gal) = 87 fl oz

7.266 Volume, gal = $(4 \text{ ft})(5 \text{ ft})(3 \text{ ft})\left(7.48 \text{ gal/ft}^3\right) = 449 \text{ gal}$

7.267 Volume, gal = $(50 \text{ ft})(20 \text{ ft})(8 \text{ ft})\left(7.48 \text{ gal/ft}^3\right)$

= 59,840 gal

7.268 Volume, gal = $(40 \text{ ft})(16 \text{ ft})(8 \text{ ft})\left(7.48 \text{ gal/ft}^3\right)$

= 38,298 gal

42 in/12 in/ft = 3.5 ft

7.269 Volume, gal = $(5 \text{ ft})(5 \text{ ft})(3.5 \text{ft})\left(7.48 \text{ gal/ft}^3\right)$

= 655 gal

2 in/12 in/ft = 0.17 ft

7.270 Volume, gal = $(40 \text{ ft})(25 \text{ ft})(9.17 \text{ ft})\left(7.48 \text{ gal/ft}^3\right)$

= 68,592 gal

CHAPTER 8

8.1   26 ft
8.2   77 ft
8.3   eccentric, segmental
8.4   flow nozzle
8.5   ultrasonic flowmeter
8.6   4,937 gal
8.7   4.57
8.8   213,904 ft^3
8.9   103 ft
8.10   8,064 lb
8.11   always constant
8.12   pressure due to the depth of water
8.13   the line that connects the piezometric surface along a pipeline
8.14   0.28 ft
8.15   254.1 ft
8.16 6.2×10^{-8}
8.17   0.86 ft
8.18   pressure energy due to the velocity of the water
8.19   a pumping condition where the size of the impeller of the pump and above the surface of the water from which the pump is running
8.20   The slope of the specific energy line

CHAPTER 9

9.1   positive-displacement
9.2   high-viscosity
9.3   positive-displacement
9.4   high
9.5   high
9.6   eye
9.7   static, dynamic
9.8   shut off
9.9 $V^2/2g$
9.10   Total head
9.11   head capacity, efficiency, horsepower demand
9.12   water
9.13   suction lift
9.14   elevation head
9.15   water hp and pump efficiency
9.16   centrifugal force
9.17   stuffing box
9.18   impeller
9.19   rings, impeller
9.20   casing

CHAPTER 10 REVIEW

10.1   a flexible piping component that absorbs thermal and/or terminal movement
10.2   fluid
10.3   fluid
10.4   connected
10.5   flow
10.6   pressure loss
10.7   increases
10.8   automatically
10.9   insulation
10.10 leakage
10.11 four times
10.12 routine preventive maintenance
10.13 12
10.14 schedule, thickness
10.15 increases
82.16 ferrous
10.17 increases
10.18 iron oxide
10.19 cast-iron
10.20 iron
10.21 corrosion
10.22 decreases
10.23 clay, concrete, plastic, glass, or wood
10.24 corrosion-proof
10.25 cement
10.26 pressed
10.27 turbulent, lower
10.28 steel
10.29 fusion
10.30 flexible
10.31 aluminum
10.32 annealed
10.33 fusion
10.34 metals, plastics

10.35 laminar flow
10.36 reinforced nonmetallic
10.37 wire-reinforced
10.38 Dacron®
10.39 diameter
10.40 flexibility
10.41 E.E.
10.42 reinforced, pressure
10.43 flexible
10.44 expansion joint
10.45 vibration dampener
10.46 plain
10.47 bends
10.48 pressure
10.49 plug
10.50 a long-radius elbow

CHAPTER 11

11.1 Na
11.2 H_2SO_4
11.3 7
11.4 base
11.5 changes
11.6 solid, liquid, gas
11.7 element
11.8 compound
11.9 periodic
11.10 solvent, solute
11.11 An atom or group of atoms that carries a positive or negative electrical charge as a result of having lost or gained one or more electrons
11.12 Colloid
11.13 Turbidity
11.14 Result of dissolved chemicals
11.15 Toxicity
11.16 Organic
11.17 0; 14
11.18 ability of water's ability to neutralize an acid
11.19 calcium and magnesium
11.20 base

CHAPTER 12

12.1 bacteria, viruses, protozoa
12.2 during rain storms
12.3 no
12.4 binary fission
12.5 spheres, rods, spirals
12.6 typhoid, cholera, gastroenteritis
12.7 amoebic dysentery, Giardiasis
12.8 cyst
12.9 host
12.10 plug screens and machinery; cause taste and odor problems

CHAPTER 13

13.1 ecosystem
13.2 benthos
13.3 periphyton
13.4 plankton
13.5 pelagic
13.6 neuston
13.7 immigration
13.8 autotrophs
13.9 lentic
13.10 dissolved oxygen solubility

CHAPTER 14

14.1 secondary maximum contaminant levels
14.2 transpiration
14.3 surface water
14.4 agriculture, municipal wastewater plants, habitat and hydrologic modifications, resource extraction, and urban runoff and storm sewers
14.5 solids content
14.6 turbidity
14.7 universal solvent
14.8 alkalinity
14.9 neutral state
14.10 lead

CHAPTER 15

15.1 muffle furnace, ceramic dishes, furnace tongs, and insulated gloves
15.2 15 min
15.3 A sample was collected all at one time. Representative of the conditions only at the time taken.
15.4 For pH, dissolved oxygen, total residual chlorine, fecal coliform, and any test by NPDES permit for grab sample.
15.5 A series of samples collected over a specified period of time in proportion to flow.
15.6 Collect from well-mixed location; clearly mark sampling points; easy location to read; no large or unusual particles; no deposits, growths, or floating materials; corrosion-resistant containers; follow safety procedures; test samples as soon as possible.
15.7 Refrigerate at 4°C
15.8 Absorption of water during cooling, contaminants, fingerprints, etc.

CHAPTER 16

16.1 cone of depression
16.2 12 in
16.3 concrete

16.4 surface water, groundwater, GUDISW

16.5 groundwater under the direct influence of surface water

16.6 easily located; softer than groundwater

16.7 the study of the properties of water and its distribution and behavior

16.8 zone of influence

16.9 GUDISW

16.10 Prevent large material from entering the intake

CHAPTER 17

17.1 Calcium (Ca) and Magnesium

17.2 Buffer

17.3 Sodium hydroxide

17.4 Chlorine Feed Rate (lbs/day)

$$= \text{Dose (mg/L)} \times \text{Flow (MGD)} \times 8.34$$
$$= (1.2 \text{ mg/L})(1.6 \text{ MGD})(8.34)$$
$$= (1.2)(1.6)(8.34)$$
$$= 16.0 \text{ lbs/day}$$

17.5 $2.4 (60 \div 25 = 2.4)$

17.6 Chlorine Dose (mg/L) − Chlorine Residual (mg/L)
$$= 1.0 \text{ mg/L} - 0.5 \text{ mg/L}$$
$$= 0.5 \text{ mg/L}$$

17.7 Residual = Dose − Demand
$$= 6.0 \text{ mg/L} - 3.3 \text{ mg/L}$$

Chlorine Residual = 2.7 mg/L

17.8 $$\text{Dose (mg/L)} = \frac{220 \text{ lbs/day Cl}_2}{3.1 \times 8.34}$$
$$= 8.5 \text{ mg/L} = 6.9 \text{ mg/L}$$

Chlorine Residual = 1.6 mg/L

17.9 A description of the soil encountered during well construction, water quantity, well casing information, and well development and testing.

17.10 dug well

17.11 disinfection residual, turbidity, coliform analysis

17.12 National Sanitation Foundation (NSF)

17.13 fit for human consumption

17.14 First, determine the required chlorine feed rate.

Feed Rate (lb/day) = Dose (mg/L) × Flow (MGD) × 8.34
$$= (0.6 \text{ mg/L})(1 \text{ MGD})(8.34)$$
$$= 5.0 \text{ lbs/day}$$

If we require 5 lbs/day of chlorine, we will require more pounds of hypochlorite because it is not 100% chlorine.

68% of the hypochlorite is available chlorine
$$68\% = 68/100 = 0.68$$

Next, $(Cl_2 \text{ Fraction})$ (Hypochlorite) = Available Chlorine

$$(0.68)(\text{x lbs/day hypochlorite}) = 5.00 \text{ lbs/day Cl}_2$$
$$\text{x lbs/day hypochlorite} = 5.00/0.68$$
$$\text{x} = 7.36 \text{ lbs/day hypochlorite}$$

17.15 public

17.16 the transport of the earth's water from one location to another

17.17 acute

17.18 reduction of pathogens to safe levels

17.19 hypochlorites

17.20 Reduce the number of pathogens to safe levels in water before the contact time is completed.

17.21 Feed Rate (lbs/day)
$$= \text{dose (mg/L)} \times \text{flow (MGD)} \times 8.34$$
$$= 0.4 \text{ (mg/L)} \times 5/3 \text{ (MGD)} \times 8.34$$
$$= 17.68 \text{ lbs/day Cl}_2$$

17.22 Residual = Dose − Demand
$$= 10 \text{ (mg/L)} - 2.6 \text{ (mg/L)}$$
$$= 7.5 \text{ mg/L}$$

17.23 Turbidity can entrap or shield microorganisms from the chlorine.

17.24 Feed Rate (lbs/day)
$$= \text{dose (mg/L)} \times \text{flow (MGD)} \times 8.34$$
$$= 0.8 \times 2.6 \times 8.34$$
$$= 17.35 \text{ lbs/day Soda Ash}$$

17.25 Given: Flow = 0.75 MGD
Shape = Circular
Size = radius = 20 ft
Depth = 10 ft
Find: Detention Time
  a. Find Tank Volume
  $$\text{Vol} = \pi r^2 h$$
  $$\text{Vol} = \pi (20 \text{ ft})^2 (10 \text{ ft})$$
  $$\text{Vol} = 12,560 \text{ ft}^3$$
  b. Flow = 0.75 MGD × 1,000,000
  $$= 750,000 \text{ gal/day}$$
  $$\text{Detention Time (h)} = \frac{12,560 \text{ ft}^3 \times 7.48 \times 24 \text{ h}}{750,000 \text{ gal/day}}$$
  Detention Time = 3.0 h

17.26 Chemical Feed Rate
$$= \text{Dose (mg/L)} \times \text{Flow (MGD)} \times (8.34)$$
$$= (35 \text{ mg/L}) \times (2 \text{ MGD}) \times (8.34)$$
$$= (70)(8.34)$$
$$= 584 \text{ lbs/day of Alum}$$

17.27 Yes

17.28 chlorine residual

17.29 a link that connects two systems and a force that causes liquids in a system to move

17.30 Moderate

17.31 negative; low

17.32 peristaltic metering pump

17.33 purchase of buffer zone around a reservoir; inspection of construction sites; public education

17.34 Given: # of Filters=3

Size (Each) = 10 ft × 7 ft

OPERATING: one out of service

Filtration Rate=280 gal/min (this is the total capacity for both filters)

Find: Filtration rate ft^2 of filter

Total Area of Filters = 10' × 7' = 70 ft^2 each

= 70' × 2' = 140 ft^2 Total

Filtration Rate

$$\frac{280 \text{ gal/min}}{140 \text{ ft}^2} = 2 \text{ gal/min/ft}^2 \text{ of filter}$$

17.35 Given: Filter Area=300 ft^2

Backwash rate=15 gal/ft^2/min

Backwash time=8 min

Find: Amount of water for backwash

Given information on per foot of filter but want to find total water needed to backwash the entire filter.

a. Find total filtration rate.

$$300 \text{ ft}^2 \times \frac{15 \text{ gal}}{\text{ft}^2/\text{min}} = 4{,}500 \text{ gal/min}$$

b. Gallons per 8-min backwash time

$$\frac{4{,}500 \text{ gal}}{\text{min}} \times 8 \text{ min} = 36{,}000 \text{ gal used}$$

Velocity = Distance Traveled ÷ Time

17.36

= 600 ft ÷ 5 min

= 120 ft/min

17.37 Safety Data Sheets (SDS)

17.38 chlorination and filtration

17.39 pump more than rated capacity

17.40 hypochlorous acid

17.41 protozoa

17.42 the removal/inactivation of the most resistant pathogens

17.43 corrosivity

17.44 turbidity, paddles speed, pH

17.45 In this problem, we want to find the velocity. Therefore, we must rearrange the general formula to solve for velocity.

V = Q/A

Given: Q=Rate of flow=11.2 cfs

A=Area in square feet

2.5 ft wide

1.4 ft deep

Find: Average Velocity

Area = (Width) × (Depth)

Step 1: = 2.5 ft × 1.4 ft

A = 3.5 ft^2

Step 2: Velocity (ft/sec)=Flow Rate (ft^3/sec)/ Area (ft^2)

$$\text{Vel} = \frac{11.2 \text{ ft}^3/\text{sec}}{3.5 \text{ ft}^2}$$

Vel = 32 ft/sec

17.46 Given: Height= 100 ft

Diameter=20 ft

Cylindrical shape

Find: Total gallons of water contained in the tank

Step 1: We must find the volume in cubic ft.

Volume = 0.785 × (Diameter)^2 × (Height)

= 0.785 × (20 ft)^2(100 ft)

= 0.785(400 ft^2)(100 ft)

= 31,400 ft^3

Step 2: Our problem asks: How many *gallons* of water will it contain?

31,400 ft^3 × 7.48 gal/ft^3 = 234,872 gal

17.47 rapid mix, flocculation, sedimentation

17.48 removal of color, suspended matter, and organics

17.49 transform soluble ions into insoluble compounds

17.50 three to four times the theoretical amount

17.51 5%

17.52 Given: Distance=1,500 ft

Time=4 min

Find: Velocity in ft/min

Step 1: Velocity=1,500 ft/4=375 ft/min

Step 2: Convert minutes to seconds

375 ft/min × 1 min/60 sec = 6.25 ft/sec

17.53 gate

17.54 achieve optimum corrosion control

17.55 50%

17.56 sodium fluoride (NaF)

17.57 mottled teeth enamel

17.58 0.75 mg/L

17.59 amount of chlorine to add for breakpoint chlorination; the correct amount of coagulant to use for proper coagulation; length of the flash mix; proper amount of mixing and settling time.

17.60 corrosion control technology
17.61 Given: Flow = 350 GPM
$\qquad$ Pipe Size = 6 in
$\qquad$ Find: Velocity (ft/sec) = Distance/Time
$\qquad$ Step 1: Convert gallons to ft^3

$$\frac{350 \text{ gal/min}}{7.48 \text{ gal/ft}^3} = 46.8 \text{ ft}^3/\text{min}$$

$\qquad$ Step 2: Find the cross-sectional area of the pipe

$$\text{Area of circle} = \pi r^2$$
$$= 3.14(3 \text{ in})(3 \text{ in})$$
$$= 28.26 \text{ in}^2$$

$\qquad$ Step 3: Convert square inches to square feet

$$\frac{28.26 \text{ in}^2}{144 \text{ in}^2/\text{ft}^2} = 0.20 \text{ ft}^2$$

$\qquad$ Step 4: Find ft/min

$$= \frac{46.8 \text{ ft}^3/\text{min}}{0.20 \text{ ft}^2} = 234 \text{ ft/min}$$

$\qquad$ Step 5: Convert minutes to seconds

$$234 \text{ ft/min} \times 1 \text{ min}/60 \text{ sec} = 3.9 \text{ ft/sec}$$

17.62 air, chlorine, or potassium permanganate
17.63 pH, alkalinity, hardness
17.64 adsorption
17.65 prior to the rapid mix basin
17.66 before the backwash, water reaches the lip of the wash water trough.
17.67 chlorine
17.68 true
17.69 true
17.70 79,269 gal
17.71 powdered activated carbon
17.72 iron and manganese
17.73 copper
17.74 soluble polyvalent cations
17.75 gains an electron in goring from the +2-oxidation state to the +3 form
17.76 bicarbonate
17.77 negative head
17.78 gravity
17.79 influent
17.80 uniform
17.81 maximize the conversion of organic carbon from the dissolved phase to the particulate phase; the removal of natural organic material; optimize the removal of DHP precursor material.
17.82 30 h
17.83 phenyl arsine oxide
17.84 Given: Surface area of pond = 20 acres

$\qquad$ Height of water collected = 2 in
$\qquad$ Find: The number of gallons collected in the reservoir after the storm.
$\qquad$ a. Convert acres to ft^2

$$20 \text{ acres} \times 43,560 \text{ ft}^2/\text{acre} = 871,200 \text{ ft}^2$$

$\qquad$ b. Convert inches to feet

$$2 \text{ in} \times \text{ft}/12 \text{ in} = 0.167 \text{ ft}$$

$\qquad$ Volume of water collected

$$= (\text{area})(\text{height})$$

$\qquad$ c.
$$= \left(871,200 \text{ ft}^2\right)(0.167 \text{ ft})$$
$$= 145,490 \text{ ft}^3$$

$\qquad$ d. Convert ft^3 to gallons

$$145,590 \text{ ft}^3 \times 7.48 \text{ gal/ft}^3 = 1,089,013 \text{ gal}$$

17.85 20.0 lbs/day Cl_2
17.86 (70 ft)(0.4) = 28 ft

$$(28 \text{ ft})(0.433) = 12.1 \text{ psi}$$

17.87 Groundwater
17.88 aeration, boiling, adsorption
17.89 Addition of powdered activated carbon
17.90 Permeability
17.91 water table
17.92 waterborne
17.93 coagulant
17.94 greater
17.95 copper sulfate
17.96 lime
17.97 disinfected
17.98 oxygen
17.99 binary fission

CHAPTER 18

18.1 the licensed operator and the responsible official
18.2 the amount of organic material in a sample that can be oxidized by a strong oxidizing agent
18.3 prevent disease, protect aquatic organisms, protect water quality
18.4 dissolved and suspended
18.5 organic indicates matter that is made up mainly of carbon, hydrogen, and oxygen and will decompose into mainly carbon dioxide and water at 550°C;

inorganic materials, such as salt, ferric chloride, iron, sand, gravel, etc.

18.6 algae, bacteria, protozoa, rotifers, virus

18.7 carbon dioxide, water, more organics, stable solids

18.8 toxic matter, inorganic dissolved solids, pathogenic organisms

18.9 raw effluent

18.10 from body wastes of humans who have disease

18.11 disease-causing

18.12 domestic waste

18.13 industrial waste

18.14 4.4%

18.15 2.3 ft

18.16 5,250-gal × 8.34 lb/gal = 43,785 lb

18.17 14,362 gal

18.18 850.7 lb/day

18.19 686 kg/day

18.20 0.121 MGD

18.21 8,477 people

18.22 9.41 lb/gal

18.23 cutter may be sharpened and/or replaced when needed. Cutter alignment must be adjusted as needed

18.24 grit is heavy inorganic matter. Sand, gravel, metal filings, egg shells, coffee grounds, etc.

18.25 0.7 fps

18.26 there is a large amount of organic matter in the gut. The aeration rate must be increased to prevent settling of the organic solids

18.27 to remove settleable and floatable solids

18.28 to remove the settleable solids formed by the biological activity

18.29 7,962 gpd/ft

18.30 stabilization pond, oxidation pond, polishing pond

18.31 settling, anaerobic digestion of settled solids, aerobic/anaerobic decomposition of dissolved and colloidal organic solids by bacteria producing stable solids and carbon dioxide, photosynthesis

18.32 products of oxygen by algae; summer effluent is high in solids (algae) and low in BOD; winter effluent is low in solids and high in BOD

18.33 eliminates wide diurnal and seasonal variation in pond dissolved oxygen

18.34 standard, high rate, roughing

18.35 increase waste rate

18.36 decrease, decrease, decrease, increase, increase

18.37 ten containers

18.38 88 days

18.39 103 cylinders; $2,823.49

18.40 4,716 lb/day

18.41 21.5 lb

18.42 27 days

18.43 64.1%

18.44 National Pollutant Discharge Elimination System

16.45 By increasing the primary sludge pumping rate or by adding dilution water

18.46 7.0 pH

18.47 either because the microorganisms have been killed or are absent

18.48 the time to do the test, 3 h versus 5 days

18.49 dark greasy

18.50 increases

18.51 temperature, pH, toxicity, waste rate, aeration tank configuration

18.52 Can function with or without dissolved oxygen. Prefer dissolved oxygen but can use chemically combined oxygen such as sulfate or nitrate.

18.53 Organic

18.54 Living organisms

18.55 Final

18.56 Colloidal

18.57 Not possible

18.58 Aerobic, facultative

18.59 Different

18.60 Reduced

18.61 Temperature

18.62 BOD

18.63 F/M

18.64 Secondary clarifier weirs

18.65 To separate and return biosolids to the aeration tank

18.66 Declining

18.67 1.5 and 2.5 mg/L

18.68 increased MLVSS concentration

18.69 decreased waste rate

18.70 decreased MCRT

18.71 concentration of aeration influent solids

18.72 compete mix is more resistant to shock loads

18.73 decrease the grit channel aeration rate

18.74 increase

18.75 floor level

18.76 $22.77

18.77 anoxic

18.78 C:N:P

18.79 Secondary

18.80 Are not

18.81 2 ft/sec

18.82 lower

18.83 chlorine residual

18.84 2 h

18.85 0.1

18.86 800 gal/ft²/day

18.87 mono-chloramine

18.88 0.2–0.5

18.89 nitrogen

18.90 decrease explosive hazard, decrease odor release, maintain temperature, collect gas

18.91 algae

18.92 dissolved solids

18.93 0.0005 ppm

18.94 Ponds have historically been used to provide detention time for wastewater to allow it to be stabilized through natural processes. Wastewater is treated by the action of bacteria (both aerobic and anaerobic), other micro- and macro-organisms, algae, and by the physical process of gravity settling. When properly designed, ponds are capable of providing the equivalent of secondary treatment for both BOD and suspended solids.

18.95

Advantages	Disadvantages
Low construction cost	Large land requirements
Low operational cost	Possible groundwater contamination from leakage
Low energy usage	Climatic conditions affect the treatment
Can accept surge loadings	Possible suspended solids problems (algae)
Low chemical usage	Possible spring odor problems (after ice-out)
Easy operation	Animal problems (muskrats, turtles, etc.)
No continuous sludge handling	Vegetation problems (rooted weeds, duckweed, algae)
	Localized sludge problems (deposition near inlet)

18.96 **Aerobic:** An aerobic pond system would have oxygen distributed throughout the entire area. This would be similar to a clean lake with anaerobic conditions occurring only in bottom sediments. This condition would probably only occur in a treatment system upon initial start-up with the pond would be filled with a clear water source, or when completely mixed with supplement air.

Anaerobic: An anaerobic pond would be devoid of all oxygen throughout the entire area. This type of pond system would only be used in special applications, usually for treating certain industrial wastes. If a normal pond system is totally anaerobic, it is organically overloaded. The only exception would be under ice cover for a fill-and-draw type facility.

Aerated: An aerated pond system would have supplemental air sources to provide dissolved oxygen. This is usually accomplished with surface mechanical aerators and mixers, or by various forms of diffusers supplied with compressed air from mechanical blowers or compressors. For equal-sized ponds, the aerated pond would provide the best treatment due to the mechanical addition of oxygen, and for a given organic loading, would require the least amount of land area.

Facultative: Most stabilization pond facilities are of this type. The pond contains an aerobic surface zone, an anaerobic bottom zone, and a transitional (facultative) zone in between. This allows aerobic organisms to function in the upper area, Anaerobic organisms in the lower and sludge area, and facultative organisms in the middle area. A facultative organism can use dissolved oxygen or combined oxygen because they can adapt to changing conditions. They can continue decomposition when the system changes form aerobic to anaerobic, or from anaerobic to aerobic.

18.97 A. Photosynthesis by algae within the pond (the main source of oxygen in most pond-type systems, especially shallow ponds in the 3–5 for depth range).

B. Diffusion of atmospheric oxygen at the pond surface with the action of the wind providing mixing of the oxygen-rich surface layer with the water below.

C. The use of compressed air systems or surface mechanical aerators.

18.98 Oxygen levels vary with depth for a number of reasons. The main reason is the relationship of the organisms with the pond. Other reasons are the physical actions within the pond and the loading to the pond.

The relationship of organisms involves the general interaction between algae and bacteria.

The algae are the main source of oxygen in a pond system. Algae growth is greatest near the surface where light penetration and photosynthesis are the greatest. Oxygen levels decrease with depth, due to less light penetration needed for photosynthesis.

The algae use carbon dioxide in the process of photosynthesis and produce oxygen. The bacteria stabilize organic matter using oxygen and produce carbon dioxide. The physical diffusion of atmospheric oxygen occurs at the surface of ponds and is mixed in the upper layers by wind action. The amount of mixing is limited, so the oxygen levels decrease with depth.

The final factor affecting oxygen levels is the organic loading to the system. If organic loadings are small, the oxygen levels will be maintained at greater depths. If organic overloading occurs, the whole pond could go anaerobic.

18.99 Dissolved oxygen meter; Winkler dissolved oxygen determination.

$$\text{Area of pond } \left(\text{in ft}^2\right) = \text{length (ft)} \times \text{Width (ft)}$$

18.100 Area of pond (in acres) = surface area $\left(\text{ft}^2\right)$

$$= 1 \text{ acre } \left(\text{ft}^2\right)$$

$$\text{Area of pond} = \text{length (ft)} \times \text{width (ft)}$$

$$= 400 \times 300$$

$$= 120,000 \text{ ft}^2$$

$$\text{Area of pond} = \text{surface area (ft}^2)$$

$$(\text{in acres}) = 1 \text{ acre } \left(\text{ft}^2\right)$$

$$= 120,000$$

$$= 43,560$$

$$= 2.75 \text{ acres}$$

CHAPTER 19

19.1	B	19.30	A
19.2	C	19.31	C
19.3	D	19.32	C
19.4	C	19.33	A
19.5	C	19.34	A
19.6	C	19.35	D
19.7	B	19.36	D
19.8	B	19.37	C
19.9	A	19.38	D
19.10	D	19.39	B
19.11	B	19.40	A
19.12	B	19.41	C
19.13	A	19.42	B
19.14	D	19.43	A
19.15	B	19.44	C
19.16	A	19.45	A
19.17	A	19.46	D
19.18	B	19.47	D
19.19	C	19.48	B
19.20	D	19.49	B
19.21	C	19.50	A
19.22	C	19.51	B
19.23	C	19.52	C
19.24	D	19.53	B
19.25	C	19.54	A
19.26	C	19.55	A
19.27	A	19.56	D
19.28	D	19.57	B
19.29	D	19.58	B
		19.59	B
		19.60	E
		19.61	D
		19.62	D
		19.63	A
		19.64	C
		19.65	D
		19.66	A
		19.67	C
		19.68	B
		19.69	D
		19.70	B
		19.71	D
		19.72	D
		19.73	C
		19.74	A
		19.75	B
		19.76	B
		19.77	D
		19.78	B
		19.79	A
		19.80	A
		19.81	D
		19.82	A
		19.83	C
		19.84	D
		19.85	B
		19.86	B

19.87	B		19.144	C
19.88	A		19.145	D
19.89	C		19.146	E
19.90	C		19.147	D
19.91	D		19.148	A
19.92	A		19.149	A
19.93	C		19.150	B
19.94	B		19.151	C
19.95	C		19.152	D
19.96	D		19.153	C
19.97	D		19.154	A
19.98	D		19.155	B
19.99	C		19.156	A
19.100	D		19.157	C
19.101	C		19.158	B
19.102	D		19.159	B
19.103	D		19.160	A
19.104	A		19.161	D
19.105	C		19.162	B
19.106	D		19.163	C
19.107	A		19.164	C
19.108	B		19.165	C
19.109	A		19.166	C
19.110	C		19.167	A
19.111	D		19.168	B
19.112	A		19.169	B
19.113	D		19.170	A
19.114	B		19.171	A
19.115	C		19.172	D
19.116	D		19.173	B
19.117	B		19.174	C
19.118	A		19.175	A
19.119	A		19.176	D
19.120	D		19.177	C
19.121	D		19.178	A
19.122	D		19.179	D
19.123	A		19.180	B
19.124	C		19.181	B
19.125	D		19.182	B
19.126	C		19.183	C
19.127	A		19.184	C
19.128	B		19.185	B
19.129	B		19.186	B
19.130	D		19.187	A
19.131	A		19.188	C
19.132	A		19.189	D
19.133	A		19.190	A
19.134	C		19.191	C
19.135	C		19.192	D
19.136	C		19.193	B
19.137	C		19.194	C
19.138	B		19.195	C
19.139	B		19.196	B
19.140	C		19.197	A
19.141	D		19.198	D
19.142	B		19.199	B
19.143	A		19.200	D

19.201	B	19.229	C
19.202	C	19.230	A
19.203	B	19.231	D
19.204	D	19.232	A
19.205	D	19.233	C
19.206	B	19.234	A
19.207	B	19.235	A
19.208	D	19.236	D
19.209	C	19.237	B
19.210	D	19.238	B
19.211	A	19.239	C
19.212	D	19.240	D
19.213	A	19.241	A
19.214	A	19.242	A
19.215	A	19.243	C
19.216	D	19.244	C
19.217	C	19.245	D
19.218	B	19.246	B
19.219	B	19.247	B
19.220	C	19.248	A
19.221	D	19.249	A
19.222	C	19.250	D
19.223	B	19.251	C
19.224	A	19.252	C
19.225	C	19.253	D
19.226	B	19.254	C
19.227	C	19.255	B
19.228	A		

Appendix B
Formulae

1. Area
   a. *Rectangular Tank*
   $$A = L \times W$$
   b. *Circular Tank*
   $$A = \pi r^2 \text{ or } A = 0.785 \text{ day}^2$$

2. Volume
   a. *Rectangular Tank*
   $$V = L \times W \times H$$
   b. *Circular Tank*
   $$V = \pi r^2 \times H \quad \text{or} \quad 0.785 \text{ day}^2 \times H$$

   Flow: Gal/day (gpd) = gal/min (gpm) $\times$ 1,440 min/day

3.   gal/day (gpd) = gal/h(gph) $\times$ 24 h/day

   Million gal/day (MGD) = (gal/day)/1,000,000

4. Dose: lbs = ppm $\times$ MG $\times$ 8.34 lb/gal

   ppm = lbs/(MG $\times$ 8.34 lb/gal)

5. Efficiency (% removal) = $\dfrac{(\text{Influent} - \text{effluent})}{\text{Influent}} \times 100$

6. Weir loading (overflow rate) = $\dfrac{\text{total gal/day}}{\text{length of weir}}$

7. Surface settling rate = $\dfrac{\text{total gal/day}}{\text{surface area of tank}}$

8. Detention time (h) = $\dfrac{\text{capacity of tank (gal)} \times 24 \text{ h/day}}{\text{flow rate (gal/day)}}$

9. Horsepower (hp) = $\dfrac{\text{gpm} \times \text{head (ft)}}{3{,}960 \times \text{total efficiency}}$

Index

Note: **Bold** page numbers refer to tables and *italic* page numbers refer to figures.

0.5 MGD capacity plant 605
1-log virus 460
1.5-MGD capacity plant 605–606
1.55-MGD capacity plant 605–606
2-log giardia 460
2-MGD capacity plant 606
2.6-MGD capacity plant 606
3-MGD capacity plant 606
5-MGD capacity plant 606
24-MGD capacity plant 606
39-MGD capacity plant 607
54-MGD capacity plant 607
67-MGD capacity plant 607

abnormal filter operations 469
aboveground equipment enclosure 32–33
A-C *see* asbestos-cement (A-C) pipe
activated biosolids 150–151
    BOD/COD loading 151
    moving averages 150–151
    process control calculations 150
activated sludge process 511
    aeration tank 567–568
    control parameters 565–566
    equipment 560
    factors affecting operation of 560
    final settling tank (clarifier) observations 568
    formation of 561
    growth curve 561
    MCRT impact on parameters 578–579
    modification 562–565
    overview of 560
    performance-controlling factors 561–562
    process control testing and sampling 568–581
    symptoms 573–574
    terminology 558–560
Activated Sludge Process Control, Part II 566
active security barriers 35, 37
adapted BOD analyzer 47
adsorption 463, 558
adsorptive processes 500
advanced wastewater treatment 511, 593–594
    biological denitrification 597
    biological nutrient removal (BNR) 601–605
    carbon adsorption 598
    chemical treatment 594
    enhanced biological nutrient removal (EBNR) 605–607
    filtration 595–597
    microscreening 595
aerated pond 535
aerated static pile (ASP) 621–622
aerated systems 525
aeration 451, 456, 459, 524, 558, 560
    for composting 620
    influent sampling 568
    pipe loop 622
    tank 560, 567–568
aerobic
    bacteria 314, 535
    conditions 511, 558
    digestion

pH adjustment 617
    process control 169
    process control calculations 616
    time 616
    volatile solids loading 616
ponds 534
respiration 537
affinity laws, centrifugal pumps 230–231
aggressive index (AI) 458
AI *see* aggressive index (AI)
air
    binding 464
    sparging 502
    to-solids ratio 167–168, 615
alarm system 33–34
alderflies (Order: Megaloptera) 362–363
algae 535–536, 538
algal control 106
alkaline stabilization 626–627
alkalinity 517, 537–538, 561, 565, 569
    determination 137–138
alternative filters 465
alternative filtration 466
aluminum sulfate 460
amendable cyanide 49
American Society for Testing Materials (ASTM) 252, 254
American Water Works Associations (AWWA) 129, 215
ammonia (NH_3) 487
ammonia nitrogen 569
ammonium 537
amoebas 570
anaerobic
    bacteria 314, 535
    conditions 511, 513
    digestion 617–618
        normal operating ranges 618, **618**
        process control 168–170
        process control calculations 618–620
    fermentation processes 534
    ponds 534
annealing 265
annunciator 33
anoxic conditions 513
anoxygenic photosynthesis 536
ANSTO *see* Australian Nuclear Science and Technology Organization (ANSTO)
anti-intrusion detection system evasion techniques 54
anti-virus programs 53
apparent color testing/analysis
    Borger Color System 429
    indicators 429
    measurement 428–429
    odor 429
    organic sources 428–429
    standard methods 429
aquatic ecosystem, biotic and abiotic factors 345
aquatic ecosystems, incremental change determinations 385–387
Archimedes' screw pump 225, *225*

area 86, 685
arithmetic average 83–84
arithmetic mean 83–84
arming station 33
arsenic (As)
    exposure 497–498
    measurement system 47
    removal technologies
        adsorptive processes 500
        alternative technologies 500–502
        granular ferric hydroxide 501–502
        membrane processes 500
        prescriptive processes 498–499
    rule 47
    standard 497
artesian aquifers 440
artificial media drying beds 632
asbestos-cement (A-C) pipe
    application 259
    composed mixture 259
    materials 259
    operations 259
    safe work practice 260
ashing 568
asiatic clam (*Corbicula fluminea*) 477–478
ASP *see* aerated static pile (ASP)
aspergillus fumigatus 623–624, **624**
ASTM *see* American Society for Testing Materials (ASTM)
atmospheric tank 231, *232*
Australian Nuclear Science and Technology Organization (ANSTO) 502
autonomous operations 22–23
average daily limitation 645
average flow rate calculation 90
average hourly limitation 645
average monthly limitation 645
average use calculation 163
average weekly limitation 645
AWWA *see* American Water Works Associations (AWWA)

Bacilli 475
Bacillus thuringiensis israelensis (Bti) 536
backflow prevention devices 34–35
backpressure 34
backsiphonage 34–35
backwashing 464, 465
    pumping rate 122
    rate 120–121
    rise rate 121
    water tank depth 122
    water volume 121–122
bacteria 475, 534
    vs. algae 535–536
    inactivation 494
bagged biosolids 620
barminution 522
barriers
    active security barriers 35
    bollards 35
    crash beam barriers 35–36
    passive security barriers 37

barriers (*Cont.*)
  portable/removable barriers 36
  wedge barriers 35
bar screen 520
baseline audit 17
BAT *see* best available technology (BAT)
bed volumes (BVs) 500
Beetles (Order: Coleoptera) 360–362
behavioral ecology 335
below-surface wedge barriers 35
belt filter 635
  press 171–173
benchmarking
  baseline audit 17
  baseline data and tracking energy use
      15–17
  create equipment inventory and demand
      and energy distribution 17–18
  field investigation 17
  potential results of 15
  process 14–15
  stage approach to 14
  steps 15
  targets 15
benthic habitat
  erosion and deposition 349
  filamentous algae 354
  macroinvertebrates
    advantages 354–355
    aquatic organisms 354
    characteristics 354
    and food web 356
    identification 355–356
    organization units 356
    running waters 356
    water quality 355
  plants and animals 354
  silt and organic materials 354
Bernoulli
  equation 205–206
  principle 203, *205*
  theorem 205
best available technology (BAT) 11
BHP *see* brake horsepower (BHP)
Big Gulch Wastewater Treatment Plant 67–68
binary fission 314
bioactive substances 7
biochemical operations 511
biochemical oxygen demand (BOD) 18, 163,
      164, 513, 531, 535, 579
  analysis 412
  BOD$_5$ 180
    calculation 412–413
    seeded 180
    test procedure 412, **412**
    unseeded 180
  BOD 7-day moving average 180–181
  description 411
  measurement 411
  organic loading 157, 562
  removal efficiency 157–158
  sampling considerations 411
  and SS removal 530–531
  and suspended solids 146
  symptom 550–551
  testing 412
biological aerosols 620
Biological Condition Gradient (BCG) 385, 388
biological contamination, monitoring sensors 47
biological denitrification 597
biological nitrification 597

biological nutrient removal (BNR) 601
  description 601–602
  operation and maintenance 604–605
  performance 603
  process 602–603
biological stability 479
biological towers 541
biomass 541
biometric finger geometry recognition 37–38
biometric hand geometry recognition 37–38
biometric security systems 37–38
biomonitoring
  advantage
    fish 383–384
    macroinvertebrates 384
    periphyton 383
  definition 383
  protocols
    fish 383–384
    macroinvertebrates 385–388
    periphyton 383
  sampling *see* sampling
biosolids 513, 607, 631
  composting 620
  digestion 168
  feed rate 172
  land application of
    process control calculations 643
    sampling and testing 643
  math 165–168
  production 163–164
  pumped/day in pounds 166
  pumping 165
  pumping time 166
  quality parameters 620
  retention time 170
  thickening 166
biosolids density index (BDI) 184–185
biosolids dewatering 170–171
  belt filter press 171–173
  filter loading 173–174
    filter yield 173–174
    percent solids recovery 174
    vacuum filter operating time 174
  flocculant dosage 173
  hydraulic loading rate 171–172
    biosolids feed rate 172
    flocculant feed rate 172–173
    solids loading rate 172
  plate and frame press 171
    net filter yield 171
    solids loading rate 172
  pressure filtration 171
  rotary vacuum filter dewatering 173
  sand drying beds 174
    biosolids withdrawal to drying beds 175
    solids loading rate 175
    total biosolids applied 174–175
  total suspended solids (TSS) 173
biosolids disposal 175
  composting biosolids 177–178
  composting calculations 177–178
  disposal cost 175–176
    crop nitrogen requirement 176
    maximum allowable applications,
        metals loading 176–177
    metals loading 176
    plant available nitrogen (PAN) 176
    site life based on metals loading 177
  land application 175
biosolids/sludge 291

biosolids volume index (BVI) 184–185
BNR *see* biological nutrient removal (BNR)
BOD *see* biochemical oxygen demand (BOD)
body feed 465
bollards 35
bored wells 444
bottom line on security 63
brake horsepower (BHP) 230
breakpoint chlorination 582
  curve 488, *488*
brine for regeneration 141–142
Broad Street Well 474
bromamine 593
bromate monitoring 495
bromide ions 481–482
bromine chloride 593
brown algae 536
Bti see Bacillus thuringiensis israelensis (Bti)
bulking 558
  agents 620
buried exterior intrusion sensors 41
BVs *see* bed volumes (BVs)
BVI *see* biosolids volume index (BVI)

CAA *see* Clean Air Act (CAA)
caddisflies (Order: Trichoptera) 359, *359*
cadmium 458
calcium carbonate (CaCO$_3$) 456, 459
calcium hardness 135–136
calcium hypochlorites 489, 492, 495
calibration procedures 454–455
  dry feeder 453–454
  positive displacement pump 454
candida parapsilosis 493
cantilever crash beams 36
cantilever gates 36
capillary water 609
carbon 540
carbonaceous BOD 513
carbon adsorption 598
  land application 599–601
  operational, observations, and
      troubleshooting 598
carbonate hardness 137
carbon dioxide 533
carbon to nitrogen ratio (C:N Ratio) 620
card identification/access/tracking systems
      38–39
card reader system 38–39
carrying capacity 344
cartridge filter system 465
cathodic protection 459
cation exchange process 457
cavitation 231
CCTV system *see* closed circuit television
      (CCTV) system
centrate clarity, poor 639
centrifugal force 524
centrifugal pumps 233
  advantages 236
  application 237
  description 233
  disadvantages 236–237
  electronic control systems 239
    flow equalization system 240
    motor controllers 240
    protective instrumentation 240
    sonar system 240
    supervisory instrumentation 240–241
    temperature detectors 240
    vibration monitors 240

modifications
    recessed impeller/vortex pumps 242, *242*
    submersible pumps 241, *241*
    turbine pumps *242*, 242–243
pump characteristics 235
    brake horsepower requirements 236
    efficiency 236
    head 235–236
pump control systems 237–238
    electrode control systems 239
    float control 238
    pneumatic controls 238–239
pump theory 235
terminology
    base plate 234
    bearings 234
    casing 234–235
centrifugation 639
centrifuge
    testing 569
    thickening 168
C/F *see* coagulation/flocculation (C/F)
C factor 210, **210**
chamber and basin volume 108
chemical
    coagulation 96
    conditioning 635
    contamination, monitoring sensors 47
    dosage calculations 92
        chlorine dosage 93–94
        dosage formula pie chart 92–93
        hypochlorite dosage 94
    feed calibration 162–163
    feeders 453–455
    feed pump 161
    feed rate 96, 158–159
    limits and testing 495
    sensors
        adapted BOD analyzer 47
        arsenic measurement system 47
        chlorine measurement system 48–49
        portable cyanide analyzer 49–50
        total organic carbon (TOC) analyzer
            47–48
    sludges 608
    solution feeder setting, mL/min 162
    solutions 452–453
    usage determination/calculation 112,
        128–129
chemical addition 452, 608
    calibration procedures 454–455
    chemical feeders 453–455
    chemical solutions 452–453
    corrosion control 457–460
    hardness treatment 456–457
    iron and manganese removal 455
        aeration 456
        ion exchange 456
        manganese greensand 456
        oxidation 455–456
        potassium permanganate oxidation 456
        precipitation 455
        sequestering 456
chemical dosing 158
    aerobic digestion process control
        calculations 168
    air-to-solids ratio 167–168
    anaerobic digestion process control
        calculations 169
    average use calculations 163
    biosolids

digestion 168
production 165
pumped/day in pounds 166
pumping 165
pumping time 166
retention time 170
thickening 166
centrifuge thickening calculations 168
CF *see* concentration factor (CF)
CFR *see* Code of Federal Regulations
    (CFR)
chemical feed calibration 162–163
chemical feed pump 161
chemical feed rate 158–159
chemical solution feeder setting, mL/min 162
chemical solutions 160
    chlorine demand 159
    chlorine residual 159
    concentration factor (CF) 167
    daily biosolids production estimation 165
    digestion time, days 169
    estimated gas production in cubic feet/
        day 170
    gravity/dissolved air flotation thickener
        calculations 166
        daily biosolids production 165
        surface loading rate 167
    hypochlorite dosage 159–160
    percent moisture reduction in digested
        biosolids 170
    percent solids 165
    percent volatile matter reduction 170
    pH adjustment 169
    primary clarifier solids production
        calculations 164
    primary solids production calculation 164
    pumping calculations 163–164
    required seed volume in gallons 169
    secondary clarifier solids production
        calculation 164–165
    secondary solids production calculation 164
    solids loading rate 167
    solids pumped per day in pounds 166
    solution chemical feeder setting, GPD 161
    stabilization 168
    volatile acids to alkalinity ratio 169–170
    volatile matter pumped per day in pounds 166
    volatile solids loading 168–169
chemical oxygen demand (COD) 147–148, 151,
    513, 517, 568
chironomidae 536
chloral hydrate 484
chloramines 487, 582
chlorination 124, 582
    application methods 495–496
    breakpoint chlorination 487–488, *488*
    breakpoint chlorination curve 488, *488*
    chlorine chemistry 487
    chlorine dosage determination 489
    chlorine generation 491–492
    chlorine terminology 486–487
    DBPs 494–495
    disinfection efficacy 493–494
    environmental hazards and safety 586
    equipment 583
    factors affecting 493
    gas chlorination 488–489
    hypochlorination 489
    measuring chlorine residual 493
    operation 583
    pathogen inactivation 493–494

process calculations 588–589
process description 583
safety and handling considerations 496
terminology 582
uses, application and doses of chlorine
    492, 493
chlorinator 584
chlorine 455–456, 491, 494, 495, 605–606
    advantages and disadvantages of 496–497
    chemistry 487
    contact time 486–487
    demand 125, 159, 486, 582, 588
    disinfection 124, 497, **497**
        chemical handling 586–587
        chlorination equipment 583
        chlorination operation 583
        chlorination process description 583
        chlorination terminology 582
        environmental hazards and safety 586
        facts and process description 582–583
        troubleshooting operation problems
            583–586
    dosage 93–94, 126, 489–491
    dose 125, 159, 484, 588
    facts 582–583
    feed rate 588
    gas 488, 492, 496, 582
    generation 491–492
    measurement system 48–49
    oxidation 627
    reaction 484
    residual 125, 159, 486
    residual test 493
    systems 583
    terminology 486–487
    uses, application and doses 492, **492**
chlorine dioxide (ClO₂) 479
chlorine pressure reducing valve (CPRV) 585
chlorine residual testing/analysis
    diethyl-*p*-phenylenediamine-ferrous
        ammonium sulfate (DPD-FAS)
        titration
        calculation 431
        equipment 431
        procedure 430
    diethyl-*p*-phenylenediamine (DPD)
        spectrophotometry
        chemicals requirement 430
        equipment requirement 430
        measurement 430
        procedure 430
    fluorides 431
    free chlorine residual 430
    plant's NPDES discharge permit 430
    total residual chlorine (TRC) 429
chlorite monitoring 495
chloropicrin 482
cholera 472–474
city of Bartlett Wastewater Treatment Plant 69
cladocerans 536
clarifier 513
Clean Air Act (CAA) 496
Clean Water Act (CWA) 27, 532
Clean Water Act Amendments 11
closed circuit television (CCTV) system 46
clumping 568
CMMS *see* computer maintenance
    management system (CMMS)
coagulation 107, 460–462
coagulation-assisted microfiltration 499
coagulation/flocculation (C/F) 498

cocci 475
COD *see* chemical oxygen demand (COD)
Code of Federal Regulations (CFR) 511
coil type vacuum filter 635
coliform 513
    bacteria 471
colloidal suspended solids 608
colorimetric methods 407
combined filter effluent monitoring 466–467
combined filter effluent reporting 467
combined residual 125, 486
combined wastewater 517
comminutor 522
communication and integration 51
    electronic controllers 51–52
    two-way radios 52
    wireless data communications 52–53
community ecology 335
community water systems (CWSs) 28
complete mix 562
complex conversions 100
    concentration to quantity 100–101
    quantity to concentration 101
    quantity to volume/flow rate 101
composite sample 179
compost 620
compost-enriched soil 623
composting biosolids 177
composting pile
    advantages and disadvantages 623
    biosolids-derived compost, malodorous
        compounds in 626
    bottom line on composting odor control 626
    composting odor problems 624–626
    construction of 622
    lime stabilization 626–627
    measuring odors 626
comprehensive performance evaluation
        (CPE) 469
computer-literate operator's work 25
computer maintenance management system
        (CMMS) 25–26
concentration 452–453
concentration factor (CF) 167, 615
conductivity testing
    description 421
    discharges 421
    equipment considerations 421–422
    measurement 421–422
    sampling 421–422
    testing 421–422
cone of depression 207
coning 558
contact
    stabilization 562–563
    tanks 583
continuous monitoring 9
control panel 33, 37
conventional activated sludge 562
conventional treatment 466
conventional water treatment model 450, *450*
conversions 84–85
    temperature conversions 85
copper sulfate dosing 106–107
    chamber and basin volume calculations 108
    coagulation 107
    detention time 108
    flocculation 107–108
corrosion
    control 457–460
    definition 457

factors affecting 458
    types 458
corrosive resistant pipe materials 459
CPE *see* comprehensive performance
        evaluation (CPE)
CPRV *see* chlorine pressure reducing valve
        (CPRV)
crash beam barriers 35–36
critical flow 212
cross-connection control procedures 9
cryptosporidium 463, 465, 466, 476, 493, 494
    basics of 320
    cryptosporidiosis 326
    life cycle 322, *324*
    named species **326**
    Surface Water Treatment Rule 326
CT factor 484
CT values 484, 487
curing air 620
current issues
    cash cows/cash dogs 4–5
    multiple-barrier concept 9–10
        wastewater operations 9–10
    paradigm shift 8–9
    sick water 6–8
cut-off head 235
CWA *see* Clean Water Act (CWA)
CWSs *see* community water systems (CWSs)
cyanobacteria 535
cyber protection devices 53
    anti-virus programs 53
    firewalls 53–54
    network intrusion hardware and software 54
    pest eradication software 53
cyber threats, control systems 58–59
cyclone degritter 524, 525

daily biosolids production estimation 165
daily discharge 645
damselflies (Order: Odonata) 363–364, *364*
Darcy-Weisbach equation 209
dark repair 591
DBP *see* disinfection by-products (DBP)
D/DBP rule *see* disinfectant/disinfection
        by-product (D/DBP) rule
dechlorination 586
declining growth phase 561
deep-well turbine pump calculation 105–106
deferrization 455
demanganization 455
density 453
    and specific gravity 197–198
dental caries 129
dental fluorosis 129
Department of Energy (DOE) 59
detection devices 33
detention time 91, 108, 113, 529, 685
dewatering
    biosolids, natural methods of 633–634
    using aquatic plants 633–634
diaphragm pump 243, *243*
diatomaceous earth filters 465
diethyl-*p*-phenylenediamine (DPD) 48,
        430, 431
    color comparator method 493
differential pressure flowmeters 216
    operating principle 216
    types
        nozzle 217
        orifice 217, *217*
        Pitot-static tube 218, *218*

Venturi 217, *217*
diffused aeration systems 560
diffused air aeration 558
diffuser 558
digital world 25–26
direct filtration 465
    alternative filters 465
    common filter problems 465–466
    filtration and compliance with turbidity
        requirements 466
        additional compliance issues 468
        IESWTR regulatory requirements
            466–467
        individual filter follow-up action 468
        individual filter self-assessment 469
        reporting and recordkeeping 467–470
        variances and exemptions 470
direct radiation measurement 42–43
discharge 200
    monitoring report 645
    permit 645
disinfectant/disinfection by-product (D/DBP)
        rule 499
disinfectant residual regulatory requirements
        484–485
disinfection 471–472, *472*
    efficacy 493–494
    methods of 484–485
    pathogen inactivation mechanisms 493
        algal growth prevention, sedimentation
            basins and filters 480
        coagulation and filtration efficiency
            improvement 480
        color removal 480
        DBP formation, minimization of 477
        disinfectants uses 477–480
        environmental effects 493–494
        iron and manganese, oxidation of
            478–479
        nuisance Asiatic clams and zebra
            mussels, control of 477–478
        re-growth prevention, distribution
            system and maintenance 479
        taste and odors removal 479
    primary concern, pathogens of 475
    pump handle removal, water treatment
        473–474
    recent waterborne disease outbreaks
        476–477
    residuals types 480
    Sherlock Holmes-type at the pump 472–474
        cholera 472–473
        1854 London, flashback to 473
    status of health information 480, **481**
    strategy selection 484
    wastewater
        symptom 584–586
disinfection by-products (DBP) 477, 494–495
    control strategies
        precursor removal 483
        source water quality control 482–483
    formation 477, 479, 480–481
        DBP precursors 481–482
        inorganic by-products and
            disinfectants 482
        organic oxidation by-products 482
        pH impacts on 482
    precursors 481–482
    rule 495
    status of health information 480, **481**
    types 480

disposal cost 175–176, 643
dissolved gases 517
dissolved organic carbon (DOC) 481
dissolved oxygen (DO) 297–298, 513, 537, 565, 568, 615
  testing
    biochemical oxygen demand (BOD) 408
    concentrations *vs.* temperature 408, **408**
    meter and probe 409–411
    oxygen 407–408
    profile 571
    sampling and equipment considerations 408
    standard methods 407
    Winkler method 408–409
dissolved solids 518
distribution arm 541
distribution system
  disinfection calculation 99–100
  integrity 9
diurnal flow variation 517
DO *see* dissolved oxygen (DO)
dobsonflies (Order: Megaloptera) 363–364
DOC *see* dissolved organic carbon (DOC)
DOE *see* Department of Energy (DOE)
domestic (sewage) wastewater 517
door frames 44
door impact test 44
Doppler type ultrasonic flowmeters 219, *219*
dosage formula pie chart 92–93
dose 685
double-cylinder locks 42
DPD *see* diethyl-*p*-phenylenediamine (DPD)
dragonflies (Order: Odonata) *363,* 363–364
drainage basin 437–438
drawdown 207
drilled wells 444
drinking water quality monitoring
  adequate control 403
  cleaning procedures
    acid wash procedures 405
    general preparation of sampling containers 405–407
  criteria 403
  good/bad
    chemical pollutants 403
    factors 392–393
    pollutants source 403
    pollution problems 403–404
    screen 404
    trends determination 404
  preparation and sampling considerations 405–407
  preservation and storage 406–407
  sample collecting, stream
    screw-cap bottles 406
  sample types 405
  state water quality standards programs 404
  water quality monitoring program 404–405
drinking water reservoirs 43
drinking water sources, assessing and protecting 9
driven wells 444
drop-arm crash beams 36
Dr. Snow's theory 473
dry feeders 453–455
dry hypochlorite rate 126–127
drying bed
  operational capacity 630–631

operational observations, problems, and troubleshooting procedures 634–635
dug wells 443
dynaSand filters 605

EBCT *see* empty bed contact time (EBCT)
EBNR *see* enhanced biological nutrient removal (EBNR)
ecology
  behavioral 335
  benthic habitat *see* benthic habitat
  categorized 335
  community 335
  description 334–335
  ecosystem *see* ecosystem ecology
  freshwater *see* freshwater, ecosystem
  history 337–338
  importance 335
  levels of organization 339
  macroinvertebrate glossary *see* Macroinvertebrate glossary
  organism's environment division 335
  population *see* Population ecology
  stream ecosystem 345–346
  stream genesis and structure *see* Stream genesis and structure
  streams, leaf processing in 336–337
  study 335–336
ecosystem ecology 335
  agroecosystem model 338
  autotrophic and heterotrophic components 339
  biotic and abiotic materials 339–340
  description 339–340
  energy flow 340–341
  food chain efficiency 341
  pattern and process *343*
  productivity 342–343
  pyramids 342
ED-50 *see* effective dilutions-50 (ED-50)
EDR *see* electrodialysis reversal (EDR)
effective dilutions-50 (ED-50) 626
efficiency 685
effluent limitations 645, 646–647
1854 London, flashback to 473
Electric Power Research Institutes (EPRI) 17
electrode control systems 239
electrodialysis reversal (EDR) 500
electronic controllers 51–52
electronic methods 407
elements and symbols 292
empty bed contact time (EBCT) 500–501
endogenous respiration 561
endotoxins 620
Energy Conservation in Wastewater Treatment Facilities: Manual of Practice No. MFD-2 18
energy grade line 208
energy head 203
enhanced biological nutrient removal (EBNR) 605–607
enhanced coagulation 499
Environmental Protection Agency (EPA) 457–458, 471, 495, 620, 633
EPA *see* Environmental Protection Agency (EPA)
EPRI *see* Electric Power Research Institutes (EPRI)
EQ slude *see* Exceptional Quality (EQ) sludge
erratic drum rotation 595
escherichia coli 476, 493

estimated return rate 575
estimating daily sludge production 610
Exceptional Quality (EQ) sludge 620
excessive sloughing 556
extended aeration 563–564
exterior doorways 44

facultative bacteria 314
facultative organisms 513
facultative pond 534–535
faucet flow estimation 178
FBI *see* Federal Bureau of Investigation (FBI)
fecal coliform 513
fecal coliform bacteria testing
  bacteriological sampling 425–426
  colony counting
    calculations 428
    description 428
    interferences 428
  description 422–423
  EPA's 2001 Total Coliform Rule 816-F-01-035 422–423
  equipment and techniques 425
  indicators 422–423
  membrane filter procedure 427–428
  MPN calculation 427
  MPN presumptive test 426–427
  multiple-tube fermentation technique 426
  sampling and equipment considerations 424–425
  serial dilution procedure 425
  sterile dilution water preparation 425
  sterilization 425
  total coliform group 422–424
  USEPA's Total Coliform Rule
    compliance 424
    requirements 423–424
Federal Bureau of Investigation (FBI) 54
Federal Water Pollution Control Act of 1972 12
feed rates 139–140
feedstock 621
fences 39–40
filamentous bacteria 559
filamentous organisms 571
filter
  backwash 97
  bed expansion 123–124
  breakthrough 465
  drum 635
  effluent turbidity 596
  flies 551
  loading 173–174
  media 635
  presses 637
    process control calculations 638
  surface clogging 596
  underdrain 541
filtration 96, 595–597
  calculations 117–118
    backwash pumping rate 122
    backwash rate 120–121
    backwash rise rate 121
    backwash water tank depth 122
    backwash water volume 121–122
    filter bed expansion 123–124
    filtration rate 119
    flow rate through a filter 118–119
    percent mud ball volume 123
    percent product water used for backwatering 123
    unit filter run volume (UFRV) 119–120

filtration (*Cont.*)
   distribution system disinfection
      calculations 99–100
   filter backwash 97
   rate 119
   rate of filtration calculation 96–97
   storage tank calculations 98–99
   types of 463–465
      diatomaceous earth filters 465
      pressure filter systems 464–465
      rapid sand filters 464
      slow sand filters 463–464
   water distribution system calculations 97
final settling tank (clarifier) observations 568
fire
   alarm systems 34
   hydrant locks 40–41
   proof doors 44
   walls 53–54
fittings
   connections type
      flanged fitting 273
      screwed fittings 273
      welded fitting 273
   function
      branch connections 272
      connecting lines 273
      direction of flow 271, 272
      materials 271–272
      sealing lines 272
      sizes of lines 272
   pipes and tubes 271
fixed bollards 36
fixed film systems 532
fixed-location sensors 46
fixed wedge barriers 35
flagellated protozoa 570
float
   cake 613
   control systems 238
floatable solids 559
flocculant
   dosage 173
   feed rate 172–173
flocculation 107–108, 461, 561
floc formation 594
floc solids 513
floc water 608
flotation thickening 613
flow 88–89, 685
   amount 216
   average flow rate calculations 90
   conversion calculations 91
   diagram, WWT process
      Big Gulch *68*
      De Pere *66*
      Sheboygan *67*
   distribution 529, 584
   equalization 525
   equalization system 240
   instantaneous flow into and out of a
      rectangular tank 89
   instantaneous flow rates 89
   rate 216
      in acre-inches day 156–157
      into a cylindrical tank 89–90
      through filter 118–119
   through pipeline 90
   velocity calculations 90
flow measurement 215–223
   devices 216

differential pressure flowmeters 216
   magnetic flowmeters 218, *218*
   positive-displacement flowmeters
      220, *221*
   ultrasonic flowmeters 219
   velocity flowmeters 219–220, *220*
   old-fashioned way 216
   open-channel 220–223
      flumes *222*, 222–223
      weirs *221*, 221–222
   traditional flow measurement 216
fluidized bed furnace 641
fluid velocity through pipeline 226
flumes *222*, 222–223
fluoridation 129–131
   calculated dosage problems 134–135
   fluoride compounds 129
      fluorosilicic acid 130
      sodium fluoride 130
      sodium fluorosilicate 130
   optimal fluoride levels 130–131
   process calculations
      calculated dosages 133–134
      fluoride feed rate 131
      percent fluoride ion in a compound
         131–132
   water fluoridation 129
fluoride 129
   compounds 129
      fluorosilicic acid 130
      sodium fluoride 130
      sodium fluorosilicate 130
   feed rate 132–133
      for saturator 133
fluorosilicic acid 130
F/M ratio *see* food-to-microorganism ratio
   (F/M ratio)
food-to-microorganism ratio (F/M ratio)
   151–152, 513, 559, 576–577
   control 577
   waste rate calculation 151–152
force 86–88
   main collection system 519
   and pressure 198–200
      effects of water under pressure 200
      hydrostatic pressure 199–200
frame grabber 38
free available chlorine 487
free residual 488
free-swimming ciliated protozoa 570
free water 608
   surface 203
freeze-assisted drying beds 633
freezing 551–552
freshwater
   ecosystem
      process 345
      species diversity 345
   source 437
friction head 201, 228
   loss 207–210
      C factor 210, **210**
      flow in pipelines 208, *208–209*
      major head loss 209
      minor head loss 210
      slope 210
full-scale chlorination treatment 478

GAC adsorption 483
Galileo's theory 8
Galvanic corrosion 458

gambusia 536
GBMSD *see* Green Bay Metropolitan
   Sewerage District (GBMSD)
gas chlorination 488–489
gastropoda (lung-breathing snails) 364
gates 36, 39
gauges
   accumulator 285
   air receiver 285–286
   double check system 282
   heat exchangers 286
   operating parameters 282
   parameter indications 282
   pressure 200
   pressure gauges 283
      bellows gauge 284
      bourdon tube 283–284
      plunger gauge 284
      spring-operated 283
   pre-start check 282
   temperature gauges
      bimetallic 284
      description 284–285
      industrial-type thermometer 284
      measurement scales 284
   vacuum breaker 285
getting well water 7
giardia 487
   chemical disinfection 323
   cysts 494
   giardiasis 320
   G. lamblia 463, 466, 476
   infection categories 321
   life cycle 322, *324*
   water supply 320
giardiasis 476
glass shatter protection films 40
global paradigm 8
Gould Biosolids Age 153
Gould Sludge Age 559
grab sample 179
granular ferric hydroxide 501–502
gravity
   collection system 519
   thickening 612–613
   velocity-controlled grit removal 523–524
green algae 536
Green Bay Metropolitan Sewerage District
   (GBMSD) 65–66
grit channel velocity calculation 144–145
grit removal 143–145, 522–524
   aeration 524
   calculations 524–525
   centrifugal force 524
   gravity/velocity-controlled grit removal
      523–524
groundwater 439–441
   aquifer 439, 441
   infiltration 516
   quality 441
   surface water, direct influence of 441
   treatment requirements 441–442
      D/DBP rule 442
      IESWT rule 442
      regulatory deadlines 442
groundwater under the direct influence
   of surface water (GUDISW) 9,
   441, 466
GUDISW *see* groundwater under the
   direct influence of surface water
   (GUDISW)

HAA concentration 484–485
Handbook of Water and Wastewater Treatment Plant Operations 10
hardness
    calculation 457
    measurement 413
    treatment 456–457
hardwired systems 34
hatch security 41
hazards 27
Hazen-Williams equation 209–213
HCL *see* hydrochloric acid (HCl)
HCN *see* hydrogen cyanide (HCN)
HDT *see* hydraulic detention time (HDT)
head 200–201
    defined 86
    friction head 201
    loss 204
    and pressure 201
    static head 201
    total dynamic head 201
    velocity head 201
Hepatitis B virus 475
Hero Hercules 7
heterotrophic plate count (HPC) 479
HGL *see* hydraulic grade line (HGL)
hidden function 10
high clarifier effluent suspended solids 550–551
high-rate trickling filters 546
hirudinea (leeches) 364
HMI *see* human Machine Interface (HMI)
HOCl *see* hypochlorous acid (HOCl)
horsepower (Hp) 229, 685
    brake horsepower (BHP) 230
    hydraulic horsepower (WHP) 229
Hp *see* horsepower (Hp)
HPC *see* heterotrophic plate count (HPC)
human Machine Interface (HMI) 55
human parasitic protozoans 475, **476**
hydrant lock 40–41
hydraulic depth 213
hydraulic detention time (HDT) 92, 157, 515–516
    in days 92
    in hours 92
    in minutes 92
hydraulic grade line (HGL) 204, 208, 214
hydraulic (water) horsepower (WHP) 229
hydraulic loading 147, 541, 552, 562
hydraulic loading rate 113, 147, 171–172, 548, 557
    biosolids feed rate 172
    flocculant feed rate 172–173
    solids loading rate 172
hydraulic machine pumps 225–226
    basic pumping calculations 226
        affinity laws, centrifugal pumps 230–231
        brake horsepower (BHP) 230
        fluid velocity through pipeline 226
        friction head 228
        horsepower 229
        net positive suction head (NPSH) 231–233
        pressure to head conversion 229
        pressure-velocity relationship 226–227
        series and parallel pumps 233
        specific speed 230
        static head 227
        total head 229

velocity head 228–229
centrifugal pumps 233
    advantages 236
    application 237
    description 233
    disadvantages 236–237
    electronic control systems 239
    modifications 241
    pump characteristics 235
    pump control systems 237–239
    pump theory 235
    terminology 234–235
    positive displacement pumps 243
        diaphragm pump 243, *243*
        peristaltic pumps 243–244
        piston/reciprocating pump 243
hydraulic radius 212–213
hydrochloric acid (HCl) 487
hydrogen cyanide (HCN) 49
hydrogen sulfide 626
hydrologic cycle 436
hydrostatic pressure 199–200
hypochlorination 489
hypochlorite
    dosage 94, 159–160
    facts 583
    operation 583
    solution feed rate 127–128
    systems 583
hypochlorous acid (HOCl) 487, 493

IDSs *see* intrusion detection systems (IDSs)
IESWT *see* interim enhanced surface water treatment (IESWT)
IESWTR *see* Interim Enhanced Surface Water Treatment Rule (IESWTR)
individual filter
    follow-up action 468
    monitoring 466
    requirements 467–468
    self-assessment
        comprehensive performance evaluation 469
        notification 469
        tier 1 violations 469–470
        tier 2 violations 470–471
industrial contributors 566
industrial hoses
    coupling
        all-metal hoses 270, 271, *271*
        low-pressure applications 269
        maintenance 271
        quick-connect 270
        quick-disconnect 270
    description 266
    nomenclature
        bend radius 267
        maintenance operations 266–267
        reinforcing materials 267
    standards, codes and sizes 268
        metallic hose 270
        nonmetallic hose *see* nonmetallic hose
industrial wastewater 516
inorganic by-products and disinfectants 482
inorganic phosphates 459
inorganic solids 518, 609
instantaneous flow into and out of rectangular tank 89
instantaneous flow rates 89
intercellular water 628
interim enhanced surface water treatment (IESWT) 437, 442, 462, 493

Interim Enhanced Surface Water Treatment Rule (IESWTR) 466–467
    regulatory requirements
        applicability 466
        combined filter effluent monitoring 466–467
        individual filter monitoring 467
interior doorways 44
interior intrusion sensors 34, 41
International Organization for Standards (ISO) 14
International Ozone Association (IOA) 485
intrusion detection systems (IDSs) 60
intrusion sensors 34, 41
invertebrates 536
IOA *see* International Ozone Association (IOA)
ion exchange 457, 500
    capacity 140
    process 457
iris recognition systems 38
iron and manganese removal techniques
    aeration 456
    ion exchange 456
    manganese greensand 456
    oxidation 455–456
    potassium permanganate oxidation 456
    precipitation 455
    sequestering 456
iron bacteria 460
iron filings 502
iron/manganese (Fe/Mn) oxidation 498–499
iron oxide coated sand 500
irrigation 599
ISO *see* International Organization for Standards (ISO)
isolated bacteria 535

jar tests 461, 572
jetted wells 444
joining metallic pipe
    bell-and-spigot joint 256, *256*
    description 256, *256*
    flanged joints 257, *257*
    screwed or threaded joints 256, *256*
    soldered and brazed joints 257
    welded joints 257, *257*
        butt-welded 257
        socket-welded 257

ladder access control 41–42
laminar flow 208, 212
LAN *see* local area network (LAN)
land application
    of biosolids 175
    operational observations, problems, and troubleshooting 599–601
    type and modes of 599
langelier saturation index (LI) 458
law of conservation of energy 203
law of continuity 202
LCRMR *see* Lead and Copper Rule Minor Revisions (LCRMR)
lead 254
    intoxication 457
lead and Copper Rule 457–458
Lead and Copper Rule Minor Revisions (LCRMR) 458
legionella pneumophila 476–477
licensure 28–30
light intensity 538

lime
  dosage calculation
    g/min 117
    lbs/day 117
    mg/L 115–117
  softening 467, 499
  stabilization 626–627
limit of technology (LOT) 603
linear gate 36
liquid
  alum 461
  bleach 492
local alarm 33
local area network (LAN) 53
local paradigm 8
locks 42
log growth phase 561
LOT *see* limit of technology (LOT)

macroinvertebrate glossary 356–357
  insect macroinvertebrates
    alderflies and dobsonflies 362–364
    beetles 360–362
    caddisflies 359, *359*
    dragonflies and damselflies *363,*
      363–364, *364*
    mayflies 357–358, *358*
    stoneflies 358, *358*
    true flies 359–360
    water strider 362
  non-insect macroinvertebrates
    gastropoda 364
    hirudinea 364
    oligochaeta 364
macroinvertebrates protocol, biomonitoring
  the biotic index
    BMWP 387
    identification keys 387
    pollution measurement 387
    principles 387
    types 387
  metrics 388
magnesium hardness 136
magnetic flowmeters 218, *218*
major head loss 209
  calculation 209
malware 53
manganese greensand 456
manhole intrusion sensors 42
manhole locks 42
manning equation 209
Manual of Water Utility Operations 451
mass balance 579
  biological process 580–581
  and measuring plant performance 101–102
  for settling tanks 102
  settling tank suspended solids 579–580
  using BOD removal 102–103
    measuring plant performance 103
    percent volatile matter reduction in
      sludge 103
    plant performance/efficiency 103
    unit process performance/efficiency 103
Master Terminal Unit (MTU) 55
math operations 81
  applied math operations
    mass balance and measuring plant
      performance 101–102
    mass balance for settling tanks 102
    mass balance using BOD removal
      102–103

area 86
arithmetic average 83–84
calculation steps 82
chemical coagulation and sedimentation 96
  calculating feed rate 96
  calculating solution strength 96
chemical dosage calculations 92
  chlorine dosage 93–94
  dosage formula pie chart 92–93
  hypochlorite dosage 94
chemical solution feeder calibration
  111–112
chemical solution feeder setting
  gpd 109
  mL/min 109
complex conversions 100
  concentration to quantity 100–101
  quantity to concentration 101
  quantity to volume or flow rate 101
conversions 84–85
  temperature conversions 84–85
detention time 91
equivalents 82–83
filtration 96
  distribution system disinfection
    calculations 99–100
  filter backwash 97
  rate of filtration calculation 96–97
  storage tank calculations 98–99
  water distribution system calculations 97
  water flow velocity 97–98
flow 93
  calculations 89–91
force 86–88
formulae 82–83
head 86–88
hydraulic detention time (HDT) 92
  in days 92
  in hours 92
  in minutes 92
median 83–84
percent removal 95
percent volatile matter reduction in sludge
  95–96
population equivalent (PE) 95
pressure 86–88
specific gravity 95
symbols 82–83
units 84–85
  milligrams per liter 85–86
volume 86
wastewater math concepts
  activated biosolids 150–151
  biosolids dewatering 170–174
  biosolids disposal 175
  chemical dosing 159–170
  preliminary treatment calculations
    142–145
  primary treatment calculations 145–146
  rotating biological contactors (RBCs)
    148–150
  solids inventory 151
  treatment ponds 156
  trickling filter calculations 147
water math concepts
  chemical usage determination 112
  chemical use calculations 128–129
  copper sulfate dosing 106–107
  dry chemical feeder calibration 111
  dry chemical feeder setting, Pound/day
    108–109

  filtration calculations 117–124
  fluoridation 129–131
  percent of solutions 110
  sedimentation calculations 112–113
  surface overflow rate 113–114
  water chlorination calculations
    124–127
  water softening 135–139
  water sources 103–104
  water storage calculations 103–104, 106
water/wastewater laboratory calculations
  biosolids density index (BDI) 184–185
  biosolids volume index (BVI) 184–185
  BOD_5 180–181
  faucet flow estimation 178
  fixed solids 183
  moles and molarity 181–182
  service line flushing time 178–179
  settleability 182–185
  settleable solids 182–183
  total solids 183
  volatile solids 183
  volatile suspended solids (VSS) 183
  wastewater suspended solids 183–184
maximum contaminant level goal (MCLG)
  466
maximum contaminant levels (MCLs) 47, 49,
  450, 458, 495, 502
maximum daily discharge 645
maximum discharge 645
mayflies (Order: Ephemeroptera) 357–358, *358*
MCC *see* motor control center (MCC)
MCLs *see* maximum contaminant levels
  (MCLs)
MCLG *see* maximum contaminant level goal
  (MCLG)
mean cell residence time (MCRT) 153, 514,
  559, 578
  control 578
  impact on parameters
    control values for 578
    mass balance 579
    recordkeeping requirements 578–579
    secondary clarifier 581
    waste quantities/requirements 578–579
    waste rate in million gallons/day 579
mean flow velocity 114
median 83–84
membrane
  filtration 483
  filtration systems 465
  processes 500
Membrane Bioreactors (MBRs) 597
Merriam-Webster's Collegiate Dictionary,
  Tenth Edition 513
metallic piping
  classification
    American Society for Testing and
      Materials (ASTM) ratings 252
    American Water Works Association
      (AWWA) ratings 252
    ASME B16.5 ratings 252
    manufacturers ratings 252
    National Fire Protection Association
      (NFPA) ratings 252
    pressure-temperature rating system 252
  description 251
  maintenance characteristics
    air binding 255
    backflow 255, *255*
    corrosion effects 255–256

expansion and flexibility 255
  isolation 255
  valve selection 255
  water hammer 255
materials characteristics
  cast-iron pipe 254
  Ductile-iron pipe 254
  metallurgy see Metallurgy
  steel pipe 254
materials selection
  capability 251
  characteristics 251
  components 251
  environmental factors 251
sizes 251–252
types
  code identification 253
  process lines 253
  service lines 253
wall thickness designations 252
metallurgy
  cast iron 254
  definition 253
  ferrous metal 253
  lead 254
  nonferrous metal 253
  pig iron 253
  piping standards 254
  steel 254
metals loading 643–645
metric ton 621
MF see microfiltration (MF)
MGD see million gallons per day (MGD)
microbial repair 59
microbiology
  bacteria 313–314
    bacilli 314
    capsules 312
    cell structure 312
    cell wall 313
    cocci 314
    cytoplasm 313
    destruction 315
    flagella 312–313
    forms 314
    growth factors 314–315
    inclusions 313
    mesosome 313
    nucleoid 313
    plasma/cytoplasmic membrane 313
    ribosomes 313
    shapes/arrangements 314, *314*
    waterborne 315
  classification 310–311
  crustaceans 316–317
  definition 309
  differentiation 312
  fungi 318
  nematodes/flatworms 318–319
  principles 309
  protozoa 315–316
  viruses 317
  water treatment 319–320
    Cryptosporidium see Cryptosporidium
    Cyclospora 327–328
    Giardia see Giardia
    helminths 328
    pathogenic protozoa 320
  water/wastewater microorganisms 309–310
microfiltration (MF) 500
microscopic examination 569

microscreening 595
milligrams per liter 85–86
million gallons per day (MGD) 92
minimum discharge 645
minor head loss 210
mixed liquor 514, 559
  color 568
  total suspended solids (TSS) 565–566
mixed liquor suspended solids (MLSS) 151,
    514, 559, 565–566, 569, 572
mixed liquor volatile suspended solids
    (MLVSS) 151–152
mixing and flocculation 462, *462*
mixis 538
MLE see Modified Ludzack-Ettinger (MLE)
    process
MLSS see mixed liquor suspended solids
    (MLSS)
MLVSS see mixed liquor volatile suspended
    solids (MLVSS)
Modified Ludzack-Ettinger (MLE) process
    564, 602
modified university of cape town (UCT)
    process 602
moisture removal 621
molarity 181–182
molecular weight cutoff (MWCO) 483
moles 181
monochloramine 477–478
motor control center (MCC) 18
motor controllers 240
MTU see Master Terminal Unit (MTU)
mud ball formation 596
muddy-bottom streams
  definition 396
  habitat assessment 398–399
  steps 396–398
multiple-barrier concept 9–10
  wastewater operations 9–10
multiple hearth furnace 641
MWCO see molecular weight cutoff (MWCO)
mycobacterium fortuitum 493

nanofiltration (NF) 500
National Environmental Policy Act
    (NEPA) 12
National Fire Protection Association (NFPA)
    ratings 252
National Infrastructure Protection Center
    (NIPC) 59
National Pollution Discharge Elimination
    System (NPDES) 11, 47, 51
National Primary Drinking Water Regulation
    (NPDWR) 469
National Science Foundation (NSF) 467
NCWs see non-community water systems
    (NCWSs)
natural organic matter (NOM) 480–481
nematodes 559
NEPA see National Environmental Policy
    Act (NEPA)
net filter yield 171
net positive suction head (NPSH) 231–233
network honeypot 54
network intrusion hardware and software 54
NF see nanofiltration (NF)
NFPA see National Fire Protection Association
    (NFPA)
NIMBY see Not-In-My-Backyard (NIMBY)
1977 Clean Water Act Amendment 12
1987 Clean Water Act 12

1989 Surface Water Treatment Rule
    (SWTR) 466
1995 Community Water Systems Survey
    reports 485
1996 Safe Drinking Water Act (SDWA)
    Amendments 466
NIPC see National Infrastructure Protection
    Center (NIPC)
nitrates measurement
  cadmium reduction method 417–418
  description 417
  electrode method 418
  sampling and equipment considerations 417
nitrification 601
nitrogen 539
  compounds 517–518
  cycle 537
  removal 537, 601–602
  requirement 644
nitrogenous oxygen demand (NOD) 514
nitrosomonas 601
NOD see nitrogenous oxygen demand (NOD)
NOM see natural organic matter (NOM)
nominal pipe size (NPS) 252
noncarbonate hardness 137
  removal of 138–139
non-community water systems (NCWSs) 28
nonmetallic hose
  description 268
  horizontal-braided hose 268
  reinforced horizontal-braided-wire hose 268
  types
    Dacron 270
    natural latex 270
    nylon 270
    pure gum 270
    silicone rubber 270
    Teflon 270
  vertical-braided hose 268, 268–269
  wire-reinforced hose 269, *269*
  wire-woven hose 269, *269*
  wrapped hose 269, *269*
nonmetallic piping
  clay pipes
    bell-and-spigot joint 258, *258*
    description 258
    unglazed (not glassy) 258
    vitrified (glass-like) 258
  concrete pipe
    advantages 259
    asbestos-cement (A-C) 259–260
    description 259
    disadvantages 259
    nonreinforced 259
    reinforced 259
    reinforced and prestressed 259
  description 257–258
  materials 258
  plastic pipe
    advantages 260–261
    description 260
    polyvinyl chloride (PVC) 261
non-uniform flow 208
normality 181–182
Not-In-My-Backyard (NIMBY) 625
nozzle 217
NPDES see National Pollution Discharge
    Elimination System (NPDES)
NPDES permits 646
  monitoring 646
  reporting 646

NPDWR *see* National Primary Drinking Water Regulation (NPDWR)
NPS *see* nominal pipe size (NPS)
NPSH *see* net positive suction head (NPSH)
NPSH available (NPSHA) calculation 231–233
NSF *see* National Science Foundation (NSF)
nutrients 561

Occupational Safety and Health Act (OHSA) 27
Occupational Safety and Health Administration (OSHA) 12, 496, 519
odors 550
oligochaeta (Family Tuificidae, Genus:Tubifex) 364
OJT *see* on-the-job training (OJT)
on-the-job training (OJT) 28
open atmospheric tank 231, *232*
open-channel flow 212–215
    calculations 212–213
    characteristics 212
        critical flow 212
        laminar and turbulent flow 212
        uniform and varied flow 212
    measurement 220
        flumes *222*, 222–223
        weirs *221*, 221–222
    parameters used 212–215
        hydraulic depth 213
        hydraulic radius 212–213
        slope 213–215
operator
    certification 28–30
    duties 27–28
optimal fluoride levels 130–131
optimizing treatment processes 9
ordering chlorine 589
organic loading 546, 562
    rate 147–150, 157, 548, 552–553, 557
organic nitrogen 495
organic oxidation by-products 482
organic solids 518, 609
organization, levels of ecology 339
orifice 217, *217*
orthophosphate measurement
    ascorbic acid method, orthophosphate 417
    phosphorus cycles 416
    phosphorus formation 416
    phosphorus testing 416
    sampling and equipment considerations 416–417
OSHA *see* Occupational Safety and Health Administration (OSHA)
OTE *see* oxygen-transfer efficiency (OTE)
outdoor equipment enclosure 32–33
over-oxidized activated sludge 571
oxidation 479
    air 621
    ditch detention time 156
    ditches 564–565
    pond 534
oxygen-transfer efficiency (OTE) 18
ozonation
    advantages and disadvantages 592
    applicability 592
    bromine chloride 593
ozone 592
    disinfection 592

package plants 563–564
PAN *see* plant available nitrogen (PAN)
PAOs *see* phosphate-accumulating organisms (PAOs)

paradigm shift 8–9
parshall flume *222*, 223
particle water 609
passive security barriers 37
pathogenic organisms 621
pathogens 609
    inactivation, environmental effects 493–494
    inactivation mechanisms 493–494
        algal growth prevention, sedimentation basins and filters 480
        coagulation and filtration efficiency improvement 480
        color removal 480
        DBP formation, minimization of 477
        disinfectants uses 477–480
        iron and manganese, oxidation of 478–479
        nuisance Asiatic clams and zebra mussels, control of 477–478
        re-growth prevention, distribution system and maintenance 479
        taste and odors removal 479
paved drying beds 631–632
peaking air 621
percent moisture reduction, digested biosolids 170
percent mud ball volume 123
percent product water, backwatering 123
percent removal 95
percent settled biosolids test 115
percent solids 165
percent strength
    of liquid solutions 110
    of mixed solutions 110–111
    of solutions 128
    using dry hypochlorite 128
percent total solids 530
percent volatile matter reduction 170, 619
    sludge 95–96, 515
perimeter intrusion sensors 34
peristaltic pumps 243–244
permits, records, and reports
    NPDES permits 646
    reporting calculations 647–648
    sampling and testing 646–647
pest eradication software 53
pH 493–494, 496, 537–538, 561, 565
    adjustment 169
    measurement
        analytical and equipment considerations 414
        color comparators 414
        meter 414
        "pocket pals" 414
PHAs *see* polyhydroxyalkanoates (PHAs)
pharmaceuticals and personal care products (PPCPs) 7
phosphate-accumulating organisms (PAOs) 602
phosphorus 518, 539
    removal 602
photooxidation 502
photorepair 591
photosynthesis 536
phragmites communis 633
physical asset monitoring and control devices
    aboveground equipment enclosures 32–33
    alarms 33–34
    backflow prevention devices 34–35
    barriers 35–37
    card identification/access/tracking systems 38–39

fences 39–40
    films for glass shatter protection 40
    fire hydrant locks 40–41
    hatch security 41
    intrusion sensors 41
    ladder access control 41–42
    locks 42
    manhole intrusion sensors 42
    manhole locks 42
    outdoor equipment enclosures 32–33
    radiation detection equipment, monitoring personnel and packages 42–43
    reservoir covers 43–44
    side-hinged door security 44–45
    valve lockout devices 45
    vent security 45–46
    visual surveillance monitoring 46
piezometer 203–204
piezometric surface 203–204
pin floc 568
pipeline flow 208, *208–209*
piping 252
    hydraulics 210–212
    networks 211–212
    standards 254
piping systems
    ancillaries *see* gauges
    definitions 246–247
    fluids *vs.* liquids 248
    maintenance
        accessories 250
        fluid flow 248–249
        hangers and supports 250, *250*
        performance 249–250
        protective features 250
        scaling 249
        temperature effects and insulation 251
    metallic *see* Metallic piping
    nonmetallic *see* nonmetallic piping
    operators responsibilities 246
    protective devices
        applications 279–280
        filter 280
        strainers 280
        traps *see* Traps
    single-line piping 246, *246*
piston/reciprocating pump 243
pitot-static tube 218, *218*
pitting 458
plain settling tanks (clarifiers) 528
plant available nitrogen (PAN) 176, 643–644
plant performance/efficiency 514–515
plant security
    bottom line on security 63
    SCADA 54–55
        adoption of technologies with known vulnerabilities 57–58
        applications in water/wastewater 56
        cyber threats to control systems 58–59
        increasing risk 57
        securing control systems 59
        security improvement steps 59–63
        vulnerabilities 56–57
    security hardware/devices 32
        communication and integration 51–53
        cyber protection devices 53–54
        physical asset monitoring and control devices 32–51
        water monitoring devices 46–51
plate and frame press 171
PLCs *see* programmable logic controllers (PLCs)

PMCL *see* proposed maximum contaminant level (PMCL)
pneumatic control systems 238–239
    for pump motor control 238–239
pneumatic pumping stations 519
point source 645
poliovirus 475
polishing pond 534
pollutant 621
polyhydroxyalkanoates (PHAs) 602
polyvinyl chloride (PVC) 254, 261, 445, 459, 621
ponding 549–550
ponds 532
    based on location 534–535
    based on type, processes occurring with 534–535
    biochemistry in 536–538
    elements of 535–536
    nutritional requirements 539–540
    physical factors 538–539
    process control calculations for 540–541
    types of 533
population ecology 335
    animal populations 344
    carrying capacity 344
    definition 343
    density 343
    distribution patterns 343
    organisms study 343
    species diversity 343
    succession
        description 344
        old-field succession 344
        primary succession or bare-rock succession 344
        secondary succession 344
population equivalent (PE) 95
population loading 158
porosity 439
portable cyanide analyzer 49–50
portable field monitors, VOCs 50
portable/removable barriers 36
portable sensors 46
positive-displacement flowmeters 220, *221*
    six common principles 220, *221*
positive displacement pumps 243, 454
    diaphragm pump 243, *243*
    peristaltic pumps 243–244
    piston/reciprocating pump 243
potable water, definitions 431
potassium permanganate 456, 479, 485
    oxidation 456
POTWs *see* publicly owned treatment works (POTWs)
pounds formula 489
PPCPs *see* pharmaceuticals and personal care products (PPCPs)
prechlorination 452, 480
precursor removal 483
preliminary treatment 142
    aerated systems 525
    chemical addition 525
    equalization 525
    grit removal 143–145, 522–524
        aeration 524
        calculations 524–525
        centrifugal force 524
        gravity/velocity-controlled grit removal 523–524
        grit channel velocity calculation 144–145

preaeration 525
preliminary treatment sampling and testing 525, **526**
    process control calculations 526–527
    screening 142, 520–521
        manually cleaned screens 521
        mechanically cleaned screens 521
        pit capacity calculations 143
        removal calculations 143–144
        removal computations 522
        safety screening 521–522
    shredding 522
prescriptive processes 498–499
pressure 86–88
    filter systems 464–465
    filtration 171
        operational observations, problems, and troubleshooting procedures 637–638
    and head 201
    to head conversion 229
    velocity relationship 226–227
pretreatment
    aeration 451
    chemical addition 452–453
        calibration procedures 454–455
        chemical feeder calibration 453–455
        chemical feeders 453–455
        chemical solutions 452–453
        corrosion control 457–460
        hardness treatment 456–457
        iron and manganese removal 455–456
        iron and manganese removal techniques 455–456
    screening 451–452
Primacy Agency 470–471
primary clarifier solids production 164
primary culprit 459
primary effluent 531
primary sludge 608
primary solids production 164
primary treatment calculations
    BOD and suspended solids removed 146
    process control calculations 145
    surface loading rate 145–146
    weir overflow rate 146
primary treatment, sedimentation
    effluent from settling tanks 531
    operator observations, process problems, and troubleshooting 528
    overview of 527–528
    problem analysis 531
    process control calculations
        BOD and SS removal 530–531
        detention time 529
        percent removal 529
        percent total solids 530
        process control calculations 529–531
        sludge pumping 530
        surface loading rate 529–530
        weir overflow rate 530
    process description 527–528
    sedimentation tanks types 528
privatization 13–14
process
    calculations 614–615
    condition *vs.* organisms present/ population **570**
    control testing and sampling 584
        aeration influent sampling 568
        aeration tank 568–572
        F/M ratio control 577

food to microorganism ratio (F/M ratio) 576–577
    process control adjustments 572–573
    process control calculations 574–576
    return-activated sludge and wasteactivated sludge 572
    settling tank 572
    settling tank effluent 572
    settling tank influent 571–572
    troubleshooting operational problems 573–574
    residuals 163–164, 607
    sidestreams 567
Process Safety Management (PSM) 496
programmable logic controllers (PLCs) 52, 55
propeller meter 219, *220*
proposed maximum contaminant level (PMCL) 458
protective instrumentation, centrifugal pumps 240
protocol analysis 54
protozoa 475, 559, 570–571
    inactivation 494
PSM *see* Process Safety Management (PSM)
publicly owned treatment works (POTWs) 4–5, 32, 474, 511
public water systems (PWSs) 32, 437, 442
pumping
    stations 519
    station wet well calculations 519–520
    water level 207
pumping calculations 226
    affinity laws, centrifugal pumps 230–231
    fluid velocity through pipeline 226
    friction head 228
    horsepower 229
        brake horsepower (BHP) 230
        hydraulic horsepower (WHP) 229
    net positive suction head (NPSH) 231–233
    pressure to head conversion 229
    pressure-velocity relationship 226–227
    series and parallel pumps 233
    specific speed 230
        suction specific speed 230
    static head 227
        static discharge head 227–228
        static suction head 227
        static suction lift 227
    total head 229
    velocity head 228–229
pure oxygen 562
purple sulfur bacteria 535
PVC *see* polyvinyl chloride (PVC)
PWSs *see* public water systems (PWSs)

quagga mussel *(Dreissena bugensis)* 478
quicklime 626

racetrack type 564
Rachel's Creek Sanitation District 18–19
Rack Load Test 44
radial bearing 234
radiation detection equipment 42–43, 46
    for monitoring personnel and packages 42–43
    for monitoring water assets 46
radiological contamination, monitoring sensors 47
rapid sand filters 464
RAS *see* return activated sludge (RAS)
RASS *see* return activated sludge solids (RASS)

rate of filtration calculation 96–97
raw sewage stabilization pond 533–534
RBCs *see* rotating biological contactors (RBCs)
recarbonation calculation 139–140
recessed impeller/vortex pumps 242, *242*
recirculation 546, 547–549, *548*
    flow 148
rectangular weir 221, *222*
recycle flow 615
red algae 536
re-engineering 13–14
remote telemetry units (RTUs) 53
removable bollards 36
reporting calculations 647–648
required MLVSS quantity 577
reservoir covers 43–44
residual solids 291
resin 140
resistance temperature devices (RTDs) 240
respiration 536–537
retention time 527
retractable bollards 36
return activated sludge (RAS) 67, 68, 559, 563, 572
return activated sludge solids (RASS) 514, 568
return sludge 560
reverse osmosis (RO) 500
Reynold's number 209
rising sludge 559
Risk Management Program (RMP) Rule 496
RMP *see* Risk Management Program (RMP) Rule
RO *see* reverse osmosis (RO)
rocky-bottom streams
    definition 392
    habitat assessment
        definition 393–394
        steps 394–396
    steps 392–393
roofed water storage tank 231, *232*
rotary drum filter 635
rotary vacuum filters
    dewatering 173
    operational observations, problems, and troubleshooting procedures 635–636
    process control calculations 636
    types of 635
rotating biological contactors (RBCs) 148, 532, 557
    cross-section and treatment system 554, *554*
    equipment 554–555
    expected performance 555
    hydraulic loading rate 149
    operation 555
    operator observations 555
    organic loading rate 149–150
    process 602
    process control calculations 148, 556–557
    process control sampling and testing 555
    soluble BOD 149
    total media area 150
    troubleshooting operational problems 555–556
rotifers 559
roughing filters 546, 547
roughness factor 228

RTDs *see* resistance temperature devices (RTDs)
RTUs *see* remote telemetry units (RTUs)

Safe Drinking Water Act (SDWA) 9, 11, 463, 469, 503
    1974 Act 27
    1996 Act 27, 28
Safe Drinking Water Act Reauthorization 9
salt for regeneration 141–142
sampling 43
    and testing 646–647
sampling, streams
    bottom line 402
    devices
        dissolved oxygen (DO) 400
        miscellaneous equipment 402
        nets 400
        plankton sampling 401
        Secchi disk 401–402
        sediment samplers (dredges) 400
        temperature monitor 400
    frequency and notes 391
    macroinvertebrate
        equipment 391–392
        muddy-bottom streams *see* muddy-bottom streams
        rocky-bottom streams *see* rocky-bottom streams
    multi-habitat approach
        cobble 390
        grid 390, 391, *391*
        sand 390
        snags 390
        submerged macrophytes 390
        transect 390
        vegetated banks 390
    planning 389
    post-sampling routine 399
    stations 390–391
sand drying beds 174
sanitary wastewater 517
saprobity system 570
SCADA *see* Supervisory Control and Data Acquisition System (SCADA)
schedule (SCH) 252
screening 142–143
    pit capacity calculations 143
    removal calculations 143–144
scum 608
    removal 529
secondary
    clarifier solids production 164–165
    sludge 608
    solids production 164
secondary maximum contaminant levels (SMCLs) 455
secondary treatment
    rotating biological contactors
        cross-section and treatment system 554, *554*
        equipment 554–555
        expected performance 555
        operation 555
        operator observations 555
        process control calculations 556–557
        process control sampling and testing 555
        troubleshooting operational problems 555–556
    treatment ponds 532–533
        based on location 534–535

    based on type, processes occurring with 534–535
    biochemistry in 536–538
    elements of 535–536
    nutritional requirements 539–540
    physical factors 538–539
    process control calculations for 540
    types of 532–533
    trickling filters
        activated sludge modifications 541
        classification 541, **546**
        cross-section of 541, *545*
        definitions 541, 546
        equipment 546–547
        general process description 547
        operator observations 548–549
        overview of 547–548
        process calculations 552–553
        process control sampling and testing 549
        standard operating procedures 547
        troubleshooting operational problems 549–552
secure control systems 59
securing doorways 44
security
    barriers 37
    hardware/devices 32
        communication and integration 51–53
        cyber protection devices 53–54
        physical asset monitoring and control devices 32–51
        water monitoring devices 46–51
    improvement steps, SCADA 59–63
    upgrading 19–20
sedimentation 291, *462*, 462–463
    calculations 112–113
        detention time 113
        tank volume calculation 113
seed volume 618–619
self-contained breathing apparatus (SCBA) 488
septic tanks 528
septum 465
sequencing batch reactor (SBR) process 602
sequestering 456
series and parallel pumps 233
service line flushing time 178–179
    composite sampling calculation 179–180
settleability 514
    test 182
settleable solids 182–183
settled biosolids volume (SBV) 182
settled sludge volume (SSV) 514, 559, 568, 571–572, 574–575
settling tank 553, 554, 572
    effluent 572
    influent 571–572
    sludge 594–595
shallow wells 443–444
Sheboygan Regional Wastewater Treatment Plant 66–67
Sherlock Holmes-type at the pump 472–474
shock load 559
shredding 522
sick water 6–8
side-hinged door security 44–45
silicates (SiO$_2$) 459
single-cylinder locks 42
slope 210, 213–215
sloughing 546

slow sand and diatomaceous earth filtration 467
slow sand filters 463–464
sludge 607
    additions 617
    blanket depth 566, 572
    bulking 568
    characteristics 608–613
    color and odor 567
    dewatering 628
        biosolids characteristics affecting 628
        characteristics affecting 628–629
    direct contact 609
    handling 607–608
    incineration 640
        incineration processes 641
        operational observations, problems,
            and troubleshooting procedures
            641–643
        process description 640
    indirect contact 610
    pathogens and vector attraction 609
    production
        in pounds/million gallons 611
        in wet tons/year 612
    pumped per day 611
    pumping 530
    pumping calculations 610
    pumping time 610–611
    removal 529
    retention time 619
    solids washout 568
    sources of 608
    stabilization
        aerobic digestion 615–616
        anaerobic digestion 617–618
        composting 620
    thickening 612–613
    withdrawal 617
sludge volume index (SVI) 154–155, 514, 559
SMI see sulfur-modified iron (SMI)
smoke detector alarm 33
SOCs see synthetic organic chemicals (SOCs)
soda ash (Na$_2$CO$_3$) 138, 459
sodium fluoride 130
sodium fluorosilicate 130
sodium hydroxide 500
sodium hypochlorite 478, 489, 491–492, 495, 496
sodium polyphosphates 459
sodium silicate 459
sodium sulfate 605
soil resistivity 459
solids 518
    concentrators 613
    loading rate 167, 172, 614–615
    sludge/biosolids handling 607–608
        background information 608
solids inventory 151
    estimating return rates from SSV$_{60}$ 154
    food-to-microorganism ratio (F/M ratio)
        151–152
        waste rate calculation 151–152
    Gould Biosolids Age 153
    mass balance
        biosolids waste 155–156
        settling tank suspended solids 155
    mean cell residence time (MCRT) 153
    oxidation ditch detention time 156
    sludge volume index (SVI) 154–155
solids measurement
    higher solids 419
    sampling and equipment considerations

precautions 419
    total solids 418–419
    variations 419
total solids 418–419
total suspended solids (TSS)
    calculations 420
    test procedure 419–420
volatile suspended solids testing
    description 420
    procedure 420
    total volatile suspended solids (TVSS)
        calculation 420–421
solids retention times (SRTs) 564
solubility 452
soluble BOD 556–557
    concentration, RBC 149
soluble cyanide 49
solution
    chemical feeder setting, GPD 161
    feeders 453
    feed system 583
    strength calculation 96
sonar system 240
source water quality control 482–483
specific capacity 208
specific energy 208
specific gravity 95, 453
specific speed 230
    suction specific speed 230
specific ultraviolet absorption (SUVA) 495
specific yield 105
spirilla 475
spot map 473
SRTs see solids retention times (SRTs)
stabilization 168
    operation, sludge dewatering 627–628
    performance factors 634
    pond processes 533
staging **3,** 3–4, 546
stalked ciliated protozoa 571
standardization methods 407
standard rate filter 547
standard temperature and pressure (STP) 196
state recordkeeping requirements 468
state reporting requirements 468
static discharge head 227–228
static head 201
static suction head 227
static suction lift 227
    from open reservoir 231, *232*
static water level 207
steady flow 208
step aeration 562
sterilization 471
Stevin's Law 197
stoneflies (Order: Plecoptera) 358, *358*
storage tank calculation 98–99
storm water 517
    runoff 516
STP see standard temperature and pressure
    (STP)
straggler floc 568
straining 463
stream ecosystem, biotic and abiotic factors
    345–346
stream genesis and structure 346–347
    adaptive changes 353
    bars, riffles and pools 350
    channels characteristics 349
    flood plain
        balanced aquarium 351

characteristics 349
    drift 352
    limiting factor 351
    organisms classification 351
    pool zone 351
    riffle zone 351
    rock measurement 352
    run zone 351
    substrate particles 352
material transport 348–349
sinuosity 349
stream current adaptations 352–353
stream profiles 349
water discharge 347–348
water-flow 347
submersible pumps 241, *241*
sulfur-modified iron (SMI) 501
supernatant withdrawal 617
Supervisory Control and Data Acquisition
    System (SCADA) 54–55
    applications in water/wastewater 56
    cyber threats to control systems 58–59
    increasing risk 57
    securing control systems 59
    security improvement steps 59–63
    technologies adoption with known
        vulnerabilities 57–58
    vulnerabilities 56–57
supervisory instrumentation 240–241
surface
    foam and scum 567
    loading rate 145–146, 614
    mounted wedge barricades 35
    settling rate 529–530, 685
surface overflow rate 113–114, 529–530
    determining lime dosage
        g/min 117
        lbs/day 117
        mg/L 115–117
    mean flow velocity 114
    percent settled biosolids 115
    weir loading rate 114–115
surface waters
    advantages/disadvantages 437
    hydrology 437–438
    intake 438–439
    quality 439
    raw water storage 438
    screening devices 439
Surface Water Treatment Rule (SWTR) 8–9,
    118, 326, 437, 463, 484
    1996 Amendments 493
suspended growth systems 532
suspended solids 518
SUVA see specific ultraviolet absorption
    (SUVA)
swarming attacks 59
swing beam design 36
swing gates 36
SWTR see Surface Water Treatment Rule
    (SWTR)
synthetic organic chemicals (SOCs) 304
system recordkeeping requirements 468

tank volume calculation 113
TCR see total coliform rule (TCR)
TDS see total dissolved solids (TDS)
temperature
    conversions 85
    detectors 240
terrorism 19, 55

tertiary treatment 594
test methods 407
   biochemical oxygen demand (BOD)
      analysis 412
      calculation 412–413
      description 411
      sampling considerations 411
      testing 412
   colorimetric 407
   dissolved oxygen (DO) 407–408
      meter and probe 409–411
      sampling and equipment
         considerations 408
      Winkler method 408–409
   electronic 407
   titrimetric 407
   visual 407
THMFP *see* THM formation potential
   (THMFP)
thermal treatment 627
THM formation potential (THMFP) 481
threshold odor number (TON) 429
thrust 200, *200*
   anchor 200, *200*
   bearing 234
   block 200
time-of-flight ultrasonic flowmeters *218,* 219
titrimetric methods 407
TKN *see* Total Kjeldahl Nitrogen (TKN)
TON *see* threshold odor number (TON)
total alkalinity
   analytical and equipment considerations
      burettes 422
      digital titrators 422
      titrators 422
   definition 422
   measurement 422
   testing 422
total chlorine residual test 125, 488, 584
total coliform rule (TCR) 479
total cyanide 49
total dissolved solids (TDS) 518
total dynamic head 201
total flow 552
total hardness 136–137
total head 229
Total Kjeldahl Nitrogen (TKN) 518, 572
total media area 150, 557
total organic carbon (TOC) analyzer 47–48,
   481, 483, 495
total quality management (TQM) 14
total residual chlorine (TRC) 429, 584
total residual limitations 583
total suspended solids (TSS) 173, 183, 518,
   536, 571, 572, 593
total trihalomethanes (TTHMs) 452, 456,
   482–483, 495
total volatile suspended solids (TVSS)
   420–421
toxicity 561–562
toxicity characteristic leaching procedure
   (TCLP) 502
toxicity monitoring/toxicity meters 51
TQM *see* total quality management (TQM)
TRC *see* total residual chlorine (TRC)
traditional flow measurement 216
traffic anomaly detection 54
transformational change 8–9
traps
   description 280–281
   maintenance 281–282

testing 282
thermostatic traps 280–281
   types 280–281
treatment ponds 156
   BOD loading 157
   BOD removal efficiency 157–158
   flow rate in acre-inches day 157
   hydraulic detention time 157
   organic loading rate 157
   parameters 156
   population loading 158
   process control calculations 156
treatment process models 64–65, *66*
   Big Gulch Wastewater Treatment Plant
      67–68
   City of Bartlett Wastewater Treatment
      Plant 69
   Green Bay Metropolitan Sewerage District
      (GBMSD) 65–66
   Sheboygan Regional Wastewater Treatment
      Plant 66–67
triangular V-notch weir 221, *222*
trickling filters 147–148
   activated sludge modifications 541
   calculations 147
   classification **546**
   cross-section of 541, *545*
   definitions 541, 546
   equipment 546–547
   general process description 547
   hydraulic loading 147
   operator observations 548–549
   organic loading rate 147–148
   overview of 547–548
   process calculations 552–553
   process control sampling and testing 549
   recirculation flow 148
   standard operating procedures 547
   troubleshooting operational problems
      549–552
troubleshooting 584–586
   wastewater ponds 541, **542–544**
true flies (Order: Diptera) 359–360
TSS *see* total suspended solids (TSS)
TTHM Stimulated Distribution System 481
tuberculation 458, 459
tubificid worms 388
tubing
   applications 265
   bending
      correct and incorrect bends 265, *265*
      spring-type benders 265
   chemical advantage 263
   connecting tubing
      cutting tubing 264
      description 264
      flared joints 264–265
      flareless joints 264–265
      soldered or compression connection
         264
      soldering tubing 264
   description 261
   fittings 261
   fittings and connections 273–274
   governing factors 267–268
   mechanical advantage
      diameter and flexibility 261
      inner-wall surfaces 263
      laminar flow 263
      threaded joints *262,* 263
      turbulent flow 263

      water hammer 263
   vs. piping
      advantages 261
      applications 261
      cost 261–262
      physical characteristics and installation
         261
      thickness 261
   types 265–266
turbidity 493
   description 414
   operations 414
   particles 415
   sampling and equipment considerations
      414–415
   Secchi disk 415
   sources 414
turbine meter 219, *220*
turbine pumps *242,* 242–243
turbulence 567
turbulent flow 208, 212
two story (Imhoff) tank 528
2002 EPA report 22
2021 Report Card for American Infrastructure **21**
two-way radios 52

UCT *see* University of Cape Town (UCT)
UF *see* ultrafiltration (UF)
UFRV *see* unit filter run volume (UFRV)
UL *see* Underwriters Laboratory (UL)
ultrafiltration (UF) 483, 500
ultrasonic flowmeters *218,* 219
   Doppler type 219, *219*
   time-of-flight *218,* 219
ultraviolet (UV) 484
   absorbance 481
   irradiation
      advantages and disadvantages 590
      applicability 591
      operation and maintenance 591–592
   lamps 590
   light tubes 590
under-oxidized activated sludge 571
Underwriters Laboratory (UL) 34
   UL 752 45
uniform flow 208, 212
United States 471, 473–475, 485, 491, 497, 532,
   624, 646
United States Environmental Protection
      Agency (USEPA) 9, 466, 467–471,
      480, 482, 487, 503, 628, 645
unit filter run volume (UFRV) 119–120
unit process performance and efficiency 515
units 84–85
   milligrams per liter 85–86
University of Cape Town (UCT) 605
U.S. Centers for Disease Control (CDC) 473
U.S. Department of Transportation 496
USEPA *see* United States Environmental
      Protection Agency (USEPA)
UV *see* ultraviolet (UV)

vacuum-assisted drying beds 632–633
vacuum collection system 519
valves
   construction 275
   definition 274
   design 274
   end connections 275, **275**
   features 274
   lockout devices 45

maintenance 279
operations
  magnetic valve 279
  pneumatic and hydraulic valve 279
trim 274, **274**
types 276
  ball valves 276
  butterfly valves 277
  check valves 277
  diaphragm valves 277–278
  gate valves 276
  needle valves 276–277
  plug valves 277
  pressure-reducing valves 279
  quick-opening valves 277
  regulating valves 278
  relief valves 278–279

variances and exemptions 471
varied flow 208, 212
velocity 459
  calculations 90
  flowmeters 219–220
  head 201, 228–229
vent security 45–46
venturi 217, *217*
vertical turbine pump calculation 105–106
vibration monitors 240
vibrios 475
viruses 475
  inactivation 494
visual methods 407
visual surveillance monitoring 46
VM *see* volatile matter (VM)
VOCs *see* volatile organic chemicals/
      compounds (VOCs)
volatile acids 617
  to alkalinity ratio 169–170, 619
volatile matter (VM) 611
volatile organic chemicals (VOCs) 305
volatile organic compounds (VOCs) 50
volatile solids loading 168–169
volatile suspended solids (VSS) 183–184
volume 86, 685
volute-casing centrifugal pump 235
VSS *see* volatile suspended solids (VSS)

WAS *see* waste-activated sludge (WAS)
WASS *see* waste activated sludge solids (WASS)
waste-activated sludge (WAS) 568, 572, 576
  flow rate 566
waste activated sludge solids (WASS) 514
waste sludge 560
wastewater characteristics
  biological characteristics and processes 518
  chemical characteristics 517–518
  physical characteristics 517
wastewater collection systems
  force main collection system 519
  gravity collection system 519
  pumping stations 519
  pumping station wet well calculations 519
  vacuum collection system 519
wastewater disinfection
  chlorine disinfection 582–586
    chemical handling 586–587
    chlorination equipment 583
    chlorination operation 583
    chlorination process description 583
    chlorination terminology 582
    environmental hazards and safety 586

facts and process description 582–583
troubleshooting operation problems
  583–586
no disinfection 593
ozonation 592–593
  advantages and disadvantages 592
  applicability 592
  bromine chloride 593
UV irradiation 589–592
  advantages and disadvantages 590
  applicability 591
  operation and maintenance 591–592
wastewater operations 9–10
wastewater operators, treatment process 511, *512*
wastewater sources
  classification 516–517
  wastewater generation 516
wastewater suspended solids 183–184
wastewater Treatment Facility (WWTP) 65, 605
  influent data profile
    Big Gulch **68**
    City of Bartlett **69**
    De Pere **65**
    Sheboygan **67**
wastewater treatment operations
  activated sludge process
    aeration tank 567–568
    control parameters 565–566
    factors affecting operation of 560
    formation of 561
    growth curve 561
    MCRT impact on parameters 578–579
    modification 562–565
    performance-controlling factors
     561–562
  advanced wastewater treatment 593–594
    biological denitrification 597
    biological nutrient removal (BNR)
     601–605
    carbon adsorption 598
    chemical treatment 594
    enhanced biological nutrient removal
     (EBNR) 605–607
    filtration 595–597
    microscreening 595
  biosolids, land application of 643
  centrifugation 639
  hydraulic detention time (HDT) 515–516
  percent volatile matter reduction, sludge 515
  permits, records, and reports
    NPDES permits 646
    reporting calculations 647–648
    sampling and testing 646–647
  plant performance/efficiency 514–515
  preliminary treatment
    aerated systems 525
    chemical addition 525
    equalization 525
    grit removal 522–524
    preaeration 525
    preliminary treatment sampling and
     testing 525, **526**
    process control calculations 526–527
    screening 520–522
    shredding 522
  primary treatment, sedimentation
    effluent from settling tanks 531
    operator observations, process
     problems, and troubleshooting 528
    overview of 527
    problem analysis 531

process control calculations 529–531
process description 527–528
sedimentation tanks types 528
rotary vacuum filtration 635–636
secondary treatment
  rotating biological contactors 554–557
  treatment ponds 532–533
  trickling filters 541, *545,* 546–553
sludge 607
  characteristics 608–613
  direct contact 609
  incineration 640–641
  indirect contact 610
  pathogens and vector attraction 609
  stabilization 615–620
  thickening 612–613
unit process performance and efficiency 515
wastewater characteristics
  biological characteristics and processes
   518
  chemical characteristics 517–518
  physical characteristics 517
wastewater collection systems
  force main collection system 519
  gravity collection system 519
  pumping stations 519
  pumping station wet well
   calculations 519
  vacuum collection system 519
wastewater disinfection
  chlorine disinfection 582–586
  no disinfection 593
  ozonation 592–593
  UV irradiation 589–592
  wastewater operators 511
wastewater sources
  classification 516–517
  wastewater generation 516
wastewater terminology and definitions
  511, 513–514
water and wastewater operations challenges
  autonomous operations 22–23
  benchmarking
    baseline audit 17
    baseline data and tracking energy use
     15–17
    create equipment inventory and
     demand and energy distribution
     17–18
    field investigation 17
    potential results of 15
    process 14–15
    12 stage approach 14
    steps 15
    targets 15
  compliance with new, changing, and
   existing regulations 11
  maintaining infrastructure 12–13
  maintaining sustainable infrastructure 22
  privatization and/or re-engineering 13–14
  Rachel's Creek Sanitation District 18–19
  sustainable water/wastewater
   infrastructure 21
  technical *vs.* professional management 20
  water/wastewater infrastructure gap 22
waterborne diseases 309, **475**
waterborne pathogens 309
  attributes of 475, **475**
water chemistry 291
  chemical parameters
    fluorides 306

water chemistry (*Cont.*)
  heavy metals 306
  nutrients 306–307
  organics 305
  SOC 305
  total dissolved solids 305–306
  VOC 305–307
  concepts
    emulsions 291–292
    ion 292
    mass concentration 292
    miscibility/solubility 291
    suspension/sediments/particles/
      solids 291
  constituents
    acids 299, **299**
    bases 299
    color 297
    DO 297–298
    inorganic matter 298
    metals 298
    organic matter 298
    pH 300–301
    salts 299
    solids 297
    sulfate concentrations 299–300
    turbidity 297
  definitions 292–293
  fundamentals
    compound substances 295–296
    elements 292
    matter 292
  measurements
    alkalinity 301
    chemicals 302–304
    disinfection 302
    hardness 301–302, **302**
    odor control/waste water treatment 302
    specific conductance 301
    taste/odor removal 303
    temperature 301
  softening
    chemical precipitation 303
    ion exchange 303–304
    recarbonation/stabilization 304
  solutions 296
water chlorination calculations 124
  calculating dry hypochlorite rate 126–127
  calculating hypochlorite solution feed rate 127–128
  calculating percent strength of solutions 128
  calculating percent strength using dry hypochlorite 128
  chlorine demand 125
  chlorine disinfection 124
  chlorine dose 125
  chlorine residual 125
water cycle 367–369
water distribution system calculation 97
Water Environment Federation (WEF) 18
water fluoridation 129
water hammer 200
water hydraulics 196
  basic concepts 196–197
  Bernoulli's equation 205–206
  Bernoulli's theorem 205
  conservation of energy 203
  density and specific gravity 197–198
  energy head 203
  flow measurement 215–223

    devices 216–220
    old-fashioned way 216
    open-channel 220–223
    traditional flow measurement 216
  force and pressure 198–200
    effects of water under pressure 200
    hydrostatic pressure 199–200
  friction head loss 207
    C factor 210, **210**
    flow in pipelines 208, *208–209*
    major head loss 209
    minor head loss 210
    slope 210
  head 200–201
    friction head 201
    loss 204
    and pressure 201
    static head 201
    total dynamic head 201
    velocity head 201
  hydraulic grade line (HGL) 204
  open-channel flow 212–215
    calculations 212–213
    characteristics 212
    parameters used 212–215
  piezometric surface 203–204
  piping hydraulics 210–212
  Stevin's Law 197
  water in motion, flow and discharge rates 201–203
    area and velocity 202–203
    pressure and velocity 203
  well hydraulics 207
  wet well hydraulics 207
water in motion, flow and discharge rates 201–203
  area and velocity 202–203
  pressure and velocity 203
water monitoring devices 46–47
  chemical sensors
    adapted BOD analyzer 47
    arsenic measurement system 47
    chlorine measurement system 48–49
    portable cyanide analyzer 49–50
    total organic carbon (TOC) analyzer 47–48
  portable field monitors to measure VOCs 50
  radiation detection equipment 46
    for monitoring water assets 46
  sensors for monitoring chemical, biological, and radiological contamination 47
  toxicity monitoring/meters 51
Water Pollution Control Act Amendments 12
water quality 367
  biological characteristics, water and wastewater
    bacteria 381
    protozoa 381
    viruses 381
    worms (helminths) 382
  characteristics 372
  chemical characteristics, wastewater
    inorganic substances 379–381
    organic substances 379
  chemical characteristics, water
    alkalinity 376
    fluoride 377
    hardness 376–377
    metals 377
    nutrients 378

    organic 377–378
    total dissolved solids 376
  Clean Water Act (CWA) 369
  monitoring sensor 46
  physical characteristics, water and wastewater
    color 373–374
    solids 372–373
    taste and odor 374–375
    temperature 375
    turbidity 373
  Safe Drinking Water Act (SDWA) 370
    consumer confidence report rule 371–372
    implementation 370–371
    maximum contaminant levels (MCLs) 370, **371**
watershed 438, *438*
water softening 135–139
  alkalinity determination 137–138
  calcium hardness 135–136
  carbonate hardness 137
  feed rates 139–140
  ion exchange capacity 140
  magnesium hardness 136
  noncarbonate hardness 137
    removal 138–139
  recarbonation calculation 139–140
  salt and brine required for regeneration 141–142
  total hardness 136–137
  treatment time calculation 141
  water treatment capacity 140–141
water solution 452
water sources 103–104
  deep-well turbine pump calculations 105–106
  specific yield 105
  vertical turbine pump calculations 105–106
  well casing disinfection 105
  well drawdown 104
  well yield 104–105
water storage calculations 103–104, 106
water strider (Order: Hemiptera) 362
water treatment capacity 140–141
water treatment operations
  arsenic removal from drinking water
    arsenic exposure 497–498
    technologies 497–502
  chlorination
    application methods 495–496
    breakpoint chlorination 487–488, *488*
    breakpoint chlorination curve 488, *488*
    chlorine chemistry 487
    chlorine dosage determination 489–491
    chlorine generation 491–492
    chlorine terminology 486–487
    DBPs 494–495
    disinfection efficacy 493–494
    factors affecting 493
    gas chlorination 488–489
    hypochlorination 489
    measuring chlorine residual 493
    pathogen inactivation 493–494
    safety and handling considerations 496
    uses, application and doses, chlorine 492
  coagulation 460–462
  CT factor 484
  current national disinfection practices 485
  DBP control strategies
    precursor removal 483

water treatment operations (*Cont.*)
    source water quality control 482–483
    DBP formation
        DBP precursors 481–482
        inorganic by-products and
            disinfectants 482
        organic oxidation by-products 482
        pH impacts on 482
    DBPs and disinfection residuals types 480
    direct filtration 465
        alternative filters 465
        common filter problems 465–466
        and compliance with turbidity
            requirements 466–471
    disinfectant residual regulatory
        requirements 484–485
    disinfection 471–472, *472*
        methods of 484–485
        pathogen inactivation, mechanisms of
            477–480
        primary concern, pathogens of 475
        pump handle removal, water treatment
            473–474
        recent waterborne disease outbreaks
            476–477
        Sherlock Holmes-type at the
            pump 472–474
        strategy selection 484
    filtration
        diatomaceous earth filters 465
        pressure filter systems 464–465
        rapid sand filters 464
        slow sand filters 463–464
    mixing and flocculation 449, 462, *462*
    pretreatment 451
        aeration 451
        chemical addition 452–453
        screening 451–452
    purpose of 450
    sedimentation *462*, 462–463
    stages of 450–451
water treatment plants (WTPs) 6
water treatment process 451
    biochemical cycles 329–332
        carbon 329–330
        nitrogen 330–331
        phosphorus 332
        sulfur 331
    biological process
        aerobic 328

        anaerobic 328
        anoxic 329
        growth cycles 329
        photosynthesis 329
    microbiology process 319–320
        Cryptosporidium see Cryptosporidium
        Giardia *see* Giardia
        helminths 328
        pathogenic protozoa 320
water under pressure 200
water/wastewater
    conveyance
        early conveyance systems 245
        fittings *see* fittings
        goal today 245
        industrial hoses *see* industrial hoses
        Kawamura (1999) 245
        Nayyar (2000) 245
        operators 246
        piping *see* piping systems
        single-line piping 246, *246*
        tubing *see* tubing
        United States Environmental Protection
            Agency (USEPA) 245
        valves *see* valves
    operators 24
        emergency responders, plant operators
            26–27
        licensure 28–30
        operator certification 28–30
        operator duties 27–28
        record setting 25
        working conditions 27–28
    setting the stage 59, 64
    treatment process models 64–65
        Big Gulch Wastewater Treatment Plant
            67–68
        City of Bartlett Wastewater Treatment
            Plant 69
        Green Bay Metropolitan Sewerage
            District (GBMSD) 65–66
        Sheboygan Regional Wastewater
            Treatment Plant 66–67
waterworks operator 449
weak acid dissociable (WAD) cyanide 49
wedge barriers 35
wedgewater system 632
wedgewire beds 632
weir 221–222
    condition 529

loading (overflow rate) 685
loading rate 114–115
overflow rate 146, 530
well casing disinfection 105
well drawdown 104
well hydraulics 207
well systems
    abandonment 447
    casing vent 444
    components of 444–445
    deep wells 444
    drop pipe 444–445
    evaluation 445–446
    grout 444
    maintenance 447
    miscellaneous components 445
    problems 447
    pumps 446
    routine operation/recordkeeping
        requirements 446
    sanitary seal 444
    screens 444
    seven-step process 443
    shallow wells 443–444
    types 443–444
    well casing 444
    well pad 444
    well site requirements 443
well yield 104–105, 207
wetwell/drywell pumping stations 519
wet well hydraulics 207
wet well pumping stations 519
white slime 556
whole effluent toxicity (WET) tests 51
Wireless Access Point (WAP) 52
wireless data communication system 52–53
Wireless Network Interface Card 52
wireless systems 34
working conditions, operators 27–28
WWTP *see* wastewater Treatment Facility
    (WWTP)

Y2K fiasco 10

zebra mussel (*Dreissena polymorpha*) 477–478
zeolite clays 457
zinc phosphates 459
zone of influence 207
zone of saturation 439
zoogleal slime 514

Printed in the United States
by Baker & Taylor Publisher Services